PROCEEDINGS OF THE 14th WORLD CONGRESS
International Federation of Automatic Control

Beijing, P.R. China, 5 - 9 July 1999

(in 18 volumes)

Edited by

Han-Fu CHEN
Chinese Academy of Sciences
Beijing, P.R.China

Dai-Zhan CHENG
Chinese Academy of Sciences
Beijing, P.R.China

and

Ji-Feng ZHANG
Chinese Academy of Sciences
Beijing, P.R.China

VOLUME C
CONTROL DESIGN

Volume Editors

A. ISIDORI (Italy) and S. ENGELL (Germany)

Published for the

INTERNATIONAL FEDERATION OF AUTOMATIC CONTROL

by

PERGAMON
An Imprint of Elsevier Science

UK	Elsevier Science Ltd, The Boulevard, Langford Lane, Kidlington, Oxford, OX5 1GB, UK
USA	Elsevier Science Inc., 660 White Plains Road, Tarrytown, New York 10591-5153, USA
JAPAN	Elsevier Science Japan, Tsunashima Building Annex, 3-20-12 Yushima, Bunkyo-ku, Tokyo 113, Japan

First edition 1999

Library of Congress Cataloging in Publication Data

A catalogue record for this book is available from the Library of Congress

British Library Cataloguing in Publication Data

A catalogue record for this book is available from the British Library

ISBN 0-08-043214 X

Printed in Great Britain

Transferred to Digital Printing 2008

14th Triennial World Congress of IFAC,
the International Federation of Automatic Control

Hosted by
The Chinese Association of Automation

CONTENTS OF VOLUMES

PREFACE

As the last triennial World Congress of the International Federation of Automatic Control (IFAC) of this century, the 14th IFAC Congress is recognized as a preeminent event for the worldwide community of control systems science and engineering. It is a great pleasure and honor for Chinese control researchers and engineers to host the Congress in Beijing.

Beijing is one of the oldest cities in the world. Its civilization can be traced back to 1000 BC when it was a small town. Since the Jin Dynasty (AD 1115), and during the Yuan, Ming and Qing Dynasties, Beijing has been the capital. It is famous partly for its many historical sites, including the Great Wall, the Forbidden City, the Summer Palace, the Ming Tombs, etc. Beijing is also famous for the development of automatic machines over many centuries, including the armillary sphere for simulating the Universe and the seismograph for detecting earthquakes, making the city one of great world centres for automation.

Beijing is also a young city; every day it shows a new face to the world. The fast pace of change in Beijing is fuelled by the rapid growth of the economy and the development of science and technology, including automation. China has vast potential for the development of automatic control. Beijing cordially welcomes you!

In the light of the outstanding efforts of our predecessors, our goal in organizing this Congress is to continue to increase the quality of the papers and of the Congress activities. We hope this Congress will be remembered as a showcase for the fruitful results on control theory and application achieved since the beginning of this century. In addition, we expect that several of the results and ideas demonstrated in this Congress will have lasting influence in the next century on the directions taken by the field of control systems engineering.

The response from the international control community to the call for papers was strong. The International Program Committee (IPC) received 2169 submitted papers, 53 invited session proposals and eight panel discussion proposals from 68 countries and regions. We were impressed by the high quality of the submissions. The technical program for the Congress covers a broad range of topics, with sessions arranged into nine congress tracks and subdivided into 45 technical areas. Reviews were conducted by all of the IFAC Technical Committees, with the help of hundreds of worldwide experts. Through this rigorous review procedure 1556 papers, one keynote speech, five plenary lectures, 47 invited session proposals and seven panel discussion proposals were accepted. They came from 59 countries and regions. The IPC is confident that the program that has been selected will provide significant contributions to each of the subject areas of the congress tracks.

There are five plenary lectures scheduled, one at the start of each of the five days of the Congress. The plenary speakers include some of the world's most outstanding authorities in various control fields. The topics of the plenary lectures include surveys, prospects and introductions to advanced control theory and applications. In selecting the plenary speakers, we have been guided by a desire to provide perspectives on theoretical topics, important applications, and other issues of common interest. It was a difficult task to select five plenary speakers among so many outstanding candidates who were nominated by our colleagues.

Following the plenary addresses, the congress program is composed of more than 1500 papers, to be presented in 214 lecture-type sessions and 47 poster-type sessions, the latter arranged in three time clusters. The paper sessions are supplemented by seven panel discussions. No distinction is made between poster papers and lecture papers in the Proceedings.

The Proceedings of the Congress are published in two forms, printed and CD-ROM. The Proceedings in its printed form consists of 18 volumes: 17 subject volumes, which contain all lecture and poster papers and a plenary volume. The plenary volume contains the keynote speech, plenary papers, panel discussions, and the table of contents, author index and subject index for the full set of subject volumes.

The CD-ROM, containing all the material in the 18 volumes, is provided to each registered participant as part of the registration material. Both the printed Proceedings and the CD-ROM will be marketed after the Congress, as the official Congress Proceedings, by the IFAC Publisher, Elsevier Science Ltd.

We expect that the reader, whether a participant in the Congress, or a researcher or practising engineer studying the material afterwards, will find the Congress publications useful, informative and helpful in their related creative works.

The Congress organizers have made every effort to create a high quality environment for the Congress, but China, as a developing country, still may seem unsatisfactory in many respects, including the language barrier. You may feel inconvenienced and you may need to be patient in communicating your needs. Imagination, plus a little spirit of adventure, may help you to have some unexpected and unforgettable experiences in this part of the world, which, in the end, makes life colorful.

Finally, we would like to thank all IPC members, all IFAC Technical Committee chairs and all reviewers for their valuable contribution to the Congress. We are also indebted to all our friends from different parts of the world who have given support, help and useful suggestions.

Han-Fu CHEN, Daizhan CHENG and Jifeng ZHANG

CONTENTS OF VOLUME C

CONTROL DESIGN
(Technical Area 2a)

14th Triennial World Congress, Beijing, P.R. China

A DESIGN OF 2-DOF CONTROL SYSTEMS BASED ON COPRIME FACTORIZATION AND LMI OPTIMIZATION

Shinichi Miyamoto * Glenn Vinnicombe **

* *Nagasaki R&D Center, Mitsubisi Heavy Industries, Ltd., 5-717-1,Fukahori-machi,Nagasaki,851-0392, JAPAN*
miyamoto@csl.ngsrdc.mhi.co.jp
** *Department of Engineering, University of Cambridge, Trumpington Street, Cambridge, CB2 1PZ, U.K.*
gv@eng.cam.ac.uk

Abstract: In this paper, a 2-DOF control system design is presented. First, control systems for reference tracking are considered, where the feedforward controllers are designed for the given feedback controller. Second, anti-windup controllers are designed for the given feedback controller. For both systems, design problems are formulated as problems of factorizing the feedback controller and they can be solved with the LMI optimization method. Then, 3-DOF control systems is proposed by combining the above two 2-DOF control systems. Last, design examples of 2-DOF control systems are given to demonstrate our design method. Copyright 1999 ©IFAC

Keywords: Control Systems, Feedforward Control, Convex Optimization, Robust Control, Tracking Systems

1. INTRODUCTION

In this paper, we consider 2-DOF (Degree Of Freedom) control system design. When the plant output is required to track the reference, a feedforward controller is often introduced to improve the response for the reference (Vidyasagar, 1985). This can be regarded as a 2-DOF control system where the response for the reference can be improved without changing the feedback properties.

It is known that when the plant has saturation nonlinearity, stability may be deteriorated (Doyle *et al.*, 1987). Anti-windup controller is introduced for the plants such that the control system remains stable. In (Miyamoto and Vinnicombe, 1996), design can be formulated as a problem of factorizing the feedback controller. This can also be regarded as a 2-DOF control system as anti-windup property can be improved without changing the feedback properties.

In this paper, we assume that the feedback controller is already designed. Then the problem of designing 2-DOF control systems can be regarded as a problem of designing an additional controller (a feedforward controller or an anti-windup controller). One approach for these problems is to factorize the controller as $C = (Q\tilde{V})^{-1}(Q\tilde{U})$ and to choose a $Q \in \mathcal{H}_\infty$ such that the control system has an appropriate property. These problems can be formulated as H_∞ standard problems, but the order of the obtained controller often becomes high.

In this paper, we propose an LMI(Linear Matrix Inequality)-based design. We show that these problems can be converted to an LMI problem by using observer-based parameterization. The controller is obtained by optimizing the observer gain for the cost function with H_∞-norm. The advantage of this method is that the order of the obtained 2-DOF controller is the same as the feedback controller.

The 2-DOF control designs for reference tracking and anti-windup will be proposed in Section 2 and Section 3, respectively. In Section 4, 3-DOF control systems which combine two 2-DOF control systems for reference tracking and anti-windup, will be described. In Section 5, design examples will be shown.

In this paper, we consider an m-input p-output n_P-state plant P and an n_C-state controller C. Their state-space representation is given as

$$P = \left[\begin{array}{c|c} A_P & B_P \\ \hline C_P & D_P \end{array}\right], \quad C = \left[\begin{array}{c|c} A_C & B_C \\ \hline C_C & D_C \end{array}\right].$$

2. REFERENCE TRACKING SYSTEM

In this section, we consider a reference tracking problem. It is assumed that the feedback controller C is already given. Then a feedforward controller is designed such that the transfer function from the reference to the output is appropriate.

2.1 *Problem Formulation*

The problem can be regarded as a problem of factorizing C with fixed-order $\tilde{U}, \tilde{V}$ (See Figure 1). We suppose that the orders of $\tilde{U}$ and $\tilde{V}$ are the same as C. Such $\tilde{U}, \tilde{V}$ can be written by introducing $L \in \mathcal{R}^{n_C \times m}$ and $Q_1 \in \mathcal{R}^{m \times m}$ as

$$[-\tilde{U}\ \tilde{V}] = \left[\begin{array}{c|cc} A_C + LC_C & -(B_C + LD_C) & L \\ \hline Q_1 C_C & -Q_1 D_C & Q_1 \end{array}\right] \quad (1)$$

for invertible Q_1 and stable $A_C + LC_C$ (Zhou *et al.*, 1995).

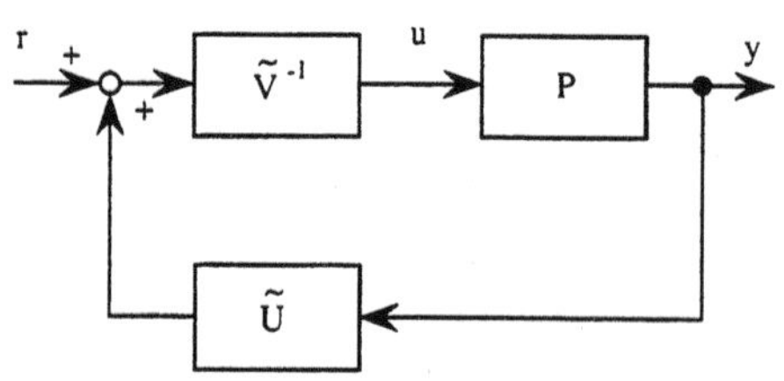

Fig. 1. 2-DOF control system $(\tilde{U}, \tilde{V})$

There are many approaches to factorize C. Here, we consider a design method based on an H_∞ mixed sensitivity problem. The complementary sensitivity T and the sensitivity S are considered for feedback control design. These transfer functions are important properties for feedback control systems (Doyle *et al.*, 1992). T shows bandwidth and robust stability for multiplicative perturbation. S shows sensivity for disturbance and tracking property. For 2-DOF control systems for reference tracking, the transfer functions for the reference are important. Here we use G_{yr} and G_{er} instead of T and S. G_{er} is the transfer function from the reference to the error $e = r - y$. We consider the following problem.

Problem 1. [Feedforward Controller Design 1] Given a plant P, a feedback controller C and weights W_1, W_2, find L, Q_1 which infimizes the norm $\left\| \begin{bmatrix} W_1 G_{yr} \\ W_2 G_{er} \end{bmatrix} \right\|_\infty$. The transfer functions G_{yr} and G_{er} are given as

$$\begin{aligned} G_{yr} &= P(I - CP)^{-1}\tilde{V}^{-1} \\ G_{er} &= I - P(I - CP)^{-1}\tilde{V}^{-1}. \end{aligned}$$

2.2 *Parameterization with Coprime Factorization*

First, we will see that the problem can be converted to an LMI optimization problem. The transfer function $G_{yr} = P(I - CP)^{-1}\tilde{V}^{-1}$ can be represented as

$$G_{yr} := \left[\begin{array}{c|c} A_T & B_T \\ \hline C_T & D_T \end{array}\right]$$

where

$$A_T = \begin{bmatrix} A_P + B_P D_C R_{PC}^{-1} C_P, & -B_P R_{CP}^{-1} C_C \\ -B_C R_{PC}^{-1} C_P, & A_C + B_C D_P R_{CP}^{-1} C_C \end{bmatrix},$$
$$B_T = \begin{bmatrix} B_P R_{CP}^{-1} Q_1^{-1} \\ (-B_C D_P R_{CP}^{-1} + L) Q_1^{-1} \end{bmatrix},$$
$$C_T = \begin{bmatrix} R_{PC}^{-1} C_P & -D_P R_{CP}^{-1} C_C \end{bmatrix},$$
$$D_T = D_P R_{CP}^{-1} Q_1^{-1},$$
$$R_{CP} = I - D_C D_P, \ R_{PC} = I - D_P D_C.$$

The transfer function $\begin{bmatrix} W_1 P(I - CP)^{-1}\tilde{V}^{-1} \\ W_2(I - P(I - CP)^{-1}\tilde{V}^{-1}) \end{bmatrix}$ is realized as

$$\begin{aligned} &\begin{bmatrix} W_1 P(I - CP)^{-1}\tilde{V}^{-1} \\ W_2(I - P(I - CP)^{-1}\tilde{V}^{-1}) \end{bmatrix} \\ &= \left[\begin{array}{ccc|c} A_T & 0 & 0 & B_T \\ B_{W1} C_T & A_{W1} & 0 & B_{W1} D_T \\ -B_{W2} C_T & 0 & A_{W2} & B_{W2}(I - D_T) \\ \hline D_{W1} C_T & C_{W1} & 0 & D_{W1} D_T \\ -D_{W2} C_T & 0 & C_{W2} & D_{W2}(I - D_T) \end{array}\right] \\ &:= \left[\begin{array}{c|c} A_{ST} & B_{ST} \\ \hline C_{ST} & D_{ST} \end{array}\right]. \end{aligned}$$

Problem 1 can be represented with LMI (Boyd *et al.*, 1994) as,

$$\inf_{Q_1, L_1, X = X^T > 0} \gamma \text{ such that}$$
$$\begin{bmatrix} A_{ST}^T X + X A_{ST} & X B_{ST} & C_{ST}^T \\ B_{ST}^T X & -\gamma I & D_{ST}^T \\ C_{ST} & D_{ST} & -\gamma I \end{bmatrix} < 0.$$

This can be modified as

$$\inf_{Q_1, L_1, X=X^T>0} \gamma \text{ such that}$$
$$\begin{bmatrix} XA_{ST}^T + A_{ST}X & B_{ST} & C_{ST}^T X \\ B_{ST}^T & -\gamma I & D_{ST}^T \\ C_{ST}X & D_{ST} & -\gamma I \end{bmatrix} < 0. \tag{2}$$

Though it is not LMI with L and Q_1, it becomes LMI with $L_1 = LQ_1^{-1}$ and Q_1^{-1}. Furthermore, since γ and X are also linear in the matrix inequality, the optimal L_1, Q_1 are obtained and the fixed-order controller is obtained with the LMI optimization method.

Next, we will see that the order of the 2-DOF controller is the same as the feedback controller. The controller is given as follows:

$$u = \tilde{V}^{-1}(\tilde{U}y + r). \tag{3}$$

The state-space representation of the controller C_2 $\left(u = C_2 \begin{bmatrix} r \\ y \end{bmatrix}\right)$, is given as follows:

$$C_2 = \left[\begin{array}{c|cc} A_C & -LQ_1^{-1} & B_C \\ \hline C_C & Q_1^{-1} & D_C \end{array}\right]. \tag{4}$$

Hence, the order of the whole controller is n_C which is the same as the feedback controller C.

Remark 2. The similar method can be applied to $\tilde{U}, \tilde{V}$ of which orders are higher than C. Assume that the controller C has stable and uncontrollable modes. Such C can be written as

$$\left[\begin{array}{c|c} \bar{A}_C & \bar{B}_C \\ \hline \bar{C}_C & \bar{D}_C \end{array}\right] = \left[\begin{array}{cc|c} A_C & 0 & B_C \\ 0 & A_Q & 0 \\ \hline C_C & C_Q & D_C \end{array}\right], \tag{5}$$

where $A_Q \in \mathcal{R}^{n_Q \times n_Q}$ is stable and uncontrollable.

Problem 3. [Feedforward Controller Design 2] Given a plant P, a feedback controller C and weights W_1, W_2, find L, Q_1 which infimize the following

$$\max \{\|W_1 P(I - CP)^{-1}\tilde{V}^{-1}\|_\infty, \\ \|W_2[I - P(I - CP)^{-1}\tilde{V}^{-1}]\|_\infty\}.$$

This can be solved because the cost function can be represented with 2 LMIs and then they can be regarded as a single large LMI. The cost function can be expected to be less conservative, but the dimension of the variable increases because 2 LMIs are considered.

The design method can be extended to other kinds of problems such as simultaneous control design and decentral control design (Miyamoto and Vinnicombe, 1997) since the method is based on the LMI representation.

3. CONTROL SYSTEM FOR SATURATION NONLINEARITY

In (Kothare *et al.*, 1994),(Miyamoto and Vinnicombe, 1996), it is shown that any anti-windup controllers can be represented with coprime factorization of $C = \tilde{V}^{-1}\tilde{U}$ (Fig. 2). This can also be regarded as a 2-DOF control design as the anti-windup property can be changed without changing the feedback property.

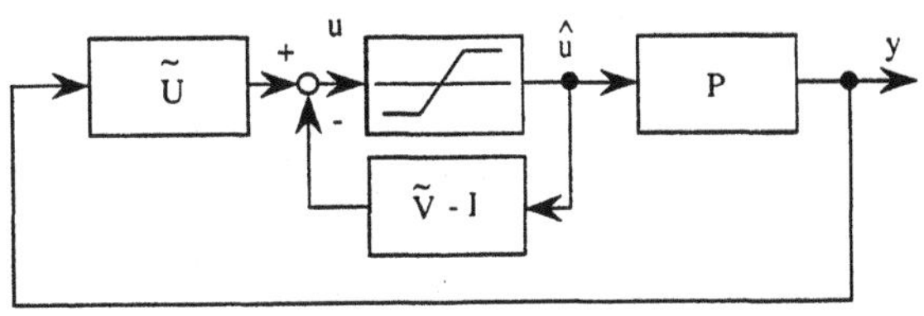

Fig. 2. Anti-windup design with CF

3.1 *Problem Formulation*

Stability and performance can be taken into account by regarding the saturation as a perturbation (Figure 3).

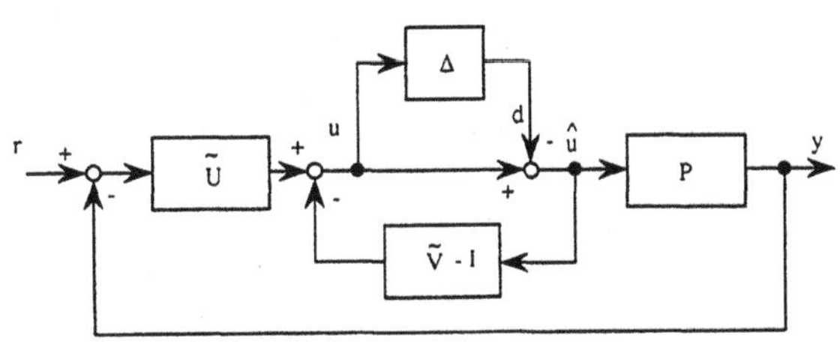

Fig. 3. Stability against saturation

As written in (Miyamoto and Vinnicombe, 1996), the following three transfer functions are important for anti-windup control design.

G_{ud}: transfer function from d to u
$G_{\hat{u}d}$: transfer function from d to $\hat{u}$
G_{yd}: transfer function from d to y

The transfer function $G_{\hat{u}d}, G_{yd}$ and G_{ud} can be written as follows:

$$\begin{aligned} G_{ud} &= -I + (I - CP)^{-1}\tilde{V}^{-1} \\ G_{\hat{u}d} &= (I - CP)^{-1}\tilde{V}^{-1} \\ G_{yd} &= P(I - CP)^{-1}\tilde{V}^{-1}. \end{aligned}$$

As perturbation satisfies $|\Delta| < 1$, stability is guaranteed if $\|G_{ud}\|_\infty < 1$. Performance when saturation occurs, is better with smaller G_{yd} and $G_{\hat{u}d}$.

In this paper, we consider the following problem (Miyamoto and Vinnicombe, 1996).

Problem 4. [Anti-windup Control Design 1]
Given a plant P, a controller C and frequency weights W_1, W_2, find $\tilde{U}, \tilde{V}$ such that

$$\inf_L \left\| \begin{bmatrix} W_1 G_{\hat{u}d} \\ W_2 G_{yd} \end{bmatrix} \right\|_\infty \text{ subject to } \|G_{ud}\|_\infty < 1.$$

3.2 *Parameterization with Coprime Factorization*

The design problem can be formulated as a problem of factorizing the controller C with fixed-order $\tilde{U}$ and $\tilde{V}$ which are written as

$$[-\tilde{U}\ \tilde{V}] = \left[\begin{array}{c|cc} A_C + LC_C & -(B_C + LD_C) & L \\ \hline C_C & -D_C & I \end{array}\right] \quad (6)$$

for stable $A_C + LC_C$ (Figure 2).

The transfer functions $G_{ud}, G_{\hat{u}d}, G_{yd}$ are realized as follows:

$$\begin{aligned} G_{ud} &= \left[\begin{array}{c|c} A_{ud} & B_{ud} \\ \hline C_{ud} & D_{ud} \end{array}\right], \\ G_{\hat{u}d} &= \left[\begin{array}{c|c} A_{ud} & B_{ud} \\ \hline C_{ud} & -I + D_{ud} \end{array}\right], \quad (7) \\ G_{yd} &= \left[\begin{array}{c|c} A_{ud} & B_{ud} \\ \hline C_{yd} & D_{yd} \end{array}\right], \end{aligned}$$

where

$$A_{ud} = \begin{bmatrix} A_P + B_P D_C R_{PC}^{-1} C_P, & -B_P R_{CP}^{-1} C_C \\ -B_C R_{PC}^{-1} C_P, & A_C + B_C D_P R_{CP}^{-1} C_C \end{bmatrix},$$

$$B_{ud} = \begin{bmatrix} B_P R_{CP}^{-1} \\ -B_C D_P R_{CP}^{-1} + L \end{bmatrix},$$

$$C_{ud} = \begin{bmatrix} D_C R_{PC}^{-1} C_P & -R_{CP}^{-1} C_C \end{bmatrix},$$

$$C_{yd} = \begin{bmatrix} R_{PC}^{-1} C_P & -D_P R_{CP}^{-1} C_C \end{bmatrix},$$

$$D_{ud} = -I + R_{CP}^{-1},\ D_{yd} = D_P R_{CP}^{-1},$$

$$R_{CP} = I - D_C D_P,\ R_{PC} = I - D_P D_C.$$

It can be seen that the parameter matrix L is only in the "B" matrix of the system. Hence, the condition $\|G_{ud}\|_\infty < 1$ can be represented with an LMI. The problem $\inf_L \left\| \begin{bmatrix} W_1 G_{\hat{u}d} \\ W_2 G_{yd} \end{bmatrix} \right\|_\infty$ can also be written with an LMI since

$$\begin{aligned} &\begin{bmatrix} W_1 (I - CP)^{-1} \tilde{V}^{-1} \\ W_2 P (I - CP)^{-1} \tilde{V}^{-1} \end{bmatrix} \\ &= \left[\begin{array}{ccc|c} A_{ud} & 0 & 0 & B_{ud} \\ B_{W1} C_{ud} & A_{W1} & 0 & B_{W1}(I - D_{ud}) \\ B_{W2} C_{yd} & 0 & A_{W2} & B_{W2} D_{yd} \\ \hline D_{W1} C_{ud} & C_{W1} & 0 & D_{W1}(I - D_{ud}) \\ D_{W2} C_{yd} & 0 & C_{W2} & D_{W2} D_{yd} \end{array}\right] \\ &:= \left[\begin{array}{c|c} A_{\hat{u}yd} & B_{\hat{u}yd} \\ \hline C_{\hat{u}yd} & D_{\hat{u}yd} \end{array}\right], \end{aligned}$$

and the parameter matrix L is only in the "B" matrix of the system. Then, the problem can be written as

$$\inf_{L, X = X^T > 0, Y = Y^T > 0} \gamma \text{ such that}$$

$$\begin{bmatrix} X A_{\hat{u}yd}^T + A_{\hat{u}yd} X & B_{\hat{u}yd} & C_{\hat{u}yd}^T X \\ B_{\hat{u}yd}^T & -\gamma I & D_{\hat{u}yd}^T \\ C_{\hat{u}yd} X & D_{\hat{u}yd} & -\gamma I \end{bmatrix} < 0,$$

$$\begin{bmatrix} Y A_{ud}^T + A_{ud} Y & B_{ud} & C_{ud}^T Y \\ B_{ud}^T & -I & D_{ud}^T \\ C_{ud} Y & D_{ud} & -I \end{bmatrix} < 0.$$

In this problem, L, X, Y and γ are linear and the LMI optimization can be applied.

The major advantage of this method is that the order of the controller is n_C which is the same as the feedback controller C, that is, no additional dynamics is necessary for an anti-windup controller. This is because the controller $C_2 \left(u = C_2 \begin{bmatrix} y \\ \hat{u} \end{bmatrix} \right)$, is given as follows

$$C_2 = \left[\begin{array}{c|cc} A_C + LC_C & B_C + LD_C & -L \\ \hline C_C & D_C & 0 \end{array}\right]. \quad (8)$$

The disadvantage of this method is that the existence of a solution is not guaranteed since the parameter space has less freedom. Note that it is always guaranteed in the design method studied in (Miyamoto and Vinnicombe, 1996).

As shown in the previous section, the control system of which degree is higher than n_P can be considered. It can also be extended to a simultaneous control design for multiple plants.

Note that the following problem where two transfer functions $G_{\hat{u}d}$ and G_{ud} are considered separately, can also be solved.

Problem 5. [Anti-windup Control Design 2]
Given a plant P, a controller C and frequency weights W_1, W_2, find L such that

$$\inf_L \max\{\|W_1 G_{\hat{u}d}\|_\infty, \|W_2 G_{yd}\|_\infty\}$$
$$\text{subject to } \|G_{ud}\|_\infty < 1.$$

4. 3-DOF CONTROL SYSTEM

In this section, we consider 3-DOF control systems. In the previous two sections, we considered two 2-DOF control systems. One is a system for feedback property and tracking property, and the other is a system for feedback property and anti-windup property. Since tracking property and anti-windup property are independent of each other, they can be combined and a 3-DOF control systems can be considered. Here, it is assumed that a feedback controller (C), a compensator for tracking ($\tilde{U}_1, \tilde{V}_1$) and an anti-windup controller ($\tilde{U}_2, \tilde{V}_2$) are already given as

$$[-\tilde{U}_1\ \tilde{V}_1] = \left[\begin{array}{c|cc} A_C + L_1 C_C & -(B_C + L_1 D_C) & L_1 \\ \hline Q_3 C_C & -Q_3 D_C & Q_3 \end{array}\right], \quad (9)$$

$$[-\tilde{U}_2\ \tilde{V}_2] = \left[\begin{array}{c|cc} A_C + L_2 C_C & -(B_C + L_2 D_C) & L_2 \\ \hline C_C & -D_C & I \end{array}\right]. \quad (10)$$

By constructing the following control input, reference tracking property and anti-windup property can be chosen independently (Figure 4),

$$u = \tilde{V}_2 \tilde{V}_1^{-1}(r + \tilde{U}_1 y) - (\tilde{V}_2 - I)\hat{u}. \quad (11)$$

The term $\tilde{U}_2$ does not appear in the above equation, but it is equivalent to $\tilde{V}_2\tilde{V}_1^{-1}\tilde{U}_1$ (Figure 4). The transfer functions $G_{yr}, G_{er}, G_{ud}, G_{\hat{u}d}$ and G_{yd} are given as follows:

$$\begin{aligned} G_{yr} &= P(I - CP)^{-1}\tilde{V}_1^{-1} \\ G_{er} &= I - P(I - CP)^{-1}\tilde{V}_1^{-1} \\ G_{ud} &= -I + (I - CP)^{-1}\tilde{V}_2^{-1} \\ G_{\hat{u}d} &= (I - CP)^{-1}\tilde{V}_2^{-1} \\ G_{yd} &= P(I - CP)^{-1}\tilde{V}_2^{-1}. \end{aligned}$$

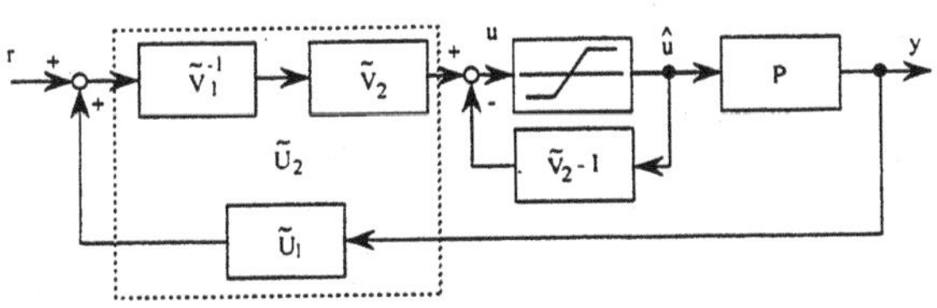

Fig. 4. 3-DOF control system with $\tilde{U}_1, \tilde{U}_2, \tilde{V}_1, \tilde{V}_2$

The state-space representation of the controller $C_3 \left(u = C_3 \begin{bmatrix} r \\ y \\ \hat{u} \end{bmatrix} \right)$, is given as

$$C_3 = \left[\begin{array}{c|ccc} A_C + L_2 C_C & (-L_1 + L_2)Q_3^{-1} & B_C + L_2 D_C & -L_2 \\ \hline C_C & Q_3^{-1} & D_C & 0 \end{array}\right] \quad (12)$$

where Q_3 is an invertible matrix. The transfer function from $\begin{bmatrix} y \\ \hat{u} \end{bmatrix}$ to u is the same as C_2 in (8) since $\tilde{U}_2 = \tilde{V}_2\tilde{V}_1^{-1}\tilde{U}_1$. The transfer function from $\begin{bmatrix} r \\ y \end{bmatrix}$ to u when $u = \hat{u}$, is the same as C_2 in (4).

5. EXAMPLE

In this section, we consider the problem stated in (Doyle *et al.*, 1987), but the plant is modified to a strictly proper plant and a loop shaping controller with a higher-order is used instead of a simple integrator. The plant P is given as

$$P = \frac{(s^2 + 2\zeta_1\omega_1 + \omega_1^2)(s^2 + 2\zeta_2\omega_1 + \omega_1^2)}{(s^2 + 2\zeta_1\omega_2 + \omega_2^2)(s^2 + 2\zeta_2\omega_2 + \omega_2^2)} \times \frac{2}{s + 10}$$

where the parameters ω_i, ζ_i are given as $\omega_1 = 0.2115$, $\omega_2 = 0.0473$, $\zeta_1 = 0.3827$, $\zeta_2 = 0.9239$.

The pre-compensator W is chosen as $W = \frac{5}{s}$. The optimal controller C_1 for the weighted plant PW is obtained with the loop shaping method (McFarlane and Glover, 1992) and the robust stability margin b_{PW,C_1} becomes 0.5529. The order of the controller $C(= WC_1)$ becomes 6th. The reference is a signal of +1 between 0 and 4 sec and -1 after 4 sec, which is the same as (Doyle *et al.*, 1987).

The response with saturation nonlinearity and without anti-windup scheme is also shown in Figure 5. It is not so easy to design an AW controller for this problem. It is shown in (Doyle *et al.*, 1987) that CAW (Conventional Antiwindp) scheme does not work for this problem.

In the following, we consider 2 kinds of 2-DOF control systems.

[Case 1: Anti-windup Control (based on CF parameterization)]

We consider Problem 5 with the same weights as Case 1. However, in this case, $\|I - G_{ud}\|_\infty$ cannot be made less than 1 with any gain L. So, we relax the constraint as $\|I - G_{ud}\|_\infty < 1.01$.

Note that the global stability is no more guaranteed in this problem. The infimum of the cost function is given as 0.0479. The response with the obtained L for the reference [+1 -1] is depicted in Figure 6. The reference is added at the plant output in Figure 2. It can be seen that it works well for this reference.

[Case 2: Feedforward Control]

We consider Problem 3 with the following weights,

$$W_1 = \frac{0.5(s + 0.1)}{0.01(s + 10)}, \quad W_2 = \frac{0.5(s + 0.1)}{s + 0.001},$$

so that the bandwidth of G_{yr} becomes lower. By solving the LMI-based method, we obtain $\tilde{U}, \tilde{V}$ which attain the infimum of 0.5374. The response for the reference [+1 -1] is depicted in Figure 7. It can be seen that the response is improved by changing the transfer function G_{yr}.

However, if the disturbance with high frequency is injected at the plant output y, the response will be deteriorated or the system may be unstable. By introducing a 3-DOF control system proposed in Section 4, it is possible to change the transfer function for the reference while maintaining the anti-windup property.

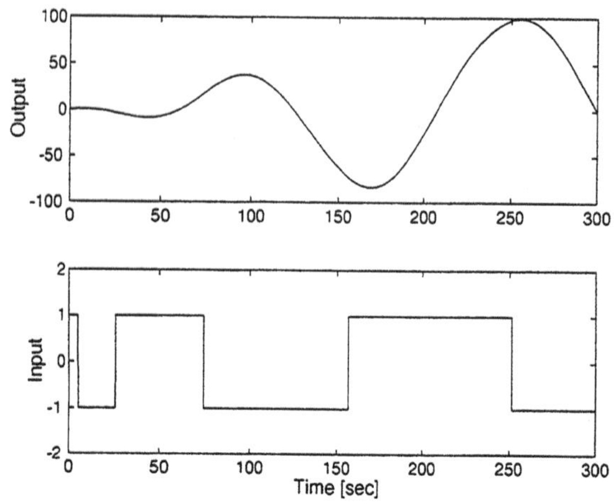

Fig. 5. Step response (without AW)

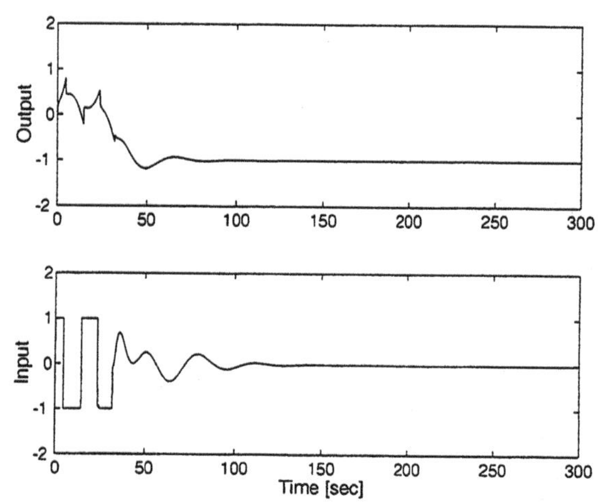

Fig. 6. Step response (Case 1)

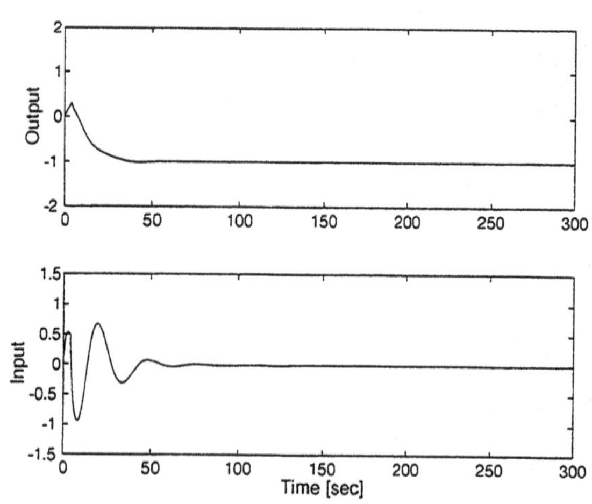

Fig. 7. Step response (Case 2)

6. SUMMARY

In this paper, we presented 2-DOF control system design. First, we considered control systems for reference tracking, where the feedforward controllers are designed for the given feedback controller. Second, anti-windup controllers were designed for the given feedback controller. For both systems, the design can be formulated as a problem of factorizing fixed-order C with $\tilde{U}, \tilde{V}$. This problem can be regarded as a problem of choosing the matrix L in the fixed-order coprime factorization and it can be solved with the LMI optimization method.

Then, we proposed 3-DOF control systems by combining the above two 2-DOF control systems. This is possible as a reference tracking property and an anti-windup property are independent. Last, design examples of 2-DOF control systems were given to demonstrate our design method.

The main advantage of the LMI-based design methods is that the order of the controller is the same as the feedback controller even for the 3-DOF control systems. Furthermore, it can be extended to a simultaneous design for multiple plants and a decentralized control design.

7. REFERENCES

Boyd, S., L.El Ghaoui, E. Feron and V. Balakrishnan (1994). *Linear Matrix Inequalities in Systems and Control Theory.* SIAM books.

Doyle, J.C., B.A. Francis and A.R. Tannenbaum (1992). *Feedback Control Theory.* Macmillan.

Doyle, J.C., R.S. Smith and D.F. Enns (1987). Control of plants with input saturation nonlinearities. In: *Proc. of ACC.* Minneapolis. pp. 1034–1039.

Kothare, M.V., P.J. Campo, M. Morari and C.N. Nett (1994). A unified framework for the study of anti-windup designs. *Automatica* **30**(12), 1869–1883.

McFarlane, D.C. and K. Glover (1992). A loop shaping design procedure using H_∞ synthesis. *IEEE Transaction on Automatic Control* **37**(6), 759–769.

Miyamoto, S. and G. Vinnicombe (1996). Robust control of plants with saturation nonlinearity based on coprime factor representations. In: *35th IEEE CDC* . Kobe. pp. 2838–2040.

Miyamoto, S. and G. Vinnicombe (1997). Fixed structure H_∞ control design for multiple plants — a coprime factorization and LMI approach. In: *4th ECC.* Brussels.

Vidyasagar, M. (1985). *Control System Synthesis: A Fatorization Approach.* MIT Press.

Zhou, K., J.C. Doyle and K. Glover (1995). *Robust and Optimal Control.* Prentice-Hall.

Sensitivity Minimization of MIMO Systems with A Stable Controller

Abdul-Wahid A. Saif
King Fahd University of Petroleum & Minerals
Systems Engineering Department
P.O.Box 1626
Dhahran 31261
Saudi Arabia
E-mail: awsaif@dpc.kfupm.edu.sa

Da-Wei Gu
University of Leicester
Department of Engineering
Leicester LE 7RH
U.K.
E-mail: dwg@sun.engg.le.ac.uk

Abstract

In this paper, the sensitivity minimization problem of multivariable linear time-invariant, biproper or strictly proper, systems using a stable controller is considered. The problem is reduced to a 1-block or 2-block $\mathcal{H}_\infty$ - optimization problem in the case of square or nonsquare systems, respectively. The solution of this optimization problem, if satisfying an upper bound constraint which depends on an initial choice of a unimodular transfer function in $\Re\mathcal{H}_\infty$, would be the Youla parameter and hence a stable stabilizing controller to the original system. A search procedure for an initial choice of an unimodular function is discussed.

Keywords: *Strong Stabilization, Sensitivity minimization, 1-Block/2-Block problems.*

1 Introduction

The strong stabilization problem (StSP) is defined as that of finding a stable controller that stabilizes a given plant. A study of strong stabilizability is mainly of practical interest for control engineers. If a plant is stabilized using a stable compensator then the resulting closed-loop stable transfer function matrix from the input disturbance to the output, or equivalently, the control signal to the output, would have the same number of right half plane zeros as the original plant, (Vidyasagar, 1985). On the other hand, stabilization using an unstable compensator always introduces additional right half plane zeros into this closed-loop transfer function matrix beyond those of the original plant. As it is known that the right half plane zeros of a system affect its ability to track reference signals and/or to reject disturbances, it is preferable to use a stable stabilizing compensator whenever possible. In addition, several other problems in reliable stabilization are related to strong stabilizability, such as the simultaneous stabilization problem.

The StSP of linear time-invariant systems was first addressed and solved for SISO systems in (Youla, *et al.*,1974), who presented a tractable condition, known as the parity interlacing property (PIP), to check whether a given plant is strongly stabilizable or not. While the PIP is an elegant result, it does not indicate how to find such a controller if it exists. Some attempts to provide such a stable controller were given in (Youla , *et al.*, 1974; Ganesh and Person, 1989; Dorato, *et al.*, 1989; Ito, *et al.*, 1993) for SISO systems using the Nevanlinna-Pick interpolation technique. References (Jacobus, *et al.*, 1990; Halevi, *et al.*, 1991; William and Bernstein, 1993; Halevi, 1994) suggested some modifications in the Riccati equations that would result in a stable controller. While the modifications are different, all these methods result in nonstandard Riccati equation(s) with solutions that are difficult to obtain and, furthermore, not even guaranteed to exist. A greater restriction on the zeros of the sought stable controller was considered in (Wei, 1990). In (Sideris and Safonov, 1985), the problem was addressed for MIMO systems where a finite dimensional optimization problem was formulated to satisfy the Pick interpolation condition. In (Saif, *et al.*, 1997), the StSP was formulated as an $\mathcal{H}_\infty$ optimization problem, while the matrix Nevanlinna-Pick interpolation technique was used to solve the StSP in (Saif, *et al.*, 1998a).

In practice, stability alone is never sufficient and a procedure just for finding a stable stabilizing controller is not of much use by itself. In this direction, the sensitivity minimization problem using stable controller was addressed in (Ito, *et al.*, 1993) for SISO systems using the Nevanlinna-Pick interpolation technique and in (Nakayama, 1995, Saif,A., 1998b,c) by tuning the free parameter of the $\mathcal{H}_\infty$ control problem to get a stable controller.

In this paper, the sensitivity minimization problem of MIMO systems using stable compensators is considered. The problem will be formulated as an $\mathcal{H}_\infty$ one-block or two-block problem. A search procedure for a unimodular transfer function matrix that is needed in this work is proposed with the help of LMI technique.

This paper is organized as follows. The statement of the problem is given in Section 2. Its formulation as a one-block or two-block problem is given in Section 3. Section 4 present a state-space procedure to search for a stable transfer matrix function whose inverse is stable such that a certain condition in the solution of the sensitivity optimization problem is satisfied. Simulation results will be reported elsewhere. The case when the system is strictly proper, is treated in Section 5. Section 6 concludes this paper.

In the paper $\Re\mathcal{H}_\infty$ denotes the set of all real rational matrices whose elements are stable and proper. Other notations will be clear from the context.

2 Statement of the Problem

Define the plant $G(s)$ and the controller $K(s)$ in the right and left coprime factorization form as

$$G = NM^{-1} = \tilde{M}^{-1}\tilde{N}, \text{and } K = -YX^{-1} = -\tilde{X}^{-1}\tilde{Y} \quad (2.1)$$

The transfer function matrices N, M, X, Y, $\tilde{N}$, $\tilde{M}$, $\tilde{X}$ and $\tilde{Y}$ are all in $\Re\mathcal{H}_\infty$ satisfying the double Bezout identity

$$\begin{bmatrix} \tilde{X} & -\tilde{Y} \\ -\tilde{N} & \tilde{M} \end{bmatrix} \begin{bmatrix} M & Y \\ N & X \end{bmatrix} = I \quad (2.2)$$

The Youla parameterization of all stabilizing controllers is given by

$$\begin{aligned} K &= -(Y - MQ)(X - NQ)^{-1} && (2.3) \\ &= -(\tilde{X} - Q\tilde{N})^{-1}(\tilde{Y} - Q\tilde{M}) && (2.4) \end{aligned}$$

where $Q(s)$ is the Youla free parameter in $\Re\mathcal{H}_\infty$. All realizable sensitivity matrices

$$S(s) = (I + GK)^{-1} \quad (2.5)$$

are parametrized by

$$S = (X - NQ)\tilde{M} \quad (2.6)$$

The transfer function matrices N,M,X,Y,$\tilde{N}$,$\tilde{M}$,$\tilde{X}$ and $\tilde{Y}$ each can be expressed in state-space form by choosing a real matrix F such that $A + BF$ is stable and a real matrix H so that $A + HC$ is stable, then

$$\begin{bmatrix} M(s) & Y(s) \\ N(s) & X(s) \end{bmatrix} := \left[\begin{array}{c|cc} A+BF & B & -H \\ \hline F & I & 0 \\ C+DF & D & I \end{array}\right]$$

$$\begin{bmatrix} \tilde{X} & -\tilde{Y} \\ -\tilde{N} & \tilde{M} \end{bmatrix} := \left[\begin{array}{c|cc} A+HC & -(B+HD) & H \\ \hline F & I & 0 \\ C & -D & I \end{array}\right]$$

The problem is to find $Q \in \Re\mathcal{H}_\infty$ that minimize the sensitivity function S given by (2.6) and make the controller K in (2.3) stable.

From (2.3), the stability of the controller is equivalent to the requirement that

$$\det(X - NQ) \neq 0 \quad \text{at} \quad \mathrm{Re}(s) > 0 \quad (2.7)$$

This problem was addressed in (Sideris and Safonov, 1985) but the procedure presented there to solve it was not tractable. In the next section, the above problem is reformulated as an H_∞- optimization problem.

3 Re-formulation and Solution of the Problem

The sensitivity optimization problem and the strong stabilization problem can be formulated as the following constrained H_∞-optimization problem:

$$\min_{Q \in \Re\mathcal{H}_\infty} \left\|(X - NQ)\tilde{M}\right\|_\infty \quad (3.1)$$

subject to

$$X - NQ = U \quad (3.2)$$

where U is an unimodular function in $\Re\mathcal{H}_\infty$.

In this section re-formulation of the above problem will be given along with its solution for two cases, square and nonsquare bi-proper plants.

3.1 Square Systems

Assume the given system is square and bi-proper, i.e. has full rank at infinity. Let the inner-outer factorization of N and $\tilde{M}$ be denoted as $N = N_i N_o$ and $\tilde{M} = \tilde{M}_o \tilde{M}_i$, then the problem is transformed to

$$\begin{aligned} &\min_{Q \in \Re\mathcal{H}_\infty} \left\|(X - N_i N_o Q)\tilde{M}_o \tilde{M}_i\right\|_\infty \\ = &\min_{Q \in \Re\mathcal{H}_\infty} \left\|X\tilde{M}_o - N_i N_o Q \tilde{M}_o\right\|_\infty \\ = &\min_{\bar{Q} \in \Re\mathcal{H}_\infty} \left\|\bar{X} - N_i \bar{Q}\right\|_\infty \quad (3.3) \end{aligned}$$

subject to

$$\bar{X} - N_i \bar{Q} = \bar{U} \quad (3.4)$$

where $\bar{X} = X\tilde{M}_o$, $\bar{Q} = N_o Q \tilde{M}_o$ and $\bar{U}$ is unimodular. This constrained optimization problem is re-formulated as follows: Denote the unimodular function $\bar{U}, \bar{U}^{-1} \in \Re\mathcal{H}_\infty$ as $\bar{U} = R + \gamma T$, where

$R \in \Re\mathcal{H}_\infty$ and T is a free unimodular function in $\Re\mathcal{H}_\infty$. If T is chosen such that

$$\|RT^{-1}\|_\infty < \gamma \tag{3.5}$$

it can be shown that $\bar{U}$ is unimodular (Saif, *et al.*, 1997). For a given $T(s)$,

$$\|(\bar{X} - N_i\bar{Q})T^{-1} - \gamma I\|_\infty \geq \|(\bar{X} - N_i\bar{Q})\,T^{-1}\|_\infty - \gamma \tag{3.6}$$

Therefore, a solution $\bar{Q} \in \Re\mathcal{H}_\infty$ of the following inequality

$$\|(\bar{X}T^{-1} - \gamma I) - N_i\bar{Q}T^{-1}\|_\infty < \gamma \tag{3.7}$$

will satisfy (3.5) and gaurantee that

$$\|(\bar{X} - N_i\bar{Q})\,T^{-1}\|_\infty \leq 2\gamma$$

and, then

$$\|\bar{X} - N_i\bar{Q}\|_\infty \leq 2\gamma\,\|T\|_\infty.$$

That will give a sub-optimal minimum for (3.3) subject to (3.4), given T is known.

In the following, the solution of (3.7) will be transformed into a constrained 1-block problem.

Since multiplication by an inner function does not change the infinity norm, the inequality (3.7) becomes

$$\min_{\bar{Q} \in \Re\mathcal{H}_\infty} \|N_i^*(\bar{X}T^{-1} - \gamma I) - \bar{Q}T^{-1}\|_\infty < \gamma \tag{3.8}$$

Taking the stable projection of $N_i^*(\bar{X}T^{-1} - \gamma I)$ as

$$N_i^*(\bar{X}T^{-1} - \gamma I) = R_s + R_u \tag{3.9}$$

where $R_s = [N_i^*(\bar{X}T^{-1} - \gamma I)]_+$ and $R_u = [N_i^*(\bar{X}T^{-1} - \gamma I)]_-$ are the stable and anti-stable (all poles are in the right half of the complex plane) projections of $N_i^*(\bar{X}T^{-1} - \gamma I)$, respectively. Then the inequality (3.8) is further reduced into

$$\min_{\widehat{Q} \in \Re\mathcal{H}_\infty} \|R_u - \widehat{Q}\|_\infty < \gamma \tag{3.10}$$

where $\widehat{Q} = \{-R_s + \bar{Q}T^{-1}\}$.

The left hand side is a 1-block model matching problem. The condition on the existence of a solution to the inequality (3.10) is known, from the Nehari's Theorem (Glover 1984), as $\gamma \geq \|\Gamma_{R_u}\|$, where $\|\Gamma_{R_u}\|$ is the Hankel norm of $R_u(-s)$. Once a solution $\widehat{Q}$ is obtained the required Q may be recovered as

$$Q = N_o^{-1}\left(R_s + \widehat{Q}\right)T\tilde{M}_o^{-1} \tag{3.11}$$

3.2 Non-Square Systems

Assume the plant $G(s) \in \Re(s)^{r\times m}$, where $r > m$. Then N_i in (3.7) is in $\Re\mathcal{H}_\infty^{r\times m}$. Let $N_i^\perp$ be the complementary inner part of N_i such that $\left[\, N_i \;\; N_i^\perp \,\right] \in \Re\mathcal{H}_\infty^{r\times r}$ is square and inner. Denote $\bar{X}_\gamma = (\bar{X}T^{-1} - \gamma I)$. Similarly as in §3.1, we have the following inequality

$$\min_{\bar{Q} \in \Re\mathcal{H}_\infty} \left\|\bar{X}_\gamma - \left[\, N_i \;\; N_i^\perp \,\right]\begin{bmatrix} \bar{Q} \\ 0 \end{bmatrix} T^{-1}\right\|_\infty < \gamma \tag{3.12}$$

which can be transformed as in the square case into

$$\begin{aligned}
&\min_{\bar{Q} \in \Re\mathcal{H}_\infty} \left\|\left[\, N_i \;\; N_i^\perp \,\right]^* \bar{X}_\gamma - \begin{bmatrix} \bar{Q}T^{-1} \\ 0 \end{bmatrix}\right\|_\infty < \gamma \\
&\min_{\bar{Q} \in \Re\mathcal{H}_\infty} \left\|\begin{Bmatrix} N_i^* \\ (N_i^\perp)^* \end{Bmatrix} \bar{X}_\gamma - \begin{bmatrix} \bar{Q}T^{-1} \\ 0 \end{bmatrix}\right\|_\infty < \gamma \\
&\min_{\bar{Q} \in \Re\mathcal{H}_\infty} \left\|\begin{Bmatrix} N_i^*\bar{X}_\gamma \\ (N_i^\perp)^*\bar{X}_\gamma \end{Bmatrix} - \begin{bmatrix} \bar{Q}T^{-1} \\ 0 \end{bmatrix}\right\|_\infty < \gamma
\end{aligned} \tag{3.13}$$

Let $R_1 = N_i^*\bar{X}_\gamma$ and $R_2 = (N_i^\perp)^*\bar{X}_\gamma$, and $\widehat{Q} = \bar{Q}T^{-1}$. Then (3.13) becomes

$$\min_{\widehat{Q} \in \Re\mathcal{H}_\infty} \left\|\begin{matrix} R_1 - \widehat{Q} \\ R_2 \end{matrix}\right\|_\infty < \gamma \tag{3.14}$$

where R_1 and R_2 are not in general anti-stable. This problem may be reduced into an ordinary 2-block problem using the following procedure (Saif *et al*, 1994). Find the stable and anti-stable projection of R_1 and the multiplicative inner/anti-stable projection of R_2. Then $R_1 = N_i^*\bar{X}_\gamma = [N_i^*\bar{X}_\gamma]_- + [N_i^*\bar{X}_\gamma]_+ = R_{1u} + R_{1s}$ and $R_2 = (N_i^\perp)^*\bar{X}_\gamma = R_{2si}R_{2u}$, where R_{2si} is stable and inner, and R_{2u} is anti-stable. Then (3.14) becomes

$$\min_{\widehat{Q} \in \Re\mathcal{H}_\infty} \left\|\begin{matrix} R_{1u} - \left(-R_{1s} + \widehat{Q}\right) \\ R_{2si}R_{2u} \end{matrix}\right\|_\infty < \gamma \tag{3.15}$$

Furthermore,

$$\min_{\widehat{Q} \in \Re\mathcal{H}_\infty} \left\|\begin{bmatrix} I & 0 \\ 0 & R_{2si}^* \end{bmatrix}\begin{bmatrix} R_{1u} - (R_{1s} + \widehat{Q}) \\ R_{2si}R_{2u} \end{bmatrix}\right\|_\infty < \gamma \tag{3.16}$$

where $R_{2si}^*R_{2si} = I$. Let $\check{Q} = \left(-R_{1s} + \widehat{Q}\right)$, then we have the following 2-block problem

$$\min_{\check{Q} \in \Re\mathcal{H}_\infty} \left\|\begin{matrix} R_{1u} - \check{Q} \\ R_{2u} \end{matrix}\right\|_\infty < \gamma \tag{3.17}$$

when solved, Q is recovered as

$$Q = N_o^{-1}\left(-R_{1s} + \check{Q}\right)T\tilde{M}_o^{-1} \tag{3.18}$$

When $r < m$, a dual problem may be obtained by minimizing the sensitivity function $S = (I + KG)^{-1}$ which can be parameterized as $S = M(\tilde{Y} - Q\tilde{N})$.

The problem that needs to be considered is a searching procedure for a unimodular function $T(s) \in \Re\mathcal{H}_\infty$ that will yield a solution for (3.10) or (3.14) and this will be addressed in the next section.

4 Searching for Unimodular Function $T(s)$

In this section a tractable procedure to search for unimodular function $T(s) \in \mathcal{RH}_\infty$, that will satisfy inequality (3.7) is proposed. The another case of (3.12) can be considered in a similar way.

Let $\bar{X}(s)$ and $T^{-1}(s)$ be written in state-space form as

$$\bar{X} = \left[\begin{array}{c|c} \bar{A} & \bar{B} \\ \hline \bar{C} & \bar{D} \end{array}\right] \qquad \text{and } T^{-1} = \left[\begin{array}{c|c} A_T & B_T \\ \hline C_T & D_T \end{array}\right] \tag{4.1}$$

Then

$$\bar{X}T^{-1} - \gamma I = \left[\begin{array}{cc|c} \bar{A} & \bar{B}C_T & \bar{B}D_T \\ 0 & A_T & B_T \\ \hline \bar{C} & \bar{D}C_T & \bar{D}D_T - \gamma I \end{array}\right]$$

Using similarity transformation it is easy to show that

$$\bar{X}T^{-1} - \gamma I = \left[\begin{array}{c|c} A_T & B_T \\ \hline -\bar{C} + \bar{D}C_T & \bar{D}D_T - \gamma I \end{array}\right] \tag{4.2}$$

$$\text{if } \left\{ \begin{array}{rcl} A_T & = & \bar{A} - \bar{B}C_T \\ B_T & = & -\bar{B}D_T \end{array} \right.$$

Here C_T is chosen to guarantee the stability of $(\bar{A} - \bar{B}C_T)$ and D_T and γ are free parameters will be chosen to satisfy the following condition

$$\|\Gamma_{R_u}\| \leq \gamma \tag{4.3}$$

A state-space representation of $R_u(s)$ will be given next.

Let N_i^* be denoted by

$$N_i^* = \left[\begin{array}{c|c} A_1 & B_1 \\ \hline C_1 & D_1 \end{array}\right] \tag{4.4}$$

then $N_i^* \left(\bar{X}T^{-1} - \gamma I\right)$ becomes as follows

$$\begin{aligned} & N_i^* \left(\bar{X}T^{-1} - \gamma I\right) \\ = & \left[\begin{array}{c|c} A_1 & B_1 \\ \hline C_1 & D_1 \end{array}\right] \left[\begin{array}{c|c} A_T & B_T \\ \hline -\bar{C} + \bar{D}C_T & \bar{D}D_T - \gamma I \end{array}\right] \\ = & \left[\begin{array}{cc|c} A_1 & B_1(-\bar{C} + \bar{D}C_T) & B_1(\bar{D}D_T - \gamma I) \\ 0 & A_T & B_T \\ \hline C_1 & D_1(-\bar{C} + \bar{D}C_T) & D_1(\bar{D}D_T - \gamma I) \end{array}\right] \end{aligned}$$

which can be written, using similarity transformation, as

$$N_i^* \left(\bar{X}T^{-1} - \gamma I\right) = R_u(s) + R_s(s)$$

where

$$R_s = \left[\begin{array}{c|c} A_T & B_T \\ \hline Y & D_1(\bar{D}D_T - \gamma I) \end{array}\right]$$

$$R_u = \left[\begin{array}{c|c} A_1 & -\gamma B_1 + \left(B_1\bar{D} + X_1\bar{B}\right)D_T \\ \hline C_1 & 0 \end{array}\right] \tag{4.5}$$

$Y = C_1X_1 + D_1(-\bar{C} + \bar{D}C_T)$ and X_1 solves

$$A_1X_1 - X_1A_T + B_1(-\bar{C} + \bar{D}C_T) = 0$$

It is obvious that R_u is a function of the free parameters γ and D_T.

The sensitivity minimization problem is reduced to find γ and D_T such that (4.3) is satisfied. At the moment, there is no procedure to find γ and D_T such that (4.3) will be satisfied except by trial-and-error. Instead, condition (4.3) may be relaxed as follows. Since

$$\|\Gamma_{R_u}\| \leq \|R_u(-s)\|_\infty \tag{4.6}$$

we may find γ and D_T such that

$$\|R_u(-s)\|_\infty \leq \gamma \tag{4.7}$$

and this is equivalent to solving the following Linear Matrix Inequality (LMI) problem (Boyd, *et al.*, 1994):

$$\min_{P = P^T, D_T, \gamma \geq 0} \gamma$$

subject to

$$\begin{pmatrix} (-A_1)P + P(-A_1)^T & PC_1^T & -\gamma B_1 + \mathcal{F}D_T \\ C_1^T P & -\gamma I & 0 \\ -\gamma B_1^T + D_T^T\mathcal{F}^T & 0 & -\gamma I \end{pmatrix} \leq 0 \tag{4.8}$$

where $\mathcal{F} = \left(B_1\bar{D} + X_1\bar{B}\right)$. Since $T^{-1}(s)$ is stable, it is easy to show that $T(s)$ is also stable.

The procedure given in this section to search for $T(s) \in \Re H_\infty$ tends to result in a high order dynamic $T(s)$. More work is needed to get a lower order one.

Remark 1 *If $T(s)$ is to be chosen as a constant, invertible matrix $T(s) = D_T$ or $T(s) = \gamma I$, the LMI in (4.8) could be modified accordingly.*

5 Strictly Proper Systems

The derivation in §3 is for bi-proper systems, i.e. systems which have full rank at infinity. When the given systems are strictly proper, the outer factorization N_o of N in equation (3.3) is not invertible and hence Q in equation (3.11) is not proper. In this section, the procedure, to minimize the sensitivity function described in §3, will be modified to tackle strictly proper systems. Assume the system $G(s)$ is given in state-space form as

$$G = \left[\begin{array}{c|c} A & B \\ \hline C & 0 \end{array}\right] \tag{5.1}$$

such that CB has full rank and has right and left coprime factorization $G = NM^{-1} = \tilde{M}^{-1}\tilde{N}$. First, the following lemma is available.

Lemma 1 *(Saif, et al., 1998) Given a strictly proper system* $N(s) = \left[\begin{array}{c|c} A_N & B_N \\ \hline C_N & 0 \end{array}\right]$, *then for given* $\alpha_o, \gamma_o > 0$, *the system* $N_{\alpha\gamma} = (\alpha_o s + \gamma_o)N$ *may be realized as*

$$N_{\alpha\gamma} = \left[\begin{array}{c|c} A_N & B_N \\ \hline C_N(\alpha_o A_N + \gamma_o I) & \alpha_o C_N B_N \end{array}\right]$$

or

$$N_{\alpha\gamma} = \left[\begin{array}{c|c} A_N & (\alpha_o A_N + \gamma_o I)B_N \\ \hline C_N & \alpha_o C_N B_N \end{array}\right]$$

■

Now $N_{\alpha\gamma}$ has IOF whose outer part is unimodular since it has full rank at infinity by definition. Define $\mathcal{T}$ as $\mathcal{T} = (\alpha_o s + \gamma_o)I$. Then the procedure is proceeded as follows. Equation (3.1) becomes

$$\begin{array}{ll} \min\limits_{Q \in \Re\mathcal{H}_\infty} & \left\|(X - NQ)\tilde{M}_o\tilde{M}_i\right\|_\infty \\ \min\limits_{Q \in \Re\mathcal{H}_\infty} & \left\|X\tilde{M}_o - NQ\tilde{M}_o\right\|_\infty \\ \min\limits_{\bar{Q} \in \Re\mathcal{H}_\infty} & \|\bar{X} - N\bar{Q}\|_\infty \end{array} \tag{5.2}$$

subject to

$$\bar{X} - N\bar{Q} = \bar{U} \tag{5.3}$$

Defining $\bar{U}$ as $\bar{U} = R + \gamma T$, and multiplying both sides of equation (5.3) by T^{-1} then

$$(\bar{X}T^{-1} - \gamma I) - N\bar{Q}T^{-1} = RT^{-1} \tag{5.4}$$

Introducing $\mathcal{T}$ as

$$(\bar{X}T^{-1} - \gamma I) - \mathcal{T}^{-1}\mathcal{T}N\bar{Q}T^{-1} = RT^{-1}$$

and finding the inner-outer factorization of $\mathcal{T}N$ as $\mathcal{T}N = \hat{N}_i\hat{N}_o$, then

$$\mathcal{T}^{-1}\left\{\mathcal{T}(\bar{X}T^{-1} - \gamma I) - \hat{N}_i\hat{N}_o\bar{Q}T^{-1}\right\} = RT^{-1}$$

As in §3.1, the constrained sensitivity minimization problem in (5.2, 5.3) will be minimized if the following problem is solved

$$\min_{\bar{Q} \in \Re\mathcal{H}_\infty} \left\|\mathcal{T}^{-1}\left\{\mathcal{T}(\bar{X}T^{-1} - \gamma I) - \hat{N}_i\hat{N}_o\bar{Q}\right\}\right\|_\infty < \gamma \tag{5.5}$$

If $\mathcal{T}$ is chosen such that $\|\mathcal{T}^{-1}\|_\infty \leq 1$, say, $\mathcal{T} = (s+1)I$, then, the above equation is reduced into

$$\min_{\bar{Q} \in \Re\mathcal{H}_\infty} \left\|\hat{N}_i^*\mathcal{T}(\bar{X}T^{-1} - \gamma I) - \hat{N}_o\bar{Q}\right\|_\infty < \gamma \tag{5.6}$$

In this equation the term $\mathcal{T}(\bar{X}T^{-1} - \gamma I)$ is not proper. To make it proper, the selection procedure for $T(s)$ given in §4 needs to be modified as follows. Let us choose $T(\infty)$ such that $\bar{X}T^{-1}\big|_{s=\infty} = \gamma I$. In this case, the state-space representation of $(\bar{X}T^{-1} - \gamma I)$ becomes

$$\begin{aligned} \bar{X}T^{-1} - \gamma I &= \left[\begin{array}{cc|c} \bar{A} & \bar{B}C_T & \bar{B}D_T \\ 0 & A_T & B_T \\ \hline \bar{C} & \bar{D}C_T & \bar{D}D_T - \gamma I \end{array}\right] \\ &= \left[\begin{array}{c|c} \bar{A} - \bar{B}C_T & -\gamma\bar{B}\bar{D}^{-1} \\ \hline -\bar{C} + \bar{D}C_T & 0 \end{array}\right] \\ &\text{if } \left\{\begin{array}{rcl} A_T &=& \bar{A} - \bar{B}C_T \\ B_T &=& -\bar{B}D_T \\ D_T &=& \gamma\bar{D}^{-1} \end{array}\right. \end{aligned} \tag{5.7}$$

Then $\mathcal{T}(\bar{X}T^{-1} - \gamma I)$ becomes

$$\begin{aligned} &\mathcal{T}(\bar{X}T^{-1} - \gamma I) \\ &= \left[\begin{array}{c|c} \bar{A} - \bar{B}C_T & -\gamma(\bar{A} - \bar{B}C_T + I)\bar{B}\bar{D}^{-1} \\ \hline -\bar{C} + \bar{D}C_T & -\gamma(-\bar{C} + \bar{D}C_T)\bar{B}\bar{D}^{-1} \end{array}\right] \end{aligned}$$

due to $\mathcal{T} = (s+1)I$ and the previous lemma. It can be shown as in §4 that

$$\begin{aligned} &N_i^*\mathcal{T}(\bar{X}T^{-1} - \gamma I) \\ &= \left[\begin{array}{c|c} A_1 & -\gamma L \\ \hline C_1 & 0 \end{array}\right] \\ &\quad + \left[\begin{array}{c|c} A_T & -\gamma(\bar{A} - \bar{B}C_T + I)\bar{B}\bar{D}^{-1} \\ \hline Z & -\gamma D_1(-\bar{C} + \bar{D}C_T)\bar{B}\bar{D}^{-1} \end{array}\right] \\ &= R_u(s) + R_s(s) \end{aligned}$$

where

$$\begin{aligned} L &= \left\{B_1(-\bar{C} + \bar{D}C_T) - X_1(\bar{A} - \bar{B}C_T + I)\right\}\bar{B}\bar{D}^{-1} \\ Z &= C_1X_1 + D_1(-\bar{C} + \bar{D}C_T) \end{aligned}$$

Then the LMI in (4.8) becomes

$$\min_{P = P^T, \gamma \geq 0} \gamma$$

subject to

$$\begin{pmatrix} (-A_1)P + P(-A_1) & PC_1^T & -\gamma L \\ C_1P & -\gamma I & 0 \\ -\gamma L^T & 0 & -\gamma I \end{pmatrix} \leq 0 \tag{5.8}$$

In the case of strictly proper systems, the unimodular function $T(s) \in \Re H_\infty$ is almost determined except for the free parameter γ that needs to be determined by solving the LMI given in (5.8).

6 Conclusion

A constructive procedure is proposed in the paper to minimize the sensitivity function with stable controller. The problem is formulated as 1-block or 2-block H_∞ optimization problem for square and nonsquare systems respectively. The successful execution of the method depends on appropriately chosen unimodular transfer function matrix $T(s)$. A procedure to choose $T(s)$ was also proposed which led to solving a Linear Matrix Inequality. The method presented here to find $T(s)$ will result in a high order dynamic $T(s)$. A new method to find simple $T(s)$ is needed.

Acknowledgment

The first author would like to thank King Fahd University of Petroleum and Minerals for its support.

References

Boyd, S., *et al.*, (1994). Linear Matrix Inequalities un Systems and Control Theory. SIAM studies in applied mathematics; Vol. 15.

Dorato, P., Hong-Bae Park and Yunzhi Li (1989). "An Algorithm for Interpolation with units in $\mathcal{H}^\infty$ with Application to Feedback Stabilization." Automatica, Vol. 25, No. 3, pp. 427-430.

Ganesh, C. and Pearson, J. B. (1989). "$\mathcal{H}^2$-Optimization with Stable Controllers", Automatica, Vol. 25, No. 4, pp. 629-634.

Glover, K. (1984). "All Optimal Hankel norm approximation of Linear Multivariable Systems and their $L_\infty-$ error bounds." Int. J. Control, Vol. 39, No. 6, pp. 1115-1193.

Halevi, Y. (1994). "Stable LQG Controllers." IEEE Trans. on Automatic Control, Vol. AC-39, No. 10, pp. 2104-2106. Also appeared in American Control Conference, 1987.

Halevi, Y. , Bernstein, D.S. and Haddad, W. (1991). " On Stable Full-Order and Reduced-Order LQG Controllers",Optimal Control Appl. & Methods, Vol. 12, pp.163-172.

Ito, H., Ohmori, H. and Sano, A. (1993). "Design of Stable Controllers Attaining Low $\mathcal{H}^\infty$ Weighted Sensitivity." IEEE Trans. on Automatic Control, Vol. AC-38, No. 3, pp. 485-488.

Jacobus, M. , Jamshidi, M. ,Abdallah, C. , Dorato, P. and Bernstein, D. (1990). "Suboptimal Strong Stabilization Using Fixed Order Dynamic Compensator", American Control Conference, pp. 2659-2660.

Nakayama, T., Ohmori, H., Sano, A. and Ito, H. (1995). "A Design of $\mathcal{H}_\infty$ Stable Controller." 3rd ECC, Rome, Italy, pp. 1830-1833.

Saif, Abdul-Wahid A., Postlethwaite, Ian,Kavranoglo, D. and Gu, Da-Wei (1994). " Simultaneous Stabilization Via Robust Stabilization Theory." IEE Control 94 International Converence, Coventry, U.K.

Saif, Abdul-Wahid A., Gu, Da-Wei and Postlethwaite, Ian (1997). " Strong Stabilization of MIMO Systems via $\mathcal{H}_\infty$-Optimization". Systems & Control Letters 32, pp. 111-120.

Saif, Abdul-Wahid A., Gu, Da-Wei and Postlethwaite, Ian (1998a). " Strong Stabilization of MIMO Systems via Unimodular Interpolation in $\mathcal{H}_\infty$".Int. J. Control, Vol. 69, No. 6, pp. 797-818.

Saif, Abdul-Wahid A., (1998b). "Parameterization of all Stable Stabilizing Controllers". UKACC International Conference on CONTROL'98, Univ. of Wales, Swansea, U.K., Sept. 1-4.

Saif, Abdul-Wahid A., (1998c). "H_∞-Stable Controllers". The 3rd Portugese International Conference, CONTROLO'98, Coimbra, Portugal, Sept. 9-11.

Sideris, A. and Safonov, M.G. (1985). " Infinity-Norm Optimization with a Stable Controller." Proceedings of American Control Conference, pp. 804-805.

Vidyasagar, M. (1985). Control system synthesis: a Factorization Approach. MIT Press, Cambridge, Massachusetts.

Wei, K. (1990). "Stabilization of linear plant via a stable compensator having no real unstable zeros." Systems & Control Letters 15, pp. 259-264.

William W. Y. and Bernstein, D. S. (1993). "$\mathcal{H}_2$-Suboptimal Stable Stabilization", Proceedings of the 32nd Conference on Decision and Control.

Youla, D.C. , Bongiorno, J.J. and Lu, C.N. (1974). "Single-loop Feedback Stabilization of Linear Multivariable Dynamic Plants." Automatic, Vol.10, pp.159-173.

A NEW APPROACH TO ROBUSTNESS AND STABILIZATION OF CONTROL SYSTEMS VIA PERTURBATION THEORY

B. T. Polyak, P. S. Shcherbakov

Institute of Control Science, Profsojuznaya 65, Moscow 117806, Russia
E-mail: boris@ipu.rssi.ru, sherba@ipu.rssi.ru, Fax: (7-095) 334-2016

Abstract: An iterative procedure of non-smooth optimization is proposed, which is aimed at moving zeros of a parameter-dependent polynomial toward a prescribed domain. The method is based on the first- and second-order approximations for zeros of perturbed polynomials and deals directly with perturbed zeros rather than coefficients; it applies to a broad class of problems that involve any differentiable dependence on the parameters. A similar method is developed for matrix families. An immediate application to the control theory leads to solution of a number of fundamental problems: Robust stability of polynomial and matrix families; Maximal degree of stability; Stabilization via low-order controllers; Simultaneous stabilization. The efficacy of the approach is illustrated by (i) Robust stability of families of polynomials and matrices and (ii) Finding a stable polynomial in a polynomial family. *Copyright © 1999 IFAC*

Keywords: Linear systems; Robust stability; Stabilization; Perturbation theory.

1. INTRODUCTION

Considered are families of parametrized polynomials $p(s,q)$ or matrices $A(q)$, where $q \in \mathbb{R}^{\ell}$ is the vector of parameters (uncertainties). Mathematically, the following two problems are posed:

RM (Robustness Margins): Let the *nominal* polynomial $p_0(s) \doteq p(s,0)$ be stable and let $\|q\| \leq \gamma$ for some vector norm $\|\cdot\|$ and $\gamma > 0$; we write $q \in Q_\gamma$, where Q_γ is a ball of radius γ in the norm $\|\cdot\|$. Find the *robustness margin*

$$\gamma_{\max} \doteq \max\{\gamma : \ p(s,q) \text{ is stable } \forall q \in Q_\gamma\}.$$

ST (Stabilization): Suppose the nominal $p(s,0)$ is unstable. Detect if there exists a q such that $p(s,q)$ is stable; if the answer is positive, find a stabilizing value of q (possibly the smallest).

The same problems are posed for matrix families. Said another way, the RM problem is: Given a *stable* element in the family, find the closest unstable element and the distance between them. Similarly, the ST problem is: Given an *unstable* element in the family, find if there exist stable elements (possibly, a closest stable).

Many classical problems in control theory reduce to the two formulations above; the RM setup relates to the robustness analysis, while the ST setup relates to design. To illustrate, suppose a plant is specified by its transfer function $P(s) = N_p(s)/D_p(s)$, where N_p and D_p are polynomials. The problem is to find if there exists a stabilizing controller in the given parametrized class of controllers $\mathcal{C} = \{C(s,q) = N_c(s,q)/D_c(s,q)\}$, where $q \in Q \subseteq \mathbb{R}^{\ell}$. In that case, the characteristic polynomial of the closed-loop system is $p(s,q) = N_p(s)N_c(s,q) + D_p(s)D_c(s,q)$, and we arrive at the ST setup.

The approach proposed is based on the perturbation theory for polynomials and matrices. Both RM and ST problems are treated on a uniform basis although there is an intrinsic difference between them. The method deals directly with perturbed zeros or eigenvalues rather than coefficients, which is more adequate and provides the method with more flexibility. The approach applies to both statements for various vector norms and types of dependence on the parameters; real and complex parameters are equally treatable as well as Hurwitz stability and Schur stability.

To the authors' knowledge, quite a few papers pursue similar approaches. In the pioneering work (Polak and Wardi, 1982), a method of nondifferentiable optimization was developed for control systems design with frequency-dependent singular value constraints and performance index. In contrast, the method in the present paper deals directly with frequency-independent eigenvalues and unconstrained problems. Polak and Wardi (1982) do not distinguish between the ST and RM problems, and their method is inapplicable to the RM setup; polynomial families are not considered as special case. Nevertheless, the works by E. Polak made a significant impact to the development of optimization-based numerical methods for design problems. In (Keel and Bhattacharyya, 1997), a linear programming (LP) approach was proposed for solving ST problems; the constraints in the LP problem were formulated in terms of the coefficients of polynomials, and the objective was to attain a target interval polynomial having *a priori* specified properties. The technique is not based on perturbation theory for zeros and eigenvalues.

2. ROBUST STABILITY OF POLYNOMIALS

2.1 Perturbed Zeros and Stability Conditions.

The takeoff point for the exposition to follow is a linear approximation to perturbed zeros of polynomials.

Proposition $\mathcal{P}$. Let $p(s,q)$ be a polynomial in s that depends on the vector $q=(q_i)_1^\ell \in \mathbb{R}^\ell$ of parameters, and $\deg p(s,q) \doteq n = \text{const}(q)$. Suppose $p(s,q)$ is differentiable with respect to q at the point $q=0$ and denote

$$\pi_i(s) = \left.\frac{\partial p(s,q)}{\partial q_i}\right|_{q=0}, \qquad i=\overline{1,\ell}. \qquad (1)$$

If $s_k \doteq s_k(0)$ is a simple zero of the *nominal* polynomial $p_0(s) \doteq p(s,0)$, then for sufficiently small q there exists a zero $s_k(q)$ of the *perturbed* polynomial $p(s,q)$ such that

$$s_k(q) = s_k - (\pi^k, q)/r_k + o(q), \qquad (2)$$

$$\pi^k = (\pi_i^k)_1^\ell, \;\; \pi_i^k = \pi_i(s_k), \;\; r_k = p_0'(s)\big|_{s=s_k}. \qquad (3)$$

These relations are a particular case of Proposition $\mathcal{M}$ (see Sec. 3); they can be obtained by straightforward expanding in Taylor series. For an *affine* polynomial family $p(s,q) = p_0(s) + \sum_{i=1}^\ell q_i p_i(s)$, we have $\pi_i(s) = p_i(s)$, and (2) yields $s_k(q) \approx s_k - (p_i(s_k), q)/p_0'(s_k)$.

Now let $p(s,0)$ be stable (unless otherwise stated, Hurwitz stability is studied: $\max_k \text{Re}\, s_k < 0$). To preserve stability of a perturbed polynomial, we require $\text{Re}\, s_k(q) < 0$ for all k. With the linear approximation $s_k(q) \approx s_k - (\pi^k, q)/r_k$ in (2), this yields first-order conditions of stability for $p(s,q)$:

$$\begin{aligned} &\max_k(\text{Re}\, s_k + (u^k, q)) < 0, \\ &u_i^k \doteq -\text{Re}(\pi_i^k/r_k), \;\; i=\overline{1,\ell}. \end{aligned} \qquad (4)$$

If Schur stability is meant, that is $\max_k |s_k| < 1$, then (4) has the form $\max_k |s_k + (u^k,q)| < 1$, $u_i^k = -\pi_i^k/r_k, i=\overline{1,\ell}$.

2.2 Robust Stability.

The RM problem formulated in Sec. 1 can be written down as follows: Find

$$\gamma_{\max} = \max\{\gamma\colon \max_k \text{Re}\, s_k(q) < 0 \,\forall q \in Q_\gamma\}. \quad (5)$$

With (4) in hand, this leads to the following estimate for $\gamma_{\max}$ based on linear approximation (2):

$$\overline{\gamma} = \max\{\gamma\colon \max_k(\text{Re}\, s_k + \gamma \max_{q\in Q_1}(u^k,q)) < 0\}. \quad (6)$$

The innermost max is computable for commonly used norms, and closed-form solutions to problem (6) are readily available:

l_∞-norm(interval uncertainties). Let $|q_i| \le 1$, then $\max_{q\in Q_1}(u^k,q) = \sum_{i=1}^\ell |u_i^k|$ and $\overline{q}_i = \text{sign}(u_i^k)$, where $\overline{q} = (\overline{q}_i)_1^\ell$ is the maximizer. From (6) it follows $\overline{\gamma} = \min_k(-\text{Re}\, s_k/\sum_{i=1}^\ell |u_i^k|)$, $\overline{q}_i = \overline{\gamma}\,\text{sign}(u_i^m)$, where $m = \text{argmin}_k(-\text{Re}\, s_k/\sum_{i=1}^\ell |u_i^k|)$.

Euclidean norm (spherical uncertainties). Let $\|q\| = (\sum_{i=1}^\ell q_i^2)^{1/2} \le 1$, then $\max_{q\in Q_1}(u^k,q) = \|u^k\|$, $\overline{q} = u^k/\|u^k\|$, and (6) gives $\overline{\gamma} = \min_k(-\text{Re}\, s_k/\|u^k\|)$ and $\overline{q} = \overline{\gamma}\, u^m/\|u^m\|$, where $m = \text{argmin}_k(-\text{Re}\, s_k/\|u^k\|)$.

Similar expressions can be obtained for other l_p norms and for the case of Schur stability.

2.3 Upper Bound for Robustness Margin.

Let us introduce the function $\eta(q) = \max_k \text{Re} s_k(q)$ and note that (5) is equivalent to finding a minimum norm solution to the problem $\eta(q) \ge 0$:

$$\text{Find} \;\; \min\|q\| \;\; \text{subject to} \;\; \eta(q) \ge 0.$$

We present an iterative method for solving this problem using linearization (4). Instead of solving (5) for $\eta(q)$, problem (6) is solved for the linearized function $\overline{\eta}(q) = \max_k(\text{Re}\, s_k + (u^k,q))$ (4) and its solution $\overline{q}$ is taken as a first-step approximation to the solution of (5). The next step is performed with $p(s,0)$ being changed for $p(s,\overline{q})$ and so forth.

Algorithm I.

1. $\widehat{q} = 0$.
2. Compute zeros s_k of $p(s,0)$ and vectors u^k according to (1), (3), (4).
3. Compute $h^k \doteq \max_{\|q\|\le 1}(u^k,q)$ and the maximizer $q^k \doteq \text{argmax}_{\|q\|\le 1}(u^k,q)$. Compute $\gamma \doteq \min_k(-\text{Re}\, s_k/h^k)$ and $m = \text{argmin}_k(-\text{Re}\, s_k/h^k)$; take q^m as the direction and γ as the step-size: $dq \doteq \gamma \cdot q^m$.
4. If $p(s,dq)$ is stable, change variables $q := q - dq$ (i.e., adopt $p(s,dq)$ as a "new" $p(s,0)$) and put $\widehat{q} := \widehat{q} + dq$. Go to step 2. Otherwise find $\alpha_{\min} = \min\{\alpha > 0 : p(s, \alpha\cdot dq)$ is unstable$\}$ and put $\widehat{q} \doteq \widehat{q} + \alpha_{\min}\cdot dq$.
5. Adopt $\overline{\gamma} \doteq \|\widehat{q}\|$ as an upper bound for the robustness margin.

At each step, a minimum norm increment dq to the vector of parameters is found that vanishes the maximal real part of linear approximation (2) to zeros of $p(s,q)$ so that the zeros are shifted toward the right-half plane iteratively in q.

The algorithm exposes a very fast convergence to a destabilizing point $\widehat{q}$, and the distance between $p(s,0)$ and $p(s,\widehat{q})$ is quite precise an estimate of $\gamma_{\max}$. The method is locally optimal and, generally speaking, it may not converge to a global solution, i.e., $\widehat{q}$ is not the closest destabilizing point, hence $\overline{\gamma} \geq \gamma_{\max}$. However, it was tested over a great variety of problems and showed a very good performance. The following "globalization" may be suggessted: Run the algorithm repeatedly from various initial points in a neighborhood of $q = 0$ to obtain "randomized" estimates of $\overline{\gamma}$; then pick the smallest solution obtained, call it $\widetilde{\gamma}$.

2.4 Examples.

Example 1: Affine uncertainty. For $q \in \mathbb{R}^2$, let $p(s,q) = p_0(s) + q_1p_1(s) + q_2p_2(s)$, $|q_i| \leq \gamma$, be an interval affine family with stable nominal $p_0(s)$. For the case of two parameters, the results are well visualized by using the D-decomposition technique (Siljiak, 1969). Figure 1 depicts the stability domain in q-space determined in this way; the larger square relates to $\overline{\gamma} = 0.0319$ obtained after the first ($q^0 = [0\ 0]$) run of the algorithm (converged after 6 steps); the smaller square relates to the refined estimate $\widetilde{\gamma} = 0.0197$. Polynomials $p_i(s)$, $i = 0,1,2$, having degrees 5, 4, 3, were generated randomly. The exact solution given by the frequency domain method in (Tsypkin and Polyak, 1991) is $\gamma_{\max} = 0.0181$.

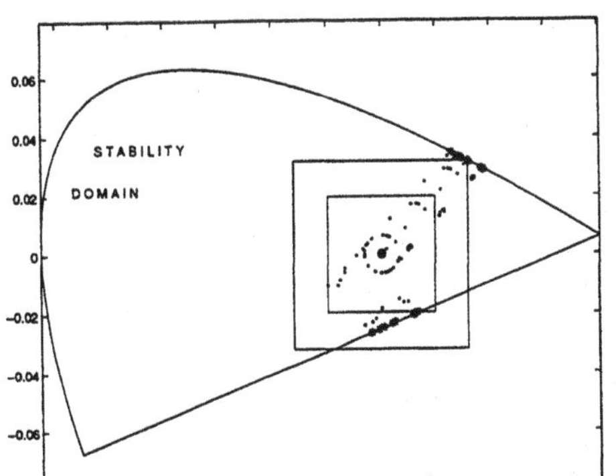

Fig. 1. Robustness margins in Example 1.

Example 2: Multilinear dependence. This example is borrowed from (De Gaston and Safonov, 1988); it reduces to the interval polynomial family

$$p(s,q) = p_0(s) + \sum_{i=1}^{3} q_ip_i(s) + q_1q_2p_{12}(s),$$

where $|q_1| \leq 0.2\gamma$, $|q_2| \leq 0.3\gamma$, $|q_3| \leq 0.1\gamma$, and

$$p_0(s) = s^4 + 20s^3 + 124s^2 + 1040s + 1600;$$
$$p_1(s) = s^3 + 16s^2 + 60s; \quad p_2(s) = s^3 + 14s^2 + 40s;$$
$$p_{12}(s) = s^2 + 10s; \quad p_3(s) = 800s + 1600,$$

with stable nominal $p_0(s)$. Application of Algorithm I with $q^0 = [0\ 0\ 0]$ gives $\overline{\gamma} = 3.4408$ after 4 iterations, while the exact solution is $\gamma_{\max} = 3.44$.

3. ROBUST STABILITY OF MATRICES

A straightforward extension of Algorithm II to the matrix case is: Given a parametrized matrix family $A(q)$, $\|q\| \leq \gamma$, such that $A(0)$ is stable, find

$$\gamma_{\max} = \max\{\gamma : A(q) \text{ is stable } \forall \|q\| \leq \gamma\}. \quad (7)$$

3.1 Perturbation Formula for Eigenvalues.

Proposition $\mathcal{M}$. Let $A(q)$ be a real $n \times n$ matrix that depends on the vector of real parameters $q = (q_i)_1^\ell$. Assume that $A(q)$ is differentiable at $q = 0$:

$$D_i \doteq \left.\frac{\partial A(q)}{\partial q_i}\right|_{q=0}, \qquad i = \overline{1,\ell}.$$

Let $\lambda_k \doteq \lambda_k(0)$ be an algebraically simple eigenvalue of $A(0)$ and let x_k and y_k denote the right and left eigenvectors of $A(0)$ associated with λ_k. Then for sufficiently small q there exists an eigenvalue $\lambda_k(q)$ of the matrix $A(q)$ such that

$$\lambda_k(q) = \lambda_k + \sum_{i=1}^{\ell} \frac{y_k^* D_i x_k}{y_k^* x_k} q_i + o(q).$$

This result is a slight modification of Theorem 6.3.12 (Horn and Johnson, 1986).

It is said that x and y are right and left eigenvectors of A associated with its eigenvalue λ if $Ax = \lambda x$ and $y^*A = \lambda y^*$, where * denotes complex conjugation. Right eigenvectors x_k can be normalized to $x_k^* x_k = 1$; left eigenvectors can be normalized to $y_k^* x_k = 1$. It is also known that $y_k^* x_m = 0$ for $k \neq m$, while $y_k^* x_k \neq 0$ for an algebraically simple eigenvalue λ_k. With this scaling, the relation above rewrites to

$$\lambda_k(q) \approx \lambda_k + (v^k, q),\ v^k = (v_i^k)_1^\ell,\ v_i^k = y_k^* D_i x_k, \quad (8)$$

which is an approximation to perturbed eigenvalues that is linear in parameters. This general form particularizes to various simple situations. For instance, $A(0) = A_0$, $D_i = A_i$ for an affine family $A(q) = A_0 + \sum_{i=1}^{\ell} q_i A_i$.

3.2 Upper Bound for Robust Stability Margin.

We introduce vectors w^k, $k = \overline{1,n}$, such that $w_i^k = \mathrm{Re}(y_k^* D_i x_k)$, where x_k, y_k, and D_i are defined in Proposition $\mathcal{M}$, and x_k and y_k are scaled. Assuming that the nominal $A(0)$ is Hurwitz and using the perturbation formula (8), we obtain linearized stability conditions for a perturbed matrix $A(q)$ similarly to (4) (with u^k being changed for w^k). With these conditions, we immediately obtain an estimate for the robustness radius $\gamma_{\max}$ by introducing some vector norm $\|\cdot\|$ on $\mathbb{R}^\ell$ and letting the perturbation vector q vary in the ball of radius γ in this norm:

$$\overline{\gamma} = \max\{\gamma : \max_k(\mathrm{Re}\lambda_k + \gamma \max_{\|q\| \leq 1}(w^k, q)) < 0\},$$
$$w_i^k = \mathrm{Re}(y_k^* D_i x_k),\ i = \overline{1,\ell},\ k = \overline{1,n}. \quad (9)$$

Note that this formulation admits for the following statement of the problem: Find the maximal value of γ retaining the stability of $A+\gamma\Delta$, where A is stable and the perturbation matrix Δ is bounded in some norm. For instance, with Frobenius matrix norm $\|A\|_F = (\sum_{i,j} a_{ij}^2)^{1/2}$ we obtain $\overline{\gamma} = \min_k(-\mathrm{Re}\lambda_k/\|W^k\|_F)$, $W_{ij}^k = \mathrm{Re}((y_i^k)^* x_j^k)$, $i,j = \overline{1,n}$. Here, x_j^k is the j-th component of the eigenvector x^k. The destabilizing uncertainty is $\widehat{\Delta} = \overline{\gamma} W^m/\|W^m\|_F$, where $m = \mathrm{argmin}_k(-\mathrm{Re}\lambda_k/\|W^k\|_F)$.

Using linearization (8)–(9) of the original problem (7) yields the same algorithm for estimating the robustness margin of a matrix family, as in the polynomial case. At each step of this procedure, eigenvalues of $A(q)$ are shifted toward the instability domain $\mathrm{Re}\,\lambda(q) \geq 0$ in the same way as were zeros of the polynomial $p(s,q)$, namely, by vanishing the maximal real part of their linear approximation. The only difference is in the coefficient vector of the underlying linear form (w^k in (9) instead of u^k in (6)).

3.4 Examples.

Example 3: Interval uncertainty. $A = A_0 + \Delta$, $|\Delta_{ij}| \leq \gamma$. In (Polyak and Panchenko, 1997) a Metzlerian matrix $A_0 \in \mathbb{R}^{40\times 40}$ was considered with $a_{ii} = -60$, $a_{ij} = 2$ for $i < j$, $a_{ij} = 1$ for $i > j$. The exact value of the interval stability radius was found to be $\gamma_{\max} = 0.0913$. Iterative method gives $\overline{\gamma} = 0.0913$ after 2 steps.

Example 4: Spherical uncertainty. $A = A_0 + \Delta$, $\|\Delta\|_F \leq \gamma$. For A_0 of general form, only lower estimates for $\gamma_{\max}$ are available in the literature, e.g., γ_L (Bhattacharyya, *et al.*, 1995), γ_B (Yemel'yanov, *et al.*, 1997). In the experiment, the nominal A_0 was taken the 10×10 matrix **-grcar(10)** from the TestMatrix Toolbox in Matlab; the following results were obtained. Lower estimates: $\gamma_L = 0.0125$; $\gamma_B = 0.0297$. Iterative method: $\overline{\gamma} = 0.1163$ after 3 steps. Direct Monte Carlo sampling over the larger radius $\gamma = 4\overline{\gamma}$ detected less than 1% of unstable samples that confirms in a sense the quality of $\overline{\gamma}$.

Examples presented in Sec. 2 and 3 are typical and constitute small part of the massive body of numerical experiments conducted with Algorithm I in the polynomial and matrix cases. The method shows excellent performance both in terms of accuracy and rate of convergence. polynomial and matrix cases.

4. STABILIZATION

This section is devoted to a polynomial family $p(s,q)$ parametrized by $q \in \mathbb{R}^\ell$ such that $\deg p(s,q) = n = \mathrm{const}\,(q)$ and $p(s,0)$ is unstable. The problem is to find if there exists a $\widehat{q}$ such that $p(s,\widehat{q})$ is stable.

4.1 Main Idea.

We use the same idea of iteratively shifting zeros of $p(s,q)$ so as to satisfy the conditions $\mathrm{Re}\,s_k(q) < 0$, $k = \overline{1,n}$. This procedure is based on the perturbation formula (2)–(3) for zeros of $p(s,q)$ and its extensions to the case of multiple roots. The essential difference between the RM and ST problems is that in order to satisfy the target condition $\max_k \mathrm{Re}\,s_k(q) \geq 0$ in the RM problem, it is sufficient to move only one zero to the right (the one with the maximal real part), whereas to satisfy the target condition $\mathrm{Re}\,s_k(q) < 0\ \forall k$ in the ST problem, *all* unstable zeros must be moved to the left-half plane. As a result, in the ST problem, the method may have to shift *several zeros at once* at some step, which is, however, not a serious obstacle and can be performed using Proposition $\mathcal{P}$. What is more important is that there might be multiple zeros among them, and this is quite often the case. This requires a more subtle account for perturbations. Indeed, the linear approximation to perturbed zeros, obtained in Proposition $\mathcal{P}$, is only valid under the assumption that the zero is simple. In Sec. 4.4, second-order conditions, which are also linear in parameters are derived. They provide a simple way to account for multiple zeros and to perform a step of the method by solving a well-defined problem of quadratic programming.

4.2 A Quadratic Programming Problem.

A step of the method: At the current point $q = 0$, linear approximations $\mathrm{Re}\,s_k(q) \approx (a^k, q) - b_k$ are built, and we solve the problem

$$\|dq\| \to \min \text{ subject to } (a^k, dq) - b_k \leq 0,\ k = \overline{1,n},$$

whose solution defines the step of the method. That is, at each step we solve the quadratic programming problem (QP)

$$\min \|dq\| \qquad \text{subject to } A \cdot dq \leq b, \qquad (10)$$

where the matrix A of constraints is composed of a^k, and the vector b is composed of b_k.

Let $\eta \doteq \max_i \mathrm{Re}\,s_i$; then the zero s_k is called *rightmost* if $\mathrm{Re}\,s_k = \eta$, i.e., it has maximal real part. Clearly, only rightmost zeros should form the constraints in the QP (10), hence, only such zeros are considered below.

4.3 Stability Conditions.

Suppose $p(s,0)$ is unstable, i.e., there exist s_k such that $\mathrm{Re}\,s_k(0) \geq 0$. We introduce accuracy $\delta > 0$, and a polynomial is considered stable if $\max_k \mathrm{Re}\,s_k < -\delta$. Let Δ_k be the perturbation of a rightmost zero s_k, caused by dq. To make this zero stable, we require $\mathrm{Re}(s_k + \Delta_k) \leq -\delta$, i.e.

$$\mathrm{Re}\,\Delta_k \leq -\mu_k, \qquad \mu_k \doteq -(\mathrm{Re}\,s_k + \delta). \qquad (11)$$

The key issue is a proper representation of perturbations Δ_k as linear functions of dq.

4.4 Linearized Conditions.

Below, we distinguish between simple and multiple rightmost zeros and derive different perturbation formulae, both linear in parameters.

First-order approximation. Assume that s_k is a simple zero, i.e., $p(s_k,0)=0$; $p'(s,0)|_{s=s_k}\neq 0$. By Proposition $\mathcal{P}$, the following linearization of condition (11) is readily available:

$$(w^k,\ dq)\leq -\mu_k,\quad w_i^k=-\mathrm{Re}\frac{\pi_i(s_k)}{p_0'(s_k)},\quad \pi_i(s)=\left.\frac{\partial p(s,q)}{\partial q_i}\right|_{q=0},\quad p_0'(s_k)=p'(s,0)|_{s=s_k}. \tag{12}$$

Each such inequality constitutes a constraint in the QP (10) with $a^k=w^k$ and $b_k=-\mu_k$. If s_k is complex, then s_k and s_k^* contribute the same constraints and one of them can be omitted. Hence, the number of constraints is equal to the number of different real parts of rightmost zeros.

Second-order approximation. Let a zero s_k have multiplicity 2, that is, $p(s_k,0)=0$; $p'(s,0)|_{s=s_k}=0$; $p''(s,0)|_{s=s_k}\neq 0$. Taylor series expansion of $p(s,q)$ at $(s_k,0)$ to the accuracy of second-order terms gives

$$\frac{1}{2}\Delta_k^2 p_0''(s_k)+\Delta_k\sum_{i=1}^{\ell}dq_i\cdot\pi_i'(s_k)+\sum_{i=1}^{\ell}dq_i\cdot\pi_i(s_k)=0,$$

$$p_0''(s_k)\doteq p''(s,0)|_{s=s_k},\quad \pi_i'(s_k)\doteq\pi_i'(s)|_{s=s_k}.$$

Denoting

$$u_i^k\doteq\pi_i'(s_k)/p_0''(s_k),\quad v_i^k\doteq 2\pi_i(s_k)/p_0''(s_k) \tag{13}$$

and

$$\alpha\doteq(u^k,\ dq),\qquad \beta\doteq(v^k,\ dq),$$

(we omit subscript k in α and β) we obtain the quadratic equation in Δ_k: $\Delta_k^2+2\alpha\Delta_k+\beta=0$ having solutions $\Delta_k^{(1,2)}=-\alpha\pm\sqrt{\alpha^2-\beta}$. Assume now that s_k is pure real and impose linear conditions on α and β ensuring the fulfillment of (11). There are two options to guarantee (11) by choosing $\Delta_k^{(1,2)}$ either real or complex.

Option 1 is to split the s_k into a pair of complex conjugate zeros $s_k^+=\widehat{s}_k+jt_k$ and $s_k^-=\widehat{s}_k-jt_k$, where $\widehat{s}_k\leq-\delta$ and $t_k\neq 0$. In that case it is sufficient to set $\alpha^2-\beta<0$ and $\alpha=\mu_k$. For instance, the conditions $\alpha=\mu_k$ and $\beta\geq\mu_k^2$ guarantee the fulfillment of (11), and we arrive at the following linear conditions on dq:

$$\begin{cases}(u^k,\ dq) & = & \mu_k,\\ (-v^k,\ dq) & \leq & -\mu_k^2.\end{cases} \tag{14}$$

Option 2 is to split s_k into 2 pure real zeros $s_k^+\leq-\delta$ and $s_k^-\leq-\delta$, for which purpose we set $\alpha\geq\mu_k$ and $\beta=\mu_k^2$ ensuring the relaxed conditions $s_k^+\leq-\delta/2$ and $s_k^-\leq-\delta$. With this choice, requirement (11) is not satisfied, although we do reduce $\mathrm{Re}\,s_k$. We thus have

$$\begin{cases}(-u^k,\ dq) & \leq & -\mu_k,\\ (v^k,\ dq) & = & \mu_k^2.\end{cases} \tag{15}$$

With both options we obtain linear constraints (14) or (15) on dq forcing the reduction of the real part of a multiple zero s_k. A procedure is developed to decide which of the two options to choose (respectively, a zero is called multiple zero of type 1 or type 2).

At each step of the method, we distinguish between the three categories of rightmost zeros: (i) simple zeros, either complex or real; (ii) multiple zeros of type 1; (iii) multiple zeros of type 2, and compose inequalities (12) or (14), (13), or (15), (13), respectively. These inequalities altogether form the constraints in the QP (10). A solution to this problem provides a step of the method such that zeros are shifted leftward. Note that every multiple zero leads to the pair of constraints (14) or (15) so that the total number of constraints is equal to the number of different real parts of rightmost zeros (accounting for their multiplicity).

Algorithm II.

1. Specify numerical constants and set $\widehat{q}=0$.
2. Calculate $\eta=\max_k \mathrm{Re}\,s_k$, where s_k are the zeros of $p(s,0)$. If $\eta\leq-\delta$, stop; $\widehat{q}$ is a stabilizing value of parameters. Otherwise, identify rightmost zeros, calculate the vectors w^k,u^k,v^k according to (12), (13), compose inequalities (12), (14), (15).
3. Solve the QP (10) with constraints (12), (14), (15) to obtain a solution dq.
4. Find $\gamma_{\min}=\arg\min_{0\leq\gamma\leq 1}\eta(\gamma)$, where $\eta(\gamma)=\max_k \mathrm{Re}\,s_k$ and s_k are zeros of $p(s,\gamma\cdot dq)$. Change variables $q:=q-\gamma_{\min}\cdot dq$ (i.e., adopt $p(s,\gamma_{\min}\cdot dq)$ as a "new" $p(s,0)$) and put $\widehat{q}:=\widehat{q}+\gamma_{\min}\cdot dq$. Go to step 2.

Algorithm II is derived under the assumptions that there are no complex multiple zeros and that pure real zeros have multiplicity not higher than 2. So, generally speaking, there is no guarantee for the algorithm to converge to a stable point. However, our rich computational experience to date shows that the violation of these assumptions is rather a pathology, while the occurrence of a double real root is an intrinsic feature of ST problems.

4.5 Examples.

Example 5: Static Output Feedback Stabilization. Given a linear system $\dot{x}=Fx+Gu$, $y=Hx$, find if there exists a stabilizing control law $y=Ku$. An example in (Anderson, *et al.*, 1975) involves a 1×2 gain matrix $K=[q_1\ q_2]$; a very lengthy derivation based on the decidability algorithms of Tarski and Seidenberg gave a stability point $\widehat{q}=[2\ 50]$. However, the problem reduces to finding a stable polynomial in the family $p(s,q)=p_0(s)+q_1p_1(s)+q_2p_2(s)$, where $p_0(s)=s^3-13s$; $p_1(s)=s^2-5s$; $p_2(s)=s+1$. The problem contains two uncertain parameters, and D-decomposition can be used to describe the whole stability domain of $p(s,q)$ (Fig. 2). To illustrate the method, we start from $q^0 = [0\ 0]$ and obtain a stabilizing point $\hat{q} = [4.7631\ \ 46.7713]$ after 20 steps.

Note that in general, static output feedback stabilization problems are NP-hard (Blondel and Tsitsiklis, 1997).

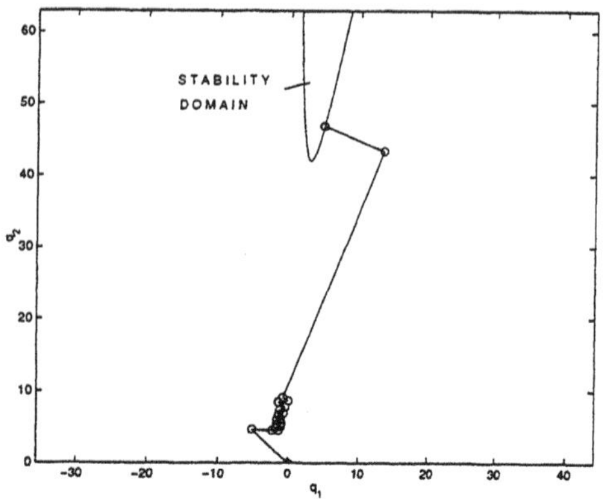

Fig. 2. Trajectory of Algorithm II.

Example 6: Stabilization via PID Controller. Given a SISO plant specified by transfer function $P(s) = 1/(s+2)(s+1)^2(s-1)$, find a stabilizing PID controller of the form $C(s,q) = q_1 + q_2 s + q_3/s$. The problem reduces to the ST formulation by considering the closed-loop characteristic polynomial $p(s,q) = s(s+2)(s+1)^2(s-1) + q_1 s + q_2 s^2 + q_3$. Application of Algorithm II with initial point $q^0 = [0\ 0\ 0]$ yields a stabilizing point $\hat{q} = [2.1793\ \ 5.5783\ \ 0.3472]$ after 24 steps.

Example 7: Matrix Stabilization. Algorithm II directly modifies to the matrix case; its behavior is exemplified by finding a stable element in the affine family $A(q) = A_0 + q_1 A_1 + q_2 A_2 + q_3 A_3$ with unstable 3×3 nominal A_0 having eigenvalues $\lambda_1 = 2$, $\lambda_{2,3} = 1 \pm j$ and perturbation matrices A_i, $i = 1,2,3$, with entries generated randomly uniformly on $[-1,\ 1]$. For $q^0 = [0\ 0\ 0]$, after 11 steps the algorithm achieves a stabilizing point $\hat{q} = [-4.5757\ \ 2.3821\ \ 7.5503]$ such that the eigenvalues of $A(\hat{q})$ are $\lambda_1 = -5.2960$ and $\lambda_{2,3} = -0.01 \pm j1.8385$.

6. CONCLUSION

The approach developed is a powerful tool for solving fundamental problems in control theory which admit no solution by existing techniques. It has a simple underlying idea and applies to a wide range of problems involving differentiable dependence on the parameters. The method is numerically stable and exposes a very good performance for a large number of test problems, both in terms of accuracy and rate of convergence.

Preliminary results are already obtained for the problem of finding a stable polynomial in an interval family (ST problem with constraints on q), finding the maximal stability degree of parameterized families, etc. No essential difficulties are expected when applying the method to robust stabilization, simultaneous stabilization, etc. For instance, under simultaneous stabilization, each of the plants contribute its own set of inequalities to the cumulative constraints in the QP problem, and the number of constraints grows, whereas the structure of the problem remains the same.

The method can be further refined by using slightly different second-order conditions (Algorithm II) and optimizing the algorithms numerically (finer step-size correction, choice of constants).

7. ACKNOWLEDGEMENTS

The work was supported by grants RFFI 96-01-00993 and INTAS 93-1034ext.

REFERENCES

Anderson, B.D.O., N.K.Bose and E.I.Jury (1975) Output feedback stabilization and related problems—solution via decision methods. *IEEE TAC*, **20**, 53–66.

Bhattacharyya, S.P., H.Chapellat and L.H.Keel (1995) *Robust control: The parametric approach.* Prentice Hall PTR, Upper Saddle River, NJ.

Blondel, V. and J.N.Tsitsiklis (1997) NP-hardness of some linear control design problems. *SIAM J. Contr. Optim.*, **35**, 2118–2127.

deGaston, R.R.E. and M.G.Safonov (1988) Exact calculation of the multiloop stability margin. *IEEE TAC*, **33**, 156–171.

Emel'yanov, S.V., S.K.Korovin and N.A.Bobylev (1997) On a particular problem in robustness theory. *Sov. Phys. Dokl.*, **357**, 747–749.

Horn, R.A. and C.R.Johnson (1986) *Matrix analysis.* Cambridge University Press, Cambridge.

Keel, L.H. and S.P.Bhattacharyya (1997) A linear programming approach to controller design. In: *Proc. 36-th CDC*, 2139–2148, San Diego, CA.

Polak, E. and Y.Wardi (1982) Nondifferentiable optimization algorithm for designing control systems having singular value inequalities. *IEEE TAC*, **18**, 267–283.

Polyak, B.T. and O.B.Panchenko (1997) A probabilistic approach to robust stability of interval matrices. *Sov. Phys. Dokl., Ser. Control Theory*, **353**, 456–458.

Siljiak, D.D. (1969) *Nonlinear systems,* Wiley, New York.

Tsypkin, Ya.Z. and B.T.Polyak (1991) Frequency domain approach to robust stability of continuous systems. In: *Systems and Control: Topics in Theory and Applications* (Kozin, F and T. Ono. (Eds.)), 389–399, MITA Press, Osaka.

14th Triennial World Congress, Beijing, P.R. China

C-2a-01-4

A DESIGN METHOD OF SIMULTANEOUS STABILIZING CONTROLLER USING FILTERED INVERSE SYSTEMS

Kou Yamada *

* *Department of Electorical and Information Engineering*
Yamagata University
4-3-16 Jonan, Yonezawa, 992-8510, Japan
Tel:+81-238-26-3326, Fax:+81-238-26-3326
Email: yamada@eie.yz.yamagata-u.ac.jp

Abstract: In this paper, we consider simultaneous stabilizing problem for multivariable systems. At first we propose a state space design procedure of simultaneous stabilizing controller based on the idea of filtered inverse systems for the case stable inverse systems exist. We expand the procedure for the case stable inverse systems exist, that is, the case of stable inverse system does not exit is also considered. Main contribution of this paper is

(1) to simplify the procedure proposed by Minto and Vidyasagar, Chen et al.
(2) to expand the procedure for the system that has unstable zeros and to show freedom of simultaneous stabilizing controller
(3) to give explicit simultaneous stabilizing controller based on state space description.

Keywords: inverse system, simultaneous stabilization, state space methods, decoupling, coprime factorization

1. INTRODUCTION

In this paper, we consider a design method of simultaneous stabilizing controller. To design simultaneous stabilization controller is the problem such that a single controller $C(s)$ is designed to stabilize all of plants $G_i(s)(i = 1, \ldots, r)$(Saeks and J.Murray, 1982). To find such $C(s)$ is called simultaneous stabilization problem. Simultaneous stabilization problem is shown in many control system design problems, for example the problem to stabilize a nonlinear system with linearized at several operating points by linear controller. Simultaneous stabilization problem is also applied to the reliable control systems for sensor failure and so on(Veilette *et al.*, 1982).

Many of papers have been considered on simultaneous stabilization problem(Saeks and J.Murray, 1982; Vidyasagar and Viswanadham, 1982; Ghosh, 1986; Wei, 1986; Minto and Vidyasagar, 1986; Veilette *et al.*, 1982; Kale *et al.*, 1990; Ghosh and Byrnes, 1983; Chen and Chow, 1993; Chen *et al.*, 1995). For single-input and single-output systems, in the case of $r = 2$, simultaneous stabilization problem was solved by Saeks and Murry(Saeks and J.Murray, 1982) by using algebraic geometric method. Vidyasager and Viswanadham solved this problem by using algebraic method(Vidyasagar and Viswanadham, 1982). But if $r > 3$, no tractable procedure was not presented.

Simultaneous stabilization problem for Multi-Input and Multi-Output systems was considered by Ghosh and Byrnes as a problem simultaneous pole placement problem(Ghosh and Byrnes, 1983) at first time. This is not attractive since applicable system is only with fixed relative degrees.

Minto and Vidyasagar solve this problem by using generalized stable derivative matrix and inverse system(Minto and Vidyasagar, 1986). Chen et al. expanded the procedure proposed by Minto and Vidyasager and parametrization of all simultaneous stabilizing controller for the systems with no unstable zeros(Chen *et al.*, 1995).

Method proposed by Minto and Vidyasagar, Chen is discussed on state space method, but computation of generalized stable derivative matrix is not on state space method. This has computational complexity. In the other hand, for the system with unstable zeros, simultaneous stabilizing controller design method was considered by Chen et al. but not sufficient.

The purpose of this paper is to propose more simple procedure built on state space method for the system with unstable zeros or with no unstable zeros than by Minto and Vidyasagar, Chen et al. by using the idea of filtered inverse system. At first, for the system with no unstable zeros, we propose a procedure to obtain simultaneous stabilizing controller. This procedure has much merit than that by Minto and Vidyasagar, Chen et al. for example simple manipulation, shorthand and easy computation. Next for the system with unstable zeros, we give a state space design procedure to obtain simultaneous stabilizing controller. For the system with unstable zeros, simultaneous stabilizing controller proposed by Chen et al. can not express freedom of simultaneous stabilizing controller, but proposed controller can be expressed.

Notations

$\mathcal{R}$	field of real numbers
$\mathcal{R}(s)$	the space of all real-rational transfer function
$\left[\begin{array}{c\|c} A & B \\ \hline C & D \end{array}\right]$	shorthand for the state space description $C(sI-A)^{-1}B+D$
RH_∞	Stable proper rational transfer matrices
$\mathcal{U}$	Unimodular matrices on RH_∞,that is if $G(s) \in \mathcal{U}$, then $G(s), G^{-1}(s) \in RH_\infty$.

2. SIMULTANEOUS STABILIZATION PROBLEM

Let us consider the control problem to stabilize all of rth number of the systems $G_i(s)(i=1,\dots,r) \in \mathcal{R}^{p\times m}(s)$ given by

$$G_i(s) = \left[\begin{array}{c|c} A_i & B_i \\ \hline C_i & 0 \end{array}\right] \tag{1}$$

by using controller $C(s)$

$$u = C(s)y. \tag{2}$$

Here $A_i \in \mathcal{R}^{n_i\times n_i}(i=1,\dots,r)$, $B_i \in \mathcal{R}^{n_i\times m}$, $C_i \in \mathcal{R}^{p\times n_i}$, $C(s) \in \mathcal{R}^{m\times p}(s)$, (A_i, B_i) is controllable, (C_i, A_i) is observable. For simple explanation, it is assumed to satisfy $m \geq p$, $rp < m+p$. If $m \geq p$, the condition $rp < m+p$ always can be hold by projecting the output y to a single input(Minto and Vidyasagar, 1986). Let us call controller $C(s)$ simultaneous stabilizing controller if $C(s)$ stabilize all of the plant defined by (1).

By using the idea of filtered inverse systems(Yoshikawa and Sugie, 1982; Yamada and Watanabe, 1992; Yamada and Watanabe, 1996a), simultaneous stabilizing controller is obtained following procedure.

Procedure 1

(1) Obtain left coprime factorization on RH_∞ of all of $G_i(s)$ as $G_i(s) = D_i^{-1}(s)N_i(s)$. From (Nett *et al.*, 1984), state space description of left coprime factorization of $G_i(s) = D_i^{-1}(s)N_i(s)$ is given by

$$\begin{cases} D_i(s) = \left[\begin{array}{c|c} A_i - H_iC & -H_i \\ \hline C_i & I_p \end{array}\right] \\ N_i(s) = \left[\begin{array}{c|c} A_i - H_iC_i & B_i \\ \hline C_i & 0_{r\times m} \end{array}\right] \end{cases} \tag{3}$$

Here K_i is appropriate size of matrix to make eigenvalues of $A_i - K_iC_i$ to be in the open left half plane.

(2) Let state space description of $M(s) \in \mathcal{R}^{rp\times(m+p)}(s)$ to be

$$M(s) = \begin{bmatrix} D_1(s) & N_1(s) \\ \vdots & \vdots \\ D_r(s) & N_r(s) \end{bmatrix} = \left[\begin{array}{c|c} A_M & B_M \\ \hline C_M & D_M \end{array}\right]. \tag{4}$$

Here, A_M, B_M, C_M,D_M is defined by

$$A_M = \begin{bmatrix} A_1 - H_1C_1 & & 0 \\ & \ddots & \\ 0 & & A_r - H_rC_r \end{bmatrix} \in \mathcal{R}^{n\times n} \tag{5}$$

$$B_M = \begin{bmatrix} -H_1 & B_1 \\ \vdots & \vdots \\ -H_r & B_r \end{bmatrix} \in \mathcal{R}^{n\times(m+p)} \tag{6}$$

$$C_M = \begin{bmatrix} C_1 & & 0 \\ & \ddots & \\ 0 & & C_r \end{bmatrix} \in \mathcal{R}^{rp\times n} \tag{7}$$

$$D_M = \begin{bmatrix} I_p & 0_{m\times p} \\ \vdots & \vdots \\ I_p & 0_{m\times p} \end{bmatrix} \in \mathcal{R}^{rp\times(m+p)} \tag{8}$$

$$n = \sum_{i=1}^{r} n_i$$

respectively.

(3) For easy explanation, it is assumed that regular matrix $T \in R^{rp \times rp}$ exists such that the system $TM(s)$ can be decoupled by state feedback control. Find such T.

(4) Obtain right inverse system $R(s) \in RH_\infty^{(m+p)\times rp}$ of $TM(s)$ and right zeroing system $S(s) \in RH_\infty^{(m+p)\times(m+p-rp)}$ of $TM(s)$ that satisfy

$$TM(s)R(s) = V(s) \tag{9}$$

$$TM(s)S(s) = 0_{rp\times(m+p-rp)} \tag{10}$$

$$V(s) = \operatorname{diag}\left\{\underbrace{1,\ldots,1}_{p}, \frac{1}{(1+s\beta)^{\alpha_{p+1}}}, \ldots, \frac{1}{(1+s\beta)^{\alpha_{rp}}}\right\} \quad (\beta > 0). \tag{11}$$

(5) Assign $U_i(s) \in \mathcal{U}^{p\times p}(i = 1,\ldots,r)$ to satisfy

$$V^{-1}(s)\begin{bmatrix} U_1(s) \\ \vdots \\ U_r(s) \end{bmatrix} = V^{-1}(s)U(s) \in RH_\infty^{rp\times p}. \tag{12}$$

(6) Parametrization of all simultaneous stabilization controller $C(s)$ is given by

$$C(s) = N_c(s)D_c^{-1}(s) \tag{13}$$

(Chen *et al.*, 1995). Here let $D_c(s)$, $N_c(s)$ be

$$\begin{bmatrix} D_c(s) \\ N_c(s) \end{bmatrix} = R(s)V^{-1}(s)U(s) + S(s)Q(s) = \begin{bmatrix} R_1(s) \\ R_2(s) \end{bmatrix} V^{-1}(s)U(s) + \begin{bmatrix} S_1(s) \\ S_s(s) \end{bmatrix} Q(s) \tag{14}$$

respectively, and $Q(s) \in RH_\infty^{(m+p-rp)\times p}$ is the system such that the direct term of $Q(s)$ is regular and the direct pass term of $R_1(s)T^{-1}V^{-1}(s)U(s) + S_1(s)Q(s)$ is regular.

To obtain simultaneous stabilizing controller by using this procedure, $R(s)$ and $S(s)$ must be stable. This requires all of zeros of $M(s)$ must be in open left half plane(Chen *et al.*, 1995). The condition that $M(s)$ must have no unstable zeros, is clear from the result in (Yoshikawa and Sugie, 1982; Yamada and Watanabe, 1992; Yamada and Watanabe, 1996a).

Remarks 1. Procedure 1 produce illusion such that arbitrary system can be simultaneously stabilizable. It is well known that the condition of existence of $U(s)$ in (5) is ascribed to strong stabilization problems(Chen *et al.*, 1995). For single-input single-output systems, the condition of existence of $U(s)$ is equivalent to p.i.p condition(Chen *et al.*, 1995).

Remarks 2. This procedure does not guarantee that N_c and D_c is coprime, but coprimeness is modulated by $Q(s)$ and coprimeness is confirmed by easy manipulation. Therefore coprimeness of N_c and D_c is not problem.

3. STATE SPACE DESIGN

From preceding section, $R(s)$ and $S(s)$ must be stable. Therefore in this section , $M(s)$ has assumed to have no unstable zeros. Under this assumption, we give explicit state space description of $V(s)$, $R(s)$, $S(s)$ in Procedure 1.

Let C_T and D_T be

$$C_T = TC_M \tag{15}$$

$$D_T = TD_M = \begin{bmatrix} I_p & 0 \\ 0 & 0 \end{bmatrix} \tag{16}$$

respectively. Factorize C_T be

$$C_T = \begin{bmatrix} C_{T1} \\ \vdots \\ C_{T(rp)} \end{bmatrix} \quad C_{Ti} \in R^{1\times m}. \tag{17}$$

Set $\alpha_i (i = 1,\ldots,rp)$ in (11), β_{ij}, Φ and X to be

$$\alpha_i = \begin{cases} 0 \quad (i = 1,\ldots,p) \\ \min\left(j | C_{Ti}A_M^{j-1}B_M \neq 0; j = 1,\ldots n\right) \\ \quad (i = p+1,\ldots,rp) \end{cases} \tag{18}$$

$$\beta_{ij} = \left\{ \begin{matrix} \alpha_i \\ j \end{matrix} \right\} \beta^{-j} \tag{19}$$

$$(i = p+1,\ldots,p; j = 1,\ldots,\alpha_i)$$

Ψ

$$= \begin{bmatrix} C_{T1} \\ \vdots \\ C_{Tp} \\ C_{T(p+1)}A_M^{\alpha_{p+1}} + \beta_{(p+1)1}C_{T(p+1)}A_M^{\alpha_{p+1}-1} \\ +\cdots+\beta_{(p+1)\alpha_{p+1}}C_{T(p+1)} \\ \vdots \\ C_{T(rp)}A_M^{\alpha_{rp}} + \beta_{(rp)1}C_{T(rp)}A_M^{\alpha_{rp}-1} \\ +\cdots+\beta_{(rp)\alpha_{rp}}C_{T(rp)} \end{bmatrix} \tag{20}$$

$$\Phi = \begin{bmatrix} I_p & 0 \\ C_{T(p+1)}A_M^{\alpha_{p+1}-1}B_M \\ \vdots \\ C_{T(rp)}A_M^{\alpha_{rp}-1}B_M \end{bmatrix} \tag{21}$$

$$X = \begin{bmatrix} I_p & & & 0 \\ & \beta_{(p+1)\alpha_{(p+1)}} & & \\ & & \ddots & \\ 0 & & & \beta_{rp\alpha_{rp}} \end{bmatrix} \tag{22}$$

respectively.

From the assumption that the system (1) can be decoupled by state feedback control,

$$\operatorname{rank}\Phi = rp \tag{23}$$

holds. $\hat{\Phi}$ and $\bar{\Phi}$ exist satisfying

$$\Phi\hat{\Phi} = I_{rp} \tag{24}$$

$$\Phi\bar{\Phi} = 0_{rp\times(m+p-rp)}. \tag{25}$$

Of all $\hat{\Phi}$ satisfying (25), by using Φ to make $A_M - B_M\hat{\Phi}\Psi$ be stable, state space description of $R(s)$ and $S(s)$ are written by

$$R(s) = \left[\begin{array}{c|c} A_M - B_M\hat{\Phi}\Psi & B_M\hat{\Phi}X \\ \hline -\hat{\Phi}\Psi & \hat{\Phi}X \end{array}\right] \tag{26}$$

$$S(s) = \left[\begin{array}{c|c} A_M - B_M\hat{\Phi}\Psi & B_M\bar{\Phi} \\ \hline -\hat{\Phi}\Psi & \bar{\Phi} \end{array}\right] \tag{27}$$

respectively. The condition there exists Φ to make $A_M - B_M\hat{\Phi}\Psi$ to be stable was considered in (Moylan, 1977; Yamada and Watanabe, 1992; Yamada and Watanabe, 1996a). From (Moylan, 1977; Yamada and Watanabe, 1992; Yamada and Watanabe, 1996a) if $M(s)$ has no unstable zeros, then such Φ always exists.

State space description of $V(s)$ is denoted by

$$V(s) = \left[\begin{array}{c|c} A_M - B_M\hat{\Phi}\Psi & B_M\hat{\Phi}X \\ \hline C_T - D_T\hat{\Phi}\Psi & D_M\hat{\Phi}X \end{array}\right]. \tag{28}$$

$R(s), S(s), V(s)$ denoted by (26), (27), (28) satisfy the conditions in previous section. To proof of this result is shown by easy numerical manipulation as (Yamada and Watanabe, 1992; Yamada and Watanabe, 1996a). Therefore proof is omitted on account of limited space.

4. EXPANSION FOR THE SYSTEM WITH UNSTABLE ZEROS

In previous section, we discuss under the assumption that $M(s)$ has no unstable zeros. In this section, Procedure 1 will be expanded to obtain simultaneous stabilizing controller for $M(s)$ with unstable zeros.

4.1 *Procedure*

Procedure to obtain simultaneous stabilizing controller for $M(s)$ with unstable zeros was considered in (Chen *et al.*, 1995). (Chen *et al.*, 1995) discusses only idea to obtain simultaneous stabilizing controller but calculating procedure. Therefore procedure in (Chen *et al.*, 1995) can not give simultaneous stabilizing controller. In this paper, we propose procedure with easy calculation of simultaneous stabilizing controller by using the structure of stable filtered inverse systems(Yamada and Watanabe, 1996b).

Procedure 2

(1) Obtain left coprime factorization on RH_∞ of all of $G_i(s)$ as $G_i(s) = D_i^{-1}(s)N_i(s)$ by using (3).
(2) Let $M(s) \in \mathcal{R}^{rp\times(m+P)}(s)$ be (4).
(3) For easy explanation, it is assumed that regular matrix $T \in R^{rp\times rp}$ exists such that the system $TM(s)$ can be decoupled by state feedback control. Find such T.
(4) Obtain right stable filtered inverse system $R(s) \in RH_\infty^{(m+p)\times rp}$ of $TM(s)$ and right stable zeroing system $S(s) \in RH_\infty^{(m+p)\times(m+p-rp)}$ of $TM(s)$ that satisfy

$$TM(s)R_{st}(s) = V(s)M_{unst}(s) \tag{29}$$

$$TM(s)S_{st}(s) = 0_{rp\times(m+p-rp)}. \tag{30}$$

Here $M_{unst}(s) \in RH_\infty^{rp\times rp}$ has zeros of all unstable zeros of $M(s)$, and direct pass term of $M_{unst}(s)$ is regular matrix.
(5) Assign $U_i(s) \in \mathcal{U}^{p\times p}(i = 1,\ldots,r)$ to satisfy (12).
(6) Simultaneous stabilizing controller $C(s)$ is given by

$$C(s) = N_c(s)D_c^{-1}(s). \tag{31}$$

Where $D_c(s)$, $N_c(s)$ are denoted by

$$\begin{bmatrix} D_c(s) \\ N_c(s) \end{bmatrix}$$
$$= R_{st}(s)M_{unst}^{-1}(s)V(s)U(s) + S(s)f^{-1}(s)Q^*(s)\,(I_p + Q(s)f(s))$$
$$= \begin{bmatrix} R_{st1}(s) \\ R_{st2}(s) \end{bmatrix} M_{unst}^{-1}(s)V(s)U(s) + \begin{bmatrix} S_{st1}(s) \\ S_{st2}(s) \end{bmatrix} f^{-1}(s)Q^*(s)\,(I + Q(s)f(s)). \quad (32)$$

$f(s)$ is setted to satisfy

$$f(s) = \prod_{i=1}^{\rho} \frac{(s-\gamma_i)}{s+\bar{\gamma}_i} \quad (33)$$

,where $\gamma_1, \ldots, \gamma_\rho$ is unstable zeros of $M_{unst}(s)$. $\{\bar{\cdot}\}$ means conjugate complex number of $\{\cdot\}$. $Q^*(s) \in RH_\infty^{rp\times p}$ holds

$$\begin{bmatrix} D_c(s) \\ N_c(s) \end{bmatrix} \in RH_\infty^{(m+p)\times p}$$

, that is

$$R_{st}(s)M_{unst}^{-1}(s)V(s)U(s) + S(s)f^{-1}(s)Q^*(s) \in RH_\infty^{(m+p)\times p}.$$

$Q(s)$ is arbitrary system such that direct pass term of $Q(s)$ is regular and $Q(s)$ makes

$$R_{st1}(s)M_{unst}^{-1}(s)V(s)U(s) + S_{st1}(s)f(s)Q^*(s)$$

to have regular direct pass term . If $M_{unst}(s)$ has simple unstable zeros, then problem to set $Q^*(s)$ is reduced to the problem of interpolation on RH_∞ to holds

$$(s-\gamma_i)\left\{R_{st}(s)M_{unst}^{-1}(s)V(s)U(s) + S(s)f^{-1}(s)Q^*(s)\right\}\Big|_{s=\gamma_i} = 0 \quad (34)$$
$$(i = 1, \ldots, \rho).$$

$C(s)$ denoted in (31) is confirmed to be simultaneous stabilizing controller as follows. Since $Q^*(s)$ satifies interpolation condition of (35),

$$(s-\gamma_i)\left\{R_{st}(s)M_{unst}^{-1}(s)V(s)U(s) + S(s)f^{-1}(s)Q^*(s)(I_p + Q(s)f(s))\right\}\Big|_{s=\gamma_i} = 0$$
$$(i = 1, \ldots, \rho) \quad (35)$$

is satisfied too. From $Q(s) \in RH_\infty$, $N_c(s), D_c(s) \in RH_\infty$ hold. (29) and (30) yield

$$M(s)\begin{bmatrix} D_c(s) \\ N_c(s) \end{bmatrix}$$
$$= M(s)R_{st}(s)M_{unst}^{-1}(s)V(s)U(s) + M(s)S(s)f^{-1}(s)Q^*(s)\,(I_p + Q(s)f(s))$$
$$= U(s). \quad (36)$$

This equation means

$$D_i(s)D_c(s) + N_i(s)N_c(s) = U_i \;(i = 1, \ldots, r). \quad (37)$$

Since $D_c(s), N_c(s) \in RH_\infty$, from (Vidyasagar, 1985), controller $C(s)$ stabilizes all of $G_i(s)(i = 1, \ldots, r)$.

When $M(s)$ has unstable zeros, class of simultaneous controller is narrower than the case that $M(s)$ has no unstable zeros. In addition $Q^*(s)$ is depend on $U(s)$. In general if $U(s)$ is changed, then $Q^*(s)$ must be changed. This limitation is exists in procedure proposed in (Chen *et al.*, 1995).

Next we will show when we change $U(s)$, $Q^*(s)$ can be obtained by easy and simple calculation. From the assumption (12), $V(s)U(s) \in RH_\infty$. $Q^*(s)$ is constructed with minimum change by

$$Q^*(s) = \hat{Q}(s)V(s)U(s). \quad (38)$$

Here $\hat{Q}(s) \in RH_\infty$ is system to satisfy interpolation condition given by

$$(s-\gamma_i)\left\{R_{st}(s)M_{unst}^{-1}(s) + S(s)f^{-1}(s)\hat{Q}(s)\right\}\Big|_{s=\gamma_i} = 0 \quad (39)$$
$$(i = 1, \ldots, \rho).$$

4.2 *State space design method*

In this subsection, we give state space design method of $V(s)$, $M_{unst}(s)$, $R_{st}(s)$, $S_{st}(s)$ that requires in Procedure 2. Procedure is obtained by using same numerical manipulation to (Yamada and Watanabe, 1996b). Therefore here is shown only result.

Set T to make $TM(s)$ to be decoupled by using state feedback control. Decide C_T, D_T, β_{ij}, Φ, Ψ to (15), (16), (20), (21), (20) respectively. Set $\hat{\Phi}$ to satisfy (24) and to satisfy $A_M - B_M\hat{\Phi}\Psi$ has no poles on the imaginary axis.Let $\bar{\Phi}$ hold (25), and let $\bar{D}_l \in \mathcal{R}^{rp\times rp}$ be appropriate regular matrix.

From the assumption, the Riccati equation

$$P(A_M - B_M\hat{\Phi}\Psi) + (A_M - B_M\hat{\Phi}\Psi)^T P - P(B_M\hat{\Phi}X\bar{D}_l^{-1})E^{-1}(B_M\hat{\Phi}X\bar{D}_l^{-1})^T P = 0 \quad (40)$$

$$E = \bar{D}_l^{-T}\bar{D}_l^{-1} \quad (41)$$

has solution $P = P^T \geq 0$. By using this solution P, let F be

$$F = -\bar{D}_l X^{-1}\Psi - E^{-1}(B_M\hat{\Phi}X\bar{D}_l^{-1})^T P \quad (42)$$

, then $A_M + B_M\hat{\Phi}X\bar{D}_l^{-1}F$ is stable.

By using this F, $\hat{\Phi}$ and $\bar{\Phi}$, state space description of $R_{st}(s)$, $S_{st}(s)$, $M_{unst}(s)$ are given by

$$R_{st}(s) = \left[\begin{array}{c|c} A_M + B_M\hat{\Phi}X\bar{D}_l^{-1}F & B_M\hat{\Phi}X\bar{D}_l^{-1}E^{-\frac{1}{2}} \\ \hline \hat{\Phi}X\bar{D}_l^{-1}F & \hat{\Phi}X\bar{D}_l^{-1}E^{-\frac{1}{2}} \end{array}\right] \tag{43}$$

$$S_{st}(s) = \left[\begin{array}{c|c} A_M + B_M\hat{\Phi}X\bar{D}_l^{-1}F & B_M\bar{\Phi} \\ \hline \bar{\Phi}F & \bar{\Phi} \end{array}\right] \tag{44}$$

$$M_{unst}(s) = \left[\begin{array}{c|c} A_M + B_M\hat{\Phi}X\bar{D}_l^{-1}F & B_M\hat{\Phi}X\bar{D}_l^{-1}E^{-\frac{1}{2}} \\ \hline F & E^{-\frac{1}{2}} \end{array}\right] \tag{45}$$

respectively. In addition state space description of $V^{-1}(s)$ is denoted by(28).

5. CONCLUSION

In this paper, we proposed a procedure to obtain simultaneous stabilizing controller by using the idea of filtered inverse systems(Yoshikawa and Sugie, 1982; Yamada and Watanabe, 1992; Yamada and Watanabe, 1996a). This procedure was simplified more than by Minto and Vidyasagar(Minto and Vidyasagar, 1986), Chen et al.(Chen and Chow, 1993; Chen *et al.*, 1995). We showed explicit state space description of simultaneous stabilizing controller. Both for the system with no unstable zeros and with unstable zeros, procedure to obtain simultaneous stabilizing controller and calculation method was proposed. For the system with unstable zeros, simultaneous stabilizing controller proposed by Chen et al.(Chen and Chow, 1993; Chen *et al.*, 1995) is not denoted by free parameter, but proposed controller is denoted by using free parameter.

The author would like to express his gratitude to Prof. K. Watanabe of Yamagata University and Prof. M. Ikeda of Osaka University and Prof. Y. Ohta of Osaka University for helpful discussions.

6. REFERENCES

Chen, H. B. and J. H. Chow (1993). Simultaneous stabilization of systems with common components. *Proc. 1993 American Control Conf.* pp. 3155–3159.

Chen, H. B., J. H. Chow, M. A. Kale and K.D.Minto (1995). Simultaneous stabilization using stable system inverseion. *Automatica* **31**, 531–542.

Ghosh, B. K. (1986). Trancendental and interpolation method in simultaneous stabilizing and simultaneous partial pole placement problem. *SIAM J. Control optimization* **24**, 1091–1109.

Ghosh, B. K. and C. I. Byrnes (1983). Simultaneous stabilization and simultaneous pole-placement by nonswitching dynamic. *IEEE Trans. on Automatic Control* **28**, 735–741.

Kale, M. A., J.H.Chow and K. D. Minto (1990). A contoller parametrization and pole-placement design for simultaneous stabilization. *Proc. 1990 Conf. American Control Conference* pp. 116–121.

Minto, K. D. and M. Vidyasagar (1986). A state space approach to simultaneous stabilization. *Control Theory And Advanced Technology* **2**, 39–64.

Moylan, P. J. (1977). Stable inversion of linear systems. *IEEE Trans. on Automatic Control* **22**, 74–78.

Nett, C. N., C. A. Jacobson and M.J.Balas (1984). A connection between state-space and doubly coprime fractional representation. *IEEE Trans. on Automatic Control* **29**, 831–832.

Saeks, R. and J.Murray (1982). Fractional representation, algebric geometry and the simultaneous stabilizationproblm. *IEEE Trans. on Automatic Control* **27**, 895–903.

Veilette, R. J., J. V. Medanic and W. R. Perkins (1982). Design of reliable control systems. *IEEE Trans. on Automatic Control* **27**, 1085–1095.

Vidyasagar, M. (1985). *Control System Synthesis: A Factrization Approach.* MIT Press. Cambridge.

Vidyasagar, M. and N. Viswanadham (1982). Algebraic design techniques for reliable stabilization. *IEEE Trans. on Automatic Control* **27**, 1085–1095.

Wei, K. (1986). The solution of a transcendental problem and its application in simultaneous stabilization problem. *IEEE Trans. on Automatic Control* **37**, 1305–1315.

Yamada, K. and K. Watanabe (1992). A state space design method of filtered inverse systems. *Trans. of the Society of Instrument and Control Engineers* **28**, 923–930(in Japanese).

Yamada, K. and K. Watanabe (1996a). A state space design method of filtered inverse systems and their application. *MTNS '96.*

Yamada, K. and K. Watanabe (1996b). A state space design method of stable filtered inverse systems and its application to h_2 suboptimal internal model control. *Proc. IFAC'96* pp. 379–382.

Yoshikawa, T. and T. Sugie (1982). Filtered inverses with application to servo systems. *Trans. of the Society of Instrument and Control Engineers* **18**, 792–799(in Japanese).

C-2a-01-5
14th Triennial World Congress, Beijing, P.R. China

WEIGHT SELECTION FOR H_∞-CONTROL USING GENETIC ALGORITHMS

Anna-Karin Christiansson* Bengt Lennartson**

* *University of Trollhättan/Uddevalla, Trollhättan, Sweden, akc@thn.htu.se*
** *Chalmers University of Technology, Gothenburgh, Sweden, bl@s2.chalmers.se*

Abstract: A method for selecting weighting functions for $\mathcal{H}_\infty$-controller design is proposed. Performance and robustness of the closed loop system are judged by treating low, mid and high frequency properties together. The method includes a genetic algorithm for treating the nonlinear optimisation. The procedure gives an objective tool for comparing different weighting function structures. Results show that an extremely simple S/KS-design may be as attractive as more complicated designs. Furthermore, comparisons show that the $\mathcal{H}_\infty$-case can further improve performance compared to the $\mathcal{H}_2$-case without losing stability. Examples are shown for both a SISO- and a MIMO-plant.

Keywords: H-infinity-control, weighting functions, performance characteristics, genetic algorithms, MIMO

1. INTRODUCTION

It is a non-trivial task to select optimal weights in $\mathcal{H}_2$- and $\mathcal{H}_\infty$-controller design. First, a suitable ***structure*** of weights is needed, and then for each structure, suitable ***parameters*** are to be chosen. In the literature, different methods for weight selection are proposed, e.g. (Green and Limebeer, 1995; Maciejowski, 1989; Skogestad and Postlethwaite, 1996). Most of them turn out to be variations of $S/KS/T$-structures, where S stands for the sensitivity function, T for the complementary sensitivity function, and KS for the control sensitivity function. To achieve a low order controller, the orders of the weighting functions has to be low. In this paper, some fairly simple weighting functions are suggested based on a systematic evaluation procedure.

The issue in $\mathcal{H}_\infty$-controller design is to minimise the norm $||G_{zw}||_\infty = \max_{w \neq 0}(||z||_2/||w||_2)$, where the performance measure is z and the exogenous input is w. In the structure according to Figure 1, the plant to control is G, the controller to design is K, and the weights used in the design procedure are W_T, W_S, W_{KS} and W_v. The transfer matrix from w to z for the so augmented system (grey boxes in the figure) can be written as

$$G_{zw} = \begin{bmatrix} -W_T T & W_T S G W_V \\ W_S S & W_S S G W_V \\ -W_{KS} K S & -W_{KS} T W_V \end{bmatrix}$$

This shows that W_T is mainly used to form the complementary sensitivity function T; W_S forms the sensitivity function S and W_{KS} forms the control sensitivity function KS. The weight W_V is introduced when it is necessary to avoid pole-zero cancellations, e.g. for highly resonant plants. Note however, that the close relationship between these functions makes it impossible to handle them separately.

This paper is organised as follows: Section 2 presents the used performance measures, recently proposed in (Lennartson and Kristiansson, 1997).

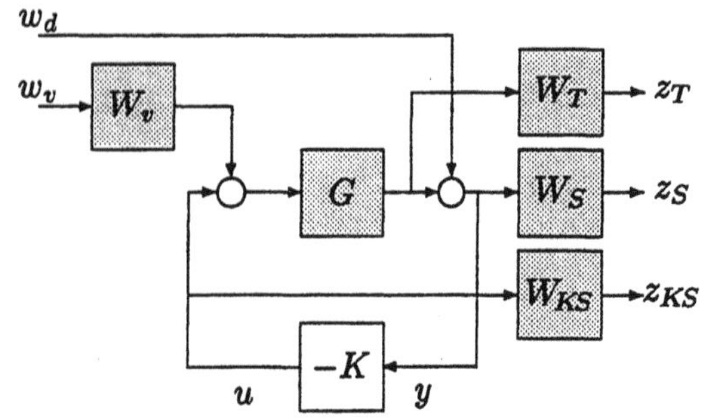

Fig. 1. Proposed $S/KS/T$-weighting function structure. Grey blocks constitutes the augmented plant.

Section 3 discusses the proposed weight structure and suitable parameters. The optimisation is carried through using a genetic algorithm presented in Section 4 and the method is applied on two plants with results shown in Sections 5 and 6. Finally, in Section 7 conclusions are drawn. The examples shown are run in a Matlab environment using some functions from the μ-toolbox.

2. EVALUATION MEASURE FOR CONTROLLER DESIGN

This section presents the evaluation measures that are used for performance and robustnes of the closed loop given a controller K. They have recently been proposed by (Lennartson and Kristiansson, 1997). The closed loop behaviour is considered in both low-, mid-, and high- frequency ranges in a unified framework.

First, in the low frequency range, integral action is of major interest. A performance measure that approximates the integral gain is J_{LF}, defined as

$$J_{LF} = \|W_{LF}GS\|_\infty \tag{1}$$

where $W_{LF} = 1/s$ in the SISO-case. A smaller J_{LF} implies a smaller output error. Thus one design issue is to minimise J_{LF}.

Second, in the mid frequency range the main issue is to ensure good stability margins. Three constraints are considered in the generalised maximum sensitivity, GM_S. These are in the SISO-case,

- S_m, the maximum value of the sensitivity function, equal to the inverse of the shortest distance for GK to the point [-1,0] in the Nyquist plot
- T_m, the maximum value of the complementary sensitivity function. It is strongly related to the phase margin
- G_m, the gain margin

In the examples below, the limits

$$S_m \leq 1.7 \qquad T_m \leq 1.3 \qquad A_m \geq 3.0 \tag{2}$$

have been used for the SISO-application. For the MIMO-application, different "directions" (for the properly scaled plant) have been treated so that they give roughly equal measures. Furthermore, in the absence of a good measure for the gain margin in the MIMO case, only the limits $S_m \leq 1.4$ and $T_m \leq 1.3$ have been considered. These turn out to be adequate for the plants investigated so far.

Third, in the high frequency range the focus is mainly on measurement noise and high frequency uncertainties in the plant description. This is taken into account through the measure J_{HF}, defined as

$$J_{HF} = \|W_{HF}KS\|_\infty \tag{3}$$

where W_{HF} is a high pass filter with unit gain above half the closed loop bandwidth. Thus, J_{HF} reflects the size of the control signal in the high frequency domain. Therefore it shall be minimised, especially when the actuators are sensitive to large and rapid control variations.

Finally, all these measures (1) - (3) are taken into account when choosing the best controller. For all controllers that pass GM_S-constraints (2), [J_{HF},J_{LF}]-values are plotted as points in a J_{LF}/J_{HF}-diagram. To evaluate which is the "best" controller among several, choose the one that gives the smallest J_{LF} given a certain J_{HF}-value. See (Lennartson and Kristiansson, 1997) for a more detailed discussion on this topic.

3. WEIGHT PROPOSALS

The weighting function selection includes two steps: first choose a ***structure*** of weights, then choose ***parameter values*** for each weight. It is worth noting, that a ***weighting function*** in this respect is a filter of appropriate size and with appropriate frequency properties.

In this work different structures have been investigated and the one presented in Figure 1 has shown to give roughly the same properties as others of equivalent complexity suggested in the literature, see e.g. (Skogestad and Postlethwaite, 1996).

Yet another design issue is to keep the controller order low. As is known, the controller order will be the same as that of the augmented system, i.e. plant G plus weights W_x. Therefore, a recommendation is to choose as many weights as possible to be constant and thus not contribute to increasing the total order. Different weights have been investigated and those with asymptotic gains in Figure 2 have shown to give best results. The weight W_S, for the sensitivity function S, is mainly introduced to provide integral action,

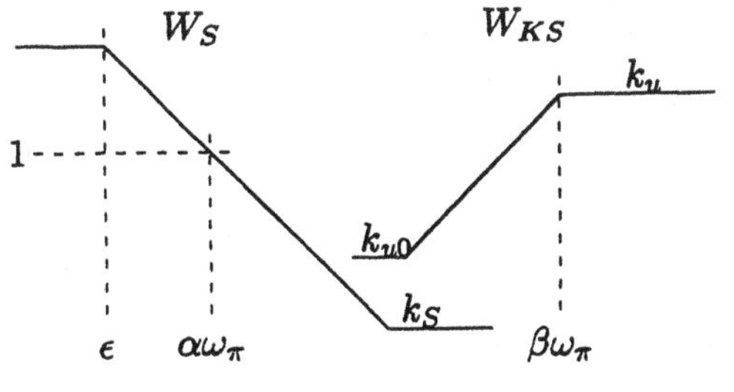

Fig. 2. Asymptotic amplitude diagrams for W_{KS}- and W_S-weight functions. w_π is the frequency where the plant phase equals $-\pi$ radians. W_T is held constant at level k_T.

and thus give small J_{LF}; ϵ and k_S are very small ($\approx 10^{-7}$). $\epsilon \neq 0$ is introduced since $\mathcal{H}_\infty$-design does not allow any poles on the imaginary axis.

The weight W_{KS}, for the control sensitivity function KS, gives a means to keep the control signal limited (J_{HF}). The reason to introduce a high pass weight (breakpoint at $\beta\omega_\pi$) is that there is no need to put large penalty on the control signal at low frequencies. The reason for not letting k_{u0} be zero, however, is that there must be some penalty on the control signal for an $\mathcal{H}_\infty$-solution to exist. Also the complementary sensitivity function T is affected by W_{KS}, since $KS = G^{-1}T$ (provided G is invertible). With constant high frequency gain for W_{KS}, there is therefore no need to introduce an extra high frequency weight W_T, if T can have the same roll-off as the plant.

Thus, α, β, k_u, k_T, k_S are tuning parameters in this case. For each set of parameter values, a continuous-time $\mathcal{H}_\infty$-controller is designed. To judge whether this controller is good in some sense, the closed loop measures (J_{HF}, J_{LF} and GM_S) are calculated as described in Section 2.

4. AUTOMATED EVALUATION PROCEDURE BASED ON A GENETIC ALGORITHM

The number of parameters introduced in the weighting functions was initially high, roughly two per weight and input/output. The number of weighting functions in the structure was initially six, three at plant input at three at its output. The problem is multivariable and includes γ-iteration in the search of smallest γ-value. It includes solving two nonlinear Riccati differential equations. The criteria for selecting the "best" controller are rather complex, see Section 2. Therefore some automatic optimisation procedure was looked for. Experiences from a number of manually optimised controllers have been implemented into a genetic algorithm, which has shown to be both powerful and flexible in the search of the best weights, see below.

4.1 *Genetic algorithm*

The concept genetic algorithm imitates the genetic evolution in real life and relies on survival of the fittest (recombination of good characteristics). Chromosomes are made up of genes and a chromosome uniquely defines the properties of a specimen. The chromosomes in the parent generation affect the properties of the next generation. The main idea when using genetic algorithms for optimisation, is to combine the "best" parents in some respect to yield even better offspring, but also to allow for random crossovers and mutations. One central point is to evaluate every specimen. Iterations are performed until desired convergence is reached (Goldberg, 1989).

This approach is applied to the weight selection in the following way.

- generate many (≈ 100) sets of design parameters, chromosomes. The genes are randomly distributed in as large a test space as is meaningful for the case in question
- design an $\mathcal{H}_\infty$-controller for each chromosome and calculate J_{HF}-, J_{LF}- and GM_S-values
- give each controller in the generation a point in a J_{HF}/J_{LF}-diagram
- for each generation find a curve of the best points passing the GM_S-constraint according to Section 2
- calculate a relative measure (r_m) for each controller based on
  - · distance to the best curve so far
  - · J_{LF}-value
  - · bandwidth ω_b (4)
- generate new chromosomes as combinations of the best ones from former generation plus more randomly distributed chromosomes

The procedure is iterated until desired convergence is reached. How the relative measure r_m is calculated in the parent generation and how it is used for generation of new chromosomes is of vital interest for the quality of the optimisation and for the convergence rate. The initial test space is also fundamental for the result. When other properties are of interest, the criteria in (4) can easily be modified. This is one of the advantages with this optimisation method. More details on this subject can be found in (Christiansson, 1998).

5. EXAMPLE 1: WEIGHT SELECTION FOR A SISO-PLANT

As an example, results from the optimisation method will first be presented applied to a SISO-plant with transfer function

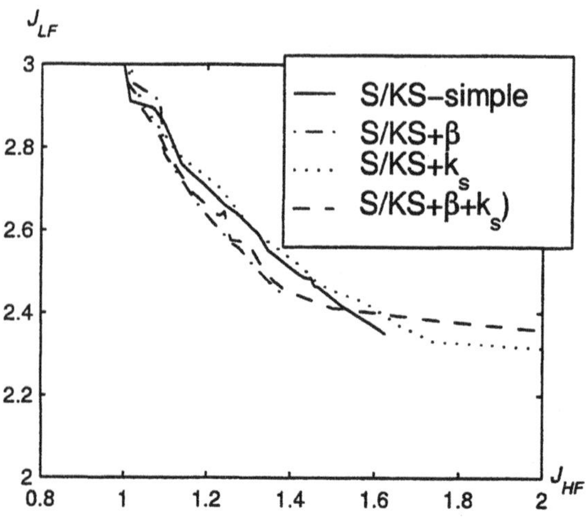

Fig. 3. SISO: J_{HF}/J_{LF}-curves for S/KS-designs. Solid curve refers to simple S/KS-design; dash-dotted: breakpoint β introduced; dotted: k_S introduced and dashed: both β and k_S introduced.

$$G(s) = \frac{e^{-s}}{(1+s)^2} \tag{5}$$

This plant has $\omega_\pi \approx 1.4$[rad/s], the frequency where the plant phase equals $-\pi$ radians. Its κ-value, defined as $\kappa = |G(j\omega_\pi)|/|G(j0)|$, is κ=0.37, see (Åström and Hägglund, 1995). The higher κ the more difficult to control ($0 < \kappa < 1$). The weighting function structure is according to Figure 1 and the tuning parameters are according to Figure 2. However, many more structures and parameter choices have been investigated, but they have not been so good.

First an extremely simple case is shown, and then one weight option (parameter) at a time is introduced to show the influence of this single parameter. Recall, that a small J_{LF}-value is desired for a given J_{HF}-value. In the graphs, different line styles refer to different cases described below, see Figure 3 for details.

5.1 *Parameter influence in S/KS-design*

The first simple S/KS-design has W_{KS} constant, $k_S = 10^{-7}$, and W_T=0. The parameters that are free for optimisation are thus α and k_u. This results in the solid J_{HF}/J_{LF}- curve in Figure 3. The bandwidth depends on J_{HF} as shown solid in Figure 4.

The role of a breakpoint in W_{KS} (β). The reason for introducing the β-parameter in W_{KS}, see Figure 2, is to allow larger control signal at lower frequencies. This results in the dash-dotted curves in Figures 3 and 4. It is clear that this breakpoint, at $\beta\omega_\pi$, for a given J_{HF} gives reduced J_{LF} and

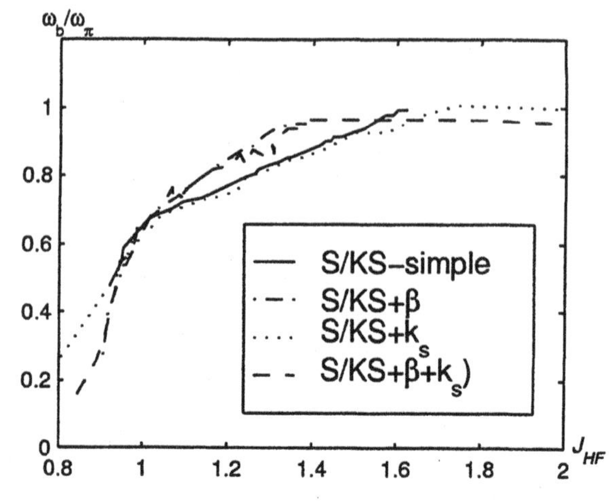

Fig. 4. SISO: Relative bandwidth, ω_b/ω_π, as a function of J_{HF} for S/KS-designs. See text in Figure 3 for more details.

increased bandwidth ω_b. The prize for this improvement is, as usual, that the resulting control signal increases at higher frequencies. Optimal value on the breakpoint parameter β turns out to be in the range from 0.5 to 1.

The role of k_S and k_{u0}. The parameters k_S and k_{u0} have also been used as tuning parameters, since they are proposed in (Skogestad and Postlethwaite, 1996). k_S should, according to them, prevent overshooting and k_{u0} should ensure W_S to dominate at lower frequencies. In this study, however, no significant improvements have been noticed and therefore the suggestion is to leave them constant and small (e.g. 10^{-7}). Increased k_S rather gives larger J_{HF}, and hence increased bandwidth.

5.2 *Comparison with $\mathcal{H}_2$-design.*

Comparisons have been made with corresponding S/KS $\mathcal{H}_2$-designs, by letting the γ-value be large, e.g. 1000. The J_{LF}/J_{HF}-curves are close to the ones from $\mathcal{H}_\infty$-design but only for smaller J_{HF} (up to $\approx$ 0.9 for this plant). For larger, but still reasonable control signals, $\mathcal{H}_2$-solutions do not even exist. Similar results apply to the bandwidth. This clearly indicates that extra control energy can be used in the $\mathcal{H}_\infty$-case to improve performance without losing stability. An $\mathcal{H}_2$-design - LQG/LTR - may give sufficient stability margins, but requires considerably higher control energy.

5.3 *S/KS/T-design.*

Since constant W_T is proposed for LTR, loop transfer recovery, in the $\mathcal{H}_2$-case, (Jovik and Lennartson, 1996; Stein and Athans, 1987), it was also investigated here. As T for the considered

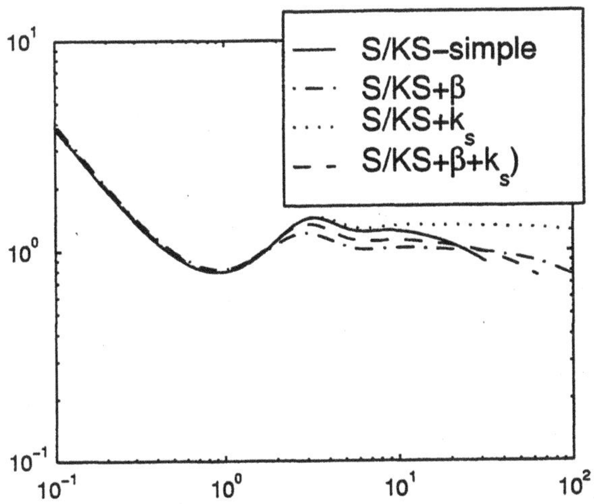

Fig. 5. Controller singular values as functions of ω [rad/s]. See Figure 3 for details .

plant model never had a peak for adequate parameter values, this made no improvement, rather the contrary. Probably this parameter is more relevant for resonant plants.

Introducing the additional weight W_V in Figure 1 will give a means to avoid pole-zero-cancellations. This is not relevant for the system under consideration. For resonant systems, however, this option is important.

5.4 *Controller singular values*

It is also interesting to study the controller singular values from the different designs. Figure 5 shows controller singular values from cases above with same J_{HF}-value, J_{HF}= 1.3. The differences are mainly in the frequency range from ω_π and upwards.

6. EXAMPLE 2: WEIGHT SELECTION FOR A MIMO JET ENGINE

Similar results as for the above presented SISO-application, have been achieved when designing weights for a simplified model of the jet engine, that is the target for the studies. This model has nine states, two inputs and two outputs. For this MIMO-application, the constraints in (4) have been modified in terms of properties in the two "directions" (due to the number of inputs/outputs). One desire was to reach as large and equal bandwidth as possible in both "directions". The results from S/KS-tuning (both with and without breakpoint in W_{KS}) are compared with results in (Härefors, 1995), where the weights are chosen by traditional frequency and time analysis. Some of those results are shown in Figures 6 and 7 and in Table 1. An "operating point" with $J_{HF}\approx$ 4.3 is chosen for the comparison. The circle

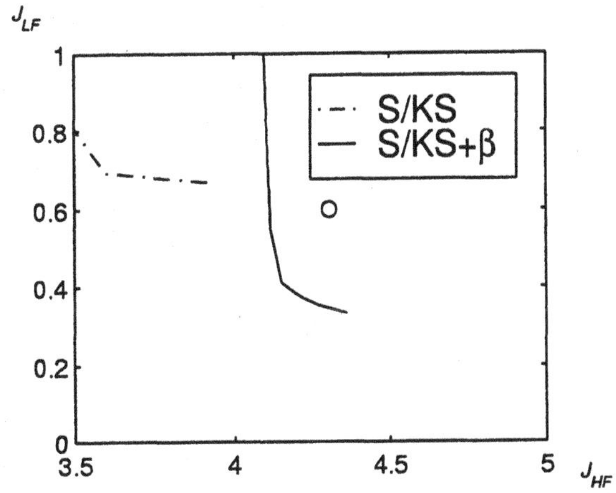

Fig. 6. $J_{HF}\,/\,J_{LF}$-curves for controlling a MIMO-jet engine, different designs. Solid curve refers to simple S/KS-design and dash-dotted when W_{KS} is high-pass. The circle is the point that stems from the original design.

in Figure 6 results from the original design. This has both larger value on J_{HF} and J_{LF} compared to the proposed S/KS-design with high pass W_{KS}. Similar improvements can be seen in terms of J_{HF} and bandwidth. With high pass W_{KS}, still higher and more equal bandwidths can be achieved in the two "directions" compared to the original design in (Härefors, 1995). High pass W_{KS} is needed to come close to the stability limit, here represented by S_m.

Worth noting is also, that this design structure has improved numerical properties in the solving of the Riccati equations involved, compared with the original one. This leads to the recommendation to go further and use this method for further work on sampled data $\mathcal{H}_\infty$design, which is the goal for the research!

7. CONCLUSIONS

The proposed procedure gives a powerful and objective tool for judging different recommended weight structures for $\mathcal{H}_\infty$-controller design. In fact, it is shown that even an extremely simple S/KS-weight structure with integral penalty on the sensitivity S and constant penalty on the control signal, may be as attractive as more complicated weight structures from performance and robustness points of view. The most important improvement from this simple scheme is found to be a high pass filter as penalty on the control signal. The breakpoint for this high pass filter gives best results somewhat below the ultimate frequency ω_π.

It is encouraging to note, that the same results have been reached for a variety of different SISO- and MIMO-plants. In fact, the procedure

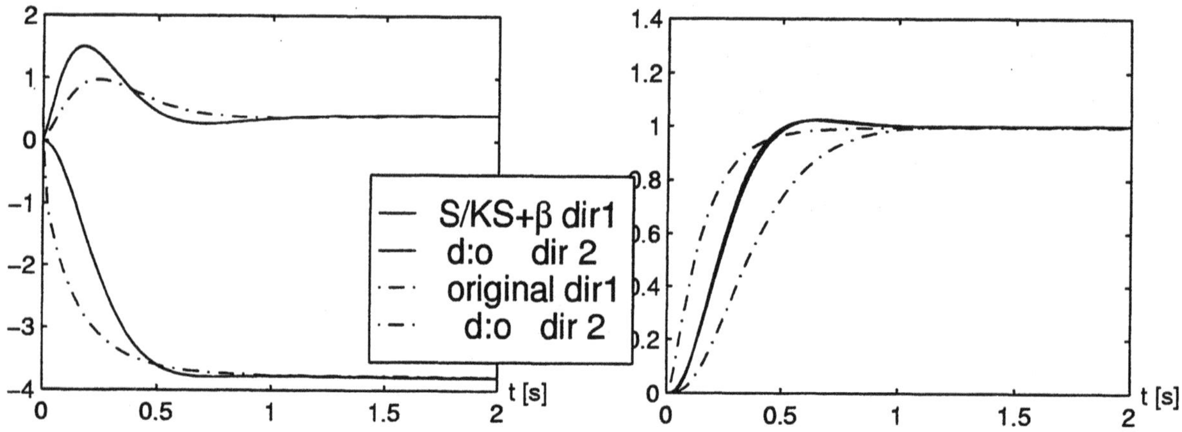

Fig. 7. Time plots from controller designs for the jet engine. Left shows control signal and right the output due to unit step in reference, one input at a time (both signals shown in the same diagram). Solid curves refer to S/KS-design with high pass W_{KS} and dash-dotted to the original design.($J_{HF} \approx 4.3$)

Table 1. Numeric values for some MIMO-designs

Design	J_{HF}	J_{LF}	ω_{b1}/ω_{b2}	S_{max}	T_{max}
"Original"	4.3	0.60	7.6/4.2	1.4	1.0
S/KS, const W_{KS}	(3.9)	(0.67)	(7.5/4.4)	(1.3)	1.0
$S/KS/T$, hp W_{KS}	4.3	0.31	8.5/6.7	1.4	1.0
S/KS, hp W_{KS}	4.3	0.31	10.7/7.6	1.4	1.0

is even more appealing in the MIMO-case, since it can easily be modified for different kinds of constraints. Compared to the $\mathcal{H}_2$-case, additional but reasonable control energy can be used in the $\mathcal{H}_\infty$-case to improve performance without losing stability. The method for comparing controller behaviour in this paper relies on a recently developed evaluation procedure for performance and robustness in all frequency regions of interest.

In the optimisation procedure, a genetic algorithm has shown to be very useful. Different types of constraints can easily be implemented into the genetic algorithm. The convergence rate and quality of the procedure mainly depend on how well the evaluation criterion can be formulated.

ACKNOWLEDGEMENTS

This work is partially financed by the Swedish aerospace research programme NFFP and run in cooperation with Volvo Aero Corporation, Sweden.

8. REFERENCES

Åström, K. J. and T. Hägglund (1995). *PID Controllers: Theory, Design and Tuning.* 2nd ed.. Instrument Society of America.

Christiansson, A-K (1998). Weight selection in $\mathcal{H}_\infty$-design. Report 1998:1. University of Trollhättan/Uddevalla.

Goldberg, D.E. (1989). *Genetic Algorithms in Search, Optimization and Mashine learning.* Adison-Wesley Publishing Company.

Green, M. and D. J. N. Limebeer (1995). *Linear Robust Control.* Prentice Hall, Englewood Cliffs, N. J.

Härefors, M. (1995). Multivariable Control for a Jet Engine. Licenciate thesis. Chalmers University of Technology.

Jovik, I. and B. Lennartson (1996). On the choice of weighting functions of an $\mathcal{H}_2/\mathcal{H}_\infty$ -design problem. Vol. C. IFAC. San Fransisco.

Lennartson, B. and B. Kristiansson (1997). Pass band and high frequency robustness for pid-control. CDC. San Diego, CA. pp. 2666–2671.

Maciejowski, J. M. (1989). *Multivariable Feedback Design.* Addison-Wesley Publishing Company.

Skogestad, S. and I. Postlethwaite (1996). *Multivariable Feedback Control, Analysis and Design.* John Wiley & Sons.

Stein, G. and M. Athans (1987). The LQG/LTR procedure for multivariable feedback control design. *IEEE Transactions on Automatic Control* **AC-32**, 105–114.

14th Triennial World Congress, Beijing, P.R. China

C-2a-01-6

LINEAR CONTROL DESIGN OF LINEAR DESCRIPTOR SYSTEMS

Peter C. Müller

Safety Control Engineering
University of Wuppertal
D-42097 Wuppertal, Germany
mueller@srm.uni-wuppertal.de

Abstract: In recent years the analysis and synthesis of control systems in descriptor form has been established. The general description of dynamical systems by differential-algebraic equations is important for many applications in mechanics and mechatronics, in electrical and electronic engineering, and in chemical engineering as well. This contribution deals with linear descriptor systems and its control design with respect to a quadratic performance criterion and based on a generalized Luenberger observer. Here, properness plays an important role whether usual design algorithms can be applied or not. The distinction of proper and non-proper systems is necessary to distinguish descriptor systems which are exclusively governed by the control input or by its higher-order time-derivatives additionally. In the unusual case of non-proper systems a quite different problem of optimal control design and observer realization has to be considered. Both cases will be solved completely.

Keywords: Descriptor systems, LQ design, regulator, observer.

1. INTRODUCTION

The investigation of dynamical processes is usually based on a mathematical model. Increasing complexity of the systems under consideration requires more and more modular modeling. A consequent modular organization of modeling simplifies the general view of the complex system and permits the utilization of known models of subsystems as well. This concept of subsystem modeling leads obviously to the use of redundant coordinates in describing the complete system behavior. Typically the systems are characterized by differential equations describing the dynamics and by algebraic equations describing the couplings among the subsystems. Therefore, the full description consists of differential-algebraic equations. Including actors and sensors to discuss control problems, then we speak on singular control systems or on descriptor systems (Luenberger, 1977). In the last decade there was an essential progress dealing with dynamic systems described by differential-algebraic equations (DAE). In control theory descriptor systems have been investigated (Dai, 1989). But still there are many unsolved problems related to the analysis, design and simulation of descriptor systems. Recent progress on parameter identification of linear mechanical descriptor systems has been presented by Schmidt (1994). New methods for analysis and control design of linear mechanical descriptor systems have been developed by Schüpphaus (1995). A profound investigation of observers for general descriptor systems has been conducted by Hou (1995). Also the stability behavior of such systems has been discussed (Bajic, 1992; Müller, 1993). Similarly, the calculus of variations and Pontryagin's maximum principal are checked for the application for descriptor systems. First attempts to the optimal control design have been presented by Cobb (1983) and Jonckheere (1988). Particularly the linear-quadratic optimal regulator for descriptor systems has been dealt with by Bender and Laub (1987). But usually the papers ask for several restrictions on the type of control problem to present some necessary optimality conditions. These restrictions are often not satisfied by real practical problems, such that the above-mentioned results cannot be applied. The main point of this lack is that in those papers it will not be distinguished between proper and non-proper descriptor systems. Properness and non-properness distinguish between the cases if the descriptor system is exclusively governed by the control input or by its higher order time-derivatives additionally. In the unusual case of non-proper systems a quite different problem of control design has to be considered (Müller, 1997). In this contribution, the general solution of the optimal control design of linear descriptor systems with respect to an infinite-horizon quadratic performance criterion and the related design of modified Luenberger observer will be presented completely.

2. PROBLEM STATEMENT

Linear time-invariant descriptor systems are presented by

$$\mathbf{E}\dot{\mathbf{x}}(t) = \mathbf{A}\mathbf{x}(t) + \mathbf{B}\mathbf{u}(t), \qquad (1)$$

$$\mathbf{y}(t) = \mathbf{C}\mathbf{x}(t) + \mathbf{D}\mathbf{u}(t) \qquad (2)$$

where $\mathbf{x}$ is an n-dimensional descriptor vector, $\mathbf{u}$ denotes the r-dimensional control input vector, and $\mathbf{y}$ characterizes the m-dimensional measurement output vector. The matrices $\mathbf{E}$, $\mathbf{A}$ are nxn-matrices, and $\mathbf{B}$, $\mathbf{C}$, $\mathbf{D}$ have dimensions nxr, mxn, mxr, respectively. The essential property of descriptor systems is that $\mathbf{E}$ is a singular matrix,

$$\operatorname{rank} \mathbf{E} < n, \qquad (3)$$

such that (1) consists of differential and algebraic equations

For the control design the quadratic performance criterion

$$J = \frac{1}{2}\int_0^\infty \begin{bmatrix} \mathbf{x} \\ \mathbf{u} \end{bmatrix}^T \begin{bmatrix} \mathbf{Q} & \mathbf{Z} \\ \mathbf{Z}^T & \mathbf{R} \end{bmatrix} \begin{bmatrix} \mathbf{x} \\ \mathbf{u} \end{bmatrix} dt \rightarrow \min_{\mathbf{u}} \qquad (4)$$

is considered where

$$\mathbf{R} > 0\,, \begin{bmatrix} \mathbf{Q} & \mathbf{Z} \\ \mathbf{Z}^T & \mathbf{R} \end{bmatrix} \geq 0 \qquad (5)$$

is assumed. The problem to be solved consists in designing a feedback control $\mathbf{u}$ which minimizes (4) having regard to the dynamic system (1).

The expected feedback control depends on the descriptor variables $\mathbf{x}$ but for the realization of the control the measurements $\mathbf{y}$ (2) are only available. Therefore, in a second step the observer problem has to be solved. Until now either singular observers are designed under the reasonable condition of "R-observability" or regular Luenberger observers can be used under the strong conditions of "R- and I-observability" (Shin and Kabamba, 1988; Dai, 1989; Hou 1995). In this paper a different solution is presented by introducing a modified Luenberger observer of reduced order where "R-observability" is only assumed.

3. SYSTEM ANALYSIS

Solving the above stated problem, regularity of the descriptor system is assumed, i.e. that the related matrix pencil $(s\mathbf{E}-\mathbf{A})$ is regular (Dai, 1989). This standard assumption is satisfied if and only if the characteristic polynomial does not vanish identically:

$$p(s) \equiv \det(s\mathbf{E} - \mathbf{A}) \not\equiv 0. \qquad (6)$$

It is well-known in matrix theory (Dai, 1989) that under this condition the matrix pencil $(s\mathbf{E}-\mathbf{A})$ is strictly equivalent to the Weierstrass-Kronecker-canonical form, i.e. there exist two regular matrices

$$\mathbf{R} = \begin{bmatrix} \mathbf{R}_1 \\ \mathbf{R}_2 \end{bmatrix}, \ \mathbf{S} = [\mathbf{S}_1 \ \mathbf{S}_2] \qquad (7)$$

such that

$$\mathbf{RES} = \begin{bmatrix} \mathbf{I}_1 & 0 \\ 0 & \mathbf{N}_k \end{bmatrix}, \ \mathbf{RAS} = \begin{bmatrix} \mathbf{A}_1 & 0 \\ 0 & \mathbf{I}_2 \end{bmatrix}, \qquad (8)$$

$$\mathbf{RB} = \begin{bmatrix} \mathbf{B}_1 \\ \mathbf{B}_2 \end{bmatrix}, \ \mathbf{CS} = [\mathbf{C}_1 \ \mathbf{C}_2]\,, \qquad (9)$$

where the identity matrices $\mathbf{I}_1$, $\mathbf{I}_2$ are of dimensions n_1, n_2 with $n_1 + n_2 = n$ and the n_2xn_2-matrix $\mathbf{N}_k$ is nilpotent of degree k ($\mathbf{N}_k^{k-1} \neq 0$, $\mathbf{N}_k^k = 0$) defining the index of the linear descriptor system. Introducing new coordinates

$$\begin{bmatrix} \mathbf{x}_1 \\ \mathbf{x}_2 \end{bmatrix} = \mathbf{S}^{-1}\mathbf{x}\,, \ \mathbf{x} = \mathbf{S}_1\mathbf{x}_1 + \mathbf{S}_2\mathbf{x}_2 \qquad (10)$$

the system is decoupled in the "slow" subsystem

$$\dot{\mathbf{x}}_1(t) = \mathbf{A}_1\mathbf{x}_1(t) + \mathbf{B}_1\mathbf{u}(t) \qquad (11)$$

and the "fast" subsystem

$$\mathbf{N}_k\dot{\mathbf{x}}_2(t) = \mathbf{x}_2(t) + \mathbf{B}_2\mathbf{u}(t). \quad (12)$$

The measurement equation yields

$$\mathbf{y}(t) = \mathbf{C}_1\mathbf{x}_1(t) + \mathbf{C}_2\mathbf{x}_2(t) + \mathbf{D}\mathbf{u}(t). \quad (13)$$

The solution of the DAE (1) is easily represented by the decomposition (11,12). The solution of the slow subsystem (11) is obtained in classical manner, the consistent solution of the fast subsystem (12) is given by (cf. Dai, 1989)

$$\mathbf{x}_2(t) = -\mathbf{B}_2\mathbf{u}(t) - \mathbf{N}_k\mathbf{B}_2\dot{\mathbf{u}}(t) - \cdots - \mathbf{N}_k^{j-1}\mathbf{B}_2\mathbf{u}^{(j-1)}(t). \quad (14)$$

It generally depends on the higher-order time-derivatives of the control input. The number $j \leq k$ is defined such that $\mathbf{N}_k^{j-1}\mathbf{B}_2 \neq 0$, $\mathbf{N}_k^j\mathbf{B}_2 = 0$. The notion of a "consistent" solution means that the algebraic constraints are satisfied everywhere, which includes consistent initial conditions. While the representation (1) does not show explicitly how consistent initial conditions look like, the Weierstrass-Kronecker-canonical representation uniquely asks for initial conditions $\mathbf{x}_{20} = \mathbf{x}_2(0)$ which satisfy (14) for $t \to 0$, where $\mathbf{x}_{10} = \mathbf{x}_1(0)$ is arbitrary.

A special feature of the solution (14) is the appearance of the time-derivatives of the control input $\mathbf{u}(t)$. In the following we distinguish the two cases where the solution depends either only on $\mathbf{u}(t)$ but not on its derivatives $\dot{\mathbf{u}}(t), \ldots, \mathbf{u}^{(j-1)}$ or on $\mathbf{u}(t)$ and its derivatives $\dot{\mathbf{u}}(t), \ldots, \mathbf{u}^{(j-1)}$ according to the general case. In the first case the system is called "proper", in the second case "non-proper" according to related proper and non-proper transfer matrix functions. The system (1) is proper if and only if in the representation (11,12) the equation

$$\mathbf{N}_k\mathbf{B}_2 = \mathbf{0} \quad (15)$$

holds (Dai, 1989; Hou, 1995).

4. LINEAR-QUADRATIC OPTIMAL CONTROL DESIGN

To solve the optimization problem (4), firstly the performance criterion is rewritten regarding the coordinate transformation (10) into the Weierstrass-Kronecker-canonical representation (11,12). Then criterion (4) yields

$$J = \frac{1}{2}\int_0^\infty \begin{bmatrix} \mathbf{x}_1 \\ \mathbf{x}_2 \\ \mathbf{u} \end{bmatrix}^T \begin{bmatrix} \mathbf{Q}_{11} & \mathbf{Q}_{12} & \mathbf{Z}_1 \\ \mathbf{Q}_{12}^T & \mathbf{Q}_{22} & \mathbf{Z}_2 \\ \mathbf{Z}_1^T & \mathbf{Z}_2^T & \mathbf{R} \end{bmatrix} \begin{bmatrix} \mathbf{x}_1 \\ \mathbf{x}_2 \\ \mathbf{u} \end{bmatrix} dt \to \min_{\mathbf{u}} \quad (16)$$

where

$$\mathbf{Q}_{ij} = \mathbf{S}_i^T\mathbf{Q}\mathbf{S}_j\ ,\ \mathbf{Z}_i = \mathbf{S}_i^T\mathbf{Z}\ ,\ i,j = 1,2. \quad (17)$$

The minimization of (16) has to be performed with respect to the slow and fast subsystems (11) and (12). According to the properness or non-properness of the descriptor system, two different optimization problems have to be considered.

4.1 Proper Descriptor Systems

If (15) holds then the solution

$$\mathbf{x}_2(t) = -\mathbf{B}_2u(t) \quad (18)$$

is used to replace the $\mathbf{x}_2$-variables in the performance criterion (16):

$$J = \frac{1}{2}\int_0^\infty \begin{bmatrix} \mathbf{x}_1 \\ \mathbf{u} \end{bmatrix}^T \begin{bmatrix} \mathbf{Q}_{11} & \mathbf{Z}_1 - \mathbf{Q}_{12}\mathbf{B}_2 \\ \mathbf{Z}_1^T - \mathbf{B}_2^T\mathbf{Q}_{12}^T & \mathbf{R} + \mathbf{B}_2^T\mathbf{Q}_{22}\mathbf{B}_2 - \mathbf{B}_2^T\mathbf{Z}_2 - \mathbf{Z}_2^T\mathbf{B}_2 \end{bmatrix} \begin{bmatrix} \mathbf{x}_1 \\ \mathbf{u} \end{bmatrix} dt \to \min_{\mathbf{u}} \quad (19)$$

For proper descriptor systems a standard linear-quadratic optimal control problem has been derived. The control of the slow subsystem (11) has to be designed with respect to the criterion (19). Therefore, the solution is obtained by the "Riccati" procedure for a system of reduced order n_1 leading to the following result.

Linear-Quadratic Optimal Control of Proper Systems. If the linear descriptor system is proper then the optimal control with respect to a quadratic performance criterion is a proportional feedback of the states of the slow subsystem according to

$$\mathbf{u}(t) = -\bar{\mathbf{R}}^{-1}(\mathbf{Z}_1^T - \mathbf{B}_2^T\mathbf{Q}_{12}^T + \mathbf{B}_1^T\mathbf{P}_1)\mathbf{x}_1(t) \quad (20)$$

where $\mathbf{P}_1$ is the unique positive (semi-)definite solution of the algebraic Riccati equation

$$\bar{\mathbf{A}}_1^T\mathbf{P}_1 + \mathbf{P}_1\bar{\mathbf{A}}_1 - \mathbf{P}_1\mathbf{B}_1\bar{\mathbf{R}}^{-1}\mathbf{B}_1^T\mathbf{P}_1 + \bar{\mathbf{Q}}_1 = \mathbf{0} \quad (21)$$

where the matrices

$$\bar{\mathbf{A}}_1 = \mathbf{A}_1 - \mathbf{B}_1\bar{\mathbf{R}}^{-1}(\mathbf{Z}_1^T - \mathbf{B}_2^T\mathbf{Q}_{12}^T), \quad (22)$$

$$\bar{\mathbf{R}} = \mathbf{R} + \mathbf{B}_2^T\mathbf{Q}_{22}\mathbf{B}_2 - (\mathbf{B}_2^T\mathbf{Z}_2 + \mathbf{Z}_2^T\mathbf{B}_2), \quad (23)$$

$$\bar{\mathbf{Q}}_1 = \mathbf{Q}_{11} - (\mathbf{Z}_1 - \mathbf{Q}_{12}\mathbf{B}_2)\bar{\mathbf{R}}^{-1}(\mathbf{Z}_1^T - \mathbf{B}_2^T\mathbf{Q}_{12}^T) \quad (24)$$

have been used.

For proper descriptor systems a standard solution of a reduced order system has been established. Therefore, the conditions for the existence of $\mathbf{P}_1 \geq \mathbf{0}$ are well known: Stabilizability of the matrix pair $(\bar{\mathbf{A}}_1, \mathbf{B}_1)$, i.e. $(\mathbf{A}_1, \mathbf{B}_1)$, and detectability of the matrix pair $(\bar{\mathbf{A}}_1, \bar{\mathbf{Q}}_1)$. The first condition can be represented by the original representation as the stabilizability of (1) (Dai, 1989).

4.2 Non-Proper Descriptor Systems

For non-proper problems an extension of state and control variables has to be carried out to deal correctly with the influence of the time-derivatives of the control input:

$$\xi_1 = \mathbf{u}\,,\ \xi_2 = \dot{\mathbf{u}}\,,\ \ldots\,,\ \xi_{j-1} = \mathbf{u}^{(j-2)}\,, \tag{25}$$

$$\mathbf{v} = \mathbf{u}^{(j-1)}. \tag{26}$$

Then (14) is replaced by

$$\begin{aligned}\mathbf{x}_2 &= -\mathbf{B}_2\xi_1 - \mathbf{N}_k\mathbf{B}_2\xi_2 - \cdots \\ &\quad -\mathbf{N}_k^{j-2}\mathbf{B}_2\xi_{j-1} - \mathbf{N}_k^{j-1}\mathbf{B}_2\mathbf{v}.\end{aligned} \tag{27}$$

Here, $\mathbf{v}$ is considered as a new control input vector. Introducing an extended state vector

$$x_e = \begin{bmatrix}\mathbf{x}_1 \\ \xi_e\end{bmatrix}\,,\ \xi_e^T = [\xi_1^T\,\xi_2^T\,\cdots\,\xi_{j-1}^T], \tag{28}$$

an extended dynamical system can be described including the dynamics of the slow subsystem (11) and of the extension (25,26):

$$\dot{\mathbf{x}}_e(t) = \mathbf{A}_e\mathbf{x}_e(t) + \mathbf{B}_e\mathbf{v}(t), \tag{29}$$

$$\mathbf{A}_e = \left[\begin{array}{c|c|cccc}\mathbf{A}_1 & \mathbf{B}_1 & 0 & \cdots & 0 \\ \hline 0 & 0 & \mathbf{I}_r & \cdots & 0 \\ & & & \ddots & \\ 0 & 0 & 0 & \cdots & \mathbf{I}_r \\ 0 & 0 & 0 & \cdots & 0\end{array}\right],\ \mathbf{B}_e = \left[\begin{array}{c}0 \\ \hline 0 \\ \vdots \\ 0 \\ \mathbf{I}_r\end{array}\right]. \tag{30}$$

It is not difficult to show that stabilizability and controllability of (29,30) coincide with stabilizability and controllability of the slow subsystem(11).

The solution (14) of the fast subsystem (12) can be written as

$$\mathbf{x}_2(t) = -[0\ \mathbf{B}_{2e}]\mathbf{x}_e(t) - \mathbf{N}_k^{j-1}\mathbf{B}_2\mathbf{v}(t) \tag{31}$$

with

$$\mathbf{B}_{2e} = [\mathbf{B}_2\ \mathbf{N}_k\mathbf{B}_2\ \cdots\ \mathbf{N}_k^{j-1}\mathbf{B}_2]. \tag{32}$$

Substituting (31) in the performance criterion (16), a linear-quadratic optimal control problem with respect to $\mathbf{x}_e$ and $\mathbf{v}$ appears. But obviously a singular control problem is obtained. There is not a regular weighting of the new control input $\mathbf{v}$. To regularize the problem it is recommended to introduce a weighting matrix $\mathbf{R}_v$. From a mathematical point of view this is reasonable to avoid mathematical problems. But also from an engineering point of view this procedure makes sense. Although the description (1) depends only on the control input $\mathbf{u}$ as usual, the detailed analysis shows the explicit dependence of the solution on the time-derivatives of $\mathbf{u}$ according to the hidden constraints. In this case, every engineer will introduce a weighting matrix with respect to the highest derivative $\mathbf{u}^{(j-1)}$ to deal with the problem reasonably, e.g. to avoid impulses in $\mathbf{u}^{(j-1)}(t)$. If this additional matrix $\mathbf{R}_v$ is introduced then the question may arise whether it is reasonable to introduce also some weightings ($\mathbf{Q}_\xi$) of the variables ξ_e. By this, finally a modified performance criterion is defined:

$$\begin{aligned}J_e &= J + \frac{1}{2}\int_0^\infty \xi_e^T\mathbf{Q}_\xi\xi_e dt + \frac{1}{2}\int_0^\infty \mathbf{v}^T\mathbf{R}_v\mathbf{v}dt \\ &= \frac{1}{2}\int_0^\infty \begin{bmatrix}\mathbf{x}_e \\ \mathbf{v}\end{bmatrix}^T \begin{bmatrix}\mathbf{Q}_e & \mathbf{Z}_e \\ \mathbf{Z}_e^T & \mathbf{R}_e\end{bmatrix}\begin{bmatrix}\mathbf{x}_e \\ \mathbf{v}\end{bmatrix} dt \\ &\rightarrow \min_{\mathbf{v}}\end{aligned} \tag{33}$$

The weighting matrices $\mathbf{Q}_e$, $\mathbf{Z}_e$, $\mathbf{R}_e$ are calculated straightforward resulting in

$$\mathbf{Q}_e = \left[\begin{array}{c|c}\mathbf{Q}_{11} & -\mathbf{Q}_{12}\mathbf{B}_{2e} + \mathbf{Z}_1\,[\mathbf{I}_r\ 0 \ldots 0] \\ \hline -\mathbf{B}_{2e}^T\mathbf{Q}_{12}^T + \begin{bmatrix}\mathbf{I}_r \\ 0 \\ \vdots \\ 0\end{bmatrix}\mathbf{Z}_1^T & \begin{array}{c}\mathbf{Q}_\xi + \mathbf{B}_{2e}^T\mathbf{Q}_{22}\mathbf{B}_{2e} - \mathbf{B}_{2e}^T\mathbf{Z}_2[\mathbf{I}_r\ 0 \ldots 0] \\ -\begin{bmatrix}\mathbf{I}_r \\ 0 \\ \vdots \\ 0\end{bmatrix}\mathbf{Z}_2^T\mathbf{B}_{2e} + \begin{bmatrix}\mathbf{I}_r \\ 0 \\ \vdots \\ 0\end{bmatrix}\mathbf{R}\,[\mathbf{I}_r\ 0 \ldots 0]\end{array}\end{array}\right] \tag{34}$$

$$\mathbf{Z}_e = \left[\begin{array}{c}-\mathbf{Q}_{12}\mathbf{N}_k^{j-1}\mathbf{B}_2 \\ \hline \mathbf{B}_{2e}^T\mathbf{Q}_{22}\mathbf{N}_k^{j-1}\mathbf{B}_2 - \begin{bmatrix}\mathbf{I}_r \\ 0 \\ \vdots \\ 0\end{bmatrix}\mathbf{Z}_2^T\mathbf{N}_k^{j-1}\mathbf{B}_2\end{array}\right] \tag{35}$$

$$\mathbf{R}_e = \mathbf{B}_2^T(\mathbf{N}_k^{j-1})^T\mathbf{Q}_{22}\mathbf{N}_k^{j-1}\mathbf{B}_2 + \mathbf{R}_v. \tag{36}$$

The optimization problem for the non-proper descriptor system is defined by (29) and (33). Therefore, now a standard linear-quadratic optimal control problem for the extended system (29) of order

$$n_e = n_1 + r(j-1) \tag{37}$$

has been obtained.

Linear-Quadratic Optimal Control of Non-Proper Systems. If the linear descriptor system is non-proper and if the performance criterion (33) has been extended, then the solution of the linear-quadratic optimal control problem is represented by

$$\mathbf{v}(t) = -\mathbf{R}_e^{-1}(\mathbf{Z}_e^T + \mathbf{B}_e^T\mathbf{P}_e)\mathbf{x}_e(t) \quad (38)$$

where $\mathbf{P}_e$ is the unique positive (semi-)definite solution of the algebraic Riccati equation

$$\bar{\mathbf{A}}_e^T\mathbf{P}_e + \mathbf{P}_e\bar{\mathbf{A}}_e - \mathbf{P}_e\mathbf{B}_e\mathbf{R}_e^{-1}\mathbf{B}_e^T\mathbf{P}_e + \bar{\mathbf{Q}}_e = \mathbf{0}, \quad (39)$$

where

$$\bar{\mathbf{A}}_e = \mathbf{A}_e - \mathbf{B}_e\mathbf{R}_e^{-1}\mathbf{Z}_e^T, \; \bar{\mathbf{Q}}_e = \mathbf{Q}_e - \mathbf{Z}_e\mathbf{R}_e^{-1}\mathbf{Z}_e^T. \quad (40)$$

Under weak conditions of $(\mathbf{A}_1, \mathbf{B}_1)$-stabilizability and $(\bar{\mathbf{A}}_e, \bar{\mathbf{Q}}_e)$-detectability the resulting closed-loop control system

$$\dot{\mathbf{x}}_e(t) = \hat{\mathbf{A}}_e\mathbf{x}_e(t) \quad (41)$$

with the system matrix

$$\hat{\mathbf{A}}_e = \mathbf{A}_e - \mathbf{B}_e\mathbf{R}_e^{-1}(\mathbf{Z}_e^T + \mathbf{B}_e^T\mathbf{P}_e) \quad (42)$$

is asymptotically stable.

Now, the original meaning of the extended variables is called back in mind. Defining the gain matrix of (38) as

$$\mathbf{R}_e^{-1}(\mathbf{Z}_e^T + \mathbf{B}_e^T\mathbf{P}_e) = [\mathbf{K}_0 \; \mathbf{K}_1 \; \cdots \; \mathbf{K}_{j-1}], \quad (43)$$

then the result (38) represents a dynamic feedback control

$$\mathbf{u}^{(j-1)}(t) + \mathbf{K}_{j-1}\mathbf{u}^{(j-2)}(t) + \cdots + \mathbf{K}_1\mathbf{u}(t) = -\mathbf{K}_0\mathbf{x}_1(t) \quad (44)$$

for the slow subsystem (11). According to the asymptotic stability of (41) the coupled system (11,44) is asymptotically stable. The dimension of the optimally controlled system is n_e (37).

5. OBSERVER DESIGN

For the realization of the dynamic feedback control (44) [or the static feedback control (20)] or, generally, for the realization of a descriptor state feedback control the estimation of the descriptor states is required. In contrary to Shin and Kabamba (1988) or Dai (1989) the observer design is based on the Weierstrass-Kronecker-canonical representation (11,12). Because the control vector $\mathbf{u}(t)$ and, in principal, also the time-derivatives $\dot{\mathbf{u}}, \ldots, \mathbf{u}^{(j-1)}(t)$ are known input signals, the observer problem is rewritten as follows:

$$\begin{aligned} \hat{\mathbf{x}}_2(t) &= \mathbf{x}_2(t) & (45)\\ \bar{\mathbf{y}}(t) &= \mathbf{y}(t) - \mathbf{C}_2\mathbf{x}_2(t) - \mathbf{D}\mathbf{u}(t). & (46)\end{aligned}$$

where $\mathbf{x}_2(t)$ is known from (14) and $\mathbf{y}(t)$ is given by (13). Altogether it remains the standard observer problem of the slow subsystem with the modified measurements (46):

$$\dot{\mathbf{x}}_1(t) = \mathbf{A}_1\mathbf{x}_1(t) + \mathbf{B}_1\mathbf{u}(t) \; , \; \bar{\mathbf{y}}(t) = \mathbf{C}_1\mathbf{x}_1(t). \quad (47)$$

An identity Luenberger observer for estimating $\hat{\mathbf{x}}_1(t)$ is given by

$$\dot{\hat{\mathbf{x}}} = \mathbf{A}_1\hat{\mathbf{x}}_1(t) + \mathbf{B}_1\mathbf{u}(t) + \mathbf{L}_1(\bar{\mathbf{y}}(t) - \hat{\bar{\mathbf{y}}}(t)). \quad (48)$$

Regarding (13,45,46) the observer will be realized by

$$\begin{aligned} \dot{\hat{\mathbf{x}}}_1(t) = \; & (\mathbf{A}_1 - \mathbf{L}_1\mathbf{C}_1)\hat{\mathbf{x}}_1(t) + (\mathbf{B}_1 - \mathbf{L}_1\mathbf{D})\mathbf{u}(t) \\ & + \mathbf{L}_1\mathbf{y}(t) + \mathbf{L}_1\mathbf{C}_2[\mathbf{B}_2\mathbf{u}(t) + \cdots \\ & + \mathbf{N}_k^{j-1}\mathbf{B}_2\mathbf{u}^{(j-1)}(t)]. \quad (49)\end{aligned}$$

Looking for the observer errors

$$\mathbf{e}_1(t) = \hat{\mathbf{x}}_1(t) - \mathbf{x}_1(t) \; , \; \mathbf{e}_2(t) = \hat{\mathbf{x}}_2(t) - \mathbf{x}_2(t), \quad (50)$$

the error dynamics

$$\dot{\mathbf{e}}_1(t) = (\mathbf{A}_1 - \mathbf{L}_1\mathbf{C}_1)\mathbf{e}_1(t) \; , \; \mathbf{e}_2(t) \equiv \mathbf{0}. \quad (51)$$

are obtained. Therefore, the following result is summarized.

Generalized Luenberger Observer. For proper and non-proper descriptor systems the descriptor states can be estimated by an extended Luenberger observer consisting of (49) and (45), which can be designed asymptotically stable if and only if the matrix pair $(\mathbf{A}_1, \mathbf{C}_1)$, i.e. the matrix triple $(\mathbf{E}, \mathbf{A}, \mathbf{C})$ is detectable.

To realize the observer (49) for non-proper descriptor systems, then $\mathbf{u}(t)$ and time-derivatives $\dot{\mathbf{u}}(t), \ldots, \mathbf{u}^{(j-1)}(t)$ have to be available. In general, this is a non-trivial problem. But in connection with the optimal control design of (44), the time-derivatives are generated within the filter (44) of the dynamic feedback. The details will be presented in the next section.

6. CLOSED-LOOP CONTROL SYSTEM

The realization of the linear-quadratic optimal regulator (20) for proper and (44) for non-proper descriptor systems requires the estimation of the variables $\mathbf{x}_1$ of the slow subsystem by an observer (49). Therefore, the optimal control is realized by replacing $\mathbf{x}_1(t)$ in (44) by $\hat{\mathbf{x}}_1$ of the observer (49). Then, using the Weierstrass-Kronecker-canonical representation of the descriptor system, the plant is described by (11) and (14)

$$\begin{aligned} \dot{\mathbf{x}}_1(t) &= \mathbf{A}_1\mathbf{x}_1(t) + \mathbf{B}_1\mathbf{u}(t), \\ \mathbf{x}_2(t) &= -\mathbf{B}_2\mathbf{u}(t) - \mathbf{N}_k\mathbf{B}_2\dot{\mathbf{u}}(t) - \cdots \\ &\quad - \mathbf{N}_k^{j-1}\mathbf{B}_2\mathbf{u}^{(j-1)}(t),\end{aligned}$$

and the optimal control according to the quadratic performance criterion (33) consists of a dynamic feedback realized by the observer (49) and by the low pass filter

$$\mathbf{u}^{(j-1)}(t) + \mathbf{K}_{j-1}\mathbf{u}^{(j-2)}(t) + \cdots + \mathbf{K}_1\mathbf{u}(t) = \mathbf{K}_0\hat{\mathbf{x}}(t). \quad (52)$$

The gain matrices $\mathbf{K}_i$, $i = 0, \ldots, j-1$, are assigned due to (43).

It follows immediately by the design steps that the algebraic separation principle of the regulator and observer design holds. Therefore, the closed-loop control system can be designed as shown above.

All over, a general procedure for the control design of linear descriptor systems has been developed independent on the index of the descriptor system and on proper or non-proper system behavior. The closed loop control system can be designed asymptotically stable under very weak controllability and observability conditions such as stabilizability and detectability of the descriptor system (1,2). The control design is based on standard numerical algorithms applied to properly defined systems of increased or reduced dimensions.

REFERENCES

Bajić, V.B. (1992). *Lyapunov's Direct Method in the Analysis of Singular Systems and Networks.* Shades Technical Publications, Hillcrest, Natal, RSA.

Bender, D.J. and A.J. Laub (1987). The linear-quadratic optimal regulator for descriptor systems. *IEEE Trans. Autom. Control* **32**, 672-688.

Cobb, D. (1983). Descriptor variable systems and optimal state regulation. *IEEE Trans. Autom. Control* **28**, 601-611.

Dai, L. (1989). *Singular Control Systems.* Lecture Notes in Control and Information Sciences **118**, Springer, Berlin-Heidelberg.

Hou, M. (1995). *Descriptor Systems: Observers and Fault Diagnosis.* Fortschr.-Ber. VDI, Reihe 8, Nr. 482, VDI, Düsseldorf.

Jonckheere, E. (1988). Variational calculus for descriptor problems. *IEEE Trans. Autom. Control* **33**, 491-495.

Luenberger, D.G. (1977). Dynamic equations in descriptor form. *IEEE Trans. Autom. Control* **22**, 312-321.

Müller, P.C. (1993). Stability of linear mechanical systems with holonomic constraints. *Appl. Mech. Rev.* **46**, no. 11, part 2, S160-S164.

Müller, P.C. (1997). Linear-quadratic optimal regulator for descriptor systems. In: *15th IMACS World Congress on Scientific Computation, Modelling and Applied Mathematics* (A. Sydow, Ed.). Vol. 5, pp. 49-54, Wissenschaft & Technik Verlag, Berlin.

Schmidt, T. (1994). *Parameterschätzung bei Mehrkörpersystemen mit Zwangsbedingungen* (Parameter Estimation of Multibody Systems with Constraints, in German). Fortschr.-Ber. VDI, Reihe 8, Nr. 433, VDI, Düsseldorf.

Schüpphaus, R. (1995). *Regelungstechnische Analyse und Synthese von Mehrkörpersystemen in Deskriptorform* (Control analysis and design of multibody systems in descriptor form, in German), Fortschr.-Ber. VDI, Reihe 8, Nr. 478, VDI, Düsseldorf.

Shin, K.-C. and P.T. Kabamba (1988). Observation and estimation in linear descriptor systems with application to constrained dynamical systems. *ASME J. Dynamic Systems, Measurement, and Control* **110**, 255-265.

14th Triennial World Congress, Beijing, P.R. China

PERFORMANCE LIMITATIONS IN THE ROBUST SERVOMECHANISM PROBLEM FOR SAMPLED LTI SYSTEMS

L. Ben Jemaa E. J. Davison

Department of Electrical and Computer Engineering, University of Toronto
10 King's College Road, Toronto, Ontario M5S 3G4 Canada
benjema@control.toronto.edu, ted@control.toronto.edu

Abstract: Fundamental limitations for error tracking/regulation are obtained for the robust servomechanism problem (RSP) for a sampled system. In studying this problem, the cheap control problem for a multi-input/multi-output discrete time system is considered, and explicit expressions are obtained for the limiting steady state solution of its associated algebraic Riccati equation (ARE), as the weight on the control energy tends to zero. Application of these results is then made to the "sampled system cheap robust servomechanism problem", and explicit expressions for the limiting costs for error tracking regulation are obtained. These limitations can be characterized completely by the number and location of the unstable transmission zeros.

Keywords: Sampled systems; cheap control problem; optimal control; performance limitations; servomechanism problem; non-minimum phase

1. INTRODUCTION

The robust servomechanism problem using a cheap control approach has been extensively studied for continuous LTI systems (Davison and Scherzinger, 1987; Qiu and Davison, 1993). In this case, for minimum phase LTI systems, perfect asymptotic tracking/disturbance rejection and arbitrarily good transient response, for a specified class of reference and disturbance signals can be achieved (Davison and Scherzinger, 1987). The robust servomechanism problem for non-minimum phase systems, however, has fundamental performance limitations for error tracking/regulation which can be characterized by the right-hand plane transmission zeros of the system (Qiu and Davison, 1993). In this paper, the robust servomechanism problem (RSP) using a cheap control approach is considered for multi-input/multi-output sampled LTI system. The cheap control problem has been extensively studied for minimum phase discrete time systems (Jemaa and Davison, 1996; Kwakernaak and Sivan, 1972; Priel and Shaked, 1983), and explicit expressions for the limiting optimal discrete cheap control and the limiting steady state solution of the associated algebraic Riccati equation (ARE), as the weight on the control energy tends to zero, have been derived (Jemaa and Davison, 1996). The multivariable discrete cheap control problem for non-minimum phase systems has not been considered to the same extent. In this paper, the cheap control problem for discrete time LTI systems is studied, and it is then used to obtain fundamental performance limitations on the behavior of the RSP for non-minimum phase sampled systems. An input/output approach has been considered by Qiu and Chen (1998) which studies similar problems.

2. DEVELOPMENT

Let S(C,A,B) be a discrete time system modeled by

$$\begin{aligned} x(k+1) &= Ax(k) + Bu(k) \\ y(k) &= Cx(k) \end{aligned} \quad (1)$$

where the control $u(k)$, the state $x(k)$, the output $y(k)$ are real vectors in $\mathbf{R}^m$, $\mathbf{R}^n$, $\mathbf{R}^r$. respectively. Consider the discrete time system $S(C,A,B)$, and let the quadratic cost to be minimized be given by

$$J_\epsilon = \sum_{k=0}^{\infty} y'(k)y(k) + \epsilon u'(k)u(k) \quad (2)$$

where ϵ is a positive scalar. Throughout this paper, we will assume that the system has an equal number of inputs and outputs ($m = r$), (A,B) is stabilizable, and (C,A) is detectable. In order to find the optimal stabilizing solution to the discrete cheap control problem which minimizes (2), it is required to solve a Riccati-type equation given by.

$$P_\epsilon - A'P_\epsilon A + A'P_\epsilon B(\epsilon I + B'P_\epsilon B)^{-1}B'P_\epsilon A = Q \quad (3)$$

where $Q = C'C$, and P_ϵ is its steady state solution. In this case, the optimal controller is given by

$$u(k) = -G_\epsilon x(k) \quad (4)$$

where G_ϵ is given by

$$G_\epsilon = (\epsilon I + B'P_\epsilon B)^{-1}B'P_\epsilon A$$

Let P be the limiting steady state solution of (3) as $\epsilon \to 0$ ($P := \lim_{\epsilon\to 0} P_\epsilon$); then if $B'PB$ is a full rank matrix, P satisfies the algebraic riccati equation (ARE):

$$P - A'PA + A'PB(B'PB)^{-1}B'PA = Q \quad (5)$$

and the limiting stabilizing optimal feedback cheap control matrix $G := \lim_{\epsilon\to 0} G_\epsilon$ is given by

$$G = (B'PB)^{-1}B'PA \quad (6)$$

It follows that the minimum achievable cost as $\epsilon \to 0$ is given by

$$J_{optimal} = x'(0)Px(0)$$

3. PRELIMINARY RESULTS

The following preliminary results are needed to solve the cheap control problem for discrete non-minimum phase systems. Let $\bar{z}$ denote the complex conjugate of z, and $F^*(z)$ denote the hermitian transpose of $F(z)$.

A stable transfer function matrix F is called inner if $F^*(\bar{z}^{-1})F(z) = I$ and if all zeros of F are located outside the unit circle. A transfer function matrix F is said to be right-invertible if F has full row rank for at least one z in the complex plane.

The following theorem shows that for any given stable state space representation (C,A,B) in which A^{-1} exists, we can find a matrix D such that the transfer function matrix $F(z) = D + C(zI-A)^{-1}B$ is an inner transfer function matrix (Jemaa and Davison, 1998).

Theorem 1. Given the discrete stable state space representation, (C,A,B) where A^{-1} exists,

I) The following two statements are equivalent:
  a) $\exists D$ and a scalar δ, such that $F^*(\bar{z}^{-1})F(z) = \delta^2 I$
  where $F = D + C(zI - A)^{-1}B$.
  b) $\exists P, Q \in \mathbf{R}^{n\times n}$ such that
    1) $P = P'$, $Q = Q'$
    2) $P = APA' + BB'$
    3) $Q = A'QA + C'C$
    4) $PQ = \delta^2 I$

II) Assume that (I,a) is true; then $\exists D$ satisfying:
  5) $D'D = (I - B'QB)$
  6) $D'C + B'QA = 0$
  7) $DB' + CPA' = 0$

Corollary 1. Let (C,A,B,D) be a balanced realization (Moore, 1981) of an inner transfer function matrix $F(z)$; then this implies that $\exists D$ such that 2) to 7) of theorem 1 are all satisfied when $P = Q = I$, $\delta = 1$.

Lemma 1. (Qiu and Davison, 1993) A non-minimum phase right invertible transfer function matrix F can always be factorized as $F = F_1F_2$ such that F_1 is inner, F_2 is minimum phase and right invertible, and such that the unstable poles of F_2 are equal to the unstable poles of F; moreover if F is strictly proper, then F_2 is strictly proper.

Lemma 2. (Qiu and Davison, 1993) Given a non-minimum phase right invertible transfer function matrix F, let $F = F_1F_2$ be the factorization described in lemma 1 Let (C_1, A_1, B_1, D_1) be a balanced realization of F_1 and let (C_2, A_2, B_2) be any stabilizable and detectable realization of F_2; then, a stabilizable and detectable realization of F is given by

$$A = \begin{bmatrix} A_1 & B_1C_2 \\ 0 & A_2 \end{bmatrix}, \; B = \begin{bmatrix} 0 \\ B_2 \end{bmatrix}, \quad (7)$$
$$C = \begin{bmatrix} C_1 & D_1C_2 \end{bmatrix}, \; D = 0$$

4. THE CHEAP CONTROL PROBLEM FOR DISCRETE SYSTEMS

In the following theorem, we will show that the limiting solution of ARE (5) for a non-minimum phase (C,A,B) system takes on a very simple form if (C,A,B) is a factorized realization as presented in lemma 2.

Theorem 2. Given a non-minimum phase transfer function matrix F, assume F is factorized as $F = F_1 F_2$ corresponding to lemma 2, and let the stabilizable and detectable realization of F be given by (C, A, B) (7). Let P_2 be the limiting stabilizing solution of the ARE (5) associated with the minimum phase system (C_2, A_2, B_2); then the limiting stabilizing solution of the ARE (5) associated with (C, A, B), is given by

$$P = \begin{bmatrix} I & 0 \\ 0 & P_2 \end{bmatrix} \tag{8}$$

5. THE ROBUST SERVOMECHANISM PROBLEM FOR SAMPLED LTI SYSTEMS

It is desired now to study the fundamental limitations associated with solving the robust servomechanism problem (RSP) Davison (1976) for sampled systems, when no penalties are imposed on the control signal energy of the corresponding cheap control performance index.

Consider the square continuous plant modeled by

$$\begin{aligned} \dot{x} &= \mathcal{A}x + \mathcal{B}(u + \omega) \\ y &= \mathcal{C}x + \eta, \qquad e = y - y_{ref} \end{aligned} \tag{9}$$

where $u \in \mathbf{R}^r$, $y \in \mathbf{R}^r$, $x \in \mathbf{R}^n$, $\omega \in \mathbf{R}^r$, $\eta \in \mathbf{R}^r$, u is the input, ω and η are constant unmeasurable disturbances, and y is the output, $y_{ref} \in \mathbf{R}^r$ is a constant set point, and e is the error in the system. The following existence results are now obtained:

Lemma 3. (Davison, 1976) There exists a solution to the RSP for (9) iff the following conditions are all satisfied:

(i) $(\mathcal{C}, \mathcal{A}, \mathcal{B})$ is stabilizable and detectable

(ii) $rank \begin{bmatrix} \mathcal{A} & \mathcal{B} \\ \mathcal{C} & 0 \end{bmatrix} = n + r$

(iii) y is measurable.

Assume now that the plant is sampled with a Zero Order Hold (ZOH), with sampling interval $h > 0$, so that the resultant sampled system is described by :

$$\begin{aligned} x_{k+1} &= Ax_k + B(u_k + \omega) \\ y_k &= Cx_k + \eta \\ e_k &= y_k - y_{ref} \end{aligned} \tag{10}$$

where $A := e^{h\mathcal{A}}$, $B := \int_0^h e^{\mathcal{A}(h-\tau)}\mathcal{B}d\tau$, $C := \mathcal{C}$, and ω, η are constant unmeasurable disturbances, and y_{ref} is a constant set-point tracking signal.

Lemma 4. (Goldenberg and Davison, 1974) There exists a solution to the RSP for (10) iff the following conditions are all satisfied:

(i) (C, A, B) is stabilizable and detectable

(ii) $rank \begin{bmatrix} A - I & B \\ C & 0 \end{bmatrix} = n + r$

(iii) y_k is measurable.

The following existence result is now obtained re a solution to the RSP for the sampled system (10):

Lemma 5. Given the continuous system (9), assume that there exists a solution to the RSP for (9), i.e. the conditions of lemma 3 are all satisfied; then for almost all $h > 0$, there exists a solution to the RSP for the sampled system (10), i.e. the conditions of lemma 4 are satisfied.

The following intermediate result is now obtained:

Lemma 6. Given the continuous LTI non-degenerate (Davison and Wang, 1974) system (9), then for almost all (Davison and Wang, 1973) plant parameters $(\mathcal{C}, \mathcal{A}, \mathcal{B})$ and almost all $h > 0$, the resultant sampled system (10) has the property that (CB) is non-singular; in particular if $(\mathcal{C}, \mathcal{A}, \mathcal{B})$ has a non-singular interactor matrix, then (CB) is non-singular for almost all $h > 0$.

It is desired now to find a "high performance " controller to solve the RSP for (10) in the presence of constant disturbances, ω and η, and constant set-points, y_{ref}. In order to achieve this goal, the following "cheap performance index " will be minimized:

$$J = h \sum_{k=o}^{\infty} \left\{ e_{k-1}' e_{k-1} + \epsilon (u_k - u_{k-1})' (u_k - u_{k-1}) \right\} \tag{11}$$

where $e_k := y_k - y_{ref}$ is the error in the system, and $\epsilon \to 0$, subject to the constraint that:

$$\begin{aligned} \begin{bmatrix} x_{k+1} - x_k \\ e_k \end{bmatrix} &= \begin{bmatrix} A & 0 \\ C & I \end{bmatrix} \begin{bmatrix} x_k - x_{k-1} \\ e_{k-1} \end{bmatrix} + \begin{bmatrix} B \\ 0 \end{bmatrix} (u_k - u_{k-1}) \\ e(k-1) &= \begin{bmatrix} 0 & I \end{bmatrix} \begin{bmatrix} x_k - x_{k-1} \\ e_{k-1} \end{bmatrix} \end{aligned} \tag{12}$$

5.1 *Robust Servomechanism Control Results for Minimum Phase Systems*

Assume now that the existence conditions for a solution to the RSP for the continuous system (9) given by lemma 3 are satisfied, and that h is chosen so that the existence conditions for a solution to the RSP for sampled system (10) given by lemma 4 are satisfied. This will be true from lemma 5 for almost all h.

Assume also that the sampled system (10) with the above choice of sampling interval h has a total of ρ transmission zeros and is minimum phase. Consider the optimal controller which solves the RSP and which minimizes J given by (11), and let J_{opt} denote the optimal "cheap control cost " as $\epsilon \to 0$ in (11). The following result is obtained:

Theorem 3. Given a sampled system (10), which has the property that there exists a solution to the

RSP and is minimum phase, consider the optimal controller which minimizes the performance index (11); then

I) If $x(0) = 0$, $y_{ref} \neq 0$, $\eta \neq 0$, and $\omega = 0$, the optimal cost J as $\epsilon \to 0$ is given by:

$$J_{opt} = h(y_{ref} - \eta)' M (y_{ref} - \eta) \quad (13)$$

where M is a constant matrix with $trace(M) = n + r - \rho$.

II) Assume in (10) that (CB) is invertible. Then
(a) if $x(0) = 0$, $y_{ref} \neq 0$, $\eta \neq 0$, and $\omega = 0$, the optimal cost J as $\epsilon \to 0$ is given by:

$$J_{opt} = h(y_{ref} - \eta)' M (y_{ref} - \eta) \quad (14)$$

where $M = 2I_r$ and $trace(M) = 2r$.
(b) if $x(0) = 0$, $y_{ref} = 0$, $\eta = 0$, and $\omega \neq 0$, the optimal cost J as $\epsilon \to 0$ is given by:

$$J_{opt} = h\omega' (CB)' (CB) \omega. \quad (15)$$

(c) The optimal controller which minimizes (11) as $\epsilon \to 0$ is given by

$$u_k - u_{k-1} = -(CB)^{-1} \{C(A+I)(x_k - x_{k-1}) - e_{k-1}\} \quad (16)$$

Remark 1. The limiting optimal performance costs, given by $trace(M)$ obtained in theorem 3 are of interest, in that they indicate the "best possible" achievable performance as measured by $J_{opt} = hE(\sum_0^\infty e'_{k-1} e_{k-1})$, where E is the expectation operator, which can be achieved. It is also to be noted that (CB) is generically invertible from lemma 5, so that the results of (II) hold generically. It is finally to be noted in theorem 3 that $J_{opt} \to 0$ as $h \to 0$, which coincides with the continuous plant RSP "perfect control" (Davison and Scherzinger, 1987) result.

5.2 *Robust Servomechanism Control Results for Non-minimum Phase Systems*

It will be assumed that the sampled system (10) has $\rho \in [1, n-1]$ transmission zeros, and is non-minimum phase with $p \in [1, \rho]$ unstable transmission zeros given by λ_i, $i = 1, 2, ..., p$.

Theorem 4. Given a sampled system (10) which has the property that there exists a solution to the RSP and is non-minimum phase, consider the optimal controller which minimizes the performance index (11); then

I) If $x(0) = 0$, $y_{ref} \neq 0$, $\eta \neq 0$, and $\omega = 0$, the optimal cost J as $\epsilon \to 0$ is given by:

$$J_{opt} = h(y_{ref} - \eta)' M (y_{ref} - \eta) \quad (17)$$

where M is a constant matrix with

$$trace(M) = n + r - \rho + \sum_{i=1}^{p} \frac{(\lambda_i + 1)}{(\lambda_i - 1)}.$$

II) Assume in (10) that (CB) is invertible. Then
(a) if $x(0) = 0$, $y_{ref} \neq 0$, $\eta \neq 0$, and $\omega = 0$, the optimal cost J as $\epsilon \to 0$ is given by:

$$J_{opt} = h(y_{ref} - \eta)' M (y_{ref} - \eta) \quad (18)$$

where M is a constant matrix with

$$trace(M) = 2r + \sum_{i=1}^{p} \frac{(\lambda_i + 1)}{(\lambda_i - 1)}$$

(b) if $x(0) = 0$, $y_{ref} = 0$, $\eta = 0$, and $\omega \neq 0$, the optimal cost J as $\epsilon \to 0$ is given by:

$$J_{opt} = h\omega' M \omega. \quad (19)$$

where M is a constant matrix given by $M = \Theta' \Theta$, where

$$\Theta = \prod_{i=p}^{1} (I_r + \frac{(\lambda_i \bar{\lambda}_i - 1)}{\bar{\lambda}_i + 1)} \bar{v}_i v_i') CB \quad (20)$$

where the vector v_i is given by

$$[u_i' \; v_i'] \begin{bmatrix} \lambda_i I - A & -B \\ -C_m^{i-1} & 0 \end{bmatrix} = 0 \quad (21)$$

and v_i is normalized,i.e. $\bar{v}_i' v_i = 1$, and where

$$C_m^0 = C$$
$$C_m^i = C_m^{i-1} - \left(\frac{\lambda_i \bar{\lambda}_i - 1}{\bar{\lambda}_i + 1} \right) \bar{v}_i u_i' (A + I).$$

The following results provides an explicit expression for result II (b) of theorem 4 for the special case of SISO systems:

Corollary 2. Given a SISO sampled system (10), which has the property that there exists a solution to the RSP and is non-minimum phase, let $x = 0$, $y_{ref} = 0$, $\eta = 0$, $\omega \neq 0$; then

$$J_{opt} = h\omega' M \omega = hM\omega^2 \quad (22)$$

where M is constant matrix given by

$$M = (CB)^2 \prod_{i=1}^{p} \lambda_i^2$$

Remark 2. In this case, the limiting optimal performance costs, given by $trace(M)$, obtained in theorem 4, indicate that not all non-minimum phase systems behave in the same way, e.g. if a sampled system has a real transmission zero $TZ = \theta > 1$ then as $\theta \to 1$ the effect of the unstable TZ becomes "arbitrarily bad", whereas if $\theta \to \infty$, then the effect of the unstable TZ is "innocent". Likewise if a sampled system has a pair of unstable complex conjugate transmission zeros $TZ = a + bj$, $a - bj$, then if $a = 1$, the effect of the unstable TZ is "innocent".

6. EXAMPLES

Four examples will be considered. The first example is a minimum phase sampled system, the second example may or may not be non-minimum phase depending on the sampling period, the third example is a non-minimum phase sampled system, and the fourth example is a non-minimum phase flexible beam system example.

Example 1. Consider the following continuous minimum phase "mass-spring" system given by:

$$\mathcal{A}=\begin{bmatrix} 0 & 1 & 0 & 0 & 0 & 0 \\ -0.2 & -0.2 & 0.1 & 0.01 & 0.1 & 0.01 \\ 0 & 0 & 0 & 1 & 0 & 0 \\ 1 & 0.1 & -1 & -0.1 & 0 & 0 \\ 0 & 0 & 0 & 0 & 0 & 1 \\ 1 & 0.1 & 0 & 0 & -1 & -0.1 \end{bmatrix},\ \mathcal{B}=\begin{bmatrix} 0 & 0 \\ 0 & 0 \\ 0 & 0 \\ 1 & 0 \\ 0 & 0 \\ 0 & 1 \end{bmatrix},$$

$$\text{and} \qquad \mathcal{C}=\begin{bmatrix} 0 & 0 & 1 & 0 & 0 & 0 \\ 0 & 0 & 0 & 0 & 1 & 0 \end{bmatrix}$$

The transmission zeros of this system are given by: $\{-0.01 \pm 0.4471i\}$. If we sample this system with $h = 0.01$, the sampled data system is minimum phase with transmission zeros given by:

$$\{0.9999 \pm 0.0045i,\ -0.9997,\ -0.9997\}$$

and has the property that (CB) is invertible. If $x(0) = 0$, $y_{ref} = [0\ 0]'$, $\eta = 0$, then it follows from (15) that the limiting optimal cost, as $\epsilon \to 0$, is given by

$$J_{opt} = 2.5 \times 10^{-11} \omega' \omega,$$

and if $x(0) = 0$, $\omega = [0\ 0]'$, then it follows from (14) that the limiting optimal cost, as $\epsilon \to 0$, is given by

$$J_{opt} = 0.02 y_{ref}' y_{ref}$$

A simulation of the resultant closed loop system, using the perfect controller (16), for the case of zero initial conditions is given in Figure 1, for the case when $\omega = [0\ 0]'$, $y_{ref} = [1\ 1]'$, and $y_{ref} = [0\ 0]'$, $\omega = [1\ 1]'$. It is seen in Fig. 1 that the "optimal" transient error costs obtained, do confirm the results obtained in (15), (14), i.e. in the case of Fig.1 (a), Fig.1(b), $J_{opt} = h\sum_0^\infty e_{k-1}' e_{k-1} = 10^{-2}(2+2) = 0.02 y_{ref}' y_{ref}$ and $J_{opt} = 10^{-2}(0 + 2(5 \times 10^{-5})^2) = 2.5 \times 10^{-11} \omega' \omega$ respectively.

Example 2. Consider a minimum phase system with the transfer function

$$T(s) = \frac{1}{(s+1)^3}$$

The corresponding sampled transfer function is non-minimum phase for $0 \le h \le 1.8399$ and as

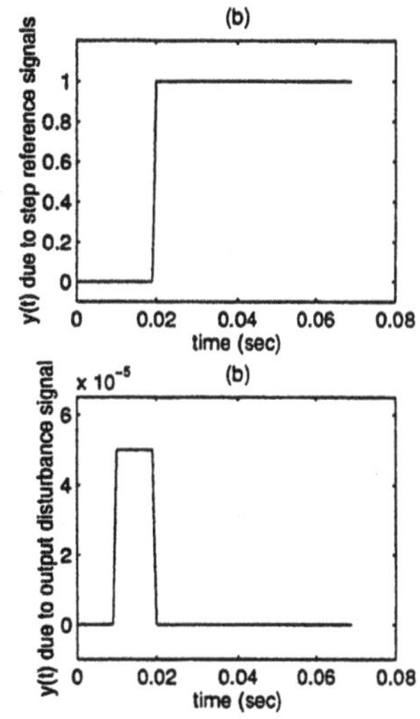

Fig. 1. Closed loop response with RSC for example 6.1: (a): $y_{ref} = [1\ 1]'$, $\omega = [0\ 0]'$, (b): $y_{ref} = [0\ 0]'$, $\omega = [1\ 1]'$.

$h \to 0$, it has an unstable zero which approaches $z_1 = -3.732$. It is of interest now to determine what are the fundamental performance limitations of this non-minimum phase system as $h \to 0$. In this case, as $h \to 0$, the term $h\frac{z_1+1}{z_1-1} \to 0$ and so $J_{opt} \to 0$ in (17)-(19). *This results coincides with the cost associated with the minimum phase continuous time system (Davison and Scherzinger, 1987). Thus the fact that the sampled system is non-minimum phase with a transmission zero* $z_1 = -3.732$ *as* $h \to 0$ *is innocent.*

Example 3. Consider a non-minimum phase system with the transfer function

$$T(s) = \frac{s-1}{s^3}$$

In this case, the sampled system is a non-minimum phase system, with (CB) invertible, and its unstable transmission zero is given by

$$z_1 = \frac{2(3+h)}{-4h + \sqrt{12h^2 + 36}}$$

As $h \to 0$, $z_1 \to 1$, and $h\frac{z_1+1}{z_1-1} \to 2$. It follows that the cost (18) *tends to* $J_{opt} = 2(y_{ref} - \eta)^2$ *which coincides with the cost associated with the non-minimum phase continuous time system (Qiu and Davison, 1993). Thus, in this case, the non-minimum phase zero is not innocent.*

Example 4. Consider the following continuous non-minimum phase "flexible beam" system modeled by the following transfer function

$$F(s) = \frac{1.65s^4 - 0.331s^3 - 576s^2 + 90.6s + 19080}{s^6 + 0.996s^5 + 463s^4 + 97.8s^3 + 12131s^2 + 8.11s}$$

The transmission zeros of this system are given by: $\{-6.0019,\ 6.1745,\ 1.7679e+01,\ -1.7651e+01\}$. If we sample this system with $h = 0.1$, the resultant sampled system is non-minimum phase, with (CB) invertible, and with the following transmission zeros

$$\{6.7603, 1.8527, 5.4903\text{e-}01, 1.5085\text{e-}01, \text{-}9.6517\text{e-}01\}.$$

If $x(0) = 0$, $y_{ref} = 0$, $\eta = 0$, then it follows from (22) that the limiting optimal cost, as $\epsilon \to 0$, is given by

$$J_{opt} = 1.7016\text{e-}04\omega^2,$$

and if $x(0) = 0$, $y_{ref} = 0$, $\omega = 0$, then it follows from (18) that the limiting optimal cost, as $\epsilon \to 0$, is given by

$$J_{opt} = 0.6692\eta'\eta$$

The closed loop system obtained by solving the cheap control problem with $\epsilon = 1\text{e} - 12$ is simulated with $x(0) = 0$, $y_{ref} = 0$ in Fig. 2, and in this case, if $\omega = 1$, $\eta = 0$, then $J_{opt} = 1.7024\text{e-}04$, and if $\eta = 1$, $\omega = 0$, then $J_{opt} = 0.6692$.

It is seen in Figure 2 that the optimal transient costs obtained, do confirm the results obtained in (22) and (18).

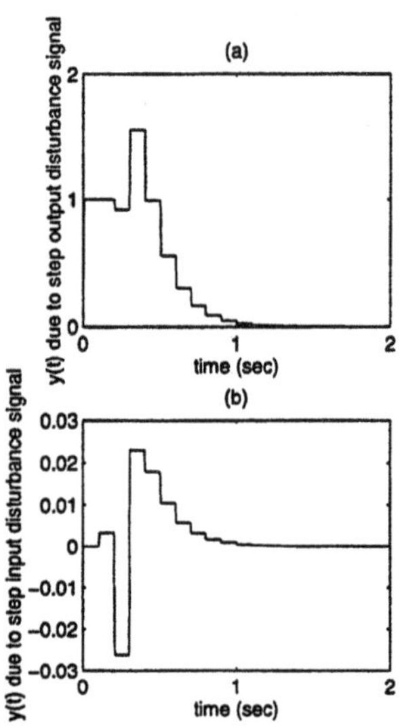

Fig. 2. Closed loop response with RSC for example 6.4 : (a): $\eta = 1$, $\omega = 0$, (b) $\eta = 0$, $\omega = 1$.

7. CONCLUSIONS

It is shown that the fundamental limitation on the closed loop performance of the robust servomechanism problem (RSP) for minimum phase sampled systems is independent of the order of the plant, and the infinite transmission zero structure of the continuous plant model, and that the fundamental limitation on the closed loop performance of the RSP for non-minimum phase sampled systems can be characterized by the number and the location of the transmission zeros of the sampled open loop system which lie outside the unit circle. These limitations can be used to evaluate an open loop system, i.e. to determine whether the system is "inherently hard to control", and to assess a given closed loop design, i.e. to determine how near the closed loop system's performance is from the best possible attainable.

REFERENCES

Davison, E. J. and Scherzinger, B. M. (1987). Perfect Control of the Robust Servomechanism Problem. *IEEE Transactions on Automatic Control*, 32(8):689–702.

Davison, E. J. (1976). The robust control of a servomechanism problem for linear time-invariant multivariable systems. *IEEE Trans. Automat. Contr.*, AC-21(1):25–34.

Davison, E. J. and Wang, S. H. (1974). Properties and calculation of transmission zeros of linear multivariable systems. *Automatica*, 10:643–658.

Davison, E. J. and Wang, S. H. (1973). Properties of LTI multivariable systems subject to arbitrary output and state feedback. *IEEE Trans. Automat. Contr.*, AC-18(1):24–32.

Goldenberg, A. and Davison, E. J. (1974). The feedforward and robust control of a general Servomechanism Problem with time lag. 8th *Annual Princeton Conf. Inf. Sci. Syst.*, pages 80–84.

Jemaa, L. B. and Davison, E. J. (1998). Performance limitations in the robust servomechanism problem for sampled LTI systems. *Systems Control Group, Dept. of Electrical and Computer Engineering, The University of Toronto, Report #9802.*

Jemaa, L. B. and Davison, E. J. (1996). The perfect robust servomechanism problem for sampled LTI systems. *Proc. of the 4th IEEE Mediterranean Sym. on New Direction in Control and Automation, Maleme Crete*, pages 298–303.

Kwakernaak, H. and Sivan, R. (1972). *Linear Optimal Control Systems.* Willy-Interscience, New York. London. Sydney. Toronto.

Moore, B. (1981). Principal component analysis in linear systems: controllability, observability, and model reduction. *IEEE Trans. Auto. Contr.*, 26(3).

Priel, B. and Shaked, U. (1983). ' cheap ' optimal control of discrete single input single output systems. *Int. J. Systems.*, 38(6):1087–1113.

Qiu, L. and Chen, J. (1998). Time domain characterizations of performance limitations of feedback control. *Submitted for publication.*

Qiu, L. and Davison, E. J. (1993). Performance limitations of non-minimum phase systems in the servomechanism problem. *Automatica*, 29(2):337–349.

DESIGN OF SAMPLED-DATA CONTROL SYSTEMS VIA THE PRINCIPLE OF MATCHING

T. Ono, T. Ishihara and H. Inooka

Graduate School of Information Sciences, Tohoku University,
Sendai, 980-8579, JAPAN.
ono@control.is.tohoku.ac.jp
ishihara@control.is.tohoku.ac.jp
inooka@control.is.tohoku.ac.jp

Abstract: The control system and its environment are said to be matched if the response of the system is acceptable for all exogenous inputs from the environment. In this paper, we give a necessary and sufficient condition for a match for sampled-data systems where the exogenous inputs belong to the class of inputs with finite energy and/or finite rate of change. The lifting technique is used to obtain the time-invariant discrete-time input-output model with the intersample behavior. It is shown that the lifted model can effectively be used to check the matching condition. A numerical example is given to illustrate the proposed design technique. *Copyright © 1999 IFAC*

Keywords: Linear control systems, Control design, Sampled-data systems, Digital control, Linear output feedback, Disturbance rejection

1. INTRODUCTION

The principle of matching (Zakian, 1991, 1996) provides a systematic procedure to design a control system which ensures the design specification for all possible exogenous inputs from the environment. For the critical control problem (Zakian, 1989), which requires that the controlled signals are always kept within the prescribed bounds, several matching conditions for practically important classes of exogenous inputs have been obtained for continuous-time control systems (Zakian, 1986, 1987, 1989, 1991, 1996; Rutland, 1992; Liu, 1990) and for discrete-time control systems (Liu, 1992).

For sampled-data (SD) control systems, Whidborne (1993) obtained the matching condition for a class of inputs with finite rate of change. But its evaluation is not easy since it is derived from a time-varying model. From the practical viewpoint, it is necessary to obtain the matching conditions for other classes of exogenous inputs and to give the efficient methods for evaluating them.

In this paper, we consider the design of SD control systems via the principle of matching. The lifting technique (Bamieh and Pearson, 1991, 1992; Yamamoto and Khargonekar, 1996) is used to obtain a time-invariant discrete-time input-output model with intersample behavior. For the class of inputs with finite energy and/or finite rate of change, we give the matching condition for critical control problems using the lifted model. The obtained condition is represented as the restriction on the impulse and/or step responses, and can easily be evaluated by a Lyapunov equation and an iterative calculation.

This paper is organized as follows. In section 2, we give the notation used in this paper, formulation of the design problem and a brief explanation of the principle of matching. In section 3, a time-invariant discrete-time input-output model of SD control systems is given by the lifting technique. Section 4 derives the matching condition from the input-output model. In section 5, we discuss efficient numerical evaluation of the matching condition. Section 6 presents a design example for a unity feedback SD control system. Conclusion is summarized in Section 7.

2. PRELIMINARIES

2.1 *Notations*

Let h be the sampling period, $\mathbb{R}$ be the set of all real numbers, $\mathbb{R}_+$ be the set of non-negative real numbers and $\mathbb{Z}_+$ be the set of non-negative integers. The space $\mathcal{C}(\mathbb{X}, \mathbb{R}^n)$ denotes the set of all piecewise continuous functions from the range $\mathbb{X}$ to $\mathbb{R}^n$. Especially, for $p \in \{2, \infty\}$, $\mathcal{L}_p(\mathbb{X}, \mathbb{R}^n)$ is the subspace of $\mathcal{C}(\mathbb{X}, \mathbb{R}^n)$ such that the following norm is finite.

$$\|x\|_{\mathcal{L}_p(\mathbb{X},\mathbb{R}^n)} = \begin{cases} \left(\int_{\mathbb{X}} \|x(t)\|^2 dt\right)^{\frac{1}{2}}, & p = 2 \\ \sup_t \|x(t)\|, & p = \infty \end{cases} \quad (1)$$

where $\|\cdot\|$ is the Euclidean norm. On the other hand, $c(\mathbb{Z}_+, \mathcal{N})$ denotes the set of all function-space valued sequences where each element lies in $\mathcal{N} := \mathcal{C}([0, h], \mathbb{R}^n)$. Especially, $\ell_p(\mathbb{Z}_+, \mathcal{K}_p)$ is the subspace of $c(\mathbb{Z}_+, \mathcal{N})$ such that each element lies in $\mathcal{K}_p := \mathcal{L}_p([0, h], \mathbb{R}^n)$ and the following norm is finite.

$$\|\hat{x}\|_{\ell_p(\mathbb{Z}_+,\mathcal{K}_p)} = \begin{cases} \left(\sum_{k=0}^{\infty} \|\hat{x}(k)\|^2_{\mathcal{K}_p}\right)^{\frac{1}{2}}, & p = 2 \\ \sup_k \|\hat{x}(k)\|_{\mathcal{K}_p}, & p = \infty \end{cases} \quad (2)$$

For simplicity, if no confusion is occurred, we omit the dimension of signal, and write $\mathcal{C}(\mathbb{R}_+)$, $\mathcal{L}_p(\mathbb{R}_+)$, $c(\mathbb{Z}_+)$ and $\ell_p(\mathbb{Z}_+)$ instead of $\mathcal{C}(\mathbb{R}_+, \mathbb{R}^n)$, $\mathcal{L}_p(\mathbb{R}_+, \mathbb{R}^n)$, $c(\mathbb{Z}_+, \mathcal{N})$ and $\ell_p(\mathbb{Z}_+, \mathcal{K}_p)$. The norms defined in (1) to (2) is written by $\|\cdot\|_{\mathcal{L}_p(\mathbb{R}_+)}$ and $\|\cdot\|_{\ell_p(\mathbb{Z}_+)}$. The symbols row$(E, i)$ and col(E, j) represent the ith row vector of matrix E and the jth column vector of E, respectively.

2.2 *Problem statement*

The system of interest is the sampled-data control system shown in Fig. 1 consisting of a continuous-time generalized plant G with the state space realization:

$$\begin{bmatrix} \dot{x}(t) \\ z(t) \\ y(t) \end{bmatrix} = \begin{bmatrix} A & B_w & B_u \\ C_z & D_w & D_u \\ C_y & D_y & 0 \end{bmatrix} \begin{bmatrix} x(t) \\ w(t) \\ u(t) \end{bmatrix}, \quad (3)$$

a discrete-time controller K with the state space realization:

$$\begin{bmatrix} \xi[k] \\ u[k] \end{bmatrix} = \begin{bmatrix} A_K & B_K \\ C_K & D_K \end{bmatrix} \begin{bmatrix} \xi[k] \\ y[k] \end{bmatrix}, \quad (4)$$

an ideal sampler S and a zero-order holder H, where $z(t)$ is a signal to be controlled of dimension

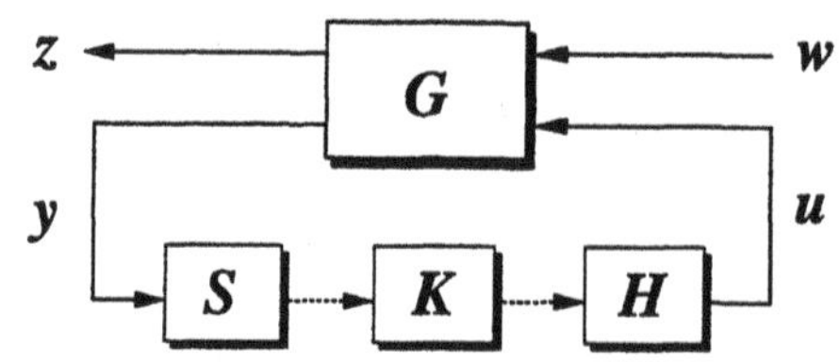

Fig. 1. Sampled-data control system

N_z and $w(t)$ is an exogenous input vector of dimension N_w. We make some assumptions on the input $w(t)$.

(A1) All of N_w elements of w are generated from different parts of a system or its environment.

(A2) For $j \in \{1, \ldots, N\}$, the element w_j belongs to the space $\mathcal{F}_2(M_j)$ which comprises all inputs such that the energy is below M_j.

$$\mathcal{F}_2(M_j) := \{ f \in \mathcal{L}_2(\mathbb{R}_+) : \|f\|_{\mathcal{L}_2(\mathbb{R}_+)} \le M_j \} \quad (5)$$

(A3) For $j \in \{N + 1, \ldots, N_w\}$, the element w_j belongs to the space $\mathcal{F}_\infty^{\text{rate}}(M_j)$ which comprises all inputs such that the absolute value of the rate of change is below M_j.

$$\mathcal{F}_\infty^{\text{rate}}(M_j) := \left\{ f \in \mathcal{L}_\infty(\mathbb{R}_+) : \begin{array}{l} f : \text{continuous} \\ f(0) = 0 \\ \|\dot{f}\|_{\mathcal{L}_\infty(\mathbb{R}_+)} \le M_j \end{array} \right\} \quad (6)$$

The space $\mathcal{F}_2(M_j)$ is a good model for transient inputs. On the other hand, the space $\mathcal{F}_\infty^{\text{rate}}(M_j)$ is a good model for persistent inputs. The space of the input w satisfying (A1) through (A3) is represented by $\mathcal{F}(M)$ in which M denotes the set $(M_1, \ldots, M_{N_w})$.

The space $\mathcal{F}_2(M_j)$ includes inputs with infinite amplitude like an impulse. Hence we make a further assumption so that $z(t)$ and $w(t)$ are bounded even if such an input is applied to the system.

(A4) col$(D_w, j) = 0$ and col$(D_y, j) = 0$ for $j \in \{1, \ldots, N\}$

In this paper, we consider the design problem of a SD control system which guarantees the critical specification that each $|z_i(t)|$ is kept below the prespecified level ε_i, that is to say,

$$\|z_i\|_\infty \le \varepsilon_i \text{ for all } i \in \{1, \ldots, N_z\} \quad (7)$$

where $\|z_i\|_\infty := \sup\{|z_i(t)| : t \in \mathbb{R}_+\}$.

2.3 *Principle of matching*

We give a brief explanation of the principle of matching. For further details, see (Zakian, 1991, 1996). The principle of matching considers the two input spaces. The first is the possible input space $\mathcal{P}$, the set of all exogenous inputs that are actually applied or likely to be applied to the control system such as reference signals, disturbances and sensor noises. This space $\mathcal{P}$ characterizes the environment. The second is the tolerable input space $\mathcal{T}$, the set of all inputs which ensure the design specification. This space $\mathcal{T}$, on the other hand, characterizes the control system. The control system and its environment are said to be matched if the tolerable input space $\mathcal{T}$ includes the possible input space $\mathcal{P}$. Thus, in the framework of the principle of matching, the control system and its environment are chosen so that $\mathcal{P} \subseteq \mathcal{T}$ is satisfied.

To get a match, the space $\mathcal{P}$ needs to be known in advance. Now we can set $\mathcal{P} = \mathcal{F}(M)$. The tolerable input space is defined from (7) as

$$\mathcal{T} = \mathcal{T}_1 \cap \mathcal{T}_2 \cap \cdots \cap \mathcal{T}_{N_z} \tag{8}$$

where $\mathcal{T}_i = \{w \in C(\mathbb{R}_+, \mathbb{R}^{N_w}) : \|z_i\|_\infty \le \varepsilon_i\}$. Let $\|z_i\|_{\text{peak}}$ be the supremum of the absolute value of z_i for any $t \in \mathbb{R}_+$ and $w \in \mathcal{F}(M)$, that is

$$\|z_i\|_{\text{peak}} := \sup\{|z_i(t,w)| : t \in \mathbb{R}_+,\ w \in \mathcal{F}(M)\} \tag{9}$$

Therefore, in the framework of the principle of matching, our present design problem is reformulated into the following problem.

$$\|z_i\|_{\text{peak}} \le \varepsilon_i \quad \text{for all } i \in \{1, \ldots, N_z\} \tag{10}$$

3. TIME-INVARIANT INPUT-OUTPUT MODEL OF SD CONTROL SYSTEMS

In this section, we give an input-output model of a SD control system via the lifting. Let L be the lifting operator that maps a function x in $C(\mathbb{R}_+)$ to a function valued sequence $\hat{x}$ in $c(\mathbb{Z}_+)$.

$$L : x \in C(\mathbb{R}_+) \longmapsto \hat{x} \in c(\mathbb{Z}_+) \tag{11}$$

It can be seen from the definitions in (1) and (2) that $\|x\|_{\mathcal{L}_2(\mathbb{R}_+)} = \|\hat{x}\|_{\ell_2(\mathbb{Z}_+)}$ and $\|x\|_{\mathcal{L}_\infty(\mathbb{R}_+)} = \|\hat{x}\|_{\ell_\infty(\mathbb{Z}_+)}$. So the lifting defined on the space $\mathcal{L}_p(\mathbb{R}_+)$ is norm-preserving isomorphism. From now on, the lifted signal of $x \in \mathcal{X} \subseteq C(\mathbb{R}_+)$ and its kth element are denoted by $\hat{x}$ and $\hat{x}(k,\theta)$ respectively, in which $\hat{x}(k,\theta) := x(kh+\theta)$, $\theta \in [0,h]$. The space of $\hat{x}$ is denoted by $\hat{\mathcal{X}}$.

It is well known that the relationship between $\hat{z}$ and $\hat{w}$ is given as the following time-invariant discrete-time model.

$$\hat{z}(k,\theta,\hat{w}) = (\hat{D}\hat{w})(k,\theta) + \sum_{l=0}^{k-1} \hat{C}(\theta)\hat{A}^{k-l-1}\hat{B}\hat{w}(l) \tag{12}$$

where matrices $\hat{A}$, $\hat{C}(\theta)$ and operators $\hat{D}$, $\hat{B}$ are of the following form.

$$\hat{A} = \begin{bmatrix} e^{Ah} + B_u(h)B_uD_KC_y & B_u(h)C_K \\ B_KC_y & A_K \end{bmatrix}$$

$$B_u(\theta) = \int_0^\theta e^{A\tau} d\tau B_u$$

$$\hat{B}\hat{w}(l) = \begin{bmatrix} \hat{B}_w\hat{w}(l) \\ \hat{B}_u\hat{w}(l) \end{bmatrix}$$

$$\hat{B}_w\hat{w}(l) = \int_0^h \left\{ e^{A(h-\tau)}B_w + B_u(h)B_uD_KD_y\delta(\tau) \right\} \times \hat{w}(l,\tau)d\tau$$

$$\hat{B}_u\hat{w}(l) = \int_0^h B_KD_y\delta(\tau)\hat{w}(l,\tau)d\tau$$

$$\hat{C}(\theta) = [\,\hat{C}_z(\theta) \quad \hat{C}_y(\theta)\,]$$

$$\hat{C}_z(\theta) = C_ze^{A\theta} + (D_u + C_zB_u(\theta))\,D_KC_y$$

$$\hat{C}_y(\theta) = (D_u + C_zB_u(\theta))\,C_K$$

$$(\hat{D}\hat{w})(k,\theta) = C_z \int_0^\theta e^{A(\theta-\tau)}B_w\hat{w}(k,\tau)d\tau + D_w\hat{w}(k,\theta) + (D_u + C_zB_u(\theta))\,D_KD_y\hat{w}(k,0)$$

and $\delta(t)$ is a Dirac delta function. From (12), the relationship between $\hat{z}_i$ and $\hat{w}_j$ can be obtained as follows.

Lemma 1. Suppose the assumptions (A2) and (A3) are satisfied for the SD control system in Fig. 1. Then $\hat{z}_i$ and $\hat{w}_j$ are related by

$$\hat{z}_i(k,\theta,\hat{w}) = \sum_{j=1}^{N_w} (T_{ij}\hat{w}_j)(k,\theta) \tag{13}$$

where

$$(T_{ij}\hat{w}_j)(k,\theta) = \begin{cases} \displaystyle\sum_{l=0}^{k} \int_0^h \hat{z}_i(k-l,\theta,\hat{\delta}_j(\tau))\hat{w}_j(l,\tau)d\tau & \text{for } j \in \{1,\ldots,N\} \\ \displaystyle\sum_{l=0}^{k} \int_0^h \hat{z}_i(k-l,\theta,\hat{1}_j(\tau))\hat{w}_j^{(1)}(l,\tau)d\tau & \text{for } j \in \{N+1,\ldots,N_w\} \end{cases} \tag{14}$$

and $\hat{w}_j^{(1)}(k,\tau) = d\hat{w}_j(k,\tau)/d\tau$, $\hat{\delta}_j(\tau)$ denotes the lifted signal of $\delta_j(\tau)$ where the jth entry is the impulse with delay τ, $\delta(t-\tau)$, and the other $N_w - 1$ entries are all zero, and $\hat{1}_j(\tau)$ denotes the lifted signal of $1_j(\tau)$ where the jth entry is the unit step with delay τ, $1(t-\tau)$, and other $N_w - 1$ entries are all zero.

4. MATCHING CONDITION FOR $\mathcal{F}(M)$

Let us define the peak norm of $\hat{z}_i$ as follows.

$$\|\hat{z}_i\|_{\text{peak}} := \sup\{|\hat{z}_i(k,\theta,\hat{w})| : k \in \mathcal{Z}_+,\ \theta \in [0,h],\ \hat{w} \in \hat{\mathcal{F}}(M)\} \tag{15}$$

It is obvious from (9) and (15) that

$$\|z_i\|_{\text{peak}} = \|\hat{z}_i\|_{\text{peak}}. \tag{16}$$

We now have the following theorem.

Theorem 1. Suppose $\hat{z}_i$ and $\hat{w}_j$ are related by (13). Then

$$\|\hat{z}_i\|_{\text{peak}} = \sup_{\theta\in[0,h]} \left(\sum_{j=1}^{N} M_j\phi_{ij}(\theta) + \sum_{j=N+1}^{N_w} M_j\psi_{ij}(\theta) \right) \tag{17}$$

where

$$\phi_{ij}(\theta) = \left(\sum_{k=0}^{\infty} \int_0^h |\hat{z}_i(k,\theta,\hat{\delta}_j(\tau))|^2 d\tau \right)^{\frac{1}{2}} \tag{18}$$

and

$$\psi_{ij}(\theta) = \sum_{k=0}^{\infty} \int_0^h |\hat{z}_i(k,\theta,\hat{1}_j(\tau))| d\tau. \tag{19}$$

Proof: From (13), (15) and (A1),

$$\|\hat{z}_i\|_{\text{peak}} = \sup_{k,\theta,\hat{w}_j} \sum_{j=1}^{N_w} |(T_{ij}\hat{w}_j)(k,\theta)| \tag{20}$$

Thus we prove the following two equations.

$$M_j\phi_{ij}(\theta) = \sup\{|(T_{ij}\hat{w}_j)(k,\theta)| : k\in\mathbb{Z}_+,\ \hat{w}_j\in\hat{\mathcal{F}}_2(M_j)\} \tag{21}$$

$$M_j\psi_{ij}(\theta) = \sup\{|(T_{ij}\hat{w}_j)(k,\theta)| : k\in\mathbb{Z}_+,\ \hat{w}_j\in\hat{\mathcal{F}}_\infty^{\text{rate}}(M_j)\} \tag{22}$$

First, we prove (21). According to the Schwartz inequality,

$$\begin{aligned} &|(T_{ij}\hat{w}_j)(k,\theta)| \\ &\le \sum_{l=0}^{k} \int_0^h |\hat{z}_i(k-l,\theta,\hat{\delta}_j(\tau))||\hat{w}_j(l,\tau)| d\tau \\ &\le \sum_{l=0}^{k} \left(\int_0^h |\hat{z}_i(k-l,\theta,\hat{\delta}_j(\tau))|^2 d\tau \right)^{\frac{1}{2}} \\ &\qquad \times \left(\int_0^h |\hat{w}_j(l,\tau)|^2 d\tau \right)^{\frac{1}{2}} \\ &\le M_j\phi_{ij}(\theta). \end{aligned} \tag{23}$$

Consider the following input $\hat{v}$ at the jth ($j\in\{1,\ldots,N\}$) input channel.

$$\hat{v}(l,\tau) = \begin{cases} M_j\hat{z}_i(k-l,\theta,\hat{\delta}_j(\tau))/N(k,\theta) \\ \quad \text{if } l\in\{0,\ldots,k\},\ \tau\in[0,h] \\ 0 \text{ otherwise} \end{cases} \tag{24}$$

$N(k,\theta)$ is of the following form.

$$N(k,\theta) = \left(\sum_{l=0}^{k} \int_0^h |\hat{z}_i(l,\theta,\hat{\delta}_j(\tau))|^2 d\tau \right)^{\frac{1}{2}} \tag{25}$$

Clearly, $\hat{v}$ belongs to $\hat{\mathcal{F}}_2(M_j)$ since $\|\hat{v}\|_{\ell_2(\mathbb{Z}_+)} = M_j$. For this input, $|(T_{ij}\hat{v})(\infty,\theta)| = M_j\phi_{ij}(\theta)$. Therefore, the equation (21) is obtained. Next we prove (22). For $\hat{w}_j\in\hat{\mathcal{F}}_\infty^{\text{rate}}(M_j)$, we have

$$\begin{aligned} &|(T_{ij}\hat{w}_j)(k,\theta)| \\ &\le \sum_{l=0}^{k} \int_0^h |\hat{z}_i(k-l,\theta,\hat{1}_j(\tau))||\hat{w}_j^{(1)}(l,\tau)| d\tau \\ &\le M_j\psi_{ij}(\theta). \end{aligned} \tag{26}$$

Consider the following $\hat{v}$ at the jth ($j\in\{N+1,\ldots,N_w\}$) input channel.

$$\hat{v}^{(1)}(l,\tau) = \begin{cases} M_j \operatorname{sgn}\{\hat{z}_i(k-l,\theta,\hat{1}_j(\tau))\} \\ \quad \text{if } l\in\{0,\ldots,k\},\ \tau\in[0,h] \\ 0 \text{ otherwise} \end{cases} \tag{27}$$

The above input $\hat{v}$ belongs to $\hat{\mathcal{F}}_\infty^{\text{rate}}(M_j)$ since $\|\hat{v}^{(1)}\|_{\ell_\infty(\mathbb{Z}_+)} = M_j$. For input $\hat{v}$, $|(T_{ij}\hat{v})(\infty,\theta)| = M_j\psi_{ij}(\theta)$. Therefore we get (22). □

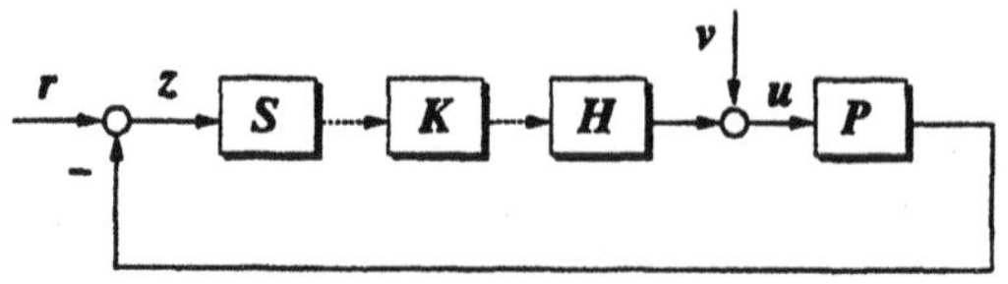

Fig. 2. Unity feedback control system

The function $\psi_{ij}(\theta)$ in (19) is generally complex to compute since it requires the integration of the response for $\hat{1}_j(\tau)$. But for a unity feedback control system shown in Fig. 2, where w_j is a reference signal and z_i is an error between an output of plant and a reference signal, $\psi_{ij}(\theta)$ can be rewritten in a simpler form.

Corollary 1. Consider the unity feedback control system shown in Fig. 2. Then $\psi_{ij}(\theta)$ in (19) is rewritten by

$$\psi_{ij}(\theta) = h\|\hat{z}_i(\theta,\hat{1}_j(0))\|_1 + \theta \tag{28}$$

where

$$\|\hat{z}_i(\theta,\hat{1}_j(0))\|_1 = \sum_{k=0}^{\infty} |\hat{z}_i(k,\theta,\hat{1}_j(0))|. \tag{29}$$

Proof: From Fig. 3, it follows that

$$\hat{z}_i(0,\theta,\hat{1}_j(\tau)) = \begin{cases} \hat{z}_i(0,\theta,\hat{1}_j(0)) & \text{for } \tau=0 \\ 1 & \text{for } \tau\in(0,\theta] \end{cases}$$

and

$$\hat{z}_i(l,\theta,\hat{1}_j(\tau)) = \begin{cases} \hat{z}_i(l,\theta,\hat{1}_j(0)) & \text{for } \tau=0 \\ \hat{z}_i(l-1,\theta,\hat{1}_j(0)) & \text{for } \tau\in(0,h]. \end{cases}$$

Thus we have

$$\begin{aligned} \phi_{ij}(\theta) &= \lim_{\alpha\to+0} \int_\alpha^\theta |\hat{z}_i(0,\theta,\hat{1}_j(\tau))| d\tau \\ &\quad + \sum_{l=1}^{\infty} \lim_{\alpha\to+0} \int_\alpha^h |\hat{z}_i(l,\theta,\hat{1}_j(\tau))| d\tau \\ &= \theta + h\sum_{l=0}^{\infty} |\hat{z}_i(l,\theta,\hat{1}_j(0))|, \end{aligned}$$

which yields (28). □

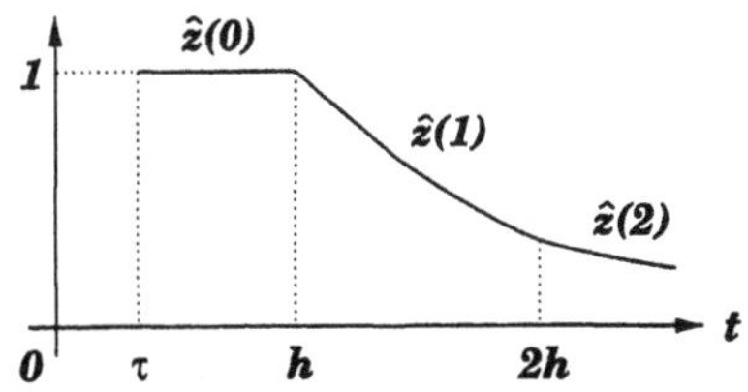

Fig. 3. Step response for a unity feedback control system

The following theorem gives the necessary and sufficient condition for a match.

Theorem 2. Define $\Lambda_i(\mathcal{F}(M))$ to be

$$\Lambda_i(\mathcal{F}(M)) = \sup_{\theta\in[0,h)} \left(\sum_{j=1}^{N} M_j\phi_{ij}(\theta) + \sum_{j=N+1}^{N_w} M_j\psi_{ij}(\theta) \right) \quad (30)$$

where $\phi_{ij}(\theta)$ is defined by (18) and $\psi_{ij}(\theta)$ by (19) or (28). Then the SD control system and its environment characterized by $\mathcal{F}(M)$ are matched if and only if

$$\Lambda_i(\mathcal{F}(M)) \le \varepsilon_i \quad \text{for all} \quad i \in \{1,\dots,N_z\}. \quad (31)$$

Remark 1. To achieve a match, $\psi_{ij}(\theta)$ needs to be finite. Thus the system must have the integral action, that is, $z_i(t) \to 0\,(t\to\infty)$ when applying the unit step at the jth ($j \in \{N+1,\dots,N_w\}$) input channel.

5. COMPUTATION OF $\phi_{IJ}(\theta)$ AND $\psi_{IJ}(\theta)$

Evaluation of the matching condition in (31) requires to compute $\phi_{ij}(\theta)$ and $\psi_{ij}(\theta)$ for $\theta \in [0,h)$. In this section, we discuss how to compute $\phi_{ij}(\theta)$ in (18) and $\psi_{ij}(\theta)$ in (28).

5.1 *Computation of* $\phi_{ij}(\theta)$

It is possible to calculate $\phi_{ij}(\theta)$ in (18) with a Lyapunov equation.

Theorem 3. For $\theta \in [0,h)$, $\phi_{ij}(\theta)$ is computed by

$$\phi_{ij}(\theta) = \left\{ C_{zi}W_j(\theta)C_{zi}' + \hat{C}_i(\theta)X_j\hat{C}_i(\theta)' \right\}^{\frac{1}{2}} \quad (32)$$

where $C_{zi} = \text{row}(C_z, i)$, $\hat{C}_i(\theta) = \text{row}(\hat{C}(\theta), i)$,

$$W_j(\theta) = \int_0^\theta e^{A(\theta-\tau)}B_{wj}B_{wj}'e^{A'(\theta-\tau)}d\tau, \quad (33)$$

$B_{wj} = \text{col}(B_w, j)$ and X_j is the solution of the following Lyapunov equation.

$$X_j = \hat{A}X_j\hat{A}' + Y_j \quad (34)$$

$$Y_j = \begin{bmatrix} W_j(h) & 0 \\ 0 & 0 \end{bmatrix} \quad (35)$$

Using the lemma on matrix exponential function (Chen and Francis, 1995), the function-valued matrix $W_j(\theta)$ in (33) is computed as follows.

Lemma 2. Define the matrix exponential function with variable $\theta \in [0,h]$ by

$$\begin{bmatrix} Q_{11}(\theta) & Q_{12}(\theta) \\ 0 & Q_{22}(\theta) \end{bmatrix} = \exp\left\{ \begin{bmatrix} A & B_{wj}B_{wj}' \\ 0 & -A' \end{bmatrix} \theta \right\}.$$

Then $W_j(\theta)$ is computed by

$$W_j(\theta) = Q_{12}(\theta)Q_{11}(\theta)'. \quad (36)$$

5.2 *Computation of* $\psi_{ij}(\theta)$

Let us consider the calculation of $\psi_{ij}(\theta)$ in (28). It requires to calculate the absolute value of $z_i(k,\theta,\hat{1}_j(0))$. The step response $z_i(k,\theta,\hat{1}_j(0))$ can be computed iteratively as follows.

Theorem 4. For $\theta \in [0,h)$, $\hat{z}_i(k,\theta,\hat{1}_j(0))$ can be iteratively computed by

$$\hat{z}_i(k,\theta,\hat{1}_j(0)) = \begin{cases} \hat{D}_{dij}(\theta) & \text{for } k=0 \\ \hat{z}_i(k-1,\theta,\hat{1}_j(0)) + \hat{C}_i(\theta)\hat{A}^{k-1}\hat{B}_{dj} & \text{for } k \in \{1,\dots,\infty\} \end{cases} \quad (37)$$

where $\hat{D}_{dij}(\theta)$ is the (i,j) entry of $\hat{D}_d(\theta)$ and $\hat{B}_{dj} = \text{col}(\hat{B}_d, j)$. The matrices $\hat{D}_d(\theta)$ and $\hat{B}_d$ are of the following form.

$$\hat{D}_d(\theta) = C_zB_w(\theta) + D_w + (D_u + C_zB_u(\theta))\,D_KD_y \quad (38)$$

$$\hat{B}_d = \begin{bmatrix} B_w(h) + B_u(h)B_uD_KD_y \\ B_KD_y \end{bmatrix} \quad (39)$$

where

$$B_w(\theta) = \int_0^\theta e^{A\tau}d\tau B_w. \quad (40)$$

6. DESIGN EXAMPLE

We show a design example of a unity feedback SD control system depicted in Fig. 2. The continuous-time plant P is the 3rd-order dynamical system with the following transfer function (Whidborne and Liu, 1993).

$$P(s) = \frac{27697}{s(s^2 + 1429s + 42653)} \quad (41)$$

The input v is a disturbance belonging to the space $\mathcal{F}_2(0.8)$ and the input r is a reference signal belonging to the space $\mathcal{F}_\infty^{\text{rate}}(1.0)$. The controlled signal z is an error between a reference signal r and an output of plant. Our purpose here is to

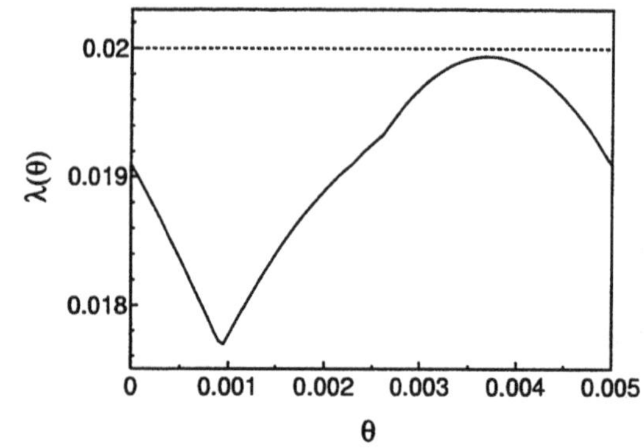

Fig. 4. Plot of $\lambda(\theta) = M_1\phi_{11}(\theta) + M_2\psi_{12}(\theta)$

get an internally stabilizing controller $K(z)$ which guarantees $||z||_{\text{peak}} \leq 0.02$.

Defining an exogenous input vector by $w = [v \ r]'$, we can obtain a generalized plant G with

$$A = \begin{bmatrix} -1429.0 & -166.6 & 0 \\ 256.0 & 0 & 0 \\ 0 & 4.0 & 0 \end{bmatrix}, \quad B_w = \begin{bmatrix} 4 & 0 \\ 0 & 0 \\ 0 & 0 \end{bmatrix},$$

$$B_u = [4 \ 0 \ 0]', \ C_z = [0 \ 0 \ -6.76],$$
$$D_w = [0 \ 1], \ D_u = 0,$$
$$C_y = [0 \ 0 \ -6.76], \ D_y = [0 \ 1].$$

The assumptions (A1) through (A4) are satisfied for above G. From (31), the sampling period h needs to be selected so that

$$M_1\phi_{11}(0) + M_2\psi_{12}(0) \leq 0.02 \tag{42}$$

is satisfied. Fortunately the left hand in (42) is independent of the controller parameters. This leads to the restriction $h \leq 0.0123$. Furthermore we restrict the controller $K(z)$ to be 2nd-order. Under these conditions, we search the controller which satisfies (31) with the numerical search method known as the moving boundaries process (e.g., Whidborne and Liu, 1993).

As a beginning, we took h to be 0.01, but we failed to get the controller which achieves a match (though there might exist such a controller). Next, taking $h = 0.005$, we could get a matching controller with transfer function

$$K(z) = 3930\frac{z^2 + 1.331z + 0.4796}{z^2 - 0.0338z - 0.4321}. \tag{43}$$

Fig. 4 shows the plot of

$$\lambda(\theta) = M_1\phi_{11}(\theta) + M_2\psi_{12}(\theta) \tag{44}$$

for above $K(z)$. It can be seen from Fig. 4 that $\Lambda(\mathcal{F}(M)) = \sup\{\lambda(\theta) : \theta \in [0, 0.005)\} = 0.01989 < 0.02$. So the controller $K(z)$ in (43) achieves a match.

7. CONCLUSION

We have discussed the matching condition for SD control systems using the lifting technique. For exogenous inputs belonging to the class of inputs with finite energy and/or finite rate of change, we show that the matching condition is represented as the restriction on the impulse and/or step responses. It can easily be evaluated with the standard numerical methods. This significant benefit owes to the lifting technique. A numerical design example was presented to show the effectiveness of the proposed design method.

References

Bamieh, B. and Pearson, J. B. (1991). A lifting technique for linear periodic systems with applications to sampled-data control. *Systems & Control Letters*, **17**, 79-88.

Bamieh, B. A. and Pearson, J. B. (1992). A general framework for linear periodic systems with applications to $\mathcal{H}^\infty$ sampled-data control. *IEEE Trans. Automatic Control*, **37**, 418-435.

Chen, T. and Francis, B. (1995). *Optimal Sampled-Data Control Systems*, Springer.

Liu, G. P. (1990). Frequency-domain approach for critical systems. *International Journal of Control*, **52**, 1507-1519.

Liu, G. P. (1992). Disturbance spaces and sup regulators in discrete time. *Systems & Control Letters*, **18**, 33-38.

Rutland, N. K. (1992). Illustration of the principle of design: vehicle speed control. *International Journal of Control*, **55**, 1319-1334.

Yamamoto, Y. and Khargonekar, P. P. (1996). Frequency response of sampled-data systems. *IEEE Trans. Automatic Control*, **41**, 166-176.

Whidborne, J. F. and Liu, G. P. (1993a). *Critical Control Systems*, Research Studies Press Ltd..

Zakian, V. (1986). A performance criterion. *International Journal of Control*, **43**, 921-931.

Zakian, V. (1987). Input spaces and output performance. *International Journal of Control*, **46**, 185-191.

Zakian, V. (1989). Critical systems and tolerable inputs. *International Journal of Control*, **49**, 1285-1289.

Zakian, V. (1991). Well matched systems. *IMA Journal of Mathematical Control & Information*, **8**, 29-38.

Zakian, V. (1996). Perspectives on the principle of matching and the method of inequalities. *International Journal of Control*, **65**, 147-175.

14th Triennial World Congress, Beijing, P.R. China

PERTURBATION ANALYSIS OF THE DISCRETE-TIME H_∞ OPTIMAL CONTROL PROBLEM

Vera Angelova Angelova *
Mihail Mihaylov Konstantinov **
Petko Hristov Petkov ***

* *Institute of Information Technologies, Bulgarian Academy of Sciences, Ak. G. Bonchev Str., Bl. 2, 1113 Sofia, Bulgaria*
e-mail: vangelova@iit.acad.bg
** *University of Architecture & Civil Engineering*
1 Hr. Smirnenski Blvd., 1421 Sofia, Bulgaria
e-mail: mmk_fte@uacg.acad.bg
*** *Department of Automatics, Technical University of Sofia*
1756 Sofia, Bulgaria
e-mail: php@mbox.digsys.bg

Abstract: The paper deals with the sensitivity of the standard H_∞ optimization problem for discrete-time linear multivariable systems. Both local and non-local bounds for the perturbation in the solution are obtained. A numerical example demonstrates the effectiveness of the estimates proposed. *Copyright © 1999 IFAC*

Keywords: Discrete-time systems, Riccati equations, Perturbation analysis, Condition numbers, Error analysis

1. INTRODUCTION

Many techniques have been derived for solving H_∞ control problems for continuous- and discrete-time systems. At the same time the numerical aspects of the H_∞ optimal control problems are not developed to a sufficient extent. In particular the numerical solution to the corresponding Riccati equations often runs into difficulties. This is due to the use of unreliable numerical algorithms or to the absence of tools for studying the sensitivity of the problem. Hence, it is necessary to derive quantitative measures for the conditioning of the corresponding matrix Riccati equations in order to get reliable bounds for the errors in the computed solution to the standard H_∞ optimization problem.

The aim of this paper is to present a perturbation analysis for the generalized discrete-time matrix Riccati equation, related to the H_∞ optimization problem for linear discrete-time control systems. For the continuous-time H_∞ optimization problem, using the approach developed in (Konstantinov, Christov and Petkov, 1986), local and non-local perturbation analysis is presented in (Konstantinov, Petkov and Christov, 1995).

The paper begins with a definition of the H_∞ optimal discrete-time control problem. Next, a perturbation analysis of the related Riccati equation is done. The problem is reformulated as an equivalent operator equation. The latter is studied applying the Banach fixed point principle which allows to establish a non-local perturbation bound for the solution. The performance of the derived bound is illustrated by a numerical example.

The following abbreviations are used: $\mathcal{R}^{n\times n}$ – the space of real $n \times n$ matrices; $\mathcal{R}^n = \mathcal{R}^{n\times 1}$; I_n – the unit $n \times n$ matrix; 0 – the zero matrix; $\mathrm{vec}(A) \in \mathcal{R}^{n^2}$ – the column-wise vector representation of $A \in \mathcal{R}^{n\times n}$; $\|.\|_2$ – the Euclidean vector norm or the spectral matrix norm; $\|.\|_F$ – the Frobenius matrix norm; $A \otimes B$ – the Kroneker product of the matrices A and B.

2. MAIN RESULTS

Consider the linear discrete-time control system

$$x(t+1) = Ax(t) + B_1u(t) + B_2w(t), \quad (1)$$

$$z(t) = Cx(t) + D_1u(t) + D_2w(t); \; t = 0, 1, 2, \ldots,$$

where $x(t) \in \mathcal{R}^n$ is the vector of state variables, $u(t) \in \mathcal{R}^m$ is the control input, $w(t) \in \mathcal{R}^l$ is the unknown disturbance, $z(t) \in \mathcal{R}^p$ is the controlled output and A, B_1, B_2, C, D_1, D_2 are constant matrices of appropriate dimensions. The H_∞ optimization problem is to find a compensator described by a static feedback law

$$u(t) = L_1x(t) + L_2w(t) \quad (2)$$

such that the closed-loop system is internally stable and for the closed loop-system the ℓ_2 induced operator norm from the disturbance w to the output z is less than 1, i.e. the H_∞ norm of the closed-loop system is less than 1. The following conditions for solvability of this problem are given in (Stoorvogel, 1992).

Let the system (A, B_1, C, D_1) has no invariant zeros on the unit circle and is left-invertible. A controller (2) exists such that the closed-loop system is internally stable and its H_∞ norm is less than 1 if and only if a symmetric matrix $X_0 \geq 0$ exists such that $V(X_0) := D_1^\top D_1 + B_1^\top X_0 B_1 > 0$ and $I_l - D_2^\top D_2 - B_2^\top X_0 B_2 + (B_2^\top X_0 B_1 + D_2^\top D_1) \times V^{-1}(X_0)(B_1^\top X_0 B_2 + D_1^\top D_2) > 0$ (this implies that the matrix

$$G(X_0, a) := D^\top D - \begin{bmatrix} 0 & 0 \\ 0 & I_l \end{bmatrix} + B^\top X_0 B,$$

where $a := (A, B, C, D)$, $B := \begin{bmatrix} B_1 & B_2 \end{bmatrix}$, $D := \begin{bmatrix} D_1 & D_2 \end{bmatrix}$, is invertible) and $X = X_0$ satisfies the generalized discrete-time matrix Riccati equation

$$X = A^\top XA + C^\top C - (A^\top XB + C^\top D) \quad (3)$$
$$\times G^{-1}(X, a)(B^\top XA + D^\top C).$$

Moreover, for

$$L_1 = -(D_1^\top D_1 + B_1^\top X_0 B_1)^{-1}(B_1^\top X_0 A + D_1^\top C)$$

$$L_2 = -(D_1^\top D_1 + B_1^\top X_0 B_1)^{-1}(B_1^\top X_0 B_2 + D_1^\top D_2)$$

the closed-loop state matrix

$$F(X_0, a) := A - BG^{-1}(X_0, a)(B^\top X_0 A + D^\top C)$$

is convergent.

Thus the problem of studying the sensitivity of the H_∞ optimization problem comes down to the perturbation analysis of the generalized discrete matrix Riccati equation (3).

The equivalent form

$$F^\top(X, a)XA - X + C^\top C \quad (4)$$
$$-(A^\top XB + C^\top D)G^{-1}(X, a)D^\top C = 0$$

of equation (3) is more convenient for perturbation analysis. The round-off errors in the solution and the parameter uncertainties in the model are represented by equivalent perturbations in the coefficient matrices:

$$M \to M + \Delta M; \; M \in \mathcal{M} := \{A, B, C, D\},$$

where the only information about ΔM is that $\|\Delta M\|_F \leq \Delta_M$ for $\Delta_M > 0$ being a given quantity. The perturbed equation is

$$F^\top(Y, a + \Delta a)Y(A + \Delta A) - Y \quad (5)$$
$$+(C + \Delta C)^\top(C + \Delta C)$$
$$-((A + \Delta A)^\top Y(B + \Delta B)$$
$$+(C + \Delta C)^\top(D + \Delta D))$$
$$\times G(Y, a + \Delta a)^{-1}(D + \Delta D)^\top(C + \Delta C) = 0,$$

where $\Delta a := (\Delta A, \Delta B, \Delta C, \Delta D)$, $\Delta B := \begin{bmatrix} \Delta B_1 & \Delta B_2 \end{bmatrix}$, $\Delta D := \begin{bmatrix} \Delta D_1 & \Delta D_2 \end{bmatrix}$. Also,

$$G(Y, a + \Delta a) = G(Y, a) + J(Y, \Delta a),$$

$$J(Y, \Delta a) := \Delta D^\top D + D^\top \Delta D + \Delta D^\top \Delta D$$
$$+\Delta B^\top YB + B^\top Y\Delta B + \Delta B^\top Y \Delta B$$

$$F^\top(Y, a + \Delta a) = (A + \Delta A)^\top$$
$$-((A + \Delta A)^\top Y(B + \Delta B)$$
$$+(C + \Delta C)^\top(D + \Delta D))$$
$$\times G^{-1}(Y, a + \Delta a)(B + \Delta B)^\top.$$

Since the Fréchet derivative of the left-hand side of (4) in X at the solution matrix X_0 is invertible, the perturbed equation (5) has an unique solution $Y = X_0 + \Delta X$ in the neighborhood of X_0.

Denote by $\Delta := [\Delta_A, \Delta_B, \Delta_C, \Delta_D]^\top \in \mathcal{R}_+^4$ the vector of absolute perturbations $\Delta_A := \|\Delta A\|_F$, ...,$\Delta_D := \|\Delta D\|_F$ in the data matrices $A, \ldots, D$.

The aim is to derive:

- A local estimate for the norm $\Delta_X := \|\Delta X\|_F$ of the perturbation ΔX as a function of Δ, which is valid for $\|\Delta\|$ asymptotically small.
- A domain $\Omega \subset \mathcal{R}_+^4$, $0 \in \Omega$, such that for each $\Delta \in \Omega$ the perturbed H_∞ control problem is solvable, i. e. equation (5) has an unique solution $Y = X_0 + \Delta X$ in the neighbourhood of X_0, whose elements are analytic functions of the elements of the perturbations.

- A function $f : \Omega \to \mathcal{R}_+$, which is analytic, non-decreasing in each of its arguments, $f(0) = 0$ and such that the perturbations in the solution satisfy $\Delta_X \le f(\Delta),\ \ \Delta \in \Omega$.

Consider first the conditioning of the matrix Riccati equation (4). Denote by $\Theta(X, a)$ the left-hand side of (4). Then, having in mind that $\Theta(X_0, a) = 0$, the perturbed equation (5) may be written as

$$\Theta(X_0 + \Delta X, a + \Delta a) = \Theta_X(\Delta X)$$
$$+ \sum_{M \in \mathcal{M}} \Theta_M(\Delta M) + N(\Delta X, \Delta a) = 0,$$

where $\Theta_Z(.) := \Theta_Z(X_0, a)(.)$ are the Fréchet derivatives of $\Theta(X, a)$ in the matrix argument Z, evaluated for $X = X_0$ and $N(\Delta X, \Delta a)$ contains second and higher order terms in ΔX, Δa. A straightforward calculation leads to

$$\Theta_X(Z) = F_0^\top Z F_0 - Z, \qquad (6)$$
$$\Theta_A(Z) = F_0^\top X_0 Z + Z^\top X_0 F_0,$$
$$\Theta_B(Z) = -F_0^\top X_0 Z G_0^{-1} S_0 - S_0^\top G_0^{-1} Z^\top X_0 F_0,$$
$$\Theta_C(Z) = (C - DG_0^{-1}S_0)^\top Z$$
$$+ Z^\top (C - DG_0^{-1}S_0),$$
$$\Theta_D(Z) = (DG_0^{-1}S_0 - C)^\top Z G_0^{-1} S_0$$
$$+ S_0^\top G_0^{-1} Z^\top (DG_0^{-1}S_0 - C),$$

where $F_0 := F(X_0, a)$, $G_0 := G(X_0, a)$ and $S_0 := S(X_0, a) = B^\top X_0 A + D^\top C$.

Since the matrix F_0 is convergent the operator $\Theta_X(.)$ is invertible and

$$\Delta X = -\sum_{M \in \mathcal{M}} \Theta_X^{-1} \circ \Theta_M(\Delta M) \qquad (7)$$
$$-\Theta_X^{-1}(N(\Delta X, \Delta a)).$$

The relation (7) gives

$$\Delta_X \le \sum_{M \in \mathcal{M}} K_M \Delta_M + O(\|\Delta\|^2), \qquad (8)$$

where the scalars

$$K_M := \|\Theta_X^{-1} \circ \Theta_M\|,\ M \in \mathcal{M}, \qquad (9)$$

are the ***absolute individual condition numbers*** of the associated with the discrete-time H_∞ optimization problem matrix Riccati equation (4) (here $\|.\|$ is the corresponding induced norm in the space of linear operators $\mathcal{R}^{n^2 \times n^2} \to \mathcal{R}^{n^2 \times n^2}$). An estimate in terms of relative perturbations follows from (8) and (9). The scalars $k_M := K_M \|M\|_F / \|X\|_F$ are the *relative condition numbers* with respect to perturbations in $M \in \mathcal{M}$.

The calculation of the condition numbers K_M is straightforward. Denote by $\Lambda_M \in \mathcal{R}^{n^2 \times n^2}$ the matrix representation of the operator $\Theta_M(.)$. Then

$$\Lambda_X := F_0^\top \otimes F_0^\top - I_{n^2},$$
$$\Lambda_A := I_n \otimes F_0^\top X_0 + (F_0^\top X_0 \otimes I_n)\Pi_n,$$
$$\Lambda_B := -S_0^\top G_0^{-1} \otimes F_0^\top X_0 - (F_0^\top X_0 \otimes S_0^\top G_0^{-1})\Pi_n,$$
$$\Lambda_C := I_n \otimes (C - DG_0^{-1}S_0)^\top$$
$$+((C - DG_0^{-1}S_0)^\top \otimes I_n)\Pi_n,$$
$$\Lambda_D := S_0^\top G_0^{-1} \otimes (DG_0^{-1}S_0 - C)^\top$$
$$+((DG_0^{-1}S_0 - C)^\top \otimes S^\top G_0^{-1})\Pi_n,$$

where $\Pi_n \in \mathcal{R}^{n^2 \times n^2}$ is the vec-permutation matrix such that $\mathrm{vec}(M^\top) = \Pi_n \mathrm{vec}(M)$. Hence

$$K_M = \|\Lambda_X^{-1} \Lambda_M\|_2,\ M \in \mathcal{M}. \qquad (10)$$

Although linear estimates of type (8), (10) are commonly used in practice, there are sharper local estimates which are homogeneous functions of $\Delta := [\Delta_1, \Delta_2, \Delta_3, \Delta_4]^\top := [\Delta_A, \Delta_B, \Delta_C, \Delta_D]^\top$. These latter estimates are obtained as follows. Let the matrices $H_i \in \mathcal{R}^{k \times s_i}$ and the quantities $\Delta_i \ge 0$ $(i = 1, \ldots, q)$ be given. Consider the problem of estimating the maximum norm of the vector

$$\alpha = \sum_{i=1}^{q} H_i \beta_i \in \mathcal{R}^k,\ \ \beta_i \in \mathcal{R}^{s_i} \qquad (11)$$

subject to the constraints $\|\beta_i\| \le \Delta_i$.

Denoting $\Delta = [\Delta_1, \ldots, \Delta_q]^\top \in \mathcal{R}_+^q$ one has

$$\|\alpha\|_2 \le \mathrm{Est}_1(\Delta, H) := \sum_{i=1}^{q} K_i \Delta_i,$$
$$H := [H_1, \ldots, H_q] \in \mathcal{R}^{k \times s},$$

where $K_i := \|H_i\|_2$ and $s := s_1 + \cdots + s_q$.

Rewriting (11) as $\alpha = H\beta$, where $\beta := [\beta_1^\top \ldots \beta_q^\top]^\top \in \mathcal{R}^s$ one also has

$$\|\alpha\|_2 \le \mathrm{Est}_2(\Delta, H) := \|H\|_2 \sqrt{\Delta_1^2 + \cdots + \Delta_q^2}.$$

Finally one gets

$$\|\alpha\|_2 \le \mathrm{Est}_3(\Delta, H) := \sqrt{\Delta^\top U \Delta} = \sqrt{\sum_{i,j=1}^{q,q} u_{ij} \Delta_i \Delta_j},$$

where $U = [u_{ij}] \in \mathcal{R}_+^{q \times q}$, $u_{ij} := \|H_i^\top H_j\|_2$.

It may be shown that $\mathrm{Est}_3(\Delta, H) \le \mathrm{Est}_1(\Delta, H)$. Nevertheless the bound Est_1 is often used since it contains the individual condition numbers K_i. In turn, the estimates Est_2 and Est_3 are alternative, i.e. depending on Δ and H both cases $\mathrm{Est}_2(\Delta, H) \le \mathrm{Est}_3(\Delta, H)$ and $\mathrm{Est}_2(\Delta, H) > \mathrm{Est}_3(\Delta, H)$ are possible. Hence the combined improved estimate

$$\|\alpha\| \le \mathrm{Est}(\Delta, H),$$

where

$$\mathrm{Est}(\Delta, H) := \min\{\mathrm{Est}_2(\Delta, H), \mathrm{Est}_3(\Delta, H)\}$$

is valid. The above results are applicable to the original problem setting $H_1 = -\Lambda_X^{-1}\Lambda_A$, $H_2 = -\Lambda_X^{-1}\Lambda_B$, $H_3 = -\Lambda_X^{-1}\Lambda_C$, $H_4 = -\Lambda_X^{-1}\Lambda_D$.

Local estimates of the type considered above in fact neglect terms of order $O(\|\Delta\|_2^2)$, i.e. they are valid only for $\|\Delta\|_2 \to 0$. Unfortunately, it is usually impossible to say, having a small but a finite perturbation Δ, whether the neglected terms are indeed neglectible. One can not even claim that the magnitude of the neglected terms is less (or of the order of) the value of the local estimate. Even for simple linear equations the local estimate may always underestimate the actual perturbation. Moreover, for some critical values of the perturbations in the coefficients the solution may not exist (or go to infinity when these critical values are approached). Nevertheless, even in such cases the local estimates still produce "a bound" for a very large or non-existent solution...

The disadvantages of the local estimates may be overcome using the techniques of non-linear perturbation analysis. As a result one gets a domain $\Omega \subset \mathcal{R}_+^4$ and a non-linear function $f : \Omega \to \mathcal{R}_+$ such that $\|\Delta X\| \leq f(\Delta)$ for all $\Delta \in \Omega$ as described later on. In connection with this two important issues must be emphasized. First, the inclusion $\Delta \in \Omega$ guarantees that the perturbed solution has a solution of the desired class (this is an independent and essential point). And second, the bound $\|\Delta X\| \leq f(\Delta)$ is rigorous, i.e. it is simply true for perturbations with $\Delta \in \Omega$ unlike the local bounds, which may be severely violated. Unfortunately, the non-local bounds may be pessimistic in some cases.

Rewrite the perturbed equation (5) as

$$\Theta(Y, a) + \Theta(Y, a + \Delta a) - \Theta(Y, a) = 0. \quad (12)$$

Note that

$$\Theta(Y, a) = \Theta(X_0, a) + \Theta_X(\Delta X) + \Gamma(\Delta X),$$

where $\Gamma(Z) := -F_0^{\mathsf{T}} Z B G_0^{-1} B^{\mathsf{T}} Z F(Y, a)$.

A straightforward calculation shows that

$$\begin{aligned}\Theta(Y, a + \Delta a) - \Theta(Y, a) &= F^{\mathsf{T}}(Y, a) Y \Delta A \\ &+ \Delta A^{\mathsf{T}} Y A + \Delta A^{\mathsf{T}} Y \Delta A \\ &+ S^{\mathsf{T}}(Y, a) G^{-1}(Y, a) J(Y, \Delta a) \\ &\times G^{-1}(Y, a + \Delta a)(S(Y, a) \\ &+ B^{\mathsf{T}} Y \Delta A) \\ &- S^{\mathsf{T}}(Y, a) G^{-1}(Y, a + \Delta a)(\Delta B^{\mathsf{T}} Y (A \Delta A) \\ &+ D^{\mathsf{T}} \Delta C + \Delta D^{\mathsf{T}} C + \Delta D^{\mathsf{T}} \Delta C) \\ &-(A^{\mathsf{T}} Y \Delta B + \Delta A^{\mathsf{T}} Y B + \Delta A^{\mathsf{T}} Y \Delta B + \Delta C^{\mathsf{T}} D + \Delta C^{\mathsf{T}} \Delta D \\ &+ C^{\mathsf{T}} \Delta D) G^{-1}(Y, a + \Delta a)((B + \Delta B)^{\mathsf{T}} Y (A + \Delta A) \\ &+ (D + \Delta D)^{\mathsf{T}} (C + \Delta C)) + C^{\mathsf{T}} \Delta C + \Delta C^{\mathsf{T}} C + \Delta C^{\mathsf{T}} \Delta C.\end{aligned}$$

To simplify the expressions the following substitutions are used: $r := \|\Theta_X^{-1}\| = \|\Lambda_X^{-1}\|_2$, $\gamma := \|G^{-1}(X_0, a)\|_2$, $a_1 := \|F(X_0, a)\|_2$, $a_0 := \|A\|_2$, $b := \|B\|_2$, $c := \|C\|_2$, $d := \|D\|_2$, $x := \|X_0\|_2$, $s_0 := \|B^{\mathsf{T}} X_0 A + D^{\mathsf{T}} C\|_2$.

Suppose that

$$\Delta_X \leq \frac{1}{\gamma b^2}. \quad (13)$$

It follows from $G^{-1}(Y, a + \Delta a) = (G(Y, a) + J(Y, \Delta a))^{-1}$ that

$$\|(G(Y, a) + J(Y, \Delta a))^{-1}\|_2 \leq \frac{\|G^{-1}(Y, a)\|_2}{1 - \|G^{-1}(Y, a)\|_2 \|J(Y, \Delta a)\|_2}.$$

Similarly the representation $G^{-1}(Y, a) = (G(X_0, a) + B^{\mathsf{T}} \Delta X B)^{-1}$ yields

$$\|G^{-1}(Y, a)\|_2 \leq \frac{\gamma}{1 - \gamma b^2 \Delta_X}.$$

One also has $F^{\mathsf{T}}(Y, a) = F_0^{\mathsf{T}} + F_0^{\mathsf{T}} \Delta X B G^{-1}(Y, a) B^{\mathsf{T}}$ and hence

$$\|F(Y, a)\|_2 \leq \frac{a_1}{1 - \gamma b^2 \Delta_X}.$$

Rewrite (12) as an operator equation $\Delta X = \Psi(\Delta X, \Delta a)$, where the operator Ψ is defined from

$$\begin{aligned}\Psi(Z, \Delta a) := -\Theta_X^{-1}(\Gamma(Z) + \Theta(X_0 + Z, a + \Delta a) \\ -\Theta(X_0 + Z, a)).\end{aligned}$$

Under some conditions on Δ there exists $\rho = f(\Delta)$ such that $\Psi(., \Delta a) : \mathcal{R}^{n\times n} \to \mathcal{R}^{n\times n}$ is contractive and maps the set

$$\mathcal{G}_\rho = \{Z : \|Z\|_F \leq \rho, Z = Z^{\mathsf{T}}, X_0 + Z \geq 0\}$$

into itself. Then according to the Banach fixed point principle there exists a unique solution $\Delta X \in \mathcal{G}_\rho$ of the operator equation. Thus $\|\Delta X\|_F \leq \rho = f(\Delta)$, which is the desired non-local bound.

Let $Z \in \mathcal{G}_\rho$. One gets

$$\begin{aligned}\|\Psi(Z, \Delta a)\|_F &\leq r(\|\Gamma(Z)\|_F + \|\Theta(X_0 + Z, a + \Delta a) \\ &\quad -\Theta(X_0 + Z, a)\|_F) \\ &\leq g(\rho, \Delta) := \gamma_0(\Delta) + \gamma_1(\Delta)\rho \\ &\quad +\gamma_2(\Delta)\rho^2 + \gamma_3(\Delta)\rho^3,\end{aligned}$$

where the quantities γ_i are given by $\gamma_i(\Delta) = r\eta(\Delta)\zeta_i(\Delta)$; $i = 0, \ldots, 3$. Here

$$\begin{aligned}\zeta_0(\Delta) &:= \pi_0(a_1 x + \nu_{XA} + x\Delta_A)\Delta_A \\ &+ \pi_0 \Delta_{CC} + \gamma \Delta_{AXB} \Delta_{BXA} \\ &+ s_0\gamma(s_0 r\gamma + \nu_{BX}\gamma r \Delta_A + \Delta_{XAB}),\end{aligned}$$

$$\zeta_1(\Delta) := \pi_0 a_1^2 \nu_{BGB} + \pi_0(a_0 + \Delta_{AA})\Delta_A$$

$$+\pi_1(\nu_{XA}+x\Delta_A)\Delta_A$$
$$-a_1 x\gamma(b^2+\varphi)\Delta_A$$
$$+s_0\gamma(b\gamma\tau+\Delta_B)\Delta_{AA}$$
$$+(s_0\varphi+a_0 b\tau)(s_0+\nu_{BX}\Delta_A)\gamma^2$$
$$-s_0 b^2\gamma^2\Delta_{XAB}-a_0 b^3\gamma^2\Delta_{XAB}$$
$$+\gamma(a_0\Delta_B+\Delta_A\Delta_{BB}-b^2\gamma\Delta_{AXB})\Delta_{BXA}$$
$$+\gamma\Delta_{AA}\Delta_{BB}\Delta_{AXB}+\pi_1\Delta_{CC},$$

$$\zeta_2(\Delta):=\pi_1\Delta_A\Delta_{AA}$$
$$+\pi_2(\nu_{XA}+x\Delta_A)\Delta_A+\pi_2\Delta_{CC}-a_1\gamma(b^2+\varphi)\Delta_A$$
$$+b\gamma^2(s_0\varphi+a_0 b\tau)\Delta_{AA}$$
$$+a_0 b\varphi\gamma^2(s_0+\nu_{BX}\Delta_A)$$
$$+b\gamma(a_0-s_0 b\gamma)\Delta_{AA}\Delta_B$$
$$+\gamma(a_0\Delta_B+\Delta_A\Delta_{BB})\Delta_{AA}\Delta_{BB}$$
$$-b^2\gamma^2(a_0\Delta_B+\Delta_A\Delta_{BB})\Delta_{BXA}$$
$$-a_1^2\nu_{BGB}\gamma(b^2+\varphi)-b^2\gamma^2\Delta_{AA}\Delta_{BB}\Delta_{AXB},$$

$$\zeta_3(\Delta):=\pi_2\Delta_A\Delta_{AA}+a_0 b^2\gamma^2(\varphi-b\Delta_B)\Delta_{AA}$$
$$-b^2\gamma^2(a_0\Delta_B+\Delta_A\Delta_{BB})\Delta_{AA}\Delta_{BB},$$

$$\eta(\Delta):=\frac{1}{\pi_0+\pi_1\rho+\pi_2\rho^2},\ \pi_0:=1-\gamma\tau,$$
$$\pi_1:=-\gamma(\varphi+b^2(1+\pi_0)),$$
$$\pi_2:=b^2\gamma^2(b^2+\varphi),$$
$$\tau:=(2d+\Delta_D)\Delta_D+x\varphi,\ \varphi:=(2b+\Delta_B)\Delta_B,$$

$$\Delta_{MM}:=\|M\|_2+\Delta_M;\ M=A,B,C,D,$$
$$\nu_{XA}:=\|X_0A\|_2,\ \nu_{BX}:=\|B^\top X_0\|_2,$$
$$\nu_{BGB}:=\|BG^{-1}(X_0,a)B^\top\|_2,$$
$$\Delta_{DC}:=c\Delta_D+d\Delta_C+\Delta_C\Delta_D,\ \Delta_{CC}:=2c\Delta_C+\Delta_C^2,$$
$$\Delta_{XAB}:=\nu_{XA}\Delta_B+x\Delta_A\Delta_B+\Delta_{DC},$$
$$\Delta_{ABX}:=\nu_{AX}\Delta_B+\nu_{XB}\Delta_A+x\Delta_A\Delta_B+\Delta_{DC},$$
$$\Delta_{BXA}:=s_0+\nu_{BX}\Delta_A+\nu_{XA}\Delta_B$$
$$+x\Delta_A\Delta_B+\Delta_{DC}.$$

In a similar way for $Y,Z\in\mathcal{G}_\rho$ the following inequality is valid

$$\|\Psi(Y,\Delta a)-\Psi(Z,\Delta a)\|_F\le g'_\rho(\rho,\Delta)\|Y-Z\|_F.$$

Hence the function $g(\rho,\Delta)$ is a Lyapunov majorant for the operator equation. Therefore there exists a closed set Ω such that for $\Delta\in\Omega$ the majorant equation

$$\rho=g(\rho,\Delta), \tag{14}$$

$g(\rho,\Delta):=\gamma_0(\Delta)+\gamma_1(\Delta)\rho+\gamma_2(\Delta)\rho^2+\gamma_3(\Delta)\rho^3$, has a root $\rho=\rho(\Delta)$ vanishing together with Δ. The boundary $\partial\Omega$ of the domain Ω is obtained by eliminating ρ from the system of equations

$$\rho=g(\rho,\Delta),\ 1=g'_\rho(\rho,\Delta),$$

which is equivalent to

$$\rho=g(\rho,\Delta):=\xi_0(\Delta)+\xi_1(\Delta)\rho \tag{15}$$
$$+\xi_2(\Delta)\rho^2+\xi_3(\Delta)\rho^3,$$
$$1=g'_\rho(\rho,\Delta)=\xi_1(\Delta)+2\xi_2(\Delta)\rho+3\xi_3(\Delta)\rho^2$$

with

$$\xi_0(\Delta):=r\zeta_0(\Delta),\ \xi_1(\Delta):=r\zeta_1(\Delta)-\pi_0+1$$
$$\xi_2(\Delta):=r\zeta_2(\Delta)-\pi_1,\ \xi_3(\Delta):=r\zeta_3(\Delta)-\pi_2.$$

For $\xi_3(\Delta)>0$ the system of equations (15) gives

$$2\xi_2^3(\Delta)+27\xi_0(\Delta)\xi_3^2(\Delta)+9\xi_2(\Delta)\xi_3(\Delta)(1-\xi_1(\Delta))=$$
$$2(\xi_2^2(\Delta)+3\xi_3(\Delta)(1-\xi_1(\Delta)))^{3/2},$$

which is equivalent to zeroing the discriminant of the polynomial $g(\rho,\Delta)$ from (14):

$$\omega_1(\Delta)=\omega_2(\Delta),$$

where

$$\omega_1(\Delta):=4\xi_0(\Delta)\xi_2^3(\Delta)+27\xi_0^2(\Delta)\xi_3^2(\Delta)$$
$$+18\xi_0(\Delta)\xi_2(\Delta)\xi_3(\Delta)(1-\xi_1(\Delta)),$$
$$\omega_2(\Delta):=\xi_2^2(\Delta)(1-\xi_1(\Delta))^2+4\xi_3(\Delta)(1-\xi_1(\Delta))^3.$$

This equality defines a hypersurface in $\mathcal{R}_+^4$,

$$\{\Delta\in\mathcal{R}_+^4:\omega_1(\Delta)=\omega_2(\Delta)\}. \tag{16}$$

The set $\partial\Omega$ consists of the hypersurface (16) and parts of the coordinate planes in $\mathcal{R}_+^4$. Thus

$$\Omega:=\{\Delta\in\mathcal{R}_+^4:\omega_1(\Delta)\le\omega_2(\Delta)\}. \tag{17}$$

Denote by $\Omega^0\subset\Omega$ the part of Ω, corresponding to the strict inequality in (17). Then for $\Delta\in\Omega^0$ the majorant equation (14) has two positive roots $\rho_1(\Delta)<\rho_2(\Delta)$ and one may choose

$$\rho=\rho_1(\Delta). \tag{18}$$

If the equality in (17) is fulfilled then the majorant equation has a double positive root $\rho_1(\Delta)=\rho_2(\Delta)=\rho(\Delta)$ and one again may use (18). Note also that for $\Delta\in\Omega^0$ the elements of the perturbation ΔP are real analytic functions of the elements of the perturbations in the coefficient matrices of the Riccati equation.

Although the inclusion $\Delta\in\Omega$ is directly checkable and the computation of $\rho_1(\Delta)$ is straightforward, one may wish to avoid the computation of the roots of the cubic equation. For this purpose the estimates may be simplified (making them slightly less sharp) by considering a new equation

$$\rho=g^0(\rho,\Delta)$$

where the function g^0 is chosen so that $g^0(\rho,\Delta)\ge g(\rho,\Delta)$ for $\Delta\in\Omega$, $\rho\in[0,\rho^0(\Delta)]$, and the interval $[0,\rho^0(\Delta)]$ contains the root $\rho_1(\Delta)$.

A function of this type is e.g. $g^0(\rho,\Delta)=\xi_0(\Delta)+\xi_1(\Delta)\rho+(\xi_2(\Delta)+\xi_3(\Delta)\rho^0(\Delta))\rho^2$ where $\rho^0(\Delta)\ge\rho_1(\Delta)$. To be sure that this inequality is fulfilled, one may choose $\rho_0(\Delta)$ as the value of ρ for which $g'_\rho(\rho,\Delta) = 1$, i.e.,

$$\rho^0(\Delta) = \frac{1-\xi_1(\Delta)}{\xi_2(\Delta)+\sqrt{\xi_2(\Delta)^2+3\xi_3(\Delta)(1-\xi_1(\Delta))}}.$$

This is possible since in Ω^0 one has $\xi_1(\Delta) < 1$.

Note that the graphs of $g(.,\Delta)$ and $g^0(.,\Delta)$ have two intersection points $(0,\xi_0(\Delta))$ and $(\rho^0(\Delta), g(\rho^0(\Delta),\Delta))$ and a joint tangent at the intersection point $(0,\xi_0(\Delta))$. This shows that the function g^0 is a good approximation of g.

Solving the equation $\rho = g^0(\rho,\Delta)$ one gets the bound

$$f(\Delta) := \rho_1^0(\Delta) = \frac{2\xi_0(\Delta)}{1-\xi_1(\Delta)+\sqrt{d(\Delta)}}, \quad (19)$$

where $d(\Delta) :=$

$$(1-\xi_1(\Delta))^2 - 4\xi_0(\Delta)(\xi_2(\Delta)+\xi_3(\Delta)\rho^0(\Delta)).$$

The expression in the right-hand side of (19) is correctly defined if

$$\xi_1(\Delta) + 2\sqrt{\xi_0(\Delta)(\xi_2(\Delta)+\xi_3(\Delta)\rho^0(\Delta))} \le 1,$$

which is equivalent to

$$\xi_1(\Delta) + 2\left(\frac{\xi_0(\Delta)}{3}(2\xi_2(\Delta) + (\xi_2(\Delta)^2+3\xi_3(\Delta)(1-\xi_1(\Delta)))^{\frac{1}{2}}\right)^{\frac{1}{2}} \le 1.$$

The last inequality is fulfilled for $\Delta \in \Omega$.

Hence one has the following non-local perturbation bound for X_0

$$\Delta_X = \|\Delta X\|_F \le \rho_1(\Delta) \le f(\Delta), \quad (20)$$

where f is defined via (19).

To illustrate the effectiveness of the estimates obtained consider the state regulator problem for a second order discrete-time system with matrices $A = VA^*V$, $B_1 = VB_1^*V$, $C = VC^*V$, $D_1 = VD_1^*V$, $D_2 = VD_2^*V$, $B_2 = VB_2^*V$, where $V = I_2 - 2vv^\top/\|v\|^2$, $v = [1,2]^\top$, $A^* = \mathrm{diag}(-1,-2)$, $B_1^* = \mathrm{diag}(1,2)$, $C^* = \mathrm{diag}(2,1)$, $B_2^* = \mathrm{diag}(1,0.1)$, $D_1^* = \mathrm{diag}(0.1,0.2)$, $D_2^* = \mathrm{diag}(0.2,0.5)$. The perturbations in the data are taken as $\Delta A = \mathrm{diag}(2,1) * 10^{-j}$, $j = 10,9,\ldots,1$, $\Delta B_1 = 0$, $\Delta C = 0$, $\Delta B_2 = 0$, $\Delta D_1 = 0$, $\Delta D_2 = 0$, which gives $\mathrm{Est} = \mathrm{Est}_1 = \mathrm{Est}_3 < \mathrm{Est}_2$. This problem was designed so as to have closed form solutions $X_0 = VX_0^*V$ and $X_0 + \Delta X$ of the unperturbed and perturbed Riccati equations respectively.

The actual relative perturbation

$$\delta_X := \|\Delta X\|_F / \|X_0\|_F$$

in the solution is estimated by the above bounds. The cases when the non-local estimate is not valid, since the existence condition $\Delta \in \Omega$ is violated, are denoted by an asterisk.

The results are shown in the table bellow.

j	δ_X	local	non-local
10	1.8758×10^{-11}	1.9655×10^{-11}	4.4180×10^{-9}
9	1.8758×10^{-10}	1.9655×10^{-10}	4.4180×10^{-8}
8	1.8758×10^{-9}	1.9655×10^{-9}	4.4181×10^{-7}
7	1.8758×10^{-8}	1.9655×10^{-8}	4.4186×10^{-6}
6	1.8758×10^{-7}	1.9655×10^{-7}	4.4236×10^{-5}
5	1.8758×10^{-6}	1.9655×10^{-6}	4.4750×10^{-4}
4	1.8758×10^{-5}	1.9655×10^{-5}	5.2722×10^{-3}
3	1.8758×10^{-4}	1.9655×10^{-4}	*
2	00019	0.0020	*
1	0.0187	0.0197	*

3. REFERENCES

Konstantinov, M., N. Christov, P. Petkov. (1986). Perturbation analysis of the continuous and discrete matrix Riccati equations. *Proc. 1986 ACC*, Seattle, **1**, 636-639.

Konstantinov, M., P. Petkov, N. Christov. (1995). Conditioning of the continuous-time H_∞ optimization problem. *Proc. 3 ECC*, Rome 1995, **1**, 613-618.

Stoorvogel, A. (1992). *The H_∞ Control Problem.* Prentice Hall, Hemel Hempstead.

Zhou, K., J. Doyle, K. Glover. (1995). *Robust and Optimal Control.* Prentice Hall, Englewood Cliffs.

14th Triennial World Congress, Beijing, P.R. China

C-2a-02-4

DESIGN OF MULTIRATE CONTROL SYSTEMS FORCED BY SINUSOIDAL EXOGENOUS SIGNALS AND SUBJECT TO PERTURBATIONS AND INSTRUMENTATION FAILURES

Arturo Locatelli and Nicola Schiavoni

Dipartimento di Elettronica e Informazione
Politecnico di Milano
Piazza L. da Vinci 32, 20133 Milano, Italy
email {locatell-schiavon}@elet.polimi.it

Abstract: This paper considers the design of digital control systems, where the measurement mechanism is nonstandard, that is, each output is measured at its own rate. It addresses the classical regulator problem, under the assumptions that the exogenous signals are of a sinusoidal type, the instrumentation is subject to failures or/and the plant parameters undergo large variations. Some sufficient conditions are given for the existence of least-order regulators able to solve the problems outlined.

Keywords: Control system, multirate, stabilization, regulation, robustness.

1. INTRODUCTION

Multirate control systems are characterized by the fact that each input is updated at its own rate and/or each output is measured at its own rate; as such, they are able to model many real-life situations. These control systems have received a great deal of attention for many years, the relevant research activity having followed two different directions (Araki, 1993). Many authors assume that the updating/measurement scheme is to be designed, whereas other authors face classical design problems for a fixed nonstandard updating/measurement scheme. This paper follows the second stream of activity and tackles the regulator problem (asymptotic stability and zero error regulation) under additional robustness constraints. First, the case where some loops open because of failures in the instrumentation (sensors and actuators) is considered; the *Reliable Regulator Problem (RRP)* requires that stability and zero-error regulation are maintained, to the maximum possible extent, even in these situations. Then, the design of a single regulator for a whole set of systems, representing various possible models of a partially uncertain plant, is considered in a second problem, called *Simultaneous Regulator Problem (SRP)*. The combined *Simultaneous Reliable Regulator Problem (SRRP)* is also considered along with some other extensions. Constructive solvability conditions are here presented sufficient to solve the above problems by means of least-order regulators, under the assumption that the plant is asymptotically stable. These regulators also supply the control systems with a generalized form of the unconditional stability property (Campo, and Morari, 1994). The results in this area are based on (Locatelli, and Schiavoni, 1993a) and details can be found in (Locatelli, and Schiavoni, 1993b) for *RRP* and (Locatelli, and Schiavoni, 1995) for *SRP*. All these papers make reference to decentralized control structures. Rather unexpectedly, the formulation itself of the problems turns out to be a little more intriguing when the regulator is centralized. This case has been treated in (Locatelli, and Schiavoni, 1996), with reference to stepwise exogenous signals, and in (Locatelli, and

Schiavoni, 1997), for polynomial ones. Here, the situation where the signals are sinusoids is dealt with both for centralized and decentralized structures, so extending the recent results reported in (Locatelli, and Schiavoni, 1998). The considered updating mechanism is standard (like in (Locatelli, and Schiavoni, 1997) and in contrast to (Locatelli, and Schiavoni, 1996)), since otherwise the problems do not admit any solution. Put in a different way, plants with infrequent output sampling are only considered. Next section presents the formal statement of *RRP* along with a constructive sufficient solvability condition. Analogously, Section 3 is entirely devoted to *SRP* and Section 4 is concerned with various extensions, among which *SRRP*. The paper ends with an example (Section 5) and some concluding remarks (Section 6). For the sake of conciseness, the proofs are omitted: they are essentially based on previous works by the authors (Locatelli, *et al.*, 1985; 1986; Schiavoni, 1983; 1990; 1991; Schiavoni, and Locatelli, 1991) and exploit time-invariant reformulations of the involved periodic systems (Meyer, and Burrus, 1975).

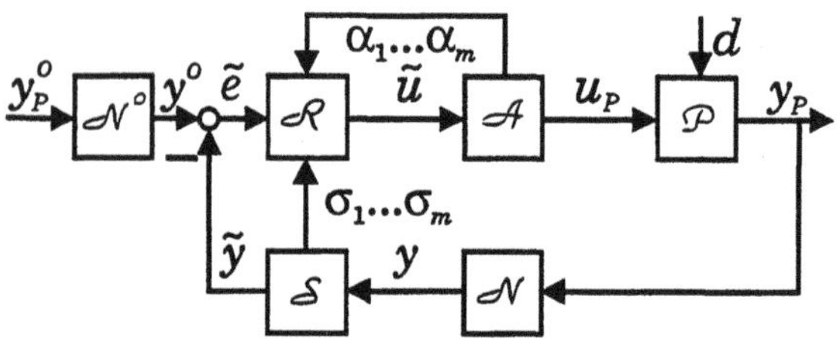

Fig. 1. The multirate control system for *RRP*.

2. RELIABLE REGULATOR PROBLEM

The control system considered in this section is represented in Fig. 1. Loosely speaking, the problem consists of designing a stabilizing controller also ensuring zero-error regulation, under sinusoidal exogenous signals, both when the instrumentation is properly working and when a failure occurs. In order to properly face plant parameters variations, the controller must incorporate a reduplication of the model of the exogenous signals, which makes it not asymptotically stable. Turning to a formal description of the control system, the plant $\mathcal{P}$ is asymptotically stable and given by

$$x_P(t+1)=A_P x_P(t)+B_P u_P(t)+E_P d(t)$$
$$y_P(t)=C_P x_P(t)+F_P d(t)$$

where $u_P \in R^m$, $x_P \in R^n$, $y_P \in R^m$, $d \in R^r$. Now assume that the i-th component y_{Pi}, $i \in M := \{1,2,\dots,m\}$, of the output vector y_P is periodically measured at times $t = kT_{Ni} + \tau_{Ni}$, k integer, T_{Ni} positive and finite integer, $\tau_{Ni} < T_{Ni}$ nonnegative integer. This measurement mechanism can be modelled as the following system $\mathcal{N}$:

$$y(t)=N(t)y_P(t) \tag{1.a}$$

$$N(t):=\mathrm{diag}\{n_i(t)\},\ n_i(t):=\begin{cases}1 & t=kT_{Ni}+\tau_{Ni}\\ 0 & t\neq kT_{Ni}+\tau_{Ni}\end{cases} \tag{1.b}$$

Observe that the output y_i is set to zero in the whole interval between two successive measures. The sensor $\mathcal{S}$ and actuator $\mathcal{A}$ are described by

$$\tilde{y}(t)=\mathrm{diag}\{\sigma_1,\sigma_2,\dots,\sigma_m\}y(t)$$
$$u_P(t)=\mathrm{diag}\{\alpha_1,\alpha_2,\dots,\alpha_m\}\tilde{u}(t)$$

respectively, where $\sigma_i=1$ ($\alpha_i=1$) if the i-th sensor $\mathcal{S}_i$ (actuator $\mathcal{A}_i$) is properly working and $\sigma_i=0$ ($\alpha_i=0$) if a failure occurs. The reference signal $y_P^o \in R^m$ for y_P and the disturbance d are

$$y_P^o(t)=Y\,\mathrm{sen}(\vartheta t),\ d(t)=D\,\mathrm{sen}(\vartheta t),\ 0<\vartheta<\pi \tag{2}$$

The regulator $\mathcal{R}$ to be designed makes its output $\tilde{u}$ to depend on the sensed error

$$\tilde{e}(t):=y^o(t)-\tilde{y}(t)$$

where y^o is the output of the system $\mathcal{N}^o$ defined by

$$y^o(t)=N(t)y_P^o(t) \tag{3}$$

Observe that the i-th component of $\tilde{e}$ equals the i-th component of the measured error

$$e(t):=y^o(t)-y(t) \tag{4}$$

only when $\mathcal{S}_i$ is properly working and the i-th component of $\tilde{u}$ equals the i-th component of u_P only when $\mathcal{A}_i$ is properly working. The regulator $\mathcal{R}$ is basically linear, that is, the relationship it establishes between $\tilde{e}$ and $\tilde{u}$ is linear for each operating condition of the multivariable actuator and sensor. However, the regulator is assumed to know such a condition, that is, it knows the value of the α_i's and the σ_i's, and, when a failure occurs in loop j, $j \in M$, it sets the effect of $\tilde{e}_j$ on $\tilde{u}$ to zero, which is equivalent to zeroing the j-th column of the transfer function of $\mathcal{R}$. The rationale behind that is simple. For as concerns a failure of the j-th sensor, the variable $\tilde{e}_j$ is different from e_j and does not contain any piece of information useful for control. On the contrary, a nonzero signal at the input of the regulator (which will include an appropriate reduplication of the model of the exogenous signals) would render impossible the zeroing of the error components different from the j-th one. With reference to a failure of the j-th actuator, observe that the j-th row of the regulator transfer function can be considered as set to zero in this case, without any effect on the behavior of the closed-loop system. Then, note that no more than $m-1$ components or combinations of the outputs can be regulated by means of $m-1$ control variables. Therefore, it is advisable to renounce to control a well specified output in order to try to regulate all the remaining ones: the choice of the j-th component of $\tilde{e}$ is anyhow arbitrary. As a convention, for any j, $j \in M$, denote by $\mathcal{R}_{0j}$ and $\mathcal{R}_{jj}$ the minimal realizations of the transfer functions obtained by the one of $\mathcal{R}$ after zeroing the j-th column and the j-th row and column, respectively. The system viewed from the regulator is T-periodic, with

$$T := \underset{i \in M}{l.c.m.}\{T_{Ni}\} \qquad (5)$$

Then, letting $M^j := M - \{j\}$, $j \in M$, and the control system error $e_P \in R^m$ be

$$e_P(t) := y_P^o(t) - y_P(t) \qquad (6)$$

the problem of this section is the following.

Reliable regulator problem (RRP) Find a regulator $\mathcal{R}$ such that:

(a) If no failures occur (i.e., $\alpha_i \sigma_i = 1$, $i \in M$)
   (i) The closed-loop system $(\mathcal{P},\mathcal{N},\mathcal{S},\mathcal{R},\mathcal{A})$ is asymptotically stable;
   (ii) The regulation constraint
   $$\lim_{t \to \infty} e_P(t) = 0$$
   holds true for all vectors Y and D and the perturbations of matrices A_P, B_P, C_P, E_P and F_P, which preserve the asymptotic stability of the closed-loop system $(\mathcal{P},\mathcal{N},\mathcal{S},\mathcal{R},\mathcal{A})$;

(b) If a failure occurs in the j-th loop (i.e., $\alpha_i \sigma_i = 1$, $i \in M^j$, $\alpha_j \sigma_j = 0$)
   (i) The closed-loop systems $(\mathcal{P},\mathcal{N},\mathcal{S},\mathcal{R}_{0j},\mathcal{A})$ and $(\mathcal{P},\mathcal{N},\mathcal{S},\mathcal{R}_{jj},\mathcal{A})$ are asymptotically stable;
   (ii) The regulation constraint
   $$\lim_{t \to \infty} e_{Pi}(t) = 0 \, , \, i \in M^j$$
   holds true for all vectors Y and D and the perturbations of matrices A_P, B_P, C_P, E_P and F_P, which preserve the asymptotic stability of the closed-loop systems $(\mathcal{P},\mathcal{N},\mathcal{S},\mathcal{R}_{0j},\mathcal{A})$ and $(\mathcal{P},\mathcal{N},\mathcal{S},\mathcal{R}_{jj},\mathcal{A})$.

It is worth noting that what is called for in (b) concerning stability is the maximum one can require; asymptotic stability of $(\mathcal{P},\mathcal{N},\mathcal{S},\mathcal{R},\mathcal{A})$ can never be achieved in failed conditions, since $\mathcal{R}$ will contain unreachable/unobservable parts which are not asymptotically stable. For any ϑ, $0 < \vartheta < \pi$, let $G_{P\vartheta} := C_P\left(e^{j\vartheta}I - A_P\right)^{-1}B_P := \{g_{P\vartheta ij}\}$; then, for any matrix Λ, denote by $\Lambda(i,j)$ the matrix obtained from Λ after deleting its i-th and j-th column; conventionally, if $i = 0$ ($j = 0$) no row (column) is deleted.

Theorem 1 If there exists a single complex $m \times m$ matrix $\boldsymbol{\Phi}$ such that all the eigenvalues λ of the matrices $G_{P\vartheta}(i,i)\boldsymbol{\Phi}(i,i)$, $G_{P\vartheta}(i,0)\boldsymbol{\Phi}(0,i)$, $i \in \{0\} \cup M$, satisfy the condition

$$\cos(\arg(\lambda) - \vartheta) < 0 \qquad (7)$$

then *RRP* admits a solution.

Observe that the nonsingularity of matrix $G_{P\vartheta}$ is a necessary condition not only for the existence of $\boldsymbol{\Phi}$ (since otherwise $\arg(\lambda)$ is not well defined for all λ), but also for the existence of a regulator satisfying part (a) of the statement of *RRP*. Whenever the condition of Theorem 1 is verified, time invariant regulators $\mathcal{R}$ solving *RRP* can be identified up to the value of their gain, which must be sufficiently small, as stated in the following theorem, where

$$\overline{N} := \sum_{t=1}^{T} N(t) \qquad (8)$$

is positive definite (see the definition of $N(t)$).

Theorem 2 For any matrix $\boldsymbol{\Phi}$ satisfying the condition of Theorem 1, there exists $\bar{\varepsilon} > 0$ such that, for any $\varepsilon \in (0, \bar{\varepsilon})$, a solution to *RRP* is provided by the regulator $\mathcal{R}$ with transfer function

$$R(z) := \frac{\varepsilon}{z^2 - 2z\cos\vartheta + 1}(K_1 + K_2 z)\overline{N}^{-1} \qquad (9)$$

$$K_1 := -(\mathrm{Re}[\boldsymbol{\Phi}]\cos\vartheta + \mathrm{Im}[\boldsymbol{\Phi}]\mathrm{sen}\,\vartheta), \; K_2 := \mathrm{Re}[\boldsymbol{\Phi}] \qquad (10)$$

The regulators of Theorem 2 are of the least possible order, as they only contain the reduplicated model of the exogenous variables necessary to steer the plant error to zero. Further, they supply the control system with a generalized form of the unconditional stability property, which is defined in (Campo, and Morari, 1994). Theorem 1 and, consequently, Theorem 2 are not completely satisfactory, as the condition of the former is not given in terms of the problem data and, in general, can be checked in an easy way for very small values of m only. However, there are special cases where it is possible to test the condition in terms of matrix $G_{P\vartheta}$ only and to give an explicit expression of $\boldsymbol{\Phi}$.

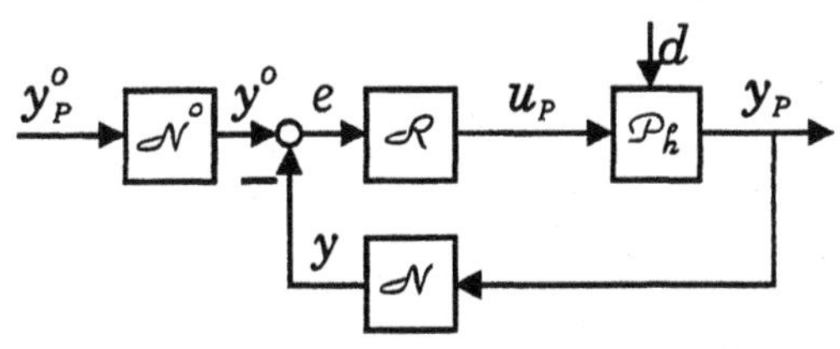

Fig. 2. The multirate control system for *SRP*.

3. SIMULTANEOUS REGULATOR PROBLEM

The control system considered in this section is represented in Fig. 2. Loosely speaking, the problem consists of designing a stabilizing controller also ensuring zero-error regulation, under sinusoidal exogenous signals, for a whole finite set of models of the plant. Again, the controller must incorporate a reduplication of the model of the exogenous signals, which makes it not asymptotically stable. The plant under control is asymptotically stable and described by the s models $\mathcal{P}_h$, $h \in S := \{1,2,...,s\}$,

$$x_P^h(t+1) = A_P^h x_P^h(t) + B_P^h u_P(t) + E_P^h d(t)$$
$$y_P(t) = C_P^h x_P^h(t) + F_P^h d(t)$$

Further, the measurement mechanism is still described by the system $\mathcal{N}$ defined in eqns. (1) and again the reference signal $y_P^o \in R^m$ for y_P and the disturbance d are as in eqn. (2). The regulator $\mathcal{R}$ for this problem is linear and makes its output u_P to depend on the measured error e (eqn. 4). The system viewed from it is T-periodic (eqn. (5)). Finally, remembering the already given definitions of y^o and e_P (eqns. (3) and (6)), the problem of this section is the following.

Simultaneous regulator problem (SRP) Find a single regulator $\mathcal{R}$ such that, for all $h \in S$,
(i) The closed-loop system $(\mathcal{P}_h, \mathcal{N}, \mathcal{R})$ is asymptotically stable;
(ii) The regulation constraint

$$\lim_{t \to \infty} e_P(t) = 0$$

holds true for all vectors Y and D and the perturbations of matrices A_P^h, B_P^h, C_P^h, E_P^h and F_P^h, which preserve the asymptotic stability of the closed-loop system $(\mathcal{P}_h, \mathcal{N}, \mathcal{R})$.

For any $\vartheta, 0 < \vartheta < \pi$, let $G_{P\vartheta}^h := C_P^h \left(e^{j\vartheta} I - A_P^h\right)^{-1} B_P^h$ $:= \{g_{P\vartheta ij}^h\}$, $h \in S$. Proceeding as in Section 2, the following results hold, where $\overline{N}$ is as in eqn. (8).

Theorem 3 If there exists a single complex $m \times m$ matrix $\boldsymbol{\Phi}$ such that all the eigenvalues λ of the matrices $G_{P\vartheta}^h \boldsymbol{\Phi}$, $h \in S$, satisfy condition (7), then *SRP* admits a solution.

Theorem 4 For any matrix $\boldsymbol{\Phi}$ satisfying the condition of Theorem 3, there exists $\bar{\varepsilon} > 0$ such that, for any $\varepsilon \in (0, \bar{\varepsilon})$, a solution to *SRP* is provided by the regulator $\mathcal{R}$ with transfer function (9), (10).

The regulators of Theorem 4 are of the least possible order, as they only contain the reduplicated model of the exogenous variables necessary to steer the plant error to zero. Further, they supply the control system with a generalized form of the unconditional stability property (Campo, and Morari, 1994).

4. EXTENSIONS

4.1 Generalized reliable regulator problem

The above stated *RRP* can be extended in a straightforward way to the case of an arbitrary number of contemporary failures, giving rise to the *Generalized Reliable Regulator Problem (GRRP)*. For any couple J_1, J_2 of elements of the power set P_M of M and any matrix $\boldsymbol{A}$, let denote by $\boldsymbol{A}(J_1, J_2)$ the matrix obtained from $\boldsymbol{A}$ after deleting the rows and columns with indices in J_1 and J_2; conventionally, if J_1 (J_2) is the empty set ϕ no row (column) is deleted. Then, a solvability condition for *GRRP* is the following.

Theorem 5 If there exists a single complex $m \times m$ matrix $\boldsymbol{\Phi}$ such that all the eigenvalues λ of the matrices $G_{P\vartheta}(J,J)\boldsymbol{\Phi}(J,J)$, $G_{P\vartheta}(J,\phi)\boldsymbol{\Phi}(\phi,J)$, $M \in P_M$, satisfy condition (7), then there exists $\bar{\varepsilon} > 0$ such that, for any $\varepsilon \in (0, \bar{\varepsilon})$, a solution to *GRRP* is provided by the regulator $\mathcal{R}$ with transfer function (9), (10).

Of course, in the *GRRP* context, the columns of $\boldsymbol{R}(z)$ corresponding to the actual failures must be zeroed. Since $M \in P_M$, the asymptotic stability of the plant is now an unavoidable assumption.

4.2 Reliable actuators

An interesting particular case of *RRP* is the one where only the sensors are subject to failures, whereas all the actuators behave properly. Under that circumstance, a necessary and sufficient solvability condition can be given, together with an explicit expression for matrix $\boldsymbol{\Phi}$. First, observe that, after a renumbering of the components of u_P (or y_P), matrix $G_{P\vartheta}(j,j)$, $i \in M$, is nonsingular, provided that $\det(G_{P\vartheta}) \neq 0$. Assuming that such a reordering has been performed, the following result holds.

Theorem 6 Assume that there are no actuator failures. Then, *RRP* has a solution if and only if $G_{P\vartheta}$ is nonsingular. Moreover, there exists $\bar{\varepsilon} > 0$ such that, for any $\varepsilon \in (0, \bar{\varepsilon})$, a solution to *RRP* is provided by the regulator $\mathcal{R}$ with transfer function (9), (10) and $\boldsymbol{\Phi} = G_{P\vartheta}^{-1}$.

The columns of $\boldsymbol{R}(z)$ corresponding to the actual failures must be zeroed, as described in Section 2. The regulator of Theorem 6 does not guarantee reliability against actuator failures, but it can be shown to ensure asymptotic stability and zeroing of $m-1$ combinations of the system error components.

4.3 Simultaneous reliable regulator problem

The *RRP* and *SRP* can be combined to give rise to the *Simultaneous Reliable Regulator Problem (SRRP)*. In this case, the conditions in the statement of the *RRP* must be fulfilled for all the models (9) of the plant and a combinations of Theorem 1-4 provides the following result.

Theorem 7 If there exists a single complex $m \times m$ matrix $\boldsymbol{\Phi}$ such that all the eigenvalues λ of the matrices $G_{P\vartheta}^h(j,j)\boldsymbol{\Phi}(j,j)$, $G_{P\vartheta}^h(j,0)\boldsymbol{\Phi}(0,j)$, $j \in M \cup \{0\}$, $h \in S$, satisfy condition (7), then there exists $\bar{\varepsilon} > 0$ such that, for any $\varepsilon \in (0,\bar{\varepsilon})$, a solution to *SRRP* is provided by the regulator $\mathcal{R}$ with transfer function (9), (10).

4.4 Decentralized control structures

The problem statements and the solutions dealt with in the preceding sections and subsections (with the exception of 4.2) can be extended in a fairly easy way to encompass decentralized constraints on the control structures. Further, some of them, relative to the design of completely decentralized regulators for whom the i-th scalar output depends on the i-th scalar input only and referred to the case of polynomial exogenous signals, have already appeared in the literature (Locatelli, and Schiavoni, 1993a; 1993b; 1995). Indeed, that situation is simpler to treat than the one considered in this paper, because an actuator failure has the effect of rendering impossible the zeroing of the corresponding error variables, but has no effect on the other errors, provided stability is preserved. Therefore, there is no need to detect the status of the actuators and it can be proven that each local regulator must know only the status of the corresponding sensor. On the contrary, in a centralized control structure, there is a coupling of all regulator inputs and outputs. Apart from the above considerations, the results for completely decentralized control structures are obtained from Theorems 1-5, 7 by just adding the conditions that matrix $\boldsymbol{\Phi}$ is diagonal. Moreover, sufficient solvability conditions for *RRP* and *SRP* can be stated in terms of the problems data under certain circumstances, which allow one to give explicit expressions for $\boldsymbol{\Phi}$. In the forthcoming theorems the symbol $(\cdot)^*$ denotes complex conjugation.

Theorem 8 If matrix $G_{P\vartheta}$ is triangular and nonsingular or diagonally dominant, then there exists $\bar{\varepsilon} > 0$ such that, for any $\varepsilon \in (0,\bar{\varepsilon})$, a completely decentralized solution to *RRP* is provided by the regulator $\mathcal{R}$ with transfer function (9), (10) and

$$\boldsymbol{\Phi} = \mathrm{diag}\left\{\frac{(g_{P\vartheta ii})^*}{|g_{P\vartheta ii}|^2}\right\} e^{j\vartheta}$$

The range of application of Theorem 8 is larger than it might appear at first glance, since there are cases where matrix $G_{P\vartheta}$ can be made diagonally dominant by premultiplying it by an appropriate diagonal matrix (Siljak, 1978). This matrix operation corresponds to applying a decentralized preregulator to the plant; then, Theorem 8 can be exploited.

Theorem 9 If matrix $G_{P\vartheta}^h$, $h \in S$, is triangular and nonsingular or diagonally dominant, and further $-\pi/2 < \arg(g_{P\vartheta ii}^h) - \arg(g_{P\vartheta ii}^1) < \pi/2$, $i \in M$, $h \in S - \{1\}$ then there exists $\bar{\varepsilon} > 0$ such that, for any $\varepsilon \in (0,\bar{\varepsilon})$, a completely decentralized solution to *SRP* is provided by the regulator $\mathcal{R}$ with transfer function (9), (10) and

$$\boldsymbol{\Phi} = \mathrm{diag}\left\{\frac{(g_{P\vartheta ii}^1)^*}{|g_{P\vartheta ii}^1|^2}\right\} e^{j\vartheta}$$

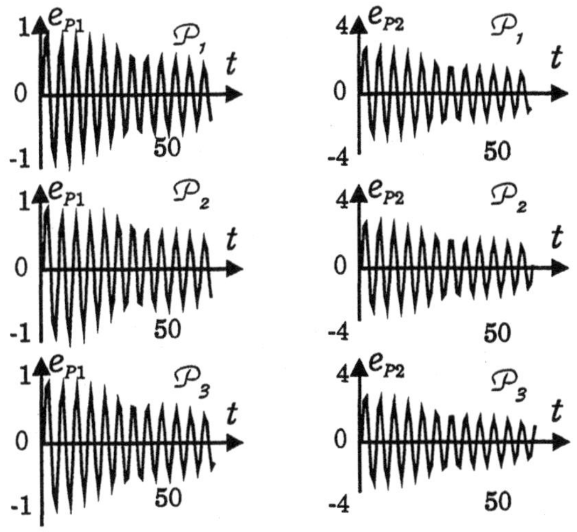

Fig. 3. Error responses to a sinusoidal reference signal.

5. ILLUSTRATIVE EXAMPLE

An example of *SRP* is now discussed. Consider the model of the binary distillation column presented in (Takamatsu, *et al.*, 1979) and assume that only the dinamical matrix is uncertain ($s = 3$), so that:

$$A_P^1 = \begin{bmatrix} .8773 & -.16288 \\ .0962 & .8413 \end{bmatrix}, A_P^2 = 1.05 A_P^1, A_P^3 = .95 A_P^1$$

$$B_P^h = -\begin{bmatrix} .02602 & .00516 \\ .0735 & .1348 \end{bmatrix},\ C_P^h = I_2,\ h = 1,2,3$$

The output components can be measured alternately every two time instants, so that $n_1(0) = n_2(1) = 0$ and $n_1(1) = n_2(0) = 1$. The angular frequency of the exogenous signals is $\vartheta = 1$. It can easily be checked that the condition of Theorem 3 is met with $\Phi = (G_{P\vartheta}^I)^{-1}$. A regulator which solves *SRP* is designed according to Theorem 4 with $\varepsilon = .1$. The resulting responses of the errors to the input $y_P^o(t) = [\sin(t)\ \ 3\sin(t)]'$ are shown in Fig. 3. Stability and zero error regulation are apparently attained, but the transients are very poorly damped.

6. CONCLUDING REMARKS

With reference to multirate control systems, the paper considered some problems concerning various robustness issues. All the proposed solutions consist of least-order small-gain regulators, which are of interest even though they supply the control systems with sluggish transients, because the dynamical behavior has not been taken into account in the problems formulation. Indeed, a typical design technique (useful when decentralization and other constraints on the regulator structure are present) is to choose a suitably parametrized class of regulators and then seek the element which is the best one in some suitable sense. The actual optimization is usually performed by means of iterative procedures, to be initialized by regulators guaranteing stability, regulation and robustness. Then, the results of the previous sections can be exploited whenever the adopted class of regulators includes those specified by eqns. (9), (10). Usually, some more dynamics are added to the simple reduplication of the internal model of the exogenous signals, which does not ensure satisfactory transients responses.

ACKNOWLEDGEMENTS

This paper has partially been supported by Politecnico di Milano (Fondo di Ricerca di Ateneo), MURST (Project Identification and Control of Industrial Systems) and CNR (CESTIA).

REFERENCES

Araki, M. (1993). Recent developments in digital control theory. *Proc. 12th IFAC World Congr.*, **9**, 251-260.

Campo, P., and M. Morari (1994). Achievable closed-loop properties of systems under decentralized control. *IEEE Trans. Automat. Contr.*, **AC-39**, 932-943.

Locatelli, A., R. Scattolini and N. Schiavoni (1985). Two problems in the design of discrete decentralised control systems subject to finite perturbations. *Proc. 4th IMA Conf. Contr. Th.*, 423-432.

Locatelli, A., R. Scattolini and N. Schiavoni (1986). On the design of reliable robust decentralized regulators for linear systems. *Large Scale Syst. Th. Appl.*, **10**, 95-113.

Locatelli, A., and N. Schiavoni (1993a). The initialization problem in the design of structurally constrained multirate digital control systems. *Proc. 2nd European Contr. Conf.*, **III**, 1709-1714.

Locatelli, A., and N. Schiavoni (1993b). Reliable regulation of multirate sampled-data systems. *Proc. 12th IFAC World Congr.*, **4**, 19-22.

Locatelli, A., and N. Schiavoni (1995). Simultaneous regulation of multirate sampled-data systems. *Proc. 1995 American Contr. Conf.*, **2**, 1503-1507.

Locatelli, A., and N. Schiavoni (1996). Design of multirate control systems subject to instrumentation failures and perturbation of the plant parameters. *Proc. 13th IFAC World Congr.*, **C**, 331-336.

Locatelli, A., and N. Schiavoni (1997). Design of multirate control systems forced by polynomial exogenous signals and subject to perturbations and instrumentation failures. *Proc. 36th IEEE Conf. Dec. Contr.*, 2043-2048.

Locatelli, A., and N. Schiavoni (1998). On the design of multirate control systems subject to perturbations and instrumentation failures. *Proc. MTNS 98*, in press.

Meyer, R. A., and C. S. Burrus (1975). A unified analysis of multirate and periodically time-varying digital filters. *IEEE Trans. Ccts. Syst.*, **CAS-21**, 162-168.

Schiavoni, N. (1983). On the simultaneous design of decentralized linear control systems. *Proc. 22th IEEE Conf. Dec. Contr.*, **2**, 529-530.

Schiavoni, N. (1990). Further results on the design of multivariable regulators under structure and robustness constraints. *Int. J. Contr.*, **51**, 669-685.

Schiavoni, N. (1991). Some new results on the design of robust decentralized regulators. *Int. J. Syst. Sc.*, **22**, 2425-2439.

Schiavoni, N., and A. Locatelli (1991). On the design of reliable digital multivariable regulators under structure constraints. *Proc. 30th IEEE Conf. Dec. Contr.*, **3**, 2822-2823.

Siljak, D. D. (1978). *Large-scale dynamic systems: stability and structure*. North Holland, New York.

Takamatsu, T., I. Hashimoto and Y. Nakai (1979). A geometric approach to multivariable control design of a distillation column. *Automatica*, **15**, 387-402.

14th Triennial World Congress, Beijing, P.R. China

C-2a-02-5

HIGH PERFORMANCE PERFECT TRACKING CONTROL BASED ON MULTIRATE FEEDFORWARD / FEEDBACK CONTROLLERS WITH GENERALIZED SAMPLING PERIODS

Hiroshi Fujimoto* Yoichi Hori* Atsuo Kawamura**

* *Department of Electrical Engineering, University of Tokyo, Tokyo, 113-8656 Japan, email: fuji@hori.t.u-tokyo.ac.jp*
** *Department of Electrical and Computer Engineering, Yokohama National University, Yokohama, 240-8501 Japan*

Abstract: In this paper, a novel perfect tracking control method based on the multirate sampling control is proposed, in which the concepts of the two-degree-of-freedom control are employed. Moreover, by generalizing the multirate input and output sampling periods, the proposed method can match the restrictions of the sampling periods caused by the hardware, and achieve higher performance. The main results of this paper are 1) the perfect tracking performance is assured in the inter-sampling points, if the input multiplicity is increased, 2) higher feedback performance is achieved, if the output multiplicity is increased. Illustrative examples of position control using a dc servo motor are presented, and simulations and experiments demonstrate the advantages of this approach. *Copyright 1999 IFAC*

Keywords: Digital control, motion control, multirate sampling control, two-degree-of-freedom, tracking control

1. INTRODUCTION

In the digital motion control system, either the point to point control or the tracking control is usually employed. The point to point controller moves the plant output from one point to another point without indicating the path. Thus, it sometimes has the large tracking error because the transient path is not concerned. In the tracking control, however, the controlled plant follows the smoothed desired trajectory. Therefore, the tracking controllers are employed for high speed and high precision servo systems.

The best tracking controller is ideally the Perfect Tracking Controller (PTC) which controls the controlled object with zero tracking error (Tomizuka, 1987). The perfect tracking can be achieved using the feedforward controller $C_1[z]$ which is realized by the inverse system of the closed-loop system $G_{cl}[z]$.

$$C_1[z] = \frac{1}{z^d G_{cl}[z]} \tag{1}$$

where d is the relative degree of $G_{cl}[z]$.

However, the discrete-time plant discretized by the zero-order-hold usually has unstable zeros (Åström *et al.*, 1984). Thus, $C_1[z]$ becomes unstable because $G_{cl}[z]$ has the unstable zeros. Therefore, in the conventional digital control systems utilizing the zero-order-holds, the perfect tracking is usually impossible.

From this viewpoint, two feedforward control methods are proposed for the discrete-time non-minimum phase plant by Tomizuka (1987). First, The Stable Pole Zero Canceling (SPZC) controller cancels all poles and stable zeros of the closed-loop system, which has both phase and gain errors caused by the uncancellable unstable zeros. Second, Zero Phase Error Tracking Controller (ZPETC) adds the factors which cancel the phase error to SPZC. However, the gain error caused by the unstable zeros has remained.

A lot of methods are proposed to compensate the gain error of ZPETC (Haack and Tomizuka, 1991; Torfs *et al.*, 1992; Gross *et al.*, 1994). However, these methods are not able to realize the perfect tracking because the zero-order-holds are assumed to be employed.

Authors have proposed a novel perfect tracking control method in use of multirate feedforward control without utilizing the zero-order-hold (Fujimoto and Kawamura, 1998). In this paper, the feedback controller is also made multirate, and the periods of the plant input and output are generalized. Those generalizations improve the closed-loop characteristics and the tracking performance in the inter-sampling points.

Kabamba (1987) and Mita *et al.* (1990) resolved the unstable zeros problems of the discrete-time plant by zero assignment in use of multirate control. However, Moore *et al.* (1993) shows that those methods have disadvantages of large overshoot and oscillation in the inter-sampling points because the control input changes back and forth very quickly. On the other hand, the proposed method never has this problem because all of the plant states are controlled along the smoothed desired trajectories.

Recently, the modern sampled-data control theories have developed, which can optimize the intersample response (e.g. Chen and Francis, 1995; Hara *et al.*, 1996). In contrast, the proposed method makes a simple and practical approach to grantee the smooth inter-sample responses by giving trajectories to all of the plant states (e.g. position and velocity) at every sampling point T.

2. GENERALIZATIONS OF THE SAMPLING PERIODS

A digital tracking control system usually has two samplers for the reference signal $r(t)$ and the output $y(t)$, and one holder on the input $u(t)$ as shown in Fig. 1. Therefore, there exist the three time periods T, T_y, and T_u which represent the period of $r(t), y(t)$, and $u(t)$, respectively. The input period T_u is generally decided by the speed of the actuator, D/A converter, or the calculation on the CPU. Moreover, the output period T_y is also determined by the speed of the sensor or the A/D converter.

Real control systems usually hold the restrictions on T_u and/or T_y. Thus, the conventional digital

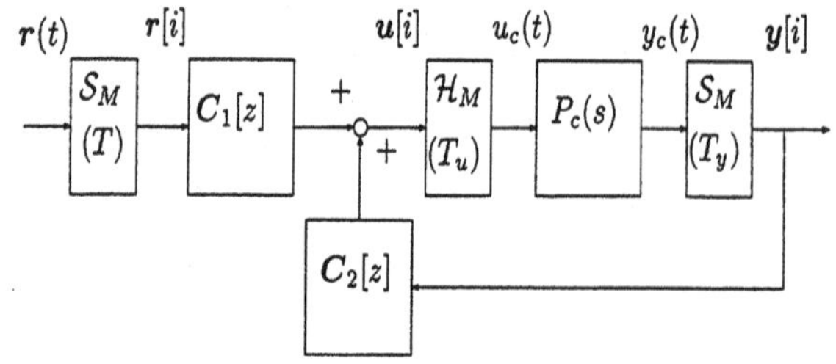

Fig. 1. Two-degree-of-freedom control system.

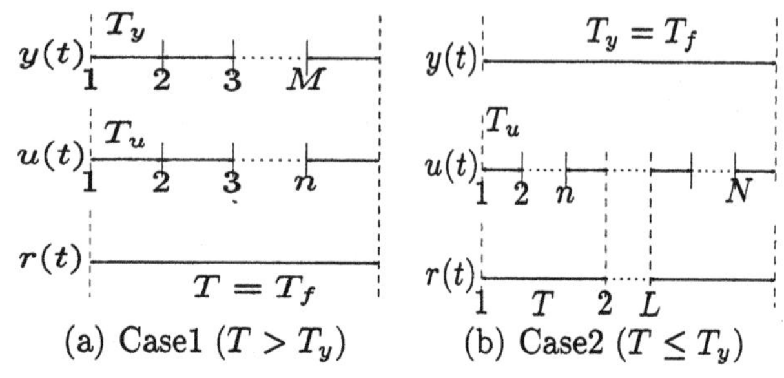

Fig. 2. Multirate sampling control.

control systems make these three periods equal to the longer period between T_u and T_y.

On the other hand, authors showed that the perfect tracking can be achieved on every sampling point T by letting $T = nT_u$, where n is the plant order (Fujimoto and Kawamura, 1998). Although the conventional methods with zero-order-hold are unable to realize the perfect tracking, the method has succeeded in it by introducing new control scheme. In spite of that, it has a problem; on the high order plant, the output sampling period T_y gets too long and the closed-loop characteristics become poor because the assumption $T_y = nT_u$ is fixed.

In this paper, high performance perfect tracking control method is proposed, by which the control system matches the hardware restrictions on the periods of the plant input holding and output sampling. In the following discussions, $T = nT_u$ is regarded as the condition for the perfect tracking, and the assumption $T_y = T$ is removed and generalized.

In this paper, the following two cases are considered, which are very ordinary in industries. First, although T_u is decided in advance by the hardware restrictions, the plant output can be detected M times during $T(= nT_u > T_y)$, as shown in Fig. 2(a). This case is referred to as case 1 in this paper. On this case, the closed-loop characteristics can be improved because much information of the plant can be detected. Second, although T_y is decided in advance, the plant input can be changed $L \times n (\triangleq N)$ during T_y, as shown in Fig. 2(b). It is also referred to as case 2. On this case, the perfect tracking can be assured L times during inter-

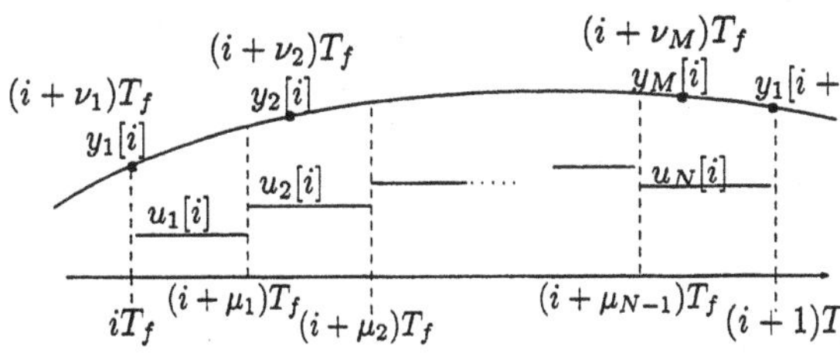

Fig. 3. Generalized multirate sampling control.

sample points of T_y. The output multiplicity M, input multiplicity N and L are positive integers.

In the two multi-period systems as shown in Fig. 2, the longer period between T and T_y is defined as the flame period T_f (Araki, 1993). If this definition is employed, case 1 and 2 can be dealt with together.

3. DESIGNS OF THE PROPOSED CONTROLLERS

In this section, the proposed perfect tracking control method is presented. For simplification, the continuous-time plant is assumed to be SISO system. The proposed method, however, can be extended to deal with the MIMO system by the same way as Fujimoto *et al.* (1999).

3.1 *Plant Discretization and Parameterization*

In order to deal with case 1 and 2 together, the multirate control scheme is represented by Fig. 3 in use of input and output multiplicity N and M. Thus, in case 1, N should be set to n. Moreover, in case 2, M also have to be set to 1.

Consider the continuous-time nth order plant described by

$$\dot{\boldsymbol{x}}(t) = \boldsymbol{A}_c\boldsymbol{x}(t) + \boldsymbol{b}_c u(t)\ ,\ \ y(t) = \boldsymbol{c}_c\boldsymbol{x}(t). \quad (2)$$

The discrete-time plant discretized by the generalized multirate sampling control (Fig. 3) becomes

$$\boldsymbol{x}[i+1] = \boldsymbol{A}\boldsymbol{x}[i] + \boldsymbol{B}\boldsymbol{u}[i]\ ,\ \ \boldsymbol{y}[i] = \boldsymbol{C}\boldsymbol{x}[i] + \boldsymbol{D}\boldsymbol{u}[i] \quad (3)$$

where $\boldsymbol{x}[i] = \boldsymbol{x}(iT)$, and where matrices $\boldsymbol{A}, \boldsymbol{B}, \boldsymbol{C}, \boldsymbol{D}$ and vectors $\boldsymbol{u}, \boldsymbol{y}$ are given by

$$\left[\begin{array}{c|c} \boldsymbol{A} & \boldsymbol{B} \\ \hline \boldsymbol{C} & \boldsymbol{D} \end{array}\right] \triangleq \left[\begin{array}{c|ccc} e^{\boldsymbol{A}_c T_f} & \boldsymbol{b}_1 & \cdots & \boldsymbol{b}_N \\ \hline \boldsymbol{c}_1 & d_{11} & \cdots & d_{1N} \\ \vdots & \vdots & & \vdots \\ \boldsymbol{c}_M & d_{M1} & \cdots & d_{MN} \end{array}\right] \quad (4)$$

$$\boldsymbol{u} \triangleq [u_1, \cdots, u_N]^T,\ \boldsymbol{y} \triangleq [y_1, \cdots, y_M]^T \quad (5)$$

$$\boldsymbol{b}_j \triangleq \int_{(1-\mu_j)T_f}^{(1-\mu_{(j-1)})T_f} e^{\boldsymbol{A}_c\tau}\boldsymbol{b}_c d\tau\ ,\ \ \boldsymbol{c}_k \triangleq \boldsymbol{c}_c e^{\boldsymbol{A}_c \nu_k T_f} \quad (6)$$

$$d_{kj} \triangleq \begin{cases} \mu_j < \nu_k : & \boldsymbol{c}_c\int_{(\nu_k-\mu_j)T_f}^{(\nu_k-\mu_{(j-1)})T_f} e^{\boldsymbol{A}_c\tau}\boldsymbol{b}_c d\tau \\ \mu_{(j-1)} < \nu_k \leq \mu_j : & \boldsymbol{c}_c\int_0^{(\nu_k-\mu_{(j-1)})T_f} e^{\boldsymbol{A}_c\tau}\boldsymbol{b}_c d\tau \\ \nu_k \leq \mu_{(j-1)} : & 0 \end{cases}$$

$$0 = \mu_0 < \mu_1 < \mu_2 < \ldots < \mu_N = 1 \quad (7)$$

$$0 \leq \nu_1 < \nu_2 < \ldots < \nu_M < 1 \quad (8)$$

where $\mu_j (j = 0, 1, \cdots, N)$ and $\nu_k (k = 1, \cdots, M)$ are the parameters for multirate sampling as shown in Fig. 3. If T_f is divided at same intervals, $\mu_j = j/N, \nu_k = (k-1)/M$.

The coprime factorization of the pulse transfer function $\boldsymbol{P}[z]$ realized by (3) is described by

$$\boldsymbol{P}[z] = \boldsymbol{N}\boldsymbol{M}^{-1} \quad (9)$$

where $\boldsymbol{M}, \boldsymbol{N} \in \boldsymbol{RH}_\infty$ (Sugie and Yoshikawa, 1986).

The proposed method employs the multirate-input control as the two-degree-of-freedom control, as shown in Fig. 1. In the figure, $\mathcal{H}_M$, $\mathcal{S}_M$ represent the multirate hold, the (multirate) sampler, respectively.

In the ideal tracking control system, the transfer characteristic (G_{yr}) from the command r to the output y is generally 1. In this paper, the feedforward controller $C_1[z]$ is considered so that the transfer characteristic from the desired state $\boldsymbol{x}_d$ to the plant state $\boldsymbol{x}$ can be $\boldsymbol{I}$.

3.2 *Design of the feedback controller* $C_2[z]$

Before the perfect tracking controller $\boldsymbol{C}_1[z]$ is designed, the $\boldsymbol{C}_2[z]$ has to be determined. Here, the given feedback controller $\boldsymbol{C}_2[z]$ must be a robust controller that let the sensitivity function $\boldsymbol{S}[z] = (\boldsymbol{I} - \boldsymbol{P}[z]\boldsymbol{C}_2[z])^{-1}$ be small enough. The reason is that the sensitivity function $\boldsymbol{S}[z]$ represents the variation of the command response characteristic $\boldsymbol{G_{yr}}$ under the variation of $\boldsymbol{P}[z]$ (Sugie and Yoshikawa, 1986).

In Fig. 1, $C_2[z]$ is described as a multirate input and multirate output controller. However, in this section, the simplest method is proposed, in which the $\boldsymbol{C}_2[z]$ is obtained from the popular single-rate controller such as a disturbance observer or H_∞ controller.

First, in case 1, T_y is set to equal to $T_u (M = n)$, and $C_{2s}[z^{\frac{1}{n}}] = \{\boldsymbol{A}_s, \boldsymbol{b}_s, \boldsymbol{c}_s, d_s\}$ is designed on sampling period T_y. Then, $C_2[z]$ moving on every T_f is obtained from $C_{2s}[z^{\frac{1}{n}}]$ moving on every $T_y = T_f/n$ by (Chen and Francis, 1995)

$$C_2[z] = \left[\begin{array}{c|cccc} \boldsymbol{A}_s^n & \boldsymbol{A}_s^{n-1}\boldsymbol{b}_s & \boldsymbol{A}_s^{n-2}\boldsymbol{b}_s & \cdots & \boldsymbol{b}_s \\ \hline \boldsymbol{c}_s & d_s & 0 & \cdots & 0 \\ \boldsymbol{c}_s\boldsymbol{A}_s & \boldsymbol{c}_s\boldsymbol{b}_s & d_s & \cdots & 0 \\ \vdots & \vdots & & & \vdots \\ \boldsymbol{c}_s\boldsymbol{A}_s^{n-1} & \boldsymbol{c}_s\boldsymbol{A}_s^{n-2}\boldsymbol{b}_s & \boldsymbol{c}_s\boldsymbol{A}_s^{n-3}\boldsymbol{b}_s & \cdots & d_s \end{array}\right]. \quad (10)$$

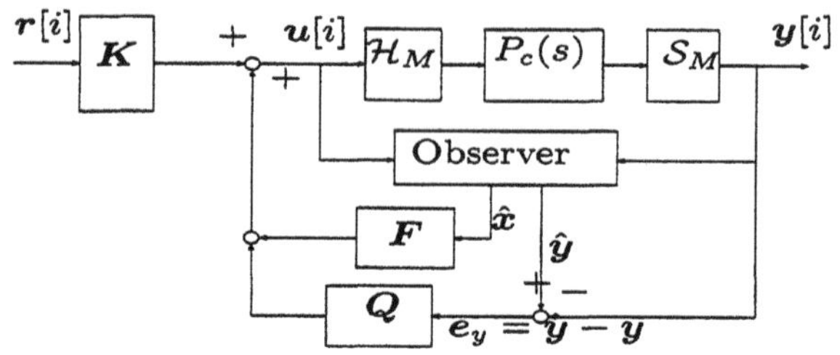

Fig. 4. Basic structure of TDOF control.

Second, in case 2, $C_{2s}[z]$ is designed on T_y. Then, the outputs of the $C_2[z]$ become the same values during one sampling period T_y as represented by

$$C_2[z] = [\,\overbrace{C_{2s}[z], \cdots, C_{2s}[z]}^{N}\,]^T. \tag{11}$$

3.3 *Design of the Perfect Tracking Controller* $C_1[z]$

In this section, the feedforward controller $C_1[z]$ is designed so that the perfect tracking can be assured at every sampling point T.

(3) described on T_f can be transferred to following equation on $T = T_f/L$.

$$\bar{x}[i+1] = \bar{A}x[i] + \bar{B}u[i] \tag{12}$$

where $q \triangleq 1/L$, and where matrices $\bar{A}, \bar{B}$ and vectors $\bar{x}$ are given by

$$\bar{x}[i+1] \triangleq \begin{bmatrix} x[i+q] \\ \vdots \\ x[i+lq] \\ \vdots \\ x[i+1] \end{bmatrix}, \quad \bar{A} \triangleq \begin{bmatrix} e^{A_c T} \\ \vdots \\ e^{A_c lT} \\ \vdots \\ e^{A_c LT} \end{bmatrix} \tag{13}$$

$$\bar{B} \triangleq \begin{bmatrix} B_L & O & \cdots & \cdots & \cdots & O \\ \vdots & \ddots & & & & O \\ B_{L-l} & \cdots & B_L & O & \cdots & O \\ \vdots & & & & \ddots & O \\ B_1 & B_2 & \cdots & \cdots & \cdots & B_L \end{bmatrix} \tag{14}$$

$$B_l = [b_{(l-1)n+1}, \cdots, b_{ln}] \quad (l = 1, \cdots, L). \tag{15}$$

From Fig. 1, the multirate control law of the proposed method is described by

$$u = C_1 r + C_2 y \tag{16}$$
$$= F\hat{x} + Qe_y + Kr \tag{17}$$

where $K, Q \in RH_\infty$ are free parameters. Therefore, Fig. 1 can be transferred to Fig. 4 (Zhou *et al.*, 1996). In this paper, K becomes a constant matrix.

Because the estimation errors of the observer become zero ($\hat{x} = x, e_y = 0$) for the nominal plant, from (12) and (17), this system is represented by

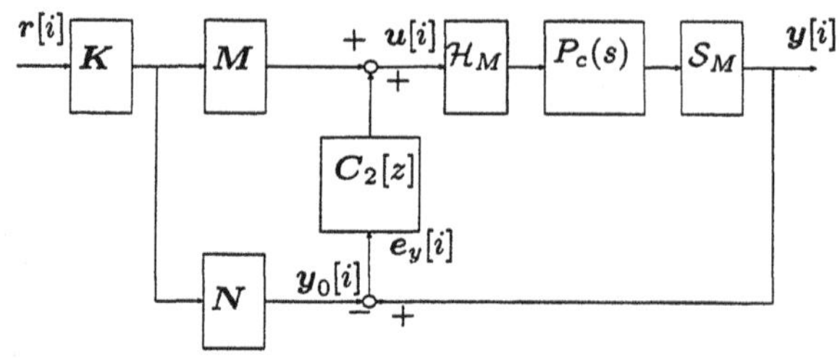

Fig. 5. Implementation of the proposed controller.

$$\bar{x}[i+1] = (\bar{A} + \bar{B}F)x[i] + \bar{B}Kr[i]. \tag{18}$$

Because non-singularity of the matrix B can be assured by $T = nT_u$ (Araki, 1993), $\bar{B}$ also becomes non-singular. Therefore, the parameters F, K can be selected so that following equations are satisfied

$$\bar{A} + \bar{B}F = O\,, \quad \bar{B}K = I. \tag{19}$$

From (19), F, K are given by

$$F = -\bar{B}^{-1}\bar{A}\,, \quad K = \bar{B}^{-1}. \tag{20}$$

Therefore, (18) is described by

$$\bar{x}[i+1] = r[i]. \tag{21}$$

Utilizing the future desired state, let the reference input be

$$r[i] = \bar{x}_d[i+1] \tag{22}$$

where $\bar{x}_d[i]$ is desired state. From (21) and (22), we find the perfect tracking $\bar{x}[i] = \bar{x}_d[i]$ can be achieved at every sampling point T.

Here, (16) and Fig. 1 can be transferred to (23) and Fig. 5 (Sugie and Yoshikawa, 1986). Therefore, the proposed controller can be simply implemented by substituting (23) for (20) and (24), and minimally realizing $[C_1[z], C_2[z]]$.

$$u = (M - C_2 N)Kr + C_2 y \tag{23}$$

$$\begin{aligned} M &= \left[\begin{array}{c|c} A+BF & B \\ \hline F & I \end{array}\right] = I + z^{-1}FB \\ N &= \left[\begin{array}{c|c} A+BF & B \\ \hline C+DF & D \end{array}\right] = D + z^{-1}(C+DF)B \end{aligned} \tag{24}$$

4. ILLUSTRATIVE EXAMPLES

In this section, the simulated and experimental results for the position tracking control system of the dc servo motor are presented, and the advantages of the proposed approach are demonstrated.

4.1 *Case2 on* $L = 1$

First, the simplest examples in case 2 on $L = 1$ are considered. The dc servo motor is described by

$$P_c(s) = \frac{K}{Js^2}. \tag{25}$$

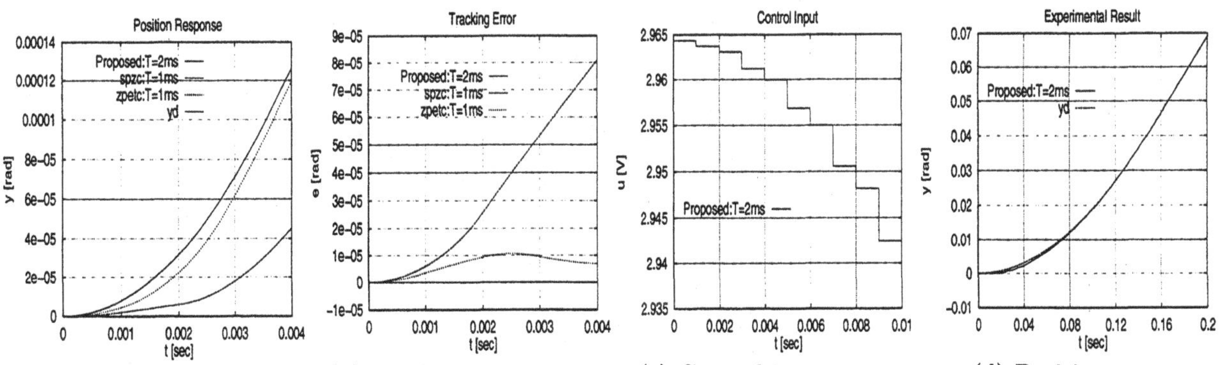

(a) Position (Simulation) (b) Tracking error (Simulation) (c) Control input (Simulation) (d) Position (Experiment)

Fig. 6. Simulation and experimental results ($T_{ref} = 0.5[\text{s}]$).

The feedback controller $C_2[z]$ is obtained from the H_∞ mixed-sensitivity problem, which includes an integrator and becomes 3rd order(Mita *et al.*, 1995). Calculating (23) and minimally realizing the obtained $C_1[z]$ and $C_2[z]$, the controller $[C_1, C_2]$ becomes 5th order system.

Simulated and experimental results under the sinusoidal desired trajectories of period $T_{ref} = 0.5[\text{s}]$ are shown in Fig. 6. Assuming $T_y = 2[\text{ms}]$, the each periods become $T = 2[\text{ms}]$ and $T_u = T_y/Ln = 1[\text{ms}]$ because this plant is 2nd order.

In the following simulations and experiment, the proposed method is compared with both SPZC and ZPETC proposed by Tomizuka (1987) at same input sampling period (T_u). Therefore, the reference and the output sampling periods (T and T_y) of proposed method are twice as long as those of SPZC and ZPETC because of $T_y = T_u = T = 1[\text{ms}]$ on them. In spite of that, the results of proposed method have better tracking performance than those methods.

Fig. 6(a)(b) show that the proposed method gives better performance both than the SPZC and the ZPETC. While the responses of the SPZC and ZPETC have large tracking errors caused by the unstable zero, those of the proposed method have zero tracking errors. The simulated time response of the control input is shown in Fig. 6(c), which indicates that the control input of the proposed method is smooth in spite of using the multirate input control. Thus, we find the proposed multirate feedforward method is very practical. Moreover, the experimental result also indicates that the proposed method has high tracking performance as shown in Fig. 6(d). Furthermore, Fig. 6(a)(b) also show that the inter-sample responses are very smooth because not only position but also velocity follow the desired trajectories at every sampling point T.

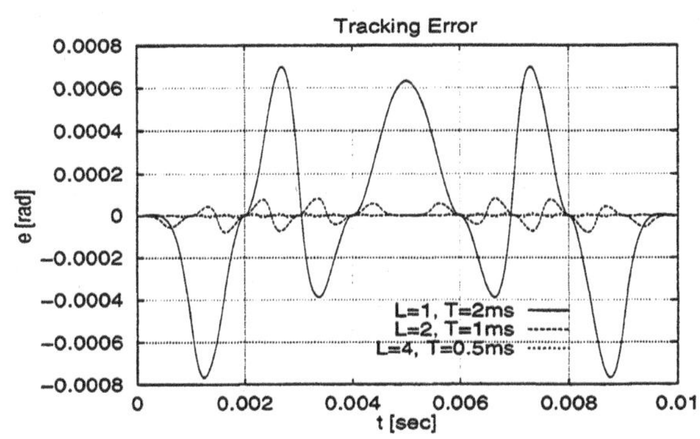

Fig. 7. Responses on $L = 1, 2, 4$ ($T_{ref} = 10[\text{ms}]$).

4.2 *Case2 on* $L = 2, 4$

Second, $T_y = 2[\text{ms}]$ is fixed for the same plant as section 4.1, and the control input is changed more frequently than $T_u = 1[\text{ms}]$.

Although in Fig. 6(b), the desired trajectories were very slow ($T_{ref} = 0.5[\text{s}]$), Fig. 7 shows the simulated results for more fast trajectories $T_{ref} = 10[\text{ms}]$.

Compared with $L = 1$, the tracking performances are improved for large input multiplicity $L = 2, 4$. The perfect tracking is also assured L times during T_y.

4.3 *Case1 on* $M = n$

Third, although $T_y = 2[\text{ms}]$ is assumed in section 4.1, the output is detected more frequently at $T_y = T_u = 1[\text{ms}]$ in this section. That is equal to case 1 on $M = n$.

In this case, the tracking performance is remained unchanged from section 4.1. However, the closed-loop characteristics can be improved because much information of the plant can be detected. Fig. 8 shows the simulated results when the plant inertia J becomes three times larger than the nominal inertia J_n. We find that the

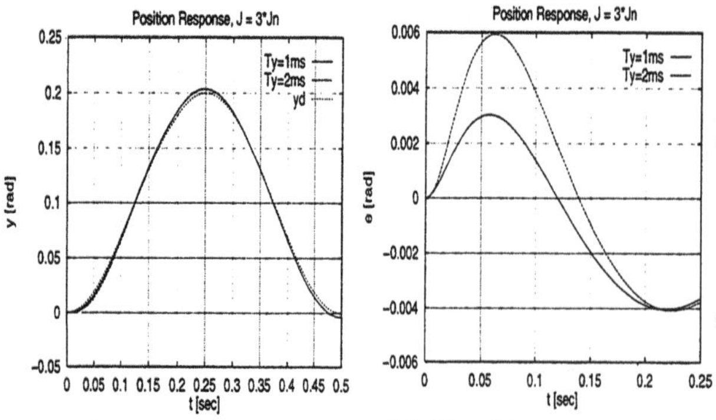

(a) Position (Simulation) (b) Tracking error (Simulation)

Fig. 8. Responses under inertia variation $J = 3J_n$ ($T_{ref} = 0.5$[s]).

high performance robustness is achieved by the H_∞ feedback controller as shown in Fig. 8(a). Moreover, Fig. 8(b) also shows that this case of $T_y = 1$[ms], $M = 2$ has higher performance robustness than the case of $T_y = 2$[ms], $M = 1$.

5. CONCLUSION

A novel perfect tracking control method in use of the multirate sampling control was proposed. Moreover, high performance perfect tracking control method was achieved, by which the control system matches the hardware restrictions on the periods of the plant input holding and output sampling. The remarkable advantages of the proposed method are 1) the tracking performance can be improved on the inter-sampling points, if the input can be changed more frequently, 2) higher feedback performance is achieved, if the output can be detected more frequently. Furthermore, illustrative examples of position control using a dc servo motor are performed, and the advantages of this approach are demonstrated.

Finally, the authors would like to note that part of this research is carried out with a subsidy of the Scientific Research Fund of the Ministry of Education.

6. REFERENCES

Araki, M. (1993). Recent developments in digital control theory. In: *12th IFAC World Congress.* Vol. 9. pp. 251–260.

Åström, K. J., P. Hangander and J. Sternby (1984). Zeros of sampled system. *Automatica* **20**(1), 31–38.

Chen, T. and B. Francis (1995). *Optimal Sampled-Data Control Systems.* Springer.

Fujimoto, H., A. Kawamura and M. Tomizuka (1999). Generalized digital redesign method for linear feedback system based on N-delay control. *IEEE/ASME Trans. Mechatronics.* (to be published).

Fujimoto, H. and A. Kawamura (1998). Perfect tracking digital motion control based on two-degree-of-freedom multirate feedforward control. In: *International Workshop on Advanced Motion Control.* pp. 322–327.

Gross, E., M. Tomizuka and W. Messner (1994). Cancellation of discrete time unstable zeros by feedforward control. *ASME Journal of Dynamics System, Measurement, and Control* **116**, 33–38.

Haack, B. and M. Tomizuka (1991). The effect of adding zeros to feedforward controllers. *ASME Journal of Dynamics System, Measurement, and Control* **113**, 6–10.

Hara, S., Y. Yamamoto and H. Fujioka (1996). Modern and classical analysis / synthesis methods in sampled-data control – a brief overview with numerical examples –. In: *Conf. on Decision and Control.* pp. 1251–1256.

Kabamba, P. T. (1987). Control of linear systems using generalized sampled-data hold functions. *IEEE Trans. Automat. Contr.* **32**(9), 772–783.

Mita, T., M. Hirata and S. B. Villas (1995). H_∞ servo controller design for plants having poles on the $j\omega$ axis. In: *Conf. on Decision and Control.* pp. 2568–2573.

Mita, T., Y. Chida, Y. Kazu and H. Numasato (1990). Two-delay robust digital control and its applications – avoiding the problem on unstable limiting zeros. *IEEE Trans. Automat. Contr.* **35**(8), 962–970.

Moore, K. L., S. P. Bhattacharyya and M. Dahleh (1993). Capabilities and limitations of multirate control schemes. *Automatica* **29**(4), 941–951.

Sugie, T. and T. Yoshikawa (1986). General solution of robust tracking problem in two-degree-of-freedom control systems. *IEEE Trans. Automat. Contr.* **31**(6), 552–554.

Tomizuka, M. (1987). Zero phase error tracking algorithm for digital control. *ASME Journal of Dynamics System, Measurement, and Control* **109**, 65–68.

Torfs, D., J. De Schutter and J. Swevers (1992). Extended bandwidth zero phase error tracking control of nonminimal phase systems. *ASME Journal of Dynamics System, Measurement, and Control* **114**, 347–351.

Zhou, K., J. Doyle and K. Glover (1996). *Robust and Optimal Control.* Prentice-Hall, Inc. (Translated into Japanese by K.Z. Liu and Z.H. Luo).

14th Triennial World Congress, Beijing, P.R. China

C-2a-02-6

HIGH PERFORMANCE TRACKING CONTROL SYSTEM UNDER MEASUREMENT CONSTRAINTS BY MULTI-RATE CONTROL

Yuping Gu and Masayoshi Tomizuka

Department of Mechanical Engineering,
University of California at Berkeley, Berkeley, CA 94720-1740, USA
gyp@me.Berkeley.edu tomizuka@me.Berkeley.edu

Abstract: This paper describes a multi-rate feedback/feedforward system applicable to the design and analysis of tracking control system under measurement constraints. In this system, the feedback controller is updated at the slow rate, while the feedforward controller is updated at the fast rate with a desired sequence, which is obtained by interpolating the slow sampled sequence of the desired trajectory. The tracking performance is evaluated by including intersample behavior of the output. Illustrative examples are given to further explain the advantages of multi-rate control.

Keywords: multi-rate control, feedforward control, digital control, tracking control

1. INTRODUCTION

Multi-rate digital control systems are those which use more than one sampling rate. Fig. 1 is a typical single-input single-output (SISO) digital control system, which has a digital controller and a controlled plant. If the controller updating rate $f_c = \frac{1}{T_c}$ is the same as the plant output sampling rate $f_s = \frac{1}{T_s}$, then it is a single-rate system. However, practical considerations may mean that it is not necessarily the best way to sample plant output and change its input at the same rate.

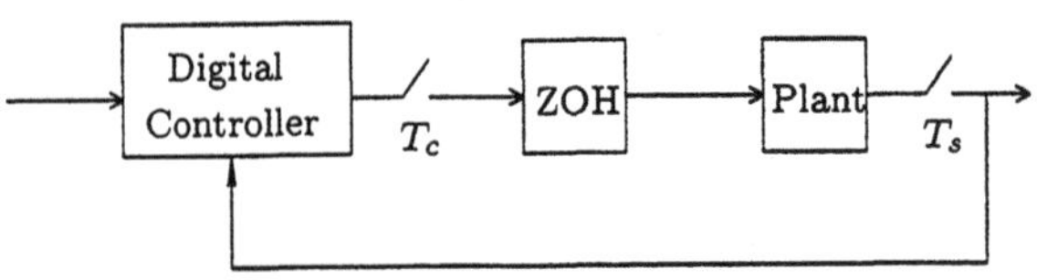

Fig. 1. Digital control system

There are many practical reasons for multi-rate digital control. For example, many disk drives use sector-servo to control the read/write head position relative to track center lines (Chiang, 1990) (Phan, 1993). In the sector-servo system, every track is evenly divided angularly into sectors to store data, and position information is written at sector boundaries. In order to utilize as much disk surface as possible for data storage, the number of sectors on each track should not be too large. This limits the measurement sampling rate for digital control. In this case, digital control with a standard zero order hold may not be appropriate, and it has been suggested to update the control input at a higher rate (Chiang, 1990). Although single-rate systems are easy to design and implement, it can be inefficient for many practical systems which provide measurements at slow sampling rates. Another reason for using multi-rate control may be that the sampling rate is constrained by the computational speed of microprocessors for digital control. We can not assume that the most powerful digital signal processors can be applied to every application.

Various multi-rate control problems presented in the literature (Moore *et al.*, 1993) can be classified into two types: (i)N-delay input control when the input is updated at N times more often than the output is sampled, and (ii) N-delay output control when the output is sampled N times more often than the input is updated. This paper presents one version of N-delay input control, the purpose of which is to improve tracking performance.

The remainder of this paper is organized as follows. In section 2, a multi-rate feedback/feedforward control system for tracking of smooth desired output signals is analyzed, and a method for analyzing the performance of single-rate systems incorporating intersample behaviors is presented. Section 3 presents illustrative examples to verify the analysis. The example shows an additional advantage of the proposed multi-rate approach, which is related to unstable zeros arising from sampling. Concluding remarks are given in section 4.

2. MULTI-RATE FEEDBACK/FEEDFORWARD CONTROL SYSTEM

2.1 *Preliminary*

Consider a single-rate control system as configured in Fig. 2, where $z_1 = e^{sT}$.

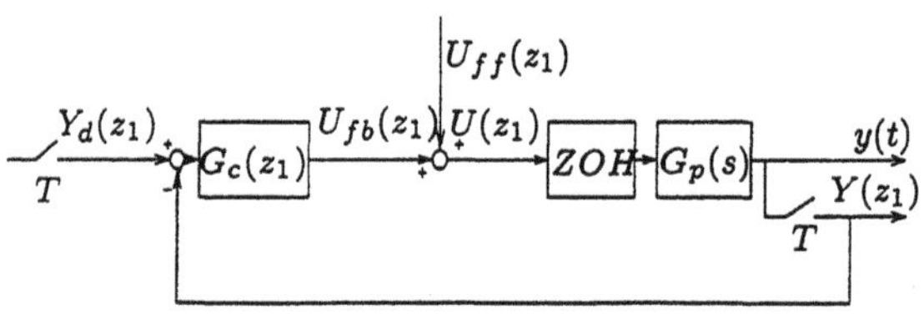

Fig. 2. Single rate feedback/feedforward system

$U_{ff}(z_1)$ represents the feedforward input, and is determined so that $y(k)$ remains close to $y_d(k)$. where $y(k) := y(kT)$ and $y_d(k)$ is a given desired sequence generated from a smooth desired trajectory $y_d(t)$.

Perfect Tracking: Assume that the closed loop system is stable. Then if the zero order hold equivalent of $G_p(s)$ does not possess any unstable zeros, $U_{ff}(z_1)$ given by

$$U_{ff}(z_1) = \{(1 - z_1^{-1})\mathcal{Z}_T[\mathcal{L}^{-1}\{\frac{G_p(s)}{s}\}]\}^{-1} Y_d(z_1) = [G_p(z_1)]^{-1} Y_d(z_1) \quad (1)$$

assures $Y(z_1) = Y_d(z_1)$, i.e. $y(k) = y_d(k)$ under zero initial condition. We call the single rate system with the feedforward control law given by (1) the basic system. This statement can be easily verified by noting

$$Y(z_1) = \frac{1}{1 + G_c(z_1)G_p(z_1)} \cdot [G_p(z_1)G_c(z_1) + G_p(z_1)G_p^{-1}(z_1)]Y_d(z_1) = Y_d(z_1)$$

Suppose that the feedforward control input can be updated at a higher rate than the feedback control input. The update rate of the feedback control input is determined by the sampling period T, i.e. $f_{fb} = \frac{1}{T}$ where f_{fb} is the update rate. Suppose that the feedforward control input is updated two times faster than the feedback control input, i.e. $f_{ff} = 2f_{fb}$. In this case, the overall control system is depicted as shown in Fig. 3. The system is a multi-rate sampling system.

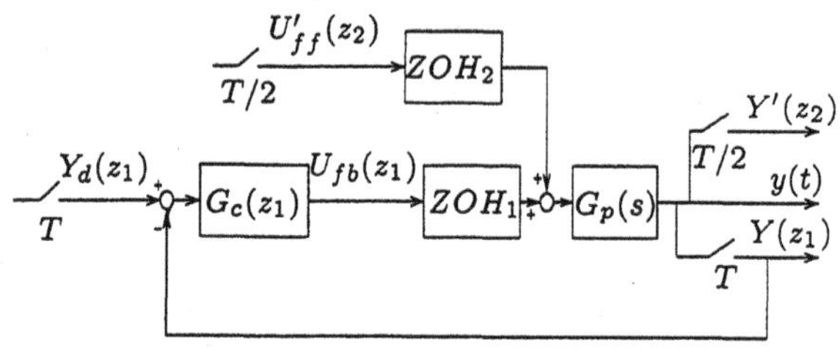

Fig. 3. Multi-rate feedback/feedforward system

In Fig. 3, ZOH_1 and ZOH_2 hold discrete time sequences $u_{fb}(k)$ and $u_{ff}(k')$ over T and $\frac{T}{2}$ sec. respectively. k denotes time for the feedback control input sequence and k' for the feedforward control input sequence. They are related to the continuous time by kT and $k'\frac{T}{2}$.

The purpose of the multi-rate control is to enhance the tracking performance of the system. Often, $y_d(k)$ is generated by sampling a continuous time signal $y_d(t)$ every T sec. The choice of the sampling period for feedback control depends on a variety of factors. It may be constrained by measurement schemes and may have to be selected low. In such situations, perfect tracking of sampled sequences may not imply even acceptable tracking of $y_d(t)$ by $y(t)$, and intersample behavior need to be considered. By updating the feedforward control input at a higher rate, the tracking performance may be improved if we include intersample behavior of the output in performance evaluation.

2.2 *Performance Evaluation of the Basic System*

To discuss possible improvement, if any, by the proposed multi-rate scheme, we must first define the tracking performance of the basic system including intersample behavior. This can be done by comparing the sampled sequence $y(k')$ (note that it is sampled twice as fast as the measurement sampler for feedback control) with a desired sequence $y_d(k')$. In Fig. 2, $Y_d(z_1)$ denotes the Z-transform of $y_d(k)$, $y_d(k)$ provides $y_d(0)$, $y_d(T)$, $y_d(2T)$, ... To examine tracking performance at higher sampling rate, we need to define $y_d(\frac{T}{2})$, $y_d(\frac{T}{3})$, ... Assuming that $y_d(k)$ is obtained by sampling a reasonably smooth continuous time signal, we define $y_d(\frac{2k+1}{2}T)$ by

$$y_d(\frac{2k+1}{2}T) = \frac{y_d(k) + y_d(k+1)}{2} \quad (2)$$

Then $y_d(k')$ is well defined for all $k' \geq 0$. Notice that $y_d(\frac{2k+1}{2}T)$ is not generated from $y_d(t)$.

We may evaluate tracking performance at the higher update rate by computing $Y'(z_2)/Y'_d(z_2)$, where $Y'(z_2)$ and $Y'_d(z_2)$ are the Z-transforms of $y(k')$ and $y_d(k')$ respectively (Oppenheim and Schafer, 1989).

Computation of $Y'_d(z_2)$

$$\begin{aligned} Y'_d(z_2) &= \sum_{k'=0}^{\infty} y_d(k') z_2^{-k'} \\ &= \sum_{k=0}^{\infty} y_d(k) z_2^{-2k} + \\ &\quad \sum_{k=0}^{\infty} \frac{y_d(k) + y_d(k+1)}{2} z_2^{-(2k+1)} \\ &= \sum_{k=0}^{\infty} y_d(k) z_2^{-2k} (\frac{z_2 + 2 + z_2^{-1}}{2}) \\ &= \frac{z_2 + 2 + z_2^{-1}}{2} Y_d(z_1)_{z_1 = z_2^2} \end{aligned} \quad (3)$$

Computation of $Y'(z_2)$

Notice that $Y(z_1)$ is obtained from $y(k') := y(\frac{T}{2}k')$, where $k' = 0, 2, 4, 6, \ldots$. We need $y(\frac{T}{2}k')$ for $k' = 1, 3, 5, \ldots$, i.e. a sequence $y(\frac{T}{2}(2k+1))$, $k = 0, 1, 2, \ldots$. Now we note

$$\begin{aligned} &\sum_{k=0}^{\infty} y(kT) z_1^{-k} \\ &= Y(z_1) \\ &= (1 - z_1^{-1}) \mathcal{Z}_T [\mathcal{L}^{-1} \{\frac{G_p(s)}{s}\}] U(z_1) \\ &= G_p(z_1) U(z_1) \\ &\sum_{k=0}^{\infty} y((k + \frac{1}{2})T) z_1^{-k} \\ &= Y_{int}(z_1) \\ &= (1 - z_1^{-1}) z_1 \mathcal{Z}_T [\mathcal{L}^{-1} \{\frac{G_p(s) e^{-\frac{T}{2}s}}{s}\}] U(z_1) \\ &= G_{pint}(z_1) U(z_1) \end{aligned}$$

then

$$\begin{aligned} &Y'(z_2) \\ &= \sum_{k=0}^{\infty} y(k') z_2^{-k'} \\ &= \sum_{k=0}^{\infty} y(2k\frac{T}{2}) z_2^{-2k} + \sum_{k=0}^{\infty} y((2k+1)\frac{T}{2}) z_2^{-(2k+1)} \\ &= Y(z_1)_{z_1 = z_2^2} + Y_{int}(z_1)_{z_1 = z_2^2} z_2^{-1} \\ &= [G_p(z_1)_{z_1 = z_2^2} + G_{pint}(z_1)_{z_1 = z_2^2} z_2^{-1}] U(z_1) \end{aligned} \quad (4)$$

$U(z_1)$ is related to $Y_d(z_1)$ by

$$\begin{aligned} U(z_1) &= \frac{1}{1 + G_p(z_1) G_c(z_1)} \cdot \\ &\quad [G_p^{-1}(z_1) Y_d(z_1) + G_c(z_1) Y_d(z_1)] \\ &= \frac{1}{G_p(z_1)} Y_d(z_1) \end{aligned} \quad (5)$$

From Eqs.(4) and (5)

$$\begin{aligned} Y'(z_2) = [G_p(z_2^2) + G_{pint}(z_2^2) z_2^{-1}] \cdot \\ \frac{1}{G_p(z_2^2)} Y_d(z_1)_{z_1 = z_2^2} \end{aligned} \quad (6)$$

From Eqs. (3) and (6), we obtain the relation,

$$\frac{Y'(z_2)}{Y'_d(z_2)} = \frac{2[G_p(z_2^2) + G_{pint}(z_2^2) z_2^{-1}]}{G_p(z_2^2)[z_2 + 2 + z_2^{-1}]} \quad (7)$$

By setting $z_2 = e^{jw\frac{T}{2}}$, we can evaluate the tracking performance of the basic system in the frequency domain including intersample behavior.

2.3 *Feedforward Control Law for Multi-rate System*

For the multi-rate system in Fig. 3, it makes intuitive sense to determine $U'_{ff}(z_2)$ by

$$\begin{aligned} U'_{ff}(z_2) &= [(1 - z_2^{-1}) \mathcal{Z}_{\frac{T}{2}} [\mathcal{L}^{-1} \{\frac{G_p(s)}{s}\}]]^{-1} Y'_d(z_2) \\ &= [G'_p(z_2)]^{-1} Y'_d(z_2) \end{aligned} \quad (8)$$

Note that $G'_p(z_2)$ is the zero order hold equivalent of $G_p(s)$ when the sampling period is $\frac{T}{2}$.

If $u_{ff}(k')$ given by Eq.(8) is applied to the plant initially at rest, $y(k')(k' = 0, 1, 2, \ldots)$ will evolve exactly as $y_d(k')$ as long as the $y_d(k')$ sequence starts with a sufficient number of zeros. This number should be larger than the relative degree of $G'_p(z_2)$. Then, the feedback controller $G_c(z_1)$ will never come into play, i.e. its output remains zero, and $y(k') = y_d(k')$, or $Y'(z_2) = Y'_d(z_2)$ is achieved. This point can be made more mathematically rigorous as described in (Gu and Tomizuka, 1999).

3. SIMULATION RESULTS

In this section, we present illustrative examples to support the analysis in the previous section. Fig. 4 shows the simulated system.

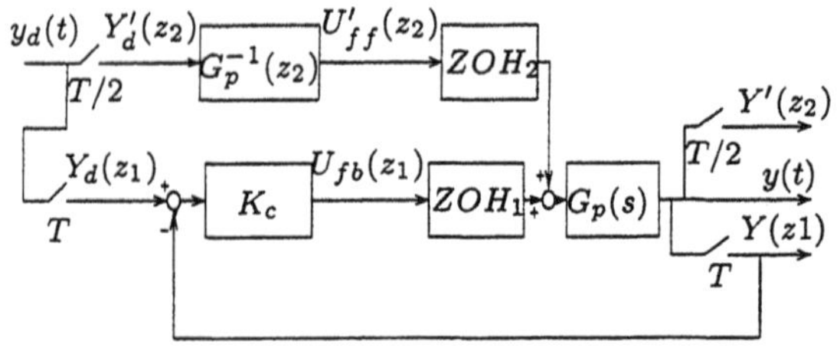

Fig. 4. Simulation of multi-rate feedback/feedforward system

In Fig. 4, the plant is $G_p(s) = \frac{1}{0.5s+1}$, the feedback controller is $K_c = 2$, and the feedforward controller $G_p^{-1}(z_2)$ is the inverse of the zero order hold equivalent of $G_p(s)$ when the sampling period is $\frac{T}{2}$, where $T = 1sec$. The desired trajectory we use is

$$y_d(t) = \begin{cases} t & 0 \le t \le 3 \\ 3 & 3 < t \le 5 \\ -2t + 13 & 5 < t \le 8 \\ -3 & 8 < t \le 10 \end{cases}$$

Figure 5 shows the tracking performance $y(t)$ of single and multi-rate system. The dotted line is the desired trajectory. In this case, $y_d(t)$ was selected such that $y_d(\frac{2k+1}{2}T)$ by Eq.(2) matches the sampled value of $y_d(t)$ at that time instance. If the feedforward control input is updated every T and the feedforward control input is based on Eq.(1), $y(k) = y_d(k)$ is attained. However, $y(t)$ does not track $y_d(t)$. Notice that within one sampling period T, the plant exhibit a first order exponential response since the zero order hold fixes the input at constant. This will make $y(t)$ deviate from $y_d(t)$ which is a straight line between two adjacent sampling instances. By applying the multi-rate feedforward control, the error is significantly reduced. It is obvious that further error reduction is possible by updating the feedforward input at even higher rates.

Figure 6 shows the tracking performance of the single rate system in the frequency domain. Equation (7) was utilized to obtain this plot. As seen in the figure, the tracking performance of the single rate system is good at low frequencies, but it is not good at high frequencies in terms of both the gain and phase characteristics. In particular, the figure shows that $y(k')$ leads $y_d(k')$ at high frequencies. This is consistent with the fact that the time plot of $y(k')$ leads that of $y_d(k')$ during two sampling instances. This figure also supports the validity of Eq.(7) as a measure of the tracking performance of the single rate system incorporating intersample behavior.

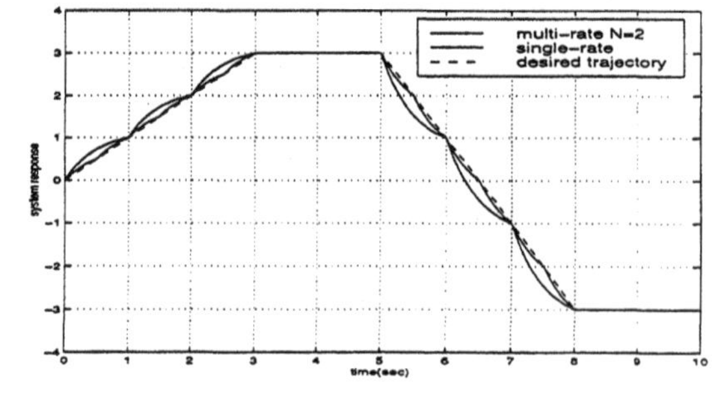

Fig. 5. Tracking performance of single and multi-rate system

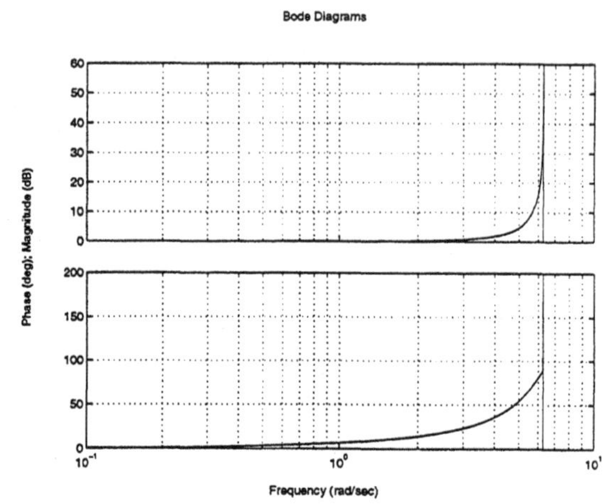

Fig. 6. System frequency response

We now consider a plant with a pure time delay and show an additional advantage of the proposed multi-rate feedforward controller: i.e. the plant is described by $G_p(s) = \frac{e^{-0.4s}}{0.5s+1}$. The desired trajectory is

$$y_d(t) = \begin{cases} 0 & 0 \le t \le 0.6 \\ \frac{5}{3}t - 1 & 0.6 < t \le 2.4 \\ 3 & 2.4 < t \le 3.6 \\ -\frac{10}{3}t + 15 & 3.6 < t \le 5.4 \\ -3 & 5.4 < t \le 6 \end{cases}$$

For single-rate control with $T = 0.6sec$,

$$G_p(z_1) = \frac{0.3297(z_1 + 1.1196)}{z_1(z_1 - 0.3012)}$$

For multi-rate control with $N = 2$, with $T/2 = 0.3sec$,

$$G_p(z_2) = \frac{0.3297(z_2 + 0.3685)}{z_2^2(z_2 - 0.5488)}$$

Note that for single-rate control, $ZPETC$ or any other approximate inverse of $G_p(z_1)$ must be used in the feedforward path since the ZOH equivalent of the plant $G_p(z_1)$ has one uncancellable zero and the overall system must be constructed as shown in Fig. 7. Notice that the reference input for the closed loop system is now filtered by $\frac{B^u(z_1^{-1})B^u(z_1)}{[B^u(1)]^2}$, which is the expected output of the plant given ZPETC as the feedforward controller (Gu and Tomizuka, 1999).

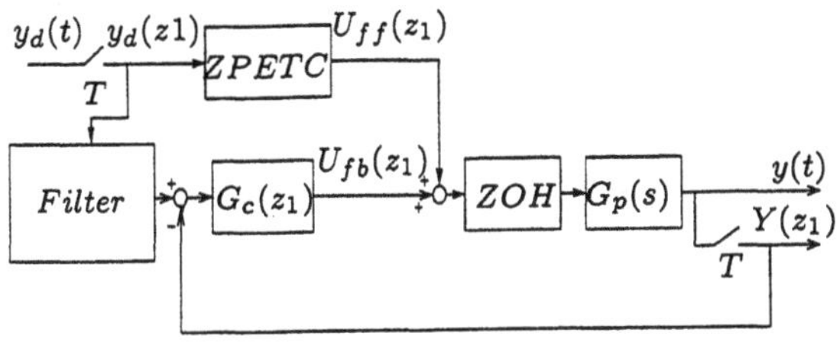

Fig. 7. Single-rate feedback/feedforward system for nonminimum phase discrete time plant

By cutting the sampling period for feedforward control into one half, the ZOH equivalent of the plant $G_p(z_2)$ does not posses any unstable zero. Thus, it becomes possible to cancel all plant poles and zeros for perfect tracking. This point is another advantage of the proposed multi-rate feedback/feedforward control scheme.

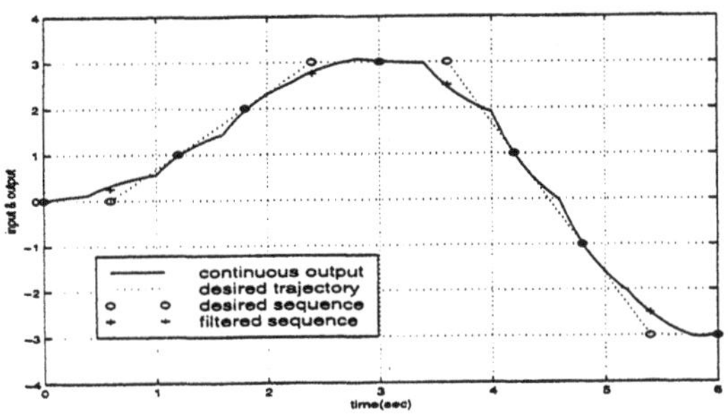

Fig. 8. Tracking performance of single-rate system

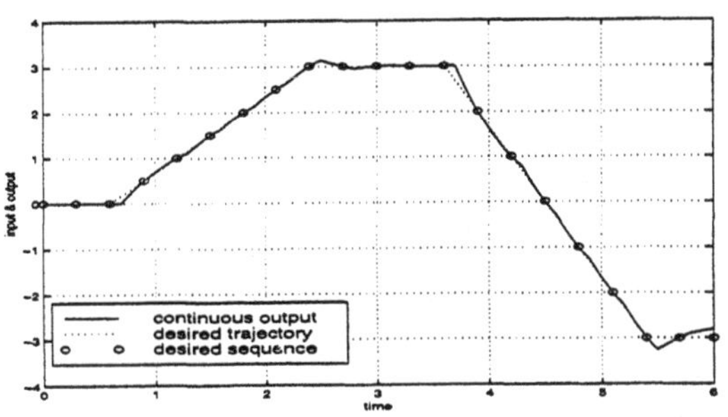

Fig. 9. Tracking performance of multi-rate system for N=2

Figure 8 and Fig. 9 show the tracking performance of single and multi-rate system respectively. The single rate system no longer provides perfect tracking even at discrete time instances. This is due to the use of ZPETC, which is an approximate inverse of $G_p(z_1)$. Since the inverse of $G_p(z_2)$ is asymptotically stable, it can be used as the feedforward controller in the multi-rate case. As shown in Fig. 9, this makes the multi-rate system achieve perfect tracking.

4. CONCLUSION

The multi-rate feedback/feedforward system proposed in this paper is applicable to the design and analysis of tracking control system under measurement constraint. For reasonably smooth continuous time desired trajectory, interpolating slowly sampled sequences, information lost by sampling can be recovered. When the sampling rate is low, it is important to include intersample behavior in performance evaluation. This paper described such a case for the design of feedforward controller. Two illustrative examples were presented to provide further insights and show the advantage of multi-rate feedforward control. The plant output remains closer to the desired output by increasing the frequency of updating the feedforward control input. In certain cases, it is possible to select the sampling period for feedforward control so that the discrete time model does not possess any unstable zero. A more general and complete version of the multi-rate feedforward controller, which includes systems with unstable inverses, is in Gu and Tomizuka (1999).

5. REFERENCES

Chiang, Wen-Wei (1990). Multi-rate state-space digital controller for sector servo systems. In: *The 29th IEEE Conference on Decision and Control.* Honolulu,HI. pp. 1902–1907.

Gu, Yuping and Masayoshi Tomizuka (1999). Multi-rate feedforward tracking control for plants with nonminimum phase discrete time models. In: *The 1999 American Control Conference.* San Diego, CA.

Moore, Kevin L., S.P. Bhattacharyya and Mohammed Dahleh (1993). Capabilities and limitations of multi-rate control schemes. *Automatica* **29**(4), 941–951.

Oppenheim, Alan V. and Ronald W. Schafer (1989). *Discrete-time signal processing.* Pretince Hill. Englewood Cliffs, N.J.

Phan, Duc T. (1993). The design and modeling of multi-rate digital control systems for disk drive application. In: *The 1993 Asian-Pacific Workshop on Advances in motion control.* Singapore. pp. 189–205.

FEEDBACK PASSIVITY OF NONLINEAR MULTIVARIABLE SYSTEMS

Hebertt Sira-Ramírez [*,1] **Miguel Rios-Bolívar** [*]

[*] *Departamento Sistemas de Control*
Universidad de Los Andes
Mérida 5101, Venezuela
Fax: +58-74-402846
email isira@ing.ula.ve, riosm@ing.ula.ve

Abstract: In this article, *projection operators* are used for obtaining a natural drift vector field decomposition that clearly exhibits the workless, the dissipative and the locally destabilizing forces acting within any nonlinear multivariable system equipped with a given positive definite storage function. A simple input coordinate transformation, which is shown to always exist, places the system into a Passivity Based Control Canonical form (PBCCF). Such a canonical form is ready for a direct application of the "energy shapping plus damping injection" controller design method.

Keywords: Multivariable systems, Nonlinear Systems, Passivity based control

1. INTRODUCTION

Passivity based control of nonlinear systems enjoys popularity due to several advantages related to controller simplicity, robustness and the physically appealing features of the approach. A seminal contribution is that of Willems (1972) in the context of ***dissipative systems***. The articles by Hill (1976, 1977) constitute also a general approach with emphasis on conditions for stability of feedback interconnected systems. A geometric approach to feedback equivalence of nonlinear passive systems was contributed by the work of Byrnes *et al* (1991).

The passivity approach has found application developments in the area of robotics, electrical machines and power electronics. The reader is referred to the works by Ortega and his coworkers (1989, 1995), and Sira-Ramírez *et al* (1997). The initial developments of passivity based control were carried out within the context of Hamiltonian systems, with a view towards applications in the control of robotic manipulator systems (Takegaki and Arimoto, 1981) General developments were later advanced, within the formulation of Lagrangian systems, in the work of Ortega *et al* (1995) and Shishkin *et al* (1996).

Nevertheless, passivity based control should not be regarded as a control technique which is suitable only in the fields of mechanical, electrical and electromechanical systems (or, *lagrangian* systems in short). Applications in the context of chemical and biochemical processes are also possible, as demonstrated by the works of Sira-Ramírez and Delgado (1997), and Sira-Ramírez and Angulo-Núñez (1997).

In this article, it is proposed a general procedure for the systematic transformation of **any** nonlinear square multivariable state space system, with nonzero desired equilibrium points and equipped with a ***given*** positive definite storage function, into a ***Passivity Based Control Canonical Form*** (PBCCF). A ***natural*** drift vector field decomposition is first obtained on the basis of a *projection*

[1] This research was supported by the National Council for Scientific and Technological Research of Venezuela (CONICIT) under Research Grant S1-95-000886

operator, induced by the given storage function of the system and the set of the control input vector fields. This decomposition is aimed at naturally exhibiting the workless, the dissipative (or beneficial) and the locally destabilizing nonlinearities acting on the uncontrolled motions of the system. A first state dependent input coordinate transformation makes the locally destabilizing drift vector field component invariant with respect to the storage function level sets. As a consequence of this partial feedback, a canonical form, addressed here as the *Passivity Based Control Canonical Form* (PBCCF), is then obtained by means of a second, nonsingular, transformation in the input space of the system. This transformation is known to *always* exist.

The PBCCF represents a further generalization of the so called *Generalized Hamiltonian canonical form*, extensively studied in the literature (van der Schaft, 1996). The proposed PBCCF differs from the Generalized Hamiltonian canonical form in that it explicitly, and naturally, includes the dissipative forces acting on the system uncontrolled dynamics. The role of the Hamiltonian function is played, not surprisingly, by the energy storage function of the system.

In the PBCCF the locally destabilizing vector field may be readily eliminated by means of partial state feedback, represented by a suitable state dependent input coordinate transformation. The resulting system is clearly shown to be output passive or strictly output passive.

2. A CANONICAL FORM FOR PASSIVITY BASED CONTROL

2.1 *Assumptions, definitions and results*

Consider the class of nonlinear multivariable *square* systems described by

$$\dot{x}(t) = f(x) + G(x)u \ ; \quad x \in \mathcal{X} \subset \mathbf{R}^n \ ;$$
$$y = h(x) \ ; \ y \in \mathcal{Y} \subset \mathbf{R}^m \quad u \in \mathcal{U} \subset \mathbf{R}^m \tag{2.1}$$

where $\mathcal{X}$ denotes the *operating region* of the system, constituted by a sufficiently large open set containing a continuum of equilibrium points, possibly parametrized by a constant control input value $u = U \in \mathcal{U}$, of the form $x = \overline{x}(U)$ and given by the solution of $f(\overline{x}) + g(\overline{x})U = 0$. In particular, for $u = 0$, we assume $f(\overline{x}) = 0$ implies $\overline{x} = 0$. However, motivated by a large class of real life systems, we are specifically interested in *nonzero* constant state equilibrium points $x = \overline{x}$, corresponding to nonzero constant control inputs $u = U$. Our results *are not* valid for the cases where the origin is the only equilibrium point of the system. The $n \times m$ matrix $G(x)$ is addressed as the *input matrix* and it is constituted by m columns, $[g_1(x), \ldots, g_m(x)]$ which are assumed to locally span an m-dimensional distribution in $\mathcal{X}$.

We assume that a smooth positive definite *storage function* $V : \mathbf{R}^n \to \mathbf{R}^+$ is given such that $V(0) = 0$.

We define, with certain abuse of notation, the quantity $L_G V(x)$ the m-dimensional *row* vector of ordered directional derivatives of $V(x)$ with respect to the set of vector fields $\{g_i(x) \ ; \ i = 1, \ldots, m\}$. In other words,

$$L_G V(x) = \frac{\partial V}{\partial x^T} G(x)$$

Similarly, by $L_G^T V(x)$, we denote the column vector $(L_G V(x))^T = G^T(x)\partial V/\partial x$. Notice then that the square of the norm of $L_G V(x)$, $||L_G V||^2$, coincides with the product $L_G V(x) L_G^T V(x)$.

Throughout the entire article, the following assumption holds valid.

Assumption 1

$$||L_G V(x)|| \neq 0 \quad \forall \, x \in \mathcal{X} \tag{2.2}$$

The condition (2.2) is usually known as the *transversality condition* and simply establishes that there is at least one vector field $g_i(x)$, among the columns of the matrix $G(x)$ which is not orthogonal to the gradient of V in $\mathcal{X}$. In other words, the control vector fields $g_i(x) \ ; \ i = 1, \ldots, m$ are not all tangential, at each x, to the storage function level sets defined, in the state space of the system, as $\{x \in \mathcal{X} \ : \ V(x) = \mathit{constant}\}$. This condition is quite familiar in sliding mode control of nonlinear systems (Sira-Ramírez, 1988) and it amounts to having a storage function which is not *vector relative degree* equals to $\{0, \ldots, 0\}$ in $\mathcal{X}$.

For each $x \in \mathcal{X}$, we define a *projection operator, along the span of the input matrix $G(x)$ onto the tangent space to the constant level sets of the energy function $V(x)$*, as the matrix $M(x)$ given by

$$M(x) = \left[I - \frac{1}{||L_G V||^2} G(x) L_G^T V(x) \frac{\partial V}{\partial x^T} \right] \tag{2.3}$$

The following proposition points out some properties of the matrix $M(x)$ which further justify the given name of "projection operator"

Proposition 1. The matrix $M(x)$ enjoys the following properties:

$$\frac{\partial V}{\partial x} \in \mathcal{K}er \ M^T(x)$$
$$M(x)(I - M(x)) = 0 \tag{2.4}$$
$$G(x) L_G^T V(x) \in \mathcal{K}er \ M(x)$$

The first property is equivalent to $\partial V/\partial x^T M(x) = 0$ and simply states that the span of $M(x)$ lies in the tangent space to the level sets of $V(x)$. The second property establishes that $M(x)$ leaves invariant any vector field, or distribution, in the tangent space to the level sets of $V(x)$. Finally, the third property establishes that there is a vector field in the distribution spanned by $G(x)$, which is projected to zero by the operator $M(x)$. This vector field is the vector obtained by mapping the m-dimensional vector $L_G^T V(x)$ through the input matrix $G(x)$. The following proposition depicts further properties of the projection matrix $M(x)$

Proposition 2. Let $f(x)$ be a smooth vector field, then the vector $M(x)f(x)$ can be written as

$$M(x)f(x) = \tilde{\mathcal{J}}(x)\frac{\partial V}{\partial x}$$

where $\tilde{\mathcal{J}}(x)$ is a skew-symmetric matrix, i.e. $\tilde{\mathcal{J}}(x) + \tilde{\mathcal{J}}^T(x) = 0$.

On the other hand the vector field $(I - M(x))f(x)$ can be written as

$$(I - M(x))f(x) = -\frac{1}{2}\tilde{\mathcal{J}}(x)\frac{\partial V}{\partial x} + \mathcal{S}(x)\frac{\partial V}{\partial x}$$

where $\mathcal{S}(x)$ is a symmetric matrix, i.e., $\mathcal{S}(x) = \mathcal{S}^T(x)$

Proof.

The first part of the proposition is straightforward from algebraic manipulations. For the proof of the second part of the proposition note that,

$$\begin{aligned}(I - M(x))f(x) &= \frac{1}{||L_G V||^2}G(x)L_G^T V(x)\frac{\partial V}{\partial x^T}f(x) \\ &= \frac{1}{||L_G V||^2}\left[G(x)L_G^T V(x)f^T(x)\right]\frac{\partial V}{\partial x} \qquad (2.5)\end{aligned}$$

The result follows from the fact that *any* square matrix, in particular,

$$N(x) = \frac{1}{||L_G V||^2}\left[G(x)L_G^T V(x)f^T(x)\right]$$

can always be written as the sum of a symmetric matrix and a skew symmetric matrix, i.e.,

$$\begin{aligned}N(x) &= 1/2(N(x) - N^T(x)) \\ &\quad +1/2(N(x) + N^T(x))\end{aligned}$$

The first summand, which is written as,

$$\begin{aligned}\frac{1}{2}(N(x) - N^T(x)) &= \frac{1}{2||L_G V||^2}\left[G(x)L_G^T V(x)f^T(x)\right. \\ &\quad \left. - f(x)L_G V(x)G^T(x)\right] \\ &= -\frac{1}{2}\tilde{\mathcal{J}}(x)\end{aligned}$$

is clearly skew-symmetric, while the second summand $(1/2)(N(x) + N^T(x))$ is symmetric. For the purposes of further reference we define the matrix $\mathcal{S}(x)$ as follows

$$\begin{aligned}\mathcal{S}(x) &= \frac{1}{2}\left[N(x) + N^T(x)\right] \\ &= \frac{1}{2||L_G V||^2}\left[G(x)L_G^T V(x)f^T(x)\right. \\ &\quad \left. + f(x)L_G V(x)G^T(x)\right]\end{aligned}$$

and the matrix $\mathcal{J}(x)$ as

$$\mathcal{J}(x) = \frac{1}{2}\tilde{\mathcal{J}}(x)$$

2.2 *Vector field decompositions through projection operators*

As a consequence of the results and definitions of the previous section, we have the following proposition,

Proposition 3. An n-dimensional smooth vector field $f(x)$ can be trivially decomposed in the following sum,

$$\begin{aligned}f(x) &= M(x)f(x) + (I - M(x))f(x) \\ &= \mathcal{J}(x)\frac{\partial V}{\partial x} + \mathcal{S}(x)\frac{\partial V}{\partial x}\end{aligned}$$

where $\mathcal{J}(x)$ is skew-symmetric and $\mathcal{S}(x)$ is symmetric.

Proof. It is obtained from algebraic manipulations.

The following lemma is a well known decomposition result for symmetric matrices,

Lemma 4. Let $\mathcal{S}(x)$ be a symmetric matrix, then $\mathcal{S}(x)$ can be decomposed as the sum of a positive semi-definite matrix $\mathcal{S}_p(x)$ and a negative semi-definite matrix $\mathcal{S}_n(x)$. If the above decomposition is not possible, then either $\mathcal{S}(x)$ is positive definite or, at least, positive semi-definite or, else, it is negative definite or, at least, negative semi-definite.

2.3 *A Hamiltonian-like canonical form for nonlinear systems*

As a corollary to the above results, a nonlinear system of the form (2.1), with a positive definite storage function $V(x)$, satisfying the transversality condition $L_G V(x) \neq 0$, can always be rewritten as

$$\begin{aligned}\dot{x} &= \mathcal{J}(x)\frac{\partial V}{\partial x} + \mathcal{S}_p(x)\frac{\partial V}{\partial x} + \mathcal{S}_n(x)\frac{\partial V}{\partial x} \\ &\quad + G(x)u \qquad (2.6)\end{aligned}$$

with $\mathcal{J}(x)$ being skew-symmetric, and $\mathcal{S}_p(x)$ being positive semi-definite and $\mathcal{S}_n(x)$ negative semi-definite. However, if $\mathcal{S}_p(x)$ is positive definite, then $\mathcal{S}_n(x)$ is zero and conversely if $\mathcal{S}_n(x)$ is negative definite then $\mathcal{S}_p(x)$ is zero.

2.4 *Feedback passivity of systems in canonical form*

Consider a nonlinear multivariable system, with a positive definite storage function $V(x)$, such that the transversality condition (2.2) is satisfied. Then, the system was shown to be expressible in the following canonical form,

$$\begin{aligned} \dot{x} &= \mathcal{J}(x)\frac{\partial V}{\partial x} + \mathcal{S}_n(x)\frac{\partial V}{\partial x} + \mathcal{S}_p(x) + G(x)u \\ y &= h(x) \end{aligned} \tag{2.7}$$

Along the solutions of the system (2.7), the time derivative of the storage function $V(x)$ is evaluated as follows,

$$\begin{aligned} \dot{V} &= \frac{\partial V}{\partial x^T}\mathcal{J}(x)\frac{\partial V}{\partial x} + \frac{\partial V}{\partial x^T}\mathcal{S}_p(x)\frac{\partial V}{\partial x} \\ &\quad + \frac{\partial V}{\partial x^T}\mathcal{S}_n(x)\frac{\partial V}{\partial x} + L_G V(x)u \\ &= \frac{\partial V}{\partial x^T}\mathcal{S}_p(x)\frac{\partial V}{\partial x} + \frac{\partial V}{\partial x^T}\mathcal{S}_n(x)\frac{\partial V}{\partial x} + L_G V(x)u \\ &\leq \frac{\partial V}{\partial x^T}\mathcal{S}_p(x)\frac{\partial V}{\partial x} + L_G V(x)u \end{aligned} \tag{2.8}$$

Consider then, the following input coordinate transformation, with v denoting a new external independent control input,

$$u = v - \frac{L_G^T V(x)}{||L_G V||^2}\left[\frac{\partial V}{\partial x^T}\mathcal{S}_p(x)\frac{\partial V}{\partial x}\right] \tag{2.9}$$

It is clear that the transformed system is given by

$$\begin{aligned} \dot{x} &= \mathcal{J}(x)\frac{\partial V}{\partial x} + \mathcal{S}_n(x)\frac{\partial V}{\partial x} \\ &\quad + \left(I - \frac{1}{||L_G V||^2}G(x)L_G^T V(x)\frac{\partial V}{\partial x^T}\right)\mathcal{S}_p(x)\frac{\partial V}{\partial x} \\ &\quad + G(x)v \\ y &= h(x) \end{aligned} \tag{2.10}$$

Notice that, as shown in the previous section, that the projected vector field given by

$$\left(I - \frac{1}{||L_G V||^2}G(x)L_G^T V(x)\frac{\partial V}{\partial x^T}\right)\mathcal{S}_p(x)\frac{\partial V}{\partial x}$$

can also be rewritten as,

$$\begin{aligned} &\frac{1}{||L_G V||^2}\left[\mathcal{S}_p(x)\frac{\partial V}{\partial x}L_G V(x)G^T(x)\right. \\ &\quad \left. - G(x)L_G^T V(x)\frac{\partial V}{\partial x^T}\mathcal{S}_p(x)\right]\frac{\partial V}{\partial x} = \mathcal{K}(x)\frac{\partial V}{\partial x} \end{aligned}$$

with $\mathcal{K}(x)$ skew-symmetric. In other words, the transformed system (2.10) is of the form,

$$\begin{aligned} \dot{x} &= \mathcal{W}(x)\frac{\partial V}{\partial x} + \mathcal{S}_n(x)\frac{\partial V}{\partial x} + G(x)v \\ y &= h(x) \end{aligned} \tag{2.11}$$

with $\mathcal{W}(x) = \mathcal{J}(x) + \mathcal{K}(x)$ being a skew symmetric matrix.

The time derivative of $V(x)$ along the controlled solutions of the transformed system, (2.11), now read as follows,

$$\begin{aligned} \dot{V}(x) &= \frac{\partial V}{\partial x^T}\mathcal{W}(x)\frac{\partial V}{\partial x} + \frac{\partial V}{\partial x^T}\mathcal{S}_n(x)\frac{\partial V}{\partial x} + L_G V(x)v \\ &\leq L_G V(x)v \end{aligned} \tag{2.12}$$

If the output map $h(x)$ satisfied the equality $L_G V(x) = h^T(x)$, the transformed system (2.11) would then be *passive* from the input vector v, towards the original output vector y. In fact, this last equality constitutes one of the well knwon Kalman-Yakubovich-Popov (KYP) conditions (Byrnes *et al*, 1991). In general, however, such an equality does not hold valid and an additional assumption must be enforced on $L_G V(x)$ in order to have the partially closed loop system passive.

The following fact is equivalent to the passivity of the transformed system (2.11) modulo a simple input coordinate transformation.

Fact

The row vector, $L_G V(x)$, can always be expressed as,

$$L_G V(x) = h^T(x)P(x) \quad \forall\, x \in \mathcal{X} \tag{2.13}$$

where $P(x)$ is an $m \times m$ matrix, which is invertible in all of $\mathcal{X}$.

It is clear that the matrix $P(x)$ always exists.

Proposition 5. Consider the control input transformation $w = P(x)v$, then, the transformed system,

$$\begin{aligned} \dot{x} &= \mathcal{W}(x)\frac{\partial V}{\partial x} + \mathcal{S}_n(x)\frac{\partial V}{\partial x} + G(x)P^{-1}(x)w \\ y &= h(x) \end{aligned} \tag{2.14}$$

is output passive whenever $\mathcal{S}_n(x)$ is negative semidefinite and it is strictly ouptut passive if $\mathcal{S}_n$(x) is strictly negative definite.

Proof

The time derivative of $V(x)$ along the solutions of the transformed system is given by

$$\dot{V}(x) = \frac{\partial V}{\partial x^T}\mathcal{W}(x)\frac{\partial V}{\partial x} + \frac{\partial V}{\partial x^T}\mathcal{S}_n(x)\frac{\partial V}{\partial x}$$
$$+L_G V(x)P^{-1}(x)w$$
$$= \frac{\partial V}{\partial x^T}\mathcal{S}_n(x)\frac{\partial V}{\partial x} + h^T(x)P(x)P^{-1}(x)w$$
$$\leq h^T(x)w = y^T w \qquad (2.15)$$

Evidently if $S_n(x)$ is negative definite, the last inequality above becomes a strict inequality. In such a case, a sufficiently small but positive constant γ can always be found such that

$$\dot{V}(x) = \frac{\partial V}{\partial x^T}\mathcal{S}_n(x)\frac{\partial V}{\partial x} + y^T w \leq y^T w - \gamma||y||^2$$

and the system is strictly output passive.

Notice that if we let $\tilde{G}(x)$ denote the transformed control input matrix $[G(x)P^{-1}(x)]$ then the system may, in fact, be written as

$$\dot{x} = \mathcal{W}(x)\frac{\partial V}{\partial x} + \mathcal{S}_n(x)\frac{\partial V}{\partial x} + \tilde{G}(x)w$$
$$y = h(x) = (P^{-1}(x))^T L_G^T V(x)$$
$$= (P^{-1}(x))^T G^T(x)\frac{\partial V}{\partial x} = \tilde{G}^T(x)\frac{\partial V}{\partial x} \qquad (2.16)$$

which, except for the damping term, $\mathcal{S}_n(x)\partial V/\partial x$, is in the same form as the *Generalized Hamiltonian systems*, widely studied in the literature (van der Schaft, 1996).

Summarizing, we have shown that, if the transversality condition (2.2) is satisfied, it is always possible to prescribe a suitable, state dependent, input coordinate transformation of the form

$$u = P^{-1}(x)w - \frac{L_G^T V(x)}{||L_G V||^2}\left[\frac{\partial V}{\partial x^T}\mathcal{S}_p(x)\frac{\partial V}{\partial x}\right] \qquad (2.17)$$

with $P(x)$ being an invertible $m \times m$ matrix satisfying (2.13), such that the given square multivariable nonlinear system (2.1) is expressible in the following *Passivity Based Control Canonical Form*

$$\dot{x} = \mathcal{W}(x)\frac{\partial V}{\partial x} + \mathcal{D}(x)\frac{\partial V}{\partial x} + \mathcal{B}(x)w$$
$$y = \mathcal{B}^T(x)\frac{\partial V}{\partial x} \qquad (2.18)$$

with $\mathcal{W}(x) + \mathcal{W}^T(x) = 0$, $D(x) = D^T(x) \leq 0$.

3. AN EXAMPLE

3.1 *An exothermic chemical reactor example*

Consider the following continuous multivariable reactor model (Viel *et al*, 1997), in which a first order and exothermic reaction $A \rightarrow B$ occurs.

$$\dot{x}_1 = -k(x_3)x_1 + \beta(u_1 - x_1)$$
$$\dot{x}_2 = k(x_3)x_1 - \beta x_2$$
$$\dot{x}_3 = \alpha k(x_3)x_1 - qx_3 + u_2$$
$$y = [x_1 \ ; \ x_3]^T \qquad (3.1)$$

where x_1 is the concentration of the reactant A and x_2 is the concentration of the product B. The variable x_3 represents the reactor temperature. $\beta > 0$ is a constant associated with the dilution rate while $\alpha > 0$ is the exothermicity of the reaction. The control input u_1 represents the concentration of the reactant A in the feed flow, denoted by x_1^{feed} when it is not considered as an input variable (Viel *et al*, 1997). The control input u_2 corresponds to a suitable, and well known, combination of the feed temperature T^{in} and the coolant temperature T_w. In fact, $u_2 = \beta T^{in} + \epsilon T_w$ where $\epsilon > 0$ being the heat transfer rate constant. The constant q is defined as $q = \beta + e > 0$. The function $k(x_3)$ is given by,

$$k(x_3) = k_0 \exp\left(-\frac{k_1}{x_3}\right) \qquad (3.2)$$

The operating region of the system is constituted by the positive orthant

$$\mathcal{X} = \{x \in R^3 \mid x_1 > 0 \ ; \ x_2 > 0 \ ; \ x_3 > 0\}$$

The constant equilibrium point of the system may be parameterized in terms of the equilibria for the variables x_1 and x_2, as follows

$$\overline{x}_3 = k_1 \frac{1}{\log(k_0) - \log(\left[\beta\frac{\overline{x}_2}{\overline{x}_1}\right])}$$
$$\overline{u}_1 = \overline{x}_1 + \overline{x}_2$$
$$\overline{u}_2 = q\overline{x}_3 - \alpha k(\overline{x}_3))$$

Consider the following positive definite *storage function* $V(x)$, given by

$$V(x) = \frac{1}{2}\left(x_1^2 + x_2^2 + x_3^2\right) \qquad (3.3)$$

In this case, $\partial V/\partial x = x$ and, both, the transversality condition and the output factorization condition are valid. These take the form

$$L_G V(x) = [x_1 \ ; \ x_3] = h^T(x) = y^T \neq 0 \qquad (3.4)$$

The time derivative of $V(x)$, along the controlled trajectories of the reactor system (3.1), is given by

$$\dot{V}(x) = -\left(k(x_3) + \beta\right)x_1^2 - \beta x_2^2 - qx_3^2$$
$$+x_1\left[\beta\, u_1 + k(x_3)x_2\right]$$
$$+x_3\left[u_2 + \alpha\, k(x_3)x_1\right] \qquad (3.5)$$

It is evident that the following input coordinate transformation makes the locally destabilizing drift vector field component tangent to the level sets of $V(x)$,

$$\begin{aligned} v_1 &= \beta u_1 + k(x_3)x_2 \\ v_2 &= u_2 + \alpha k(x_3)x_1 \end{aligned} \qquad (3.6)$$

The proposed input coordinate transformation also renders the system state variables x_1, x_3, as passive outputs without any further transformations. Indeed, letting $y^T = [x_1 \; x_3]$ and $v = [v_1 \; v_2]^T$, we have, from (3.5) and (3.6), that, for all x, the following *infinitesimal form* of the basic passivity inequality is satisfied,

$$\dot{V} \leq x_1 v_1 + x_3 v_2 = y^T v$$

In new input coordinates, (v_1, v_2), the system may be rewritten in matrix form as

$$\begin{bmatrix} \dot{x}_1 \\ \dot{x}_2 \\ \dot{x}_3 \end{bmatrix} = \begin{bmatrix} 0 & -k(x_3) & 0 \\ k(x_3) & 0 & 0 \\ 0 & 0 & 0 \end{bmatrix} \frac{\partial V}{\partial x}$$
$$+ \begin{bmatrix} -k(x_3) - \beta & 0 & 0 \\ 0 & -\beta & 0 \\ 0 & 0 & -q \end{bmatrix} \frac{\partial V}{\partial x} + \begin{bmatrix} 1 & 0 \\ 0 & 0 \\ 0 & 1 \end{bmatrix} \begin{bmatrix} v_1 \\ v_2 \end{bmatrix}$$
$$y = \begin{bmatrix} 1 & 0 \\ 0 & 0 \\ 0 & 1 \end{bmatrix}^T \frac{\partial V}{\partial x} \qquad (3.7)$$

4. CONCLUSIONS

In this article, we have proposed a general canonical form which is suitable for the systematic derivation of passivity based regulation schemes for nonlinear multivariable systems with nonzero equilibrium points.

The proposed canonical form entitles obtaining a natural drift vector field decomposition based on the use of a projection operator, entirely defined on the basis of the storage function of the system and of the control input matrix. The decomposition clearly exhibits the well known *workless forces*, the *damping forces* and the locally *destabilizing forces* comprising the system dynamics.

5. REFERENCES

Byrnes, Ch. I., A. Isidori and J. C. Willems(1991), "Passivity, feedback equivalence and the global stabilization of minimum phase nonlinear systems" *IEEE Transactions on Automatic Control* **36**, pp. 1228-1240.

Hill, D. and P. Moylan (1976), "The stability of nonlinear dissipative systems" *IEEE Transactions on Automatic Control* **21**, pp. 708-711.

Hill, D. and P. Moylan (1977), "Stability results for nonlinear feedback systems" *Automatica* **13**, pp. 377-382.

Ortega, R., A. Loria, R. Kelly and L. Praly (1995), " On passivity based output feedback global stabilization of Euler-Lagrange systems" *International Journal of Robust and Nonolinear Control*, **5**, pp. 313-324.

Ortega, R., P.J. Nicklasson and G. Espinosa(1995), "On speed control of induction motors" in *Proceedings of the American Control Conference*,pp. 3521-3525 Seattle, Washington.

Ortega, R. and M. Spong (1989), "Adaptive motion control of rigid robots: a tutorial" *Automatica*, **25**, pp. 877-888.

Shishkin, S., R. Ortega, D. Hill and A. Loria (1996), "On output feedback stabilization of Euler-Lagrange systems with nondissipative forces", *Systems & Control Letters*, **27**, pp. 315-324.

Sira-Ramírez, H. (1988), "Differential geometric methods in variable sturcture control" *International Journal of Control* **48**, pp 1359-1390.

Sira-Ramírez, H. and M. I. Angulo-Núñez (1997), "Passivity based control of nonlinear chemical processes" *International Journal of Control*, **68**, pp. 971-996.

Sira-Ramírez, H. and M. Delgado " Passivity Based Mechatronic Control of Nonlinear Continuous Processess" in *Advances in Control* C. T. Leondes (Editor), (to appear).

Sira-Ramírez, H., R. Ortega, R. A. Pérez-Moreno and M. García- Esteban (1997), "Passivity based controllers for the stabilization of dc-to-dc power converters," *Automatica*, **33**, pp. 499-513.

Takegaki, M. and S. Arimoto (1981), "A new feedback method for dynamic control of manipulators" *ASME J. Dynamic Systems, Measurement and Control*, **102**, pp. 119-125.

van der Schaft, A. (1996) *L_2 and Passivity Techniques in Nonlinear Control*, Lecture Notes in Control and Information Systems, **218**, Springer-Verlag, London.

Viel, F., F. Jadot and G. Bastin (1997), "Global Stabilization of Exothermic Chemical Reactors under Input Constraints", *Automatica*, **33**, No. 8, pp. 1437–1448.

Willems, J. C. (1972), "Dissipative dynamical systems, part I and II" *Archives of Rational Mechanics and Analysis*, **45**, pp. 321-393.

14th Triennial World Congress, Beijing, P.R. China

C-2a-03-2

FEEDBACK STABILISATION WITH POSITIVE CONTROL OF A CLASS OF DISSIPATIVE MASS-BALANCE SYSTEMS

Georges Bastin* and **Laurent Praly**◇

* *Centre for Systems Engineering and Applied Mechanics (CESAME) - UCL*
Av. G. Lemaitre 4, B-1348 Louvain-La-Neuve, Belgium
Fax : +3210472380 ; e-mail : bastin@auto.ucl.ac.be
◇*Ecole des Mines de Paris (CAS)*
34, Rue Saint Honoré
77305 Fontainebleau Cedex, France

Abstract. In many process control applications, the system under consideration is *positive.* This means that both the state variables and the control input are physically constrained to remain non-negative along the system trajectories. For such systems, the design of state feedback controllers makes sense only if the control function is guaranteed to provide a non-negative value at each time instant. The purpose of this paper is to present a positive control law for the feedback stabilisation of a class of positive mass-balance systems which are dissipative but can nevertheless be globally unstable. The approach is illustrated with an application to the control of an industrial grinding circuit. *Copyright ©1999IFAC*

Keywords : positive control systems, stabilisation, dissipative, mass balance.

1. INTRODUCTION

In many practical applications of control engineering, the dynamical system under consideration is *positive.* This means that both the state variables and the control input are physically constrained to remain non-negative along the system trajectories as stated in the following definition :

Definition. Positive System. A control system $\dot{x} = f(x,u) \;\; x \in \mathbb{R}^n, \;\; u \in \mathbb{R}$ is positive if

$$\left.\begin{array}{l} x(0) \in \mathbb{R}^n_+ \\ u(t) \in \mathbb{R}_+ \;\; \forall t \geq 0 \end{array}\right\} \Rightarrow x(t) \in \mathbb{R}^n_+ \;\; \forall t \geq 0.$$

(**Notation.** The set of non-negative real numbers is denoted as usual $\mathbb{R}_+ = \{a \in \mathbb{R}, a \geq 0\}$. For any integer n, the set $\mathbb{R}^n_+$ is called the "non-negative orthant". Similarly the set of positive real numbers is denoted $\mathbb{P} = \{a \in \mathbb{R}, a > 0\}$ and $\mathbb{P}^n$ is called the "positive orthant".)
For such systems, it is an evidence that the design of state feedback controllers makes sense only if the control function is guaranteed to provide a non-negative value at each time instant.

The purpose of the present paper is to design a positive control law for the feedback stabilisation of a class of positive mass-balance systems which are described in Section 2. These systems are dissipative but can nevertheless be globally unstable. In Section 3, a positive control law is proposed in order to achieve global output stabilization with state boundedness in the positive orthant. The controlled output has a clear physical meaning : it is the total mass contained in the system. The approach is illustrated with an application to a compartmental model of an industrial grinding circuit in Section 4. Some final comments are given in Section 5.

2. POSITIVE MASS BALANCE SYSTEMS

We consider a class of single-input mass balance dynamical systems described by a state equation of the form :

$$\dot{x} = f(x) - Ax + bu \tag{1}$$

with state $x \in \mathbb{R}^n_+$ and control input $u \in \mathbb{R}_+$, under the following conditions:

C1. The function

$$f(x) = (f_1(x), f_2(x), \ldots, f_n(x))^T : \mathbb{R}^n_+ \to \mathbb{R}^n$$

is continuous

C2. $f(0) = 0$

C3. $f_i(x) \geq 0 \;\; \forall x \in \mathbb{R}^n_+$ with $x_i = 0$

C4. $\sum_{i=1}^n f_i(x) = 0 \;\; \forall x \in \mathbb{R}^n_+$

C5. The matrix A is diagonal

$$A = \text{ diag}(a_1, a_2, \ldots, a_n)$$

with $a_i \geq 0 \;\; \forall i$ and $a_i > 0$ for at least one i.

C6. The vector $b = (b_1, b_2, \ldots, b_n)^T$ has non negative entries and at least one of them is positive : $b_i \geq 0 \;\; \forall i$ and $b_i > 0$ for at least one i.

C7. The unforced system $\dot{x} = f(x) - Ax$ with output $y = \sum_{i=1}^n a_i x_i$ is zero state detectable in the sense of [5] (Definition 2.27).

System (1) with conditions (C1 - C7) is representative of a wide class of mass balance systems of interest in industrial and environmental applications. Typical examples are compartmental systems (see e.g. [2]), chemical or biological stirred tank reactors (see e.g. [1]), or Lotka-Volterra ecological systems (see e.g. [3]).

In these systems :

1. The state variables x_i represent the amount of mass of various species involved in the system. The state vector x is often called the *composition* because it represents the distribution of mass among the various species.

2. The functions $f_i(x)$ represent various mass transformation effects like :

   - exchange of material between compartments in compartmental systems
   - kinetic transformations in chemical and biological reactors
   - interactions (predation, competition for food, ...) between biological species in ecological systems.

3. The terms $a_i x_i$ represent an outflow of material leaving the system (withdrawal, excretion, mortality of living organisms, etc ...)

4. The terms $b_i u$ represent an inflow of material injected into the system from the outside (like feeding of reactants or nutrients for instance).

The total mass contained in the system is

$$M(x) = \sum_{i=1}^n x_i$$

In all cases, Condition C4 expresses that the system is *mass conservative* in the sense that the involved transformations preserve the mass balance inside the system. This is easily seen if we consider the special case of system (1) without inflows ($u = 0$) and without outflows ($a_i = 0, \forall i$). Then the system reduces to $\dot{x} = f(x)$ and $dM(x)/dt = 0$ which shows that the total mass is indeed conserved.

Under conditions C1-C7, the mass balance system (1) has the following properties.

Properties

1. The system is *positive.* Indeed, if $x_i = 0$, then $\dot{x}_i = f_i(x) + b_i u \geq 0$ from conditions C3 and C6.

2. If $u = 0$ (no inflow), the unforced system $\dot{x} = f(x) - Ax$ is *dissipative* in the sense that the total mass $M(x)$ decreases along the system trajectories, because from Condition C4 we have $dM(x)/dt = -\sum_{i=1}^n a_i x_i \leq 0$.

3. If $u = 0$ (no inflow), the origin $x = 0$ is a globally asymptotically stable equilibrium of the unforced system $\dot{x} = f(x) - Ax$. This follows readily from Condition C7 and Lasalle's Theorem.

3. CONTROL DESIGN FOR GLOBAL STABILISATION

Although the system is dissipative when the control input u is zero (no inflow), it can nevertheless be *globally unstable* when there is a non zero inflow $u(t) > 0$ which is the normal mode of operation in practical applications. The symptom of this instability is an unbounded accumulation of mass inside the system. An example will be given in the application section of the paper. This obviously makes the problem of feedback stabilisation of mass balance systems in the positive orthant relevant and sensible. One way of approaching the problem is to consider that the control objective is to globally stabilize the total mass $M(x)$ at a given positive set point $M^* > 0$ in order to prevent the unbounded mass accumulation. This control objective may be achieved with the following *positive* control law :

$$u(x) = \max(0, \tilde{u}(x)) \tag{2}$$

$$\tilde{u}(x) = \left(\sum_{i=1}^n b_i\right)^{-1} \left(\sum_{i=1}^n a_i x_i + \lambda(M^* - M(x))\right) \tag{3}$$

where λ is an arbitrary positive design parameter. The set $\Omega = \{x : M(x) = M^* \text{ and } x \in \mathbb{R}_+^n\}$ (which is a portion of hyperplane) is called an "iso-mass", because it is the set of all compositions x corresponding to the same total mass M^*.

The control law (2) is a saturated input-output feedback linearisation with the total mass $y = M(x)$ as regulated output. The stabilizing properties of this control law are given in the following theorem.

Theorem. For the closed loop system (1)-(2) with arbitrary initial conditions in the non-negative orthant $x(0) \in \mathbb{R}_+^n$:

(i) the iso-mass Ω is positively invariant

(ii) the state $x(t)$ is bounded for all $t > 0$ and converges to the iso-mass Ω.

Proof. Along the closed loop trajectories, we have :

$$\frac{dM(x)}{dt} = -\sum_{i=1}^{n} a_i x_i + \left(\sum_{i=1}^{n} b_i\right) u(x)$$

(i) if $x \in \Omega$, then $M(x) = M^*$ and

$$u(x) = \tilde{u}(x) = \left(\sum_{i=1}^{n} b_i\right)^{-1} \left(\sum_{i=1}^{n} a_i x_i\right)$$

hence $\dot{M}(x) = 0$ which proves that Ω is positively invariant.

(ii) if $x \neq \Omega$, then $u(x) = \begin{cases} 0 & \text{if } \tilde{u}(x) < 0 \\ \tilde{u}(x) & \text{if } \tilde{u}(x) \geq 0 \end{cases}$

Suppose that $\tilde{u}(x) < 0$ and $u(x) = 0$, then necessarily $M(x) > M^*$.

Consider the Lyapunov function candidate $V = \frac{1}{2}(M^* - M(x))^2$. We have :

$$\begin{aligned} \dot{V} &= -(M^* - M(x))\frac{dM(x)}{dt} \\ &= (M^* - M(x))\left(\sum_{i=1}^{n} a_i x_i\right) \leq 0 \end{aligned}$$

Suppose that $u(x) = \tilde{u}(x) \geq 0$, then

$$\dot{V} = -\lambda(M^* - M(x))^2 \leq 0$$

If $\dot{V} = 0$ then either $x \in \Omega$ which is a positively invariant set of the closed loop (see above)

$$\text{or } x \in \{x : \sum_{i=1}^{n} a_i x_i = 0 \text{ and } M(x) > M^*\}$$

which does not contain any invariant set from condition C7. The result then follows from Lasalle's theorem. ■

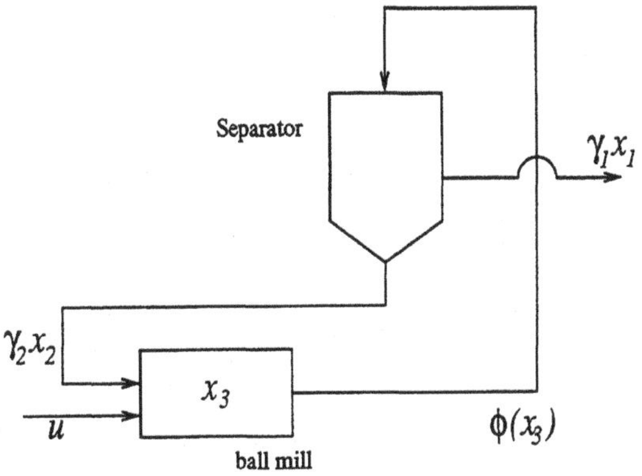

Figure 1: An industrial ball mill

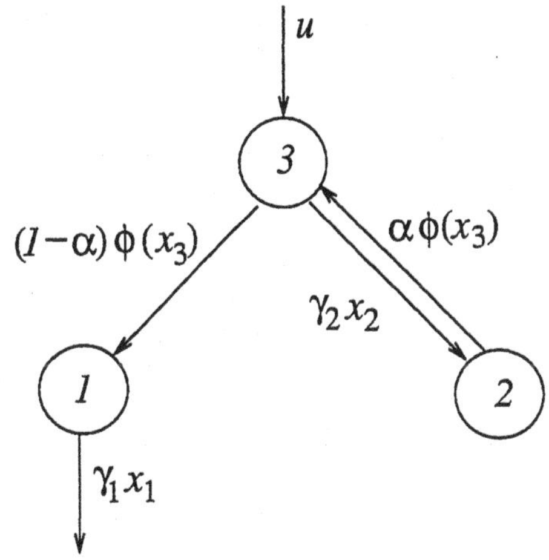

Figure 2: Compartmental structure of the system

4. APPLICATION TO A COMPARTMENTAL SYSTEM

A schematic lay-out of an industrial grinding circuit used in cement industries is depicted in Fig.1.

It is made up of the interconnection of a ball mill and a separator as shown in the figure. The ball mill is fed with raw material. After grinding, the milled material is introduced in a separator where the finished product is separated from the oversize particles which are recycled to the ball mill. A simple dynamical model has been proposed (see [4]) for this system under the form of a compartmental system with three compartments as represented in Fig.2.

The corresponding state space model is as follows :

$$\begin{aligned} \dot{x}_1 &= -\gamma_1 x_1 + (1-\alpha)\phi(x_3) \\ \dot{x}_2 &= -\gamma_2 x_2 + \alpha\phi(x_3) \\ \dot{x}_3 &= \gamma_2 x_2 - \phi(x_3) + u \end{aligned}$$

with the following notations and definitions :

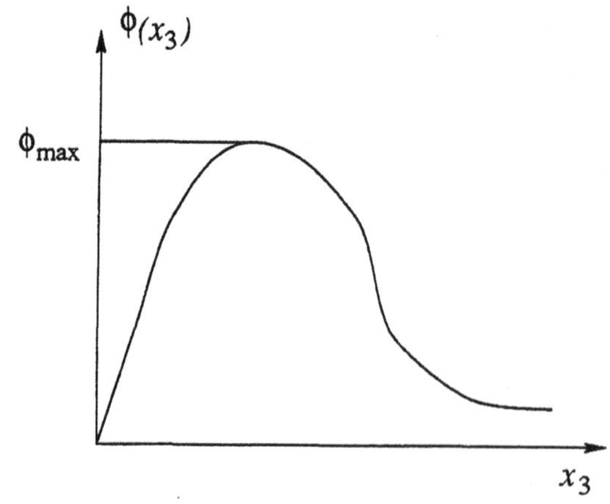

Figure 3: The grinding function

x_1	=	amount of finished product in the separator
x_2	=	amount of oversize particles in the separator
x_3	=	amount of material in the ball mill
u	=	feeding rate
$\gamma_1 x_1$	=	outflow rate of finished product
$\gamma_2 x_2$	=	flowrate of recycled product
$\phi(x_3)$	=	grinding function

The parameter α is the separation constant of the separator ($0 < \alpha < 1$). The grinding function $\phi(x_3)$ is non monotonic as represented in Fig. 3. This model is readily seen to be a special case of the general mass-balance model (1) with the following definitions :

$$\begin{array}{ll} f_1(x) = (1-\alpha)\phi(x_3) & a_1 = \gamma_1 \\ f_2(x) = -\gamma_2 x_2 + \alpha\phi(x_3) & a_2 = 0 \\ f_3(x) = \gamma_2 x_2 - \phi(x_3) & a_3 = 0 \end{array}$$

It is easy to check that Conditions C1 to C6 are satisfied. The following sequence of equalities for the unforced system (with $u = 0$) shows that condition C7 is also satisfied :

$$\begin{aligned} x_1 = 0 \quad &\Longrightarrow \quad \dot{x}_1 = 0 \Longrightarrow x_3 = 0 \\ &\Longrightarrow \quad \dot{x}_3 = 0 \Longrightarrow x_2 = 0 \end{aligned}$$

When the control input is constant $u = \bar{u}$ (constant) > 0, the global instability of the system appears if the state is initialised in the set D defined by the following inequalities (see Fig. 4) :

$$D \left\{ \begin{array}{c} (1-\alpha)\phi(x_3) < \gamma_1 x_1 < \bar{u} \\ \alpha\phi(x_3) < \gamma_2 x_2 \\ \partial\phi/\partial x_3 < 0 \end{array} \right.$$

Indeed, it can be shown that this set D is positively invariant and if $x(0) \in D$ then $x_1 \to 0$ $x_2 \to 0$ $x_3 \to \infty$. This means that there is an irreversible accumulation of material in the mill with a decrease of the production to zero. In the jargon of cement industries, this is called *mill plugging*. In practice, the state may be lead to the set D by intermittent disturbances like variations of hardness of the raw material.

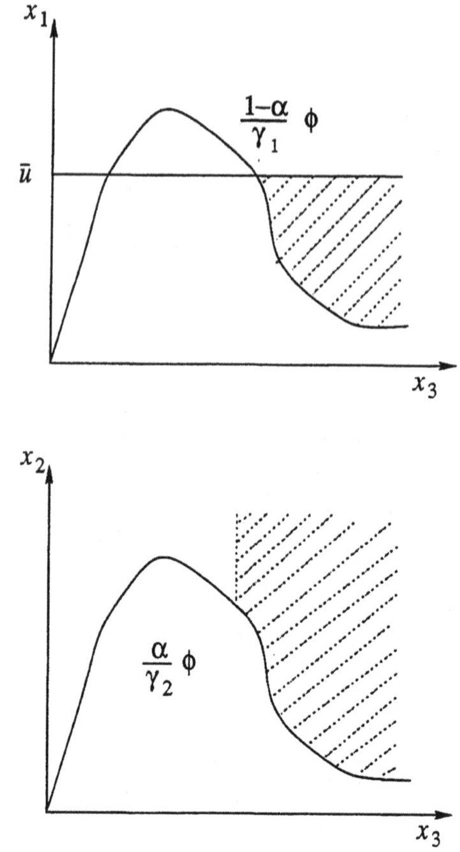

Figure 4: The stability invariant set D

The iso-mass is $M(x) = x_1 + x_2 + x_3$ and we have $\dot{M}(x) = -\gamma_1 x_1 + u$. The control law is written :

$$u(x) = \max(0, \tilde{u}(x))$$

with :

$$\begin{aligned} \tilde{u}(x) &= \gamma_1 x_1 + \lambda(M^* - M(x)) \\ &= \lambda M^* + (\gamma_1 - \lambda)x_1 - \lambda x_2 - \lambda x_3 \end{aligned}$$

It is interesting to analyse the behaviour of the system in the invariant set Ω which is in fact the behaviour of the zero dynamics (computed for the constant regulated output $y = M(x) = x_1 + x_2 + x_3 = M^*$) :

$$\left\{ \begin{array}{ll} \dot{x}_2 = & -\gamma_2 x_2 + \alpha\phi(x_3) \\ \dot{x}_3 = & \gamma_1(M^* - x_2 - x_3) + \gamma_2 x_2 - \phi(x_3) \end{array} \right.$$

The equilibria of the zero-dynamics must satisfy the following relation :

$$\underbrace{\left(\frac{(\gamma_1 - \gamma_2)}{\gamma_2}\alpha + 1\right)\phi(x_3) + \gamma_1 x_3}_{\psi(x_3)} = \gamma_1 M^*$$

The zero-dynamics have a unique equilibrium in Ω if the following inequality is satisfied :

$$\frac{\partial\psi}{\partial x_3} > 0 \Longrightarrow \frac{\partial\phi}{\partial x_3} > -\frac{\gamma_1\gamma_2}{\alpha\gamma_1 + (1-\alpha)\gamma_2}$$

If $\gamma_2 \geq \gamma_1$, this unique equilibrium is easily shown to be globally asymptotically stable in Ω by using the Bendixsson theorem (e.g. [6]).

In this case, it follows that the feedback controller is able not only to prevent the mill from plugging by regulating the total mass at an arbitrary set point, but also to stabilise the system at a unique equilibrium which is globally asymptotically stable in its domain of physical existence (the positive orthant).

5. FINAL COMMENTS

1. The controller (2) proposed in this paper has an interesting robustness property. Indeed it is fully independent from $f(x)$ which represents the internal transformations of the system. This means that the feedback stabilisation is robust against a full modelling uncertainty regarding $f(x)$ provided it satisfies conditions C1 to C4. This is quite important because in many practical applications, $f(x)$ is precisely the most uncertain term of the model.

2. The mass-conservative condition C4 is critical for our result. It is interesting to note that the positive models encountered in the literature are not always mass conservative, even when they describe mass transformations in chemical or biological systems. An archetype is the classical Lotka prey-predator model :

$$\begin{aligned} \dot{x}_1 &= -x_1x_2 + k_1x_1 \\ \dot{x}_2 &= x_1x_2 - k_2x_2 \end{aligned}$$

   If Condition C5 is satisfied, this model cannot verify Condition C4 and is therefore not mass conservative. The underlying reason is that the term k_1x_1 represents a creation of preys "ex-nihilo", under the assumption that the resources for the development of preys are unlimited.

3. Condition C7 which guarantees the dissipativity of the unforced system is also critical for our result. It implies that, in absence of feeding ($u = 0$), there is a natural "wash-out" of the material contained in the system. Besides the fact that it is a common property in many practical applications, it must be emphasized that, without natural dissipativity, there is no hope to globally stabilise the total mass $M(x)$ at an *arbitrary* set point.

References

[1] Bastin G. and J. Lévine (1993), "On state accessibility in reaction systems", IEEE Transactions in Automatic Control, Vol. 38(5), pp. 733 - 742.

[2] Godfrey K. (1983), "Compartmental models and their applications", Academic Press.

[3] Luenberger D. (1979), "Introduction to dynamic systems", Wiley.

[4] Magni L., G. Bastin and V. Wertz (1999), "Multivariable nonlinear predictive control of cement mills", to appear in IEEE Transaction on Control Technology.

[5] Sepulchre R., M. Jankovic, P. Kokotovic (1997), "Constructive nonlinear control", Springer Verlag.

[6] Vidyasagar M. (1993), "Nonlinear systems analysis", Prentice Hall.

14th Triennial World Congress, Beijing, P.R. China

C-2a-03-3

FINITE SPECTRUM ASSIGNMENT FOR NONLINEAR SYSTEMS WITH NON-COMMENSURATE DELAYS

Toshiki Oguchi * **Atsushi Watanabe** *
Takayoshi Nakamizo **

* *Dept. of Precision Engineering , Tokyo Metropolitan University*
1-1, Minami-Ohsawa, Hachioji-shi, Tokyo 192-0397 JAPAN
oguchi@control.prec.metro-u.ac.jp
** *Dept. of Mechanical Engineering, Ashikaga Inst. of Technology*
268-1, Omaechou, Ashikaga-shi, Tochigi 326-8558 JAPAN

Abstract: In this paper single input nonlinear systems with non-commensurate time-delays in state variables are considered. The objective is to design a coordinates transformation and a feedback which yield a finite dimensional linear controllable closed-loop system in a new coordinates, whose spectrum is located at an arbitrarily preassigned set of n points in the complex plane. An algebraic design procedure for this problem is proposed here. The proposed method is related to the exact linearization technique for systems with a finite dimension, and recognized as the extension of the finite spectrum assignment of linear retarded systems.

Keywords: Nonlinear systems, Time-delay, Delay compensation, Stabilization, Differential geometric methods, Pole assignment

1. INTRODUCTION

The differential geometric linearization methods are recognized as useful tools for designing nonlinear control systems (Isidori, 1995; Nijmeijer and van der Schaft, 1990). While this approach is essentially not applicable to systems with an infinite number of dimensions, there have been several attempts (Kravaris and Wright, 1989; Huang and Wang, 1992; Oguchi *et al.*, 1994*a*; Henson and Seborg, 1994; Oguchi *et al.*, 1994*b*; Oguchi, 1995; Oguchi and Watanabe, 1998) to extend the applicability of the linearization methods to time delay systems. They are based on state prediction or decoupling of delayed state variables. Although the former deals with systems with time delays at inputs and the latter is applied to retarded systems which have time delays coupled with state variables, their applicability is limited to narrow class of systems. On the other hand, the finite spectrum assignment (Morse, 1976; Sontag, 1976; Lee and Żak, 1982; Manitius and Olbrot, 1979) is a useful controller design method for retarded linear systems. Since this method uses a linear coordinates transformation, it can not be directly applied to nonlinear systems.

This paper is to attempt to apply the finite spectrum assignment method to nonlinear systems with non-commensurate time-delay in the state variables by combining the method with the differential geometric linearization. We introduce extensions of the Lie derivative and the Lie bracket to nonlinear differential-difference control systems for this purpose. By use of these extensions we derive a nonlinear coordinates transformation with delayed-state variables and a state feedback law with delayed-state variables which transform nonlinear retarded systems into a finite dimensional linear systems. A sufficient condition for solvability of this problem is also considered. Furthermore we also show that the coordinates transformation is invertible under the obtained condition. Though the idea of finite spectrum assignment for nonlinear systems with commensurate time-delay in state variables can be found in a paper (Oguchi *et al.*, 1997), the systems treated in this paper are of more general form than that of the paper (Oguchi *et al.*, 1997)

2. PRELIMINARY

Throughout this paper, we consider a single input nonlinear system,

$$\begin{aligned} \dot{x} &= f(x(t), x(t-\ell_{f_1}), \ldots, x(t-\ell_{f_k})) \\ &\quad + g(x(t))u(t), \qquad (1) \\ x(t) &= \psi(t), t \in [-\ell_{f_k}, 0], \quad x(0) = \psi(0) = x_0, \end{aligned}$$

where $\ell_{f_i}, 1 \le i \le k$ are fixed time delays such as $\ell_{f_1} \le \ell_{f_2} \le \cdots \le \ell_{f_k}$, the $n \times 1$ vector $x(t)$ is the state at time t, and the scalar u is the control and $\psi : \Re^+ \to \Re^n$ is a continuous vector valued function specifying the initial state of the system. It is assumed that $f(\cdot)$ and $g(x)$ are real smooth vector fields with $f(x_0, x_0, \cdots, x_0) = 0$. Thus system (1) is a dynamical model with any length of time delays in the state.

We begin with some mathematical definitions for the nonlinear systems with time-delays in the state described by equation (1).

First, we introduce extensions of the Lie derivative and the Lie bracket into nonlinear difference-differential equations. For the case of commensurate time-delay, the delayed state derivative has been already proposed (Oguchi, 1995). By a simple extension, we define a delayed state derivative with non-commensurate time-delays.

Definition 1. **(delayed state derivative)** *Let* $\bar{f}(x,\sigma) \equiv f(x(t), x(t-\ell_{f_1}), \ldots, x(t-\ell_{f_k}))$ *be an* n*-vector whose components are functions of* $x(t)$ *and* $x(t-\ell_{f_i})$, $1 \le i \le k$, $\bar{\phi}(x,\sigma) \equiv \phi(x(t), x(t-\ell_{\phi_1}), \ldots, x(t-\ell_{\phi_m}))$, $1 \le i \le m$, *a real-valued function of* $x(t)$ *and* $x(t-\ell_{\phi_i})$, $1 \le i \le m$, *and* ℓ_{f_i} *and* ℓ_{ϕ_i} *positive time delays. Then the derivative of* $\bar{\phi}(x,\sigma)$ *along* $\bar{f}(x,\sigma)$ *is defined by*

$$L_{\bar{f}}\bar{\phi}(x,\sigma) \stackrel{\text{def}}{=} \frac{\partial \bar{\phi}}{\partial x(t)}\bar{f} + \sum_{i=1}^{m} \frac{\partial \bar{\phi}}{\partial \sigma_{\phi_i} x}\sigma_{\phi_i}\bar{f}(x,\sigma),$$

where $\sigma_{f_i} : x(t) \to x(t-\ell_{f_i})$ *and* $\sigma_{\phi_i} : x(t) \to x(t-\ell_{\phi_i})$ *be the pure delay operators. We call this operator the "delayed-state derivative".*

Since $L_{\bar{f}}\bar{\phi}(x,\sigma)$ is a real-valued function with time delays, repeated delayed-state derivatives can be defined. For $\gamma \ge 2$, the general form of higher order derivatives can be denoted by

$$L_{\bar{f}}^{\gamma+1}\bar{\phi}(x,\sigma) = L_{\bar{f}}L_{\bar{f}}^{\gamma}\bar{\phi}(x,\sigma).$$

For the case of all $\ell_* = 0$, since $x(t)$ is identified with a point x of $\Re^n$, this operator coincides with the conventional Lie derivative.

Next, we define an extension of the Lie bracket similarly.

Definition 2. **(delayed state bracket)** *Let* $\bar{f}(x,\sigma) \equiv f(x(t), x(t-\ell_{f_1}), \ldots, x(t-\ell_{f_m}))$, $\bar{g}(x,\sigma) \equiv g(x(t), x(t-\ell_{g_1}), \ldots, x(t-\ell_{g_p}))$ *be vectors whose components are functions of* $x(t)$ *and* $x(t-\ell_*)$. *Then the delayed-state bracket of* $\bar{f}(x,\sigma)$ *and* $g(x,\sigma)$ *is defined by*

$$[\bar{f},\bar{g}](x,\sigma) \stackrel{\text{def}}{=} \sum_{i=0}^{p} \frac{\partial \bar{g}}{\partial \sigma_{g_i} x}(\sigma_{g_i}\bar{f}) - \sum_{j=0}^{m} \frac{\partial \bar{f}}{\partial \sigma_{f_j} x}(\sigma_{f_j}\bar{g}),$$

where $\sigma_{f_0} : x(t) \to x(t)$ *and* $\sigma_{g_0} : x(t) \to x(t)$. $\sigma_{g_i} : x(t) \to x(t-\ell_{g_i})$, $1 \le i \le k$ *are the pure delay operators.*

Repeated bracketing of $\bar{g}(x,\sigma)$ *with* $\bar{f}(x,\sigma)$ *can be written down as follows.*

$$\begin{aligned} [\bar{f},[\bar{f},\bar{g}]](x,\sigma) &= \sum_{i=0}^{p}\sum_{j=0}^{m} \frac{\partial[\bar{f},\bar{g}]}{\partial \sigma_{g_i}\sigma_{f_j} x}(\sigma_{g_i}\sigma_{f_j}\bar{f}) \\ &\quad - \sum_{j=0}^{m} \frac{\partial \bar{f}}{\partial \sigma_{f_j} x}(\sigma_{f_j}[\bar{f},\bar{g}]) \end{aligned}$$

Then we denote such operations recursively by

$$\begin{aligned} &ad_{\bar{f}}^{0}\bar{g}(x,\sigma) \stackrel{\text{def}}{=} \bar{g}(x,\sigma), \\ &ad_{\bar{f}}^{i}\bar{g}(x,\sigma) \stackrel{\text{def}}{=} [\bar{f}, ad_{\bar{f}}^{i-1}\bar{g}(x,\sigma)] \quad \text{for } i \ge 1. \quad (2) \end{aligned}$$

If $\ell = 0$, this operator coincides with the conventional Lie bracket.

The delayed state derivative and the delayed state bracket satisfy the following relation.

$$L_{ad_{\bar{f}}\bar{g}}\bar{\phi}(x,\sigma) = L_{\bar{f}}L_{\bar{g}}\bar{\phi}(x,\sigma) - L_{\bar{g}}L_{\bar{f}}\bar{\phi}(x,\sigma), \qquad (3)$$

where $\bar{\phi}$ is a real-valued function defined by def 1. This equation corresponds to Leibniz's formula for the Lie bracket and the Lie derivative:

$$L_{ad_f g}\varphi(x) = L_f L_g \varphi(x) - L_g L_f \varphi(x). \qquad (4)$$

Using relation (3), we obtain the following theorem.

Theorem 3. Given $\bar{f}, \bar{g}$ *and* $\bar{\phi}$ *defined above, the relations*

$$\begin{cases} L_{\bar{g}} L_{\bar{f}}^{i} \bar{\phi} = 0, & 0 \le i \le \rho - 1 \\ L_{\bar{g}} L_{\bar{f}}^{\rho} \bar{\phi} \ne 0 \end{cases} \tag{5}$$

are equivalent to

$$\begin{cases} L_{ad_{\bar{f}}^{i} \bar{g}} \bar{\phi} = 0, & 0 \le i \le \rho - 1, \\ L_{ad_{\bar{f}}^{\rho} \bar{g}} \bar{\phi} = (-1)^{\rho} L_{\bar{g}} L_{\bar{f}}^{\rho} \bar{\phi} \ne 0. \end{cases} \tag{6}$$

The relation (3) and the above theorem are proved by a same procedure as the proof of a similar results for the Lie bracket and the Lie derivative (Isidori, 1995).

3. MAIN RESULTS

3.1 *Finite Spectrum Assignment for Retarded Nonlinear Systems*

By introducing the pure delay operator σ_{f_i} : $x(t) \to x(t - \ell_{f_i})$, $i = 1, \ldots, k$, we rewrite system (1) as

$$\begin{aligned} \dot{x}(t) &= f(x(t), \sigma_{f_1} x(t), \ldots, \sigma_{f_k} x(t)) + g(x(t))u(t) \\ &= \bar{f}(x, \sigma) + g(x)u(t). \end{aligned} \tag{7}$$

The main problem here is designing a feedback and a coordinates transformation that transform the retarded nonlinear system to a finite-dimensional linear controllable system. It can be formally stated by the following definition.

Definition 4. **(A finite spectrum assignment problem for single input retarded nonlinear systems)**
Let a plant of nonlinear system with time delays in state variables (7) be given. Then, if possible, find a feedback with delayed-state variables,

$$u(t) = \alpha(x, \sigma) \tag{8}$$

and a coordinates transformation with delayed-state variables, $z(t) = \Phi(x, \sigma)$, such that the resulting closed loop system

$$\dot{x}(t) = \bar{f}(x, \sigma) + g(x)\alpha(x, \sigma) \tag{9}$$

is linear in the coordinates $z(t) = \Phi(x, \sigma)$, and has spectrums at arbitrarily preassigned self-conjugate set of n points λ_i, $i = 1, \ldots, n$, in the complex plane,i.e. it is denoted by $\dot{z}(t) = Az(t)$ for some suitable matrix $A \in \Re^{n \times n}$ which has totally n eigenvalues at λ_i, $i = 1, \ldots, n$.

For this problem, we obtain the following theorem.

Theorem 5. The finite spectrum assignment problem for retarded nonlinear systems described by (7) is solvable if there exists a $C^{\infty}-$class real-valued function $\phi(x)$ of the state $x(t)$ satisfying the following conditions:

(1) $L_g L_{\bar{f}}^{i} \phi(x, \sigma) = 0$ *for* $i = 0, 1, \cdots, n-2$

(2) $\frac{\partial L_{\bar{f}}^{n-1} \phi(x,\sigma)}{\partial x(t)} g(x, \sigma) \ne 0$

(3) $\frac{\partial L_{\bar{f}}^{n-1} \phi(x,\sigma)}{\partial \sigma_{f_{j_1}} \cdots \sigma_{f_{j_{n-2}}} x} \sigma_{f_{j_1}} \cdots \sigma_{f_{j_{n-2}}} g(x, \sigma) = 0$
for $1 \le^{\forall} j_1 \le k, 0 \le^{\forall} j_2, \ldots, ^{\forall} j_{n-2} \le k$,
where $\sigma_{f_0} : x(t) \to x(t)$,

for any $\sigma_{f_{j_1}} \cdots \sigma_{f_{j_k}} x \in \Re^n$.

When these conditions are satisfied and the preassigned poles are given by $\lambda_1, \ldots, \lambda_n$, the required feedback law is given by

$$u(t) = \frac{-L_{\bar{f}}^{n} \phi(x, \sigma) + v(t)}{L_g L_{\bar{f}}^{n-1} \phi(x, \sigma)} \tag{10}$$

and the required coordinates transformation by

$$z(t) = \begin{pmatrix} \phi(x) \\ L_{\bar{f}} \phi(x, \sigma) \\ \vdots \\ L_{\bar{f}}^{n-1} \phi(x, \sigma) \end{pmatrix}, \tag{11}$$

where $v = \sum_{i=1}^{n} a_i z_i(t)$. Here a_i are the coefficients of the desired characteristic polynomial of the closed loop system $(s - \lambda_1)(s - \lambda_2) \cdots (s - \lambda_n) = s^n + a_n s^{n-1} + \cdots + a_0$.

Furthermore there exists an open neighborhood of x_0 such that the coordinates transformation (11) is a diffeomorphism under the assumption that the past values of the state are known.

PROOF. We assume that $n \ge 3$ and a real-valued function $\phi(x)$ satisfies conditions *1*, *2* and *3*. By differentiating the function $\phi(x) \equiv z_1(t)$ with respect to time t, we obtain

$$\begin{aligned} \dot{z}_1(t) &\equiv \frac{\partial \phi(x)}{\partial x(t)} \frac{dx(t)}{dt} \\ &= L_{\bar{f}} \phi(x, \sigma) + L_g \phi(x) u(t). \end{aligned} \tag{12}$$

By condition *1*, the function ϕ satisfies $L_g \phi(x) = 0$. Therefore we obtain

$$\dot{z}_1(t) = L_{\bar{f}} \phi(x, \sigma) \equiv z_2(t). \tag{13}$$

Repeating differentiation, we obtain

$$\dot{z}_2(t) = L_{\bar{f}}^2 \phi(x, \sigma) + L_{gu} L_{\bar{f}} \phi(x, \sigma). \tag{14}$$

If $L_g L_{\bar{f}}^{i} \phi(x, \sigma) = 0$, $L_{gu} L_{\bar{f}}^{i} \phi(x, \sigma) = 0$. Therefore we obtain $\dot{z}_2(t) = L_{\bar{f}}^2 \phi(x, \sigma) \equiv z_3(t)$. By repeating the same operations n times, we obtain

$$\dot{z}_n(t) = L_{\bar{f}}^{n} \phi(x, \sigma) + L_{gu} L_{\bar{f}}^{n-1} \phi(x, \sigma). \tag{15}$$

The second term on the right-hand side of the above equation can be decomposed as follows.

$$L_{gu}L_{\bar{f}}^{n-1}\phi(x,\sigma) = \frac{\partial L_{\bar{f}}^{n-1}\phi(x,\sigma)}{\partial x(t)}g(x,\sigma)u(t)$$
$$+\sum_{j_1=1}^{k}\sum_{j_2=0}^{k}\cdots\sum_{j_{n-1}=0}^{k}\frac{\partial L_{\bar{f}}^{n-1}\phi(x,\sigma)}{\partial\sigma_{f_{j_1}}\sigma_{f_{j_2}}\cdots\sigma_{f_{j_{n-1}}}x}$$
$$\times\sigma_{f_{j_1}}\sigma_{f_{j_2}}\cdots\sigma_{f_{j_{n-1}}}\{g(x,\sigma)u(t)\} \quad (16)$$

By conditions *2* and *3*, all the terms but the first term in the right hand side vanish, and we have

$$L_gL_{\bar{f}}^{n-1}\phi(x,\sigma) = \frac{\partial L_{\bar{f}}^{n-1}\phi(x,\sigma)}{\partial x(t)}g(x,\sigma)u(t)$$
$$= L_g\bar{f}\phi(x,\sigma)u(t). \quad (17)$$

By substituting equation (17) in equation (15), the n-times differentiation of $\phi(x)$ with respect to t becomes as follows.

$$\dot{z}_n(t) = L_{\bar{f}}^n\phi(x,\sigma) + L_gL_{\bar{f}}\phi(x,\sigma)u(t) \quad (18)$$

Therefore if we choose coordinates transformation (11) and feedback law (10), system (7) is transformed to a finite-dimensional linear system whose spectrum is placed at the preassigned points in the complex plane, $\dot{z}(t) = Az(t)$, where matrix $A \in \Re^{n\times n}$ is given by

$$A = \begin{pmatrix} 0 & 1 & \cdots & 0 \\ \vdots & \ddots & \ddots & \vdots \\ 0 & \cdots & 0 & 1 \\ -a_1 & -a_2 & \cdots & -a_n \end{pmatrix} \quad (19)$$

Next we prove that there exists an open neighborhood of x_0 such that the coordinates transformation (11) is a diffeomorphism under the assumption that the past values of the state are known.

Let $C(x,\sigma)$ be a matrix defined by

$$C(x,\sigma) = \left(g \quad ad_{\bar{f}}g \quad \cdots \quad ad_{\bar{f}}^{n-1}g\right). \quad (20)$$

Furthermore we define the $n \times n$ matrix $M(x,\sigma)$ as

$$M(x,\sigma) \equiv \frac{\partial\Phi(x,\sigma)}{\partial x(t)}C(x,\sigma) \quad (21)$$

and the (i,j)–th elements $M_{(i,j)}(x,\sigma)$ of $M(x,\sigma)$ can be described by

$$M_{(i,j)}(x,\sigma) = <\frac{\partial L_{\bar{f}}^{i-1}\phi(x,\sigma)}{\partial x}, ad_{\bar{f}}^{j-1}g(x,\sigma)>$$

where $<\cdot,\cdot>$ is the inner product notation.

While, by Theorem 3, we obtain

$$L_{ad_{\bar{f}}^{j-1}g}L_{\bar{f}}^{n-j}\phi(x,\sigma) = (-1)^{j-1}L_gL_{\bar{f}}^{j-1}L_{\bar{f}}^{n-j}\phi(x,\sigma)$$
$$= (-1)^{j-1}L_gL_{\bar{f}}^{n-1}\phi(x,\sigma)$$
$$\equiv (-1)^{j-1}\rho(x,\sigma) \quad (22)$$

for $j = 1,\cdots,n$, where $L_gL_{\bar{f}}^{n-1}\phi(x,\sigma) \equiv \rho(x,\sigma)$. Here under condition *3*, we have

$$L_{ad_{\bar{f}}^{n-i}g}L_{\bar{f}}^{i-1}\phi(x,\sigma) = \frac{\partial L_{\bar{f}}^{i-1}\phi(x,\sigma)}{\partial x}ad_{\bar{f}}^{n-i}g(x,\sigma).$$

Therefore by conditions *1*,*2* and *3*, $M(x,\sigma)$ can be rewritten as follows.

$$M(x,\sigma) = \begin{pmatrix} 0 & \cdots & L_{ad_{\bar{f}}^{n-1}g}\phi \\ \vdots & & * \\ L_gL_{\bar{f}}^{n-1}\phi & * & * \end{pmatrix}$$
$$= \begin{pmatrix} 0 & \cdots & (-1)^{n-1}\rho(x,\sigma) \\ \vdots & & * \\ \rho(x,\sigma) & * & * \end{pmatrix}.$$

By condition *1*, $\rho(x,\sigma)$ does not vanish even if all the delayed-state variables are equal to zero. This means that $M(x,\sigma)$ is nonsingular for any $x(t)$ in a neighborhood of the origin, and in turn we can deduce that $\frac{\partial\Phi(x,\sigma)}{\partial x(t)}$ is also nonsingular. As the result, from the inverse function theorem there exists a open neighborhood of x_0 such that coordinates transformation (11) with the known past value of the state is diffeomorphism. This completes the proof. □

Next, we consider the condition for the existence of such a function $\phi(x)$ that satisfies the conditions of Theorem 5. By Theorem 3, condition 1 is equivalent to the following condition.

$$L_{ad_{\bar{f}}^ig}\phi(x,\sigma) = 0 \text{ for } i = 0,1,\ldots,n-2 \quad (23)$$

Conditions 2 and 3 mean that

$$L_gL_{\bar{f}}^{n-1}\phi(x,\sigma) = (-1)^{n-1}\frac{\partial\phi}{\partial x}ad_{\bar{f}}^{n-1}g$$
$$= \frac{\partial L_{\bar{f}}^{n-1}\phi(x,\sigma)}{\partial x(t)}g(x,\sigma)$$
$$\equiv \xi_0(x,\sigma) \neq 0 \quad (24)$$

By equations (23) and (24), conditions 1, 2 and 3 can be replaced by the following conditions **(i)**, **(ii)** and **(iii)** respectively.

(i) $L_{ad_{\bar{f}}^ig}\phi(x,\sigma) = 0$ for $i = 0,1,\ldots,n-2$

(ii) $L_{ad_{\bar{f}}^{n-1}g}\phi(x,\sigma) \neq 0$

(iii) $L_{ad_{\bar{f}}^{n-1}g}\phi(x,\sigma) = (-1)^{n-1}\frac{\partial L_{\bar{f}}^{n-1}\phi(x,\sigma)}{\partial x(t)}g(x,\sigma)$

An obvious sufficient condition for existence of $\phi(x)$ satisfying conditions **(i)** and **(ii)** is given as follows.

Theorem 6. There exists a real-valued function $\phi(x)$ defined in a neighborhood U of x_0 satisfying conditions (i) *and* (ii) *if*

a) *the matrix* $\{g, ad_{\bar{f}}g, \ldots, ad_{\bar{f}}^{n-1}g\}(x,\sigma)$ *has rank n for any* σ,
b) $ad_{\bar{f}}^{i}g(x,\sigma) = ad_{\bar{f}}^{i}g(x,0)$ *for* $i = 0, 1, \ldots, n-2$ *and the distribution* $\mathcal{D} = span\{g, ad_{\bar{f}}g, \ldots, ad_{\bar{f}}^{n-2}g\}$ *is involutive.*

3.2 *Replacing condition 3 of Theorem 5*

By extending the class of feedback law (8), condition 3 of Theorem 5 can be replaced by a looser condition.

Now we consider such a type of feedback as $u(t) = \beta(x, u, \sigma)$ instead of equation (8), and assume that the function $\phi(x)$ satisfies conditions *1* and *2* of Theorem 5 but does not satisfy condition *3*. Then instead of equation (10), a feedback law $u(t)$ which transforms system (7) into a finite-dimensional linear system must satisfies the following equation.

$$\frac{\partial L_{\bar{f}}^{n-1}\phi(x,\sigma)}{\partial x(t)} g(x,\sigma)u(t) = -\sum_{i=0}^{n-1} a_i L_{\bar{f}}^{i}\phi(x,\sigma) - \sum_{j_1=1}^{k}\sum_{j_2=0}^{k}\cdots\sum_{j_{n-1}=0}^{k} \frac{\partial L_{\bar{f}}^{n-1}\phi(x,\sigma)}{\partial\sigma_{f_{j_1}}\sigma_{f_{j_2}}\cdots\sigma_{f_{j_{n-1}}}x} \times\sigma_{f_{j_1}}\sigma_{f_{j_2}}\cdots\sigma_{f_{j_{n-1}}}\{g(x,\sigma)u(t)\} - L_{\bar{f}}^{n}\phi(x,\sigma).$$

Since condition *2* is satisfied, equation (10) can be rewritten as

$$u(t) = \alpha(x,\sigma) - \sum_{j_1=1}^{k}\sum_{j_2=0}^{k}\cdots\sum_{j_{n-1}=0}^{k} \eta_{j_1\cdots j_{n-1}} u(t - \mathcal{L}_{j_1\cdots j_{n-1}}). \tag{25}$$

where $\mathcal{L}_{j_1\cdots j_{n-1}} = \sum_{i=1}^{n-1}\ell_{f_{j_i}}$ and $\ell_{f_0} = 0$. In this equation, η_j and $\alpha(x,\sigma)$ are scalar-valued functions defined as follows.

$$\eta_{j_1\cdots j_{n-1}} \equiv \frac{\xi_{j_1\cdots j_{n-1}}(x,\sigma)}{\xi_0(x,\sigma)},$$

$$\alpha(x,\sigma) = -\frac{L_{\bar{f}}^{n}\phi(x,\sigma) + \sum_{i=0}^{n-1} a_i L_{\bar{f}}^{i}\phi(x,\sigma)}{\xi_0(x,\sigma)},$$

where we define ξ_i by

$$\xi_{j_1\cdots j_{n-1}}(x,\sigma) \equiv \frac{\partial L_{\bar{f}}^{n-1}\phi(x,\sigma)}{\partial\sigma_{f_{j_1}}\cdots\sigma_{f_{j_{n-1}}}x} g(x(t - \mathcal{L}_{j_1\cdots j_{n-1}})),$$

and ξ_0 by $\xi_0(x,\sigma) \equiv \frac{\partial L_{\bar{f}}^{n-1}\phi(x,\sigma)}{\partial x} g(x)$. Equation (25) means that the input $u(t)$ has a dynamics, which is required that if condition *3* does not hold,

> *(3') input dynamics (25) is stable around the origin in z-coordinates.*

While condition *3* requires that a feedback $u(t)$ has no dynamics, condition *3'* requires only that even if $u(t)$ has a dynamics, it is stable around the origin. Therefore condition *3'* is less restrictive than condition *3* of Theorem5.

4. EXAMPLE

(Example 1.) We consider an unstable system,

$$\dot{x} = \begin{pmatrix} x_2 + x_1(t-\ell_1) \\ x_3 + x_1\cdot x_2(t-\ell_2) \\ x_3^2 \end{pmatrix} + \begin{pmatrix} 0 \\ 0 \\ 1 \end{pmatrix} u(t). \tag{26}$$

Now, if we choose the function $\phi(x) = x_1(t)$, which satisfies conditions *1*, *2* and *3* of Theorem 5. By choosing the coordinates transformation

$$z(t) = \begin{pmatrix} x_1 \\ x_2 + \sigma_1 x_1 \\ x_3 + x_1\cdot\sigma_2 x_2 + \sigma_1 x_2 + \sigma_1^2 x_1 \end{pmatrix}$$

and the feedback law

$$u(t) = -x_3^2 - (x_2 + \sigma_1 x_1)\sigma_2 x_2 - \sigma_1 x_3 - x_1(\sigma_2 x_3 + \sigma_2 x_1\cdot\sigma_2^2 x_2) - \sigma_1 x_1\cdot\sigma_1\sigma_2 x_2 - \sigma_1^2 x_2 - \sigma_1^3 x_1 + v,$$

we obtain a finite spectrum assigned system

$$\dot{z}(t) = \begin{pmatrix} 0 & 1 & 0 \\ 0 & 0 & 1 \\ -a_0 & -a_1 & -a_2 \end{pmatrix} z(t), \tag{27}$$

where $v(t) = -a_0 z_1(t) - a_1 z_2(t) - a_2 z_3(t)$. We consider the regulation problem to the origin in z-coordinates. The initial conditions of the state variables x are given by $x(\tau) = 0$ for $\tau < 0$ and $x(0) = (1.0, 0.5, 0.4)^T$ and the lengths of the time-delays are $\ell_1 = 1.0$[sec] and $\ell_2 = 0.7$[sec]. A simulation result in Fig.1 shows that the state $z(t)$ converges to 0 in z-coordinates.

(Example 2.) Consider an unstable system

$$\dot{x} = \begin{pmatrix} x_1\cdot x_1(t-\ell_1) + x_2 + \frac{1}{2}x_2(t-\ell_2) \\ x_1 + x_2 \end{pmatrix} + (0 \quad 1)^T u.$$

In this case, if we choose the function $\phi(x) = x_1(t)$, a finite spectrum assignment can be accomplished though $\phi(x)$ satisfies condition 1 and 2 of

Theorem 5 but does not satisfy condition 3. In fact, applying a coordinates transformation

$$z(t) = \begin{pmatrix} x_1 \\ x_1 \cdot \sigma_{f_1} x_1 + x_2 + \frac{1}{2}\sigma_{f_2} x_2 \end{pmatrix}$$

and a feedback $u(t) = \frac{1}{2}u(t - \ell_2) + \alpha(x, \sigma)$, where

$$\begin{aligned} \alpha(x,\sigma) = &-(x_1 \cdot \sigma_{f_1} x_1 + x_1 + \frac{1}{2}\sigma_{f_2} x_2) \cdot \sigma_{f_1} x_1 \\ &- x_1(\sigma_{f_1} x_1 \cdot \sigma_{f_1}^2 x_1 + \sigma_{f_1} x_2 + \frac{1}{2}\sigma_{f_1}\sigma_{f_2} x_2) \\ &- (x_1 + x_2 + \frac{1}{2}(\sigma_{f_2} x_1 + \sigma_{f_2} x_2) + v(t), \end{aligned}$$

we obtain a finite spectrum assigned system

$$\dot{z}(t) = \begin{pmatrix} 0 & 1 \\ -a_0 & -a_1 \end{pmatrix} z(t), \qquad (28)$$

where $\sigma_{f_1} x(t) = x(t - \ell_1)$, $\sigma_{f_2} x(t) = x(t - \ell_2)$ and $v(t) = -a_0 z_1(t) - a_1 z_2(t)$. The initial conditions of the state variables x are given by $x(\tau) = 0$ for $\tau < 0$ and $x(0) = (0.5, 0.4)^T$, and the lengths of the time-delays are $\ell_1 = 0.7$[sec] and $\ell_2 = 1.0$[sec]. Simulation result Fig.2 shows that this system is stabilized through our proposed procedure.

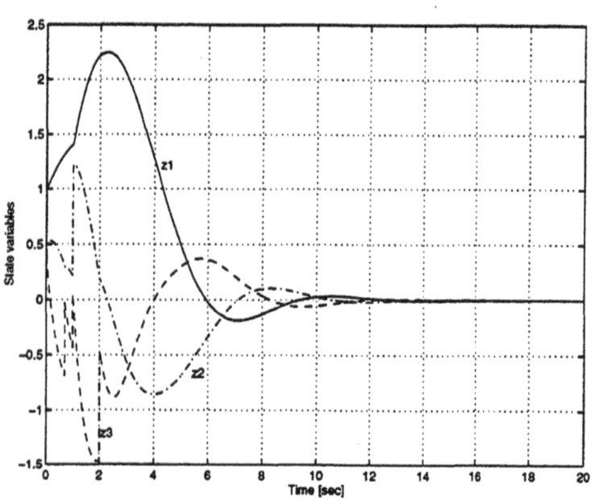

Fig. 1. Example 1 (z-coordinates)

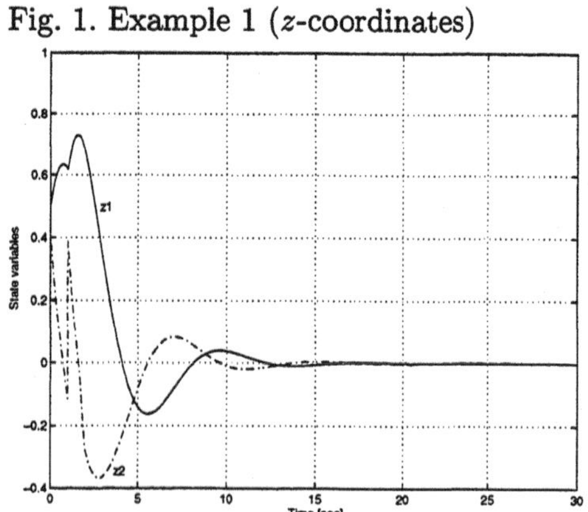

Fig. 2. Example 2 (z-coordinates)

5. CONCLUSION

In this paper, we considered the finite spectrum assignment for nonlinear systems with non-commensurate time-delays in state variables. The proposed method is related with both differential geometric linearization of finite dimensional nonlinear systems and the algebraic method of finite spectrum assignment for $\mathcal{R}$-controllable retarded linear systems (Morse, 1976; Sontag, 1976).

6. REFERENCES

Henson, M. A. and D. E. Seborg (1994). Time delay compensation for nonlinear processes. *Ind. Eng. Chem. Res.* **33**(6), 1493–1500.

Huang, H.-P. and G.-B. Wang (1992). Dead-time compensation for nonlinear processes with disturbances. *Int. J. Syst. Science* **23**(10), 1761–1776.

Isidori, A. (1995). *Nonlinear Control Systems -An Introduction.* 3rd ed.. Springer-Verlag.

Kravaris, C. and R. A. Wright (1989). Deadtime compensation for nonlinear processes. *AIChE J.* **35**(9), 1535–1542.

Lee, E. B. and S. H. Żak (1982). On spectrum placement for linear time invariant delay systems. *IEEE Trans. Automat. Contr.* **AC-27**(2), 446–449.

Manitius, A. Z. and A. W. Olbrot (1979). Finite spectrum assignment problem for systems with delays. *IEEE Trans. Automat. Contr.* **AC-24**(4), 541–553.

Morse, A. S. (1976). Ring models for delay-differential systems. *Automatica* **12**, 529–531.

Nijmeijer, H. and A. van der Schaft (1990). *Nonlinear Dynamical Control Systems.* Springer-Verlag.

Oguchi, T. (1995). Control Systems Design of Nonlinear Systems with Time Delays by using Differential Geometry (in Japanese). Ph.D. dissertation. Tokyo Metropolitan Univ.

Oguchi, T., A. Watanabe and S. Kawata (1994*a*). Exact linearization of nonlinear systems with input time delay with model based state prediction. In: *Proc. ASCC.* Tokyo. pp. 353–356.

Oguchi, T., A. Watanabe and S. Kawata (1994*b*). Input-output linearization of multi-variable nonlinear systems with input time delays. *Trans. JSME (Vol.C)(in Japanese)* **60**(578), 3431–3439.

Oguchi, T., A. Watanabe and T. Nakamizo (1997). A finite pole assignment procedure for retarded nonlinear systems. In: *Proc. ACC.* Albuquerque. pp. 2682–2686.

Oguchi, T. and A. Watanabe (1998). Input-output linearization of nonlinear systems with time-delays in state variables. *Int. J. Syst. Science* **29**(6), 573–578.

Sontag, E. D. (1976). Linear systems over commutative rings: A survey. *Ricerche di Automatica* **7**(1), 1–34.

14th Triennial World Congress, Beijing, P.R. China

STATE AND OUTPUT FEEDBACK CONTROLLER DESIGN FOR ABSOLUTE STABILIZATION PROBLEMS *

Lei Guo, Chun-Bo Feng and Shiji Song

Research Institute of Automation, Southeast University, Nanjing, 210096, P.R. China.Email:leiguo@seu.edu.cn

Abstract: The absolute stabilization problem for Lur'e systems with multiple nonlinear loops is discussed based on state-space approach in this paper. Solvability conditions are presented to design nonlinear controllers such that the closed-loop system is absolutely stable via the algebraic matrix inequality (AMI) approach. It is shown that feedback controllers exist if and only if a class of special multilinear matrix inequalities (MLMIs) are solvable. Also, the AMI-based design method obtained in this paper is simplified to be computationally feasible and tractable. The approach can be generalized to deal with other problems such as H^2, H^∞ and dissipation control problem. *Copyright © 1999 IFAC*

Keywords: nonlinear systems, decentralized control, robust control.

1 Introduction

Although the importance of the stabilization control problem for nonlinear systems has been recognized for a long time, researches are not very sufficient even for the Lur'e system, which is a kind of typical nonlinear systems (See, e.g., Knobloch, *et al.*, 1993; Narendra and Taylor, 1973 and references therein). Recently, Savkin and Petersen (1995) considered an extension of Popov criterion in state space. A necessary and sufficient condition involved in two parameter-dependent algebraic Riccati equation (ARE) (with a coupled constraints) was presented for this property. However, the Lur'e system under consideration in Savkin and Petersen (1995) was confined to have only one nonlinear loop and some assumptions made on the Lur'e system were redundant.

In this paper we consider the absolute stabilization control problems for the generalized plants described by the Lur'e system with multiple nonlinearities. The objective is to design controllers such that the closed-loop system are absolutely stable. The Lur'e control vector is considered as a measurable output. Thus, unlike the linear feedback case, the control law given in this paper is nonlinear and the former case can be dealt with as a special one. Based on LMI approach in control theory (Boyd, *et al.*, 1994), both state feedback and output feedback cases are discussed and the solvability conditions are reduced to multilinear matrix inequalities (MLMIs). If the number of the nonlinear loops turns to be one, results obtained in this note are sufficient and necessary in the sense of existence of *Lyapunov* function.

The controlled plant G is described as

$$\left\{\begin{array}{lll} \dot{x} & = & Ax \quad +B_p p \quad +B_u u \\ y & = & C_y x \\ q & = & C_q x \end{array}\right. , \qquad (1)$$

where $p^T = (p_1(t), \cdots, p_{n_p}(t)), q^T = (q_1(t), \cdots, q_{n_p}(t))$, $p_i(t) = \phi_i(q_i(t)), i = 1, 2, \cdots, n_p$. $\phi_i(\sigma)$ satisfies the following sector condition

$$0 \leq \sigma\phi_i(\sigma) \leq l_i\sigma^2. \qquad (2)$$

And $x \in R^n, u \in R^{n_u}$ and $y \in R^{n_y}$, represent state, control input, and measurable output, respectively. $p \in R^{n_p}$ is Lur'e control vector, while $q \in R^{n_p}$ is the signal for control, which relates p and x via a sector bound function. Suppose coefficient matrices in (1) have appropriate dimensions. Denote $L := diag\{l_1, \cdots, l_{n_p}\}$ to describe the bound of the nonlinear sector.

Consider the following free system

$$\left\{\begin{array}{lll} \dot{x} & = & Ax \quad +B_p p \\ q & = & C_q x \end{array}\right. , (1)'$$

The absolute stability of (1)' can be converted to the existence of the Lyapunov function in the following form

$$V(x) = x^T P x + 2\sum_{i=1}^{n_p} \lambda_i \int_0^{C_{i,q}x} \phi_i(\sigma) d\sigma, \quad (3)$$

where $P > 0, \Lambda := diag\{\lambda_i, \cdots, \lambda_{n_p}\} \geq 0$, and $C_q^T := (C_{1,q}^T, \cdots, C_{n_p,q}^T)$. It is well known that the existence of $V(x)$ in the form of (3) is equivalent to the famous Popov criterion (Savkin and Peterson, 1995).

The following Lemma reduced the absolute stability for Lur'e system to the solvability condition of a certain LMI.

Lemma 1 (Absolutely Stable Condition): For Lur'e system $(1)'$, *if there exist* $P > 0, \Lambda \geq 0$, *and* $T := diag\{t_1, \cdots, t_{n_p}\} \geq 0$ *satisfying*

$$\begin{bmatrix} A^T P + PA & M_1 \\ M_1^T & \Lambda C_q B_p + B_p^T C_q^T \Lambda - 2T \end{bmatrix} < 0, \qquad (4)$$

where

$$M_1 := PB_p + A^T C_q^T \Lambda + C_q^T LT,$$

then $(1)'$ *is absolutely stable.*

Lemma 1 follows from the statement of §8.2 of Boyd, *et al.* (1994) with a little difference.

Remark 1: When $n_p = 1$, the conditions (4) in the above Lemma are equivalent to the existence of Lyapunov function. When $n_p > 1$, the results are conservative. Rapoport (1987) devised a way of determining if there exists a Lyapunov function of the form (3) nonconservatively by generating appropriate LMIs.

Absolute Stabilization Problem: To design controllers such that the closed-loop Lur'e system is absolutely stable.

2 Main Results

2.1 State feedback Case

Suppose the Lur'e control signal p is measurable and consider feedback law $u = Kx + K_p p$.

Thus, u is nonlinear since it is related to x via a nonlinear function. When p is not available, its gain K_p can be set to be 0 and the corresponding results for linear feedback controllers can be given.

The closed-loop system becomes

$$\begin{cases} \dot{x} = \bar{A}x + \bar{B}_p p \\ q = C_q x \end{cases}, \quad (5)$$

where $\bar{A} = A + B_u K$, $\bar{B}_p = B_p + B_u K_p$. Our goal is to find K and K_p such that (5) is absolutely stable.

Denote $Q := P^{-1}$ and $KQ = R$. Applying Lemma 1 to the closed-loop system, the inequality (4) corresponding to the closed loop system (5) can be rewritten to be the following form via matrix transformation.

$$\begin{bmatrix} R^T B_u^T + QA^T + AQ + B_u R & M_{sf1} \\ M_{sf1}^T & M_{sf2} \end{bmatrix} < 0, \quad (6)$$

where,

$$M_{sf1} := B_p + B_u K_p + R^T B_u^T C_q^T \Lambda + QA^T C_q^T \Lambda + QC_q^T LT,$$

$$M_{sf2} := \Lambda C_q (B_p + B_u K_p) + (B_p + B_u K_p)^T C_q^T \Lambda - 2T$$

Hence, we have the following result for absolute stabilization.

Theorem 1: If there exist Λ,T,$Q(>0)$, R and K_p satisfying (6), then there exists an state feedback nonlinear controller $u = Kx + K_p p$ $(K = RQ^{-1})$ such that the closed-loop system is absolutely stable.

Proof is omitted for brevity.

Remark 2: (6) is a bilinear matrix inequality (BLMI) of K_p,Q,R, T and Λ. For BLMI, Goh, et al., presented local and global optimization methods in (Goh, *et al.*,1994a, 1994b). If the parameter T (related to $\mathcal{S}$-procedure) and Λ of the Lyapunov function (3) is given, the condition is exactly a linear matrix inequality (LMI) for K_p,Q and R.

2.2 Dynamic Output Feedback Case

Suppose that Σ_C is a nonlinear dynamical output feedback controller with order of n_c, whose state equation has the following form

$$\begin{cases} \dot{x}_c = A_c x_c + B_c y + B_{cp} p \\ u = C_c x_c + D_c y + D_{cp} p \end{cases}. \quad (7)$$

Combine (7) with the plant (1), then the closed-loop system Σ_{cl} is described by

$$\begin{cases} \dot{\bar{x}} = \bar{A}\bar{x} + \bar{B}_p p \\ q = \bar{C}_q \bar{x} \end{cases}, \quad (8)$$

where

$$\bar{A} := \hat{A} + \hat{B}_u G \hat{C}_y, \bar{C}_q = \begin{bmatrix} C_q & 0 \end{bmatrix},$$
$$\bar{B}_p = \hat{B}_p + \hat{B}_u \hat{D}_p = \begin{bmatrix} B_p + B_u D_{cp} \\ B_{cp} \end{bmatrix}, \quad (9)$$

and

$$\begin{bmatrix} \hat{A} & \hat{B}_u \\ \hat{C}_y & G^T \end{bmatrix} := \left[\begin{array}{cc|cc} A & 0 & B_u & 0 \\ 0 & 0 & 0 & I_{n_c} \\ \hline C_y & 0 & D_c^T & B_c^T \\ 0 & I_{n_c} & C_c^T & A_c^T \end{array}\right],$$
$$\hat{B}_p = \begin{bmatrix} B_p \\ 0 \end{bmatrix}, \hat{D}_p = \begin{bmatrix} D_{cp} \\ B_{cp} \end{bmatrix}. \quad (10)$$

Our objective is to find G and $\hat{D}_p$ such that (8) is absolutely stable.

Take the coefficient matrices of $\bar{G}$ in place of those in (4) correspondingly. Hence, (4) implies

$$\Omega + BGC + (BGC)^T < 0, \quad (11)$$

where

$$B := \begin{bmatrix} P\hat{B}_u \\ \Lambda\bar{C}_q\hat{B}_u \end{bmatrix}, C := \begin{bmatrix} \hat{C}_y & 0 \end{bmatrix},$$
$$\Omega := \begin{bmatrix} \hat{A}^T P + P\hat{A} & \hat{M}_1 \\ \hat{M}_1^T & \hat{M}_2 \end{bmatrix},$$

and

$$\hat{M}_1 = P\hat{B}_p + \hat{A}^T\hat{C}_q^T\Lambda + \Lambda\hat{B}_u\hat{D}_p + \hat{C}_q^T LT,$$

$$\begin{aligned}\hat{M}_2 = \; & \Lambda\bar{C}_q\hat{B}_p + \hat{B}_p^T\bar{C}_q^T\Lambda + \Lambda\bar{C}_p\hat{B}_u\hat{D}_p \\ & +\hat{D}_p^T\hat{B}_u^T\bar{C}_p^T\Lambda - 2T\end{aligned}.$$

Set $Q := P^{-1}$ and note that

$$B^{\perp} = \begin{bmatrix} \hat{B}_u \\ 0 \end{bmatrix}^{\perp} \begin{bmatrix} Q & 0 \\ -\Lambda\bar{C}_q Q & I \end{bmatrix},$$

and

$$\begin{aligned}\begin{bmatrix} \hat{C}_y^T \\ 0 \end{bmatrix}^{\perp} &= \left[\begin{array}{ccc} C_y^{T\perp} & 0 & 0 \\ \hline 0 & 0 & I \end{array}\right], \\ \begin{bmatrix} \hat{B}_u \\ 0 \end{bmatrix}^{\perp} &= \left[\begin{array}{ccc} B_u^{\perp} & 0 & 0 \\ \hline 0 & 0 & I \end{array}\right]\end{aligned}, \qquad (12)$$

where $B^{\perp}$ is defined as $B^{\perp}B = 0$ and $B^{\perp}B^{\perp T} > I$. Partition P to be a block matrix compatible with (10),

$$P := \begin{bmatrix} X & P_{12} \\ P_{12}^T & P_{22} \end{bmatrix}, Y := (X - P_{12}P_{22}^{-1}P_{12}^T)^{-1}. \qquad (13)$$

In this case $Q = P^{-1} = \begin{bmatrix} Y & * \\ * & * \end{bmatrix}$ and

$$X - Y^{-1} = P_{12}P_{22}^{-1}P_{12}^T. \qquad (14)$$

It can be shown that Σ_{cl} is absolutely stable if and only if the following two formulas hold (see Boyd, *et al.*, 1994 for details.)

$$\Pi_1 \begin{bmatrix} A^T X + XA & M_1 \\ M_1^T & M_2 \end{bmatrix} \Pi_1^T < 0, \qquad (15)$$

$$\Pi_2 \begin{bmatrix} YA^T + AY & N_1 \\ N_1^T & N_2 \end{bmatrix} \Pi_2^T < 0, \qquad (16)$$

where

$$\Pi_1 := \begin{bmatrix} C_y^{T\perp} & 0 \\ 0 & I \end{bmatrix}, \Pi_2 := \begin{bmatrix} B_u^{\perp} & 0 \\ 0 & I \end{bmatrix},$$

and

$$\begin{aligned} M_1 = \; & XB_p + A^T C_q^T\Lambda + C_q^T LT \\ & +XB_u D_{cp} + P_{12}B_{cp}, \\ M_2 = \; & \Lambda C_q B_p + B_p^T C_q^T \Lambda + \Lambda C_q B_u D_{cp} \\ & +D_{cp}^T B_u^T C_q^T \Lambda - 2T, \\ N_1 = \; & B_p - A^T Y C_q^T \Lambda + Y C_q^T LT + B_u D_{cp}, \\ N_2 = \; & -\Lambda C_q Y C_q^T LT - TLC_q Y C_q^T \Lambda - 2T. \end{aligned} \qquad (17)$$

The following condition results from the decomposition of (13)

$$\begin{bmatrix} X & I \\ I & Y \end{bmatrix} \geq 0, \qquad (18)$$

Theorem 2: If there exist Λ,T,$X > 0$,$Y > 0$,P_{12}, B_{cp} and D_{cp} satisfying (15), (16), (18) and the constraint (14) for an arbitrary $P_{22} > 0$, then there exists an nonlinear output feedback controller Σ_C such that the closed-loop system is absolutely stable. In this case, if we suppose the column dimension of P_{12} is r, the order of Σ_C is $n_c = r$.

Proof: Based on the above derivation, it is shown that the conditions in Theorem 2 hold if and only if (4) (in Lemma 1) holds when the coefficient matrices of Σ_{cl} are substituted to (4). Hence, Σ_{cl} is absolutely stable. According to (10), (13) and (14), it is easy to show that n_c is equal to the dimension of P_{22}, i.e., r. *Q.E.D.*

2.3 On Computation and Design Procedure

As to the constraint (14), we present a theorem in the following to show that this constraint can be removed without loss of generality.

Noting that P_{12} and B_{cp} appeared in (15) have the form of $P_{12}B_{cp}$, if we define $P_0 = P_{12}B_{cp}$, Theorem 2 can be simplified as follows.

Theorem 3: If there exist $\Lambda, T, X > 0, Y > 0, P_0$ *and* D_{cp} *satisfying (15), (16) and (18), then there exists an output feedback controller* Σ_C *such that the closed-loop system is absolutely stable.*

Proof is omitted for brevity.

Although the constraint is removed by Theorem 3, the (15) is bilinear for the controller parameter D_{cp} and X. In the following, we consider a more tractable case, where D_{cp} in (7) is set to 0.

Thus, the submatrices in (15) and (16) can be simplified and (15) becomes linear for X, Λ, T and P_0. In addition, a static condition that should be satisfied is

$$\Lambda C_q B_p + B_p^T C_q^T \Lambda - 2T < 0. \qquad (19)$$

The design method based on iteration algorithm can be presented correspondingly.

(i) Solve LMI (19) to get a pair of Λ and T;

(ii) Solve LMI (15) to get a pair of $X > 0$ and P_0, if (15) has no solutions for X and P_0, turn to step (i);

(iii) Solve LMI (16) and (18) to get $Y > 0$, if (16) and (18) have no solutions for Y, turn to step (i);

(iv) If $X - Y^{-1} > 0$ (otherwise, choose a appropriate e such that $(X + eI, Y)$ satisfies (15), (16) and (18) as well as $X + eI - Y_{-1} > 0$), denote $B_{cp} = P_0$, $P_{12} = I$ and $P_{22} = (X - Y^{-1})^{-1}$. Thus, P is obtained;

(v) Solve G via LMI (11). Thus, the nonlinear controller is obtained.

Since each step in the above procedure is only involved in convex optimization of LMIs, for which there has been a Toolbox in Matlab, this design method is feasible and traceable.

In fact, the absolute stabilization problem stated in this paper is similar to the problem discussed by Savkin and Petersen (1995), if only linear feedback controllers are considered. Three improvements have been made in this paper: First, the Lur'e system considered here has multiple nonlinear loops instead of single one. Second, the Lur'e control vector p is used for the feedback law and hence the controllers become nonlinear. Third, some assumptions such as controllability of (A, B_p), observablity of (A, C_q), and $B_u^T C_q^T C_q B_u > 0$ supposed in Savkin and peterson (1995) are removed here. If we confined the controllers to be linear, the results can be simplified as follows.

It is well known that Lur'e systems can be considered as linear systems with nonlinear uncertainties. Some robust control methods for example, the approach discussed by Xie, *et al.*, (1992) can deal with Lur'e systems. However, a direct treatment on Lur'e systems may describe their properties much more explicitly. Moreover, when the Lur'e control variable p is needed for feedback controllers, the robust results for linear systems may be unavailable.

3 Conclusion

This paper discuss the absolute stabilization problem problems for Lur'e systems with multiple nonlinearities. Solvability conditions are presented in term of multilinear matrix inequalities (MLMIs) for both state feedback and output feedback cases. Feasible design algorithms obtained in the paper are involved in optimization of LMI or BLMI, which have been studied by Boyd, *et al.*, (1994), Goh, *et al.*, (1994a,1994b) more sufficiently. The approach presented in this note can be used to deal with other problems for Lur'e systems related to LMIs such as L_2-gain control, passive control and other dissipative control problems.

References

Knobloch, H. W., A. Isidori and D. Flockerzi. (1993). *Topics in control theory.* Switzerland, Birkhauser Verlag.

Narendra, K. S. and J. H. Taylor. (1973). *Frequency domain criteria for absolute stability.* New York, Academic Press, 1973.

Savkin, A., V. and I. R. Petersen. (1995). A method for robust stabilization related to the Popov stability criterion. *Int. J. Control,* 62(5):1105-1115.

Boyd, S., L. EL Ghaoui, E. Feron, and V. Balakrishnan. (1994). *Linear matrix inequality in system and control theory.* Philadelphia, SIAM Studies in Applied Mathematics, Vol.15.

Goh, K., C., L. Turan, M. G. Safonov, et al. (1994a). Bilinear matrix inequality properties and computational methods. In:*Proc. of Conf. on ACC,* 1994, 850-855.

Goh, K. C., M. G. Safonov and G. P. Papavassilopoulos. (1994b). A global optimization approach for the BMI problem. In:*Proc. of Conf. on CDC,* 2009-2014.

Rapoport, L.B. (1987). Absolute stability of control systems with several nonlinear stationary elements. *Automation and Remote Control,* 623-630.

Xie, L., M. Fu and C. de Souza (1992). H^∞ control and quadratic stabilization of systems with parameter uncertainties via output feedback. *IEEE Trans. on AC.,* AC-37:1253-1260.

C-2a-03-5
14th Triennial World Congress, Beijing, P.R. China

HYBRID CONTROL OF SANDWICH SYSTEMS WITH NONSMOOTH NONLINEARITIES

Gang Tao
Department of Electrical Engineering
University of Virginia
Charlottesville, VA 22903, USA
E-mail: gt9s@virginia.edu

Abstract: This paper presents two discrete/continuous-time hybrid control schemes using different feedback signals for control of a class of sandwich systems with a nonsmooth nonlinear block such as a dead-zone between two linear time-invariant dynamic blocks. The proposed control schemes employ an inner-loop discrete-time feedback design and an outer-loop continuous-time feedback design, combined with a dead-zone inverse, to cancel the dead-zone effect for improving output tracking. The paper also presents two switching backlash compensation control schemes for a gear-train sandwich system with backlash: one for known parameters and the other for unknown parameters, for improving system tracking performance.

Keywords: Adaptive control, backlash, dead-zone, hybrid control, nonlinearity inverse, sandwich system, tracking.

1 Introduction

Nonsmooth nonlinearities such as dead-zones, backlash, hysteresis are common in actuators and sensors of control systems, such as mechanical connections, hydraulic servo-valves and electric servomotors [1] - [5], and even appear in biomedical systems [6]. Usually poorly known, they often severely limit system performance. It has been a challenging task to control such nonlinear systems. The study of adaptive control for systems with unknown dead-zones at input was initiated by Recker et al. [7] where an adaptive scheme was proposed with full state measurement, and extended with indirect adaptive control designs [8]. A direct adaptive inverse control approach has been developed for systems with an unknown dead-zone, backlash, hysteresis or piecewise-linearity at the input or output of a linear part [9] or a nonlinear part [10]. Those results and others [11], [12], [13] employing adaptive nonlinearity inverses for control of two-block systems with input nonlinearities indicate that adaptive inverse controllers significantly improve system tracking performance.

In this paper, a more general problem is considered: control of sandwich nonlinear systems, especially, those with dead-zone, backlash, hysteresis or piecewise-linearities. Control schemes are presented for sandwich nonlinear systems with a nonsmooth nonlinear characteristic such as a dead-zone, backlash, or hysteresis between two linear time-invariant dynamic blocks and for a sandwich system which has a feedback structure with backlash. The proposed control schemes all have nonlinearity compensation features designed for improving system tracking performance.

2 Control of Three-Block Sandwich Systems

Consider the plant with a nonlinear block $N(\cdot)$ between two linear blocks $G_i(s)$, $G_o(s)$:

$$\begin{aligned} y(t) &= G_o(s)[z](t), \; z(t) = N(u(t)) \\ u(t) &= G_i(s)[v](t) \end{aligned} \tag{2.1}$$

where s is the differentiation operator: $s[x](t) = \dot{x}(t)$, or the Laplace transform operator as the case may be, $v(t)$ is the accessible control input, $y(t)$ is the measured output, $G_o(s)$ and $G_i(s)$ are both rational transfer functions with their denominator degrees being n_o and n_i respectively, and $N(\cdot)$ is a nonsmooth characteristic such as a dead-zone, backlash, hysteresis or piecewise-linear element.

An example of such sandwich plants is a hydraulic actuator with a pilot valve. The block $G_i(s)$ represents the dynamics of the pilot valve piston whose overlapping on the fluid orifice causes a dead-zone represented by $N(\cdot)$. The actuator piston with load, as represented by $G_o(s)$, is controlled by the flow through the orifice.

To cancel the nonlinearity between two linear dynamic blocks, two new control schemes consisting of an inner-outer loop feedback structure incorporated with inverses for the nonlinearity $N(\cdot)$ are presented, for the two cases:

(i) the signal $u(t)$ is measured while $z(t)$ is not;
(ii) the signal $z(t)$ is measured while $u(t)$ is not.

As a comparison, control schemes were developed in [9] for the discrete-time sandwich plant with a linear dynamics $G(D)$, an input nonlinearity $N_i(\cdot)$, and an output nonlinearity $N_o(\cdot)$:

$$\begin{aligned} y(k) &= N_o(z(k)),\ z(k) = G(D)[u](k) \\ u(k) &= N_i(v(k)) \end{aligned} \quad (2.2)$$

where D is the advance operator: $D[x](k) = x(k+1)$), or the z-transform operator as the case may be. It should be noted that, unlike the sandwich plant (2.2) which can be modeled in discrete time, the plant (2.1), operating in continuous time, can not be modeled with G_i, $N(\cdot)$ and G_o all in discrete time. For case (i), only G_i can be modeled in discrete time, while for case (ii), only G_i and $N(\cdot)$ can be modeled in discrete time.

2.1 Design with $u(t)$ Measured

When the signal $u(t)$ is measured, the control scheme is designed to have an inner-loop feedback law, an inverse $NI(\cdot)$ [9] for the nonlinearity $N(\cdot)$, and an outer-loop feedback law.

The inner-loop feedback law is a discrete-time model reference controller based on a discrete-time version of the linear block $G_i(s)$:

$$v(k) = \theta_{11}^T \omega_{11}(k) + \theta_{12}^T \omega_{12}(k) + \theta_{13} w_s(k) \quad (2.3)$$

where $w_s(k)$ is the sampled version of $w(t)$, and

$$\omega_{11}(k) = a_\lambda(D)[v](k),\ \omega_{12}(k) = b_\lambda(D)[u_s](k)$$

with $a_\lambda(D) = (D^{-n_i+1}, \ldots, D^{-1})^T$, $b_\lambda(D) = (a_\lambda^T(D), 1)^T$ being vector filters and $u_s(k)$ being the sampled version of $u(t)$. This inner-loop controller is to cancel the linear part $G_i(s)$ whose relative degree is n_i^* to ensure the matching

$$u(k) = w_s(k - n_i^*) = w(k - n_i^*).$$

The block $NI(\cdot)$ is a continuous-time inverse for the nonlinearity $N(\cdot)$:

$$w(t) = NI(u_d(t)) \quad (2.4)$$

which is to achieve the delayed inversion of $u_d(t)$ at the sampling instant k:

$$\begin{aligned} z(k) &= N(u(k)) = N(w(k - n_i^*)) \\ &= N(NI(u_d(k - n_i^*))) = u_d(k - n_i^*). \end{aligned} \quad (2.5)$$

The outer feedback loop is a continuous-time model reference controller based on a continuous-time transfer function $G_o(s)$:

$$u_d(t) = \theta_{21}^T \omega_{21}(t) + \theta_{22}^T \omega_{22}(t) + \theta_{23} r(t) + f(u_d(t)) \quad (2.6)$$

where $r(t)$ is a bounded reference signal,

$$\omega_{21}(t) = \frac{a(s)}{\Lambda(s)}[u_d](t),\ \omega_{22}(t) = \frac{b(s)}{\Lambda(s)}[y](t)$$

$$a(s) = (1, s, \ldots, s^{n_o-2})^T,\ b(s) = (1, s, \ldots, s^{n_o-1})^T$$

and $\Lambda(s)$ being a stable polynomial of degree $n_o - 1$, and $f(\cdot)$ is a functional (see [14], [15]) to compensate for the time delay $n_i^* T_s$ caused by the inner-loop controller and the inverse $NI(\cdot)$, where $T_s > 0$ is the sampling interval used for the inner-loop controller. The outer-loop controller is to reduce the tracking error $y(t) - y_m(t)$ for the reference signal $y_m(t)$ generated from a continuous-time reference model:

$$y_m(t) = \frac{1}{P_m(s)}[\bar{r}](t),\ \bar{r}(t) = r(t - n_i^* T_s) \quad (2.7)$$

for a stable polynomial $P_m(s)$ of degree n_o^* equal to the relative degree of $G_o(s)$.

2.2 Design with $z(t)$ Measured

When the signal $z(t)$ is measured, the control scheme consists of an inner-loop feedback control law combined with a pair of identical inverses $NI(\cdot)$ for the nonlinearity $N(\cdot)$, and an outer-loop feedback control law, as shown in Figure 1.

The inner feedback control loop with two discrete-time inverses of $N(\cdot)$ is

$$v(k) = \theta_{11}^T \omega_{11}(k) + \theta_{12}^T \omega_{12}(k) + \theta_{13} w(k) \quad (2.8)$$

$$w(k) = NI(u_{ds}(k)) \quad (2.9)$$

$$\bar{z}(k) = NI(z_s(k)) \quad (2.10)$$

where

$$\omega_{11}(k) = a_\lambda(D)[v](k),\ \omega_{12}(k) = b_\lambda(D)[\bar{z}](k)$$

with $u_{ds}(k)$, $z_s(k)$ being the sampled $u_d(t)$, $z(t)$.

The linear feedback controller (2.8) is a discrete-time model reference controller based on a discrete-time model of $G_i(s)$, which, with the help

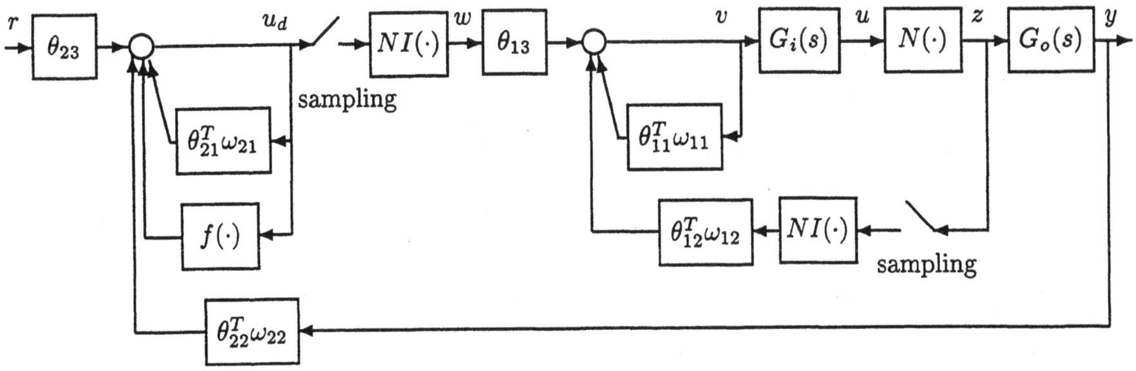

Figure 1: Control system with $z(t)$ measured.

of the two discrete-time inverses (2.9) and (2.10), is to cancel both the linear and nonlinear parts $G_i(s)$, $N(\cdot)$ to ensure the matching

$$z(k) = u_d(k - n_i^*).$$

The outer-loop design is a continuous-time model reference controller designed for $G_o(s)$:

$$u_d(t) = \theta_{21}^T\omega_{21}(t) + \theta_{22}^T\omega_{22}(t) + \theta_{23}r(t) + f(u_d(t)) \tag{2.11}$$

where $r(t)$ is a bounded reference signal,

$$\omega_{21}(t) = \frac{a(s)}{\Lambda(s)}[u_d](t),\ \omega_{22}(t) = \frac{b(s)}{\Lambda(s)}[y](t)$$

and $f(\cdot)$ is a functional to compensate for the time delay $n_i^*T_s$ caused by the inner-loop feedback controller with nonlinearity inverses. Similar to the controller (2.6), this outer-loop controller is for the plant output $y(t)$ to track the continuous-time reference signal $y_m(t)$ in (2.7).

2.3 Performance Analysis

For the case when the signal $u(t)$ is measured, the inner controller (2.3) plus the inverse (2.4) leads to the delayed inversion of $u_d(t)$ at $t = k$, that is, (2.5) holds. If $u_d(t)$ is taken as the new input, then the relationship from $u_d(t)$ to the plant output $y(t)$ is approximately equal to $G_o(s)$ plus an input delay $\tau = n_i^*T_s$. For such a linear system with input delay, the model reference controller structure (2.6), with the additional term $f(\cdot)$ for delay compensation [15], [16], is applicable. For the control design for the case when the signal $z(t)$ is measured, the situation is similar.

The detailed analysis and simulation study are not included here. Some issues of interest in this direction include the development of adaptive versions of the control schemes (2.3) - (2.7), (2.8) - (2.11), with adaptive inverses $\widehat{NI}(\cdot)$, for $G_i(s)$ and/or $G_o(s)$ known or unknown, and $N(\cdot)$ unknown, and the associated closed-loop stability and performance evaluation.

Control of the sandwich plant (2.1) is an important open problem. The novel ideas in the control schemes (2.3) - (2.7) and (2.8) - (2.11) are the hybrid designs which employ a discrete-time inner-loop feedback controller structure and a continuous-time outer-loop feedback controller structure combined with inverses for the nonlinear block, and can cancel the sandwich nonlinearity between two linear blocks.

3 Control of A Sandwich System with Feedback

As an example of sandwich plants with feedback connections, a gear-train system is considered (as that in a gun-turret system) which has backlash and flexibility [16], [17], [18].

Let $\theta_l(t)$ be the controlled output (angular displacement of the driven gear) and $\theta_m(t)$ be the measured intermediate variable (angular displacement of the driving gear). The impact force on gear teeth due to backlash and flexibility is

$$u_f = k\delta,\ \delta = DZ(\theta_m - \theta_l) \tag{3.1}$$

where k is the stiffness constant of the gear teeth and $DZ(\cdot)$ is the dead-zone characteristic

$$DZ(\theta_m - \theta_l) = \begin{cases} \theta_m - \theta_l - b & \text{if } \theta_m - \theta_l \geq b \\ 0 & \text{if } -b < \theta_m - \theta_l < b \\ \theta_m - \theta_l + b & \text{if } \theta_m - \theta_l \leq -b \end{cases}$$

with $b > 0$, which models the backlash effects.

The gear-train sandwich system with feedback connection is shown in Figure 2 and described by

$$J_m\ddot{\theta}_m + b_m\dot{\theta}_m = u - u_f \tag{3.2}$$

$$J_l\ddot{\theta}_l + b_l\dot{\theta}_l = u_f \tag{3.3}$$

where $u(t)$ is the control input (torque) to be designed, the impact force $u_f(t)$ is defined in (3.1), and $b_l\dot{\theta}_l$ and $b_m\dot{\theta}_m$, with $b_l \geq 0$ and $b_m \geq 0$, represent viscous friction forces.

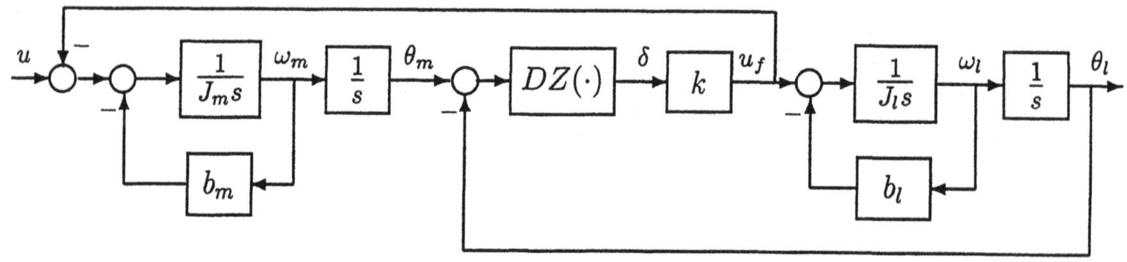

Figure 2: Gear-train sandwich system model.

The control objective is to design a feedback control $u(t)$ such that the closed-loop system is stable and the system output $\theta_l(t)$ tracks a given reference signal $\theta_d(t)$ as close as possible. Obviously, the key feature of a desirable design is its ability to cancel the backlash effects.

At any given time, the system operates in either of the two phases: impact phase (when two gears are in contact) and backlash phase (when two gears are not in contact). The backlash width b characterizes the "size" of the backlash effect on system tracking performance. To develop backlash compensation control schemes, the following two assumptions are made:

(A1) the knowledge when the system operates in the backlash phase is available;

(A2) the backlash width parameter b is known.

3.1 Designs for Rigid Gears with Known Parameters

In this section, an inverse control design is presented for the rigid gears case ($k = \infty$) with system parameters J_m, J_l, b_m and b_l known.

For $k = \infty$, the system (3.2) - (3.3) has a simpler form: when in the *impact phase*, that is, when $\theta_m(t) - \theta_l = b$, $\dot{\theta}_m(t) - \dot{\theta}_l(t) = 0$, or $\theta_m(t) - \theta_l(t) = -b$, $\dot{\theta}_m(t) - \dot{\theta}_l = 0$, it becomes

$$J_0\ddot{\theta}_m + b_0\dot{\theta}_m = u, \; J_0 = J_l + J_m, \; b_0 = b_l + b_m \tag{3.4}$$

when in the *backlash phase*, that is, for $\theta_m(t)-\theta_l = b$, $\dot{\theta}_m(t)-\dot{\theta}_l < 0$, or $\theta_m(t)-\theta_l = -b$, $\dot{\theta}_m(t)-\dot{\theta}_l > 0$, or $-b < \theta_m(t) - \theta_l < b$, it becomes

$$J_m\ddot{\theta}_m + b_m\dot{\theta}_m = u, \; J_l\ddot{\theta}_l + b_l\dot{\theta}_l = 0. \tag{3.5}$$

3.1.1 Impact-Phase Control Design

For the system (3.4) in the *impact phase*, the control signal is

$$\begin{aligned} u(t) &= b_0\dot{\theta}_m + J_0(\ddot{\theta}_d(t) - k_1(\dot{\theta}_m(t) - \dot{\theta}_d(t)) \\ &\quad -k_2(\theta_m(t) - \theta_d(t) + \theta_i)) \end{aligned} \tag{3.6}$$

where $\theta_d(t)$ is the desired output, $k_1 > 0$, $k_2 > 0$ are design parameters, $\theta_i = -b$ if $\theta_m(t) - \theta_l = b$, or $\theta_i = b$ if $\theta_m(t) - \theta_l = -b$.

From the condition: $\ddot{\theta}_m(t) = \ddot{\theta}_l(t)$ and $\dot{\theta}_m(t) = \dot{\theta}_l(t)$, in the *impact phase*, the controller (3.6) results in the desired error equation

$$\ddot{\theta}_l - \ddot{\theta}_d + k_1(\dot{\theta}_l - \dot{\theta}_d) + k_2(\theta_l - \theta_d)) = 0. \tag{3.7}$$

3.1.2 Backlash-Phase Control Design

For the system (3.5) in the *backlash phase*, the control signal is

$$\begin{aligned} u(t) &= b_m\dot{\theta}_m + J_m(\ddot{\theta}_l(t) - k_3(\dot{\theta}_m(t) - \dot{\theta}_l(t)) \\ &\quad -k_4(\theta_m(t) - \theta_l(t) + \theta_b)) \end{aligned} \tag{3.8}$$

where $k_3 > 0$, $k_4 > 0$ are design parameters, and $\theta_b = -b - \epsilon$ if $\theta_l(t) - \theta_d(t) < 0$, or $\theta_b = 0$ if $\theta_l(t) - \theta_d(t) = 0$, or $\theta_b = b + \epsilon$ if $\theta_l(t) - \theta_d(t) > 0$ for a small design parameter $\epsilon > 0$. The controller (3.8) results in the error equation

$$\ddot{\theta}_m - \ddot{\theta}_l + k_3(\dot{\theta}_m - \dot{\theta}_l) + k_4(\theta_m - \theta_l + \theta_b)) = 0 \tag{3.9}$$

which implies that $\lim_{t\to\infty}(\theta_m(t) - \theta_l(t) + \theta_b) = 0$ if the closed-loop system stays in the backlash phase. Thus, the backlash phase control $u(t)$ in (3.8) makes θ_m traverse through the backlash, ready for the next desired move if the tracking error $\theta_l(t) - \theta_d(t) \neq 0$, or leads to $\theta_m(t) = \theta_l(t)$ if $\theta_l(t) - \theta_d(t) = 0$. The use of $\epsilon > 0$ in (3.8) is to ensure that $\theta_m(t)$ reaches $\theta_l(t) - b$ or $\theta_l(t) + b$ in a finite time which can be made smaller with larger k_3 and k_4, leading to a faster error system (3.9).

3.1.3 Discussion

The choice of k_i, $i = 1,2,3,4$, in (3.6) and (3.8), determines the dynamics of the error equations (3.7) and (3.9), that is, the closed-loop system performance. It is important that the choice of k_3 and k_4 results in a fast dynamics of the error system (3.8) so that the convergence of $\theta_m(t)$ to $\theta_l(t) - \theta_b$ is fast. The proposed control schemes employ an inverse for the gear-train backlash nonlinearity—the use of θ_i in (3.6) and θ_b in (3.8) is to compensate for the unwanted backlash effects.

It is interesting to note that the signal $\ddot{\theta}_l(t)$ for implementing (3.8) can be obtained from (3.5): $\ddot{\theta}_l(t) = -\frac{b_l}{J_l}\dot{\theta}_l(t)$, as it is in the backlash phase.

To examine the system stability, in the impact phase, the closed-loop system is (3.6) which is obviously asymptotically stable with the choice of $k_1 > 0$ and $k_2 > 0$. When the system is in backlash phase, the motion of the driven gear is $J_l\ddot{\theta}_l + b_l\dot{\theta}_l = 0$. If $b_l > 0$, then the solution $\theta_l(t)$ of this dynamic equation is bounded if the initial conditions are bounded. Hence, the closed-loop error system (3.9) for the backlash phase is also asymptotically stable. If $b_l = 0$, that is, if no friction is present, then the solution $\theta_l(t)$ may grow linearly in the time t. However, with $\epsilon > 0$ used in the control (3.8) and the choice of $k_3 > 0$ and $k_4 > 0$ large for a fast dynamic error system (3.9), $\theta_m - \theta_l(t) + \theta_0$ ($\theta_0 = b$ or $\theta_0 = -b$) reaches zero at a finite time to make the gears leave the backlash phase, so that $\theta_l(t)$ still remains bounded.

3.2 Designs for Rigid Gears with Unknown Parameters

When the system parameters J_m, J_l, b_m, and b_l are all unknown, the following adaptive control scheme is proposed for the rigid gear case.

3.2.1 Impact-Phase Adaptive Design

When the system is in the *impact phase*, its dynamic equation is in (3.4), with J_0 and b_0 unknown and $\theta_m(t) - \theta_l = b$ or $\theta_m(t) - \theta_l(t) = -b$. To develop an adaptive control scheme for (3.4), the error signal is defined as

$$e(t) = \theta_m(t) + \theta_i - \theta_d(t)$$

where $\theta_d(t)$ is the desired system output such that $\theta_d(t)$, $\dot{\theta}_d(t)$ and $\ddot{\theta}_d(t)$ are all bounded and available, and $\theta_i = -b$ if $\theta_m(t) - \theta_l = b$, or $\theta_i = b$ if $\theta_m(t) - \theta_l = -b$ (that is, $e(t) = \theta_l(t) - \theta_d(t)$ is in fact the tracking error between the system output $\theta_l(t)$ and the desired output $\theta_d(t)$). A frequently used technique [19] is to define

$$s(t) = \dot{e}(t) + \lambda_1 e(t),\; v(t) = \dot{\theta}_d(t) - \lambda_1 e(t),\; \lambda_1 > 0.$$

Let $\hat{J}_0(t)$ and $\hat{b}_0(t)$ be the estimates of J_0 and b_0. The control input is chosen as

$$u(t) = \hat{J}_0(t)\dot{v}(t) + \hat{b}_0(t)(s(t)+v(t)) - \lambda_2 s(t),\; \lambda_2 > 0$$

and the adaptive law to update $\hat{p}_0 = (\hat{J}_0, \hat{b}_0)^T$ is

$$\dot{\hat{p}}_0(t) = -\Gamma_0 s(t)\zeta(t),\; \Gamma_0 = \Gamma_0^T > 0$$

$$\zeta(t) = (\dot{v}(t), s(t) + v(t))^T.$$

This adaptive design turns out desirable:

$$V(s, \hat{p}_0 - p_0) = s^2 J_0 + (\hat{p}_0 - p_0)^T\Gamma_0^{-1}(\hat{p}_0 - p_0)$$

$$\dot{V} = -2\lambda_2 s^2 \le 0.$$

3.2.2 Backlash-Phase Adaptive Design

When the system is in the *backlash phase*, its dynamic equations are in (3.5). The motion of $J_m\ddot{\theta}_m + b_m\dot{\theta}_m = u$ is to be controlled so that $\theta_m(t)$ is to track $\theta_l(t) - \theta_b$, where $\theta_b = -b - \epsilon$ if $\theta_l(t) - \theta_d(t) < 0$, or $\theta_b = 0$ if $\theta_l(t) - \theta_d(t) = 0$, or $\theta_b = b + \epsilon$ if $\theta_l(t) - \theta_d(t) > 0$ for a small design parameter $\epsilon > 0$. The unknown parameters are J_m and b_m, and the tracking error is

$$e_m(t) = \theta_m(t) - \theta_l(t) - \theta_b.$$

To design a control signal $u(t)$ for (3.5), it is assumed that $\theta_l(t)$, $\dot{\theta}_l(t)$ and $\ddot{\theta}_l(t)$ are all bounded and available for measurement, during the time interval when the system is operating in the backlash phase. Such a boundedness condition can be met since $J_l\ddot{\theta}_l + b_l\dot{\theta}_l = 0$ is a linear system.

To proceed, define

$$\begin{aligned} s_m(t) &= \dot{e}_m(t) + \lambda_3 e_m(t) \\ v_m(t) &= \dot{\theta}_l(t) - \lambda_3 e_m(t),\; \lambda_3 > 0 \end{aligned} \quad (3.10)$$

and let $\hat{J}_m(t)$ and $\hat{b}_m(t)$ be the estimates of J_m and b_m. The control is chosen as

$$\begin{aligned} u(t) &= \hat{J}_m(t)\dot{v}_m(t) + \hat{b}_m(t)(s_m(t) + v_m(t)) \\ &\quad -\lambda_4 s_m(t),\; \lambda_4 > 0 \end{aligned} \quad (3.11)$$

and the update law for $\hat{p}_m = (\hat{J}_m, \hat{b}_m)^T$ is

$$\dot{\hat{p}}_m(t) = -\Gamma_m s_m(t)\zeta_m(t),\; \Gamma_m = \Gamma_m^T > 0$$

$$\zeta_m(t) = (\dot{v}_m(t), s_m(t) + v_m(t))^T.$$

This adaptive design is also desirable:

$$V(s_m, \hat{p}_m - p_m) = s_m^2 J_m + (\hat{p}_m - p_m)^T\Gamma_m^{-1}(\hat{p}_m - p_m)$$

$$\dot{V} = -2\lambda_4 s_m^2 \le 0.$$

3.3 Discussion

Thus far, two control schemes with backlash compensation are proposed for control of systems with backlash between two linear blocks. In the present control problem, the most desirable function of a backlash compensation controller is its ability to cancel the backlash effects using a control signal not directly reaching the backlash. To invert the backlash through a dynamic block representing the driving gear dynamics, the fastness of the system response during the backlash phase is crucial. If the backlash phase control gains k_3 and k_4 in (3.7) (or λ_3 and λ_4 in (3.10) and (3.11)) are large, then the system response in the backlash phase can be made fast so that the cancellation of backlash can be achieved at a fast rate.

4 Conclusions

In this paper, a class of open control problems have been addressed: control of sandwich systems with nonsmooth nonlinearities. Such problems are important because many control systems have sandwich structures and dead-zone, backlash and hysteresis are common practical nonlinearities in control systems. Several designs were presented for control of three-block sandwich systems with a nonsmooth nonlinear characteristic such as dead-zone, backlash, or hysteresis between two linear dynamic blocks and for a feedback sandwich system with backlash. These designs take into account the nonsmooth nonlinearities and provide special functions to compensate the undesirable nonlinearity effects so that the system tracking performance can be improved despite the presence of nonlinearities. The nonlinearity compensation techniques used in the developed control schemes are the discrete-continuous time hybrid designs in Section 2 and the two-phase switching designs in Section 3. Encouraged by these preliminary results, more research is currently under way on the related control issues in this direction.

Acknowledgements

This research was supported by the National Science Foundation under grant ECS-9619363. The author would like to thank Professor Petar Kokotović and Dr. Kenan Ezal of University of California for their helpful comments on this work.

References

[1] John G. Truxal, *Control Engineers' Handbook*, McGraw-Hill, New York, 1958.

[2] H. E. Merritt, *Hydraulic Control Systems*, John Wiley & Sons, Inc., New York, 1967.

[3] A. Netushil, *Theory of Automatic Control*, Mir Publishers, Moscow, 1973.

[4] M. A. Krasnoselskii and A. V. Pokrovskii, *Systems with Hysteresis*, Springer-Verlag, Berlin, 1983.

[5] I. D. Mayergoyz, *Mathematical Models of Hysteresis*, Springer-Verlag, Berlin, 1991.

[6] M. S. Hatwell, B. J. Oderkerk, C. A. Sacher and G. F. Inbar, "The development of a model reference adaptive controller to control the knee joint of paraplegics," *IEEE Transactions on Automatic Control*, vol. AC-36, no. 6, pp. 683-691, 1991.

[7] D. Recker, P. Kokotović, D. Rhode and J. Winkelman, "Adaptive nonlinear control of systems containing a dead-zone," *Proceedings of the 30th IEEE Conference on Decision and Control*, pp. 2111-2115, Brighton, England, 1991.

[8] D. Recker, *Adaptive Control of Systems Containing Piecewise Linear Nonlinearities*, Ph. D., Thesis, University of Illinois, Urbana, IL, 1993.

[9] G. Tao and P. V. Kokotović, *Adaptive Control of Systems with Actuator and Sensor Nonlinearities*, John Wiley & Sons, 1996.

[10] M. Tian and G. Tao, "Adaptive dead-zone compensation for output-feedback canonical systems," *International Journal of Control*, vol. 67, no. 5, pp. 791–812, 1997.

[11] J.-H. Kim, J.-H. Park, S.-W. Lee, and E. K. P. Chong, "A two-layered fuzzy logic controller for systems with dead-zones," *IEEE Trans. on Industrial Electronics*, vol. 41, no. 2, pp. 155-162, April 1994.

[12] A. J. Kurdila and G. Webb, "Compensation for distributed hysteresis operators in active structural systems," *Journal of Guidance, Control, and Dynamics*, vol. 20, no. 6, November-December, 1997.

[13] F. L. Lewis, K. Liu, R. R. Selmic and Li-Xin Wang, "Adaptive fuzzy logic compensation of actuator dead-zones," *Journal of Robot Systems*, vol. 14, no. 6, pp. 501-511, 1997.

[14] R. Ortega and R. Lozano, " Globally stable adaptive controller for systems with delay," *Int. Journal Control*, vol. 47, no. 1, pp. 17 - 23, 1988.

[15] G. Tao, "Model reference adaptive control of multivariable plants with delays," *International Journal of Control*, vol. 55, no. 2, pp. 393-414, 1992.

[16] H. G. Kwatny, C. Teolis and G. Blankenship, "Robust, adaptive control of systems with discontinuous nonlinearities," Techno-Sciences, Inc., Lanham, MD, Phase I Report, 1995.

[17] J.-H. Yang and L.-C. Fu, "Nonlinear adaptive control for manipulator system with gear backlash," *Proceedings of the 35th IEEE Conference on Decision and Control*, pp. 4369-4374, Kobe, Japan, 1996.

[18] K. Ezal, P. V. Kokotović and G. Tao, "Optimal control of tracking systems with backlash and flexibility," *Proceedings of the 36th IEEE Conference on Decision and Control*, pp. 1749–1754, San Diego, CA, 1997.

[19] J. Slotine and W. Li, "On the adaptive control of robot manipulators," *International Journal of Robotics Research*, vol. 6, no. 3, 1987, pp. 49-59.

14th Triennial World Congress, Beijing, P.R. China

C-2a-03-6

INVERSE OPTIMAL STABILIZATION FOR FEEDFORWARD NONLINEAR SYSTEMS[1]

Tarek Ahmed-Ali*, Frédéric Mazenc and Françoise Lamnabhi-Lagarrigue***

*Laboratoire des Signaux et Systèmes, C.N.R.S.
SUPELEC
91192 Gif-sur-Yvette, France
ahmed@lss.supelec.fr
lamnabhi@lss.supelec.fr

**Process Systems Engineering Centre
Imperial College of Science Technology and Medecine
London SW 7 2AZ, UK
f.mazenc@ic.ac.uk

Abstract: In this paper is solved the inverse optimal stabilization problem for nonlinear nonaffine control discrete-time systems which are globally stable when uncontrolled. Stabilizing feedback laws and cost functionals are given by explicit formulas. We apply the result to feedforward systems. *Copyright © 1999 IFAC*

Keywords: Inverse optimality, Stabilization, Feedforward systems

1. Introduction

Inverse optimality is a notion introduced a few years ago by Moylan and Anderson [11] and studied also later by Freeman and Kokotović [1], [2]. It arises from the wish to have for nonlinear systems the nice robustness properties of the stabilizing feedbacks obtained via the standard optimization results without having to solve the Hamilton-Jacobi-Bellman equation which is not always a feasible task. In the inverse optimal approach, a stabilizing feedback is designed first and then it is shown to be optimal for a cost functional of the form

$$\int_0^\infty \left(l(x) + u^\top R(x)u\right) dt$$

where l and R are positive functions. In other words, the functions $l(x)$ and $R(x)$ are *a posteriori* determined from the data of a particular stabilizing feedback, rather than *a priori* chosen by the designer i.e. regardless of any stabilizing feedback. Some results in this new part of the nonlinear control theory were also given by Sepulchre, Janković and Kokotović [3, Section 3.5] for systems satisfying the conditions of the Jurdjevic-Quinn or Artstein's theorem and by Krstić and Li [5], [6]. So far, to the best knowledge of our, all the works on this subject are dealing with continuous-time systems. The purpose of the present paper is to address the problem of inverse optimal stabilization of general classes of nonlinear discrete-time systems. More precisely, its aim is twofold. In a first part, we show that the stabilizing feedbacks designed in [8, Appendix A] and in [9, Section 2] for nonlinear discrete-time systems

$$x_{i+1} = f(x_i) + g(x_i, u_i)u_i$$

which are globally stable when $u = 0$ also minimizes cost functionals of the form

$$J = \sum_{i=0}^{\infty}(\alpha(x_i, u_i) + \beta(x_i, u_i)|u_i|^2).$$

[1] Work partially supported by the European Commission's Training and Mobility of Researchers (TMR) Contract # ERBFMRX-CT970137

In a second part, we apply the results of the first to a particular class of discrete-time feedforward systems i.e. systems which admit a representation of the form:

$$\begin{aligned} z_{i+1} &= \mathcal{F}(z_i) + \psi(z_i, \xi_i) + g_1(z_i, \xi_i, u_i)u_i \\ \xi_{i+1} &= a(\xi_i) + g_1(z_i, \xi_i, u_i)u_i \end{aligned}$$

where $z_{i+1} = \mathcal{F}(z_i)$ is globally stable, $\xi_{i+1} = a(\xi_i)$ is globally asymptotically stable and locally exponentially stable and $\psi(z, 0) = 0$ for all z. These systems have been studied for the first time by A. Teel in [12]. Many extentions have followed: see [8, 3, 10]. A discrete-time version of the main result of [8] is proved in [9]. In the present paper, we present a discrete-time version of the main result of [3]. Let us mention that the study of this class of systems is, from a practical point of view, appealing. On the one hand, many physical systems are described by feedforward equations (Cart-pendulum system, Ball and Beam with friction term, PVTOL) and, on the other hand the technique provides us with bounded feedbacks for null-controllable systems.

Preliminaries

1. Throughout the paper we assume that the functions encountered are sufficiently smooth.

2. We denote by χ_i the solution of the discrete-time system:

$$\chi_{i+1} = \mathcal{H}(\chi_i) \tag{1}$$

with the initial condition $\chi_0 = \chi$.

3. For a sequence χ_i solution of (1) and a function $\mathcal{V}$ we denote by $\Delta\mathcal{V}$ the term $\mathcal{V}(\chi_{i+1}) - \mathcal{V}(\chi_i)$.

4. A function $\mathcal{V}$ is positive definite if

$$\mathcal{V}(\chi) > 0, \forall \chi \neq 0 .$$

2. Inverse optimal control

In this section, we study the inverse problem of optimal stabilization for discrete-time nonlinear systems

$$x_{i+1} = f(x_i) + g(x_i, u_i)u_i \tag{2}$$

with $f(0) = 0$ and $x_i \in \mathbf{R}^n$ and $u \in \mathbf{R}$ and introduce the following assumptions:

H1. There exists a proper, positive definite function V and such that $V(f(x)) - V(x) \leq 0$.

H1'. There exists a positive definite function V and such that $V(f(x)) - V(x) \leq 0$.

H2. The sets

$$\Omega = \{x \in \mathbf{R}^n : V(f^{i+1}(x)) = V(f^i(x))\}$$

$$S = \{x \in \mathbf{R}^n : \frac{\partial V}{\partial \alpha}|_{\alpha = f^{i+1}(x)} g(f^i(x), 0) = 0\}$$
$$i = 0, 1, 2, \ldots$$

are such that

$$\Omega \bigcap S = \{0\}.$$

Remark. The systems (2) is a single-input system. Generalizing our results to multi-input systems can be proved but it turns out that the proofs are then much more intricate.

2.1. Globally asymptotically stabilizing feedback

Let us recall a result which is an immediate consequence of [8, Lemma II.4] or of the feedback design of [9, Section2].

Theorem 2.1 *Consider the discrete-time systems (2). Assume that Assumptions* ***H1*** *and* ***H2*** *are satisfied. Then for all function $\mu(x) > 0$, there exists a smooth function $\phi(x)$ such that the following feedback control*

$$\overline{u}(x) = -\phi(x)h(x,0), \qquad 0 < \phi(x) \leq \mu(x) \tag{3}$$

$$h(x,u) = \int_0^1 \frac{\partial V}{\partial \alpha}|_{\alpha = f(x)+g(x,u)u\theta} g(x,u)d\theta \tag{4}$$

globally asymptotically stabilizes the system (2).

Theorem 2.2 *Consider the discrete-time systems (2). Assume that Assumptions* ***H1'*** *and* ***H2*** *are satisfied. Then, for all function $\mu(x) > 0$ such that there exists a smooth function $\phi(x)$ such that all the solutions of (2) in closed-loop with the following feedback control*

$$\overline{u}(x) = -\phi(x)h(x,0), \qquad 0 < \phi(x) \leq \mu(x) \tag{5}$$

$$h(x,u) = \int_0^1 \frac{\partial V}{\partial \alpha}|_{\alpha = f(x)+g(x,u)u\theta} g(x,u)d\theta \tag{6}$$

are bounded, the system (2) is globally asymptotically stabilizes by the feedbacks (5).

Discussion of Theorem 2.1 and Theorem 2.2.

i) The main advantage of the feedbacks (3), (5) is that, contrary to those proposed in [7], there are given by explicit formulas and not as the implicit solutions of nonlinear algebraic equations which do not necessarily admit a solution.

ii) Theorem 2.1 is a discrete-time nonaffine version of the Jurdjevic-Quinn theorem [4]: Assumption **H1** guarantees the global stability of (2) with $u = 0$ and the technical Assumption **H2** guarantees that a detectability property which allows to conclude by invoking the LaSalle invariance principle with arbitrarily small feedbacks is satisfied.

iii) The advantage of Theorem 2.2 is clear: it does not require the knowledge of a proper function V.

2.2. Optimal criterium design for continuous-time systems

Let us consider

$$\begin{aligned} \dot{x} &= f(x) + g(x,u)u \\ h(x,u) &= \frac{\partial V}{\partial x}(x)g(x,u) \end{aligned}$$

Definition 1: The *inverse optimal stabilization problem* for discrete-time systems (2) is solvable if there exist positive real-valued functions $\alpha(x,u)$ and $\beta(x,u)$ such that there exists a feedback law $u(x)$ which globally asymptotically stabilizes (2) and at the same time minimizes the cost functional

$$J = \int_0^\infty (\alpha(x,u) + \beta(x,u)u^2)dt.$$

Let us state our main result.

Theorem 2.3 *Let us consider the class of systems (2). Assume that the assumptions of Theorem 2.1 or Theorem 2.2 are satisfied. Then, for all function $\mu(x) > 0$ the inverse optimal stabilization problem is solved by the following control law*

$$\overline{u}(x) = -\phi(x)h(x,0), \qquad 0 < \phi(x) \leq \mu(x) \tag{7}$$

with $h(x,u)$ given in (4) and the cost functional

$$J = \int_0^\infty (\alpha(x,u) + \beta(x,u)u^2)dt \tag{8}$$

with

$$\alpha(x,u) = W(x) + \frac{1}{2}\rho(u)\phi(x)h(x,0)^2 \tag{9}$$

$$\beta(x,u) = k(x,u) + \frac{\rho(u)}{2\phi(x)} + \frac{\rho(u)-1}{u}h(x,0) \tag{10}$$

where $\rho(u)$ is a strictly positive function given by an explicit formula and such that $\rho(0) = 1$ and k is defined by

$$-h(x,u) = -h(x,0) + k(x,u)u\ . \tag{11}$$

Proof. Theorem 2.1 or Theorem 2.2 provides us with a globally asymptotically stabilizing feedback $\overline{u}$ of the form (3) or (5). Let us consider the following criterium

$$S = \int_0^\infty \left[W(x) - \frac{\partial V}{\partial x}(x)g(x,u)u + \frac{\rho(u)}{2\phi(x)}(u - \overline{u}(x))^2\right] dt\ . \tag{12}$$

Let us observe that, since ρ is a positive function, $u = \overline{u}(x)$ minimizes S.

Moreover

$$\begin{aligned} S &= \int_0^\infty (W(x) - \frac{\partial V}{\partial x}(x)g(x,u)u \\ &+ \frac{\rho(u)}{2\phi(x)}(u^2 - 2u\overline{u}(x) + \overline{u}(x)^2))dt \\ &= \int_0^\infty (W(x) + \frac{\rho(u)}{2\phi(x)}\overline{u}(x)^2 - h(x,u)u \\ &- \frac{\rho(u)}{\phi(x)}u\overline{u}(x) + \frac{\rho(u)}{2\phi(x)}u^2)dt\ . \end{aligned}$$

Next, we rewrite S as:

$$\begin{aligned} S &= \int_0^\infty W(x) + \frac{1}{2}\rho(u)\phi(x)h(x,0)^2 \\ &+ (\rho(u)-1)uh(x,0) + (k(x,u) + \frac{\rho(u)}{2\phi(x)})u^2)dt\ . \end{aligned}$$

At last, we get:

$$S = \int_0^\infty \left[\alpha(x,u) + \beta(x,u)u^2\right] dt \,. \tag{13}$$

with α and β defined respectively in (9) and (10).

Assumption **H1** guarantees that α is positive. The proof is completed by using the following Lemma:

Lemma: It is possible to determine $\phi(x)$ and $\rho(u)$ such that $\beta(x,u)$ is a strictly positive function.

Proof: Since ρ is smooth and such that $\rho(0) = 1$, the function $\left|\frac{1-\rho(u)}{u}\right|$ is bounded on a neighborhood of the origin. So, we may determine a function $\phi(x)$ such that, for all x, u

$$\frac{\rho(u)}{4\phi(x)} - h(x,0)\frac{1-\rho(u)}{u} \geq 0 \,.$$

Next, let us prove that we may determine $\phi(x)$ and $\rho(u)$ so that

$$k(x,u) + \frac{\rho(u)}{4\phi(x)} \geq 0. \tag{14}$$

One can always determine explicit formulas of positive functions m, l such that $|k(x,u)| \leq l(x)m(u)$ and $m(0) \leq 1$. So, (14) holds if

$$\frac{1}{\phi(x)} \geq 4l(x)\frac{1}{\rho(u)}m(u).$$

which is satisfied with $\phi(x) \leq \frac{1}{4l(x)+1}$ and $\rho(u) \geq m(u)$. This allows us to conclude our proof.

2.3. *Optimal criterium design for discrete-time systems*

Definition 1: The *inverse optimal stabilization problem* for discrete-time systems (2) is solvable if there exist positive real-valued functions $\alpha(x,u)$ and $\beta(x,u)$ such that there exists a feedback law $u(x)$ which globally asymptotically stabilizes (2) and at the same time minimizes the cost functional

$$J = \sum_{i=0}^{\infty}(\alpha(x_i,u_i) + \beta(x_i,u_i)u_i^2).$$

Let us state our main result.

Theorem 2.4 *Let us consider the class of systems (2). Assume that the assumptions of Theorem 2.1 or Theorem 2.2 are satisfied. Then, for all function $\mu(x) > 0$ the inverse optimal stabilization problem is solved by the following control law*

$$\overline{u}(x) = -\phi(x)h(x,0), \qquad 0 < \phi(x) \leq \mu(x) \tag{15}$$

with $h(x,u)$ given in (4) and the cost functional

$$J = \sum_{i=0}^{\infty}(\alpha(x_i,u_i) + \beta(x_i,u_i)u_i^2) \tag{16}$$

with

$$\alpha(x,u) = V(x) - V(f(x)) + \frac{1}{2}\phi(x)\rho(u)h(x,0)^2 \tag{17}$$

$$\beta(x,u) = k(x,u) + \frac{\rho(u)}{2\phi(x)} - h(x,0)\frac{1-\rho(u)}{u} \tag{18}$$

where $\rho(u)$ is a strictly positive function given by an explicit formula and such that $\rho(0) = 1$ and k is defined by

$$-h(x,u) = -h(x,0) + k(x,u)u \,. \tag{19}$$

Proof. Theorem 2.1 or Theorem 2.2 provides us with a globally asymptotically stabilizing feedback $\overline{u}$ of the form (3) or (5). Let us consider the following criterium

$$\begin{aligned} S &= -\sum_{i=0}^{\infty}(V(x_{i+1}) - V(x_i)) + \sum_{i=0}^{\infty}\frac{\rho(u_i)}{2\phi(x_i)}(u_i - \overline{u}(x_i))^2 \\ &= -\sum_{i=0}^{\infty}(V(f(x_i)) - V(x_i) + h(x_i,u_i)u_i) \\ &+ \sum_{i=0}^{\infty}\frac{\rho(u_i)}{2\phi(x_i)}(u_i - \overline{u}(x_i))^2 \,. \end{aligned}$$

Let us observe that, since ρ is a positive function, $u = \overline{u}(x)$ minimizes S. Next, we rewrite S as:

$$\begin{aligned} S &= -\sum_{i=0}^{\infty}(V(f(x_i)) - V(x_i)) - \sum_{i=0}^{\infty}h(x_i,u_i)u_i \\ &+ \sum_{i=0}^{\infty}\left[\frac{\rho(u_i)}{2\phi(x_i)}u_i^2 - \frac{\rho(u_i)}{\phi(x_i)}\overline{u}(x_i)u_i + \frac{\rho(u_i)}{2\phi(x_i)}\overline{u}(x_i)^2\right] \,. \end{aligned}$$

Using (19), we get:

$$\begin{aligned} S &= -\sum_{i=0}^{\infty}(V(f(x_i)) - V(x_i)) \\ &+ \sum_{i=0}^{\infty}k(x_i,u_i)u_i^2 - \sum_{i=0}^{\infty}h(x_i,0)u_i \\ &+ \sum_{i=0}^{\infty}\left[\frac{\rho(u_i)}{2\phi(x_i)}u_i^2 + h(x_i,0)\rho(u_i)u_i + \frac{\rho(u_i)}{2\phi(x_i)}\overline{u}(x_i)^2\right] \,. \end{aligned}$$

Regrouping the terms differently, we obtain:

$$\begin{aligned} S &= \sum_{i=0}^{\infty}\left[(V(x_i) - V(f(x_i))) + \frac{1}{2}\phi(x_i)\rho(u_i)h(x_i,0)^2\right] \\ &+ \sum_{i=0}^{\infty}\left[k(x_i,u_i) + \frac{\rho(u_i)}{2\phi(x_i)} - h(x_i,0)\frac{1-\rho(u_i)}{u_i}\right]u_i^2 \,. \end{aligned}$$

Therefore

$$S = \sum_{i=0}^{\infty}\alpha(x_i,u_i) + \sum_{i=0}^{\infty}\beta(x_i,u_i)u_i^2$$

with α and β defined respectively in (17) and (18). Assumption **H1** guarantees that α is positive. The proof is completed by using the previous Lemma.

3. Feedforward discrete-time nonlinear systems

In this section, we particularize the results of Section 2 to the class of the discrete-time feedforward systems i.e. systems having the following representation:

$$\begin{aligned} z_{i+1} &= \mathcal{F}(z_i) + \psi(z_i,\xi_i) + g_1(z_i,\xi_i,u_i)u_i \\ \xi_{i+1} &= a(\xi_i) + g_2(z_i,\xi_i,u_i)u_i \end{aligned} \tag{20}$$

with $z_i \in \mathbf{R}^{n_z}$, $\xi_i \in \mathbf{R}^{n_\xi}$. We introduce the following assumptions

A1. There exists a proper, positive definite function W_1 which is zero at the origin and such that $W_1(\mathcal{F}(z)) - W_1(z) \leq 0$.

A2. There exist a function W_2 positive definite radially unbounded zero at the origin and a function ν positive definite and zero at the origin such that

$$W_2(a(\xi)) - W_2(\xi) \leq -\nu(\xi).$$

Moreover both W_2 and ν are lower bounded on a neighborhood of the origin by a positive definite quadratic function.

A3. There exist two differentiable positive functions γ_0 and γ_1 zero at the origin and such that

$$|\psi(z,\xi)| \leq \gamma_0(\xi) + \gamma_1(\xi)W_1(z)$$

A4. The following inequality is satisfied:

$$\left|\frac{\partial W_1}{\partial \alpha}\right| \leq 1\,.$$

Discussion of the assumptions.

• The Assumptions **A1** to **A4** are the standard assumptions of the forwarding and the family of systems (20) is slightly larger than the one studied in [9] since it is not required on $z_{i+1} = \mathcal{F}(z_i)$ to be linear.

• As an immediate consequence of **A2**, we have that the system $\xi_{i+1} = a(\xi_i)$ is globally asymptotically stable and locally exponentially stable.

• If is known a function $\mathcal{W}_1$ such that

$$\left|\frac{\partial \mathcal{W}_1}{\partial \alpha}\right| \leq \mathcal{L}(\mathcal{W}_1)$$

where $\mathcal{L}$ is a positive function such that $\frac{1}{1+\mathcal{L}} \notin L^1$, then Assumption **A4** is satisfied by

$$W_1 = \int_0^{\mathcal{W}_1} \frac{1}{\mathcal{L}(s)+1} ds.$$

If $\mathcal{W}_1$ is a quadratic form, the corresponding function W_1 satisfies a linear growth property and Assumption **A3** is a linear growth assumption imposed on the coupling term.

In order to derive inverse optimal controls from the result of the previous part, we first prove that Assumptions **A1** to **A4** ensure the Lyapunov stability of the free system associated with (20). We construct a candidate Lyapunov function depending on the given functions W_1 and W_2.

The construction we adopt mimics the one proposed in [3] (see also [10]) for continuous time feedforward systems. An alternative construction is given in [9].

3.1. Stability of the uncontrolled system

Let us introduce the notations: $x = (z, \xi)^\top$,

$$f(z,\xi) = \begin{pmatrix} \mathcal{F}(z) + \psi(z,\xi) \\ a(\xi) \end{pmatrix},$$

$$g(z,\xi,u) = \begin{pmatrix} g_1(z,\xi,u) \\ g_2(z,\xi,u) \end{pmatrix}.$$

Theorem 3.1 *Assume that the Assumptions **A1** to **A4** are satisfied by the system (20). Then this system is Lyapunov stable when $u = 0$ and there exists a function zero at the origin, positive definite and of the form*

$$V(x) = W_1(z) + \Phi(z,\xi) + W_2(\xi) \tag{21}$$

such that $\Delta V \leq 0$ when $u = 0$.

Remark. One may easily deduce from the forthcoming proof that for all real-valued functions k, l of class $\mathcal{K}^\infty$ there exists a cross-term Φ_{kl} such that $V_{kl} = l(W_1) + \Phi_{kl} + k(W_2)$ is proper, positive definite zero at zero and such that $\Delta V_{kl} \leq 0$.

Proof. First, let us determine a cross term function Φ such that the candidate Lyapunov function given in (21) satisfies $\Delta V \leq 0$. The expression of the variation ΔV is

$$\begin{aligned}\Delta V &= W_1(\mathcal{F}(z_i) + \psi(z_i,\xi_i)) - W_1(z_i) \\ &\quad + \Phi(z_{i+1},\xi_{i+1}) - \Phi(z_i,\xi_i) + W_2(a(\xi_i)) - W_2(\xi_i)\,.\end{aligned}$$

Since

$$\begin{aligned}W_1(\mathcal{F}(z_i) &+ \psi(z_i,\xi_i)) - W_1(\mathcal{F}(z_i)) = \\ &\int_0^1 \left(\frac{\partial W_1}{\partial \alpha}|_{\alpha=\mathcal{F}(z_i)+\psi(z_i,\xi_i)\theta}\psi(z_i,\xi_i)\right) d\theta\end{aligned}$$

Assumption **A1** implies that ΔV is negative if

$$\begin{aligned}\Phi(z_{i+1},\xi_{i+1}) &- \Phi(z_i,\xi_i) = \\ &- \int_0^1 \left(\frac{\partial W_1}{\partial \alpha}|_{\alpha=\mathcal{F}(z_i)+\psi(z_i,\xi_i)\theta}\psi(z_i,\xi_i)\right) d\theta.\end{aligned}$$

Let us denote the right hand side of this expression by $-q(z_i,\xi_i)$. It straightforwardly follows from (22) that

$$\Phi(z,\xi) = \sum_{i=0}^{\infty} q(z_i,\xi_i) \tag{22}$$

provided that the right hand side of (22) is a well-defined function.

The next step of the proof consists of showing that this power series converges. From the definition of $q(z,\xi)$ and Assumptions **A3** and **A4** we successively obtain:

$$\begin{aligned}|q(z,\xi)| &\leq \left|\int_0^1 \frac{\partial W_1}{\partial \alpha}|_{\alpha=\mathcal{F}(z)+\psi(z,\xi)\theta}\psi(z,\xi)d\theta\right| \\ &\leq |\psi(z,\xi)| \leq \gamma_0(\xi) + \gamma_1(\xi)W_1(z)\,.\end{aligned}$$

From Assumption **A2**, we deduce that there exists a positive and increasing function $\tilde{\gamma}$ zero at the origin and a strictly positive real number r such that:

$$|q(z_i,\xi_i)| \leq \tilde{\gamma}(\xi)e^{-ri}[W_1(z_i) + 1]\,. \tag{23}$$

So, to conclude we have to prove that the sequence z_i is bounded. This can be easily seen as follows:

Using Assumptions **A3** and **A4**, we deduce that

$$\begin{aligned}W_1(z_{i+1}) - W_1(z_i) &\leq W_1(\mathcal{F}(z_i) + \psi(z_i,\xi_i)) - W_1(\mathcal{F}(z_i)) \\ &\leq \gamma_0(\xi_i) + \gamma_1(\xi_i)W_1(z_i)\,.\end{aligned}$$

Using Assumption **A2**, we deduce that there exist a function Γ_1 smooth, positive, zero at zero and $r > 0$ such that:

$$\frac{W_1(z_{i+1}) + 1}{W_1(z_i) + 1} \leq 1 + \Gamma_1(\xi)e^{-ri}\,.$$

It follows that

$$\ln(W_1(z_{i+1}) + 1) \leq \ln(W_1(z) + 1) + \sum_{j=0}^{i} \ln\left[1 + \Gamma_1(\xi)e^{-rj}\right]$$

which implies that there exists Γ_2 such that for all integer l,

$$W_1(z_l) \leq (W_1(z)+1)\Gamma_2(\xi) . \quad (24)$$

To show that V defined in (21) is a positive function, consider

$$\begin{aligned} W_1(z)+\Phi(z,\xi) &= W_1(z)+\sum_{i=0}^{\infty}[W_1(\mathcal{F}(z_i)+\psi(z_i,\xi_i)) \\ &- W_1(\mathcal{F}(z_i))] \\ &= W_1(z)+\sum_{i=0}^{\infty}[W_1(z_{i+1})-W_1(\mathcal{F}(z_i))] . \end{aligned}$$

For all integer $J>0$, we have

$$\begin{aligned} W_1(z)+\Phi(z,\xi) &= \sum_{i=0}^{J}[W_1(z_i)-W_1(\mathcal{F}(z_i))]+W_1(z_{J+1}) \\ &+ \sum_{i=J+1}^{\infty}[W_1(z_{i+1})-W_1(\mathcal{F}(z_i))] . \end{aligned}$$

According to **A1**, the term $\sum_{i=0}^{J}[W_1(z_i)-W_1(\mathcal{F}(z_i))]$ is positive. On the other hand, using Assumption **A3** and **A4**, one can prove easily that:

$$\lim_{J\to+\infty}\sum_{i=J+1}^{\infty}[W_1(z_{i+1})-W_1(\mathcal{F}(z_i))] = 0 .$$

It follows that $W_1(z)+\Phi(z,\xi)$ is positive. Since $\Phi(z,0)=0$ for all z, it straightforwardly follows that $V(z,\xi)$ is a positive definite function.

Remark. It is worth noting that surprisingly, Assumptions **A1** to **A4** do not guarantee that V is radially unbounded whereas similar assumptions for continuous-time systems ensure that the corresponding function V is radially unbounded.

3.2. Inverse optimal controls for feedforward systems

Let us state an inverse optimal result for feedforward systems.

Corollary: Assume that the system (20) satisfies the Assumptions **A1** to **A4**. Assume that the cross term of the function V provided by Theorem 3.1 is continuously differentiable and that Assumption **H2** is satisfied. Then for this system the inverse optimal stabilization problem is solvable.

Proof. This result is an immediate consequence of Theorem 2.4 and Theorem 3.1 when V is a proper function. When V is not a proper function then Assumption **H1'** and not Assumption **H1** is satisfied. To apply Theorem 2.2, we have to prove that there exist feedbacks of the form (3) which do not destabilize the system. This can be done by using the arguments similar to those invoked above in the proof of Theorem 3.1.

REFERENCES

[1] R.A. Freeman and P.V. Kokotovic: *Inverse optimality in robust stabilization*, SIAM Journal on Control and Optimization, vol. 34, pp. 1365–1391, 1996.

[2] R.A. Freeman and P.V. Kokotovic: *Robust Nonlinear Control* Design, Birkhauser, Boston, 1996.

[3] M.J. Jankovic, R. Sepulchre, P.V. Kokotovic: *Global stabilization of an enlarged class of cascade nonlinear systems.* IEEE Trans. on Automatic Control, vol.41, no.12, pp.1723-1735, 1996.

[4] V. Jurdjevic, J.P. Quinn : *Controllability and stability.* J. Differential Equations, vol.4 , pp. 381-389, 1978.

[5] M. Krstić and Zhong-Hua Li: *Inverse optimal design of input-to-state stabilizing nonlinear controllers*, Preprint 1997.

[6] Zhong-Hua Li and M. Krstić: *Optimal design of adaptive tracking controllers for nonlinear systems*, Proc. ACC, Albuquerque, New Mexico, 1997.

[7] Wei Lin: *Further results on global stabilization of discrete-time nonlinear systems.* Systems & Control Letters, 29, 51–59, 1996.

[8] F. Mazenc, L. Praly: *Adding an integration and Global asymptotic stabilization of feedforward systems.* IEEE Trans. on Automatic Control, vol.41, no.11, pp.1559-1578, 1996.

[9] F. Mazenc and H. Nijmeijer: *Forwarding in discrete-time nonlinear systems.* To appear in the International Journal of control.

[10] F. Mazenc, R. Sepulchre, M. Jankovic: *Lyapunov functions for stable cascades and applications to global stabilization.* 36th CDC conference, San Diego, 1997.

[11] P.J. Moylan, B.D. Anderson: *Nonlinear regulator theory and an inverse optimal control problem.* IEEE Trans. Automatic Control, vol. 18, pp. 460-465, 1973.

[12] A. Teel: *Feedback Stabilization: Nonlinear Solutions to Inherently Nonlinear Problems.* PhD Dissertation, University of California, Berkeley, 1992.

14th Triennial World Congress, Beijing, P.R. China

Vibrational Control of an Underactuated Mechanical System: Control Design through Averaging Analysis

Keum-Shik Hong*, Hyun-Chul Sohn, and Kyo-Il Lee†**

** School of Mechanical Engineering, Pusan National University, San 30, Changjeon-dong, Kumjeong-ku, Pusan, 609-735, Korea. Tel.: +82-51-510-2454, Fax.: +82-51-514-0685, Email: kshong@hyowon.pusan.ac.kr*
*** Graduate College, Pusan National University, Korea*
Email: hcson@hyowon.pusan.ac.kr
† School of Mechanical and Aerospace Engineering, Seoul National University, San 56-1, Shinlim-dong, Kwanak-ku, Seoul, 151-742, Korea.

Abstract: An open loop vibrational control for an underactuated mechanical system with amplitude and frequency modulation is investigated. Because there is no direct control input to an unactuated joint, the dynamic coupling between actuated and unactuated joints is utilized for controlling unactuated joints. The actuated joints are first of all positioned to their desired locations. Then, periodic oscillatory inputs are applied to the actuated joints to move the remaining unactuated joints to their target positions. The amplitudes and frequencies of the vibrations introduced are determined through averaging analysis. A systematic way of obtaining an averaged system for the underactuated system via a coordinate transformation is developed. In the event of actuator failure, a manipulator in zero gravity space can be controlled in this way. A control design example of the planar 2R manipulator with a free joint with no brake is illustrated. Simulation and experimental results are provided. *Copyright © 1999 IFAC*

Keywords: Averaging, Feedback Linearization, Open Loop Control, Underactuated Manipulator, Vibrational Control

1. INTRODUCTION

An underactuated mechanical system refers to the system with fewer number of actuators than the degrees of freedom that the system possesses. Therefore, manipulators with passive or free joints can become underactuated systems since the number of control inputs is smaller than the number of generalized coordinates or the dimension of the configuration space. Recently, the focus of the control of underactuated systems has been the reduction of the number of actuators and/or sensors, and the improvement of the reliability, by a fault-tolerant design, of fully-actuated manipulators which are working in hazardous areas. It is particularly important for a space robot working in outer space to have the ability to control the failed joints in the event of actuator failure.

Recently, scholastic papers which applied periodic oscillations to control the manipulators with free joints have appeared (Nakamura et al., 1997; Suzuki and Nakamura, 1997). Note that the active joint variables appearing in the unactuated joint's dynamics can be considered as varying system parameters. Therefore, periodic motion of the active joint acts as a parametric vibrational control for controlling the unactuated joints. Suzuki and

Nakamura (1997) investigated an oscillatory control based on Poincare map analysis. De Luca et al. (1997) also investigated a constructive open loop control which involves nilpotent approximation and iterative steps.

Vibrational control is an open loop control technique which utilizes high frequency zero mean vibrations to modify, in a desired manner, the behavior of dynamical systems. Motivated by the stabilization of an inverted pendulum by a fast vertical oscillation of the support point, the vibrational control theory for nonlinear systems matured in the middle of 1980's by Bellman *et al.* (1986). The theory has also been extended to parabolic partial differential equations (Bentsman and Hong, 1993), and functional differential equations (Bentsman et al., 1991).

The method of averaging is an asymptotic method which permits the analysis of the dynamic behavior of a time-varying system via a time-invariant (averaged) system. The time-invariant system is obtained by averaging the right hand side of the original time-varying system that is in the standard form.

In this paper, once all active joints reach their set points, periodic inputs are applied to an active joint for the purpose of positioning the remaining unactuated joints to their target positions via the dynamic coupling between actuated and unactuated joints. Since the input is periodic, the active joint returns back to its original position in each period. The design issue now becomes how to move the unactuated joints to their set positions. Since the system is time-varying, a direct procedure for designing control input is not readily available. However, the averaging theory guarantees that the stability properties of an original time-varying system and an averaged system are the same under certain conditions. Therefore, the input magnitudes and frequencies of vibrations are calculated by analyzing the trajectories of an averaged system instead of the original system.

2. VIBRATIONAL CONTROL

Consider a k degrees of freedom open-loop mechanism with joint variables $q_1, \cdots, q_k$. It is assumed that each joint has a single degree of freedom and only $m < k$ joints are active. And the remaining $l = k - m$ joints are unactuated. It is assumed that all joint variables, either actuated or unactuated, can be measured. Using the Lagrange method, one can derive the equations of motion of the system, and rearrange the equations so that the coordinates for actuated joints are grouped in $q_1 \in R^m$ and the coordinates for unactuated joints are grouped in $q_2 \in R^l$. A partial feedback linearization of the system yields (Hong et al., 1998).

$$\ddot{q}_1 = u \tag{1}$$

$$M_{22}(q)\ddot{q}_2 + C_2(q,\dot{q}) + G_2(q) = -M_{21}(q)u \tag{2}$$

Define the state vector as $x = [q_2 \quad \dot{q}_2]^T \in R^n$, where $n = 2l$. Then, the state equation of (2) becomes

$$\dot{x} = \begin{bmatrix} x_2 \\ -M_{22}^{-1}(x_1,q_1)\{C_2(x_1,x_2,q_1,\dot{q}_1) \\ \qquad + G_2(x_1,q_1) + M_{21}(x_1,q_1)\ddot{q}_1\} \end{bmatrix} \tag{3}$$

$$\equiv X(x;q_1,\dot{q}_1,\ddot{q}_1) \quad \equiv X(x,\lambda)$$

where $X : R^n \to R^n$. q_1, $\dot{q}_1$ and $\ddot{q}_1$ are considered as system parameters. The symbol λ is introduced to emphasize that there exists one selective parameter in which vibrations can be introduced. It is remarked that only a subset of $\{q_1, \dot{q}_1, \ddot{q}_1\}$ can appear in (3) and it is dependant on the structure of the underactuated systems considered. λ is taken as the second highest derivative in the set of $\{q_1, \dot{q}_1, \ddot{q}_1\}$. For example, if only q_1 and $\dot{q}_1$ appear, then $\lambda = q_1$. Assuming that all q_1, $\dot{q}_1$, $\ddot{q}_1$ exist in (3), i.e. $\lambda(t)$ is assumed to be $\dot{q}_1(t)$, an oscillatory input is introduced into (3) as follows

$$\lambda(t) \to \lambda_0 + \alpha f(\omega t) \tag{4}$$

where λ_0 is a constant, and $\alpha f(\omega t)$ is a zero mean T-periodic function in which α and ω denote its amplitude and frequency. The following relations also hold

$$\ddot{q}_1 = \alpha\omega f'(\omega t), \text{ and } q_1 = \frac{\alpha}{\omega}F(\omega t) \tag{5}$$

where f' and F are the derivative and anti-derivative of f, respectively. Substituting (4) and (5) into (3) yields

$$\dot{x} = X(x, \lambda_0 + \alpha f(\omega t))$$

$$= \begin{bmatrix} x_2 \\ -M_{22}^{-1}\left(x_1, \frac{\alpha}{\omega}F(\omega t)\right)\begin{Bmatrix} C_2(x_1,x_2,\frac{\alpha}{\omega}F(\omega t), \lambda_0 + \alpha f(\omega t)) + \\ G_2(x_1,\frac{\alpha}{\omega}F(\omega t)) + \alpha\omega M_{21}(x_1,\frac{\alpha}{\omega}F(\omega t))f'(\omega t) \end{Bmatrix} \end{bmatrix} \tag{6}$$

Assume that (6) is decomposed into two parts as follows

$$\dot{x} = X_0(x, \omega t, 1/\omega) + \omega X_1(x, \omega t) \tag{7}$$

The second term consists of the terms that contain the derivatives of $f(\omega t)$. The decomposition of (7) is always possible since the second highest derivative had been chosen as a vibratile parameter in (3b). The second term in (7) is now utilized as a generating equation of the coordinate transformation, which transforms (7) into the standard form of averaging. Therefore, the generating equation takes the form

$$\dot{\xi} = X_1(\xi, t) \tag{8}$$

Let $h(t,c): R \times R^n \rightarrow R^n$ be a general solution of (13) which is T-periodic. Note that $c \in R^n$ can be uniquely determined once initial conditions $\xi(t_0) \in \Omega \subset R^n$ are provided. Introduction of a new variable $p(t)$ using the general solution $h(t,c)$ acts as a coordinate transformation between $x(t)$ and $p(t)$ as follows

$$x(t) = h(\omega t, p(t)) \tag{9}$$

Differentiating both sides of (9) with respect to t and utilizing (7), the following is obtained

$$\dot{p}(t) = \left[\frac{\partial h(\omega t, p(t))}{\partial p}\right]^{-1} X_0\big(h(\omega t, p(t)), \omega t, 1/\omega\big) \tag{10}$$

In slow time scale such that $\tau = \omega t$ with $z(\tau) = p(t)$ and $\varepsilon \equiv 1/\omega$, the following standard form is obtained

$$\dot{z}(\tau) = \varepsilon\left[\frac{\partial h(\tau, z(\tau))}{\partial z}\right]^{-1} X_0\big(h(\tau, z(\tau)), \tau, \varepsilon\big). \tag{11}$$

3. AVERAGING

Consider differential equations of the form

$$\dot{z} = \varepsilon g(z,t,\varepsilon), \quad z(0) = z_0 \tag{12}$$

where $z \in R^n$, $t \geq 0$, $0 < \varepsilon \leq \varepsilon_0$, and g is piece-wise continuous in t. Note that for small ε, the time variation of z is slow as compared to the time variation of g. (12) is called the standard form of averaging, in which ε appears in front of g.

Theorem 1: Consider system (3). There exists a parametric vibration such that system (3), with the vibrations, can be transformed into the standard form of (12).

Define the mean value of $g(z,t,0)$ as follows

$$g_{av} = \lim_{T\rightarrow\infty} \frac{1}{T} \int_{t_0}^{t_0+T} g(z,t,0)\, dt \tag{13}$$

It is assumed that the limit in (13) exists uniformly in t_0 and z. Then, the averaged system associated with (12) is defined as

$$\dot{y} = \varepsilon g_{av}(y), \quad y(0) = z_0 \tag{14}$$

Theorem 2 (Basic Averaging Theorem: Sastry and Bodson, 1989, p.172): Let (12) and (14) satisfy the following assumptions: i) $z = 0$ is an equlibrium point of (12), ii) $g(z,t,\varepsilon)$ is locally Lipschitz continuous in both x and ε, iii) $y = 0$ is an equil ibrium point of (14) iv) $g_{av}(y)$ is locally Lipschitz continuous in y, v) $e(z,t) \equiv g(z,t,0) - g_{av}(z)$ is piecewise continuous in t, has bounded and continuous partial derivatives in z, and $e(0,t) = 0$, for all $t \geq 0$. Moreover, $e(z,t)$ and $\partial e(z,t)/\partial z$ have zero mean values. Then, there exists a continuous and strictly increasing function $\varphi(\varepsilon)$ with $\varphi(0) = 0$ such that for a given $T \geq 0$

$$\| z(t) - y(t) \| \leq \varphi(\varepsilon) b_T \tag{15}$$

for some $b_T \geq 0$, $\varepsilon_T > 0$ and for all $t \in [0,\ T/\varepsilon]$ and $\varepsilon \leq \varepsilon_T$.

Theorem 3 (Exponential Stability Theorem: Sastry and Bodson, 1989, p.173): Let (12) and (14) satisfy assumptions i)~v) in Theorem 2. Assume further that (14) satisfies vi) $g_{av}(y)$ has continuous and bounded first partial derivatives in y, vii) $y = 0$ is an exponentially stable equilibrium point. Then, $z = 0$ of (12) is exponentially stable for ε sufficiently small.

Remark 1: Theorem 2 establishes that the trajectories of (12) and (14) are arbitrarily close on intervals $[0,\ T/\varepsilon]$, when ε is sufficiently small. The error is in the order of $\varphi(\varepsilon)$. Although the intervals $[0,\ T/\varepsilon]$ are unbounded, Theorem 2 does not allow us to compare the stability of the two systems. On the other hand, Theorem 3 states a local exponential stability result. It is further noted that the original system will be globally exponentially stable if the averaged system is globally exponentially stable, and provided that all assumptions are valid globally.

4. A PLANAR 2R MANIPULATOR

In this section, as an example of an underactuated mechanical system, and utilizing a planar 2R manipulator with a free joint, the control system design procedure in Section 3 is illustrated step by step. As shown in the sequel, linear control theory is not applicable in this case since the linearized system at any operating point is not controllable.

Fig. 1 shows an experimental 2R manipulator, in which the second joint is free, with parameter values tabulated below. Using the Lagrange equation, the following equations of motion are obtained.

$$M_{11}(\theta_2)\ddot{\theta}_1 + M_{12}(\theta_2)\ddot{\theta}_2 + C_1(\theta_2,\dot{\theta}_1,\dot{\theta}_2) = \tau \tag{16}$$

$$M_{12}(\theta_2)\ddot{\theta}_1 + M_{22}(\theta_2)\ddot{\theta}_2 + C_2(\theta_2,\dot{\theta}_1,\dot{\theta}_2) = 0$$

where all coefficients and their values are found in (Hong et al., 1998; Suzuki and Nakamura, 1997). Note that the gravity term does not appear in the equations, and check that the linearized equation at an operating point $(\theta_1,\theta_2,\dot{\theta}_1,\dot{\theta}_2) = (\theta_1,\theta_2,0,0)$ is not controllable.

The partially linearized system, which corresponds to (1) and (2), is

$$\ddot{\theta}_1 = u \tag{17a}$$

$$\ddot{\theta}_2 = -(1+d\cos\theta_2)\ddot{\theta}_1 - d\dot{\theta}_1^2\sin\theta_2 \quad (17b)$$

where $d = m_2 l_1 l_{2c}/(m_2 l_{2c}^2 + I_2)$ is a constant. Now assume that the active joint θ_1 has been positioned at its desired location with an appropriate control action. For instance,

$$u = \ddot{\theta}_{1d} + k_v(\dot{\theta}_{1d} - \dot{\theta}_1) + k_p(\theta_{1d} - \theta_1) \quad (18)$$

would suffice, in which k_p and k_v are position and velocity gains. While the active joint is steered to its target position, the position of the free joint will be arbitrary. Now, restricting our control task to the free joint, observe that $\ddot{\theta}_1$ and $\dot{\theta}_1^2$ are parameters in θ_2-dynamics (17b). When the state variables are defined as $x_1 = \theta_2, x_2 = \dot{\theta}_2$, the state equation of (17b) becomes

$$\dot{x}_1 = x_2, \quad (19)$$

$$\dot{x}_2 = -(1+d\cos x_1)\ddot{\theta}_1 - d\sin x_1(\dot{\theta}_1)^2.$$

Introduce the vibrations in Section 3 as follows

$$\lambda(t) = \dot{\theta}_1(t) \rightarrow 0 + \alpha\sin\omega t \quad (20)$$

Hence, $\lambda_0 = 0$ and $\alpha f(\omega t) = \alpha\sin\omega t$ are used. Substituting (20) into (19) yields

$$\begin{bmatrix}\dot{x}_1\\ \dot{x}_2\end{bmatrix} = \begin{bmatrix}x_2\\ -d\sin x_1(\alpha\sin\omega t)^2\end{bmatrix} + \omega\begin{bmatrix}0\\ -\alpha\cos\omega t(1+d\cos x_1)\end{bmatrix} \quad (21)$$

Note that (21) is in the form of (7). Therefore, the generating equation of (8) takes the form

$$\begin{bmatrix}\dot{\xi}_1\\ \dot{\xi}_2\end{bmatrix} = \begin{bmatrix}0\\ -\alpha\cos t(1+d\cos\xi_1)\end{bmatrix} \quad (22)$$

A general solution of (22) is given as

$$h(t,c) = \begin{bmatrix}h_1(t,c)\\ h_2(t,c)\end{bmatrix} = \begin{bmatrix}c_1\\ c_2 - \alpha(1+d\cos c_1)\sin t\end{bmatrix} \quad (23)$$

where c_1 and c_2 are constants to be determined from initial conditions. A coordinate transformation which will convert (21) into the standard form of averaging is defined via (23) as

$$\begin{bmatrix}x_1(t)\\ x_2(t)\end{bmatrix} = \begin{bmatrix}p_1(t)\\ p_2(t) - \alpha(1+d\cos p_1(t))\sin t\end{bmatrix} \quad (24)$$

Therefore, differentiating (24) with respect to time and utilizing (21) yields

$$\dot{p}_1(t) = p_2 - \alpha(1+d\cos p_1)\sin\omega t, \quad (25)$$

$$\dot{p}_2(t) = d\alpha\sin p_1\sin\omega t(-p_2 + d\alpha\cos p_1\sin\omega t).$$

Note that (25) takes the form of (10) in Section 3. In slow time $\tau = \omega t$, with $p(t) = z(\tau)$ and $\varepsilon = 1/\omega$, the following standard form is obtained

$$\dot{z}_1(\tau) = \varepsilon[z_2 - \alpha(1+d\cos z_1)\sin\tau], \quad (26)$$

$$\dot{z}_2(\tau) = \varepsilon d\alpha\sin z_1\sin\tau(-z_2 + d\alpha\cos z_1\sin\tau).$$

Finally, by applying the definition of averaging, the following averaged system is obtained.

$$\dot{y}_1 = \varepsilon y_2, \quad y_1(0) = x_{10} \quad (27)$$

$$\dot{y}_2 = \varepsilon\frac{d^2\alpha^2}{4}\sin 2y_1, \quad y_2(0) = x_{20}$$

The control system design is now based upon (27) instead of (21). It is noted that the initial conditions of the averaged system are not, in general, the same as those of the original system. It is also noted that a sufficiently small ε can always be obtained since ε was defined by $1/\omega$.

. *Control System Design*

The trajectories of transformed system (25), the phase portrait of averaged system (27), and the trajectories of the averaged system with different α's are found in (Hong et al. 1998). It is observed that an arbitrary point in the state space can be reached by modulating α. It is also noted that in the 2R manipulator case changing ω does not influence the trajectories of the averaged system. However, by increasing ω the amplitude of $\theta_1(t)$ gets smaller and the Poincare map becomes dense. Therefore, more precise movement can be achieved by increasing ω.

Let (y_1, y_2) be a present state and (y_{1d}, y_{2d}) be a desired state. Integration of (27) from (y_1, y_2) to (y_{1d}, y_{2d}) yields the following relationship

$$\frac{d^2\alpha^2}{2}\cos^2 y_1 + y_2^2 = \frac{d^2\alpha^2}{2}\cos^2 y_{1d} + y_{2d}^2 \quad (28)$$

Therefore, the amplitude of a vibration that steers from (y_1, y_2) to (y_{1d}, y_{2d}) is calculated as follows.

$$\alpha = \sqrt{\frac{2(y_2^2 - y_{2d}^2)}{d^2(\cos^2 y_{1d} - \cos^2 y_1)}} \quad (29)$$

Four different patterns are sketched in Fig. 2.

. Example: Pattern #1

(a) Step 1: Set initial amplitude and frequency as $\alpha = \alpha_1$ and $\omega = \omega_1$. As an illustration, Fig. 3 shows the trajectories of the steering problem from $(\theta_2, \dot{\theta}_2) = (0.5, 0)$ to $(\pi/4, 0)$ and following Pattern #1. These are computer-simulated results, and small circles indicate the Poincare map. The initial amplitude and frequency are $\alpha = 1.3$ and $\omega = 6\pi$. (b) Step 2: If $\theta_2(t)$ becomes $\pi/2$, then change α to α_{max}. In Fig. 6, $\alpha_{max} = 1.67$ is used. (c) Step 3: If $\theta_2(t)$ gets less than its previous value, then change α to, α_3 where $\alpha_3 \leq \alpha_1$. In Fig. 6, $\alpha_3 = 1.3$ is used. (d) Step 4: If $\theta_2(t)$ gets to $\pi/2$ for the second time, then vary α sequentially in each oscillation according to (29) (cruising quarter). (e) Step 5: Finally, if $|\theta_2(t) - \theta_{2d}| < \delta$, where δ is an error bound, then switch ω to a higher frequency ω_2. In Fig. 3, $\omega_2 = 30\pi$ is used. Similar analyses can be made on Pattern #2, #3, and #4.

5. EXPERIMENTS

The experimental setup is composed of a 2R

mechanism, a data acquisition board, an interface board that includes a servo controller and two signal conditioners, and a PC equipped with a digital signal process card. The actuator at the first joint is a high speed DC coreless motor with a nominal voltage of 12V and a power consumption of 6W (Maker: Maxon, Model: 2326). A planetary gear with a gear ratio of 32.1:1 is attached to the head of the motor. A servo controller regulates the input current to the motor, in which the PWM method with a carrier frequency of 20KHz is used. The rotational angles of both joints are measured by potentiometers. The second link is directly attached to the potentiometer. The damping effect of the potentiometer is negligible. The measured signals from the potentiometers are filtered by lowpass filters with a cutoff frequency of 10Hz before entering the DSP board. The time delay occurring at the lowpass filters is less than 10 μ sec. The control algorithm is executed at a DSP board of dSPACE DS1102 with the TI chip TMS320C31. In order to compare simulation and experimental results, the control algorithms are first coded with MATLAB/SIMULINK. Then, after converting them to C codes with Real-Time Workshop, they are downloaded to DS1102 for the experiments.

Fig. 4 shows a vibrationally controlled motion of the free joint $\theta_2(t)$, in which the initial angle and the target angles were 0.4 radian and $\pi/4$ radian, respectively. Fig. 5 shows the control input that was applied to obtain the motion in Fig. 4. At 2 seconds, initial vibrations with amplitude 1.3 and frequency 6π were applied. At around 4 seconds, the amplitude was changed to 1.67, and then 1.3 following Pattern #1. The tiny oscillations appearing after 7 seconds indicate that the frequency has been increased to $\omega_2 = 30\pi$. Experimental results agree well with simulation results.

6. CONCLUSIONS

An open loop vibrational control for underactuated mechanical systems has been investigated in this paper. The control strategy to move an unactuated joint lies in utilizing the dynamic coupling, which occurs when the actuated joint is oscillating, between the actuated and unactuated parts. With the introduction of vibrations, the whole system becomes time-varying. The stability analysis of the time-varying system is carried out using averaging analysis. The averaging method was extended to the system with the derivatives and anti-derivatives of vibrations. A systematic way of obtaining an averaged system, via the generating equation for an underactuated system, was developed. To illustrate the design procedure, a 2R planar manipulator with a free joint was demonstrated. The calculation procedure of the amplitudes and frequencies of vibrations was demonstrated. In zero gravity outer space and using this strategy, a manipulator with a failed joint can be controlled. The control strategy in this paper provides a viable tool when the conventional control scheme is not applicable and/or actuator failure occurs.

ACKNOWLEDGEMENTS

This work was supported in part by the Korea Science and Engineering Foundation under KOSEF Grant 971-0920-137-2.

REFERENCES

Bellman, R., Bentsman, J., and Meerkov, S. M. (1986a). Vibrational Control of Nonlinear Systems: Vibrational Stabilizability. *IEEE Trans. Automatic Control*, **Vol. 31**, pp. 710-716.

Bellman, R., Bentsman, J., and Meerkov, S. M. (1986b). Vibrational Control of Nonlinear Systems: Vibrational Controllability and Transient Behavior. *IEEE Trans. Automatic Control*, **Vol. 31**, pp. 717-724.

Bentsman, J., and Hong, K. S. (1993). Transient Behavior Analysis of Vibrationally Controlled Nonlinear Parabolic Systems. *IEEE Trans. Automatic Control*, **Vol. 38**, pp. 1603-1607.

Bentsman, J., Hong, K. S., and Fakhfakh, J. (1991). Vibrational Control of Nonlinear Time Lag Systems: Vibrational Stabilization and Transient Behavior. *Automatica*, **Vol. 27**, No. 3, pp. 491-500.

Hong, K. S., and Bentsmann, J. (1994). Application of Averaging Method for Integro-Differential Equations to Model Reference Adaptive Control of Parabolic Systems. *Automatica*, **Vol. 30**, No. 9, pp.1415-1419.

Hong, K. S., Lee, K. R., and Lee, K. I. (1998). Vibrational Control of Underactuated Mechanical Systems: Control Design through Averaging Analysis. *Pro. American Control Conference*, Philadelphia, DA, pp. 3482-3486.

De Luca, A., Mattone, R., and Oriolo, G. (1997). Stabilization of Underactuated Robot: Theory and Experiments for a Planar 2R Manipulator. *Proceedings IEEE Int. Conf. on Robotics and Automation*, pp. 3274-3280.

Nakamura, Y., Suzuki, T., and Koinuma, M. (1997). Nonlinear Behavior and Control of a Nonholonomic Free-Joint Manipulator. *IEEE Trans. Robotics and Automation*, **Vol. 13**, No. 6, pp. 853-862.

Sastry, S., and Bodson, M. (1989). *Adaptive Control: Stability, Convergence, and Robustness*, Prentice-Hall, Inc., Englewood Cliffs, NJ.

Suzuki, T., and Nakamura, Y. (1997). Control of Manipulators with Free-Joints via the Averaging Method. *Proceedings IEEE Int. Conf. on Robotics and Automation*, pp. 2998-3005.

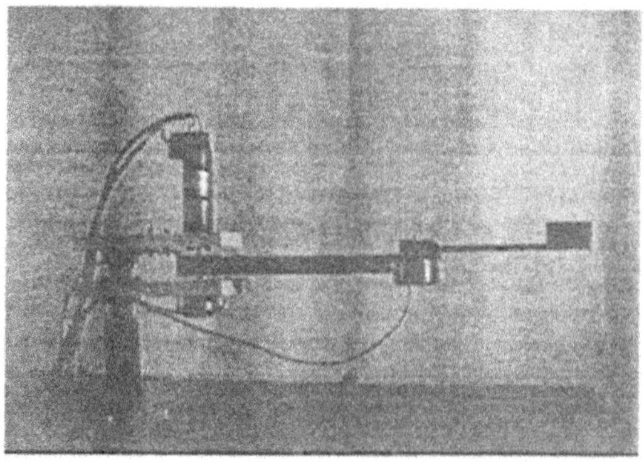

Parameters	link 1	link 2
mass(kg)	200	255
Inertia(10^{-3} kgm^2)	3.097	3.499
Length(10^{-3} m)	185	135
Distance to mass center (10^{-3} m)	103	60

Fig. 1. Experimental 2R manipulator with a free joint

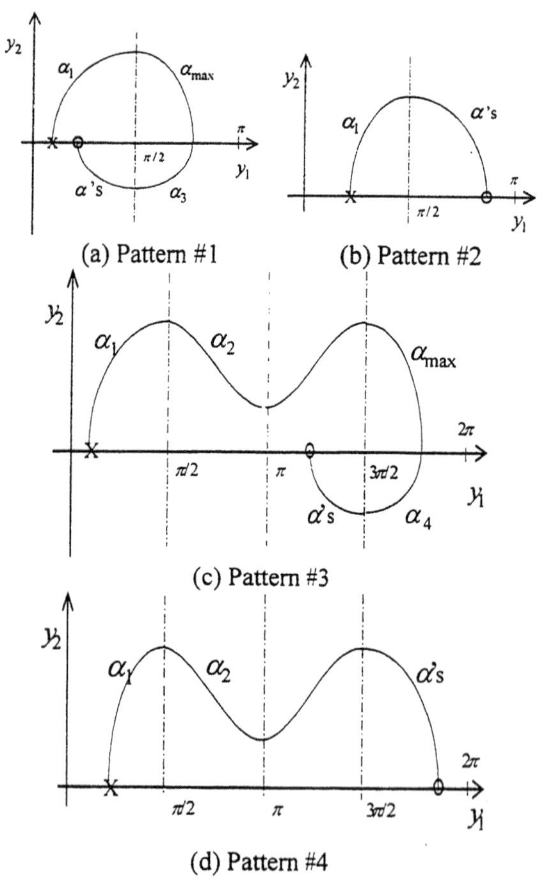

Fig. 2. Control strategies (×: Initial states, O: Target states)

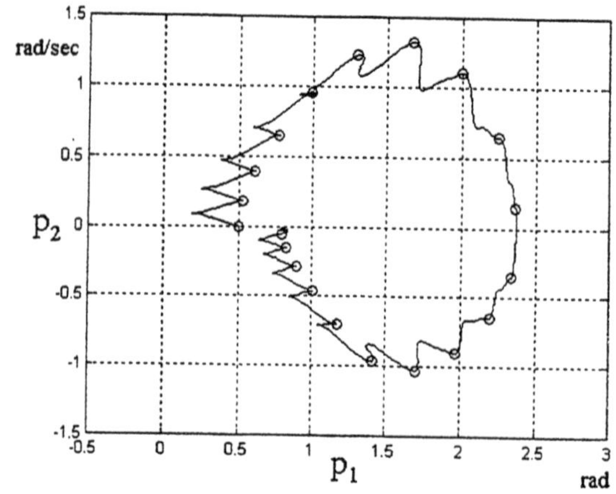

Fig. 3. Computer simulation of a steering problem from initial angle θ_2=0.5 rad to target angle $\theta_2=\pi/4$ rad following Pattern #1 (α_1=1.3, α_{max}=1.67, α_3=1.3, $\omega_1=6\pi$, $\omega_2=30\pi$)

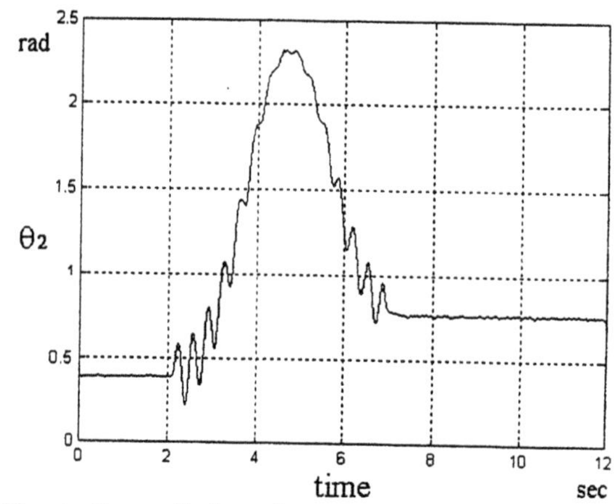

Fig. 4. Controlled motion of the free joint following Pattern #1 (experimental results)

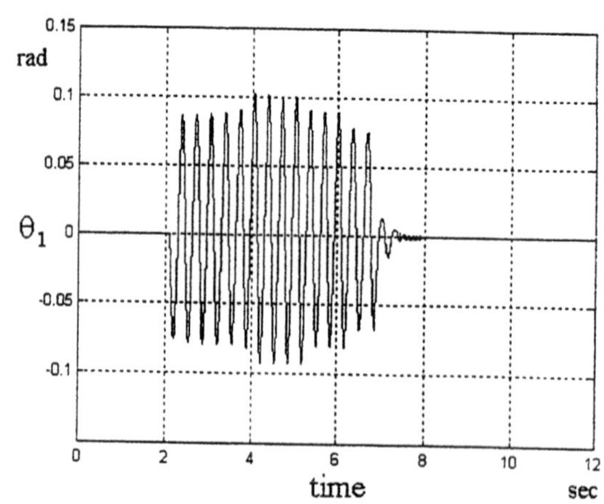

Fig. 5. Control input for obtaining Fig. 8: α_1=1.3, α_{max}=1.67, α_3=1.3, $\omega_1=6\pi$, $\omega_2=30\pi$

14th Triennial World Congress, Beijing, P.R. China

VIBRATION CONTROL OF A FLEXIBLE BEAM WITH COLLOCATED PIEZOELECTRIC SENSORS AND ACTUATORS

Yong LI

Beijing Institute of Control Engineering, Chinese Academy of Space Technology
Beijing 100080, P. R. CHINA
E-mail: liy@public.east.cn.net

Abstract: In this paper, vibration control of a flexible cantilevered beam with collocated piezoelectric sensors and actuators is studied. A dynamic governing equation of motion for the smart cantilevered beam is derived by applying Hamilton principle. Linear feedback controls for the smart beam are presented. By using LaSall's invariant principle in infinite dimensional space and linear semigroup theory, it is shown that implementation of these controls result in vibration suppression.

Keywords: Active control, distributed-parameter systems, mechanical systems, stabilizability, stabilizing controllers.

1. INTRODUCTION

Within the past decade, there has been an increasing interest in the development of smart structure or intelligent structures for space applications. New generation space structures tend to have relatively low modal frequencies. Moreover, these modes tend to be densely spaced. Quite often, the bandwidth of both control forces imparted to the structure and disturbances acting on it encompass many of these modes. This makes it difficult, for example, to slew a large spacecraft without exciting some vibrational modes. To fulfill the requirement of precise pointing accuracy for large spacecraft, much effort has been directed toward looking for smart structure, i.e., structures with highly integrated sensors and actuators, which can be used to change the mechanical properties of the structure. As of now, piezoelectric materials are most often employed as sensors or actuators for smart structure applications.

Research on smart structure systems using piezoelectric materials was first undertaken by Bailey and Hubbard (1985). By utilizing a uniform layer of the material bounded to one face of a cantilevered beam as the actuator, they implemented a control strategy using Lyapunov's direct method, and showed that vibrational mode of the beam could be controlled based upon the measurement of the angular velocity at the beam's tip. Burke et. al. (1988) showed that spatially-varying piezoelectric film actuator distributions can be applied to control all vibrational modes of flexible beams with nearly arbitrary boundary condition. Piezoelectric actuators were also used as elements of intelligent structures by Crawly and de Luis (1987). It is well known that piezoelectric materials, unlike other actuators, have the advantage of also acting as sensors. From the viewpoint of stability, collocated sensors and actuators are advantage, while in applications, maybe non-collocated

[1] The author is supported by the National Natural Science Foundation of China.

sensors and actuators give better performance. In this paper, however, only collocated sensors and actuators will be considered.

As is evident from previous studies, numberous researchers have used piezoelectric materials only to focus on designing of the control law and demonstrating the stability and performance of the proposed control algorithm by simulation and experimental results without severe theoretical analysis for system stability.

It is well known that the flexible beam is an infinite dimensional system. In order to faithfully measure and control the system without using a truncated model, there is a need in designing control algorithms directly from the partial differential equation model to avoid spillover. In this paper, beam structure have been selected since most aerospace structures consist of beam bending members as the primary load-carrying elements. A dynamic governing equation of motion for a smart cantilevered beam is derived by applying Hamilton principle. We present some linear feedback control laws for the smart beam and show that implementation of these control laws result in vibration suppression by using LaSall's invariant principle in infinite dimensional space and linear semigroup theory.

2. SYSTEM MODELS

Consider a structure that consists of a cantilevered beam, collocated m actuators and m sensors, which are the layers made of piezoelectric ceramic (PZT) and piezoelectric polymer polyvinylidene fluoride (PVDF) materials, respectively, attached to both sides of the beam. Let h stands for the thickness of the different layers of the composite beam. The subscripts s, b and a denote sensor, beam and actuator respectively.

2.1 Actuators Models

Piezoelectric actuators are made based on the constitutive property of the PZT material. When control voltages $V^{(i)}(x,t)$ is applied to the ith PZT material actuator, $1 \leq i \leq m$ the induced strain $\varepsilon_p^{(i)}$ in the ith PZT material actuator is given by

$$\varepsilon_p^{(i)}(x,t) = \frac{d_{31}}{h_a} V^{(i)}(x,t), \tag{1}$$

where d_{31} is the piezoelectric field and strain field constant.

Suppose the bonding between actuator layers and the beam is perfect, i.e., there is no shear lag layer effects on the smart beam. The induced strain ε_p has two effects on the beam. One effect is that it introduces a longitudinal strain to insure a force equilibrium along the axial. The second effect is that the net force in each layer acts through the moment arm with the bending moment which is introduced as the control mechanism . The bending moment produced from the ith PZT actuator can be expressed as (Bailey and Hubbard 1985):

$$M^{(i)}(x,t) = k_a \varepsilon_p^{(i)}(x,t), \tag{2}$$

where k_a is a constant depending on the geometry and materials of the beam.

For the development work, we assume that

$$V^{(i)}(x,t) = c_1 V^{(i)}(t) F^{(i)}(x), \tag{3}$$

where function $F^{(i)}(x)$, being the local width of the electrodes covering both sides of the ith actuator, is the weight function or shape function of the ith actuator. $F^{(i)}(l) = F_x^{(i)}(l) = 0$ for $1 \leq i \leq m-1$, $V^{(i)}(t)$ is the input voltage to the ith PZT actuator layer as the control variable. c_1 is a positive constant.

Substituting equation (1), (3) into (2) yields

$$M^{(i)}(x,t) = cV^{(i)}(t) F^{(i)}(x), \tag{4}$$

where $c = c_1 k_a \frac{d_{31}}{h_a}$. Equation (4) describe relationship between the applied voltages and actuation of the ith PZT actuators.

2.2 Sensors Models

Piezoelectric polymer polyvinylidene fluoride (PVDF) is also strain sensitive and relies on the applied strain to produce electrical charge. This process is the reverse to piezoelectric actuation. The amount of electrical charge is proportional to the amount of strain induced by the structure.

For the case that sensors and actuators are collocated, we have that the function $F^{(i)}(x)$ is also the weight function or shape function of the ith sensor. Then the output of the ith sensor is given by

$$I^{(i)}(t) = k_s \int_0^l F^{(i)}(x) u_{xxt}(x,t) dx, \tag{5}$$

where $u(x,t)$ denotes the transverse displacement of the beam at time t and position x along the beam length direction. k_s is a constant determined by the sensor geometry, material and the capacitance between the electrodes of both side of the sensor surfaces. The subscripts $(\cdot)_x$ and $(\cdot)_t$ denote partial differentiation.

2.3 System Model

We use the Euler-Bernoulli beam model to describe the dynamical behavior of the beam. Applying the Euler-Bernoulli beam theory, the kinetic energy T and potential energy V of the structure including the piezoelectric actuators and sensors are expressed as

$$T = \frac{1}{2}\int_0^l \rho u_t^2(x,t)dx,$$

$$V = \frac{1}{2}\int_0^l \frac{1}{EI}[EIu_{xx}(x,t) - \sum_{i=1}^{m} M^{(i)}(x,t)]^2 dx,$$

where l is the length of the beam, EI is the effective bending stiffness of the smart flexible beam and ρ is the mass per length of the composite beam.

By applying the Hamilton principle, the governing equation of motion for transverse vibration $u(x,t)$ and associated boundary conditions are obtained as follows:

$$\rho u_{tt}(x,t)+EIu_{xxxx}(x,t)-c\sum_{i=1}^{m} V^{(i)}(t)F_{xx}^{(i)}(x) = 0, \quad (6)$$

$$u(0,t) = 0, \qquad u_x(0,t) = 0,$$
$$EIu_{xx}(l,t) = cV^{(m}(t)F^{(m)}(l),$$
$$EIu_{xxx}(l,t) = cV^{(m}(t)F_x^{(m)}(l).$$

3. CONTROL LAW: DESIGN AND ANALYSIS

Our goal is to suppress the structural vibration of the system described by equation (6) using input voltages $V^{(1)}(t), V^{(2)}(t), ...V^{(m)}(t)$ to PZT actuators. In order to complete the mission, we design the following control law:

$$V^{(i)}(t) = -k_i I^{(i)}(t), \quad i = 1,2,...,m \quad (7)$$

where $k_i (i = 1,2,...,m)$ are positive constants.

Define the function space $\mathcal{H}$ as

$$\mathcal{H} = \{(u,v)^T | u \in H_0^2(0,l), v \in L^2(0,l)\},$$

where the spaces $L^2(0,l)$ and $H_0^k(0,l)$ are defined as follows

$$L^2(0,l) = \{f : [0,l] \to R | \int_0^l f^2 dx < \infty\},$$

$$H_0^k(0,l) = \{f \in L^2(0,l)|$$
$$f, f', ..., f^{(k)} \in L^2(0,l), f(0) = f'(0) = 0\},$$

for all positive integer k.

In $\mathcal{H}$, we define the energy inner product as follows:

$$\langle z, \hat{z}\rangle_{\mathcal{H}} = \frac{1}{2}\int_0^l [\rho u_t \hat{u}_t + EIu_{xx}\hat{u}_{xx}]dx, \quad (8)$$

where $z = (u,u_t)^T \in \mathcal{H}$, $\hat{z} = (\hat{u},\hat{u}_t)^T \in \mathcal{H}$. The corresponding energy norm is given by

$$(\|z\|_{\mathcal{H}})^2 = \frac{1}{2}\int_0^l (\rho u_t^2 + EIu_{xx}^2)dx. \quad (9)$$

We note that $\mathcal{H}$ together with the energy inner product derived by (8) becomes a Hilbert space (Mogül, 1991).

From (6) and (7), we see that the closed-loop system can be written as follows:

$$\dot{z} = \mathcal{A}z, \quad (10)$$

where $z = (u,u_t)^T$. The linear operator $\mathcal{A} : \mathcal{H} \to \mathcal{H}$ is defined as

$$\mathcal{A}z = \Big(u_t, \quad -\frac{EI}{\rho}u_{xxxx} - \frac{ck_s}{\rho} \cdot \sum_{i=1}^{m} k_i F_{xx}^{(i)}(x)\int_0^l F^{(i)}(x)u_{xxt}dx\Big)^T.$$

The domain of the operator $\mathcal{A}$ is defined as

$$D(\mathcal{A}) := \{(u,v)^T | u \in H_0^4(0,l), v \in H_0^2(0,l),$$

$$EIu_{xx}(l) = -ck_m k_s F^{(m)}(l)\int_0^l F^{(m)}(x)u_{xxt}dx,$$

$$EIu_{xxx}(l) = -ck_m k_s F_x^{(m)}(l)\int_0^l F^{(m)}(x)u_{xxt}dx\}.$$

The energy function associated with (10) is defined by

$$E(t) = \|z(t)\|_{\mathcal{H}}^2 = \frac{1}{2}\int_0^l (\rho u_t^2 + EIu_{xx}^2)dx \quad (11)$$

Lemma 1. Consider the system given by (10). Then for $t \geq 0$, $E(t)$ is a nonincreasing function of time t along the classical solution of equation (10).

Proof: Differentiating (11) with respect to t and using equation (10), we obtain that

$$\dot{E}(t) = \int_0^l (\rho u_t u_{tt} + EIu_{xx}u_{xxt})dx$$

$$= \int_0^l [u_t(-EIu_{xxxx} - ck_s\sum_{i=1}^{m} k_i F_{xx}^{(i)}(x) \cdot \int_0^l F^{(i)}(\tau)u_{\tau\tau t}d\tau) + EIu_{xx}u_{xxt}]dx$$

$$= EI(-\int_0^l u_t du_{xxx} + \int_0^l u_{xx}u_{xxt}dx)$$

$$-ck_s\sum_{i=1}^{m}k_i\int_0^l F^{(i)}(x)u_{xxt}dx\int_0^l u_t dF_x^{(i)}$$

$$= EI(-u_tu_{xxx}|_0^l+u_{xt}u_{xx}|_0^l)$$

$$-ck_s\sum_{i=1}^{m}k_i\int_0^l F^{(i)}(x)u_{xxt}dx$$

$$\cdot[u_tF_x^{(i)}|_0^l - u_{xt}F^{(i)}|_0^l + \int_0^l F^{(i)}(x)u_{xxt}dx]$$

$$= -ck_s\sum_{i=1}^{m}k_i[\int_0^l F^{(i)}(x)u_{xxt}dx]^2 \leq 0. \tag{12}$$

Thus, we conclude that for $t \geq 0$, $E(t)$ is a nonincreasing function of t along the classical solution of equation (10).

□

Define an operator A on $L^2(0,l)$ by

$$A\varphi(x) = \varphi''''(x),$$

$$D(A) = \{\varphi \in L^2(0,l)|$$
$$\varphi(0) = \varphi'(0) = \varphi''(l) = \varphi'''(l) = 0\}$$

It is well known that A is a self-adjoint and positive definite operator. Let $\{\lambda_n, \varphi_n(x)\}_{n=1}^{\infty}$ be the eigenpairs of operator A. Then we have the following (Luo, et. al. 1997)

i) $\lambda_n = \beta_n^4$ with β_n satisfying

$$1 + \cos\beta_n l\cosh\beta_n l = 0,$$

$\beta_n > 0$, and $\beta_n \to \infty$ as $n \to \infty$.

ii)

$$\varphi_n(x) = (\cosh\beta_n x - \cos\beta_n x)$$
$$-\frac{\cosh\beta_n l + \cos\beta_n l}{\sinh\beta_n l + \sin\beta_n l}(\sinh\beta_n x - \sin\beta_n x)$$

forms an orthogonal basis on $L^2(0,l)$.

For the shape functions $F^{(i)}(x)$ (i=1, 2, ..., m), we have the following assumption

Assumption 1

$$F^{(i)}(x) \in H^2(0,l) = \{f|f, f', f'' \in L^2(0,l)\},$$

and for any $n, n = 1, 2, ...,$ there exists an $i, 1 \leq i \leq m$ such that

$$\int_0^l F^{(i)}(x)\varphi_n''(x)dx \neq 0 \tag{13}$$

Lemma 2. Consider the system given by (10). Suppose Assumption 1 holds. Let S denote the subspace of $\mathcal{H}$ defined by

$$S := \{z(t) \in \mathcal{H}|dE(t)/dt = 0, \quad t \geq 0\}.$$

Then $z(\cdot)$, the classical solution of equation (10) which lies in S, satisfies $z(t) \equiv 0, t \geq 0$.

Proof: Let $z = (u, u_t)^\tau$ be the classical solution of equation (10) which lies in S. Then, from (12), we have

$$\begin{cases} \rho u_{tt}(x,t) + EIu_{xxxx}(x,t) = 0, \\ u(0,t) = u_x(0,t) = u_{xx}(l,t) = u_{xxx}(l,t) = 0 \\ \int_0^l F^{(i)}(x)u_{xxt}(x,t)dx = 0, \qquad i = 1,2,...,m \end{cases} \tag{14}$$

Define

$$A_1 = \begin{bmatrix} 0 & I \\ -\frac{EI}{\rho}A & 0 \end{bmatrix},$$

The domain of the linear operator A_1 is

$$D(A_1) := \{(u, u_t)^\tau | u \in H_0^4(0,l),$$
$$u_t \in H_0^2(0,l), u_{xx}(l) = u_{xxx}(l) = 0\}.$$

Then the system (14) can be rewritten as follows:

$$\dot{z}(t) = A_1 z(t), \quad z(0) = z_0 \tag{15}$$

with $\int_0^l F^{(i)}(x)u_{xxt}(x,t)dx = 0, i = 1,2,...,m,$ where $z = (u, u_t)^\tau$.

It is well known that A_1 generates an exponentially decaying C_0 semigroup $T_1(t)$ on $\mathcal{H}$. For $\Phi = (\xi, \eta)^\tau \in \mathcal{H}$, denote $\alpha_n = \langle \xi, \varphi_n \rangle_{\mathcal{H}}$, $\beta_n = \langle \eta, \varphi_n \rangle_{\mathcal{H}}$, then we have

$$T_1(t)\Phi = \sum_{n=1}^{\infty}$$

$$\begin{bmatrix} (\alpha_n\cos\sqrt{\frac{EI}{\rho}\lambda_n}t + \frac{\beta_n}{\sqrt{\frac{EI}{\rho}\lambda_n}}\sin\sqrt{\frac{EI}{\rho}\lambda_n}t)\varphi_n \\ (\beta_n\cos\sqrt{\frac{EI}{\rho}\lambda_n}t - \sqrt{\frac{EI}{\rho}\lambda_n}\alpha_n\sin\sqrt{\frac{EI}{\rho}\lambda_n}t)\varphi_n \end{bmatrix}$$

Then, from (15), we deduce that

$$u(x,t) = \sum_{n=1}^{\infty}(\langle u(x,0), \varphi_n(x)\rangle_{\mathcal{H}}\cos\sqrt{\frac{EI}{\rho}\lambda_n}t$$

$$+\frac{\langle u_t(x,0), \varphi_n(x)\rangle_{\mathcal{H}}}{\sqrt{\frac{EI}{\rho}\lambda_n}}\sin\sqrt{\frac{EI}{\rho}\lambda_n}t)\varphi_n(x),$$

$$u_t(x,t) = \sum_{n=1}^{\infty}(\langle u_t(x,0), \varphi_n(x)\rangle_{\mathcal{H}}\cos\sqrt{\frac{EI}{\rho}\lambda_n}t$$

$$-\sqrt{\frac{EI}{\rho}}\lambda_n\langle u(x,0),\varphi_n(x)\rangle_{\mathcal{H}}\sin\sqrt{\frac{EI}{\rho}}\lambda_n t)\varphi_n(x).$$

This implies that for any $i, i = 1, 2, ..., m$

$$0 = \int_0^l F^{(i)}(x)u_{xxt}(x,t)dx$$

$$= \int_0^l F^{(i)}(x)\sum_{n=1}^{\infty}(\langle u_t(x,0),\varphi_n(x)\rangle_{\mathcal{H}}\cos\sqrt{\frac{EI}{\rho}}\lambda_n t$$

$$-\sqrt{\frac{EI}{\rho}}\lambda_n\langle u(x,0),\varphi_n(x)\rangle_{\mathcal{H}}\sin\sqrt{\frac{EI}{\rho}}\lambda_n t)\varphi_n''(x)dx$$

$$= \sum_{n=1}^{\infty}(\int_0^l F^{(i)}(x)\varphi_n''(x)dx\langle u_t(x,0),\varphi_n(x)\rangle_{\mathcal{H}}$$

$$\cdot\cos\sqrt{\frac{EI}{\rho}}\lambda_n t\cdot$$

$$-\sqrt{\frac{EI}{\rho}}\lambda_n\int_0^l F^{(i)}(x)\varphi_n''(x)dx\langle u(x,0),\varphi_n(x)\rangle_{\mathcal{H}}$$

$$\cdot\sin\sqrt{\frac{EI}{\rho}}\lambda_n t),$$

which means that

$$\int_0^l F^{(i)}(x)\varphi_n''(x)dx\langle u_t(x,0),\varphi_n(x)\rangle_{\mathcal{H}}$$

and

$$-\sqrt{\frac{EI}{\rho}}\lambda_n\int_0^l F^{(i)}(x)\varphi_n''(x)dx\langle u(x,0),\varphi_n(x)\rangle_{\mathcal{H}}$$

are the Fourier coefficients of the uniformly almost periodic function 0. Hence for any $n \geq 1$ and $i \in 1, 2, ..., m$

$$\int_0^l F^{(i)}(x)\varphi_n''(x)dx\langle u(x,0),\varphi_n(x)\rangle_{\mathcal{H}}$$

$$= -\sqrt{\frac{EI}{\rho}}\lambda_n\int_0^l F^{(i)}(x)\varphi_n''(x)dx\langle u_t(x,0),\varphi_n(x)\rangle_{\mathcal{H}}$$

$$= 0$$

But from Assumption 1, we know that for any $n = 1, 2, ...$, there exists an $i, 1 \leq i \leq m$ such that $\int_0^l F^{(i)}(x)\varphi_n''(x)dx \neq 0$. Then we have

$$\langle u(x,0),\varphi_n(x)\rangle_{\mathcal{H}} = \langle u_t(x,0),\varphi_n(x)\rangle_{\mathcal{H}} = 0,$$

which implies $u(x,0) = u_t(x,0) = 0$ and $E(0) = 0$. Thus we have $z(t) \equiv 0$ for $t \geq 0$.

□

From Lemma 1, by the Sobolev embedding theorem and the argument similar to that used in Guo, et. al. 1995, we can obtain the following lemma

Lemma 3. The operator $\mathcal{A} : D(\mathcal{A}) \subset \mathcal{H} \to \mathcal{H}$ generates a C_0 semigroup of contraction on $\mathcal{H}$ with $(\lambda I - \mathcal{A})^{-1}$ being a compact operator for $\lambda > 0$.

By lemma1-lemma 3, using LaSalle's invariance principle (Hale 1969) and linear semigroup theory (Pazy 1983), we immediately have the following result.

Theorem 1. Consider the system given by (10). Then for $z(0) \in D(\mathcal{A})$, equation (10) has a unique classical solution $z(t)$. Moreover, if Assumption 1 holds, we have

$$\lim_{t\to\infty}||z(t)||^2_{\mathcal{H}} = 0. \tag{16}$$

Consider two special case of $i = 1$, we have the following corollarys

Corollary 1. Consider the system (10) with $i = 1, F^{(1)}(x) \equiv 1$. Then for $z(0) \in D(\mathcal{A})$, the system has a unique classical solution $z(t)$, which satisfies

$$\lim_{t\to\infty}||z(t)||^2_{\mathcal{H}} = 0.$$

Proof: By Theorem 1, we see that we only need to show Assumption 1 holds for the case $i = 1, F^{(1)}(x) \equiv 1$. In this case, we have $F^{(1)}(x) \in H^2(0,l)$, and

$$\int_0^l F^{(1)}(x)\varphi_n''(x)dx = \int_0^l \varphi_n''(x)dx = \varphi_n'(l) - \varphi_n'(0)$$

$$= \beta_n[(\sinh\beta_n l + \sin\beta_n l)$$

$$-\frac{\cosh\beta_n l + \cos\beta_n l}{\sinh\beta_n l + \sin\beta_n l}(\cosh\beta_n l - \cos\beta_n l)]$$

$$= \frac{2\beta_n\sinh\beta_n l\sin\beta_n l}{\sinh\beta_n l + \sin\beta_n l}$$

Since $\beta_n > 0, \beta_n l > 0$, it follows that $\sinh\beta_n l > 0, \sinh\beta_n l + \sin\beta_n l > 0$ and $\cosh\beta_n l > 1$ which together with $\cosh\beta_n l\cos\beta_n l = -1$ implies that $|\cos\beta_n l| < 1$, this means $\sin\beta_n l \neq 0$. Thus, we obtain

$$\beta_n\sinh\beta_n l\sin\beta_n l \neq 0, \qquad n = 1, 2, ...,$$

i. e., Assumption 1 holds for $i = 1, F^{(1)}(x) \equiv 1$.

□

Corollary 2. Consider the system (10) with $i = 1, F^{(1)}(x) = l - x$. Then for $z(0) \in D(\mathcal{A})$, the system has a unique classical solution $z(t)$, which satisfies

$$\lim_{t\to\infty}||z(t)||^2_{\mathcal{H}} = 0.$$

Proof: Similarly, we only need to show Assumption 1 holds for the case $F(x) = l - x$. In this case, we have $F^{(1)}(x) \in H^2(0, l)$, and

$$\int_0^l F^{(1)}(x)\varphi_n''(x)dx = \int_0^l (l-x)\varphi_n''(x)dx$$

$$= (l-x)\varphi_n'(x)|_0^l + \int_0^l \varphi_n'(x)dx = \varphi_n(l)$$

$$= (\cosh\beta_n l - \cos\beta_n l)$$

$$-\frac{\cosh\beta_n l + \cos\beta_n l}{\sinh\beta_n l + \sin\beta_n l}(\sinh\beta_n l - \sin\beta_n l)$$

$$= 2\frac{\cosh\beta_n l \sin\beta_n l - \sinh\beta_n l \cos\beta_n l}{\sinh\beta_n l + \sin\beta_n l}$$

Since $\cosh\beta_n l \cos\beta_n l = -1$ and $\beta_n l > 0$, it follows that $\cosh\beta_n l > 0$ and $\cos\beta_n l < 0$, which implies that $\beta_n l \in (\frac{\pi}{2} + 2k\pi, \frac{3\pi}{2} + 2k\pi)$ for some nonegative integer k. Then we have $\cosh\beta_n l > \frac{e^{\pi/2} + e^{-\pi/2}}{2} > \sqrt{2}$, which, together with $\cosh\beta_n l \cos\beta_n l = -1$, means $-\sqrt{2}/2 < \cos\beta_n l < 0$. Thus we obtain $|\frac{\sin\beta_n l}{\cos\beta_n l}| > 1$, which together with the fact that $0 < \frac{\sinh\beta_n l}{\cosh\beta_n l} < 1$ follows that

$$\frac{\sin\beta_n l}{\cos\beta_n l} - \frac{\sinh\beta_n l}{\cosh\beta_n l} \neq 0,$$

i. e.,

$$\cosh\beta_n l \sin\beta_n l - \sinh\beta_n l \cos\beta_n l \neq 0.$$

Thus, Assumption 1 holds for $F^{(1)}(x) = l - x$.

□

4. CONCLUSIONS

In this paper, an active distributed damper for a cantilever beam is designed and evaluated. The distributed-parameter active elements are piezoelectric sensors and actuators. Linear feedback controls for the smart beam are developed using Lyapunov's second method. Based upon LaSall's invariant principle in infinite dimensional space and linear semigroup theory, it is shown that implementation of these control algorithms result in vibration suppression.

REFERENCES

Bailey, T., and Hubbard, J.E., 1985, Distributed piezoelectric polymer active vibration control of a cantilever beam. *Journal of Guidance, Control, and Dynamics*, **8**, 605-611.

Burke, S. E., and Hubbard, J.E., 1988, Distributed actuator design for flexible beams. *Automatica*, **24**, 619-627.

Chen, G., Delfour, M. C., Krall, A. M., and Payre, G., 1987, Modelling, stabilization and control of serially connected beams. *SIAM Journal of Control and Optimization*, **25**, 526-546.

Crawley, E. F., and de Luis, J., 1987, Use of piezoelectric actuator as elements of intelligent structures. *AIAA Journal*, **25**, 1373-1385.

Guo, B. Z., and Song, Q., 1995, Tracking control of a flexible beam by nonlinear boundary feedback. *J. of Applied Mathematics and Stochastic Analysis*, **8**, 47-58.

Hale, J. K., 1969, Dynamical systems and stability. *J. Math. Anal. Appl.*, **26**, 39-59.

Luo, Z. H., and Guo, B. Z., 1997, Shear force feedback control of a single-link flexible robot with a revolute joint. *IEEE trans. on Autom. Contr.*, **42**, 53-65.

Morgül, Ö., 1991, Orientation and stabilization of a flexible beam attached to a rigid body: planar motion. *IEEE trans. on Autom. Contr.*, **36**, 953-962.

Pazy, A., 1983, *Semigroups of Linear Operators and applications to Partial Differential Equation* (New York, Springer–Verlag).

14th Triennial World Congress, Beijing, P.R. China

ENERGY SHAPING CONTROL METHOD FOR ROBOTIC FORCE/POSITION REGULATION AND MOTION CONTROL

Natsuo Tanaka * **Masayuki Fujita** *,**

* *School of Information Science JAIST, Tatsunokuchi Ishikawa, 923-1211, Japan, E-mail: natsuo@jaist.ac.jp*
** *Faculty of Engineering Kanazawa Univ., Kanazawa Ishikawa, 920-8667, Japan, E-mail: fujita@t.kanazawa-u.ac.jp*

Abstract: In this paper, an adaptive $\mathcal{H}_\infty$ SP–D (Saturated Proportional and Differential) control method is introduced for robot manipulators interacting with a compliant environment. The proposed control method is able to achieve both tracking control on a contact surface of the environment and force/position regulation along a vertical direction of the surface simultaneously. Furthermore, disturbance attenuation is revealed by exploiting an $\mathcal{L}_2$-gain of the manipulator system, and some experimental results show effectiveness of the proposed method. The proposed SP–D control method is derived from a novel concept, that is, an energy shaping of the manipulator system. It is one of the energy based control methods such as the Lyapunov-based control and the passivity-based control. In other words, the proposed method is also fundamental and essential for robot control theory. Contribution of this article is to investigate some abilities of the energy shaping control method for manipulators in contact with the environment.

Keywords: energy shaping, SP–D control, force/position regulation, tracking control, robot manipulator

1. INTRODUCTION

Energy based control scheme for robot manipulators such as the Lyapunov-based control and the passivity-based control has attracted much attention (Canudas de Wit *et al.*, 1996; Arimoto, 1996). The control scheme is based on the principle of energy conservation. In recent years, interesting concepts such as quasi-natural potential (Arimoto, 1996) and energy shaping (Kelly *et al.*, 1996) are introduced with respect to the energy based control scheme. These concepts derive a SP–D (Saturated Proportional and Differential) control method.

In previous researches (Takegaki and Arimoto, 1981; Kelly, 1997), PD control method with gravity compensation is shown to guarantee the global asymptotic stability in a set-point control problem. Generally, physical parameters of the manipulator dynamics have some kinds of uncertainties. Thus, it is difficult to avoid effects of offset errors. In order to remove the offset errors, PID control method is effective for the set-point control problem. Unfortunately, the PID control method can only local asymptotic stability of the manipulator system (Arimoto and Miyazaki, 1984). However, a new SP–ID (Saturated Proportional, Integral and Differential) control method which is a PID type control is presented (Arimoto, 1994). The SP–ID control method based on the mentioned SP–D control can ensure the global asymptotic stability even if the manipulator system is subjected to the parametric uncertainties. One of some reasons for the global asymptotic stability is to exploit not a quadratic potential energy but a shaped potential energy for the Lyapunov function. It is worth mentioning that the energy shaping control method is a fundamental and essential of the the energy based control theories.

One of our projects is to reinforce the robot control theories with respect to the energy–shaping control method, further to evaluate the control method by applying to actual manipulators practically. In our previous work (Tanaka and Fujita, 1997; Fujita and Tanaka, 1997), a robust SP–D control method which ensures the globally uniformly ultimate boundedness is proposed. The previous work contributed to investigate robustness of the energy–shaping control method against the parametric uncertainties in the tracking control problems. Then, a nominal SP–D force/position regulator is introduced for the manipulator interacting with a compliant environment (Tanaka and Fujita, 1998). However, the regulator requires a stiffness coefficient of the environment precisely. Further, it is not necessary to be robust against the parametric uncertainties.

This paper presents an adaptive $\mathcal{H}_\infty$ SP–D control method which is able to achieve the tracking control and the force/position regulation simultaneously. The tracking control is conducted on a contact surface of the compliant environment, the force/position regulation is along the vertical axis of the surface. By using the adaptive control scheme, the proposed control method becomes robust against the parametric uncertainties. Furthermore, robustness against force disturbances exerted on a tip of the manipulators is analyzed by exploiting the $\mathcal{L}_2$-gain of the manipulator system. Some experimental results on the actual manipulator confirm the above analysis. Main contribution of this work is an investigation of the energy–shaping control method with respect to the robustness against both the parametric uncertainties and the force disturbances.

2. PRELIMINARIES

2.1 *manipulator dynamics*

In this work, a non-redundant rigid manipulator with $n(\leq 3)$ joints is considered. A dynamical model of the manipulator in contact with a compliant environment can be written as follows.

$$M(q)\ddot{q} + C(q,\dot{q})\dot{q} + g(q) = \tau + J(q)^T(f + w) \quad (1)$$

where $q \in \mathbb{R}^n$, $\tau \in \mathbb{R}^n$ and $f \in \mathbb{R}^n$ denote a vector of joint angles, torques and contact forces exerted to an end-effector of the manipulator, respectively. The matrix $J(q) \in \mathbb{R}^{n\times n}$ means the robot Jacobian, $w \in \mathbb{R}^n$ represents a vector of force disturbances. The inertia matrix $M(q) \in \mathbb{R}^{n\times n}$ is positive symmetric for any q and the term $C(q,\dot{q})\dot{q} \in \mathbb{R}^n$ represents a vector of the Coriolis and centrifugal forces. It is worth mentioning that $\dot{M}(q) - 2C(q,\dot{q})$ is skew-symmetric for any $q, \dot{q}$. The term $g(q) \in \mathbb{R}^n$ corresponds to the gravitational forces. Choose an adequate regressor $Y(q,\dot{q},\dot{q},\ddot{q}) \in \mathbb{R}^{n\times m}$, the manipulator dynamics can be simplified as follows (Canudas de Wit *et al.*, 1996).

$$M(q)\ddot{q} + C(q,\dot{q})\dot{q} + g(q) = Y(q,\dot{q},\dot{q},\ddot{q})\theta$$

where $\theta \in \mathbb{R}^m$ represents a vector of physical parameters containing mass, length, inertia momentum of each link.

One of elementary tasks for the manipulator is to press the compliant environment. In other words, the end-effector located at a contact position $p \in \mathbb{R}^n$ is pressed by the force f from the environment. The environment is modeled as follows.

$$f = K_e(p - p_0) \quad (2)$$

where $K_e := \mathrm{diag}\{k_{ex}, k_{ey}, k_{ez}\}$ denotes a stiffness coefficient matrix of the environment. Subscript x of the k_{ex} represents a vertical direction to the contact surface, and subscript y, z mean horizontal components on the surface. The vector $p_0 \in \mathbb{R}^n$ is a position at rest. In this article, tracking control on the contact surface (on the y-z surface) and force/position regulation along the vertical direction are discussed. Thus, $k_{ey}, k_{ez} = 0$ are defined. It is supposed that the contact between the manipulator and the environment is not lost. Further, there exist no friction forces on the contact surface. Therefore, the contact forces become $f := [f_x, 0, 0]^T$, desired contact forces $f_d := [f_{dx}, 0, 0]^T$ (f_{dx} : const.) is defined. Similarly, desired positions $p_d := [p_{dx}, p_{dy}, p_{dz}]^T$, desired velocities $\dot{p}_d := [0, \dot{p}_{dy}, \dot{p}_{dz}]^T$ are given. It is worth noticing that the desired position p_{dx} is related to the desired force f_{dx} with respect to the environment model (2). Consequently, the desired position p_{dx} should obey the desired force f_{dx}. The other desired positions p_{dy}, p_{dz} and desired velocities $\dot{p}_{dy}, \dot{p}_{dz}$ are given in order to achieve the desired trajectory on the contact surface.

With respect to the force/position regulation, a potential energy of the manipulator system is composed of the positional errors $\tilde{p} := p - p_d$ and the force errors $\tilde{f} := f - f_d$. These errors have a linear relation which is obtained from (2).

$$\tilde{f} = K_e\tilde{p} \quad (3)$$

Since of k_{ey} and k_{ez} are zero, the above relation is simplified as follows.

$$\tilde{f}_x = k_{ex}\tilde{p}_x \quad (4)$$

2.2 *properties of shaped potential energy*

In this subsection, a novel potential energy which derives the SP–D control method will be introduced. A key idea of this work is to combine the positional errors $\tilde{p}$ and each integration of the force errors $\tilde{f}$ as follows.

$$\xi := \tilde{p} + \Lambda_f \int_0^t \tilde{f}(\tau)d\tau \tag{5}$$

where $\Lambda_f \in \mathbb{R}^{n\times n}$ is a diagonal matrix with all positive elements. Furthermore, the following auxiliary variable s is also important idea.

$$s := J^{-1}\left\{\dot{\xi} + \Lambda\varphi(\xi)\right\} \tag{6}$$

where $\varphi(\xi) := [\varphi(\xi_1), \ldots, \varphi(\xi_n)] \in \mathbb{R}^n$, and $\Lambda \in \mathbb{R}^{n\times n}$ is also a diagonal matrix with all positive elements.

Both nonlinear saturation function $\varphi(\cdot) \in C^1$ (C^n : n times continuously differentiable) and the potential energy $U(\cdot) \in C^2$ with respect to ξ have following properties.

$$\text{(P1)}\ \varphi(0) = 0, \quad \varphi(\xi_i) \neq 0\ (\xi \neq 0)$$

$$\text{(P2)} \begin{cases} \dfrac{\partial\varphi(\xi_i)}{\partial\xi_i} > 0\ (|\xi_i| \leq \delta_i) \\ \dfrac{\partial\varphi(\xi_i)}{\partial\xi_i} = 0\ (|\xi_i| > \delta_i) \end{cases}$$

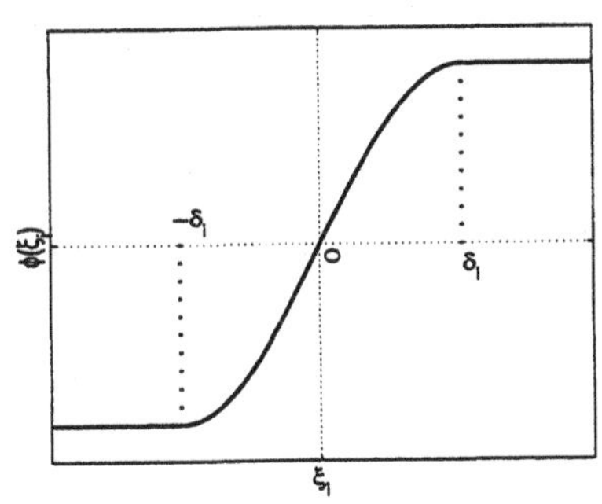

Fig. 1. Saturation function

Instead of the above property (P2),

$$\text{(P2)}'\ \frac{\partial\varphi(\xi_i)}{\partial\xi_i} > 0$$

can be available. That is, the derivative function $\varphi(\xi_i)$ has a monotone increasing property at $-\delta_i < \xi_i < \delta_i$.

Furthermore,

$$\text{(P3)}\ U(\mathbf{0}) = 0, \quad U(\xi) > 0\ (\xi \neq 0)$$

$$\text{(P4)}\ \frac{\partial^T U(\xi)}{\partial\xi} = K_p\varphi(\xi)$$

where K_p represents a positive diagonal matrix which is related to feedback gains, and $\frac{\partial}{\partial\xi} = [\frac{\partial}{\partial\xi_1}, \ldots, \frac{\partial}{\partial\xi_n}]$.

3. ADAPTIVE $\mathcal{H}_\infty$ SP–D CONTROL METHOD

If the end-effector of the manipulator is changed, parameters such as mass, inertia momentum and length of the last link is changed consequently. That is, the physical parameter θ is also changed. Therefore, an adaptive control method is exploited for changing nominal parameter $\hat{\theta}$, and gives robustness against the parametric uncertainties.

Based on the SP–D control method, a new adaptive $\mathcal{H}_\infty$ SP–D control law is proposed for the motion tracking and the force/position regulation.

$$\tau = Y\hat{\theta} - K_v s - J^T K_p \varphi(\xi) - J^T f \tag{7}$$

$$\dot{\hat{\theta}} = -\Gamma^{-1} Y^T s \tag{8}$$

where, gains $K_v \in \mathbb{R}^{n\times n}$, $K_p \in \mathbb{R}^{n\times n}$, $\Gamma \in \mathbb{R}^{m\times m}$ are diagonal matrices with all positive elements, respectively. Virtual reference trajectories are defined as follows.

$$\dot{q}_s := J^{-1}\left\{\dot{p}_d - \Lambda_f \tilde{f} - \Lambda\varphi(\xi)\right\} \tag{9}$$

Then, the auxiliary variable s has the following relation.

$$s = \dot{q} - \dot{q}_s \tag{10}$$

In fact, substituting the virtual reference $\dot{q}_s$ into the above relation, the auxiliary variable s is obtained by using (5). The first term $Y\hat{\theta}$ of the control law (7) compensates the manipulator dynamics, the second term $K_v s$ and the third term $J^T K_p \varphi(\xi)$ feedback errors. The last term $J^T f$ corresponds to a torque vector of the contact force f which is measured by a force/torque sensor on the tip.

Substituting the control law (7) into the manipulator dynamics (1),

$$M\dot{s} + (C + K_v)s + J^T K_p \varphi(\xi) + Y\tilde{\theta} = J^T w \tag{11}$$

is obtained. The vector $\tilde{\theta} := \theta - \hat{\theta}$ denotes the parametric uncertainties. Define the state variable $x := [s^T, \xi^T, \tilde{\theta}^T]^T$, the following closed-loop system Σ_{wz} is derived.

$$\Sigma_{wz} \begin{cases} \dot{x} = f(x) + g(x)w \\ z = h(x) \end{cases} \tag{12}$$

where

$$f(x) = \begin{bmatrix} -M^{-1}\{(C+K_v)s + J^T K_p \varphi(\xi) + Y\tilde{\theta}\} \\ Js - \Lambda\varphi(\xi) \\ \Gamma^{-1} Y^T s \end{bmatrix}$$

$$g(x) = \begin{bmatrix} M^{-1} J^T \\ 0 \\ 0 \end{bmatrix}, \quad h(x) = \begin{bmatrix} \dot{\xi} \\ \varphi(\xi) \end{bmatrix}$$

Further, the variable z means a pricing vector containing the positional errors $\tilde{p}$, the velocity errors $\dot{\tilde{p}}$, the force errors $\tilde{f}$ and the integration of the force errors.

4. PERFORMANCE ANALYSIS BY USING $\mathcal{L}_2$-NORM

The trajectory tracking on the contact surface and the force/position regulation along the vertical direction can be shown to be achieved as follows.

Firstly, the internal stability is shown in case of the force disturbances $w = 0$. The following Lyapunov function candidate is defined.

$$V(x) := \frac{1}{2}s^T M s + U(\xi) + \frac{1}{2}\tilde{\theta}^T \Gamma \tilde{\theta} \qquad (13)$$

The function V is positive definite since of the positive matrices M , Γ and the property (P3). The second term U in the right-hand side is the shaped potential energy with respect to ξ. It is worth noting that the energy is not quadratic form necessarily.

Along the solution trajectory of the closed-loop system (11), the derivative function is derived as follows.

$$\dot{V}(x) = -s^T K_v s - \varphi(\xi)^T \Lambda^T K_p \varphi(\xi) \leq 0 \quad (14)$$

The variables s and ξ are uniformly continuous, and the function φ is continuous. Thus, $s \to 0$ and $\xi \to 0$ are ensured by using the Barbalat's lemma. The adaptation law (8) ensures that the parametric uncertainties $\tilde{\theta}$ become to be constant. Therefore internal stability of the closed-loop system is guaranteed.

Remark 1. On the contact surface, the force errors $\tilde{f}_y$ and $\tilde{f}_z$ are zero since of the assumption about the friction forces. Consequently, $\tilde{p}_y(=\xi_y)$, $\tilde{p}_z(=\xi_z)$, $\dot{\tilde{p}}_y(=\dot{\xi}_y)$, $\dot{\tilde{p}}_z(=\dot{\xi}_z) \to 0$ as $t \to \infty$. The trajectory tracking on the contact surface is achieved.

Furthermore, the force/position regulation is also achieved. The force error and the positional error satisfy the following equalities.

$$\tilde{p}_x = -\lambda_{fx}\int_0^t \tilde{f}_x(\tau)d\tau, \quad \dot{\tilde{p}}_x = -\lambda_{fx}\tilde{f}_x$$

Using the relation of the force error and the position error (4),

$$\dot{y} = -k_{ex}\lambda_{fx} y \qquad (15)$$

is obtained, where the variable y is defined as follows.

$$y := \left[\int_0^t \tilde{f}_x d\tau,\ \tilde{f}_x\right]^T$$

Therefore, given the adequate reference which satisfies the equation (4), the force/position regulation is achieved. □

Secondly, the $\mathcal{L}_2$ stability of the closed-loop system Σ_{wz} is discussed in case of $w \neq 0$. The derivative Lyapunov function is derived as follows.

$$\dot{V}(x) = (Js)^T w - s^T K_v s - \varphi(\xi)^T \Lambda^T K_p \varphi(\xi) \qquad (16)$$

On the other side,

$$\dot{V}(x) = \frac{\partial V(x)}{\partial x} f(x) + \frac{\partial V(x)}{\partial x} g(x) w \qquad (17)$$

is satisfied clearly. With respect to the disturbance w, the following equation are satisfied identically.

$$\frac{\partial V(x)}{\partial x} f(x) = -s^T K_v s - \varphi(\xi)^T \Lambda^T K_p \varphi(\xi) \quad (18)$$

$$\frac{\partial V(x)}{\partial x} g(x) = (Js)^T \qquad (19)$$

By using the above equations, the Lyapunov function V satisfies the following Hamilton-Jacobi inequality (van der Schaft, 1996).

$$\frac{\partial V}{\partial x} f(x) + \frac{1}{2\gamma^2}\frac{\partial V}{\partial x} g(x) g(x)^T \frac{\partial^T V}{\partial x} + \frac{1}{2} h(x)^T h(x) \leq 0 \quad (20)$$

where γ is a positive constant. In fact, substitute the equations (18) and (19) into the left-hand side of the Hamilton-Jacobi inequality.

$$\begin{aligned}\frac{\partial V}{\partial x} f(x) &+ \frac{1}{2\gamma^2}\frac{\partial V}{\partial x} g(x) g(x)^T \frac{\partial^T V}{\partial x} + \frac{1}{2} h(x)^T h(x) \\ &= -s^T K_v s - \varphi(\xi)^T \Lambda^T K_p \varphi(\xi) \\ &\quad + \frac{1}{2\gamma^2} s^T (J^T J) s + \frac{1}{2}\|z\|^2\end{aligned}$$

is obtained. Substitute the following inequality

$$\begin{aligned}\|z\|^2 &= \|Js - \Lambda\varphi(\xi)\|^2 + \|\varphi(\xi)\|^2 \\ &\leq 2\lambda_{\max}(J^T J)\|s\|^2 + \sum_{i=1}^{n}(2\lambda_i^2 + 1)\varphi(\xi_i)^2\end{aligned}$$

into the last term,

$$\begin{aligned}\frac{\partial V}{\partial x} f(x) &+ \frac{1}{2\gamma^2}\frac{\partial V}{\partial x} g(x) g(x)^T \frac{\partial^T V}{\partial x} + \frac{1}{2} h(x)^T h(x) \\ &\leq -\left\{\lambda_{\max}(K_v) - \left(\frac{1}{2\gamma^2} + 1\right)\lambda_{\max}(J^T J)\right\}\|s\|^2 \\ &\quad - \sum_{i=1}^{n}\lambda_i\left\{k_{pi} - \lambda_i - \frac{1}{2\lambda_i}\right\}\varphi(\xi_i)^2 \quad (21)\end{aligned}$$

is derived.

Theorem 2. Given a constant $\varepsilon > 0$. Define the gains K_v and K_p such as

(i) $\lambda_{\min}(\boldsymbol{K}_v) \geq \left(\frac{1}{2\gamma^2}+1\right)\lambda_{\max}(\boldsymbol{J}^T\boldsymbol{J})$

(ii) $k_{pi} \geq \lambda_i + \frac{1}{2\lambda_i} \qquad (i=1,\ldots,n)$.

Satisfied $V(\boldsymbol{x}(\boldsymbol{0})) \leq \varepsilon$, the $\mathcal{L}_2$-gain of the closed loop system Σ_{wz} satisfies

$$||\Sigma_{wz}||_{i2} := \sup_{\boldsymbol{w}\in\mathcal{L}_2^n} \frac{||z||_2}{||\boldsymbol{w}||_2} \leq \left(\gamma^2 + \frac{2\varepsilon}{||\boldsymbol{w}||_2^2}\right)^{\frac{1}{2}}$$

with respect to the force disturbances $\boldsymbol{w} \in \mathcal{L}_2^n$, where the symbols $||\cdot||_2$ and $||\cdot||_{i2}$ mean the $\mathcal{L}_2$ norm and the induced $\mathcal{L}_2$-gain respectively.

Proof: By the conditions with respect to $\boldsymbol{K}_v$ and $\boldsymbol{K}_p$, the Hamilton-Jacobi inequality (20) is satisfied. (See the inequality (21).) From the equation (17), the following $\dot{V}$ is derived.

$$\dot{V}(\boldsymbol{x}) = -\frac{1}{2}\gamma^2\left|\left|\boldsymbol{w} - \frac{1}{\gamma^2}g(\boldsymbol{x})^T\frac{\partial^T V}{\partial \boldsymbol{x}}\right|\right|^2 + \frac{1}{2}\gamma^2||\boldsymbol{w}||^2 + \frac{\partial V}{\partial \boldsymbol{x}}f(\boldsymbol{x}) + \frac{1}{2\gamma^2}\frac{\partial V}{\partial \boldsymbol{x}}g(\boldsymbol{x})g(\boldsymbol{x})^T\frac{\partial^T V}{\partial \boldsymbol{x}}$$

Substitute (20) into the third and the fourth terms.

$$\begin{aligned}\dot{V}(\boldsymbol{x}(t)) &\leq \frac{1}{2}\gamma^2||\boldsymbol{w}||^2 - \frac{1}{2}||z||^2 \\ &\quad -\frac{1}{2}\gamma^2\left|\left|\boldsymbol{w} - \frac{1}{\gamma^2}g(\boldsymbol{x})^T\frac{\partial^T V}{\partial \boldsymbol{x}}\right|\right|^2 \\ &\leq v(\boldsymbol{w}, \boldsymbol{z})\end{aligned}$$

where

$$v(\boldsymbol{w}, \boldsymbol{z}) := \frac{1}{2}\gamma^2||\boldsymbol{w}||^2 - \frac{1}{2}||z||^2.$$

The function v denotes the energy supply rate, *i.e.*, $\mathcal{L}_2$-gain supply rate (van der Schaft, 1996). In other words, the V function means a storage function. Integrate the above inequality,

$$\int_0^t ||z(\tau)||^2 d\tau \leq \gamma^2 \int_0^t ||\boldsymbol{w}||^2 d\tau + 2V(\boldsymbol{x}(0)) \tag{22}$$

is obtained. Substitute the condition $V(\boldsymbol{x}(0)) \leq \varepsilon$ into the last term,

$$||\Sigma_{wz}||_{i2} \leq \left(\gamma^2 + \frac{2\varepsilon}{||\boldsymbol{w}||_2^2}\right)^{\frac{1}{2}} \tag{23}$$

is shown. □

5. EXPERIMENTS

The proposed adaptive $\mathcal{H}_\infty$ control law is confirmed by some experiments with two-link direct drive manipulator system. (See Fig. 2.) This experimental system has two directions, axis X and axis Y. The trajectory tracking and the force/position regulation is conducted along the axis X and the axis Y, respectively. Initial positions of the end-effector is set $\boldsymbol{p} = [0.34, -0.05]^T$ m. The desired force $f_{dx} = 29.0$ N is given along the axis X, and the desired position $p_{dx} = 0.35$ m is dependently is defined. These values f_{dx} and p_{dx} are measured experimentally. On the contact surface, that is, along the axis Y, the desired trajectory is given smoothly. The desired trajectory achieves $y_d = 0.0$ m at time $t = 5.0$ s. In order to confirm the boundedness of the $\mathcal{L}_2$-gain $||\Sigma_{wz}||_{i2}$, the force disturbance $\boldsymbol{w} = [-10.0, 0.0]^T$ N is subjected at time $6.0 \leq t < 7.0$. The following gains are defined to satisfy the conditions in Theorem 2.

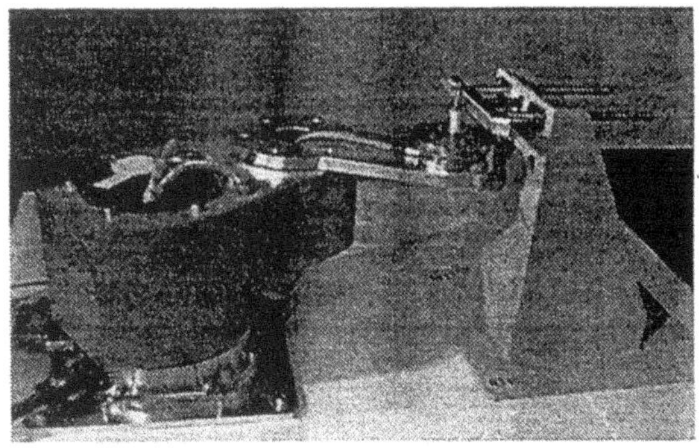

Fig. 2. DD manipulator system

$$\boldsymbol{K}_v = \text{diag}\{1.0, 1.0\}, \quad \boldsymbol{K}_p = \text{diag}\{10, 70\},$$
$$\boldsymbol{\Lambda} = \text{diag}\{0.2, 0.2\}, \quad \boldsymbol{\Lambda}_f = \text{diag}\{0.1, 0.05\},$$
$$\boldsymbol{\Gamma} = \text{diag}\{1.0, 1.0\}$$

Experimental results are shown in Fig.3, Fig.4 and Fig.5. Figure 3 and Figure 4 show the position and the velocities of the end-effector along the axis X and the axis Y, respectively. Figure 5 shows the contact force along the axis X. Solid line means the experimental results, and dotted line represents the desired values. Even though the force disturbance $\boldsymbol{w}$ affected the position $\boldsymbol{p}$ and the velocity $\dot{\boldsymbol{p}}$ and the contact force f_X, the objectives, the force/position regulation along the axis X and the motion control along the axis Y, are achieved. Further, the $\mathcal{L}_2$-gain $||\Sigma_{wz}||_{i2} \leq 0.46$ was calculated from $||\boldsymbol{w}||_2$ and $||z||_2$ and confirmed to be robust against the disturbance $\boldsymbol{w}$. It means that disturbance attenuation was confirmed.

6. CONCLUSIONS

This article presented the adaptive $\mathcal{H}_\infty$ SP–D control method for robot manipulators. The proposed control method was obtained by exploiting the energy–shaping control with respect to the potential energy, and achieved the trajectory tracking and the force/position regulation simultaneously. The proposed control method was inheriting a

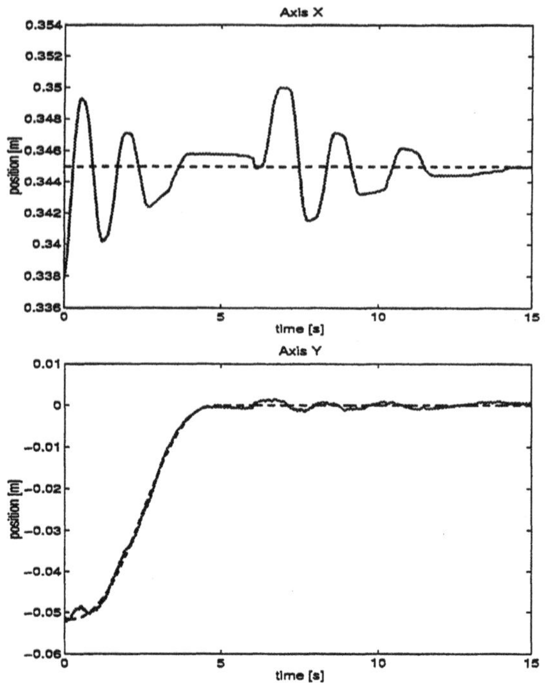

Fig. 3. Positions along the axis X and the axis Y

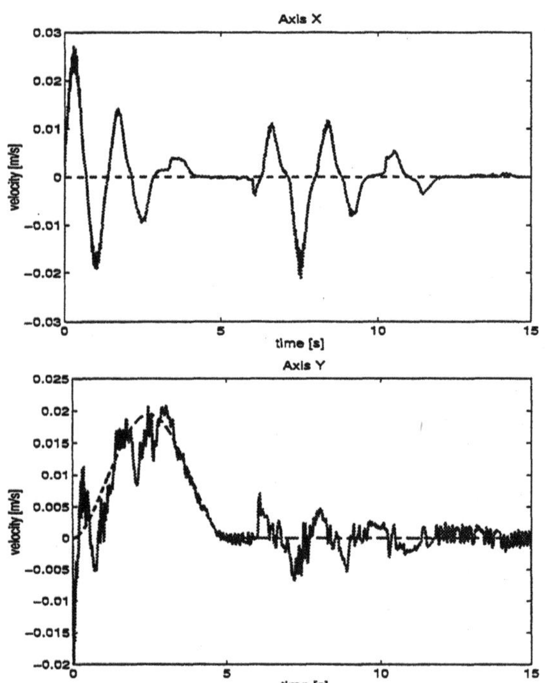

Fig. 4. Velocities along the axis X and the axis Y

simple structure by extending the previous SP–D control method. In other words, it has the advantage of implementing the control method. In fact, the proposed control method is shown to be effective for actual manipulators experimentally.

Main objective of our projects is investigation of the abilities with respect to the SP–D control method for manipulators. This work contributed to clarify the robustness of the SP–D control method in both the tracking and the force/position control problems.

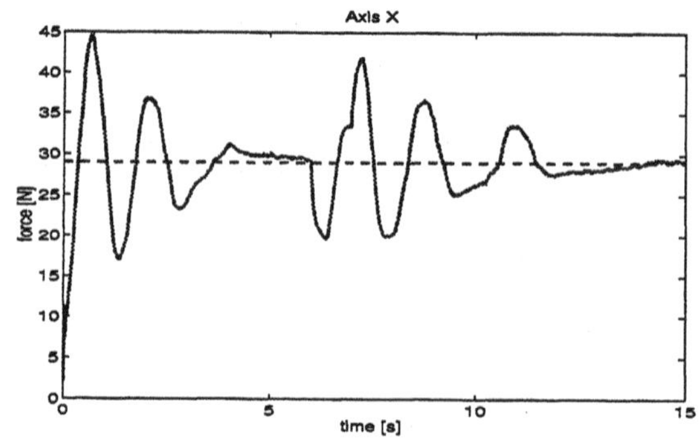

Fig. 5. Contact force along the axis X

REFERENCES

Arimoto, S. (1994). A class of quasi-natural potentials and hyper-stable pid servo-loops for nonlinear robotic systems. *Trans. of the Society of Instrument and Control Engineerings* **30**(9), 1005–1012.

Arimoto, S. (1996). *Control Theory of Non-Linear Mechanical Systems.* Oxford University Press. New York.

Arimoto, S. and F. Miyazaki (1984). Stability and robustness of pid feedback control for robot manipulators of sensory capability. In: *Robotics Research: The First International Symposium* (M. Brady and R. Paul, Eds.). pp. 783–799. MIT Press. London.

Canudas de Wit, C., B. Siciliano and Eds. G. Bastin (1996). *Theory of Robot Control.* Springer-Verlag. London.

Fujita, M. and N. Tanaka (1997). A new type of robust tracking control of robot manipulators based on generalized sp–d control scheme. *Proc. 1997 IEEE Int. Conf. on Robotics and Automation* pp. 2371–2376.

Kelly, R. (1997). Pd control with desired gravity compensation of robotic manipulators: A review. *The Int. J. of Robotics Research* **16**(5), 660–672.

Kelly, R., V. Santibáñes and F. Reyes (1996). On saturated-proportional derivative feedback with adaptive gravity compensation of robot manipulators. *Int. J. of Adaptive Control and Signal Processing* **10**(4), 465–479.

Takegaki, M. and S. Arimoto (1981). A new feedback method for dynamic control of manipulators. *J. of Dynamic Systems, Measurement, and Control* **102**, 119–125.

Tanaka, N. and M. Fujita (1997). A new type of robust tracking sp–d control for manipulators. *Trans. on IEEJ* **117**(10), 1529–1534.

Tanaka, N. and M. Fujita (1998). Torque disturbance attenuation and position/force tracking based on sp–d control method for manipulators (in japanese). *Trans. on IEEJ* **118**(2), 271–276.

van der Schaft, A. (1996). *$\mathcal{L}_2$-Gain and Passivity Techniques in Nonlinear Control.* Springer-Verlag. London.

14th Triennial World Congress, Beijing, P.R. China

MIXED H_2/H_∞ CONTROL OF FLEXIBLE STRUCTURES [1]

D. P. de Farias M. C. de Oliveira J. C. Geromel

LAC-DT / School of Electrical and Computer Engineering, UNICAMP, C.P. 6101, 13081-970, Campinas, SP, Brazil,
e-mail: geromel@dt.fee.unicamp.br

Abstract: This paper addresses the design of full order linear dynamic output feedback controllers for flexible structures. Unstructured H_∞ uncertainty models are introduced for systems in modal coordinates and in reduced order form. Then a controller is designed in order to minimize a given H_2 performance function while keeping the maximum supported H_∞ perturbation bellow some appropriate level. To solve this problem we develop an algorithm able to provide local optimal solutions to optimization problems with convex constraints and non-convex but differentiable objective functions. A controller design procedure based on a trade-off curve is proposed and a simple example is solved, providing a comparison between the proposed method and the usual minimization of an upper bound to the H_2 norm. The method is applied to a flexible structure model.

Keywords: H_∞ control, Quadratic control, Numerical methods.

1. INTRODUCTION

The control of flexible structures poses several challenges that may be very difficult to handle by standard synthesis procedures. The dynamics of such systems are usually high order and present many closely spaced low frequency modes. Finite-element methods which are the usual source of mathematical models of real world flexible structures are known to provide models which may not accurately describe high order modes dynamics. Theoretical flexible structure models are infinite dimensional and must be preprocessed by some truncation or order reduction procedure before being used in controller design. All such features ask for the introduction of some kind of uncertainty modeling in the design process so that the obtained controllers may succeed when applied to the real physical structure. Furthermore, in view of the high order of the plant dynamics the controller must be able to handle only a reduced set of measurements, which in practice discards the use of a state feedback structure.

The most powerful procedures available in the literature to the design of controllers for uncertain systems are based either on H_∞ or on quadratic stability theory. Several results are able to handle the H_∞ controller design under output feedback (Colaneri et al. (1997)), but most of the available procedures deal only with unstructured uncertainty. However, the particular form of flexible structure models leads in general to a highly structured uncertainty description, which may render the available designs very conservative. Despite this fact, many authors (see for instance Boulet et al. (1997)) have been successful in providing uncertainty models that work well in an unstructured H_∞ design framework. With respect to performance, this unstructured framework does not easily admit the introduction of additional H_∞ performance constraints without increasing conservativeness, which generally asks for some

[1] This work has been supported in part by grants from "Fundação de Amparo à Pesquisa do Estado de São Paulo - FAPESP" and "Conselho Nacional de Desenvolvimento Científico e Tecnológico - CNPq" - Brazil.

compromise solution. Some authors try to reduce conservativeness with a H_∞/μ synthesis approach (Smith et al. (1994)). In order to try to overcome some of these difficulties we propose to address this problem on a mixed H_2/H_∞ framework. The approach involves three steps: a) setup of an H_∞ unstructured uncertainty model to cope with plant uncertainty, b) numerical solution to a mixed control problem whose objective is the minimization of an H_2 performance cost subject to the H_∞ constraint developed in step a, and c) development of a controller design procedure.

2. ROBUST H_∞ DESIGN

Flexible structure models are usually described by the state space equations

$$\dot{x} = A_n x + B_n u \quad (1)$$

$$y = C_n x \quad (2)$$

where

$$A_n = \begin{bmatrix} 0 & I \\ -\Lambda & -\Gamma \end{bmatrix},\ B_n = \begin{bmatrix} 0 \\ B \end{bmatrix},\ C_n = \begin{bmatrix} E & F \end{bmatrix}$$

Furthermore, the matrices $B \in R^{n\times m}$, $E, F \in R^{r\times n}$, and the diagonal matrices Γ and $\Lambda \in R^{n\times n}$ with entries $\Gamma_{i,i} = 2\xi_i\omega_i$, $\Lambda_{i,i} = \omega_i^2$ are given in the modal coordinates. In Boulet et al. (1997) a left-coprime factorization is applied to the Laplace transform of the system in (1)-(2) and a structured perturbation model is developed from real parametric uncertainty bounds along with a particular weighting function. Although the nature of the perturbation model is taken to be highly structured the actual design is performed as if the perturbation were unstructured, by finding a standard full-order output feedback controller that satisfies a γ level constraint in the closed-loop H_∞ norm. Compared with the available structured uncertainty design methods, this approach has a much simpler computational solution.

In this paper we introduce a normalized variational uncertainty model which is much simpler and naive than, for instance, the model described in Boulet et al. (1997), but which provides enough room for the control engineer to change some of the available parameters in order to adequately tune a controller. In this model the uncertainty description is less realistic, in the sense that it is not directly drawn from bounds in the real parameter range of variations, but is much simpler. Since it does not involve frequency dependent scalings or weighting functions it may be changed without increasing the order of the generalized plant. The basic idea is to induce in the state space model (1)-(2) matrix perturbations in the form $A = A_n + \Delta_A$, $B = B_n + \Delta_B$, $C = C_n + \Delta_C$ where

$$\Delta_A = \begin{bmatrix} 0 & 0 \\ -\Delta_1\Lambda & -\Delta_1\Gamma \end{bmatrix},\ \Delta_B = \begin{bmatrix} 0 \\ \alpha\Delta_2 B \end{bmatrix},\quad (3)$$

$$\Delta_C = \begin{bmatrix} \alpha\Delta_3 E & \alpha\Delta_3 F \end{bmatrix} \quad (4)$$

Introducing the usual four port model

$$\dot{x} = Ax + B_1 w_1 + B_2 u \quad (5)$$

$$y = C_2 x + D_{21} w_1 \quad (6)$$

$$z_1 = C_1 x + D_{12} u \quad (7)$$

with $A = A_n$, $B_2 = B_n$, $C_2 = C_n$,

$$B_1 = \begin{bmatrix} 0 & 0 & 0 \\ I & I & 0 \end{bmatrix},\ C_1 = \begin{bmatrix} -\Lambda & -\Gamma \\ 0 & 0 \\ E & F \end{bmatrix},$$

$$D_{12} = \alpha \begin{bmatrix} 0 & B' & 0 \end{bmatrix}',\ D_{21} = \alpha \begin{bmatrix} 0 & 0 & I \end{bmatrix}$$

we can ensure that if the H_∞ norm of the closed-loop transfer function from w_1 to z_1 is less than γ, then the system is robust to unstructured perturbations in the form $w_1 = \Delta z_1$, $||\Delta||_\infty < \gamma^{-1}$. In this case it is also immediate to conclude that (1)-(2) is robustly stable with respect to (3)-(4) since these structured perturbations are such that $\Delta = \text{block diag}\{\Delta_1, \Delta_2, \Delta_3\}$, $||\Delta||_\infty < \gamma^{-1}$. It is important to remark that, on the contrary of Boulet et al. (1997), we consider perturbations on both B_n and C_n matrices. It is also easy to see that the resulting robust design will be such that $||\Delta_i||_\infty < (\alpha\gamma)^{-1}$, $i = 2, 3$ so that setting $\alpha > 1$ we may lower the requirement of robustness face to uncertainties in the input and output matrices without changing the importance of robustness face to the dynamic uncertainty matrix Δ_1.

If the flexible structure model is preprocessed by some order reduction procedure, as for instance, balanced truncation (see Moore (1981)), the diagonal structure of the modal matrices may be lost, which makes difficult to fit real uncertainty bounds into a reduced order unstructured uncertainty framework. However, a simple reduced order normalized variational uncertainty model is available yet. For instance, it is possible to work with the uncertainty model

$$A_\Delta = \Delta_1 A_{nr},\ B_\Delta = \alpha\Delta_2 B_{nr},\ C_\Delta = \alpha\Delta_3 C_{nr} \quad (8)$$

where the matrices subscripted by r are obtained from (1)-(2) by some order reduction procedure.

3. H_2 PERFORMANCE AND THE MIXED H_2/H_∞ CONTROL PROBLEM

In the analysis of flexible structures the time response to an impulsive input is of great value in the comprehension of the system behavior. So, it is natural to consider as the performance index to be minimized in controller synthesis the H_2 norm. More specifically, and denoting by $T(z, w; s)$ the transfer function from the input w to the output

z, our objective is to design a controller for the six port generalized plant

$$\dot{x} = Ax + B_0 w_0 + B_1 w_1 + B_2 u \quad (9)$$

$$y = C_2 x + D_{20} w_0 + D_{21} w_1 \quad (10)$$

$$z_0 = C_0 x + D_0 u \quad (11)$$

$$z_1 = C_1 x + D_{12} u \quad (12)$$

that minimizes the H_2 norm of the closed loop transfer function from w_0 to z_0 and keeps the H_∞ norm of the transfer function from w_1 to z_1 under some prescribed level, i.e., to find a solution to the mixed H_2/H_∞ control problem

$$\min \{||T(z_0, w_0; s)||_2 \ : \ ||T(z_1, w_1; s)||_\infty < \gamma\} \quad (13)$$

4. NUMERICAL SOLUTION TO THE MIXED H_2/H_∞ CONTROL PROBLEM

Until today, there is no numerical algorithm in the literature that can find an optimal solution (not even a local one) to the control problem stated in the previous section. Imposing extra assumptions, for instance $w = w_0 = w_1$, some existing methods are able to find a sub-optimal solution based on the minimization of an upper bound to the objective function. This procedure, as will be illustrated in the examples, may produce poor solutions, which do not present a sound decrease in $||T(z_0, w_0; s)||_2$. In this section we propose a numerical algorithm that is able to find a local optimal solution to the mixed H_2/H_∞ problem without any additional assumption. The algorithm, generically described bellow, may also be applied to other optimization problems formulated in the same form. Introducing suitable changes of variables (see section 4.1), it is possible to rewrite the mixed H_2/H_∞ problem (13) in the form

$$\min_x \{f(x) \ : \ x \in \mathcal{X}\} \quad (14)$$

where $f(.)$ is a non-convex yet differentiable function and $\mathcal{X}$ is a convex set described by a Linear Matrix Inequality (LMI), that is, $\mathcal{X} = \{x \mid F(x) = F(x)' > 0\}$ where $F(\cdot)$ is an affine matrix operator. Assuming that x_k is a feasible point, i.e. $F(x_k) > 0$, we define the following convex optimization problem

$$\min_z \{< \nabla f(x_k), z > \ : \ F(z) > 0\} \quad (15)$$

Letting z_k be its optimal solution and keeping in mind that x_k is a feasible solution we have that $< \nabla f(x_k), z_k - x_k > \leq 0$. Therefore, defining $d_k := z_k - x_k$, two different situations are possible:

1) $< \nabla f(x_k), d_k > < 0$: since the constraint set is convex, the point $x_{k+1} = x_k + \alpha d_k$ will always be feasible for $0 \leq \alpha \leq 1$. Furthermore, there is always an $\alpha > 0$ small enough so that $f(x_{k+1}) < f(x_k)$. Thus, by solving the problem

$$\min_{0 \leq \alpha \leq 1} f(x_k + \alpha d_k) \quad (16)$$

by unidimensional search on the direction d_k we may simultaneously decrease the objective function and find a new feasible point.
2) $< \nabla f(x_k), d_k > = 0$: with no loss of generality, we can consider $z_k = x_k$. Furthermore, it is possible to write the optimality conditions for problem (15) and show that they are indeed the necessary optimality conditions for problem (14), which implies that x_k is a local optimum to (14).

Hence, the following numerical method converges to a local optimal solution to problem (14):

1) Let x_0 be such that $F(x_0) > 0$. If there is no such x_0, problem (14) has no solution. Set $k = 0$.
2) Solve the convex optimization problem (15) and get its optimal solution z_k. Let $d_k := z_k - x_k$ and $\theta_k := < \nabla f(x_k), d_k >$.
3) If $\theta_k \approx 0$, x_k is a local optimal solution, then stop. Otherwise go to step 4.
4) Solve the unidimensional search problem (16) and get its optimal solution α_k. Let $x_{k+1} := x_k + \alpha_k d_k$, $k \leftarrow k + 1$ and go to step 2.

In order to accelerate convergence and decrease the computational effort it is possible to replace the direction d_k in step 2 by $d_k = -H(x_k)^{-1} \nabla f(x_k)$, where $H(\cdot)$ is the positive definite Hessian matrix of a convex barrier function associated to the constraints, in this case the usual logarithmic barrier $b(z) = -\log\det[F(z)]$. In the linear case this method is known as the *affine scaling algorithm.* It is interesting to notice that setting $H(x_k) = I$ the algorithm becomes a pure steepest-descent algorithm and, since the constraints are not taken into account when choosing the descent direction, d_k reduces to the gradient of the unconstrained problem. Our experience with this accelerated algorithm shows that the replacement of d_k by the affine scaling direction is effective in providing a significant reduction in the computational work required to find a local optimal and generally does not alter the optimality degree of the obtained solution when compared with the original step 2.

4.1 *Solving the mixed H_2/H_∞ control problem*

In order to be solved by the algorithm introduced above it is necessary to rewrite the mixed H_2/H_∞ control problem (13) in the form (14). With respect to the generalized plant (9)-(12) we shall consider the following dynamic output feedback controller with the same order as that of the plant

$$\dot{x}_c = A_c x_c + B_c y \quad (17)$$

$$u = C_c x_c \quad (18)$$

With no loss of generality, we impose $D'_{12}D_{12} = I$ and $D_{21}D'_{21} = I$. It is possible to devise two approaches to reduce this control problem to the form (14). The first one makes use of the well known observer based parametrization. So, given the following particular controller structure

$$A_c = A_\infty + B_{2\infty}K + L_\infty C_2,\ B_c = -L_\infty,\ C_c = K$$

where A_∞, $B_{2\infty}$ and L_∞ may be found in Colaneri et al. (1997), the following are equivalent:

a) $\|T(z_1, w_1; s)\|_\infty < \gamma$
b) There exists (X, L) such that $X = X' > 0$,

$$\begin{bmatrix} A_\infty X + XA'_\infty + B_{2\infty}L + L'B'_{2\infty} + L_\infty L'_\infty & (\bullet)' \\ C_1X + D_{12}L & -\gamma^2 I \end{bmatrix} < 0$$

is feasible and $K = LX^{-1}$.

For a proof of that equivalence see Colaneri et al. (1997). Furthermore, notice that the constraint set in *b)* is an LMI, and since every gain K corresponding to a pair $(L, X) \in \mathcal{X}$ robustly stabilizes the system, through the substitution of K in the controller structure it follows that the norm $\|T(z_0, w_0; s)\|_2^2$ may be expressed as a function of L and X. So, considering $\mathcal{X}$ as the LMI given above and calculating the function $f(X, L) = \|T(z_0, w_0; s)\|_2^2$ it is possible to rewrite the observer based output feedback control problem in the form (14) as desired. Notice that in the case $w_0 \neq w_1$ it may be shown that the obtained solution may be conservative since we are implicitly assuming that the closed loop observability Grammian has a very particular structure.

As a second and more general approach it is possible to use the output feedback controller parametrization as described, for instance, in Scherer (1995). In this case, all controller matrices A_c, B_c and C_c shall be determined by optimization procedures and no special structure is assumed in the H_2 norm calculation. The associated inequalities may be found in the references and are omitted for brevity.

5. CONTROLLER DESIGN PROCEDURE

With the tools developed in the previous sections we are able to develop a mixed H_2/H_∞ controller design procedure which provides the designer with the ability to select a controller that represents an adequate compromise between robustness and performance. So, with respect to the plant (9)-(12) we start determining $\gamma_{\min} = \min \|T(z_1, w_1; s)\|_\infty$. Then, starting from a value of $\gamma = \gamma_0$ slightly above $\gamma_{\min}$ we find a feasible initial controller, run the algorithm of section 4 and get a local optimal solution to the mixed H_2/H_∞ control problem for some controller parametrization. Notice that if the value of γ_0 is sufficiently close to $\gamma_{\min}$ the norm of the obtained mixed controller will be approximately the H_2 norm of $T(z_0, w_0; s)$ calculated for the minimal H_∞ controller. Then setting $k = 1$ and observing that the algorithm in section 4 always provides a local optimal solution to the mixed H_2/H_∞ problem we relax the value of $\gamma_k > \gamma_{k-1}$ and run the algorithm taking as an initial solution the optimal solution associated to γ_{k-1}. In this way we may ensure that the following descent property always hold

$$\|T(z_0, w_0; s, \gamma_k)\|_2 \le \|T(z_0, w_0; s, \gamma_{k-1})\|_2$$

where the dependency of γ emphasizes the fact that these norms are evaluated for the optimal controllers obtained as explained. In this manner we may draw a trade-off curve between robustness (the H_∞ norm) and performance (the H_2 norm) and adequately select a controller that satisfies the project specifications. A summary of this design procedure is given bellow:

1) Determine $\gamma_{\min}$.
2) Set $\gamma_0 = (1 + \epsilon)\gamma_{\min}$, $\epsilon > 0$, and determine a local optimal solution to the mixed H_2/H_∞ control problem. Set $k = 1$.
3) Starting from the mixed H_2/H_∞ local optimal solution associated to the value of γ_{k-1}, set $\gamma_k > \gamma_{k-1}$ and apply the algorithm in section 4 to determine the current local optimal solution.
4) If $\gamma_k \ge \gamma_{\max} > \gamma_0$, draw the trade-off curve and select the controller that most adequately fits the design specifications. Otherwise increment k and go back to step 3.

This design method is illustrated by the following simple numerical example with matrices

$$A = \begin{bmatrix} 0 & 1 \\ 2 & 1 \end{bmatrix},\ B_0 = B_1 = \begin{bmatrix} 0 \\ 0 \end{bmatrix},\ B_2 = \begin{bmatrix} 0 \\ 1 \end{bmatrix},$$

$$C_0 = \begin{bmatrix} -1 & 0 \end{bmatrix},\ C_1 = \begin{bmatrix} 0 & 0 \end{bmatrix},\ C_2 = \begin{bmatrix} -1 & 1 \end{bmatrix},$$

$$D_0 = 1,\ D_{12} = 1,\ D_{20} = D_{21} = 1$$

Notice that $w = w_0 = w_1$ and we will be able to compare the optimal H_2/H_∞ controller with the central controller and with that obtained from the minimization of an upper bound to the objective function (Baeyens and Khargonekar (1994)). Four designs were tried, namely, the central H_∞ controller labeled **C**, the suboptimal minimization of an upper bound to the H_2 norm labeled **S** and the application of the proposed algorithm for the two possible output feedback controller parametrizations discussed in section 4.1. For this simple example, both parametrizations provided virtually the same result in terms of the minimized H_2 norm and, for this reason, they are uniquely labeled as **O**. The trade-off curves can be seen in Figure 1. For low values of γ, we observe that the controller **S** provides poor performance, even worse than that of the central controller **C**. For large values of γ the situation is reversed, which can be easily understood, since the minimized

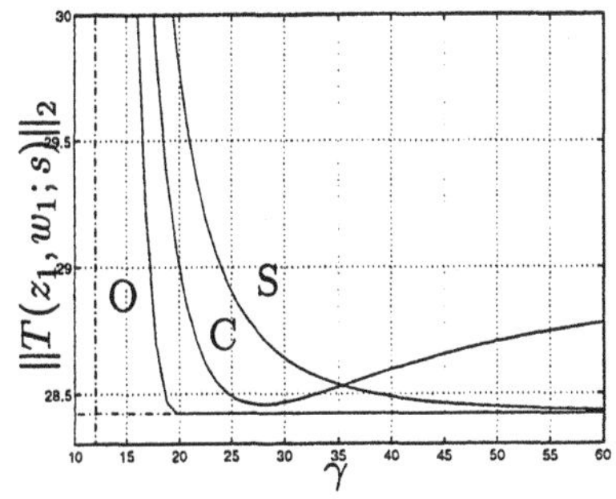

Fig. 1. Trade-Off curves

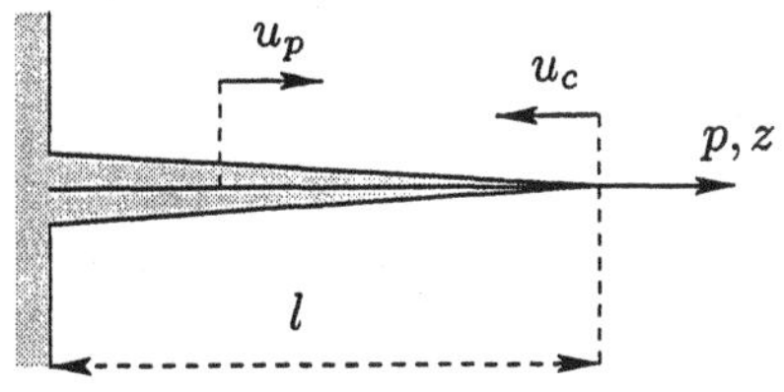

Fig. 2. Variable cross section rod

upper bound becomes closer to the actual H_2 norm and the suboptimal controller tends to the optimal H_2 controller. Denoting by $\gamma_2 = 19.7$ the $\|T(z_1, w; s)\|_\infty$ calculated with the H_2 optimal controller, we see that the controller **O** has the best performance and indeed matches the H_2 minimal norm.

6. FLEXIBLE STRUCTURE DESIGN

In this section we perform some numerical experiments that illustrate the features of the design methods proposed in this paper applied to a typical flexible structure model. The flexible structures (see Meirovitch (1986, 1967)) we consider have infinite dimensional linear models in the form (1)-(2) with natural frequencies $\omega_i, i = 1, \ldots, \infty$, to be determined as the solution of given algebraic equations. The example we consider here is an homogeneous variable cross section rod with one free end as illustrated in Figure 2. In this figure both perturbation (u_p) and control (u_c) forces are assumed to act axially and z represents the axial displacement. Considering the linear mass density $M_0 = 1kg/m$, the product of the modulus of elasticity and the maximum cross section area $EA_o = 1N$ and the length $l = 10m$ we determine the undamped linear model parameters. We consider that the control input u_c is placed at $p = l$ and that the perturbation input u_p is placed at $p = 0.63l$. These inputs define the input matrices B_p and B_c. The control structure is collocated, that means, the measured output y is also taken at $p = l$. We assume that both displacement and velocity at $p = l$ are being measured.

For practical purposes we truncate the given infinite dimensional models by taking only the first 12 modes and consider the corresponding state space models in the form (1)-(2), with 24 states, as our *reference* model. Damping is added by

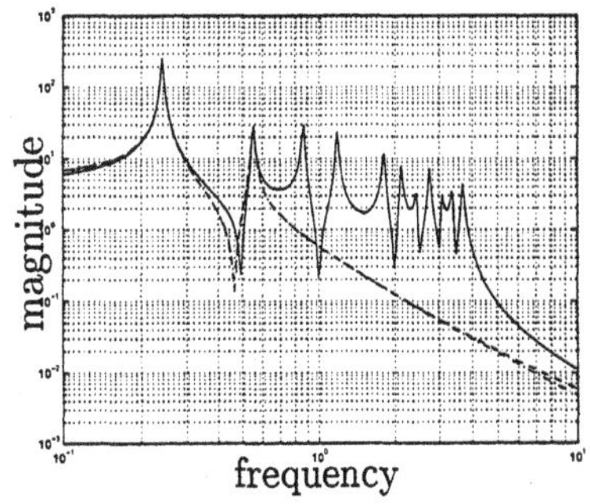

Fig. 3. Open loop Bode diagrams

taking $\xi = 0.01$ for all modes. With respect to the matrices in the generalized plant (9)-(12) we define the following augmented perturbation input $w_0 = \begin{bmatrix} u_p & u_n \end{bmatrix}'$ where u_n is a measurement noise, and define matrices B_0 and D_{20} as $B_0 = \begin{bmatrix} B_p & 0 \end{bmatrix}$ and $D_{20} = \begin{bmatrix} 0 & I \end{bmatrix}$. The control input is taken as $u = u_c$ so that $B_2 = B_c$. The open loop Bode diagram between the displacement and the perturbation input u_p is shown in Figure 3 as a solid line.

In order to illustrate the proposed design procedure we solve two reduced order output feedback control problems by designing controllers for a reduced order model of our reference plant. We try to find a controller for the reduced plant that is also effective with respect to the original high order reference plant by introducing robustness considerations in the design via the mixed H_2/H_∞ control problem.

In the first case we reduce the order of the plant using balanced-truncation (Moore (1981)) with $n_r = 4$. Robustness is introduced according to the model (8) with $\alpha = 10$. In this way we are mainly considering uncertainty in the dynamic rather than in the input and output matrices. Performance in the H_2 setup is measured by the cost output z_0, defined by $C_0 = \begin{bmatrix} I & 0 \end{bmatrix}'$ and $D_0 = \begin{bmatrix} 0 & 1 \end{bmatrix}'$, with respect to the augmented perturbation input w_0. Notice that the input associated to robustness (w_1) is quite different from the perturbation input (w_0). The Bode diagram of the obtained reduced plant (dashed line) can be seen and compared with the full order response (solid line) in Figure 3. Following the steps given in section 5 we have set $\gamma_0 = 74.89$. It is interesting to notice that in this case the optimal H_2 controller do not stabilizes the given reference model. Then, we have run the algorithm and plotted the design trade-off curve seen in Figure 4 for values of γ between γ_0 and $\gamma_{\max} = 8\gamma_0 \approx 600$. In this figure the solid line corresponds to the closed loop H_2 norm with the local optimal mixed H_2/H_∞ controller while the dashed line shows the same function calculated with the central controller. Notice that for large values of γ the value of $\|T(z_0, w_0; s)\|_2$ approaches that of the optimal H_2 controller, represented in the figure as a dotted lower bound line. In this sense the design proce-

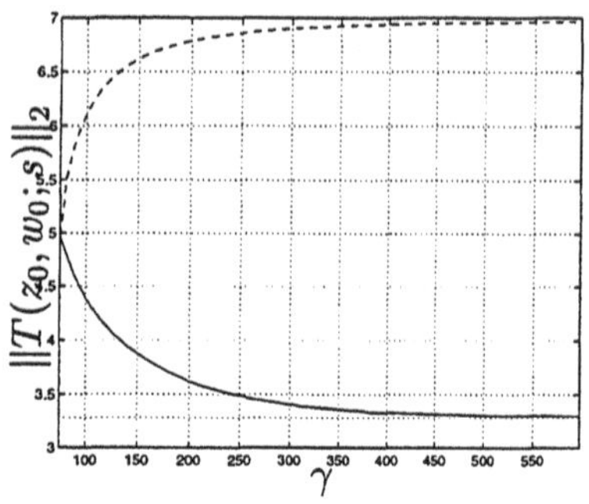

Fig. 4. Trade-off curve

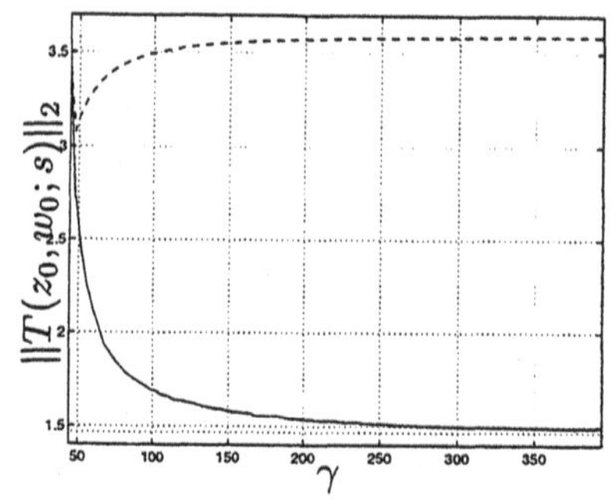

Fig. 6. Trade-off curve

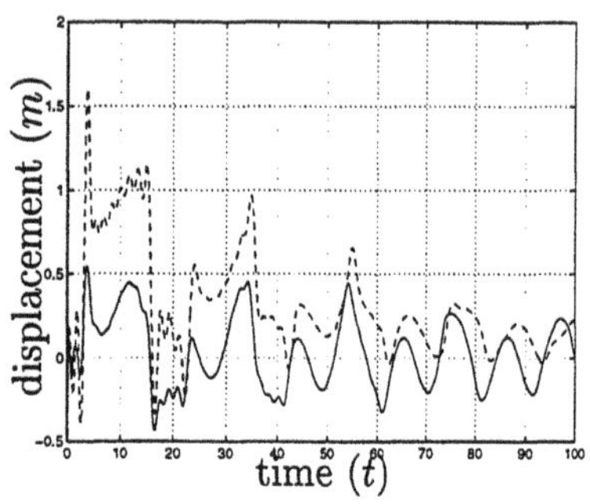

Fig. 5. Closed loop impulsive response

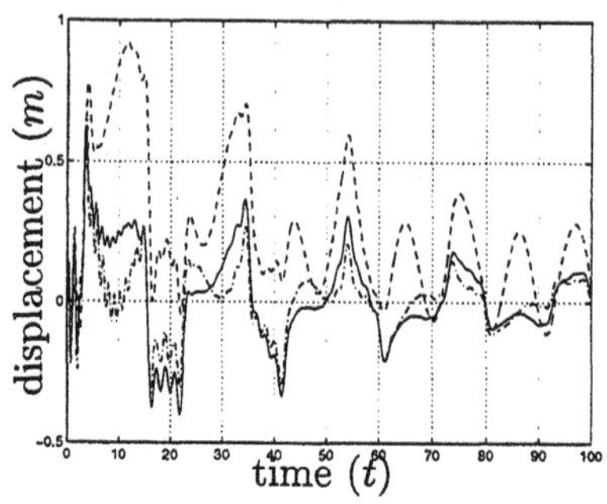

Fig. 7. Closed loop impulsive response

dure was able to provide an H_2 cost that is near optimal. Also remarkable is the fact that, in the considered interval, the value of $||T(z_0, w_0; s)||_2$ associated with the central H_∞ controller never decreases. To finish the design we have selected a value of $\gamma^* = 271.5 \approx 0.5 \times 557.3$. For our purpose, that is to stabilize the reference model this level of robustness is enough and in terms of performance the $||T(z_0, w_0; s)||_2$ is just 5% larger than the one of the optimal H_2 norm controller, which does not stabilizes the reference model. Closed loop controller performance may be evaluated by looking at Figure 5 the responses of the reference model to an impulse at u_p are plotted for the optimal mixed H_2/H_∞ controller (solid) and for the central H_∞ controller (dashed).

In the second case we illustrate the application of the more elaborate uncertainty description (3)-(4) by reducing the order of the plant by truncation. We retain the first 2 modes of the considered flexible structure model which provide a state space representation with order $n_r = 4$. The Bode diagram of the obtained truncated plant (dash-dot) can be compared with the plots in Figures 3. With the same settings as in the previous design, we have found $\gamma_{\min} = 43.78$ and set $\gamma_0 = 43.82$. Surprisingly enough, the minimal H_2 norm controller is able to stabilize the reference model. The trade-off curves are in Figure 6. As a compromise solution we have chosen $\gamma^* = 78.87 \approx 0.2 \times 386$. The optimal mixed H_2/H_∞ cost $||T(z_0, w_0; s)||_\infty = 1.81$ is about 23% larger than the optimal H_2 cost. The impulse responses of the optimal mixed controller (solid) and the optimal H_∞ (dashed) are in Figure 7 along with the impulse response of the optimal H_2 controller (dash-dot).

REFERENCES

E. Baeyens and P. P. Khargonekar. Some examples in mixed H_2/H_∞ control. In *Proceedings of the American Control Conference*, Baltimore, Maryland, 1994.

Benoit Boulet, Bruce A. Francis, Peter C. Hugues, and Tony Hong. Uncertainty modeling and experiments in H_∞ control of large flexible space structures. *IEEE Transactions on Control Systems Technology*, 5(5):504–519, 1997.

Patrizio Colaneri, José C. Geromel, and Arturo Locatelli. *Control Theory and Design : An $RH_2 - RH_\infty$ viewpoint.* Academic Press, 1997.

L. Meirovitch. *Analytical Methods in Vibrations.* The MacMillian Co., New York, NY, 1967.

L. Meirovitch. *Elements of Vibrational Analysis.* McGraw Hill, New York, NY, 1986.

B. C. Moore. Principal components analysis in linear systems: controlability, observability and model reduction. *IEEE Transactions on Automatic Control*, 26(1), 1981.

Carsten Scherer. Mixed H_2/H_∞ control for linear parametrically varying systems. In *Proceedings of the IEEE Conference on Decision and Control*, pages 3182–3187, 1995.

R. S. Smith, C. C. Chu, and J. L. Fanson. The design of H_∞ controllers for an experimental noncollocated flexible structure problem. *IEEE Transactions on Control Systems Technology*, 2: 101–109, 1994.

Active Vibration Suppression of Plate By the Use of Piezoceramic Sensors and Actuators

Yaxi Shen, Abdollah Homaifar, Da Chen

yaxi@ncat.edu, homaifar@ncat.edu, dachen@ncat.edu
NASA Center for Aerospace Research
Department of Electrical Engineering
North Carolina A&T State University
Greensboro, NC 27411, USA

Abstract: This work presents an active vibration control system of a thin test plate. Software simulation are realized using different control methodologies, such as localized control with Hybrid-fuzzy PD controller and State feedback controller. Both are shown to attenuate vibration at resonance frequencies. The results confirm that these two kinds of control methods are reliable in active damping of the vibration.

Keywords: Modeling, Active control, Fuzzy logic, Pole assignment.

I. INTRODUCTION

A typical spacecraft consists of a series of two-dimensional thin metal panels (Figure 1) that are clamped around their individual boundaries by a continuous metal frame. Avionic equipment are mounted to these panels and can be exposed to excessive vibration due to large acoustic emanations during launch. These vibrations are characterized by a large number of resonances over a wide frequency band. Electrically actuated piezoceramic elements mounted on the plate can create mechanical forces and moments to attenuate the acoustic vibrations and minimize the g-loads transmitted to the avionics equipment (Falangas, et al., 1994). Figure 2 illustrates the method by which a voltage applied across a piezoceramic patch can create a plate deformation. The application of an electric field along the polarization direction of the piezoceramic element forces it to expand along the directions perpendicular to the electric field, producing a local strain on the plate surface. On the other hand, if the electric field is opposite of the polarization direction, the piezoceramic will contract in the direction perpendicular to the electric field. This expansion or contraction of piezoceramic elements bonded on a plate generates local moments on the plate surface. If the applied voltage is kept at a low level (well below the depoling level), the relationship between the local moment and piezoceramic element voltage is linear. A number of these elements mounted to the plate can deform the plate in the vicinity of the expected plate maximum deflections. These maximum deflections occur at discrete (natural) frequencies and are at distinct positions along the plate. We have already conducted an experiment using piezoceramic actuators and accelerometers mounted on a rectangular plate with an active damping controller implemented on a personal computer. Appropriate closed-loop controller (dSPACE controller) were designed and the quantitative amount of active attenuation was measured (Shen, et al., 1998). Preliminary test results indicated that these technique can lower the loads on payloads and permit a reduction of the size, weight and cost of the structure.

In this paper, we specially want to introduce Hybrid fuzzy-PD control and state-feedback control, and compare their performance through simulation. The methodology developed in section 2 is applicable to a plate with multiple attached piezoceramic patchs. Section 3 discusses several possible closed-loop control architectures to attenuate the vibration disturbance on the plate. Section 4 presents some results of active vibration suppression when the control architectures were tested (Crawley and Luis, 1986).

Fig. 1. Experiment plate configuration

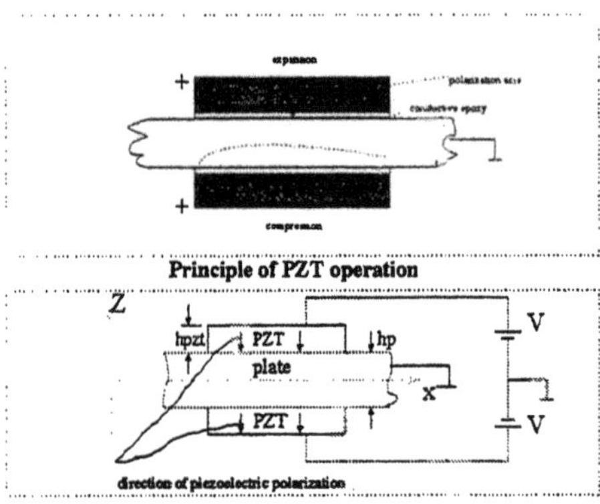

Fig. 2. Principles of piezoceramic patch operation

II. MODEL EQUATIONS OF PLATE

The structure was modeled from experimental data as well as analytically from first principles. The complete derivation of equations of the motion of the plate can be found in (Nagle, et al.,1996). Here, we only present a synopsis. According to the force and moment analysis, we can get the equation of the motion for the plate with multiple PZT patches from the equilibrium equations of force and moments (if we assume that the density of the plate is approximately the same as that of the PZT, and the thickness of the PZT is negligible compared with that of the plate):

$$\begin{aligned} & D\{\tfrac{\partial^4 w}{\partial x^4} + 2\tfrac{\partial^4 w}{\partial x^2 \partial y^2} + \tfrac{\partial^4 w}{\partial y^4} \\ & +\alpha\psi \sum_{p=1}^{M}[\tfrac{\partial^2}{\partial x^2}\left(\chi_p\left(\tfrac{\partial^2 w}{\partial x^2} + v\tfrac{\partial^2 w}{\partial y^2}\right)\right) \\ & +2(1-v)\tfrac{\partial^2}{\partial x \partial y}\left(\chi_p \tfrac{\partial^2 w}{\partial x \partial y}\right) \\ & +\tfrac{\partial^2}{\partial y^2}\left(\chi_p\left(\tfrac{\partial^2 w}{\partial y^2}+\nu\tfrac{\partial^2 w}{\partial x^2}\right)\right)]\}+\rho_p h_p \ddot{w} \\ & = \sum_{q=1}^{N} F_q + \tfrac{Y_{pz}}{1-\nu}\sum_{p=1}^{M} h_t d_{31}\left(\tfrac{\partial^2}{\partial x^2}+\nu\tfrac{\partial^2}{\partial y^2}\right)(\chi_p V_p) \end{aligned} \tag{1}$$

where $D = Y_p h_p^3/12(1-\nu_p^2), \alpha = 8\xi^3 + 12\xi^2 + 6\xi, \xi = h_{pz}/h_p$, h_{pz}, h_p are the thickness of the PZT and the plate, h_t is total thickness of the plate and the PZT patches, w is the displacement of vibration for the plate, $\psi = Y_{pz}/Y_p$, Y_{pz}, Y_p are Young's modulus of the PZT patches and the plate, ν is the Poisson's ratio, $\chi_p = 1$ or 0, when the PZT pair p is present or not, the subscript p denotes the patch pair number, M is the total number of patch pairs, F_q is the qth lateral force applied to plate, ρ_p is the density of the plate, d_{31} is the charge coefficient (Elishakoff, 1977).

Consider the free vibration problem ($V_p = 0$ and $F_q = 0$). Assumed $W(x,y)$ to be separable in the x and y directions (Elishakoff, 1977):

$$w(x,y,t) = W(x,y)e^{jwt}, W(x,y) = X(x)Y(y) \tag{2}$$

Assume that the plate is homogeneous such that the elastic effect of PZT patches is distributed over the entire plate, an approximate solution for the eigenvalues and eignfunctions of equation with the appropriate boundary conditions are given by

$$\omega = \left(\frac{D}{\rho_p}\right)^{1/2}\left[\left(\frac{\bar{r}\pi}{a}\right)^2 + \left(\frac{\bar{s}\pi}{b}\right)^2\right] \tag{3}$$

$$\begin{aligned} X_{rs}(x) &= \cosh\frac{\theta\pi x}{a} - \cos\frac{\bar{r}\pi x}{a} \\ & -\frac{C_1 - c_1}{S_1 - s_1\theta/\bar{r}}\left(\frac{\theta\pi x}{a} - \frac{\theta}{\bar{r}}\sin\frac{\bar{r}\pi x}{a}\right) \end{aligned} \tag{4}$$

$$\begin{aligned} Y_{rs}(y) &= \cosh\frac{\alpha\pi x}{b} - \cos\frac{\bar{s}\pi x}{b} \\ & -\frac{C_2 - c_2}{S_2 - s_2\alpha/\bar{s}}\left(\frac{\alpha\pi x}{b} - \frac{\alpha}{\bar{s}}\sin\frac{\bar{s}\pi x}{b}\right) \end{aligned} \tag{5}$$

$$\Psi_{rs}(x,y) = X_{rs}(x)Y_{rs}(y) \tag{6}$$

where a and b are the plate length along the x-axis and y-axis. $C_1, c_1, C_2, c_2, S_1, s_1, S_2, s_2, \theta, \alpha, \delta$ are defined as

$$\begin{aligned} & c_1 = \cos\bar{r}\pi, s_1 = \sin\bar{r}\pi, \\ & C_1 = \cosh\theta\pi, S_1 = \sinh\theta\pi \\ & c_2 = \cos\bar{s}\pi, s_2 = \sin\bar{s}\pi, \\ & C_2 = \cosh\alpha\pi, S_2 = \sinh\alpha\pi \\ & \theta = \sqrt{\tfrac{2\bar{s}^2}{\delta^2} + \bar{r}^2}, \alpha = \sqrt{\tfrac{2\bar{r}^2}{\delta^2} + \bar{s}^2}, \delta = \tfrac{a}{b} \end{aligned} \tag{7}$$

For a plate clamped all around $\bar{r}$ and $\bar{s}$ which correspond to mode r in the x-direction and mode s in the y direction are the root of the following transcendental equations

$$1 - c_1C_1 + \frac{\bar{s}\delta}{\bar{r}}\left[2 + \frac{\bar{r}^2}{\bar{s}^2\delta^2}\right]^{1/2} s_1S_1 = 0 \tag{8}$$

$$1 - c_2C_2 + \frac{\bar{r}}{\bar{s}\delta}\left[2 + \frac{\bar{s}^2\delta^2}{\bar{r}^2}\right]^{1/2} s_2S_2 = 0 \tag{9}$$

For a simply-supported plate , $\bar{r} = r, \bar{s} = s$. After applying orthogonality conditions and expressing the solution as

$$w(x,y,t) = \sum_{j=1}^{\infty}\Psi_j(x,y)\eta_j(t) \tag{10}$$

where Ψ_j is the eigenfunction of free vibration, η is the generalized displacement, we obtain the following infinite-dimensional system of equations for forced vibration

$$\ddot{\eta}_j + 2\zeta_j\dot{\eta}_j + \omega_j^2\eta_j = \sum_{q=1}^{N} F_{jq} + \sum_{p=1}^{M} Q_{jp}V_p \qquad (11)$$

where F_{jq} is the force applied at position q, V_p is the voltage applied to PZT p, j is the mode number. If the first n modes are selected for experiment, the differential equation of the plate system is given by

$$\begin{aligned} &\tilde{M}_{n\times n}\ddot{\eta}_{n\times 1} + \tilde{C}_{n\times n}\dot{\eta} + \tilde{K}_{n\times n}\eta \qquad (12) \\ = \;& \tilde{F}_{n\times 1}F(t) + \tilde{D}_{n\times 5}V_{5\times 1} \end{aligned}$$

where $\tilde{M}$ is mass matrix, $\tilde{C}$ is damping ratio matrix, and $\tilde{K}$ is spring stiffness matrix. According to the above derivation, we can represent the system using the state-space equations

$$\begin{aligned} \frac{d}{dt}\begin{bmatrix} x_1 \\ x_2 \end{bmatrix} &= \begin{bmatrix} 0 & I \\ -\tilde{K} & -\tilde{C} \end{bmatrix}\begin{bmatrix} x_1 \\ x_2 \end{bmatrix} \qquad (13) \\ &+ \begin{bmatrix} 0 \\ \tilde{D} \end{bmatrix} V + \begin{bmatrix} 0 \\ \tilde{F} \end{bmatrix} F(t) \end{aligned}$$

Our analysis mainly concentrated on the first three mode-shapes. The reasons are:

- Lower-frequency modes have much higher vibration amplitudes than the higher-frequency modes. These modes are the main contributor to vibration (Falangas, et al.,1994).
- The actuator/sensor collocation guarantees that the system is positive real and stable regardless of the gain value at the lower-frequency mode. This guarantee does not apply to higher-frequency modes because the collocation principle doesn't apply to a mode whose wavelength is comparable to the size of the PZT actuator. In addition, phase lags and computation delays can drive some higher frequency modes unstable.

III. CONTROL SYSTEM DESIGN

The control objectives are (1) to reduce vibrations at a given frequency or a set of frequencies, (2) to control vibration without "spillover" into uncontrolled modes, (3) to be able to change the frequencies to be controlled without redesign, and (4) at same time to remain robust in face of model uncertainty. We prefer to use localized control for system configuration (Figure 3)(Dworak, et al., 1992). An accelerometer is used as feedback to a collocated PZT actuator. The concept in this configuration is simple since no coordinate transformation to decouple modes is utilized. Furthermore, it is more suitable for application where there is not sufficient knowledge about the structural model. In this paper, we consider the following two methods.

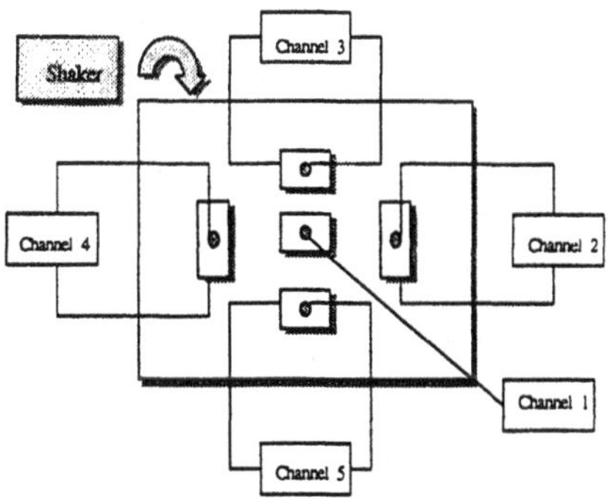

Fig. 3. Localized control

TABLE I
FUZZY RULE TABLE

dE/dt \ ERROR	NL	NS	ZE	PS	PL
NL	L	ML	M	MS	S
NS	ML	M	MS	MS	S
ZE	M	MS	S	MS	M
PS	MS	MS	MS	M	ML
PL	S	MS	M	ML	L

Kp

dE/dt \ ERROR	NL	NS	ZE	PS	PL
NL	L	ML	M	MS	MS
NS	ML	M	MS	MS	S
ZE	M	MS	S	S	M
PS	ML	S	MS	M	ML
PL	M	MS	M	ML	L

Kd

Fuzzy-PD controller: A hybrid fuzzy-PD controller based on regular PD controller is designed and simulated using MATLAB. We select the displacement error and its time derivative as input and individual proportional, and derivative coefficients of the PD controller as outputs for the fuzzy inference system. Appropriate membership functions associated with fuzzy sets are constructed to ensure a smooth transition between neighboring fuzzy sets, and a set of fuzzy rules is developed (Figure 4) (Homaifar, et al., 1994). To design the rule base we utilize our knowledge about the system under control, as well as the existing relationship between P and D coefficients and the performance of the plate system. Table I reflects the developed rule base for our simulation where NL, NS, ZE, PS, and PL stand for Negative large, Negative medium, Zero, Positive small and Positive large, L, ML, M, MS and S stand for Large, Medium large, Medium, Medium small and Small respectively. For example, the meaning of two rules from the rule tables are

If Error is Negative Large and Errordot is Positive Large, then K_p is Small and K_d is Medium.
If Error is Positive Small and Errordot is Zero, then K_p is Medium Small and K_d is Small.

Figure 5 and 6 give the relationship between inputs and outputs of fuzzy inference system. Figure 7 reflects the

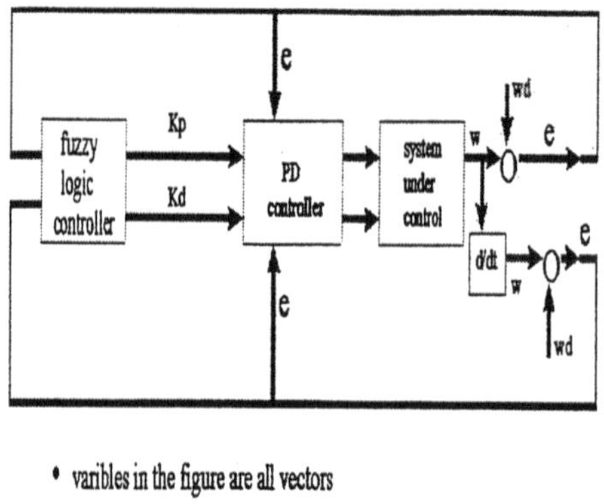

Fig. 4. Hybrid-fuzzy PD control system

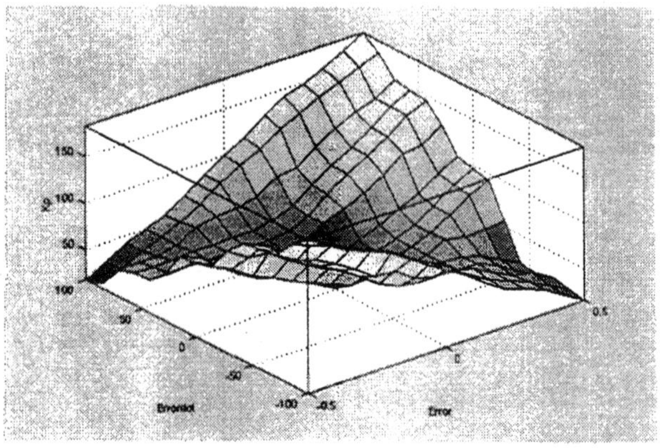

Fig. 5. The relationship betweent two input fuzzy sets and the output fuzzy set Kp

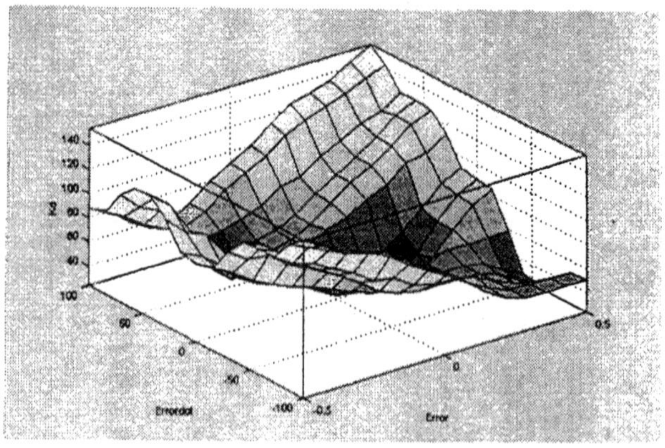

Fig. 6. The relationship betweent two input fuzzy sets and the output fuzzy set Kd

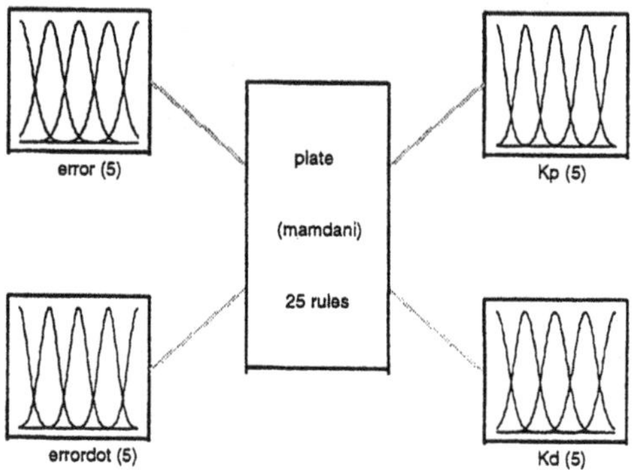

Fig. 7. Fuzzy inference system

names and characterizing parameters (i.e. e, $\frac{de}{dt}$ and K_p, K_d) of each fuzzy set (Homaifar, et al., 1994).

Stability analysis of output feedback: Our control methods are based on the idea of symmetric output feedback. In order to investigate the stability of the closed-loop, consider the following equations of motion for the first 8 modes (Junkins and Kim, 1993).

$$\begin{aligned} &\tilde{M}_{8\times8}\ddot{\eta}_{8\times1} + \tilde{C}_{8\times8}\dot{\eta} + \tilde{K}_{8\times8}\eta \\ = \ &\tilde{F}_{8\times1}F(t) + \tilde{D}_{8\times5}V_{5\times1} \end{aligned} \quad (14)$$

where V is control vector, $\tilde{M}$ is 8×8 positive definite symmetric mass matrix, $\tilde{C}$ is 8×8 positive semidefinite symmetric structural damping matrix, $\tilde{K}$ is 8×8 positive semidefinite symmetric stiffness matrix, and $\tilde{D}$ is 8×5 control influence matrix. We introduce the following symmetric output feedback form of the control law (collocated sensors and actuators):

$$\begin{aligned} V &= -(G_1 y + G_2 \dot{y}), y = D^T\eta \\ &\Rightarrow \ V = -(G_1 D^T \eta + G_2 D^T \dot{\eta}) \end{aligned} \quad (15)$$

G_1, G_2 are fully populated, $m \times m$ positive definite symmetric gain matrices. Based on the assumption that the sensors and actuators are perfectly linear and instantaneous, the closed-loop equations can be written as

$$M\ddot{\eta} + (C + DG_2D^T)\dot{\eta} + (K + C + DG_1D^T)\eta = 0 \quad (16)$$

Since G_i are positive definite, then the control induced stiffness and damping perturbations DG_iD^T which is symmetric positive semidefinite. We have shown that for all members of this special class of output feedback controllers, asymptotic stability of closed-loop system is guaranteed. An important consequence of the stability analysis is that the stability of a controllable symmetric system (having symmetric positive definite output feedback) is maintained:

- *Regardless of the number of modes retained in the reduced order model*
- *Regardless of inaccuracies in the parameter values used in the design*

Since G1 and G2 are being chosen positive definite, then M, $K + DG_1D^T$, $C + DG_2D^T$ are always positive definite regardless of full order or reduced order modal are used. Thus, by using the symmetric output feedback controller, the potential instability problems due to spillover can be completely avoided and the parameters do not have to be known accurately to guarantee stability. This closed-loop stability analysis leads us to the conclusion that if the system is controllable, the closed loop system is at least asymptotically stable so long as the gain matrices are chosen as positive definite matrices.

State-feedback Controller using Pole-placement: The plate system is given by

$$\begin{aligned} \dot{x} &= Ax + Bu + Gw \\ y &= Cx + v \end{aligned} \quad (17)$$

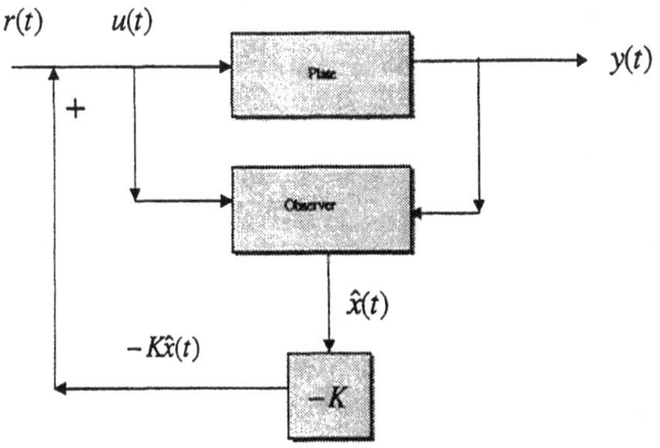

Fig. 8. State-feedback controller combined with the observer

where $x(t) \in R^{16}$ is the state, $u(t) \in R^5$ is the control input, $w(t)$ is the process noise from a shaker, and $v(t)$ is the measurement noise. The full state-feedback control

$$u = -Kx + r \tag{18}$$

is designed, with $r(t)$ being the reference input command.. However, the control law above cannot be implemented since not all the state usually are available as measurements.. So we designed an observer to observe the states

$$\dot{\hat{x}} = (A - LC)\hat{x} + Bu + Ly \tag{19}$$

Then based on the estimated states $\hat{x}(t)$, the control law is givn by

$$u = -K\hat{x} + r \tag{20}$$

The closed-loop structure using this controller is shown in Figure 8.. Define the estimation error as

$$\tilde{x}(t) = x(t) - \hat{x}(t) \tag{21}$$

we may write the control law as

$$u = -Kx + K\tilde{x} + r \tag{22}$$

when used in equation (17), yields

$$\dot{x} = (A - BK)x + BK\tilde{x} + Br + Gw \tag{23}$$

combine it with the error dynamics

$$\dot{\tilde{x}} = (A - LC)\tilde{x} + Gw - Lv \tag{24}$$

we can get the following augmented system

$$\begin{aligned} \begin{bmatrix} \dot{x} \\ \dot{\tilde{x}} \end{bmatrix} &= \begin{bmatrix} A - BK & BK \\ 0 & A - LC \end{bmatrix} \begin{bmatrix} x \\ \tilde{x} \end{bmatrix} \\ &+ \begin{bmatrix} B \\ 0 \end{bmatrix} r + \begin{bmatrix} G \\ G \end{bmatrix} w - \begin{bmatrix} 0 \\ L \end{bmatrix} v \\ y &= \begin{bmatrix} C & 0 \end{bmatrix} \begin{bmatrix} x \\ \tilde{x} \end{bmatrix} + v \end{aligned} \tag{25}$$

Since the augmented system is block triangular, the closed-loop characteristic equation is

$$\Delta(s) = \mid sI - (A - BK) \mid \cdot \mid sI - (A - LC) \mid = 0 \tag{26}$$

That is, the closed-loop poles are nothing but the plant poles that result by choosing K, and the desired observer poles that result by choosing L. The state feedback gain K and observer gain L can be selected separately for desired closed-loop behavior according to separation principle. Because the system's settling-time or rising time has a relationship with the closed-loop poles, so we can place the poles at desired locations according to required criteria. The observer poles (i.e., those of (A-LC)) should be chosen faster than the desired closed-loop plant poles (i.e., those of (A-BK)) for good closed-loop behavior (Lewis and Syrmos, 1995).

IV. SIMULATION RESULTS & CONCLUSIONS

The rectangular plate we use is .05" thick aluminum with edges of 28" and 21". The frame of the structure clamps the plate between two 1" square aluminum members. The frame is supported at the corners with legs of angle aluminum which are anchored to a benchop, preventing any rigid body motion. We use accelerometer (PCB SN 1722) to detect the signal. The signal then flows to a band-pass filter which is implemented using Simulink on PC, the sampling time is 0.1ms. All control algorithms and A/D, D/A signal conversions are realized through dSPACE environment. The signal finally passes through a PCB Piezotronics 790 series power amplifier before arriving at the actuators. The result of real-time vibration control using rate feedback controller is shown in (Shen et al., 1998). The simulation result of open-loop uncontrolled system with Guassian noise disturbance is shown in Figure 9. The simulation result of vibration control using hybrid fuzzy-PD controller is shown in Figure 10. The simulation result of vibration control using state-feedback controller is shown in Figure 11. From the results of the experiment, we can get the following comparision. Given the same initial states, process noise (Guassian noise) from the shaker, it is found that State-feedback controller has shorter setting-time and smaller steady-state error than fuzzy-PD controller for Gussian noise suppression. This is only basic comparison. Our future objective is to compare them in real-time vibration control.

This paper has described the modeling of a controlled plate to be used for the comparison of various control methodologies. Localized control configuration using Hybrid-fuzzy PD controller and State-feedback controller were shown to attenuate vibration at several natural frequencies. Stability analysis of output feedback using this plate model predict superior performance of Fuzzy-PD controller for a broad band disturbance. A simulation result confirmed that Hybrid fuzzy-PD controller and State-feedback controller are reliable in active damping the vibration.

ACKNOWLEDGMENT

We would like to thank Eric A. Wheater, Ph.D candidate of the Structures Group of NASA Core at North Carolina A&T State University for his assistance and involvement. This work is partially funded by grants from

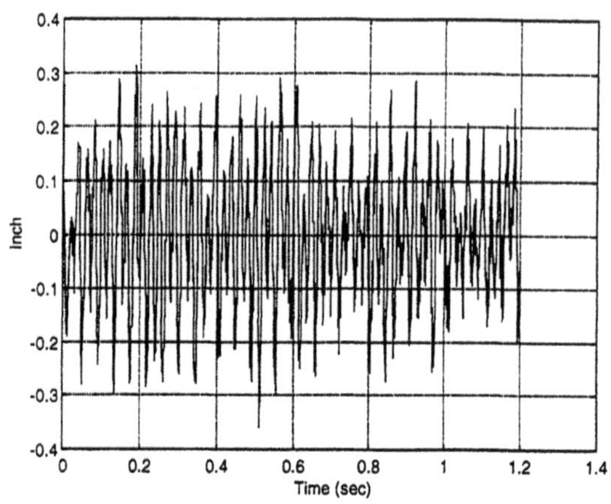

Fig. 9. The simulation result of the open-loop system under Guassian noise disturbance.

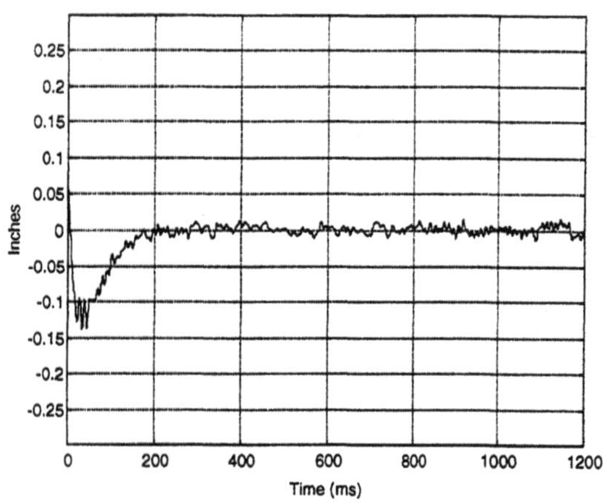

Fig. 10. The simulation of the closed-loop system using hybrid-fuzzy PD controller under Guassian noise at the first mode

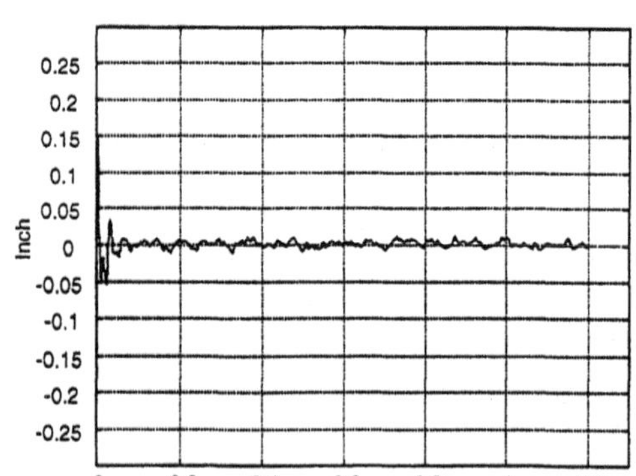

Fig. 11. The simulation result of the closed-loop system using state-feedback controller under Gassian noise at the first mode.

the NASA Center of Aerospace at North Carolina A&T State University under grant number NAGW-2924.

V. REFERENCE

Crawley, E.F. and J.de Luis (1986), "Experimental Verification of Distributed Piezoelectric Actuators for Use in Precision Space Structures," *Proc. 27th AIAA Struct. Dyn. Conf.*, pp. 116-124.

Dworak, J. A., E.T. Falangas, S. Koshigoe, and G.T. Tsheng (1992), "Vibro-acoustic Control Using PZT Actuators," *presented at the 3rd Int. conf. Adaptive structures* in Nov.92, San Diego, CA, pp. 529-545.

Elishakoff, I. (1977), "Random Vibrations of Orthopic Plates Clamped or Simply Supported All Round ," *Acta Mechanica*, Vol. 28, pp. 165-176.

Falangas, E.T., J.A. Dworak, and S. Koshigoe (1994), "Controlling Plate Vibrations Using Piezoelectric Actuators," *IEEE Control System*, pp. 34-41.

Homaifar, A., B. Sayyarrodsari, and W.E. Snyder (1994), "A Theoretical Justification for Nonlinear Controlling Property of a class of Fuzzy Logic Controllers," *Third IEEE International Conference on Fuzzy Systems.* pp. 382-386.

Junkins, J. L. and Y. Kim (1993), "Introduction to Dynamics and Control of Flexible Structures," AIAA, Education Series, Washington, DC, 1993.

Lewis, F.L. and V.L. Syrmos (1995), "*Optimal Control*," Second Edition, John Wiley & Sons, Inc., New York, 1995.

Nagle, J. C., A. Homaifar, A. Abdelnaser, and M. Bikdash, "Modeling and Control of a Thin Plate," *Proceedings of the 1st National student conference*, The National Alliance of NASA University Research Centers at Minority Institutions, Editors : E. O. Daso, and S., Mebanee, TRI Press, March 31, April 2, 1996, pp. 362-365.

Shen, Y., A. Homaifar, M. Bikdash, and A. Naser, "Real-Time Active Vibration Control Using Piezoelectric Actuators in Plate Structures," *Proceedings of 3nd NASA-URC conference*, The National Alliance of NASA University Research Centers at Minority Institutions, Editors: Coleman, T. L., white, B. and Goodman, S.,TRI Press, Feb. 21-26, 1998, pp. 342-347.

FEASIBLE TRAJECTORIES OF LINEAR DYNAMIC SYSTEMS WITH INEQUALITY CONSTRAINTS USING HIGHER-ORDER REPRESENTATIONS

Sunil K. Agrawal* Nadeem Faiz* Richard M. Murray**

* *Department of Mechanical Engineering, University of Delaware, Newark, DE, USA. agrawal@me.udel.edu*
** *Control and Dynamical Systems, California Institute of Technology, Pasadena, CA, USA. murray@cds.caltech.edu*

Abstract: This paper proposes a method to determine trajectories of dynamic systems that steer between two end points while satisfying linear inequality constraints arising from limits on states and inputs. The method exploits the structure of the dynamic system to eliminate the state equations. The linear inequalities on inputs and states are replaced by a finite set of linear inequalities on the mode coefficients and changing the problem of trajectory generation to finding a convex polytope on the mode coefficients. The procedure is demonstrated in an example and verified experimentally.

Keywords: Real-time planning, higher-order forms, flatness based planning, convex programming

1. INTRODUCTION

For linear dynamic systems, the equations of motion are typically written as

$$\dot{\bar{x}} = \bar{A}\bar{x} + \bar{B}u \qquad (1)$$

where $\bar{A}$ is a $(n \times n)$ matrix, $\bar{x}$ is a $(n \times 1)$ state vector, $\bar{B}$ is a $(n \times m)$ matrix, and u is a $(m \times 1)$ control vector. We assume here that the controllability index of $(\bar{A}, \bar{B})$ pair is p, an integer equal to $\frac{n}{m}$ (Kailath, 1980). The statement of the problem is to determine feasible trajectories $u(t)$ and $\bar{x}(t)$ that will steer the system from an initial state $\bar{x}_0$ at time t_0 to a final state $\bar{x}_f$ at t_f, while satisfying inequality constraints of the form

$$C\bar{x} + Du + e \leq 0 \qquad (2)$$

where C and D are respectively $(l \times n)$ and $(l \times m)$ matrices and e is a $(l \times 1)$ constant vector. These constraints may arise from different physical phenomena such as actuator limits, state limits, or from a complex interaction between these two.

Among the methods used for real-time trajectory determination of dynamic systems, higher-order forms (or, flat forms) have proved to be computationally quite attractive. Some new real-time optimal planning schemes have used the higher-order form with direct methods ((Nieuwstadt and Murray, 1995), (Nieuwstadt and Murray, 1998)). Indirect methods together with higher-order forms have also been used for optimal planning using results from higher-order variational calculus ((Agrawal and Veeraklaew, 1996), (Agrawal and Faiz, 1998)). Some recent studies on real-time planning have used homotopy methods (Hauser and Meyer, 1998). However, none of these studies considered the planning problem in the presence of inequality constraints which is the subject of this paper. The motivation for real-time motion planning comes from applications such as missiles, uninhabited combat air vehicles (UCAVs) and pilot-less airplanes.

2. SYSTEM HIGHER-ORDER REPRESENTATION

On applying a linear transformation $\bar{x} = Tx$, $(\bar{A}, \bar{B})$ transforms to (A, B) where $\bar{A}T = TA$ and $\bar{B} = TB$. On choosing the transformation matrix $T = [\bar{B}\ \bar{A}\bar{B}\ \bar{A}^2\bar{B}\ ...\ \bar{A}^{p-1}\bar{B}]$, (this matrix is nonsingular because of the controllability index property of the system) the system matrices A and B take the following forms:

$$A = \begin{bmatrix} 0 & 0 & 0 & ... & E_1 \\ I_m & 0 & 0 & ... & E_2 \\ 0 & I_m & 0 & ... & E_3 \\ \vdots & \vdots & \ddots & \ddots & \vdots \\ 0 & 0 & ... & I_m & E_p \end{bmatrix}, \quad B = \begin{bmatrix} I_m \\ 0 \\ 0 \\ \vdots \\ 0 \end{bmatrix} \quad (3)$$

where E_i are $(m \times m)$ matrices, I_m is an m-dimensional identity matrix, and 0 are $(m \times m)$ zero matrices, $[E_1, E_2 \ldots E_p]^T = T^{-1}\bar{A}^p\bar{B}$. From this new form of A and B, the control vector u and the state vector x can be written as higher derivatives of y, an m-dimensional vector consisting of the last m entries of x:

$$\begin{bmatrix} u \\ x_1 \\ \vdots \\ x_{p-1} \\ x_p \end{bmatrix} = \begin{bmatrix} -E_1 & -E_2 & ... & -E_p & I_m \\ -E_2 & -E_3 & ... & I_m & 0 \\ \vdots & \vdots\ \vdots & & \vdots & \vdots \\ -E_p & I_m & ... & 0 & 0 \\ I_m & 0 & ... & 0 & 0 \end{bmatrix} \begin{bmatrix} y \\ y^{(1)} \\ \vdots \\ y^{(p-1)} \\ y^{(p)} \end{bmatrix} \quad (4)$$

where x_i are the m-dimensional vectors partitioned from x and $y^{(p-1)}$ represents the $p-1$th derivative of y. From the given boundary conditions of $\bar{x}$ at t_0 and t_f, the boundary conditions on the higher derivatives y, $\dot{y}$,..., $y^{(p-1)}$ can be computed using Eqs. (4) at t_0 and t_f.

2.1 *Constraint Representation*

On substituting $\bar{x} = Tx$ in the constraint $C\bar{x} + Du + e \leq 0$, the inequality constraints can be written in a symbolic form:

$$M_0\, y + ... + M_{p-1}y^{(p-1)} + M_p y^{(p)} + e \leq 0 \quad (5)$$

where M_i are $(l \times m)$ matrices obtained by appropriately multiplying the terms in the constraint equation $C\bar{x} + Du + e \leq 0$. A salient feature of Eq. (5) is that it has the state Eqs. (1) explicitly embedded in it. In other words, a feasible trajectory of Eq. (5) that satisfies the boundary conditions of y and its derivatives is also a feasible solution of the state equations (1) and constraints (2).

3. CHARACTERIZATION OF THE SOLUTION

As outlined in the last section, the statement of the problem is to find a solution $y(t)$ over the time domain t_0 and t_f such that it satisfies Eq. (5) and boundary conditions of y, $y^{(1)}$,...,$y^{(p-1)}$ at t_0 and t_f. In this paper, the solution $y(t)$ is characterized as

$$y(t) = \Phi_0(t) + \sum_{j=1}^{k} a_j\phi_j(t) \quad (6)$$

where $\Phi_0(t)$ are m-dimensional vector functions that satisfy all boundary conditions of the problem, $\phi_j(t)$ are scalars that satisfy the boundary conditions homogeneously, and a_j are m-dimensional vectors of mode coefficients (Fletcher, 1984). With this form, $y(t)$ satisfies all boundary conditions of the problem regardless of the values of a_j.

On substituting $y(t)$ from Eq. (6) in Eq. (??), the form for the inequalities becomes

$$\sum_{i=0}^{p}\sum_{j=1}^{k} M_i a_j \phi_j^{(i)}(t) + \sum_{i=0}^{p} M_i \Phi_0^{(i)}(t) + e \leq 0 \quad (7)$$

This equation should be valid at all points in the domain $[t_0, t_f]$, i.e., it represents an infinite number of constraints on the mode coefficients a_1,...,a_k. It must be noted that these constraints are linear in the elements of the mode coefficients a_j and nonlinear in time. In the rest of this section, two schemes are considered to satisfy Eq. (7) within the domain.

3.1 *Collocation Scheme*

In the first scheme, a finite number of collocation points is selected within the domain of interest. At each of the collocation points, the constraint functions and a finite number of their derivatives are bounded to ensure satisfaction of Eq. (7). In order to describe this method intuitively, consider a single inequality $|g(t)| < c$. Given $g(t)$, $g(t+\Delta T)$ can be approximated by

$$g(t + \Delta T) = g(t) + g'(t)\Delta T + ...g^{(r)}(t)\frac{(\Delta T)^r}{r!} \quad (8)$$

where $0 \leq \Delta T < 1$ and terms of the order $(\Delta T)^{r+1}$ and higher are considered negligible. If the functions $g^{(i)}(t)$ satisfy appropriate bounds, i.e., $|g^{(i)}(t)| \leq g_i$, $i = 0,...,r$, $g(t + \Delta T)$ is bounded in the neighborhood of t in the following way:

$$|g(t + \Delta T)| \leq g_0 + g_1\Delta T + ...g_r\frac{(\Delta T)^r}{r!} \quad (9)$$

As a result, given the original bound c for $g(t)$ and a choice for derivative bounds g_i, $i = 1, ..., r$, a modified bound g_0 can be determined

$$g_0 = c - [g_1 \Delta T + ... + g_r \frac{(\Delta T)^r}{r!}] \quad (10)$$

with the following interpretation: If $|g(t)| \leq g_0$ and its derivatives are bounded according to $|g^{(i)}(t)| \leq g_i$, $i = 1, ..., r$, then $|g(t + \Delta T)| \leq c$ in the neighborhood of t under the specific assumptions on terms of order $(\Delta T)^{r+1}$ and higher.

Using this reasoning, for the system described in Eq. (7), $N + 2$ equally spaced collocation points $t_0, t_1, ..., t_N, t_f$ are chosen such that $t_0 < t_1 < t_2 < ... < t_N < t_f$ and $0 \leq \Delta T < 1$ is the spacing between the collocation points. At a collocation point, for every constraint, a bound on its r derivatives is specified and an appropriate bound on the constraint is computed using Eq. (10). For the l inequality constraints of Eq. (7), on bounding their r derivatives at $N + 2$ collocation points, results in a total of $(N+2)(r+1)l$ linear inequalities on the km elements of the mode coefficients a_1,...,a_k. In general, these linear inequalities enclose a convex polyhedral region in the coefficient space which represent the feasible trajectories of the dynamic system within the characterization of Eq. (6).

3.2 *Conservative Scheme*

In this scheme, the elements of the vector $\Phi_0(t)$ and the mode functions $\phi_j(t)$ are chosen prudently to come up with inequalities on the elements of a_1,...,a_k that satisfy Eq. (7) at every point in the domain of interest. The method also utilizes the property of polynomial functions $1, t, t^2, ..., t^k$ to be positive in the domain $0 \leq t_0 < t \leq t_f$ and for their derivatives to be closed within the set. Some special linear combinations of polynomial functions, e.g., Chebyshev's polynomials, also share this property.

In order to get an intuitive understanding of the method, consider a single linear inequality $g(t) \leq c$. Let the function $g(t)$ be parameterized using suitable basis functions so that the constraint is written in the following form,

$$g(t) - c = \alpha_1 \gamma_1(t) + ... + \alpha_k \gamma_k(t) \leq 0 \quad (11)$$

where α_i are linear functions of the parameters used to express $g(t)$ and $\gamma_i(t)$ are functions of time. Now, if each $\gamma_i(t)$ possessed the property that it is non-negative over $[t_0, t_f]$, a sufficient condition for the inequality (11) to hold in the domain of interest is that the coefficients $\alpha_i \leq 0$. It must be noted that the basis functions that are used to parameterize $g(t)$ need not be the same

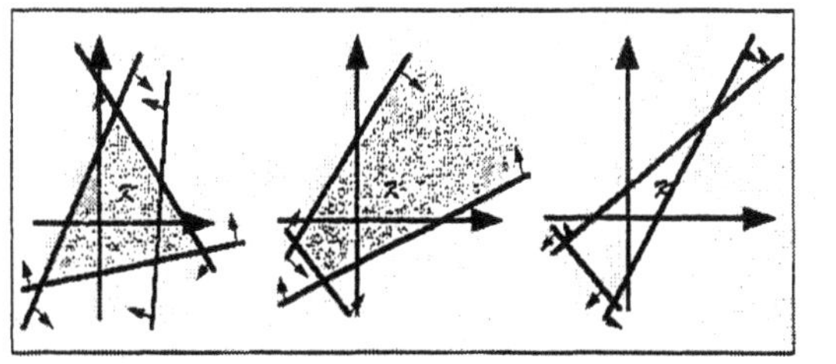

Fig. 1. Convex sets $\mathcal{K}$ in two variables with (i) consistent feasible region, (ii) unbounded feasible region, (iii) inconsistent feasible region.

as $\gamma_i(t)$. For example, one could use Chebyshev's polynomials to express $g(t)$ while select $\gamma_i(t)$ to be the simple polynomials $1, t, t^2, ..., t^k$.

Using this reasoning, with proper choice of $\Phi_0(t)$ and $\phi_j(t)$, the component equations of (7) are written in the form of Eq. (11). Then, appropriate linear inequalities are imposed on the elements of a_1,...,a_k to guarantee satisfaction of the constraints everywhere in the domain. These linear inequalities again enclose a convex polyhedral region in the coefficient space which represent the feasible trajectories of the dynamic system. The computational aspects and limitation of both methods will be discussed in the next section.

4. LINEAR INEQUALITIES AND CONVEX SETS

From the two schemes described in Section 3, it is clear that finding a solution of Eq. (7) is equivalent to characterizing the convex feasible region bounded by I linear inequalities on the elements of a_1,...,a_k in $\mathcal{R}^{mk}$. In general, the forms of the inequalities arising from the collocation scheme of Section 3.1 and conservative scheme of Section 3.2 will be different. However, the inequalities are still linear in the two cases. This section briefly outlines the computational aspects of characterizing the convex feasible region based on the procedures described in (Solodovnikov, 1980). Detailed descriptions of convex optimization are available in (Boyd and Vandenberghe, 1996).

Every constraint partitions $\mathcal{R}^{mk}$ in a half space. Let the intersection of these half spaces be denoted by a convex set $\mathcal{K}$. It is possible for $\mathcal{K}$ to be (i) empty, if the inequalities are inconsistent, (ii) a bounded polytope, or (iii) an unbounded convex region. These possibilities are illustrated for a 2-dimensional space in Fig. 1. It is implicitly assumed in this section that the system of inequalities is consistent, i.e., the convex set $\mathcal{K}$ is non-empty.

In order to mathematically characterize the feasible region $\mathcal{K}$, one needs to define some terminologies and mathematical procedures. In general, a system of linear inequalities is non homogeneous.

i.e., contains constant terms. We label such a system as $\mathcal{N}(I)$. Correspondingly, one can define a system of homogeneous inequalities, i.e., all constant terms are set to zero, labeled as $\mathcal{H}(I)$. Also, one defines $\mathcal{N}(E)$ corresponding to $\mathcal{N}(I)$ and $\mathcal{H}(E)$ corresponding to $\mathcal{H}(I)$, where all inequalities are replaced by equalities. The characteristics of $\mathcal{K}$ depend on the properties of these four systems $\mathcal{N}(I)$, $\mathcal{H}(I)$, $\mathcal{N}(E)$, and $\mathcal{H}(E)$.

A system of inequalities $\mathcal{N}(I)$ is called *normal* if the corresponding system of homogeneous equations $\mathcal{H}(E)$ has only null solution. We will first consider the case where the system is normal and later extend these ideas to systems which are not normal. A normal system of inequalities $\mathcal{N}(I)$ possesses *vertices*, which are defined as points of $\mathcal{K}$ which are not interior points of any line segment lying entirely within $\mathcal{K}$.

In order to find the vertices of $\mathcal{K}$, from the system $\mathcal{N}(E)$ consisting of I equations, we solve all combinations of mk equations which have unique solution points. In general, there are ${}^{I}C_{mk}$ choices of such combinations. From these solutions, we discard those which do not satisfy the original inequalities $\mathcal{N}(I)$. Let us assume that this process results in a set of points $P_1,...,P_H$. These H points are the vertices of $\mathcal{K}$. Using these H points, a *convex hull* $\langle P_1, P_2, ..., P_H\rangle$ is defined as follows:

$$\langle P_1, P_2, ..., P_H\rangle = s_1 P_1 + s_2 P_2 + ... + s_H P_H$$
$$s1,\ s2,\ ..., s_l \geq 0 \ \ \&\ \ s_1 + s_2 + ... + s_l = 1 \quad (12)$$

In order to fully characterize $\mathcal{K}$, one needs to check the existence of a *polyhedral cone*. This is done by solving all combinations of $mk-1$ equations from $\mathcal{H}(E)$. There are ${}^{I}C_{mk-1}$ such choices and corresponding to each, we pick a solution and its negative and check if $\mathcal{H}(I)$ is satisfied by any of these two points. All points that satisfy $\mathcal{H}(I)$ through this procedure are retained and are labeled as $N_1,...,N_C$. The convex cone is defined as $(N_1, ..., N_C)$, where

$$(N_1, N_2, ..., N_C) = r_1 N_1 + r_2 N_2 + ... + r_C N_C$$
$$r1,\ r2,\ ..., r_C \geq 0 \quad (13)$$

Proposition 1: If the system of linear inequalities $\mathcal{N}(I)$ is normal and the convex hull of its vertices, as defined in Eq. (12), is $\langle P_1, P_2, ..., P_H\rangle$ and the convex cone, as defined in Eq. (13), is $(N_1, N_2, ..., N_C)$, then the domain $\mathcal{K}$ has the following analytical form

$$\mathcal{K} = \langle P_1, P_2, ..., P_H\rangle + (N_1, N_2, ..., N_C) \quad (14)$$

From this characterization, if $(N_1, N_2, ..., N_C)$ is empty, the domain $\mathcal{K}$ is a convex bounded polytope. If $\langle P_1, P_2, ..., P_H\rangle$ is empty, the domain $\mathcal{K}$ is a convex cone. In general, $\mathcal{K}$ is a semi bounded convex region. If both $(N_1, N_2, ..., N_C)$ and $\langle P_1, P_2, ..., P_H\rangle$ are empty, $\mathcal{K}$ is empty.

If the system of inequalities $\mathcal{N}(I)$ is *not normal*, the corresponding system of homogeneous equations $\mathcal{H}(E)$ contains a higher dimensional subspace $\mathcal{L}$ besides the origin. For such a case, $\mathcal{K}$ has the following form:

Proposition 2: If the system of linear inequalities $\mathcal{N}(I)$ is not normal, then

$$\mathcal{K} = \mathcal{K}_r + \mathcal{L} \quad (15)$$

where $\mathcal{K}_r$ is the convex region enclosed by a reduced set of inequalities $\mathcal{N}_r(\mathcal{I})$ which is obtained from $\mathcal{N}(I)$ by setting an appropriate number of variables to zero that equals the dimension of $\mathcal{L}$.

4.1 *Real-time Computations*

From the steps of Section 3, the trajectory determination problem was reduced to finding a solution of Eq. (7). The schemes of Sections 3.1 and 3.2 further modified the problem to finding intersection of I half spaces characterized by $\mathcal{N}(I)$. From the point of view of determination of a feasible trajectory for the system, any solution of the mode coefficients that satisfies $\mathcal{N}(I)$ is acceptable. However, our approach in this paper is to characterize the entire set of feasible solutions $\mathcal{K}$ enclosed by $\mathcal{N}(I)$ using convexity theory so that the feasible trajectory can be selected in a judicious way.

In general, the inequalities $\mathcal{N}(I)$ arising from the procedures of last section are normal and $I > mk$. From the characterization of $\mathcal{K}$, the main computation steps are to find the vertices P_i of the convex hull and N_i of the convex cone. In order to compute the vertices P_i, one requires solving ${}^{I}C_{mk}$ sets of mk linear equations in mk variables, followed by I inequality checks to determine feasibility with respect to $\mathcal{N}(I)$. In the computation of vertices N_i, one requires solving ${}^{I}C_{mk-1}$ sets of $mk-1$ linear equations in $mk-1$ variables, followed by $2I$ inequality checks to determine feasibility with respect to $\mathcal{H}(I)$. These solutions must be done in real-time as end-conditions may change continuously.

A significant computational effort goes into solution of the vertices P_i as end conditions change, since ${}^{I}C_{mk}$ sets of mk linear equations in mk variables are solved. A similar computational burden is faced in the computation of N_i. However, it is important to note a property stated in the following proposition:

Proposition 3: If the only difference in two trajectory computations is a new set of end conditions then the set $\mathcal{N}(I)$ in the two computations

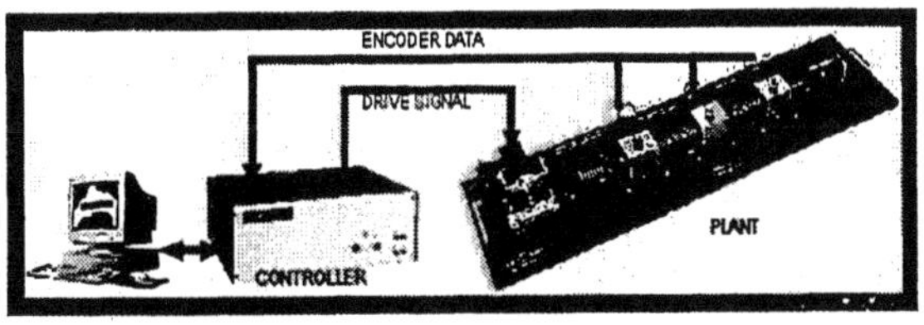

Fig. 2. A rectilinear three mass-spring-damper system.

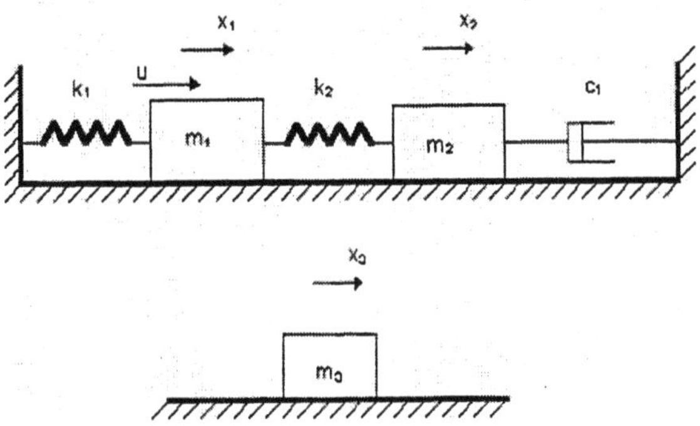

Fig. 3. A schematic model of the three mass-spring-damper system.

differ only in their homogeneous terms. As a consequence, to compute P_i, it is no longer necessary to solve in real-time ${}^{I}C_{mk}$ sets of mk linear equations in mk variables. Instead, the corresponding matrix inverses may be computed off-line and stored. In real-time, a single matrix-vector computation is required to determine a P_i. Also, due to the above mentioned property of $\mathcal{N}(I)$, $\mathcal{H}(I)$ in the two computations are identical. As a result, the computation of N_i needs to be performed only once even when the boundary conditions are changed.

As a consequence of the above observations, the real-time trajectory generation becomes even more attractive as boundary conditions are changed. The computations do not require matrix inversions or solution of simultaneous linear equations in real-time, only matrix-vector multiplications.

5. EXAMPLE

This example is motivated from an experiment of a rectilinear spring-mass-damper system shown in Fig. 2. The system is driven by a single motor acting on mass 1. Mass 3 is physically disconnected from the other masses. It is moved within its limits by an operator, its motion being followed by mass 2. This experiment is designed to demonstrate real-time planning and on-line implementation during situations such as "pursuit-evasion".

A schematic of this system is shown in Fig. 3. It consists of two degrees-of-freedom and a single input, i.e., $n = 4$ and $m = 1$. The equations of motion for this system are:

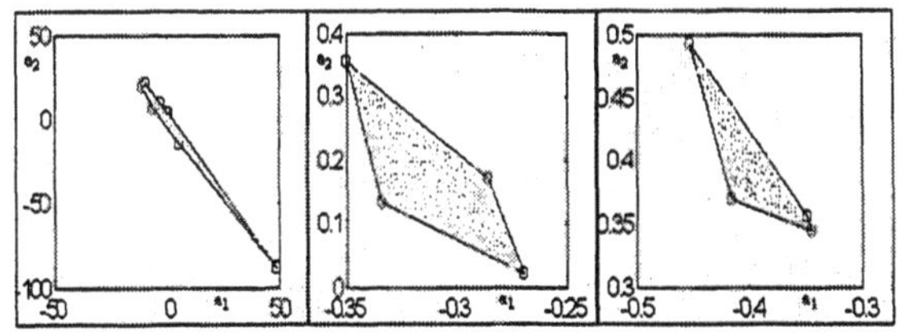

Fig. 4. Feasible regions with collocation schemes: (i) 6 collocation points, (ii) 6 collocation points with bounded first derivatives, (iii) 6 collocation points with bounded first and second derivatives.

$$\begin{aligned} m_1 x_1^{(2)} &= -k_2(x_1 - x_2) - k_1 x_1 + u \\ m_2 x_2^{(2)} &= k_2(x_1 - x_2) - c x_2^{(1)} \end{aligned} \quad (16)$$

From the structure, these equations can be rewritten in the following higher-order form without change of coordinates:

$$x_2 = y \quad (17)$$

$$x_1 = \frac{m_2}{k_2} y^{(2)} + \frac{c}{k_2} y^{(1)} + y \quad (18)$$

$$\begin{aligned} u = {} & \frac{m_1 m_2}{k_2} y^{(4)} + \frac{m_1 c_2}{k_2} y^{(3)} + \frac{c_2(k_1 + k_2)}{k_2} y^{(1)} + \\ & [m_1 + \frac{m_2(k_1 + k_2)}{k_1}] y^{(2)} + k_1 y \end{aligned} \quad (19)$$

From the physical limitations, the motion of the system must satisfy the following 2-sided constraints: $|x_1| \leq 3\text{cm}$, $|x_2| \leq 3\text{cm}$, and $|u| \leq 60\text{Nm}$. On substituting from Eqs. (19), we get

$$\begin{aligned} |y| &\leq 0.03 \\ |\frac{m_2}{k_2} y^{(2)} + \frac{c}{k_2} y^{(1)} + y| &\leq 0.03 \\ |\frac{m_1 m_2}{k_2} y^{(4)} + \frac{m_1 c_2}{k_2} y^{(3)} + k_1 y| &\leq 60 \end{aligned} \quad (20)$$

The boundary conditions at t_0 on x_1, $\dot{x}_1$, x_2, $\dot{x}_2$ are known using the encoder readings. At t_f, the boundary conditions on x_2 and $\dot{x}_2$ are obtained from sampled position and rate of slider 3 using a specific heuristic. As a result, four boundary conditions on $y^{(3)}$, $y^{(2)}$, $y^{(1)}$, y at t_0 and two boundary conditions on $y^{(1)}$ and y at t_f are known. The statement of the problem is to find a feasible trajectory for the dynamic system that satisfies Eqs. (20) and the appropriate boundary conditions on derivatives of y.

First, we want to note that due to the 2-sided nature of the constraints (20), the conservative scheme can not be used. This scheme gives a set of inconsistent inequalities. As a result, the collocation scheme is used to develop the feasible solution selected to have the following form:

$$y(t) = \phi_0(t) + a_1\phi_1(t) + a_2\phi_2(t) \quad (21)$$

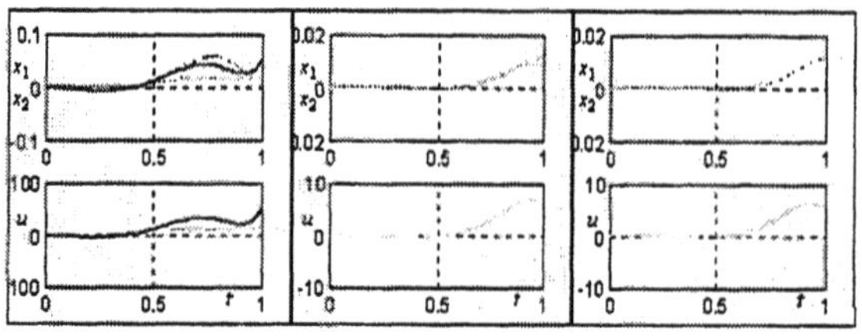

Fig. 5. Constraint satisfaction with a vertex point and the centroid of Fig. 4: (i) 6 collocation points, (ii) 6 collocation points with bounded first derivatives, (iii) 6 collocation points with bounded first and second derivatives.

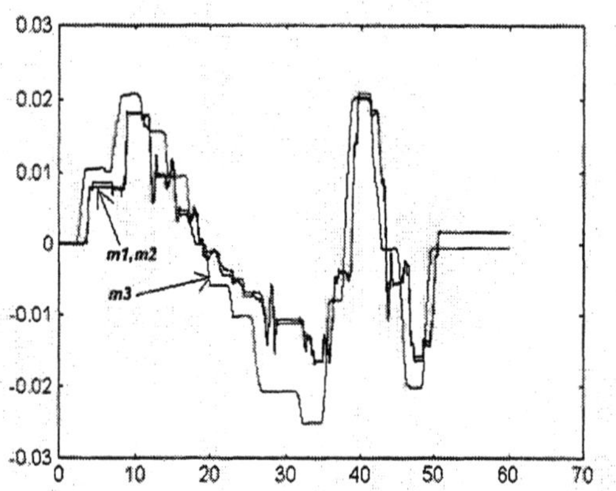

Fig. 6. A typical 'pursuit-evasion' experiment where mass 3 is moved around and mass 2 follows mass 3 while satisfying the constraints.

where $\phi_0(t) = b_0 + b_1 t + b_2 t^2 + b_3 t^3 + b_4 t^3(1-t) + b_5 t^3(1-t)^2$, $\phi_1(t) = t^4(1-t)^2$, and $\phi_2 = t^5(1-t)^2$, assuming $t_0 = 0$ and $t_f = 1$. The coefficients b_i are determined from the six boundary conditions specified on the problem. Due to the forms of $\phi_1(t)$ and $\phi_2(t)$, $y(t)$ is admissible regardless of the values for a_1 and a_2. For a representative set of boundary conditions, $(x_2, \dot{x}_2, x_1, \dot{x}_1)^T|_0 = (0.0, -0.0015, 0.0, 0.0)^T$ and $(x_2, \dot{x}_2)^T|_1 = (0.011, 0.0015)^T$, variants of the algorithm was implemented in simulation using six equally spaced collocation points.

Fig. 4 shows the feasible regions under varying specifications. From the figure, one observes that the size of the feasible region reduced progressively as further bounds on the derivatives of the constraints are added. For each case, trajectories corresponding to a vertex point and centroid of the feasible region were checked for constraint satisfaction and are plotted in Fig. 5. From these plots, we notice that as higher derivatives of the constraints are bounded, the trajectories satisfy the constraints throughout the domain.

In the hardware implementation, an exponential trajectory tracker is designed to follow $y_d(t)$. The form of the input $u(t)$ is

$$u(t) = \beta_4 y_d^{(4)}(t) + \beta_3 y_d^{(3)}(t) + \ldots \beta_1 y_d^{(1)}(t) + \beta_0 y_d(t) + \gamma_3 e^{(3)}(t) + \ldots \gamma_1 e^{(1)}(t) + \gamma_0 e(t) \quad (22)$$

where $e(t) = y_d(t) - y(t)$, β_i are the coefficients of $y^{(i)}$ in the higher-order model of the system, and γ_i are the feedback gains selected to make the error dynamics stable. Figure 6 shows the typical results of an experiment where mass 3 is moved over a length of time and mass 2 follows the motion of mass 3 while satisfying the constraints.

6. CONCLUSION

This paper presented a methodology to generate feasible trajectories of linear dynamic systems expressed in higher-order forms in the presence of inequality constraints. On representing the feasible solution by basis functions, the feasible trajectory determination problem was changed to the problem of finding a convex polytope bounded by linear constraints. The method was successfully implemented in hardware to simulate a 'pursuit-evasion' maneuver with a linear spring-mass-damper system. We believe that these methodologies can be extended for real-time planning of nonlinear systems such as air vehicles and missiles.

7. REFERENCES

Agrawal, S.K. and N. Faiz (1998). A new efficient method for optimization of a class of nonlinear systems without lagrange multipliers. *Journal of Optimization Theory and Applications* **97**(1), 11-28.

Agrawal, S.K. and T. Veeraklaew (1996). A higher-order method for dynamic optimization of a class of linear time-invariant dynamic systems. *Journal of Dynamic Systems, Measurements, and Control, Transactions of ASME* **118**(4), 786-791.

Boyd, S. and L. Vandenberghe (1996). *Convex Optimization: EE364 Lecture Notes.*

Fletcher, C.A.J. (1984). *Computational Galerkin Methods.* Springer Verlag.

Hauser, J. and D.G. Meyer (1998). Trajectory morphing for nonlinear systems. *Proceedings of the American Control Conference, Philadelphia, Pennsylvania, 1998* pp. 2065-2070.

Kailath, T. (1980). *Linear Systems.* Prentice-Hall, Inc.

Nieuwstadt, M.J. and R.M. Murray (1995). Approximate trajectory generation for differentially flat systems with zero dynamics. *Proceedings pf the 34th IEEE Conference on Decision and Control.*

Nieuwstadt, M.J. and R.M. Murray (1998). Rapid hover to forward flight transitions for a thrust vectored aircraft.

Solodovnikov, A.S. (1980). *System of Linear Inequalities.* University of Chicago press, Chicago.

Robust Controller Design for Uncertain Time-delay Systems Containing Saturating Actuators

Hongye Su*, James Lam, Jian Chu***

*Institute of Industrial Process Control, Zhejiang University
Hangzhou 310027, P.R.China
Email: (H) hysu@public.hz.zj.cn (O) hysu@iipc.zju.edu.cn
**Department of Mechanical Engineering, The University of Hong Kong
Pokfulam Road, Hong Kong

Abstract: The delay-dependent robust stabilization problem for a class of uncertain linear time-delay systems containing saturating actuator is considered in this paper. The uncertain time-delay systems under consideration are described by state differential equations with time-varying unknown-but-bounded uncertain parameters and delayed state, the delay is assumed to be constant bounded but unknown. A new sufficient condition for the existence of memoryless robust stabilizing control law is derived based on linear matrix inequality technique. The results depend on the size of delay and are given in terms of several linear matrix inequalities (LMIs). Numerical example is presented to illustrate the obtained results.

Keywords: Uncertain time-delay systems, robust stabilization, saturating actuator, state feedback, Linear Matrix Inequality (LMI)

1. INTRODUCTION

In many practical situations, some non-linearities introduced by the actuator dynamics, and one of the common non-linearities is the saturation. Generally, the physical limitations of the actuator are unavoidable in the operation of driving the actuator by the signals emitted from the designed controllers, thus causing actuator saturation. The latter not only deteriorates the control system performance, but can also lead to undesirable stability effects. If such non-linearity saturation is not taken into account in the control system design, an integral wind up or limit cycle may occur (see for example Krikelis 1980, Glattfelder *et al.* 1983, Kosut *et al.* 1983, and the references therein). The stability analysis and stabilization of linear systems with saturating actuator have been investigated by many researchers in the past few years. The stability analysis of a SISO system with saturating actuator was studied by Glattfelder et al. (1983) and Krikelis (1980). In those papers, Popov's criterion is employed to analyse the stability of the saturating system via PI control. Krikelis et al. (1984) has studied the tracking problem of a system subject to actuator saturation and integrator wind up. They have derived a region where asymptotic stability is guaranteed. Chen *et al.* (1988a) studied the stability analysis problem for the multivariable case and obtained sufficient and bounded input bounded output stability conditions using the time domain approach. This results was extended to the case of linear systems with delayed state and sufficient conditions for state feedback stabilization given in Klai *et al.* (1994) and Tissir *et al.* (1992), which are independent of the size of delay. Chen *et al.* (1988b) considered the problem of the stabilization of the saturating time-delay system with state feedback and sampled-state feedback, delay-dependent and delay-independent conditions, expressed in terms of norm or matrix measures, are derived to ensure the stability of the closed-loop saturating time-delay system under control. Upper bounds on time delay which guarantee the asymptotic stability of the closed-loop system via a state feedback control law are given in Su *et al.* (1991) using a Razumikhin approach. The case of a linear system with delayed-input is treated in Shen *et al.* (1989), where sufficient conditions are given for stabilization via state or output feedback. Recently, special interest has been devoted to the robust stabilization problem for uncertain time-delay systems containing saturating actuator. For example, sufficient conditions for output stabilization, independent of the size of delay, are given in Chou *et al.* (1989) using the concept of matrix measure and the comparison theory for an uncertain linear time-delay system with single state delay and saturating actuator in time domain. Niculescu *et al.* (1996) studied the problem of robust stabilization for a class of uncertain linear time-delay systems containing a saturating actuator by using Razumikhin approach. A delay-dependent criteria for robust stabilization via linear memoryless state feedback control law was obtained and the upper bound on time-delay is given such that the uncertain system is robustly stabilizable. But the obtained results are also given in terms of the solution of algebraic Riccati equation and still need the tuning of scalar and positive definite symmetric matrix.

In this paper, we extend the linear matrix inequality (LMI) method to time-varying uncertain linear dynamic systems with saturating actuator and unknown-but-bounded time-delay in state variable. It is assumed that perfect information of plant states is available for feedback. The synthesis problem addressed is to design a memoryless state feedback control law such that the closed-loop system is asymptotically stable for all admissible uncertainties. Based on linear matrix inequality (LMI) approach, a new sufficient condition for the existence of such a control law is presented in terms of several linear matrix inequalities (LMIs). The obtained design approach in this paper is depend on the size of delay and does not involve any tuning of parameters, as in the case with the robust stability and stabilization method (Li *et al.* 1996), and it can be computed effectively (Boyd *et al.* 1994).

2.SYSTEM DESCRIPTION AND DEFINITIONS

Consider the following uncertain time-delay systems described by

$$\begin{aligned} \dot{x}(t) &= (A+\Delta A(t))x(t) + (A_1+\Delta A_1(t))x(t-d) \\ &\quad + (B+\Delta B(t))u'(t), \\ u'(t) &= Sat(u(t)), \\ Sat(u(t)) &= [sat(u_1(t)) \ \cdots \ sat(u_m(t))] \\ x(t) &= \phi(t), \ t\in[-\tau,\ 0] \end{aligned} \tag{1}$$

where $x(t)\in R^n$ is the state vector, $u(t)\in R^m$ is the control input vector to the actuator (emitted from the designed controller), $u'(t)\in R^m$ is the control input vector to the plant; $A\in R^{n\times n}$, $A_1\in R^{n\times n}$, $B\in R^{n\times m}$ are known constant matrices. The matrices $\Delta A(\cdot)$, $\Delta A_1(\cdot)$ and $\Delta B(\cdot)$ are real-valued continuous matrix functions representing time-varying parameter uncertainties in the system model with appropriate dimensions. The saturation function is defined by (shown in Fig. 1)

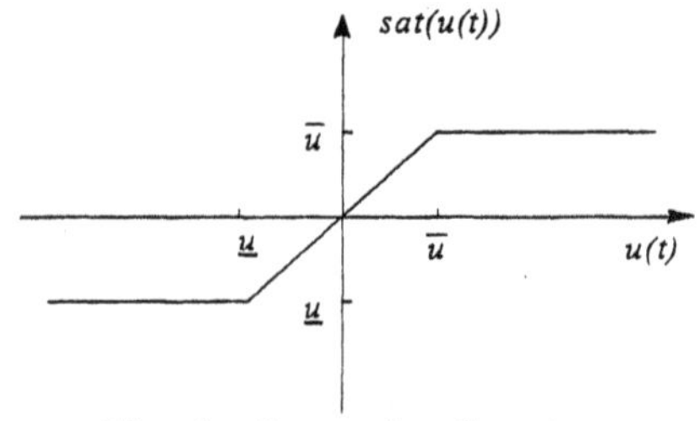

Fig. 1 Saturation function

$$sat(u_i(t)) = \begin{cases} \underline{u}_i, & u_i(t) \le \underline{u}_i < 0 \\ u_i(t), & \underline{u}_i \le u_i(t) < \bar{u}_i \\ \bar{u}_i, & u_i(t) \ge \bar{u}_i > 0 \end{cases}$$

d is unknown scalars denote the delays in the state and control, respectively, and it is assumed that there exist positive number τ such that

$$0 \le d \le \tau \tag{2}$$

holds for all t; $\phi(t)$ is a smooth vector-valued continuous initial function defined in the Banach space $C^n[-\tau,\ 0]$ of smooth functions

$$\psi\colon [-\tau,\ 0] \mapsto R^n \text{ with } \|\psi\|_\infty := \sup_{-\tau\le\eta\le0} \|\psi(\eta)\|$$

In this paper, the admissible uncertainties are assumed to be of the form

$$\begin{aligned} &\Delta A(t) = H_1F_1(t)E_1,\ \Delta A_1(t) = H_2F_2(t)E_2, \\ &\Delta B(t) = H_3F_3(t)E_3 \end{aligned} \tag{3}$$

where $F_i(t)\in R^{s_i\times q_i}$, $i=1,\ 2,\ 3$ are unknown real time varying matrices with Lebesgue measurable elements, satisfying

$$F_i^T(t)F_i(t) \le I,\ i=1,\ 2,\ 3 \tag{4}$$

and H_i, E_i, $i=1,\ 2,\ 3$ are known real constant matrices with appropriate dimensions which characterize how the uncertain parameters in $F_i(t)$, $i=1,\ 2,\ 3$ enter the nominal matrices A, A_1 and B. To facilitate further development, we introduce

$$\begin{aligned} &A(t) = A+\Delta A(t),\quad A_1(t) = A_1+\Delta A_1(t), \\ &B(t) = B+\Delta B(t) \end{aligned} \tag{5}$$

Throughout this paper, we shall use the following concept of robust stabilization for the uncertain time-delay system of the form (1)~(3).

Definition 1: The class of uncertain time-delay systems (1)~(3) is said to be robustly stable if the trivial solution $x(t)\equiv 0$ of the functional differential equation associated to (1) with $u(t)\equiv 0$ is asymptotically stable for all admissible uncertainties $\Delta A(t)$ and $\Delta A_1(t)$. The class of uncertain time-delay systems (1)~(3) is said to be robustly stabilizable if there exists a static linear state feedback control law $u(t)=Kx(t)$ such that the resulting closed-loop system is robustly stable.

In this paper, we shall develop delay-dependent

conditions for robust stability of the uncertain time-delay system with saturating actuator (1)~(3) and the correspondent delay-dependent robust stabilizing control law design approach. More specifically, our objective is to determine bounds for the time-delay which ensure that the system (1)~(3) is robustly stabilizable. Suitable memoryless state feedback control law will be developed. It will be shown in the sequel that the above problems can be solved by using linear matrix inequalities.

We introduce some useful lemmas which will be essential for the proofs in the next section; see, e.g. Wang *et al.* (1992).

Lemma 1: Given any positive definite symmetric matrix R, vector $X(t)$, $Y(t)$ and matrix $F(t)$ with appropriate dimensions, and satisfies $F^T(t)F(t) \le I$, then

$$2X^T(t)F(t)Y(t) \le X^T(t)RX(t) + Y(t)^T R^{-1} Y(t)$$

Lemma 2: Let A, D and E be real constant matrices with appropriate dimensions, and matrix $F(t)$ satisfies $F^T(t)F(t) \le I$. Then we have:

(a) For any scalar $\varepsilon > 0$,

$$DF(t)E + E^T F^T(t) D^T \le \varepsilon DD^T + \varepsilon^{-1} E^T E$$

(b) For any matrix $P > 0$ and scalar $\varepsilon > 0$ such that $\varepsilon I - EPE^T > 0$, then

$$(A + DF(t)E)P(A + DF(t)E)^T \le APA^T + APE^T\left(\varepsilon I - EPE^T\right)^{-1} EPA^T + \varepsilon DD^T$$

(c) For any matrix $P>0$ and scalar $\varepsilon > 0$ such that $P - \varepsilon DD^T > 0$, then

$$(A + DF(t)E)^T P^{-1}(A + DF(t)E) \le A^T\left(P - \varepsilon DD^T\right)^{-1} A + \varepsilon^{-1} E^T E$$

3. ROBUST STABILIZING CONTROL LAW SYNTHESIS

In this section, we shall give a new robust stabilization sufficient condition and correspondent synthesis algorithm for uncertain linear time-delay systems with saturating actuator. Upper bound τ on the time delay d is given such that the uncertain linear time-delay system is robustly stabilizable for any time delay $0 \le d \le \tau$.

The approach presented have makes use of the Lyapunov theorem and LMI method. The difficulty of this stability study comes from the nonlinearity given by the saturating actuator.

The main result is derived as follow:

Theorem 1: For the uncertain linear time-delay system (1)~(3), given scalar τ satisfying (2), this uncertain linear time-delay system is robustly stabilizable if there exist positive definite symmetric matrices $X > 0$, $T_{1j} > 0, j = 1,2,3$; $Q > 0$; matrix Y, and positive scalars, $\alpha_i > 0, i = 1,2,3$; $\beta_j > 0, j = 1,2,3$; $\varepsilon > 0$ and $\sigma > 0$ satisfying the following linear matrix inequalities (LMIs),

$$\begin{bmatrix} Q - BB^T - \varepsilon H_3 H_3^T & BE_3^T \\ E_3 B^T & \varepsilon I - E_3 E_3^T \end{bmatrix} \ge 0$$

$$\begin{bmatrix} X & XA^T & XE_1^{\ T} \\ AX & T_{11} - \beta_1 H_1 H_1^T & 0 \\ E_1 X & 0 & \beta_1 \end{bmatrix} \ge 0$$

$$\begin{bmatrix} X & XA_1^T & XE_2^{\ T} \\ A_1 X & T_{12} - \beta_2 H_2 H_2^T & 0 \\ E_2 X & 0 & \beta_2 \end{bmatrix} \ge 0$$

$$\begin{bmatrix} X & Y^T B^T & Y^T E_3^T \\ B\,Y & T_{13} - \beta_3 H_3 H_3^T & 0 \\ E_3 Y & 0 & \beta_3 \end{bmatrix} \ge 0$$

$$\begin{bmatrix} S & M_1 & M_2 & M_3 & M_4 \\ M_1^T & -N_1 & 0 & 0 & 0 \\ M_2^T & 0 & -N_2 & 0 & 0 \\ M_3^T & 0 & 0 & -N_3 & 0 \\ M_4^T & 0 & 0 & 0 & -N_4 \end{bmatrix} < 0 \quad (6)$$

where

$$S = AX + XA^T + A_1 X + XA_1^T + BY + Y^T B^T + \sum_{i=1}^{3} \alpha_i H_i H_i^T + Q + 3\tau X + \tau(A_1 T A_1^T + \sigma H_2 H_2^{\ T})$$

$$M_1 = \left[XE_1^T \quad XE_2^T\right], \quad N_1 = diag(\alpha_1 I, \alpha_2 I)$$

$$M_2 = Y^T E_3^T, \quad N_2 = \alpha_3 I$$

$$M_3 = \tau A_1 T E_2^T, \quad N_3 = \tau(\sigma I - E_2 T E_2^T)$$

$$M_4 = \left[Y^T \quad \tau Y^T\right], \quad N_4 = diag(I, \ \tau I)$$

Moreover, a suitable robustly stabilizing control law is given by

$$u(t) = 2YX^{-1}x(t) \quad (7)$$

Proof: For the uncertain linear time-delay system (1)~(3), introduce the control law $u(t) = 2Kx(t)$, where the control law gain matrix $K \in R^{m \times n}$ is to be found, the closed-loop system is

$$\dot{x}(t) = [A(t) + B(t)K]x(t) + A_1(t)x(t-d) + B(t)\eta(t) \quad (8)$$

$$x(t) = \phi(t), \quad t \in [-\tau, 0]$$

where $\eta(t) = Sat(2Kx(t)) - Kx(t)$. Obviously, vector function $\eta(t)$ satisfies the following inequality

$$\eta^T(t)\eta(t) \le x^T(t) K^T K x(t) \quad (9)$$

Using the Leibniz-Newton formula and the system (8) can be written as

$$\dot{x}(t)=[A(t)+B(t)K+A_1(t)]x(t)$$
$$-\int_{-d}^{0}A_1(t)\{[A(t+s)+B(t+s)K]x(t+s)+B(t)\eta(t)$$
$$+A_1(t+s)x(t-d+s)+B(t+s)\eta(t+s)\}ds$$
$$x(t)=\varphi(t),\quad t\in[-2\tau,\ 0] \tag{10}$$

where $\varphi(t)$ is a smooth vector-valued continuous initial function.

Obviously, the asymptotic stability of system (10) can assure the asymptotic stability of closed-loop system (8). Let the following Lyapunov functional candidate for the system (10)

$$V(x(t),t)=x^T(t)Px(t)+\int_{-d}^{0}[\int_{t+s}^{t}x^T(\theta)R_1x(\theta)d\theta$$
$$+\int_{t-d+s}^{t}x^T(\theta)R_2x(\theta)d\theta]ds \tag{11}$$

where $P,\ R_1,R_2$ are positive definite symmetric matrices. The derivative of $V(x(t),t)$ along the trajectory of the system (10) with respect to time t is given by

$$\dot{V}(x(t),t)=x^T(t)\{[A(t)+B(t)K+A_1(t)]^TP$$
$$+P[A(t)+B(t)K+A_1(t)]\}x(t) \tag{12}$$
$$+f(x(t),t)+g(x(t),t)$$

where

$$f(x(t),t)=dx^T(t)R_1x(t)+dx^T(t)R_2x(t)$$
$$-\int_{-d}^{0}x^T(t+s)R_1x(t+s)ds$$
$$-\int_{-d}^{0}x^T(t-d+s)R_2x(t-d+s)ds$$
$$g(x(t),t)=2x^T(t)PB(t)\eta(t)$$
$$-2x^T(t)P\{\int_{-d}^{0}A_1(t)[(A(t+s)+B(t+s)K)x(t+s)$$
$$+A_1(t+s)x(t-d+s)+B(t+s)\eta(t+s)]ds\}$$

By using Lemma 1 to (12), we have

$$\dot{V}(x(t),t)\le x^T(t)W_1x(t)+f(x(t),t)+g(x(t),t) \tag{13}$$

where

$$W_1=A^TP+PA+A_1^TP+PA_1+K^TB^TP+PBK$$
$$+P(\sum_{i=1}^{3}\alpha_iH_iH_i^T)P+\sum_{i=1}^{2}\alpha_i^{-1}E_i^TE_i$$
$$+\alpha_3^{-1}K^TE_3^TE_3K$$

and

$$g(x(t),t)\le x^T(t)PB(t)B^T(t)Px(t)+x^T(t)K^TKx(t)$$
$$+\sum_{j=1}^{3}\int_{-d}^{0}x^T(t)PA_1(t)T_{1j}A_1^T(t)Px(t)ds$$
$$+\int_{-d}^{0}x^T(t+s)A^T(t+s)T_{11}^{-1}A(t+s)x(t+s)ds$$
$$+\int_{-d}^{0}x^T(t-d+s)A_1^T(t+s)T_{12}^{-1}A_1(t+s)x(t-d+s)d$$
$$+\int_{-d}^{0}x^T(t+s)K^TB^T(t+s)T_{13}^{-1}B(t+s)Kx(t+s)d$$
$$+\int_{-d}^{0}x^T(t+s)K^TKx(t+s)ds \tag{14}$$
$$+\int_{-d}^{0}x^T(t)PA_1(t)B(t+s)B^T(t+s)A_1^T(t)Px(t)ds$$

Assume that there exist scalar $\varepsilon>0$ and positive definite symmetric matrix $Q>0$ such that the following inequality are satisfied

$$BB^T+BE_3^T(\varepsilon I-E_3E_3^T)^{-1}E_3B^T+\varepsilon H_3H_3^T\le Q \tag{15}$$

where

$$\varepsilon I-E_3E_3^T>0 \tag{16}$$

Then using Lemma 2(b), we have

$$B(t)B^T(t)\le Q \tag{17}$$
$$B(t+s)B^T(t+s)\le Q \tag{18}$$

Let $T=Q+\sum_{j=1}^{3}T_{1j}$ and assume that there also exists scalar $\sigma>0$ such that the following inequality is satisfied

$$\sigma I-E_2TE_2^T>0 \tag{19}$$

Then using Lemma 2(b), we have

$$A_1(t)TA_1^T(t)\le A_1TA_1^T+\sigma H_2H_2^T$$
$$+A_1TE_2^T(\sigma I-E_2TE_2^T)^{-1}E_2TA_1^T \tag{20}$$

Assume that there exist scalars $\beta_j>0,\ j=1,2,3$ such that the following inequalities are satisfied

$$A^T(T_{11}-\beta_1H_1H_1^T)^{-1}A+\beta_1^{-1}E_1^TE_1\le P \tag{21}$$
$$A_1^T(T_{12}-\beta_2H_2H_2^T)^{-1}A_1+\beta_2^{-1}E_2^TE_2\le P \tag{22}$$
$$K^TB^T(T_{13}-\beta_3H_3H_3^T)^{-1}BK+\beta_3^{-1}K^TE_3^TE_3K\le P \tag{23}$$

where

$$T_{1j}-\beta_jH_jH_j^T>0,\ j=1,2,3 \tag{24}$$

Then using Lemma 2(c), we have

$$A^T(t+s)T_{11}^{-1}A(t+s)\le P,\ \forall t\ge 0$$
$$A_1^T(t+s)T_{12}^{-1}A_1(t+s)\le P,\ \forall t\ge 0$$
$$K^TB^T(t+s)T_{13}^{-1}B(t+s)K\le P,\ \forall t\ge 0 \tag{25}$$

Then applying (17)(18), (20) and (25) to (13) and let

$$R_1=2P+K^TK,\ R_2=P \tag{26}$$

we can get

$$\dot{V}(x(t),t)\le x^T(t)[W_1+PW_2P+W_3]x(t)$$
$$=x^T(t)W_4x(t) \tag{27}$$

where

$$W_2 = Q + \tau[A_1 T A_1^T + A_1 T E_2^T (\sigma I - E_2 T E_2^T)^{-1} E_2 T A_1^T + \sigma H_2 H_2^T] \quad (28)$$

$$W_3 = (1+\tau) K^T K + 3\tau P \quad (29)$$

$$\begin{aligned} W_4 = {} & A^T P + PA + A_1^T P + PA_1 + K^T B^T P + PBK \\ & + P(\sum_{i=1}^{3} \alpha_i H_i H_i^T) P + \sum_{i=1}^{2} \alpha_i^{-1} E_i^T E_i \\ & + \alpha_3^{-1} K^T E_3^T E_3 K + (1+\tau) K^T K + 3\tau P \\ & + P[Q + \tau(A_1 T A_1^T \\ & + A_1 T E_2^T (\sigma I - E_2 T E_2^T)^{-1} E_2 T A_1^T + \sigma H_2 H_2^T)]P \end{aligned} \quad (30)$$

Then if for some scalar $\tau > 0$, there exist positive definite symmetric matrices P, $T_{1j}, j = 1,2,3$, Q; matrix K and positive scalars $\alpha_i > 0, i = 1,2,3;$ $\beta_j > 0, j = 1,2,3;$ $\varepsilon > 0$ and $\sigma > 0$ satisfying the inequalities (15)(16), (19), (21)~(24) and $W_4 < 0$, then for any $0 \le d \le \tau$, we have

$$\dot{V}(x(t),t) \le -\alpha \|x(t)\|^2$$

where $\alpha = -\lambda_{\max}(W_4) > 0$ and $\lambda_{\max}(W_4)$ denotes the maximum eigenvalue of matrix W_4. Therefore, it follows from Lyapunov theorem and *Definition 1* that the system (10) is asymptotically stable for any $0 \le d \le \tau$ and for any admissible uncertainties $F_i(t)$ satisfying (4). This implies that the closed-loop system (8) is robustly stable and original uncertain time-delay system (1)~(3) is robustly stabilizable for any $0 \le d \le \tau$ and for any admissible uncertainties $F_i(t)$ satisfying (4).

Also introducing the new variable X, let

$$X = P^{-1} \quad (31)$$

and let $W_5 = X W_4 X$, equation (30) can be rewritten as

$$\begin{aligned} W_5 = {} & AX + XA^T + A_1 X + XA_1^T + BKX \\ & + XK^T B^T + \sum_{i=1}^{3} \alpha_i H_i H_i^T + Q \\ & + \tau(A_1 T A_1^T + A_1 T E_2^T (\sigma I - E_2 T E_2^T)^{-1} E_2 T A_1^T + \sigma H_2 H_2^T) \quad (32) \\ & + \sum_{i=1}^{2} \alpha_i^{-1} X E_i^T E_i X + \alpha_3^{-1} X K^T E_3^T E_3 K X \\ & + (1+\tau) X K^T K X + 3\tau X \end{aligned}$$

Let $Y = KX$ and using schur complements, we obtain that inequalities (15)(16), (19), (21)~(24) and $W_5 < 0$ are equivalent to those LMIs in (6). □

Remark 1: Theorem 1 provides a delay-dependent condition for robust stabilizability of the uncertain linear time-delay system (1)~(3) with constrained input (saturating actuator) and the correspondent delay-dependent linear memoryless state feedback control law synthesis approach based on LMI technique. Since this robust stability criteria and robustly stabilizing control law is dependent of the size of the time-delays, in general, it is expected to be less conservative than the delay-independent robust stability criteria and robust stabilizing control law design methods of Chou *et al.* (1989). In contrast with the results in Niculescu *et al.* (1996), which developed delay-dependent robust stabilization method for uncertain time-delay systems with constrained input in terms of the solution of Riccati equations, the obtained results in Theorem 1 is given in terms of the solutions of linear matrix inequalities and does not need any tuning of parameters and it can be calculated very effectively by using interior point algorithms.

The upper bound τ for time delay d which ensure that the uncertain linear time-delay system (1)~(3) is robustly stabilizable for any $0 \le d \le \tau$ can be determined by solving the following quasi-convex optimization problem,

maximize τ

Subject to LMIs (6) and

$X > 0$, $T_{1j} > 0, j = 1,2,3$; $Q > 0$;

matrix Y, and positive scalars, $\alpha_i > 0, i = 1,2,3;$

$\beta_j > 0, j = 1,2,3;$ $\varepsilon > 0$ and $\sigma > 0$

4. EXAMPLE

To illustrate the effectiveness of the previous obtained results in this paper, a numerical examples is studied.

Consider an uncertain time-delay system (1)~(3), whose actuator is saturated at level ± and whose dynamic are described as follows,

$$A = \begin{bmatrix} -2 & 0 \\ 1 & -3 \end{bmatrix}, A_1 = \begin{bmatrix} -1 & 0 \\ -0.8 & -1 \end{bmatrix},$$

$$B = \begin{bmatrix} 1 & 2 \\ -1 & 4 \end{bmatrix}, H_i = \begin{bmatrix} 0.2 & 0 \\ 0 & 0.2 \end{bmatrix},$$

$$E_i = \begin{bmatrix} 1 & 0 \\ 0 & 1 \end{bmatrix}, \quad i=1,2,3$$

and $F_i^T(t) F_i(t) \le I$, $i = 1, 2, 3$. Applying Theorem 1 to this uncertain time-delay system, it is found, using the software package LMI Lab, that this system is robustly stabilizable for any time-delay $d \le 0.284$. If there is not any uncertainties in this system, i.e. $H_i = 0$ and $E_i = 0$, $i = 1,2,3$, the system is robustly stabilizable for any time-delay $d \le 0.6153$ by using the software package LMI Lab.

5. CONCLUSION

This paper deals with the problem of robust stabilization synthesis for a class of uncertain linear time-delay systems with saturating actuator. Delay-dependent LMI based methods of designing linear memoryless state feedback control laws has been developed. A new criteria of robustly stabilizable for uncertain time-delay systems is given in terms of linear matrix inequality (LMI). A numerical example shows that the presented method is feasible and less conservative than reported methods.

6. ACKNOWLEDGMENT

This work is supported by the National Natural Science Foundation of China (NSFC: 69604006), the Zhejiang Provincial Natural Science Foundation of China.

REFERENCES

Boyd, S., L. Tl Ghaoui, E. Feron and Balakrishnan, V. (1994), *Linear Matrix Inequalities in Systems and Control Theory*, SIAM, Philadelphia.

Chen, B.S. and Wang S.S. (1988a), The Stability of Feedback Control with Nonlinear Saturating Actuator: Time Domain Approach. *IEEE Trans. Auto. Control*, AC-33, pp. 483-487.

Chen, B.S., Wang S.S. and Lu, H.-C. (1988b), Stabilization of Time-delay Systems Containing Saturating Actuators. *Int. J. Control*, 47, pp. 867-881.

Chen, B.S. and Lu, H.-C. (1988c), State Estimation of Large-scale Systems. *Int. J. Control*, 47(6), p. 1613.

Chou. J.H., Horng, I.-R. and Chen, B.S. (1989), Dynamic Feedback Compensator for Uncertain Time-delay Systems Containing Saturating Actuator. *Int. J. Control*, 49, pp. 961-968.

Glattfelder, A.H. and Schaufelberger, W. (1983), Stability Analysis of Single Loop Control Systems with Saturation and Antireset-windup Circuits. *IEEE Trans. Auto. Control*, AC-28, pp. 1074-1080.

Hale, J. (1971), *Theory of Functional Differential Equations*. Spring-Verlag, New York.

Klai, M., Tarbouriech, S. and Burgat, C. (1994), Some Independent Time-delay Stabilization of Linear Systems with Saturating Controls. Proc. of IEE Contr.'94, Coventry, U.K., pp. 1358-1363.

Kosut, R.L. (1983), Design of Linear Systems with Saturating Linear Control Bounded State. *IEEE Trans. Auto. Control*, AC-28, pp. 121-124.

Krikelis, N.J. (1980), State Feedback Integral Control with Intelligent Integrator. *Int. J. Control*, 32, pp. 465-473.

Krikelis, N.J. and Barkas, S.K. (1984), Design of Tracking Systems Subject to Actuator Saturation and Integrator Wind-up. *Int. J. Control*, 39(4), p. 667.

Li, X. and C. E. de Souza (1996a), Robust Stabilization and H^∞ Control of Uncertain Linear Time-delay Systems. Proc. of 13th Triennial World Congress of IFAC, San Francisco, USA, June 30-July 5, pp113-118.

Liu, W., Chitour, Y. and Sontag, E. 1(996), On Finite-gain Stabilizability of Linear Systems Subject to Input Saturation. *SIAM J. Contr. And Optimization*, 34(4), pp. 1190-1219.

Mori, T., Fukuma, N. and Kuwahara, M. (1981), Simple Stability Criteria for Single and Composite Linear Systems with Time Delays. Int. J. Control, 34, pp. 1175-1184.

Niculescu, S.-I., Dion, J.-M. and Dugard, L. (1996), Robust Stabilization for Uncertain Time-delay Systems Containing Saturating Actuators, *IEEE Trans. Auto. Control*, AC-41, pp. 742-747.

Shen, J.-C. and Kung, F.-C. (1989), Stabilization of Input-delay Systems with Saturating Actuators. *Int. J. Control*, 50, pp. 1667-1680.

Su, Hongye, Wang, Jingcheng, Yu, Li and Chu, Jian (1997a), Robust Memoryless H^∞ Controller Design for A Class of Time-varying Uncertain Linear Time-delay Systems. Proc. ACC'97, Albuquerque NM, USA, Vol. 6, 3662-3663..

Su, Hongye, Chu, Jian, Wang, Jingcheng and Wang, Shuqing (1997b), Robust Stabilizing Control for Linear Time-varying Uncertain Time-delay Systems with Dynamic Output Feedback Controller. Proc. ADCHEM'97, Banff, Canada, 149-154.

Su, Hongye, Chu, Jian (1998), Robust Stabilizing Control for Uncertain Time-delay Systems Contain Saturating Actuators, IFAC Workshop on LTDC '98, Grenoble, France, 7-12.

Su, T.-J., Liu, P.L. and Tsay, J.T. (1991), Stabilization of Delay-dependent for Saturating Actuator Systems. Proc. of 30th Conference Decision Control, Brighton, U.K., pp2891-2892

Tissir, E. and Hmamed, A. (1996), Further Results on Stability of $\dot{x}(t) = Ax(t) + Bx(t-\tau)$. *Automatica*, 32(12), pp. 1723-1726.

Wang, W.J. and Chen B.S. (1988), Stability of Large-Scale Systems with Saturating Actuators. *Int. J. Control*, 47(3), p. 827.

Wang, Y., Xie, L. and de Souza, C.E. (1992), Robust Control of a Class of uncertain nonlinear systems. *Syst. & Contr. Letter*, 19, pp. 139-149.

Yu, Li, Chu, Jian and Su, Hongye (1996), Robust Memoryless H^∞ Controller Design for Linear Time-delay Systems with Norm-Bounded Time-varying Uncertainty. *Automatica*, 32(12), pp. 1759-1762.

14th Triennial World Congress, Beijing, P.R. China

C-2a-05-2

SLIDING MODE CONTROL OF TWO-MASS POSITIONING SYSTEMS

Yu-Feng Li, Bengt Eriksson and Jan Wikander

Department of Machine Design, Mechatronics Lab, Royal Institute of Technology, 100 44 Stockholm, Sweden, E-mail: yufeng@damek.kth.se

Abstract: A variety of servo controlled processes used in industry can be modelled as two-mass systems. The point-to-point and path following positioning control of such systems is highly affected by uncertainties due to varying compliance and nonlinear friction. This paper presents a design of the sliding mode control for this type of systems. The paper shows that the difficulty of choosing the switching gain for stability can be solved by the combination of a disturbance observer; and the robustness can be guaranteed during both reaching and convergence phases by the use of a nonlinear time-shifting sliding surface. Experiments are also performed to verify the feasibility of the proposed controller.

Keywords: two-mass systems, point-to-point control, nonlinear sliding surface, disturbance observer

1. INTRODUCTION

Variable structure control (VSC) provides a systematic approach to the problem of maintaining stability and consistent performance in the face of modelling imprecision. Sliding mode control (SMC) as a subclass of VSC is characterized by a discontinuous control action which changes structure upon reaching a set of predetermined switch surfaces. It allows the trade-offs between modelling detail and achievable control performance to be quantified in a simple fashion (Slotine and Li, 1991). Once the sliding motion occurs, the system has invariant properties which make the motion independent of certain system parameter variations and disturbances. Thus the system performance can be completely determined by the design of the sliding manifold.

Conventional SMC methods have the following common features:

- SMC usually requires the knowledge of the upper bounds of each uncertainty in the model and the disturbance on the process. It is not always possible to know these bounds in practice. As the result, an over-conservative gain is selected in order to satisfy the stability condition;
- The sliding manifold may be composed of several submanifolds, and the control laws depend highly on the specified sliding surfaces, i.e. the change of each sliding manifold requires redesign of the corresponding controller;
- It is known that sliding motion occurs only after the system reaches the surfaces. However, most of the conventional sliding manifolds typically do not consider system constraints during the reaching phase. In fact, SMC may be very sensitive to parameter variation or unmodeled dynamics.

Several papers have been published with the consideration of each of the above individual problems. An example is Ghosh-Roy *et al.* (1997) who apply the concept of perturbation to remedy the problem of selecting the switching gain. Stepanenko and Su (1993) proposed a nonlinear sliding manifold, which provides adaptability to the parameters variation and thus gives a controller which learns the bounds of uncertainties rather than uses the upper bounds of them. Choi *et. al.* (1993) use moving linear switching surfaces for taking the initial condition into account during the reaching phase.

The purpose of this paper is to present a design of an SMC for a flexible link two-mass positioning system. With the combination of a nonlinear time-varying sliding surface (NTSS) and on-line disturbance estimation, the control is robust to uncertainties due to unmodeled dynamics and nonlinear friction. The positioning error is shown to be less than the measurement resolution. The following sections are organized as follows: section 2 presents conventional SMC and SMC in combination with a disturbance observer (DOB). The Lyapunov method is used for the stability analysis. The design of a time-shifting nonlinear sliding surface is given in section 3. In section 4, experimental results are presented to examine the performance of the proposed controller. Conclusions are given in section 5.

2. SLIDING MODE CONTROL WITH DISTURBANCE OBSERVER

A variety of servo controlled processes used in industry can be modelled as two-mass systems and often the measurement is only performed at the motor side, as shown in Fig. 1. The mathematical expression of the model can simply be written as:

$$J_m\ddot{\theta}_m + f_m\dot{\theta}_m = K_m u_m - T_d \quad (1)$$

$$T_d = k_t(\theta_m - \theta_l) + b(\dot{\theta}_m - \dot{\theta}_l) + T_f + T_{ext} \quad (2)$$

where θ_m, θ_l is the motor and the load angles, K_m, the motor torque constant, J_m, the equivalent moment of inertia of the motor, f_m, the viscous friction. T_d is the disturbance which contains friction T_f, external low frequency disturbances T_{ext} and also the torque due to torsion of the flexible linkage, i.e. $k_t(\theta_m - \theta_l) + b(\dot{\theta}_m - \dot{\theta}_l)$, where k_t is the stiffness of the flexible shaft and b the damping coefficient. Since the viscous friction is typically time-varying and not known in detail, it is incorporated into the disturbance T_d. By taking the parameter uncertainty into account, the dynamic system can then simply be written as:

$$\ddot{\theta} = (k + \Delta k)u(t) + d(t) \quad (3)$$

and the nominal model used for design of the control law is

$$\ddot{\theta} = ku(t) \quad (4)$$

where, θ represents the motor angle and $k = K_m/J_m$. Disturbances and modelling uncertainties are expressed by $d(t) = -T_d/J_m$. Δk contains the uncertainties of inertia and torque constant. Note that all the subscripts in (1) are omitted from now on for the simplification of reading.

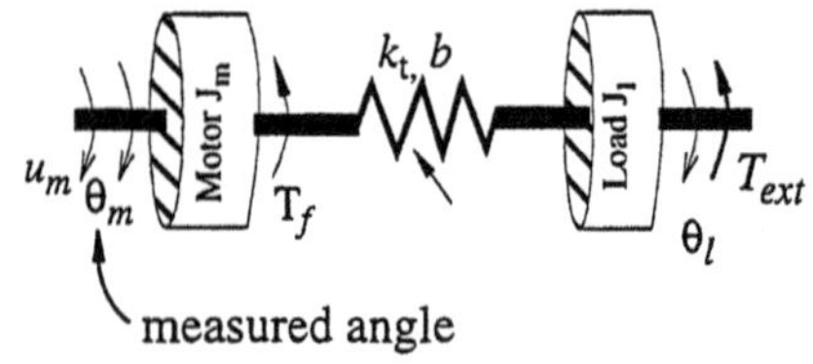

Fig.1. Two- mass systems

2.1 Conventional SMC

The control task considered here is restricted only to point-to-point positioning, i.e., to drive θ to the desired position θ_d with required accuracy. To accomplish this, the control law must be designed to assure sliding motion. In general, a sliding surface can be defined as

$$s = \dot{e} - v(e) \quad (5)$$

then, $$\dot{s} = \ddot{e} - v'(e)\dot{e} \quad (6)$$

where $e = \theta - \theta_d$, is the position error. $v(e) \in R$ is a function to be designed and $v'(e) = \frac{d}{de}v(e)$. A common approach of SMC design is to consider the Lyapunov function candidate

$$V = \frac{1}{2}s^2 \quad (7)$$

which must satisfy the attractive condition of sliding mode, i.e

$$\dot{V} = s\dot{s} < 0 \quad (8)$$

One method is to choose s dynamics as:

$$\dot{s} = -K_1 s - K_2 \operatorname{sgn}(s) \quad (9)$$

where $K_1>0$, $K_2>0$ are two positive real numbers. Since only point-to-point control is considered, θ_d is a constant to be the desired position, which implies

$$\dot{e} = \dot{\theta} - \dot{\theta}_d = \dot{\theta} \quad (10)$$

Hence, from the nominal model (4) and eqns (5) to (10), the control law can be derived as

$$u(t) = -\frac{1}{k}(K_1 s + K_2 \operatorname{sgn}(s) - v'(e)\dot{\theta}) \quad (11)$$

Substituting the control law of eqn (11) in the dynamic system (3), the actual s dynamics becomes

$$\dot{s} = -K_1 s - K_2 \operatorname{sgn}(s) + \Delta k u(t) + d(t) \quad (12)$$

To satisfy the condition (8), K_2 must be chosen as

$$K_2 \geq |\Delta k u(t) + d(t)| \quad (13)$$

This implies that the assumptions on bounds are needed in order to make the control law feasible. In general, the bound of parameter uncertainty, $|\Delta k|$, is usually not difficult to know. However, the assumption on the bounded disturbance $d(t)$ is rather restrictive as long as unmodeled dynamics are concerned. Generally speaking, the unmodeled dynamics is a function of the system states and may grow beyond any constant bound if the system becomes unstable. Hence, the selection of the switching gain K_2 becomes a key problem. The attractive condition requires that K_2 must be larger than an upper bound of disturbances including uncertainties. On the contrary,

K_2 together with K_1 should also be limited to a certain value in order not to excite the unmodeled dynamics.

2.2 SMC design with DOB

To avoid the above difficulty of selecting K_2, one solution is to introduce a disturbance observer (DOB). By considering all unknown terms as a single quantity defined by

$$D(t) = \Delta k u(t) + d(t) \quad (14)$$

then a reasonable estimation of $D(t)$ can be derived from eqn (3),

$$\hat{D}(t) = \hat{\ddot{\theta}}(t) - k u(t) \quad (15)$$

where $\hat{\ddot{\theta}}$ is the estimated system acceleration evaluated from the second derivative of position. Note that $\hat{D}$ can hardly be used directly due to high quantization level. Therefore, the estimated disturbance must be filtered as

$$\hat{D}_{flt}(t) = F(t)\hat{D}(t) \quad (16)$$

before it is added to the control that resulted in eqn (11). $F(t)$ is a filter designed according to the required control bandwidth. It is worth to mention here that the relatively high sampling frequency is required for applying this method, since a high cut-off frequency of $F(t)$ is needed in order to keep the delay introduced by the filter small enough so that it does not cause instability. Thus, the new control law becomes:

$$u(t) = -\frac{1}{k}(K_1 s + K_2 \operatorname{sgn}(s) - v'(e)\dot{\theta} + \hat{D}_{flt}) \quad (17)$$

which yields the dynamics for s

$$\dot{s} = -K_1 s - K_2 \operatorname{sgn}(s) + D - \hat{D}_{flt} \quad (18)$$

and the choice of the switching gain K_2 becomes

$$K_2 \geq |D_e| = |D - \hat{D}_{flt}| \quad (19)$$

Now the control requires only the estimation error upper bounds $|\hat{D}_e|$ instead of the disturbance upper bounds themselves. Thus, the required switching gain is much lower than the gain resulting from the conventional SMC design. This is an obvious advantage of using DOB in SMC over the conventional SMC.

Note that the term $K_2\operatorname{sgn}(s)$ leads to undesirable chattering in practice. Therefore, in a real implementation the sgn(s) is usually replaced by a saturation function as follows:

$$sat(s) = \begin{cases} s/\rho, & \text{if } |s| \leq \rho \\ \operatorname{sgn}(s), & \text{if } |s| > \rho \end{cases} \quad (20)$$

i.e. by introducing a thin boundary layer ρ neighbouring the switching surface, the control becomes continuous inside the boundary layer, thus eliminating chattering. ρ is a trade off between control bandwidth and required control precision.

2.3 Integral action

Since the *sgn(s)* term is replaced by a saturation function, the system can not exactly reach the predetermined sliding surface, but only reach its boundary as described above. In practice, the boundary layer may possibly not be chosen thin enough to provide the desired accuracy. Hence, an integrator may also be added in order to eliminate the steady state error when the control is inside the boundary layer. In order to avoid the potential cause of instability due to saturation effects. The integration is stopped as long as the control signal is saturated. Let

$$K(r) = k_i r \quad (21)$$

where, $\dot{r} = s$, and K_i is defined as

$$k_i = \begin{cases} k_i, & |u(t)| < u_{max} \\ 0, & |u(t)| \geq u_{max} \end{cases} \quad (22)$$

u_{max} is the maximum allowed control signal. The final control law is then

$$u(t) = -\frac{1}{k}(K_1 s + K_2 sat(s) + K(r) - v'(e)\dot{\theta} + \hat{D}_{flt}) \quad (23)$$

3. SLIDING SURFACE DESIGN

This section presents the design of a nonlinear time-shifting sliding surface (NTSS) by taking into account initial conditions and system operation constraints.

3.1 Nonlinear sliding surface

In order to demonstrate the feasibility of the proposed approach, the nonlinear function $v(e)$ in eqn (5) is chosen as follows

$$v(e) = -\omega_p S(e) \quad (24)$$

$$S(e) = \tanh(ce) \quad (25)$$

where ω_p and c are the two positive design parameters. The function $v(e)$ has the following properties:

$$v(-e) = -v(e), \qquad |v(e)| \leq \omega_p \quad (26)$$

and with a bounded first derivative

$$|v'(e)| = \frac{\omega_p c}{\cosh^2(ce)} \leq \omega_p c \quad (27)$$

By substituting $v(e)$ in eqn (5), the sliding surface becomes

$$s = \dot{\theta} + \omega_p \tanh(ce) \quad (28)$$

When the system states are in the vicinity of the sliding surface, one has that

$$s \approx 0 \rightarrow \dot{\theta} = -\omega_p \tanh(ce) \quad (29)$$

3.2 Time-shifting sliding surface

The qualities of the closed loop system with the process states on, or close to, the sliding surface are independent of parameter variations in the process and to external and internal disturbances. However, before the process has reached the sliding surface, the properties of the controller are the same as those of a nonlinear PID controller, with the position error as controller input. The magnitude of the control signal thus depends strongly on the desired positioning distance, i.e. a large desired position will require a high control input to the process. For the true nonlinear and low damped process without a load side sensor, the large control input will cause an undesired oscillatory motion. Also, because of the high order of the true process, (four or higher), compared with the second order nominal process, the closed loop system will even become unstable with the desired controller gains and with a desired position exceeding a certain level. This can be solved by, either letting the surface parameters be a function of position error or by designing the reference position as a function of time. The first method will be used in this paper and is described below.

A moving switching surface proposed by Choi, *et al.* (1993) can be used to accomplish this task. By defining a shifting sliding surface for the Nominal model (3) as

$$s_i = \dot{\theta}(t) + \omega_i S(e), \quad i = 0,1,\ldots,p \quad (30)$$

Next, define the predetermined surface as s_p *and* the initial surface as s_0 which goes through the given initial conditions with the appropriate ω_0. The motion is performed by updating the parameter

$$\omega_{i+1} = \omega_i + \alpha(\tau), \quad i = 0,1,\ldots,p-1 \quad (31)$$

where, τ is a finite time during which the system stays on surface s_i. The shifted surfaces are shown in Fig. 2.

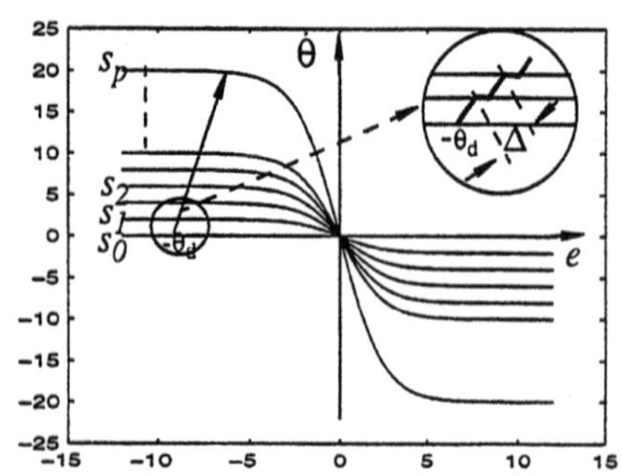

Fig.2. Construction of the shifting sliding surface as a function of *e*

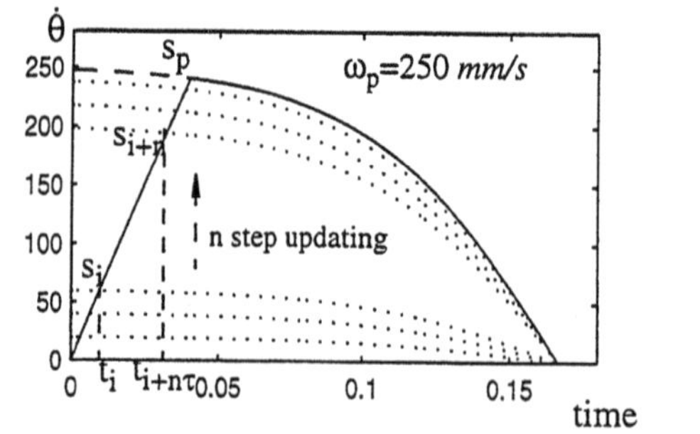

Fig.3. Construction of the shifting sliding surface as a function of time.

The updating will be stopped when the surface s_i reaches s_p i.e

$$s_p = \dot{\theta}(t_p) + \omega_p S(e) \quad (32)$$

The shifting distance between each surface is

$$\Delta = \alpha(\tau) S(e) \quad (33)$$

Without loss of generality, the initial point can be assumed to be at $(\theta(t_0), \dot{\theta}(t_0)) = (0, 0)$, which implies $e(t_0) = -\theta_d$, and $s_0 = 0$. With such a time-shifting sliding surface and controller (23), the sliding condition (8) is still satisfied. The proof is referred to Choi, *et al.* (1993).

Eqn (29) gives the velocity when the system is in the region of sliding motion. Using the above modification, Fig. 3 shows exactly the desired velocity profile for the SMC of the point-to-point control system.

One significant advantage of choosing such a sliding surface is that it is easy to take into account the constraints on maximum acceleration and velocity related to the actuator constraints. In fact, ω_p equals approximately the maximum velocity ω_{max}, i.e.

$$\omega_p \approx \omega_{max} \quad (34)$$

and the maximum acceleration a_{max} is for the two phases, given by

1. Reaching phase, if the shifting parameter $\alpha(\tau)$ is defined by

$$\alpha(\tau) = a_0 \tau \quad (35)$$

then, $$a_0 \approx a_{max} \quad (36)$$

2. Converging phase, the acceleration is actually adjusted by the parameter *c*. The maximum acceleration is calculated as follows:

   When the system is in the neighbourhood of the sliding mode, taking a derivative on both sides of eqn (29) with respect to time and using eqn (10) will turn out to be

$$\ddot{\theta} \approx \frac{\omega_p^2 c \tanh(ce)}{\cosh^2(ce)} \quad (37)$$

It can also be found that when $\tanh(ce) = 1/\sqrt{3}$, $\ddot{\theta}$ reaches its maximum value. Thus, the parameter c should be chosen as

$$c \le \frac{3\sqrt{3}}{2}(a_{max}/\omega_{max}^2) \quad (38)$$

4. EXPERIMENTAL EVALUATION

4.1 The test equipment

The main outline of the test equipment is shown in Fig. 4. and the model parameters in Table 1. A more detailed description can be found in Li, *et al.* (1998). The main control problems are:

- Varying stiffness of the ball-screw 5-10 *Nm/rad.*
- Position dependent friction 10-30% of maximum motor torque.
- No load side position sensor
- Unmodeled resonance frequencies due to other flexible components close to the desired control bandwidth.

TABLE 1. Parameters of theY-axis of SMR

Motor inertia: $J_m = 1.38 \times 10^{-4}$ (kgm^2)
Load inertia: $J_l = 7.45 \times 10^{-4}$ (kgm^2)
Damping coefficient: $d = 0.02$ *(Nm/rad/s)*
Stiffness: $k_t = 5 \sim 10$ *(Nm/rad)*
Torque constant: $K_m = 0.356$ *(Nm/A)*,
Screw pitch: p=0.040 *m*

It is already known from previous experiments that by using only a conventional velocity PI controller, it is difficult to achieve the required control goal. The nonlinear controller used in Li, *et al.* (1998) gives almost equivalent performance comparing with the results presented here, however lacking the theoretical explanation.

The nominal model for the SMR used in the proposed controller is

$$\ddot{y} = ku(t) \quad (39)$$

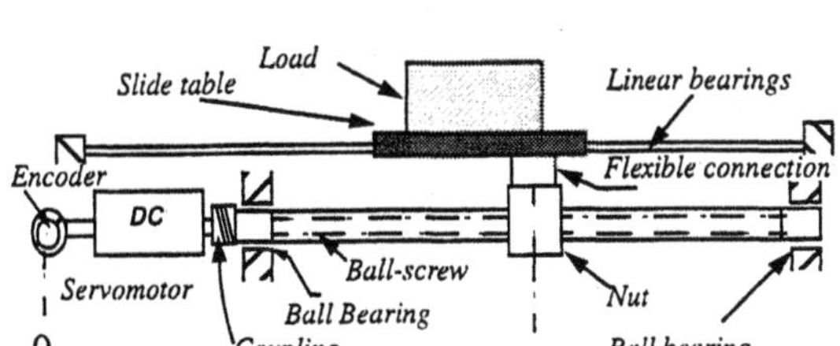

Fig.4. The Y-axis of the MYDATA surface mount robot (SMR)

where, $u(t)$ is input motor current, $y = \frac{p}{2\pi}\theta$ is the linear position transformed from motor angle θ, p is the pitch of the screw. Hence,

$$k = (p/2\pi)K_m/J_m \quad (40)$$

Since the desired positioning distance y_d may vary because of the purpose of SMR, the maximum velocity ω_{max} is determined by y_d according to a triangular velocity profile, as in the following eqn (42). Considering the motor as an ideal torques source, the triangular can be regarded as an ideal time optimal velocity profile refer to the nominal model (39), as it has the shortest arrival time t_f under a constant acceleration. It is easy to calculate t_f,

$$t_f = 2\sqrt{y_d/a_{max}} \quad (41)$$

Therefore ω_{max} is defined by

$$\omega_{max} = a_{max}t_f/2 = \sqrt{a_{max}y_d} \quad (42)$$

Normally ω_{max} is below the maximum velocity since the typical positioning distances y_d of the SMR is relatively short. In case ω_{max} exceeds the limitation, then the value of the maximum velocity is used. Thus, the predetermined sliding surface is chosen as

$$s = \dot{y} + \omega_{max}\tanh(ce) \quad (43)$$

where $e = y - y_d$, with c from eqn (38) as:

$$c = \frac{3\sqrt{3}}{2y_d} \quad (44)$$

The shifting parameter in eqn (35) is chosen such that $a_0 = a_{max} = 6.25\ m/s^2$, $\tau = T_s = 0.4$ ms, where T_s is the sampling time used in the experiments. A first order filter is applied for filtering the estimated disturbance. The Laplace transform of the filter is $F(s) = 1/(0.002s + 1)$. The remaining parameters are chosen as K_1=K_2=1.62, k_i=324, $\rho = 0.5x10^{-3}$ *m* and $y_d = 0.01$ *m* positioning is performed in each experiment.

4.2 The experimental results

Fig. 5, and Fig. 6 show the experimental results of SMC using NTSS without DOB and with DOB respectively. It can be seen that in the first case, the unmodeled dynamics is excited and large chattering occurs. The second case shows very good performance. The controlled system reaches the required accuracy +-4 μm with the total positioning time 0.18 *s*, and goes into +-2 μm with 0.23 *s*. It is also interesting to investigate the first reaching time to the desired position. From Fig. 7 it can be seen that the SMC with DOB needs a little longer time, 0.14 *s*, to reach the position, compared to 0.08 *s* in the case without DOB,

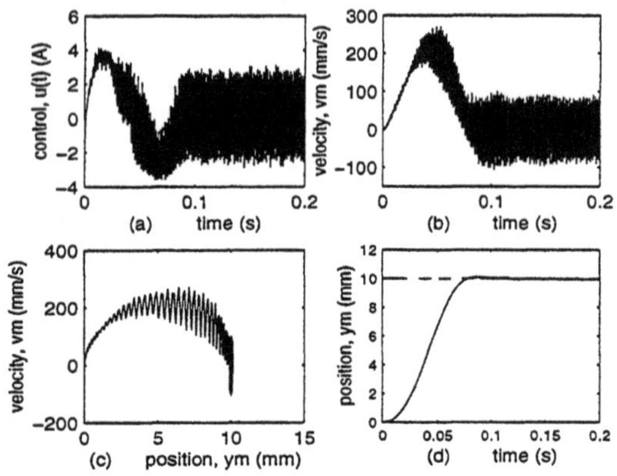

Fig.5. Conventional SMC without disturbance observer; (a) control input, (b) velocity output, (c) phase plant, (d) position output.

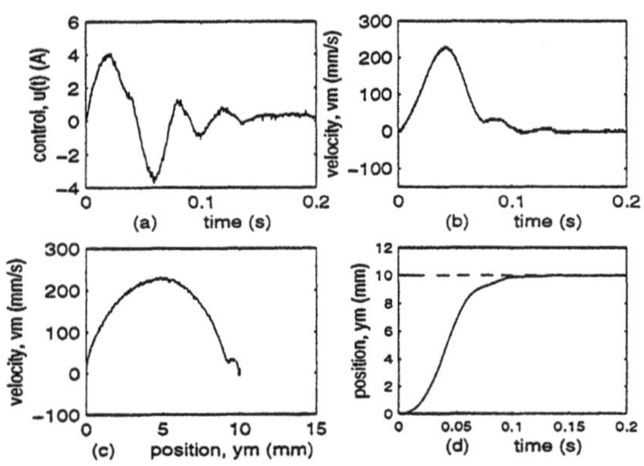

Fig.6. SMC with disturbance observer; (a) control input, (b) velocity output, (c) phase plant, (d) position output.

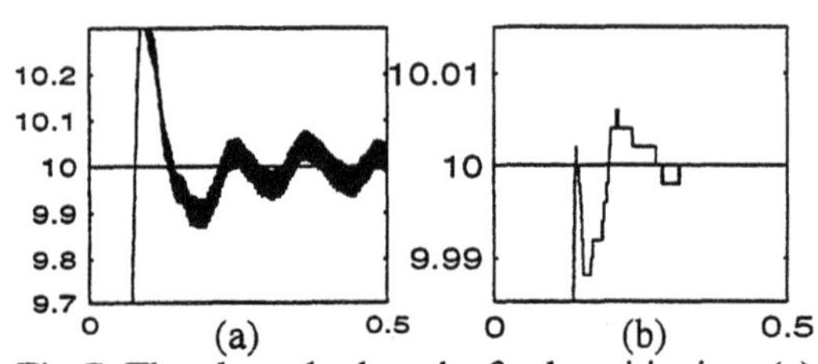

Fig.7. The closer look at the final positioning, (a) without DOB; (b) with DOB.

which equals the ideal positioning time t_f calculated by eqn (41) for a 0.01 *m* positioning distance.

Remarks:

K_1 and K_2 are experimentally determined for this particular system, the values are limited in order to avoid exciting the high frequency modes.

In the control law (23), the term $v'(e)\theta$ is actually an additional velocity feedback, the feedback gain is a function of position error. Therefore, if the value of $v(e)$ is fixed and incorporated into the first term K_1, this will not violate the attractive condition of the sliding mode.

The use of NTSS forces the system to always operate in the neighbouring region of either sub-sliding or sliding manifold. Since the final steady-state error is eliminated by the integral term, one can actually increase the boundary parameter ρ, bringing the controlled system inside the boundary layer for almost all of the time. Increasing ρ actually increases the robustness of the controller, however, it may slow down the convergence to the final position. The DOB also delays the reaching to the desired position which has been seen from the experimental results. This is the price needed to pay for the robustness. In fact, if ρ is chosen large enough, then the switching term in the control law (23) can be omitted, this yields a continuous but nonlinear control which is very similar to the one used in Li, *et al.* (1998), i.e, the control reduces to a decoupled PI controller in the velocity loop.

5. CONCLUSIONS

The SMC approach to control of uncertain dynamical systems has been studied for a long time. However, many of the results are only limited to simulations. On the other hand, when a conventional SMC is applied to real controlled processes, undesired vibrations may occur because of disturbance and actuator limitations, especially when there are flexible components in the processes. This paper presents a SMC design, by using a DOB to deal with the internal disturbance caused by flexibility and other external disturbances, and by selecting a NTSS to take into account the constraints of the process related to the actuator limitations. Experiments are performed at different position of the SMR with the varying stiffness of the ball-screw and position dependent friction. The results show that the proposed controller is less sensitive to the dynamic uncertainties and it has strong robustness regarding both stability and performance.

6. REFERENCE

Choi, S-B., C-C. Cheong and D-W. Park (1993). Moving switching surfaces for robust control of second-order variable structure systems. *International Journal of Control*, **Vol.** 58, **No.** 1, pp 229-245.

Ghosh-Roy, R. J. Mousra and N. Olgac (1997). Implementation of sliding mode control using the concept of perturbation. *Mechatronics* **Vol.** 7, **No.** 8, pp. 723-736.

Li, Y.-F., B. Eriksson and J. Wikander (1998). Robust position control for a DC servomechanism subject to friction and compliance. *Proceeding of the 6th UK Mechatronics Forum International Conference*, Skovde, Sweden. pp117-122.

Slotine, J. E., W. Li (1991). *Applied Nonlinear Control.* Prentice-Hall, Inc, Inglewood Cliffs, New Jersey

Stepanenko, Y and C-Y. Su (1993). Variable structure control of robot manipulators with nonlinear sliding manifold. *International Journal of Control*, **Vol.** 58, **No.** 2, pp 285-300.

14th Triennial World Congress, Beijing, P.R. China

C-2a-05-3

FURTHER RESULTS ON TRACKING CONTROL OF NONHOLONOMIC CHAINED SYSTEMS

Zhong-Ping Jiang* Henk Nijmeijer,[1]**

* *Department of Electrical Engineering, Polytechnic University, Six Metrotech Center, Brooklyn, NY 11201, U.S.A.*
** *Faculty of Mathematical Sciences, Dept of Systems, Signals and Control, University of Twente, P.O. Box 217, 7500 AE Enschede, The Netherlands.*

Abstract: In this paper we deal with the state-feedback tracking control problem for a class of nonholonomic dynamic systems in chained form. By means of the recursive technique proposed in our recent work, we first show that the semiglobal exponential tracking can be easily obtained for systems in chained form. Then, we show that, whenever the control inputs are subject to saturation constraints, a semiglobal asymptotic tracking controller can be found on the basis of Lyapunov redesign and backstepping techniques. *Copyright © 1999 IFAC*

Keywords: Nonlinear control; mechanical systems; tracking; saturation control.

1. INTRODUCTION

Recent years have witnessed a lot of research activities on the control and stabilization of nonholonomic dynamical systems – see the survey paper (Kolmanovsky and McClamroch 1995) and references cited there for practical motivations and various novel approaches. One of the technical burdens for the use of traditional control design methods to solve the stabilization problem is that there is no (time-invariant) continuous state-feedback, because Brockett's necessary condition (Brockett 1983) is not met by such systems. The time-varying feedback appears to be a useful approach to get around the impossibility of stabilizing a nonholonomic system by means of (time-invariant) continuous state-feedback. More recently, the problem of tracking control for nonholonomic dynamical systems has also received attention – see, for instance, (Fierro and Lewis 1995), (Fliess *et al.* 1995), (Kanayama *et al.* 1990), (Murray *et al.* 1992)–(Sørdalen 1993) as well as (Jiang and Nijmeijer 1997*b*)–(Jiang and Nijmeijer 1999).

The purpose of the paper is to continue our papers (Jiang and Nijmeijer 1997*a*, Jiang and Nijmeijer 1999) and to show that the tracking methodology proposed in those papers can be extended to estabilish semiglobal exponential tracking and semiglobal asymptotic tracking with input saturation. As in (Jiang and Nijmeijer 1997*a*, Jiang and Nijmeijer 1999), we focus on a class of nonholonomic systems in chained form, or extended chained form. As we will see below, the goals are achieved through a step-by-step control design procedure and the stability proof is rather simple as in (Jiang and Nijmeijer 1997*a*, Jiang and Nijmeijer 1999) invoking advanced Lyapunov stability theory (Khalil 1996).

Section 2 states the problem and recalls briefly the control method in our recent papers (Jiang and Nijmeijer 1997*a*, Jiang and Nijmeijer 1999). Section 3 gives a result on semiglobal exponential

[1] Corresponding Author. This author is also with the Faculty of Mechanical Engineering, Technical University of Eindhoven, P.O. Box 513, 5600 MB Eindhoven, The Netherlands. h.nijmeijer@math.utwente.nl

stability. Section 4 proposes a modified version of the control design scheme in Section 2 in order to design saturated tracking control laws for our class of nonholonomic systems in chained form. Finally, we conclude the paper in Section 5.

2. PRIOR RESULTS AND PROBLEM STATEMENT

We consider nonholonomic mechanical systems which can be transformed into the following chained form (cf. (Murray *et al.* 1992))

$$\begin{aligned} \dot{x}_1 &= u_1 \\ \dot{x}_2 &= u_2 \\ \dot{x}_3 &= x_2 u_1 \\ &\vdots \\ \dot{x}_n &= x_{n-1} u_1 \end{aligned} \qquad (1)$$

where $x = (x_1, \ldots, x_n) \in \mathbb{R}^n$ is the state, $u = (u_1, u_2) \in \mathbb{R}^2$ is the control input.

We assume, like in (Jiang and Nijmeijer 1997*b*, Jiang and Nijmeijer 1997*a*, Jiang and Nijmeijer 1999), that the reference trajectory $x_d = (x_{1d}, \ldots, x_{nd})$ is generated by a system in chained form:

$$\begin{aligned} \dot{x}_{1d} &= u_{1d} \\ \dot{x}_{2d} &= u_{2d} \\ \dot{x}_{id} &= x_{(i-1)d} u_{1d}\,, \quad 3 \le i \le n \end{aligned} \qquad (2)$$

where $u_d = (u_{1d}, u_{2d})$ is the time-varying reference control input.

Based on this model reference control, several solutions were given in (Jiang and Nijmeijer 1997*a*, Jiang and Nijmeijer 1999) to the following semi-global tracking problem for a class of nonlinear systems in chained form (1).

Definition 1. The *tracking control problem* is said to be *semi-globally* solvable for the system (1) if, given *any* compact set $S \in \mathbb{R}^n$ containing the origin, we can design appropriate Lipschitz continuous time-varying state-feedback controllers of the form

$$u_1 = \mu_1(t, x)\,, \quad u_2 = \mu_2(t, x) \qquad (3)$$

such that, for any initial tracking error $x_e(0) = x(0) - x_d(0)$ in S, all the signals of the closed loop system (1)-(3) are uniformly bounded over $[0, \infty)$. Furthermore, the tracking errors $x_e(t) := x(t) - x_d(t)$ satisfy

$$\lim_{t \to +\infty} |x(t) - x_d(t)| = 0\,. \qquad (4)$$

It should be noted that the *global* tracking problem was solved in our earlier paper (Jiang and Nijmeijer 1997*b*) (see also (Samson and Ait-Abderrahim 1991)) for a two-wheeled mobile robot which kinematic model can be put into a third-order system in chained form (1) via diffeomorphism and feedback.

The purpose of this paper is to address the semiglobal tracking problem whenever the control inputs are subject to saturation constraints:

$$|u_1(t)| \le u_{1\max}\,, \qquad |u_2(t)| \le u_{2\max} \qquad (5)$$

where $u_{1\max} > \sup_{t \ge 0} |u_{1d}(t)|$ and $u_{2\max} > \sup_{t \ge 0} |u_{2d}(t)|$ are given arbitrary constant real numbers.

Before approaching this practically important saturation issue, we first summarize how the semiglobal tracking problem without input saturation was solved in (Jiang and Nijmeijer 1997*a*, Jiang and Nijmeijer 1999) by means of a recursive technique.

Step a : Let $x_e = x - x_d$ and introduce the global change of state coordinates

$$\begin{aligned} y_i &= x_{(n-i+1)e} - x_{n-i} x_{1e}\,, \quad 1 \le i \le n-2 \\ y_{n-1} &= x_{2e} \\ y_n &= x_{1e}\,, \end{aligned} \qquad (6)$$

so that the tracking error x_e-dynamics is transformed into:

$$\begin{aligned} \dot{y}_1 &= u_{1d} y_2 - x_{n-2}(u_1 - u_{1d}) y_n \\ &\vdots \\ \dot{y}_{n-3} &= u_{1d} y_{n-2} - x_2 (u_1 - u_{1d}) y_n \\ \dot{y}_{n-2} &= u_{1d} y_{n-1} - u_2 y_n \\ \dot{y}_{n-1} &= u_2 - u_{2d} \\ \dot{y}_n &= u_1 - u_{1d} \end{aligned} \qquad (7)$$

Clearly, for *any* vector $(x_{2d}, \ldots, x_{(n-1)d})$, the tracking error x_e is zero if and only if $y = (y_1, \ldots, y_n)$ is equal to zero. With this in mind, the state feedback semiglobal tracking problem for the system (1) has been translated into a state feedback semiglobal *stabilization* problem for the transformed time-varying system (7).

Step b : Using the currently popular integrator backstepping methodology (cf. (M. Krstić and Kokotović 1995)), we find a new change of state coordinates

$$\begin{aligned} \bar{y}_1 &= y_1\,, \quad \bar{y}_n = y_n\,, \quad 2 \le i \le n-1 \\ \bar{y}_i &= y_i - \alpha_{i-1}(y_1, \ldots, y_{i-1}, \phi_1, \ldots, \phi_{i-1}) \end{aligned} \qquad (8)$$

where the functions ϕ_i's and α_i's are defined by

$$\phi_1(u_{1d}) = u_{1d}^{2\ell+1} , \tag{9}$$

$$\phi_i = \frac{1}{u_{1d}}\frac{d}{dt}\phi_{i-1}(u_{1d}, \ldots, u_{1d}^{(i-2)})) \tag{10}$$

$$2 \le i \le n-2 ,$$

$$\alpha_1(y_1, \phi_1) = -c_1\phi_1(u_{1d})y_1 , \tag{11}$$

$$\alpha_i = -c_i\phi_1\overline{y}_i - \overline{y}_{i-1} + \sum_{k=1}^{i-1}\frac{\partial\alpha_{i-1}}{\partial\phi_k}\phi_{k+1} + \sum_{k=1}^{i-1}\frac{\partial\alpha_{i-1}}{\partial y_k}y_{k+1} \tag{12}$$

with $\ell \ge n-3$ a nonnegative integer and the c_i's positive real numbers.

In the new coordinates, the system (7) reads as

$$\begin{aligned}
\dot{\overline{y}}_1 &= -u_{1d}^{2\ell+2}\overline{y}_1 + u_{1d}\overline{y}_2 \\
&\quad -x_{n-2}(u_1 - u_{1d})\overline{y}_n \\
&\vdots \\
\dot{\overline{y}}_i &= -u_{1d}^{2\ell+2}\overline{y}_i - u_{1d}\overline{y}_{i-1} + u_{1d}\overline{y}_{i+1} \\
&\quad - \left[x_{n-i-1} - \sum_{k=1}^{i-1}\frac{\partial\alpha_{i-1}}{\partial y_k}x_{n-k-1}\right] \\
&\quad \times(u_1 - u_{1d})\overline{y}_n \qquad (13) \\
&\vdots \\
\dot{\overline{y}}_{n-2} &= -u_{1d}^{2\ell+2}\overline{y}_{n-2} - u_{1d}\overline{y}_{n-3} \\
&\quad + u_{1d}\overline{y}_{n-1} - u_2\overline{y}_n \\
&\quad + \sum_{k=1}^{n-3}\frac{\partial\alpha_{n-3}}{\partial y_k}x_{n-k-1}(u_1 - u_{1d})\overline{y}_n \\
\dot{\overline{y}}_{n-1} &= u_2 - u_{2d} - \dot{\alpha}_{n-2} \\
\dot{\overline{y}}_n &= u_1 - u_{1d}
\end{aligned}$$

It has been shown in (Jiang and Nijmeijer 1997*a*, Jiang and Nijmeijer 1999) that the tracking problem is semiglobally solved for the system (1) if the following tracking controllers are chosen:

$$u_2 = u_{2d} - c_{n-1}\overline{y}_{n-1} - u_{1d}\overline{y}_{n-2} + \sum_{i=1}^{n-2}\frac{\partial\alpha_{n-2}}{\partial\phi_i}\dot{\phi}_i + \sum_{i=1}^{n-2}\frac{\partial\alpha_{n-2}}{\partial y_i}u_{1d}y_{i+1} \tag{14}$$

$$u_1 = u_{1d} + (\lambda - \Delta_1)^{-1}\left[-c_n\overline{y}_n + \left(\overline{y}_{n-2} - \overline{y}_{n-1}\frac{\partial\alpha_{n-2}}{\partial y_{n-2}}\right)u_2\right] \tag{15}$$

where c_{n-1}, c_n, $\lambda > 0$ and Δ_1, a smooth function of $(y_1, \ldots, y_{n-1}, u_{1d}, \ldots, u_{1d}^{(n-3)})$. More precisely, the following result can be established.

Proposition 1. **(Jiang and Nijmeijer 1997*a*)** Assume that $x_{id}(t)$ $(2 \le i \le n-1)$, $u_d(t)$ and its derivatives $u_{1d}^{(i)}(t)$ up to order $n-2$ are bounded on $[0, \infty)$. If $u_{1d}(t)$ does not go to zero as $t \to +\infty$, given any compact neighborhood S of the origin in $\mathbb{R}^n$, we set $c_i > 0$ for all $1 \le i \le n$ and can find a sufficiently large $\lambda > 0$ so that, for any initial tracking conditions $x_e(0)$ in S, the solutions of the resulting closed-loop system (7), (14) and (15) and the tracking errors $x_e(t)$ are uniformly bounded. In particular,

$$\lim_{t\to+\infty}|x(t) - x_d(t)| = 0 \tag{16}$$

It was shown in (Jiang and Nijmeijer 1999) that the conditions of Proposition 1 on the reference states and inputs can be weakened to guarantee semiglobal asymptotic tracking. It was also shown in (Jiang and Nijmeijer 1997*a*, Jiang and Nijmeijer 1999) that exponential tracking is achieved under the extra condition that $\liminf_{t\to+\infty}|u_{1d}(t)| > 0$. Indeed, we show next that a *semiglobal* exponential tracking result can be ensured under the same condition.

3. SEMIGLOBAL EXPONENTIAL TRACKING

As mentioned, the semiglobal asymptotic tracking and local exponential tracking results were based on the systematic control design scheme which we recalled in Section 2. We show that the same design procedure implies a semiglobal exponential tracking result if we choose the design parameters c_i's appropriately. More specifically,

Proposition 2. Under the conditions of Proposition 1, if the reference input u_{1d} satisfies the property that $\liminf_{t\to\infty}|u_{1d}(t)| > 0$, then given any compact neighborhood S of the origin in $\mathbb{R}^n$, we can pick a sufficiently large $\lambda > 0$ such that, for any initial tracking conditions $x_e(t_0)$ in S, the tracking errors $x_e(t)$ satisfy

$$|x_e(t)| \le k_1e^{-k_2(t-t_0)}|x_e(t_0)| , \tag{17}$$

$\forall t \ge t_0 \ge 0$, where k_1 and k_2 are positive real numbers (which may depend on S).

Proof. Consider the Lyapunov function candidate

$$V(\overline{y}) = \frac{1}{2}\overline{y}_1^2 + \cdots + \frac{1}{2}\overline{y}_{n-1}^2 + \frac{1}{2}\overline{y}_n^2 \tag{18}$$

where $\overline{y} := (\overline{y}_1, \ldots, \overline{y}_{n-1}, \overline{y}_n)$.

It is easy to verify that the time derivative of V along the solutions of (13), (14) and (15) satisfies

$$\dot{V} \leq -\sum_{i=1}^{n-2} c_i \overline{y}_i^2 u_{1d}^{2\ell+2} - c_{n-1}\overline{y}_{n-1}^2 - c_n \overline{y}_n^2 \quad (19)$$

By assumptions, there exist a constant $l_u > 0$ and a finite time instant $t^* \geq 0$ such that $|u_{1d}(t)| \geq l_u$ for all $t \geq t^*$. Define a positive real number σ by

$$\sigma = \min\{c_1 l_u^{2\ell+2}, \ldots, c_{n-2} l_u^{2\ell+2}, c_{n-1}, c_n\}$$

Then, (19) implies

$$\dot{V}(t, \overline{y}) \leq -2\sigma V(\overline{y}), \quad \forall t \geq t^* \quad (20)$$

Hence, it follows from Gronwall's lemma (Khalil 1996) together with (19) that, for any pair of time instants $t \geq t_0 \geq 0$,

$$V(\overline{y}(t)) \leq e^{2\sigma t^*} e^{-2\sigma(t-t_0)} V(\overline{y}(t_0)) \quad (21)$$

In other words, there exist two positive constants k_3 and k_4 such that, $\forall t \geq t_0 \geq 0$

$$|\overline{y}(t)| \leq k_3 e^{-k_4(t-t_0)} |\overline{y}(t_0)| \quad (22)$$

It is important to note that this inequality is true for all solutions $x_e(t)$ initiated from any given compact set S, which are bounded because of Proposition 1 in (Jiang and Nijmeijer 1997*a*).

By the definition of $\overline{y}$ and y in (8) to (12) and (6) respectively, it follows using a recursive argument that there exist two constants k_1 and k_2 (which may depend on S) such that for every initial tracking error $x_e(t_0)$ in S, the corresponding solution $x_e(t)$ satisfies $(t \geq t_0 \geq 0)$

$$|x_e(t)| \leq k_1 e^{-k_2(t-t_0)} |x_e(t_0)| \quad (23)$$

The proof of Proposition 2 is therefore completed.

We close this section by another result on semiglobal exponential tracking which is obtained via a simpler approach and is of independent interest.

In this case, we assume that the sign of $u_{1d}(t)$ is constant and known for all $t \geq 0$. Without loss of generality, we assume that $u_{1d}(t) \geq 0$ and $\liminf_{t\to\infty} u_{1d}(t) > 0$.

Let $(a_1, \ldots, a_{n-1})$ be a vector so that

$$A = \begin{pmatrix} 0 & 1 & 0 & \cdots & 0 \\ 0 & 0 & 1 & \cdots & 0 \\ \vdots & \vdots & \vdots & \vdots & \vdots \\ 0 & 0 & 0 & \cdots & 1 \\ a_1 & a_2 & a_3 & \cdots & a_{n-1} \end{pmatrix} \quad (24)$$

is an $(n-1) \times (n-1)$ stable matrix.

Let $P > 0$ be the solution to the Lyapunov matrix equation

$$PA + A^T P = -I \quad (25)$$

Choose

$$u_2 = u_{2d} + u_{1d} \sum_{i=1}^{n-1} a_i y_i \quad (26)$$

$$u_1 = u_{1d} - \left(\lambda - \sum_{i=1}^{n-1}\sum_{j=2}^{n-2} b_{ij} y_i x_j\right)^{-1} \times \quad (27)$$

$$\left(-c_n y_n - \sum_{i=1}^{n-1} b_{i1} y_i u_2\right) \quad (28)$$

with $c_n > 0$, $b_{ij} \in \mathbb{R}$. Differentiating the function

$$W = (y_1, \ldots, y_{n-1}) P (y_1, \ldots, y_{n-1})^T + \frac{\lambda}{2} y_n^2$$

with respect to the solutions of (7) yields

$$\dot{W} = -u_{1d} |(y_1, \ldots, y_{n-1})|^2 - c_n y_n^2 \quad (29)$$

Finally, the same semiglobal exponential tracking result can be similarly proved as in Proposition 2.

Proposition 3. Assume that $x_{id}(t)$ $(2 \leq i \leq n-1)$, $u_d(t)$ and its derivative $\dot{u}_{1d}(t)$ are bounded on $[0, \infty)$. If $u_{1d}(t) \geq 0$ for all $t \geq 0$ and $\liminf_{t\to\infty} u_{1d}(t) > 0$, then given any compact neighborhood S of the origin in $\mathbb{R}^n$, we can pick a sufficiently large $\lambda > 0$ such that, for any initial tracking conditions $x_e(t_0)$ in S, the tracking errors $x_e(t)$ satisfy

$$|x_e(t)| \leq k_5 e^{-k_6(t-t_0)} |x_e(t_0)|, \quad (30)$$

$\forall t \geq t_0 \geq 0$, where k_5 and k_6 are positive real numbers (which may depend on S).

Remark 1. Note that u_2 in (26) is a linear feedback control law. It is also of interest to note that the semiglobal exponential convergence of the tracking errors $x_e(t)$ can be even guaranteed by application of a *linear* feedback law of the type

$$u_1 = u_{1d} - c_n y_n \quad (31)$$

4. BOUNDED STATE-FEEDBACK TRACKING

The main purpose of this section is to show how to modify the control design procedure in Section 2 so that a bounded state-feedback law can be designed to achieve semiglobal asymptotic tracking. As in (Jiang and Nijmeijer 1997*a*, Jiang and Nijmeijer 1999), such a semiglobal asymptotic tracking controller with saturation constraint (5) is found using a stepwise scheme based on the transformed system (7), which we formulate below.

Step 1 : We begin with the y_1-subsystem of (7) with y_2 viewed as the virtual control input. Let $z_1 = y_1$ and $V_1 := \frac{1}{2}z_1^2$. The time derivative of V_1 along the solutions of (7) is:

$$\dot{V}_1 = z_1 y_2 u_{1d} - z_1 x_{n-2}(u_1 - u_{1d})y_n \quad (32)$$

This leads us to choose a virtual control function like $\alpha_1 = 0$ and introduce a new variable $z_2 = y_2 - \alpha_1$. Consider the function

$$W_1 = \varepsilon_1 \log(1 + V_1(z_1)) \quad (33)$$

where $\varepsilon_1 > 0$ is a design parameter.

Then, (32) gives

$$\dot{W}_1 = \frac{\varepsilon_1 z_1}{1 + V_1(z_1)}(z_2 u_{1d} - x_{n-2}(u_1 - u_{1d})y_n)$$

Step 2 : Consider the (y_1, y_2)-subsystem of (7) with y_3 viewed as the virtual control input. Let $V_2 := W_1 + \frac{1}{2}z_2^2$. Differentiating V_2 with respect to the time yields

$$\dot{V}_2 = \left(y_3 + \frac{\varepsilon_1 z_1}{1 + V_1(z_1)}\right) z_2 u_{1d} - y_n(u_1 - u_{1d}) \times \left(\frac{\varepsilon_1 z_1 x_{n-2}}{1 + V_1(z_1)} + y_2 x_{n-3}\right)$$

This leads us to choose a virtual control function $y_3 = \alpha_2(y_1)$ and a new variable as

$$\alpha_2 = -\frac{\varepsilon_1 z_1}{1 + V_1(z_1)}, \quad z_3 = y_3 - \alpha_2(y_1)$$

As a consequence, given any $\varepsilon_2 > 0$, the time derivative of the positive definite and proper function

$$W_2 = \varepsilon_2 \log(1 + V_2(z_1, z_2)) \quad (34)$$

satisfies

$$\dot{W}_2 = \frac{\varepsilon_2 z_2 z_3 u_{1d}}{1 + V_2(z_1, z_2)} - \left(\frac{\varepsilon_2 z_2 x_{n-3}}{1 + V_2(z_1, z_2)} + \frac{\varepsilon_1 \varepsilon_2 z_1 x_{n-2}}{(1 + V_1(z_1))(1 + V_2(z_1, z_2))}\right)(u_1 - u_{1d})y_n$$

Step $n-1$: As a direct application of backstepping (cf. (Jiang and Nijmeijer 1999)), we obtain a function

$$V_{n-1} = W_{n-2}(z_1, \ldots, z_{n-2}) + \frac{1}{2}z_{n-1}^2 \quad (35)$$

which satisfies

$$\dot{V}_{n-1} = z_{n-1}\left(u_2 - u_{2d} + \frac{\varepsilon_{n-2} z_{n-2} u_{1d}}{1 + V_{n-2}(Z_{n-2})} - \sum_{i=1}^{n-2} \frac{\partial \alpha_{n-2}}{\partial y_i} u_{1d} y_{i+1}\right) \quad (36)$$
$$-\Delta_2(Z_2, \widetilde{x})(u_1 - u_{1d})y_n + \left(z_{n-1}\frac{\partial \alpha_{n-2}}{\partial y_{n-2}} - \frac{\varepsilon_{n-2} z_{n-2}}{1 + V_{n-2}(Z_{n-2})}\right) u_2 y_n$$

where $\widetilde{x} = (x_2, \ldots, x_{n-2})$.

Therefore, by choosing the control law u_2 as

$$u_2 = u_{2d} - f_1(z_{n-1}) - \frac{\varepsilon_{n-2} z_{n-2} u_{1d}}{1 + V_{n-2}(Z_{n-2})} + \sum_{i=1}^{n-2} \frac{\partial \alpha_{n-2}}{\partial y_i} u_{1d} y_{i+1} \quad (37)$$

where $f_1 : \mathbb{R} \to \mathbb{R}$ is any bounded, "first-sector third-sector" function (i.e. $f_1(0) = 0$ and $z f_1(z) > 0$ for all $z \neq 0$.), (36) yields:

$$\dot{V}_{n-1} = -z_{n-1} f_1(z_{n-1}) + \left(z_{n-1}\frac{\partial \alpha_{n-2}}{\partial y_{n-2}} - \frac{\varepsilon_{n-2} z_{n-2}}{1 + V_{n-2}(Z_{n-2})}\right) u_2 y_n - \Delta_2(Z_2, \widetilde{x})(u_1 - u_{1d})y_n \quad (38)$$

Step n : Now, consider the Lyapunov function for the whole system (7)

$$V_n = V_{n-1}(z_1, \ldots, z_{n-1}) + \frac{\lambda}{2}z_n^2 \quad (39)$$

where $\lambda > 0$ is a design parameter.

As in (Jiang and Nijmeijer 1997*a*, Jiang and Nijmeijer 1999), whenever choosing the tracking controller u_1 as

$$u_1 = u_{1d} + (\lambda - \Delta_2)^{-1}\left[-f_2(z_n)u_2 \cdot - \left(z_{n-1}\frac{\partial \alpha_{n-2}}{\partial y_{n-2}} - \frac{\varepsilon_{n-2} z_{n-2}}{1 + V_{n-2}(Z_{n-2})}\right)\right] \quad (40)$$

with f_2 arbitrary bounded, "first-sector third-sector" function, the time derivative of V_n along the solutions of (7) gives

$$\dot{V}_n = -z_{n-1} f_1(z_{n-1}) - z_n f_2(z_n) \quad (41)$$

Finally, we can prove the following result using the same type of arguments as in (Jiang and Nijmeijer 1999, Proof of Prop. 1).

Proposition 4. Assume that $x_{id}(t)$ $(2 \leq i \leq n-1)$, $u_d(t)$ and $\dot{u}_{1d}(t)$ are bounded for all $t \in [0, \infty)$. If $u_{1d}(t)$ does not go to zero as $t \to +\infty$, the semiglobal asymptotic tracking problem is solvable for system (7) subject to input saturation (5). More precisely, given any compact neighborhood S of the origin in $\mathbb{R}^n$, we can find suitable design functions f_1, f_2 and tune the design parameters λ, ε_1 to ε_{n-2} such that, for any initial tracking conditions $x_e(0)$ in S, the input constraint (5) holds while the solutions of the resulting closed-loop system (7), (37) and (40) satisfy (5) and the tracking errors $x_e(t)$ converge to zero as $t \to \infty$.

Remark 2. As in (Jiang and Nijmeijer 1999), exponential stability of the closed-loop system can be guaranteed provided the reference input u_d satisfies the conditions of (Jiang and Nijmeijer 1999, Corollary 2). It is also worth noting that extension to a simplified dynamic model as introduced in (Jiang and Nijmeijer 1999) is rather straightforward.

5. CONCLUSION

In this paper we have shown that recent results in our work (Jiang and Nijmeijer 1997*a*, Jiang and Nijmeijer 1999) can be easily extended to the case of semiglobal exponential tracking and tracking under input constraints. As before, the desired tracking control laws have been designed on the basis of a recursive control design procedure. The semiglobal results presented in this paper are meaningful from a practical point of view because chainability is in general not global for nonlinear mechanical systems with nonholonomic constraints. Nevertheless, our results have shown the boundedness property for the solutions in any (bounded) domain where we have chainability. Last but not the least it is more or less direct to extend our results in this paper to systems in extended chained form, or dynamic model.

6. REFERENCES

Brockett, R. W. (1983). Asymptotic stability and feedback stabilization. *Differential Geometric Control Theory,* R.W. Brockett, R.S. Millman and H.J. Sussmann, eds. pp. 181–191.

Fierro, R. and F. L. Lewis (1995). Control of a nonholonomic mobile robot: backstepping kinematics into dynamics. *Proc. 34th IEEE Conf. Dec. Control* pp. 3805–3810.

Fliess, M., J. Levine, P. Martin and P. Rouchon (1995). Design of trajectory stabilizing feedback for driftless flat systems. *Proc. 3rd ECC* pp. 1882–1887.

Jiang, Z. P. and H. Nijmeijer (1997*a*). Backstepping-based tracking control of nonholonomic chained systems. *ECC'97,* Paper No. 672.

Jiang, Z. P. and H. Nijmeijer (1997*b*). Tracking control of mobile robots: a case study in backstepping. *Automatica* **33**, 1393–1399.

Jiang, Z. P. and H. Nijmeijer (1999). A recursive technique for tracking control of nonholonomic systems in chained form. *IEEE Trans. Automat. Control.*

Kanayama, Y., Y. Kimura, F. Miyazaki and T. Noguchi (1990). A stable tracking control scheme for an autonomous mobile robot. *Proc. IEEE 1990 Int. Conf. on Robotics and Automation* pp. 384–389.

Khalil, H. K. (1996). *Nonlinear Systems.* 2nd Ed., Prentice Hall. Upper Saddle River.

Kolmanovsky, I. and N. H. McClamroch (1995). Developments in nonholonomic control systems. *IEEE Control Systems Magazine* **15**, 20–36.

M. Krstić, I. Kanellakopoulos and P. V. Kokotović (1995). *Nonlinear and Adaptive Control Design.* John Wiley & Sons. New York.

Murray, R. M., G. Walsh and S. Sastry (1992). Stabilization and tracking for nonholonomic control systems using time-varying state feedback. *IFAC Nonlinear Control Systems Design* pp. 109–114.

Samson, C. and K. Ait-Abderrahim (1991). Feedback control of a nonholonomic wheeled cart in cartesian space. *Proc. 1991 IEEE Int. Conf. Robotics and Automation* pp. 1136–1141.

Sørdalen, O. J. (1993). *Feedback Control of Nonholonomic Mobile Robots.* Ph.D. thesis, Norwegian Institute of Technology.

STATE AND OUTPUT FEEDBACK DESIGN FOR ROBUST TRACKING OF LINEAR SYSTEMS WITH RATE LIMITED ACTUATORS

Zongli Lin * **Meir Pachter** ** **Siva Banda** *** **Yacov Shamash** ****

* *Department of Electrical Engineering, University of Virginia, Charlottesville, VA 22903, USA; Email: zl5y@virginia.edu*
** *Department of EECS, Air Force Institute of Technology, Wright-Patterson AFB, OH 45433, USA*
*** *Flight Dynamics Dir.(WL/FIGC), Write Laboratory, Wright-Patterson AFB, OH 45433, USA*
**** *College of Engineering and Applied Science, SUNY at Stony Brook, Stony Brook, NY 11794-2200, USA.*

Abstract: A design technique (Lin *et al.*, 1997*b*) recently proposed for stabilization of a linear system with rate-limited actuators is utilized to design feedback laws that cause the system output to track a desired command signal. This design technique combines two design techniques recently developed for linear systems with position limited actuators, piecewise-linear LQ control (Wredenhagen and Belanger, 1994) and low-and-high gain feedback (Saberi *et al.*, 1996), and hence takes advantage of both design techniques, while avoiding their disadvantages. In the case that only the output is available for feedback, the performance of the state feedback law is preserved by the use of a fast observer. An open-loop exponentially unstable fighter aircraft is used to demonstrate the effectiveness of the proposed control design method.

Keywords: actuator rate saturation, piecewise-linear LQ control, low-and-high gain feedback, tracking, flight control

1. INTRODUCTION

Every physical actuator is subject to constraints. These constraints include both position and rate saturation. In the past few years there has been much interest concerning the stabilization of linear systems with position saturating actuators, resulting in several promising design techniques. In this paper, we will recourse to the low-and-high gain (LHG) design technique (Saberi *et al.*, 1996) and the piecewise linear LQ control (PLC) design technique (Wredenhagen and Belanger, 1994). Design techniques that are conceptually similar to the PLC design, the high performance bounded control (HPB), the scheduled low gain feedback and scheduled low-and-high gain feedback were recently developed in (Miotto *et al.*, 1996), (Megretski, 1996) and (Lin, 1998), respectively. Additional related work might be found in the above mentioned papers and in (Bernstein and Michel, 1995).

Recently, a method of designing stabilizing feedback control laws for linear systems taking into account the effect of actuator rate saturation was proposed in (Lin *et al.*, 1997*b*). The proposed feedback design, referred to as the combined PLC/LHG state feedback design method, combines the two recently developed design techniques, the PLC and the LHG design techniques, both originally proposed for linear systems with actuator position saturation. It inherits the advantages of both design techniques, while avoiding their disadvantages. Thus, exact knowledge of the dynamics of the actuators will not be needed and the actuator disturbances can be rejected.

The objective of this paper is to utilize the combined PLC/LHG state feedback design technique developed in (Lin *et al.*, 1997*b*) for stabilization to design state feedback laws that achieve robust tracking performance. The signals to be tracked are modeled as the state of a reference system and the tracking control problem is formulated as an output regulation problem ((Francis, 1984), (Lin *et al.*, 1996)). In the case that only the output is available for feedback, the performance of the state feedback laws is preserved by the use of a fast observer. This is a generalization of (Lin *et al.*, 1997*a*), a preliminary version of this paper, where only state feedback result is presented. It is also noted here that the PLC design method as originally proposed in (Wredenhagen and Belanger, 1994) is only for state feedback. Although the LHG output feedback design is given in (Saberi *et al.*, 1996), its design and analysis do not carry through in the combined PLC/LHG design, since the closed-loop system under the combined PLC/LHG state feedback law is discontinuous. Indeed, discontinuity in the PLC/LHG state feedback laws is the primary source of difficulty in its implementation with observer state.

2. PROBLEM FORMULATION

Consider the linear dynamical system

$$\begin{cases} \dot{x} = Ax + Bv, \quad x(0) \in \mathcal{X} \subset \mathbb{R}^n \\ \dot{v} = \mathrm{sat}_\Delta(-Tv + Tu + d), \\ \qquad\qquad v(0) \in \mathcal{V} \subset \mathbb{R}^m \\ \dot{w} = Sw, \quad w(0) \in \mathcal{W} \subset \mathbb{R}^s \\ e = Cx + Qw \\ y = Ex \end{cases} \tag{1}$$

where the first equation describes the plant, with state $x \in \mathbb{R}^n$, input $u \in \mathbb{R}^m$ and output $y \in \mathbb{R}^q$, the second equation represents actuator dynamics with state $v \in \mathbb{R}^m$ and is subject to disturbance $d \in \mathbb{R}^m$ and rate saturation $\mathrm{sat}_\Delta(\cdot)$, the third equation defines the reference signal $-Qw \in \mathbb{R}^p$ to be tracked by the plant output $Cx \in \mathbb{R}^p$. For $\Delta = (\Delta_1, \Delta_2, \cdots, \Delta_m)$, $\Delta_i > 0$, the actuator rate saturation function $\mathrm{sat}_\Delta : \mathbb{R}^m \to \mathbb{R}^m$ is the standard saturation function, i.e., $\mathrm{sat}_\Delta(v) = [\mathrm{sat}_{\Delta_1}(v_1), \mathrm{sat}_{\Delta_2}(v_2), \cdots, \mathrm{sat}_{\Delta_m}(v_m)]$, $\mathrm{sat}_{\Delta_i}(v_i) = \mathrm{sign}(v_i)\min\{\Delta_i, |v_i|\}$, the positive definite diagonal matrices $T = \mathrm{diag}(\tau_1, \tau_2, \cdots, \tau_m)$ represents the "time constants" of the actuators. Finally, all three sets $\mathcal{X}$, $\mathcal{V}$ and $\mathcal{W}$ are bounded.

Assumption 1. The pair (A, B) is stabilizable;

Assumption 2. The eigenvalues of S have nonnegative real parts.

Assumption 3. The disturbance is uniformly bounded by a known (arbitrarily large) constant D, i.e., $\|d(t)\| \leq D$, $\forall t \geq 0$.

Assumption 4. The pair (A, E) is observable.

Definition 1. The data $(D, \mathcal{Z}_\infty, \mathcal{E}_\infty)$ is said to be *admissible for state feedback [output feedback]* if D is a nonnegative real number, and $\mathcal{Z}_\infty$ and $\mathcal{E}_\infty$ are respectively a subset of $\mathbb{R}^{n+m}$ $[\mathbb{R}^{2n+m}]$ and $\mathbb{R}^p$ both containing the origin as an interior point.

Problem 2. [State feedback robust tracking problem.] Given the data $(D, \mathcal{Z}_\infty, \mathcal{E}_\infty)$, admissible for state feedback, find, if possible, a state feedback law $u = F(x, v, w)$, such that the closed-loop system satisfies,

(1) every trajectory of the system
$$\begin{cases} \dot{x} = Ax + Bv \\ \dot{v} = \mathrm{sat}_\Delta(-Tv + TF(x, v, 0) + d) \end{cases} \tag{2}$$
starting from $\mathcal{X} \times \mathcal{V}$ enters $\mathcal{Z}_\infty$ in a finite time and remains in it thereafter.

(2) for all $x(0) \in \mathcal{X}$, $v(0) \in \mathcal{V}$ and $w(0) \in \mathcal{W}$, the solution of the closed-loop system is such that $e(t)$ enters the set $\mathcal{E}_\infty$ in a finite time and remains in it thereafter.

Problem 3. [Output feedback robust tracking problem.] Given the data $(D, \mathcal{Z}_\infty, \mathcal{E}_\infty)$, admissible for output feedback, find, if possible, an output feedback law
$$\begin{cases} \dot{\psi} = \alpha(\psi, y), \quad \psi(0) \in \mathcal{X} \subset \mathbb{R}^n \\ u = F(\psi, v, w) \end{cases}$$
such that the closed-loop system satisfies

(1) every trajectory of the system
$$\begin{cases} \dot{x} = Ax + Bv \\ \dot{v} = \mathrm{sat}_\Delta(-Tv + TF(\psi, v, 0) + d) \\ \dot{\psi} = \alpha(\psi, Ex) \end{cases}$$
starting from $\mathcal{X} \times V \times \mathcal{X}$ enters $\mathcal{Z}_\infty$ in a finite time and remains in it thereafter.

(2) for all $x(0), \psi(0) \in \mathcal{X}$, $v(0) \in \mathcal{V}$ and $w(0) \in \mathcal{W}$, the solution of the closed-loop system is such that $e(t)$ enters the set $\mathcal{E}_\infty$ in a finite time and remains in it thereafter.

3. A COMBINED PLC/LHG DESIGN ALGORITHM

This section recalls from (Lin *et al.*, 1997*b*) a design algorithm that is a combination of the piecewise linear LQ control (Wredenhagen and Belanger, 1994) and the low-and-high gain feedback (Saberi *et al.*, 1996).

3.1 *Piecewise Linear LQ Control Design*

Consider the linear dynamical system subject to actuator position saturation,

$$\dot{x} = Ax + B\mathrm{sat}_\Delta(u), \; x(0) \in \mathcal{X} \subset \mathbb{R}^n, \; u \in \mathbb{R}^m \tag{3}$$

where the saturation function $\mathrm{sat}_\Delta : \mathbb{R}^m \to \mathbb{R}^m$ is as defined in Section 2, the pair (A, B) is assumed to be stabilizable, and $\mathcal{X}$ is bounded.

The PLC design is based on the following LQ algebraic Riccati equation,

$$A'P + PA - PBR^{-1}B'P + I = 0 \quad (4)$$

where $R = \text{diag}(\epsilon) = \text{diag}(\epsilon_1, \epsilon_2, \cdots, \epsilon_m)$, $\epsilon_i > 0$, are the design parameters to be chosen later.

Key to the PLC scheme is the notion of invariant sets. A nonempty subset of ε in $\mathbb{R}^n$ is positively invariant if for a dynamical system and for any initial condition $x(0) \in \varepsilon$, $x(t) \in \varepsilon$ for all $t \geq 0$. For the closed-loop system comprising system (3) and the LQ control $u = -R^{-1}B'Px$, simple Lyapunov analysis shows that the Lyapunov level set

$$\varepsilon(P, \rho) = \{x : x'Px \leq \rho\}, \ \forall \rho > 0$$

is an invariant set, provided that saturation does not occur for all $x \in \varepsilon(P, \rho)$. To avoid the saturation from occurring, while fully utilizing the available control capacity, for a given ρ, $\epsilon = (\epsilon_1, \epsilon_2, \cdots, \epsilon_m)$ will be chosen to be the smallest such that

$$|u_i| = \left|\frac{1}{\epsilon_i}B_i'Px\right| \leq \Delta_i, \ \forall x \in \varepsilon(P, \rho)$$

where B_i is the ith column of matrix B and u_i is the ith element of u. The existence and uniqueness of such an ϵ are established, and an algorithm for computing such an ϵ is also given, in (Wredenhagen and Belanger, 1994). More specifically, it is shown through the existence of a unique fixed point that the following iteration converges from any initial value to the desired value of ϵ,

$$\epsilon_{n+1} = \sqrt{\rho}\Phi(\epsilon_n) \quad (5)$$

where $\Phi(\epsilon) = [\phi_1(\epsilon), \phi_2(\epsilon), \cdots, \phi_m(\epsilon)]'$, and for each $i = 1$ to m,

$$\phi_i(\epsilon) = \frac{1}{\Delta_i}\sqrt{B_i'P(\epsilon)B_i}$$

The aim of the PLC scheme is to increase the state feedback gain piecewisely while adhering to actuator bounds as the trajectories converge towards the origin. This is achieved by constructing nested level sets, $\varepsilon_0, \varepsilon_1, \cdots, \varepsilon_N$, in such a way that the trajectories traverse successively the surface of each ε_i and the control law is switched to higher and higher gains as each surface is crossed.

The procedure for designing a PLC law is as follows. Given the set of initial conditions $\mathcal{X} \subset \mathbb{R}^n$, choose an initial level set ε_0 as,

$$\varepsilon_0 = \inf_{\rho}\{\varepsilon(P, \rho) : \mathcal{X} \subset \varepsilon(P, \rho)\} \quad (6)$$

We denote the value of ρ associated with ε_0 as ρ_0, and the corresponding values of ϵ, R and P as ϵ_0, R_0 and P_0 respectively. A simple approach to determining ε_0 and ρ_0 can also be found in (Wredenhagen and Belanger, 1994). More specifically, it is shown that the size of ε_0 grows monotonically as the parameter ρ grows. Hence, ε_0 and ρ_0 can be determined by a simple iteration procedure. Note that, as explained in (Wredenhagen and Belanger, 1994), increasing ρ indefinitely for exponentially unstable A will not result in an ε_0 that grows without bound.

To determine the inner level sets ε_i's, choose successively smaller ρ_i where $\rho_{i+1} < \rho_i$ for each $i = 1, 2, \cdots, N$. A simple choice of such ρ_i's is the geometric sequence of the form

$$\rho_i = \rho_0(\Delta\rho)^i, \ i = 0, 1, 2, \cdots, N$$

where the ρ-reduction' factor $\Delta\rho \in (0, 1)$. (Consequently, the values of ϵ, R and P associated with each of these ρ_i's are denoted as ϵ_i, R_i and P_i respectively.) For a discussion on the choice of N and $\Delta\rho$, see (Wredenhagen and Belanger, 1994).

As shown in (Wredenhagen and Belanger, 1994), a critical property of such a sequence of level sets ε_i is that they are strictly nested in the sense that ε_{i+1} is strictly inside ε_i for each $i = 0$ to $N - 1$.

3.2 *Low-and-High Gain Feedback Design*

Consider the linear dynamical system subject to actuator position saturation, input additive disturbances and uncertainties,

$$\dot{x} = Ax + B\text{sat}_{\Delta}(u + f(x) + d), \qquad x(0) \in \mathcal{X} \subset \mathbb{R}^n, u \in \mathbb{R}^m \quad (7)$$

where the saturation function $\text{sat}_{\Delta} : \mathbb{R}^m \to \mathbb{R}^m$ is as defined in Section 2, the locally Lipschitz function $f : \mathbb{R}^n \to \mathbb{R}^m$ represents the input additive plant uncertainties and d the input-additive disturbance. The LHG feedback design for this system is given as follows. First, the level set ε_0 is determined as in the PLC design. Correspondingly, a state feedback law with possibly low feedback gain is determined as,

$$u_L = -R_0^{-1}B'P_0x$$

A high gain state feedback is then constructed as,

$$u_H = -kR_0^{-1}B'P_0x, \ k \geq 0$$

The final low-and-high gain state feedback is then given by a simple addition of u_L and u_H, viz.,

$$u = -(1 + k)R_0^{-1}B'P_0x, \ k \geq 0$$

Here the design parameter k is referred to as the high gain parameter. As demonstrated in (Saberi *et al.*, 1996) the freedom in choosing the value of this high gain parameter can be utilized to achieve robust stabilization in the presence of input additive plant uncertainties $f(x)$ and input-additive disturbance rejection. Moreover, the transient speed for the states not in the range space of $B'P_0$ will increase as the value of k

3.3 *Combined PLC/LHG Feedback Design*

This subsection presents the proposed combined PLC/LHG state feedback design for linear systems subject to actuator rate saturation (1).

Step 1. Choose a pre-feedback

$$u = v + \bar{u} \tag{8}$$

Let $\tilde{x} = [x', v']'$. Then the system (1) under the above pre-feedback is given by,

$$\dot{\tilde{x}} = \tilde{A}\tilde{x} + \tilde{B}\mathrm{sat}_\Delta(T\bar{u}+d),\ \tilde{x}(0) \in \mathcal{X}\times\mathcal{V} \subset \mathbb{R}^{n+m} \tag{9}$$

where

$$\tilde{A} = \begin{bmatrix} A & B \\ 0 & 0 \end{bmatrix},\quad \tilde{B} = \begin{bmatrix} 0 \\ I \end{bmatrix} \tag{10}$$

Assumption 1, i.e., the pair (A, B) is stabilizable, implies that $(\tilde{A}, \tilde{B})$ is stabilizable.

Step 2. Apply the PLC state feedback design algorithm to system (9), and obtain a sequence of nested level sets $\varepsilon_0, \varepsilon_1, \cdots, \varepsilon_N$ (and correspondingly, the parameters $\epsilon_0, \epsilon_1, \cdots, \epsilon_N$) and a piecewise linear state feedback law,

$$\bar{u} = \begin{cases} \bar{u}_i = -T^{-1}\tilde{R}_i^{-1}\tilde{B}'\tilde{P}_i\tilde{x}, \\ \quad \text{for } \tilde{x} \in \varepsilon_i \setminus \varepsilon_{i+1},\ i = 0 \text{ to } N-1 \\ \bar{u}_N = -T^{-1}\tilde{R}_N^{-1}\tilde{B}'\tilde{P}_N\tilde{x}, \quad \text{for } \tilde{x} \in \varepsilon_N \end{cases} \tag{11}$$

Step 3. Design the LHG state feedback based on the PLC feedback law (11) and obtain the following combined final PLC/LHG feedback law,

$$u = \begin{cases} u_i = v - (k+1)T^{-1}\tilde{R}_i^{-1}\tilde{B}'\tilde{P}_i\tilde{x}, \\ \quad \text{for } \tilde{x} \in \varepsilon_i \setminus \varepsilon_{i+1},\ i = 0 \text{ to } N-1 \\ u_N = v - (k+1)T^{-1}\tilde{R}_N^{-1}\tilde{B}'\tilde{P}_N\tilde{x}, \\ \quad \text{for } \tilde{x} \in \varepsilon_N \end{cases} \tag{12}$$

where $k \geq 0$ is the high gain parameter.

4. STATE FEEDBACK DESIGN FOR ROBUST TRACKING

Step 1. Carry out Step 1 of the combined PLC/LHG feedback design of Section 3.3 as follows. Choose a pre-feedback

$$u = v + \bar{u} \tag{13}$$

Let $\tilde{x} = [x', v']'$, Then system (1) under the above pre-feedback becomes,

$$\dot{\tilde{x}} = \tilde{A}\tilde{x} + \tilde{B}\mathrm{sat}_\Delta(T\bar{u} + d) \tag{14}$$

where $\tilde{A}$ and $\tilde{B}$ are given by (10). Also let $\tilde{x}(0) \in \Xi = (\mathcal{X} \times \mathcal{V}) \cup \{\tilde{x} - \Pi w : \tilde{x} \in \mathcal{X} \times \mathcal{V}, w \in \mathcal{W}\}$.

Step 2. Carry out Step 2 of the combined PLC/LHG feedback design of Section 3.3 using δ_i instead of Δ_i and Ξ instead of $\mathcal{X} \times \mathcal{V}$, and obtain a sequence of nested level sets $\varepsilon_0, \varepsilon_1, \cdots, \varepsilon_N$ (and correspondingly, the parameters $\epsilon_0, \epsilon_1, \cdots, \epsilon_N$) and a piecewise linear feedback law,

$$\bar{u} = \begin{cases} \bar{u}_i = -T^{-1}\tilde{R}_i^{-1}\tilde{B}'\tilde{P}_i\tilde{x}, \\ \quad \text{for } \tilde{x} \in \varepsilon_i \setminus \varepsilon_{i+1},\ i = 0 \text{ to } N-1 \\ \bar{u}_N = -T^{-1}\tilde{R}_N^{-1}\tilde{B}'\tilde{P}_N\tilde{x}, \quad \text{for } \tilde{x} \in \varepsilon_N \end{cases} \tag{15}$$

Step 3. Design the final combined PLC/LHG tracking control laws as,

$$u = \begin{cases} u_i = v - (k+1)T^{-1}\tilde{R}_i^{-1}\tilde{B}'\tilde{P}_i\tilde{x} \\ \quad +T^{-1}[(k+1)\tilde{B}'\tilde{R}^{-1}\tilde{P}_i\Pi + \Gamma]w, \\ \quad \text{for } \tilde{x} \in \varepsilon_i \setminus \varepsilon_{i+1},\ i = 0 \text{ to } N-1 \\ u_N = v - (k+1)T^{-1}\tilde{R}_N^{-1}\tilde{B}'\tilde{P}_N\tilde{x} \\ \quad +T^{-1}[(k+1)\tilde{B}'\tilde{R}_N^{-1}\tilde{P}_N\Pi + \Gamma]w \\ \quad \text{for } \tilde{x} \in \varepsilon_N \end{cases} \tag{16}$$

where $k \geq 0$ is the high gain parameter.

Theorem 4. Let Assumptions 1-3 hold, and given the data $(D, \mathcal{W}_\infty, \mathcal{E}_\infty)$ admissible for state feedback, then Problem 2 can be solved by (16) for sufficiently large k, if there exist matrices Π and Γ such that

(1) they satisfy the linear matrix equations,

$$\begin{cases} \Pi S = \tilde{A}\Pi + \tilde{B}\Gamma \\ \tilde{C}\Pi + Q = 0 \end{cases} \tag{17}$$

where $\tilde{A}$ and $\tilde{B}$ are as given by (10), and $\tilde{C} = [\,C \quad 0_{p\times m}\,]$.

(2) there exists a $\delta = [\delta_1, \delta_2, \cdots, \delta_m]$, $\delta_i > 0$, such that $|\Gamma_i w(t)| \leq \Delta_i - \delta_i$ for all $w(0) \in \mathcal{W}$ and all $t \geq 0$, where Γ_i is the ith row of Γ.

5. OUTPUT FEEDBACK DESIGN FOR ROBUST TRACKING

Step 1: Fast Observer Design. The family of fast observers, parameterized in ℓ is given by,

$$\dot{\hat{x}} = A\hat{x} - L(\ell)(y - C\hat{x}) + Bv,\ \hat{x}(0) \in \mathcal{X} \tag{18}$$

where $L(\ell)$, $\ell > 0$ is chosen such that $\lambda(A + L(\ell)C) = \ell\Lambda$ and Λ is any set of n complex numbers, with negative real parts and close under complex conjugation.

Step 2: State Feedback Design. Carry out the state feedback design as in the proof of Theorem 4. In the design, instead of letting $\Xi = (\mathcal{X}\times\mathcal{V}) \cup \{\tilde{x} - \Pi w : \tilde{x} \in \mathcal{X}\times\mathcal{V}, w \in \mathcal{W}\}$, let Ξ be any bounded set such that $(\mathcal{X}\times\mathcal{V}) \cup \{\tilde{x} - \Pi w : \tilde{x} \in \mathcal{X}\times\mathcal{V}, w \in \mathcal{W}\}$ is strictly inside it. Let the resulting state feedback laws be given by,

$$u = \begin{cases} u_i = v - (k+1)T^{-1}\tilde{R}_i^{-1}\tilde{B}'\tilde{P}_i\tilde{x} \\ \quad +T^{-1}[(k+1)\tilde{B}'\tilde{R}^{-1}\tilde{P}_i\Pi + \Gamma]w, \\ \quad \text{for } \tilde{x} \in \varepsilon_i \setminus \varepsilon_{i+1},\ i = 0 \text{ to } N-1 \\ u_N = v - (k+1)T^{-1}\tilde{R}_N^{-1}\tilde{B}'\tilde{P}_N\tilde{x} \\ +T^{-1}[(k+1)\tilde{B}'\tilde{R}_N^{-1}\tilde{P}_N\Pi + \Gamma]w, \\ \quad \text{for } \tilde{x} \in \varepsilon_N \end{cases} \tag{19}$$

where

$$\varepsilon_i = \{\tilde{x} \in \mathbb{R}^{n+m} : \tilde{x}' P_i \tilde{x} \leq \rho_i\},\ i = 0, 1, \cdots, N$$

is the sequence of strictly nested level sets and $k \geq 0$ is the high gain parameter.

Step 3: Output Feedback Laws. The final output feedback laws are obtained by implementing the state feedback law as obtained in Step 1 with the fast observer as obtained in Step 2. The resulting output feedback law is given by,

$$u=\begin{cases} u_i=v-(k+1)T^{-1}\tilde{R}_i^{-1}\tilde{B}'\tilde{P}_i\hat{\tilde{x}} \\ \quad +T^{-1}[(k+1)\tilde{B}'\tilde{R}^{-1}\tilde{P}_i\Pi+\Gamma]w, \\ \quad \text{for } \hat{\tilde{x}} \in \varepsilon_i\backslash\varepsilon_{i+1}, i=0 \text{ to } N-1 \\ u_N=v-(k+1)T^{-1}\tilde{R}_N^{-1}\tilde{B}'\tilde{P}_N\hat{\tilde{x}} \\ +T^{-1}[(k+1)\tilde{B}'\tilde{R}_N^{-1}\tilde{P}_N\Pi+\Gamma]w, \\ \quad \text{for } \hat{\tilde{x}} \in \varepsilon_N \end{cases} \quad (20)$$

where $\hat{\tilde{x}} = [\hat{x}', v']'$.

Theorem 5. Let Assumptions 1-3 and 4 hold, and given the data $(D, \mathcal{W}_\infty, \mathcal{E}_\infty)$ admissible for output feedback, then Problem 3 can be solved by (20) for sufficiently large k, if there exist matrices Π and Γ such that

(1) they satisfy the linear matrix equations,

$$\begin{cases} \Pi S = \tilde{A}\Pi + \tilde{B}\Gamma \\ \tilde{C}\Pi + Q = 0 \end{cases} \quad (21)$$

where $\tilde{A}$ and $\tilde{B}$ are as given by (10), and $\tilde{C} = [\,C \quad 0_{p\times m}\,]$.

(2) there exists a $\delta = [\delta_1, \delta_2, \cdots, \delta_m]$, $\delta_i > 0$, such that $|\Gamma_i w(t)| \leq \Delta_i - \delta_i$ for all $w(0) \in \mathcal{W}$ and all $t \geq 0$, where Γ_i is the ith row of Γ.

6. ROBUST TRACKING OF AN F-16 AIRCRAFT

In this section, the applicability and effectiveness of the proposed combined PLC/LHG design algorithm is demonstrated with an F-16 fighter aircraft derivative. At the flight condition corresponding to an altitude of 10,000 feet and a Mach number of 0.7, the second order pitch plane dynamics (short period) of this aircraft is given by (Pachter *et al.*, 1995),

$$\begin{cases} \begin{bmatrix}\dot{\alpha}\\ \dot{q}\end{bmatrix}=\begin{bmatrix}-1.15 & 0.994\\ 3.724 & -1.26\end{bmatrix}\begin{bmatrix}\alpha\\ q\end{bmatrix}+\begin{bmatrix}-0.177\\ -19.5\end{bmatrix}\delta, \\ \qquad |\alpha(0)| \leq 0.1\text{rad}, |q(0)| \leq 0.5\text{rad/sec} \\ \dot{\delta}=\text{sat}_1(-T\delta+Tu+d), \quad |\delta(0)| \leq 0.1\text{rad} \\ \dot{w}=\begin{bmatrix}0 & 1\\ -1 & 0\end{bmatrix}w, \ |w_i(0)|\leq 0.5\text{rad/sec}, i=1,2 \\ e=[0 \quad 1]\begin{bmatrix}\alpha\\ q\end{bmatrix}+[0 \quad 1]w \\ y=[1 \quad 0]\begin{bmatrix}\alpha\\ q\end{bmatrix} \end{cases}$$

where α, q and δ are respectively the trim value of the angle of attack, pitch rate and stabilizer deflection. Here the actuator is rate limited to ± 1.0 rad/sec. The constants $T = 20$ sec, which corresponds to an actuator bandwidth of 20 rad/sec.

In what follows, some simulation results are shown for a family of state and output feedback laws that are designed according to the algorithms of the previous two sections.

Figs. 1 and 2 are simulation results for the closed-loop system under the state feedback laws. They show good utilization of the available actuator rate capacity, and that the degree of tracking accuracy in the presence of actuator disturbance increases as the value of the parameter k increases.

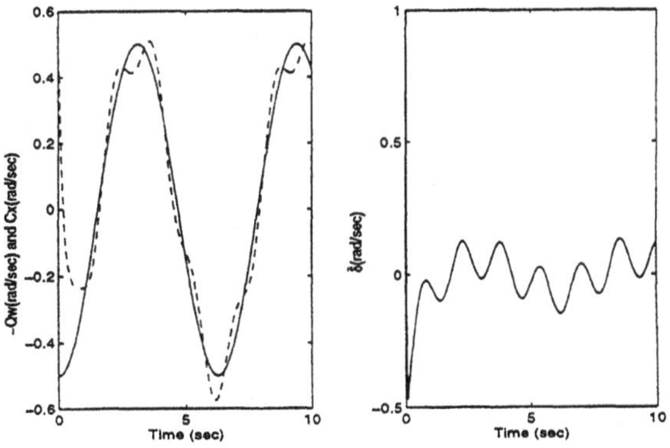

Fig. 1. Combined PLC/LHG design ($N = 5$, $k = 40$). $(\alpha(0), q(0), \delta(0), w_1(0), w_2(0)) = (-0.1, 0.5, 0.1, 0, 0.5)$ and $d = 5\sin(4t+3)$. The left plot: the plant output Cx (dashed line) and the reference signal $-Qw$ (solid line). The right plot: the actuator rate $\dot{\delta}$.

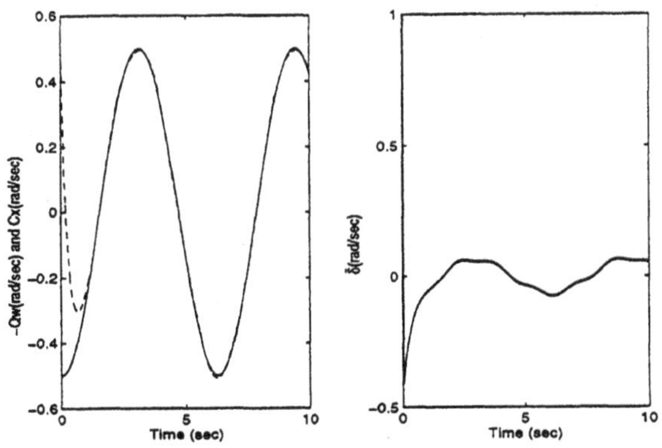

Fig. 2. Combined PLC/LHG design ($N = 5$, $k = 400$). $(\alpha(0), q(0), \delta(0), w_1(0), w_2(0)) = (-0.1, 0.5, 0.1, 0, 0.5)$ and $d = 5\sin(4t+3)$. The left plot: the plant output Cx (dashed line) and the reference signal $-Qw$ (solid line). The right plot: the actuator rate $\dot{\delta}$.

We next compare the performance of the combined PLC/LHG design with both the PLC and the LHG design. To compare with the PLC laws, we set $k = 0$ in (16). The simulation (Fig. 3) shows that in the presence of the same disturbance, the aircraft becomes unstable. To compare with the LHG feedback law, we set $N = 0$ in (16). The simulation (Fig. 4) shows that the combined PLC/LHG feedback law results in much better transient performance. By increasing the value of N, the transience performance can be further improved.

Shown in Fig. 5 are some simulation results for the closed-loop system under the output feedback laws. In the simulation, $(\alpha(0), q(0), \delta(0), w_1(0), w_2(0)) = (-0.1, 0.5, 0.1, 0, 0.5)$, $d(t) = 5\sin(4t+3)$, which are the same as in the simulations in the

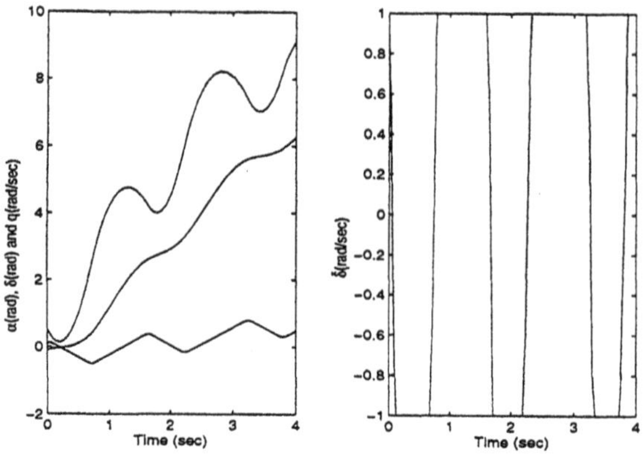

Fig. 3. PLC Design ($N = 5$, $k = 0$). $(\alpha(0), q(0), \delta(0), w_1(0), w_2(0)) = (-0.1, 0.5, 0.1, 0, 0.5)$ and $d = 5\sin(4t+3)$. The left plot: states of the plant and the actuator. The right plot: the actuator rate $\dot{\delta}$.

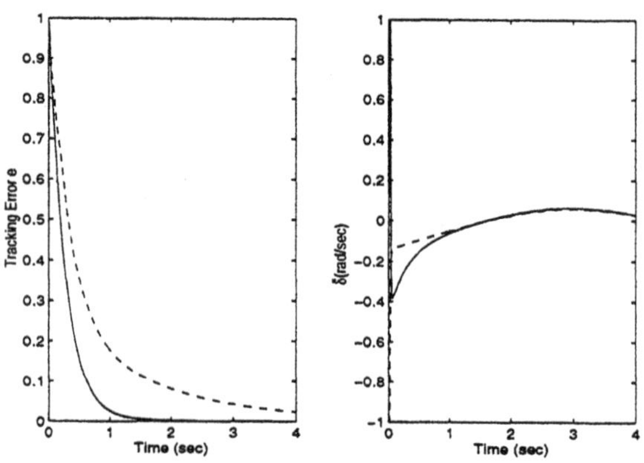

Fig. 4. Transience performance: PLC/LHG ($N = 5, k = 40$, solid lines) vs. LHG ($N = 0, k = 40$, dashed lines). $(\alpha(0), q(0), \delta(0), w_1(0), w_2(0)) = (-0.1, 0.5, 0.1, 0, 0.5)$ and $d = 0$. The left plot: tracking error e. The right plot: the actuator rate $\dot{\delta}$.

state feedback case, and $(\hat{x}_1(0), \hat{x}_2(0)) = (0, 0)$. With $k = 400$, $\ell = 1000$, these simulations results are almost the same as those shown in Fig. 2.

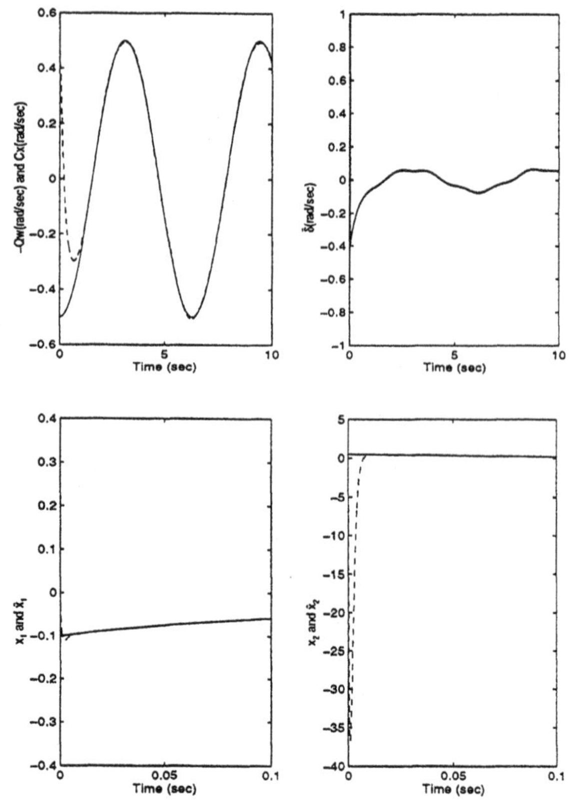

Fig. 5. PLC Design ($N = 5$, $k = 400$). $(\alpha(0), q(0), \delta(0), w_1(0), w_2(0), \hat{x}_1(0), \hat{x}_2(0)) = (-0.1, 0.5, 0.1, 0, 0.5, 0, 0)$ and $d = 5\sin(4t + 3)$. The upper left plot: the plant output Cx (dashed line) and the reference signal $-Qw$ (solid line). The upper right plot: the actuator rate $\dot{\delta}$. The lower plots: plant states $x(t)$ (solid line) and their estimates $\hat{x}$ (dashed line).

7. REFERENCES

Bernstein, D.S. and A.N. Michel (1995). A chronological bibliography on saturating actuators. *Int. J. Robust and Nonlinear Control* **5**, 375–380.

Francis, B. (1984). The linear multivariable regulator problem. *SIAM J. Control Optimization* **15**, 486–505.

Lin, Z. (1998). Global control of linear systems with saturating actuators. *Automatica* **34**, 897–905.

Lin, Z., A. Stoorvogel and A. Saberi (1996). Output regulation for linear systems subject to input saturation. *Automatica* **32**, 29–47.

Lin, Z., M. Pachter, S. Banda and Y. Shamash (1997*a*). Feedback design for robust tracking of linear with rate limited actuators. *Proceedings of 1997 AIAA GNC* pp. 780–788.

Lin, Z., M. Pachter, S. Banda and Y. Shamash (1997*b*). Stabilizing feedback design for linear systems with rate limited actuators. In: *Control of Uncertain Systems with Bounded Inputs* (S. Tarbouriech and Germain Garcia, Eds.). pp. 173–186. Springer-Verlag. London.

Megretski, A. (1996). Output feedback stabilization with saturated control: making the input-output map l_2-bounded. *Proc. 13th World Congress of IFAC* **D**, 435–440.

Miotto, P., J.M. Shewchun, E. Feron and J.D. Paduano (1996). High performance bounded control synthesis with application to the f18 harv. *Proc. of the AIAA GNC.*

Pachter, M., P.R. Chandler and M. Mears (1995). Reconfigurable tracking control with saturation. *AIAA Journal of Guidance, Control and Dynamics* **18**, 1016–1022.

Saberi, A., Z. Lin and A.R. Teel (1996). Control of linear systems with saturating actuators. *IEEE Trans. Auto. Contr.* **41**, 368–378.

Wredenhagen, G.F. and P.R. Belanger (1994). Piecewise-linear lq control for systems with input constraints. *Automatica* **30**, 403–416.

14th Triennial World Congress, Beijing, P.R. China

C-2a-05-5

A LEARNING APPROACH TO STICK-SLIP FRICTION COMPENSATION

Seong-Il Cho and In-Joong Ha *

* *School of Electrical Engineering, Seoul National University, san 56-1 Shinrim-Dong, Kwanak-Ku, Seoul, 151-742, Korea*
e-mail:ijha@asri.snu.ac.kr

Abstract: This paper describes a novel learning control scheme for tracking periodic-trajectories in mechanical systems with friction. It is based on the fact that the solution of the closed-loop system tends to be periodic in steady state. When the closed-loop system reaches the steady state, the proposed learning control scheme updates the control input. By doing this iteratively, the proposed learning control scheme eventually can drive the tracking error to zero. Neither the information of the system mass nor the parametric model for friction is required for successful tracking. In particular, the proposed learning control scheme is computationally quite simple. Furthermore, its generality is well supported through rigorous convergence analysis.

Keywords: friction, periodic, tracking, steady-state, learning

1. INTRODUCTION

In motion control systems, friction is a primary source of disturbance, which can degrade control performance significantly. In this context, various friction compensation methods have been considered by many authors (Armstrong-Hélouvry *et al.*, 1994). One of such friction compensation methods is the model-based friction compensation approach taken in (Walrath, 1984; Canudas de wit *et al.*, 1987; Friedland and Park, 1992; Baril and Gutman, 1997). It seems well suited for the case that an accurate parametric model for friction is available. In reality, however, it is often hard to obtain an accurate parametric model for friction. For this reason, a model-free design methodology was taken in (Du and Nair, 1998), in which state-dependent parasitic effects such as friction were approximated by Gaussian neural networks. However, its main drawback is computational complexity.

On the other hand, it is well known that a learning approach is quite effective in the case that *a priori* knowledge of the system model is limited. In fact, learning control schemes have been popularly used in robotic manipulators and CNC machines (Arimoto *et al.*, 1985; Bondi *et al.*, 1988; Hara *et al.*, 1988; Tung *et al.*, 1993; Ha *et al.*, 1998). Furthermore, the feasibility of a learning control scheme for friction compensation has been well demonstrated in (Tung *et al.*, 1993) through experiments using a repetitive control scheme. The learning control schemes are usually computationally simple and hence can be implemented at cheap cost. Instead, iterative operations along with some memory are required.

In this paper, we also take a learning approach to friction compensation. Our learning control scheme extends basically the work in (Ha *et al.*, 1998) to the problem of periodic-trajectory tracking in mechanical systems with friction. In particular, our learning control scheme differs fundamentally from the repetitive control schemes (Hara *et al.*, 1988; Tung *et al.*, 1993) and the conventional iterative learning schemes (Arimoto *et al.*, 1985; Bondi *et al.*, 1988) in that its update scheme is based on the fact that the solution of the closed-loop system tends to be periodic in steady state. Compared to other friction compensation methods, ours has the following merits. First, the proposed learning control scheme is computationally quite simple. Second, it is shown through a mathematically rigorous analysis on discontinuous nonlinear systems that the tracking error eventually converges to zero in a few number of iterations but neither the accurate information of the system mass nor the parametric model for friction is required for the convergence of the tracking error.

In our development, the usual Euclidean norm of a vector $x \in \mathcal{R}^n$ is denoted by $||x||$. On the other hand, a function $f : [0, \infty) \rightarrow \mathcal{R}$ is called T-periodic if $f(t+T) = f(t)$, $\forall t \geq 0$, for a constant $T > 0$. Finally, the inner product and the norm on $L_2[0,\ T]$ are denoted by $\langle f|g\rangle \triangleq \int_0^T f(t)g(t)dt$ and $||f||_T \triangleq \langle f|f\rangle^{1/2}$, respectively.

2. PRELIMINARY RESULT

The dynamics of the mechanical system with friction considered in this paper are described by the following 2nd-order differential equation.

$$m\ddot{x}(t) \ + \ F(\dot{x}(t), u(t)) \ = \ u(t), \qquad (1)$$

where m is a mass, $x(t) \in \mathcal{R}$ is the position, F is the friction force, and $u(t) \in \mathcal{R}$ is the control input. Here, m and F are unknown. However, the structure of the friction force is assumed to be *a priori* known as follows (Lee and Kim, 1995).

$$F(\dot{x}, u) = F_s(\dot{x}) + F_0(\dot{x}, u). \qquad (2)$$

Here, the function F_s represents the friction force at non-zero velocity and is assumed to be modeled as follows.

$$F_s(\dot{x}) \ = B\dot{x} \ + \ F_n(\dot{x}), \qquad (3)$$

where B is the viscous friction coefficient and $F_n(\dot{x})$ is the nonlinear function of $\dot{x}$ which is continuous except at zero velocity. The typical form of the function F_s is shown in Fig.1.

On the other hand, the function F_0 represents the friction force at zero velocity and its mathematical model is given by

$$F_0(\dot{x}, u) \ \triangleq \ \begin{cases} 0 & \text{if } \dot{x} \neq 0 \\ \lambda(u) & \text{if } \dot{x} = 0 \end{cases}, \qquad (4)$$

where

$$\lambda(u) \triangleq \begin{cases} F_s(0+) & \text{if } u \ \geq F_s(0+) > \ 0 \\ u & \text{if } F_s(0-) < \ u \ < F_s(0+) \\ F_s(0-) & \text{if } u \ \leq F_s(0-) < \ 0 \end{cases}. \qquad (5)$$

For mathematical tractability, we assume that the function F_s in (3) has the following properties,

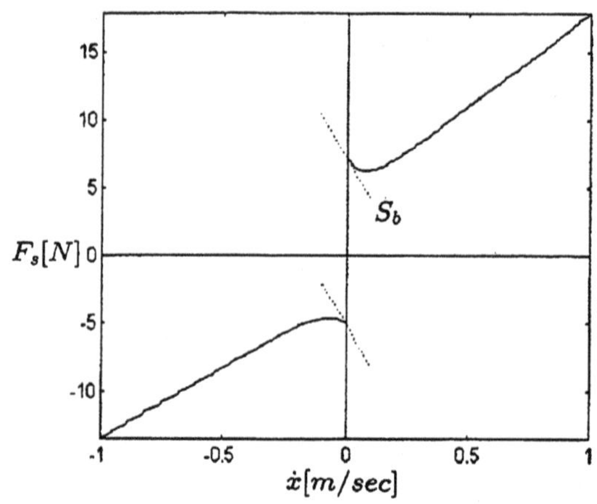

Fig. 1. Friction Curve.

which have been accepted commonly in the prior literature.

$$\frac{F_s(\dot{x}_1) - F_s(\dot{x}_2)}{\dot{x}_1 - \dot{x}_2} \geq S_b, \qquad \text{if} \quad \dot{x}_1 \neq \dot{x}_2, \quad (6)$$

$$|F_n(\dot{x})| \ \leq \ F_b,\ \forall \dot{x} \in \mathcal{R} \text{ and } F_n(0) = 0, \ (7)$$

where S_b and F_b are constants. In fact, many empirical models including the Gaussian model and the Lorentzian model (Armstrong-Hélouvry *et al.*, 1994) can be easily shown to satisfy the above assumptions.

Our tracking controller consists of the well-known proportional-plus-derivative (PD) controller and a feedforward input, say u^f, for error correction as follows.

$$u(t)=K_p(x^d(t)-x(t))+K_d(\dot{x}^d(t)-\dot{x}(t))+ u^f(t), \ (8)$$

where K_p, K_d are positive constants. Here, the function $x^d : [0, \infty) \rightarrow \mathcal{R}$ represents the desired trajectory which is assumed to be T-periodic and twice continuously differentiable. On the other hand, the function $u^f : [0, \infty) \rightarrow \mathcal{R}$ is updated iteratively, as described later. The following Theorem 1 shows that as $t \rightarrow \infty$, any solution of the closed-loop system given by (1) and (8) approaches a T-periodic solution.

Theorem 1. (Existence of Steady-State Oscillation) Suppose that u^f in (8) is continuous and T-periodic. If the controller gains K_p, K_d in (8) are chosen so as to satisfy

$$K_p > 0,\ K_d + S_b > 0, \qquad (9)$$

then any solution, say $z(t) \triangleq [x(t)\ \dot{x}(t)]^T \in \mathcal{R}^2$, of the closed-loop system given by (1) and (8) converges to an absolutely continuous T-periodic solution, say $\bar{z}(t) \triangleq [\bar{x}(t)\ \dot{\bar{x}}(t)]^T \in \mathcal{R}^2$ such that

$$\lim_{t\rightarrow\infty} ||z(t) - \bar{z}(t)|| = 0. \qquad (10)$$

□

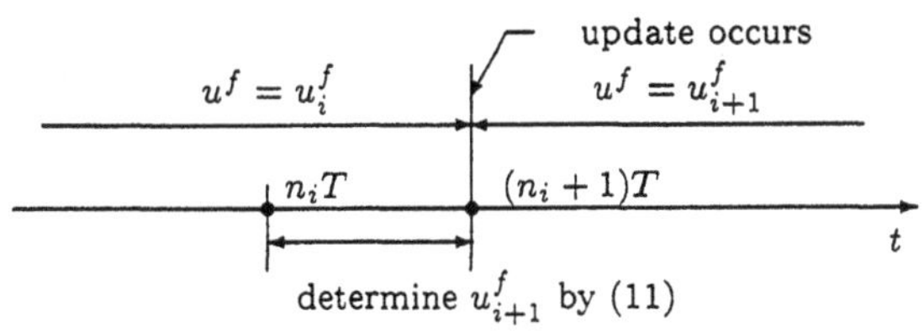

Fig. 2. Timing diagram of the update scheme.

In what follows, the T-periodic solution $\bar{z}$ will be called *steady-state oscillation.* Theorem 1 is very important since it provides the theoretical foundation for our learning approach. Nonetheless, the proof of Theorem 1 is omitted here because of limited space but will be presented elsewhere.

3. MAIN RESULT

To describe precisely the update scheme for u^f in (8), we need some notational convention. The subscript i is used to denote the iteration number, while u_i^f represents the *ith* update of the function u^f. Correspondingly, the subscript i is attached to all variables that are affected by letting $u^f = u_i^f$ in (8). The n_i are nonnegative integers. Note that we need to define the function u^f only for one period $[0, T]$ since it is T-periodic. Finally, the tracking error is denoted by $e(t) \triangleq x^d(t) - x(t)$. Now, we are ready to describe the iterative update scheme for u^f.

Step 1) Set i=0. Let $u^f(\bar{t}) = u_0^f(\bar{t}) \triangleq 0, \forall \bar{t} \in [0, T]$.

Step 2) Wait until the closed-loop system given by (1) and (8) exhibits the steady-state oscillation. Then, store the time history of the velocity error $\dot{e}_i$ for one period, say, $[n_iT, (n_i+1)T]$.

Step 3) Let $\bar{e}_i(\bar{t}) \triangleq e_i(\bar{t} + n_iT)$ and $\dot{\bar{e}}_i(\bar{t}) \triangleq \dot{e}_i(\bar{t} + n_iT)$ for all $\bar{t} \in [0, T]$. Determine u_{i+1}^f using

$$u_{i+1}^f(\bar{t}) \triangleq u_i^f(\bar{t}) + \gamma \dot{\bar{e}}_i(\bar{t}) + \frac{K_p}{T}\int_0^T \bar{e}_i(\tau)d\tau, \forall \bar{t} \in [0, T], \quad (11)$$

where γ is a constant satisfying

$$0 < \gamma < 2(S_b + K_d). \quad (12)$$

Step 4) At $t = (n_i + 1)T$, let $u^f = u_{i+1}^f$. Then, increase i by 1 and jump to *Step 2.*

The timing diagram of the proposed update scheme is depicted in Fig.2. Note that the function u^f is not updated continuously with respect to time. Instead, it is switched once at each iteration from the present waveform to a new waveform after the closed-loop system with $u^f = u_i^f$ reaches the steady state.

As can be seen from (11), there are two error correction terms in the update of u^f. The one is a T-periodic signal proportional to the derivative of the steady-state tracking error. As can be seen from the proof of the following Theorem 2, it plays a key role in driving the tracking error to a constant (which may not be zero) as the iteration number increases. The other is a constant proportional to the time-average value of the steady-state tracking error. As can be seen below, it is to assure that the tracking error goes necessarily to zero as the iteration number increases.

Now, we are ready to state our main result.

Theorem 2. (Convergence of Tracking Error) Suppose that the feedforward input u^f in (8) is updated according to *Step 1) - Step 4).* Suppose further that the set

$$\mathcal{N} \triangleq \{t \in [0, T] | \dot{x}^d(t) = 0\} \text{ is of measure zero.} \quad (13)$$

Then,

$$\lim_{i \to \infty} \bar{e}_i(t) = 0, \ \forall t \in [0, T]. \quad (14)$$

□

Proof: First, we prove that

$$\lim_{i \to \infty} \|\dot{\bar{e}}_i\|_T = 0. \quad (15)$$

Define a T-periodic function u^d by

$$u^d(t) \triangleq m\ddot{x}^d(t) + F_s(\dot{x}^d(t)). \quad (16)$$

Let $\bar{u}^d$ and $\bar{u}^f$ be the time-average components of u^d and u^f, respectively. Let $\tilde{u}^d \triangleq u^d - \bar{u}^d$ and $\tilde{u}^f \triangleq u^f - \bar{u}^f$. Then, we can rewrite (11) as follows.

$$\bar{u}_{i+1}^f = \bar{u}_i^f + \frac{K_p}{T}\int_0^T \bar{e}_i(\tau)d\tau, \tilde{u}_{i+1}^f(t) = \tilde{u}_i^f(t) + \gamma \dot{\bar{e}}_i(t). \quad (17)$$

Define new variables ϕ and ξ by

$$\phi_i \triangleq \bar{u}^d - \bar{u}_i^f, \xi_i(t) \triangleq \tilde{u}^d(t) - \tilde{u}_i^f(t). \quad (18)$$

By (17) and (18), we can obtain that

$$\xi_{i+1}(t) = \xi_i(t) - \gamma \dot{\bar{e}}_i(t), \forall t \in [0, T], \phi_{i+1} = \phi_i - K_p \bar{E}_i, \quad (19)$$

where $\bar{E}_i \triangleq \frac{1}{T}\int_0^T \bar{e}_i(\tau)d\tau$. Hence, we have

$$\|\xi_{i+1}\|_T^2 = \|\xi_i\|_T^2 + \gamma^2\|\dot{\bar{e}}_i\|_T^2 - 2\gamma\langle \xi_i | \dot{\bar{e}}_i \rangle. \quad (20)$$

On the other hand, we can obtain the following dynamic equation for the tracking error in steady state from (1), (8), (16), and (18).

$$m\ddot{\bar{e}}_i + F_s(\dot{x}^d) - F(\dot{x}^d - \dot{\bar{e}}_i, K_p\bar{e}_i + K_d\dot{\bar{e}}_i + u^d - \phi_i - \xi_i) + K_d\dot{\bar{e}}_i + K_p\bar{e}_i = \phi_i + \xi_i. \quad (21)$$

Since ϕ_i is a constant, this can be written as

$$m(\ddot{\bar{e}}_i - \frac{\ddot{\phi}_i}{K_p}) + \Delta F_i + K_d(\dot{\bar{e}}_i - \frac{\dot{\phi}_i}{K_p}) + K_p(\bar{e}_i - \frac{\phi_i}{K_p}) = \xi_i, \quad (22)$$

where $\Delta F_i \triangleq F_s(\dot{x}^d) - F(\dot{x}^d - (\dot{\bar{e}}_i - \frac{\dot{\phi}_i}{K_p}), K_p(\bar{e}_i - \frac{\phi_i}{K_p}) + K_d(\dot{\bar{e}}_i - \frac{\dot{\phi}_i}{K_p}) + u^d - \xi_i)$. Furthermore, we can easily show using (6) and Theorem 1 that

$$\langle \dot{\bar{e}}_i | \bar{e}_i \rangle = \langle \dot{\bar{e}}_i | \ddot{\bar{e}}_i \rangle = 0. \quad (23)$$

$$(\dot{x}_1 - \dot{x}_2)\left[F_s(\dot{x}_1) - F(\dot{x}_2, u) + K_d\dot{x}_1 - K_d\dot{x}_2\right] \geq (S_b + K_d)|\dot{x}_1 - \dot{x}_2|^2, \forall \dot{x}_1, \dot{x}_2, u \in \mathcal{R}. \quad (24)$$

By (22)-(24), we can find the following inequality for the last term in the Right-Hand Side (RHS) of (20).

$$\langle \xi_i | \dot{\bar{e}}_i \rangle = \langle m(\ddot{\bar{e}}_i - \frac{\ddot{\phi}_i}{K_p}) + \Delta F_i + K_d(\dot{\bar{e}}_i - \frac{\dot{\phi}_i}{K_p}) + K_p(\bar{e}_i - \frac{\phi_i}{K_p}) | \dot{\bar{e}}_i - \frac{\dot{\phi}_i}{K_p} \rangle$$
$$= \langle \Delta F_i + K_d(\dot{\bar{e}}_i - \frac{\dot{\phi}_i}{K_p}) | \dot{\bar{e}}_i - \frac{\dot{\phi}_i}{K_p} \rangle \geq (S_b + K_d)\|\dot{\bar{e}}_i\|_T^2. \quad (25)$$

Now, the following inequality is the direct consequence of (12), (20), and (25).

$$\|\xi_{i+1}\|_T^2 - \|\xi_i\|_T^2 \leq -\gamma\{2(S_b + K_d) - \gamma\}\|\dot{\bar{e}}_i\|_T^2 \leq 0, \forall i = 0, 1, 2, \cdots. (26)$$

By induction using this fact, we can show that

$$\gamma\{2(S_b + K_d) - \gamma\} \sum_{i=0}^{\infty} \|\dot{\bar{e}}_i\|_T^2 \leq \|\xi_0\|_T^2 < \infty. \quad (27)$$

This clearly implies (15).

It is shown in the Appendix that if (15) holds, then

$$\lim_{i\to\infty} \frac{1}{T} \int_0^T \left(\bar{F}_i(\tau) - \bar{F}_{i-1}(\tau)\right) d\tau = 0, \quad (28)$$

where $\bar{F}_i(\tau) \triangleq F(\dot{\bar{x}}_i(\tau), K_p\bar{e}_i(\tau) + K_d\dot{\bar{e}}_i(\tau) + u^d(\tau) - \phi_i - \xi_i(\tau))$. Now, we prove the desired property in (14) using this fact. For this aim, we first integrate both sides of (21) over [0, T]. As the result, we have

$$\frac{1}{T} \int_0^T (F_s(\dot{x}^d(\tau)) - \bar{F}_i(\tau))d\tau + K_p\bar{E}_i = \frac{1}{T} \int_0^T (\phi_i + \xi_i(\tau))d\tau = \phi_i. \quad (29)$$

By this along with (19),

$$\phi_{i+1} = \frac{1}{T} \int_0^T (F_s(\dot{x}^d(\tau)) - \bar{F}_i(\tau))d\tau. \quad (30)$$

Therefore, we have

$$K_p\bar{E}_i = \frac{1}{T} \int_0^T (\bar{F}_i(\tau) - \bar{F}_{i-1}(\tau))d\tau. \quad (31)$$

By (28), this implies that

$$\lim_{i\to\infty} \bar{E}_i = 0. \quad (32)$$

For two arbitrary constants $t, t_0 \in [0,\ T]$, we define the function $\hat{E}_i : \mathcal{R}^2 \to \mathcal{R}$ by $\hat{E}_i(t, t_0) \triangleq \int_{t_0}^t \dot{\bar{e}}_i(\tau)d\tau$. By the definitions of $\bar{E}_i$ and $\hat{E}_i$, we then have

$$\bar{E}_i = \frac{1}{T} \int_0^T \hat{E}_i(\tau, t_0)d\tau + \bar{e}_i(t_0). \quad (33)$$

By (15) along with the Schwarz Inequality, we also have

$$\lim_{i\to\infty} \hat{E}_i(t, t_0) = 0, \quad \forall t, t_0 \in [0\ \ T]. \quad (34)$$

Now, our claim in (14) is the immediate consequence of (32), (33), and (34).

Note that many dynamical tracking problems such as circular end milling in CNC machines can satisfy the condition in (13) although set-point tracking problems may not.

4. SIMULATION RESULTS

For our simulation purpose, the function F_s in (2) was assumed to be expressed by

$$F_s(\dot{x}) \triangleq F_s^+(\dot{x})\nu(\dot{x}) + F_s^-(\dot{x})[1 - \nu(\dot{x})], \quad (35)$$

where $\nu(\dot{x}) \triangleq \begin{cases} 1 \text{ if } \dot{x} > 0 \\ 0 \text{ if } \dot{x} < 0 \end{cases}$. $F_s^+(\dot{x}) \triangleq B^+\dot{x} + F_c^+ + (F_z^+ - F_c^+)e^{-|\dot{x}/\dot{x}_s^+|}$, and $F_s^-(\dot{x}) \triangleq B^-\dot{x} + F_c^- + (F_z^- - F_c^-)e^{-|\dot{x}/\dot{x}_s^-|}$, where $B^+ = 13.7[N\sec/m]$, $B^- = 10.5[N\sec/m]$, $F_c^+ = 4.2[N]$, $F_c^- = -3.0[N]$, $F_z^+ = 7.3[N]$, $F_z^- = -5.0[N]$, $\dot{x}_s^+ = 0.07[m/\sec]$, and $\dot{x}_s^- = 0.09[m/\sec]$. The friction curve corresponding to the above data can be plotted graphically as shown in Fig.1. Observe from Fig.1 that the asymmetry of friction as well as the Stribeck effect has been considered fully. On the other hand, we chose x^d, m, K_p, K_d, and γ as follows.

$$x^d(t) = sin(\pi t)[m], m = 1.0[kg],$$
$$K_p = 120, K_d = 80, \gamma = 70. \quad (36)$$

To verify Theorem 1 first, we have investigated the dynamic behavior of the closed-loop system given by (1) and (8). The simulation results for three different initial conditions given in Fig.3 clearly show that the closed-loop system tends to oscillate in steady state. To verify Theorem 2 next, u^f in (8) was updated according to *Step 1) - Step 4)*. To confirm that the closed-loop system reaches the steady state at each iteration, the time interval when the update occurs was chosen sufficiently large to be 8 [sec]. The simulation results are shown in Fig.4, where the encircled number ① stands for the *ith* iteration. Fig.4 shows that the proposed learning control scheme drives $\bar{e}$ and $\dot{\bar{e}}$ to almost zero in six iterations.

5. CONCLUSION

To our best knowledge, this is the first result to present a mathematically rigorous analysis on iterative learning control of discontinuous nonlinear systems. Nonetheless, the proposed learning control scheme has some drawbacks to resolve in further research. First, it is necessary to perform different trials for each of new desired trajectories. Second, the convergence rate may be slow since each update occurs only after the closed-loop system reaches the steady state. Finally, the proposed update scheme requires some memory to store the feedforward input for one period.

Nonetheless, the proposed scheme can be implemented easily at low cost. Furthermore, neither the information of the system mass nor the parametric model for friction is required for successful tracking. As a matter of fact, we have shown that the generality and feasibility of the proposed scheme are well supported through rigorous convergence analysis and simulation results. Hence, it should find a wide range of applications in industries.

Appendix A

Proof of (28)

Define the subsets $\mathcal{O}$, $\mathcal{A}_i$, $\mathcal{B}_i$, $\mathcal{C}_i$ of the interval $[0,T]$ as follows. $\mathcal{O} \triangleq \{t \in [0,T] | \lim_{i\to\infty} \dot{\bar{x}}_i(t) \neq \dot{x}^d(t)\}$, $\mathcal{A}_i \triangleq \{t \in [0,T] | \dot{\bar{x}}_i(t)\dot{\bar{x}}_{i-1}(t) = 0\} - \mathcal{O}$, $\mathcal{B}_i \triangleq \{t \in [0,T] | \dot{\bar{x}}_i(t)\dot{\bar{x}}_{i-1}(t) < 0\} - \mathcal{O}$, and $\mathcal{C}_i \triangleq \{t \in [0,T] | \dot{\bar{x}}_i(t)\dot{\bar{x}}_{i-1}(t) > 0\} - \mathcal{O}$. Then, $[0,T] = \mathcal{O} \bigcup \mathcal{A}_i \bigcup \mathcal{B}_i \bigcup \mathcal{C}_i$, $\forall i = 1,2,3,\cdots$. Therefore, we can write

$$\int_0^T (\bar{F}_i - \bar{F}_{i-1})d\tau = \int_{\mathcal{O}} (\bar{F}_i - \bar{F}_{i-1})d\tau + \int_{\mathcal{A}_i} (\bar{F}_i - \bar{F}_{i-1})d\tau$$
$$+ \int_{\mathcal{B}_i} (\bar{F}_i - \bar{F}_{i-1})d\tau + \int_{\mathcal{C}_i} (\bar{F}_i - \bar{F}_{i-1})d\tau. \quad (A.1)$$

Firstly, we evaluate the first two terms in the RHS of (A.1). In what follows, $\mu(\mathcal{S})$ denotes the Lebesgue measure of a set $\mathcal{S}$ (Royden, 1989). Note from (13) and (15) that $\mu(\mathcal{O}) = 0$ and $\lim_{i\to\infty} \mu(\mathcal{A}_i) = 0$. Therefore, we have

$$\lim_{i\to\infty} \{ \int_{\mathcal{O}} (\bar{F}_i(\tau) - \bar{F}_{i-1}(\tau))d\tau$$
$$+ \int_{\mathcal{A}_i} (\bar{F}_i(\tau) - \bar{F}_{i-1}(\tau))d\tau \} = 0. \quad (A.2)$$

Secondly, we evaluate the third term in the RHS of (A.1). Choose an arbitrary constant $\epsilon > 0$. Then, define the sets $\mathcal{P}_\epsilon$, $\mathcal{N}_\epsilon$, and $\mathcal{S}_\epsilon$ as follows. $\mathcal{P}_\epsilon \triangleq \{t \in [0,T] | \dot{x}^d(t) \geq \epsilon\}$, $\mathcal{N}_\epsilon \triangleq \{t \in [0,T] | \dot{x}^d(t) \leq -\epsilon\}$, and $\mathcal{S}_\epsilon \triangleq \{t \in [0,T] || \dot{x}^d(t)| < \epsilon\}$. By Proposition 3.24 in (Royden, 1989) along with (15), there is a subset $\mathcal{E}$ of $[0,T]$ and an integer N such that

$$\mu(\mathcal{E}) < \epsilon, |\dot{\bar{x}}_{i-1} - \dot{x}^d| < \epsilon, \forall t \in [0,T] - \mathcal{E}, \forall i \geq N. \quad (A.3)$$

Furthermore, it follows from (A.3) and the definition of $\mathcal{B}_i$ that if $i \geq N$, $t \in \mathcal{B}_i$ implies that $t \notin (\mathcal{P}_\epsilon - \mathcal{E}) \bigcup (\mathcal{N}_\epsilon - \mathcal{E})$ and hence that $t \in \mathcal{S}_\epsilon \cup \mathcal{E}$. Therefore, we conclude that

$$\mathcal{B}_i \subset \mathcal{S}_\epsilon \cup \mathcal{E}, \ \forall i \geq N. \quad (A.4)$$

On the other hand, we see from (13) that $\lim_{\epsilon\to 0} \mu(\mathcal{S}_\epsilon) = 0$. By (A.3) and (A.4), this implies that $\lim_{i\to\infty} \mu(\mathcal{B}_i) = 0$. Therefore, we have

$$\lim_{i\to\infty} \int_{\mathcal{B}_i} (\bar{F}_i(\tau) - \bar{F}_{i-1}(\tau))d\tau = 0. \quad (A.5)$$

Finally, we evaluate the last term in the RHS of (A.1). Note that the above function $\bar{F}_i - \bar{F}_{i-1}$ is continuous at each $t \in \mathcal{C}_i$. Therefore, for any $\epsilon > 0$, there exists an integer N such that $|\bar{F}_i(t) - \bar{F}_{i-1}(t)| < \epsilon$, $\forall t \in \mathcal{C}_i$, $\forall i \geq N$. This implies that

$$\lim_{i\to\infty} \int_{\mathcal{C}_i} (\bar{F}_i(\tau) - \bar{F}_{i-1}(\tau))d\tau = 0. \quad (A.6)$$

Now, (28) is the immediate consequence of (A.2), (A.5), and (A.6).

REFERENCES

Arimoto, S., S. Kawamura, F. Miyazaki and S. Tamaki (1985). Learning control theory for dynamical systems. *Proc. of IEEE 24th CDC* **3**, 1375–1380.

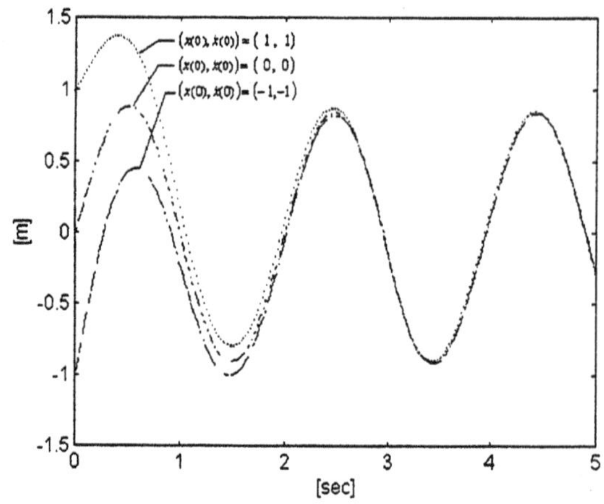

(a) Steady-state response of position x

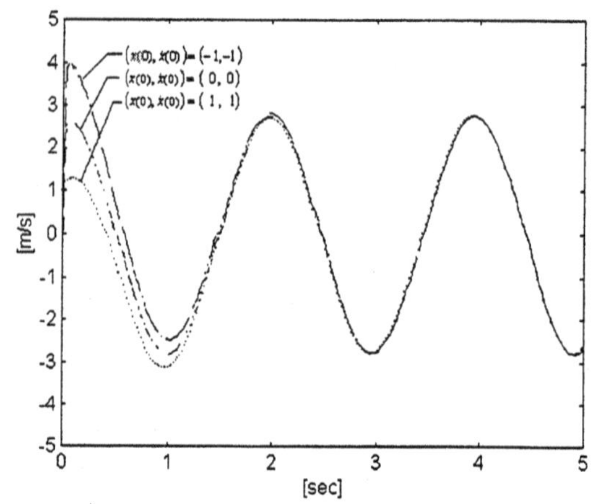

(b) Steady-state response of velocity $\dot{x}$

Solid line: $(x(0), \dot{x}(0)) = (0,0)$, Dashed line: $(x(0), \dot{x}(0)) = (-1,-1)$, Dot-Dashed line: $(x(0), \dot{x}(0)) = (1,1)$

Fig. 3. Steady-state characteristics of the closed-loop system($u^f \equiv 0$)

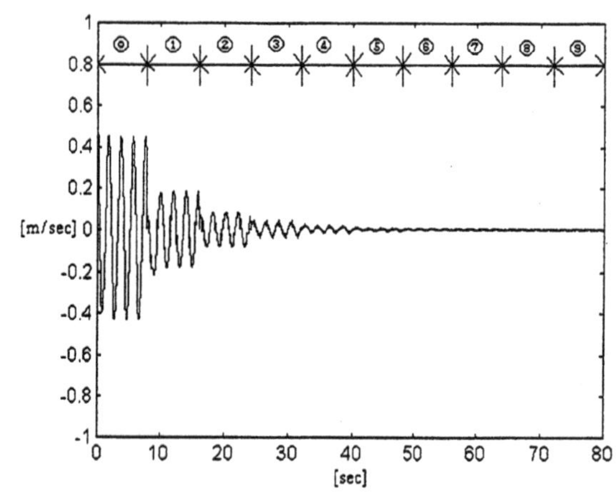

(a) Time history of velocity error $\dot{e}$

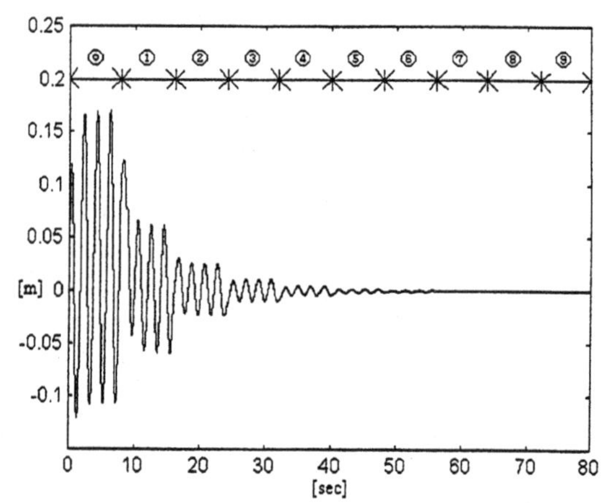

(b) Time history of position error e

Fig. 4. Effect of the iterative update scheme

Armstrong-Hélouvry, B., P. Dupont and Canudas de Wit (1994). A survey of models, analysis tools and compensation methods for the control of machines with friction. *Automatica* **30**, 1083–1138.

Baril, C. G. and P. Gutman (1997). Performance enhancing adaptive friction compensation for uncertain systems. *IEEE Control Systems Technology* **5**, 466–479.

Bondi, P., G. Casalino and L. Gambardella (1988). On the iterative learning control theory for robot manipulators. *IEEE J. Robotics and Automation* **4**, 14–22.

Canudas de Wit, C., K. J. Åström and K. Braun (1987). Adaptive friction compensation in dc-motor drives. *IEEE J. Robotics and Automation* **3**, 681–685.

Du, H. and S. S. Nair (1998). Low velocity friction compensation. *IEEE Control Systems Magazine* **18**, 61–69.

Friedland, B. and Y. J. Park (1992). On adaptive friction compensation. *IEEE Trans. Automat. Contr.* **37**, 1609–1612.

Hara, S., Y. Yamamoto, T. Omata and M.Makano (1988). Repetitive control systems: A new type servo system for periodic exogeneous signals. *IEEE Trans. Automat. Contr.* **33**, 659–667.

Ha, I. J., S. H. Han and Y. H. Kim (1998). Iterative identification of state-dependent disturbance torque for high precision velocity control of servo motors. *IEEE Trans. Automat. Contr.* **43**, 724–729.

Lee, S. W. and J. H. Kim (1995). Robust adaptive stick-slip friction compensation. *IEEE Industrial Electronics* **42**, 474–479.

Royden, H. L. (1989). *Real Analysis*. Vol. 3rd. Macmillan Publishing Company.

Tung, E. D., G. Anwar and M. Tomizuka (1993). Low velocity friction compensation and feedforward solution based on repetitive control. *ASME J. Dynamic Syst., Meas. and Contr.* **115**, 279–284.

Walrath, C. D. (1984). Adaptive bearing friction compensation based on recent knowledge of dynamic friction. *Automatica* **6**, 717–727.

14th Triennial World Congress, Beijing, P.R. China

GLOBAL PREDICTIVE PI-CONTROL OF ANCBI PLANTS UNDER POSITIONAL AND RATE ACTUATOR SATURATIONS

David Angeli * Alessandro Casavola ** Edoardo Mosca *

* *Dip. Sistemi e Informatica, Università di Firenze, Italy*
email: angeli,mosca@dsi.unifi.it
** *Dip. Elettronica, Informatica e Sistemistica,*
Università della Calabria, Italy
email: casavola@si.deis.unical.it

Abstract: A receding-horizon strategy is presented for set-point tracking and global stabilization of ANCBI plants in the presence of actuators with position and rate saturation constraints. The algorithm, providing an integral action, is able to handle any piecewise constant disturbance acting on the system output and state.

Keywords: Predictive control, PI controllers, Constraint satisfaction

1. INTRODUCTION

Dynamic system control in the presence of input saturations and/or state-related constraints has been a significant topic of automatic control since its origin. Despite the efforts and the various approaches, only in the beginning of the nineties the class of dynamical linear systems which can be controlled to zero with arbitrarily small controls has been characterized, (Sussmann *et al.*, 1994) . These are the so called ANCBI (Asymptotically-Null-Controllable-with-Bounded-Inputs) systems.

While positional actuator saturation has been addressed in the recent literature, e.g. see (Zheng and Morari, 1994), and (Alvarez-Ramirez and Suarez, 1996), few design techniques are currently available to deal jointly with actuator positional and rate saturations. Actuator rate saturation often presents a serious challenge in many control engineering fields, such as flight control (see (Dornheim, 1992) and (Lenorovitz, 1990)). In particular, it is known (Berg *et al.*, 1996) that actuator rate saturation can induce a considerable destabilizing effect due to phase-lag. The presence of joint constraints on both input magnitude and increments was recently considered in (Lin, 1997) showing that ANCBI systems are (no matter what the amplitude of the constraints is) semi-globally stabilizable through linear feedback . For other contributions to the topic see also (Lin *et al.*, 1997) and (Tyan and Bernstein, 1997). However, all of the mentioned approaches, deal only with the stabilization issue and little or no indication is given on how to optimize the performance of the closed-loop system.

Hereafter, we present a predictive control law which ensures global exponential stability, set-point tracking and rejection of constant disturbances for ANCBI systems subject to position and rate input constraints. The work extends the predictive controller presented in (Casavola *et al.*, 1998), where only one type of constraints is considered.

2. PROBLEM FORMULATION

Consider the following discrete-time linear dynamic square system

$$\begin{cases} x(t+1) = \Phi x(t) + Gu(t) + \tilde{G}\xi \\ y(t) = Hx(t) + \zeta \end{cases} \quad (1)$$

with: $t \in \mathbb{Z}_{-1} := \{-1, 0, 1, \cdots\}$; state $x(t) \in \mathbb{R}^n$; input $u(t) \in \mathbb{R}^m$; output $y(t) \in \mathbb{R}^m$; and unknown constant disturbances $\xi \in \mathbb{R}^p$ and $\zeta \in \mathbb{R}^m$. Further, let $r \in \mathbb{R}^m$ be the constant set-point to be tracked by the output and

$$\varepsilon(t) := y(t) - r, \tag{2}$$

denote the tracking error. The following assumption is adopted throughout

$$(H, \Phi, G) \text{ stabilizable and detectable} \tag{3}$$

The plant input $u(t)$ and its increments $\delta u(t) := u(t) - u(t-1)$, $\forall t \in \mathbb{Z}_+ := \{0, 1, \ldots\}$ are subject to the following saturation constraints

$$u(t) \in \Omega(U^-, U^+), \quad \forall t \in \mathbb{Z}_{-1} \tag{4}$$

$$\delta u(t) \in \Omega(\Delta^-, \Delta^+), \quad \forall t \in \mathbb{Z}_+ \tag{5}$$

where $\Omega(U^-, U^+) \subset \mathbb{R}^m$ and $\Omega(\Delta^-, \Delta^+) \subset \mathbb{R}^m$ represent closed parallelotopic regions

$$\Omega(U^-, U^+) := \{u \in \mathbb{R}^m : U^- \le u \le U^+\} \tag{6}$$

$$\Omega(\Delta^-, \Delta^+) := \{\delta u \in \mathbb{R}^m : \Delta^- \le \delta u \le \Delta^+\} \tag{7}$$

where $U^- := [U_1^-, \ldots, U_m^-]'$, $U^+ := [U_1^+, \ldots, U_m^+]'$ with U_i^- and U_i^+ extended reals, $U_i^- < 0 < U_i^+$, $i \in \underline{m} := \{1, 2, ..., m\}$, and similarly for $\Delta^- := [\Delta_1^-, \ldots, \Delta_m^-]'$ and $\Delta^+ := [\Delta_1^+, \ldots, \Delta_m^+]'$. Notice that the vector inequalities in (6) and (7) are to be interpreted in a component-wise sense.

The aim is to find a possibly nonlinear feedback control which globally exponentially stabilizes (1) subject to (4)-(5) and possibly yields an offset-free asymptotic tracking of constant set-points r irrespective of disturbances acting on the system.

A classic approach adopted to asymptotically reject constant disturbances is to enforce an "integral action" from ε to u. According to the internal model principle this can be done by providing an integrator in the feedback loop. The related design can be carried out by resorting to the following *incremental model* of (1)

$$\begin{cases} \chi(t+1) = \mathcal{A}\,\chi(t) + \mathcal{B}\,\delta u(t) \\ \varepsilon(t) = \mathcal{C}\,\chi(t) \end{cases} \tag{8}$$

where $t \in \mathbb{Z}_+, \chi(t) := [\delta x'(t) \;\; \varepsilon'(t-1)]'$, $\delta x(t) := x(t) - x(t-1)$ and

$$\mathcal{A} = \begin{bmatrix} \Phi & 0 \\ H & \mathrm{I}_m \end{bmatrix} \quad \mathcal{B} = \begin{bmatrix} G \\ 0_m \end{bmatrix} \quad \mathcal{C} = \begin{bmatrix} H & \mathrm{I}_m \end{bmatrix} \tag{9}$$

Because the exogenous vectors ξ, ζ and r do not show up in (8), a linear state-feedback law $\delta u(t) = \mathcal{F}\chi(t)$, which stabilizes (8) ensures an offset-free steady-state tracking error. As for (8), direct application of PBH rank tests shows that detectability of (H, Φ) implies detectability of $(\mathcal{C}, \mathcal{A})$, whereas $(\mathcal{A}, \mathcal{B})$ results stabilizable if and only if (Φ, G) is such and

$$\det \begin{bmatrix} \mathrm{I}_n - \Phi & G \\ H & 0_m \end{bmatrix} \neq 0 \tag{10}$$

The latter is a necessary and sufficient condition for the existence of a stabilizing linear state-feedback law with integral action for system (1) under (3) (Davison, 1976).

It is worth pointing out that, if $(I_n - \Phi)$ is non-singular, a non-zero steady-state value of the input, viz. $u(\infty) \neq 0_m$, is required to compensate for the effect of constant disturbances, and yield an offset-free tracking error $\varepsilon(\infty) = 0$. Further, if $\delta u(t) = \mathcal{F}\chi(t)$ is stabilizing, $u(\infty)$, which only depends on the system matrices and exogenous signals ξ, ζ and r, turns out to be unaffected by $\mathcal{F}$. To see this, rewrite (1) in steady-state for $\varepsilon(\infty) = 0$ to find

$$\begin{bmatrix} \mathrm{I}_n - \Phi & G \\ -H & 0_m \end{bmatrix} \begin{bmatrix} x(\infty) \\ -u(\infty) \end{bmatrix} = \begin{bmatrix} \tilde{G}\xi \\ \zeta - r \end{bmatrix} \tag{11}$$

Then, exploiting (10), (11) yields

$$u(\infty) = \mathcal{L}\begin{bmatrix} \tilde{G}\xi \\ \zeta - r \end{bmatrix}, \quad \mathcal{L} = -\Gamma_2,$$

$$\Gamma = \begin{bmatrix} \mathrm{I}_n - \Phi & G \\ -H & 0_m \end{bmatrix}^{-1} = \begin{bmatrix} \Gamma_1 \\ \cdots \\ \Gamma_2 \end{bmatrix} \begin{matrix} \}n \\ \\ \}m \end{matrix} \tag{12}$$

This shows that $u(\infty)$ is unaffected by $\mathcal{F}$. We next reexpress $\mathcal{L}$ in a different form. As $\mathcal{F}$ stabilizes (8), it is not difficult to see that

$$u(\infty) = u(t-1) + \mathcal{F}(\mathrm{I}_{n+m} - \mathcal{A}_{\mathcal{F}})^{-1}\chi(t) \tag{13}$$

where $\mathcal{A}_{\mathcal{F}} := (\mathcal{A} + \mathcal{B}\mathcal{F})$. Since

$$\chi(t) = \begin{bmatrix} (\Phi - \mathrm{I}_n)x(t-1) + Gu(t-1) + \tilde{G}\xi \\ Hx(t-1) + \zeta - r \end{bmatrix}$$

equating (12) and (13), we find

$$\mathcal{L} = \mathcal{F}(I_{n+m} - \mathcal{A}_{\mathcal{F}})^{-1} \tag{14}$$

As a consequence, being $\mathcal{L}$ unaffected by $\mathcal{F}$, the matrix on the R.H.S. of (14) does not depend on $\mathcal{F}$ either. By looking at (13), we conclude that $u(\infty) = u(t-1) + \mathcal{L}\chi(t)$, $\forall t \in \mathbb{Z}_+$. Let

$$\tilde{u}(t) := u(t) - u(\infty), \ \forall t \in \mathbb{Z}_{-1} \tag{15}$$

Note that $\tilde{u}(t)$ is the displacement of $u(t)$ from the steady-state input $u(\infty)$. Then,

$$\delta u(t) = \tilde{u}(t) + \mathcal{L}\,\chi(t) = \mathcal{F}\chi(t), \tag{16}$$

Because $\mathcal{L}\chi(t)$ is unaffected by $\mathcal{F}$, $\tilde{u}(t)$ can substitute $\delta u(t)$ in (8) and considered as another possible manipulable input.

This choice is a convenient one in order to enable one to take care of both the positional and the incremental input constraints (4) and, respectively, (5). Accordingly, (8), rewritten in terms of $\tilde{u}(t)$, becomes

$$\begin{cases} \chi(t+1) = \tilde{\mathcal{A}}\,\chi(t) + \mathcal{B}\,\tilde{u}(t) \\ \varepsilon(t) = \mathcal{C}\chi(t) \end{cases} \tag{17}$$

where $t \in \mathbb{Z}_+$ and $\tilde{u}(-1) = -\mathcal{L}\chi(0)$ and

$$\tilde{\mathcal{A}} := \mathcal{A} + \mathcal{B}\mathcal{L} \tag{18}$$

Further, it is possible to show that under (3) and (10)

$$(\mathcal{C}, \tilde{\mathcal{A}}, \mathcal{B}) \text{ is stabilizable and detectable.} \tag{19}$$

The constraints (4)-(5) are now reexpressed in terms of $\tilde{u}$ and $\delta\tilde{u}(t)$ relatively to the new model (17). This can be done by observing that $\delta u(t) = \delta\tilde{u}(t)$ by (15). Then, (4) and (5) are equivalent to

$$\tilde{u}(t) \in \Omega(U^- - u(\infty), U^+ - u(\infty)), \tag{20}$$

$$\delta\tilde{u}(t) \in \Omega(\Delta^-, \Delta^+), \tag{21}$$

provided that $\tilde{u}(-1) = -\mathcal{L}\chi(0)$ belongs to $\Omega(U^- - u(\infty), U^+ - u(\infty))$. Basically the original set-point tracking problem under constraints (4)-(5) and constant unknown disturbances has been converted into a disturbance-free regulation problem under equivalent constraints.

3. ADMISSIBLE FEEDBACK DESIGN PROBLEM

For reasons which will be clearer in the next Section , we assume temporarily that

$$u(\infty) \in \Omega(U^-, U^+) \tag{22}$$

$$\tilde{u}(-1) \in \Omega(\alpha\Delta^-, \alpha\Delta^+),\ \alpha \in (0,1) \tag{23}$$

Inclusion (22) amounts to assuming that ξ, ζ and r are jointly within the range of the input signal, whereas (23) that $u(-1)$ is close to $u(\infty)$.

Given $\chi(0)$, our goal is to find a linear state-feedback law

$$\tilde{u}(t) = \mathcal{F}\chi(t),\ \forall t \in \mathbb{Z}_+ \tag{24}$$

which stabilizes (17) subject to (20) and (21). To this end, we recall some properties of LQ regulation (LQR) relevant for our subsequent developments. Consider, relatively to (17), the following quadratic performance index

$$J_\rho(\chi(0), \tilde{u}(\cdot)) = \sum_{t=0}^{+\infty} \|\varepsilon(t)\|^2_{\Psi_\varepsilon} + \rho\|\tilde{u}(t)\|^2_{\Psi_{\tilde{u}}} \tag{25}$$

where $\rho \in (0, +\infty)$, $\Psi_\varepsilon = \Psi'_\varepsilon > 0$, $\Psi_{\tilde{u}} = \Psi'_{\tilde{u}} > 0$ and $\|v\|^2_\Psi := v'\Psi v$. It is well known that, under (19), the LQ control $\tilde{u}_\rho(t) = \mathcal{F}_\rho\,\chi(t)$

$$\mathcal{F}_\rho = -(\rho\Psi_{\tilde{u}} + \mathcal{B}'\mathcal{P}_\rho\mathcal{B})^{-1}\mathcal{B}'\mathcal{P}_\rho\tilde{\mathcal{A}} \tag{26}$$

asymptotically stabilizes (17) and yields the minimum cost

$$\min_{\tilde{u}(\cdot)} J_\rho(\chi(0), \tilde{u}(\cdot)) = \chi'(0)\mathcal{P}_\rho\chi(0) \tag{27}$$

In (26), $\mathcal{P}_\rho = \mathcal{P}'_\rho \geq 0$ is the unique positive semidefinite solution of the corresponding algebraic Riccati equation (ARE). Observe further that

$$\mathcal{P}_\rho = \mathcal{W}^{\tilde{u}}_\rho + \mathcal{W}^\varepsilon_\rho \tag{28}$$

if $\mathcal{W}^{\tilde{u}}_\rho = \mathcal{W}^{\tilde{u}'}_\rho \geq 0$ and $\mathcal{W}^\varepsilon_\rho = \mathcal{W}^{\varepsilon'}_\rho \geq 0$ are the reachability and, respectively, the observability Gramian, solutions of the Lyapunov equations

$$\begin{aligned} \mathcal{W}^{\tilde{u}}_\rho &= \tilde{\mathcal{A}}'_\rho\mathcal{W}^{\tilde{u}}_\rho\tilde{\mathcal{A}}_\rho + \rho\mathcal{F}'_\rho\Psi_{\tilde{u}}\mathcal{F}_\rho \\ \mathcal{W}^\varepsilon_\rho &= \tilde{\mathcal{A}}'_\rho\mathcal{W}^\varepsilon_\rho\tilde{\mathcal{A}}_\rho + \mathcal{C}'\Psi_\varepsilon\mathcal{C} \end{aligned} \tag{29}$$

with $\tilde{\mathcal{A}}_\rho := (\tilde{\mathcal{A}} + \mathcal{B}\mathcal{F}_\rho)$.

Consider next, the following ρ-parameterized family of ellipsoidal sets

$$\mathcal{C}(P_\rho, \sigma^2) := \{v \in \mathbb{R}^n : v'P_\rho v \leq \sigma^2\} \tag{30}$$

with σ^2 a positive real. A key property is that $\mathcal{C}(\mathcal{P}_\rho, \sigma^2)$ is positively invariant, viz. if the state-feedback (26) is used, $\chi(0) \in \mathcal{C}(\mathcal{P}_\rho, \sigma^2)$ implies that $\chi(t) \in \mathcal{C}(\mathcal{P}_\rho, \sigma^2)$, $t \in \mathbb{Z}_+$. In fact, $\chi(t) = \mathcal{A}^t_\rho\chi(0)$, $\forall t \in \mathbb{Z}_+$, along the optimal trajectory and (26) together with the monotonicity property of the (ARE) imply that

$$\mathcal{A}'_\rho\mathcal{P}_\rho\mathcal{A}_\rho - \mathcal{P}_\rho \leq 0 \tag{31}$$

It can be similarly seen that the ellipsoidal sets $\mathcal{C}(\mathcal{W}^{\tilde{u}}_\rho, \sigma^2)$ and $\mathcal{C}(\mathcal{W}^\varepsilon_\rho, \sigma^2)$ are also positively invariant for (17) and (26).

The foregoing invariance properties can be used so as to determine $\mathcal{F}_\rho$ which give rise to feasible trajectories for given $\chi(0)$. To this end, from (27) and (28) it follows that

$$\sum_{t=0}^{+\infty} \|\tilde{u}_\rho(t)\|^2_{\Psi_{\tilde{u}}} = \chi(0)'\,\bar{\mathcal{W}}^{\tilde{u}}_\rho\,\chi(0) \tag{32}$$

where $\bar{\mathcal{W}}^{\tilde{u}}_\rho := \frac{\mathcal{W}^{\tilde{u}}_\rho}{\rho}$. In a similar way, it can be found that

$$\sum_{t=1}^{+\infty} \|\delta\tilde{u}_\rho(t)\|^2_{\Psi_{\tilde{u}}} = \chi(0)'\bar{\mathcal{V}}_\rho\chi(0) \tag{33}$$

where $\bar{\mathcal{V}}_\rho = (\tilde{\mathcal{A}}_\rho - \mathrm{I})'\bar{\mathcal{W}}^{\tilde{u}}_\rho(\tilde{\mathcal{A}}_\rho - \mathrm{I})$.

Positional and incremental constraints (20) and (21) are jointly satisfied by the state-feedback (26) if there exists a $\rho > 0$ such that

$$\chi(0) \in \mathcal{C}(\bar{\mathcal{W}}_\rho^{\bar{u}}, w^2) \cap \mathcal{C}(\bar{\mathcal{V}}_\rho, \underline{\lambda}(\Psi_u)\Delta^2) \quad (34)$$

where $\Delta := \min_{i \in \underline{m}}\{-\Delta_i, \Delta_i\}$ and $w^2 := \underline{\lambda}(\Psi_u)\min\{v^2, (1-\alpha)^2\Delta^2\}$, with α as in (23). It can be shown that the selection rule (34) ensures satisfaction of (4) and (5). This is simple to be seen for (4). Relatively to (5), we have

$$\begin{aligned} |\delta u_i(t)| &= |u_i(\infty) + \{\mathcal{F}_\rho\chi(t)\}_i - u_i(t-1)| \\ &\leq \quad |u_i(\infty) - u_i(t-1)| + |\{\mathcal{F}_\rho\chi(t)\}_i| \\ &\leq \quad \min(-\Delta_i^-, \Delta_i^+) \end{aligned}$$

where α is specified in (23).

In order to determine conditions under which existence of ρ satisfying (34) can be ensured, we consider some monotonicity properties of $\mathcal{P}_\rho$, $\mathcal{W}_\rho^{\bar{u}}$ and $\mathcal{W}_\rho^{\varepsilon}$, whose proof can be found in (Casavola *at al.*, 1998).

Lemma 1 - Let (19) be fulfilled and ρ_1 and ρ_2 two positive reals. Then, $\rho_2 > \rho_1$ implies

$$\mathcal{P}_{\rho_2} > \mathcal{P}_{\rho_1} \quad \frac{\mathcal{P}_{\rho_2}}{\rho_2} < \frac{\mathcal{P}_{\rho_1}}{\rho_1} \quad (35)$$

(and analogous inequalities for $\mathcal{W}_\rho$).Further,

$$\lim_{\rho \to \infty} \frac{\mathcal{P}_\rho}{\rho} = \bar{\mathcal{P}}_\infty = 0_{n \times n} \quad (36)$$

provided that the system (17) is Asymptotically Null-Controllable with Bounded Inputs (ANCBI). A consequence of Lemma 1 is that large enough ρ satisfying (34) can always be found for any $\chi(0) \in \mathbb{R}^{n+m}$ provided that (17) be ANCBI. Consequently,

$$\begin{aligned} \rho = \min\{\rho \geq 0 : \chi(0) \in \mathcal{C}(\bar{\mathcal{W}}_\rho^{\bar{u}}, w^2) \\ \cap\, \mathcal{C}(\bar{\mathcal{V}}_\rho, \underline{\lambda}(\Psi_u)\Delta^2)\} \end{aligned} \quad (37)$$

can be only solved approximately. The above discussion can be summed up as follows.

Lemma 2 - Let (19), (22) and (23) be fulfilled. Let ρ be selected either via (37). Then, provided that (17) is ANCBI,

$$u(t) = \mathcal{F}_\rho\chi(t) + u(\infty) \quad (38)$$

is a globally asymptotically stable state-feedback control law for (17) which ensures constraints fulfilment for all $t \in \mathbb{Z}_+$ and for any $\chi(0) \in \mathbb{R}^{n+m}$ and offset-free asymptotic tracking of the constant reference r. If, on the contrary (17) is not ANCBI, the above properties hold true provided that (37) admit a solution, viz. $\chi(0) \in \mathcal{C}(\bar{\mathcal{W}}_\infty^{\bar{u}}, z^2)\} \bigcap \mathcal{C}(\bar{\mathcal{V}}_\infty, \underline{\lambda}(\Psi_u)\Delta^2)$

4. REFERENCE GOVERNOR DESIGN PROBLEM

An issue to be further elaborated hereafter is the fact that disturbances and set-point r need not yield an $u(\infty)$ satifying (22). This makes in general set-point tracking incompatible with arbitrary ξ, ζ and r. Under these circustances, our approach is to restore global problem solvability by appropriately modifying r so as to make $u(\infty)$ admissible. More specifically, we shall describe a reference governor unit (RG) which on-line transform r in a sequence $\hat{r}(t)$ of set-point targets converging in finite time to a constant $\hat{r}$ so that the related $\hat{u}_\infty(t)$, viz. the value of $u(\infty)$ for $r = \hat{r}(t)$, settles in finite time on an admissible value $\hat{u}(\infty)$ corresponding to $\hat{r}$.

Firstly, we must guarantee that, for all ξ and ζ, there exists $\hat{r}$ such that the corresponding steady-state value of the control input $\hat{u}(\infty)$ satisfy the positional constraint (4).

Lemma 3 - Let (10) be satisfied. Then, for all $\xi \in \mathbb{R}^p$ and $\zeta \in \mathbb{R}^m$, there exist set-points $r \in \mathbb{R}^m$ such that $u(\infty) \in \Omega(U^-, U^+)$ if and only if

$$\text{span}\left\{\mathcal{L}\begin{bmatrix}\tilde{G}\\0\end{bmatrix}\right\} \subseteq \text{span}\left\{\mathcal{L}\begin{bmatrix}0_{n\times m}\\ \mathrm{I}_m\end{bmatrix}\right\} \quad (39)$$

Then, under (39), let $\hat{r}$ be an appropriate approximation of r. Further, define $\hat{u}(\infty)$ as the steady-state input corresponding to $y(\infty) = \hat{r}$. By (12), we have

$$\hat{u}(\infty) = u(\infty) + \mathcal{L}\mathcal{D}\,(\hat{r} - r) \quad (40)$$

where $\mathcal{D} = [0_{m\times n} \;\; -\mathrm{I}_m]'$. Further, let $\delta > 0$, with $\delta \leq \alpha\Delta$, be an arbitrarily small constant introduced in order to strictly fulfill the constraint (20), viz. $u(t) \in \Omega(U^- + \bar{\delta}, U^+ - \bar{\delta})$, $\bar{\delta} := [\delta, \delta, ..., \delta]' \in \mathbb{R}^m$, with $(U_i^- + \delta) < 0$ and $(U_i^+ - \delta) > 0$. Then, the best admissible approximation of r is given by

$$\begin{aligned} \hat{r} = \quad & \arg\min \|r - v\|^2 \\ & u(\infty) + \mathcal{L}\mathcal{D}\,(v - r) \in \\ & \in \Omega(U^- + \bar{\delta}, U^+ - \bar{\delta}) \end{aligned} \quad (41)$$

We remark, as specified in Lemma 1, that (41) always admits a solution for all $\xi \in \mathbb{R}^p$ if and only if (39) is satisfied. Further, at each time t, let us consider the following optimization problem for some constant $\alpha \in (1/2, 1)$

$$\begin{aligned} \hat{u}_\infty(t) = \quad & \arg\min \|u - \hat{u}_\infty\|^2 \\ & u \in \mathbb{R}^m \\ & u \in \Omega(U^- + \bar{\delta}, U^+ - \bar{\delta}) \\ & u - u(t-1) \in \Omega(\alpha\Delta^-, \alpha\Delta^+) \end{aligned} \quad (42)$$

The above problems defines a sequence of steady-state values which hopefully converges to $\hat{u}(\infty)$. This is required because it is not possible in general to satisfy istantaneously $u(-1) - \hat{u}(\infty) \in$

$\Omega(\alpha\Delta^-, \alpha\Delta^+)$ but one can hope to do that in successive steps. For reasons of space we omit the proofs of the following lemma 4 and 5, which we defer to a forthcoming paper.

Lemma 4 - Assume $u(-1) \in \Omega(U^-, U^+)$. Then, the optimization problem (42) is always solvable and the filtering action performed has a finite stopping time, viz. there exists a t_o such that

$$\hat{u}_\infty(t) = \hat{u}(\infty), \quad \forall t \geq t_o \qquad (43)$$

The combined use of the RG (41) and (42) actions plus LQ state-feedback law (38) gives rise to the following overall control action

$$u(t) = \mathcal{F}_\rho \hat{\chi}(t) + \hat{u}_\infty(t) \qquad (44)$$

where $\hat{\chi}(t) := [\delta x'(t)\ \hat{\varepsilon}'(t-1)]', \hat{\varepsilon}(t) := y(t) - \hat{r}$

Lemma 5 - Let (19) and (39) be fulfilled and $u(-1) \in \Omega(U^-, U^+)$. Let ρ be selected via (37) on the base of $\hat{\chi}(0)$. Then, provided that (17) is ANCBI, (44) is a globally asymptotically stable state-feedback control law for (17) which ensures constraints fulfilment for all $t \in \mathbb{Z}_+$ and for any $\hat{\chi}(0) \in \mathbb{R}^{n+m}$ and offset-free asymptotic tracking of the constant reference $\hat{r}$. If, on the contrary (17) is not ANCBI, the above properties hold true provided that (37) admit a solution, viz. $\hat{\chi}(0) \in \mathcal{C}(\bar{\mathcal{W}}^{\tilde{u}}_\infty, z^2)\} \bigcap \mathcal{C}(\bar{\mathcal{V}}_\infty, \underline{\lambda}(\Psi_u)\Delta^2)$.

5. STABILITY AND RECEDING HORIZON IMPLEMENTATION

In order to use the above procedure in real-time, by following a receding horizon strategy, it is convenient to denote by $v(t+k|t)$, $k \in \mathbb{Z}_+$, the prediction on v at time $t+k$ based on the information available at time t. By definition we set $v(t-1|t) := v(t-1)$. Accordingly, we define v-increment predictions as $\delta v(t+k|t) := v(t+k|t) - v(t+k-1|t)$, $k \in \mathbb{Z}_+$. Notice that, accordingly to the receding-horizon mode of operation, $v(t|t) = v(t)$ and $\delta v(t|t) = \delta v(t)$. Using the foregoing notation, our prediction model is

$$\begin{aligned} \chi(t+k+1|t) &= \mathcal{A}\,\chi(t+k|t) + \mathcal{B}\,\delta u(t+k|t) \\ \varepsilon(t+k|t) &= \mathcal{C}\chi(t+k|t) \end{aligned}$$

where the disturbances and the reference signal, along the predictions, have been considered constant and equal to $\xi(t)$, $\zeta(t)$ and $r(t)$. Assume now that $\delta u(t+k|t) = \mathcal{F}\chi(t+k|t)$, $k \in \mathbb{Z}_+$, for some stabilizing feedback matrix $\mathcal{F}$. Then, in analogy to (17), we adopt the following prediction model related to $\tilde{u}(t+k|t) := u(t+k|t) - u(\infty|t)$,

$$\begin{aligned} \chi(t+k+1|t) &= \tilde{\mathcal{A}}\,\chi(t+k|t) + \mathcal{B}\,\tilde{u}(t+k|t) \\ \varepsilon(t+k|t) &= \mathcal{C}\chi(t+k|t) \end{aligned}$$

where $u(\infty|t) := \lim_{k\to+\infty} u(t+k|t) = u(t-1) + \mathcal{L}\chi(t)$. Based on the previous model, and considering the RG device, we can rewrite the performance index (25) and the ρ-selection rule (34). The only difference is that this time $\alpha \in (1/2, 1)$. The overall algorithm consists of the following steps.

- **Main Loop (generic step $t \geq 0$)**
  (1) Measure $y(t)$ and $\delta x(t)$;
  (2) Compute $u(\infty) = u(t-1) + \mathcal{L}\chi(t)$;
  (3) Compute $\hat{r}(t)$ as in (41) with r replaced by $r(t)$;
  (4) Compute $\hat{u}(\infty) = u(\infty) + \mathcal{LD}(\hat{r}(t) - r(t))$;
  (5) Compute $\hat{u}_\infty(t)$ as in (42);
  (6) Set $\hat{\chi}(t) = [\delta x'(t), (y(t-1) - \hat{r}(t-1))']'$;
  (7) Compute $\rho(t)$ as in (34);
  (8) Compute the matrix $\mathcal{F}_{\rho(t)}$;
  (9) Feed $u(t) = \hat{u}_\infty(t) + \mathcal{F}_{\rho(t)}\hat{\chi}(t)$ into the system;
  (10) Go to 1 with t replaced by $t+1$.

It is easy to recognize that the above algorithm always gives rise to control inputs $u(t)$ which satisfy both positional and incremental constraints (4) and (5) no matter what the reference signal $r(t)$ and the disturbance $\xi(t)$ and $\zeta(t)$ are.
Lemma 6 - Let (3), (10) and (39) be satisfied, $u(-1) \in \Omega(U^-, U^+)$ and (1) be ANCBI. Then, then the control action

$$u(t) = \mathcal{F}_{\rho(t)}\hat{\chi}(t) + \hat{u}_\infty(t) \qquad (45)$$

with $\hat{u}_\infty(t)$ determined by the reference governor strategy (42) and $\rho(t)$ selected via (34) globally stabilizes system (8) provided that there exists a finite time $\bar{t}$ after that the disturbances and the reference become constant.
The action of the reference governor is illustrated by taking into account the following plant

$$Y(s) = \frac{1}{\left(s^2 + \frac{1}{4}\right)(s+1)(s+2)}\left[U(s) + \Xi(s)\right]$$

We sampled the system with $T_s = 0.2s$.. We set: $|U^-| = U^+ = 50$; $|\Delta^-| = \Delta^+ = 10$; $\Psi_\varepsilon = \Psi_{\tilde{u}} = 1$; $\underline{\rho} = 0.1$; $\alpha = 0.51$ and $\delta = 5$. The corresponding simulations are shown inf Fig. 1. The procedure involves some conservatism as it emerges looking at the contraints which are almost never active and, as a consequence, at the rather unsatisfactory performance obtained during the transient. Most of this conservatism is due to the additional constraints required in order to guarantee a converging filtering action on $u(\infty)$. As a remedy we introduce N initial "free moves" in the virtual control, before closing it onto a constant stabilizing feedback gain. Because only the first sample of the optimal virtual command is applied at each time step, a much more efficient exploitation of the input range is obtained, leading, very quickly

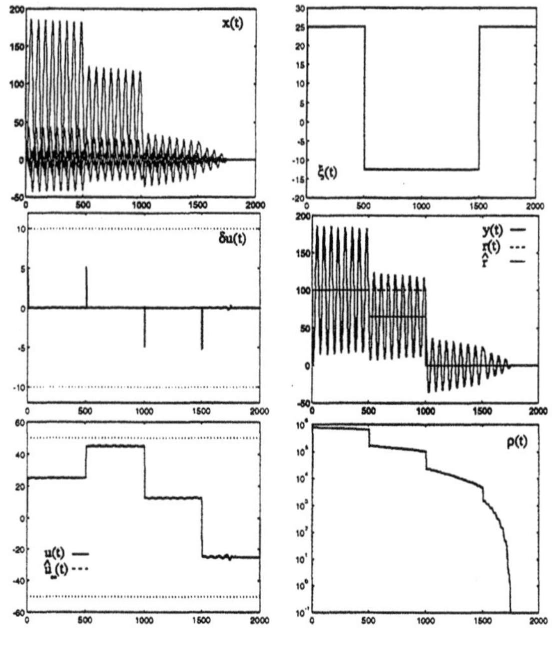

Fig. 1. Simulation under constraints

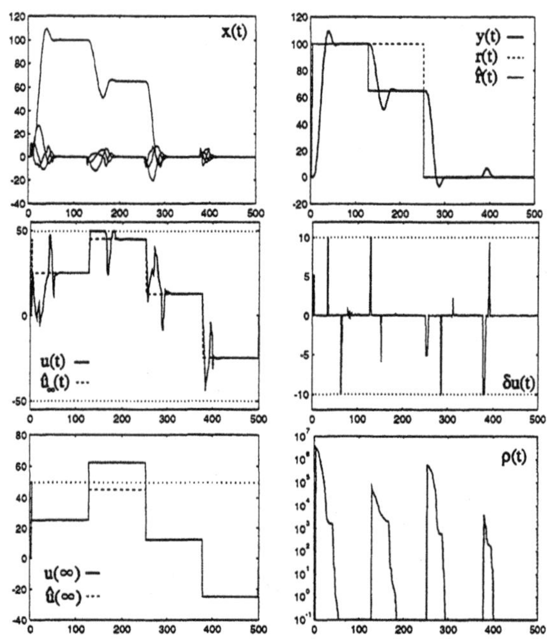

Fig. 2. Simulation with $N = 4$

as N grows, to high tracking performance compatibly with constraints; we defer the details of how this can be done to a forthcoming paper. Fig. 2 show the behaviour of the overall scheme with $N = 4$. As it is possible to verify, thanks to the free control moves, both constraints become active. The resulting closed-loop dynamics are clearly improved.

6. CONCLUSIONS

This paper has addressed the problem of stabilizing a discrete-time linear plant subject to positional and incremental input saturations. It has been proved that, under this strategy, global asymptotic stability and offset-free tracking of admissible set-point references is achieved for ANCBI plants without constraints violation. Feasibility and stability results are achieved independently of the control horizon N which is a free design knobs.

REFERENCES

J. Alvarez-Ramirez and R. Suarez, "Global Stabilization of discrete-time linear systems with bounded inputs", *Int. Journ. of Adapt. Contr. and Sign. Proc.*, Vol. 10, pp. 409-416, 1996.

J. M. Berg, K. D. Hammett, C. A. Schwartz, S. S. Banda, "An analysis of the destabilizing effect of daisy chained rate-limited actuators" , *IEEE Trans. Control Sys. Tech.*, Vol. 4, No. 2, 1996

A. Casavola, M. Giannelli and E. Mosca, "Globally stabilizing predictive regulation of input-saturated linear systems", *Proc. American Control Conference 1998*, Philadelphia, 1998.

E. J. Davison, "The steady-state invertibility and feedforward control of linear time-invariant systems", *IEEE Trans. on Automatic Control*, Vol. 21, pp. 529-534, 1976.

M. A. Dornheim, "Report pinpoints factors leading to YF-22 crash", *Aviation Week Space Technology*, pp. 53-54, Nov. 9, 1992.

J. M. Lenorovitz, "Gripen control problems resolved through in-flight, ground simulations", *Aviation Week Space Technology*, pp. 74-75, June 18, 1990.

Z. Lin, "Semi-global exponential stabilization of linear discrete-time systems subject to input saturation via linear feedbacks.", *System & Control Letters*, Vol. 24, pp. 125-132, 1995.

Z. Lin, M. Pachter, S. Banda, Y. Shamash, "Stabilizing Feedback Design for Linear Systems with Rate Limited Actuators", *Control of Uncertain Systems with Bounded Inputs*, Lect. Notes in Control and Information Sciences, Vol. 227, pp. 173-186, Springer, 1997.

Z. Lin, "Semi-global stabilization of linear systems with position and rate-limited actuators.", *System & Control Letters*, Vol. 30, pp. 1-11, 1997.

H. J. Sussmann, E. D. Sontag and Y. Yang, "A general result on the stabilization of linear systems using bounded control", *IEEE Trans. Aut. Control.*, Vol. 39, pp. 2411-2424, 1994.

F. Tyan and D. S. Bernstein, "Dynamic output feedback compensation for linear systems with independent amplitude and rate saturations", *International Journal of Control*, Vol. 67, N. 1, pp. 89-116, 1997.

A. Zheng and M. Morari, " Global Stabilization of linear discrete-time systems with bounded controls - A model predictive control approach", *Proc. ACC 94 Contr. Conf.*, Baltimore, 1994.

14th Triennial World Congress, Beijing, P.R. China

C-2a-06-1

DESIGN OF STABILIZING CONTROL LAWS FOR SECOND-ORDER SWITCHED SYSTEMS

Xuping Xu and Panos J. Antsaklis

Department of Electrical Engineering,
University of Notre Dame,
Notre Dame, IN 46556, USA
{xxu, antsaklis.1}@nd.edu

Abstract: In this paper, we design asymptotically stabilizing switching control laws for switched systems consisting of several second-order LTI subsystems with unstable foci. Switching is needed for the stabilization of a system consisting of several subsystems if each of its subsystems is unstable. We first study switched systems consisting of two subsystems with unstable foci and conic switching stabilizing control laws are derived. Then the result is extended to several subsystems. In particular, necessary and sufficient conditions for the asymptotic stabilizability of a switched system are derived. If the switched system is asymptotically stabilizable, an asymptotically stabilizing switching law can also be obtained. Finally, we briefly compare the method derived here to a matrix inequality based design method. *Copyright © 1999 IFAC*

Keywords: Switching theory; Hybrid; Stabilization; Control system synthesis;

1. INTRODUCTION

A switched system is a system that consists of several subsystems and a (switching) law that specifies which subsystem dynamics will be followed by the system trajectory at each instant of time. Switchings may be time-driven (i.e., a switching happens at specific time instants) or event-driven (i.e., a switching happens when some internal or external event takes place). One issue of major concern is the stability issue.

There are many papers on stability analysis and design of switched systems (Branicky, 1994*a*; Branicky, 1994*b*; Guckenheimer, 1995; Johansson and Rantzer, 1996; Peleties and DeCarlo, 1991). Several authors adopt multiple candidate Lyapunov functions for analysis (Branicky, 1994*a*; Peleties and DeCarlo, 1991). In (Pettersson and Lennartson, 1996; Johansson and Rantzer, 1996), Linear Matrix Inequality (LMI) problems are formulated for the stability analysis of switched systems.

In this paper, we focus on the design of stabilizing switching control laws. We establish necessary and sufficient conditions for the asymptotic stabilizability of a special kind of switched systems, namely, switched systems consisting of second-order LTI subsystems with unstable foci. We will design switching control laws for this kind of switched systems by studying the vector fields of the subsystems. The effectiveness of our approach for this kind of switched systems will be shown in examples.

The outline of the paper is as follows. In Section 2, we look into the asymptotic stabilizability problem for switched systems with two subsystems. In Section 3, the method in Section 2 is extended to the case of several subsystems. In Section 4, examples illustrate our method, which is then compared to a matrix inequality method. Section 5 concludes the paper.

2. TWO SUBSYSTEMS

In this section, we will design stabilizing switching laws for switched systems consisting of two second-order LTI subsystems with unstable foci. By a LTI subsystem with unstable focus, we mean that the origin of $\mathbb{R}^2$ plane is a unstable focus for the subsystem (Khalil, 1996).

We will study the vector fields of both subsystems and give an intuitive way of obtaining a stabilizing switching control law. We shall say that a subsystem is of clockwise(counterclockwise) direction if starting from any nonzero initial point in the phase plane its trajectory is a spiral around the origin in the clockwise(counterclockwise) direction. In the following, we will first consider switched systems with subsystems of the same direction and then consider switched systems with subsystems of opposite directions. For more information on the direction of second order LTI systems with focus, please refer to (Xu and Antsaklis, 1998).

2.1 *Two Subsystems of the Same Direction*

We will consider the following switched systems,

$$\dot{x}(t) = A_1 x(t),\ \dot{x}(t) = A_2 x(t), \quad (1)$$

whose subsystems are both with unstable foci and are both of the clockwise direction. For switched systems whose subsystems both are of counterclockwise direction, all the following discussion can be applied in a completely analogous manner.

Let $x = (x_1, x_2)^T$ be a nonzero point on $\mathbb{R}^2$ plane, and let

$$f_1 = A_1 x = (a_1, a_2)^T,\ f_2 = A_2 x = (a_3, a_4)^T \quad (2)$$

We can view x, f_1 and f_2 as vectors in $\mathbb{R}^2$ and we denote $\theta_i, i = 1, 2$ to be the angle from x to f_i measured counterclockwise with respect to x (θ_i is confined to $-\pi \le \theta_i < \pi$). So in this case, $-\pi \le \theta_i \le 0$ (Figure 1(a)). Since if it were otherwise, assume that f_i (here we use f_1 for illustration) is on the other side of x as shown in Figure 1(b), then in sufficiently small elapsed time dt, the trajectory will travel in the counterclockwise direction, which contradicts our assumption.

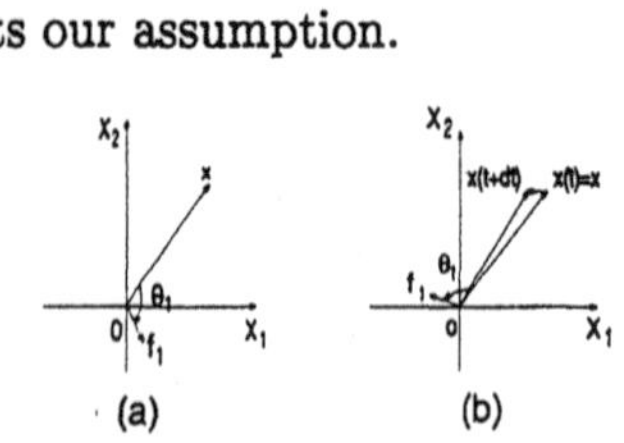

Fig. 1. (a) The angle θ_1 from x to f_1. (b) f_1 is on the other side of vector x.

Note: In the following, we will take $a_1x_1 + a_2x_2$ as the dot product of vectors f_1 and x.

Fig. 2. Dot Product.

And we will also take $a_1x_2 - a_2x_1$ as the dot product of vectors f_1 and $y = (x_2, -x_1)^T$. In this way, we can provide a geometrical interpretation for the above expressions (refer to Figure 2), i.e., $a_1x_1 + a_2x_2 = x^T f_1 = \|x\|_2 \|f_1\|_2 \cos\theta_1$ and $a_1x_2 - a_2x_1 = y^T f_1 = \|y\|_2\|f_1\|_2 \cos(\theta_1 + \frac{\pi}{2}) = \|x\|_2\|f_1\|_2 \cos(\theta_1 + \frac{\pi}{2})$. Similar argument also applies to f_2.

Now we design the switching law that asymptotically stabilizes the switched system. In other words, by our switching law, we want to drive the trajectory closer and closer to the origin, i.e., $\|x(t)\|_2 \to 0$ as $t \to 0$. In the following, we try to design such a switching law by always choosing a subsystem which has the potential to drive the trajectory closer to the origin.

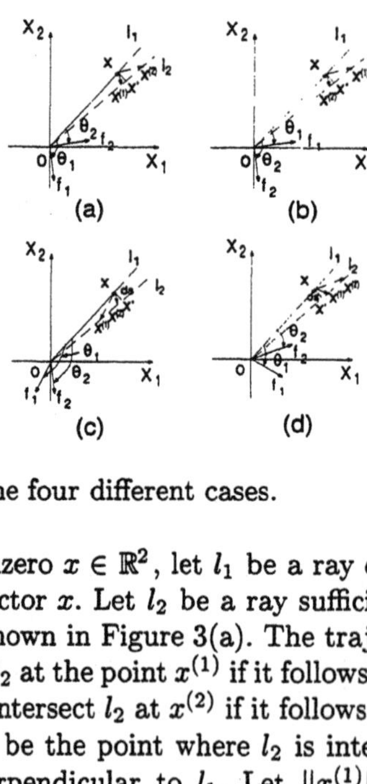

Fig. 3. The four different cases.

For a nonzero $x \in \mathbb{R}^2$, let l_1 be a ray determined by the vector x. Let l_2 be a ray sufficiently close to l_1 as shown in Figure 3(a). The trajectory will intersect l_2 at the point $x^{(1)}$ if it follows subsystem 1, it will intersect l_2 at $x^{(2)}$ if it follows subsystem 2. Let x^* be the point where l_2 is intersected by a line perpendicular to l_1. Let $\|x^{(1)}\|_2$, $\|x^{(2)}\|_2$ be the distances from $x^{(1)}$, $x^{(2)}$ to the origin, respectively. And let $ds = \|x^* - x\|_2$ (Figure 3). Also we define the following conic regions

$$E_{is} = \{x| -\pi \le \theta_i(f_i) \le -\frac{\pi}{2} \text{ or } \frac{\pi}{2} \le \theta_i(f_i) < \pi\}$$
$$= \{x|x^T f_i(x) = x^T A_i x \le 0\},\ i = 1, 2 \quad (3)$$
$$E_{iu} = \{x| -\frac{\pi}{2} \le \theta_i(f_i) \le \frac{\pi}{2}\}$$
$$= \{x|x^T f_i(x) = x^T A_i x \ge 0\},\ i = 1, 2. \quad (4)$$

Now we try to associate a better subsystem to x according to the followings.

Case 1: Assume that $x \in E_{1s} \cap E_{2u}$. In this case, as shown in Figure 3(a), $\|x^{(1)}\|_2 \le \|x^*\|_2 \le \|x^{(2)}\|_2$ when ds is sufficiently small. So we can choose subsystem 1 in order to drive the trajectory closer to the origin. We denote the conic region $E_{1s} \cap E_{2u}$ as Ω_1.

Case 2: Assume that $x \in E_{1u} \cap E_{2s}$. In this case, as shown in Figure 3(b), we can choose subsystem 2 by a similar argument as in Case 1. We denote the conic region $E_{1u} \cap E_{2s}$ as Ω_2.

Case 3: Assume that $x \in E_{1s} \cap E_{2s}$. In this case, as shown in Figure 3(c), consider $\frac{\|x\|_2 - \|x^{(1)}\|_2}{\|x\|_2 - \|x^{(2)}\|_2}$ as $ds \to 0$. Let dt_1 and dt_2 be the time the system would take to intersect l_2 along the vector f_1 and f_2, respectively. When ds is sufficiently small, it is not difficult to see that we approximately have

$$dt_1 = \frac{ds}{\|f_1\|_2 \cos(\theta_1 + \frac{\pi}{2})} = \frac{ds}{\frac{(a_1x_2 - a_2x_1)}{\sqrt{x_2^2 + x_1^2}}}, \tag{5}$$

$$dt_2 = \frac{ds}{\|f_2\|_2 \cos(\theta_2 + \frac{\pi}{2})} = \frac{ds}{\frac{(a_3x_2 - a_4x_1)}{\sqrt{x_2^2 + x_1^2}}}. \tag{6}$$

By (5) and (6) and using L'Hospital Rule, we obtain

$$\lim_{ds \to 0} \frac{\|x\|_2 - \|x^{(1)}\|_2}{\|x\|_2 - \|x^{(2)}\|_2}$$
$$= \lim_{ds \to 0} \frac{\sqrt{x_1^2 + x_2^2} - \sqrt{(x_1 + a_1 dt_1)^2 + (x_2 + a_2 dt_1)^2}}{\sqrt{x_1^2 + x_2^2} - \sqrt{(x_1 + a_3 dt_2)^2 + (x_2 + a_4 dt_2)^2}}$$
$$= \frac{(a_1x_1 + a_2x_2)(a_3x_2 - a_4x_1)}{(a_1x_2 - a_2x_1)(a_3x_1 + a_4x_2)} \tag{7}$$
$$= \frac{\cos\theta_1 \cos(\theta_2 + \frac{\pi}{2})}{\cos(\theta_1 + \frac{\pi}{2}) \cos\theta_2} \tag{8}$$

In view of the above, we see that if

$$\lim_{ds \to 0} \frac{\|x\|_2 - \|x^{(1)}\|_2}{\|x\|_2 - \|x^{(2)}\|_2} \ge 1, \tag{9}$$

or, by simplification using (7) (notice $(a_1x_2 - a_2x_1)(a_3x_1 + a_4x_2) < 0$ in this case), if $a_2a_3 - a_1a_4 \le 0$, then we will have $\|x^{(1)}\|_2 \le \|x^{(2)}\|_2$ for sufficiently small ds. In this case, we can choose subsystem 1. And we denote the conic region $E_{1s} \cap E_{2s} \cap \{x | a_2a_3 - a_1a_4 \le 0\}$ as Ω_3.

Using (8), (9) can also be translated into the condition $\sin(\theta_1 - \theta_2) \le 0$, in other words, $\theta_1 \le \theta_2$. So in this case, if $\theta_1 \le \theta_2$, we choose subsystem 1. Similarly, if $a_2a_3 - a_1a_4 \ge 0$, i.e., $\theta_1 \ge \theta_2$, we choose subsystem 2 and we denote the conic region $E_{1s} \cap E_{2s} \cap \{x | a_2a_3 - a_1a_4 \ge 0\}$ as Ω_4.

Case 4: Assume that $x \in E_{1u} \cap E_{2u}$. In this case, as shown in Figure 3(d), consider $\frac{\|x^{(1)}\|_2 - \|x\|_2}{\|x^{(2)}\|_2 - \|x\|_2}$ as $ds \to 0$. In a manner analogous to the derivation of (8) above, we obtain

$$\lim_{ds \to 0} \frac{\|x^{(1)}\|_2 - \|x\|_2}{\|x^{(2)}\|_2 - \|x\|_2}$$
$$= \frac{(a_1x_1 + a_2x_2)(a_3x_2 - a_4x_1)}{(a_1x_2 - a_2x_1)(a_3x_1 + a_4x_2)} \tag{10}$$
$$= \frac{\cos\theta_1 \cos(\theta_2 + \frac{\pi}{2})}{\cos(\theta_1 + \frac{\pi}{2}) \cos\theta_2} \tag{11}$$

Then as in Case 3, we see that if

$$\lim_{ds \to 0} \frac{\|x^{(1)}\|_2 - \|x\|_2}{\|x^{(2)}\|_2 - \|x\|_2} \le 1, \tag{12}$$

or, by simplification (notice $(a_1x_2 - a_2x_1)(a_3x_1 + a_4x_2) > 0$ in this case), if $a_2a_3 - a_1a_4 \le 0$, then we will have $\|x^{(1)}\|_2 \le \|x^{(2)}\|_2$ for sufficiently small ds, so we can choose subsystem 1. And we denote the conic region $E_{1u} \cap E_{2u} \cap \{x | a_2a_3 - a_1a_4 \le 0\}$ as Ω_5.

Similarly to Case 3, this condition can also be translated into the condition $\theta_1 \le \theta_2$. So in this case, if $\theta_1 \le \theta_2$, we choose subsystem 1. Similarly, if $a_2a_3 - a_1a_4 \ge 0$, i.e., $\theta_1 \ge \theta_2$, we choose subsystem 2 and we denote the conic region $E_{1u} \cap E_{2u} \cap \{x | a_2a_3 - a_1a_4 \ge 0\}$ as Ω_6. □

Notice that Ω_i, $i = 1, 2, \cdots, 6$, are all conic regions that partition the $\mathbb{R}^2$ plane. We can associate with each region a subsystem according to the above discussion. *In essence, we associate with each point on $\mathbb{R}^2$ a subsystem i whose θ_i is smaller.* This partition of the $\mathbb{R}^2$ plane is of particular importance here. In the following, we call this switching law (that uses the partition and the associated subsystems) the **conic switching law**. We will show that by using the conic switching law, we can decide whether the system (1) is asymptotically stabilizable or not. If (1) is asymptotically stabilizable, the partition and associated subsystems also provide a stabilizing switching control law.

Note: If at a point x and $\theta_1 = \theta_2$, then we associate with x the subsystem whose valid conic region is to be entered.

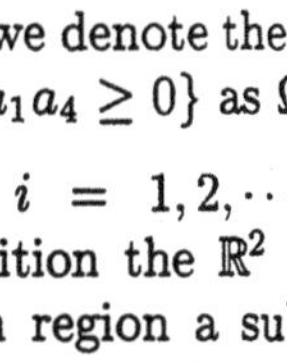

Fig. 4. The two rays and related points.

Consider the system (1). Let l_1 and l_2 be two rays that go through the origin and are in the same conic region of one Ω_i. Suppose l_2 is on the clockwise side of l_1 in the conic region and the conic region with angle from l_1 to l_2 is inside Ω_i. Suppose x_0 is on l_1. Let $x^{(1)}$ be the point on l_2 where the trajectory of the system intersects l_2 for the first time if the system evolves according to the subsystem associated with Ω_i in the conic switching law. Let $x^{(2)}$ be the point on l_2 where the trajectory of the system intersects l_2 for the first time if the system evolves according to the other subsystem. Figure 4 shows the two rays and corresponding points. For the above-mentioned rays and points, we have the following lemma.

Lemma 1. If l_1 and l_2 are in the same conic region Ω_i, $i = 1, 2, \cdots, 6$, then $\|x^{(1)}\|_2 \leq \|x^{(2)}\|_2$.

Proof: See (Xu and Antsaklis, 1998). □

In view of the above lemma, we can prove the following theorem (Xu and Antsaklis, 1998).

Theorem 2. Let l_1 be a ray that goes through the origin. Let x_0 be on l_1. If x^* is the point on l_1 where the trajectory intersects l_1 for the first time after leaving x_0, when the switched system evolves according to the conic switching law, and if x_1 is the point on l_1 where the trajectory intersects l_1 for the first time after leaving x_0, when the switched system evolves according some arbitrary switching law, then we have $\|x^*\|_2 \leq \|x_1\|_2$.

Theorem 2 implies the following theorem for the necessary and sufficient conditions of the asymptotic stabilizability of switched system (1) with subsystems of the same direction.

Theorem 3. The switched system (1) with subsystems of the same direction is asymptotically stabilizable if and only if $\|x^*\|_2 < \|x_0\|_2$ by the conic switching law, where x^* and x_0 are the same as in Theorem 2.

2.2 *Two Subsystems of Opposite Directions*

We will consider the following switched system,

$$\dot{x}(t) = A_1 x(t), \; \dot{x}(t) = A_2 x(t), \qquad (13)$$

where both subsystems are with unstable foci, and subsystem 1 is of clockwise direction while subsystem 2 is of counterclockwise direction. Figure 5 shows the angle θ_1 ($-\pi \leq \theta_1 \leq 0$) and θ_2 ($0 \leq \theta_2 < \pi$) in this case.

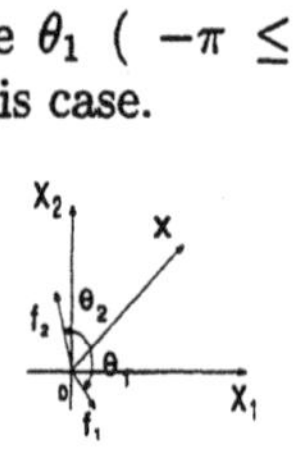

Fig. 5. The angle θ_i from x to f_i, $i = 1, 2$.

We would like to consider a switching law that asymptotically stabilizes the switched system. By our switching law, we want to lead the trajectory closer and closer to the origin, i.e., $\|x(t)\|_2 \to 0$ as $t \to 0$. In the following, we will design a switching law by choosing a subsystem which has the potential to drive the trajectory closer to the origin. We first define the following conic regions

$$\Omega_1 = E_{1s} \cap E_{2s}, \qquad (14)$$

$$\Omega_2 = E_{1u} \cap E_{2u}, \qquad (15)$$

$$\Omega_3 = E_{1s} \cap E_{2u} \cap \{x | a_2 a_3 - a_1 a_4 \geq 0\}, \qquad (16)$$

$$\Omega_4 = E_{1s} \cap E_{2u} \cap \{x | a_2 a_3 - a_1 a_4 \leq 0\}, \qquad (17)$$

$$\Omega_5 = E_{1u} \cap E_{2s} \cap \{x | a_2 a_3 - a_1 a_4 \geq 0\}, \qquad (18)$$

$$\Omega_6 = E_{1u} \cap E_{2s} \cap \{x | a_2 a_3 - a_1 a_4 \leq 0\}, \qquad (19)$$

where E_{is}, E_{iu} are as defined in (3) and (4). Notice, some Ω_i's in this case are different from Ω_i's in the previous subsection.

Case 1: If $Int(\Omega_1) \neq \emptyset$, it is clear that any trajectory of the switched system that is totally inside Ω_1 is stable (Figure 6(a)). Now if the initial point of the switched system is outside Ω_1, we can follow the trajectory of any one subsystem to go into Ω_1. Then we can use the **conic switching law**, i.e., *switch to another subsystem upon intersecting the boundary of* Ω_1, as this will give us a trajectory that is totally inside Ω_1 and it is bounded. If there exists $x \in \Omega_1$, such that $|\theta_i(f_i(x))| \neq \frac{\pi}{2}$ for some $i = 1$, or 2, then the switched system is asymptotically stablizable by the conic switching law.

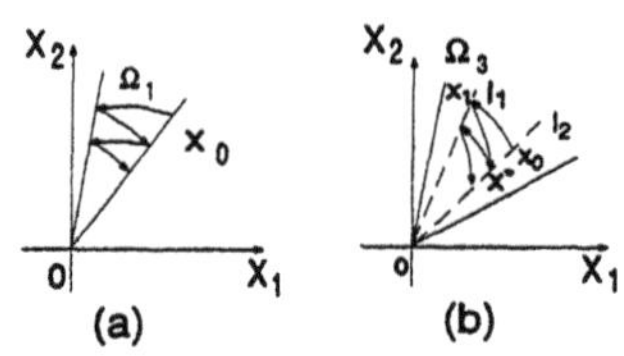

Fig. 6. (a) Case 1:$Int(\Omega_1) \neq \emptyset$. (b) Case 2:$Int(\Omega_3) \neq \emptyset$.

Case 2: Let $Int(\Omega_3) \neq \emptyset$. In this case, assume l_1 and l_2 be two rays that go through the origin and are in the same conic region Ω_3, and that the conic region with angle from l_1 to l_2 is inside Ω_3 (Figure 6(b)). Suppose l_2 is on the clockwise side of l_1 in the conic region and suppose x_0 is on l_2. Then we have the following Lemma.

Lemma 4. Let the switched system follow subsystem 2 until the trajectory intersects l_1 for the first time at x_1 and then let the system switch to subsystem 1 and evolve according to subsystem 1. Suppose x^* is a point on l_2 where the trajectory intersects l_2 for the first time after the switching. Then we have $\|x^*\|_2 \leq \|x_0\|_2$.

Proof: See (Xu and Antsaklis, 1998). □

According to Lemma 4, the **conic switching law** (i.e., *switch to another subsystem upon intersecting the boundary of* Ω_3 so as to keep the trajectory inside Ω_3 (Figure 6(b))) will give us a trajectory that is totally inside Ω_3 so it is bounded. If there exists $x \in \Omega_3$, such that $|\theta_1(f_1)| + |\theta_2(f_2)| \neq \pi$ at x, then the switched system is asymptotically stablizable by the conic switching law.

Case 3: Let $Int(\Omega_5) \neq \emptyset$. Then similar argument as in Case 2 can be applied and we find that the system can be stabilized by conic switching law to keep the trajectory inside Ω_5. □

The following theorem shows us that the system can only be stabilized for the above-mentioned three cases.

Theorem 5. The switched system (13) with two subsystems of opposite directions is asymptotically stabilizable if and only if $Int(\Omega_1) \cup Int(\Omega_3) \cup Int(\Omega_5) \neq \emptyset$ and there exists $x \in Int(\Omega_1) \cup Int(\Omega_3) \cup Int(\Omega_5)$ such that $|\theta_1(f_1)| + |\theta_2(f_2)| \neq \pi$ at x.

Proof: See (Xu and Antsaklis, 1998). □

3. SEVERAL SUBSYSTEMS

Now we study stabilizing switch laws for switched systems with several second-order LTI subsystems with unstable foci

$$\dot{x}(t) = A_i x(t),\ i = 1, 2, \cdots, N. \qquad (20)$$

3.1 *All Subsystems of the Same Direction*

Assume that all the N subsystems are of the clockwise direction. Notice that from the discussion in subsection 2.1, we find the conic switching law by associating with each point on $\mathbb{R}^2$ a subsystem i with smaller θ_i. As an extension of the conic switching law there, we obtain the **conic switching law** by associating with each point on $\mathbb{R}^2$ a subsystem i where θ_i is the smallest.

Similar argument as in section 2.1 can be made for several subsystems to show the following theorem.

Theorem 6. Let l_1 be a ray that goes through the origin. Let x_0 be on l_1 and consider the conic switching law. Let x^* on l_1 be the point where the trajectory intersects l_1 for the first time after leaving x_0. The switched system (20) is asymptotically stabilizable if and only if $||x^*||_2 < ||x_0||_2$.

3.2 *Not All Subsystems of the Same Direction*

Assume that $K(K > 0)$ subsystems are of clockwise direction and $M(M > 0)$ subsystems are of counterclockwise direction $(K + M = N)$. Combining the previous results, the following theorem can be obtained.

Theorem 7. The switched system (20) with $K(K > 0)$ subsystems $S_1^-, \cdots, S_K^-$ of clockwise direction and $M(M > 0)$ subsystems $S_1^+, \cdots, S_M^+$ of counterclockwise direction is asymptotically stabilizable if and only if at least one of the following three conditions holds:

(1) The switched system consisting of $S_1^-, \cdots, S_K^-$ is asymptotically stabilizable.
(2) The switched system consisting of $S_1^+, \cdots, S_M^+$ is asymptotically stabilizable.
(3) There exist i and j with $1 \leq i \leq K$ and $1 \leq j \leq M$ such that the switched system consisting of two subsystems S_i^- and S_j^+ is asymptotically stabilizable.

4. EXAMPLES AND COMPARISON TO MATRIX INEQUALITY APPROACH

4.1 *Examples*

In the following examples, the conic switching laws are used.

Example 8. Consider the switched system (1) with two subsystems of the same direction with

$$A_1 = \begin{bmatrix} 1 & 13 \\ -2 & 3 \end{bmatrix},\ A_2 = \begin{bmatrix} -1 & 2 \\ -10 & 3 \end{bmatrix}.$$

It can be shown that it is asymptotically stabilizable using the conic switching law, though each subsystem has unstable foci. The regions for subsystems are specified and a stabilizing control law is derived. Figure 7 shows the regions for subsystems, the trajectory and the time domain responses of the switched system.

Example 9. Consider the switched system (13) with two subsystems of different directions with

$$A_1 = \begin{bmatrix} -2 & 52 \\ -8 & 6 \end{bmatrix},\ A_2 = \begin{bmatrix} 11 & -10 \\ 50 & -9 \end{bmatrix}.$$

It is asymptotically stabilizable since the region $Int(\Omega_1)$ is nonempty. Figure 8 shows the conic regions, the trajectory and the time domain responses of the switched system.

4.2 *Comparison to a Matrix Inequality Approach*

In (Pettersson and Lennartson, 1996), Pettersson introduced a LMI based stability analysis method. Note that, by some modification of the LMI approach, the design of stabilizing control laws can be formulated as matrix inequality problems (Xu and Antsaklis, 1998). Yet because these problems typically are complicated and may be nonlinear in some variables, they are generally very difficult to solve. And if we cannot obtain a solution

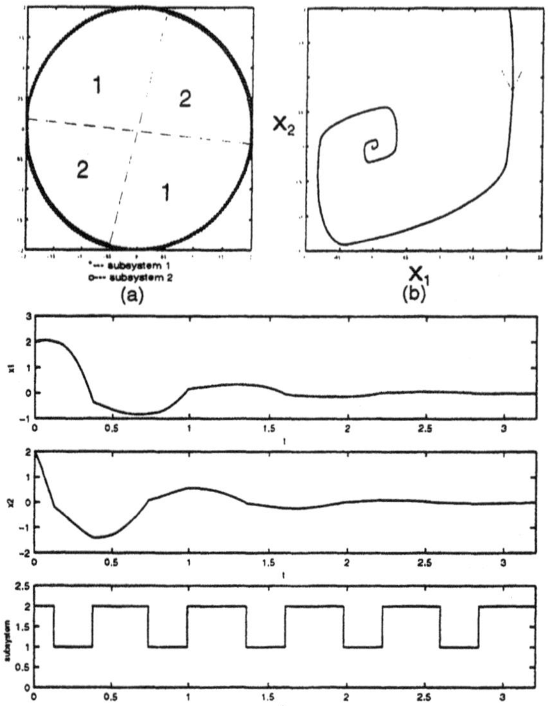

Fig. 7. Example 8: (a) The regions for subsystems. (b) The trajectory of the system. (c) Time domain responses. ($x_0 = (2,2)^T$)

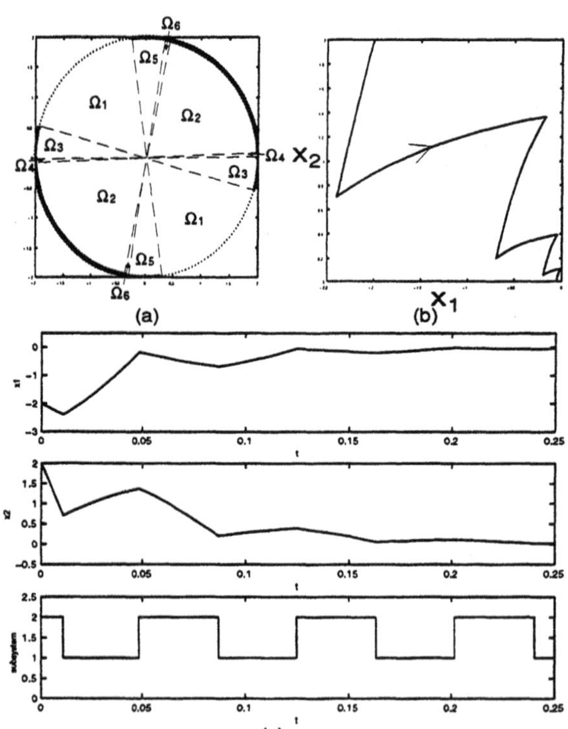

Fig. 8. Example 9: (a) The conic regions. (b) The trajectory of the system. (c) Time domain responses. ($x_0 = (-2,2)^T$)

for the problem, we may not know whether the switched system is stabilizable or not. Even for second-order LTI switched systems discussed in this paper, the matrix inequality problem may not be easy to solve. One may need to associate multiple candidate Lyapunov functions for each subsystem and hope to find a solution.

Therefore, although the matrix inequality formulation for control design seems to be general for n-th order LTI switched systems, they are difficult to solve even for second-order switched systems. Our method in this paper can solve the design problem efficiently for second-order switched systems.

5. CONCLUSION

In this paper, we look into the problem of finding asymptotically stabilizing switching control laws for switched systems consisting of several second-order LTI subsystems with unstable foci. Necessary and sufficient conditions were given for the asymptotic stabilizability of such switched systems. Also a design procedure for deriving such stabilizing control laws was presented. For the general second-order LTI switched systems with subsystems having unstable nodes or saddle points instead of unstable foci, the design of stabilizing switching control laws poses further research questions and it is currently under investigation.

6. REFERENCES

Branicky, M.S. (1994*a*). Analysis of continuous switching systems: Theory and examples. In: *Proc. of ACC.* Baltimore. pp. 3110–3114.

Branicky, M.S. (1994*b*). Stability of switched and hybrid systems. In: *Proc. of 33rd CDC.* pp. 3498–3503.

Guckenheimer, J. (1995). A robust hybrid stabilization strategy for equilibria. *IEEE Trans. on AC* **40**(2), 321–326.

Johansson, M. and A. Rantzer (1996). Computation of piecewise quadratic lyapunov functions for hybrid systems. Technical Report ISRN LUTFD2/TFRT-7549-SE. Department of Automatic Control, Lund Institute of Technology.

Khalil, H.K. (1996). *Nonlinear Systems.* Prentice-Hall.

Peleties, P. and R. DeCarlo (1991). Asymptotic stability of m-switched systems using lyapunov-like functions. In: *Proc. of ACC.* Boston. pp. 1679–1684.

Pettersson, S. and B. Lennartson (1996). Stability and robustness for hybrid systems. In: *Proc. of CDC.* Kobe. pp. 1202–1207.

Xu, X. and P. J. Antsaklis (1998). Stabilization of second-order switched systems. ISIS Techical Report. University of Notre Dame. (`http://www.nd.edu/~isis/tech.html`).

14th Triennial World Congress, Beijing, P.R. China

C-2a-06-2

APPLICATION OF LYAPUNOV FUNCTION METHOD TO THE DESIGN OF VARIABLE STRUCTURE CONTROL FOR A CLASS OF MEASURE DIFFERENTIAL SYSTEMS WITH IMPULSIVE DISTURBANCES*

Dong YUE, Shifan XU

College of Information & Electrical Engineering, China University of Mining & echnology
Xuzhou, Jiangsu, China (dongy@cumt.edu.cn)

Abstract: It is different from the usually considered systems with norm-bounded uncertainties that the effects of impulsive disturbances always exist in the measure differential systems with impulsive disturbances(MDSID). In the present paper, the Lyapunov function will be used to the design of variable structure control for the MDSID. It will be shown that the given control can counteract the effect of both norm-bounded uncertainties and impulsive disturbances and ensure the asymptotic stability of closed-loop systems. Finally, a simulation result is given to illustrate the effectiveness of the scheme. *Copyright © 1999 IFAC*

Key words: measure differential systems; variable structure control; Lyapunov function; switching function.

1. INTRODUCTION

Many practical systems occur impulsive phenomena, such as different games, biological phenomena, economic problems and some other mechanical systems. These impulsive phenomena can be described by measure differential equations (Das, P.C, et al., 1971; Xu Yuantong, 1988). Different from the usually considered uncertain systems, the effects of impulsive disturbances always exist in the measure differential systems with impulsive disturbances (MDSID). Obviously, the impulsive disturbances are not norm-bounded, thus the previous results of robust control can not be directly used for MDSID.

Recently, much attention has been paid to the study of sliding mode control due to more strong robustness to the parameter uncertainties and external disturbances than some other controls. Moreover, the sliding mode control theory has been also extensively used in the control synthesis of real systems, such as robotics systems and large scale structure systems (Gao Wei-bing, 1990; Dong Yue, 1995). It is well known that, however, the uncertainties of systems are required to be norm-bounded for the previous design of variable structure control. Obviously, the norm-bounded conditions of uncertainties can not be suitable for MDSID because of the existence of impulsive disturbance. Hence, the natural problem arising is whether the usual variable structure systems theory can also be used for the design of variable structure control for MDSID. It is still an open problem now.

The present paper will consider the design problem of variable structure control for MDSID. It will be noted that, under some restrictive conditions, a variable structure control with different switching energy can be developed for MDSID. And the given control can ensure both the realization of sliding mode in a finite time and the asymptotic stability of closed loop system. Finally, to illustrate the effectiveness of the scheme, a simulation result is given.

2. SYSTEM MODEL

Consider the following measure differential systems with norm-bounded uncertainties and impulsive disturbances

$$Dx(t) = (A+\Delta A)x(t) + Bu(t) + Bd(t) + Gx(t)Dw(t) \tag{1}$$

where $x(t) \in R^n$ is the state vector, $u(t) \in R^m$ is the input vector. A,B are matrices of suitable dimensions. Dx, Dw denote the distributional derivative of functions x,w, respectively. $w: R_+ \to R$ is a function of right continuous and bounded variation.

Definition (Xu Yuantong, 1988): The function $w: R_+ \to R$ is thought to belong to the class S^*, if

I>. w is right continuous and bounded variation on every compact subinterval of R_+.

II>. The discontinuous points $t_1 < t_2 < \ldots\ldots$ are isolated, and $t_k \to \infty$ as $k \to \infty$.

III>. w is differentiable on the subinterval $[t_k, t_{k+1})(k = 1,2,\ldots)$ of R_+, where the derivative of w at point t_k denotes the right derivative.

Suppose that $w \in S^*$ and has the following form

$$w = t + \sum_{k=1}^{\infty} \alpha_k H_k(t)$$

where

$$H_k(t) = \begin{cases} 0, & t < t_k \\ 1, & t \geq t_k \end{cases}$$

α_k are some constants. Hence,

$$Dw(t) = 1 + \sum_{k=1}^{\infty} \alpha_k \delta_k(t_k)$$

where $\delta_k(t_k)$ stand for the Dirac measure, $t_k - t_{k+1} \geq \delta > 0$.

In (1), $\Delta A, G$ denote the parameter uncertain matrices, d(t) is the external disturbance. They satisfy

Assumption 1: External disturbance d(t) is bounded, i.e., $\rho > 0$ exists such that

$$\|d(t)\| \leq \rho.$$

Assumption 2: The uncertain matrices $\Delta A, G$ can be decomposed into the following form

$$\Delta A = \Delta A_1 + \Delta A_2$$
$$G = G_1 + G_2$$

where $\Delta A_1, G_1$ satisfy

$$\Delta A_1^T \Delta A_1 \leq A_1, \quad G_1^T G_1 \leq G_2$$

where A_1, G_2 are two semi-positive matrices. $\Delta A_2, G_2$ satisfy the matching conditions, that is, there exist two matrices E_1, E_2 such that

$$\Delta A_2 = BE_1, \quad G_2 = BE_2.$$

3. DESIGN OF VARIABLE STRUCTURE CONTROL

To design the control, we first consider the following Riccati equation as

$$(A+\gamma I)^T P + P(A+\gamma I) - 2PBR^{-1}B^T P + \gamma^{-1}\|P\|(A_1 + G_0) + Q = 0 \tag{3}$$

where R, Q are positive matrices, $\gamma > 0$ is a constant. If there exists a positive definite solution P to equation (3) for some R, Q and γ, we can then design a variable structure control of form

$$u(t) = -R^{-1}B^T Px(t) - \xi(t)\operatorname{sgn}(B^T Px) \tag{4}$$

where $\xi(t) > \rho + \xi + (\|E_1\| + \|E_2\|\|x(t)\|), \xi > 0$, *sgn* denotes the sign function, and if let $B^T = [B_1 \ldots\ldots B_m]^T$, then

$$\operatorname{sgn}(B^T Px) = [\operatorname{sgn}(B_1^T Px) \ldots\ldots \operatorname{sgn}(B_m^T Px)]^T$$

Next, we will analyze the asymptotic behavior of system (1) under the control (4). For this purpose, we construct the Lyapunov function candidate as

$$V(t) = x^T(t)Px(t)$$

For $t \in [t_k, t_{k+1})$,

$$\dot{V}(t) = 2x^T P[A + \Delta A_1 + G_1 - BR^{-1}B^T P]x$$
$$+ 2x^T P[(\Delta A_2 + G_2)x - \xi(t)B\operatorname{sgn}(B^T Px) + Bd(t)]$$

Note that

$$2x^T P\Delta A_1 x \leq \gamma x^T Px + \gamma^{-1}x^T \Delta A_1^T P\Delta A_1 x$$
$$\leq \gamma x^T Px + \gamma^{-1}\|P\|x^T A_1 x$$

$$2x^T PG_1 x \leq \gamma x^T Px + \gamma^{-1}\|P\|x^T G_0 x$$

Hence, one can obtain

$$\dot{V}(t) \leq x^T[(A+\gamma I)P + P(A+\gamma I) - 2PBR^{-1}B^T P$$
$$+ \gamma^{-1}\|P\|(A_1 + G_0)]x + 2[\rho + (\|E_1\|$$
$$+ \|E_2\|)\|x\|]|B^T Px| - 2\xi(t)|B^T Px|$$

In terms of the Riccati equation one yields

$$\dot{V}(t) \le -x^T Qx - \xi|B^T Px|$$
$$\le -x^T Qx \le -\frac{\lambda_{\min}(Q)}{\lambda_{\max}(P)}V(t)$$

and further obtains

$$V(t) \le e^{-\lambda_1(t-t_k)}V(t_k), \qquad t \in [t_k, t_{k+1}) \qquad (5)$$

where $\lambda_1 = \lambda_{\min}(Q)\lambda_{\max}^{-1}(P)$.

From system (1), one can obtain

$$x(t_k) - x(t_{k-h}) = \int_{t_{k-h}}^{t_k} ([A + \Delta A]x(t) + Bu(t) + Bd(t))dt$$
$$+ \int_{t_{k-h}}^{t_k} Gxdw(t)$$

Letting $h \to 0$, one gets

$$x(t_k) - x(t_{k-0}) = \alpha_k Gx(t_k)$$

If $I - \alpha_k G$ is nonsingular, then in terms of above equality one yields

$$x(t_k) = [I - \alpha_k G]^{-1} x(t_{k-0}) \stackrel{\Delta}{=} G_k x(t_{k-0}) \qquad (6)$$

From (5) and (6), it is ease to see that

$$V(t_k) = x^T(t_k)Px(_k)$$
$$= x^T(t_{k-0})G_k^T PG_k x(t_{k-0})$$
$$\le \frac{\lambda_{\max}(P)\lambda_{\max}(G_k^T G_k)}{\lambda_{\min}(P)} x^T(t_{k-0})Px(t_{k-0})$$
$$\stackrel{\Delta}{=} \lambda_2^k V(t_{k-0})$$

For $\forall t \in R_+$, there exists some N>0 such that $t_N \le t < t_{N+1}$. Hence,

$$V(t) \le e^{-\lambda_1(t-t_N)}V(t_N)$$
$$\le e^{-\lambda_1(t-t_N)}\lambda_2^N V(t_{N-0}) \le \lambda_2^N e^{-\lambda_1(t-t_{N-1})}V(t_{N-1})$$

Further, by inductive method, one can see that

$$V(t) \le \prod_{k=0}^{N} \lambda_2^k e^{-\lambda_1(t-t_0)}V(t_0) \qquad (7)$$

If $\lambda_3 > 0$ exists such that

$$\prod_{k=0}^{N} \lambda_2^k \le e^{\lambda_3(t_N - t_0)} \qquad (8)$$

Substituting (8) into (7) yields

$$V(t) \le e^{-(\lambda_1-\lambda_3)(t-t_0)}V(t_0)$$

Theorem 1: Suppose that for $\gamma > 0, R > 0, Q > 0$, there exists a positive solution P to Riccati equation (3), $G_k = [I - \alpha_k G]^{-1} (k = 1,2,\ldots)$ exist and satisfy

$$\prod_{k=0}^{N} \lambda_2^k \le e^{\lambda_3(t_N - t_0)}$$

$\lambda_3 > 0, \lambda_1 > \lambda_3$, then under the control (4), the trajectories of system (1) are asymptotically stable.

Proof: The conclusion of theorem 1 can be easily induced by the above discussion.

4. EXISTENCE OF SLIDING MODE MOTION

To study the sliding mode motion, we must first design the switching function(also called sliding mode function). In fact, the switching function in (4) has already been developed as follows

$$s(t) = B^T Px(t)$$

Next, it will be shown that, under the control (4), the sliding mode motion can be realized in a finite time. In other word, the trajectories of closed loop of system (1) can reach the switching function $s(t)=0$ in a finite time.

From the expression of $s(t)$, one can see that

$$Dx(t) = B^T PDx(t)$$
$$= B^T P(A + \Delta A)x(t) - B^T PBR^{-1}B^T Px(t)$$
$$- B^T PB\xi(t)\,\mathrm{sgn}(s) + B^T PBd(t) + B^T PGx(t)Dw(t) \qquad (9)$$

Constructing Lyapunov function candidate as

$$L(t) = s^T (B^T PB)^{-1} s$$

where $|B^T PB$ is supposed to be nonsingular. In fact, if B is of full rank, $B^T PB$ is therefore inevitable.

For $t \in [t_k, t_{k+1})$,

$$\dot{L}(t) = 2s^T (B^T PB)^{-1}[B^T P(A + \Delta A) - B^T PBR^{-1}B^T P$$
$$+ B^T PG]x(t) - 2s^T \xi(t)\,\mathrm{sgn}(s) + 2s^T d(t)$$
$$\le 2\|(B^T PB)^{-1}\|\|B^T P(A + \Delta A - BR^{-1}B^T PG)\|\|x(t)\||s|$$
$$- 2(\xi(t) - \rho)|s| \qquad (10)$$

By theorem 1, it can be seen that $x(t) \to 0$ as $t \to \infty$. Hence, there exists N such that, for $K > N$ and $t \in [t_k, t_{k+1})$,

$$\|x(t)\| \le \frac{\xi}{2\|(B^T PB)^{-1}\|\|B^T P(A + \Delta A - B^T R^{-1}B^T PG)\|}$$

From (10) one can get

$$\dot{L}(t) \le -\xi|s|, \quad t \in [t_k, t_{k+1})$$

Note that

$$\frac{1}{\sqrt{m}}\lambda_{\min}^{1/2}(B^T PB)^{-1}|s| \le \lambda_{\min}^{1/2}(B^T PB)^{-1}(s^T s)^{1/2}$$

$$\leq L^{1/2}(t) = [s^T (B^T PB)^{-1} s]^{1/2}$$

$$\leq \lambda_{\max}^{1/2} (B^T PB)^{-1} (s^T s)^{-1}$$

$$\leq \lambda_{\max}^{1/2} (B^T PB)^{-1} |s| \tag{11}$$

Hence,

$$\dot{L}(t) \leq -\xi \lambda_{\max}^{1/2} (B^T PB)^{-1} L^{1/2}(t), \quad t \in [t_k, t_{k+1}),$$

It is easy to see that

$$L^{1/2}(t) - L^{1/2}(t_k) \leq \frac{-\xi \lambda_{\max}^{-\frac{1}{2}} (B^T PB)^{-1} (t - t_k)}{2}$$

Letting $t \to t_{k+1-0}$, one can have

$$L^{1/2}(t_{k+1-0}) - L^{1/2}(t_k) \leq -\frac{\xi \delta \lambda_{\max}^{-\frac{1}{2}} (B^T PB)^{-1}}{2}$$

and thus,

$$L^{1/2}(t_{k+1-0}) \leq L^{1/2}(t_k) - \frac{\xi \delta \lambda_{\max}^{-\frac{1}{2}} (B^T PB)^{-1}}{2}$$

$$\leq \lambda_{\max}^{1/2} (B^T PB)^{-1} |s(t_k)| - \frac{\xi \delta \lambda_{\max}^{-\frac{1}{2}} (B^T PB)^{-1}}{2}$$

Further, using inequality (11), one can see that

$$|s(t_{k+1-0})| \leq \sqrt{m} [\frac{\lambda_{\max}(B^T PB)^{-1}}{\lambda_{\min}(B^T PB)^{-1}}]^{1/2} |s(t_k)|$$

$$- \frac{\sqrt{m} \xi \delta \lambda_{\max}^{-\frac{1}{2}} (B^T PB)^{-1}}{2 \lambda_{\min}(B^T PB)^{-1}}$$

Defining

$$k_1 = [\frac{\lambda_{\max}(B^T PB)^{-1}}{\lambda_{\min}(B^T PB)^{-1}}]^{1/2}, \quad k_2 = \frac{\xi \delta \lambda_{\max}^{-\frac{1}{2}} (B^T PB)^{-1}}{2 \lambda_{\min}(B^T PB)^{-1}}.$$

Thus, one can yield

$$|s(t_{k+1-0})| \leq k_1 \sqrt{m} |s(t_k)| - k_2 \sqrt{m}$$

From (9), one gets

$$s(t_{k+1}) - s(t_{k+1-0}) = B^T PGx(t_{k+1}) \alpha_k$$

and thus has

$$|s(t_{k+1})| \leq |s(t_{k+1-0})| + |\alpha_k| \|B^T PG\| \|x(t_{k+1})\|$$

$$\leq |s(t_{k+1-0})| + \alpha \|B^T PG\| \|x(t_{k+1})\| \tag{12}$$

$\alpha = \sup\{|\alpha_k|, k = 1,2,......\}$.

By theorem 1, it is easy to see that, for k large enough,

$$\|x(t_{k+1})\| \leq \frac{k_2 \sqrt{m}}{2\alpha \|B^T PG\|} \tag{13}$$

Substituting (13) into (12) yields

$$|s(t_{k+1})| \leq |s(t_{k+1-0})| + \frac{k_2 \sqrt{m}}{2}$$

$$\leq k_1 \sqrt{m} |s(t_k)| - \frac{k_2 \sqrt{m}}{2}$$

In terms of the inductive method, it can be seen that

$$|s(t_{k+1})| \leq (k_1 \sqrt{m})^q |s(t_{k-q})| - \frac{k_2}{2} \{k_1^{q-1} (\sqrt{m})^q$$

$$+......+\sqrt{m}\}$$

$$\leq (k_1 \sqrt{m})^q |s(t_{k-q})| - \frac{k_2 k_1^{q-1} (\sqrt{m})^q}{2} \{1 + \frac{1}{k_1 \sqrt{m}}$$

Choosing $k = 2q$, if q is large enough, then

$$1 + \frac{1}{k_1 \sqrt{m}} +......+ \frac{1}{(k_1 \sqrt{m})^q} \geq \frac{k_1 \sqrt{m}}{1 + k_1 \sqrt{m}}$$

Hence, for q large enough, one has

$$|s(t_{2q+1})| \leq (k_1 \sqrt{m})^q |s(t_q)| - \frac{k_2 \sqrt{m} (k_1 \sqrt{m})^q}{1 + k_1 \sqrt{m}} \tag{14}$$

Since $s(t) = Cx(t)$, thus, it can be easily seen by theorem 1 that, for q large enough,

$$|s(t_q)| \leq \beta. \tag{15}$$

where $\beta > 0$ is a arbitrarily chosen constant.

Substituting (15) into (14) obtains

$$|s(t_{2q+1})| \leq (k_1 \sqrt{m})^q \{\beta - \frac{k_2 \sqrt{m}}{4(1 + k_1 \sqrt{m})}\} \tag{16}$$

If choosing $\xi > \frac{8 \beta \lambda_{\min} (B^T PB)^{-1} (1 + k_1 \sqrt{m})}{\delta \lambda_{\min}^{-\frac{1}{2}} (B^T PB)^{-1} \sqrt{m}}$, then

$$k_2 > \frac{4\beta(1 + k_1 \sqrt{m})}{\sqrt{m}} \tag{17}$$

Further, from (16) and (17), one can obtain

$$|s(t_{2q+1})| < 0 \tag{18}$$

Theorem 2: Suppose that B is of full rank and the conditions of theorem 1 hold,

$$\xi > \frac{8 \beta \lambda_{\min} (B^T PB)^{-1} (1 + k_1 \sqrt{m})}{\delta \lambda_{\min}^{-\frac{1}{2}} (B^T PB)^{-1} \sqrt{m}},$$

then, under the control (4), the trajectories of system (1) can reach the switching function $s(t) = B^T Px$ in a finite time and remain on it .

proof: Obviously, (18) implies that the switching

function *s(t)* is reachable in a finite time. If there exist $t_p (p = 1,2,...)$, which satisfy $t_p \to \infty \;\; as \;\; p \to \infty$, such that

$$\left|s(t_p)\right| > 0 \tag{19}$$

It has been shown from (18) that, when q is large enough,

$$\left|s(t_{2q+1})\right| < \sigma .$$

Suppose that q_0 is so large that for $q > q_0$,

$$\left|s(t_{2q+1})\right| < 0 .$$

Obviously, if p is chosen large enough, then there exist $q_1 > q_0$ such that $t_p \in [t_{2q_1+1}, t_{2q_1+2})$. It has been shown from above discussion that

$$\dot{L}(t) \le -\xi|s|, \qquad t \in [t_{2q_1+1}, t_{2q_1+2})$$

and

$$\left|s(t_{2q_1+1})\right| < 0 .$$

They imply that

$$\left|s(t\;)\right| < 0, \quad t \in [t_{2q_1+1}, t_{2q_1+2})$$

Hence, one can obtain that

$$\left|s(t_p)\right| < 0$$

This contradict (19).

Remark 2: By theorem 2, it is known that the choice of switching energy is dependent on $k_1, m \;\; and \;\; \beta$, where β is a constant which can be arbitrarily chosen. However, it must be indicated that, if β is chosen smaller, the reaching time of switching function will become long.

5. EXAMPLE

Consider the following system

$$Dx(t) = (\begin{bmatrix} -2 & 1 \\ 0 & -1 \end{bmatrix} + \begin{bmatrix} \alpha & 0 \\ \alpha & 0 \end{bmatrix})x(t)$$
$$+ \begin{bmatrix} 0.5 & 0 \\ 0 & 0.5 \end{bmatrix} x(t)Dw(t) + \begin{bmatrix} 0 \\ 1 \end{bmatrix} u(t) + \begin{bmatrix} 0 \\ 1 \end{bmatrix} d(t) \tag{20}$$

where $x(t) \in R^{2}, |\alpha| \le 0.5, |d(t)| \le 1, \delta = 0.1$.

It can be easily seen that, if choosing $R = I_2, Q = \begin{bmatrix} 2.5 & -1 \\ -1 & 3.5 \end{bmatrix}, \gamma = 1,$ then $P = I_2$ is a solution to Riccati equation (3). Design the control as

$$u(t) = -x_2(t) - \delta(t)\,\mathrm{sgn}(x(t)),$$

where $\delta(t) > 1 + 2\|x(t)\|$. Fig-1 and Fig-2 stand for the response of system (20) with the initial conditions $x_1(0) = 2$ and $x_2(0) = 1$ under the above control.

6. CONCLUSION

It is the first time to study the design problem of variable structure control for a class of measure differential systems with impulsive disturbances. The variable structure control was proposed by using Lyapunov function method. In this paper, we also considered the reaching problem of the sliding mode. It has been proved that the given control can ensure the realization of sliding mode motion and the asymptotic stability of closed loop system.

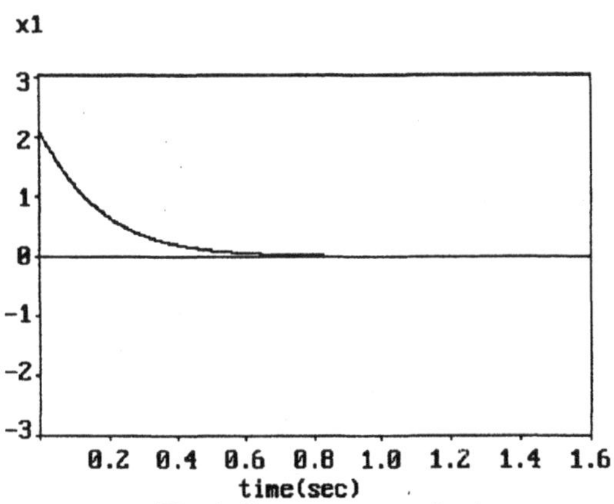

Fig.1 The response of x1

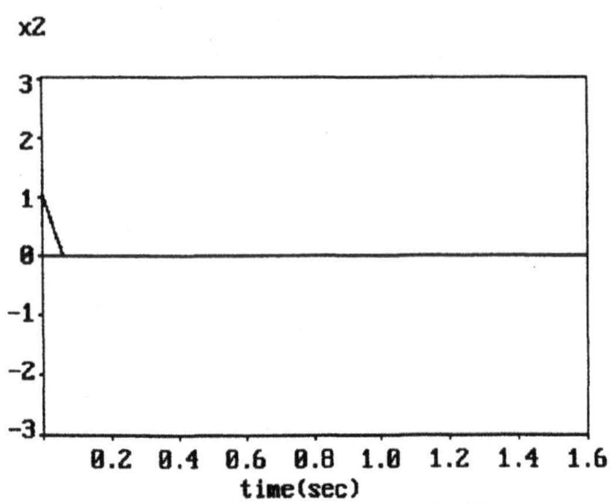

Fig.2 The response of x2

REFERENCES

Das, P.C and P.R.Sharma (1971). On Optimal Controls for Measure Delay-differential Equations. *SIAM.J.Control*, **Vol.9**, 43-61.

Xu Yuan-tong (1988). *Functional Differential*

Equations and Measure Differential Equations, Zhongshan University Press.

Gao Wei-bing (1990). *The Basis of Variable Structure Control Theory,* Academic Press. China.

.Dong Yue (1995). Variable Structure Control of Distributed Parameter Systems with Unbounded Disturbances, *Control Theory and Application*, **No.3.**

---

* *This work is supported by National Science Foundation of China (69874042)*

14th Triennial World Congress, Beijing, P.R. China

C-2a-06-3

SLIDING MODE CONTROLLER-OBSERVER SYSTEMS WITH UNMATCHED UNCERTAINTY

Ali Jafari Koshkouei and **Alan S. I. Zinober**

Applied Mathematics Department,
The University of Sheffield, Sheffield S10 2TN, UK
email: {a.jafari, a.zinober}@sheffield.ac.uk

Abstract: This paper presents sufficient conditions for the sliding mode control of a system with disturbance input. The behaviour of the sliding dynamics and a method for the design of asymptotically stable sliding observers for linear multivariable systems are studied. The sliding observer design ensures that, in the presence of unmatched uncertainty, the estimated state nearly approaches the actual state. Certain sufficient conditions should be satisfied for the asymptotic stability of the error system. *Copyright © 1999 IFAC*

Keywords: Sliding mode control, observers, uncertain linear systems, stabilization methods, variable structure control, feedforward compensation, control design.

1. INTRODUCTION

Sliding mode observers, as well as sliding mode controllers, are known for their robustness and insensitivity with respect to unknown parameter variations. The fundamental difference between a sliding mode observer and other observer designs is that the sliding mode observer is usually discontinuous and the state error trajectories move onto a specified attractive hyperplane. Robustness, insensitivity properties and simplicity of design make sliding observers a powerful approach.

Walcott and Żak (1987) discussed the state observation of nonlinear dynamic systems with bounded nonlinearities/uncertainties. They advocated an observer design method using Lyapunov and min-max methods. Their approach requires the matching condition and is linked to the strictly positive real condition.

Sira-Ramírez and Spurgeon (1994) discussed matching conditions of the sliding mode observer for linear systems and also studied the generalized observer canonical form. Koshkouei and Zinober (1995) have described methods for designing an asymptotically stable observer, the existence of the sliding mode and the stability of state reconstruction systems of MIMO linear systems including disturbance input. They also studied an observer for a SISO linear system with unmatched uncertainty and presented certain conditions for the stability of the system (1998).

Dorling and Zinober (1983) compared the full and reduced order observers with the Utkin observer (1992). They reported some difficulty in the selection of an appropriate constant switched gain to ensure that the sliding mode occurs, and discussed the elimination of chattering. However, the unmatched uncertainty was shown to affect the ideal dynamics prescribed by the chosen sliding surface.

In Section 2 sufficient conditions for the existence of the sliding mode for a linear system with disturbance input are proposed. In Section 3 an approach for observer design for a system which does *not* satisfy the *matching condition*, is developed. Sufficient conditions for the existence of the sliding mode for the reconstruction error system are proposed to ensure ultimate boundedness or asymptotic stability of the error system. When there is unmatched uncertainty, the system stability may not be achieved. However, a region may exist in which the state error trajectory converges to the sliding surface within a finite time, and thereafter remains on this surface to the origin. In this case, the disturbance rejection problem for the sliding system may not be completely satisfied, but when the sliding mode occurs, the state trajectory moves within a neighbourhood of the sliding surface to the origin. An example illustrating the results is presented in Section 4.

In this paper $\sigma_M(\cdot)$, $\sigma_m(\cdot)$ $\lambda_{\max}(\cdot)$ and $\lambda_{\min}(\cdot)$ refer to the largest singular value, the smallest singular value, the largest and the smallest eigenvalue of $(\cdot)$, respectively. Also p.d., p.d.s. and u.p.d.s. stand for positive definite, positive definite symmetric and unique positive definite symmetric.

2. SUFFICIENT SLIDING MODE CONDITIONS FOR SYSTEMS WITH UNMATCHED UNCERTAINTY

Consider the time-invariant system

$$\dot{x}(t) = Ax(t) + Bu(t) + \Gamma\xi(t) \tag{1}$$

$$y(t) = Cx(t) \tag{2}$$

where $x \in \mathbb{R}^n$ is the state variable, $A \in \mathbb{R}^{n\times n}$, $B \in \mathbb{R}^{n\times m}$ is full rank, $u \in \mathbb{R}^m$ is the control, $C \in \mathbb{R}^{m\times n}$ such that CB is non-singular matrix, $y \in \mathbb{R}^m$ is the "output" required to implement sliding mode control, $\Gamma \in \mathbb{R}^{n\times m}$ is the perturbation input map and $\xi \in \mathbb{R}^m$ the bounded disturbance input, i.e. there exists a positive real number M such that $\|\xi\| \leq M$. The real number M is chosen as small as possible and, if $\sup\|\xi(t)\|$ is known, $M = \sup\|\xi(t)\|$. It is assumed that (A, B) is completely controllable and (A, C) completely observable.

The sliding surface is $y = Cx = 0$. The ideal sliding mode occurs if there exist a finite time t_s such that

$$y = Cx = 0 \qquad t \geq t_s \tag{3}$$

where the time t_s is the time when the sliding mode is attained. The equivalent control is the theoretical effective linear control of the system during the sliding mode and is given by

$$u_{eq} = -(CB)^{-1}(CAx + C\Gamma\xi) \tag{4}$$

Choose the control

$$u = -(CB)^{-1}(CAx + K\text{sgn } y) \tag{5}$$

where sgn indicates the signum function, and the matrix $K - C\Gamma M$ is a p.d. matrix. For simplicity, consider the matrix K to be a diagonal matrix with positive real entries such that

$$M\sigma_M(C\Gamma) < \sigma_m(K) \tag{6}$$

Then the output signal is given by

$$\dot{y} = C\Gamma\xi - K\text{sgn } y \tag{7}$$

and, for $y = \begin{bmatrix} y_1 & y_2 & \dots & y_m \end{bmatrix}^T$, the reaching condition of the sliding mode is

$$\dot{y}_i y_i < 0, \qquad \forall i \quad 1 \leq i \leq m \tag{8}$$

Let $C\Gamma\xi = \begin{bmatrix} \gamma_1 & \gamma_2 & \dots & \gamma_m \end{bmatrix}^T$. If (7) is satisfied, then

$$\dot{y}_i = \gamma_i - k_i \text{ sgn } y_i \qquad \forall i,\ 1 \leq i \leq m$$

where t_{is} is the time to reach the surface $y_i = 0$. Therefore, (8) implies

$$\dot{y}_i \text{ sgn } y_i = \gamma_i \text{ sgn } y_i - k_i < 0$$

Hence, a sufficient condition for the state trajectories to converge to the surface $y_i = 0$ is

$$\gamma_i \text{ sgn } y_i < k_i$$

which holds if $|\gamma_i| < k_i$ is satisfied.

For any $1 \leq i \leq m$, there is a real number η_i such that $|\gamma_i| \leq \eta_i < k_i$ and $\eta_i \leq \|C\Gamma\|M$. Hence the i-th output y_i satisfies

$$|y_i| \begin{cases} \geq (\eta_i - k_i)(t - t_{is}) & \text{if } t < t_{is} \\ = 0 & \text{if } t \geq t_{is} \end{cases}$$

Then

$$t_{is} \leq \frac{|y_i(0)|}{k_i - \eta_i}, \qquad \forall i \quad 1 \leq i \leq m$$

Assume the condition (6) is true, then

$$t_s = \max_{1\leq i\leq m} t_{is} \leq \max_{1\leq i\leq m} \frac{y_i(0)}{k_i - \eta_i} \leq \frac{\|y(0)\|}{\sigma_m(K) - M\sigma_M(C\Gamma)}$$

3. SLIDING MODE OBSERVER DESIGN

Sliding observers potentially offer advantages similar to those of sliding controllers, in particular, inherent robustness to parametric uncertainty and straightforward application to important classes of systems. A robust sliding observer for the system (1)-(2) may be selected as

$$\dot{\hat{x}} = A\hat{x} + Bu + H(y - \hat{y}) + \Lambda v \tag{9}$$

$$\hat{y} = C\hat{x} \tag{10}$$

where $\hat{x}$ is the state estimate, $H \in \mathbb{R}^{n\times m}$ the observer gain, $v \in \mathbb{R}^m$ an external discontinuous feedforward compensation signal and $\Lambda \in \mathbb{R}^{n\times m}$ the feedforward injection map such that $C\Lambda$ is a nonsingular matrix. The state reconstruction error is defined as $e = x - \hat{x}$. Subtracting (1) from (9) gives the dynamical reconstruction error system

$$\dot{e} = (A - HC)e + \Gamma\xi - \Lambda v \quad (11)$$

$$e_y = Ce \quad (12)$$

where $e_y = y - \hat{y}$ is the output reconstruction error.

The ideal sliding mode for the system (11)-(12) satisfies $e_y = 0$, $\dot{e}_y = 0$ (Utkin, 1992). The virtual equivalent feedforward input is given by

$$v_{eq} = (C\Lambda)^{-1}(CAe + C\Gamma\xi) \quad (13)$$

Substituting (13) in the state reconstruction error system (11) gives the reduced order system

$$\dot{e} = (I - \Lambda(C\Lambda)^{-1}C)Ae + (I - \Lambda(C\Lambda)^{-1}C)\Gamma\xi \quad (14)$$

with m of the eigenvalues (14) zero and the remaining $n - m$ eigenvalues to be assigned. The reduced order system is independent of the disturbance input signal if there exists an $m \times m$ matrix D such that

$$\Gamma = \Lambda D \quad (15)$$

which is the matching condition. One needs to obtain H, Λ and v so that the stability of the observer system is preserved. The observer gain H can be found in two ways; pole assignment methods and the LQ method.

In the LQ method, the algebraic Riccati equation (ARE)

$$AP + PA^T - PC^TR^{-1}CP = -Q \quad (16)$$

with Q, R are arbitrary semi-p.d.s. and p.d.s. matrices respectively, has a u.p.d.s. matrix solution P. Then $A - HC$ is stable with

$$H = PC^TR^{-1} \quad (17)$$

To ensure that the state approaches and crosses the sliding surface, v should be a discontinuous function. Let P_f be the u.p.d.s. solution of the Lyapunov equation

$$(A - HC)P_f + P_f(A - HC)^T = -Q_f \quad (18)$$

where Q_f is an arbitrary p.d.s. matrix. Consider the discontinuous feedforward input

$$v = W\frac{Ce}{||Ce||} \quad (19)$$

where W is an $m \times m$ diagonal p.d. matrix with

$$\lambda_{\min}(W) \geq M||D||\frac{\lambda_{\max}(CP_f^{-1}C^T)}{\lambda_{\min}(CP_f{}^{-1}C^T)} \quad (20)$$

Assume the condition (15) is satisfied. It is desired to find conditions such that $\lim_{t\to\infty} e(t) = 0$. Set

$$\Lambda = P_f^{-1}C^TW^{-1} \quad (21)$$

Note that $C\Lambda$ is a nonsingular matrix and W is a p.d. matrix. Therefore $C\Lambda W$ is nonsingular and

$$\lambda_{\min}(C\Lambda W) = \lambda_{\min}(CP_f^{-1}C^T) \neq 0$$

The quadratic stability of the reconstruction error system is guaranteed by (20) and (21). A Lyapunov function candidate for (11) is

$$V(e) = e^TP_fe \quad (22)$$

If $Ce \neq 0$, then

$$\begin{aligned}\dot{V} &\leq -e^TQ_fe + 2\,||e^TC^T||\,(||W^{-1}D||M - 1)\\ &\leq -e^TQ_fe + 2\,||e^TC^T||\left(\frac{1}{\lambda_{\min}(W)}||D||M - 1\right)\\ &< 0 \end{aligned} \quad (23)$$

since

$$\lambda_{\min}(W) \geq M||D||\frac{\lambda_{\max}(CP_f^{-1}C^T)}{\lambda_{\min}(CP_f{}^{-1}C^T)} \geq M||D||$$

If $Ce = 0$, $v = v_{eq}$ and

$$\begin{aligned}\dot{V} &= -e^TQ_fe + 2e^TC^TW^{-1}D\xi - 2e^TC^TW^{-1}v_{eq}\\ &= -e^TQ_fe < 0\end{aligned}$$

A condition for the existence of the sliding mode is now established. Since the system is stable, the convergent sliding mode exists, i.e. $\lim_{t\to\infty} e_y(t) = 0$. A region, the so-called sliding region, is now found so that, after a finite time, the state error trajectory lies in the vicinity of the sliding surface and inside the region. Consider

$$\begin{aligned}e_y^T\dot{e}_y &\leq ||Ce||\;||C(A - HC)e|| + ||e^TC^T|| \times\\ &\qquad \lambda_{\max}(CP_f^{-1}C^T)||W^{-1}D||M -\\ &\qquad e^TC^T\lambda_{\min}\left(CP_f^{-1}C^T\right)\frac{Ce}{||Ce||}\\ &\leq ||Ce||\left[||C(A - HC)||.||e|| + \lambda_{\max}(CP_f^{-1}C^T) \times\right.\\ &\qquad \left.||W^{-1}D||M - \lambda_{\min}(CP_f^{-1}C^T)\right]\end{aligned} \quad (24)$$

A sufficient condition for the sliding mode is that the right-hand side of (24) be nonpositive. Thus

$$\begin{aligned}||e|| &\leq \frac{\lambda_{\min}(CP_f^{-1}C^T) - \lambda_{\max}\left(CP_f^{-1}C^T\right)||W^{-1}D||M}{\sigma_M(C(A - HC))}\\ &= r_s\end{aligned} \quad (25)$$

or

$$\begin{aligned}||e|| &\leq \frac{\lambda_{\min}(CP_f^{-1}C^T) - \lambda_{\max}(CP_f^{-1}C^T)||W^{-1}D||M}{\sigma_M(C)\sigma_M(A - HC)}\\ &= \hat{r}_s\end{aligned} \quad (26)$$

3.1 *Sliding Error System with Unmatched Uncertainty*

If the error system in the presence of uncertainty does not satisfy the matching condition, the error system stability is not generally guaranteed.

However, if some conditions on the uncertainty such as boundedness are available, an asymptotic observer can be achieved (Koshkouei and Zinober, 1998). The behaviour of the system depends upon the norm of a matrix which it is named the 'unmatched uncertainty matrix' (Koshkouei and Zinober, 1998). The norm of this matrix, ϵ, is called the 'unmatched uncertainty distance'. If the 'unmatched uncertainty distance' is zero, the matching condition is completely satisfied. Let $D \in \mathbb{R}^{m\times m}$ be a matrix such that

$$\left[0_{m\times(n-m)}\ I_m\right]\Gamma = \left[0_{m\times(n-m)}\ I_m\right]\Lambda D \quad (27)$$

Set $E = \Gamma - \Lambda D$ and $||E|| = \epsilon$.

Definition 1. For the system (1), the uncertainty $\Gamma\xi$ is said to satisfy a matching condition with ϵ-approximation (or ϵ-matching condition) if there exists a matrix D such that

(*i*) condition (27) holds;
(*ii*) $||E|| = ||\Gamma - \Lambda D|| = \epsilon$.

$\Gamma = \Lambda D + E$ where $\epsilon = ||E||$. ΛD is the matched uncertainty matrix and E is the unmatched uncertainty matrix of the matrix Γ. So the unmatched distance is $||E|| = \epsilon$. If $\epsilon \neq 0$, the error system (11) may no longer be generally asymptotically stable. However, if a condition on the disturbance input like the convergence of the norm of disturbance input signal to zero, or the norm of the disturbance signal is bounded by the norm of the output error holds, the system is asymptotically stable. Otherwise, only ultimate boundedness results. So the state error trajectory enters a region centred on the origin and thereafter remains within this region.

Definition 2. The solution $e(t; e_0, t_0)$ of (11) is

(*i*) *ultimately bounded*, with respect to a compact set $S \subset \mathbb{R}^n$, if there exists a nonnegative time $T(t_0, e_0, S)$ such that for all $t \geq t_0 + T(t_0, e_0, S)$, $e(t) \in S$.
(*ii*) *uniformly ultimately bounded* with respect to a compact set $S \subset \mathbb{R}^n$, if $T(e_0, S)$, possibly dependent on e_0 but not on t_0, defined as (*i*) is independent of t_0, i.e. there exists a nonnegative time $T(e_0, S)$ such that for all $t \geq t_0 + T(e_0, S)$, $e(t) \in S$.

Consider the Lyapunov function (22). Then, similarly to (23), one obtains

$$\begin{aligned}\dot{V} &\leq -||e||^2\lambda_{\min}(Q_f) + 2||e||||P_fE||M \\ &= -||e||\{(1-\theta)\lambda_{\min}(Q_f)||e|| + \\ &\qquad \theta\lambda_{\min}(Q_f)||e|| - 2||P_fE||M\}\end{aligned}$$

where $0 < \theta \leq 1$. So, if

$$||e|| \geq \frac{2||P_fE||M}{\theta\lambda_{\min}(Q_f)}, \quad (28)$$

$\dot{V} < 0$. Since V is a monotonically decreasing function on the outside of the set

$$\Omega_E = \left\{e \in \mathbb{R}^n \mid ||e|| \leq \frac{2||P_fE||M}{\theta\lambda_{\min}(Q_f)} = r_\Omega\right\} \quad (29)$$

the maximum value of V on the compact set Ω_E is attained for $||e|| = \dfrac{2||P_fE||M}{\theta\lambda_{\min}(Q_f)}$, and the state error trajectory enters the ellipsoid

$$\mathscr{E}_E = \{e \in \mathbb{R}^n \mid V(e(t)) \leq r^2\} \quad (30)$$

where

$$r = \frac{2M||P_fE||\sqrt{\lambda_{\max}(P)}}{\theta\lambda_{\min}(Q_f)}$$

The set $\mathscr{E}_E$ is closed and bounded. So according to the Heine-Borel theorem the set $\mathscr{E}_E$ is a compact set. The solution of the error system is uniformly ultimately bounded with ultimate boundedness ratio

$$b_u = \frac{2||P_fE||M}{\theta\lambda_{\min}(Q_f)}\sqrt{\frac{\lambda_{\max}(P_f)}{\lambda_{\min}(P_f)}} \quad (31)$$

and ultimate boundedness radius

$$r_u = \inf\{r \in \mathbb{R} \mid \mathscr{E}_E \subset B_r\}$$

where B_r is a ball with radius r centred on the origin. Note that all trajectories starting inside $\mathscr{E}_E$, remain within this set for all future time, and all trajectories starting outside $\mathscr{E}_E$ enter this compact set within a finite time and remain inside thereafter. Hence $e(t)$ is bounded by

$$||e(t)|| \leq \max\{||e(0)||, b_u\} \quad (32)$$

If $||e(0)||$ is sufficiently small, then $||e(t)|| \leq b_u$ for all $t \geq 0$.

Explicit bounds on $e(t)$ and $e(t)^TP_fe(t)$ are now obtained, and it is then shown that the state trajectory enters the set $\mathscr{E}_E$ in a finite time and remains inside $\mathscr{E}_E$ thereafter. Consider

$$\begin{aligned}\frac{d(e^T(t)P_fe(t)e^{\mu t})}{dt} &= \frac{d\left(e^T(t)P_fe(t)\right)}{dt}e^{\mu t} + \\ &\qquad \mu e^T(t)P_fe(t)e^{\mu t} \\ &\leq \left(-||e(t)||^2\lambda_{\min}(Q_f) + 2||e(t)||||P_fE||M\right) \\ &\qquad \times e^{\mu t} + \lambda_{\min}(Q_f)||e(t)||^2e^{\mu t} \\ &= 2||e(t)||.||P_fE||Me^{\mu t} \quad (33)\end{aligned}$$

where $\mu = \lambda_{\min}(Q_f)/\lambda_{\max}(P_f)$. Integrating both sides (33) over the interval $[0, t]$ and multiplying both sides by $e^{-\mu t}$ gives

$$e^TP_fe \leq e^T(0)P_fe(0)e^{-\mu t} + 2b_u||P_fE||Me^{-\mu t}\int_0^t e^{\mu\tau}d\tau$$

since $||e|| \leq b_u$. Therefore

$$e^T P_f e \leq e^T(0) P_f e(0) e^{-\mu t} + 2 b_u M \|P_f E\| \times (1 - e^{-\mu t}) \lambda_{\max}(P_f) / \lambda_{\min}(Q_f) \quad (34)$$

Since after a finite time

$$\|e\| \leq 2\|P_f E\| M / \theta \lambda_{\min}(Q_f)$$

equation (34) shows that the state error $e(t)$ converges to the compact set $\mathscr{E}_E$ (30), i.e.

$$\lim_{t \to \infty} e^T P_f e \leq 4M^2 \|P_f E\|^2 \frac{\lambda_{\max}(P_f)}{\theta \lambda_{\min}^2(Q_f)} \sqrt{\frac{\lambda_{\max}(P_f)}{\lambda_{\min}(P_f)}}$$

and then

$$\lim_{t \to \infty} d(e(t), \mathscr{E}_E) = 0 \quad (35)$$

where d denotes the Euclidean metric on $\mathbb{R}^n$ and

$$d(e(t), \mathscr{E}_E) = \inf_{\alpha \in \mathscr{E}_E} d(e(t), \alpha))$$

The reaching time to $\mathscr{E}_E$ is finite, otherwise for all $t > 0$, $\dot{V}(e(t)) < 0$ and $V(e(t))$ tends to zero asymptotically. The result (35) shows that the state error trajectory enters $\mathscr{E}_E$ at finite time t_e and after this time remains inside $\mathscr{E}_E$. Hence, boundedness is guaranteed in the presence of a bounded disturbance with possibly unknown bound. But the size of $\mathscr{E}_E$ cannot be estimated *a priori* if no bound on the disturbance input is given.

Equation (34) also yields

$$\|e\| \leq \|e(0)\| e^{-\frac{1}{2}\mu t} \sqrt{\frac{\lambda_{\max}(P_f)}{\lambda_{\min}(P_f)}} + \frac{2M\|P_f E\|}{\sqrt{\theta} \lambda_{\min}(Q_f)} \times \sqrt{1 - e^{-\mu t}} \left(\frac{\lambda_{\max}(P_f)}{\lambda_{\min}(P_f)} \right)^{3/4}$$

When $t \to \infty$

$$\|e\| \leq \frac{2M\|P_f E\|}{\sqrt{\theta} \lambda_{\min}(Q_f)} \left(\frac{\lambda_{\max}(P_f)}{\lambda_{\min}(P_f)} \right)^{3/4} = \sqrt{\theta} \left(\frac{\lambda_{\max}(P_f)}{\lambda_{\min}(P_f)} \right)^{1/4} b_u = \hat{b}_u$$

So

$$\hat{b}_u \begin{cases} > b_u & \text{if } 1 \geq \theta > \eta \\ = b_u & \text{if } \theta = \eta \\ < b_u & \text{if } \theta < \eta \end{cases}, \quad (36)$$

where $\eta = \sqrt{\dfrac{\lambda_{\min}(P_f)}{\lambda_{\max}(P_f)}}$. On the other hand, from (29),

$$\hat{b}_u \begin{cases} > r_\Omega & \text{if } 1 \geq \theta > \eta^3 \\ = r_\Omega & \text{if } \theta = \eta^3 \\ < r_\Omega & \text{if } \theta < \eta^3 \end{cases} \quad (37)$$

However, $b_u \geq r_\Omega$ and the equality holds if $P = pI$ $(p > 0)$. Although Ω_E is not the smallest ultimately bounded set, when Ω_E is a small neighbourhood about the origin, the concept of uniform ultimate boundedness is tantamount to 'practical' asymptotic stability.

Note that, for suitable performance, one should choose $\theta = 1$ or certainly close to 1.

Remark 3. When the matching condition is not satisfied ($\epsilon \neq 0$), the gain matrix W should be chosen so that

$$\lambda_{\min}(W) \geq \frac{M\|D\|\lambda_{\max}(CP_f^{-1}C^T) + M\|CE\|}{\lambda_{\min}(CP_f{}^{-1}C^T)} \quad (38)$$

In this case the sliding region is

$$\mathscr{S}_r = \{e \in \mathbb{R}^n \mid \|e\| \leq r_2\} \quad (39)$$

where

$$r_2 \leq r_s - \frac{M\|CE\|}{\sigma_M(C(A - HC))} \quad (40)$$

with r_s as defined in (25).

4. EXAMPLE

The example below illustrates our results regarding the sliding mode, stability of the error system and observer design. Consider the system

$$\dot{x} = \begin{bmatrix} -1 & 2.25 & -0.4 \\ 0 & -1.50 & -2.0 \\ 0 & 0.50 & -1.0 \end{bmatrix} x + \begin{bmatrix} 0.30 & 0 \\ 0 & 1 \\ 0.12 & 0 \end{bmatrix} u + \begin{bmatrix} -0.0007 & 0.0015 \\ 0.0103 & 0.0041 \\ 0.0188 & 0.0188 \end{bmatrix} \xi$$

$$y = \begin{bmatrix} 0.2 & 0.4 & 0.89 \\ 0.2 & 0 & 1.00 \end{bmatrix} x$$

Suppose ξ is a bounded random signal satisfying $\|\xi\| < 0.1$. Using the LQ method with $R = I_2$ and $Q = I_3$, from (17)

$$H = \begin{bmatrix} 0.4697 & 0.1938 \\ 0.1936 & -0.0470 \\ 0.2004 & 0.3083 \end{bmatrix}$$

The eigenvalues of $A - HC$ are $-1.7043 \pm 1.0365i$, -0.7881. Let $Q_f = 5I_3$. Take $W = \text{diag}\,(2.5, 2.5)$. From (21)

$$\Lambda = \begin{bmatrix} -0.0117 & 0.0244 \\ 0.1709 & 0.0676 \\ 0.3133 & 0.3136 \end{bmatrix}$$

The reduced order error system is independent of ξ since $\Gamma = 0.06\Lambda$. In fact $D = 0.06I_2$. So the error system is quadratically stable. The minimum eigenvalue of $C\Lambda W$ is 0.0510. From (25) $r_s = 0.0156$. Simulation results are presented in Fig. 1.

Alternatively the eigenvalues of $A - HC$ can be specified using pole assignment technique. Suppose the desired eigenvalues of $A - HC$ are

$-1,\ -1 \pm 0.15i$. The existence of H is guaranteed by the observability of (A, C). Using the MATLAB Control Toolbox

$$H = \begin{bmatrix} 5.7656 & -5.5452 \\ -1.3486 & -0.7121 \\ 1.2296 & -1.0990 \end{bmatrix}$$

For $Q_f = 3I_3$, equation (21) gives

$$\Lambda = \begin{bmatrix} 0.0400 & 0.0568 \\ 0.0991 & 0.0020 \\ 0.2407 & 0.2658 \end{bmatrix}$$

So

$$D = \begin{bmatrix} 0.1044 & 0.0407 \\ -0.0238 & 0.0339 \end{bmatrix}$$

and then

$$E = \Gamma - \Lambda D = \begin{bmatrix} -0.0035 & -0.0021 \\ 0 & 0 \\ 0 & 0 \end{bmatrix}$$

with norm $||E|| = \epsilon = 0.0041$. So the matching condition is not satisfied. However, the state error trajectory enters a small sphere about the origin with radius

$$r_\Omega = 2M||P_f E||/\lambda_{\min}(Q_f) = 0.00039$$

and $b_u = 0.00047$. So the stability of the error system is nearly achieved. The sliding region radius (39) is $r_2 = 0.0328$. Simulation results are presented in Fig.2.

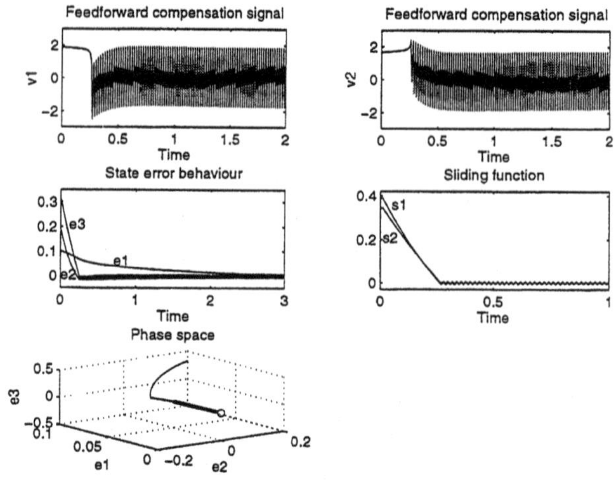

Fig. 1. Responses for the LQ method.

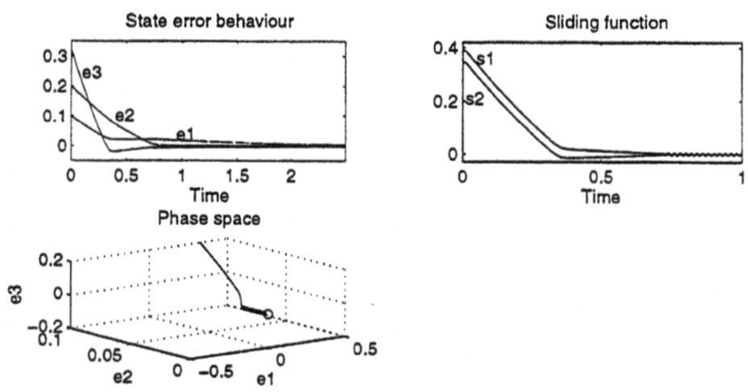

Fig. 2. Responses of for the pole assignment technique.

5. CONCLUSIONS

The existence of the sliding mode guarantees that the state trajectories converge to a sliding surface at a finite time and then move along the surface to the origin. However, the system may generally not be stable. For the system to be asymptotically stable, some further conditions are needed. For the system with unmatched uncertainty some relaxed sliding conditions to impose asymptotic stability have been studied .

A sliding observer design method has been proposed such that the estimated state nearly approaches the actual state. Certain sufficient conditions should be satisfied for the asymptotic stability of the system. These conditions are limitations on the uncertainty input. Otherwise, this method ensures only that the estimated state error approximately tends to the actual state. The accuracy of this approximation is given by (31); the state error trajectory enters a certain set in finite time and remains within the set thereafter. The sliding mode occurs after a finite time. So there exists a finite time after which the state trajectories remain in the 'sliding region' (39) and move to the origin along the sliding surface.

References

Dorling, C. M. and A. S. I. Zinober (1983). A Comparative study of the sensitivity of observers, *Proc. First IASTED Symposium on Applied Control and Identification*, Copenhagen, 6.32–6.37.

Koshkouei, A. J. and A. S. I. Zinober (1995). Sliding mode state observers for multivariable systems, *Proc. 34th IEEE Conf. Decision and Control*, New Orleans, 2115-2120.

Koshkouei, A. J. and A. S. I. Zinober (1998). Sliding mode controller-observer design for SISO linear systems, *Int. J. Systems Science*, **12**, 1363–1373.

Sira-Ramírez, H. and S. K. Spurgeon (1994). On the robust design of sliding mode observers for linear systems, *Systems & Control Letters*, **23**, 9–14.

Utkin, V. I. (1992). *Sliding Modes in Control and Optimization*, Springer-Verlag, Berlin.

Walcott, B. L. and S. H. Żak (1987). State observation of nonlinear uncertain dynamical systems, *IEEE Trans. Automat. Control*, **32**, 166–170.

C-2a-06-4
14th Triennial World Congress, Beijing, P.R. China

A NEW APPROACH TO CONTROL OF DYNAMIC SYSTEMS WITH UNILATERAL CONSTRAINTS

Boris M. Miller and Joseph Bentsman

Institute for Information Transmission Problems GSP-4, *B. Karetny Per.* 19, 101447 *Moscow, Russia*

Department of Mechanical and Industrial Engineering University of Illinois at Urbana-Champaign 1206 *W. Green St. Urbana IL* 61801 *USA*

Abstract: The problem of control for dynamic systems subjected to unilateral constraints is considered. The new approach to the description of generalized (discontinuous) solution based on the idea of discontinuos time transformation was suggesed. This approach gives the opportunity to describe the generalized solution by the differential equtions with a measure and to reduce the original optimization problem to some problem of nonsmooth optimization. As a result the optimality conditions in the special form of maximum principle can be obtained. *Copyright © 1999 IFAC*

Keywords: dynamic systems, unilateral constraints, impulses, optimal control, generalized solutions

1. INTRODUCTION

The dynamic systems with constraints can be frequently treated as systems characterized by continuos and discrete behavior. The motion of such systems can be divided into regular and singular parts, i.e., continuous and jumping, respectively. These systems are very typical for various mechanical applications, where the discrete-continuous modes of motions could arise because of shocks and friction. There has been a significant progress in this area, including the development of the rigorous mathematical framework for the description of these systems and preliminary formulations of the procedures for synthesis of control laws for them. However, the common mathematical feature of these class of systems is the presence of singularities, which manifest themselves in: discontinuities and nonsmoothness in system motion, jumps in system dimension, the lack of the continuous dependence on initial conditions and nonuniqiueness of solution of equations of motion (see, for example, (Brogliato, 1996; Jean and Moreau, 1991).

Traditionally the control of such systems has been exerted either during the nonsingular phase of the system motion or during the singularity phase, which was induced by the control action itself and did not naturally exist in the system (Miller, 1993; Miller 1996). However, in dynamic systems with constraints the impulsive actions arise due to the interaction of the system with obstacles, therefore, they are of a type of impulsive feedback form. Theory of impulsive controls, which was developed mainly for open loop controls, cannot be directly applied to systems, where impulsive actions arise, when the system achieves the boundary of some set (obstacle). It is also known that unilateral constrains must be weakly violated at the points of impact giving rise to fine spatial structure of the system trajectory in the vicinity of the point of jump. Due to these fine structures on one hand and the slow dynamics of the system to be controlled on the other hand the resulting controlled system possesses the multiscale time-space dynamics. The proper tool for the description of the resulting system is then the multi-resolution time-space analysis such as that provided by a wavelet decomposition or more specifically by a discontinuous time transformation method in impulsive control (Miller, 1993; Miller 1996). The new approach is to consider a jump as a result of some "fictitious motion" along the paths of some auxiliary system, which provides a model of "fast motion" and describes the jump, arising in the motion of dynamic system, in terms of some shift operator. This approach bases on the representation of robust hybrid system, which was obtained in (Miller, 1997), where discrete-continuos systems are treated as systems with impulsive inputs. However, if we consider these systems as ones with impulsive actions in feedback form, it becomes necessary to find a more general mathematical framework, than for standard problems with impulse controls. The main approach is based on the results obtained in (Miller, 1998) and (Bentsman and Miller, 1998), where the problem statement and results concerning the generalized solution representation and the existence of the optimal generalized solution were given. The main goal of this paper is to obtain the necessary optimality conditions in the form of generalized maximum principle.

2. PROBLEM STATEMENT

Consider the evolution of discrete-continuous dynamical system with unilateral constraint, whose behavior be described on some interval $[0,T]$ by variable $X(t) \in R^n$, which satisfies the differential equation

$$\dot{X}(t) = F(X(t), u(t)), \tag{1}$$

with given initial condition $X(0) = x_0 \in R^n$. The actions of unilateral constraint of the form $G(X(t)) \leq 0$, where $G(\cdot)$ is continuos and continuously differentiable, will be described by following intermediate conditions

$$X(\tau_i) = X(\tau_i-) + \Psi(X(\tau_i-)), \tag{2}$$

which are given for some sequence of instants $\{\tau_i, i = 0, ..., N\}, N \leq \infty$, satisfying the recurrence conditions

$$\tau_0 = 0$$

$$\tau_i = \begin{cases} \inf\{\tau_{i-1} < t \leq T : G(X(t-) = 0\}, \\ \infty, \quad \text{if the appropriate set is empty.} \end{cases} \tag{3}$$

In equation (2) $X(\tau_i-) = \lim_{t\uparrow\tau_i} X(t)$, and τ_i is the sequence of instants when the system states can change discontinuously on the boundary of constraint.

So, the state of system changes continuously in halfintervals $[0,\tau_1)$, ... $[\tau_{i-1},\tau_i)$, ... $[\tau_N,T]$, and undergoes a sudden change at every instant τ_i, whose value, due to equation (2), depends on the state preceding the jump.

We suppose that control variable in (1)

$$u \in U \subset R^m, \tag{4}$$

where U is some compact set, and function $F(X,u)$ be continuous with respect to all variables and continuously differentiable with respect to X. To be sure that solution of (1) is continuable to the right we need some additional assumptions concerning the functions $\Psi(X)$ and $G(X)$. So we suppose that $X(\tau_i) = X(\tau_i-) + \Psi(X(\tau_i-))$ is the result of the action of shift operator along the paths of differential equation

$$\dot{y}(s) = B(y(s)), \quad s \in [0,\infty) \tag{5}$$

with initial condition $y(0) = X(\tau_i-)$.

Therefore, if $\Phi(x,s)$ is the general solution of (5) with initial condition $y(0) = x$, then

$$X(\tau_i) = X(\tau_i-) + \Psi(X(\tau_i-)) = \Phi(X(\tau_i-), s^*(X(\tau_i-)), \tag{6}$$

where

$$s^*(X(\tau_i)) = \inf\{s > 0 : G(\Phi(X(\tau_i-), s) = 0\}, \quad (7)$$

and on the interval $(0, s^*(X(\tau_i))$ we have the relation $G(\Phi(X(\tau_i-), s) > 0$. We assume also that $B(y)$ be continuously differentiable with respect to y.

The typical path of the hybrid system is shown in figure 1, where τ is the instant of the hitting the boundary of the set $G(X) \leq 0$. The solid line shows the real motion of the system, which is processed in ordinary time scale. The fictitious motion, which corresponds to the fast phase of the impulsive control action is shown by dashed line.

The optimization problem to be considered is the minimization of performance criterion

$$J[X(\cdot), u(\cdot)] = \phi_0(X(T)) \quad (8)$$

with some continuously differentiable function ϕ.

The description of jump by some shift operator looks like rather artificial, however, all discrete-continuous systems that are stable (or robust) with respect to an approximation procedure of impulsive input admit such jump representation (Miller, 1997). The examples of the paths representation for mechanical systems with elastic and non-elastic shocks can be found in (Miller, 1998), (Bentsman and Miller, 1998).

3. EXISTENCE THEOREM AND PROBLEM REDUCTION

Following the line of paper (Miller, 1998) define on the interval $[0, T_1]$ the auxiliary system

$$\dot{y}(s) = \alpha(s)F(y(s), u(\eta(s))) + (1 - \alpha(s))B(y(s)),$$

$$\dot{\eta}(s) = \alpha(s) \quad (9)$$

with initial conditions

$$y(0) = x_0, \quad \eta(0) = 0.$$

Consider the following auxiliary problem for system (9) with controls $\{\alpha, u_1\}$, satisfying

$$\alpha(s) \in [0, 1],$$
$$u_1(s) \in U \quad \text{a. e.} \quad \text{on} \quad [0, T_1], \quad (10)$$

and such that $y(\cdot)$ satisfies (11) and $\eta(T_1) = T$, and such that the integral constraints

$$\int_0^{T_1} \alpha(s)G^+(y(s))ds = 0,$$
$$\int_0^{T_1} (1 - \alpha(s))G^-(y(s))ds = 0, \quad (11)$$

where

$$G^+(y) = \max\{G(y), 0\}, \quad G^-(y) = \min\{G(y), 0\}.$$

The problem will be considered on nonfixed interval $[0, T_1]$, such that $T_1 < \infty : \eta(T_1) = T$ with performance criterion

$$J'\{y(\cdot), \alpha(\cdot), u_1(\cdot)\} = \phi_0(y(T_1)) \to \min,$$

where ϕ_0 is the same as in (8).

Theorem 1. Suppose that the set $F(X, U)$ be convex for any $X \in R^n$, the set of admissible controls of auxiliary problem is non empty, and the set of admissible T_1, such that $\eta(T_1) = T$ is uniformly bounded. Then the auxiliary problem has the optimal solution $\{y^0(\cdot), \eta^0(\cdot), \alpha^0(\cdot), u_1^0(\cdot)\}$ and the optimal generalized solution of the original problem satisfies the equation

$$dX^0(t) = F(X^0(t), u^0(t))dt + B(X^0(t))d\mu^c(dt) +$$
$$\sum_{\tau \leq t} [\Phi(X^0(\tau-), \mu(\{\tau\})) - X^0(\tau-)]\delta(t - \tau)dt, \quad (12)$$

satisfying the constraint

$$G(X^0(t)) \leq 0, \quad \text{for any} \quad t \in [0, T]. \quad (13)$$

with

$$u^0(t) = u_1^0(\Gamma^0(t), \quad \mu\{[0, t]\} = \int_0^{\Gamma^0(t)} (1 - \alpha^0(s))ds,$$

where

$$\Gamma^0(t) = \inf\{s : \eta^0(s) > t\}, \quad X^0(t) = y^0(\Gamma^0(t)).$$

4. NECESSARY OPTIMALITY CONDITIONS

The auxiliary problem belongs to a class of nonsmooth optimization problems due to the nondifferentiability of functions G^+ and G^-. However, by applying the methods recently obtained for

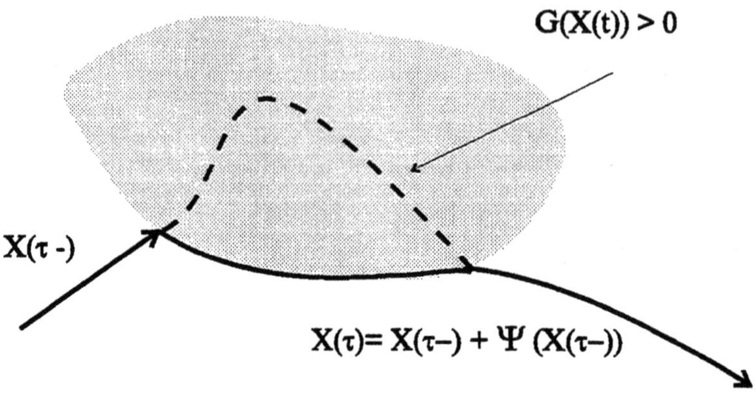

Figure 1: Path of discrete-continuous system

nonsmooth problems (see, for example (Clarke, 1983), it becomes possible to derive necessary optimality conditions in the form of generalized maximum principle.

To formulate the result introduce the notation

$$\mathcal{H}_0(\psi_x, \psi_\eta, \psi_{\mu,1}, x, u) =$$

$$< \psi_x, F(x,u) > + \psi_\eta + \psi_{\mu,1} G^+(x),$$

$$\mathcal{H}_1(\psi_x, \psi_{\mu,2}, x) =$$

$$< \psi_x, B(x) > + \psi_{\mu,2} G^-(x),$$

and define the subsets D_μ, D^c_μ, and D^d_μ, which are supports of measures $\mu(dt)$, $\mu^c(dt)$, and $\mu^d(dt)$, respectively.

Theorem 2. Suppose that $\{X^0(\cdot), u^0(\cdot), \mu(\cdot)\}$ be the optimal path, control and corresponding measure, localized on the set $\{t : G(X^0(t) = 0\}$ in the problem (1)-(8). Then there exist the set of elements that are non equal to zero simultaneously:

1. the set of constants $\lambda_0 \geq 0$, $\psi^0_\eta, \psi^0_{\mu,1}, \psi^0_{\mu,2}$;

2. the nonnegative measures $\tilde{\lambda}_1(dt)$ and $\tilde{\lambda}_2(dt)$ which are localized on the set $\{t : G(X^0(t) = 0\}$ and being absolutely continuos with respect to Lebesgue measure and measure $\mu^c(dt)$, respectively;

3. the set of nonnegative measures $\lambda_2^\tau(ds)$, defined for every $\tau \in D^d_\mu$ on the corresponding interval $[0, \mu\{\tau\}]$, being absolutely continuos with respect to the Lebesgue measure and localized on the sets $\{\omega \in [0, \mu\{\tau\}] : G(\Phi(X^0(\tau-), \omega) = 0)\}$;

4. adjoint variable $p^0_x(\cdot)$, which satisfies the system of equations

$$\begin{aligned} p^0_x(t) &= -\lambda_0 \varphi'_{0_x}(X(T)) + \\ &\int_t^T < p^0_x(\tau), F'_x(X^0(\tau), u^0(\tau)) > d\tau + \\ &\int_t^T < p^0_x(\tau), B'_x(X^0(\tau)) > d\mu^c(\tau) + \\ &\int_t^T \psi^0_{\mu,1} G'_x(X^0(\tau)) d\tilde{\lambda}_1(\tau) + \\ &\int_t^T \psi^0_{\mu,2} G'_x(X^0(\tau)) d\tilde{\lambda}_2(\tau) + \\ &\sum_{\tau > t, \tau \in D^d_\mu} \Delta p^0_x(\tau), \end{aligned} \tag{14}$$

where $\Delta p^0_x(\tau)$ be defined by relations

$$\begin{aligned} \Delta p^0_x(\tau) &= p^0_x(\tau) - p^0_x(\tau-) = \\ & p_{\tau,x}(\mu\{\tau\}) - p_{\tau,x}(0), \end{aligned} \tag{15}$$

and $p_{\tau,x}(s)$ satisfies on the interval $[0, \mu\{\tau\}]$

the following equation

$$p_{\tau,x}(s) = p_x^0(\tau)+$$

$$\int_s^{\mu\{\tau\}} < p_{\tau,x}(\omega), B_x^{'}(\Phi(X^0(\tau-),\omega)) > d\omega +$$

$$\int_s^{\mu\{\tau\}} \psi_{\mu,2} G_x^{'}(\Phi(X^0(\tau-),\omega) d\tilde{\lambda}_2^{\tau}(\omega). \tag{16}$$

such that the following inequalities (*generalized maximum principle*) are valid on the appropriate subsets, namely:

$$0 = \mathcal{H}_0(p_x^0(t), \psi_\eta^0, 0, X^0(t), u^0(t)) =$$

$$\max_{u \in U} \mathcal{H}_0(p_x^0(t), \psi_\eta^0, 0, X^0(t), u) \geq \tag{17}$$

$$\mathcal{H}_1(p_x^0(t), \psi_{\mu,2}, X^0(t)),$$

which is valid a.e. on the set $\{t \in [0,T] \setminus D_\mu\}$;

$$\max_{u \in U} \mathcal{H}_0(p_y^0(t), \psi_\eta^0, 0, X^0(t), u) \leq \tag{18}$$

$$\mathcal{H}_1(p_x^0(t), 0, X^0(t)) = 0$$

a.e. with respect to measure $\mu(dt)$ on the set $\{t \in D_\mu^c\}$;
and

$$\max_{u \in U} \mathcal{H}_0(p_{\tau,x}(\omega), \psi_\eta^0, \psi_{\mu,1}, \Phi(X^0(\tau-),\omega), u) \leq$$

$$\mathcal{H}_1(p_{\tau,x}(s), 0, \Phi(X^0(\tau-),\omega)) = 0, \tag{19}$$

for variables satisfying (16), $\tau \in D_\mu^d$.

5. CONCLUSION

So the new approach to the description of dynamic systems with unilateral constraints had been developed. The main achievement is the new concept of generalized solution, which can be obtained by time substitution in the solution of standard system, described by ordinary differential equations. This approach give the opportunity to prove the existence theorem and to reduce the originally singular control problem to a regular one and to derive the optimality conditions.

REFERENCES

Bentsman J. and B. M. Miller (1998). Control of dynamic systems with unilateral constraints and differential equations with measures. In: *Proceedings of 4th IFAC Workshop on Nonlinear Control Systems Design,* NOLCOS'98. **2**. 411–416. University of Twente, Enschede, The Netherlands.

Brogliato B. (1996). *Nonsmooth Impact Mechanics. Models, Dynamics and Control.* Lecture Notes in Control and Information Sciences. **220**. Springer-Verlag.

Clarke F. H. (1983). *Optimization and Nonsmooth Analysis,* John Wiley & Sons, New York, Chichester, Brisbane, Toronto, Singapore.

Jean M. and J.J. Moreau (1991). Dynamics of elastic or rigid bodies with frictional contact: numerical methods. *Publications of Laboratory of Mechanics and Acoustics* Marseille, April, **124**

Miller B. M. (1993). Method of discontinuous time change in problems of control for impulse and discrete-continuous systems. *Automation and Remote Control.* **54. No 12**. 1727–1750.

Miller B. M. (1996). The generalized solutions of nonlinear optimization problems with impulse control. *SIAM J. Control and Optimization.* **34, No 4**. 1420–1440.

Miller B. M. (1997). Representation of robust and non-robust solutions of nonlinear discrete-continuous systems. In: *Proceedings of International Workshop on Hybrid and Real Time Systems.* HART'97. Ed.: Oded Maler. Grenoble, March 28–28, France.

Miller B. M. (1998). Optimization of generalized solutions of nonlinear hybrid (discrete-continuous) systems. In: *Proceedings of First International Workshop on Hybrid Systems: Computation and Control.* HSCC'98. Eds.: Tomas Henzinger and Sankar Sastry. Lecture Notes in Computer Science. **1386**. 334–345. Berkeley, California, USA.

14th Triennial World Congress, Beijing, P.R. China

C-2a-06-5

OUTPUT FEEDBACK VARIABLE STRUCTURE CONTROL DESIGN FOR SINGULAR SYSTEMS

Ma Shuping Cheng Zhaolin

Department of Mathematics, Shandong University, Jinan 250100, P.R. China.
E-mail:Chengzha@jn-public.sd.cninfo.net

Abstract: In this paper, the problem of output feedback variable structure control design for multivariable linear singular systems is discussed. An algorithm of switching surface design and stability analysis of sliding mode motion are given, the output feedback variable structure control design and global asymptotic stability analysis of closed-loop are given. The control design given in this paper assures that closed-loop trajectory converges to zero with arbitrary exponent speed. *Copyright © 1999 IFAC*

Keywords: variable structure control; output feedback; linear singular systems; matrix spectrum analysis; Lyapunov method.

1. INTRODUCTION

Variable structure control system design is a nonlinear control method with control switch developed in Russia about fourty years ago (Utkin, 1978), because it has robustness to disturbance, in recent years, it has received considerable attention. There have been many mature works about variable structure control for multivariable linear systems, such as Bondarev, *et al.*(1985), EL-Khazali and Decarlo (1993, 1995). Two ways were used for control design in these works, which are full-order state feedback via observer and output feedback. Comparing these two ways, the former is easy in theory, but need observer dynamic, which complicates control design in practice. In last decade, many works focused on output feedback. As for the problem of variable structure control for multivariable linear singular systems, since the structure of singular systems is complicated, and the system state trajectory always contains impulsive behaviour, there are more difficulties need to be conquered for singular systems than for linear systems, so at present, the work is in the begining, there are not many representative works, such as Kawaji and Taha (1994), in these papers, the control structure all used the full-order state feedback with observer.

In this paper, the problem of output feedback variable structure control design for multivariable linear singular systems is discussed, there is no report about this question as the author known, and controller using the output feedback accords further with the practical demands.

2. PRELIMINARIES

This section gives some results of the linear singular systems

$$\begin{aligned} E\dot{x} &= Ax + Bu, \qquad Ex(0^-) = \alpha_0 \\ y &= Cx \end{aligned} \tag{1}$$

where $E, A \in R^{n\times n}$, $B \in R^{n\times r}$, E is singular, and the pencil $sE - A$ is regular, i.e. $\det[sE - A] \not\equiv 0$. Furthermore, we assume that the system (1) contains impulse mode, which is

$$\text{degree det } (sE - A) < \text{rank}E. \quad (2)$$

It is well known, for the system (1), there exist nonsingular matrices $P, Q \in R^{n\times n}$ such that

$$\begin{aligned} PEQ &= \begin{bmatrix} I_{n_1} & & \\ & J & \\ & & 0 \end{bmatrix} \\ PAQ &= \begin{bmatrix} A_1 & & \\ & I_{n_2} & \\ & & I_{n_3} \end{bmatrix} \\ PB &= [B_1^T \ B_2^T \ B_3^T]^T \\ CQ &= [C_1 \ C_2 \ C_3] \\ \tilde{x} &= [x_1^T \ x_2^T \ x_3^T]^T, \ x = Q\tilde{x} \end{aligned} \quad (3)$$

where $A_1 \in R^{n_1\times n_1}$, $I_{ni} \in R^{n_i\times n_i}$ is a unit matrix, $B_i \in R^{n_i\times r}$, $x_i \in R^{n_i}$, $i = 1,2,3$, $J \in R^{n_2\times n_2}$ is a nilpotent Jordan matrix with nilpotent index μ, the order of each Jordan block of J is not less than 2. Call the following system

$$\dot{x}_1 = A_1x_1 + B_1u, \quad x_1(0) = \alpha_{01} \quad (4a)$$

$$J\dot{x}_2 = x_2 + B_2u, \quad Jx_2(0^-) = \alpha_{02}, \quad x_3 = -B_3u \quad (4b)$$

$$y = C_1x_1 + C_2x_2 + C_3x_3 \quad (4c)$$

$$\alpha_{01} = [I_{n_1} \ 0 \ 0]P\alpha_0, \quad \alpha_{02} = [0 \ I_{n2} \ 0]P\alpha_0 \quad (4d)$$

as the Weierstrass canonical form of system (1). Noticing that the state trajectory of (4a), (4b) is:

$$\begin{aligned} x_1(t) &= e^{A_1t}x_1(0) + \int_0^t e^{A_1(t-\tau)}B_1u(\tau)d\tau \\ x_2(t) &= -\sum_{i=0}^{\mu-1} J^iB_2u^{(i)}(t) - (\sum_{i=0}^{\mu-2} \delta^{(i)}(t)J^{i+1}) \\ &\quad (x_2(0^-) + \sum_{i=0}^{\mu-2} J^iB_2u^{(i)}(0)) \end{aligned} \quad (5)$$

where $u^{(i)}(t)$ and $\delta^{(i)}(t)$ are the ith-order derivative of $u(t)$ and Dirac-δ function, i.e $\delta(t)$, respectively. Then, to assure that there is no impulsive behaviour in the system state trajectory, the initial values of the control function and its derivatives must and only must satisfy the following consistent condition:

$$Jx_2(0^-) = -\sum_{i=0}^{\mu-2} J^{i+1}B_2u^{(i)}(0). \quad (6)$$

Denoting the control function set satisfying (6) as U, we have:

Lemma 1 (see Chen, et al.1996). For any given $Jx_2(0^-)$, U is nonempty, that is (6) has solution $u^{(i)}(0)$, $i = 0, 1, \cdots, \mu - 2$, if and only if:

$$\text{rank}[B_2 \ \ JB_2 \cdots J^{\mu-1}B_2] = n_2 \quad (7)$$

Lemma 2 (see Chen, et al.1996). Matrix $[B_2 \ \ JB_2 \ \cdots \ J^{\mu-1}B_2]$ is of full column rank if and only if $J^{\mu-1}B_2$ is of full column rank.

Lemma 3 (see Chen, et al.1996). For any given $Jx_2(0^-)$, equation (8) has the solution $u^{(i)}(0)$, $i = 0, 1, \cdots, \mu - 2$, if and only if:
1) rank $[B_2 \ \ JB_2 \ \cdots J^{\mu-1}B_2] = n_2$;
2) $J^{\mu-1}B_2$ is of full column rank.

Remark. The meaning of lemma 3 lies in: it points out that the no impulse state trajectory and the control of this kind of systems can be described by the following standard state-space linear system:

$$\begin{bmatrix} \dot{x}_1 \\ \dot{z} \end{bmatrix} = \begin{bmatrix} A_1 & G \\ 0 & D \end{bmatrix} \begin{bmatrix} x_1 \\ z \end{bmatrix} + \begin{bmatrix} 0 \\ F \end{bmatrix} v \quad (8)$$

$$\begin{bmatrix} x_1(0) \\ z(0) \end{bmatrix} = \begin{bmatrix} \alpha_{01} \\ H\alpha_{02} \end{bmatrix} \quad (9)$$

$$x_2(t) = -[B_2 \ \ JB_2 \ \cdots J^{\mu-2}B_2 \ \ J^{\mu-1}B_2] \begin{bmatrix} z(t) \\ v(t) \end{bmatrix} \quad (10)$$

$$x_3(t) = -B_3Lz(t) \quad (11)$$

where

$$\begin{aligned} &H = -(H_1^TH_1)^{-1}H_1^T \\ &H_1 = [JB_2 \ \ J^2B_2 \ \cdots J^{\mu-1}B_2] \\ &G = [B_1 \ 0 \ \cdots 0], \quad G \in R^{n_1\times(\mu-1)r} \\ &F = [0 \ \cdots \ 0 \ \ I_r]^T, \quad F \in R^{(\mu-1)r\times r} \\ &L = [I_r \ 0 \ \cdots \ 0], \quad L \in R^{r\times(\mu-1)r} \\ &D = \begin{bmatrix} 0 & I_r & & \\ & 0 & \ddots & \\ & & \ddots & I_r \\ & & & 0 \end{bmatrix}, \\ &D \in R^{(\mu-1)r\times(\mu-1)r} \\ &z(t) = [u^T(t) \cdots (u^{(\mu-2)}(t))^T]^T \\ &v(t) = u^{(\mu-1)}(t) \end{aligned} \quad (12)$$

where α_{01}, α_{02} are shown as (4d). The above discussion gives us a fine method to discuss the no impulse state trajectory and the control of singular system (1).

3. OUTPUT FEEDBACK VARIABLE STRUCTURE CONTROL DESIGN

This section discusses the output feedback variable structure control design for singular systems

$$\begin{aligned} E\dot{x} &= Ax + Bu, \quad Ex(0^-) = \alpha_0 \\ y &= CEx \end{aligned} \quad (13)$$

where $E, A \in R^{n\times n}$, $B \in R^{n\times r}$, $C \in R^{m\times n}$, $x \in R^n, u \in R^r, y \in R^m$ are the state vector, the control vector and the output vector respectively. For relating simplicity, we use the Weierstrass canonical form of singular systems as model, i.e. E, A, B, C are Weierstrass canonical form (3). Here, we assume that the outputs are linear combinations for dynamic part of the system, i.e. Ex. Obviously, this assumption makes the system contains direct information related to the dynamic part. To realize our designing, we have assumptions for the system as follows:

Assumption 1. System (13) is impulse controllable.

Assumption 2. Matrices $J^{\mu-1}B_2$, $C_2J^{\mu-1}B_2$, and $\begin{bmatrix} A_1 & B_1 \\ C_1 & -C_2JB_2 \end{bmatrix}$ are all of full column rank.

Assumption 3. Matrix $[C_1\ C_2JB_2 \cdots C_2J^{\mu-1}B_2]$ has full row rank.

Assumption 4. The rank of matrix $[C_1\ C_2JB_2 \cdots C_2J^{\mu-2}B_2]$ is $m-r$.

Assumption 5. The input, output, and state dimensions of system (13) satisfy the following inequality:

$$n_1 + n_2 \leq m + 2r - 1 \tag{14}$$

Assumption 6. System (13) is R-controllable, R-observable.

Where parameters $A_1, J, B_i, C_i, i = 1, 2$, are shown as (3).

From assumption 1 and the provision about $J^{\mu-1}B_2$ being of full column rank in assumption 2, we can obtain that the no impulse state trajectory and the control of singular systems can completely be described by standard state-space linear systems, so the problem of variable structure control for the system (13) can transform into the problem of variable structure control for standard state-space linear system, whose state equation is (8)-(12), output equation is

$$y = [C_1 - C_2JB_2 \cdots - C_2J^{\mu-1}B_2]\begin{bmatrix} x_1 \\ z \end{bmatrix} \tag{15}$$

Denoting the former as P_1, the latter as P_2. Obviously, solving P_2 can obtain P_1. For this, we begin to solve problem P_2.

3.1 Switching surface design and stability for sliding mode motion

We choose the switching surface $\{[x_1^T\ z^T]^T : \sigma(y) = 0\}$ of problem P_2 as $\sigma(y) = Sy$. So the switching surface is

$$\pi = \{[x_1^T\ z^T]^T : \sigma(y) = Sy = 0\} \tag{16}$$

where $S \in R^{r\times m}$ is of full row rank matrix, and the choosing of S shoud assure that the switching surface π is a $(n_1 + (\mu-2)r)$ dimensional linear subspace. To obtain the sliding mode motion equation on π, we use the following equation to partition the state vector in (8) again:

$$z = [z_1^T\ z_2^T]^T,\ x^* = [x_1^T\ z_1^T]^T.$$

$$z_1 = [u^T\ (u^{(1)})^T \cdots (u^{(\mu-3)})^T]^T,\ z_2 = u^{(\mu-2)} \tag{17}$$

and rewrite (8)-(9) and (15) as:

$$\begin{bmatrix} \dot{x}^* \\ \dot{z}_2 \end{bmatrix} = \begin{bmatrix} A_{11} & A_{12} \\ 0 & 0 \end{bmatrix}\begin{bmatrix} x^* \\ z_2 \end{bmatrix} + \begin{bmatrix} 0 \\ I_r \end{bmatrix} v \tag{18}$$

$$\begin{bmatrix} x^*(0) \\ z_2(0) \end{bmatrix} = \begin{bmatrix} \beta_1 \\ \beta_2 \end{bmatrix} \tag{19}$$

$$y = [C_1^0\ C_2^0]\begin{bmatrix} x^* \\ z_2 \end{bmatrix} \tag{20}$$

where

$$A_{11} = \begin{bmatrix} A_1 & B_1 & 0 & \cdots & 0 \\ & 0 & I_r & & \vdots \\ & & \ddots & \ddots & \\ & & & \ddots & I_r \\ & & & & 0 \end{bmatrix}$$

$$A_{12} = \begin{bmatrix} 0, & \vdots & 0 & I_r \end{bmatrix}^T$$

$$\beta_1 = [I_{n_1+(\mu-2)r}\ 0]\begin{bmatrix} \alpha_{01} \\ H\alpha_{02} \end{bmatrix}$$

$$\beta_2 = [0 \cdots 0\ I_r]\begin{bmatrix} \alpha_{01} \\ H\alpha_{02} \end{bmatrix}$$

$$C_1^0 = [C_1\ - C_2JB_2 \cdots - C_2J^{\mu-2}B_2]$$

$$C_2^0 = -C_2J^{\mu-1}B_2 \tag{21}$$

$A_{11} \in R^{(n_1+(\mu-2)r)\times(n_1+(\mu-2)r)}$, $\beta_2 \in R^r$, $A_{12} \in R^{(n_1+(\mu-2)r)\times r}$, $\beta_1 \in R^{n_1+(\mu-2)r}$. Noticing that the sliding mode motion on π satisfies

$$\sigma(y(t)) = Sy(t) \equiv 0 \tag{22}$$

so on the premise that SC_2^0 has full rank, from (18) and (20) yields:

$$z_2 = -(SC_2^0)^{-1}SC_1^0x^* \tag{23}$$

$$\dot{x}^* = (A_{11} - A_{12}(SC_2^0)^{-1}SC_1^0)x^* \tag{24}$$

$$y = (I_m - C_2^0(SC_2^0)^{-1})S)C_1^0x^* \tag{25}$$

$$v_{eq} = -(SC_2^0)^{-1}SC_1^0(A_{11}x^* + A_{12}z_2). \tag{26}$$

In which, (24) is called the sliding mode motion equation on switching surface π, (25) is the output equation of sliding mode motion, (26) is the equivalent control. The first step in the following is choosing S such that the coefficient matrix $A_{11} - A_{12}(SC_2^0)^{-1}SC_1^0$ of the sliding mode equation is stable matrix. For this we introduce the following discussion.

Lemma 4. If assumptions 2, 6 hold, then linear system $[C_1^0, A_{11}, A_{12}]$ is state controllable and observable.

Lemma 5 (see Kimura,1975). Assume that linear system is $[C, A, B]$, in which A, B, C are constant matrices, $A \in R^{n\times n}$, $B \in R^{n\times r}$, $C \in R^{m\times n}$. If the system is state controllable and observable, the input, output, and state dimensions satisfy Davison-Kimura condition: $n \le m+r-1$, then for n arbitrary given numbers, $\alpha_1, \cdots, \alpha_n$ (complex numbers apear as conjugate pairs), and a small enough positive number ε, there exists feedback matrix $K \in R^{r\times m}$ such that the eigenvalues of closed-loop system via output feedback are all located in $\varepsilon-$domain of $\alpha, i = 1, 2, \cdots, n$.

The computation for feedback matrix K in lemma 5 can see Kimura(1975).

Obviously, the coefficient matrix of sliding mode motion equation (24), i.e. $A_{11} - A_{12}(SC_2^0)^{-1}SC_1^0$, can be looked as the coefficient matrix of closed-loop of system $[C_1^0, A_{11}, A_{12}]$ via output feedback, where the output feedback matrix is

$$K = (SC_2^0)^{-1}S. \qquad (27)$$

Notice that K satisfies

$$KC_2^0 = I_r \qquad (28)$$

namely that K can not arbitrarily be choosed. Analyse equation (28), we know that the general solution to (28) can be written as:

$$K = (C_2^0)^+ + PM \qquad (29)$$

where

$$(C_2^0)^+ = ((C_2^0)^\tau C_2^0)^{-1}(C_2^0)^\tau \qquad (30)$$

is the generalized inverse of C_2^0, $P \in R^{r\times(m-r)}$ is an arbitrary parameter matrix, $M \in R^{(m-r)\times m}$ is of full row rank, and M satisfies

$$MC_2^0 = 0 \qquad (31).$$

Substituting of (29), (27) into (24) yields:

$$\dot{x}^* = (\hat{A}_{11} - A_{12}P\hat{C}_1)x^* \qquad (32)$$

in which

$$\hat{A}_{11} = A_{11} - A_{12}(C_2^0)^+C_1^0, \quad \hat{C}_1 = MC_1^0. \quad (33)$$

Notice that the asymptotic stability of sliding mode motion is determined by the eigenvalues distribution of matrix $\hat{A}_{11} - A_{12}P\hat{C}_1$. In the following we first study state controllability and observability of the system $[\hat{C}_1, \hat{A}_{11}, A_{12}]$. Observing the rank of C_1^0, C_2^0 and the null-space of $(C_1^0)^\tau$, $(C_2^0)^\tau$, from assumptions 2, 4 and (21), yields:

$$\text{rank}C_1^0 = m - r, \text{rank}C_2^0 = r, \text{rank}[C_1^0\ C_2^0] = m$$

$$M_1C_1^0 = 0, \quad MC_2^0 = 0 \qquad (34)$$

where M is shown as (31), $M_1 \in R^{r\times m}$, $M \in R^{(m-r)\times m}$ are all full row rank matrices. More further, there are:

Lemma 6. Matrix $\begin{bmatrix} M \\ M_1 \end{bmatrix}$ is full rank square matrix, in which M, M_1 are shown as (34).

Lemma 7. Linear system $[\hat{C}_1, \hat{A}_{11}, A_{12}]$ is state controllable and observable and satisfies Davison-Kimura condition, in which $\hat{C}_1, \hat{A}_{11}, A_{12}$ are shown as (33), (21).

Lemma 8. For any given positive number α, there must exist $P \in R^{r\times(m-r)}$ such that the eigenvalues $\lambda_i, i = 1, 2, \cdots, n_1 + (\mu-2)r$ of $\hat{A}_{11} - A_{12}P\hat{C}_1$ satisfy: 1) λ_i are distinct real; 2) $\lambda_i < -\alpha$.

Lemma 9. If K is shown as (29), $S \in R^{r\times m}$, then S is a solution to (27) if and only if S can be shown as:

$$S = LK \qquad (35)$$

where L is an arbitrary $r \times r$ full rank square matrix.

The proofs of Lemma 6-9 are easy, in lemma 8, the computation of matrix P can see Kimura(1975).

We conclude the above disscusion as theorem.

Theorem 1. The following algorithm gives the design of switching surface, and assures that the sliding mode motion trajectory $x^*(t)$ converges to zero with exponent speed:

$$\|x^*(t)\| \le m_0e^{-\alpha t}, \quad \forall t \ge 0 \qquad (36)$$

where α is an arbitrary given positive number, m_0 is positive real, and depends on α and initial values of the state trajectory, is independent of $t \ge 0$.

Algorithm. 1) For arbitrary given positive number α, according to follows:

$$\alpha_{i+1} < \alpha_i - 1, \ i = 1, 2, \cdots, n_1 + (\mu-2)r - 1$$
$$\alpha_1 < -(\alpha + 1) \qquad (37)$$

we assign $n_1 + (\mu - 2)r$ real numbers α_i, and by lemma 8, using the Kimura algorithm to find matrix $P \in R^{r\times(m-r)}$ such that the eigenvalues $\lambda_1, \lambda_2, \cdots, \lambda_{n_1+(\mu-2)r}$ of $\hat{A}_{11} - A_{12}P\hat{C}_1$ satisfy:

$$|\lambda_i - \alpha_i| < \frac{1}{2}, \quad i = 1, 2, \cdots, n_1 + (\mu-2)r. \quad (38)$$

2) According to (30), (31), find $(C_2^0)^+$ and M , from (29), find K.
3)According to (35), find S, particularly can let $S = K$.

3.2 Output feedback variable structure control design and global asymptotic stability of closed-loop

In this section, we discuss the design of output feedback variable structure controller. For generality, the output feedback variable structure control law takes the form

$$v = F_0 y + F_1 \cdot \text{sgn}\sigma, \quad \forall \sigma(y) \neq 0 \tag{39}$$

where

$$\text{sgn}\sigma = [\text{sgn}\sigma_1 \ \text{sgn}\sigma_2 \ \cdots \ \text{sgn}\sigma_r]^T \tag{40}$$

σ_i is the ith vector of $\sigma(y) = Sy$, $i = 1, 2, \cdots, r$, $F_0 \in R^{r \times m}$, $F_1 \in R^{r \times r}$ are the feedback gain matrix waiting decided. Obviously, the key to control design is to define the domain of F_0 and F_1. For this, we introduce coordinate transformation as follows:

$$\begin{bmatrix} q \\ \sigma \end{bmatrix} = \begin{bmatrix} T^{-1} & 0 \\ SC_1^0 & SC_2^0 \end{bmatrix} \begin{bmatrix} x^* \\ z_2 \end{bmatrix} \tag{41}$$

where $T \in R^{(n_1+(\mu-2)r)\times(n_1+(\mu-2)r)}$ satisfies

$$T^{-1}(A_{11} - A_{12}(SC_2^0)^{-1}SC_1^0)T = \begin{bmatrix} \lambda_1 & & \\ & \ddots & \\ & & \lambda_{n_1+(\mu-2)r} \end{bmatrix} \tag{42}$$

It is easy to know that in the coordinate transformation (41), the system (18) and the output (20) transform as:

$$\begin{bmatrix} \dot{q} \\ \dot{\sigma} \end{bmatrix} = \begin{bmatrix} A_{11n} & A_{12n} \\ A_{21n} & A_{22n} \end{bmatrix} \begin{bmatrix} q \\ \sigma \end{bmatrix} + \begin{bmatrix} 0 \\ B_{2n} \end{bmatrix} v \tag{43}$$

$$y = [C_{1n} \ C_{2n}] \begin{bmatrix} q \\ \sigma \end{bmatrix} \tag{44}$$

where

$$\begin{aligned} A_{11n} &= \begin{bmatrix} \lambda_1 & & \\ & \ddots & \\ & & \lambda_{n_1+(\mu-2)r} \end{bmatrix} \\ A_{12n} &= T^{-1}A_{12}(SC_2^0)^{-1} \\ A_{21n} &= SC_1^0(A_{11} - A_{12}(SC_2^0)^{-1}SC_1^0)T \\ A_{22n} &= SC_1^0 A_{12}(SC_2^0)^{-1} \\ B_{2n} &= SC_2^0 \\ C_{1n} &= (I_m - C_2^0(SC_2^0)^{-1}S)C_1^0 T \\ C_{2n} &= C_2^0(SC_2^0)^{-1}. \end{aligned} \tag{45}$$

Next, use Lyapunov method to discuss global asymptotic stability for the closed-loop system

$$\begin{bmatrix} \dot{q} \\ \dot{\sigma} \end{bmatrix} = \begin{bmatrix} A_{11n} & A_{12n} \\ A_{21n} + B_{2n}F_0C_{1n} & A_{22n} + B_{2n}F_0C_{2n} \end{bmatrix} \begin{bmatrix} q \\ \sigma \end{bmatrix} + \begin{bmatrix} 0 \\ B_{2n}F_1 \end{bmatrix} \text{sgn}\sigma \tag{46}$$

of system (43),(44) via the output feedback (39), and define the domain of coefficients F_0 and F_1. We introduce quadratic Lyapunov function:

$$V(q, \sigma) = \frac{1}{2}(||q||^2 + ||\sigma||^2) \tag{47}$$

take

$$F_1 = -(SC_2^0)^{-1} \cdot \beta_0 \tag{48}$$

$$F_0 = -(SC_2^0)^{-1}(A_{22n} + K_0)S \tag{49}$$

$$K_0 = k_0 I_r, k_0 \geq 1 + \frac{\rho^2}{|\lambda_1|} \tag{50}$$

$$\rho = ||A_{12n}|| + ||A_{21n}|| \tag{51}$$

analysing $\dot{V}$:

$$\begin{aligned} \dot{V} &= q^T\dot{q} + \sigma^T\dot{\sigma} = q^T(A_{11n}q + A_{12n}\sigma) \\ &+\sigma^T((A_{21n} + B_{2n}F_0C_{1n})q \\ &+(A_{22n} + B_{2n}F_0C_{2n})\sigma) + \sigma^T B_{2n}F_1 \cdot \text{sgn}\sigma \\ &\leq -(|\lambda_1|||q||^2 - \rho||q||||\sigma|| + k_0||\sigma||^2) - \beta_0||\sigma|| \\ &= -\frac{3}{4}|\lambda_1|||q||^2 - (\frac{\sqrt{|\lambda_1|}}{2}||q|| - \frac{\rho}{\sqrt{|\lambda_1|}}||\sigma||)^2 \\ &-(k_0 - \frac{\rho^2}{|\lambda_1|})||\sigma||^2 - \beta_0||\sigma|| \\ &\leq -(\frac{3}{4}|\lambda_1|||q||^2 + ||\sigma||^2) - \beta_0||\sigma|| < 0 \\ &\forall ||\sigma||^2 + ||q||^2 \neq 0 \end{aligned} \tag{52}$$

Using Lyapunov theorem, we know immediatly that closed-loop system (46) is globally asymptotically stable. Above results are concluded as theorem, that is

Theorem 2. Assume that the open-loop system is shown as (18)-(21), the output feedback variable structure control is shown as (39), coefficients F_0, F_1 are shown as (48)-(51). Then the closed-loop system is globally asymptotically stable.

In the following, observing the possibility for that the state trajectory of the closed-loop system reaches the switching surface in finite time, and realizes sliding mode motion on it forced by the control (39), we have:

Theorem 3. The closed-loop system related in theroem 2 must reach the switching surface in finite time, and realize sliding mode motion on it.

Now, we pay attention back to system (13), and observe question P_1. Noticing that control $u(t)$ is the solution of the differential equation (8) satisfying initial condition (9), so

$$z(t) = e^{Dt} + z(0)\int_0^t e^{D(t-\tau)}Fv(\tau)d\tau \qquad (53)$$

compute e^{Dt}, we have

$$\begin{aligned} u(t) &= u(0) + tu^{(1)}(0) + \frac{t^2}{2!}u^{(2)}(0) \\ &+ \cdots + \frac{t^{\mu-2}}{(\mu-2)!}u^{(\mu-2)}(0) \\ &+ \frac{1}{(\mu-2)!}\int_0^t (t-\tau)^{\mu-2}\cdot v(\tau)d\tau \quad (54) \end{aligned}$$

where $v(\tau)$ is shown as (39),(48)-(51) Analyse asymptotic properties for the state trajectory $x(t) = [x_1^T(t)\ x_2^T(t)\ x_3^T(t)]^T$ of system (13) forced by $u(t)$. Noticing (10),(11),(17), and noticing that $x^*(t)$ reaches the switching surface π in finite time, and realizes stable sliding mode motion on it with the exponent speed (36), then

$$\|x(t)\| \leq m^* e^{-\alpha t}, \quad \forall t \geq 0 \qquad (55)$$

where α is shown as (36), m^* is positive real number, and depends on the initial value $Ex(0^-)$ and α, is independent of $t \geq 0$. Concluding above results, we have:

Theorem 4: If system (13) satisfies assumptions 1-6, then the system state trajectory drived by output feedback control law (54) converges to zero with assigned exponent speed.

Example. Assume that singular system is shown as (13), in which

$$E = \begin{bmatrix} 1 & 0 & 0 \\ 0 & 0 & 1 \\ 0 & 0 & 0 \end{bmatrix}, \quad A = \begin{bmatrix} 1 & 0 & 0 \\ 0 & 1 & 0 \\ 0 & 0 & 1 \end{bmatrix},$$
$$B = \begin{bmatrix} 1 \\ 0 \\ 1 \end{bmatrix}, \quad C = \begin{bmatrix} 0 & 1 & 0 \\ 1 & 0 & 1 \end{bmatrix},$$
$$Ex(0^-) = [\ 0\ \ 1\ \ 0\ ]^T.$$

by calculating, above system satisfies assumptions 1-6, and take $P = 3$, then

$$S = K = [-1\ \ 3],$$

Take

$$K_0 = 26, \quad F_0 = [29\ \ -87], \quad \beta_0 = 1, \quad F_1 = -1$$

then obtain control law:

$$\begin{aligned} &u = -1 + \textstyle\int_0^t v(\tau)d\tau \\ &v = [29\ \ -87]y - \text{sgn}\sigma, \quad \forall \sigma \neq 0 \\ &\sigma = [-1\ \ 3]y. \end{aligned}$$

4. CONCLUSIONS

This paper has discussed the problem of output feedback variable structure control design for multivariable linear singular systems. Given: 1. the design of switching surface and stability analysis of sliding mode motion; 2. the output feedback variable structure control design and global asymptotic stability analysis of closed-loop system. The design of control given in this paper guarantees that the closed-loop state trajectory converges to zero with arbitrary assigned exponent speed.

An important difference between singular systems and linear systems is: to assure that the state trajectory of singular systems has no impulsive behaviour, the initial values of the control and up to the $\mu-2$th-order derivative must satisfy consistent condition (6). So the control design for singular systems, which we can freely design is the $\mu-1$th-order derivative of the control, but not the control inputs, that is what the discussion is based on in this paper.

REFERENCES

Bondarev, A.G, S. A. Bondarev, N. E. Kosbyrera and V.I. Utkin(1985). Sliding modes in systems with asymptotic state observers. *Automatica*, **21**, 6-11.

Dai, L. (1989). *Singular Control Systems.* Springer-Verlag.

Davision, E.J. and S.H. Wang (1975). On pole assignment in linear multivariable systems using output feedback. *IEEE Trans. Autom. Control*, **AC-20**, 516-518.

EL-Khazali, R. and R. Decarlo (1993). Output feedback variable structure control using dynamic conpensation for linear systems. Proceedings of American Control Conf., San Francisco, 954-958.

EL-Khazali, R. and R. Decarlo (1995). Output feedback variable structure control design. *Automatica*, **31**, 805-816.

Kawaji, S. and E.Z. Taha(1994). Hyperplane design in variable structure of descriptor systems. Proceeding of the 33rd IEEE Conf. Decision and Control, Florida, USA.

Kimura, H. (1975). Pole assignment by gain output feedback. *IEEE Trans. Autom. Control*, **AC-20**, 509-516.

Utkin, V.I.(1978). *Sliding Modesand Their Application in Variable Structure Systems.* MIR, Moscow.

Chen,Y., S. Ma and Z. Cheng(1996). Singular optimal control problem of linear singular systems with linear-quadratic cost. Proceedings of Chinese Control Conference, Qingdao, 5-9.

14th Triennial World Congress, Beijing, P.R. China

A DESIGN OF A STRONGLY STABLE SELF-TUNING CONTROLLER USING COPRIME FACTORIZATION APPROACH

Akira Inoue*, Akira Yanou and Yoichi Hirashima

* *Department of Systems Engineering, Faculty of Engineering, Okayama University, Okayama, 700–8530, Japan*
Phone +81-86-251-8233, Fax +81-86-251-8233
E-mail inoue@suri.it.okayama-u.ac.jp

Abstract: This paper proposes a new self-tuning controller having new design parameters. In selecting the design parameters, the controller gives a strongly stable self-tuning controller, that is, the closed-loop system is not only stable, but also the controller itself is stable. The controller consists of an extended minimum variance controller and a parameter identification law. The extended minimum variance controller is extended by including newly introduced design parameters into the generalized minimum variance controller given by Clarke and others. The parameters are introduced by applying the coprime factorization approach and comparing Youla parametrization of stabilizing compensators with the minimum variance controller.

Keywords: minimum variance control, self-tuning controller, strongly stable, coprime factorization, adaptive control

1. INTRODUCTION

To control plants with uncertainty, self-tuning controllers are widely applied in industry(Astrom and Wittenmark, 1995). Self-tuning controller is proposed by Astrom and Wittenmark(Astrom and Wittenmark, 1973) first, which is designed to minimize the variance of the plant output. Clarke and others(Clarke. D. W. *et al.*, 1979) generalized the self-tuning controller to minimize the variance of a generalized output instead of minimizing the variance of the plant output and the generalized self-tuning controller can be applied to a wider class of plants such as, unstable plants or non-minimum phase plants.

After them, many design schemes are proposed (Astrom and Wittenmark, 1995). But they all based on the concept of Clarke and others' method. Still, there remain two problems on the self-tuning controllers: First one is that the polynomials defining the generalized output is chosen to place the poles of the closed-loop system and once these polynomials are determined then the poles of the compensator are determined and could not be assigned independently to the closed-loop poles. So, it is impossible to design a stable compensator in addition to the closed-loop stability, that is, to design a strongly stable controller. The second is that the transfer function from noise to output is also fixed by the polynomials defining the generalized output and the poles of the transfer function could not be designed.

This paper proposes a method to solve these problems by using coprime factorization approach. Using coprimely factorized rational functions to express a controlled system, the most general form (Youla parametrization) for stabilizing compensator is obtained(Vidyasagar, 1985). Using the general form of compensators, the authors have proposed a unified design method of Model Reference Adaptive Control (MRAC)(Inoue and Masuda, 1993). In this paper, the same coprime factorization approach is applied to the design of self-tuning controller. In comparing the self-tuning controller to Youla parametrization of stabilizing compensators, we introduce new design parameters in self-tuning controller. By choosing newly introduced parameters, we can design the poles of the compensator and the transfer function from noise to output and we can design a strongly stable controller.

Notations: z^{-1} denotes backward shift operator: $z^{-1}y(t) = y(t-1)$. A polynomial function and a rational function are distinguished by $A[z^{-1}]$ and $A(z^{-1})$.

2. PROBLEM STATEMENT AND GENERALIZED MINIMUM VARIANCE CONTROL LAW

Consider a single-input single-output system given by the difference equation

$$A[z^{-1}]y(t) = z^{-k_m}B[z^{-1}]u(t) + C[z^{-1}]\xi(t) \quad (1)$$
$$t = 0, 1, 2, \cdots$$

where $u(t)$ is the input, $y(t)$ is the output, k_m is the time delay, $\xi(t)$ is a white Gaussian noise with zero mean. $A[z^{-1}], B[z^{-1}]$ and $C[z^{-1}]$ are polynomials;

$$\begin{aligned} A[z^{-1}] &= 1 + a_1 z^{-1} + \cdots a_n z^{-n} \\ B[z^{-1}] &= b_0 + b_1 z^{-1} + \cdots b_m z^{-m} \\ C[z^{-1}] &= 1 + c_1 z^{-1} + \cdots c_l z^{-l} \end{aligned} \quad (2)$$

On the system (1), we put the following assumptions

Assumption 1. The degrees n, m, l of $A[z^{-1}]$, $B[z^{-1}]$, $C[z^{-1}]$ and the time delay k_m are known.

Assumption 2. The coefficients $a_1, \cdots, a_n, b_0, \cdots, b_m, c_1, \cdots, c_l$ of $A[z^{-1}]$, $B[z^{-1}]$, $C[z^{-1}]$ are unknown.

Assumption 3. The polynomials $A[z^{-1}]$ and $B[z^{-1}]$, $A[z^{-1}]$ and $C[z^{-1}]$ are coprime.

Assumption 4. The polynomial $C[z^{-1}]$ is stable

The control objective is that the output $y(t)$ has a desirable response to the reference input $w(t)$. To accomplish this objective, the generalized minimum variance control given by Clarke and others(Clarke. D. W. *et al.*, 1979) minimizes the following variance of a generalized output

$$J = E[\Phi(t + k_m)^2] \quad (3)$$

where $\Phi(t + k_m)$ is the generalized output;

$$\Phi(t + k_m) = P[z^{-1}]y(t + k_m) + Q[z^{-1}]u(t) - R[z^{-1}]w(t) \quad (4)$$

and $P[z^{-1}], Q[z^{-1}]$ and $R[z^{-1}]$ are polynomials given by a controller designer with degrees of n_p, n_q, n_r. Usually these polynomials are selected to obtain stable desirable closed-loop poles.

In this section and the next section, the coefficients of $A[z^{-1}], B[z^{-1}]$ and $C[z^{-1}]$ are assumed to be known, and the generalized minimum variance control law(Clarke. D. W. *et al.*, 1979) and an extended minimum variance control law are given.

In the generalized minimum variance control law, first, the following three Diophantine equations are solved;

$$P[z^{-1}]C[z^{-1}] = A[z^{-1}]E[z^{-1}] + z^{-k_m}F[z^{-1}] \quad (5)$$
$$G[z^{-1}] = E[z^{-1}]B[z^{-1}] + C[z^{-1}]Q[z^{-1}] \quad (6)$$
$$T[z^{-1}] = P[z^{-1}]B[z^{-1}] + Q[z^{-1}]A[z^{-1}] \quad (7)$$

where $E[z^{-1}]$ is a monic polynomial with degree $k_m - 1$ and $F[z^{-1}]$ is a polynomial with degree n_1;

$$n_1 = \max\{n - 1, n_p + l - k_m\} \quad (8)$$

Then, the generalized output $\Phi(t + k_m)$ and its prediction $\hat{\Phi}(t + k_m|t)$ are given(Clarke. D. W. *et al.*, 1979) by

$$\Phi(t + k_m) = \hat{\Phi}(t + k_m|t) + E[z^{-1}]\xi(t + k_m) \quad (9)$$
$$\hat{\Phi}(t + k_m|t) = (F[z^{-1}]y(t) + G[z^{-1}]u(t) - C[z^{-1}]R[z^{-1}]w(t))/C[z^{-1}] \quad (10)$$

Since the estimate $\hat{\Phi}(t + k_m|t)$ and the noise term $E[z^{-1}]\xi(t + k_m)$ have no correlation with each other, the control $u(t)$ to minimize the variance J is obtained by choosing $u(t)$ to make

$$\hat{\Phi}(t + k_m|t) = 0 \quad (11)$$

Making the numerator of eq (10) equal to 0, the control law is

$$u(t) = \frac{C[z^{-1}]R[z^{-1}]}{G[z^{-1}]}w(t) - \frac{F[z^{-1}]}{G[z^{-1}]}y(t) \quad (12)$$

Substituting the control (12) into the system (1), we get the closed-loop system

$$y(t) = \frac{z^{-k_m}B[z^{-1}]R[z^{-1}]}{T[z^{-1}]}w(t) + \frac{G[z^{-1}]}{T[z^{-1}]}\xi(t) \quad (13)$$

The control law (12) which is derived by Clarke and others(Clarke. D. W. *et al.*, 1979) as the generalized minimum variance control has two problems:

Problem 1. The polynomials $P[z^{-1}], Q[z^{-1}]$ and $R[z^{-1}]$ defining the generalized output $\Phi(t)$ are determined by controller designers to get the stable closed-loop characteristics, $T[z^{-1}]$. Then, the characteristic polynomial $G[z^{-1}]$ of the control law (12) is uniquely fixed and can not be designed independently to $T[z^{-1}]$.

Problem 2. The effect of the noise on the output is given by the transfer function $G[z^{-1}]/T[z^{-1}]$ and is also determined by the polynomials $P[z^{-1}]$, $Q[z^{-1}]$ and $R[z^{-1}]$.

In order to solve these problems, we will introduce new design polynomials in the generalized minimum variance controller by using the coprime factorization approach and comparing the most general stabilizing compensator, Youla parametrization form, to the control law (12).

3. EXTENDED MINIMUM VARIANCE CONTROL LAW USING COPRIME FACTORIZATION APPROACH

This section extends the generalized minimum variance control law in order to include design polynomials under the assumption that the coefficients of $A[z^{-1}], B[z^{-1}]$ and $C[z^{-1}]$ are known. For the coprime factorization approach, consider the family of stable rational functions:

$$RH_\infty = \{G(z^{-1}) = \frac{G_n[z^{-1}]}{G_d[z^{-1}]}, \quad G_d[z^{-1}] : \text{stable polynomial}\} \quad (14)$$

The transfer function is expressed by a ratio of rational function in RH_∞,

$$G(z^{-1}) = N(z^{-1})/D(z^{-1}) \quad (15)$$

where $N(z^{-1}), D(z^{-1})$ are rational functions in RH_∞ and are coprime in each other. Then, all the stabilizing two-degree-of-freedom compensator is given in Youla parametrization form(Vidyasagar, 1985):

$$u(t) = C_1(z^{-1})w(t) - C_2(z^{-1})y(t) \quad (16)$$

$$C_1(z^{-1}) = (Y(z^{-1}) - U(z^{-1})N(z^{-1}))^{-1}K(z^{-1}) \quad (17)$$

$$C_2(z^{-1}) = (Y(z^{-1}) - U(z^{-1})N(z^{-1}))^{-1} \times(X(z^{-1}) + U(z^{-1})D(z^{-1})) \quad (18)$$

where $U(z^{-1})$ and $K(z^{-1})$ are rational functions in RH_∞ and are design parameters. $X(z^{-1})$ and $Y(z^{-1})$ are also in RH_∞ and the solutions of the following Bezout equations;

$$X(z^{-1})N(z^{-1}) + Y(z^{-1})D(z^{-1}) = 1 \quad (19)$$

If the polynomials $P[z^{-1}]$, $Q[z^{-1}]$ and $R[z^{-1}]$ are chosen for $T[z^{-1}]$ to be stable, comparing transfer function (1) to (15), we can choose $N(z^{-1})$ and $D(z^{-1})$ as

$$N(z^{-1}) = z^{-k_m}B[z^{-1}]/T[z^{-1}] \quad (20)$$

$$D(z^{-1}) = A[z^{-1}]/T[z^{-1}] \quad (21)$$

And $N(z^{-1}),\ D(z^{-1}) \in RH_\infty$. Substituting (20) and (21) into Bezout equation (19) and comparing the equation to Diophantine equation (5), we get the solution of Bezout equation $X(z^{-1})$, $Y(z^{-1})$ as

$$X(z^{-1}) = F[z^{-1}]/C[z^{-1}]$$

$$Y(z^{-1}) = G[z^{-1}]/C[z^{-1}] \quad (22)$$

Then the control law (12) is obtained from the Youla parametrization (16) by selecting the design parameters as

$$K(z^{-1}) = R[z^{-1}], \quad U(z^{-1}) = 0 \quad (23)$$

In order to extend the control law (12), instead of choosing $U(z^{-1})$ as in (23), we keep $U(z^{-1})$ as a newly introduced design parameter for the control law (12). To simplify the description of the control law, using new two design polynomials $U_n[z^{-1}]$ and stable $U_d[z^{-1}]$, we express the rational function $U(z^{-1})$ in the form of

$$U(z^{-1}) = \frac{U_n[z^{-1}]}{U_d[z^{-1}]}T[z^{-1}] \quad (24)$$

Then substituting (20)~(24) into (17) and (18), we get

$$C_1(z^{-1}) = (U_d[z^{-1}]G[z^{-1}] - U_n[z^{-1}]z^{-km}C[z^{-1}] \cdot B[z^{-1}])^{-1}U_d[z^{-1}]C[z^{-1}]R[z^{-1}] \quad (25)$$

$$C_2(z^{-1}) = (U_d[z^{-1}]G[z^{-1}] - U_n[z^{-1}]z^{-km}C[z^{-1}] \cdot B[z^{-1}])^{-1}(U_d[z^{-1}]F[z^{-1}] + U_n[z^{-1}]C[z^{-1}]A[z^{-1}]) \quad (26)$$

Then using $C_1(z^{-1})$ and $C_2(z^{-1})$ of (25) and (26) in (16), a newly extended minimum variance control law is

$$(U_d[z^{-1}]G[z^{-1}] - U_n[z^{-1}]z^{-k_m}C[z^{-1}]B[z^{-1}])u(t) = -(U_d[z^{-1}]F[z^{-1}] + U_n[z^{-1}]C[z^{-1}]A[z^{-1}])y(t) + U_d[z^{-1}]C[z^{-1}]R[z^{-1}]w(t) \quad (27)$$

To calculate this control, we express the polynomial operating on $u(t)$ in the left-hand side of (27) by the leading term $u_{d_0}g_0$ and the rest of the polynomial;

$$U_d[z^{-1}]G[z^{-1}] - U_n[z^{-1}]z^{-k_m}C[z^{-1}]B[z^{-1}] \triangleq u_{d_0}g_0 + z^{-1}G'[z^{-1}] \quad (28)$$

Using this expression, the control law (27) is calculated by

$$u(t) = -\frac{1}{u_{d_0} g_0}(U_d[z^{-1}]F[z^{-1}] + U_n[z^{-1}]C[z^{-1}] \cdot A[z^{-1}])y(t) - \frac{1}{u_{d_0} g_0}G'[z^{-1}]u(t-1) + \frac{1}{u_{d_0} g_0}U_d[z^{-1}]C[z^{-1}]R[z^{-1}]w(t) \quad (29)$$

Theorem 1. Using the control law (27) or (29), the followings hold;

(i) The closed-loop system is given by

$$y(t) = \frac{z^{-k_m}B[z^{-1}]}{T[z^{-1}]}R[z^{-1}]w(t) + (Y(z^{-1}) - U(z^{-1})N(z^{-1}))\frac{C[z^{-1}]}{T[z^{-1}]}\xi(t) \quad (30)$$

(ii) The generalized output (4) is

$$\Phi(t+k_m) = (z^{k_m}E[z^{-1}] - \frac{U_n[z^{-1}]}{U_d[z^{-1}]}C[z^{-1}])\xi(t) \quad (31)$$

Since the noise $\xi(t)$ is white, from (31), the optimal prediction of $\Phi(t+k_m)$ is

$$\hat{\Phi}(t+k_m|t) = 0 \quad (32)$$

that is, the control objective (11) holds,

(iii) From (27), the poles of the compensator are the roots of equation,

$$U_d[z^{-1}]G[z^{-1}] - U_n[z^{-1}]z^{-k_m}C[z^{-1}]B[z^{-1}] = 0 \quad (33)$$

Remark 1. (i) From (30), it is clear that the transfer function from the reference input $w(t)$ to the output is equal to the one by the control law of Clarke and others(Clarke. D. W. *et al.*, 1979), and is independent of the choice of the design parameter $U_d[z^{-1}]$ and $U_n[z^{-1}]$.

(ii) The effect of the noise $\xi(t)$ on the output $y(t)$ may be reduced by choosing the parameter $U(z^{-1})$ in (30), whereas the effect of noise by the control law (12) could not be changed in (13).

(iii) The poles of the compensator can be designed by selecting polynomials $U_d[z^{-1}]$ and $U_n[z^{-1}]$ in (33), not changing the poles of the closed-loop system from the roots of $T[z^{-1}] = 0$. We may design a strongly stable controller, that is, a controller by which both of the closed-loop system and the controller itself are stable, by sequentially designing first $P[z^{-1}]$, $Q[z^{-1}]$ and $R[z^{-1}]$ for the closed-loop poles (the roots of $T[z^{-1}] = 0$) to be stable, second $U_n[z^{-1}]$ and $U_d[z^{-1}]$ for the controller poles (the roots of (33)) to be stable.

Proof. Equations (30) and (31) are straightforwardly derived from (5)~(7) and (15)~(29). □

4. SELF-TUNING CONTROLLER BY USING THE EXTENDED MINIMUM VARIANCE CONTROL

In this section, we assume that only the nominal values of the coefficients of the system (1) are known and the true or varying values are unknown, and propose a self-tuning controller by adding a parameter identification law to the extended minimum variance controller (29). First choose the design parameter $U_d[z^{-1}]$ and $U_n[z^{-1}]$ so as that the poles of controller, that is, the roots of equation (33) are stable using the nominal values of $A[z^{-1}]$, $B[z^{-1}]$ and $C[z^{-1}]$. Then the self-tuning controller consists of the following 3 steps calculated in each sampling period.

Step 1. To identify the coefficients, the next recursive least square method is used;

$$\hat{\theta}(t) = \hat{\theta}(t-1) + \frac{\Gamma(t-1)\psi(t-1)}{1+\psi^T(t-1)\Gamma(t-1)\psi(t-1)}\varepsilon(t) \quad (34)$$

$$\Gamma(t) = \Gamma(t-1) - \frac{\lambda\Gamma(t-1)\psi(t-1)\psi^T(t-1)\Gamma(t-1)}{1+\lambda\psi^T(t-1)\Gamma(t-1)\psi(t-1)} \quad (35)$$

$$\Gamma(0) = \alpha I,\ 0 < \alpha < \infty$$

$$\varepsilon(t) = y(t) - \hat{\theta}^T(t-1)\psi(t-1) \quad (36)$$

$$\eta(t) = y(t) - \hat{\theta}^T(t)\psi(t-1) \quad (37)$$

$$\hat{\theta}(t) = [\hat{a}_1(t), \cdots, \hat{a}_n(t), \hat{b}_0(t), \cdots, \hat{b}_m(t), \hat{c}_1(t), \cdots, \hat{c}_l(t)]^T \quad (38)$$

$$\psi(t) = [-y(t-1), \cdots, -y(t-n), u(t-k_m), \cdots, u(t-k_m-m), \eta(t-1), \cdots, \eta(t-l)]^T \quad (39)$$

where $\hat{a}_1(t), \cdots, \hat{c}_l(t)$ are the identified value, λ $(0 < \lambda \leq 1)$ is a forgetting factor, $\varepsilon(t)$ and $\eta(t)$ are a priori and a posteriori estimation error.

Step 2. Obtain the estimated polynomials $\hat{A}[t:z^{-1}]$, $\hat{B}[t:z^{-1}]$ and $\hat{C}[t:z^{-1}]$ by substituting the identified coefficients $\hat{a}_1(t), \cdots, \hat{c}_l(t)$ into polynomials (2). Then calculate the solutions $\hat{E}[t:z^{-1}]$, $\hat{F}[t:z^{-1}]$, $\hat{G}[t:z^{-1}]$ and $\hat{T}[t:z^{-1}]$ of Diophantine equations (5)~(7) using the estimates $\hat{A}[t:z^{-1}]$, $\hat{B}[t:z^{-1}]$ and $\hat{C}[t:z^{-1}]$ instead of $A[z^{-1}]$, $B[z^{-1}]$ and $C[z^{-1}]$.

Step 3. Calculate the control input $u(t)$ from (27) using $\hat{A}[t:z^{-1}]$, $\hat{B}[t:z^{-1}]$, $\hat{C}[t:z^{-1}]$, $\hat{F}[t:z^{-1}]$ and $\hat{G}[t:z^{-1}]$.

As for the convergence of the generalized output $\Phi(t)$, the following theorem holds.

Theorem 2. If the assumptions [A.1]~[A.4] and the following assumptions hold,

Assumption 5. Transfer function $C[z^{-1}] - \frac{\lambda}{2}$ is strongly positive real.

Assumption 6. The signal $\psi(t)$ used in parameter identification satisfies PE(Persistently Spanning) condition; that is, the symmetric matrix

$$R = \lim_{N\to\infty} \frac{1}{N} \sum_{t=1}^{N} \psi(t-1)\psi^T(t-1) \quad (40)$$

is positive definite,

then the error of the generalized output;

$$e(t) = \Phi(t) - (E[z^{-1}] - \frac{U_n[z^{-1}]}{U_d[z^{-1}]} C[z^{-1}]z^{-k_m})\xi(t) \quad (41)$$

converges to zero *a.s.*(almost surely, converge except on a set having probability zero)

$$\lim_{t\to\infty} e(t) = 0 \quad a.s. \quad (42)$$

Proof. Since the signal $\psi(t)$ satisfies PE condition and the transfer function $C[z^{-1}] - \frac{\lambda}{2}$ is strongly positive real, the error of parameter identification $\tilde{\theta}(t) = \theta - \hat{\theta}(t)$ and the error of output estimation, $z(t) = \eta(t) - \xi(t)$ converge to zero *a.s.*;

$$\lim_{t\to\infty} \tilde{\theta}(t) = 0, \quad \lim_{t\to\infty} z(t) = 0 \quad a.s. \quad (43)$$

This statement can be proved in a similar way by (Goodwin and Sin, 1984). The equation (37) is rewritten as

$$\begin{aligned} \hat{A}[t : z^{-1}]y(t) &= \hat{B}[t : z^{-1}]u(t - k_m) \\ &\quad + \hat{C}[t : z^{-1}]\eta(t) \end{aligned} \quad (44)$$

Then the error of the generalized output is given by

$$\begin{aligned} e(t) &= (\hat{E}[t : z^{-1}] - \hat{C}[t : z^{-1}]\frac{U_n[z^{-1}]}{U_d[z^{-1}]} z^{-k_m})z(t) \\ &\quad - \{(\hat{E}[t : z^{-1}] - E[z^{-1}]) - (\hat{C}[t : z^{-1}] \\ &\quad - C[z^{-1}])\frac{U_n[z^{-1}]}{U_d[z^{-1}]} z^{-k_m}\}\xi(t) \end{aligned} \quad (45)$$

Using (43) to (45), we get the convergence of $e(t)$. □

5. EXAMPLE

Consider a system and a generalized output

$$\begin{aligned} &(1 + 0.6z^{-1} + 0.7z^{-2})y(t) = \\ &\qquad z^{-1}(0.5 - 1.5z^{-1})u(t) + \xi(t) \quad (46) \\ &\Phi(t+1) = y(t+1) + 1.5u(t) - z^{-2} \cdot (-0.15)w(t) \quad (47) \end{aligned}$$

The reference input $w(t)$ is a rectangular wave with amplitude 1.0 and period of 100 steps.

The control law (12) by Clarke and others(Clarke. D. W. *et al.*, 1979) is

$$u(t) = \frac{-0.15z^{-2}}{1 - 1.5z^{-1}} w(t) - \frac{-0.6 - 0.7z^{-1}}{1 - 1.5z^{-1}} y(t) \quad (48)$$

and the closed-loop system is

$$\begin{aligned} y(t) &= \frac{-0.075z^{-3} + 0.225z^{-4}}{1 - 1.2z^{-1} + 0.35z^{-2}} w(t) \\ &\quad + \frac{1 - 1.5z^{-1}}{1 - 1.2z^{-1} + 0.35z^{-2}} \xi(t) \end{aligned} \quad (49)$$

The poles of the closed-loop system are $z^{-1} = 1/0.5,\ 1/0.7$ and stable. But the pole of the controller is $z^{-1} = 1/1.5$ and unstable, so this controller is not strongly stable.

Selecting the design parameters in the extended minimum variance controller (29) as $U_d = 1.0$, $U_n = 0.6$, the controller is

$$\begin{aligned} u(t) &= \frac{-0.15z^{-2}}{1 - 1.8z^{-1} + 0.9z^{-2}} w(t) \\ &\quad - \frac{-0.34z^{-1} + 0.42z^{-2}}{1 - 1.8z^{-1} + 0.9z^{-2}} y(t) \end{aligned} \quad (50)$$

and the closed-loop system is

$$\begin{aligned} y(t) &= \frac{-0.075z^{-3} + 0.225z^{-4}}{1 - 1.2z^{-1} + 0.35z^{-2}} w(t) + \\ &\frac{1 - 3z^{-1} + 3.05z^{-2} - 0.525z^{-3}}{1 - 2.4z^{-1} + 2.14z^{-2} - 0.84z^{-3} + 0.1225z^{-4}} \xi(t) \end{aligned} \quad (51)$$

The poles of the closed-loop system by this controller from $w(t)$ to $y(t)$ are not changed from the poles by the controller (48) and the poles of the controller are $z^{-1} = 1/(0.9 \pm 0.3i)$, $|z^{-1}| = 1/0.9487$ and are improved to be stable, that is, the controller is strongly stable. The effect of noise on the output $y(t)$ is also changed from (49) to (51).

Assuming the coefficients of the plant (46) are unknown and the variance of noise is $\sigma = 0.0036$, computer simulations are conducted. In the simulations, the least square identification law (34)~(39) with the forgetting factor $\lambda = 0.99$ is used

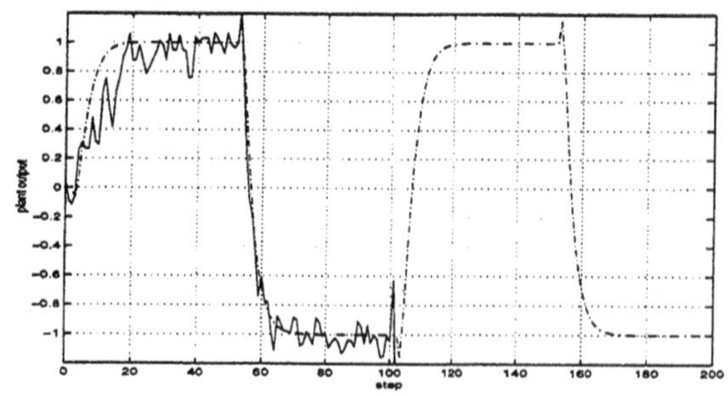

Fig. 1. Control result by STC method by Clarke and others

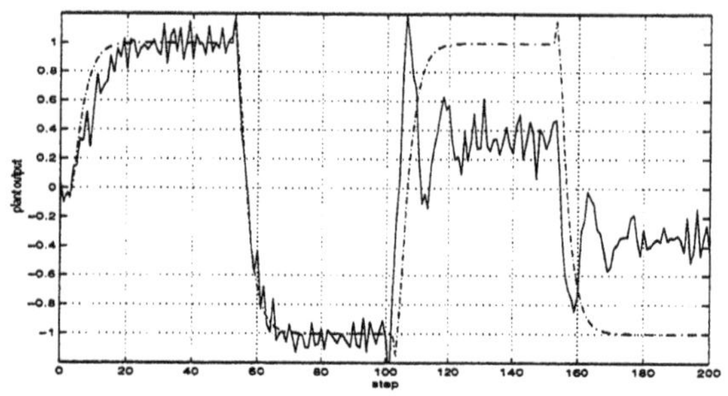

Fig. 2. Control result by STC parametrized method proposed in this paper

and the initial value of the estimated covariance matrix $\Gamma(0)$ is $\alpha = 1.0$ and the initial values of identified coefficients are set to be equal to 0.5 of the true values.

Simulation results are shown in Fig.1 for the case of using the controller (48) by Clarke and others(Clarke. D. W. *et al.*, 1979), and Fig.2 with the controller (50) proposed in this paper. In both figures, dotted lines show the output without noise and are same in Fig.1 and Fig.2. This fact indicates that the output response to reference input is not changed by introducing the design parameters $U_n[z^{-1}]$ and $U_d[z^{-1}]$.

In the simulations, the solid lines give the output responses with noise and in the simulation at step $t = 100$, the feedback loop is cut. Fig.1 shows the output by the controller (48) is divergent, whereas, the output in Fig.2 by the controller (50) stays bounded.

Fig.3 shows parameter identification processes in the case with noise and feedback break(the solid line in Fig.2).

6. CONCLUSION

In this paper, the generalized minimum variance control by Clarke and others(Clarke. D. W. *et al.*, 1979) is extended to an extended minimum variance controller including new design polynomials by using coprime factorization approach and comparing the most general two-degree-of-freedom compensator in Youla parametrization form. Without changing the response of output to

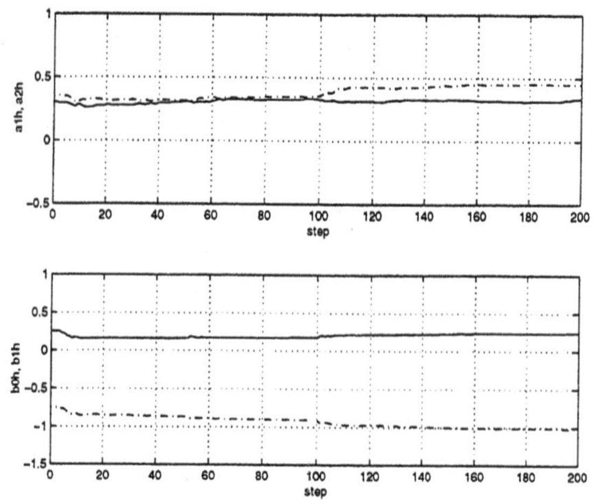

Fig. 3. Parameter identification processes in the case of the solid line in Fig.2

reference input, the poles of the compensator and the effect of noise on the output can be changed by selecting the newly introduced design polynomials in the proposed controller.

By designing the poles of the controller, a strongly stable controller is obtained.

Also proposed a self-tuning controller for systems with unknown coefficients by adding coefficient identification law to the extended minimum variance controller and proved the convergence of the controlled output.

7. REFERENCES

Astrom, K. J. and B. Wittenmark (1973). On self-tuning regulators. *Automatica* **9**, 185–199.

Astrom, K. J. and B. Wittenmark (1995). *Adaptive Control.* 2nd ed.. Addison-Wesley.

Clarke. D. W., M. A. D. Phil, P. J. Gawthrop et al. (1979). Self-tuning control. *Proc. IEE* **126, 6**, 633–640.

Goodwin, G. C. and K. S. Sin (1984). *Adaptive Filtering Prediction and Control.* Prentice Hall. New Jersey.

Inoue, A., V. T. Kroumov and S. Masuda (1993). A unified design method for model reference approach. In: *Preprints of the IFAC World Congress 1993.* Vol. 2. Sydney. pp. 83–88.

Vidyasagar, M. (1985). *Control System Synthesis: A Factorization Approach.* The MIT Press. Cambridge.

TUNING ROBUST MODEL PREDICTIVE CONTROLLERS USING LQG/LTR

C. Rowe *,[1] J.M. Maciejowski *

* *Cambridge University Engineering Department, Trumpington St., Cambridge, CB2 1PZ, U.K.*

Abstract.
In this paper, a tuning procedure for constrained model predictive control (MPC) is proposed. This takes advantage of recent extensions to the theory of MPC, which allow the use of infinite horizons in the presence of hard constraints. This allows the use of the LQG/LTR approach to tuning MPC. In this procedure, a Kalman filter is first designed by adjusting the values of two covariance matrices until its return ratio is satisfactory. This is typically assessed in the frequency domain, by inspecting the 'loop shape'. The weighting matrices of the linear quadratic regulator are then adjusted in a prescribed manner to obtain the same loop shape for the return ratio of the plant/controller combination, assuming that all constraints are inactive. The procedure is illustrated by an example based on an isothermal continuous stirred-tank reactor.

Keywords. Model Predictive Control, Linear Quadratic Gaussian, Loop Transfer Recovery, Return Ratio, Kalman Filter

1. INTRODUCTION

MPC was originally formulated using finite impulse or step response models, and later using transfer functions, however, it is possible toformulate it entirely using a state-space setting (Lee and Yu, 1994). Such a setting presents a much wider range of design decisions than the original formulations, primarily because the possibilities for disturbance modelling and state observer design are much greater. In (Lee and Yu, 1994) this problem was tackled by suggesting a 'standard model' for disturbances, which consisted of an independent disturbance added to each plant output, this disturbance being generated by passing white noise through an integrator followed by a first-order low-pass filter. It was also assumed that there is white measurement noise on each output, again independent of noise on the other outputs. This leaves the designer to choose only the time constant of the filter, the amplitude of the disturbance, and the signal-to-noise ratio for each output. This disturbance model is very well suited to many process control applications. (Lee and Yu, 1994) pointed out some of these limitations.

Once the 'standard' disturbance model proposed in (Lee and Yu, 1994) is abandoned, the tuning of MPC becomes very difficult, because there are so many parameters which the designer can vary. In (Bitmead *et al.*, 1990) it was suggested that infinite horizons should be used for MPC, in which case it becomes essentially equivalent to classical

[1] e-mail: {car24, jmm}@eng.cam.ac.uk, tel:+44-1223-332732

LQG control (Kwakernaak and Sivan, 1972). This allows the use of the tuning procedure known as *Loop Transfer Recovery*, or LTR (Doyle and Stein, 1981; Stein and Athans, 1987; Maciejowski, 1985), which has the advantage that only two of the four LQG 'parameters' need to be tuned, while the other two are chosen in a prescribed manner. This makes the LQG design methodology much easier to use, because the tuning is greatly simplified. But the approach proposed in (Bitmead *et al.*, 1990) was confined to MPC without constraints. In (Rawlings and Muske, 1993) it was shown how the problem of MPC *with* constraints could be formulated and solved, while using an infinite 'prediction horizon', though retaining a finite 'control horizon'. Subsequently it was shown in (Scokaert and Rawlings, 1996) how both horizons could be made infinite in constrained MPC. This made it possible to formulate the constrained MPC problem in a form which was essentially equivalent to LQG, and therefore one in which the LTR tuning procedure could be employed.

In this paper we show the details involved in doing this. Section 2 summarises the LQG/LTR theory and tuning procedure, while section 3 summarises the relevant MPC theory. Then section 4 shows how the LTR procedure can be applied to constrained MPC, and this is illustrated by an example in section 5.

Also, our tuning procedure assumes operation of the control scheme in conditions when no constraints are active, since it exploits the linear nature of the controller under such conditions. This seems justified if it is expected that this will be the usual mode of closed-loop operation. Alternatively, if the expectation is that 'normal' operation will be at (or close to) the intersection of some constraints, then the same procedure can be applied under the assumption that a fixed set of constraints is active, since that assumption again leads to a fixed, linear controller. But we do not give details of this in this paper.

2. LQG/LOOP TRANSFER RECOVERY

Assume the plant to be controlled may be represented by the following linear time invariant discrete time state space model:

$$x_{k+1} = Ax_k + B_u u_k + w_k \quad (1)$$
$$y_k = Cx_k + v_k \quad (2)$$

where $x_k \in \mathbb{R}^n$, $u_k \in \mathbb{R}^m$ and $y_k \in \mathbb{R}^p$ are the state, input and output vectors respectively. w_k and v_k are the process noise and the measurement noise respectively and are assumed to be zero-mean Gaussian stochastic processes which are uncorrelated in time . The usual assumptions of stabilizability and detectability of the model are made.

The covariance matrices of the noise processes are given by:

$$\mathbb{E}\{ww^T\} = W \geq 0, \quad \mathbb{E}\{vv^T\} = V > 0 \quad (3)$$
$$\mathbb{E}\{wv^T\} = 0 \quad (4)$$

The LQG problem is to find the sequence of control inputs $\{u_k, u_{k+2}, u_{k+3}, \cdots\}$ to minimize the following

$$\mathbb{J} = \lim_{n\to\infty} \mathbb{E}\Big\{\sum_{k=0}^{n} x_k^T Q x_k + u_k^T R u_k\Big\} \quad (5)$$

where the weighting matrices $Q = Q^T \geq 0$ and $R = R^T > 0$.

As is well known, the separation principle allows this problem to be solved using the optimal state estimator (Kalman Filter), in series with the optimal deterministic LQ regulator.

In discrete time there is a distinction between a Kalman predictor (KP) and a Kalman filter (KF). The predictor produces the best linear estimate in a least squares sense, of x_k given input signals $\{u_{k-1}, u_{k-2}, \cdots, u_0\}$ and output data $\{y_{k-1}, y_{k-2}, \cdots, y_0\}$. The filter updates the linear estimate produced by the predictor using data available at time k. The "filter form" is given by the following equations:

$$\hat{x}_{k/k} = \hat{x}_{k/k-1} + K_f^f(y_k - C\hat{x}_{k/k-1}) \quad (6)$$
$$\hat{x}_{k+1/k} = A\hat{x}_{k/k} + Bu_k \quad (7)$$

where

$$K_f^f = PC^T(CPC^T + V)^{-1} \quad (8)$$

and P is the positive semi-definite solution of the algebraic Riccati equation. provided that $(A, W^{1/2})$ is stabilizable and where the W and V are the covariance matrices defined in (3).

The predictor equation can be written as:

$$\hat{x}_{k+1/k} = A\hat{x}_{k/k-1} + Bu_k + K_f^p(y_k - C\hat{x}_{k/k-1}) \quad (9)$$

where

$$K_f^p = AK_f^f \quad (10)$$

Once the state estimates are obtained from the optimal state estimator, the linear quadratic regulator

(LQR) can then be used to solve the deterministic control problem. The optimal control sequence for the regulator as predicted at time k, is

$$u_{j/k} = -K_c \hat{x}_{j/k} \quad (11)$$
$$\hat{x}_{j/k} = A\hat{x}_{j-1/k} + Bu_{j-1/k} \quad (j > k) \quad (12)$$
$$K_c = (B^T \Sigma B + R)^{-1} B^T \Sigma A \quad (13)$$

and

$$\Sigma = A^T \Sigma A - A^T \Sigma B (B^T \Sigma B + R)^{-1} B^T \Sigma A + Q \quad (14)$$

Σ is the maximal solution of the algebraic Riccati equation provided that $(A, Q^{1/2})$ is detectable.

It should be noted that equation (11) only applies when state estimation is very fast and thus the Kalman filter can be used. Otherwise, the Kalman predictor is used for state estimation and thus equation (11) becomes

$$u_{j/k} = -K_c \hat{x}_{j/k-1} \quad (15)$$

2.1 *Loop Transfer Recovery*

LTR is a popular procedure for LQG design and was orignally formulated to recover the excellent robustness properties of the continuous time LQR or its dual, the KF (Doyle and Stein, 1981). The procedure may also be used for discrete-time systems even though they do not possess all the attractive properties of continuous time systems such as infinite gain margin. The following procedure is outlined in (Maciejowski, 1985) assuming that the 'filtering form' of the observer can be used.

The model is augmented if necessary, and the W and V matrices of the noise model are adjusted, until some characteristic of the KF such as the open loop return ratio,

$$\Phi(z) = C(zI - A)^{-1} K_f^p \quad (16)$$

is satisfactory. The compensator transfer function and the plant transfer function are given by $H_f(z)$ and $G(z)$ respectively and are as given below:

$$H_f(z) = zK_c[zI - (I - K_f^f C)(A - BK_c)]^{-1} K_f^f \quad (17)$$

$$G(z) = C(zI - A)^{-1} B \quad (18)$$

In general the transfer function $\Phi(z)$ is different from $G(z)H_f(z)$ but it is shown in (Maciejowski, 1985) that total asymptotic recovery of $\Phi(z)$ by $G(z)H_f(z)$ is obtained for minimum phase plants provided that $det(CB) \neq 0$. For minimum phase plants with $det(CB) \neq 0$ the values assigned to the weighting matrices are as shown below:

$$Q = C^T C \quad (19)$$
$$R = 0 \quad (20)$$

It is also shown that even though asymptotic recovery cannot be guaranteed for non-minimum phase plants, in most instances recovery can be achieved over the closed loop bandwidth provided it is made "small" enough. The achievable bandwidth is determined by the location of the non-minimum phase zero and further details are given in (Maciejowski, 1985). In this case in order to ensure internal stability the input weighting matrix, R cannot be assigned a value of zero but if $Q = C^T C$, and if R is made small enough then a 'reasonable' level of recovery may be attained. It is further pointed out that recovery cannot be guaranteed if the KP is used, but again approximate recovery is often obtained.

3. INFINITE HORIZON MPC

In recent times, much work has been done on infinite horizon MPC which incorporates both state and input constraints. In (Rawlings and Muske, 1993) a scheme is proposed which utilizes an infinite prediction horizon with a finite number of control moves in the MPC objective function. This guarantees closed loop stability even when the constraints are entered into the algorithm (subject to feasability).

In (Scokaert and Rawlings, 1996) a similar scheme was proposed which allows the control horizon m to go to infinity. This is outlined below.

The objective over the finite horizon is given by:

$$\mathbb{J}(\hat{x}_k, \pi^n) = \min_{\pi^n} \sum_{j=k}^{n} \hat{x}_{j/k}^T Q \hat{x}_{j/k} + \Delta u_{j/k}^T R \Delta u_{j/k} \quad (21)$$

Subject to constraints defined in equation (23).

where $Q \geq 0$ and $R > 0$ are symmetric weighting matrices such that $(Q^{1/2}, A)$ is detectable.

It is shown that there exists some n such that the optimal solution for the constrained finite horizon regulator is identical to the optimal solution of the constrained infinite horizon regulator defined by

$$\lim_{n \to \infty} \mathbb{J}(\hat{x}_k, \pi^n) \quad (22)$$

$$\text{subject to} \begin{cases} H\hat{x}_{j/k} \leq h & k < j, \\ Du_{j/k} \leq d & k \leq j, \\ F\Delta u_{j/k} \leq f & k \leq j. \end{cases} \quad (23)$$

A procedure is outlined to iteratively determine this value of n.

4. APPLICATION TO MPC

The above theoretical developments can now be combined into one unifying framework and applied to the MPC problem. This will be done by casting the problem in a Linear Quadratic Gaussian framework as follows.

Assume that a stabilizable and detectable linearized model of the plant exists and is given by the following LTI state-space equation

$$x_{k+1}^p = A^p\, x_k^p + B^p u_k + B_d^p\, \xi_k \quad (24)$$

$$y_k = C^p\, x_k^p + \xi_k + v_k \quad (25)$$

Assume a linear model of the disturbance exists and is given by

$$x_{k+1}^d = A^d\, x_k^d + B^d q_k \quad (26)$$

$$\xi_k = C^d\, x_k^d \quad (27)$$

where A^d has all its eigenvalues in the unit disc and q_k is white noise with $\mathbb{E}\{q_k q_k^T\} > 0$. In the DMC formulation the eigenvalues of A^d are on the unit disc itself so it would be necessary to place these slightly inside the unit disc in order for a stabilizing Ricatti solution to exist.

In order to put (5) into the standard form of (21), the state vector is augmented with the input vector so the state space model of the augmented system now becomes:

$$\begin{bmatrix} x_{k+1}^p \\ x_{k+1}^d \\ u_k \end{bmatrix} = \begin{bmatrix} A^p & 0 & B^p \\ 0 & A^d & 0 \\ 0 & 0 & I \end{bmatrix} \begin{bmatrix} x_k^p \\ x_k^d \\ u_{k-1} \end{bmatrix} + \begin{bmatrix} B^p \\ 0 \\ I \end{bmatrix} \Delta u_k + \begin{bmatrix} B_d^p & 0 \\ 0 & B^d \\ 0 & 0 \end{bmatrix} \begin{bmatrix} \xi_k \\ q_k \end{bmatrix} \quad (28)$$

$$y_k = \begin{bmatrix} C^p & C^d & 0 \end{bmatrix} \begin{bmatrix} x_k \\ x_k^d \\ u_{k-1} \end{bmatrix} + v_k \quad (29)$$

Thus, the augmented state space system can be written compactly as follows.

$$\tilde{x}_{k+1} = \tilde{A}\tilde{x}_k + \tilde{B}\Delta u_k + w_k \quad (30)$$

$$y_k = \tilde{C}x_k + v_k \quad (31)$$

The augmented system is subject to uncorrelated process noise w_k and measurement noise v_k which we assume to be zero-mean Gaussian noises with covariance matrices W and V respectively.

However it should be observed from the $\tilde{A}$ matrix given by (28) that there is an identity matrix which comes from augmenting the state with the input vector and so there are poles on the unit circle which are uncontrollable. It would not be possible to obtain a stabilizing observer. To overcome this problem these poles are placed slightly inside the unit circle.

This augmented systems is now in the standard LQG form of equations (1) to (5). However, in the MPC formulation the deviations between the output and a reference signal is penalised and so modifications will have to be made to account for this. To solve the infinite horizon optimization problem would require a knowledge of all future values of the reference signal. However since the prediction horizon will eventually be limited to some finite value, a state model of the reference may be used with an artificial dimension N_r. This state model of the reference is also used to augment the system as outlined in (Kwakernaak and Sivan, 1972).

The LQG problem then becomes that of finding the sequence of control inputs, $\{\Delta u_k, \Delta u_{k+2}, \Delta u_{k+3}, \cdots\}$ to minimize the following

$$\mathbb{J} = \lim_{n\to\infty} \mathbb{E}\left\{ \sum_{j=k}^{n} \chi_{j/k}^T\, \tilde{Q}\chi_{j/k} + \Delta\, u_{j/k}^T\, \tilde{R}\Delta u_{j/k} \right\} \quad (32)$$

where χ is the state vector formed from augmenting the state $\tilde{x_k}$ with the state of the reference model.

The LTR procedure now involves designing the KF by adjusting the values of the covariance matrices to shape the frequency responses given by (16) or (??), making the appropriate substitutions.

After the values of the covariance matrices are decided upon then the weighting matrices of the regulator will be obtained by loop recovery techniques as outlined earlier.

Finally, for implementation of MPC with constraints, the value of the prediction horizon n is chosen according to the guidelines in (Rawlings and Muske, 1993).

5. EXAMPLE

To demonstrate the tuning procedure, a simple illustrative example will be presented here. The isothermal continuous stirred-tank reactor (CSTR) which uses the dilution rate as the manipulated variable is considered (Sitsu and Bequette, 1995). The control objective is to maintain the concentration of B (x_2) at a set point of 1. The specified constraint is that x_2 should be no greater than 1.5 and the input be no greater than 150. The model is linearized around the desired operating point and then discretized at a sampling time of 0.002. For this sampling time the system is non-minimum phase with a zero at $z = 1.6906$. The necessary loop saping was ten performed.

The following graphs show recovery for different values of R for two different crossover frequencies after loop shaping was used to obtain values for W and V and Q was assigned the value specified in loop recovery theory.

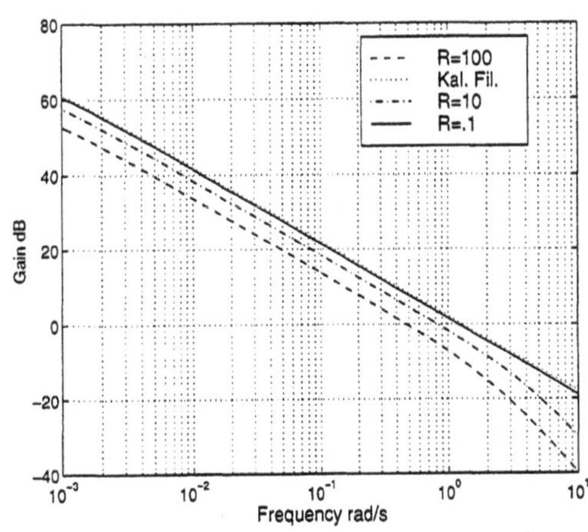

Fig. 1. Recovery of Kalman Filter Return Ratio at $\omega_c = 1\ radh^{-1}$

These two graphs illustrate that bandwidth will determine the value of R for which recovery is achieved and thus to ensure good performance the bandwidth has to be carefully chosen. The step response graphs illustrate this. The value of the prediction horizon n, used for these simulations was chosen according to the guidelines given by (Rawlings and Muske, 1993). The value of n was 7 for this problem.

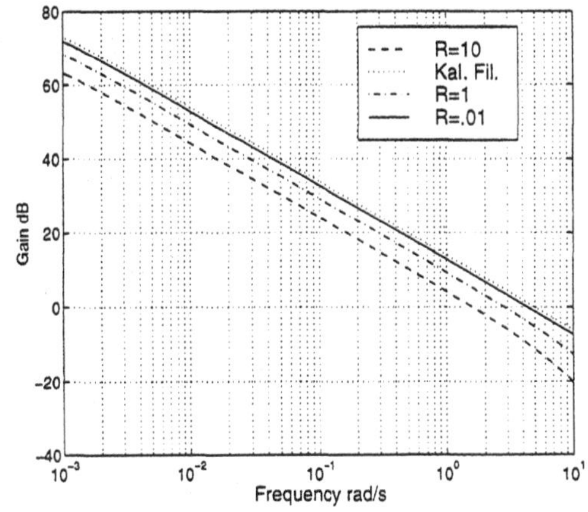

Fig. 2. Recovery of Kalman Filter Return Ratio at $\omega_c = 4\ radh^{-1}$

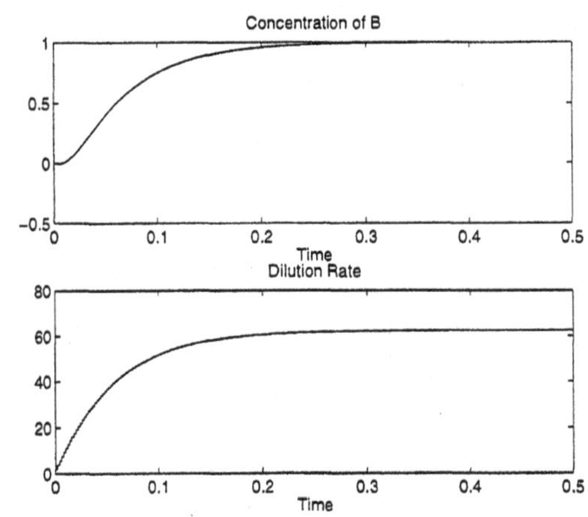

Fig. 3. Step Reponse for $\omega_c = 1\ radh^{-1}$ and $R = .1$

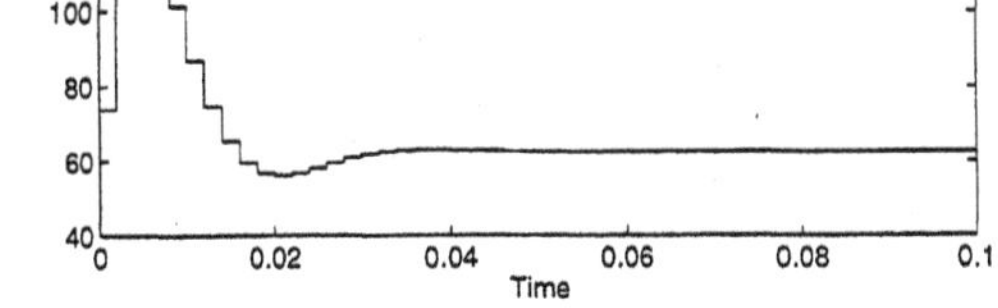

Fig. 4. Step Reponse for $\omega_c = 4\ radh^{-1}$ and $R = .01$

A simulation was done with the nonlinear plant with the constraints included.

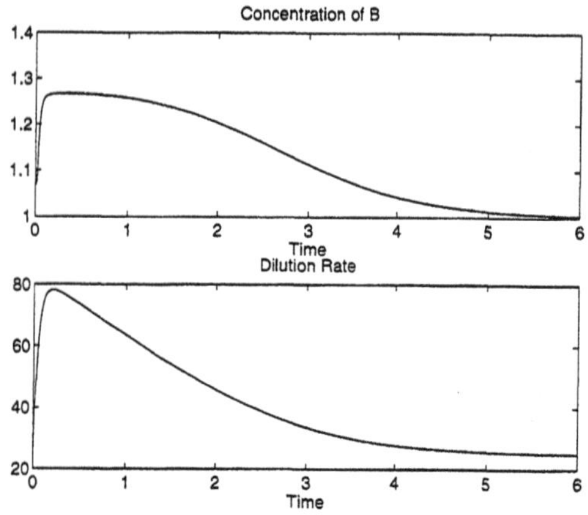

Fig. 5. Simulation with Nonlinear Plant with $x_0 = [2.8; 1.1]$

A large step disturbance was placed on x_{1f} (500% nominalvalue) which caused output constraints to be violated. The following graph shows the result.

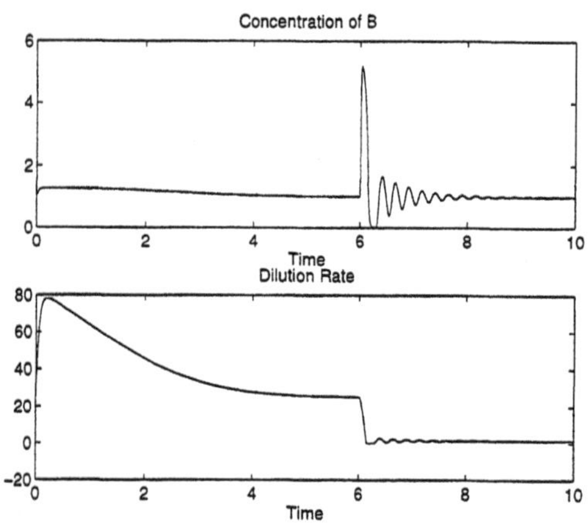

Fig. 6. Simulation with Nonlinear Plant with disturbance that causes violation of output constraint at $t = 6$

6. CONCLUSION

This paper has addressed the issue of tuning infinite horizon MPC with constraints, under the assumption that constraints are inactive most of the time using LQG/LTR. It is shown in (Lee and Yu, 1994) that one of the advantages of developing quantitative tuning rules based on robust performance, using the parameters of the the state estimator, is that robustness may be obtained even in the presence of active output constraints. Our example is consistent with this: Figure 6 shows that, even with violations of output constraints, stability was maintained.

In the case where there will be significant input uncertainity it would be necessary to first design the state-feedback return ratio using the weighting matrices and then recover the loop shape by adjusting the Kalman filter parameters. Perfect recovery cannot be obtained for this procedure, but a useful degree of recovery is often obtained in practice (Maciejowski, 1985).

7. REFERENCES

Bitmead, Robert R., Michel Gevers and Vincent Wertz (1990). *Adaptive Optimal Control - The Thinking Man's GPC.* Prentice Hall International.

Doyle, J.C. and G. Stein (1981). Multivariable feedback design:concept for a classical/modern synthesis. *IEEE Transactions on Automatic Control* **AC-26**, 4–16.

Kwakernaak, H. and R. Sivan (1972). *Linear Optimal Control Systems.* John Wiley and Sons.

Lee, J.H. and Z.H. Yu (1994). Tuning of Model Predictive Controllers for Robust Performance. *Computeres in Chemical Engineering* **18**(1), 15–37.

Maciejowski, J.M. (1985). Asymptotic Recovery for Discrete-time Systems. *I.E.E.E Transactions on Automatic Control* **30**, 602–605.

Maciejowski, J.M. (1989). *Multivariable Feedback Design.* Addison-Wesley Publishing Company.

Rawlings, James B. and Kenneth R. Muske (1993). Stability of Constrained Receding Horizon Control. *I.E.E.E Transactions on Automatic Control* **38**, 1512–1516.

Scokaert, Pierre O.M. and James B. Rawlings (1996). Infinite Horizon Linear Quadratic Control with Constraints. *Automatica* **23**(2), 109–114.

Sitsu, Phani B. and Wayne Bequette (1995). Model Predictive Control of Processes with Input Multiplicities. *Chemical Engineering Science* **6**, 921–936.

Stein, Gunter and Michael Athans (1987). The LQG/LTR Procedure for Multivariable Feedback Control Design. *I.E.E.E Transactions on Automatic Control* **32**(2), 105–113.

14th Triennial World Congress, Beijing, P.R. China

C-2a-07-2

MULTIVARIABLE FGPC CONTROLLER AND ITS PERFORMANCE ANALYSIS[1]

Zhang Huaguang He Xiqin Meng Xiangping

Department of Automatic Control, Northeast University
Shenyang, 110006, P.R.China.
E-mail: hgzhang@mail.neu.edu.cn.

Zeungnam Bien

Department of Electrical Engineering, KAIST
Taejon, 305-701, Korea
E-mail: zbien@ee.kaist.ac.kr

Abstract: In this paper, the dynamic Takagi-Sugeno (T-S) fuzzy model is used, and an on-line identification algorithm about its parameters and structures is presented . A multivariable fuzzy generalized predictive control (FGPC) approach is put forward based on the identified fuzzy model by means of Clark's principle of single-variable generalized predictive control, some of whose performances are analysed in detail. *Copyright © 1999 IFAC*

Keywords: Multivariable system, Fuzzy models, Predictive control, Intervals, Stability analysis, Performance analysis.

1. INTRODUCTION

Most of the existing control design methodologies are based on a mathematical model of the system under consideration.Therefore,system modelling is the primary issue in the design of a control system.If the dynamical system is highly nonlinear and complicated,it is usually needed to (partially) linearize the system in order to proceed its controller design (Oliver,1997) .However, for system with shifting or multiple operating conditions, the traditional mathematical model may not be sufficient to represent the dynamic behaviour.

The T-S fuzzy model is based on a fuzzy partition of input space and in each fuzzy subspace a linear input/output relation is formed (Takagi and Sugeno,1985).The output of fuzzy reasoning is given by the aggregation of the values inferred by some implications. To construct a T-S fuzzy model for a complicated nonlinear system,we first linearize the system around the operating conditions of interest and then combine these linearized systems by fuzzy If-Then rules. If there is any human intervention or

[1] This work was supported by the national natural science foundation of China and the special doctoral branch foundation of C.E.C

decision making involved in the process under consideration,T-S fuzzy model is also capable of incorporating the process by using linguistic If-Then rules.

Generalized Predictive Control or GPC is a novel long-range predictive controller (Clarke et al.,1987). Its feature lies in the long-range predictive to plant future outputs, i.e.,when the near future control u(t-1) at each sample-instant t is asserted, all the future output responses in some horizon that is significantly affected by the current control have been concerned with. By long-range predictive, GPC can be used to control some complex plants with either varying dead--time or if the plant model is overparameterized. Therefore, it is with strong robustness.

The stability analysis of the fuzzy control systems is a very difficult work. A puzzle problem is how to construct suitable Lyapunov function V(x) when using the Lyapunov direct method (Jeffrey et al., 1997). M.E.Sezer presented a method to check the stability by the Interval Matrices, but the research result was limited to a Morishima-type interval matrices (Sezer and Siljak,1970).

In our paper, a new effective method to use the interval matrices is given and can also be used in other complex discrete system directly. In this paper, it starts with the derivation of the fast multivariable fuzzy identification algorithm about T-S model. Then, a multivariable GPC algorithm is put forward and a novel fuzzy GPC control (FGPC) system is configurated. Finally, some performances, such as stability, steady state error, etc., have been analysed in detail.

Notations

$\rho(A)$ is the spectral radius of the matrix A.

$|A|$ is modulus of matrix A, i.e.,matrix with modulus elements of A.

$\lambda(A)$ is the eigenvalues of matrix A.

$A \le B$ denotes element-by-element inequality, i.e., $a_{ij} \le b_{ij}$, where A={a_{ij}}, B={b_{ij}}.

2. IDENTIFICATION OF FUZZY MODEL

A discrete time fuzzy model for a MISO system with m inputs may be expressed by n fuzzy rules. A fuzzy rule is of the following form

$$\begin{aligned}
R^i: &\textit{if}\ \ y(t)\,is\,A_1^i,\ \ y(t-1)\,is\,A_2^i,\ \ldots\ldots \\
&y(t-n_y)\,is\,A_{n_y+1}^i,\ \ u_1(t)\,is\,A_{n_y+2}^i,\ldots \\
&u(t-n_1)\,is\,A_{n_y+n_1+2}^i,\ldots, \\
&u_p(t-n_p)\,is\,A_{n_y+n_1+\ldots+n_p+p+1}^i
\end{aligned}$$

$$\begin{aligned}
\text{then}\ \ y^i(t+1) &= P_1^i y(t) + P_2^i y(t-1) + \ldots\ldots + \\
&P_{n_y+1}^i y(t-n_y) + P_{n_y+2}^i u_1(t) + \ldots \\
&+ P_{n_y+n_1+\ldots+n_p+p+1} u_p(t-n_p)
\end{aligned} \tag{1}$$

where R^i means the i-th rule, A_j^i is a fuzzy subset of a fuzzy variable, y^i the output from the i-th rule , and P_j^i a consequent parameter, $u_1(.),\ldots,u_p(.)$ are input variables, y(.) is an output variable.

For convenience, let:

$$\begin{aligned}
x_1(t) &= y(t) \\
x_2(t) &= y(t-1) \\
&\vdots \\
x_{n_y+1}(t) &= y(t-n_y) \\
x_{n_y+2}(t) &= u_1(t) \\
&\vdots \\
x_m(t) &= u_p(t-n_p)
\end{aligned}$$

where $m = n_y + 1 + \sum_{i=1}^{p} (n_i + 1)$

Then getting Eqn. (2) from Eqn. (1)

$$\begin{aligned}
R^i:\ \ &\textit{if}\ x_1\,is\,A_1^i,\ \ x_2\,is\,A_2^i,\ldots\ldots\ x_m\,is\,A_m^i, \\
&\textit{then}\ y^i(t+1) = P_1^i x_1 + P_2^i x_2 + \ldots\ldots + P_m^i x_m
\end{aligned} \tag{2}$$

where x_j is generalized input variable.The membership function of a fuzzy variable, expressed by premise parameters, will be of a convex type and formed by straight lines.

Given an input vector ($x_{10}, x_{20}, \ldots, x_{m0}$), the output $\hat{y}$ can be inferred by taking the weighted average of the y^i, i=1,2,...,n.

$$\hat{y} = \sum_{i=1}^{n} G^i y^i(t+1) \Big/ \sum_{i=1}^{n} G^i \tag{3}$$

where n is the number of fuzzy rules, y^i is obtained from the input vector by the equation of the i-th rule, and the weight G^i means the truth value of the premise of the i-th rule for the input calculated as

$$G^i = \prod_{i=1}^{m} A_j^i(x_{j0}) \tag{4}$$

where the symbol $\prod$ denotes a fuzzy operator. In this sequel, the product operator is used.

The identification algorithm of the fuzzy model consists of structure and parameters identification. The outline of the algorithm is shown in Figure 1. As the limit of the paper,the details are omited.

3. THE DESIGN OF THE GENERALIZED PREDICTIVE CONTROLLER BASED ON FUZZY MODEL

3.1 A transformation of the fuzzy model

Intuitively, since a fuzzy model is essentially a nonlinear model, the model-based design of a controller seems to be very difficult. However it can be noticed the fact that the consequent of a fuzzy process law is a linear equation so here try to use linear control theory to design the controller.

When the process has only one output and m inputs, Eqn. (5) can be infered by Eqs. (2--4)

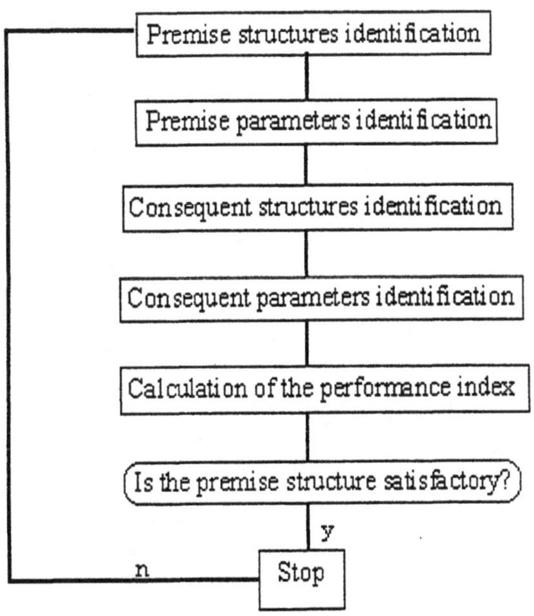

Fig. 1. The outline of the algorithm

$$y(t+1) = \frac{1}{\sum_{i=1}^{n} G^i} [\sum_{i=1}^{n} G^i P_1^i y(t) + G^i P_2^i y(t-1) + ... +$$

$$G^i P^i_{n_y+1} y(t-n_y) + G^i P^i_{n_y+2} u_1(t) + ..$$

$$+ G^i P^i_{n_y+n_1+2} u_1(t-n_1) + ... +$$

$$G^i P^i_{n_y+n_1+...+n_p+p+1} u_p(t-n_p) =$$

$$a_1 y(t) + a_2 y(t-1) + ... + a_{n_y+1} y(t-n_y) + b_1^1 u_1(t)$$

$$+ b_1^2 u_1(t-1) + ... + b_1^{n_1} u_1(t-n_1) + b_2^1 u_2(t)$$

$$+ b_2^2 u_2(t-1) + ... + b_2^{n2} u_2(t-n_2) + ... + b_m^1 u_m(t)$$

$$+ b_m^2 u_m(t-1) + ... + b_m^{nm} u_m(t-n_m) \qquad (5)$$

As the process has m output and m inputs, which can be regarded as m MISO--system, the multivariable plant can be described as:

$$A(q^{-1})\underline{Y}(t) = B(q^{-1})\underline{u}(t-1) \qquad (6)$$

where $\underline{Y}(t), \underline{u}(t-1)$ are mx1 output, input; $A(q^{-1})$, $B(q^{-1})$ are two matrix polynomials of the backward operator q^{-1}.

Eqn. (5) is a nonlinear equation, but can be seen as ordinary linear equation at each sample-instant.

When an equation or a set of equations as (5) has been formed the controller can be designed by any linear control method. In this paper, the GPC is used to design the controller.

3.2 Output prediction of multivariable plant

Suppose the multivariable plant described by the model

$$A(q^{-1})\underline{Y}(t) = B(q^{-1})\underline{u}(t-1) + \underline{\xi}(t) \big/ \Delta \qquad (7)$$

Where, $\underline{\xi}(t)$ is uncorrelated noise vector respectively and

$$A(q^{-1}) = I + A_1 q^{-1} + ... + A_{na} q^{-na}$$

$$B(q^{-1}) = B_0 + B_1 q^{-1} + ... + B_{nb} q^{-n_b}$$

$$\Delta = 1 - q^{-1}$$

$$(A_i, B_i \in R^{m \times m}) \qquad (8)$$

It can be got from Eqn.(7) that

$$A(q^{-1})\Delta\underline{Y}(t) = B(q^{-1})\Delta\underline{u}(t-1) + \underline{\xi}(t) \qquad (9)$$

In order to predicting future plant output, let's define the matrix polynomials $\Phi, \Phi^{[k]}, B^{[k]}$ as follows

$$A(q^{-1})\Delta = I - q^{-1}\Phi$$

$$\Phi^{[k]} = q^k [\Phi - \sum_{j=0}^{k-1} \Phi_j q^{-j}]$$

$$B^{[k]} = q^k [B - \sum_{j=0}^{k-1} B_j q^{-j}] \quad (k = 1,2,...) \qquad (10)$$

where, $\Phi, B, \Phi^{[k]}, B^{[k]}$ are matrix polynomials of q^{-1}; Φ_j, B_j denote the coefficient matrix of q^{-j} in the polynomials Φ, B.

Based on Eqs.(9)(10), the future 1-step ahead output is as:

$$\underline{Y}(t+1) = \Phi\underline{Y}(t) + B\Delta\underline{u}(t) + \underline{\xi}(t+1) =$$
$$\Phi\underline{Y}(t) + B^{[1]}\Delta\underline{u}(t-1) + B_0\Delta\underline{u}(t) + \underline{\xi}(t+1)$$
$$= Y_0\underline{Y}(t) + U_0\Delta\underline{u}(t-1) + H_0\Delta\underline{u}(t) + \underline{\xi}(t+1) \quad (11)$$

where, $Y_0 = \Phi, U_0 = B^{[1]}, H_0 = B_0, \underline{\xi}(t+1)$ is a unmeasurable component in the future so that the 1-step prediction is clearly

$$\underline{Y}^*(t+1) = \underline{f}(t+1) + H_0\Delta\underline{u}(t) \quad (12)$$

where, $\underline{f}(t+1)$ is the predictive information which can be estimated at time t and denoted as

$$\underline{f}(t+1) = Y_0\underline{Y}(t) + U_0\Delta\underline{u}(t-1) \quad (13)$$

Working in a similar manner, the k-step ahead prediction can be written as:

$$\underline{Y}^*(t+k) = \underline{f}(t+k) + H_0\Delta\underline{u}(t+k-1) + H_1\Delta\underline{u}(t+k-2) + \ldots + H_{k-1}\Delta\underline{u}(t) \quad (14)$$

$$\underline{f}(t+k) = Y_{k-1}\underline{Y}(t) + U_{k-1}\Delta\underline{u}(t-1) \quad (15)$$

where, the matrix polynomials Y_{k-1}, U_{k-1} and the matrix H_{k-1} can be calculated by the following recursive relationship:

$$Y_{k-1} = \Phi^{[k-1]} + \sum_{j=0}^{k-2}\Phi_j Y_{k-2-j}$$
$$H_{k-1} = B_{k-1} + \sum_{j=0}^{k-2}\Phi_j H_{k-2-j}$$
$$U_{k-1} = B^{[k]} + \sum_{j=0}^{k-2}\Phi_j U_{k-2-j} \quad (16)$$

Eqn.(14) indicates that the output prediction of multivariable plant consists of two parts: one being $\underline{f}(t+k)$ estimated at time t (determined by Eqs.(9)(10), sometimes called free response part of the prediction); the other depending on future control actions yet to be determined.

3.3 The predictive control law for multivariable system

By selecting the future control inputs, the objective of the multivariable predictive control law is to drive each output prediction "close" to its relevant reference trajectory in optimal sense, which can be realized from the minimization of the following quadratic cost criterion:

$$J = E\left\{ \sum_{j=0}^{N}[\underline{Y}(t+j)-\underline{Y}_r(t+j)]^\tau[\underline{Y}(t+j)-\underline{Y}_r(t+j)] + \sum_{j=0}^{N_u-1}\Delta\underline{u}^\tau(t+j)\beta(j)\Delta\underline{u}(t+j) \right\} \quad (17)$$

where, N, Nu are the prediction horizon and the control increment horizon respectively of multivariable plant; $\beta(j)$ is the weighting sequence of control increment, which is in general taken as a constant matrix; $\underline{Y}_r(t+j)$ is the vector of reference trajectory at future time instant j and denoted by

$$\underline{Y}_r(t+j) = [\underline{Y}_{r1}(t+j), \underline{Y}_{r2}(t+j), \ldots \underline{Y}_{rm}(t+j)]^\tau \quad (18)$$

where, $\underline{Y}_{ri}(t+j)$ (i=1,2, ...,m) is i-th reference sequence at time instant t, which is obtainable from the simple first-order lag model:

$$\underline{Y}_{ri}(t+j) = \alpha_i\underline{Y}_{ri}(t+j-1) + (1-\alpha_i)\omega_i \quad (19)$$

where, ω_i is i-th output set-point of multivariable system; α_i stands for the adjustable parameter of i-th reference trajectory, $0 \le \alpha_i \le 1$

The predictive control law can be easily obtained by minimizing the cost function Eqn.(17)

$$\Delta\underline{U} = [H^\tau H + \lambda I]^{-1}H^\tau(\underline{Y}_r - F) \quad (20)$$

where, $\underline{Y}_r = [\underline{Y}_r^\tau(t+1), \underline{Y}_r^\tau(t+2), \ldots \underline{Y}_r^\tau(t+N)]^\tau$ denotes the reference trajectory vector of dimension mN; $\underline{F} = [\underline{f}^\tau(t+1), \underline{f}^\tau(t+2), \ldots \underline{f}^\tau(t+N)]^\tau$ stands for output prediction vector of dimension mN, which can be estimated at time t;
$\Delta\underline{U} = [\Delta\underline{u}^\tau(t), \Delta\underline{u}^\tau(t+1), \ldots, \Delta\underline{u}^\tau(t+N_u-1)]^\tau$ is the control increment vector of dimension mNu; H is a matrix of dimension mNxmNu given by

$$H = \begin{bmatrix} H_0 & \cdots & \cdots & \cdots & 0 \\ H_1 & H_0 & \cdots & \cdots & 0 \\ H_2 & H_1 & H_0 & \cdots & 0 \\ \cdots & \cdots & \cdots & \cdots & \cdots \\ H_{N-1} & H_{N-2} & \cdots & \cdots & H_{N-N_u} \end{bmatrix}$$
$$(H_i \in R^{m\times m}) \quad (21)$$

β is a weighting matrix of dimension mNuxmNu given by

$$\beta = diag\{\beta(0), \beta(1), \cdots, \beta(N_u-1)\}$$
$$\beta(i) \in R^{m\times m} \quad (22)$$

Mark the former m rows of $[H^\tau H + \lambda I]^{-1}H^\tau$ in Eqn. (20) by a submatrix R_g, i.e.,

$$R_g = \left[R_{g1} \quad R_{g2} \quad \cdots \quad R_{gN} \right] \tag{23}$$

where R_{gi} is a matrix of dimension mxm.

Therefore it can be obtained from Eqn. (20) that

$$\Delta\underline{u}(t) = R_g \left[\underline{Y}_r - \underline{F} \right] = \sum_{i=1}^{N} R_{gi} \underline{Y}_r(t+i) - \sum_{i=1}^{N} R_{gi} \underline{f}(t+i) \tag{24}$$

From Eqn. (15), having

$$\Delta\underline{u}(t) = R(q)\underline{Y}_r(t) - R_y(q^{-1})\underline{Y}(t) - R_u(q^{-1})\Delta\underline{u}(t-1) \tag{25}$$

where

$$\Delta\underline{u}(t) = \underline{u}(t) - \underline{u}(t-1)$$
$$R(q) = \sum_{i=1}^{N} R_{gi} q^i$$
$$R_y(q^{-1}) = \sum_{i=1}^{N} R_{gi} Y_{i-1}$$
$$R_u(q^{-1}) = \sum_{i=1}^{N} R_{gi} U_{i-1} \tag{26}$$

4.THE STABILITY ANALYSIS OF THE FUZZY GENERALIZED PREDICTIVE CONTROL SYSTEM

From Eqn.(25), it can be deduced that

$$\underline{u}(t) = R_{yr}(q^{-1})\underline{Y}_r(t) - R_{yy}(q^{-1})\underline{Y}(t) \tag{27}$$

where $R_{yr}(q^{-1})$ and $R_{yy}(q^{-1})$ are two matrix polynomials. Then Eqn.(6) can be expressed as

$$\left[A(q^{-1}) + B(q^{-1}) R_{yy}(q^{-1}) q^{-1} \right] \underline{Y}(t) = B(q^{-1}) R_{yr}(q^{-1}) \underline{Y}_r(t-1) \tag{28}$$

Eqn.(28) can be transformated as following state space equations:

$$\underline{X}(k+1) = A_s \underline{X}(k) + B_s \underline{Y}_r(k)$$
$$\underline{Y}(k) = C_s \underline{X}(k) \tag{29}$$

where A_s is nonsinqular matrix with an appropriate dimension.

Definition 1. A family of interval matrices is defined as:

$$\Lambda(\underline{A}, \overline{A}) = \left\{ A \in R^{n \times n}; \underline{A} \le A \le \overline{A} \right\} \tag{30}$$

where $\underline{A} = (\underline{a}_{ij})$ *and* $\overline{A} = (\overline{a}_{ij})$ are constant matrices with dimension $n \times n$. It is believed that a family Λ is stable (Schur or Hurwitz stable) if every $A \in \Lambda$ is stable.

Theorem 1. A system

$$X_{k+1} = (E+F)X_k$$
$$(|F| \le U) \tag{31}$$

is stable if

$$\rho(|E| + U) < 1 \tag{32}$$

The proof can be found in the reference by Dai (1993).

Theorem 2. If $|U| \le \varepsilon |E|$ $(\varepsilon < 1)$ then the system

$$X_{k+1} = (E+U)X_k \tag{33}$$

is stable iff

$$\varepsilon < \frac{1-\rho(|E|)}{\rho(|E|)} \equiv \delta_R \tag{34}$$

Proof.

(i) Sufficiency

From Theorem 1 and with $|U| \le \varepsilon |E|$, the system (33) is stable if $\rho\ (E + \varepsilon E) \le \rho\ (|E| + |\varepsilon E|) = (1+\varepsilon)\rho(|E|) < 1$, i.e., $\varepsilon < \delta_R$

(ii) Necessity

(a) If E>0, then $E + U \ge E(1-\varepsilon) > 0$. Thus (33) is stable iff $\rho(E+U) < 1$, i.e., $\varepsilon < \delta_R$.

(b) If E<0, then the system (33) is stable with $|U| < \varepsilon |E|$

iff

$$\rho(E+U) = \rho(-(E+U)) \le \rho(-E + \varepsilon|E|)$$
$$= (1+\varepsilon)\rho(-E) = (1+\varepsilon)\rho(|E|) < 1,$$

i.e., $\varepsilon < \delta_R$

Corollary. In (29), A_s usually lies:

$$A_s \in \left[\underline{A} \quad \overline{A} \right] \tag{35}$$

where $\overline{A}$ is matrix whose elements consist of respective maximum values, $\underline{A}$ is the one of minimum values.

Let

$$E = \frac{\overline{A} + \underline{A}}{2}$$

$$U = \frac{|(\bar{A} - \underline{A})|}{2} \qquad (36)$$

Then the stability of the FGPC system can be checked by Theorem 2.

The algorithm of the on-line designing of the stability:
1. Calculating and save the $\bar{A}$ and $\underline{A}$ at each sample instant.

2. If $t \geq t_c$ (t_c is a predetermined constant)

$$E = \frac{1}{2}(\bar{A} + \underline{A})$$

$$U = \frac{1}{2}|(\bar{A} - \underline{A})|$$

$$\rho(|E|) = \max \lambda(|E|)$$

$$\delta_R = \frac{1 - \rho(|E|)}{\rho(|E|)}$$

otherwise continue.

3. $\varepsilon = \max \dfrac{|U_{ij}|}{|E_{ij}|}, \quad 1 \leq i, j \leq n$

If $\varepsilon < \delta_R$, turn to 1,

otherwise, take a close look to the system and proper tune to the parameters of the controller.

As the calculations of $\rho(|E|)$ and δ_R are more complex than ε, a larger constant t_c should be chosen properly.

5. CONCLUSIONS

A multivariable FGPC approach has be put forward by means of the principle of Clark's GPC in this paper. The stability of the closed-loop system can be checked and guaranteed.

The FGPC is also with the advantage that the design procedure and the tuning of the controller parameters are simple to understand and to implement. It can effectively control a multivariable nonlinear plant, especially with large time--delay.

There exists some problems which haven't be solved so far, such as how to improve the proposed algorithm to reduce the time required, so that it can be used for faster dynamic system.

REFERENCES

Clark, D.W., C. Mohtadi, and P.S. Toffs. (1987). Generalized Predictive Control-Par t I: the Basic Algorithm, *Automatica,* **Vol.23**, pp:137-148.

Dai,C.G.(1993). *Linear algebra in control systems,* Southeastern university publisher (in Chinese).

Jeffrey T. S.,etc.(1997). Direct adaptive fuzzy control for a class of discrete-time systems. *Proceedings of the ACC*, pp:1814-1818.

Oliver H.,et al. (1997). Non-linear system identification and predictive control of a heat exchanger, *Proceedings of the ACC*, pp:3294-3298.

Sezer, M.E and D.D. Siljak, (1994). Stability of interval matrices, *IEEE Trans on AC*, **Vol.39**, pp:368-371.

Takagi,T. and M. Sugeno, (1985). Fuzzy identification of systems and its application to modelling and control, *IEEE Trans on SMC*, **Vol.15**, pp:116-133.

14th Triennial World Congress, Beijing, P.R. China

C-2a-07-3

CONSTRAINED LINEAR MODEL-BASED PREDICTIVE CONTROL WITH AN INFINITE CONTROL AND PREDICTION HORIZON

H. H. J. Bloemen and T. J. J. van den Boom

Department of Information Technology and Systems
Systems and Control Engineering Group
P.O. Box 5031, NL-2600 GA Delft, The Netherlands
Tel. (+31) 15 2782087 Fax. (+31) 15 2786679
Email: H.Bloemen@its.tudelft.nl

Abstract: This paper proposes a new linear Model-based Predictive Control (MPC) strategy for which nominal closed loop stability is guaranteed. The new strategy consists of a combination of regular and feedback MPC resulting in a strategy with an infinite control and prediction horizon. Both the inputs up to a certain switching horizon and the state feedback (used after this switching horizon) are optimized in one convex optimization problem based on linear matrix inequalities (LMI's). It can be proven that other well known MPC strategies are special cases of the proposed strategy using a special weighting in the performance index.

Keywords: Constraints, Predictive control, Stability analysis, State feedback

1. INTRODUCTION

Initially a finite control and prediction horizon was used for Model-based Predictive Control (MPC) strategies. Without additional constraints nominal closed loop stability can in general not be proven for such a strategy. Stability analysis for unconstrained (no constraints on the input or output) MPC strategies have for example been performed in (Scattolini and Bittanti, 1990) and for an unconstrained state-feedback strategy in (Kwon *et al.*, 1983).

But in practice input and output constraints have to be accounted for in MPC strategies. For example these constraints can arise from equipment limitations or safety demands (van den Boom, 1996). Because constraint handling is an important aspect of MPC techniques, strategies that guarantee nominal closed loop stability of constrained MPC problems are more interesting from a practical point of view. Stability analysis for constrained MPC strategies have for example been performed in (Zafiriou, 1990), (Zheng and Morari, 1993) and (Zheng and Morari, 1995). Constrained MPC strategies reported in literature that achieve nominal closed loop stability are for example:

- finite prediction and control horizon MPC using an end point constraint (Clarke and Scattolini, 1991), (Mosca and Zhang, 1992)
- infinite prediction horizon, finite control horizon MPC (Rawlings and Muske, 1993), (Scokaert, 1997)
- infinite prediction and control horizon MPC using a state feedback law (Kothare, 1997)
- infinite prediction and control horizon MPC where at a certain time step (switching horizon) a switch is being made to an a priori determined state feedback law (Scokaert and Rawlings, 1996), (Lee *et al.*, 1997)

In the end point constraint MPC strategy (Clarke and Scattolini, 1991), (Mosca and Zhang, 1992) the output is forced to equal the reference value at the end of the horizon. This equality constraint restricts the performance of the controller because some of the degrees of freedom (the control inputs) have to be used to satisfy this equality constraint. Additionally an end point constraint increases the risk of infeasibility (Scokaert, 1997).

In (Rawlings and Muske, 1993) the prediction horizon is extended to infinity and the control horizon is kept finite. For simplicity assume the reference value and the control input after the control horizon to be equal to zero. Then the quadratic performance index of the MPC problem is finite if and only if the unstable modes equal zero at the end of the control horizon. The stable modes will eventually converge to zero (open loop dynamics, because the control equals zero) resulting in a finite performance index. This performance index can then serve as a Lyapunov function which proves nominal closed loop stability. To satisfy output constraints over the entire infinite prediction horizon, in (Rawlings and Muske, 1993) a constraint horizon is derived which guarantees the constraints to be satisfied over the entire prediction horizon. The end point constraint for the unstable modes is a necessary condition for the feasibility of this MPC problem and does not restrict the performance of the infinite prediction horizon finite control horizon MPC strategy (Scokaert, 1997). However the control input after the control horizon is restricted to equal zero which may restrict the performance of the controller.

In (Kothare, 1997) both the prediction and control horizon are infinite. In order to obtain a finite dimensional optimization problem a state feedback law is used to compute the control input in stead of using the control input at every interval as a degree of freedom. To guarantee nominal closed loop stability a Lyapunov function is minimized which is an upper bound for the performance index. The major drawback of this method is the conservative constraint handling of the input and output constraints, which may restrict the performance of the controller.

In (Scokaert and Rawlings, 1996) regular and feedback MPC are combined. The inputs up to a certain switching horizon are used for optimizing a performance index. At the switching horizon the input follows from an a priori determined state feedback. The constraints for the feedback part are transformed to an inequality constraint in the sense that the state at this switching instant should belong to a certain region of the state space for which the input and state constraints are satisfied for the infinite horizon, given the state feedback. This inequality constraint is evaluated after optimization. When this constraint is violated, the optimization is repeated for an increased switching horizon, leading to an iterative approach. In (Lee *et al.*, 1997) regular and feedback MPC are combined in a similar way. However, in (Lee *et al.*, 1997) constraints are used for a finite terminal weighting matrix, based on a given state feedback, that ensure satisfaction of the input and state constraints for the infinite horizon. These constraints are implemented in the optimization thus avoiding an iterative approach. In order to reduce the flexibilities that are present in choosing the finite terminal weighting matrix an additional optimization problem is stated to enlarge the region of attraction of the controller.

In this paper the method of (Lee *et al.*, 1997) is extended by means of both optimizing the inputs up to the switching horizon and the state feedback. One could see this as increasing the freedom of choosing the finite terminal weighting matrix. This freedom is used to minimize the costs associated with the state at the switching instant and the finite terminal weighting matrix. These costs can be seen as an upper bound for the performance index from the switching instant up to infinity, see (Kothare, 1997).

Because optimizing the state feedback gives additional degrees of freedom compared to the algorithm of (Lee *et al.*, 1997), the feasibility properties of the optimization problem are improved.

Nominal closed loop stability of this strategy can be guaranteed as outlined in section 3. In section 4 it is shown that the regular MPC strategy and the MPC strategies that use an end point constraint (for all the states or only for the unstable modes) are a special case of the strategy proposed in this paper.

2. PROPOSITION OF THE SWITCHING MPC STRATEGY

Consider a controllable linear discrete time state space model:

$$\mathbf{x}(k+1|k) = \mathbf{A}\mathbf{x}(k|k) + \mathbf{B}\mathbf{u}(k|k) \qquad (1)$$

$$\mathbf{y}(k|k) = \mathbf{C}\mathbf{x}(k|k) \qquad (2)$$

with $\mathbf{x}(k+1|k)$ representing the state at time $k+1$ calculated at time k, $\mathbf{u}$ representing the input and $\mathbf{y}$ representing the output. The objective is to bring the system to a steady state value $(\mathbf{y_{ss}}, \mathbf{u_{ss}})$. For simplicity assume this equilibrium to be the origin (when this equilibrium $(\mathbf{y_{ss}}, \mathbf{u_{ss}}) \neq (\mathbf{0}, \mathbf{0})$, a state transformation can be applied to yield a shifted system with equilibrium $(\mathbf{y}^*_{\mathbf{ss}}, \mathbf{u}^*_{\mathbf{ss}}) = (\mathbf{0}, \mathbf{0})$).

Define the (quadratic) performance index, calculated at time k, to be equal to:

$$J(k) = J_1(k) + J_2(k) \tag{3}$$

$$J_1(k) = \sum_{i=1}^{H_s} ||\mathbf{x}(k+i|k)||^2_{\mathbf{Q_1}} + ||\mathbf{u}(k+i-1|k)||^2_{\mathbf{R_1}}$$

$$J_2(k) = ||\mathbf{x}(k+H_s|k)||^2_{\mathbf{P}}$$

in which $\mathbf{Q_1}$ is a positive semi-definite weighting matrix and $\mathbf{R_1}$ and $\mathbf{P}$ are positive definite weighting matrices. The goal is to minimize the performance index without violating the input and output constraints which should hold for the infinite horizon:

$$|\mathbf{u}(k+i-1|k)| \leq \begin{bmatrix} u_{max,1} \\ \vdots \\ u_{max,m} \end{bmatrix}, \text{for } i = 1 \ldots \infty \tag{4}$$

$$|\mathbf{y}(k+i|k)| \leq \begin{bmatrix} y_{max,1} \\ \vdots \\ y_{max,p} \end{bmatrix}, \text{for } i = 1 \ldots \infty \tag{5}$$

In which m equals the number of inputs and p equals the number of outputs.

Optimizing the performance index (equation 3) can be stated as one LMI-based optimization problem. Derivation of the LMI's can be split into two steps:
step 1: the costs associated with $J_1(k)$
step 2: the costs associated with $J_2(k)$
The LMI's of both steps can be combined, resulting in one optimization problem.

2.1 *step 1*

The prediction of $\mathbf{x}$ and $\mathbf{y}$ up to horizon H_s can be written as (see (Kinnaert, 1989) for example):

$$\tilde{\mathbf{x}} = \tilde{\mathbf{A}}\mathbf{x}(k) + \tilde{\mathbf{B}}\tilde{\mathbf{u}} \tag{6}$$

$$\tilde{\mathbf{y}} = \tilde{\mathbf{C}}\tilde{\mathbf{x}} \tag{7}$$

with $\tilde{\mathbf{x}}$, $\tilde{\mathbf{u}}$ and $\tilde{\mathbf{y}}$ vectors containing $\mathbf{x}(k+i|k)$, $\mathbf{u}(k+i-1|k)$ and $\mathbf{y}(k+i|k)$ for $i = 1 \ldots H_s$ respectively, and $\tilde{\mathbf{A}}$, $\tilde{\mathbf{B}}$ and $\tilde{\mathbf{C}}$ prediction matrices. Now $J_1(k)$ can be written as:

$$\begin{aligned} J_1(k) &= \tilde{\mathbf{x}}^T\tilde{\mathbf{Q}}_1\tilde{\mathbf{x}} + \tilde{\mathbf{u}}^T\tilde{\mathbf{R}}_1\tilde{\mathbf{u}} \\ &= constant + 2\tilde{\mathbf{u}}^T\tilde{\mathbf{B}}^T\tilde{\mathbf{Q}}_1\tilde{\mathbf{A}}\mathbf{x}(k) \\ &\quad + \tilde{\mathbf{u}}^T(\tilde{\mathbf{B}}^T\tilde{\mathbf{Q}}_1\tilde{\mathbf{B}} + \tilde{\mathbf{R}}_1)\tilde{\mathbf{u}} \end{aligned} \tag{8}$$

using equation 6. $\tilde{\mathbf{Q}}_1$ and $\tilde{\mathbf{R}}_1$ are block diagonal matrices which (i,i)-th blocks are $\mathbf{Q_1}$ and $\mathbf{R_1}$ respectively.

Minimizing $J_1(k)$ can be stated as a LMI based optimization problem by noting that minimizing $J_1(k)$ (equation 8) is equivalent to minimizing η_1 under the constraint:

$$2\tilde{\mathbf{u}}^T\tilde{\mathbf{B}}^T\tilde{\mathbf{Q}}_1\tilde{\mathbf{A}}\mathbf{x}(k) + \tilde{\mathbf{u}}^T(\tilde{\mathbf{B}}^T\tilde{\mathbf{Q}}_1\tilde{\mathbf{B}} + \tilde{\mathbf{R}}_1)\tilde{\mathbf{u}} \leq \eta_1 \tag{9}$$

Using Schur complements (Kothare, 1997) equation 9 can be converted to a LMI in η_1 and $\tilde{\mathbf{u}}$, under the assumption that $\tilde{\mathbf{B}}^T\tilde{\mathbf{Q}}_1\tilde{\mathbf{B}} + \tilde{\mathbf{R}}_1$ is positive definite. This assumption is satisfied because $\mathbf{R_1}$ is positive definite and $\mathbf{Q_1}$ is positive semi-definite.

Next the input and output constraints (equations 4 and 5) have to be represented as LMI's. Because these constraints are linear this conversion is straightforward.

2.2 *step 2*

From H_s till infinity a feedback control law is used

$$\mathbf{u}(k+i|k) = \mathbf{F}\mathbf{x}(k+i|k), \quad \text{for } i \geq H_s \tag{10}$$

The objective is to find a stabilizing control law $\mathbf{F}$ that minimizes a performance index over a horizon H_s up to infinity, similar to $J_1(k)$. This is done by defining a Lyapunov function V for the system. Take, according to (Kothare, 1997):

$$V(\mathbf{x}) = \mathbf{x}^T\mathbf{P}\mathbf{x} \tag{11}$$

with $\mathbf{P}$ positive definite, and assume that for $i \geq H_s$:

$$\begin{aligned} &V(\mathbf{x}(k+i+1|k)) - V(\mathbf{x}(k+i|k)) \\ &\leq -(||\mathbf{x}(k+1+i|k)||^2_{\mathbf{Q_2}} + ||\mathbf{u}(k+i|k)||^2_{\mathbf{R_2}}) \end{aligned} \tag{12}$$

Normally $\mathbf{Q_2}$ and $\mathbf{R_2}$ are equal to $\mathbf{Q_1}$ and $\mathbf{R_1}$ respectively. Summation of equation 12 for $i = H_s \ldots \infty$ shows $V(\mathbf{x}(k+H_s|k)) = J_2(k)$ is an upper bound for a performance index similar to $J_1(k)$ expressed over a horizon H_s up to infinity.

In order to compute the feedback law the upper bound of the performance index, $J_2(k)$, is minimized. Minimizing $J_2(k)$ can be formulated as a LMI based optimization problem by minimizing η_2 under the constraint:

$$\mathbf{x}(k+H_s|k)^T\mathbf{S}^{-1}\mathbf{x}(k+H_s|k) \leq 1 \tag{13}$$

with $\mathbf{S} = \eta_2\mathbf{P}^{-1}$. Equation 13 can be stated as a LMI in $\tilde{\mathbf{u}}$ ($\mathbf{x}(k+H_s|k)$ is affine in $\tilde{\mathbf{u}}$) and $\mathbf{S}$. V also has to satisfy inequality 12. This leads to the constraint (using 1, 10 and 11):

$$\begin{aligned} \mathbf{x}(k+i|k)^T \{ &(\mathbf{A}+\mathbf{BF})^T\mathbf{P}(\mathbf{A}+\mathbf{BF}) - \mathbf{P} \\ &+(\mathbf{A}+\mathbf{BF})^T\mathbf{Q_2}(\mathbf{A}+\mathbf{BF}) \\ &+\mathbf{F}^T\mathbf{R_2}\mathbf{F} \}\, \mathbf{x}(k+i|k) \leq 0 \end{aligned} \tag{14}$$

for all $i \geq H_s$. Substituting $\mathbf{P} = \eta_2\mathbf{S}^{-1}$ and $\mathbf{F} = \mathbf{Y}\mathbf{S}^{-1}$, followed by pre and post multiplication by $\mathbf{S}$, and division by η_2 this can be written as a LMI in $\mathbf{Y}$, $\mathbf{S}$ and η_2 for the matrix expression between the curly brackets (see (Kothare, 1997) and (Lee *et al.*, 1997) for example).

Regarding step 2, input and output constraints should be satisfied from H_s till infinity. The constraints on the j-th input (equation 4) can be expressed as (using 10):

$$\max_{i \geq H_s} |\mathbf{u}_j(k+i|k)|^2$$
$$= \max_{i \geq H_s} |\mathbf{Y}_j \mathbf{S}^{-1} \mathbf{x}(k+i|k)|^2 \leq u^2_{max,j}$$

where the subscript j denotes the j-th row. Because $\varepsilon = \{\mathbf{z}|\mathbf{z}^T\mathbf{S}^{-1}\mathbf{z} \leq 1\}$, equation 13, is an invariant ellipsoid for the predicted states (Kothare, 1997) (Boyd *et al.*, 1994), $\mathbf{x}(k+H_s|k) \in \varepsilon$ implies $\mathbf{x}(k+i|k) \in \varepsilon$ for all $i \geq H_s$. This means:

$$\max_{i \geq H_s} |\mathbf{Y}_j \mathbf{S}^{-1} \mathbf{x}(k+i|k)|^2 \leq \max_{\mathbf{z} \in \varepsilon} |\mathbf{Y}_j \mathbf{S}^{-1} \mathbf{z}|^2$$

The j-th input constraint is satisfied if (using the invariant ellipsoid ε):

$$\mathbf{z}^T \left\{ \mathbf{S}^{-1}(\mathbf{Y}_j)^T \frac{1}{u^2_{max,j}} \mathbf{Y}_j \mathbf{S}^{-1} - \mathbf{S}^{-1} \right\} \mathbf{z} \leq 0 \quad (15)$$

After pre and post multiplication by $\mathbf{S}$ this can be written as a LMI in $\mathbf{Y}$ and $\mathbf{S}$ for the matrix expression between the curly brackets. Constraints on the output (see equation 5) can be expressed as LMI's similarly.

Note that the constraint handling for the input and output constraints is conservative when applying the feedback law.

2.3 *Combination*

The switching MPC problem now consists of:

$$\min_{\tilde{\mathbf{u}},\mathbf{F}} \eta_1 + \eta_2 \quad (16)$$

given the constraints, equations: 9, 4, 5, 13, 14, 15.

3. STABILITY

For the combined MPC strategy nominal closed loop stability can be proven using the following theorem:

Theorem 1. Under the assumption that $\mathbf{R}_1$ and $\mathbf{R}_2$ are positive definite and $\mathbf{Q}_1$ and $\mathbf{Q}_2$ are positive semi definite matrices, satisfying:

$$\mathbf{R}_2 \geq \mathbf{R}_1 \quad (17)$$
$$\mathbf{Q}_2 \geq \mathbf{Q}_1 \quad (18)$$

the switching MPC strategy (see section 2.3) guarantees nominal closed loop stability for a controllable plant once a feasible solution of the MPC problem is found.

proof Assume a feasible solution at time $t = k$ exists, then at all times $t > k$ a feasible solution exists, because at every time $t = t+1$ the shifted solution of time $t = t$ can be applied which is thus feasible. Assume at time $t = l = k+1$ we use the shifted solution of time $t = k$. Consider now the performance indices $J(k)$ and $J(l)$ given the solution for the input ($\tilde{\mathbf{u}}^*$ and $\mathbf{F}^*$) at time k. $J_1(l)$ can be written as:

$$J_1(l) = J_1(k) - (||\mathbf{x}(k+1|k)||^2_{\mathbf{Q}_1} + ||\mathbf{u}(k|k)||^2_{\mathbf{R}_1})$$
$$+ ||\mathbf{x}(k+H_s+1|k)||^2_{\mathbf{Q}_1} + ||\mathbf{u}(k+H_s|k)||^2_{\mathbf{R}_1}$$

Using equation 12 it can be derived that:

$$J_2(l) = V(\mathbf{x}(k+H_s+1|k)) \leq J_2(k)$$
$$- (||\mathbf{x}(k+H_s+1|k)||^2_{\mathbf{Q}_2} + ||\mathbf{u}(k+H_s|k)||^2_{\mathbf{R}_2})$$

Given $\mathbf{Q}_2 \geq \mathbf{Q}_1$ and $\mathbf{R}_2 \geq \mathbf{R}_1$ above expressions and equation 3 lead to:

$$J(l) \leq J(k) \quad (19)$$

Using the receding horizon strategy, at time $l = k+1$ a new minimization of $J(l)$ is performed. This convex optimization always yields a global optimum $J(l) \leq J(k)$.

This means $J(i)$ is a Lyapunov function, which proves nominal closed loop stability of the new MPC strategy (section 2.3) when the assumptions in theorem 1 are satisfied.

4. RELATION TO OTHER MPC STRATEGIES

Matrices $\mathbf{Q}_2$ and $\mathbf{R}_2$ may be chosen differently from $\mathbf{Q}_1$ and $\mathbf{R}_1$. These extra tuning parameters can be used to relate the switching MPC strategy (section 2.3) to the regular and end point constraint MPC strategies, resulting in the next theorem:

Theorem 2. Given an unconstrained system (without input and output constraints) the regular, finite horizon MPC strategy, the end point constraint MPC strategy (Clarke and Scattolini, 1991), (Mosca and Zhang, 1992) and the infinite prediction horizon finite control horizon MPC strategy (end point constraint only for the unstable modes) (Rawlings and Muske, 1993), (Scokaert, 1997) all are special cases of the switching MPC strategy proposed in this article (section 2.3).

proof For simplicity assume that the weighting matrices are related according to:

$$\mathbf{Q_2} = w_Q * \mathbf{Q_1}$$
$$\mathbf{R_2} = w_R * \mathbf{R_2}$$

in which w_Q and w_R are scalar variables. Using special values of w_Q and w_R it can be proven that the other MPC strategies all are special cases of the switching MPC strategy:

* $w_Q \to 0, w_R \to 0$. This means that no effort is made to minimize the performance index after horizon H_s, so this corresponds to the regular, finite horizon MPC strategy. Stability can not be proven for this case because w_Q and w_R need to be larger than one for the proof of theorem 1.
* $w_Q \to \infty, w_R \to \infty$. This means the weighting of the input $\mathbf{u}$ in the performance index will be infinitely large for times $H_s \ldots \infty$ and the weighting of $\mathbf{x}$ in the performance index will be infinitely large for times $H_s + 1 \ldots \infty$. So the switching MPC strategy will aim at bringing $\mathbf{u}$ to zero at the end of horizon H_s and bringing $\mathbf{x}$ to zero at the end of horizon $H_s + 1$. Because the control from time $H_s \ldots H_s + 1$ goes to zero, $\mathbf{x}$ can only become zero at time $H_s + 1$ when $\mathbf{x}$ at time H_s belongs to the kernel of $\mathbf{A}$. This corresponds to the end point constraint MPC strategy with a prediction horizon H_s+1 and a control horizon H_s. When $\mathbf{A}$ is non singular the switching MPC strategy corresponds to the end point constraint MPC strategy with a control horizon equal to the prediction horizon. Stability can still be proven for this case because w_Q and $w_R \geq 1$, which is in agreement with (Clarke and Scattolini, 1991) and (Mosca and Zhang, 1992).
* $w_Q = 1, w_R = \infty$. This means the switching MPC strategy aims at bringing the control after H_s to zero. In that case finiteness of the performance index over the infinite horizon can only be accomplished by bringing the unstable modes to zero at horizon H_s, thus resulting in the infinite prediction horizon finite control horizon MPC strategy. Stability can still be proven for this case because w_Q and $w_R \geq 1$, which is in agreement with (Rawlings and Muske, 1993) and (Scokaert, 1997).

5. SIMULATIONS

For simulation example 3.13 of (Soeterboek, 1992) has been used, which has two unstable modes:

$$\mathbf{A} = \begin{bmatrix} 2.4936 & -1.2130 & 0.4176 \\ 2.0000 & 0 & 0 \\ 0 & 1.0000 & 0 \end{bmatrix}$$
$$\mathbf{B} = \begin{bmatrix} 0.2500 \\ 0 \\ 0 \end{bmatrix}$$
$$\mathbf{C} = \begin{bmatrix} 0.0684 & 0.1288 & 0.0313 \end{bmatrix}$$

The symmetric positive definite weighting matrices in the performance index are chosen to be:

$$\mathbf{Q_1} = \mathbf{Q_2} = \begin{bmatrix} 3 & 2 & 1 \\ 2 & 3 & 2 \\ 1 & 2 & 3 \end{bmatrix}, \mathbf{R_1} = \mathbf{R_2} = \begin{bmatrix} 1 \end{bmatrix}$$

For the simulation $\mathbf{x}(0) = [1\ 1\ 1]^T$ is used.

The MPC strategy proposed in this paper is compared for situations where the state feedback can be optimized (this paper) and for situations where it is fixed (Scokaert and Rawlings, 1996), (Lee *et al.*, 1997). In case the feedback is fixed the optimal LQ regulator is chosen. The controllers are implemented in a receding horizon mode. In table 1 and 2 the value of the actually achieved performance index (*costs*) over the horizon $t = 0$ to $t = 10$ is given for the different control strategies:

$$costs = \sum_{i=1}^{10} ||\mathbf{x}(k+i|k)||^2_{\mathbf{Q_1}} + ||\mathbf{u}(k+i-1|k)||^2_{\mathbf{R_1}}$$

When the optimization problem is infeasible this is indicated by *infeas*. The effect of optimizing the state feedback is compared for different input constraints.

Table 1. Optimized state feedback

	costs		
$\|u\| \leq$	$H_s = 1$	$H_s = 2$	$H_s = 3$
∞	88.585	88.585	88.585
2.5	126.733	126.615	126.615
1.5	193.928	189.976	189.976

Table 2. LQ regulator

	costs		
$\|u\| \leq$	$H_s = 1$	$H_s = 2$	$H_s = 3$
∞	88.585	88.585	88.585
2.5	*infeas*	126.615	126.615
1.5	*infeas*	*infeas*	189.976

From tables 1 and 2 it is clear that optimizing the state feedback gives better feasibility properties than using an a priori determined state feedback. When optimizing the state feedback the problem remains feasible for small values of H_s whereas the associated *costs* only increase by a small amount.

For large horizons or for the unconstrained case, there is no difference in the methods. This can be explained by the fact that in these cases the feedback is not restricted by the input and output constraints because $\mathbf{x}(k + H_s|k)$ is then already close to zero. When input and output constraints are not active (Kothare, 1997) proves that the optimized state feedback results in the optimal LQ regulator.

6. CONCLUSION

In this paper a new Model-based Predictive Controller is proposed which is an extension of the one presented in (Lee *et al.*, 1997). At a certain switching instant the input follows from a state feedback. This state feedback is calculated by minimizing an upperbound for the performance index over the horizon from this switching instant up to infinity. Optimizing this state feedback results in better feasibility properties, compared to taking a fixed feedback. Linear Matrix Inequalities can be derived which guarantee input and output constraints to be satisfied over the infinite horizon. For the proposed strategy nominal closed loop stability can be guaranteed.

This method does not require some of the modes to go to zero in finite time, as is the case for end point constraint MPC strategies (for all the modes (Clarke and Scattolini, 1991), (Mosca and Zhang, 1992) or only the unstable modes (Rawlings and Muske, 1993), (Scokaert, 1997)). In stead of such an end point constraint the states are constrained by an invariant ellipsoid $\mathbf{x}(k + H_s|k)^T\mathbf{S}^{-1}\mathbf{x}(k + H_s|k) \leq 1$, which is less restrictive (Lee *et al.*, 1997).

Using different weighting matrices in the expressions for J_1 and J_2 it can be shown that the regular finite horizon MPC and the end point constraint MPC strategies all are special cases of the switching MPC strategy in the unconstrained case.

Acknowledgments- The authors would like to thank one of the reviewers for his helpful comment. This research was supported by the Dutch Technology Foundation (STW) under project number DEL55.3891

7. REFERENCES

Boyd, S., L. E. Ghaoui, E. Feron and V. Balakrishnan (1994). *Linear matrix inequalities in system and control theory.* SIAM, Philadelphia.

Clarke, D. W. and R. Scattolini (1991). Constrained receding-horizon predictive control. *IEE Proceedings-D* **138**(4), 347–354.

Kinnaert, M. (1989). Adaptive generalized predictive controller for MIMO systems. *Int. J. Control* **50**(1), 161–172.

Kothare, M.V. (1997). Control of systems subject to constraints. PhD thesis. California Institute of Technology Pasadena.

Kwon, W. H., A. M. Bruckstein and T. Kailath (1983). Stabilizing state-feedback design via the moving horizon method. *International Journal of Control* **37**(3), 631–643.

Lee, Jae-Won, Wook Hyun Kwon and Jinhoon Choi (1997). On stability of constrained receding horizon control with finite terminal weighting matrix. In: *ECC'97, European Control Conference, Brussels, Belgium.*

Mosca, Edoardo and Jingxin Zhang (1992). Stable redesign of predictive control. *Automatica* **28**(6), 1229–1233.

Rawlings, J.B. and K.R. Muske (1993). The stability of constrained receding horizon control. *IEEE transactions on automatic control* **38**(10), 1512–1516.

Scattolini, Riccardo and Sergio Bittanti (1990). On the choice of the horizon in long-range predictive control; some simple criteria. *Automatica* **26**(5), 915–917.

Scokaert, Pierre O. M. (1997). Infinite horizon generalized predictive control. *International Journal of control* **66**(1), 161–175.

Scokaert, Pierre O.M. and James B. Rawlings (1996). Infinite horizon linear quadratic control with constraints. In: *13th Triennial World Congress, San Francisco, USA*. IFAC. pp. 109–114.

Soeterboek, Ronald (1992). *Predictive control: a unified approach.* Prentice Hall International.

van den Boom, Ton J. J. (1996). Model based predictive control, status and perspectives. In: *CESA/IMACS, Lille, France.* pp. 1–12.

Zafiriou, E. (1990). Robust model predictive control of processes with hard constraints. *Computers chem. Engng.* **14**(4/5), 359–371.

Zheng, Alex and Manfred Morari (1995). Stability of Model Predictive Control with Mixed Constraints. *IEEE Transactions on automatic control* **40**(10), 1818–1823.

Zheng, Zhi Q. and Manfred Morari (1993). Robust stability of constrained model predictive control. In: *American Control Conference, San Francisco, California.* pp. 379–383.

14th Triennial World Congress, Beijing, P.R. China

A GENERALISED PREDICTIVE CONTROLLER FOR LINEAR TIME-VARYING SYSTEMS

Z. Li* and H. Unbehauen**

**School of Electrical, Computer and Telecommunication Engineering*
University of Wollongong, Australia, zheng@elec.uow.edu.au
***Laboratory of Control Engineering, Ruhr-University Bochum*
D-44780 Bochum, Germany, unbehauen@esr.ruhr-uni-bochum.de

Abstract: This paper studies the problem of predictive control of time-varying linear systems and develops a generalised predictive controller based on a standard time-varying autoregressive moving average model. The controller allows various control and output horizons and is applicable to a large class of time-varying linear systems. It is an extension of the previous results on the minimum variance prediction and control of time-varying linear systems based on pseudocommutation. *Copyright © 1999 IFAC*

Keywords: Adaptive control, minimum variance control, generalised minimum variance control, generalised predictive control, time-varying systems.

1. INTRODUCTION

The generalised predictive controller (GPC) for linear time-invariant (LTI) systems (Clarke, Mohtadi and Tuffs, 1987) extends naturally the minimum variance controller (Astrom, 1970) and the generalised minimum variance controller (Clarke and Gawthrop, 1975) by increasing the control and output horizons. This flexibility in the control and output horizons makes the GPC applicable to more complex systems. Both computer simulation and industrial experiments indicate that the GPC is relatively insensitive to slowly time-varying dead-time and model orders and, hence, is a very useful controller for adaptive control of slowly time-varying stochastic systems.

As the minimum variance controller and the generalised minimum variance controller the GPC is a predictor based controller, which depends on the minimum variance prediction of the plant output over the whole output horizon. Unlike the LQ optimal controllers in the state space approach the GPC was developed in a transfer function framework and cannot be easily extended from the LTI case to linear time-varying (LTV) systems because of noncommutativity of time-varying transfer operators with respect to multiplication.

Due to the unavailability of a GPC for LTV systems the LTI model based GPC is often applied for adaptive control of slowly time-varying systems. In order to adapt to unforeseeable changes in the plant dynamics the GPC is redesigned during each sampling period based on the newly updated plant parameter estimates as if the plant were LTI. Apparently, this method relies on the robustness of the GPC in withstanding variations in the plant parameters and works only in the case where the variation speed of the plant parameters are sufficiently slow. The purpose of this paper is to develop a new GPC based on an LTV plant model for control of fast time-varying systems. However, this LTV GPC can also achieve better performance even in the case where the plant parameters are slowly time-varying because variations in the plant parameters degrades the tracking performance of the LTI model based GPC.

Based on a newly developed pseudocommutation method (Li, Evans and Wittenmark, 1997) the minimum variance controller and the generalised minimum variance controller have been extended from the LTI case to LTV systems (Li and Evans, 1995 and 1997). Using the pseudocommutation method and based on an LTV controlled autoregressive moving average model (CARMA) this paper develops a new GPC for adaptive control of stochastic systems with rapidly time-varying parameters.

The remainder of this paper is organised as the following: Section 2 describes the LTV plants to be controlled and explains the generalised predictive control cost. Section 3 develops the LTV GPC followed by a computer simulation example presented by Section 4. Finally, Section 5 concludes the paper.

2. LTV PLANT AND CONTROL OBJECTIVE

The following standard LTV CARMA model is considered.

$$A(k,q^{-1})y(k+1)=B(k,q^{-1})u(k)$$
$$+C(k,q^{-1})w(k+1) \tag{1}$$

where $\{u(k)\}$ and $\{y(k)\}$ are the plant input and output sequence and $\{w(k)\}$ is a zero mean, independent and possibly nonstationary Gaussian sequence. The variance of the disturbance $\{w(t)\}$ is assumed to be unknown and uniformly bounded. In the CARMA model

$$A(k,q^{-1})=1+a_1(k)q^{-1}+a_2(k)q^{-2}+\ldots+a_n(k)q^{-n}$$
$$B(k,q^{-1})=b_0(k)+b_1(k)q^{-1}+\ldots+b_m(k)q^{-m}$$
$$C(k,q^{-1})=1+c_1(k)q^{-1}+c_2(k)q^{-2}+\ldots+c_h(k)q^{-h} \tag{2}$$

are polynomials in the one-step-delay operator q^{-1} and are referred to as moving average operators (MAOs). The following assumptions are made about the CARMA model.

A1) The LTV autoregressive operator (ARO) $C^{-1}(k,q^{-1})$ is exponentially stable (Li, Evans and Wittenmark, 1997).

A2) All the coefficients in the LTV MAO's $A(k,q^{-1})$, $B(k,q^{-1})$ and $C(k,q^{-1})$ are uniformly bounded.

A3) The determinants of the left time-varying Sylvester matrices (Li, Evans and Wittenmark, 1997) of the LTV MAO pairs $[A(k,q^{-1}), C(k,q^{-1})]$ and $[B(k,q^{-1}), C(k,q^{-1})]$ are uniformly bounded away from zero.

Clearly, the assumptions A1) and A2) are natural extensions of those made by the LTI GPC from the LTI case to LTV plants. The assumption A3) is additional and implies that the LTV CARMAR model can be written as the following LTV MACAR model (Li and Evans, 1997)

$$\begin{bmatrix} 0 & \tilde{D}(k,q^{-1}) \\ \tilde{A}(k,q^{-1}) & -\tilde{B}(k,q^{-1}) \end{bmatrix} \begin{bmatrix} x(k+1) \\ z(k) \end{bmatrix} = \begin{bmatrix} u(k) \\ w(k+1) \end{bmatrix} \tag{3}$$

$$y(k+1)=[\tilde{C}(k,q^{-1})\ 0]\begin{bmatrix} x(k+1) \\ z(k) \end{bmatrix}$$

where the LTV MAO's have the forms

$$\tilde{A}(k,q^{-1})=1+\tilde{a}_1(k)q^{-1}+\tilde{a}_2(k)q^{-2}+\ldots\tilde{a}_n(k)q^{-n}$$
$$\tilde{B}(k,q^{-1})=\tilde{b}_0+\tilde{b}_1(k)q^{-1}+\tilde{b}_2(k)q^{-2}+\ldots\tilde{b}_m(k)q^{-m} \tag{4}$$
$$\tilde{C}(k,q^{-1})=1+\tilde{c}_1(k)q^{-1}+\tilde{c}_2(k)q^{-2}+\ldots\tilde{c}_h(k)q^{-h}$$
$$\tilde{D}(k,q^{-1})=1+\tilde{d}_1(k)q^{-1}+\tilde{d}_2(k)q^{-2}+\ldots\tilde{d}_h(k)q^{-h}$$

They are determined using the following left pseudocommutation (Li, Evans and Wittenmark, 1997)

$$A(k,q^{-1})\tilde{C}(k,q^{-1})=C(k,q^{-1})\tilde{A}(k,q^{-1}) \tag{5}$$
$$B(k,q^{-1})\tilde{D}(k,q^{-1})=C(k,q^{-1})\tilde{B}(k,q^{-1})$$

The LTV ARO's $\tilde{C}^{-1}(k,q^{-1})$ and $\tilde{D}^{-1}(k,q^{-1})$ are exponentially stable because of exponential stability of the LTV ARO $C^{-1}(k,q^{-1})$ (Li, Evans and Wittenmark, 1997).

The first matrix equation in the MACAR model (3) describes plant dynamics using the unmeasurable internal states $x(k)$ and $z(k)$ and is referred to as a state equation. The second equation relates the plant output with the internal states and is referred to as an output equation. The difference between the LTV CARMA model and the LTV MACAR model is in the order of the moving average and autoregression operation. The MACAR model does the autoregression first and the moving average second while the CARMAR model does them in a reverse order. This property will be used in the next section for avoiding the noncommutativity of the LTV MAO's for the design of the LTV GPC.

Given a uniformly bounded reference sequence $\{s(k)\}$ the objective is to design a controller which minimises the following generalised output tracking error variance conditioned on the plant input and output data available at the current time k.

$$J(k)=E\sum_{i=1}^{N}\{\rho(k+i)[y(k+i)-s(k+i)]^2 + \lambda(k+i-1)u^2(k+i-1)\} \qquad (6)$$

where $\rho(k+i)$ and $\lambda(k+i)$ are positive weightings.

3. PREDICTIVE CONTROL

As in the LTI case the LTV GPC is based on a set of minimum variance predictors for the plant output over a number of steps along the output horizon. Letting N be the output horizon these predictors are developed using the LTV MAO's defined by the following equations

$$\tilde{C}(k,q^{-1}) = F_i(k,q^{-1})\tilde{A}(k,q^{-1}) + G_i(k,q^{-1})q^{-i-1} \qquad (7)$$

where $i=0, 1, ..., N-1$ and the orders of the MAO's $F_i(k, q^{-1})$ and $G_i(k, q^{-1})$ are i and $max(h-i-1, n-1)$. Comparing the coefficients for the same power of q^{-1} in the above equation we have

$$F_i(k, q^{-1})=1+f_1(k)q^{-1}+...+f_i(k)q^{-i} \qquad (8)$$

where

$$f_1(k) = \tilde{c}_1(k) - \tilde{a}_1(k)$$

$$f_2(k) = \tilde{c}_2(k) - f_1(k)\tilde{a}_1(k-1) - \tilde{a}_2(k)$$

...

$$\Phi = \begin{bmatrix} h_{00}(k) & 0 & 0 & \dots & 0 \\ h_{11}(k+1) & h_{10}(k+1) & 0 & \dots & \dots \\ \dots & \dots & \dots & \dots & \dots \\ \dots & \dots & \dots & \dots & 0 \\ h_{(N-1)(N-1)}(k+N-1) & h_{(N-1)(N-2)}(k+N-1) & \dots & \dots & h_{(N-1)0}(k+N-1) \end{bmatrix}$$

$$f_i(k) = \tilde{c}_i(k) - f_{i-1}(k)\tilde{a}_1(k-i+1) - f_{i-2}(k)\tilde{a}_2(k-i+2) - ... - f_1(k)\tilde{a}_{i-1}(k-1) - \tilde{a}_i(k) \qquad (9)$$

with $\tilde{a}_j(k)=0$ if $j>n$; $\tilde{c}_j(k)=0$ if $j>h$. Note that $\tilde{a}_j(.)$ is delayed by the one-step-delay operator q^{-1}. Substituting (7) into the output equation of the MACAR model (3) and noting the state equation it follows that

$$y(k+1)=H_0(k,q^{-1})z(k)+G_0(k,q^{-1})x(k) +F_0(k,q^{-1})w(k+1)$$

$$y(k+2)=H_1(k+1,q^{-1})z(k+1) +G_1(k+1,q^{-1})x(k)+F_1(k+1,q^{-1})w(k+2)$$

...

$$y(k+N)=H_{N-1}(k+N-1,q^{-1})z(k+N-1) +G_{N-1}(k+N-1,q^{-1})x(k) +F_{N-1}(k+N-1,q^{-1})w(k+N) \qquad (10)$$

where the LTV MAO's relating the internal state $z(k)$ to the plant output is given by

$$H_i(k+i,q^{-1})=F_i(k+i,q^{-1})\tilde{B}(k+i,q^{-1}) =h_{i0}(k+i)+h_{i1}(k+i)q^{-1}+...+h_{i(m+i)}(k+i)q^{-m-i} \qquad (11)$$

Letting

$$t_i(k)=[H_i(k,q^{-1})-h_{i0}(k)-h_{i1}(k)q^{-1}- ...-h_{ii}(k)q^{-i}]z(k)+G_i(k)x(k-i) \qquad (12)$$

we predict the plant output using

$$\hat{Y}=\Phi Z+T \qquad (13)$$

where

$$\hat{Y}=[\hat{y}(k+1)\ \ \hat{y}(k+2)\ \ ...\ \ \hat{y}(k+N)]^{\tau}$$

$$Z=[z(k)\ z(k+1)\ ...\ z(k+N-1)]^{\tau}$$

$$T=[t_0(k)\ t_1(k+1)\ ...\ t_{N-1}(k+N-1)]^{\tau} \qquad (14)$$

with the superscript τ denoting matrix transpose. In the predictor the vector Z consists of unmeasurable internal states. In order to relate Z to the plant input we define

$$v_i(k) = [\tilde{D}(k,q^{-1}) - 1 - \tilde{d}_1(k) - ... - \tilde{d}_i(k)]z(k) \qquad (15)$$

where $\tilde{d}_j(k)=0$ if $j>h$. Noting the first equation in the state equation of the MACAR model (3) we have

$$\Theta Z+V=U \tag{16}$$

where

$$U=[u(k)\ u(k+1)\ \dots\ u(k+N-1)]^\tau$$

$$V=[v_0(k)\ v_1(k+1)\ \dots\ v_{N-1}(k+N-1)]^\tau$$

$$\Theta=\begin{bmatrix} 1 & 0 & 0 & \dots & 0 \\ \tilde{d}_1(k+1) & 1 & 0 & \dots & \dots \\ \tilde{d}_2(k+2) & \tilde{d}_1(k+2) & 1 & \dots & \dots \\ \dots & \dots & \dots & \dots & 0 \\ \tilde{d}_{N-1}(k+N-1) & \tilde{d}_{N-2}(k+N-1) & \dots & \dots & 1 \end{bmatrix} \tag{17}$$

It follows that

$$Z=DU+\Psi \tag{18}$$

where

$$D=\Theta^{1}$$

$$\Psi=-\Theta^{1}V \tag{19}$$

Thus, the equations in (10) can be rewritten as

$$Y=\Phi DU+\Phi\Psi+T+FW \tag{20}$$

where

$$Y=[y(k+1)\ y(k+2)\ \dots\ y(k+N)]^\tau$$

$$W=[w(k+1)\ w(k+2)\ \dots\ w(k+N)]^\tau$$

$$F=\begin{bmatrix} 1 & 0 & 0 & \dots & 0 \\ f_1(k+1) & 1 & 0 & \dots & \dots \\ f_2(k+2) & f_1(k+2) & 1 & \dots & 0 \\ \dots & \dots & \dots & \dots & 0 \\ f_{N-1}(k+N-1) & f_{N-2}(k+N-1) & \dots & \dots & 1 \end{bmatrix} \tag{21}$$

Using the matrix notation the generalised tracking error variance (6) has the form

$$J(k,N)=E[(\hat{Y}+FW-S)^\tau P(\Phi DU+\Phi\Psi+T$$
$$+FW-S)+U^\tau RU]$$
$$=(\Phi DU+\Phi\Psi+T-S)^\tau P(\Phi DU+\Phi\Psi+T-S)$$
$$+U^\tau RU+F^\tau E(W^\tau W)F \tag{22}$$

where

$$S=[s(k+1)\ s(k+2)\ \dots\ s(k+N)]^\tau$$

$$P=diag(\rho(k+1),\ \rho(k+2),\ \dots,\ \rho(k+N))$$

$$R=diag(\lambda(k),\ \lambda(k+1),\ \dots,\ \lambda(k+N-1)) \tag{23}$$

with *diag* denoting the diagonal matrices. Zeroing the derivative of $J(k, N)$ with respect to U results in

$$D^\tau\Phi^\tau P(\Phi DU+\Phi\Psi+T-S)+RU=0 \tag{24}$$

Thus, the control variables are given by

$$U=(D^\tau\Phi^\tau P\Phi D+R)^{-1}D^\tau\Phi^\tau P(S-\Phi\Psi-T) \tag{25}$$

where T and Ψ consist of the unmeasurable internal variables. Noting the LTV MACAR model (3) we estimate the internal states using

$$\tilde{C}(k,q^{-1})\hat{x}(k)=y(k)$$
$$\tilde{D}(k,q^{-1})\hat{z}(k)=u(k) \tag{26}$$

where $\hat{x}(k)$ and $\hat{z}(k)$ are the estimates of $x(k)$ and $z(k)$ based on the plant output and input up to and including time k. Because the exponential stability of the LTV ARO's $\tilde{C}^{-1}(k,q^{-1})$ and $\tilde{D}^{-1}(k,q^{-1})$ the estimates will converge exponentially to their true values. Substituting the internal states in T and Ψ with their estimates we have

$$\hat{U}=(D^\tau\Phi^\tau P\Phi D+R)^{-1}D^\tau\Phi^\tau P(S-\Phi\hat{\Psi}-\hat{T}) \tag{27}$$

where $\hat{T}$, $\hat{U}$ and $\hat{\Psi}$ are estimates of T, U and Ψ. These estimates will also converge exponentially to their real values because T, U and Ψ are linear functions of the noise. Consequently, any error caused by using the estimates in the GPC (27) will decay exponentially to zero depending on initial conditions if the closed-loop system is exponentially stable.

It is unnecessary to compute all the control variables in (27) because the GPC uses only

$$\hat{u}(k)=\gamma(k)(S-\Phi\hat{\Psi}-\hat{T}) \tag{28}$$

where $\gamma(k)$ is the first row of $(D^\tau\Phi^\tau P\Phi D+R)^{-1}D^\tau\Phi^\tau P$. The inversion of the matrix $D^\tau\Phi^\tau P\Phi D+R$ is always possible if P or R is chosen to be uniformly positive definite.

The computation of the LTV minimum variance predictors can be carried out as follows. Comparing equation i with equation i-1 in (7) results in

$$[F_i(k,q^{-1}) - F_{i-1}(k,q^{-1})]\tilde{A}(k,q^{-1}) + [G_i(k,q^{-1})q^{-1} \quad (29)$$

$$- G_{i-1}(k,q^{-1})]q^{-i} = 0$$

It follows that

$$G_i(k,q^{-1})q^{-1} = G_{i-1}(k,q^{-1}) - f_i(k)\tilde{A}(k-i,q^{-1}) \quad (30)$$

Letting

$$G_i(k, q^{-1}) = g_{i0}(k) + g_{i1}(k)q^{-1} + \ldots + g_{ih_i}q^{-h_i} \quad (31)$$

and comparing the coefficients of the equation (30) for the same power of q^{-1} we have

$$\begin{aligned} f_i(k) &= g_{(i-1)0}(k) \\ g_{i(j-1)}(k) &= g_{(i-1)j}(k) - g_{(i-1)0}(k)\tilde{a}_j(k-i) \end{aligned} \quad (32)$$

where $j=1, 2, \ldots, h_i$, $\tilde{a}_j(k) = 0$ if $j>n$ and $g_{(i-1)j}(k)=0$ if $j>h_{i-1}$. The above equations can be used to computer recursively the LTV MAO's $F_i(k, q^{-1})$ and $G_i(k, q^{-1})$, $i= 1, 2, \ldots, N-1$. For initial values to start with we let $i=0$ in the equation (7) and have

$$g_{0(i-1)}(k) = \tilde{c}_i(k) - \tilde{a}_i(k) \quad (33)$$

where $i=1, 2, \ldots, max(h, n)$, $\tilde{a}_j(k) = 0$ if $j>n$ and $\tilde{c}_j(k) = 0$ if $j>h$.

4. SIMULATION

Considering the second order LTV plant given by

$$y(k+1)+a(k)\, y(k)=u(k)+b(k)u(k-1)$$

$$+w(k+1)+c(k)\, w(k); \quad (34)$$

where $w(k)\sim N(0, 1)$ is an independent stationary Gaussian noise with zero mean and unit variance. The plant parameters are

$$a(k) = 1 + 0.5|\cos(0.05\pi k)|$$

$$b(k) = 4.6 + 2.3\cos(0.2\pi k) \quad (35)$$

$$c(k) = 0.9\frac{k}{k+1}$$

The reference output is a zero mean square wave and the weighting matrices are given by

$$P = \begin{bmatrix} 1 & 0 \\ 0 & 1 \end{bmatrix}; \quad R = \begin{bmatrix} 0.1 & 0 \\ 0 & 10 \end{bmatrix} \quad (36)$$

The LTV ARO $C^{-1}(k,q^{-1})$ is exponentially stable because $|c(k)|$ is uniformly less than the unit. However, both the LTV ARO's $A^{-1}(k,q^{-1})$ and $B^{-1}(k,q^{-1})$ are exponentially unstable because both $|a(k)|$ and $|b(k)|$ are uniformly greater than the unit. The plant parameters (35) are rapidly time varying and the plant is difficult to control using any LTI model based method. Simulation results in Figure 1 show that the plant output follows the reference and the closed-loop system is stable.

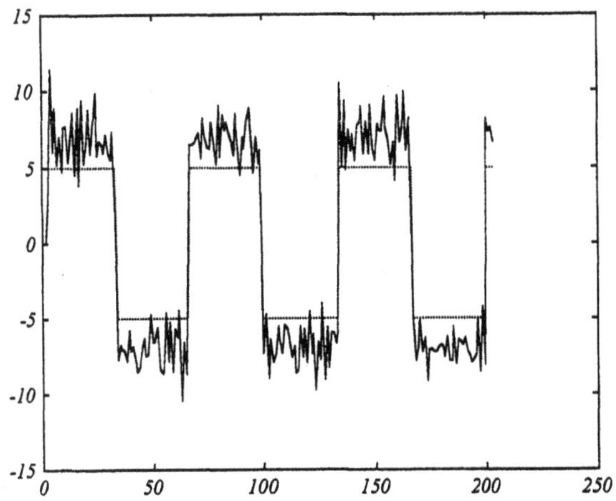

Fig. 1 Simulation results for the time-varying GPC. The dashed line represents the reference output and the solid line represents the plant output.

5. CONCLUSION

An LTV GPC is developed for LTV systems described using the LTV CARMA model. It meets the generalised predictive control objective and can be applied to a large class of LTV systems. In comparison with the LTI GPC the LTV GPC is developed under an additional assumption of left pseudocommutativity for the LTV CARMA model. Compared with the LTV MVC and the LTV GMVC the LTV GPC has the flexibility to use various control and output horizons for robustness against uncertainties in the plant dynamics.

REFERENCES

Astrom, K. J. (1970). *Introduction to Stochastic Control Theory.* New York: Academic Press.

Clarke, D. W. and P. J. Gawthrop (1975). Self-tuning controller. *Proc. IEE Part. D,* vol. 122, pp929-934.

Clarke, D. W., C. Mohtadi and P. S. Tuffs (1987) Generalised predictive control - Part I. The basic algorithm, *Automatica,* vol 23, pp. 137-148.

Clarke, D. W., C. Mohtadi and P. S. Tuffs (1987) Generalised predictive control - Part II. Extensions and Interpretations, *Automatica,* vol 23, pp. 149-160.

Li, Z. and R.J. Evans (1995). Generalised minimum variance control of time-varying systems. *Proceedings of the third European Control Conference,* Roma, Italy, vol.1, pp1301.

Li, Z. and R. J. Evans (1997). Minimum variance control of linear time-varying systems, *Automatica vol. 33,* pp1531-1537.

Li, Z., R. J. Evans and B. Wittenmark (1997). Minimum variance prediction for linear time-varying systems, *Automatica vol. 33,* pp607-618.

MULTI-RATE DYNAMIC MATRIX CONTROL (DMC) FOR MULTIVARIABLE SYSTEMS

Lifang Zhou, Hao Dong and Jixin Qian

Institute of Industrial Process Control, National Lab for Industrial Technical, Zhejiang University, Hangzhou, 310027, P. R. China
E-mail: lfzhou@iipc.zju.edu.cn

Abstract: in this paper, a predictive control method based on DMC is developed for multirate multivariable systems. The intersample behavior is considered and the optimization is performed as soon as there is a new measured output. By using the technique of FIACS[1] (Flexible Advanced Control Software), the effect of "thin" system met in this method can be eliminated. Simulation results indicate that the method is effective with zero steady-state error and the intersample behavior is not degraded. The closed-loop stability criteria are also derived in the Appendix.

Keywords: predictive control, multirate, multivariable systems, steady-state errors, stability criteria.

1. INTRODUCTION

In the control field, the predictive control is developed quickly for it's characteristic of predictive model, rolling optimization and error compensation. With the increase of complexity of the industry process, it is inadequate for the model predictive control (MPC) resorting to the single rate. So the Multi-rate MPC technique has received more and more attention in the past decade, especially in the following facts (Colaneri, *et al.*, 1990):

(1) One or more sensor require a significant time before they supply the measurement of the corresponding output variables, while the value of other output components are available without time and rate limitation.

(2) The same sensor is periodically switched to measure different plant variables at different times, or all the plant output are measured at the same time and rate, but the measured values are

[1]FIACS is the product of Zhejiang Fractal Inc., supported by "95" National Project.

transmitted to the control processing unit sequentially.

(3) Technological constraints prevent one from computing and /or updating the control components at any sampling period, while the output measurement is available with no rate or time limitations.

It's then natural to resort to multirate model based predictive control, where different sampling periods are permitted for different variables along with time shifts and delays among the measurements.

Recently several attempts have been made to develop new types of digital controllers that include periodically time-varying mechanisms. In multirate predictive control system, the case is often studied, in which the outputs are measured at different sampling rates and output measurements are available less frequently than inputs that are performed at the same sampling rate (Park, et al., 1993; Masahiro Ohshima, et al.). In order to eliminate the effect of model mismatch and disturbances, the self-tuning mechanism has been employed in some papers (Scattolini, 1988; Carini, *et al.*, 1990), the model parameters are modified as soon as the output is available. Most of these researches are based on the parameter model.

In this paper, a method based on DMC is developed for n inputs and n outputs systems in which the sampling rates of outputs and holding periods of inputs are all different, but the sampling rate of the ith output y_i is the same as the holding period of the ith input u_i, which is described by Mituhiko Araki and Koichiro Yamamoto (1986). Any kinds of system can be transformed into the above case. Simulation results indicate that this method is effective with zero steady-state error and the intersample behavior is not degraded. The closed-loop stability condition is also derived in the Appendix.

2. MULTIRATE MULTIVARIABLE PREDICTIVE CONTROL

Consider the n inputs and n outputs system, it is assumed that the sampling rate of the ith output y_i is T_i (i=1... n), the greatest common measure is τ_0, which is regarded as base rate of the system. For simplicity of notation, τ_0 is assumed to be 1 and T_i is considered as an integer. Giving the following assumption:

1) The holding period of the ith input u_i is the same as the sampling rate of the ith output y_i.
2) The step response models are known at the base rate τ_0.

The multi-rate sampling system is shown in Fig 1.

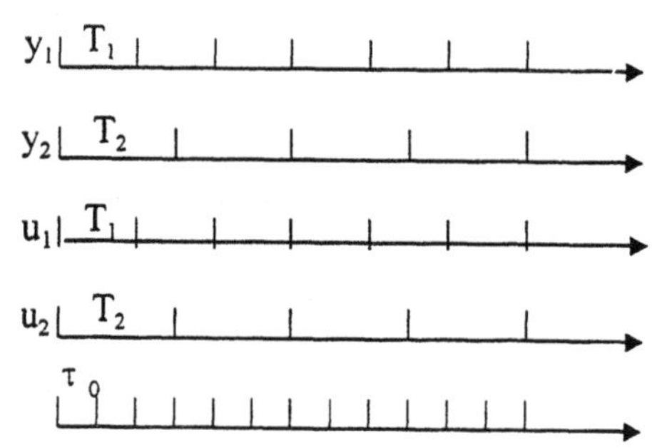

Fig. 1. Multi-rate sampling system

The least common multiple of $\{T_i\}$ is denoted by T_0, that is

$$T_0 = \text{l.c.m}\,(T_1, T_2, \ldots, T_n) \qquad (1)$$

2.1 The Predictive Model

Assume the model time horizon is N, the step response coefficients of the kth output y_k to the ith input u_i based on the base rate τ_0 are $\{a_1^{ki}, a_2^{ki}, \ldots, a_N^{ki}\}$. Choose the predictive horizon to be P and the control horizon to be M, the predicted value of the qth output at k+j time can be written:

$$y_{m,q}(k+j/k) = y_{0,q}(k+j/k) + \sum_{i=1}^{n}\sum_{l=1}^{j} a_l^{qi}\Delta u_i(k+j-l) \qquad (2)$$

for $\Delta u_i(k+d) = 0$ (d≥M),

q=1,2,...,n, j=1,2,...,P

Where $y_{m,q}(k+j/k)$ denotes the value of the qth output y_q at time k+j, which is predicted at time k.

$y_{0,q}(k+j/k)$ means the initial value of the qth output y_q at time k+j, $\Delta u_i(k+j-l)$ is the ith control increment at time k+j-l. Then the output equation can be rewritten in vector form

$$Y_m(k+1) = A_0\Delta U(k-1) + A\Delta U(k) \qquad (3)$$

$$Y_m(k+1) = \left[Y_{m,1}^T(k+1), Y_{m,2}^T(k+1), \cdots, Y_{m,n}^T(k+1)\right]^T$$

$$\Delta U(k) = \left[\Delta U_1^T(k), \Delta U_2^T(k), \cdots, \Delta U_n^T(k)\right]^T$$

$$\Delta U(k-1) = \left[\Delta U_1^T(k-1), \Delta U_2^T(k-1), \cdots, \Delta U_n^T(k-1)\right]^T$$

$$Y_{m,i}(k+1) = \left[y_{m,i}(k+1/k), \cdots, y_{m,i}(k+P/k)\right]^T$$

$$\Delta U_j(k) = \left[\Delta u_j(k), \cdots, \Delta u_j(k+M-1)\right]^T$$

$$\Delta U_j(k-1) = \left[\Delta u_j(k-N), \cdots, \Delta u_j(k-1)\right]^T$$

The matrixes A and A_0 consists of matrix A_{ij} (i=1...n, j=1...n) respectively, where A_{ij} and A_{0ij} are

$$A_{ij} = \begin{bmatrix} a_1^{ij} & & & \\ a_2^{ij} & a_1^{ij} & & \\ \vdots & \vdots & \ddots & \\ a_P^{ij} & a_{P-1}^{ij} & \cdots & a_{P-M+1}^{ij} \end{bmatrix}_{P\times M}$$

$$A_{0,ij} = \begin{bmatrix} \hat{a}_N^{ij} & \hat{a}_N^{ij} & \hat{a}_{N-1}^{ij} & \cdots & \hat{a}_2^{ij} \\ \hat{a}_N^{ij} & \hat{a}_N^{ij} & \hat{a}_N^{ij} & \cdots & \hat{a}_3^{ij} \\ \vdots & \vdots & \vdots & \ddots & \vdots \\ \hat{a}_N^{ij} & \hat{a}_N^{ij} & \hat{a}_N^{ij} & \cdots & \hat{a}_{P+1}^{ij} \end{bmatrix}$$

Because the control process is often complex and the disturbance is unmeasured, it is difficult to obtain the precise model. The modeling mismatch is always existing. In order to eliminate the effect of it, the predicted output equation should be modified using the error between the model outputs and the system outputs at current time. But at some moment, due to the different sampling rates, some outputs can not be measured, and thus can not be used to modify the predictive outputs. A method has been proposed to solve this problem in the paper (Masahiro Ohshima, *et al.*, 1994). That is using the most recent sampling output y_i (k*) to replace the current output y_i (k) if it has not the measured value at time k. Then the prediction equation is described as follows:

$$Y_p(k+1) = A_0\Delta U(k-1) + A\Delta U(k) + HE(k^*) \qquad (4)$$

where

$$E(k^*) = \left[y_1(k^*) - y_{m,1}(k^*) \quad \cdots \quad y_n(k^*) - y_{m,n}(k^*)\right]^T$$

$$H = diag\{H_i\},$$

$$H_i = \left[h_{i1} \quad h_{i2} \quad \cdots \quad h_{iP}\right]^T, \quad i=1,\ldots,n$$

H_i is the weighing factor vector. Thus we can predict the outputs response over the prediction horizon [(k+1), (k+P)] resorting to the model equation (4)

In order to avoid the jump of setpoints, usually the reference trajectory is given by

$$y_{R,l}(k+j/k) = (\alpha_l)^j y_l(k) + (1-(\alpha_l)^j)r_l^* \qquad (5)$$

Where $y_{R,l}(k+j/k)$ denotes the reference output of y_l at time k+j , r_l^* is the setpoint value of the output y_l , α_l is the tuning parameter in the reference trajectory of the lth output.

2.2 Control Strategy

In the predictive control, rolling optimization is an important characteristic. Appropriate inputs are selected so that the results predicted outputs over the predictive horizon would follow the reference trajectories as closely as possible. In other words, the manipulated variables are determined to minimize the following quadratic objective function:

$$J_p = [Y_p(k+1) - Y_R(k+1)]^T Q[Y_p(k+1) - Y_R(k+1)] + \Delta U^T(k)\lambda\Delta U(k) \qquad (6)$$

Substituting (4) into (6), J_p can be rewritten as:

$$J_p' = \Delta U^T(k) W \Delta U(k) + 2T^T(k)\Delta U(k) \qquad (7)$$

where

$$W = A^T Q A + \lambda I$$

$$T(k) = A^T Q[A_0\Delta U(k-1) + HE(k^*) - Y_R(k+1)]$$

Q and λ are weighting matrixes for predicted error and control moves respectively. Because the inputs are performed at different rates, the following constraints should be added to the quadratic objective function:

If $j\tau_0 \neq sT_i$ (s is integer) then $\Delta u_i(j) = 0$ (i=1...n)
After calculating the inputs, only the first changes in the inputs $\Delta u_i(k)$ (i=1...n) are performed

2.3 The Determination of Optimization Time

In many papers, the optimization is done at the next time (k+1) or at the time (k+T_0) after it is finished at time k. In the case described in the paper, it is possible that the outputs are not measured at time k+1. So it is unnecessary to do the optimization at time k+1. If we choose the time k+T_0 as the next optimization time, much information will be lost. In order to make full use of the information and reduce the burden of the computer, the next optimization will be done as soon as there is at least a new measured output since the time k. That is the optimization time is k+t, t is calculated as:

$$t = \min_j \{ j \mid (k+j) \bmod T_i = 0, i \in [1,n] \} \quad (8)$$

3. SIMULATION

Consider the following 2 inputs 2 outputs system:

$$\begin{bmatrix} y_1 \\ y_2 \end{bmatrix} = \begin{bmatrix} \dfrac{e^{-8s}}{15s+1} & \dfrac{e^{-2s}}{8s+1} \\ -\dfrac{e^{-4s}}{5s+1} & \dfrac{1}{10s+1} \end{bmatrix} \begin{bmatrix} u_1 \\ u_2 \end{bmatrix} \quad (9)$$

Assume the sampling rates of the output y_1 and y_2 are 2, 3 respectively, the holding periods of the input u_1 and u_2 are also 2, 3. The initial values of outputs are 0 and the setpoints are 2 and 1, respectively. The control horizon is 6 and the predictive horizon is 12. The matrix Q and λ are defined as follows:

$Q=\mathrm{diag}(1,\ldots,1)_{24*24}$ $\lambda=\mathrm{diag}(1,\ldots,1)_{12*12}$

Two cases are studied in the simulation: one is no model mismatch, the other is that model mismatch is existing. The simulation results are described in Fig 2 (a)(no model mismatch) and Fig 2(b)(model mismatch). In the case of model mismatch, the model is expressed as (9) and the system transfer function is described as

$$\begin{bmatrix} y_1 \\ y_2 \end{bmatrix} = \begin{bmatrix} \dfrac{0.5e^{-8s}}{15s+1} & \dfrac{1.5e^{-2s}}{8s+1} \\ -\dfrac{2e^{-4s}}{5s+1} & \dfrac{0.5}{10s+1} \end{bmatrix} \begin{bmatrix} u_1 \\ u_2 \end{bmatrix} \quad (10)$$

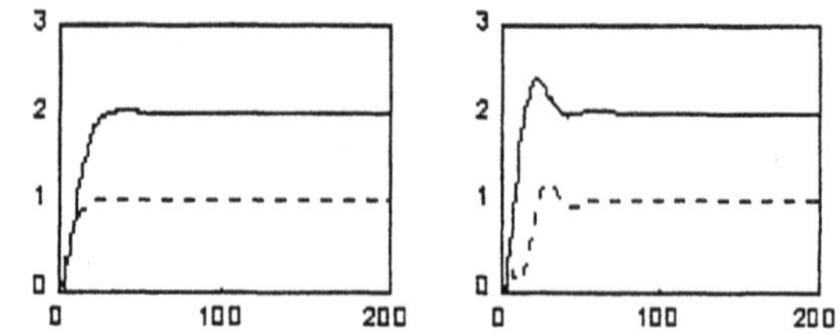

(a) no model mismatch (b) model mismatch

Fig 2. The step response curves of the multi-rate DMC system

Fig 3 shows the changes of the inputs using the method presented in this paper when the model is match or mismatch.

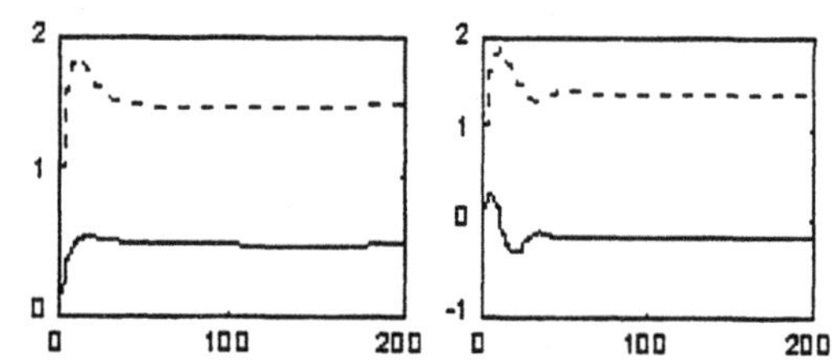

(a) no model mismatch (b) model mismatch

Fig. 3 The curves of control variables

From the Fig 2,3, we can obtain that the responses resorting to the method presented in this paper have little overshoot and short transient time while there are no model mismatch. If the model mismatch is existing, from the Fig 2 (b), it is showed that the control system in this paper can still track the change of setpoints with little oscillation. Fig 3(a) indicates the control actions at the beginning have a jump because of the step change of setpoints, and then the control increment is small.

The above analysis indicates that the method proposed in this paper is effective. The intersample behavior of the transient response is not degraded as it presented in paper (Moore, 1993). Then the following remarks can be obtained:

(1) For n inputs and n outputs multirate system, some of the outputs can not be measured at time k, in this case, we adopt the method proposed in

the paper (Masahiro Ohshima, et al., 1994]

(2) In order to calculate the control inputs, we use not only the outputs at the sampling point but also the intersample information. This can improve the dynamic performance of system, avoiding the degradation of intersample behavior.

(3) Because the holding periods of inputs are different, the constraints to the quadratic objective function are also different. But the constraints changed periodically.

(4) Using this method, the "thin" system is appeared. For the "thin" system, the free degree is less than zero, and then it is difficult to obtain good control effect. In order to solve this problem, we use the technique of FIACS, The effect of the "thin" system can be eliminated.

4. CONCLUSION

In this paper, we study the multi-rate multivariable predictive control in which the sampling rates of the outputs are different and the inputs are also performed at different rates. Using this method, the information that is known can be taken full use and the control actions are modified in time. The simulation indicates the method is effective with zero steady-state error and little overshoot. Although this is only one of the specious cases in the multi-rate control, this method can be extended to general cases. The study of the problem of "thin" system and model modification will be continued.

APPENDIX

Close-loop stability condition:

Consider the n inputs and n outputs system presented in this paper. The performance index is (7). Choose $\Delta u_i(k+j)$ which is not satisfied with the above condition to form the vector $\Delta\overline{U}(k)$

$$\Delta\overline{U}(k) = [\Delta u_1(k+l_1), \cdots, \Delta u_1(k+(n_1'-1)T_1+l_1, \cdots, \Delta u_n(k+(n_n'-1)T_n+l_n)]^T$$

where

$$l_i = \min\{l_i \mid (k+l_i) \bmod T_i = 0\}$$

$$n_i' = INT(\frac{M}{T_i}) \quad i=1\ldots n$$

Then $\Delta\overline{U}(k)$ is expressed

$$\Delta\overline{U}(k) = C_1(k)\Delta U(k) \tag{11}$$

(7) can be rewritten as

$$J_P' = \Delta\overline{U}^T(k)\overline{W}\Delta\overline{U}(k) + 2\overline{T}^T(k)\Delta\overline{U}(k) \tag{12}$$

Where

$$\overline{W} = C_1(k)WC_1^T(k) \qquad \overline{T}^T(k) = T^T(k)C_1^T(k) \tag{13}$$

let $\partial J_P' / \partial\Delta\overline{U}(k) = 0$. then

$$\Delta\overline{U}(k) = -\overline{W}^{-1}(k)\overline{T}(k) \tag{14}$$

Substituting (11), (13) into (14)

$$\Delta U(k) = B(k)[Y_R(k+1) - A_0\Delta U(k-1) - HE(k^*)] \tag{15}$$

Where

$$B(k) = C_1^T(k)[C_1(k)WC_1^T(k)]^{-1}C_1(k)A^TQ$$

then $\Delta u_i(k)$ is

$$\Delta u_i(k) = d_{(i-1)M+1}B(k)[Y_R(k+1) - A_0\Delta U(k-1) - HE(k^*)] \tag{16}$$

Using the same method, the input $\Delta u_i(k+j)$ at time (k+j) can be calculated:

$$\Delta u_i(k+j) = d_{(i-1)M+1}B(k+j)[Y_R(k+j+1) - A_0\Delta U(k+j-1) - HE((k+j)^*)] \tag{17}$$

By grouping the variables in a frame period T_0 into vectors, we can express the multirate system as a single-rate-sampled system. So we define the following vectors:

$$\widetilde{Y}(k) = [y_1(k), y_1(k+1), \cdots, y_1(k+T_0-1), y_2(k), \cdots, y_n(k+T_0-1)]^T$$

$$\widetilde{Y}_m(k) = [y_{m,1}(k), y_{m,1}(k+1), \cdots, y_{m,1}(k+T_0-1), y_{m,2}(k), \cdots, y_{m,n}(k+T_0-1)]^T$$

$$\Delta\widetilde{U}(k) = [\Delta u_1(k), \Delta u_1(k+1), \cdots, \Delta u_1(k+T_0-1), \Delta u_2(k), \cdots, \Delta u_n(k+T_0-1)]^T$$

$$\Delta\widetilde{U}(k-1) = [\Delta u_1(k-T_0), \Delta u_1(k-T_0+1), \cdots, \Delta u_1(k-1), \Delta u_2(k-T_0), \cdots, \Delta u_n(k-1)]^T$$

thus, we can obtain

$$A_0\Delta U(k+j-1) = \overline{A}_{0,0}(k+j)\Delta\widetilde{U}(k) + \cdots + \overline{A}_{0,s}(k+j)\Delta\widetilde{U}(k-s) \quad (18)$$

$$HE[(k+j)^*] = HC_2(k+j)(\widetilde{Y}(k) - \widetilde{Y}_m(k)) \quad (19)$$

$$Y_R(k+j+1) = C_3(k+j)\widetilde{Y}(k) + C_4(k+j)r^* \quad (20)$$

substituting (18),(19),(20) into (17)

$$\Delta u_i(k+j) = D_1(i,k+j)\widetilde{Y}(k) + D_2(i,k+j)\widetilde{Y}_m(k) + D_3(i,k+j)r^* - (F_0(i,k+j)\Delta\widetilde{U}(k) + F_1(i,k+j)\Delta\widetilde{U}(k-1) + \cdots + F_s(i,k+j)\Delta\widetilde{U}(k-s) \quad (21)$$

where:

$$D_1(i,k+j) = d_{(i-1)M+1}B(k+j)(C_3(k+j) - HC_2(k+j))$$

$$D_2(i,k+j) = d_{(i-1)M+1}B(k+j)HC_2(k+j)$$

$$D_3(i,k+j) = d_{(i-1)M+1}B(k+j)C_4(k+j)$$

$$F_q(i,k+j) = d_{(i-1)M+1}B(k+j)\overline{A}_{0,q}(k+j)$$

$i=1,\ldots,n \qquad j=0,\ldots,T_0-1$

Using the delay operator, q^{-1}, the inputs $\Delta\widetilde{U}(k)$ is expressed in vector form as

$$(I + F(q^{-1}))\Delta\widetilde{U}(k) = D_1\widetilde{Y}(k) + D_2\widetilde{Y}_m(k) + D_3 r^* \quad (22)$$

Where

$$D_l = [D_l(1,k), D_l(1,k+1), \cdots D_l(1,k+T_0-1), D_l(2,k), \cdots, D_l(n,k+T_0-1)]^T$$

$$F_l = [F_l(1,k), F_l(1,k+1), \cdots, F_l(1,k+T_0-1), F_l(2,k), \cdots, F_l(n,k+T_0-1)]^T$$

$$F(q^{-1}) = F_0 + F_1 q^{-1} + \cdots + F_s q^{-s}$$

The transfer functions of the process and the model are described as following respectively:

$$\widetilde{Y}(k) = G_p(q^{-1})\widetilde{U}(k), \widetilde{Y}_m(k) = G_m(q^{-1})\widetilde{U}(k) \quad (23)$$

where

$$\widetilde{U}(k) = [u_1(k), u_1(k+1), \cdots, u_1(k+T_0-1), u_2(k), \cdots, u_n(k+T_0-1)]^T$$

then $\Delta\widetilde{U}(k)$ is expressed by $\widetilde{U}(k)$ and $\widetilde{U}(k-1)$

$$\Delta\widetilde{U}(k) = \Psi_1\widetilde{U}(k) - \Psi_2\widetilde{U}(k-1) \quad (24)$$

(22) is rewritten as:

$$\{(I + F(q^{-1}))(\Psi_1 - \Psi_2 q^{-1}) - D_1 G_p(q^{-1}) - D_2 G_m(q^{-1})\}\widetilde{U}(k) = D_3 r^* \quad (25)$$

Therefore, the characteristic equation is given by

$$\det\{(I + F(q^{-1}))(\Psi_1 - \Psi_2 q^{-1}) - D_1 G_p(q^{-1}) - D_2 G_m(q^{-1})\} = 0 \quad (26)$$

When all the roots of the (26) are located at the inside of the unit circle, the control system is stable.

REFERENCE

Colaneri P., R. Scattolini, and N. Schiavoni, (1990). *Stabilization of Multirate Sampling-data Linear System*, Automatic, **v26, n3**, 377-380

Park I O, MS and J H Oh (1993). *Multi-rate generalized predictive control for multi-variable systems*, IMechE, Part 1, **v207**, 253-260.

Scattolini Riccardo (1988). *Self-tuning control of systems with infrequent and delayed output sampling*, IEE Proc., Part D, **v135, n4**, 213-219.

Mituhiko Araki, Hoichiro Yamamoto (1986). *Multivariable Multirate Sampled-Data Systems: State-Space Description, Transfer Characteristics, and Nyquist Criterion*, IEEE Trans. Automat. Contr., **v AC-31, n2**, 145-154.

Carini P., R. Micheli and R. Scattolini (1990). *Multirate self-tuning predictive control with application to binary distillation column*, Int.J.Systems Sci., **v21, n1**, 51-64.

Masahiro Ohshima, et al., (1994). *Multirate multivariable model predictive control and its application to a polymerization reactor*, INT. J. CONTROL, **v59, n3**, 731-742.

Kevin L. Moore, (1993). *Capabilities and limitation of multirate Control Schemes*, Automatic, **v29, n4**, 941-951.

14th Triennial World Congress, Beijing, P.R. China

C-2a-07-6

SIMPLE PREFILTER DESIGN IN GPC FOR A WIDE CLASS OF INDUSTRIAL PROCESSES

Julio E. Normey-Rico * **Carlos Bordons** ** **and Eduardo F. Camacho** **

* *Dpto. Automação e Sistemas, Univ. Fed. de Sta Catarina, Brazil*
** *Depto. Ing. Sist. y Automática, Univ. de Sevilla, Spain*
E-mails: {julio, bordons, eduardo}@cartuja.us.es

Abstract: This paper presents a simple design method for improving the robustness of the GPC using the T polynomial as a filter. Although different methods have been proposed in literature, the one presented here not only proposes a choice of the polynomial, but differs from the others in the way that the polynomial is used to filter the predictions. The method is valid for a wide range of processes commonly found in industry, those that can be described by means of a static gain, time constant and dead-time and has special interest when the process has uncertainties in the dead-time.

Keywords: Predictive control, Generalized predictive control, Robustness, Delay compensation, Process control.

1. INTRODUCTION

As is well known, Generalized Predictive Control (GPC) uses a CARIMA model to calculate the predictions along the horizon (Clarke *et al.*, 1987). This model has the following form:

$$A(z^{-1})y(t) = z^{-d}B(z^{-1})u(t-1) + \frac{T(z^{-1})e(t)}{\Delta} \quad (1)$$

$$A(z^{-1}) = 1 + a_1 z^{-1} + a_2 z^{-2} + \ldots + a_{na} z^{-na}$$
$$B(z^{-1}) = b_0 + b_1 z^{-1} + b_2 z^{-2} + \ldots + b_{nb} z^{-nb}$$
$$T(z^{-1}) = t_0 + t_1 z^{-1} + t_2 z^{-2} + \ldots + t_{nt} z^{-nt}$$

$y(t)$ is the plant output, $u(t)$ is the control action, $\Delta = 1 - z^{-1}$ and d is the dead-time of the system.

In this model the T-polynomial represents the stochastic characteristics of the noise and the disturbances. The main objective of the use of $T(z^{-1})$ is to reduce the effect of the noise and disturbances on the predicted output without affecting the nominal setpoint tracking. On the other hand, low frequency disturbances can be removed by the Δ operator that appears in the prediction.

However, in industrial practice it is difficult to estimate the real characteristics of the noise, so $T(z^{-1})$ is rarely estimated but best treated as a filter. If T is appropriately chosen, the roll-off of the filter can attenuate the prediction error caused by model mismatch, which is particulary important at high frequencies. Notice that dead-time uncertainties are one of the most characteristic high-frequency unmodelled dynamics, and have a dangerous influence on the closed-loop stability since they can deteriorate the system phase margin.

The T polynomial seems to be a good solution, but its correct choice is a problem that has not completely been solved, although its effect on the robustness of the closed loop system has been analyzed in several papers (Clarke *et al.*, 1987; Clarke and Mothadi, 1989; Robinson and Clarke, 1991; Yoon and Clarke, 1995). In these papers the authors deal with the correct tuning of

[1] Work supported in part by CICYT-Spain contracts TAP96-0884 and TAP98-0541 and CAPES-BRASIL Contract BEXO448/95-6

the prefilter in order to increase some robustness indices in certain particular cases of GPCs. A systematic two-step design for the GPC taking into account performance and robustness is presented in (Ansay and Wertz, 1997). But despite these efforts, the selection of T is still an open field.

In this paper the T polynomial is used as a filter in order to improve the robustness of the GPC, but in a different way as in the referred works. The proposed method makes the controller coefficients independent of T, making the controller tuning and the robustness improvement independent. Then the tuning of the prefilter is very simple and can be performed taking into account both the disturbance rejection performance and robustness of the closed-loop. This latter property is very important because it permits the tuning of the controller in two steps; first the GPC parameters are chosen to attempt the nominal set-point performance specifications and then the filter is chosen for the disturbance rejection and robust performance conditions. Also the proposed method is simpler than the T tuning method presented in (Ansay and Wertz, 1997).

There are several claims that in practice low-order models coupled with dead-times are sufficient for most purposes (Chien and Fruehauf, 1990). In process control, reduced-order modelling is often employed where a low-order transfer function plus a dead-time is used to represent the dynamic behaviuor of a high-order process (Tan *et al.*, 1996). The method described in this paper has been developed for processes that are described by a simple model consisting of a static gain, time constant and dead-time. As the model is low order, high-frequency uncertainties are easy to be found, being the dead-time uncertainties the most dangerous ones, since they cause a considerable robustness loss (Camacho and Bordons, 1999).

The paper is organized as follows: A closed-loop analysis of the GPC, where an equivalent structure of the controller is derived, is presented in section 2. In section 3 the proposed structure of the prefiltered GPC is presented and the T tuning method is given. Section 4 is dedicated to presenting some simulation results and finally the conclusions and perspectives of the work are presented in section 5.

2. CLOSED-LOOP ANALYSIS OF THE GPC

The GPC algorithm (Clarke *et al.*, 1987) consists of applying a control sequence that minimizes a multistage cost function of the form

$$J(N_1,N_2,N_3) = \sum_{j=N_1}^{N_2} [\hat{y}(t+j \mid t) - w(t+j)]^2 +$$

$$+\sum_{j=1}^{N_3} \lambda(j)[\Delta u(t+j-1)]^2 \qquad (2)$$

where N_1 and N_2 are the minimum and maximum prediction horizons, N_3 is the control horizon, $\lambda(j)$ is the control weighting sequence, $w(t+j)$ is a future set-point or reference sequence, $\Delta u(t)$ is the incremental control action and $\hat{y}(t+j \mid t)$ is the optimum j-step ahead prediction of the system output on data up to time t computed using the CARIMA model of equation (1). For the analysis of this paper it will be considered that $A(z^{-1}) = 1 + a_1 z^{-1}$ $(na = 1)$, $B(z^{-1}) = b_0$ $(nb = 0)$ and that the process always exhibits a dead-time greater than one sample $(d > 0)$. Without loss of generallity and as the process exhibits a dead-time d, the horizons are chosen as $N_1 = d+1$, $N_2 = d+N$, $N_3 = N$ and $\lambda(j) = \lambda$.

To compute the optimal prediction $\hat{y}(t+j \mid t)$ the following Diophantine equation is used (for the sake of clarity the dependency on z^{-1} is omitted from now on):

$$T = E_j \triangle A + z^{-j} F_j \qquad (3)$$

where the polynomials E_j (degree of $E_j = j-1$) and F_j (degree of $F_j = na = 1$) can be obtained by dividing T by $\tilde{A} = \Delta A$ until the remainder of the division can be factorized as $z^{-j}F_j$.

Using (1) and (3), it follows that:

$$T\hat{y}(t+j \mid t) = E_j B \triangle u(t+j-d-1) +$$

$$+F_j y(t) + T E_j e(t+j) \qquad (4)$$

Taking the expected value of the previous equation and assuming that the expected values of the future errors are zero, the optimal prediction is given by:

$$T\hat{y}(t+j \mid t) = E_j B \triangle u(t+j-d-1) + F_j y(t) \qquad (5)$$

For this paper it is interesting to separate the computation of the predictions in two sets: from $t+1$ to $t+d$ as:

$$T\hat{y}(t+j \mid t) = E_j B \triangle u(t+j-d-1) + F_j y(t) \qquad (6)$$

with $j = 1...d$ and from $t+d+1$ to $t+d+N$ as:

$$T\hat{y}(t+d+j \mid t) = E_j B \triangle u(t+j-1) +$$

$$+F_j \hat{y}(t+d) \qquad (7)$$

with $j = 1...N$. Note that for the model considered in this paper $F_j = f_{j0} + f_{j1}z^{-1}$.

Solving (7) it is possible to write:

$$\hat{\mathbf{y}}_+ = \mathbf{S}_1 \mathbf{u} + \mathbf{S}_2 \hat{\mathbf{y}}_- \qquad (8)$$

where $\hat{\mathbf{y}}_+ = (\hat{y}(t+d+1 \mid t)\ \hat{y}(t+d+2 \mid t) \dots \hat{y}(t+d+N \mid t))^T$, $\mathbf{u}$ is the vector of future control action increments and $\hat{\mathbf{y}}_- = [\hat{y}(t+d \mid t)\ \hat{y}(t+d-1 \mid t)]^T$. $\mathbf{S}_1$ and $\mathbf{S}_2$ are constant matrices of dimension $N \times N$ and $N \times 2$ respectively and their coefficients are functions of the coefficients of the polynomials T, E_j and F_j (Camacho and Bordons, 1999). If the obtained predictions are introduced in the cost function, $J(N)$ is a function of $\hat{\mathbf{y}}_-$, $\mathbf{u}$ and the reference sequence. Minimizing $J(N)$ with respect to the future control actions and keeping only $\Delta u(t)$ it is possible to write (Camacho and Bordons, 1999):

$$\Delta u(t) = l_{y1}\hat{y}(t+d \mid t) + l_{y2}\hat{y}(t+d-1 \mid t) + \sum_{i=1}^{N} k_i w(t+d+i) \quad (9)$$

where l_{yi} and k_i are functions of the model parameters b_0 and a_1 and of the controller parameters N, λ and T. The control scheme is shown in figure 1.

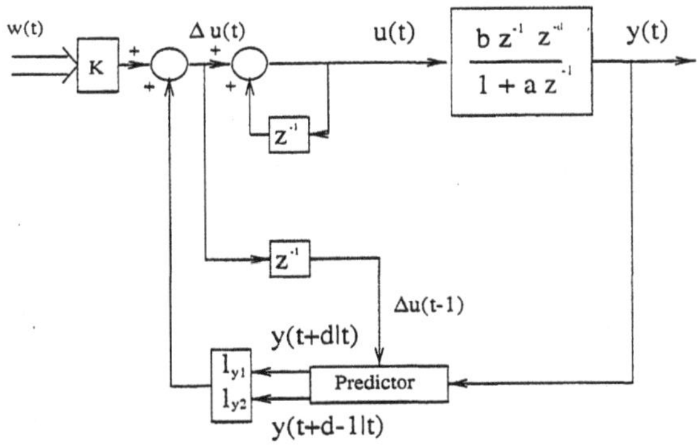

Fig. 1. Control Scheme of the GPC

The predictions in (9) can be computed using the solution of (6) as follows. First consider the solution for $j = 1$:

$$T = E_1 \triangle A + z^{-1}F_1 \quad (10)$$

where $E_1 = 1$ and $F_1 = z(T - \tilde{A})$. Using $\tilde{B} = \Delta B$ the prediction is:

$$T\hat{y}(t+1 \mid t) = \tilde{B}u(t-d) + z(T-\tilde{A})y(t) = \tilde{B}z^{-d}u(t) + z(T-\tilde{A})y(t) \quad (11)$$

Using (11) the prediction $\hat{y}(t+2 \mid t)$ can be computed as:

$$T\hat{y}(t+2 \mid t) = \tilde{B}z^{-d}u(t+1) + z(T-\tilde{A})\hat{y}(t+1 \mid t) \quad (12)$$

Multiplying (12) by T and using (11):

$$T^2\hat{y}(t+2 \mid t) = \tilde{B}z^{-d}\left[T + (T-\tilde{A})\right]u(t+1) + z^2(T-\tilde{A})^2 y(t) \quad (13)$$

Using the same procedure till $t+d$, it can be obtained that:

$$\hat{y}(t+d \mid t) = \tilde{B}\frac{z^{-d}}{T}\left[\frac{1-(1-\tilde{A}/T)^d}{\tilde{A}/T}\right]u(t+d-1) + z^d(1-\tilde{A}/T)^d y(t) \quad (14)$$

considering the plant model $P(z) = \frac{Bz^{-1-d}}{A}$ and $R(z) = z^d(1-\tilde{A}/T)^d$, it follows that:

$$\hat{y}(t+d \mid t) = P(z^d - R)u(t) + Ry(t) = = Pz^d u(t) + R(y(t) - Pu(t)) \quad (15)$$

Using this equation the final control scheme for the GPC can be drawn as in figure 2 where the relation between the control action and the predictions and references has been expressed as a two degree of freedom primary controller.

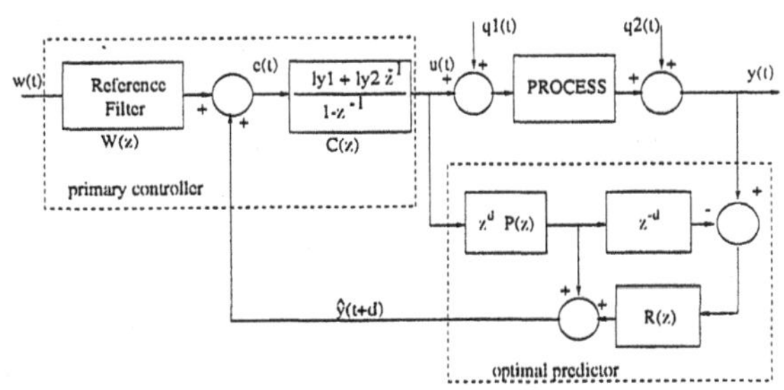

Fig. 2. Classical representation of the GPC

As can be seen only parameter T of the controller affects the computation of the prediction $\hat{y}(t+d \mid t)$ while all parameters (N, λ and T) affect the primary controller. This structure will be used in the next section to analyze the robustness of the GPC.

3. A SIMPLE DESIGN METHOD OF THE T-POLYNOMIAL

The study of the robustness of the GPC is made using the block diagram of figure 2. The plant will be represented by a transfer function P and unstructured uncertainties will be considered, that is: $P = P_n + DP$ with P_n the nominal plant. Also $P = Gz^{-d}$ where G represents the plant without the dead-time and G_n is its nominal value.

$W(z)$ and $C(z)$ represent the two-degree of freedom primary controller in figure 2. To evaluate the closed-loop performance the transfer function between the reference and the output is computed for the nominal case:

$$y(t) = \frac{WCP_n}{1+CG_n}r(t) \quad (16)$$

that is, as expected, d is eliminated from the characteristic equation. Note that R does not appear in (16).

The norm-bound uncertainty region for $DP(e^{j\omega})$ ($\Delta P(\omega)$) is computed in order to maintain closed-loop stability $\forall\omega \in (0,\pi)$ (Morari and Zafiriou, 1989):

$$| DP(e^{j\omega}) | \leq \Delta P(\omega) = \frac{| 1 + C(e^{j\omega})G_n(e^{j\omega}) |}{| C(e^{j\omega})R(e^{j\omega}) |} \quad (17)$$

As can be seen, after the definition of C and W the filter R could be used to improve the robustness of the controller. As R is a function of T, this polynomial could be chosen in order to obtain a low pass filter in R. In this way the proposed approach shows clearly the relation between the robustness of the GPC and T. However the selection of T is not straightforward because also C and W have $T-$dependence. Note that T is used to compute the predictions from $t+d+1$ to $t+d+N$ and so the final coefficients of the control law (l_i and k_i) depend on the coefficients of T. This is the principal reason for the not trivial methods presented in the literature to design T (Yoon and Clarke, 1995; Ansay and Wertz, 1997). As pointed out in (Yoon and Clarke, 1995) increasing the low pass characteristics of T do not necessary improve the robustness of the GPC. From the block diagram of the GPC shown in figure 2 these properties could be easily seen.

But this problem can be solved using the method proposed in this paper, that is, a design procedure for T that allows the independence between performance and robustness with a very simple method. First note that if T is only used to compute the optimal predictions from $t+1$ to $t+d$ and an unitary polynomial $T = 1$ is used to compute the predictions from $t+d+1$ to $t+d+N$ the dependence of C and W with T is eliminated since the first d predictions do not appear in the cost function and therefore they do not affect the controller coefficients. On the other hand R is a function of T and could be used to improve the robustness without changing the closed-loop relation (16) which defines the nominal performance of the controller. Now the tuning procedure can be done in two independent steps: (i) compute N and λ for closed loop performance and (ii) compute T in order to define the low pass characteristics of R. This two step design procedure relates the proposed controller with the internal model control IMC (Morari and Zafiriou, 1989) where a stabilizing controller is computed in a first step and in a second step a filter is included to improve the robustness. The same idea is also used in (Normey-Rico *et al.*, 1997) where the robustness of an industrial dead-time compensator controller is increased with a low pass filter.

As has been mentioned, the filter R could be used to improve the robustness of the system at the desired frequency region without modifying the reference to output nominal performance. However, the disturbance rejection performance of the system is affected by R as is the case in others $T-$design methods for the GPC (Yoon and Clarke, 1995; Ansay and Wertz, 1997). But in this controller the tuning of the filter in order to attempt a compromise between robustness and disturbance rejection is straightforward. Note that the closed-loop transfer function between the input disturbance q_1 and the control action u can be used as a messurement of the disturbance rejection qualities of the system. In general, for a good disturbance rejection performance $| u/q_1 |$ must be close to one for all the frequencies in the desired bandwidth:

$$\left|\frac{u(e^{j\omega})}{q_1(e^{j\omega})}\right| = \left|\frac{C(e^{j\omega})P_n(e^{j\omega})R(e^{j\omega})}{1 + C(e^{j\omega})G_n(e^{j\omega})}\right| = 1 \quad (18)$$

On the other hand, if we compute the multiplicative uncertainty norm boundary of the system, it follows that:

$$\delta P(\omega) = \frac{\Delta P(\omega)}{| P_n(e^{j\omega}) |} = \frac{| 1 + C(e^{j\omega})G_n(e^{j\omega}) |}{| C(e^{j\omega})P_n(e^{j\omega})R(e^{j\omega}) |} = \left|\frac{q_1(e^{j\omega})}{u(e^{j\omega})}\right| \quad (19)$$

That is, the higher disturbance rejection performance gives lower robustness. Notice that similar results can be obtained for the output disturbance $q_2(t)$.

As, in general, the model uncertainties are dominant at high frequencies, R must be chosen in order to increase the value of δP at those frequencies, but maintaining the unitary gain of u/q_1 for the frequencies above ω_0, that is, there is a compromisse between robustness and disturbance rejection.

To compute the polynomial T the expression of R is used: $R(z) = z^d(1 - \tilde{A}/T)^d$. Note that as $\tilde{A}(1) = 0$, $R(1) = 1$ for every $T \neq 0$. For the type of stable plant considered here and in order to eliminate the dependence between R and A the following T is proposed:

$$T = (1 + a_1 z^{-1})(1 - a_t z^{-1}) \qquad a_t > 0 \quad (20)$$

Thus R is given by:

$$R = \left(\frac{1 - a_t}{1 - a_t z^{-1}}\right)^d \qquad a_t > 0 \quad (21)$$

that is, R is a $d-$order low pass filter. The value of a_t is chosen having an estimation of the uncertainties and d automatically compensates the effect of the dead-time in the controller.

As was analyzed in (Camacho and Bordons, 1999), for the type of processes analyzed in this paper, the deterioration in the performance caused by the uncertainties in d is higher for higher values of the nominal dead-time. If no filter is used, a simple error of one sample could cause unstability if the nominal dead-time is 10 or higher. Because of that, the use of T is necessary in real applications when the process exhibits significant dead time. With the normal procedure used in GPC the design of T is not easy and in general high order filters are nedded to stabilize the closed-loop (Yoon and Clarke, 1995). For the method proposed here a simple T is used that could stabilize process with long dead-times and high frequency uncertainties. To show the effect of the proposed T two different dead-times and different errors in d are considered for the same gain and pole of the plant when considering $\lambda = 0.8$ and $N = 15$ that allow for a good nominal setpoint response. In both cases the same $T = A(1-0.8z^{-1})$ is used and the norm-bound uncertainty is compared to the modelling error and to the norm-bound uncertainty of the non-filtered GPC. In figure 3.a the nominal dead-time is $d = 2$ and the real dead-time is $d_p = 3$ and in figure 3.b the nominal dead-time is $d = 7$ and the real dead-time is $d_p = 9$. Note that while for the small dead-time case the GPC with $T = 1$ is stable even in the presence of the error in the estimation of d, in the other case the system is not robust stable. On the other hand the proposed filter with the same tuning gives a stable closed-loop in both cases. Note that the price to pay for increasing robustness is a smaller bandwidth for disturbance rejection as can be seen in figure 4.

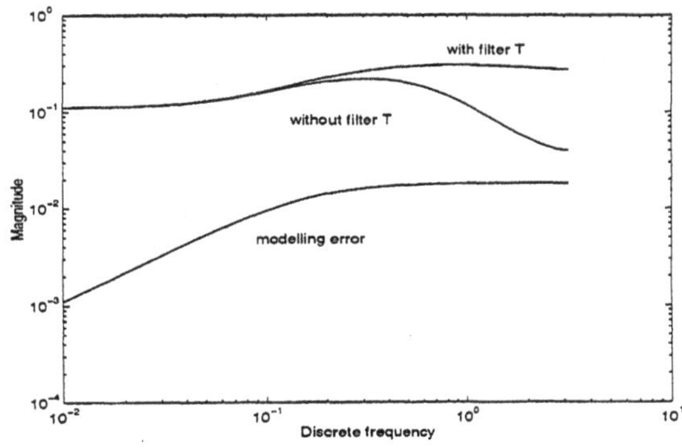

(a)

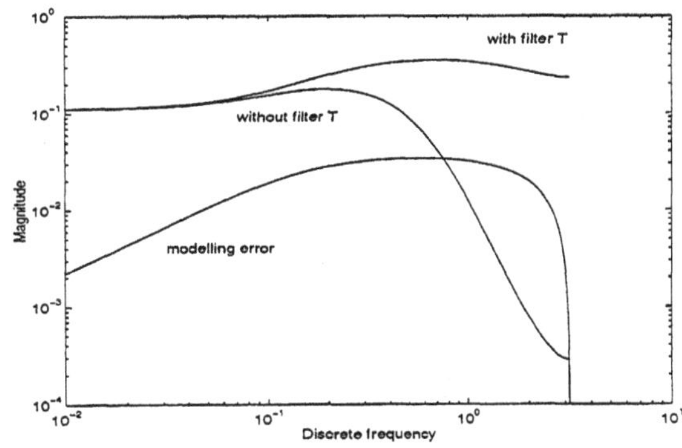

(b)

Fig. 3. Modelling error and additive norm-bound uncertainty for the GPC with filter T and without filter T for small dead-time (a) and long dead-time (b).

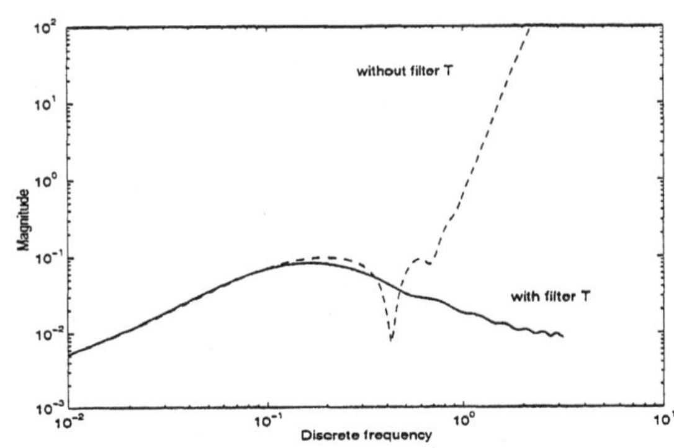

Fig. 4. Modulus of the disturbance to output transfer fuction for the long dead-time case

4. A SIMULATION CASE STUDY

In order to illustrate properties of the proposed design method of the T polynomial in the GPC, a simulation example is presented that corresponds to the temperature control in a heat exchanger. The model corresponds to a pilot plant at the "Departamento de Ingeniería de Sistemas y Automática de la Universidad de Sevilla" that is described with details in (Normey-Rico *et al.*, 1997). The model of the plant, obtained with the "step test" identification method is given by:

$$P(s) = \frac{k}{1+s\tau}e^{-t_d}$$

where the nominal values of the parameters are: $t_d = 14s, k = 0.11$ and $\tau = 6s$. By performing several identification tests with different operation conditions it was possible to note an uncertainty in the values of the plant parameters. The gain k could vary between 0.08 and 0.13, the dead-time between 11 and 18 seconds and τ between 5.7 and 6.3 seconds.

To show the performance and robustness of the proposed method for the tuning of the T−polynomial in the GPC two simulation tests were performed with the same control parameters $N = 15$, $\lambda = 1$ and the sampling time $T_s = 1s$. In the first simulation a GPC with filter $T = 1$ is compared to a GPC with filter $T = A(1-0.8z^{-1})$. The closed loop behaviour of both control systems when parameter uncertainties are considered

$(t_d = 18s, k = 0.12$ and $\tau = 5.7s)$ is shown in figure 5.a. Note that the GPC with filter $T = 1$ is unstable. In the test a step change in the reference is performed at $t = 0$ samples. To analyse the improvment in the robustness of the controller the modelling error $\mid DP \mid$ and the norm-bound uncertainty ΔP are shown in figure 5.b. Note that the proposed low order filter allows a robust stable behaviour of the controller.

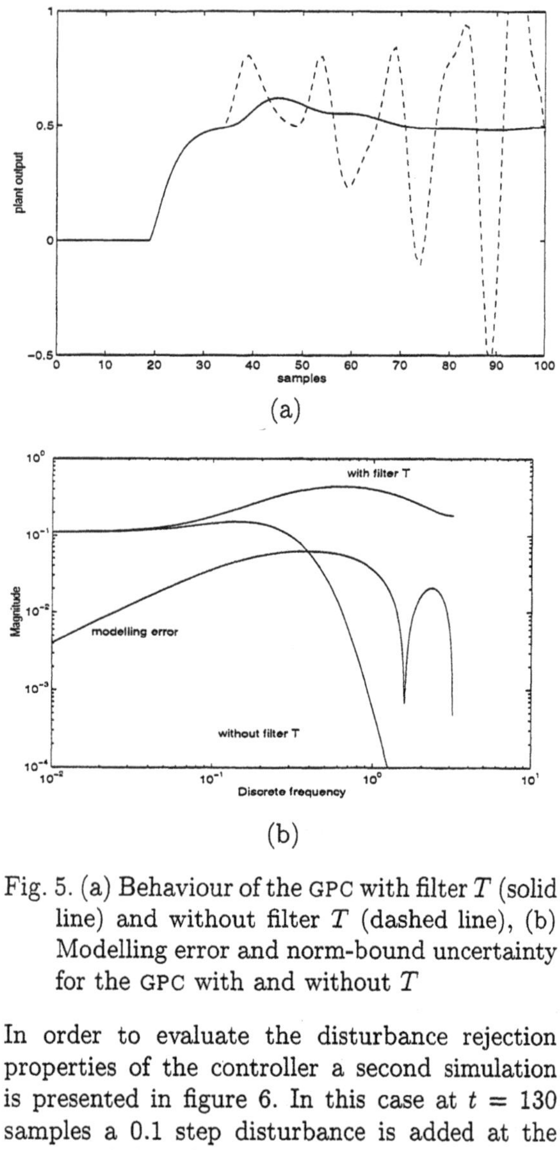

Fig. 5. (a) Behaviour of the GPC with filter T (solid line) and without filter T (dashed line), (b) Modelling error and norm-bound uncertainty for the GPC with and without T

In order to evaluate the disturbance rejection properties of the controller a second simulation is presented in figure 6. In this case at $t = 130$ samples a 0.1 step disturbance is added at the output of the plant and at $t = 210$ samples a 0.3 step disturbance is added at the input of the plant. Note that in both cases the controller has a good performance showing the good compromise between robustness and disturbance rejection.

5. CONCLUSIONS

This work has presented a simple and effective method to design the prefilter in a generalized predictive controller that can be applied to a wide class of industrial processes. The proposed method allows a two-step controller design: first computing the GPC parameters in order to obtain

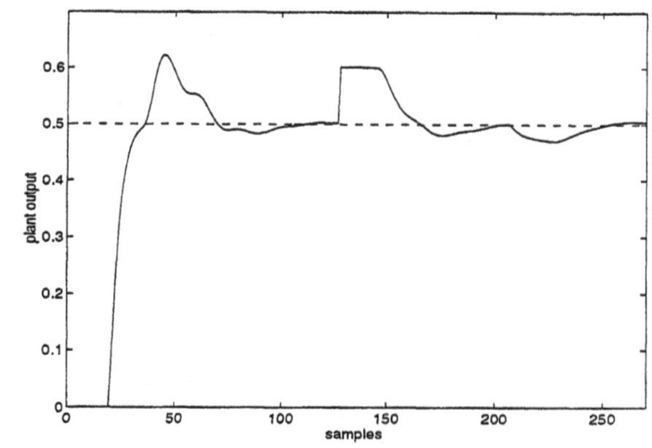

Fig. 6. Behaviour of the GPC with the proposed filter T for set-point and disturbance responces

a desired closed loop performance, and then defining a simple T polynomial in order to attempt some robust and disturbance rejection specifications. Some simulation results confirm the good qualities of the proposed controller and show that the method is simpler than the others proposed in literature.

6. REFERENCES

Ansay, P. and V. Wertz (1997). Model uncertainties in gpc: a systematic two-step design. In: *Proc. of the ECC 97.* Bruxelles.

Camacho, E.F. and C. Bordons (1999). *Model Predictive Control.* Springer Verlag.

Chien, I.L. and P.S. Fruehauf (1990). Consider imc tuning to improve performance. *Chem. Eng. Progress* **10**, 33–.

Clarke, D.W. and C. Mothadi (1989). Properties of generalized predictive control. *Automatica* **25**(6), 859–875.

Clarke, D.W., C. Mothadi and P.S. Tuffs (1987). Generalized Predictive Control. Part I The Basic Algorithm and Part II Extensions and Interpretations. *Automatica* **23**(2), 137–160.

Morari, M. and E. Zafiriou (1989). *Robust Process Control.* Prentice Hall.

Normey-Rico, J.E., C. Bordons and E.F. Camacho (1997). Improving the robustness of dead-time compensating PI controllers. *Control Engineering Practice* **5**(6), 801–810.

Robinson, T. and D. Clarke (1991). Robustness effects of a prefilter in receding-horizon predictive control. *IEE-D* **138**, 2–8.

Tan, K.K., Q. G. Wang, T. H. Lee and Q. Bi (1996). New approach to analysis and design of smith predictor controllers. *AIChE journal* **42**(6), 1793–1797.

Yoon, T. and D. Clarke (1995). Observer design in receding-horizon control. *Int. Journal of Control* **2**, 171–191.

14th Triennial World Congress, Beijing, P.R. China

STUDY ON THE ROBUST CONTROL DESIGN OF MEASURE DIFFERENTIAL SYSTEMS*

Dong YUE, Xiaoping MA, Shifan XU

College of Information & Electrical Engineering, China University of Mining & Technology, Xuzhou, Jiangsu, China (dongy@cumt.edu.cn)

Abstract: In this paper, we propose a robust control based on the observer for a class of measure differential systems with time delay. To implement the control, we only need the output signal of the systems. In the finial section, a numerical example is given to illustrate the effectiveness of proposed scheme. *Copyright © 1999 IFAC*

Key words: observer, measure differential systems, delay, stability.

1. INTRODUCTION

In practical systems, it is difficult to measure the signal of state variable of the system if not impossible. Hence, what we first consider is to utilize the output signal of the system. For the linear systems with continuous solution (LCS), a lot of papers have been reported to investigate the control design based on the output of the system (Choi.H.H, et al., 1996; Barmish.B.R, 1983). And the result has been extended to study the robust control design for uncertain systems, such as uncertain linear systems, uncertain singularly perturbed systems, uncertain distributed parameter systems, and so on.

Since the impulse always exists in the practical systems, such different games, biological phenomena, economic problems and some other mechanical systems, the study of measure differential system has been a subject of intense research in past few years (Yue.D, et al., 1997a, 1997b; Deo.S.G, et al., 1982; Xu Y , 1988; Das.P.C, et al., 1971). (Yue D, et al., 1997a) proposed a robust control scheme for a class of measure differential system with constant time delay. To analyze the stability of closed-loop, the Lyapunov functional method was employed for the first time in the study of control of measure differential systems. To deal with the stability of closed-loop with time-varying delay, (Yue.D, et al., 1997b) proposed a analysis method based on a new differential inequality with impulse and delay. This method has been used to study the robust control of measure differential system with time-varying delay, and also extended.

While the given controllers implemented by using state variable have been shown to have strong robustness for some uncertainties, the requirement for the available measurement of state variable restricts its wild use.

In this paper, using the output signal of the system, we give a design method of robust control for a class of measure differential system with constant time delay. In section 2, preliminaries are first given to describe the type and basic properties of the system. In section 3, the design procedure of the control is represented. To illustrate the effectiveness of the proposed scheme, a numerical example is given in section 4. Conclusion will be shown in the finial section.

2. PRELIMINARIES

In this paper, we consider the following measure differential system with parameter uncertainties and delay

$$Dx(t) = (A + \Delta A)x(t) + A_1 x(t-\gamma) + Bu(t) \qquad (1)$$
$$x(\theta) = x_{t_0}, \qquad \theta \in [t_0 - r, t_0]$$
$$y(t) = Cx(t)$$

where $x(t) \in R^n$ is the state vector, $u(t) \in R^m$ is the input vector, $y(t) \in R^p$. A, ΔA, A_1, B are the known and unknown parameter matrices of appropriate dimensions. γ denote the delay and satisfy $0 < \gamma \le \sigma$. . $Dx, D\omega$ denote the distributional derivative of functions x, ω, respectively. $\omega: R_+ \to R$ is a function of right continuous and bounded variation.

Definition 1: The function $\omega: R_+ \to R$ is thought to belong to the class S^*, if

I>. w is right continuous and bounded variation on every compact subinterval of R_+.

II>. The discontinuous points $t_1 < t_2 < \ldots\ldots$ are isolated, and $t_k \to \infty$ as $k \to \infty$.

III>. w is differentiable on the subinterval $[t_k, t_{k+1})(k = 1,2,\ldots)$ of R_+, where the derivative of w at point t_k denotes the right derivative.

Suppose that $w \in S^*$ and has the following form

$$w = t + \sum_{k=1}^{\infty} \alpha_k H_k(t)$$

where $H_k(t) = \begin{cases} 0, & t < t_k \\ 1, & t \ge t_k \end{cases}$, α_k are some constants.

Hence, $Dw(t) = 1 + \sum_{k=1}^{\infty} \alpha_k \delta_k(t_k)$, where $\delta_k(t_k)$ stand for the Dirac measure, $t_k - t_{k+1} \ge \delta\sigma > 0$.

Assumption: ΔA and A_1 satisfy the following relation, that is,

$$\Delta A^T \Delta A \le \delta_1 I, \quad A_1^T A_1 \le \delta_2 I$$

3. DESIGN OF CONTOL BASED ON OBSERVER

As we know, in many practical systems, the measurement of state variables is very difficult if not impossible to do. In the case of linear system with continuous solutions, a widely used method is to construct the observer-based control, which is also called compensator. This method has been extended to study the design of robust control for uncertain dynamical system with delay or not, such as uncertain linear systems, uncertain singularly perturbed systems, uncertain distributed parameter systems, and so on. Next, we will use this notion to study the design problem of robust control for the system given above.

Constructing a observer based on the output of system (1) as

$$\dot{z} = Az + L(y - Cz) + Bu(t) \qquad (2)$$

where L is a constant matrix. As mentioned above, the solutions of system (1) are always discontinuous, thus, the impulsive signal is contained in the input of system (2) because of the relation $y = Cx$. Fortunately, the output signal from system (2) is continuous even though the discontinuity of the input signal.

Based on the output signal of system (2), we design a class of linear memoryless controller as

$$u(t) = -R^{-1}B^T P_1 z(t) \qquad (3)$$

where R and P are two constant matrices to be given below.

Incorporating the systems (1), (2) and the control (3) into a complex system, we obtain

$$Dx(t) = [(A + \Delta A)x(t) + A_1 x(t-r) + Bu(t)]D\omega$$
$$\dot{z}(t) = Az(t) + L(y - Cz) + Bu(t) \qquad (4)$$
$$u(t) = -R^{-1}B^T P_1 z(t)$$

which can be rewritten as

$$Dx(t) = [(A + \Delta A)x(t) + A_1 x(t-r) - BR^{-1}B^T P_1 z(t)]D\omega \qquad (5)_1$$
$$\dot{z}(t) = Az(t) + L(y - Cz) - BR^{-1}B^T P_1 z(t) \qquad (5)_2$$

Let $e = z - x$, the time derivative of e with respect to time t yields

$$\dot{e}(t) = (A - LC)e(t) - \Delta A x(t) - A_1 x(t-r) \qquad (6)$$

Moreover, using the state transformation $e = z - x$ and for $t \in [t_{k-1}, t_k)$, system $(5)_1$ can be rewritten as

$$\dot{x}(t) = (A + \Delta A - BR^{-1}B^T P_1)x(t) - BR^{-1}B^T P_1 e(t) \qquad (7)$$

Defining another state transformation as

$$\bar{e} = e^{\alpha t} e, \quad \bar{x} = e^{\alpha t} x \qquad (8)$$

Substituting (8) into (6) and (7), we get

$$\dot{\bar{e}}(t) = (A - LC)\bar{e}(t) - \Delta A \bar{x}(t) - e^{\alpha r} A_1 \bar{x}(t-r)$$
$$\dot{\bar{x}}(t) = (A + \alpha I - BR^{-1}B^T P_1 + \Delta A)\bar{x}(t) \qquad (9)$$
$$+ e^{\alpha r} A_1 \bar{x}(t-r) - BR^{-1}B^T P_1 \bar{e}(t)$$

Taking the Lyapunov function as

$$V(t) = \bar{x}^T P_1 \bar{x} + \bar{e}^T P_2 \bar{e} + \int_{t-r}^{t} \bar{x}(\tau)^T T \bar{x}(\tau) d\tau \qquad (10)$$

where P_1, P_2 are two positive definite matrices.

Diferenting $V(t)$ with respect to time t yields

$$\dot V(t) = 2\bar x^T P_1[(A+\alpha I - BR^{-1}B^T P_1)\bar x + \Delta A\bar x + e^{\alpha r}A_1\bar x(t-r) - BR^{-1}B^T P_1\bar e$$
$$+2\bar e^T P_2[(A-LC)\bar e - \Delta A\bar x - e^{\alpha r}A_1\bar x(t-r)$$
$$+\bar x^T T\bar x + \bar x^T T\bar x(t-r)$$
$$-2P_1BR^{-1}B^T P_1]\bar x + \bar x^T[\delta_1 P_1^2 + \delta_2 e^{\alpha r}P_1^2 + I]\bar x$$
$$\le \bar x^T[(A+\alpha I)^T P_1 + P_1(A+\alpha I)$$
$$+\bar x^T(t-r)\bar x(t-r) + \bar x^T P_1BR^{-1}B^T P_1\bar x$$
$$+\bar e^T P_1BR^{-1}B^T P_1\bar e + \bar e^T[(A-LC)^T P_2$$
$$+P_2(A-LC)]\bar e + d^{-1}\delta_2\bar e^T P_2^2\bar e + d\bar x^T\bar x$$
$$+d^{-1}e^{\alpha r}\delta_2\bar e^T P_2^2\bar e + d\bar x(t-r)^T\bar x(t-r)$$
$$+\bar x^T T\bar x + \bar x(t-r)^T T\bar x(t-r)$$

Let $T=(1+d)I$, we can obtain

$$\dot V(t) \le \bar x^T[(A+\alpha I)^T P_1 + P_1(A+\alpha I) - P_1BR^{-1}B^T P_1$$
$$+(\delta_1 + e^{\alpha r}\delta_2)P_1^2 + 2(1+d)I]\bar x$$
$$+\bar e^T[(A-LC)^T P_2 + P_2(A-LC)$$
$$+P_1BR^{-1}B^T P_1 + d^{-1}(\delta_1 + e^{\alpha r}\delta_2)P_2^2]\bar e$$
$$t\in[t_{k-1},t_k)$$

If we can choose a constant d such that two positive definite matrices P_1 and P_2 exist for the following two matrix equations, that is,

$$(A+\alpha I)^T P_1 + P_1(A+\alpha I) - P_1BR^{-1}B^T P_1 + (\delta_1 + e^{\alpha r}\delta_2)P_1^2 + 2(1+d)I < 0$$
$$(A-LC)^T P_2 + P_2(A-LC) + P_1BR^{-1}B^T P_1 + d^{-1}(\delta_1 + e^{\alpha r}\delta_2)P_2^2 < 0$$

then we can have

$$\dot V(t) < 0,\quad t\in[t_{k-1},t_k)$$

From the above differential inequality we obtain

$$V(t) \le V(t_{k-1}),\quad t\in[t_{k-1},t_k)$$

Further, in terms of (10), we can know that

$$V(t)\le V(t_{k-1})$$
$$\le (\lambda_{\max}(P_1)+1+d)\|\bar e_{t_{k-1}}\|^2 + \lambda_{\max}(P_2)\|\bar x_{t_{k-1}}\|^2$$
$$\le \lambda_2(\|\bar e_{t_{k-1}}\|^2 + \|\bar x_{t_{k-1}}\|^2),$$
$$t\in[t_{k-1},t_k) \tag{11}$$

where $\lambda_2 = \max(\lambda_{\max}(P_1)+1+d, \lambda_{\max}(P_2))$.
Since

$$V(t) \ge \lambda_{\min}(P_1)\|\bar e\|^2 + \lambda_{\min}(P_2)\|\bar x\|^2 = \lambda_1(\|\bar e\|^2 + \|\bar x\|^2)$$

where $\lambda_1 = \min(\lambda_1(P_1), \lambda_{\min}(P_2))$, it is clear that

$$\|\bar e(t)\|^2 + \|\bar x(t)\|^2 \le \bar\lambda(\|\bar e_{t_{k-1}}\|^2 + \|\bar x_{t_{k-1}}\|^2)$$

where $\bar\lambda = \lambda_2/\lambda_1$.

In addition, replacing $\bar e(t)$ and $\bar x(t)$ by $e(t)$ and $x(t)$, we obtain the following relation

$$\|e(t)\|^2 + \|x(t)\|^2 \le \bar\lambda e^{-2\alpha(t-t_{k-1})}(\|e_{t_{k-1}}\|^2 + \|x_{t_{k-1}}\|^2),$$
$$t\in[t_{k-1},t_k). \tag{12}$$

Since the impulse of the solution appears in the point t_{k-1}, thus the change between $x(t_{k-1})$ and $x(t_{k-1-})$ is discontinuous, so is for $e(t_{k-1})$ and $e(t_{k-1-})$. To estimate the range of change of $x(t)$ and $e(t)$ from t_{k-1-} to t_{k-1}, we integrate the both side of system (5)₁ from t_{k-1-} to t_{k-1}

$$x(t_{k-1}) - x(t_{k-1-}) = \int_{t_{k-1-}}^{t_{k-1}}[(A+\Delta A)x(t) + A_1x(t-r) + BKz(t)]D\omega(t)$$
$$= \alpha_k(A+\Delta A)x(t_{k-1}) + \alpha_kA_1x(t_{k-1}-r) + \alpha_kBKz(t_{k-1}) \tag{13}$$

where $K = -R^{-1}B^T P_1$.
Since $z(t)$ is continuous, thus, $z(t_{k-1}) = z(t_{k-1-})$.
From (13), we can obtain

$$x(t_{k-1}) = \alpha_k(A+\Delta A)x(t_{k-1}) + (I+\alpha_kBK)x(t_{k-1-}) + \alpha_kBKe(t_{k-1-}) + \alpha_kA_1x(t_{k-1}-r)$$

Similarly, we have

$$e(t_{k-1}) = \alpha_k(A+\Delta A)x(t_{k-1}) + \alpha_kBKx(t_{k-1-}) + (I+\alpha_kBK)e(t_{k-1-}) - \alpha_kA_1x(t_{k-1}-r)$$

Representing the above equations in vector form as

$$\begin{bmatrix} x(t_{k-1}) \\ e(t_{k-1}) \end{bmatrix} = \begin{bmatrix} \alpha_k(A+\Delta A) & 0 \\ \alpha_k(A+\Delta A) & 0 \end{bmatrix}\begin{bmatrix} x(t_{k-1}) \\ e(t_{k-1}) \end{bmatrix} + \begin{bmatrix} I+\alpha_kBK & \alpha_kBK \\ \alpha_kBK & I+\alpha_kBK \end{bmatrix}\begin{bmatrix} x(t_{k-1-}) \\ e(t_{k-1-}) \end{bmatrix} + \begin{bmatrix} \alpha_kA_1 & 0 \\ -\alpha_kA_1 & 0 \end{bmatrix}\begin{bmatrix} x(t_{k-1}-r) \\ e(t_{k-1}-r) \end{bmatrix} \tag{14}$$

Defining two matrices as

$$H_k = \begin{bmatrix} I-\alpha_k(A+\Delta A) & 0 \\ -\alpha_k(A+\Delta A) & I \end{bmatrix}$$

$$P_k = \begin{bmatrix} I+\alpha_kBK & \alpha_kBK \\ \alpha_kBK & I+\alpha_kBK \end{bmatrix}, M_k = \begin{bmatrix} \alpha_kA_1 & 0 \\ -\alpha_kA_1 & 0 \end{bmatrix}$$

Clearly, if $I-\alpha_k(A+\Delta A)$ is invertible, H_k is also invetible. It is assumed in the following that $I-\alpha_k(A+\Delta A)$ is invetible. So, (14) can be represented as

$$\begin{bmatrix} x(t_{k-1}) \\ e(t_{k-1}) \end{bmatrix} = H_k^{-1} P_k \begin{bmatrix} x(t_{k-1-}) \\ e(t_{k-1-}) \end{bmatrix} + H_k^{-1} M_k \begin{bmatrix} x(t_{k-1}-r) \\ e(t_{k-1}-r) \end{bmatrix}$$

Taking the Euclidean norm of $x(t)$ and $e(t)$ yields

$$\|e(t_{k-1})\|^2 + \|x(t_{k-1})\|^2$$
$$\le (\gamma_{1k} + \gamma_{2k} e^{\alpha r}) e^{-\alpha(t_{k-1}-t_{k-2})} (\|e_{t_{k-2}}\|^2 + \|x_{t_{k-2}}\|^2) \quad (15)$$

where $\gamma_{1k} = \|H_k^{-1} P_k\|$, $\gamma_{2k} = \|H_k^{-1} M_k\|$.

In terms of (12) and (15) and using the induction method, we can prove that

$$\|e(t)\|^2 + \|x(t)\|^2$$
$$\le (\bar{\lambda})^k \gamma_k \ldots \gamma_k e^{-2\alpha(t-t_0)} (\|e_{t_0}\|^2 + \|x_{t_0}\|^2)$$
$$t \in [t_0, t_k) \quad (16)$$

where $\gamma_k = \gamma_{1k} + e^{\alpha r} \gamma_{2k}$, the constant k can be take value arbitrarily.

Theorem 1: Supposing the following

The assumption 1 hold and a constant d can be chosen such that there are two positive definite matrices P_1 and P_2 satisfying the Riccati matrix inequalities

$$(A + \alpha I)^T P_1 + P_1 (A + \alpha I) - P_1 B R^{-1} B^T P_1$$
$$+ (\delta_1 + e^{\alpha r} \delta_2) P_1^2 + 2(1+d) I < 0$$
$$(A - LC)^T P_2 + P_2 (A - LC)$$
$$+ P_1 B R^{-1} B^T P_1 + d^{-1} (\delta_1 + e^{\alpha r} \delta_2) P_2^2 < 0$$

then the solutions of system (1) under the control (3) are exponentially stable, if

<i>. $I - \alpha_k (A + \Delta A)$ $(k = 1,2,\ldots)$ are invertible.

<ii>. Defining

$$\bar{\lambda} = \frac{\max(\lambda_{\max}(P_1) + 1 + d, \lambda_{\max}(P_2))}{\min(\lambda_{\min}(P_1), \lambda_{\min}(P_2))},$$

H_k and P_k as given above, $q = \sup_k \{\max(1, \gamma_k)\}$,

then $\delta > \dfrac{\ln(\bar{\lambda} q)}{2\alpha\sigma}$.

Proof: From (16) and using the conditions of theorem, we can estimate the following inequality

$$\|e(t)\|^2 + \|x(t)\|^2 \le e^{-(2\alpha - \frac{\ln(\bar{\lambda} q)}{\delta\sigma})(t-t_0)} (\|e_{t_0}\|^2 + \|x_{t_0}\|^2)$$

which can complete the proof because of the condition <ii>.

Q.E.D

In the condition <i> of theorem 1, we need the invertibility of $I - \alpha_k (A + \Delta A)$. Next, a sufficient condition satisfying this property is given.

Theorem 2: If $I - \alpha_k A$ is invertible, and $\delta_1 |\alpha_k| \|(I - \alpha_k A)^{-1}\| < 1$, then $I - \alpha_k (A + \Delta A)$ is invertible.

4. EXAMPLE

Consider the two-dimensional system with constant delay as

$$Dx(t) = [(\begin{bmatrix} 1 & 2 \\ -1 & -1 \end{bmatrix} + \begin{bmatrix} \alpha_1 & 0 \\ 0 & \alpha_1 \end{bmatrix}) x(t)$$
$$+ \begin{bmatrix} \beta_1 & 0 \\ 0 & \beta_1 \end{bmatrix} x(t-0.2) + \begin{bmatrix} 0 \\ 1 \end{bmatrix} u(t)] D\omega(t)$$
$$x(\theta) = x_0, \quad \theta \in [t_0 - 0.2, t_0]$$
$$y(t) = \begin{bmatrix} 1 & 1 \end{bmatrix} x(t)$$

where

$$\alpha_k = \begin{cases} 1, & k = 2n-1 \\ -1, & k = 2n \end{cases}, \quad (n = 1,2,\ldots), \quad |\alpha_1| \le 0.1,$$
$$|\beta_1| \le 0.1, \quad \delta > 7$$

To design a observer-based controller for stabilizing the above system with uncertainty and time delay, we take $\alpha = 1$, $d = 5$ and $R = 0.01$. Solving P_1, P_2 and L yields

$$P_1 = \begin{bmatrix} 41.4229 & -1.1334 \\ -1.1334 & 0.3683 \end{bmatrix},$$
$$P_2 = \begin{bmatrix} 170.9239 & -171.5794 \\ -171.5794 & 174.3739 \end{bmatrix}, \quad L = \begin{bmatrix} -34 \\ 49 \end{bmatrix}$$

Constructing a observer as

$$\dot{z}(t) = \begin{bmatrix} 1 & 2 \\ -1 & -1 \end{bmatrix} z(t) + \begin{bmatrix} -34 \\ 49 \end{bmatrix} (y(t) - \begin{bmatrix} 1 & 1 \end{bmatrix} z(t))$$
$$+ \begin{bmatrix} 0 \\ 1 \end{bmatrix} u(t)$$

and finally designing the control based on the output of the above observer as

$$u(t) = \begin{bmatrix} -113.3388 & 36.8263 \end{bmatrix} z(t)$$

5. CONCLUSION

In this paper, we studied the stabilizing control problem for a class of measure differential systems with parameter uncertainty and time delay. Combining linear control theory for systems with solution, we proposed a class of observer-based robust control for measure differential systems with parameter uncertainty and time delay. Combining. We show that the observer-based control theory is also suitable for measure differential systems if the impulse behavior of the system satisfies some

REFERENCE

Choi.H.H and M.J.Chung (1996). Observer-based H_∞ Control Design for State Delayed Linear Systems, *Automatica,* **Vol.32, No.**7, 1073-1075

Barmish.B.R (1983). A New Class of Stability Controllers for Uncertain Dynamical Systems, *SIAM Journal of Control and Optimization,* **Vol.21**, 153-157.

Yue D and S.F.Xu (1997a). Robust Stabilization for a Class of Measure Differential Systems with Delays: Riccati-Like Approach, *Proceedings of IEEE SMC'97*, **Patr.5**, 380-384

Yue D and S.F.Xu (1997b). Robust Control for a Class of Measure Differential Systems with Delays, *Advances in Modeling & Analysis,* **Vol.50, No.2**, 1-14

Deo.S.G and S.G.Pandit (1982). *Differential Systems Involving Impulsive* ,Lecture Notes, Berlin; Springer-Verlag.

Xu Y (1988). *Functional Differential Equation and Measure Differential Equation*, Zhongshan University Press.

Das.P.C and P.R.Sharmn (1971). On Optimal Controls for Measure Delay Differential Equation" *SIAM.J.Control*, **9**, 43-61.

* *This work is supported by National Science Foundation of China(69874042)*

Copyright © 1999 IFAC
14th Triennial World Congress, Beijing, P.R. China

A MINIMAX DESIGN OF ROBUST I-PD CONTROLLER FOR TIME-DELAY SYSTEMS WITH PARAMETRIC UNCERTAINTY

K. Hirata * Y. Yanase ** T. Katayama ** T. Kawabe ***

* *Department of Electrical and Electronic Systems*
Osaka Prefecture University, Sakai, Osaka 599-8531, Japan
kent@ecs.ees.osakafu-u.ac.jp
** *Department of Applied Mathematics and Physics*
Graduate School of Informatics, Kyoto University
Kyoto 606-8501, Japan, katayama@kuamp.kyoto-u.ac.jp
*** *Institute of Information Sciences and Electronics*
University of Tsukuba, Tsukuba, Ibaraki 305-8573, Japan
kawabe@is.tsukuba.ac.jp

Abstract: This paper presents a design method of robust I-PD controller for SISO plants with a time-delay and parametric uncertainty by extending our earlier approach. It is assumed that the adjustable parameters of the I-PD controller are chosen so as to minimize the generalized integral of squared error (ISE) maximized by the plant parameters belonging to a given bounded set. Thus the design problem is formulated as a minimax optimization problem, which is solved by a genetic algorithm (GA). A novel feature of the present paper lies in the exact computation of the generalized ISE for time-delay systems without using Padé approximant. Numerical examples show the effectiveness of the present approach.

Keywords: Robust I-PD controller, Minimax optimization, Time-delay systems, Exact evaluation of ISE, Genetic algorithm

1. INTRODUCTION

There has been a renewed interest in proportional-integral-derivative (PID) controllers, since it has become easier to implement PID-based algorithms due to the recent development of digital process controllers. In fact, over 90% of industrial control problems are solved by PID controllers (or by their variants) in spite of the simple structures (Åström and Hägglund 1995).

Since industrial plants are too complex to obtain precise dynamics, the PID controllers must be robust for a wide range of operating conditions. There have been developed some works for robust PID tunings methods taking into account robust closed-loop stability and robust performance in the presence of plant uncertainty. For example, the IMC based method of designing PID controller is considered in Morari and Zafiriou (1989), where a low-pass filter is introduced to achieve robust stability and robust performance under nonparametric uncertainty. In Chen *et.al.* (1995), the mixed H_2/H_∞ optimization is employed to design a robust PID controller in the presence of nonparametric uncertainty, where the problem of minimizing the integral of squared error (ISE) subject to robust stability constraint is solved by using a genetic algorithm (GA).

Motivated by a minimax approach of designing PID controller under parametric uncertainty due to Lu (1992a, 1992b), we have also developed a GA based method of designing robust I-PD controller (see Fig. 1) for time-delay systems with parametric uncertainty, where the time-delay is replaced by its Padé approximant (Kawabe and Katayama 1994, 1996, 1997). More specifically, the I-PD parameters are adjusted so as to minimize performance criterion maximized by plant parameters with bounded uncertainty. It is shown in Kawabe and Katayama (1994) that a minimax design based on the ISE leads to a very large overshoot in the optimal closed-loop response. A large overshoot has also been observed in earlier PID tunings based on the ISE criterion (see e.g. Smith and Corripio (1985)). In order to circumvent this difficulty, we have applied the minimax design technique to a generalized ISE that includes a penalty on the derivative of the control variable (Kawabe and Katayama 1996, 1997). It is shown that the use of the generalized ISE cost function, which has been used in the LQ regulator based design of servomechanism, is very effective in shaping the closed-loop responses by adjusting a weighting parameter. The use of Padé approximants is, however, a drawback of (Kawabe and Katayama 1994, 1996, 1997), since the stability criterion with Padé approximants for a time-delay element may yield erroneous robust stability conditions and the evaluation of the (generalized) ISE is not exact (Marshall *et.al.* 1992).

In this paper, we present a design method based on the exact evaluation of control performance for time-delay systems using the technique developed in Marshall *et.al.* (1992), in which the PID parameter settings are extensively studied based on the ISE. The performance criterion and the optimization technique of this paper are the same as those of Kawabe and Katayama (1996, 1997). The robust stability of the closed-loop system with a time delay is determined by checking the range of time-delay in the characteristic equation. Furthermore, we give a robust stability criterion based on zero exclusion principle (Kogan 1995) for a polytope of quasi-polynomials[1]. This criterion gives a necessary and sufficient condition for stability of the polytope of quasi-polynomials, and is conveniently used in the present problem as well as in problems of adjusting parameters of controllers with a fixed configuration.

2. PROBLEM FORMULATION

2.1 *Minimax Optimization Problem*

Consider the SISO control system shown in Fig. 1, where r is the step input, u the manipulated variable, y the measured variable, e the error, and K_i, T_i, T_d, $\gamma \in \mathbf{R}$ are parameters of the controller.

[1] The quasi-polynomial is a polynomial in one complex variable and exponential powers of the variable.

We assume that the plant model is described by $P(s)e^{-hs}$, where $P(s)$ is a strictly proper rational function with parametric uncertainties and h is a delay.

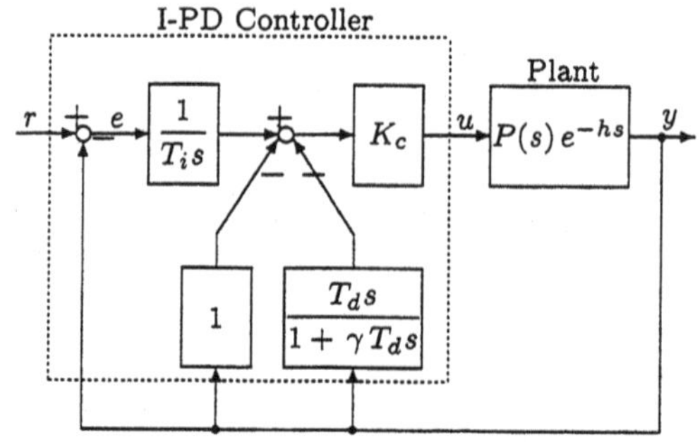

Fig. 1. Block diagram of control system

Let θ be the parameter vector of $P(s)$ and belong to a bounded set $\Theta = \{\theta \,|\, \theta_l \le \theta \le \theta_u\}$. Let $q := (K_c, T_i, T_d)^T$ be the I-PD controller parameter vector belonging to a bounded set $Q = \{q \,|\, q_l \le q \le q_u\} \subset \mathbf{R}^3$, which is the range of adjustable I-PD parameters. It is assumed that Θ and Q are given a priori.

We define the generalized ISE (Kawabe and Katayama 1996, 1997) as

$$J(q,\theta) := \int_0^\infty \left\{e^2(t) + \rho \dot{u}^2(t)\right\} dt, \qquad (1)$$

where $\rho \ge 0$ is an adjustable parameter. This performance index, employed in the LQ regulator based design of servomechanism, is a weighted sum of the ISE and the integral of squared derivative of the manipulated variable. Hence, this performance index not only evaluates the ISE but also gives a penalty to the velocity of manipulated variable.

The robust I-PD controller design based on the generalized ISE is formulated as a minimax optimization problem: $\min_{q\in Q} \max_{\theta\in\Theta} J(q,\theta)$, under the constraint that q stabilizes the closed-loop system for all $\theta \in \Theta$. The designed controller guarantees the robust stability and the best performance under worst case scenario.

2.2 *Description of the Closed-Loop System*

From the Parseval's theorem, (1) is written as

$$J(q,\theta) = \frac{1}{2\pi j}\left\{ \int_{-j\infty}^{+j\infty} e(s)e(-s)\, ds + \rho \int_{-j\infty}^{+j\infty} s\,u(s)\,[-s\,u(-s)]\, ds \right\}, \quad (2)$$

where $e(s)$ and $u(s)$ are the Laplace transforms of $e(t)$ and $u(t)$, respectively. Let $P(s) = N(s)/D(s)$, and define $A(s)$, $B_e(s)$, $C(s)$, $D_e(s)$ and $B_u(s)$ as

$$\begin{aligned} A(s) &= T_i s(1+\gamma T_d s)D(s), \\ B_e(s) &= T_i(1+\gamma T_d s)D(s), \\ C(s) &= K_c\left\{(1+\gamma)T_iT_ds^2+(T_i+\gamma T_d)s+1\right\}N(s), \\ D_e(s) &= K_cT_i\left\{(1+\gamma)T_ds+1\right\}N(s), \\ B_u(s) &= K_c(1+\gamma T_d s)D(s). \end{aligned} \quad (3)$$

Then from Fig. 1, the error $e(s)$ for the step input $r(s) = 1/s$ and the derivative of the manipulated variable are given by

$$e(s) = \frac{B_e(s) + D_e(s)e^{-hs}}{A(s) + C(s)e^{-hs}}, \quad (4)$$

$$su(s) = \frac{B_u(s)}{A(s) + C(s)e^{-hs}}, \quad (5)$$

respectively. The characteristic equation of the closed-loop system is given by

$$f(s) := A(s) + C(s)\,e^{-hs} = 0. \quad (6)$$

Since the plant is strictly proper, the degree of $A(s)$ is always greater than that of $C(s)$, so that this system is a retarded system.

3. ROBUST STABILITY CRITERION FOR TIME-DELAY SYSTEMS

3.1 *Stability Criterion*

We first consider the stability criterion of the quasi-polynomial (6) when the coefficients of $A(s)$ and $C(s)$ are fixed. Let us regard h as a variable and write the characteristic equation (6) as

$$f(s,h) := A(s) + C(s)e^{-hs} = 0. \quad (7)$$

Since a common factor of $A(s)$ and $C(s)$ is a root of (7) for any $h \geq 0$, we assume that $A(s)$ and $C(s)$ have no common factors. A procedure of finding the stability regions for h consists of four steps (Marshall *et.al.* 1992).

Step 1: Examine the stability of (7) at $h = 0$.

Step 2: Consider infinitesimally small positive h. For retarded systems, all the new roots appear in the left half-plane.

Step 3: Determine the positive zeros[2] of the polynomial

$$W(\omega^2) := A(j\omega)A(-j\omega) - C(j\omega)C(-j\omega), \quad (8)$$

and compute the corresponding h. Then analyze the behavior of the roots in the neighborhood of h.

Step 4: Check the stability region of h by sorting destabilizing and stabilizing critical value of h. If a given time-delay h is in the stability region, the closed-loop system is stable. Otherwise, the system is unstable.

3.2 *Robust Stability of time-delay systems*

Next we consider the robust stability criterion against parametric uncertainties. Namely, we assume that $N(s)$ and $D(s)$ in (3) are the (real) interval polynomials. Let N denote the number of vertices of the hyperrectangle Θ in the parameter space and

$$f^i(s) = A^i(s) + C^i(s)e^{-hs}, i = 1, \cdots, N, \quad (9)$$

be the quasi-polynomials corresponding to each vertex. Then the characteristic quasi-polynomial is generated by the convex combination of these quasi-polynomials:

$$f(s,k) := \sum_{i=1}^{r} k_i f^i(s), \quad (10)$$

$$s.t. \sum_{i=1}^{N} k_i = 1,\ 0 \leq k_i \leq 1,\ i = 1, \cdots, N.$$

Let $k := (k_1, \cdots, k_N)^T$ and define the convex polyhedron K, the quasi-polynomial family S, and the value set S_ω as

$$\begin{aligned} K &:= \left\{ k \;\middle|\; \sum_{i=1}^{N} k_i = 1,\ 0 \leq k_i \leq 1 \right\}, \\ S &:= \{f(s,k) \mid k \in K\}, \\ S_\omega &:= \{f(j\omega,k) \mid k \in K\}. \end{aligned} \quad (11)$$

The shape of S_ω is a polygonal region and each quasi-polynomial segment corresponding to the boundary of S_ω is called edge. From the edge theorem (Fu *et.al.* 1989) and the zero exclusion principle (Kogan 1995), we see that the quasi-polynomial family S is stable if and only if the boundary of S_ω does not contain or pass through the origin for all $\omega \geq 0$. Since the boundary of S_ω is the value set of the segment quasi-polynomial corresponding to two generating points of S, we consider an edge connecting $f^1(s)$ and $f^2(s)$. Define the quasi-polynomial segment

$$f(s,\lambda) := (1-\lambda)f^1(s) + \lambda f^2(s), \quad (12)$$

with $\lambda \in [0,1]$. From (9) we can express (12) as

$$f(s,\lambda) = A(s,\lambda) + C(s,\lambda)\,e^{-hs}, \quad (13)$$

[2] Each zero indicates the point where the root loci against h touch or cross the imaginary axis. As is obvious from (8), such locations are independent of delay h. Another key fact is that the number of zeros is finite, although (7) has infinitely many solutions.

where $A(s,\lambda)$ and $C(s,\lambda)$ are defined as

$$A(s,\lambda) := (1-\lambda)A^1(s) + \lambda A^2(s),$$
$$C(s,\lambda) := (1-\lambda)C^1(s) + \lambda C^2(s).$$

A stability criterion of the quasi-polynomial segment (12) is given as follows:

Theorem 3.1 Given $\lambda \in [0,1]$, let ω_λ be

$$\omega_\lambda = \sup\{\omega \mid A(j\omega,\lambda)A(-j\omega,\lambda) - C(j\omega,\lambda)C(-j\omega,\lambda) = 0\}.$$

Also let $\bar{\omega}$ be

$$\bar{\omega} = \sup\{\omega_\lambda \mid \lambda \in [0,1]\}.$$

Then, the quasi-polynomial segment (12) contains or passes through the origin for $\omega \geq 0$ if and only if there exist $\omega \in [0,\bar{\omega}]$ satisfying the following condition [3]

$$\begin{aligned} &\mathrm{Re}[f^1(j\omega)]\,\mathrm{Im}[f^2(j\omega)] - \\ &\qquad \mathrm{Re}[f^2(j\omega)]\,\mathrm{Im}[f^1(j\omega)] = 0, \\ &\mathrm{Re}[f^1(j\omega)]\,\mathrm{Re}[f^2(j\omega)] \leq 0, \\ &\mathrm{Im}[f^1(j\omega)]\,\mathrm{Im}[f^2(j\omega)] \leq 0, \end{aligned} \tag{14}$$

We can summarize the procedure examining stability of the quasi-polynomial family S as follows:

Step 1: Examine the stability of one quasi-polynomial in S. If the quasi-polynomial is stable, go to Step 2. If not, S is not stable.

Step 2: Check whether $0 \notin S_{\omega_0}$ for one $\omega_0 \geq 0$. If $0 \notin S_{\omega_0}$, go to Step 3. If not, S is not stable.

Step 3: For each edge, check the existence of $\omega \geq 0$ such that (14) holds. If there exist such an ω on at least one edge, S is not stable. If not, S is stable.

4. EVALUATION OF ISE FOR TIME-DELAY SYSTEMS

By using the technique of Marshall *et.al.* (1992), we develop a method of computing the ISE of the form

$$J = \int_0^\infty x^2(t)\,dt = \frac{1}{2\pi j}\int_{-j\infty}^{+j\infty} x(s)x(-s)ds, \tag{15}$$

where $x(s)$ is expressed as

$$x(s) = \frac{B(s) + D(s)e^{-hs}}{A(s) + C(s)e^{-hs}}. \tag{16}$$

[3] In contrast to the polynomial family case, the first equation in (14) is not a polynomial but a nonlinear function of ω. Therefore we must perform a line search instead of solving polynomial roots. However, since the interval to be searched is finite, such a computation is tractable.

It is assumed that the system is stable, so that all the poles of (16) lie in the left half-plane. For the notational convenience, we suppress the Laplace variable s of each polynomial and denote its paraconjugate by $\bar{(\cdot)}$, for example, we write $A := A(s)$ and $\bar{A} := A(-s)$. Then the integrand of (15) is expressed as

$$x(s)x(-s) = \frac{B + De^{-hs}}{A + Ce^{-hs}} \cdot \frac{\bar{B} + \bar{D}e^{hs}}{\bar{A} + \bar{C}e^{hs}}. \tag{17}$$

It can be shown that (17) is additively decomposed as

$$\begin{aligned} x(s)x(-s) &= \frac{\alpha + \beta e^{-hs}}{A + Ce^{-hs}} + \frac{\bar{\alpha} + \bar{\beta}e^{hs}}{\bar{A} + \bar{C}e^{hs}} \\ &=: H(s) + H(-s), \end{aligned} \tag{18}$$

where

$$\begin{aligned} \alpha &= \frac{A(B\bar{B} + D\bar{D}) - 2BC\bar{D}}{2(A\bar{A} - C\bar{C})}, \\ \beta &= \frac{2AD\bar{B} - C(B\bar{B} + D\bar{D})}{2(A\bar{A} - C\bar{C})}. \end{aligned} \tag{19}$$

To compute (15), we consider the contour Γ shown in Fig. 2. Denote the segments on the imaginary axis and the semi-circle as Γ_1 and Γ_2, respectively. If $A\bar{A} - C\bar{C} = 0$ has roots on the positive/negative imaginary axis, add infinitesimally small indentation to Γ_1 to the left/right around the root. Denote the integral of $H(s)$ along Γ as I, i.e.,

$$I = \frac{1}{2\pi j}\int_\Gamma \frac{\alpha + \beta e^{-hs}}{A + Ce^{-hs}}ds. \tag{20}$$

Since the system is stable, $A + Ce^{-hs} = 0$ has no root inside Γ. Then the singularities of the integrand of (20) inside Γ are limited to the roots of $A\bar{A} - C\bar{C} = 0$, which are denoted by s_i. It is crucial that $\{s_i\}$ is a finite set. It follows from the residue theorem that

$$I = -\sum_i \mathrm{Res}\left[\frac{\alpha + \beta e^{-hs}}{A + Ce^{-hs}}, s_i\right], \tag{21}$$

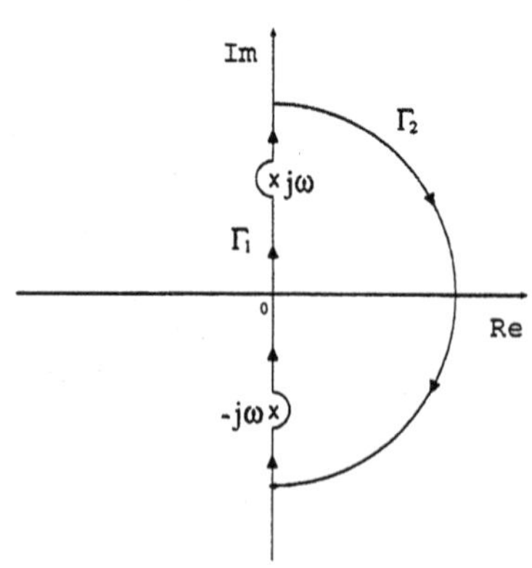

Fig. 2. Contour Γ

where Res $[f(s), s_i]$ denotes the residues of $f(s)$ at s_i. If s_i is a singularity of $H(s)$, $-s_i$ is also a singularity of $H(-s)$. With respect to residues, we have

$$\mathrm{Res}\,[H(s), s_i] = -\mathrm{Res}\,[H(-s), -s_i]\,. \quad (22)$$

It follows from (22) that the integral of $H(-s)$ along Γ is equal to I. Since $x(s)$ has no singularity on the imaginary axis, we can replace the integral path of (18) by Γ_1, i.e.,

$$J = \frac{1}{2\pi j} \int_{\Gamma_1} \{H(s) + H(-s)\}\, ds.$$

As the radius of Γ_2 goes infinity, the integral along Γ_2 becomes zero and hence we can replace Γ_1 by Γ. Thus we obtain $J = 2I$.

5. OPTIMIZATION METHOD

As stated in Section 2, the design of robust I-PD controller is reduced to a minimax optimization problem:

$$\min_{q \in Q} \max_{\theta \in \Theta} J(q, \theta), \quad (23)$$

where the closed-loop system is stable for all $\theta \in \Theta$. It should be noted that the generalized ISE $J(q, \theta)$ does not have saddle point in this case. Since there are many local optimal solutions for the controller parameters, we use a genetic searching algorithm to find a global solution.

5.1 *Algorithm of Minimax Optimization Problem*

We assume that $q := (K_c, T_i, T_d) \in Q$. For computational purpose, let $Q_d := \{q_1, \cdots, q_N\}$ be a discrete approximation of the set Q. Then the design algorithm of a robust I-PD parameters is summarized as follows.

Step 1: Check robust stability of q_i for all $\theta \in \Theta$, where q_i is generated by the GA operations described below.

Step 2: If the closed-loop system with q_i is robustly stable, compute

$$M_i = \max_{\theta \in \Theta} J(q_i, \theta), \quad (24)$$

by using (21) and the GA. If the system is not robustly stable, set $M_i = \infty$.

Step 3: If the GA algorithm stops, go to Step 4. Otherwise, go to Step 1.

Step 4: Let the minimum of M_i be M_{io}. Then the corresponding q_{io} yields the minimax robust I-PD controller.

5.2 *Structure of Genetic Algorithm*

The GA is one of the powerful tools for the noncovex optimization. To design this iterative random search algorithm, we must define some problem-dependent algorithm elements:

Representation

In the present algorithm, we adopt the Gray-code string to represent candidate solutions. Each of the I-PD parameters K_c, T_i, T_d is represented by a binary string. Three such strings are concatenated into a binary string, which defines a point in the parameter space to be searched by the GA algorithm.

Fitness function

The linear scaling method is used as a fitness function, i.e., the fitness is simply computed from the order of the cost function value of each individual among the population.

Genetic operators

The genetic operators used in the present algorithm are uniform crossover, bit mutation and the roulette wheel selection. The uniform crossover generates two offspring by exchanging a predefined number of alternate subsections between two parent strings. The recombination operator, mutation, is implemented as altering bit values at randomly selected string positions.

6. DESIGN EXAMPLES

In this section, we show a numerical example of robust I-PD controller design. The range of adjustable I-PD parameters and γ are given by $K_c \in [0.1, 15.9]$, $T_i \in [1.0, 100.0]$, $T_d \in [1.0, 100.0]$ and $\gamma = 0.1$. The parameters of the GA are given by population: 500, number of individual: 100, mutation rate: 0.2.

Example 6.1 We consider the second-order plant

$$G_3(s) = \frac{K}{T_2 s^2 + T_1 s + 1} e^{-2s}, \quad (25)$$

where the parameters belong to the set $K \in [0.8, 1.2]$, $T_1 \in [6.0, 20.0]$, $T_2 \in [225, 625]$. Table 1 shows the minimizing solution for nominal plant with $\rho = 0$ and the minimax solutions for $\rho = 0.0$, 1.0, 10.0. Fig. 3 shows the step responses for the worst case scenario with $\rho = 10$ where the solid

Table 1: Robust I-PD controller design

ρ	K_c	T_i	T_d	K	T_1	T_2	J
nominal	15.9	6.90	8.30	1.0	12.0	400	10.2
0.0	15.9	17.4	5.44	0.8	6.0	625	15.5
1.0	9.30	19.9	9.03	0.8	6.0	625	20.9
10.0	5.45	23.0	13.0	0.8	20.0	625	28.6

line and the dashed line are the responses designed by the present method and by the earlier method (Kawabe and Katayama 1996, 1997), respectively. We see that the closed-loop response designed by

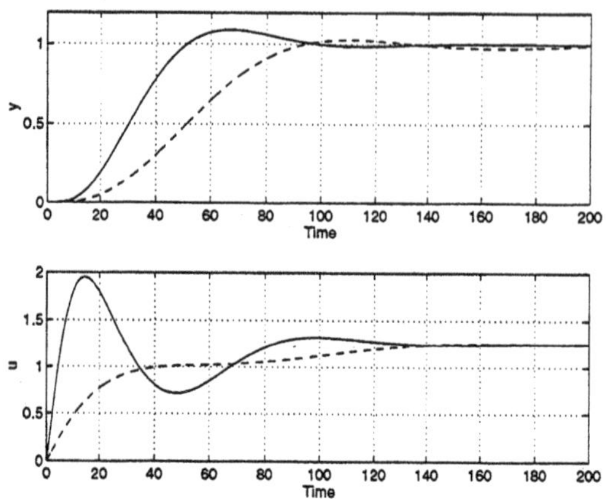

Fig. 3. Step responses of closed-loop system for the worse case scenario

the present method is much better than the one designed by our earlier method based on Padé approximant. The improvement of the closed-loop performance is due to the exact treatment of the time-delay element. As shown in Fig. 4, the robust stability region of the I-PD parameters for $T_i = 27$ is much larger than that estimated by using the Padé approximant. Thus under the Padé approximation, the region of adjustable parameters becomes much smaller than the actual feasible region. □

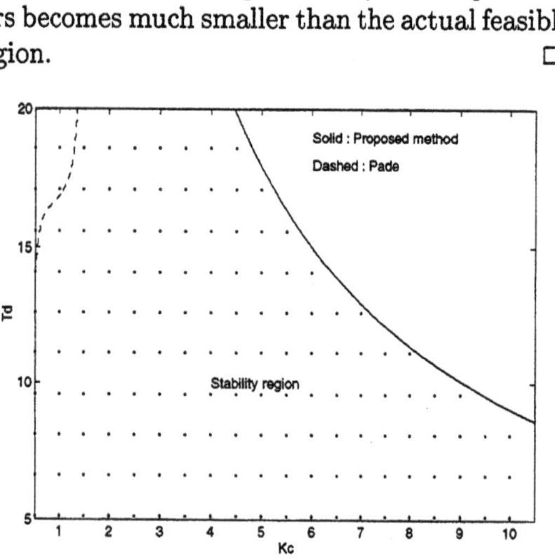

Fig. 4. The stability region of I-PD parameters for $T_i = 27$

7. CONCLUSION

We have developed a GA based design method of robust I-PD controller for time-delay systems with parametric uncertainty. A novelty of the present design method over the earlier ones is the exact evaluation of control performance and the robust stability criterion. The numerical example shows that

- The robust I-PD controller guarantees the robust stability and the best performance under worst case scenario.
- We can easily regulate the overshoot and oscillatory variation in the closed-loop responses by using the generalized ISE.
- The control performance of present design is better than the earlier results based on the Padé approximant and the Kharitonov's theorem.

8. REFERENCES

Åström, K. J., and Hägglund, T.(1995). *PID Control – Theory, Design and Tuning, 2nd ed.* Instrument Society of America.

Chen, B. S., Cheng, Y. M., and Lee, C. H.(1995). A genetic approach to mixed H_2/H_∞ optimal PID control. *IEEE Control Systems Magazine*, **15**, 51-60.

Fu, M., Olbrot, A. W., and Polis, M. P. (1989). Robust stability for time-delay systems: The edge theorem and graphical tests. *IEEE Trans. Automat. Contr.*, **34**, 813-820.

Kawabe, T., and Katayama, T. (1994). A minimax design of robust I-PD controller for system with time-delay. *Proc. 1st Asian Control Conference*, Tokyo, July, **3**, pp. 495-498.

Kawabe, T., and Katayama, T. (1996) A minimax design of robust I-PD controller based on a generalized integral-squared-error. *Trans. SICE*, **32**, 1226-1233 (in Japanese).

Kawabe, T., and Katayama, T. (1997). A minimax approach to design of robust I-PD controller. *Proc. 2nd Asian Control Conference*, Seoul, July,**I**, pp. 661-664.

Kogan, J. (1995). *Robust Stability and Convexity.* Springer-Verlag.

Lu, J. P. (1992a). Robust process control on Honeywell TDC3000. Preprints, *Honeywell Advanced Control Symposium.*

Lu, J. P. (1992b). A MIMO robust controller on the Honeywell TDC3000. *Proc. 19th Annual Honeywell Industrial Automation and Control, User's Group Symposium.*

Marshall, J.E., Górecki, H., Korytowski, A., and Walton, K. (1992). *Time-Delay Systems: Stability and Performance Criteria with Applications.* Ellis Horwood.

Morari, M., and Zafiriou, E. (1989) *Robust Process Control.* Prentice-Hall.

Smith, C. A., and Corripio, A. B. (1985). *Principles and Practice of Automatic Process Control.* Wiley.

14th Triennial World Congress, Beijing, P.R. China

CONTROL OF SOME DISTRIBUTED DELAY SYSTEMS USING GENERALIZED POPOV THEORY

D. Ivanescu* S.-I. Niculescu J.M. Dion* L. Dugard***

* *Laboratoire d'Automatique de Grenoble, INPG-CNRS (UMR5528), ENSIEG-BP 46, 38402 Saint Martin d'Hères Cedex, FRANCE, Fax : (33) 4.76.82.63.88, email : dugard@lag.ensieg.inpg.fr*
** *HEUDYASIC, Université de Technologie de Compiègne, BP 529, 60205, Compiègne, FRANCE, Fax: (33) 3.44.23.44.77 email: silviu@hds.utc.fr*

Abstract: The paper deals with the generalized Popov theory applied to systems with distributed time delay. Sufficient conditions for stabilizing this class of delayed systems as well as for γ - attenuation achievement are given in terms of algebraic properties of a Popov system via a Liapunov - Krasovskii functional. The considered approach is new in the context of distributed linear-delay systems and gives some interesting interpretations of H^∞ memoryless control problems in terms of Popov triplets and associated objects. The approach is illustrated via numerical examples.

Keywords: Time-Delay system, distributed systems, H^∞ control, Riccati theory, Kalman-Popov-Yakubovich system.

1. INTRODUCTION

In the last decade a lot of attention has been paid to the stability and stabilization problems of uncertain systems with delayed state involving unknown constant or time-varying delay (see, for instance, Kolmanovskii and Nosov 1986, Hale and Verduyn Lunel 1991, Kojima *et al.* 1993) and the references therein)

For the simplest case in which the delayed argument appears only in the state variable, a Liapunov - Krasovskii approach or a Liapunov - Razumikhin like theory was developed (Cheres *et al.*, 1989; Xie and de Souza, 1993; Lee *et al.*, 1994).

Let us mention that some results have been reported about $\mathcal{H}_\infty$ analysis and control of delay systems. For examples, memoryless $\mathcal{H}_\infty$ control problem has been investigated by Lee *et al.* where a frequency domain approach is adopted for linear constant delay systems without uncertainty. Niculescu *et al.* 1995, deal with robust $\mathcal{H}_\infty$ memoryless control problem for linear systems with time-varying state delay and norm-bounded time-varying uncertainty which appears in the state matrices and input matrix of the state equation.

But the real systems are often distributed parameters ones so it is of interest to study such a class of systems. However few works have been done in this field (see for example Dugard and Verriest *eds* , 1997 and the references therein)

In this paper we consider the control problem of a large class of distributed time-delay systems using the generalized Popov theory (see Ionescu *et al.* , 1998). Our objectives are to achieve simultaneously both closed-loop stability and disturbance attenuation for systems described in terms of linear retarded functional differential equations. This paper provides an unified approach to the tackled problems. Further extensions can be easily done through the framework of the generalized Popov theory. The development is essentially based on the 'generalized' Riccati theory presented by Ionescu, Oara and Weiss 1998 (and the references therein) which is an extension of the famous Popov's positivity theory to the indefinite sign case, usually encountered in game-theory situations.Some comments on the delay-independent stability of linear time-delay systems have been considered in Ionescu *et al.* , 1997. To be more specific, our results are based on the necessary and sufficient condition for the existence of the stabilizing solution to an adequate Kalman - Yakubovich - Popov system of indefinite sign. This approach, combined with the Krasovskii theory for time - delay systems, leads to explicit representation formulae. The delay system class considered includes discrete and distributed delay terms and can be seen as a special case of infinite-dimensional systems described by distributional convolution equations. Note that the results presented here extend previous results obtained by some of the same authors(Dion *et al.* 1998, Niculescu *et al.* 1998).

2. PROBLEM STATEMENT AND PRELIMINARY RESULTS

Consider the following state-delayed system including discrete and distributed delays:

$$\dot{x}(t) = Ax(t) + A_1 x_t(-\tau_1) \int_0^{\tau_2} A_2(\theta)x_t(-\theta)d\theta + B_1 u_1(t) + B_2 u_2(t) \quad (1)$$

where x_t represents the translation operator $x_t(\theta) = x(t+\theta)$, and $A_2(\theta)$ is a piecewise continuous matrix function, $x(t) \in \mathbf{R}^n$ is the state, $u_1(t) \in \mathbf{R}^{m_1}$, $u_2(t) \in \mathbf{R}^{m_2}$ are the disturbance and control inputs, A, A_1, B_i $i = 1, 2$ are constant matrices of appropriate dimensions.

We are interested in finding a memoryless controller

$$u_2(t) = F_2 x(t) \quad (2)$$

that simultaneously stabilizes the sytem (1) and achieves the γ - attenuation property, i.e., $\|T_{y_1 u_1}\| < \gamma$ where $T_{y_1 u_1}$ is the L^2-linear bounded input-output operator defined by the closed-loop configuration obtained by coupling (1) and (2).

Unlike the techniques previously mentioned, our interest is directed towards the tools offered by the generalized Popov theory. The interest of such approach is twofold : first to extend the Popov theory developed in the linear case to delay systems and second to provide some alternative to handle control problems for delay case.

First,we define a general quadratic cost function which corresponds to the disturbance attenuation requirement. We show that the stabilization problem is solvable if the corresponding Kalman-Popov-Yakubovich system has a stabilizing solution and a stabilizing controller can be derived from the partition. If further conditions related to the general quadratic cost function are satisfied, then the stabilizing controller also achieves disturbance attenuation requirement.

In the sequel, we will drop the explicit time dependence of $x(t)$, $u_1(t)$ and $u_2(t)$ on t for brevity. In this section several basic notions and results concerning the general Riccati theory are presented.

Definition 1. Call $\Sigma = (A, B; P)$ where $A \in \mathbf{R}^{n\times n}$, $B \in \mathbf{R}^{n\times m}$ and

$$P = \begin{bmatrix} Q & L \\ L^T & R \end{bmatrix} = P^T \in \mathbf{R}^{(n+m)\times(n+m)}$$

a Popov triplet.

Frequently, we will use the extensive notation $\Sigma = (A, B; Q, L, R)$.

Let $\Sigma = (A, B; Q, L, R)$ be a Popov triplet and let

$$J = \begin{bmatrix} -I_{m_1} & \\ & I_{m_2} \end{bmatrix}, \quad m_1 + m_2 = m \quad (3)$$

be an arbitrary sign matrix. Associated with Σ the following object:

The Kalman-Popov-Yakubovich system in J form (KPYS(Σ, J))

The following nonlinear system with unknown X, V, W:

$$\begin{aligned} R &= V^T J V \\ L + XB &= W^T J V \\ Q + A^T X + XA &= W^T J W \end{aligned} \quad (4)$$

is usually denoted as the KPYS(Σ, J).

Definition 2. Any triplet (X, V, W) for which (4) is fulfilled and in addition $X = X^T$, V is nonsingular and of lower-left block triangular form

$$V = \begin{bmatrix} V_{11} & 0 \\ V_{21} & V_{22} \end{bmatrix} \quad (5)$$

partitioned in accordance with J in (3) and $A + BF$ is exponentially stable for

$$F = -V^{-1}W, \quad (6)$$

called the stabilizing feedback gain, is called a stabilizing solution to the KPYS(Σ, J).

3. MAIN RESULTS

In a first step, let us consider the following linear distributed time-delay system:

$$\begin{cases} \dot{x} = Ax + A_1 x_t(-\tau_1) + \\ \quad + \int_0^{\tau_2} A_2(\theta) x_t(-\theta) d\theta + Bu, \\ x = \phi \text{ on } [-\tau, 0] \end{cases} \quad (7)$$

where $x \in \mathbf{R}^n$ is the state, $u \in \mathbf{R}^m$ is the input, $A, A_1 \in \mathbf{R}^{n\times n}$, $B \in \mathbf{R}^{n\times m}$, $A_2(\theta)$ is a piecewise continuous function, $\tau = max\{\tau_1, \tau_2\}$ is the delay and ϕ is any continuous n-valued function on $[-\tau, 0]$.

Let $\Sigma = (A, B; Q, L, R)$ be a Popov triplet where the entries A and B coincide with A and B in (7). Consider the *extended* time-varying Popov triplet associated to (7):

$$\Sigma_v = (A, [A_1 \;\; A_2(\cdot) \;\; B]; Q, [0 \;\; 0 \;\; L], \begin{bmatrix} R_{d1} & 0 & 0 \\ 0 & R_{d2}(\cdot) & 0 \\ 0 & 0 & R \end{bmatrix}), \quad (8)$$

where $R_{d1} \in \mathbf{R}^{n\times n}$ and R_{d2} is a continuous time-varying function with some sign constraints. Such extended Popov triplet allows to reduce the control problem of a time-varying delay system to a time-varying system free of delay. For the sake of simplicity we shall not address such problem here. Note also that if A_2 is a constant matrix, one recovers the *time-invariant* Popov triplet used in Niculescu *et al.* 1998, with R_{d2} a symmetric and strictly negative-definite matrix.

Consider also the *extended* time-invariant Popov triplet

$$\Sigma_e = (A, [A_1 \;\; M^{\frac{1}{2}}(\tau_2) \;\; B]; Q, [0 \;\; 0 \;\; L], \begin{bmatrix} R_{d1} & 0 & 0 \\ 0 & -I_n & 0 \\ 0 & 0 & R \end{bmatrix}), \quad (9)$$

where

$$M(\tau_2) = \int_0^{\tau_2} A_2(\theta) Y^{-1}(\theta) A_2(\theta)^T d\theta \quad (10)$$

for some $Y(\cdot)$ continuous time-varying function, (seen as a *parameter*) .

Let also the following (extended) sign matrix

$$J_e = \left[\begin{array}{c|c} -I_{2n} & \\ \hline & J \end{array}\right] = \left[\begin{array}{c|cc} -I_{2n} & & \\ \hline & -I_{m_1} & \\ & & I_{m_2} \end{array}\right], \quad (11)$$

where $(m_1 + m_2 = m)$, be considered. Let B, L and R be partitioned in accordance with J in (11):

$$B = [B_1 \;\; B_2] \; L = [L_1 \;\; L_2] \; R = \begin{bmatrix} R_{11} & R_{12} \\ R_{12}^T & R_{22} \end{bmatrix} \quad (12)$$

The basic result of this section is

Theorem 1. Assume that the KPYS(Σ_e, J_e) given by 9 and 11 has a stabilizing solution (X, V_e, W_e). Let the stabilizing feedback gain F_e be partitioned in accordance with J_e in (11), that is,

$$F_e = -V_e^{-1} W_e = \begin{bmatrix} F_d \\ F_1 \\ F_2 \end{bmatrix} \quad (13)$$

Let also u be split in accordance with B in (12):

$$u = \begin{bmatrix} u_1 \\ u_2 \end{bmatrix} \begin{matrix} m_1 \\ m_2 \end{matrix}.$$

Let S be a symmetric and positive-definite matrix and ε a positive scalar. Assume further that

$$X \geq 0 \quad (14)$$

$$R_{11} < 0 \quad (15)$$

$$R_{d2}(\tau_2) < 0 \quad (16)$$

$$Q + L_2 F_2 + F_2^T L_2^T + \\ + F_2^T R_{22} F_2 + R_{d1} + R_{d2}(0) > 0 \quad (17)$$

where

$$R_{d2}(\theta) = R_2 + \int_\theta^{\tau_2} Y(\theta) d\theta, \quad \theta \in [0, \tau_2] \quad (18)$$

Then the state feedback

$$u_2 = F_2 x \quad (19)$$

stabilizes (7), i.e.,

$$\begin{cases} \dot{x} = \tilde{A}x + A_1 x_t(-\tau_1) + \int_0^{\tau_2} A_2(\theta) x_t(-\theta) d\theta \\ x = \phi \text{ on } [-\tau, 0] \end{cases} \quad (20)$$

defines an exponentially stable solution for all ϕ. Here $\tilde{A} = A + B_2 F_2$.

The proof is given in *the full version of the paper* and makes use of the following Liapunov-Krasovkii functional:

$$V(x_t) = x^T(t) X x(t) + \int_{t-\tau_1}^{t} x^T(\theta)(-R_{d1}) x(\theta) d\theta \\ + \int_{t-\tau_2}^{t} x^T(\theta)(-R_{d2}(t-\theta)) x(\theta) d\theta, \quad (21)$$

where $X = X^T \geq 0$ and $R_{d1} = R_{d1}^T < 0$ are given before; the time-varying matrix function $R_{d2}(\cdot)$ is given in (18). Note that since the inequality (16) is satisfied, it follows that $-R_{d2}(\xi)$ is a symmetric and positive-definite matrix for each $\xi \in [0, \tau_2]$.

A natural consequence of this Theorem is:

Corollary 1. If all the conditions in the statement of Theorem 1 hold, then

$$\begin{cases} \dot{x} = \tilde{A}x + A_1 x_t(-\tau_1) + \\ \quad + \int_0^{\tau_2} A_2(t-\theta) x_t(-\theta) d\theta + B_1 u_1, \\ x = 0 \text{ on } [-\tau, 0] \end{cases} \quad (22)$$

defines a linear bounded input-state operator from L_+^{2,m_1} into $L_+^{2,n}$.

Let $\hat{Q}$ be any $n \times n$ symmetric matrix. Let $\Sigma = (A, B; \hat{Q}, L, R)$ be the Popov triplet constructed with $\hat{Q}$ and with entries of Σ_e. Associate with Σ the "usual" Popov index:

$$J_\Sigma(\phi, u) =< \begin{bmatrix} x \\ u \end{bmatrix}, \begin{bmatrix} \hat{Q} & L \\ L^T & R \end{bmatrix} \begin{bmatrix} x \\ u \end{bmatrix} > \quad (23)$$

where $(x, u) \in L_+^{2,n} \times L_+^{2,m}$ and x and u are linked via (7) for some ϕ.

Then we have (see the full version of the paper):

Proposition 1. Let us consider a symmetric matrix $\hat{Q}$ satisfying

$$\begin{bmatrix} \hat{Q} & L_2 \\ L_2^T & R_{22} \end{bmatrix} \geq 0 \quad (24)$$

Assume also that all the conditions in the statement of Theorem 1 hold except (17) which is modified as

$$Q + R_{d1} + R_{d2}(0) > \hat{Q} \quad (25)$$

If the controller (19) stabilizes the delay system and $\phi = 0$, i.e., (7) becomes (22), then there exists $\bar{\zeta} > 0$ such that

$$J_\Sigma(0, u_1) \leq -\bar{\zeta} \|u_1 - F_1 x\|_2^2, \quad \forall u_1 \in L_+^{2,m_1} \quad (26)$$

where

$$J_\Sigma(0, u_1) := J_\Sigma(0, u)|_{u_2 = F_2 x}. \quad (27)$$

Proposition 2. Assume that all conditions in the statement of Theorem 1 hold. Assume additionally that

$$\bar{Q} + R_{d1} + R_{d2}(0) > 0 \quad (28)$$

where

$$\bar{Q} := Q + LF + F^T L^T + F^T R F$$

Then

$$\begin{cases} \dot{x} = \tilde{A}x + A_1 x_t(-\tau_1) + \\ \quad + \int_0^{\tau_2} A_2(\theta) x_t(-\theta) d\theta + B_1 u_1 \\ v_1 = -F_1 x \qquad + u_1 \end{cases} \quad (29)$$

(with $x = 0$ on $[-\tau, 0]$) defines a linear boundedly invertible operator on L_+^{2,m_1}.

Using all the results presented before, we shall state and prove the main result of this paper.

Theorem 2. Let (7) together with the quadratic cost defined by the right-hand side of (23) be given. For arbitrary m_1, m_2 such that $m_1 + m_2 = m$, let B, L and R be partitioned as in (12).

Assume that there exists three $n \times n$ symmetric matrices $S > 0$, Q, R_{d1} and a positive scalar ε such that the KPYS(Σ_e, J_e), where Σ_e and J_e are defined by (9) and (11), respectively, has a stabilizing solution (X, V_e, W_e) and let the stabilizing feedback F_e be partitioned in accordance with (13).

Let us consider a symmetric matrix $\hat{Q}$ and assume also that the following conditions all hold:

1). $X \geq 0$

2). $\begin{bmatrix} \hat{Q} & L_2 \\ L_2^T & R_{22} \end{bmatrix} \geq 0$

3). $R_{11} < 0$

4). $R_{d2}(\tau_2) < 0$

5). $Q + R_{d1} + R_{d2}(0) > \hat{Q}$

6). $\bar{Q} + R_{d1} + R_{d2}(0) > 0$

where

$$\bar{Q} = Q + LF + F^T L^T + F^T R F$$

$$R_{d2}(0) = R_2 + \int_0^{\tau_2} Y(\theta) d\theta \quad \theta \in [0, \tau_2].$$

Then

a. $u_2 = F_2 x$ stabilizes (7)

b. There exists $c_0 > 0$ such that

$$J_\Sigma(0, u_1) \leq -c_0 \|u_1\|_2^2 \quad \forall u_1 \in L_+^{2,m_1}$$

where $J_\Sigma(0, u_1)$ has been defined by (27), (23).

4. $\mathcal{H}^\infty$-CONTROL

In this section the theory developed in Section 3 will be applied for solving the following $\mathcal{H}^\infty$-control problem formulated for state-delayed systems. Such a problem is stated as follows.

Let the system

$$\begin{cases} \dot{x} = Ax \; + A_1 x_t(-\tau_1) + \int_0^{\tau_2} A_2(\theta) x_t(-\theta) d\theta + \\ +B_1 u_1 \; + B_2 u_2 \\ y_1 = C_1 x \qquad + D_{11} u_1 + D_{12} u_2 \end{cases} \quad (30)$$

(where $x = 0$ on $[-\tau, 0]$) be given. Here x is the state (in the usual sense), u_1 and u_2 are the disturbance and control inputs, respectively, and y_1 is the output to be controlled. The state x is assumed to be accessible for measurement. We are looking for a state feedback law

$$u_2 = F_2 x \quad (31)$$

which stabilizes (30) and achieves γ-attenuation property for the closed-loop system, i.e., there exists $c_0 > 0$ such that

$$-\gamma^2 \|u_1\|_2^2 + \|y_1\|_2^2 \leq -c_0 \|u_1\|_2^2 \quad \forall u_1 \in L_+^{2,m_1} \quad (32)$$

or equivalently the system

$$\begin{cases} \dot{x} = (A + B_2 F_2)x \; + A_1 x_t(-\tau_1) + \\ \quad + \int_0^{\tau_2} A_2(\theta) x_t(-\theta) d\theta + B_1 u_1 \\ y_1 = (C_1 + D_{12} F_2)x + D_{11} u_1 \end{cases} \quad (33)$$

(where $x = 0$ on $[-\tau, 0]$) defines a γ-strictly contractive input-output map. Here γ is a prescribed tolerance for the attenuation level. Then we have:

Corollary 2. Assume that there exist two $n \times n$ symmetric matrices Q and R_d such that all the conditions of Theorem 2 hold with respect to the following particular data:

$$\begin{aligned} & \quad\; m_1 \; m_2 \\ & B = [B_1 \; B_2] \\ & \hat{Q} = C_1^T C_1, \; L = [L_1 \;\; L_2] = C_1^T [D_{11} \;\; D_{12}] \quad (34) \\ & R = \begin{bmatrix} R_{11} & R_{12} \\ R_{12}^T & R_{22} \end{bmatrix} = \begin{bmatrix} -\gamma^2 I + D_{11}^T D_{11} & D_{11}^T D_{12} \\ D_{12}^T D_{11} & D_{12}^T D_{12} \end{bmatrix} \end{aligned}$$

Then for F_2 given in Theorem 2, $u_2 = F_2 x$ is a solution to the H^∞-control problem stated above.

5. SOME UNCERTAINTY CASES

Let us consider the following uncertain state-delayed system:

$$\dot{x} = (A + \Delta A)x + (A_1 + \Delta A_1)x(t - \tau_1) + \\ + \int_0^{\tau_2} A_2(\theta) x_t(-\theta) d\theta + Bu, \quad (35)$$

Let $\Sigma = (A, B; Q, L, R)$ be a Popov triplet where the entries A and B coincide with A and B in 7. The uncertainty $[\Delta A \;\; \Delta A_1]$ satisfies the following norm-bounded conditions $\forall t \geq t_0$:

$$\Delta A = DF(t)E \quad F^T(t)F(t) \leq I, \quad (36)$$

$$\Delta A_1 = D_1 F_1(t) E_1 \quad F_1^T(t) F_1(t) \leq I, \quad (37)$$

with known matrices $D \in \Re^{n \times i}$, $D_1 \in \Re^{n \times i_1}$. The matrices E and E_1 are given weighting matrices.

We can obtain an equivalent for the theorem 2 (uncertainty case), using an apropriate extended Popov triplet:
Let $R_d \in \mathbf{R}^{n \times n}$ and consider the *extended* Popov triplet for the system (35) - (37):

$$\Sigma_e = (A, [A_1 \;\; D \;\; D_1 \;\; M(\tau_2)^{1/2} \;\; B]; \\ Q, [0 \;\; 0 \;\; 0 \;\; 0 \;\; L], \begin{bmatrix} R_{dp} & 0 \\ 0 & R \end{bmatrix}), \quad (38)$$

$$\text{with } R_{dp} = \begin{bmatrix} R_d & 0 & 0 & 0 \\ 0 & -I_1 & 0 & 0 \\ 0 & 0 & -I_d & 0 \\ 0 & 0 & 0 & -I \end{bmatrix}$$

6. AN EXAMPLE

In this section, two numerical examples are presented for comparing these stability criteria using a general form for $R_{d2}(\theta)$ with the existing ones given in Niculescu *et al.* 1998 with a particular function. First let us calculate a maximal size of the distributed delay τ_2 for a fixed γ. It should be pointed out that there are few results in the literature addressed to this problem. (Dugard and Verriest eds, 1997). Let the following unstable distributed state-delayed system

$$\dot{x} = x + \sigma^{-\tau} x + \int_0^{\tau_2} \theta x_t(-\theta) d\theta + u_1 + u_2 \\ y_1 = x \qquad\qquad + u_2 \quad (39)$$

be given. Here x, u_1, u_2 and y_1 are all scalars. The problem is to find a memoryless controller

$$u_2 = F_2 x \quad (40)$$

that achieves simultaneously closed-loop stability and γ-attenuation. The prescribed tolerance is

$$\gamma = 2 \quad (41)$$

The input data are

$$A = 1, \;\; A_1 = 1, \;\; B_1 = 1, \;\; B_2 = 1 \\ C_1 = 1, \qquad D_{11} = 0, \;\; D_{12} = 1 \quad (42)$$

as directly follows from (39). Choose $R_d = -10$. Then the extended Popov triplet (see (9)) is

$$\Sigma_e = (1, [1\ M(\tau_2)^{\frac{1}{2}}\ 1\ 1]; \quad Q, [0\ 0\ 0\ 1], \begin{bmatrix} -10 & 0 & 0 & 0 \\ 0 & -1 & 0 & 0 \\ 0 & 0 & -4 & 0 \\ 0 & 0 & 0 & 1 \end{bmatrix}) \qquad (43)$$

We consider two values for $Y(\theta)$. In the first case, we take non-linear form $Y(\theta) = \theta^2 + 1$ and with (10) we find that $M(\tau_2) = \tau_2 - arctg(\tau_2)$. But (43) is equivalent to the algebraic Riccati equation (ARE) associated with Σ_e, that is,

$$0 = A^T X + XA + Q - ([0\ 0\ L_1\ L_2] + X[A_1\ M(\tau_2)^{\frac{1}{2}}\ B_1\ B_2]) \begin{bmatrix} R_d & 0 & 0 & 0 \\ 0 & -I & 0 & 0 \\ 0 & 0 & R_{11} & R_{12} \\ 0 & 0 & R_{12}^T & R_{22} \end{bmatrix}^{-1} \left(\begin{bmatrix} 0 \\ 0 \\ L_1^T \\ L_2^T \end{bmatrix} + \begin{bmatrix} A_1^T \\ M(\tau_2)^{\frac{1}{2}} \\ B_1^T \\ B_2^T \end{bmatrix} X \right)$$

After computations we find the stabilizing solution:

$$X = \sqrt{\frac{Q-1}{0.65 - M(\tau_2)}} \qquad (44)$$

which is greater than 0 when $Q > 1$, and we can still find a solution for $\tau_{2_{max}} = 1.69$. Let us verify the conditions stated in Theorem 2. The first condition is fulfilled when $Q > 1$ as (44) shows. For the next ones, we obtain

$$R_{d2}(\tau_2) < 0, \quad Q - 1 > -R_{d1} - R_{d2}(0)$$

There are verified for all $R_2 < 0$. Let $R_2 = -0.1$ and we obtain $Q > 7.87$ Therefore all of the conditions stated in Theorem 23 hold for $Q > 7.87$ and consequently

$$u_2 = -(1 + \sqrt{\frac{Q-1}{0.65 - M}})x$$

is the desired feedback law. For the second case, we consider $A_2(\theta) = \theta$ and $Y(\theta) = \theta$. With these values we obtain $M(\tau_2) = \frac{\tau_2^2}{2}$. This time we find $\tau_{2_{max}} = 1.1$, which is a more conservative one. And all of the conditions stated in Theorem 23 hold for $Q > 10.45$

7. CONCLUSIONS

An extension of the generalized Popov theory to the case of delay system with discrete and distributed delays has been done. Our interest has been focused on the memoryless controller design for $\mathcal{H}^\infty$-control problem. As future research direction we suggest the investigation of observer-based compensation technique.

REFERENCES

Cheres, E., S. Gutman and Z. J. Palmor, "Robust stabilization of uncertain dynamic system including state delay", *IEEE Trans. Automat. Control*, vol. 34, pp. 1199–1203, 1989.

..., "Control of distributed varying delay systems using generalized Popov theory" *Internal Note , LAG - 98* .

Dion J.-M., L. Dugard, D. Ivanescu, S. I. Niculescu, V. Ionescu, "Robust H_∞ Control of Time-Delay Systems: A Generalized Popov Theory Approach" in *Perspectives in Control Theory and applications*, Springer - Verlag, June 1998.

Hale, J.K. and S. M. Verduyn Lunel, *Introduction to Functional Differential Equations*, Applied Math. Sciences, vol. **99**, Springer-Verlag, 1991.

Ionescu, V., C. Oară and M. Weiss, *Generalized Riccati Theory*, John Wiley & Sons, 1998.

Ionescu, V., S. I. Niculescu and H. Woerdeman, "On $\mathcal{L}_2$ memoryless control of time-delay systems," *Proc. 36th IEEE Conf. Dec. Contr.*, San Diego, USA, December 1997.

Ionescu, V., S.-I. Niculescu, J.-M. Dion, L. Dugard, H. Li, "Generalized Popov Theory Applied To State-Delayed Systems," *Proc. IFAC Conf. System Structure and Control*, Nantes , July 1998.

Kojima, A., K. Uchida, and E. Shimemura, "Robust stabilization of uncertain time delay systems via combined internal-external approach," *IEEE Trans. Automat. Control*, vol. **38**, pp. 373-378, 1993.

Kolmanovskii, V. B.,and V. R. Nosov, *Stability of Functional Differential Equations*, Mathematics in Science and Eng., vol. **180**, Academic Press, New York, 1986.

Lee, J. H., S. W. Kim and W. H. Kwon, "Memoryless H_∞ controllers for state delayed systems," *IEEE Trans. Automat. Contr.*, **39**, pp. 159-162, 1994.

Niculescu, S. I., C. E. de Souza, J. -M. Dion and L. Dugard, "Robust $\mathcal{H}_\infty$ memoryless control for uncertain linear systems with time-varying delay," *3rd European Contr. Conf.*, Rome, Italy, pp. 1814-1818, 1995.

Niculescu, S. I., V. Ionescu, D. Ivănescu, J. M. Dion and L. Dugard, "On generalized Popov theory for delay systems," *6th IEEE Mediterranean Conf.*, Sardaigne, Italy, June 1998.

Stability and Control of Time-Delay Systems (L. Dugard and E. I. Verriest, Eds.), LNCIS, Springer-Verlag, London, **228**

Xie, L. and C. E. de Souza, "Robust stabilization and disturbance attenuation for uncertain delay system," *Proc. 2nd European Contr. Conf.*, Groningen, The Netherlands, pp. 667-672, 1993.

14th Triennial World Congress, Beijing, P.R. China

C-2a-08-4

ROBUST H_∞ CONTROLLER SYNTHESIS FOR UNCERTAIN SYSTEMS WITH MULTIPLE TIME-VARYING DELAYS: AN LMI APPROACH

Qing-Long HAN and Driss MEHDI

LAII—ESIP, 40, Av. du Recteur Pineau, 86022 Poitiers Cedex, France.
(QLHAN, Driss.Mehdi)@esip.univ-poitiers.fr, Fax: (33) 05 49 45 40 34

Abstract: This paper is concerned with the robust H_∞ control problem for linear uncertain systems with multiple time-varying delays. Based on the linear matrix inequality (LMI) approach, we develop a method for synthesizing a robust H_∞ dynamic output feedback control law which guarantees the quadratic stability of the closed-loop system and reduces, to a prescribed level, the effect of the disturbance input on the controlled output. A sufficient condition for the existence of a robust H_∞ controller of any order is proposed in terms of three LMIs. One can easily design a robust H_∞ controller by solving the three LMIs numerically very efficiently via convex and quasi-convex optimization techniques. *Copyright ©1999 IFAC*

Keywords: Uncertain systems; delay; robust controller; quadratic stabilizability; output feedback; linear matrix inequality (LMI).

1. INTRODUCTION

H_∞ control of linear systems without time-delay has been considered and fairly complete solutions for linear systems without time-delay have been presented since late 1980s (see Francis, 1987; Green and Limebeer, 1995; Kimura, 1997 and references therein), but H_∞ control for linear time-delay systems has not been fully investigated. Lee *et al.* (1994) have extented the state-space H_∞ controller design method proposed by Petersen (1987a) to such systems. Based on a symmetric positive definite solution of a modified algebraic Riccati equation, Ge *et al.* (1996) and Yu *et al.* (1996) have proposed robust memoryless H_∞ state feedback controllers which robustly stabilize uncertain time-delay systems independently of time-delays, and guarantees an H_∞ norm bound constraint on disturbance attenuation for all admissible uncertainties. Choi and Chung (1995; 1996) have considered the systems with delayed input and further presented a design for observer-based controller.

In many control problems, the design constraint have a simple reformulation in terms of LMI. This is hardly surprising, given that LMIs are directive byproducts of Lyapunov-based criteria, and that Lyapunov techniques play a central role in the analysis and control of linear systems. See Boyd *et al.* (1994) for a survey of LMIs in control theory. One advantage of the LMI approach is that it does not need any generalized plant and another is that the controller order can be fixed by imposing a rank condition on positive definite solutions to the LMI existence conditions. Recently, existence conditions for the general H_∞ control problem have been given in terms of three LMIs and all H_∞ controller are parameterized explicity in state space (Iwasaki and Skelton ,1994; Gahinet and Apkarian, 1994;. Gahinet, 1996). Based on an LMI approach, Gu (1994) has been concerned with H_∞ control of systems under norm bounded uncertainties in all system matrices, which extended the work derived by Xie *et al.* (1992a, b). Kokame *et al.* (1998) have extended the results of Gu (1994) to linear time-delay systems. Li and de Souza (1997a, b) have presented the delay-dependent robust stability and stabilization of linear uncertain time-delay systems. Niculescu (1998) has considered a H_∞ memoryless control problem with supplementary α-stability constraints for a class of linear systems with delayed state and disturbance inputs and a sufficient condition for feasibility is also given in terms of LMI's. Jeung *et al.* (1996) have presented a robust output controller design method for linear uncertain systems with time-varying delays in both state and control input but not considered the disturbance attenuation with precribed level. Based on an LMI approach and the bounded real lemma for time-delay systems, Choi and Chung (1997) have developed an output feedback H_∞ control design for nominal linear single constant time-delay systems but they have not considered parameter uncertainties and time-varying delays. The above-mentioned references in this paragraph do not involve multiple time-varying delays which appear in both states and control inputs.

This paper investigates the robust H_∞ control problem of linear unertain systems with multiple time-varying delays. The uncertaintites are assumed to be time-varying and norm-bounded and appear in all the system matrices. The time-delays are time-varying bounded and their time derivative are less than one. The objective is to design a robust H_∞ dynamic output feedback controller which guarantees the quadratic stability of the closed-loop systems and reduces the effect of the disturbance input on the controlled output to a prescribed level. This paper is organized as follows. Section 2 states briefly the H_∞ control problem. In Sections 3 and 4, based on the LMI approach, we develop a robust dynamic output feedback control design method. Sufficient conditions for the existence of robust H_∞ controller of any order is presented in terms of LMIs. Finally, Section 5 gives conclusion and remarks.

Notations: The following notations will be used throughout this paper.

We let $\boldsymbol{R} = (-\infty, \infty)$, $\boldsymbol{R}^+ = [0, \infty)$, $\boldsymbol{R}^n$ by the real n-dimensional linear vector space over the reals with the standard Euclidean norm denoted by $\|\cdot\|$. Matrix I denotes the identity matrix of appropriate dimension. The notation $W > 0$ ($W < 0$) means that the matrix W is a symmetric positive (negative) definite matrix. Furthermore, for $\bar{d}_{max} \geq 0$, let $\mathcal{C}[-\bar{d}_{max}, 0]$ denote the *Banach space* of continuous functions φ : $[-\bar{d}_{max}, 0] \rightarrow \boldsymbol{R}^n$ with $\|\varphi\| = \sup_{-\bar{d}_{max} \leq \eta \leq 0} \|\varphi(\eta)\|$. If $x : [-\bar{d}_{max}, T) \rightarrow \boldsymbol{R}$ is continuous and $T > 0$, then for fixed $t \in [0, T)$, x_t denotes the restriction of x to the interval $[t - \bar{d}_{max}, t]$ translated to $[-\bar{d}_{max}, 0]$; *i. e.*, $x_t \in \mathcal{C}[-\bar{d}_{max}, 0]$ and $x_t(\eta) = x(t+\eta)$, $-\bar{d}_{max} \leq \eta \leq 0$.

In the sequel, note that "O.F." stands for "output feedback".

2. PROBLEM STATEMENT AND DEFINITIONS

Let $A_i(t) = A_i + \Delta A_i(t)$ for $(i = 0, 1, \ldots, m_a)$ and $B_j(t) = B_j + \Delta B_j(t)$ for $(j = 0, 1, \ldots, m_b)$, and $G(t) = G + \Delta G(t)$, $C(t) = C + \Delta C(t)$, $D(t) = D + \Delta D(t)$ and $q_{ai}(t) = x(t - d_{ai}(t))$ for $(i = 1, 2, \ldots, m_a)$ and consider the uncertain linear dynamic system with multiple time-varying delays

$$\begin{aligned} \dot{x}(t) &= A_o(t)x(t) + B_o(t)u(t) + G(t)\omega(t) \\ &\quad + \sum_{i=1}^{m_a} A_i(t)q_{ai}(t) + \sum_{j=1}^{m_b} B_j(t)u(t - d_{bj}(t)) \\ z(t) &= \begin{bmatrix} C(t)x(t) \\ D(t)u(t) \end{bmatrix} \\ y(t) &= C_o x(t) \end{aligned} \quad (1)$$

where $x(t) \in \boldsymbol{R}^n$ is the state, $u(t) \in \boldsymbol{R}^m$ is the control input, $y(t) \in \boldsymbol{R}^l$ is the measured output, $\omega(t) \in \boldsymbol{R}^p$ is the disturbance input which belongs to $L_2[0, \infty)$, $z(t) \in \boldsymbol{R}^q$ is the controlled output, A_i for $(i = 0, 1, \ldots, m_a)$, B_j for $(j = 0, 1, \ldots, m_b)$, G, C, C_o and D are known constant real matrices with appropriate dimensions. $\Delta A_i(t)$ for $(i = 0, 1, \ldots, m_a)$, $\Delta B_j(t)$ for $(j = 0, 1, \ldots, m_b)$, $\Delta G(t)$, $\Delta C(t)$, and $\Delta D(t)$ are real-valued continuous matrix functions representing time-varying parameter uncertainties in the system model, which are assumed to be of the following structure:

$$\begin{aligned} \Delta A_i(t) &= H_{ai}F(t)E_{ai}, \ \Delta G(t) = H_g F(t) E_g, \\ \Delta B_j(t) &= H_{bj}F(t)E_{bj}, \ \Delta C(t) = H_c F(t) E_c, \\ \Delta D(t) &= H_d F(t) E_d \end{aligned}$$

where H_{ai}, E_{ai} $(i = 0, 1, \ldots, m_a)$, H_{bj}, E_{bj} $(j = 0, 1, \ldots, m_b)$, H_g, H_c, H_d, E_g, E_c and E_d are known constant real matrices with appropriate dimensions, and $F(t)$ is an unknown real time-varying matrix with Lebesgue measurable elements satisfying

$$\|F(t)\| \leq 1, \ \forall t \geq 0 \quad (2)$$

The time-varying delays $d_{ai}(t) \in \boldsymbol{R}$, $(i = 1, 2, \ldots, m_a)$, $d_{bj}(t) \in \boldsymbol{R}$, $(j = 1, 2, \ldots, m_b)$, are continuous differential functions satisfying

$$\begin{aligned} 0 \leq d_{ai}(t) \leq \bar{d}_{ai}, \ \dot{d}_{ai}(t) \leq \beta_{ai} < 1 \\ 0 \leq d_{bj}(t) \leq \bar{d}_{bj}, \ \dot{d}_{bj}(t) \leq \beta_{bj} < 1 \end{aligned}$$

where $\bar{d}_{ai}$ and $\bar{d}_{bj}$ are the upper bounds of the time-varying delays $d_{ai}(t)$ and $d_{bj}(t)$ and β_{ai} and β_{bj} are the upper bounds of the time derivative of the time-varying delays $d_{ai}(t)$ and $d_{bj}(t)$, for $(i = 1, 2, \ldots, m_a)$ and $(j = 1, 2, \ldots, m_b)$ whenever it applies.

Let $\bar{d}_{max} = \max_{i,j} \{\bar{d}_{ai}, \bar{d}_{bj}\}$, $(i = 1, 2, \ldots, m_a; j = 1, 2, \ldots, m_b)$. The real-valued continuous vector initial function is given as

$$x(t) = \varphi(t), \ t \in [-\bar{d}_{max}, 0] \quad (3)$$

Assumption 1: *The triple (A_o, B_o, C_o) is stabilizable and detectable.*

Remark 1: Note that Assumption 1, which is equivalent to the stabilizability via linear dynamic output feedback of the system (1) in the absence of time-delays and parameter uncertainties, is a necessary condition for the existence of a stabilizing linear dynamic output feedback control law for the system (1).

Let the controller for system (1) be of the form of the following dynamic output feedback compensator:

$$\begin{aligned} \dot{x}_k(t) &= A_k x_k(t) + B_k y(t) \\ u(t) &= C_k x_k(t) + D_k y(t) \end{aligned} \quad (4)$$

where $A_k \in \boldsymbol{R}^{n_k x n_k}$, $B_k \in \boldsymbol{R}^{n_k x l}$, $C_k \in \boldsymbol{R}^{m x n_k}$ and $D_k \in \boldsymbol{R}^{m x l}$ and (4) is a dynamic compensator of order n_k, $0 \leq n_k \leq n$. The extreme case $n_k = 0$ represents a static gain output feedback.

By defining the initial condition of $x_k(t)$ on the interval $[-\bar{d}_{max}, 0]$ the same as that of $x(t)$ and applying the control law (4) to system (1), we have the closed-loop system

$$\dot{\hat{x}}(t) = \hat{A}_o(t)\hat{x}(t) + \sum_{j=1}^{m_b} \hat{B}_j(t)[I \ \ 0] K \hat{C}_o \hat{q}_{bj}(t)$$

$$+\hat{G}(t)\omega(t)+\sum_{i=1}^{m_a}\hat{A}_i(t)q_{ai}(t)$$

$$z(t) = \left(\hat{C}(t)+\hat{D}(t)[I\ \ 0]K\hat{C}_o\right)\hat{x}(t) \quad (5)$$

$$\hat{x}(t) = \hat{\varphi}(t),\ t\in[-\bar{d}_{max},0]$$

with $\hat{q}_{bj}(t)=\hat{x}(t-d_{bj}(t))$ and

$$\hat{A}_o(t) = \begin{bmatrix} A_o(t)+B_o(t)D_kC_o & B_o(t)C_k \\ B_kC_o & A_k \end{bmatrix},$$

$$\hat{B}_j(t) = \begin{bmatrix} B_j(t) \\ 0 \end{bmatrix},\ \hat{G}(t)=\begin{bmatrix} G(t) \\ 0 \end{bmatrix}$$

$$\hat{A}_i(t) = \begin{bmatrix} A_i(t) \\ 0 \end{bmatrix},\ \hat{C}(t)=\begin{bmatrix} C(t) & 0 \\ 0 & 0 \end{bmatrix}$$

$$\hat{x}(t) = \begin{bmatrix} x(t) \\ x_k(t) \end{bmatrix},\ \hat{\varphi}(t)=\begin{bmatrix} \varphi(t) \\ \varphi(t) \end{bmatrix},$$

$$\hat{C}_o = \begin{bmatrix} C_o & 0 \\ 0 & I \end{bmatrix}, K=\begin{bmatrix} D_k & C_k \\ B_k & A_k \end{bmatrix}$$

for $(i=1,\ldots,m_a)$ and $(j=1,\ldots,m_b)$ whenever it applies.

Note that the single variable K includs all the parameters of the controller.

The objective of this paper is to find a controller of the form (4) for system (1) such that the resulting closed-loop system (5) is quadratically stable with a given H_∞-norm constraint on disturbance attenuation for all admissible uncertainties that satisfy $\|F(t)\|\le 1$. The following definitions are firstly introduced.

Definition 1: *The closed-loop system (5) with $\omega(t)=0$ is said to be quadratically stable, if there exist matrices $\hat{P}>0$, $\hat{R}>0$ and $\hat{S}>0$ and a constant $\hat{\alpha}>0$ such that for any given admissible uncertainty $F(t)$, the Lyapunov derivative $L(\hat{x}_t,t)$ corresponding to the closed-loop system (5) and the Lyapunov-Krasovskii functional*

$$V(\hat{x}_t,t)=\hat{x}^T(t)\hat{P}\hat{x}(t)+\sum_{i=1}^{m_a}\int_{t-d_{ai}(t)}^{t}x^T(\tau)\hat{R}x(\tau)d\tau$$

$$+\sum_{j=1}^{m_b}\int_{t-d_{bj}(t)}^{t}\hat{x}^T(\tau)\hat{C}_o^TK^T[I\ \ 0]^T\hat{S}[I\ \ 0]K\hat{C}_o\hat{x}(\tau)d\tau$$

satisfies the bound

$$L(\hat{x}_t,t)=\hat{x}^T(t)\left\{\hat{A}_o^T(t)\hat{P}+\hat{P}\hat{A}_o(t)+m_a[I\ \ 0]^T\hat{R}[I\ \ 0]\right.$$
$$\left.+m_b\hat{C}_o^TK[I\ 0]^T\hat{S}[I\ 0]K\hat{C}_o\right\}\hat{x}(t)$$
$$-\sum_{i=1}^{m_a}(1-\dot{d}_{ai}(t))q_{ai}^T(t)\hat{R}q_{ai}(t)+2\sum_{i=1}^{m_a}\hat{x}^T(t)\hat{P}\hat{A}_i(t)q_{ai}(t)$$
$$-\sum_{j=1}^{m_b}(1-\dot{d}_{bj}(t))\zeta_j^T(t)\hat{S}\zeta_j(t)$$
$$+2\sum_{j=1}^{m_b}\hat{x}^T(t)\hat{P}\hat{B}_j(t)\zeta_j(t)\le-\hat{\alpha}\|\hat{x}(t)\|^2$$

for all $(t,\hat{x}_t)\in R^+\times C[-\bar{d}_{max},0]$. and $\zeta_j(t)=[I\ 0]K\hat{C}_o\hat{q}_{bj}(t)$, for $j=1,\ldots,m_b$. In this case, system (1) is said to be quadratically stabilizable by dynamic output feedback (abbreviated as QSDOF).

Definition 2: *For a given $\gamma>0$, the uncertain time-delay system described by (1) is said to be quadratically stabilizable by dynamic output feedback with an H_∞-norm bound $\gamma>0$ (abbreviated as QSDOF-H_∞-γ) if there exists a dynamic output feedback (4) such that for any admissible parameter uncertainty, the following conditions are satisfied:*

(i) *the resulting closed-loop system is quadratically stable;*

(ii) *subject to the assumption of the zero initial condition, the controlled output $z(t)$ satisfies*

$$\|z(t)\|_2\le\gamma\|\omega(t)\|_2 \quad (6)$$

In this case, the control law (4) is said to be a robust H_∞ dynamic output feedback control law for the system (1).

3. ROBUST DYNAMIC O.F. CONTROLLER

By Letting

$$\hat{R}_{ao}=m_a\hat{R}+\epsilon_{ao}^{-1}E_{ao}^TE_{ao}$$
$$\hat{R}_{ai}=(1-\beta_{ai})\hat{R}-\epsilon_{ai}^{-i}E_{ai}^TE_{ai},\ (i=1,2,\ldots,\ m_a)$$
$$\hat{S}_{bo}=m_b\hat{S}+\epsilon_{bo}^{-1}E_{bo}^TE_{bo}$$
$$\hat{S}_{bj}=(1-\beta_{bj})\hat{S}-\epsilon_{bj}^{-i}E_{bj}^TE_{bj},\ (j=1,2,\ldots,\ m_b)$$
$$[\hat{H}^T\ \ I]^T[\hat{H}^T\ \ I]=\begin{bmatrix} H^* & 0 \\ 0 & 0 \end{bmatrix}$$
$$H^*=\sum_{i=0}^{m_a}\epsilon_{ai}H_{ai}H_{ai}^T+\sum_{j=1}^{m_b}\epsilon_{bj}H_{bj}H_{bj}^T$$
$$\aleph_o=\hat{A}_o^T\hat{P}+\hat{P}\hat{A}_o+[I\ \ 0]^T\hat{R}_{ao}[I\ \ 0] \quad (7)$$
$$\aleph=\aleph_o+\hat{C}_o^TK^T\hat{B}_o^T\hat{P}+\hat{P}\hat{B}_oK\hat{C}_o+$$
$$\hat{C}_o^TK^T[I\ \ 0]^T\hat{S}_{bo}[I\ \ 0]K\hat{C}_o+\hat{P}[\hat{H}^T\ \ 0]^T[\hat{H}^T\ \ 0]\hat{P}$$

one can derive the following condition

$$L(\hat{x}_t,t)\le\hat{\xi}^T(t)\hat{Q}(t)\hat{\xi}(t) \quad (8)$$

where $\hat{Q}(t)$ is given by (17) and

$$\hat{\xi}(t)=\left[\hat{x}^T(t)\ q_{a1}^T(t)\ldots\ q_{am_a}^T(t)\ \zeta_1^T(t)\ \ldots\ \zeta_{m_b}^T(t)\right]^T$$

If the matrix $\hat{Q}(t)$ is negative definite, *i.e.* $\hat{Q}(t)<0$, then the closed-loop system (5) is quadratically stable. The existence conditions of robust dynamic output feedback controllers for (1) are obtained with the following theorem.

Theorem 1: *Consider the system (1) with $\omega(t)=0$. Let W_1 and $[W_2^T\ \ W_3^T]^T$ be any orthogonal bases of the null spaces of C_o and $[B_o^T\ \ I]$. If there exist matrices $X>0$ and $Y>0$ satisfying (18) and (19) and*

$$\begin{bmatrix} Y & I \\ I & X \end{bmatrix}\ge 0 \quad (9)$$

for some matrices $\hat{R} > 0$ and $\hat{S} > 0$ and scalars $\epsilon_{ai} > 0$ $(i = 0, 1, \ldots, m_a)$, $\epsilon_{bj} > 0$ $(j = 0, 1, \ldots, m_b)$, then there exists a robust dynamic output feedback controller for system (1), therefore the uncertain time-delay system (1) is QSDOF. Moreover, if

$$\mathrm{Rank}(I - XY) = k < n \tag{10}$$

for some matrices $X > 0$ and $Y > 0$ satisfying (9), (18) and (19), then there exists a robust dynamic output feedback controller of order k.

Remark 2: Let $C_o^{\perp}$ be the orthogonal complement of C_o. One choice of the bases of the null spaces of C_o and $[B_o^T \;\; I]$ is $W_1 = C_o^{\perp}$ and $[W_2^T \;\; W_3^T]^T = [I \;\; -B_o]^T$, respectively. When we consider system (1) with only a single time-delay in both state and input, Theorem 1 reduces to the main result of Jeung *et al.* (1996). Note that the matrix $\begin{bmatrix} C_{\perp}^T & 0 \\ 0 & I \end{bmatrix}$ in Eq. (16) of Jeung *et al.* (1996) should be $\begin{bmatrix} C_{\perp} & 0 \\ 0 & I \end{bmatrix}$.

Remark 3: If $k = Rank(I - XY) = 0$, then we set $\hat{P} = X > 0$. Otherwise, using $UV^T = I - XY$ and $\hat{P}\begin{bmatrix} Y & I \\ V^T & 0 \end{bmatrix} = \begin{bmatrix} I & X \\ 0 & U^T \end{bmatrix}$, $V \in \mathbf{R}^{nxn_k}$, $U \in \mathbf{R}^{nxn_k}$ we obtain the unique solution $\hat{P}$. It should be noted that the existence of a solution is assured if $Y > 0$ and U has full column rank (Gahinet and Apkarian, 1994). The solution $\hat{P}$ should satisfy

$$\hat{Q}_o^*(t) + \Sigma^T K^T \Gamma + \Gamma K \Sigma \; < \; 0 \tag{11}$$

which is an LMI in the control parameter K with

$$\Sigma := [\hat{C}_o \; 0 \; \ldots \; 0] \tag{12}$$

$$\Gamma := [\hat{B}_o^T \; 0 \; \ldots \; 0 \; [I \;\; 0]^T]\mathrm{diag}(\hat{P}, I, \ldots, I) \tag{13}$$

Once $\hat{P}$ is obtained, one can numerically synthesize a robust dynamic output controller in the form (4) for given $\hat{P}$ by applying a convex optimization algorithm. Alternatively, one can use more efficient and explicit schemes. All control parameters satisfying (4) can be parameterized explicitly as in Iwasaki and Skelton (1994) and Gahinet and Apkarian, (1994). Gahinet (1996) has given an improved scheme by which the controller formulas are easy to analyze and particularly well suited for efficient and numerically stable implementation.

4. ROBUST H_∞ DYNAMIC O.F. CONTROLLER

In this section, we give the result stating the QSDOF-$H_\infty - \gamma$ for system (1).

Theorem 2: *Let W_1 and $[W_2^T \;\; W_3^T]^T$ be any orthogonal bases of the null spaces of C_o and $[B_o^T \;\; I]$. For a given constant $\gamma > 0$, the system (1) is QSDOF-H_∞-γ, If there exist two matrices $X > 0$ and $Y > 0$ satisfying LMI's conditions given by (20) and (21) and*

$$\begin{bmatrix} Y & I \\ I & X \end{bmatrix} \geq 0 \tag{14}$$

for appropriate scalars $\epsilon_{ai} > 0$ for $(i = 0, 1, \ldots, m_a)$, $\epsilon_{bj} > 0$ for $(j = 0, 1, \ldots, m_b)$, $\epsilon_g > 0$, $\epsilon_c > 0$, $\epsilon_d > 0$ such that $(\epsilon_c^{-1} I - H_c^T H_c) > 0$ and $(\epsilon_d^{-1} I - H_d^T H_d) > 0$ and symmetric positive definite matrices $\hat{R} > 0$ and $\hat{S} > 0$. Moreover, if

$$\mathrm{Rank}(I - XY) = k < n \tag{15}$$

for some matrices $X > 0$ and $Y > 0$ satisfying (14), (20) and (21), then there exists a robust H_∞ dynamic output feedback controller of order k.

Remark 4: Note that Theorem 2 provides a condition for dynamic output feedback stabilization with an H_∞-norm bound γ of system (1) in terms of the solvability of linear matrix inequalities and does not present the computation of the robust controller itself, but the existence of a robust controller. The computation of a robust dynamic output feedback controller can be carried out using a procedure similar to that of the controller design in Remark 3. The proposed stabilization method has the advantage that it can be implemented numerically very efficiently using recent developed algorithms for solving LMIs, see, *e.g.* Boyd *et al.* (1994) and Nesterov and Nemirovsky (1994).

5. CONCLUSION AND REMARKS

This paper has investigated the robust H_∞ synthesis problem for time-varying uncertain linear systems with multiple time-varying delays under the specific framework of quadratic stabilizability. As a particular approach of the Lyapunov-type design, it may be conservative for robust analysis and synthesis since a fixed deterministic Lyapunov-Krasovskii functional is needed to account for all uncertainties. Nevertheless, the technique is effective for dealing with time-varying parameter uncertainty.

We have presented sufficient conditions for the existence of robust H_∞ dynamic output feedback controller of any order in terms of three LMIs. The control parameters can be parameterized explicitly in state space. The proposed method is an extension of the work of Gahinet and Apkarian (1994) and Iwasaki and Skelton (1994) to the time-varying uncertain linear systems with multiple time-varying delays. The robust H_∞ dynamic output feedback controller can been constructed using the corresponding feasible solution of the LMIs.

ACKNOWLEDGEMENTS. This work was supported by La Région Poitou Charentes, France (Grant No. 97/RPC-R-137).

REFERENCES

Boyd, S., L. El-Ghaoui, , E. Feron and V. Balakrishnan (1994). *Linear Matrix Inequalities in System and Control Theory.* SIAM studies in Applied Mathematics, vol. 15, SIAM, PA.

Choi, H. H. and M. J. Chung (1995). Memoryless H_∞ controller design for linear systems with delayed state and control. *Automatica*, **31**, 919-919; (1996). Observer-based H_∞ controller design for state delayed linear systems. *Automatica*, **32**, 1073-1075; (1997). An LMI approach to H_∞ controller design for linear time delay systems. *Automatica*, **33**, 737-739.

Francis, B. A. (1987). *A Course in H_∞ Control.* Berlin: Springer.

Gahinet, P. (1996). Explicit controller formulas for LMI-based H_∞ synthesis. *Automatica*, **32**, 1007-1014.

Gahinet, P. and A. J. Apkarian (1994). A linear matrix inequality approach to H_∞ control. *Int. J. Robust Nonlinear Control*, **4**, 421-448.

Ge, J.-H., P. M. Frank and C.-F. Lin (1996). Robust H_∞ state feedback control for linear systems with state delay and parameter uncertainty. *Automatica*, **32**, 1183-1185.

Green, M. and D. J. Limbeer (1995). *Linear Robust Control.* Prentice-Hall, Englewood Cliffs, NJ.

Gu, K. (1994). H_∞ control of systems under norm bounded uncertainties in all systems matrices. *IEEE Trans. on Automatic Control*, **AC-39**, 1320-1322.

Hale, H. and S. M. Verduyn Lunel (1991). Introduction to Functional Differential Equations. Applied Math. Sciences, Vol. 99, New York: Springer-Verlag.

Iwasaki, T. and R. E. Skelton (1994). All controller for the general H_∞ control problem: LMI existence conditions and state-space formulas. *Automatica*, **30**, 1307-1317.

Jeung, E. T., D. C. Oh, J. H. Kim and H. B. Park (1996). Robust controller design for uncertain systems with time delays: LMI approach. *Automatica*, **32**, 1229-1231.

Kimura, H. (1997). *Chain-Scattering Approach to H_∞-Control.* Birkhauser, Boston.

Kokame, H., H. Kobayashi and T. Mori (1998). Robust H_∞ performance for linear delay- differential systems with time-varying uncertainties. *IEEE Trans. on Automatic Control*, **AC-43**, 223-226.

Lee, J. H., S. W. Kim and W. H. Kwon (1994). Memoryless H_∞ controller for state delayed systems.*IEEE Trans. on Automatic Control*, **AC-39**, 159-162.

Li, X. and C. E. de Souza (1997a). Delay-dependent robust stability and stabilization of uncertain linear delay systems: A linear matrix inequality approach. *IEEE Trans. on Automatic Control*, **AC-42**, 1144-1148; (1997b). Criteria for robust stability and stabilization of uncertain linear systems with state delay. *Automatica*, **33**, 1657-1662.

Niculescu, S.-I. (1998). H_∞ memoryless control with an α-stability constraint for time-delay systems: an LMI approach. *IEEE Trans. on Automatic Control*, **AC-43**, 739-743.

Peterson, I. R. (1987a). Disturbance attenuation and H_∞ optimization: a design method based on the algebraic Riccati equation. *IEEE Trans. on Automatic Control*, **AC-32**, 427-429; (1987b). A stabilization algorithm for a class of uncertain linear systems. *Systems & Control Letters*, **8**, 351-357.

Xie, L. and C. E. de Souza (1992a). Robust H_∞ Control for linear systems with norm-bounded time-varying uncertainty. *IEEE Trans. on Automatic Control*, **AC-37**, 1188- 1191.

Xie L., M. Fu, and C. E. de Souza (1992b). H_∞ Control and quadratic stabilization of systems with parameter uncertainty via output feedback. *IEEE Trans. on Automatic Control*, **AC-37**, 1253-1256.

Yu, L., J. Chu and H. Su (1996). Robust memoryless H_∞ controller design for linear time- delay systems with norm-bounded time-varying uncertainty. *Automatica*, **32**, 1759-1762.

Appendix

$$\hat{Q}_o^*(t) = \begin{bmatrix} \aleph_o & \hat{P}\hat{A}_1 & \dots & \hat{P}\hat{A}_{m_a} & \hat{P}\hat{B}_1 & \dots & \hat{P}\hat{B}_{m_b} & \hat{P}[\hat{H}^T\ 0]^T & 0 \\ \hat{A}_1^T\hat{P} & -\hat{R}_{a1} & \dots & 0 & 0 & \dots & 0 & 0 & 0 \\ \vdots & \vdots & \vdots & \vdots & \vdots & \vdots & \vdots & \vdots & \vdots \\ \hat{A}_{m_a}^T\hat{P} & 0 & \dots & -\hat{R}_{am_a} & 0 & \dots & 0 & 0 & 0 \\ \hat{B}_1^T\hat{P} & 0 & \dots & 0 & -\hat{S}_{b1} & \dots & 0 & 0 & 0 \\ \vdots & \vdots & \vdots & \vdots & \vdots & \vdots & \vdots & \vdots & \vdots \\ \hat{B}_{m_b}^T\hat{P} & 0 & \dots & 0 & 0 & \dots & -\hat{S}_{bm_b} & 0 & 0 \\ [\hat{H}^T\ 0]\hat{P} & 0 & \dots & 0 & 0 & \dots & 0 & -I & 0 \\ 0 & 0 & \dots & 0 & 0 & \dots & 0 & 0 & -\hat{S}_{bo}^{-1} \end{bmatrix} \quad (16)$$

$$\hat{Q}(t) = \quad (17)$$
$$\begin{bmatrix} \aleph & \hat{P}\hat{A}_1 & \dots & \hat{P}\hat{A}_{m_a} & \hat{P}\hat{B}_1 & \dots & \hat{P}\hat{B}_{m_b} \\ \hat{A}_1^T\hat{P} & -\hat{R}_{a1} & \dots & 0 & 0 & \dots & 0 \\ \vdots & \vdots & \vdots & \vdots & \vdots & \dots & \vdots \\ \hat{A}_{m_a}^T\hat{P} & 0 & \dots & -\hat{R}_{am_a} & 0 & \dots & 0 \\ \hat{B}_1^T\hat{P} & 0 & \dots & 0 & -\hat{S}_{b1} & \dots & 0 \\ \vdots & \vdots & \vdots & \vdots & \vdots & \dots & \vdots \\ \hat{B}_{m_b}^T\hat{P} & 0 & \dots & 0 & 0 & \dots & -\hat{S}_{bm_b} \end{bmatrix}$$

$$W_4 = \left[\begin{array}{c|c} W_1 & 0 \\ \hline 0 & I \end{array}\right], \quad W_5 = \left[\begin{array}{c|c} W_2 & 0 \\ W_3 & 0 \\ \hline 0 & I \end{array}\right],$$

$$W_g = -\gamma^2 I + \epsilon_g^{-1} E_g^T E_g, \quad W_y = Y A_o^T + A_o Y, \quad \hat{R}_{wao} = \hat{R}_{ao} + C^T H_c \left(\epsilon_c^{-1} I - H_c^T H_c\right)^{-1} H_c^T C + \epsilon_c^{-1} E_c^T E_c$$

$$W_x = A_o^T X + X A_o + \hat{R}_{ao}, \quad [\hat{H}_\omega^T \;\; 0]^T [\hat{H}_\omega^T \;\; 0] = [\hat{H}^T \;\; 0]^T [\hat{H}^T \;\; 0] + \begin{bmatrix} \epsilon_g H_g H_g^T & 0 \\ 0 & 0 \end{bmatrix}$$

$$\hat{S}_{\omega bo} = \hat{S}_{bo} + D^T (I + H_d(\epsilon_d^{-1} I - H_d^T H_d)^{-1} H_d^T) D + \epsilon_d^{-1} E_d^T E_d, \quad W_{x\omega} = A_o^T X + X A_o + \hat{R}_{wao}$$

$$W_4^T \left[\begin{array}{c|ccccccc} W_x & XA_1 & \cdots & XA_{m_a} & XB_1 & \cdots & XB_{m_b} & XH \\ \hline A_1^T X & -\hat{R}_{a1} & \cdots & 0 & 0 & \cdots & 0 & 0 \\ \vdots & \vdots & \vdots & \vdots & \vdots & \vdots & \vdots & \vdots \\ A_{m_a}^T X & 0 & \cdots & -\hat{R}_{am_a} & 0 & \cdots & 0 & 0 \\ B_1^T X & 0 & \cdots & 0 & -\hat{S}_{b1} & \cdots & 0 & 0 \\ \vdots & \vdots & \vdots & \vdots & \vdots & \vdots & \vdots & \vdots \\ B_{m_b}^T X & 0 & \cdots & 0 & 0 & \cdots & -\hat{S}_{bm_b} & 0 \\ H^T X & 0 & \cdots & 0 & 0 & \cdots & 0 & -I \end{array}\right] W_4 < 0 \quad (18)$$

$$W_5^T \left[\begin{array}{cc|cccccccc} W_y & 0 & Y & A_1 & \cdots & A_{m_a} & B_1 & \cdots & B_{m_b} & H \\ 0 & -\hat{S}_{bo}^{-1} & 0 & 0 & \cdots & 0 & 0 & \cdots & 0 & 0 \\ \hline Y & 0 & -\hat{R}_{ao}^{-1} & 0 & \cdots & 0 & \cdots & 0 & 0 & 0 \\ A_1^T & 0 & 0 & -\hat{R}_{a1} & \cdots & 0 & \cdots & 0 & 0 & 0 \\ \vdots & \vdots & \vdots & \vdots & \vdots & \vdots & \vdots & \vdots & \vdots & \vdots \\ A_{m_a}^T & 0 & 0 & 0 & \cdots & -\hat{R}_{am_a} & 0 & \cdots & 0 & 0 \\ B_1^T & 0 & 0 & 0 & \cdots & 0 & -\hat{S}_{b1} & \cdots & 0 & 0 \\ \vdots & \vdots & \vdots & \vdots & \vdots & \vdots & \vdots & \vdots & \vdots & \vdots \\ B_{m_b}^T & 0 & 0 & 0 & \cdots & 0 & 0 & \cdots & -\hat{S}_{bm_b} & 0 \\ H^T & 0 & 0 & 0 & \cdots & 0 & 0 & \cdots & 0 & -I \end{array}\right] W_5 < 0 \quad (19)$$

$$W_4^T \left[\begin{array}{c|ccccccccc} W_{x\omega} & XA_1 & \cdots & XA_{m_a} & XB_1 & \cdots & XB_{m_b} & C^T & XG & XH_\omega \\ \hline A_1^T X & -\hat{R}_{a1} & \cdots & 0 & 0 & \cdots & 0 & 0 & 0 & 0 \\ \vdots & \vdots & \vdots & \vdots & \vdots & \vdots & \vdots & \vdots & \vdots & \vdots \\ A_{m_a}^T X & 0 & \cdots & -\hat{R}_{am_a} & 0 & \cdots & 0 & 0 & 0 & 0 \\ B_1^T X & 0 & \cdots & 0 & -\hat{S}_{b1} & \cdots & 0 & 0 & 0 & 0 \\ \vdots & \vdots & \vdots & \vdots & \vdots & \vdots & \vdots & \vdots & \vdots & \vdots \\ B_{m_b}^T X & 0 & \cdots & 0 & 0 & \cdots & -\hat{S}_{bm_b} & 0 & 0 & 0 \\ C & 0 & \cdots & 0 & 0 & \cdots & 0 & -I & 0 & 0 \\ G^T X & 0 & \cdots & 0 & 0 & \cdots & 0 & 0 & W_g & 0 \\ H_\omega^T X & 0 & \cdots & 0 & 0 & \cdots & 0 & 0 & 0 & -I \end{array}\right] W_4 < 0 \quad (20)$$

$$W_5^T \left[\begin{array}{cc|ccccccccccc} W_y & 0 & Y & A_1 & \cdots & A_{m_a} & B_1 & \cdots & B_{m_b} & YC^T & G & H_\omega \\ 0 & -\hat{S}_{\omega bo}^{-1} & 0 & 0 & \cdots & 0 & 0 & \cdots & 0 & 0 & 0 & 0 \\ \hline Y & 0 & -\hat{R}_{wao}^{-1} & 0 & \cdots & 0 & 0 & \cdots & 0 & 0 & 0 & 0 \\ A_1^T & 0 & 0 & -\hat{R}_{a1} & \cdots & 0 & 0 & \cdots & 0 & 0 & 0 & 0 \\ \vdots & \vdots & \vdots & \vdots & \vdots & \vdots & \vdots & \vdots & \vdots & \vdots & \vdots & \vdots \\ A_{m_a}^T & 0 & 0 & 0 & \cdots & -\hat{R}_{am_a} & 0 & \cdots & 0 & 0 & 0 & 0 \\ B_1^T & 0 & 0 & 0 & \cdots & 0 & -\hat{S}_{b1} & \cdots & 0 & 0 & 0 & 0 \\ \vdots & \vdots & \vdots & \vdots & \vdots & \vdots & \vdots & \vdots & \vdots & \vdots & \vdots & \vdots \\ B_{m_b}^T & 0 & 0 & 0 & \cdots & 0 & 0 & \cdots & -\hat{S}_{bm_b} & 0 & 0 & 0 \\ CY & 0 & 0 & 0 & \cdots & 0 & 0 & \cdots & 0 & -I & 0 & 0 \\ G^T & 0 & 0 & 0 & \cdots & 0 & 0 & \cdots & 0 & 0 & W_g & 0 \\ H_\omega^T & 0 & 0 & 0 & \cdots & 0 & 0 & \cdots & 0 & 0 & 0 & -I \end{array}\right] W_5 < 0 \quad (21)$$

CONTROL OF INTEGRATOR PROCESSES WITH LONG DEAD-TIME

Yu-Chu Tian[1]

Institute of Industrial Process Control, Zhejiang Univ, Hangzhou 310027, P R China

Furong Gao

Dept of Chem Eng, Hong Kong Univ of Sci & Tech, Kowloon, Hong Kong

Moses O. Tadé and Jinyu Tang

School of Chem Eng, Curtin Univ of Tech, GPO Box U 1987, Perth, WA 6845, Australia

Abstract: A novel control scheme with a simple design procedure is proposed for integrator processes with long dead-time. This scheme contains three controllers configured for different objectives. With this scheme, the set-point response and load response of the closed-loop system are separated from each other, and can then be independently designed to simultaneously obtain fast set-point tracking and efficient load rejection. Simulations show that the proposed scheme also has good robustness to variations in process dynamics compared to the existing schemes. *Copyright © 1999 IFAC*

Keywords: integrators, time delay, process control, control schemes, control design.

1. INTRODUCTION

Non-self-regulating integrator processes are common in chemical industry. A typical example of such processes is a liquid-level system with a pump attached to the outflow line. The initial stage responses of the processes with large time constant to a step change are similar to those of an integrator process with dead-time. This type of process can thus be approximated by an integrator plus dead-time model for control purpose. Chien and Fruehauf (1990) showed, through the control of an industrial distillation column, that this approximation gives good closed-loop system performance.

The control of integrator processes is thus important. The conventional proportional-integrator-derivative (PID) control gives a satisfactory performance for an integrator process without dead-time or with small dead-time, but the control of an integrator process with long dead-time is difficult.

The Smith predictor is commonly used for compensating dead-time (Smith, 1957). But it cannot be applied to an integrator process with long dead-time as it results in a steady-state offset for a load disturbance (Hang and Wong, 1979; Watannabe and Ito, 1981). Watanabe and Ito (1981) modified the Smith predictor to remove this steady-state offset.

[1] Current address: School of Chemical engineering, Curtin University of Technology, GPO Box U 1987, Perth, WA 6845, Australia. Phone: +61-8-9266 3776. Fax: +61-8-9266 3554. E-mail: tiany@che.curtin.edu.au.

The resulting set-point response is, however, too slow, and the steady-state offset cannot be eliminated if there is an estimating error in the process dead-time (Astrom, Hang and Lim, 1994; Matausek and Micic, 1996; Zhang znd Sun, 1996). Astrom, Hang and Lim (AHL) (1994) proposed a new Smith predictor to overcome these drawbacks. It allows a separate design of the closed-loop set-point and load responses. The controller tuning of the scheme is, however, rather complicated and there lacks a systematic tuning method. This motivated Matausek and Micic (MM) (1996) and Zhang and Sun (ZS) (1996) to extend the AHL scheme. Both MM and ZS schemes contain clear design procedures. MM scheme has a simple structure with only two proportional controllers. However, it has lost the separation nature of the closed-loop set-point and load responses. ZS scheme retains the good nature of the AHL scheme, allowing a separate design of the closed-loop set-point and load responses. But it contains a positive feedback loop that is a potential instability source, leading to limited system robustness, as will be demonstrated later.

This paper proposes a novel control scheme with a clear design procedure for integrator processes with long dead-time. The scheme contains three controllers and retains the separation nature of the closed-loop set-point and load responses. It will shown to have has fast set-point tracking, efficient load rejection, and good robustness.

2. PRELIMINARIES

Consider the proportional control of an integrator process with delay, as shown in figure 1. The symbols R_e and L are set-point and load disturbance, respectively. The process dynamics are governed by an integrator-plus-delay transfer function, $P(s) = G_p(s)\, e^{-ds}$, where $G_p = K_p / s$, d is time delay. y is process output. From figure 1, we have the closed-loop transfer function $H_{0r}(s)$ relating R_e to y as:

$$H_{0r}(s) = \frac{Y(s)}{R_e(s)} = \frac{(K_0 K_p / s)e^{-ds}}{1 + \dfrac{K_0 K_p}{s} e^{-ds}} = G_{0r}(s)e^{-ds}, \quad (1)$$

and the closed-loop transfer function $H_{0l}(s)$ relating L to y as:

$$H_{0l}(s) = \frac{Y(s)}{L(s)} = \frac{(K_p / s)e^{-ds}}{1 + \dfrac{K_0 K_p}{s} e^{-ds}} = G_{0l}(s)e^{-ds}. \quad (2)$$

It is clear that the original unstable pole $s = 0$ has been removed. If the closed-loop system is stable, then $H_{0r}(s) \to 1$ and $H_{0l}(s) \to 1/K_0$ as $s \to 0$, implying that the proportional controller K_0 can not eliminate steady-state offset.

The characteristic equation of the closed-loop system can be written as:

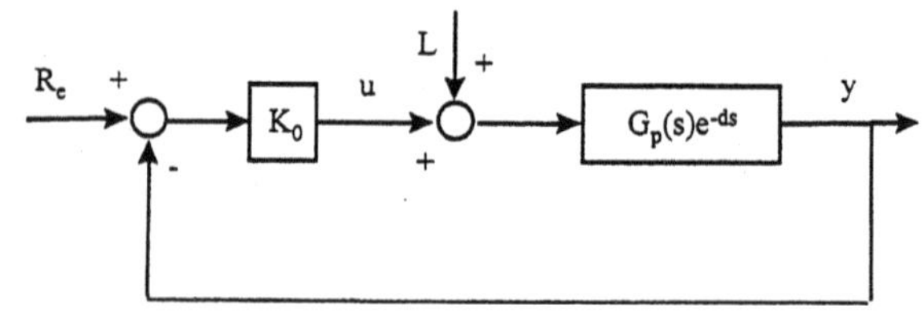

Figure 1. Proportional control of an integrator process with delay.

$$1 + W_0(s) = 0, W_0(s) = \frac{K_0 K_p}{s} e^{-ds}. \quad (3)$$

This characteristic equation determines the stability of the closed-loop system. From this equation, the phase margin, Φ_0, of the control system can be obtained from the following relations:

$$|W_0(s)| = K_0 K_p / \omega = 1, \quad (4a)$$

$$\Phi_0 = \pi + \arg\{W_0(s)\} = \pi / 2 - d\omega. \quad (4b)$$

Letting $\Phi_0 = 0$ yields the ultimate values, K_{0u}, of K_0. From the above equation, it follows:

$$K_{0u} = \pi / (2K_p d). \quad (5)$$

K_0 is recommended to be:

$$K_0 = 1/(2K_p d), \quad (6)$$

as in Matausek and Micic (1996), the resulting phase margin $\Phi_0 \approx 61^\circ$. In general, the value of K_0 decreases as Φ_0 increases. A larger Φ_0 gives better system stability but slower responses in set-point and load. In practical computation of K_0, the unknown parameters K_p and d should be respectively replaced by their estimates K_p^* and d^* in equation (6).

The applications of the above results will be discussed after the new control scheme is introduced.

3. CONTROL SCHEME

The proposed control scheme for integrator processes with long dead-time is shown in figure 2. In addition to the notations of figure 1, R denotes set-point in this figure. The scheme contains three separate controllers for different objectives. The basic proportional feedback controller K_0 discussed in the last section is employed to pre-stabilize the open-loop unstable integrator process. The controller $G_r(s)$ and is configured for set-point tracking. The controller $G_c(s)$ is used for load rejection. The three controllers, K_0, $G_r(s)$ and $G_c(s)$, are therefore referred to as pre-stabilizer, set-point controller and load controller, respectively. The scheme also contains a pre-stabilized process model, $G_{0r}^*(s)e^{-d^*s}$, an estimate of $G_{0r}(s)e^{-ds}$ in equation (1). The superscript * in figure 2 indicates that the corresponding parameter (or variable) is an estimate.

Figure 2 can be simplified into figure 3 using equations (1) and (2). The original integrator process, the pre-stabilizer and the corresponding unity feedback loop inside the dotted line box in figure 2 can then be treated as an extended process once K_0 is determined. This extended process is referred to as the pre-stabilized process here. The transfer functions of the pre-stabilized process in set-point path and in load path are described by equations (1) and (2), respectively. The set-point controller, $G_r(s)$, and the load controller, $G_c(s)$, are then synthesized for the pre-stabilized process. This explains why the pre-stabilized process model, $G_{0r}^*(s)e^{-d^*s}$, is used for set-point tracking in our control scheme.

From figure 3, we have the transfer function from R to y as:

$$H_r(s) = \frac{Y(s)}{R(s)}$$
$$= \frac{G_r(s)G_{0r}(s)e^{-ds}}{1+G_r(s)G_{0r}^*(s)} \cdot \frac{1+G_c(s)G_{0r}^*(s)e^{-d^*s}}{1+G_c(s)G_{0r}(s)e^{-ds}}, \quad (7)$$

and the transfer function from L and y as:

$$H_l(s) = \frac{Y(s)}{L(s)} = \frac{G_{0l}(s)}{1+G_c(s)G_{0r}(s)e^{-ds}}e^{-ds}. \quad (8)$$

It is seen from equation (8) that $H_l(s)$ is determined by the load controller, $G_c(s)$, and is independent of the set-point controller, $G_r(s)$. Equation (7) shows that $H_r(s)$ depends on $G_c(s)$ as well as $G_r(s)$. If the process dynamics are well modeled, i.e., $K_p^* \approx K_p$, $d^* \approx d$ and consequently $G_{0r}^*(s) \approx G_{0r}(s)$, equation (7) can be simplified into:

$$H_r(s) = \frac{Y(s)}{R(s)} \approx \frac{G_r(s)G_{0r}^*(s)}{1+G_r(s)G_{0r}^*(s)}e^{-ds}. \quad (9)$$

$H_r(s)$ thus becomes independent of $G_c(s)$, and is determined only by the set-point controller, $G_r(s)$.

An important feature of the proposed scheme can be drawn. The set-point and load responses of the control system are decoupled from each other with a good process model. The set-point and load controllers can thus be designed separately. With a well tuned pre-stabilizer, a good set-point tracking can be obtained by designing the set-point controller and a good load rejection performance can be gotten by synthesizing the load controller.

4. CONTROLLER DESIGN AND TUNING

The tuning of the pre-stabilizer K_0 is considered first. It is seen that the part inside the dotted line box in figure 2 is the same as figure 1. According to the discussions in section 2 for figure 1, K_0 can be computed from the relations of equation (4) for a given Φ_0. A relatively large Φ_0 can be chosen in the scheme since K_0 addresses the pre-stabilization of the unstable integrator process. K_0 is recommended to be computed from equation (6) with $\Phi_0 \approx 61°$.

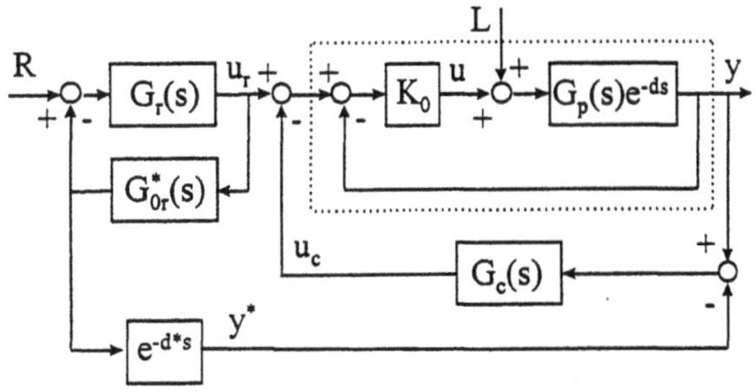

Figure 2. The proposed control scheme.

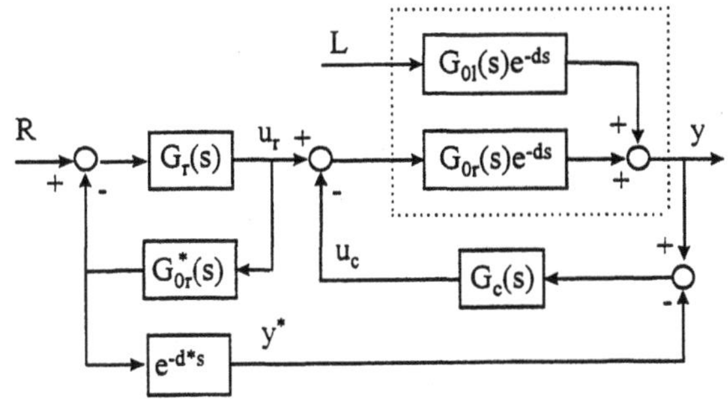

Figure 3. A simplified block diagram of figure 2.

The design of the set-point controller is simple and straightforward. The direct synthesis method is adopted to obtain a desired set-point response. Taking into account equation (9), suppose that the desired transfer function from R to y, $H_{rd}(s)$, is a first-order plus delay form of:

$$H_{rd}(s) = G_{rd}(s)e^{-ds} = \frac{1}{T_{re}s+1}e^{-ds}, \quad (10)$$

where T_{re} is the desired time constant. The set-point controller can thus be designed as for a process without delay. Letting $H_r(s)$ of equation (9) equal to $H_{rd}(s)$ of equation (10) results in:

$$G_r(s) = \frac{1}{G_{0r}^*} \cdot \frac{G_{rd}(s)}{1-G_{rd}(s)}. \quad (11)$$

Taking into account equations (1) and (6) and replacing the unknown parameters K_p and d respectively by their estimates, K_p^* and d^*, it follows that:

$$G_r(s) = \frac{2d^*}{T_{re}}\left(1+\frac{1}{2d^*s}e^{-d^*s}\right). \quad (12)$$

The resulting set-point controller, $G_r(s)$, can be viewed as a nonlinear PI controller with a controller gain of $K_r = 2d^*/T_{re}$ and a variable integral time of $T_{ri} = 2d^*/e^{-d^*s}$. It is seen that the integral action of the controller works for low-frequency signals and vanishes for high-frequency signals.

The design of the load controller, $G_c(s)$, is now considered. Similar to the set-point controller, $G_r(s)$, a nonlinear PID control law is proposed for $G_c(s)$:

$$G_c(s) = K_c\left(1+\frac{1}{T_{ci}s}\right)(1+T_{cd}s), \quad (13)$$

Substituting this control law and the expressions of $G_{0l}(s)$ and $G_{0r}(s)$ into equation (8) yields:

$$H_l(s) = \frac{(K_p/s)e^{-ds}}{\left(1+\frac{K_0K_p}{s}e^{-ds}\right)+\frac{K_cK_0K_p}{s}\left(1+\frac{1}{T_{ci}s}\right)(1+T_{cd}s)e^{-ds}} \quad (14)$$

If the integral time T_{ci} is designed to be:

$$T_{ci} = \frac{1}{K_0K_pe^{-ds}} = 2d/e^{-ds}, \quad (15)$$

equation (14) can then be simplified into:

$$H_l(s) = \frac{(K_p/s)e^{-ds}}{\left(1+\frac{K_0K_p}{s}e^{-ds}\right)\left[1+\frac{K_cK_0K_p}{s}(1+T_{cd}s)e^{-ds}\right]} \quad (16)$$

Two controller parameters K_c and T_{cd} are tuned based on the analysis of the characteristic equation of the closed-loop transfer function (16). The similar idea can be found in Matausek and Micic (1996). This characteristic equation is expressed by:

$$\left(1+\frac{K_0K_p}{s}e^{-ds}\right)\left[1+\frac{K_cK_0K_p}{s}(1+T_{cd}s)e^{-ds}\right] = 0. \quad (17)$$

The first factor of the left hand of the characteristic equation has been analyzed in section 2 (preliminaries) and is independent of the controller parameters K_c and T_{cd}. The remaining factor of the characteristic equation reads:

$$1+W_c(s) = 0,\ W_c(s) = \frac{K_cK_0K_p}{s}(1+T_{cd}s)e^{-ds}. \quad (18)$$

Consider a unity feedback system with the transfer function $W_c(s)$ in its forward path. The characteristic equation of this imaginary system is expressed by equation (18). The corresponding phase margin Φ_c, rather than the phase margin of the control system in figure 2 or 3, is determined by:

$$|W_c(j\omega)| = \frac{K_cK_0K_p\sqrt{1+(T_{cd}\omega)^2}}{\omega} = 1, \quad (19a)$$

$$\Phi_c = \pi + \arg\{W_c(j\omega)\} = \frac{\pi}{2} + \tan^{-1}(T_{cd}\omega) - d\omega. \quad (19b)$$

There are three unknown parameters in the relations of equation (19), i.e., K_c, T_{cd} and ω. These parameters cannot be derived from the above two independent equations. In this paper, a heuristical relation is proposed for determining the derivative time T_{cd} as:

$$T_{cd} = a\,d^*, \quad (20)$$

where α is a constant coefficient. For any given derivative time T_{cd}, the controller gain K_c can be obtained from the relations of equation (19) with a specified phase margin Φ_c. This papers recommends taking α = 1/3 ~ 1/2 and $\Phi_c = 60°$ for dominant time delay processes. For non-dominant time delay cases, larger values of α and Φ_c are suggested.

In summary, the controller design procedure can be divided into the following three steps:

1). Determine K_0 by equation (6) to pre-stabilize the process;
2) Determine $G_r(s)$, a nonlinear PI controller, by equation (12) with a desired set-point tracking performance; and
3). Determine $G_c(s)$, a nonlinear PID controller, by equations (13), (15), (19) and (20) for load rejection, where α = 1/3 ~ 1/2 and Φ_c = 60° are recommended for dominant time delay processes.

Steps 2) and 3) are independent of each other. Note that the estimates of the system parameters rather than the unavailable actual values are used in the design procedure.

With the designed pre-stabilizer K_0, the set-point controller $G_r(s)$ and the load controller $G_c(s)$, we have $H_r(s) \to 1$ and $H_l(s) \to 0$ as $s \to 0$. These relations imply that the closed-loop system will track the set-point without steady offset for constant changes in set-point and load.

5. NUMERICAL EXPERIMENTS

5.1 Process Descriptions and Controller Settings

To illustrate the design procedure and the effectiveness of the proposed control scheme, consider the following two integrator processes:

$$P_1(s) = \frac{1}{s}e^{-5s}, \quad (21)$$

$$P_2(s) = \frac{1}{s(s+1)(0.5s+1)(0.2s+1)(0.1s+1)}e^{-5s}. \quad (22)$$

The first process (21) that has been studied in Astrom, Hang and Lim (1994), Matausek and Micic (1996), and Zhang and Sun (1996) is adopted so that the simulation results could be compared based on the same standard. The second process (22) was introduced by Matausek and Micic (1996) to test the applicability of their control scheme to high-order integrator processes. The process is approximated by a pure integrator plus delay model in controller design, where the estimate of the model time delay is 6.7.

The controller settings and the corresponding Φ_c for the above two processes are shown in Table 1, where the values of T_{re} are taken as in Matausek and Micic (1996) and Zhang and Sun (1996). For the second process with high-order dynamics, simulations show that the satisfactory value of the adjustable parameter λ_2 of ZS scheme is about 5.

In the simulations, a unit step set-point change is introduced at time t = 0, and a step load change L = −0.1 is introduced at time t = 70, as in Astrom, Hang and Lim (1994), Matausek and Micic (1996), and

Table 1. Controller settings.

$G_p^* e^{-d^* s}$	K_0	$G_{0r}^*(s)$	T_{re}	$G_r(s)$	$G_c(s)$	Figures
$\frac{1}{s}e^{-5s}$	$\frac{1}{10}$	$\frac{1}{10s+e^{-5s}}$	$\frac{1}{0.6}$	$6\left(1+\frac{1}{10s}e^{-5s}\right)$	$1.76\left(1+\frac{1}{10s}e^{-5s}\right)(1+2.5s)$	4 - 7
$\frac{1}{s}e^{-6.7s}$	$\frac{1}{13.4}$	$\frac{1}{13.4s+e^{-6.7s}}$	1.7	$\frac{13.4}{1.7}\left(1+\frac{1}{13.4s}e^{-6.7s}\right)$	$1.68\left(1+\frac{1}{13.4s}e^{-6.7s}\right)(1+3s)$	8 - 10

Zhang and Sun (1996). Three control schemes, i.e., MM scheme (Matausek and Micic, 1996), ZS scheme (Zhang and Sun, 1996) and the proposed scheme, are compared in terms of performances of set-point tracking, load rejection and robustness. Matlab/Simulink software is used in the simulations.

5.2 The First Process

The simulation results of the first process (21) are shown in figures 4 through 7. With a good process model, figure 4 shows that all three control schemes have almost the same set-point tracking response and similar load responses. Due to the separation nature of the set-point tracking response and the load rejection response, these two responses can be optimized simultaneously with the proposed scheme and ZS scheme, but not with MM scheme.

In addition to fast set-point tracking and efficient load rejection, the proposed scheme also has good robustness compared to the other two schemes. Figure 5 shows the responses of the three schemes for the first process with a –25% estimating error in time delay, i.e., $d^* = 5$ and $d = 6.25$. It is seen that all three schemes have similar load responses, while MM and ZS schemes are more oscillatory than the proposed scheme in set-point tracking.

Figure 6 depicts the responses of the three schemes for the first process with a 25% estimating error in time delay, i.e., $d^* = 5$ and $d = 3.75$. ZS scheme behaves with large oscillations. The performance of MM scheme is acceptable in spite of its slightly oscillatory set-point tracking response. The proposed scheme outperforms the other two schemes.

The performances of MM and ZS schemes deteriorate rapidly as the estimating error in time delay increases. But the proposed scheme works well even for a large estimating error in time delay. Figure 7 illustrates the responses of MM and the proposed schemes for the first process with a 50% estimation error in time delay. MM scheme becomes unstable. The performance of the proposed scheme is still acceptable. The response of ZS scheme is not plotted because the scheme is unstable with very large oscillations.

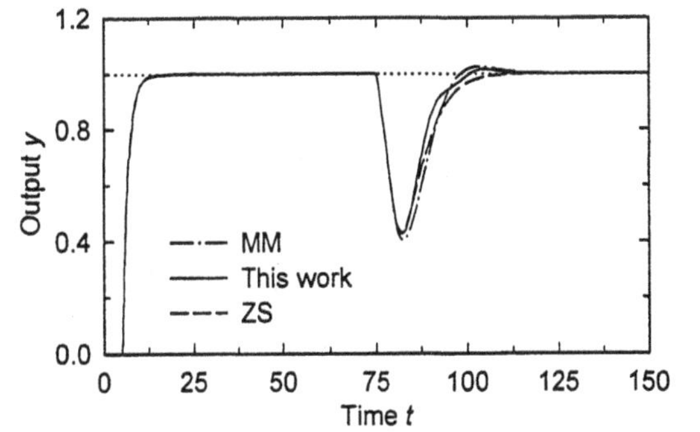

Figure 4. Closed-loop system responses of the first process with a good process model.

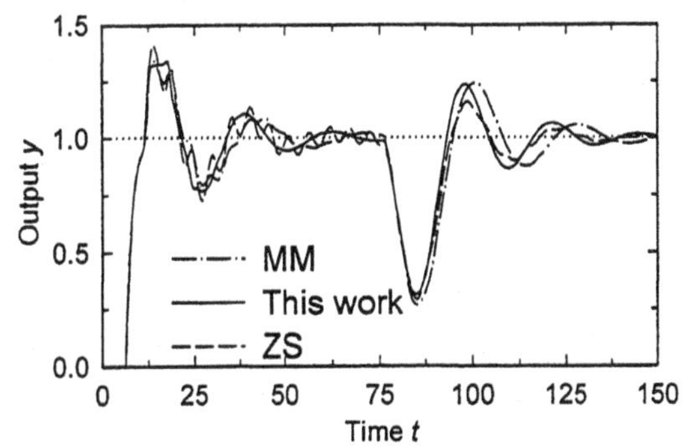

Figure 5. Closed-loop system responses of the first process with a –25% estimating error in time delay ($d^* = 5$ and $d = 6.25$).

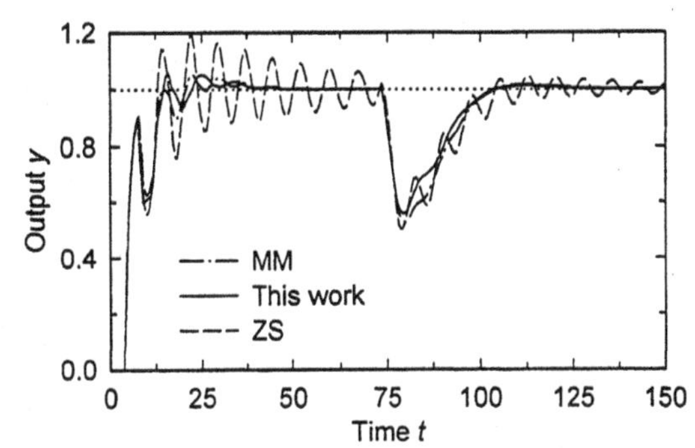

Figure 6. Closed-loop system responses of the first process with a 25% estimating error in time delay ($d^* = 5, d = 3.75$).

5.3 The Second Process

The simulation results for the second process (22) are summarized in figures 8 through 10. Figure 8 shows the responses of the three schemes for the process with a good process model. It can be observed here that the three schemes have comparable responses in both set-point and load in the nominal case.

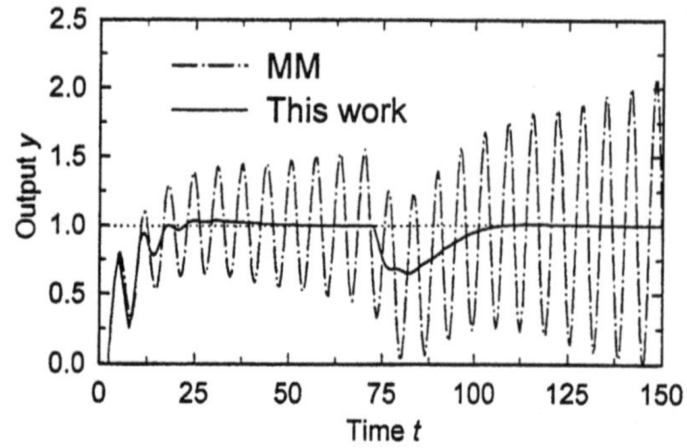

Figure 7. Closed-loop system responses of the first process with a 50% estimating error in time delay ($d^* = 5, d = 2.5$).

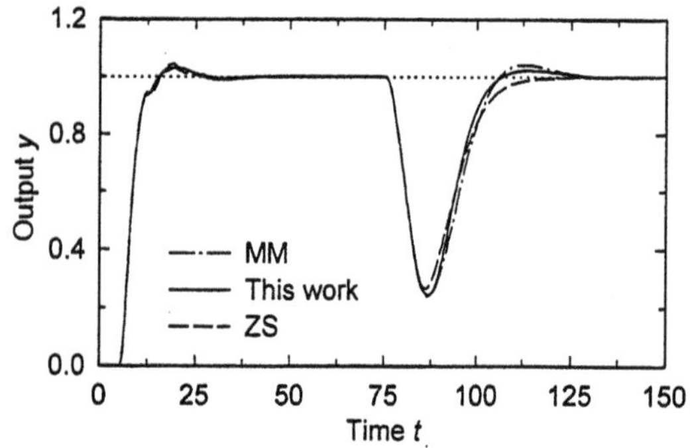

Figure 8. Closed-loop system responses of the second process with a good process model.

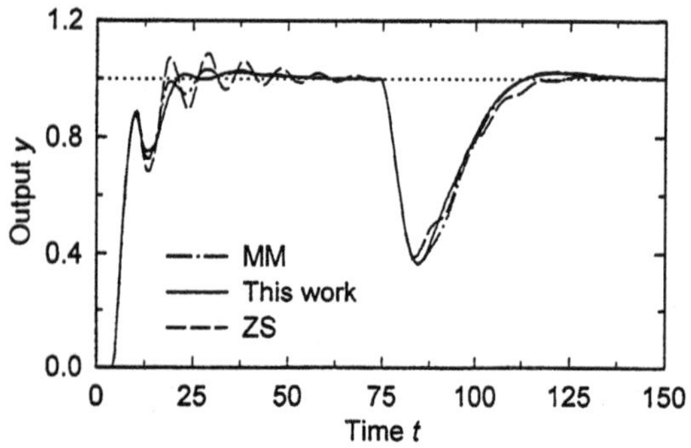

Figure 9. Closed-loop system responses of the second process at when d deviates to 3.75 from its normal value of 5.

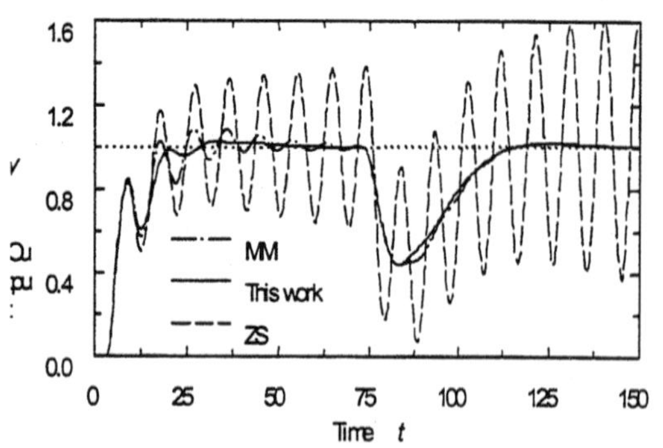

Figure 10. Closed-loop system responses of the second process at when d deviates to 3 from its normal value of 5.

Now consider the case where the process time delay deviates to 3.75 from its normal value of 5, i.e., with a 25% error in time delay being introduced. The responses of MM and ZS schemes become oscillatory, but the response of the proposed scheme continues to be smooth with an acceptable performance, as shown in figure 9. Figure 10 illustrates the simulation results in the case where the process delay changed to 3 from its normal value of 5, i.e., with an introduction of 40% error in time delay. ZS scheme becomes unstable and MM scheme behaves with strong oscillations in set-point tracking. Again, the proposed scheme turns out to be superior to the other two schemes, indicating the good robustness of the proposed scheme.

6. CONCLUSION

A novel control scheme with a clear and simple design procedure has been proposed for integrator processes with long dead-time. It contains three separate controllers. Each of these controllers has a clear physical interpretation: The pre-stabilizer, a proportional controller, is used to pre-stabilize the open-loop unstable integrator process; the set-point controller with nonlinear proportional-integral action addresses set-point tracking; and the load controller with nonlinear proportional-integral-derivative action is configured for load rejection. The set-point controller and the load controller can be synthesized independently with the separation nature of the set-point and the load responses. Simulations have shown that the proposed scheme has fast set-point tracking, efficient load rejection and good robustness to variations in process dynamics.

REFERENCES

Chien I.-L. and Fruehauf P. S. (1990). Consider IMC tuning to improve controller performance. *Chem Eng Prog*, **86(10)**, 33-41.

Smith, O. J. M. (1957). Closer control of loops with dead time. *Chem Eng Prog*, **53(5)**, 217-219.

Hang C. C. and Wong F. S. (1979). Modified Smith predictors for the control of processes with dead time. *Proc ISA Annual Conf*, 33-34.

Watannabe K. and Ito M. (1981). A process-model control for linear systems with delay. *IEEE Trans on Automat Control*, **26(6)**, 1261-1266.

Astrom K. J, Hang C. C. and Lim B. C. (1994). A new Smith predictor for controlling a process with an integrator and long dead-time. *IEEE Trans on Automat Control*, **39(2)**, 343-345.

Matausek M. R. and Micic A. D. (1996). A modified Smith predictor for controlling a process with an integrator and long dead-time. *IEEE Trans on Automat Control*, **41(8)**, 1199-1203.

Zhang W. D. and Sun Y. X. (1996). Modified Smith predictor for controlling integrator/time delay processes. *Ind Eng Chem Res.* **35**, 2769-2772.

14th Triennial World Congress, Beijing, P.R. China

C-2a-08-6

A NEW PI-PD SMITH PREDICTOR FOR CONTROL OF PROCESSES WITH LONG DEAD TIME

Ibrahim Kaya   Derek P. Atherton

School of Engineering, University of Sussex,
Falmer, Brighton BN1 9QT UK.
Email: d.p.atherton@sussex.ac.uk
Email: i.kaya@sussex.ac.uk

Abstract: Good control of processes with long dead time is often achieved using a Smith predictor configuration. Typically a PI or PID controller is used, however, it is shown in this paper that for some situations improved setpoint and disturbance responses can be obtained if a PI-PD controller is used. Several methods are possible for selecting the parameters of the PI-PD controller but when the plant transfer function has no zeros the use of the standard forms provides a simple algebraic approach, and also reveals why difficulties may be encountered if a PID controller is used. Some examples are given to show the value of the approach presented.

Keywords: Predictive control; Integral performance indices; Delay compensation; PID controllers; Disturbance rejection

1. INTRODUCTION

Plants with long time-delays can often not be controlled effectively using a simple PID controller. The main reason for this is that the additional phase lag contributed by the time delay tends to destabilize the closed loop system. The stability problem can be solved by decreasing the controller gain. However, in this case the response obtained is very sluggish.

The Smith predictor, shown in Fig 1, is well known as an effective dead-time compensator for a stable process with long time-delays (Smith, 1959). The closed loop transfer function for the system of Fig 1 is given by

$$T(s) = \frac{C(s)}{R(s)} = \frac{G_c(s)G(s)e^{-\theta s}}{1 + G_c(s)[G_m(s) + G_e(s)]} \quad (1)$$

where $G_e(s) = G(s)e^{-\theta s} - G_m(s)e^{-\theta_m s}$. $G_m(s)e^{-\theta_m s}$, $G(s)e^{-\theta s}$ and $G_c(s)$ are respectively the plant's dynamic model, and the transfer functions of the plant and the controller, which is usually a PI or a PID controller.

The stability of the Smith predictor is affected by the accuracy with which the model represents the plant. Based on the assumption that the model used matches perfectly the plant dynamics, $G_e(s) = 0$, and the closed loop transfer function of eqn. (1) reduces to

$$T_o(s) = \frac{G_c(s)G_m(s)e^{-\theta_m s}}{1 + G_c(s)G_m(s)}. \quad (2)$$

According to eqn. (2), the parameters of the primary controller, $G_c(s)$, which is typically taken as PI or PID, may be determined using a model of the delay free part of the plant.

Many possible approaches for determining or tuning the parameters of an appropriate controller, $G_c(s)$, have been given in the literature and recent contributions include Hägglund (1992), Watanabe and Ito (1981), Åström, *et al.* and Majhi and Atherton (1998). An interesting result has also

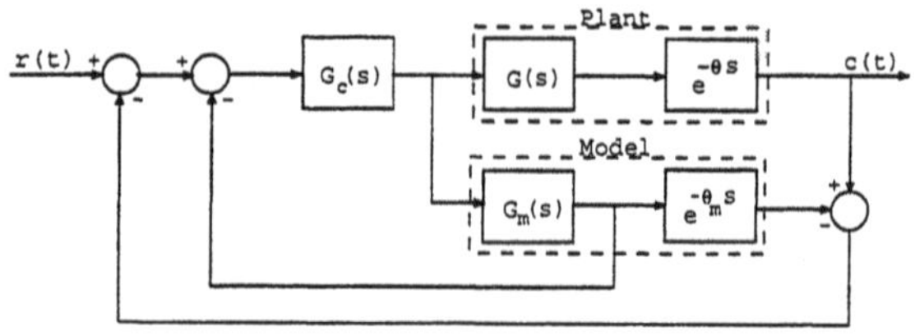

Fig. 1. The Smith predictor control scheme

been provided recently by Mataušek and Micić (1996) who have shown that by adding an additional controller the disturbance response of the Smith predictor can be controlled independently of the setpoint response.

In this paper a new control strategy for the Smith predictor is presented which replaces the conventional controller by a PI-PD structure, a controller which has been shown to significantly outperform a PID controller in some standard single input-single output systems (Atherton and Boz, 1998). The disturbance rejection controller suggested in Mataušek and Micić (1996) is also included. It is also shown that, when the plant transfer function is of relatively low order, the parameters of the PI-PD controller can be suitable chosen using the standard forms method which is a simple algebraic approach to control system design (Atherton and Boz, 1998).

The paper is organized as follows. In section 2, design using integral performance criteria is briefly reviewed and the standard forms for integral squared time error (ISTE) are given. Section 3, presents the new PI-PD Smith predictor controller and then outlines the design procedure for several simple process plant transfer function models. Examples are given in section 4, which by comparision with previous studies clearly show the advantages of the new design approach presented. Finally, conclusions are given in section 5.

2. INTEGRAL PERFORMANCE CRITERIA

In this section a very brief review of integral performance criteria is given, since the approach can be used to produce standard forms used for the controller designs used in the next section.

The use of integral performance indices for control system design is well known. Many text books (Dorf, and Bishop, 1995; Chen, 1993) include short sections devoted to the procedure. When integral performance criteria were first suggested in the early 1950s, digital computers were in their infancy and evaluations could take a long computation time.

For linear systems, the ISE can be evaluated efficiently on digital computers using the s-domain approach with Astrom's recursive algorithm (Åström, 1970). Thus for

$$J_o = \int_0^\infty [e^2(t)]dt \qquad (3)$$

the s-domain solution is given by

$$J_o = \frac{1}{2\pi j}\int_0^\infty E(s)E(-s)ds \qquad (4)$$

where $E(s) = B(s)/A(s)$, and A and B are polynomials with real coefficents, given by

$$A(s) = a_0 s^m + a_1 s^{m-1} + ... + a_{m-1}s + a_m$$

$$B(s) = b_1 s^{m-1} + ... + b_{m-1}s + b_m$$

Criteria of the form $J_n = \int_0^\infty (t^n e)^2 dt$ can also be evaluated (Zhuang and Atherton, 1993) using this approach, since $L[tf(t)] = \frac{-d}{ds}F(s)$, where L denotes the Laplace transform and $L[f(t)] = F(s)$.

Using this approach it is possible to obtain the optimal parameters of a closed loop transfer function which will provide a minimum value of the ISE. Tables of such all pole transfer functions were given many years ago (Graham and Lathrop, 1953) but are of little use in design, because even with an all pole plant transfer function the addition of a typical controller produces a closed loop transfer function with a zero. Results with a single zero were also given so that the feedback loop would follow a ramp input with zero steady-state error but these expressions are not appropriate for step response design. For a closed loop transfer function with one zero it is easy to present results for these optimum transfer functions as the position of the zero varies. These results are given in Atherton and Boz (1998) but the appropriate curves for the ones to be made use of in this paper are shown here. If the normalized closed loop transfer function is

$$T_n(s) = \frac{c_1 s + 1}{s^n + d_{n-1}s^{n-1} + ... + d_1 s + 1} \qquad (5)$$

then for a minimum ISTE the optimum values of the d's as functions of c_1 are shown in Figs. 2 and 3 for $T_2(s)$ and $T_3(s)$, respectively.

3. CONTROLLER DESIGN

The proposed Smith predictor is shown in Fig 4, where G_{c1} is a PI controller, G_{c2} is a PD, or P where it is appropriate, controller and K_o is the disturbance controller introduced in (Matašek

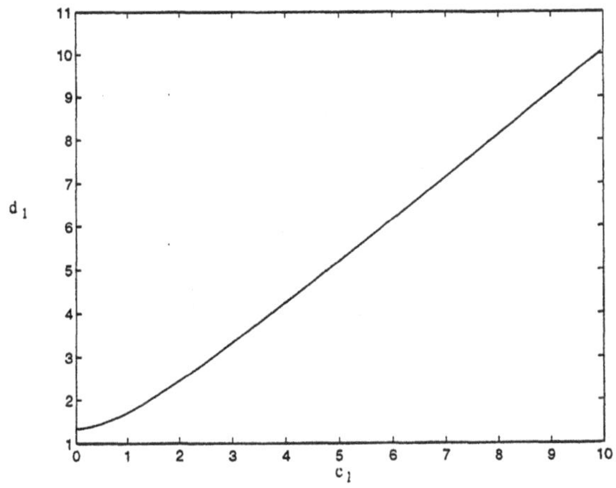

Fig. 2. Optimum values of d_1

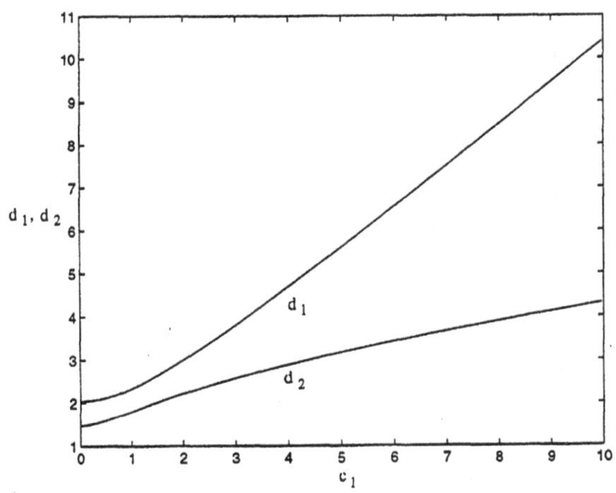

Fig. 3. Optimum values of d_1 and d_2

and Micić, 1996). Assuming exact matching between the process and the model parameters then the setpoint and disturbance responses are given by

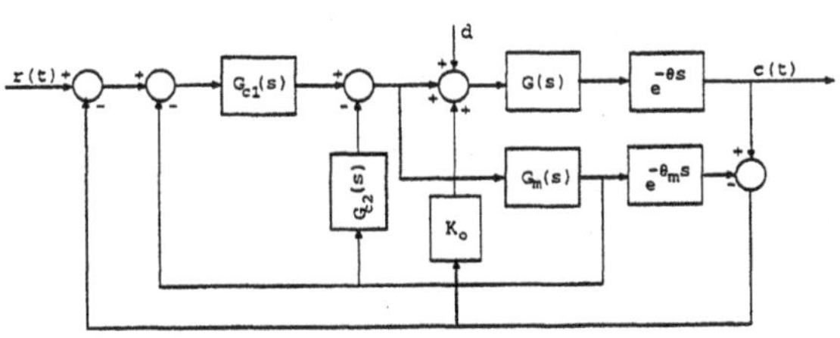

Fig. 4. The proposed Smith predictor control scheme

$$C(s) = T_r(s)R(s) + T_d(s)D(s) \tag{6}$$

where

$$T_r(s) = \frac{G_m G_{c1} e^{-\theta_m s}}{1 + G_m(G_{c1} + G_{c2})} \tag{7}$$

$$T_d(s) = \frac{G_m[1 + G_m(G_{c2} + G_{c1} - G_{c1}e^{-\theta_m s})]e^{-\theta_m s}}{[1 + G_m(G_{c1} + G_{c2})][1 + K_o G_m e^{-\theta_m s}]} \tag{8}$$

The transfer function for the setpoint response, given by eqn. (7), reveals that the parameters of the main controllers, G_{c1} and G_{c2}, may be determined using a model of the delay free part of the plant. Also it is seen that only the disturbance response is affected by the gain K_o. It has been shown (Watanabe and Ito, 1981) that the original Smith predictor gives a steady-state error under disturbances for integrating processes . That is why, the use of the gain K_o was proposed in Mataušek and Micić (1996) to improve the disturbance rejection of integrating processes. Here, it has been adopted in the proposed method, again primarily to improve disturbance rejection for integrating processes.

In the following, it is shown how standard forms are used in order to obtain the controller parameters for several specific plants.

1. If the process can be modelled by a first order plus dead-time transfer function given by

$$G_m(s) = \frac{K_m}{s+a} e^{-\theta_m s} \tag{9}$$

then the controllers are given by

$$G_{c1}(s) = K_c(1 + \frac{1}{T_i s}) \tag{10}$$

and

$$G_{c2}(s) = K_f \tag{11}$$

Using the delay free part of eqn. (7), then the resulting closed loop transfer function, $T(s)$, is given by

$$T(s) = \frac{K_m K_c (T_i s + 1)}{T_i s^2 + (a + K_c + K_f) K_m T_i s + K_m K_c} \tag{12}$$

Rearranging eqn. (12)

$$T(s_n) = \frac{c_1 s_n + 1}{s_n^2 + d_1 s_n + 1} \tag{13}$$

is obtained, where the normalized Laplace complex variable $s_n = s(T_i/KK_c)^{1/2} = s/\alpha$, which means the response of the system will be faster than the normalized response by a factor of α, where

$$\alpha = (K_m K_c / T_i)^{1/2} \tag{14}$$

and c_1 and d_1 in eqn. (13) are given by

$$c_1 = \alpha T_i \tag{15}$$

$$d_1 = \frac{a + (K_c + K_f)K_m}{\alpha} \tag{16}$$

In principle the time scale, α, can be selected by the choice of K_c, c_1 by the choice of T_i and

d_1 by the choice of K_f. In practice, K_c will be constrained, possibly to limit the initial value of the control effort, so that the choice of K_c and T_i may involve a trade off between the values chosen for α and c_1.

2. If the process can be modelled by a second order plus dead-time transfer function with complex poles which is given by

$$G(s) = \frac{K_m}{s^2 + as + b} e^{-\theta_m s} \quad (17)$$

then the controller, G_{c1}, in the forward loop is still a PI controller given by eqn. (10). However, the controller, G_{c2}, in the inner feedback loop is now a PD controller, given by

$$G_{c2} = T_d s + K_f \quad (18)$$

Following the same procedure as in case 1, the closed loop transfer function can be put in the normalized form

$$T(s_n) = \frac{c_1 s_n + 1}{s^3 + d_2 s_n^2 + d_1 s_n + 1} \quad (19)$$

where

$$c_1 = \alpha T_i \quad (20)$$

$$d_2 = \frac{a + K_m T_d}{\alpha} \quad (21)$$

$$d_1 = \frac{b + (K_c + K_f) K_m}{\alpha^2} \quad (22)$$

and

$$\alpha = (K_m K_c / T_i)^{1/3} \quad (23)$$

In this case α is again selected by the choice of K_c and c_1 by the choice of T_i, d_1 by the choice of T_d and d_2 by the choice of K_f. Also the choice of K_c and T_i may involve a trade off between the values chosen for α and c_1.

3. An integrating process plus dead-time with the following transfer function is considered.

$$G(s) = \frac{K_m}{s} e^{-\theta_m s}. \quad (24)$$

This is a special case of eqn. (9) with the constant a equal to zero. In this case the controllers G_{c1} and G_{c2} are again given by the transfer functions of case 1. It is easy to show that the standard closed loop transfer function is again given by eqn. (13) and also the constants c_1 and α are given by the same equations as in case 1. However, d_1 is now given by

$$d_1 = \frac{K_m (K_c + K_f)}{\alpha} \quad (25)$$

The choice of α, c_1 and d_1 is made according to the discussions in case 1.

With the above controllers the system will give a good response to setpoint change. However, it will give a steady-state error when a disturbance is introduced to the system. and to avoid this the gain K_o can be used.

The choice of K_o was studied in Mataušek and Micić (1996) when the proceses has the form of $K_m e^{-\theta_m}/s$ and it was found appropriate to use

$$K_o = \frac{1}{2 K_m \theta_m} \quad (26)$$

The reader may refer to (Mataušek and Micić, 1996) for more details on the derivation of K_o. It should be pointed out that this approach to finding the value of the gain K_o could also be used for higher order processes. In this case θ_m in eqn. (26) is replaced by $\theta_m + T_d$ where T_d is the sum of the time constants.

4. A first order plus integrator plus dead-time with the transfer function

$$G(s) = \frac{K_m}{s(s + a)} e^{-\theta_m s} \quad (27)$$

is a special case of 2 with $b = 0$. Thus the eqns. (20)-(23) are used with $b = 0$ in eqn. (22).

The choice of α, c_1, d_1 and d_2 is made according to the discussions in case 2. The gain K_o, given by eqn. (26), may again be used for disturbance rejection.

4. EXAMPLES

In this section several examples are given to illustrate the use of the method. The method is compared with the methods of both Hägglund and Mataušek for processes which can be modelled by a first order plus dead-time and integrator plus dead-time transfer functions respectively. Also, the control of a second order plus dead-time transfer function with complex poles and a long dead time is given using the proposed method and it is compared with a Smith predictor using a PID controller as the main controller.

4.1 *Example 1*

A FOPDT transfer function given by $e^{-10s}/(10s+1)$ is considered. It should be noticed that the time constant is quite large. K_c was limited to 1.00 and T_i was chosen as 0.25 leading to $\alpha = 0.6324$ and $c_1 = 0.1581$. For the ISTE integral performance criterion $d_1 = 1.3563$ which gives the controller parameter $K_f = 6.5772$. The results are compared with the initial tuning for a PI controller suggested by Hagglund for which $K_c = 1$ and

$T_i = 10$. The response of the system to the proposed and Hägglund's method, when the starting tuning parameters are used, is given in Fig. 5. At time $t = 60s$ a disturbance with magnitude of -0.1 is introduced to the system. The proposed method gives better results for setpoint response, although there is not a significant difference in disturbance rejection. The main reason the proposed method gives better results is because due to its large time constant its response resembles more an integrating process rather than a non-integrating one.

4.2 *Example 2*

A SOPDT transfer function given by $e^{-15s}/(s^2 + 0.2s + 1)$ is considered. K_c was again limited to 1.00, with the choice of $T_i = 0.50$. These two values give $\alpha = 1.26$ and $c_1 = 0.63$. The ISTE optimum values of d_2 and d_1, are, from Fig. 4, 1.644 and 2.163 respectively. These values can be achieved with 1.8714 and 1.4334 for T_d and K_f respectively, from eqns. (20) and (21). The response to setpoint change is given in Fig. 6 for the PI-PD controller and also shown for comparision are the results for a PID controller with parameters 1.00, 0.198, 5.065 for the gain, integral and derivative time constants respectively which were obtained first by limiting $K_c = 1.00$, then by using an ISTE optimization program to get the other two parameters. The proposed method is seen to be superior to using a PID controller, since the plant's complex poles are near to the imaginary axis.

4.3 *Example 3*

A pure integrator plus long dead-time process given by e^{-5s}/s, which was given in Mataušek and Micić (1996), is considered. As in Mataušek and Micić (1996) K_o was taken equal to 0.1. K_c was limited to 1.00 and T_i was chosen as 0.25. Following the same procedure as before $\alpha = 2.00$, $c_1 = 0.50$ and $d_1 = 1.451$ giving $K_f = 1.903$. In Mataušek and Micić (1996) a proportional only, controller with gain 0.6, is used. The responses to a unit setpoint and a disturbance change, which is of magnitude -0.1 at $t = 20s$, are given in Fig. 7. In this case only a small improvement over Matausek's approach is obtained but as the next example shows a much better improvement may be possible for other forms of plants.Also, a disturbance having a magnitude of -0.1 was introduced at $t = 70s$. A little improvement is achieved in both setpoint and disturbance response. However, in the following example it is shown that the proposed method can be more advantageous if the process again has a large time constant which means that the process almost has a double integrator action.

4.4 *Example 4*

Consider $G(s) = e^{-6.665s}/s(10s + 1)$, where the plant has both an integrator and a relatively large time constant. With $K_c = 0.50$ and $T_i = 2.00$ and again using the ISTE standard form approach one obtains $T_d = 3.757$ and $K_f = 1.336$. Using eqn. (25) $K_o = 0.03$. For the approach suggested by Mataušek and Micić (1996) the proportional controller has a gain of $K_r = 0.1$ and $K_o = 0.03$ as before. The responses of the system to a unit setpoint and -0.1 unit disturbance at $t = 100s$ are shown in Fig. 8. The far superior response of the PI-PD controlled system is now clearly evident.

5. CONCLUSIONS

A new approach to control processes, both integrating and non-integrating, with long dead-times based on a Smith predictor concept and a PIPD controller has been introduced. It is shown that the proposed method can be advantageous when the process has a large time constant, with or without an integrator, and for processes with 'poorly' located poles, i.e ligthly damped. Several procedures for obtaining the parameters of the PI-PD controllers are possible but one of the simplest approaches, which is used in this paper, is to employ ISTE standard forms as this enables the design to be completed using simple algebra.

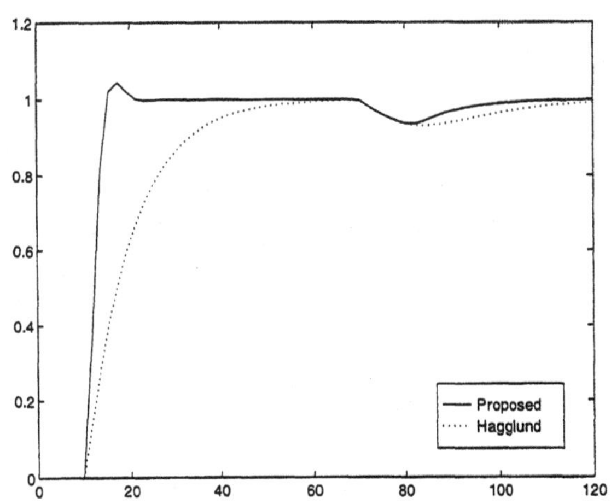

Fig. 5. Step responses for example 1

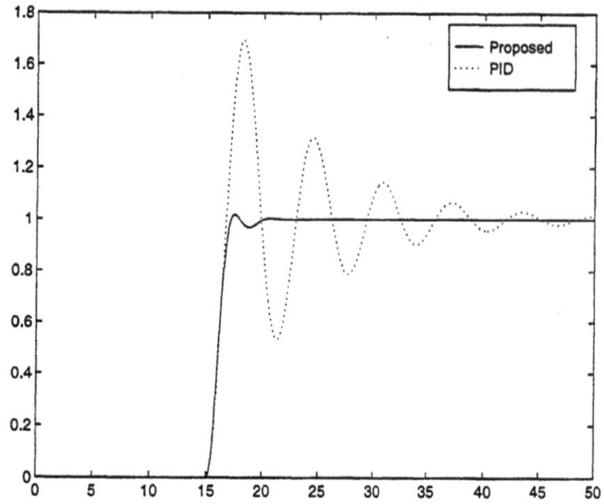

Fig. 6. Step responses for example 2

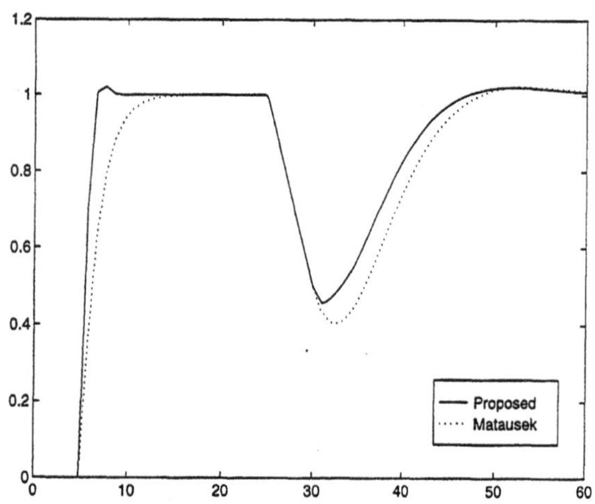

Fig. 7. Step responses for example 3

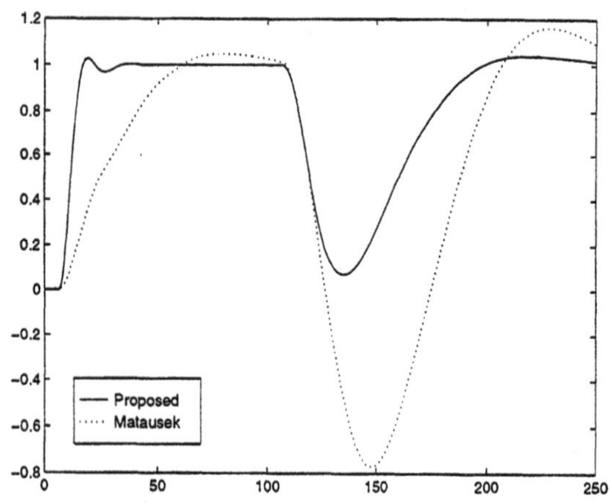

Fig. 8. Step responses for example 4

6. REFERENCES

Åström, K.J. (1970) *Introduction to Stochastic Control Theory*, Academic Press Inc..

Åström, K.J. and C.C. Hang, and B.C. Lim (1994) A new Smith predictor for controlling a process with an integrator and long dead-time. *IEEE Trans. Automat. Contr.*, **39**, 343–345.

Atherton, D.P. and A.F. Boz (1998) Using standard forms for controller design. *UKACC International Conf. on Control'98*, **2**, 1066–1071.

Chen, C.T. (1993) *Analog and Digital Control System Design*, Saunders College.

Dorf, R.C. and R.H. Bishop (1995) *Modern Control Systems*, Addison-Wesley, Reading.

Graham D. and R.C. Lathrop (1953) The synthesis of optimum response: criteria and standart forms, II, *Trans. AIEE*, 273–288.

Hägglund T. (1992) A predictive PI controller for processes with long dead-time delay, *IEEE Control Syst. Mag.*, **12**, 57–60.

Majhi, S. and D.P. Atherton (1998) A new Smith predictor and controller for unstable and integrating processes with time delay, *IEEE Conf. on CDC'98*, 1341–1345.

Matausek M.R. and A.D. Micić (1996) A modified Smith predictor for controlling a process with an integrator and long dead-time, *IEEE Trans. Automat. Contr.*, **41**, 1199–1203.

Smith O.J. (1959) A controller to overcome dead time, *ISA J.*, **6**, 28–33.

Watanabe K., M. Ito (1981) A process-model control for linear systems with delay, *IEEE Trans. Automat. Contr.*, **AC-26**, 1261–1266.

Zhuang M. and D.P. Atherton (1993) Tuning PID controllers with integral performance criteria, In: *Matlab Toolboxes and Applications*, 131–144, Peter Peregrinus.

14th Triennial World Congress, Beijing, P.R. China

A DESIGN METHOD OF LOW SENSITIVITY CONTROL SYSTEMS–SIMULTANEOUS DESIGN OF CONTROLLER AND PLANT–

Kou Yamada *,[1]

* *Department of Electorical and Information Engineering*
Yamagata University
4-3-16 Jonan, Yonezawa, 992-8510, Japan
Tel:+81-238-26-3326, Fax:+81-238-26-3326
Email: yamada@eie.yz.yamagata-u.ac.jp

Abstract: In this paper, we propose a new low sensitivity control system design method for the system with uncertainty using special feature that robust stability condition does not depend on the number of poles of the plant and that of the nominal plant. This method is built based on the idea of simultaneous design of the controller and the nominal plant parameter from frequency response data. This procedure is constructed by set membership identification method and nonlinear optimization problem. Since it is not necessary to equalize the number of unstable poles of nominal plant to that of the plant, to design robustly stable systems, it is sufficient only to check norm condition to depend on sensitivity function to hold robust stability condition. Therefore in this framework, we can construct the control system with low sensitivity characteristics and robust stability. *Copyright © 1999 IFAC*

Keywords: robust stability, the relative degree, low sensitivity control, multiplicative uncertainty, varying number of unstable poles, simultaneous design of controller and nominal plant

1. INTRODUCTION

In this paper, we consider a control system design method of robustly stable low sensitivity control systems. Several papers have been considered on robust stabilization problem(Doyle and Stein, 1981; Doyle *et al.*, 1982; Chen and Desoer, 1982; Verma *et al.*, 1986; Glover and Doyle, 1988; Doyle *et al.*, 1989). Doyle and Stein built the basis for this problem(Doyle and Stein, 1981), and the condition for the multiplicative uncertainty and additive uncertainty was shown. Chen and Deseor gave the complete proof of the result by Doyle and Stein (Chen and Desoer, 1982). Kimura (Kimura, 1984) was considered the robust stabilizability problem for single input and single output systems. Vidyasagar and Kimura (Vidyasagar and Kimura, 1986) expanded for multiple input and multiple output plant.

According to (Doyle and Stein, 1981; Doyle *et al.*, 1982; Chen and Desoer, 1982), in order to keep the stability for the large uncertainty, the complementary sensitivity function must be small value. To make the complementary sensitivity function small, brings the control systems lower performance in the meaning of disturbance attenuation property and so on. Therefore, we must make the sensitivity function small to produce the control system with high disturbance attenuation property. Since the sensitivity function and the complementary function are equal to 1, we can

[1] Supported by the Ministry of Education, Science, Sports and Culture, Grant-in-Aid for Encouragement of Young Scientists, 10750330.

not obtain either low sensitivity or high robust stability characteristics, although low sensitivity does not always make the system unstable. Maeda et al. considered this problem as an infinite gain margin one(Maeda and Vidyasagar, 1986; Maeda and Vidyasagar, 1985). Nogami et al. clarified the condition that hi-gain controller does not make the system unstable and also proposed a design method(Nogami *et al.*, 1986). In (Doyle *et al.*, 1992; Yamada, 1998; Miyasato, 1997), low sensitivity control system with robust stability characteristics can be constructed by considering appropriate class of uncertainty. Yamada showed the necessary and sufficient robust stability condition did not depend on the number of poles of the plant and that of the nominal plant with same uncertainty to (Doyle *et al.*, 1992; Yamada, 1998; Miyasato, 1997). Therefore we can construct low sensitivity control systems without information of the number of unstable poles by using result of (Yamada, 1998). Control design method based on the result in (Yamada, 1998) is discussed only in (Yamada, 1998) based on H_∞ control system design using linear matrix inequality. Control system design method is required to be considered using special feature that robust stability condition does not depend on the number of poles of the plant and that of the nominal plant.

In this paper, we propose a new control system design method for the system with uncertainty discussed in (Yamada, 1998) using special feature that robust stability condition does not depend on the number of poles of the plant and that of the nominal plant. This method is built on the idea of simultaneous design of controller and nominal plant from frequency response data. Since it is not necessary to equalize the number of unstable poles of nominal plant to that of unstable poles of the plant, to design robustly stable systems, it is sufficient to check norm condition to depend on sensitivity function. If sensitivity function is smaller value, then the robust stability condition is easy to be satisfied. Therefore in this framework, we can construct the control system with low sensitivity characteristics and robust stability from frequency response data.

NOTATIONS

R	set of real number
C	set of complex number
$R(s)$	set of real coefficient rational function of s
RH_∞	Set of stable proper real coefficient rational function
$\lVert\cdot\rVert_\infty$	H_∞ norm
$\Re\{\cdot\}$	real part of $\{\cdot\} \in C$
$\Im\{\cdot\}$	imaginary part of $\{\cdot\} \in C$
$0_{n\times m}$	zero matrix with nth row and mth columns

2. ROBUST STABILITY

In this section, we summarize robust stability problem discussed in (Yamada, 1998) and clarify the purpose of this paper. Let us consider the system in Fig. 1 . Here, $G(s) \in R(s)$ is single-input and single-output strictly proper system, $C(s) \in R(s)$ is controller, $r$is reference input, y is output and d is disturbance. $G(s)$ is assumed to have no zeros in closed right half plane, that is

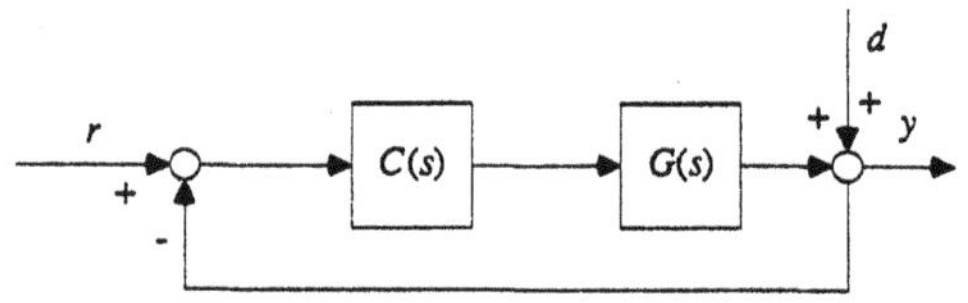

Fig. 1. Control system

$$G(s) \neq 0 \ \ \forall \Re s \geq 0 \tag{1}$$

is satisfied. Nominal plant of $G(s)$ is denoted by

$$G_m(s) = \left[\begin{array}{c|c} A_m & B_m \\ \hline C_m & 0 \end{array}\right] \in R(s),$$

and assumed to have no zeros in closed right half plane, that is

$$G_m(s) \neq 0 \ \ \forall \Re s \geq 0 \tag{2}$$

holds.

Uncertainty between $G(s)$ and $G_m(s)$ is denoted by $\Delta(s)$ and relation $G(s)$ and $G_m(s)$ is denoted by

$$G(s) = G_m(s)\,(1 + \Delta(s))\,. \tag{3}$$

Yamada(1998) clarified the necessary and sufficient robust stability condition for the system with the uncertainty defined as follows.

Definition 1. If $G(s)$ satifies following expressions, $G(s)$ is called an element of set $\Omega(G_m(s), W(s))$.

- Both $G(s)$ and $G_m(s)$ have no zeros in the closed right half plane.

$$\left|\frac{\Delta(j\omega)}{1+\Delta(j\omega)}\right| < |W(j\omega)| \ \forall\omega \in R \qquad (4)$$

Here $W(s)$ is stable rational transfer function.

□

Remarks 1. Note that the set $\Omega(G_m(s), W(s))$ include the case that the number of unstable poles of $G(s)$ is not equal to that of G_m.

Robust stability condition of all $G(s)$ of set Ω $(G_m(s),\ W(s))$ is summarized in next theorem (Yamada, 1998).

Theorem 1. Let us assumed that $C(s)$ is a stabilizing controller of $G_m(s)$. Necessary and sufficient condition that $C(s)$ is a robust stabilizing controller for set of $\Omega(G_m(s), W(s))$ is given by

$$\left\|\frac{1}{1+G_m(s)C(s)}W(s)\right\|_\infty = \|S(s)W(s)\|_\infty \le 1 \qquad (5)$$

∎

This theorem does not require the number of unstable poles of $G(s)$ as a well-known number. Therefore when the number of unstable poles of $G(s)$ is uncertain, Theorem 1 is useful. For mechanical system, the plant has no zeros in the closed right half plane from the structure. In addition the number of unstable poles is not decided only from the frequency response data. Therefore for mechanical systems, Theorem 1 is useful to design control system only using frequency response data. The purpose of this paper is to give a control design method for this problem.

3. PROBLEM FORMULATION

Let us consider following problem. Let the relative degree of $G(s)$ is m a known number. We denote $G(s)$ by

$$G(s) = G_m(s,\theta)\left(1+\Delta(s)\right). \qquad (6)$$

Here $G_m(s,\theta)$ and θ are defined by

$$G_m(s,\theta) = \frac{b(s,\theta)}{a(s,\theta)} = \frac{b_{n-m}s^{n-m} + b_{n-m-1}s^{n-m-1} + \cdots + b_0}{s^n + a_{n-1}s^{n-1} + \cdots + a_0} \qquad (7)$$

$$\theta = \left[\, a_{n-1} \cdots a_0 \ b_{n-m} \cdots b_0 \,\right]^T \qquad (8)$$

respectively. By using closed loop identification for unstable systems(Hansen and Franklin, 1988), we have already obtained Nth number of frequency response data

$$G(j\omega_i) \ \ i = 1, 2, \ldots, N.$$

Problem to be considered is to find controller $C(s)$

$$C(s,\theta_c) = \frac{q(s,\theta_c)}{p(s,\theta_c)} = \frac{q_n s^n + q_{n-1}s^{n-1} + \cdots + q_0}{s^n + p_{n-1}s^{n-1} + \cdots + p_0} \qquad (9)$$

and θ

$$\theta_c = \left[\, p_{n-1} \cdots p_0 \ q_n \cdots q_0 \,\right]^T \qquad (10)$$

to satisfy both performance specification

$$\left\|\frac{1}{1+C(s)G_m(s)}W_p(s)\right\|_\infty = \|S(s)W_p(s)\|_\infty \le 1 \qquad (11)$$

and robust stability condition (5).

4. SIMULTANEOUS DESIGN PROCEDURE

In this section, we propose a design procedure for the problem defined in preceding section.

Basic lines to built procedure are following points.

- Poles of controller is set to be equal to zeroes of nominal plant. This contributes to decrease the number of design parameters. Poles of nominal plant and zeros of controller are set to satisfy specifications (5). From the assumption that both $G(s)$ and $G_m(s)$ have no zeros in the closed right half plane, the unstable poles and unstable zeros cancellation between $G(s)$ and $G_m(s)$ do not occur.
- Stability of sensitivity function $S(s)$ is guaranteed by solving optimal problem subject to condition based on Routh-Hurwits criteria. This guarantee robust stability of the system.
- Performance specification (11) is certified in final step of procedure. If performance specification is not satisfied, weighting function is changed to holds performance specification.

A procedure to give a solution defined by preceding section is summarized below.

Procedure 1

(1) Preparation

(a) By using curve fitting method (Pintelon *et al.*, 1994; Sagara *et al.*, 1981), $b(s,\theta)$, that is zeros polynomial of plant $G(s)$, is prepared in advance under the condition that $b(s,\theta)$ has no zeros in closed right half plane.

(b) Let $b(s,\theta)$ be $b(s,\theta)=K\bar{b}(s,\theta)$. $p(s,\theta_c)$ is settled to be

$$p(s,\theta_c)=\bar{b}(s,\theta). \quad (12)$$

Here $\bar{b}(s,\theta)$ is monic polynomial.

(c) Let $\bar{q}(s)$ be $\bar{q}(s)=Kq(s)$. We have

$$G_m(s,\theta)C(s,\theta_c)=\frac{\bar{q}(s,\theta_c)}{a(s,\theta)} \quad (13)$$

$$\bar{q}(s) = \bar{q}_n s^n+\bar{q}_{n-1}s^{n-1}+\cdots+\bar{q}_0. \quad (14)$$

Then problem to be defined in preceding section is equivalent to following problem. Find design parameter $\bar{q}(s,\theta_c)$, $a(s,\theta)$ to satisfy the condition

$$\left|\frac{1-\dfrac{G_m(j\omega_i,\theta)}{G(j\omega_i)}}{1+G_m(j\omega_i,\theta)C(j\omega_i,\theta_c)}\right| = \left|\frac{p(j\omega_i,\theta_c)a(j\omega_i,\theta)-\dfrac{b(j\omega_i,\theta)p(j\omega_i,\theta_c)}{G(j\omega_i)}}{p(j\omega_i,\theta_c)a(j\omega_i,\theta)+q(j\omega_i,\theta_c)b(j\omega_i,\theta)}\right| = \left|\frac{a(j\omega_i,\theta)-\dfrac{b(j\omega_i,\theta)}{G(j\omega_i)}}{a(j\omega_i,\theta)+\bar{q}(j\omega_i,\theta_c)}\right| \le\gamma<1 \quad (15)$$

correspond to (5).

(2) Set K, γ_1 and $W_k(j\omega_i)$ be $k=1$, $\gamma_1=1$ and

$$W_k(j\omega_i)=1 \quad i=1,\ldots,N \quad (16)$$

, respectively. $\hat{\theta}$ is denoted by

$$\hat{\theta}=\left[\,a_{n-1}\cdots a_0\ \bar{q}_{n-m}\cdots\bar{q}_0\,\right]^T. \quad (17)$$

(3) Set of F_k is defined by

$$F_k = \left\{\hat{\theta}\,\middle|\,\left|\frac{W(j\omega_i)\left\{a(j\omega_i,\theta)-\dfrac{b(j\omega_i)}{G(j\omega_i)}\right\}}{a(j\omega_i,\theta)+\bar{q}(j\omega_i,\theta_c)}\right|\le\gamma_k\right\} \quad i=1,\ldots,N. \quad (18)$$

Find parameter to be included in set of F_i subject to restricting condition on set of $\hat{\theta}$

$$\hat{\theta}^T\Gamma_i\hat{\theta}+2\Psi_i\theta+\Phi_i\le 0 \quad (19)$$

$$\begin{aligned}\Gamma_i =& -\gamma_k^2\Re\left\{[\zeta_{n-1},\zeta_{n-m}]^T\right\}\Re\{[\zeta_{n-1},\zeta_{n-m}]\} \\ &-\gamma_k^2\Im\left\{[\zeta_{n-1},\zeta_{n-m}]^T\right\}\Im\{[\zeta_{n-1},\zeta_{n-m}]\} \\ &+\Re\left\{[W(j\omega_i)\zeta_{n-1},0]^T\right\} \Re\{[W(j\omega_i)\zeta_{n-1},0]\} \\ &+\Im\left\{[W(j\omega_i)\zeta_{n-1},0]^T\right\} \Im\{[W(j\omega_i)\zeta_{n-1},0]\}\end{aligned} \quad (20)$$

$$\zeta_m=\left[\,(j\omega_i)^m\cdots j\omega_i\ 1\,\right] \quad (21)$$

$$\begin{aligned}\Psi_i =& -\gamma_k^2\Re\{(j\omega_i)^n[\zeta_{n-1},\zeta_{n-m}]\} \\ &-\gamma_k^2\Im\{(j\omega_i)^n[\zeta_{n-1},\zeta_{n-m}]\} \\ &+\Re\left\{\left\{(j\omega_i)^{(n-m)}-\frac{b(j\omega_i)}{G(j\omega_i)}\right\} [W_k^2(j\omega_i)\zeta_{n-1},0]\right\} \\ &+\Im\left\{\left\{(j\omega_i)^{(n-m)}-\frac{b(j\omega_i)}{G(j\omega_i)}\right\} [W_k^2(j\omega_i)\zeta_{n-1},0]\right\}\end{aligned} \quad (22)$$

$$\Phi_i = -(\omega_i)^{2n} + \left|W_k(j\omega_i)\left\{(j\omega_i)^{(n-m)}-\frac{b(j\omega_i)}{G(j\omega_i)}\right\}\right|^2 \quad (23)$$

and that of condition of stability of sensitivity function

$$l_i\begin{bmatrix} I_{m-1} & 0 & 0 \\ 0 & I_{n-m+1} & I_{n-m+1}\end{bmatrix}\hat{\theta}>0 \quad i=1,\ldots,n \quad (24)$$

$$H_\tau=\begin{vmatrix} h_{n-1} & h_{n-3} & \cdots & h_{n-(2\tau-1)} \\ 1 & h_{n-2} & \cdots & h_{n-(2\tau-2)} \\ 0 & h_{n-1} & \cdots & h_{n-(2\tau-3)} \\ 0 & 1 & \cdots & h_{n-(2\tau-4)} \\ \vdots & \cdots & \cdots & \vdots \\ 0 & \cdots & \cdots & h_{n-\tau}\end{vmatrix}>0 \quad \tau=1,\ldots,n \quad (25)$$

$$h_\tau=l_{n-\tau}\begin{bmatrix} I_{m-1} & 0 & 0 \\ 0 & I_{n-m+1} & I_{n-m+1}\end{bmatrix}\hat{\theta} \quad (26)$$

$$l_i=\left[\,0_{1\times(i-1)}\ 1\ 0_{1\times(n-i)}\,\right] \quad (27)$$

by using nonlinear optimization method. (25) requires to guarantee that the sensitivity function $S(s)$ is stable.

(4) If F_k is null, increase γ_k. Obtain minimum γ_k and $\hat{\theta}$ to satisfy F_k is not null set. Next fix $a(s,\theta)$ and $b(s,\theta)$ and find minimum γ_k and $\hat{\theta}$ to satisfy F_k is not null set under the condition that design parameters are $p(s,\theta_c)$ and $q(s,\theta_c)$ in the same way of (2) and (3) of this procedure.

(5) Final step is to check that frequency response of sensitivity function $S(s)$ holds performance specification. Draw the frequency response of sensitivity function $S(s)$ using obtained parameter $\hat{\theta}$. If performance specification is satisfied, then set $q(s)$ and $a(s)$ from that $\hat{\theta}$ else add 1 to k and change the value $W_k(j\omega_i)$ under the condition

$$W_k(j\omega_i) \geq 1$$

and return to (2).

Remarks 2. In above procedure, pole zero cancellation between $C(s,\theta_c)$ and $G_m(s,\theta)$ does not happen. Because (25) can not be satisfied if unstable pole zero cancellation between $C(s,\theta_c)$ and $G_m(s,\theta)$ is occurred.

5. ROBUST PERFORMANCE DESIGN

In this section, we will expand the Procedure 1 to construct control systems with robust performance characteristics and robust stability. Here the system is called the system having the robust performance characteristics, if $C(s)$ satisfies

$$\left\| \frac{1}{1+C(s)G(s)} W_p(s) \right\|_\infty \leq 1 \qquad (28)$$

for all $G(s)$ included in $\Omega(G_m(s), W(s))$.

Considerable problem is summarized as follows. Problem is to find controller $C(s)$

$$C(s,\theta_c) = \frac{q(s,\theta_c)}{p(s,\theta_c)} = \frac{q_n s^n + q_{n-1}s^{n-1} + \cdots + q_0}{s^n + p_{n-1}s^{n-1} + \cdots + p_0} \qquad (29)$$

and θ

$$\theta_c = \left[p_{n-1} \cdots p_0 \; q_n \cdots q_0 \right]^T \qquad (30)$$

to satisfy both performance specification

$$\left\| \frac{1}{1+C(s)G(s)} W_p(s) \right\|_\infty \leq 1 \qquad (31)$$

and robust stability condition

$$\left\| \frac{1}{1+G_m(s)C(s)} W(s) \right\|_\infty = \|S(s)W(s)\|_\infty \leq 1 \qquad (32)$$

Note that if the controller $C(s)$ holds (31), (32) is always satisfied. Therefore only to satisfy robust performance characteristics, the problem is solved by changing (15) and F_k of Procedure 1. But the stability of sensitivity function $S(s)$ is not clear. To satisfy that the sensitivity function $S(s)$ is stable, Procedure 1 is changed as follows.

In the case of considering robust performance problem, it is sufficient only to change final step of Procedure 1 below.

(5) Final step is to check that frequency response of robust performance specification given by

$$\left| \frac{1}{1+C(j\omega_i,\theta_c)G(j\omega_i)} W_p(j\omega_i) \right| < 1 \; \forall \omega_i \qquad (33)$$
$$i = 1, \ldots, N.$$

Draw the frequency response of sensitivity function $S(s)$ using obtained parameter $\hat{\theta}$. IF performance specification is hold, then set $q(s)$ and $a(s)$ from that $\hat{\theta}$ else add 1 to k and change $W_k(j\omega_i) \geq 1$ under $W_k(j\omega_i) \geq 1$ and return to (2).

6. CONCLUSION

In this paper, we proposed a new control system design method for the system with uncertainty discussed by (Yamada, 1998) using special feature that robust stability condition does not depend on the number of poles of the plant and that of the nominal plant. Until now control system design is performed in the following three steps

(1) at first step, system identification is applied to the system.
(2) at second step, nominal plant and upper bound of uncertainty is determined from the data obtained by preceding step
(3) at second step, by using nominal plant and upper bound of uncertainty, robust stabilizing controller is constructed

Proposed method is considered to unite second step and third step of above procedure by using frequency response data. Namely this method is built on the idea of simultaneous design of controller and nominal plant from frequency response data. This procedure is built by using the idea of set membership identification method and nonlinear optimization problem. Since it is not necessary to equalize the number of unstable poles of nominal plant to that of unstable poles of the plant, to design robustly stable systems, it is sufficient only to check norm condition to depend on sensitivity function. If sensitivity function is smaller value, then the robust stability condition is easy to be satisfied. Therefore in this framework, we can construct the control system with low sensitivity characteristics and robust stability only using frequency response data.

ACKNOWLEDGEMENTS

The author would like to express his gratitude to Prof. K. Watanabe of Yamagata University, Prof. M. Ikeda of Osaka University, Prof. Y. Ohta of Osaka University and Prof. H. Fujioka of Kyoto University and for helpful discussions.

7. REFERENCES

Chen, M. J. and C. A. Desoer (1982). Necessary and sufficient condition for robust stability of linear distributed feedback systems. *International Journal of Control* **35**, 255–267.

Doyle, J. C. and G. Stein (1981). Multivariable feedback design: concepts for a classical modern synthesis. *IEEE Trans. on Automatic Cotrol* **26**, 4–16.

Doyle, J. C., B. Francis and A. Tannenbaum (1992). *Feedback control theory.* Macmillan Publishing. New York.

Doyle, J. C., J. E. Wall and G. Stein (1982). Performance and robustness analysis for structured uncertainty. *Proc. 21st IEEE conf. CDC* pp. 629–636.

Doyle, J. C., K. Glover, P. P. Khargonekar and B. A. Francis (1989). State-space solution to standard h_2 and h_∞ control problems. *IEEE Trans. Automatic Control* **34**, 831–847.

Glover, K. and J. C. Doyle (1988). State-space formulae for all stabilizing controllers that satisfy an h_∞ norm bound and relations to risk sensitivity. *Systems and Control Letters* **11**, 167–172.

Hansen, F. and G. Franklin (1988). On a fractional representation approach to closed loop experiment design. *Proc. of 1988 American Control Conference* pp. 1319–1320.

Kimura, H. (1984). Robust stabilizability for a class of transfer functions. *IEEE Trans. on Automatic Control* **29**, 788–793.

Maeda, H. and M. Vidyasagar (1985). Design of multivariable feedback systems with infinite gain margin and decoupling. *Systems and Control Letters* **6**, 127–130.

Maeda, H. and M. Vidyasagar (1986). Infinite gain margin problem in multivariable feedback systems. *Automatica* **22**, 131–133.

Miyasato, Y. (1997). Robustness for adaptive control and backstepping. *SICE seminor Adaptation and tuning on control* pp. 25–33(in Japanese).

Nogami, H., H. Maeda, M. Vidyadagar and S. Kodama (1986). Design of high gain feedback system with robust stability. *Trans. of SICE* **22**, 1014–1020(in Japanese).

Pintelon,
T., P. Guillaume, Y. Tolain, J. Schoukens and H. VVanhamme (1994). Parametric identification of transfer functions in frequency domain–a survey. *IEEE Trans. on Automatic Control* **39**, 2245–2260.

Sagara, S., K. Akizuki, T. Nakamizo and T. Katayama (1981). *System Identification.* The society of instrument and control engineers. Tokyo.

Verma, M. S., J. W. Helton and E. A. Jonckheere (1986). Robust stabilization of a family of plants with varying number of right half plane poles. *Proc. 1986 American Control Conference* pp. 1827–1832.

Vidyasagar, M. and H. Kimura (1986). Robust controllers for uncertain linear multivariable systems. *Automatica* **22**, 85–64.

Yamada, K. (1998). Robust stabilization for the plant with varying number of unstable poles and low sensitivity characteristics. *Proc. 1998 American Control Conference* pp. 2050–2054.

14th Triennial World Congress, Beijing, P.R. China

C-2a-09-2

DESIGN OF MULTIVARIABLE PID CONTROLLERS ON FREQUENCY DOMAIN BASED ON PARTIAL MODEL MATCHING

Zenta Iwai, Yohei Shimada, Ikuro Mizumoto and Mingcong Deng

Department of Mechanical Engineering and Materials Science
Kumamoto University
2-39-1 Kurokami, Kumamoto 8608555, Japan
E-mail:iwai@gpo.kumamoto-u.ac.jp

Abstract: This paper deals with a design method of multivariable PID controllers based on partial model matching on frequency domain. Stability of the resulting control system is also considered. In this method, the PID controller parameters are determined from the optimal solution of a constrained performance index which satisfies Rosenbrock's stability theorem. The design scheme is also examined from the view point of improving the tracking performance and robustness. The effectiveness of the proposed method is confirmed through a numerical example. *Copyright ©1999 IFAC*

Keywords: Multivariable PID control, frequency domain, partial model matching, optimization.

1. INTRODUCTION

Classical PID controllers are still used in most industrial control system. The popularity and wide acceptability of the method are due to its structural simplicity and robustness. However, many of PID controller design techniques are restricted on SISO case because of the complexity in MIMO case. Generally speaking, it might be better to know detailed model of controlled plant such as state-space model or transfer function matrices when we design MIMO PID control system. However, it is difficult to know such exact plant models in real application problems. As a countermeasure to this disadvantage, some methods for tuning MIMO PID controllers on the frequency domain, which use frequency response of controlled plant only at a few points, have been considered by several researchers (Luyben, 1986; Palmor, *et al.*, 1995; Ho, *et al.*, 1996; Wang, *et al.*, 1996).

Recently, Egashira and Iwai (1997) have also proposed a method for tuning of MIMO PID controllers on the frequency domain based on partial model matching (*Källén*, 1992). In this method, PID parameters are calculated so as to minimize the relative errors between the loop transfer function of control system and reference model transfer function at a few frequency points. The main advantage of this method is that it requires frequency response matrices of a process only at a few points *a priori*. Hence it attracts many interests from the practical point of view. The method, however, has a drawback that the stability of resulting control system can not basically be taken into consideration in the tuning procedure.

As an extension, this paper discusses a method for tuning of multivariable PID controllers based on partial model matching on some frequency range by taking into account the stability of control system. Here, multivariable PID controller parameters are given as an optimal solution by an optimization technique concerning a constrained performance with inequalities. These inequality constraints are constructed so as to satisfy the stability criteria of control system based on Rosenbrock's stability theorem. Further, this paper considers how to choose reference model and fitting frequencies so as to attain better tracking performance and robustness of control system with disturbances. The effectiveness of the proposed method is illustrated by a numerical example.

2. PROBLEM SETUP

Consider the following MIMO plant:

$$G(s) = [g_{ij}(s)]_{i,\ j\ =\ 1,\dots,\ m} \tag{1}$$

where, $G(s)$ is stable and proper. PID controller matrix is given by

$$C(s) = [c_{ij}(s)]_{i,\ j\ =\ 1,\dots,\ m} \tag{2}$$

$$c_{ij}(s) = k_{P_{ij}} + \frac{k_{I_{ij}}}{s} + \frac{k_{D_{ij}} s}{1+\gamma_i s} \tag{3}$$

where, $k_{P_{ij}}$, $k_{I_{ij}}$ and $k_{D_{ij}}$ are proportional, integral and derivative gains, respectively. It is assumed that the derivative part is filtered by a first-order filter with small time constant γ_i. The open-loop transfer function of control system is defined as

$$\begin{aligned} Q(s) &= G(s)C(s) = [q_{ij}(s)]_{i,\ j\ =\ 1,\dots,\ m} \\ &= \left[\sum_{p=1}^{m} g_{ip}(s) c_{pj}(s)\right]_{i,\ j\ =\ 1,\dots,\ m} \end{aligned} \tag{4}$$

The desired closed-loop model system is given by

$$\begin{aligned} G_{Mcl}(s) &= (I_m + G_M(s))^{-1} G_M(s) \\ &= \mathrm{diag}\left[\frac{g_{Mi}(s)}{1+g_{Mi}(s)}\right]_{i,\ =\ 1,\dots,\ m} \end{aligned} \tag{5}$$

where, $G_M(s)$ is the desired open-loop transfer function matrix, and defined as

$$G_M(s) = \mathrm{diag}\,[g_{Mi}(s)]_{i,\ =\ 1,\dots,\ m} \tag{6}$$

It is noted that it is not necessary to choose $G_M(s)$ as diagonal form. However, diagonal form will be taken in most cases from practical reasons. The design objectives are: (1) to design the parameters of PID controller such that the actual open-loop frequency response $G(j\omega)C(j\omega)$ matches to the pre-chosen desired open-loop frequency response $G_M(j\omega)$ which satisfy conditions of feedback stability, and (2), to choose $G_M(j\omega)$ and fitting frequencies

$$\begin{aligned} &\Omega_i = [\ \omega_{i_1},\ \omega_{i_2},\ \dots,\ \omega_{iN_i}\ ] \\ &0 < \omega_{i_k} < \infty, \quad i = 1,\dots,m \end{aligned} \tag{7}$$

so as to obtain well tracking performance for command inputs and robustness against disturbances.

3. PARAMETER TUNING OF MIMO PID CONTROLLER

3.1 Introduction of the loss function

Consider the following relative open-loop model error:

$$\varepsilon_{ii}(j\omega_{i_k}) = \frac{g_{Mi}(j\omega_{i_k}) - \sum_{p=1}^{m} g_{ip}(j\omega_{i_k}) c_{pi}(j\omega_{i_k})}{g_{Mi}(j\omega_{i_k})} \tag{8}$$

$$\begin{aligned} \varepsilon_{qi}(j\omega_{i_k}) &= \frac{\sum_{p=1}^{m} g_{qp}(j\omega_{i_k}) c_{pi}(j\omega_{i_k})}{g_{Mq}(j\omega_{i_k})}, \quad q \neq i \\ &\omega_{i_k} \in \Omega_i; \quad i = 1,\dots,m \end{aligned} \tag{9}$$

$\varepsilon_{ii}(j\omega_{i_k})$ and $\varepsilon_{qi}(j\omega_{i_k})$ can be represented as

$$\varepsilon_{ii}(j\omega_{i_k}) = 1 - \Phi_{ii}^T(j\omega_{i_k})\Theta_i \tag{10}$$

$$\varepsilon_{qi}(j\omega_{i_k}) = \Phi_{qi}^T(j\omega_{i_k})\Theta_i \tag{11}$$

where

$$\begin{aligned} \Phi_{ii}(j\omega_{i_k}) = \Big[&A_{i1}(j\omega_{i_k}), \dots, \frac{j\omega_{i_k} A_{i1}(j\omega_{i_k})}{1+j\omega_{i_k}\gamma_i} \\ &, \dots, A_{im}(j\omega_{i_k}), \dots, \frac{j\omega_{i_k} A_{im}(j\omega_{i_k})}{1+j\omega_{i_k}\gamma_i}\Big]^T \end{aligned} \tag{12}$$

$$\begin{aligned} \Theta_i &= [k_{P_{i1}}, k_{I_{i1}}, k_{D_{i1}}, \dots,\ k_{P_{im}}, k_{I_{im}}, k_{D_{im}}]^T \\ i &= 1,\dots,m \end{aligned} \tag{13}$$

$$A_{ip}(j\omega_{i_k}) = \frac{g_{ip}(j\omega_{i_k})}{g_{Mi}(j\omega_{i_k})}, \quad p = 1,2,\dots,m \tag{14}$$

$$\begin{aligned} \Phi_{qi}(j\omega_{i_k}) = \Big[&A_{q1}(j\omega_{i_k}), \dots, \frac{j\omega_{i_k} A_{q1}(j\omega_{i_k})}{1+j\omega_{i_k}\gamma_i} \\ &, \dots, A_{qm}(j\omega_{i_k}), \dots, \frac{j\omega_{i_k} A_{qm}(j\omega_{i_k})}{1+j\omega_{i_k}\gamma_i}\Big]^T \end{aligned} \tag{15}$$

$$A_{qp}(j\omega_{i_k}) = \frac{g_{qp}(j\omega_{i_k})}{g_{Mq}(j\omega_{i_k})}, p = 1,\dots,m \tag{16}$$

Then the following loss function can be defined.

$$\begin{aligned} J_i(\Theta_i) &= \sum_{k=1}^{N_i} \left|1 - \Phi_{ii}^T(j\omega_{i_k})\Theta_i\right|^2 \\ &+ \sum_{k=1}^{N_i} \left|1 - \Phi_{ii}^T(-j\omega_{i_k})\Theta_i\right|^2 \\ &+ \sum_{\substack{q=1 \\ q\neq i}}^{m} \left\{ \sum_{k=1}^{N_i} \left|\Phi_{qi}^T(j\omega_{i_k})\Theta_i\right|^2 + \sum_{k=1}^{N_i} \left|\Phi_{qi}^T(-j\omega_{i_k})\Theta_i\right|^2 \right\} \\ &= 2\sum_{i=1}^{m} N_i - 4\sum_{k=1}^{N_i} \mathrm{Re}\left[\Phi_{ii}^T(j\omega_{i_k})\right]\Theta_i \\ &+ 2\Theta_i^T \left\{ \sum_{q=1}^{m}\sum_{k=1}^{N_i} \mathrm{Re}\left[\bar{\Phi}_{qi}(j\omega_{i_k})\Phi_{qi}^T(j\omega_{i_k})\right] \right\} \Theta_i \end{aligned} \tag{17}$$

3.2 Stability of the closed-loop system

By minimizing the loss function (17), we can basically determine the PID parameters Θ_i. However, these parameters do not necessarily guarantee the stability of the system. To achieve the stability, we will take the following strategy. First, we introduce the stable diagonal reference model (6), where each $g_{Mi}(j\omega)$ is chosen so that its Nyquist plot always look at $(-1, j0)$ on the left-hand(stable) side. Then the PID parameters tuning problem can be derived as follows: They must minimize the loss function (17) under the restriction of stability. To guarantee the stability, we add the following conditions to the optimization problem of (17)(Rosenbrock's stability criterium).

[Condition 1.] Nyquist plot of $g_{ii}(s)$, $i = 1, \ldots, m$ always looks at $(-1, j0)$ on their left-hand side.

[Condition 2.] The i-th column (or row) Gershgorin band does not include $(-1, j0)$ for $i = 1, \ldots, m$.

3.3 Selection of frequencies on constraints condition

Selection of the set Ω_i remains unsolved. This problem is examined on the following from the view point of stability constraints.

Diagonal part: Let ω_{Mc_i} be phase cross-over frequency of $g_{Mi}(j\omega)$. Further assume $g_{Mi}(j\omega_{Mc_i})$ locates between $(-1, 0)$. Then $g_{ii}(j\omega)$ is chosen such that the $g_{ii}(j\omega_{Mc_i})$ locates in the circle (center: $g_{Mi}(j\omega_{Mc_i})$, radius: $r_i(j\omega_{Mc_i})$), where r_i is chosen such that the circle does not include $(-1, j0)$. If, $g_{ii}(j\omega_{Mc_i})$ satisfies the above stated condition, then, we can expect the increase of possibility of the satisfaction of condition 1.

Non-diagonal part: Theoretically, Condition 2 is satisfied if the return difference matrix: $I_m + Q(j\omega)$ is diagonally dominant. However, In many practical cases, this nature is governed by the frequency characteristics at neighborhood of phase cross-over frequency of the diagonal element. Hence, we select some finite set

$$\Omega_{ni} = [\ \omega_{ni_1}, \omega_{ni_2}, \ldots, \omega_{ni_p}\ ]$$

which includes ω_{Mc_i}, and give the condition such that the circle (center: $|1 + g_{ii}(j\omega_{ni_k})|$, radius: $\sum_{r=1} r \neq i^m |g_{ri}(j\omega_{ni_k})|$) does not include $(-1, j0)$ for every $\omega_{ni} \in \Omega_{ni}$.

3.4 Constraints and optimization

Based on the above-stated conditions, we can derive the following constraint conditions for both diagonal and non-diagonal parts.

(1) Constraint condition for diagonal elements.

$$c_{di}(\Theta_i) \leq 0 \tag{18}$$

where

$$c_{di}(\Theta_i) = 1 - \delta_i^2 - 2\mathrm{Re}\left[\Phi_{ii}^T(j\omega_{Mc_i})\right]\Theta_i$$

$$+ \ \Theta_i^T \mathrm{Re}\left[\Phi_{ii}(j\omega_{Mc_i})\Phi_{ii}^*(j\omega_{Mc_i})\right]\Theta_i \tag{19}$$

and δ_i are chosen so that the circle do not include $(-1, j0)$. That is,

$$\delta_i\,|g_{Mi}(j\omega_{Mc_i})| < |1 + g_{Mi}(j\omega_{Mc_i})| \tag{20}$$

(2) Constraint condition for non-diagonal elements.

$$c_{ni}(\Theta_i) \leq 0 \tag{21}$$

$$c_{ni}(\Theta_i) = [c_{ni_1}(\Theta_i), \ldots, c_{ni_P}(\Theta_i)]^T \tag{22}$$

where

$$\begin{aligned} -c_{ni_k}(\Theta_i) &= 1 + 2\mathrm{Re}\left[\Psi_{ii}^T(j\omega_{ni_k})\right]\Theta_i \\ &+ \Theta_i^T \Bigg\{ \mathrm{Re}\left[\Psi_{ii}(j\omega_{ni_k})\Psi_{ii}^*(j\omega_{ni_k})\right] \\ &- \alpha_i \Bigg(\sum_{\substack{q=1 \\ q\neq i}}^{m} \mathrm{Re}\left[\Psi_{qi}(j\omega_{ni_k})\Psi_{qi}^*(j\omega_{ni_k})\right] \\ &+ 2\mathrm{diag}\Big[\sum_{\substack{q=1 \\ q\neq i}}^{m-1} \sum_{\substack{r=q+1 \\ r\neq i}}^{m} \left\|\Psi_{qi}^T(j\omega_{ni_k})\right\|_2 \\ &\times \left\|\Psi_{ri}^T(j\omega_{ni_k})\right\|_2\Big]\Bigg)\Bigg\}\Theta_i \end{aligned} \tag{23}$$

$$\Psi_{ii}(j\omega) = \left[g_{i1}(j\omega), \frac{g_{i1}(j\omega)}{j\omega}, \frac{j\omega g_{i1}(j\omega)}{1 + j\omega\gamma_i}, \ldots, g_{im}(j\omega), \frac{g_{im}(j\omega)}{j\omega}, \frac{j\omega g_{im}(j\omega)}{1 + j\omega\gamma_i}\right]^T \tag{24}$$

$$\Psi_{qi}(j\omega) = \left[g_{q1}(j\omega), \frac{g_{q1}(j\omega)}{j\omega}, \frac{j\omega g_{q1}(j\omega)}{1 + j\omega\gamma_i}, \ldots, g_{qm}(j\omega), \frac{g_{qm}(j\omega)}{j\omega}, \frac{j\omega g_{qm}(j\omega)}{1 + j\omega\gamma_i}\right]^T \tag{25}$$

(3) Optimization problem.

As a result, the optimal choice of PID parameters can be attained by solving the following optimization problem.

$$\min_{\Theta_i} \{J(\Theta_i)\} \text{ subject to } c_{di}(\Theta_i) \leq 0, c_{ni}(\Theta_i) \leq 0 \tag{26}$$

In our examples, we solved the problem by using successive quadratic programing method in Matlab Optimization Toolbox.

4. A CONSIDERATION ON THE SELECTION OF REFERENCE MODEL AND FITTING FREQUENCIES

Reference model $G_M(j\omega)$ in (6) and fitting frequencies Ω_i can be chosen arbitrary as far as the resulting control system becomes stable. However, in practice, the accuracy of plant-model matching and stability of the control system are strongly affected by the choice of $G_M(j\omega)$ and Ω_i. In the following, a guideline concerning the design of reference model and the choice of fitting frequencies is considered.

4.1 Parameters selection of reference model
It is obvious that the matching accuracy between $G_M(j\omega)$ and $G(j\omega)C(j\omega)$ will be restricted by the dynamic characteristics of $G_M(j\omega)$ and $G(j\omega)C(j\omega)$. That is, if there exists a considerable characteristics deference between $G_M(j\omega)$ and $G(j\omega)C(j\omega)$, it is seems to be difficult to attain good matching between $G_M(j\omega)$ and $G(j\omega)C(j\omega)$.
However, we can choose a $G_M(j\omega)$ which will give well-posed loss function under the assumption that the phase crossover frequency of the plant $G(j\omega)$ is known. Note that, we can obtain phase crossover frequency of the plant by using, for example, the method given by *Åström* and *Hägglund* (1984).
Consider the following closed-loop reference model of which diagonal plants are given by 3rd order model:

$$G_{Mcl}(s) = \mathrm{diag}\,[g_{Mcl_i}(s)]_{i=1,\dots,m} = \mathrm{diag}\left[\frac{\omega_{M0_i}^3}{s^3+\zeta_i\omega_{M0_i}s^2+\zeta_i\omega_{M0_i}^2 s+\omega_{M0_i}^3}\right] \tag{27}$$

From (5),

$$G_M(s) = \mathrm{diag}\,[g_{M_i}(s)]_{i=1,\dots,m} = \mathrm{diag}\left[\frac{\omega_{M0_i}^3}{s^3+\zeta_i\omega_{M0_i}s^2+\zeta_i\omega_{M0_i}^2 s}\right] \tag{28}$$

Then the phase crossover frequency ω_{Mc_i} of $g_{M_i}(j\omega)$ is written as

$$\omega_{Mc_i} = \sqrt{\zeta_i}\omega_{M0_i}, \quad i = 1,\dots,m \tag{29}$$

From (4), we have

$$q_{ii}(s) = \sum_{k=1}^{m} g_{ik}(s)c_{ki}(s) \tag{30}$$

From the practical point of view, it is desirable that the matching is attained on the wide range of frequency. Taking into consideration above stated fact, we will choose ω_{M0_i} from (29) and following relation:

$$\omega_{M0_i} \in [\frac{1}{\sqrt{\zeta_i}}\min\{\omega_{c_{ij}}\}, \frac{1}{\sqrt{\zeta_i}}\max\{\omega_{c_{ij}}\}] \quad i = 1,\dots,m, j = 1,\dots,m \tag{31}$$

where $\omega_{c_{ij}}$ are the phase crossover frequencies of $g_{ij}(j\omega)$. It is noted that we can also determine $\omega_{c_{ij}}$ from experiments.
Remark 1. For example, when $\zeta_i = 2$, the $g_{M_i}(s)$ becomes a Butterworth model; when $\zeta_i = 3$, $g_{M_i}(s)$ becomes a Binomial model.

4.2 Selection of fitting frequencies
A necessary condition for solvability in the PID controller design is that the number of fitting frequencies N_i, $i = 1,\dots,m$ satisfies the relation (*Källén*, 1992):

$$N_i \geq \frac{3m}{2} \tag{32}$$

Generally speaking, the increase of N_i results in the smoothness of the matching over the range of matching frequency. However, at the same time, it causes the decrease of matching ability in the local region and the increase of computing time. In process control, to obtain good response at a steady state from our experiencing it is enough to choose fitting frequencies over a low frequency band with minimal number of N_i. Further, it is better to spread fitting frequencies logarithmically and eventually over the low frequency band.

5. SIMULATION

In this section, the effectiveness of the proposed method is confirmed through the following 2-input/2-output plant.

$$y(t) = G(s)u(t) + d(t) \tag{33}$$

$$G(s) = \begin{bmatrix} \frac{0.5}{(0.1s+1)^2(0.2s+1)^2} & \frac{4}{(0.1s+1)(0.2s+1)^2} \\ \frac{1}{(0.1s+1)(0.2s+1)^2} & \frac{2.4}{(0.1s+1)(0.2s+1)^2(0.5s+1)} \end{bmatrix}$$

where $d(t)$ is a disturbance which satisfies

$$d(t) = [\, d_1(t),\ d_2(t)\,]^T \tag{34}$$

$$d_1(t) = \begin{cases} 0.2, 10 \leq t \leq 20 \text{ [sec]}, \\ 0, \textit{otherwise}. \end{cases} \tag{35}$$

$$d_2(t) = \begin{cases} -0.2, 40 \leq t \leq 50 \text{ [sec]}, \\ 0, \textit{otherwise}. \end{cases} \tag{36}$$

Further assume the following phase crossover frequencies are known

$$\omega_{c11} = 7.0711 \text{ [rad/sec]}, \omega_{c12}=11.1803 \text{ [rad/sec]}$$
$$\omega_{c21} = 11.1803 \text{ [rad/sec]}, \omega_{c22}=4.7673 \text{ [rad/sec]} \tag{37}$$

According to the proposed design scheme, we introduce reference model $G_M(s)(i=1,2)$ as

$$G_{Mcl}(s)=\text{diag}\left[\frac{\omega_{MO_i}^3}{s^3+3\omega_{MO_i}s^2+3\omega_{MO_i}^2 s+\omega_{MO_i}^3}\right]$$

$$G_M(s)=\text{diag}\left[\frac{\omega_{MO_i}^3}{s^3+3\omega_{MO_i}s^2+3\omega_{MO_i}^2 s}\right]$$

$$\omega_{MO_1}=5.0, \omega_{MO_2}=4.0[\text{rad/sec}] \quad (38)$$

where the phase crossover frequencies of $G_M(s)$ are

$$\omega_{Mc_1}=8.663, \omega_{Mc_2}=6.930[\text{rad/sec}] \quad (39)$$

and the fitting frequencies Ω_i is given as

$$\Omega_1=[0.09, 0.3, 1.0], \Omega_2=[0.09, 0.3, 1.0][\text{rad/sec}]$$

The reference inputs are chosen such that

$$u_{M_1}(t) = 1, \quad t \geq 0 \quad (40)$$

$$u_{M_2}(t) = \begin{cases} 0, t < 30 \text{ [sec]}, \\ 1, t \geq 30 \text{ [sec]}. \end{cases} \quad (41)$$

The design parameters of diagonal and non-diagonal constraint functions are given as

$$\omega_{d1}=8.7, \omega_{d2}=6.9, [\text{rad/sec}], \delta_1=1.0, \delta_2=1.0 \quad (42)$$

$$\Omega_{n1}=[2.9841, 4.263, 6.09, 8.7, 12.4286, 17.7551]$$
$$\Omega_{n2}=[2.3667, 3.381, 4.83, 6.9, 9.8571, 14.0861]$$

$$\alpha_1=1.1, \quad \alpha_2=1.1 \quad (43)$$

The parameters for derivative terms of the controller $C(s)$ are designed as

$$\gamma_1=\gamma_2=0.1 \quad (44)$$

In this case, the parameters of PID controller can be obtained as follows.

$$\begin{aligned}
[k_{P_{11}}, k_{I_{11}}, k_{D_{11}}] &= [0.4466, -1.3736, -0.2014] \\
[k_{P_{12}}, k_{I_{12}}, k_{D_{12}}] &= [0.0906, 1.8741, 0.0891] \\
[k_{P_{21}}, k_{I_{21}}, k_{D_{21}}] &= [0.0603, 0.5833, 0.0232] \\
[k_{P_{22}}, k_{I_{22}}, k_{D_{22}}] &= [0.0082, -0.2310, -0.0046]
\end{aligned} \quad (45)$$

Simulation results are shown in Fig. 1-3. Fig. 1 shows the effect of constraint condition. Fig. 2 shows the diagonal dominance of $G(j\omega)C(j\omega)$ and Fig. 3 is an example of time response.

6. CONCLUSIONS

A design method of multivariable PID controllers based on partial model matching in frequency domain was proposed. Stability, tracking performance and robustness of the method were discussed. The effectiveness of the proposed method was confirmed through a numerical example.

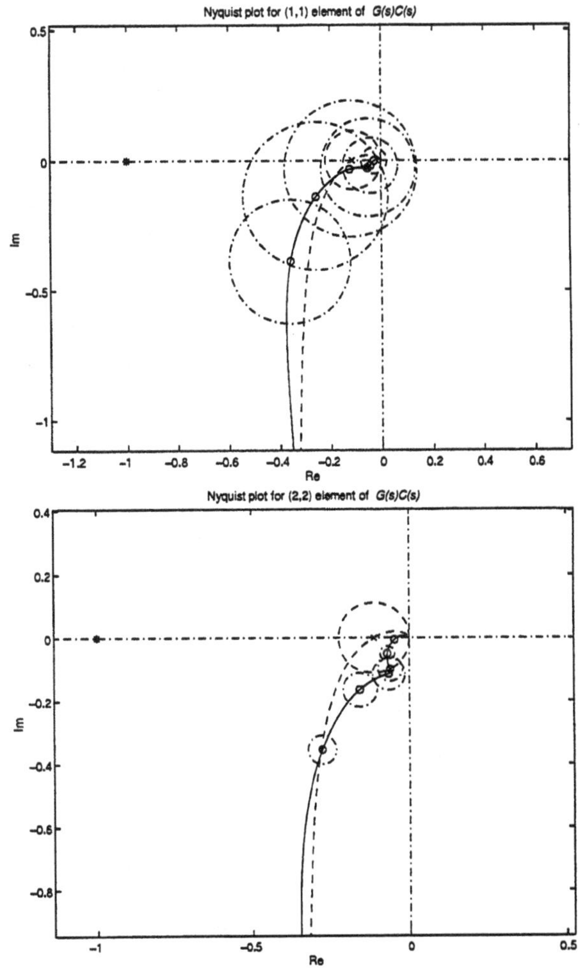

Fig. 1 Nyquist plots for $G(j\omega)C(j\omega)$ and $G_M(j\omega)$

———: Nyquist plot for the ith diagonal element of $G(j\omega)C(j\omega)$; — —: Nyquist plot for the ith diagonal element of $G_M(j\omega)$; ×: Values of $G(j\omega)C(j\omega)$ at diagonal constraint frequencies; o: Values of $G(j\omega)C(j\omega)$ and $G_M(j\omega)$ at non-diagonal constraint frequencies; · — (circle): Critical disk of Gerschgorin band at non-diagonal constraint frequencies; — — (circle): Boundary of the diagonal constraints

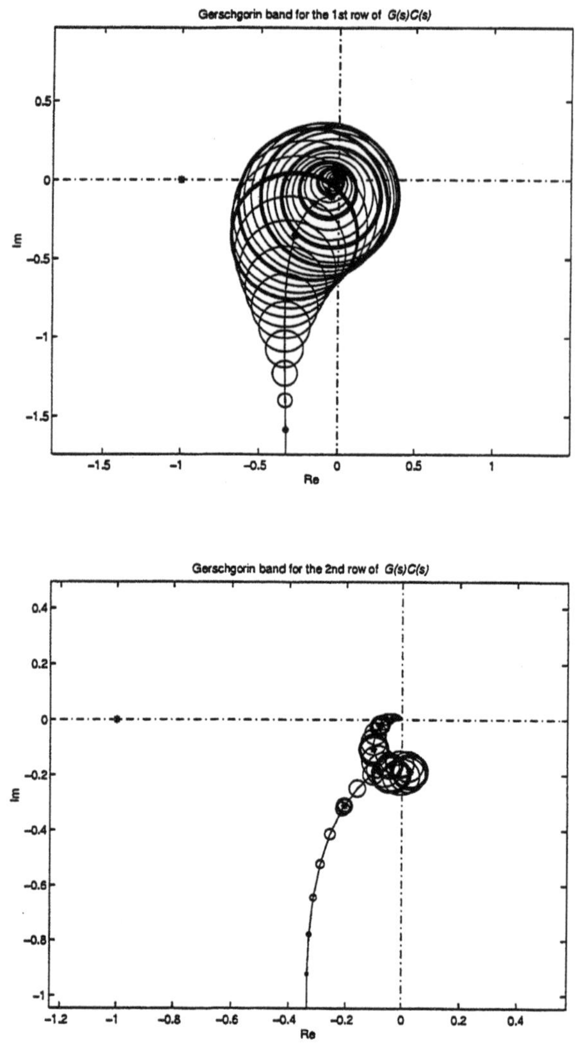

Fig. 2 Row Gerschgorin bands for $G(j\omega)C(j\omega)$

$*$: Values of $G(j\omega)C(j\omega)$ at non-diagonal constraint frequencies

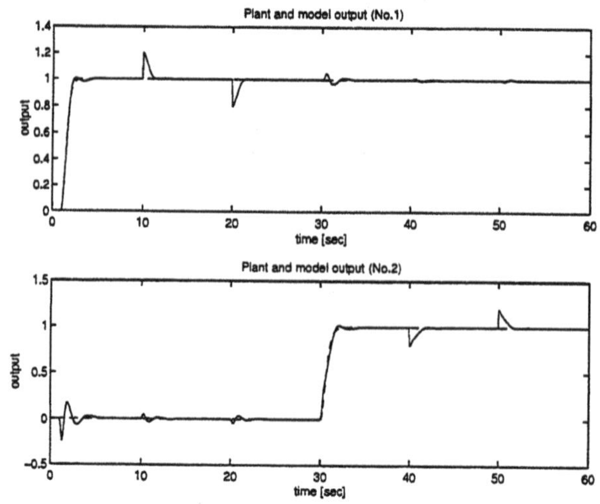

Fig. 3 Plant and model outputs

——— : Plant outputs

— — : Model outputs

REFERENCES

Åström, K.J. and T. Hägglund (1984). Automatic tuning of simple regulators with specifications on phase and amplitude margins. *Automatica*, **20**, pp. 645-651.

Egashira, T. and Z. Iwai (1996). Tuning of a multivariable PID controller on frequency domain and its application. *Trans. of JSME*, **63**, pp. 4264-4271(in Japanese).

Ho, H.K., T. H. Lee and O. P. Gan (1996). Tuning of multi-loop PID controllers based on gain and phase margins specifications. *Proc. 13th Triennial World Congress, San Francisco, USA*, **M**, pp. 211-216.

Källén, P.O. (1992). *Frequency Domain Adaptive Control*, Licentiate thesis TFRT-3211, Department of Automatic Control, Lund Institute of Technology, Lund, Sweden.

Luyben,W.L. (1986). Simple method for SISO controllers in multivariable systems. *Ind. Eng. Chem. Process Des. Dev.*, **25**, pp. 654-660.

Palmor,Z.J., Y. Halivi and N. Krasney (1995). Automatic tuning of decentralized PID controllers for TITO processes. *Automatica*, **31**, pp. 1001-1010.

Rosenbrock, H.H. (1972). *Computer-Aided Control System Design*, Academic Press.

Wang, Q.G., C.C. Hang and B. A. Zou (1996). A frequency response approach to autotuning of multivariable PID controllers. *Proc. 13th Triennial World Congress, San Francisco, USA*, **M**, pp. 295-330.

OPTIMAL COMBINATION OF IDENTIFICATION AND CONTROL FOR BOUNDED-NOISE ARX SYSTEMS

Andrzej Królikowski

Poznań University of Technology, Department of Control, Robotics and Computer Science, 60-965 Poznań, ul.Piotrowo 3A, POLAND
akrol@omega.ar-kari.put.poznan.pl

Abstract. An optimal combination of identification and control for linear bounded input/ bounded noise discrete-time ARX system is considered. Various configurations of identyfying/controlling sequences are investigated. A second-order control system is simulated. *Copyright © 1999 IFAC*

Key Words. Sequential identification/control. Bounded input. Bounded noise.

1 INTRODUCTION

The sequential identification and control can be considered as an alternative approach to the simultaneous identification and control (self-tuning). There are some reasons to use this approach for control, see for example [1]. This problem has also been considered in [2] for the case of stochastic noise systems. In this paper, an optimal combination of identification and control for bounded noise ARX systems is considered. The input applied in both identification and control periods is also assumed to be bounded in amplitude. Two identification algorithms [3],[4] are taken as a base for a design of identifying sequences, and the tracking is taken as a control objective.
The physical plant is assumed to be represented by a model of the ARX type

$$A(q^{-1})y_t = q^{-1}B(q^{-1})u_t + \eta_t \qquad (1)$$

where A, B, are polynomials in the backward shift operator q^{-1}, i.e.

$$A(q^{-1}) = 1 + a_1 q^{-1} + \dots + a_{na} q^{-na}$$

$$B(q^{-1}) = b_0 + b_1 q^{-1} + \dots + b_{nb} q^{-nb}$$

y_t is the output, and u_t is the control input bounded in amplitude

$$| u_t | \leq \alpha \qquad (2)$$

Different models can be used for η_t but as a result, the disturbance η_t is considered to be bounded with known upperbound δ_t

$$| \eta_t | \leq \delta_t \qquad (3)$$

2 IDENTIFICATION ALGORITHMS

For parameter identification of bounded-noise model (1), the methods EW-RLS [3] or FHMV [4] are used.
The model (1) can be represented in the regressor form

$$y_t = \theta^T \varphi_t + \eta_t \qquad (4)$$

where $\theta^T = (a_1, \cdots, a_{na}, b_0, \cdots, b_{nb})$ and

$$\varphi_t^T = (-y_{t-1}, \cdots, -y_{t-na}, u_{t-1}, \cdots, u_{t-nb-1}) \qquad (5)$$

2.1. The EW-RLS algorithm

The exponentially weighted recursive least squares (EW-RLS) identification algorithm derived in [3] for model (4) has a form

$$\hat{\theta}_t = \hat{\theta}_{t-1} + \frac{d_t P_{t-1} \varphi_t}{\gamma_t} (| \epsilon_t | - | \delta_t |) sgn(\epsilon_t) \qquad (6)$$

$$P_t = \frac{1}{\lambda}[P_{t-1} - \frac{d_t P_{t-1} \varphi_t \varphi_t^T P_{t-1}}{\gamma_t}(1 - | \frac{\delta_t}{\epsilon_t} |)] \qquad (7)$$

where $0 < \lambda \leq 1$, $\gamma_t = \varphi_t^T P_{t-1}\varphi_t$, $\epsilon_t = y_t - \hat{\theta}_{t-1}^T\varphi_t$ and

$$d_t = \begin{cases} 0 & \text{if } \gamma_t = 0 \text{ or } |\epsilon_t| < |\delta_t| \\ 1 & \text{otherwise} \end{cases} \qquad (8)$$

2.2. The FHMV algorithm

The minimal volume recursive algorithm of Fogel-Huang (FHMV) [4] is given by

$$\hat{\theta}_t = \hat{\theta}_{t-1} + \frac{q_t}{\delta_t^2} Z_t\varphi_t\epsilon_t \qquad (9)$$

$$Z_t = P_{t-1} - \frac{q_t P_{t-1}\varphi_t\varphi_t^T P_{t-1}}{\delta_t^2 + q_t\gamma_t} \qquad (10)$$

where $P_t = z_t Z_t$, and

$$z_t = 1 + q_t - \frac{q_t}{\delta_t^2 + q_t\gamma_t}\epsilon_t^2$$

The way of computing the coefficients q_t is given as follows [4]

$$q_t = \begin{cases} 0 & \text{if } \alpha_2^2 - 4\alpha_1\alpha_3 < 0 \\ & \text{or } -\alpha_2 + \sqrt{\alpha_2^2 - 4\alpha_1\alpha_3} \leq 0 \\ \frac{-\alpha_2 + \sqrt{\alpha_2^2 - 4\alpha_1\alpha_3}}{2\alpha_1} & \text{otherwise} \end{cases}$$

where $\alpha_1 = (n-1)R_t^2$, $\alpha_2 = R_t(2n - 1 - R_t^2 + \epsilon_t^2)$, $\alpha_3 = n(1 - \epsilon_t^2) - R_t$, $R_t = \varphi_t^T P_{t-1}\varphi_t$ and $n = n_a + n_b + 1$.

2.3. Design of identifying sequences

The goal of the input identifying sequence $\{u_t^I\}$ is twofold: to speed up the achievement of identification accuracy given by some measure, i.e. to obtain a given accuracy in the shortest possibly time, or to achieve, in a given identification time, the highest possibly accuracy determined by some measure.

2.3.1. Input sequence design for EW-RLS

The identyfying input sequence is proposed to cause the steepest descent of $detP_t$ which is proportional to the volume of ellipsoid associated with parameter estimates error. From (7), it follows that

$$\max_{|u_{t-1}| \leq \alpha} \frac{detP_{t-1}}{detP_t} \equiv \max_{|u_{t-1}| \leq \alpha} \gamma_t \qquad (11)$$

Taking into consideration

$$\gamma_t = s_{t-1}u_{t-1}^2 + t_{t-1}u_{t-1} + w_{t-1}$$

the maximization (11) yields

$$u_{t-1}^I = \alpha\ sgn\ t_{t-1} \qquad (12)$$

where

$$t_{t-1} = p_{t-1}^{na+1,1}y_{t-1} + p_{t-1}^{na+1,2}y_{t-2} + \cdots +$$

$$p_{t-1}^{na+1,na}y_{t-1} + p_{t-1}^{1,na+1}y_{t-1} + p_{t-1}^{2,na+1}y_{t-2} + \cdots +$$

$$p_{t-1}^{na,na+1}y_{t-na+1}p_{t-1}^{na+2,na+1}u_{t-1} + \cdots +$$

$$p_{t-1}^{na+nb,na+1}u_{t-na+1} + p_{t-1}^{na+1,na+2}u_{t-1} + \cdots +$$

$$+p_{t-1}^{na+1,na+nb}u_{t-na+1}$$

where $P_{t-1} = \{p_{t-1}^{k,l}\}$ for $k, l = 1, \cdots, na + nb$.

2.3.2. Input sequence design for FHMV

From Z_t and P_t (10), it follows that

$$P_t = z_t(I - \frac{q_t P_{t-1}\varphi_t\varphi_t^T}{1 + q_t\gamma_t})P_{t-1} \qquad (13)$$

to give

$$\frac{detP_t}{detP_{t-1}} = \frac{z_t^{na+nb}}{1 + q_t\gamma_t} = \frac{1}{1 + q_t\gamma_t}(1 + q_t - \frac{q_t\epsilon_t^2}{\delta_t^2 + q_t\gamma_t})^{na+nb} \qquad (14)$$

So, again the identyfying input has to be chosen as

$$\max_{|u_{t-1}| \leq \alpha} \frac{detP_{t-1}}{detP_t} \equiv \min_{|u_{t-1}| \leq \alpha} \frac{detP_t}{detP_{t-1}} \equiv \max_{|u_{t-1}| \leq \alpha} \gamma_t \qquad (15)$$

yielding the similar result as (12).

3 TRACKING CONTROL PROBLEM

The tracking controller has a form [5]

$$u_t = \hat{\theta}_t^{*T}\varphi_t^{'} \qquad (16)$$

where the controller parameters $\hat{\theta}_t^*$ are estimates of θ^* defined below and

$$\varphi_t^{'T} = (-y_t, \cdots, -y_{t-na+1}, -u_{t-1}, \cdots, -u_{t-nb}, r_{t+1}) \qquad (17)$$

where r_t is a known reference signal to be followed by the output y_t. Here it holds

$$y_{t+1} = \alpha(q^{-1})y_t + \beta(q^{-1})u_t + \nu_t \qquad (18)$$

where $\alpha(q^{-1}) = G(q^{-1})$, $\beta(q^{-1}) = F(q^{-1})B(q^{-1})$, $|\nu_t| \leq \nu_0$ while $F(q^{-1}), G(q^{-1})$ can be determined from diofantine equation

$$F(q^{-1})A(q^{-1}) + q^{-1}G(q^{-1}) = 1 \qquad (19)$$

with $deg\alpha = na - 1$, $deg\beta = nb$. Here

$$\theta^* = (\theta_y^{*T}, \theta_u^{*T}, \theta_r^*)^T \qquad (20)$$

where

$$\theta_y^{*T} = (\alpha_0^{'}, \cdots, \alpha_{na-1}^{'})$$

$$\theta_u^{*T} = (\beta_1^{'}, \cdots, \beta_{nb}^{'}), \quad \theta_r^* = \frac{1}{\beta_0}$$

with

$$\alpha_i^{'} = \frac{\alpha_i}{\beta_0}, \quad i = 1, \cdots, na-1$$

$$\beta_j^{'} = \frac{\beta_j}{\beta_0}, \quad j = 1, \cdots, nb, \quad \beta_0^{'} = \beta_0 = b_0$$

Equation (18) can be written as

$$u_t = \theta^{*T}\varphi_t - \bar{\nu}_t \tag{21}$$

where

$$\varphi_t^T = (-y_t, \cdots, -y_{t-na+1}, -u_{t-1}, \cdots, -u_{t-nb}, y_{t+1}) \tag{22}$$

and $|\ \bar{\nu}_t\ | \leq \frac{\nu_a}{b_0}$. Looking at (16),(21) and (22) one can see the tracking and disturbance rejecting mechanism of the controller (16). The bounded control applied to the system is then

$$u_t^C = sat(u_t; \alpha) \tag{23}$$

4 COMBINATION OF IDENTIFICATION AND CONTROL

The combination of identification and control consists of consecutive identification in open-loop configuration, when the bounded identifying sequence $\{u_t^I\}$ is applied to the system, and the control in closed-loop configuration, when the control sequence $\{u_t^C\}$is applied. Identification and control have to be achieved within the time interval $[0, T]$ where T is given. The identification of system parameters lasts during the period $[0, T_I]$. So, the identification period T_I and the amplitude α of an identyfying signal are the only two experiment parameters considered. A controller obtained from the parameter estimates is used during the control time T_C, so $T = T_I + T_C$. The impact of identification accuracy measured by $detP_t$ on the control performance measured by the sum of tracking errors can be analyzed in order to find an optimal trade-off. To this end, various configurations of sequential combination of EW-RLS and FHMV identification together with tracking control are simulated.

5 SIMULATIONS

Consider the model (4) with numerical parameter values $a_1 = -1.8, a_2 = 0.9, b_0 = 1.0, b_1 = 0.5$, where η_t was taken as a truncated zero mean normal variable with variance $\sigma_\eta^2 = 0.2$, and $\delta_t = \delta = 0.4$. System parameters were identified using EW-RLS and FHMV algorithms with $\hat{\theta}_0 = 0$, $P_0 = 100I$, and $\lambda = 0.95$.
The identyfying input is taken as in (12). For the above example, we have for t_{t-1}

$$t_{t-1} = p_{t-1}^{3,1}y_{t-1} + p_{t-1}^{3,2}y_{t-2} + p_{t-1}^{1,3}y_{t-1} + p_{t-1}^{2,3}y_{t-2} + p_{t-1}^{4,3}u_{t-2} + p_{t-1}^{3,4}u_{t-2}$$

The control input is taken as in (23) where for the considered example we have

$$u_t = \frac{1}{b_0}(a_1 y_t + a_2 y_{t-1} - b_1 u_{t-1} + r_{t+1})$$

and

$$\alpha_0^{'} = \frac{-a_1}{b_0}, \quad \alpha_1^{'} = \frac{-a_2}{b_0}, \quad \beta_1^{'} = \frac{b_1}{b_0}, \quad \theta_r^* = \frac{1}{b_0}$$

The reference signal was taken as $r_t = 0$ in the identification period, i.e. for $t = 1, \cdots, T_I$, while in the control period $r_t = 4$, i.e. for $t = T_I + 1, \cdots, T$. The optimal combination of consecutive identification and control periods is considered regarding the cost $J = \frac{1}{T}\sum_{t=1}^{T} e_t^2$, where the tracking error $e_t = r_t - y_t$, and $T = 50$. For optimization, the sample cost $\bar{J}$ averaged over 30 realizations was taken into consideration. The obtained results concerning optimal identification periods T_I^{opt} are shown in Figs.1,2,3 for the EW-RLS method with $\alpha = 4, 6, 8$, where $T_I^{opt} = 5, 6, 9$, respectively. The corresponding plots for the FHMV method are given in Figs.4,5,6, where $T_I^{opt} = 10, 10, 9$, respectively. Plot of signals u_t^I, u_t^C, y_t is shown in Fig.7 for $\alpha = 4$ and $T_I^{opt} = 5$ when the EW-RLS method is used. In Fig.8, the corresponding plot is shown for $T_I^{opt} = 10$ when the FHMV method is used. The decreasing rate of $\bar{J}_t$ corresponding to Figs.1,2,3 with optimal T_I^{opt} is given in Fig.9 for the EW-RLS method. The corresponding plots of $\bar{J}_t$ for the FHMV method are shown in Fig.10. Additionally, the plots of estimates for EW-RLS and FHMV methods are presented in Figs.11,12, respectively for the identification period $T_I = 10$.

6 CONCLUSIONS

An optimal combination of identification and control problem is formulated and corresponding algorithms are proposed for an ARX model with bounded disturbance. A second-order example is simulated showing that the EW-RLS identification algorithm converges faster than the FHMV algorithm. The obtained results show an optimal interplay between T_I and T_C regarding the cost $\bar{J}$ for a given α, where the identification period T_I^{opt} is shorter for the EW-RLS method than for the FHMV method. It can also be seen that T_I^{opt} depends on the imposed constraint α. An interesting, however to be expected, fact is that when the constraint α is enough large, i.e. when the control signal does not saturate, then the obtained costs $\bar{J}$ for both methods are almost equal with the same T_I^{opt}, see the case $\alpha = 8, T_I^{opt} = 9$.

REFERENCES

1. Franke B.: Optimal Combination of Sequential Identification and Control, Ph.D.Thesis, Technical University of Enschede, 1993.

2. Królikowski A.: Application of input signal design in system identification for adaptive control, *Int. J. Syst.Sci.*, Vol.17, No.2, 1986, pp.305-318.

3. Canudas de Wit C, Carrillo J.: A modified EW-RLS algorithm for systems with bounded disturbances, *Automatica*, Vol.20, No.3, 1990, pp.599-606.

4. Fogel E, Huang Y.F.: On the value of information in system identification - bounded noise case, *Automatica*, Vol.18, No.2, 1982, pp.229-238.

5. Annaswamy A.M., Karason S.P.: Discrete time adaptive control in the presence of input constraints, *Automatica*, Vol.31, No.10, 1995, pp.1421-1431.

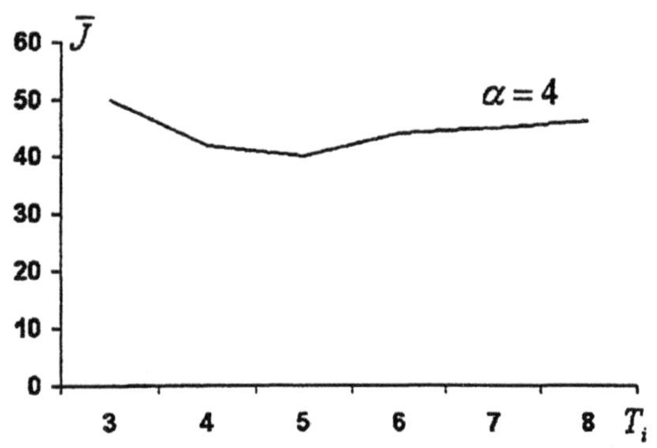

Figure 1: Optimization results, $T_I^{opt} = 5$

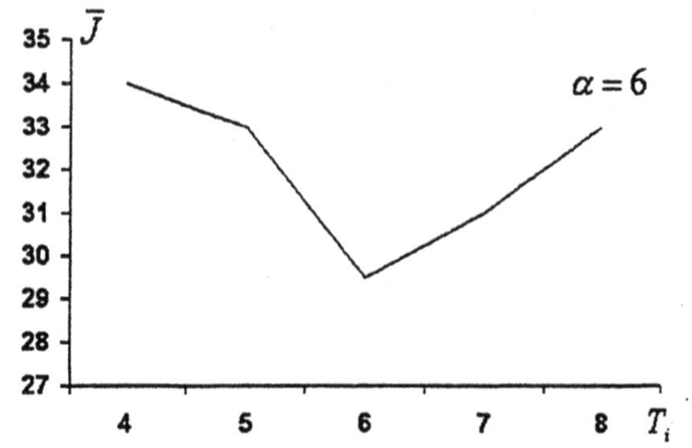

Figure 2: Optimization results, $T_I^{opt} = 6$

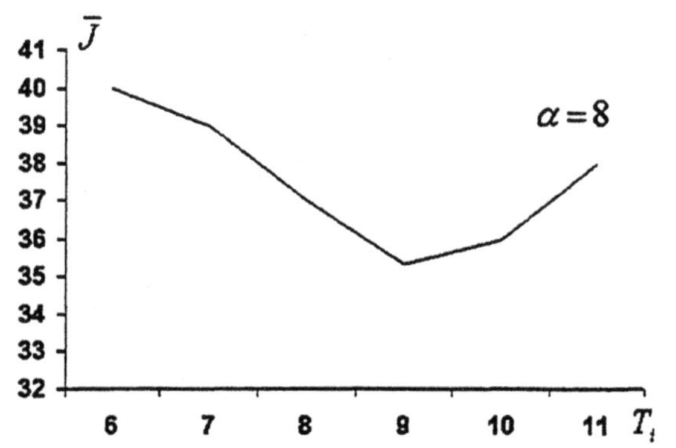

Figure 3: Optimization results, $T_I^{opt} = 9$

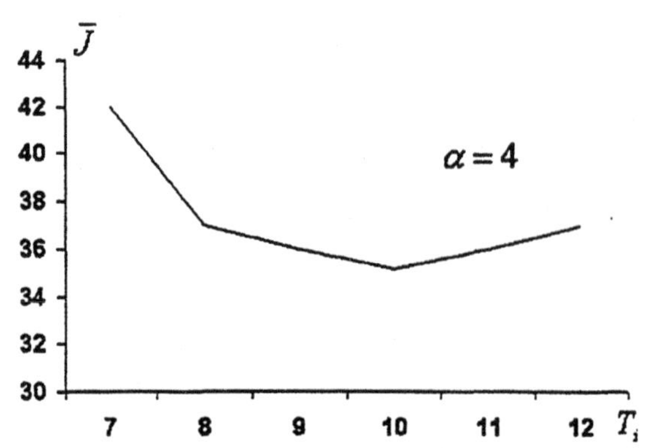

Figure 4: Optimization results, $T_I^{opt} = 10$

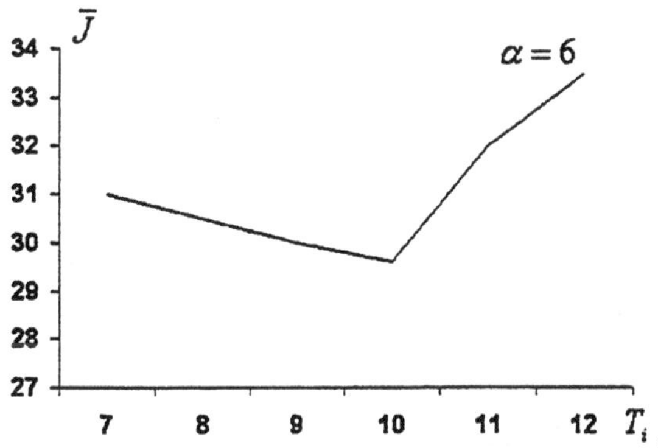

Figure 5: Optimization results, $T_I^{opt} = 10$

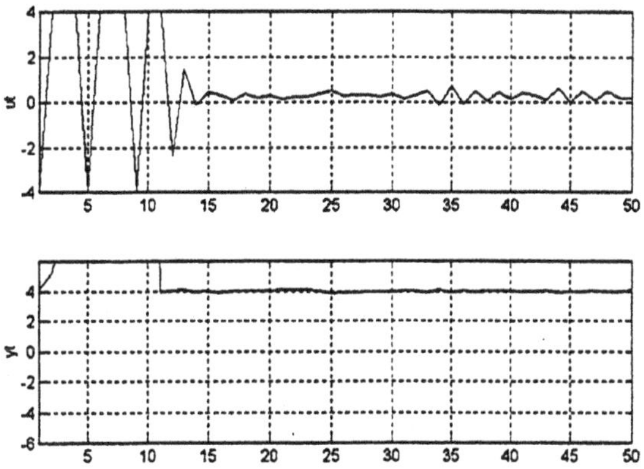

Figure 8: Identification/control: $\alpha = 4, T_I^{opt} = 10$

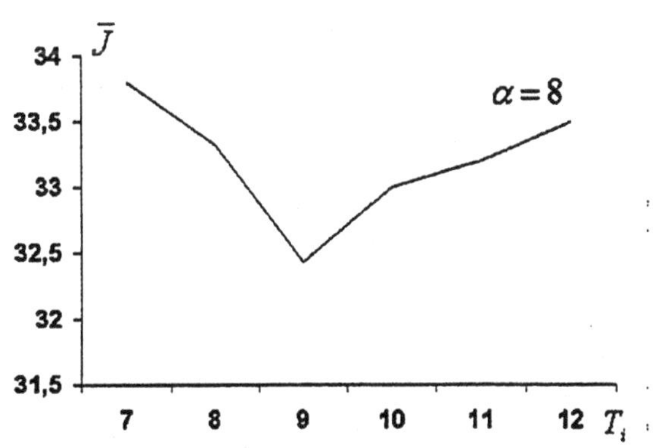

Figure 6: Optimization results, $T_I^{opt} = 9$

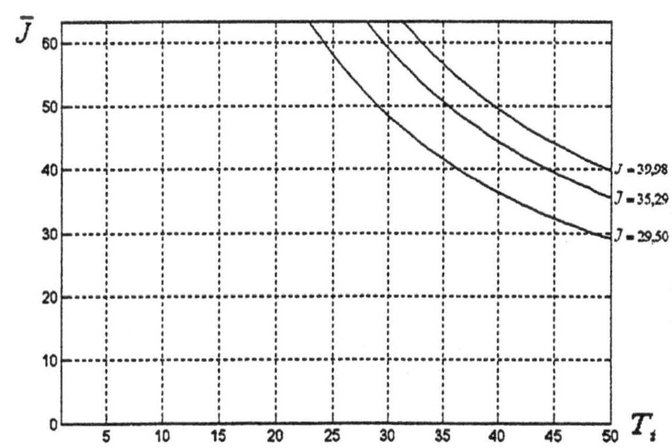

Figure 9: Sample cost $\bar{J}$

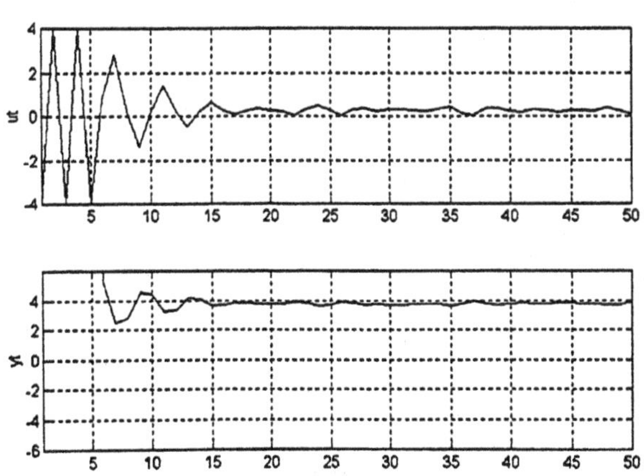

Figure 7: Identification/control: $\alpha = 4, T_I^{opt} = 5$

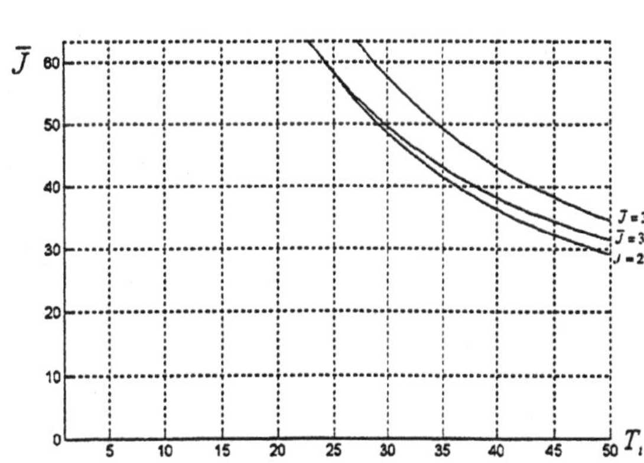

Figure 10: Sample cost $\bar{J}$

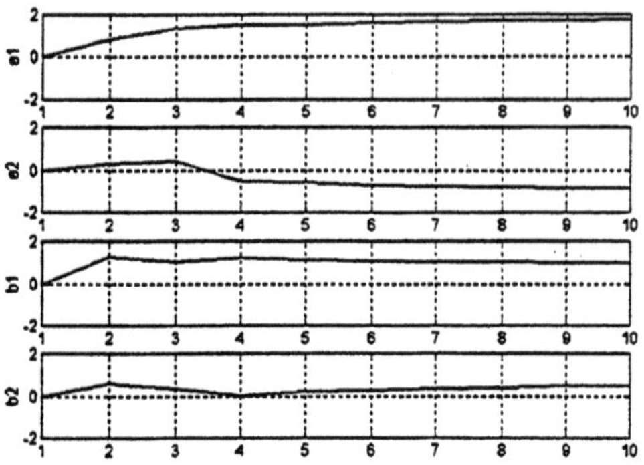

Figure 11: Estimates for EW-RLS method

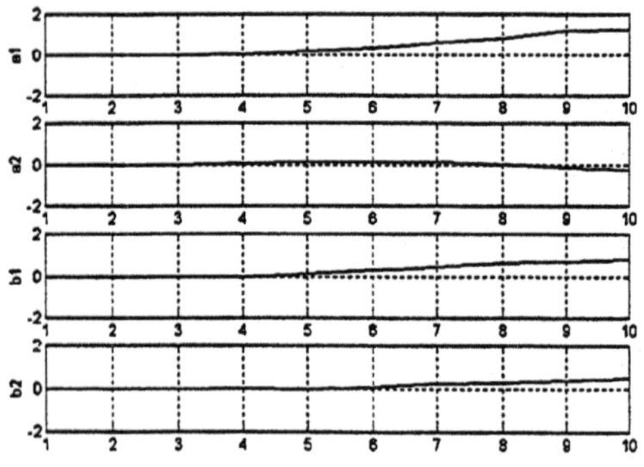

Figure 12: Estimates for FHMV method

14th Triennial World Congress, Beijing, P.R. China

C-2a-09-4

ON IDENTIFICATION FOR ROBUST CONTROL

Huipin Zhang *, **A.C. Antoulas** *

* *Department of Electrical and Computer Engineering*
Rice University
Houston, Texas 77251-1892, USA

Abstract: Given measured data we propose a model consisting of a linear, time-invariant system affected by norm-bounded perturbation. In this paper we examine the special case of (real) parameter uncertainty. The plants belonging to the resulting uncertain family form a convex set. A robust control method based on strict-positive realness results can be applied. This method allows the computation of a controller which maximizes the stability margin. Lower and upper bounds of the optimal stability margin are obtained in terms of different measures of the uncertainty. The approach depends on two key parameters: the bound of the perturbation, and the input used to obtain the data. It turns out that the size of the uncertain family can be reduced by intersecting the model families obtained through the use of different inputs.

Keywords: Identification, convex model family, norm-bounded perturbations, linear fractional representations, ellipsoids, robust performance, LMIs

1. INTRODUCTION

Robust control theory assumes the existence of a nominal plant together with uncertainty, often taken to be norm bounded. A number of approaches to system identification have been proposed in the past decade which attempt to fit this mold. One of them belongs to the general category of *set membership identification* which aims at characterizing the membership set of the parameters consistent with the data, the model structure and the perturbation characteristics. Set membership identification including error bounded identification has been studied in the literature; see e.g. (Norton, 1987), (Bai *et al.*, 1997), (Antoulas and Anderson, 1999).

We propose an approach to robust identification which seeks to provide a nominal model together with a norm-bounded perturbation. More precisely, the proposed approach can be cast into the set membership framework. It is assumed that measured input/output data is available; with some a priori knowledge, an upper bound ε on the norm of the perturbation which affects the data is deduced. Furthermore, in this paper we make the assumption that the noise is *low*, i.e. condition (19) given in corollary 3.1 is satisfied. With these ingredients, an uncertain family of plants is obtained; this family is *convex*, in particular, an ellipsoid in parameter space. Consequently, this family is given in terms of a linear fraction, the parameter being a contraction, which lends itself well to the application of robust control techniques. The robust control problem for this family with real parameter uncertainty, makes use of results developed for positive real functions; it becomes tractable since it can be converted into a quasi-convex optimization problem with linear matrix inequality (LMI) constraints. Easily computable bounds for the maximal stability margin, in terms of various measures of the uncertainty (e.g. the radius or the root-mean-square radius of the ellipsoid) are derived.

It turns out that both ε and the input function used to generate the measurements are important parameters of our approach. The issue of reducing the uncertainty by intersecting the ellipsoids obtained for different inputs, is addressed.

The rest of the paper is organized as follows. In section 2 the problem is formulated, followed by the convex parameterization of the uncertain family in section 3. The problem of robust performance is discussed in the following section. Section 5 addresses the issue of improving the robust performance when multiple sets of data record are available. The paper concludes with an illustrative example (section 6). Note that for reasons of length proofs are omitted.

2. PROBLEM FORMULATION

Consider the following input-output data records:

$$\hat{w}_t = \begin{pmatrix} \hat{u}_t \\ \hat{y}_t \end{pmatrix} \in \mathbb{R}^2, \; t = 0, 1, \; \cdots, \; N, \quad (1)$$

generated by some discrete-time dynamical system. In this paper we seek to model the data as generated by a *linear, time-invariant, discrete-time* system together with an *additive* perturbation. This implies that the data can be decomposed in the form:

$$\hat{w}_t = w_t + \tilde{w}_t = \begin{pmatrix} u_t \\ y_t \end{pmatrix} + \begin{pmatrix} \tilde{u}_t \\ \tilde{y}_t \end{pmatrix}, \quad (2)$$

where w_t is generated by a linear time-invariant system and $\tilde{w}_t$ is the perturbation.

To motivate the introduction of the tools used throughout the paper, assume that the data record is produced by a linear, time-invariant, discrete-time system. Consider a model which is compatible with (i.e. may have generated) the data w_t:

$$(p(\sigma) \quad q(\sigma))w_t = 0 \quad (3)$$

for appropriate values of t, where σ denotes the forward shift operator, and

$$p(\xi) := p_0 + p_1\xi + \; \cdots \; + p_m\xi^m, \quad p_i \in \mathbb{R},$$
$$q(\xi) := q_0 + q_1\xi + \; \cdots \; + q_n\xi^n, \quad q_i \in \mathbb{R},$$

where m, n, $(m \leq n)$, are two specified nonnegative integers. Also, the model is usually represented by its *transfer function:* $H(z) = -\frac{p(z)}{q(z)}$, or its *coefficient vector*

$$x := (p_0 \;\; p_1 \;\; \cdots \;\; p_m \;\; q_0 \;\; q_1 \;\; \cdots \;\; q_n)^T. \; (4)$$

Equivalently, (3) can be written as a difference equation: $\sum_{i=0}^{m} p_i u_{k+i} + \sum_{j=0}^{n} p_j u_{k+j} = 0$, for $k = 0, 1, \ldots, s$, where $s := N - n + 1$. In order to write these equations in matrix form, we define the *input matrix* U and *output matrix* Y

$$U = [u_{i+j}]_{i,j=0}^{m,s-1}, \text{ and } Y = [y_{i+j}]_{i,j=0}^{n,s-1}, \quad (5)$$

which are Hankel matrices. The **data matrix** is:

$$\mathcal{M} := \begin{bmatrix} U \\ Y \end{bmatrix} \in \mathbb{R}^{r \times s}, \; r := m + n + 2 \quad (6)$$

It readily follows from the above difference equations that the coefficient vector x in (4) is in the *left kernel* of the data matrix $\mathcal{M}$, i.e.,

$$x^T \mathcal{M} = 0. \quad (7)$$

The perturbed data matrix $\hat{\mathcal{M}}$, and the perturbation matrix $\tilde{\mathcal{M}}$ are defined accordingly. We thus have

$$\hat{\mathcal{M}} = \mathcal{M} + \tilde{\mathcal{M}} \quad (8)$$

Motivated by (7), we will assume that $\mathcal{M}$ has a non-trivial left kernel. Furthermore, we will assume that an *upper bound* on the norm of the perturbation $\tilde{\mathcal{M}}$ is a priori known:

$$\| \tilde{\mathcal{M}} \|_2 \; \leq \; \varepsilon. \quad (9)$$

The problem we will consider is to characterize all models compatible with (8) under the assumption (9).

Remarks. (a) The determination of ε in several different situations is discussed in (Zhang and Antoulas, 1998). **(b)** The coefficient vector x in (4) completely determines the model. In our formulation, all coefficients (including the highest q_n) are allowed to vary. In the next section we will show how to parameterize the family of uncertain models (denoted below by $\mathcal{F}$) under the normalization constraint $q_n = 1$. ■

We are now ready to describe the family of uncertain models consistent with the data (1), the model structure (3) and the perturbation characteristic (9):

$$\bar{\mathcal{F}} := \{x \in \mathbb{R}^r; \; \exists \mathcal{M}, \ni x^T\mathcal{M} = 0 \; \text{ and } \; \|\hat{\mathcal{M}} - \mathcal{M}\|_2 \leq \varepsilon\}.$$

Assume that a decomposition as in (8) exists, and $x \in \mathbb{R}^r$ is an element in the left kernel of $\mathcal{M}$. Then

$$\frac{\|x^T\hat{\mathcal{M}}\|_2}{\|x\|_2} = \frac{\|x^T(\mathcal{M} + \tilde{\mathcal{M}})\|_2}{\|x\|_2} = \frac{\|x^T\tilde{\mathcal{M}}\|_2}{\|x\|_2} \leq \varepsilon.$$

This leads to an equivalent description of the model family $\bar{\mathcal{F}}$:

$$\bar{\mathcal{F}} = \left\{ x \in \mathbb{R}^r; \; \mu(x, \hat{\mathcal{M}}) := \frac{\|x^T\hat{\mathcal{M}}\|_2}{\|x\|_2} \leq \varepsilon \right\} \quad (10)$$

where $\mu(x, \hat{\mathcal{M}})$ is called the *misfit function* between the uncertain model x and $\hat{\mathcal{M}}$. In (Antoulas, 1997) it is shown that in the special case of *low* noise (see 3.1), $\bar{\mathcal{F}}$ is a convex cone.

Our aim below is to describe the this family with the normalization constraint: $q_n = 1$, i.e. the denominator polynomial of the transfer function is normalized to one. In order to describe the resulting family $\mathcal{F}$, we proceed by partitioning $\hat{\mathcal{M}}$ as follows:

$$\hat{\mathcal{M}} = \begin{bmatrix} \hat{\mathcal{M}}_1 \\ \hat{m}_2^T \end{bmatrix}, \; \hat{m}_2^T : \text{ last row of } \hat{\mathcal{M}}, \quad (11)$$

and define

$$\begin{aligned} A &:= \hat{\mathcal{M}}_1 \hat{\mathcal{M}}_1^T - \varepsilon^2 I_{r-1} \\ b &:= \hat{\mathcal{M}}_1 \hat{m}_2 \\ c &:= \hat{m}_2^T \hat{m}_2 - \varepsilon^2 \end{aligned} \quad (12)$$

or equivalently,

$$\begin{bmatrix} A & b \\ b^T & c \end{bmatrix} := \hat{\mathcal{M}}\hat{\mathcal{M}}^T - \varepsilon^2 I_r$$

Now, define the set:

$$\mathcal{E} := \{\theta \in \mathbb{R}^{r-1};\ \theta^T A \theta + 2b^T\theta + c \le 0\} \quad (13)$$

The *normalized* model family $\mathcal{F}$ can be expressed as follows:

$$\mathcal{F} := \left\{ x = \begin{bmatrix} \theta \\ 1 \end{bmatrix} \in \mathbb{R}^r;\ \theta \in \mathcal{E} \right\} \quad (14)$$

The set $\mathcal{E}$ is called the *uncertain parameter set.* Thus the uncertain model family $\mathcal{F}$ is fully determined by the uncertain parameter set $\mathcal{E}$.

3. CHARACTERIZATION OF THE MODEL FAMILY

In this section we will characterize the model family $\mathcal{F}$, or equivalently, the uncertain parameter set $\mathcal{E}$ by studying the quadratic inequality given in (13). It turns out that under a low noise assumption, $\mathcal{E}$ is an ellipsoid.

Lemma 3.1. Let $A \in \mathbb{R}^{(r-1)\times(r-1)}, b \in \mathbb{R}^{r-1}$, $c \in \mathbb{R}$ be as defined in (12). Suppose A has the following eigenvalue decomposition

$$A = U \begin{bmatrix} D_2 & \\ & -D_1 \end{bmatrix} U^T \quad (15)$$

where U is an orthogonal matrix and D_1, D_2 are two diagonal positive definite matrices

$$D_1 = \begin{bmatrix} \beta_k^{-2} & & \\ & \ddots & \\ & & \beta_1^{-2} \end{bmatrix}, \; D_2 = \begin{bmatrix} \alpha_{r-1}^{-2} & & \\ & \ddots & \\ & & \alpha_{k+1}^{-2} \end{bmatrix}$$

with $\alpha_i > 0$, $\beta_j > 0$. Then θ is a vector satisfying the quadratic inequality (13) if and only if there holds

$$\theta = -A^{-1}b + \sum_{i=1}^{k} \beta_i \gamma_i u_i + \sum_{i=k+1}^{r-1} \alpha_i \delta_i u_i, \quad (16)$$

where u_i are columns of the matrix U, $U = [u_{r-1} \ \cdots \ u_2 \ u_1]$, and the parameters δ_i and γ_i satisfy

$$\sum_{i=k+1}^{r-1} \delta_i^2 - \sum_{i=1}^{k} \gamma_i^2 \le b^T A^{-1} b - c \quad (17)$$

Next we will make use of this result to characterize the model family $\mathcal{F}$ in terms of (14) and (13). Assume the singular value decomposition (SVD) of the matrix $\hat{\mathcal{M}}_1$ is given by

$$\hat{\mathcal{M}}_1 = U \begin{bmatrix} \sigma_{r-1} & & 0 \\ & \ddots & \vdots \\ & \sigma_1 & 0 \end{bmatrix} V^T,$$

where $U \in \mathbb{R}^{(r-1)\times(r-1)}, V \in \mathbb{R}^{s\times s}$ are orthogonal matrices and $U = [u_{r-1} \ \cdots \ u_1]$ (u_i's are left singular vectors of $\hat{\mathcal{M}}_1$), $0 \le \sigma_1 \le \cdots \le \sigma_{r-1}$. In the case when $\sigma_{k+1} > \varepsilon > \sigma_k$, the eigenvalue decomposition (15) can be easily derived for A in (12), and

$$\beta_i = \frac{1}{\sqrt{\varepsilon^2 - \sigma_i^2}}, \quad \alpha_i = \frac{1}{\sqrt{\sigma_i^2 - \varepsilon^2}} \quad (18)$$

According to the above lemma, we thus have the following

Theorem 3.1. If $\sigma_{k+1} > \varepsilon > \sigma_k$, then $\theta \in \mathcal{F}$ if and only if θ is given by (16), where the parameters δ_i and γ_i satisfy the condition (17). If, furthermore, ε is different from any singular value of the data matrix $\hat{\mathcal{M}}$, it follows that $b^T A^{-1} b - c \ne 0$, and α_i, β_i in (18) can be replaced by

$$\beta_i = \sqrt{\frac{|b^T A^{-1} b - c|}{\varepsilon^2 - \sigma_i^2}}, \quad \alpha_i = \sqrt{\frac{|b^T A^{-1} b - c|}{\sigma_i^2 - \varepsilon^2}}$$

Condition (17) can be accordingly replaced by

$$\sum_{i=k+1}^{r-1} \delta_i^2 - \sum_{i=1}^{k} \gamma_i^2 \le \operatorname{sign}(b^T A^{-1} b - c)$$

The following is the special case of *low noise.*

Corollary 3.1. Assuming that the condition

$$\sigma_{\min}(\hat{\mathcal{M}}) < \varepsilon < \sigma_1 \quad (19)$$

is satisfied, A is positive definite and $b^T A^{-1} b - c > 0$; consequently, the set $\mathcal{E}$ is an ellipsoid with center $\theta_c = -A^{-1}b$ and half axis lengths

$$\alpha_i = \sqrt{\frac{b^T A^{-1} b - c}{\sigma_i^2 - \varepsilon^2}},\ i = 1, \ldots, r-1 \quad (20)$$

Furthermore, $x \in \mathcal{F}$ if and only if $x = [\theta^T \ \ 1]^T$, and θ has the form:

$$\theta = -A^{-1}b + \sum_{i=1}^{r-1} \alpha_i \delta_i u_i \quad (21)$$

where $\delta = (\delta_1 \ \cdots \ \delta_{r-1})$ is a contraction: $\delta_1^2 + \cdots + \delta_{r-1}^2 \leq 1$.

From the vectors u_i and x_c we can define the polynomials $n_i(z)$, $d_i(z)$, $n_c(z)$, $d_c(z)$:

$$\begin{bmatrix} -n_i(z) \\ d_i(z) \end{bmatrix} := \begin{bmatrix} 1 \ z \ \cdots \ z^m \ 0 \ 0 \ \cdots \ \ 0 \\ 0 \ 0 \ \cdots \ \ 0 \ \ 1 \ z \ \cdots \ z^{n-1} \end{bmatrix} u_i$$

$$\begin{bmatrix} -n_c(z) \\ d_c(z) - z^n \end{bmatrix} := \begin{bmatrix} 1 \ z \ \cdots \ z^m \ 0 \ 0 \ \cdots \ \ 0 \\ 0 \ 0 \ \cdots \ \ 0 \ \ 1 \ z \ \cdots \ z^{n-1} \end{bmatrix} x_c$$

which determine the uncertain models. Then the transfer function of the corresponding uncertain model has the general parameterization:

$$\frac{n_\delta(z)}{d_\delta(z)} = \frac{n_c(z) + \sum_{i=1}^{r-1} \alpha_i \delta_i n_i(z)}{d_c(z) + \sum_{i=1}^{r-1} \alpha_i \delta_i d_i(z)}. \quad (22)$$

Remarks. **(a)** The semiaxis lengths α_i increase with increasing ε (upper bound on the norm of the perturbation). For a fixed ε the largest semiaxis length, or the radius of the ellipsoid is α_1, which is determined by the smallest singular value σ_1 of $\hat{\mathcal{M}}_1$. Moreover, the larger the smallest singular value, the smaller the radius of the ellipsoid. **(b)** In the case $\varepsilon \geq \sigma_1$, the characterization of the uncertain family is more involved. In this paper we will concentrate on the low noise case (19). **(c)** The transfer function (22) is a linear fraction in the real parameter δ, i.e., δ appears linearly in the numerator and denominator. **(d)** Many algorithms have been developed to exactly or approximately characterize the set $\mathcal{E}$ under the assumption that the measurements are affected by bounded noise. These methods are known as *error-bounded estimation*. For details, see e.g., (Norton, 1987), (Bai *et al.*, 1997). ■

4. ROBUST PERFORMANCE

In this section we will briefly consider the robust synthesis problem. The uncertain plant is given by (22), where δ is a *real* parameter uncertainty. Using ideas first developed in (Anderson *et al.*, 1990), and subsequently used in (Rantzer and Megretski, 1994), all controllers that simultaneously stabilize the uncertain model family, can be obtained by solving a quasi-convex optimization problem.

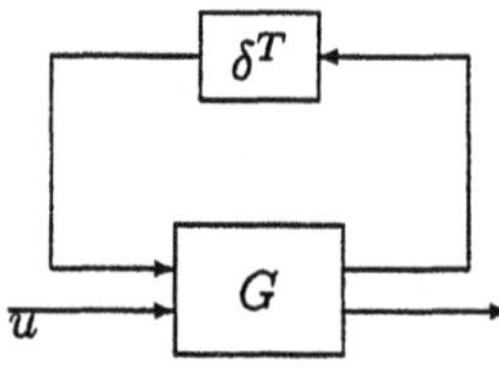

Fig. 1. Block diagram of the uncertain plant, with δ constant, $||\delta|| \leq 1$.

From (22), the transfer function of every model in the family $\mathcal{F}$ can be expressed compactly in the form:

$$G_\delta(z) = \frac{\tilde{N}(z) + \delta^T \tilde{N}_\delta(z)}{\tilde{M}(z) + \delta^T \tilde{M}_\delta(z)} \quad (23)$$

where $\tilde{M}$, $\tilde{N}$, $\tilde{M}_\delta$, $\tilde{N}_\delta$ are all known stable transfer functions (the first two are scalar, and the last two are column vectors of size $r-1$); furthermore $\tilde{M}$ and $\tilde{N}$ are assumed to be coprime, and δ is a real contractive vector of size $r-1$, representing the uncertainty. The uncertain plants with linear fractional transfer function (23) can be represented as in figure 1, where the transfer function G is determined by $\tilde{M}$, $\tilde{N}$, $\tilde{M}_\delta$ and $\tilde{N}_\delta$. The controllers that stabilize the family of uncertain models (23) have been characterized using convex parameterization in (Rantzer and Megretski, 1994).

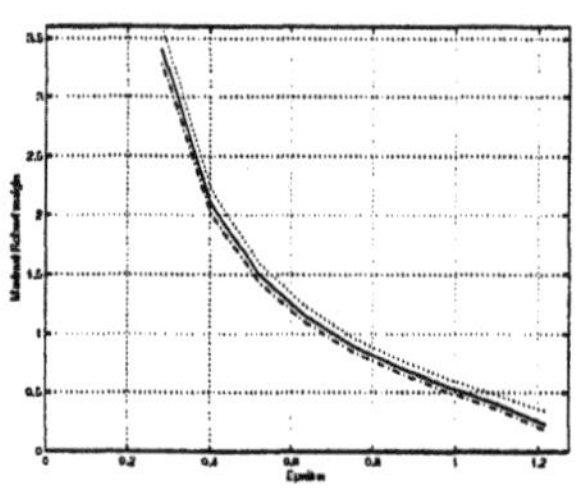

Fig. 2. Maximal stability margin together with upper and lower bounds, as a function of ε.

The advantage of this convex parameterization of the robust controllers is that the *robust performance problem* can be stated in terms of quasi-convex optimization. The *maximal stability margin* of the problem can thus be computed by solving an infinite-dimensional optimization problem. It turns out that Ritz approximation methods for solving such problems can be used. The key idea is to approximate the solution by solving a sequence of finite-dimensional optimization problems with Linear Matrix Inequality (LMI) constraints (Boyd *et al.*, 1994). In figure 2 we show (solid curve) the relationship between the maximal stability margin and the parameter ε, for a typical case. In this figure we also show the upper bound (dotted curve) and lower bound (dashdot curve) expressed in terms of the smallest semiaxis length and the radius of the ellipsoid $\mathcal{E}$ (Zhang and Antoulas, 1998). As expected, the maximal stability margin decreases with increasing perturbation bound ε. In other words, *robust performance is inversely proportional to the size of the uncertainty.*

5. UNCERTAINTY AND MULTIPLE DATA SETS

In this section we assume that the system generating the data is linear time-invariant or time-varying. The problem we wish to address is how to improve the robust performance when multiple sets of data are available or can be obtained. First let us consider the problem of determining whether or not the given input-output data record is consistent with *one* of the models in the family $\mathcal{F}$. Let the data (1) be given, together with an upper bound $\varepsilon \geq \| \tilde{\mathcal{M}} \|$ on the norm of the perturbation. We notice that the family $\mathcal{F}$ contains the model x if and only if $\mu(x, \hat{\mathcal{M}}) \leq \varepsilon$, where $\hat{\mathcal{M}}$ is the data matrix constructed in (6). More specifically, assume the misfit function between the true model x_0 and the perturbation data matrix $\tilde{\mathcal{M}}$ is ε_0. If ε is smaller than ε_0: $\mu(x_0, \hat{\mathcal{M}}) = \varepsilon_0 > \varepsilon$, then every model in the family $\mathcal{F}$ is consistent with the data record. On the other hand, if $\varepsilon \geq \varepsilon_0$, every model in the family $\mathcal{F}$ is not consistent with the data record.

The above considerations can be used to compute the uncertain model family corresponding to *multiple* sets of data. Let the sets of input-output data $\hat{w}_{ki}$, $i = 0, 1, \cdots, N_k$, and the error bound ε, be available. The corresponding uncertain model family $\mathcal{F}$ is the intersection of the uncertain model families $\mathcal{F}_k$. The strategy of intersecting the model families has the advantage that it reduces the size of the uncertain model family, thus improving, in principle, the robust performance; for details on the best performance given several measurement sets, see ((A.C. Antoulas and Iwasaki, 1999)). The membership set corresponding to multiple sets of data, although still convex, is no longer described in a simple way; one usually approximates the membership set by an ellipsoid in some optimal sense. Efficient algorithms have been developed (Boyd *et al.*, 1994) for computing various suboptimal outer approximations of the intersection of ellipsoids, by formulating the problem as a convex optimization problem with linear matrix inequality (LMI) constraints.

6. EXAMPLE

In this section we will use a time-varying mass-spring-damper system (Chen, 1997) shown in figure 3, to illustrate the results developed above. To design a robust controller for the time-varying system and analyze its performance, we need to model the input-output data as generated by a linear, time-invariant system, with additive perturbation.

Under the assumption that upper and lower bounds for the time-varying parameters are a priori known, the uncertain model family is characterized in terms of an ellipsoid. Furthermore, multiple sets of input-output data will be considered; as already mentioned, the optimal ellipsoidal bounding of the intersection of the different ellipsoids determined by the data, can be computed using convex optimization techniques. The robust performance of the controller is thus improved.

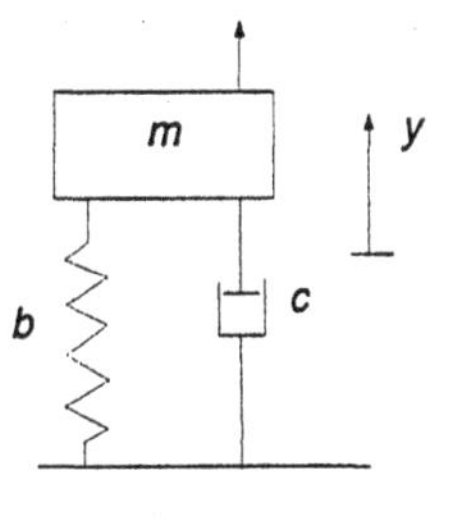

Fig. 3. Mass-spring-damper system.

Let the equation of motion be:

$$\frac{d}{dt}\left(m(t)\frac{dy}{dt}\right) + d(t)\frac{dy}{dt} + b(t)y(t) = f(t)$$

This can be written as

$$\frac{d^2y(t)}{dt^2} + \frac{c(t)}{m(t)}\frac{dy(t)}{dt} + \frac{b(t)}{m(t)}y(t) = \frac{f(t)}{m(t)} \qquad (24)$$

The parameters m, b and c are assumed to be time-varying mass, spring, and damping coefficients:

$$m(t) = m_0 + a_m \sin(\omega_m t)$$
$$b(t) = b_0 + a_b \sin(\omega_b t)$$
$$c(t) = c_0 + a_c \sin(\omega_c t)$$

where m_0, b_0, c_0 are the nominal values, ω_m, ω_b, ω_c are the frequencies, and a_m, a_b, a_c are the amplitudes. By choosing a sampling time τ, we can discretize equation (24):

$$\frac{d^2y}{dt^2}(t) \approx \frac{y(t+2\tau) - 2y(t+\tau) + y(t)}{\tau^2}$$
$$\frac{dy}{dt}(t) \approx \frac{y(t+\tau) - y(t)}{\tau}$$

For some initial time t_0, we define $t_k = t_0 + k\tau$, and $y_k = y(t_k)$, $f_k = f(t_k)$, $m_k = m(t_k)$, $b_k = b(t_k)$, $c_k = c(t_k)$. Equation (24) becomes $y_{k+2} + q_{1k}y_{k+1} + q_{0k}y_k = p_{0k}f_k$, where $p_{0k} = \frac{1}{m_k}\tau^2$, $q_{0k} = \frac{b_k}{m_k}\tau^2 - \frac{c_k}{m_k}\tau + 1$, $q_{1k} = \frac{c_k}{m_k}\tau - 2$. The numerical values are: $m_0 = 5$, $b_0 = 3$, $c_0 = 7$, $a_m = 0.45$, $a_b = 0.35$, $a_c = 0.4$, $\omega_m = 0.25\pi$, $\omega_b = 0.115\pi$, $\omega_c = 0.105\pi$, $\tau = 0.01$, $t_0 = 0$. We can obtain the locus of the point $\theta_k = (-p_{0k}, q_{0k}, q_{1k})$ with discrete time k by computing the values of the parameters p_{0k}, q_{0k}, and q_{1k}. The input has length $N = 100$ and unit energy; arbitrary initial values $y_0 = y(t_0)$, $y_1 = y(t_0 + \tau)$ are assumed; we can thus recursively compute a set of input-output data.

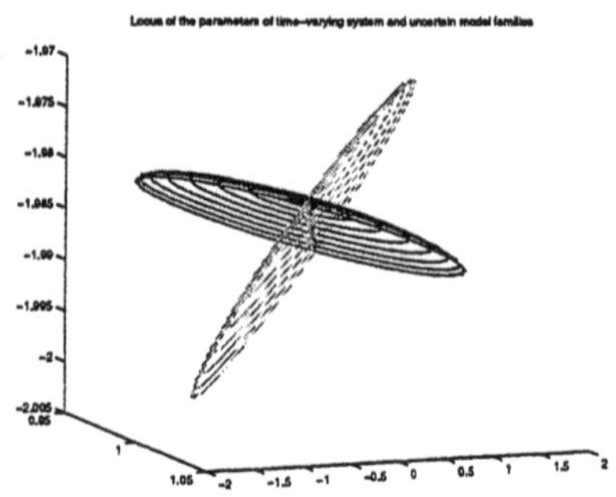

Fig. 4. The locus of the parameters of the time-varying system included in the intersection of two uncertain parameter sets (ellipsoids).

The next step is to model the resulting input-output data. as generated by a linear time-invariant, discrete-time system $y_{k+2} + q_1 y_{k+1} + q_0 y_k = p_0 f_k$, $k = 0, 1, \cdots$, where p_0, q_0, q_1 are constants. The data matrix $\hat{\mathcal{M}}$ (6) is formed. The perturbation bound ε has to be chosen large enough so that the resulting model family $\mathcal{F}$ contains the locus of the parameters θ_k. To achieve this, the locus will be put into a rectangular box; its vertices are determined by the lower bound $\theta_* = [u_* \ v_* \ w_*]^T$ and the upper bound $\theta^* = [u^* \ v^* \ w^*]^T$ of the points θ_k:

$$\theta_* \le \theta_k \le \theta^*.$$

The results of the computation are $\theta_* = [-0.0022 \ 0.8623 \ -1.8689]^T$ and $\theta^* = [-0.0018 \ 0.8691 \ -1.8374]^T$. The eight vertices $\theta = [u \ v \ w]^T$ of the box are thus determined, where $u = u_*$ or u^*, $v = v_*$ or v^*, $w = w_*$ or w^*.

ε has to be bigger than or equal to the misfit function between uncertain model $x = [\theta \ 1]^T$ and data matrix $\hat{\mathcal{M}}$, i.e.,

$$\varepsilon \ge \max \mu(x, \hat{\mathcal{M}}) = 0.064,$$

where θ represents the vertices of the box. For each chosen ε which is bigger than 0.064 and smaller than $\sigma_1 = 0.835$, the uncertain model family is an ellipsoid containing the locus of the points θ_k; see figure 4 (it can be shown that $\hat{\sigma}_1 = .053$, which is smaller than $\max \mu = .064$ as expected). For each ε between 0.064 and 0.853, the optimal robust controller that stabililizes the ellipsoid can be obtained; furthermore, the optimal performance can be computed, using the method introduced in section 4. Notice that compared to the ellipsoids, the locus is actually small. However, the intersection of the two ellipsoids is a rather tight cover of the parameter locus. Finally, in figure 2, the relation between the maximal stability margin and the size of the uncertainty, the quantity ε is depicted, where the curve indicates that better optimal performance can be achieved for a smaller ε.

7. EPILOGUE

In this paper we have proposed a method to model the given sets of measured data by means of a linear, time-invariant system together with a norm-bounded perturbation. We explored a robust control method for the resulting uncertain model family, and given several bounds on the maximal stability margin in terms of various measures of the uncertainty. The role played by the bound of the perturbation and the input (excitation) function are examined. An important conclusion is that intersection of model families obtained for different input functions, provide a way of reducing the uncertainty. Finally, an example is provided to illustrate the results. As this example indicates, the size of ε is not as important as the fact that the uncertainty can be reduced by intersecting the uncertain families.

8. REFERENCES

A.C. Antoulas, S. Gugercin and T. Iwasaki (1999). Further results in identification for robust control. *Technical Report.*

Anderson, B.D.O., S. Dasgupta, P. Khargonekar, F.J. Kraus and M. Mansour (1990). Robust strict positive realness: characterization and construction. *IEEE Trans. on Circuits and Systems* **37, No. 7**, 869–876.

Antoulas, A.C. (1997). On modeling for robust control. *Proc. IFAC SysId 1997* pp. 1707–1709.

Antoulas, A.C. and B.D.O. Anderson (1999). On the choice of inputs in identification for robust control. *Automatica.*

Bai, E.W., R. Tempo and K. Nagpal (1997). Recursive system identification in the presence of noise and model uncertainties. *System and Control Letters* **32**, 57–62.

Boyd, S., L. Ghaoui, E. Feron and V. Balakrishnan (1994). *Linear Matrix Inequalities in System and Control Theory.* SIAM Studies in Applied Mathematics V.15. Philadelphia.

Chen, J. (1997). Frequency-domain test for validation of linear fractional uncertain models. *Automatica* **42**, 748–760.

Norton, J.P. (1987). Identification and application of bounded-parameter models. *Automatica* **23**, 497–507.

Rantzer, A. and A. Megretski (1994). A convex parameterization of robustly stabilizing controllers. *IEEE Trans. on Automatic Control* **39**, 1802–1808.

Zhang, Huipin and A.C. Antoulas (1998). System identification for robust control. *Submitted to Automatica.*

14th Triennial World Congress, Beijing, P.R. China

C-2a-09-5

CO-OPERATIVE CONTROL OF MULTI-INPUT AND SINGLE-OUTPUT PROCESSES: PROBLEM FORMULATION AND SOLUTION

Qing-Guo Wang [1] Yong Zhang Wen-Jian Cai Qiang Bi Chang-Chieh Hang

National University of Singapore, Department of Electrical Engineering, Singapore 119260
[1] *Email: elewqg@nus.edu.sg*

Abstract: In control engineering, a process variable is normally regulated by a process manipulated variable. Due to physical limitations, any manipulated variable will have constraints on its amplitude. A practical problem frequently encountered in industry but less addressed by researchers is the output unreachability under input saturation: for a large disturbance or set-point change, the process variable may never reach the set-point even when the manipulated variable has been driven to saturation. The process variable can be brought back to the set-point only by activating auxiliary manipulated variables. However, timing of activation remains an open problem. A common practice adopted in industry is to wait for the process variable to reach a new steady state and then take action accordingly. This is time-consuming and may cause serious loss of productivity. In this paper, a new strategy based on disturbance estimation and output prediction is presented. It can detect unreachability early and activate the auxiliary variables timely so that the process variable will return to the set-point much faster. Simulation and real-time testing on a pilot mini-HVAC system show the effectiveness of the proposed strategy.

Keywords: Multi-input systems, co-operative control, manipulated variable constraint, disturbance, estimation, predictive control.

1. INTRODUCTION

Industrial systems usually have a large number of control loops. A loop's process variable (or process output) is often affected by other loops and disturbances in one way or another, but it can normally be brought back to its set-point quickly by regulating the loop's manipulated variable (or process input). Due to physical limitations, it is sometimes possible that a manipulated variable has reached its limit or saturation, while the corresponding process variable still cannot reach its set-point at the steady state. For ease of reference, this phenomenon is here called the output unreachability problem under input saturation. This state can be released only by activating some other manipulated variables which also affect the output and are here called the auxiliary manipulated variables.

Consider the example of reactor temperature control (Luyben, 1990). The reactor is cooled by both cooling water flowing through a jacket surrounding the reactor and condensing vapor that boils off the reactor in a heat exchanger that is cooled by a refrigerant. In practice, the nominal flow rates of cooling water and refrigerant are determined in advance, and the water is utilized as the primary manipulated variable, while the refrigerant acts as the auxiliary manipulated variable and is

[1] Author to whom all correspondence should be addressed.

normally kept constant. If there is a disturbance or set-point change, the steady state may deviate from the previous designed one. There exists a situation where the effect is so large that the temperature cannot return to the set-point even when the cooling water valve hits its limitation. In such a case, the auxiliary manipulated variable, i.e., flow of refrigerant, has to be adjusted to make the temperature steady state reach the desired value.

These is just one of industrial examples which have output unreachability. Such cases necessitate changes in the set-point of auxiliary manipulated variables. Shinskey (1981) proposed a "valve-position control" for a blend composition processes. In his control strategy, the maximum valve opening position is sent to a valve-position controller (VPC) by a high-signal selector. It then changes the set-point of the auxiliary manipulated variable to keep all valves from exceeding a preset limit. His algorithm is to wait for the process variable to reach its steady state, then regulate an auxiliary manipulated variable iteratively with trial-and-error until the steady state error becomes zero. The main drawbacks of the method are time consuming and the results largely depend on engineer's experience.

In this paper, a new control strategy, referred to as co-operative control, is proposed to systematically solve the problem. A criterion and a corresponding algorithm for activation of auxiliary manipulated variables are developed based on disturbance estimation and output steady state prediction. Compared with the conventional strategy, the new approach results in much shorter settling time. In addition, times of change required for auxiliary manipulated variables are reduced such that possible impact on the main manipulated variable are alleviated.

The paper is organized as follows. Problem formulation and preliminaries are presented in Section 2. Disturbance estimation is discussed in Section 3. An algorithm for activation of the auxiliary manipulated variable is discussed in Section 4. Simulation and real-time test on a mini-HVAC system are given in Section 5. Conclusion is in Section 6.

2. PROBLEM FORMULATION AND PRELIMINARIES

For simplicity, only two-input and single-output processes will be here addressed. The results can be easily extended to other multi-input single-output processes. A control system which can represent three examples in Introduction is shown in Fig. 1, where y is the single process output, u_1 and u_2 are the two inputs to the process, and r_1 is the set-point for y. Normally, the output y is regulated by u_1 in the indicated feedback loop in response to any disturbance or set-point change, while u_2 is kept constant. u_1 is thus viewed as the primary process input and u_2 as the auxiliary input. Sometimes, there is a control loop for u_2, too, and G_2 should then be considered as its closed-loop transfer function and u_2 as its set-point. As a result, system in Fig.1 is used for both cases. The process may have many disturbances. For a linear system, they can be collectively represented by one disturbance w entering the process at the output.

In normal operations, the control system is designed such that for any moderate set-point or load disturbance changes, there exists the primary manipulated variable $u_1(t)$ within its working range, $[u_{\min}, u_{\max}]$, which can regulate the process output adequately to achieve a zero output steady state error, i.e., $y(\infty) = r_1(\infty)$, while u_2 is kept unchanged during the entire transient. However, it is highly possible, as mentioned in Introduction, that under a large set-point or load disturbance change, there is no $u_1(t) \in [u_{\min}, u_{\max}]$ such that $y(\infty) = r_1(\infty)$, if u_2 is unchanged, leading to the output unreachability under input saturation. The objective of the present paper is to detect such a phenomenon at an early stage and activate the auxiliary manipulated variable u_2 properly and timely to restore the output reachability, i.e., $y(\infty) = r_1(\infty)$, with a change in u_2.

Assume throughout the paper that the primary manipulated variable u_1 is constrained by $u_1(t) \in [u_{\min}, u_{\max}]$ for all t, where $u_{\min}$ and $u_{\max}$ are given, and that the system is stable and has an integral control. These assumptions imply that for step inputs, the output steady state error is always zero provided that $u_1(\infty)$ exists and is strictly within the constraint $[u_{\min}, u_{\max}]$. The concern of the paper is to know if $y(\infty) = r_1(\infty)$ is achievable under various conditions. This problem naturally requires a steady state analysis of the system. It follows from Fig.1 that at the steady state, there holds

$$y(\infty) = G_1(0)u_1(\infty) + G_2(0)u_2 - w(\infty). \quad (1)$$

But $u_1(\infty)$ is limited by

$$u_{\min} \leqslant u_1(\infty) \leqslant u_{\max}. \quad (2)$$

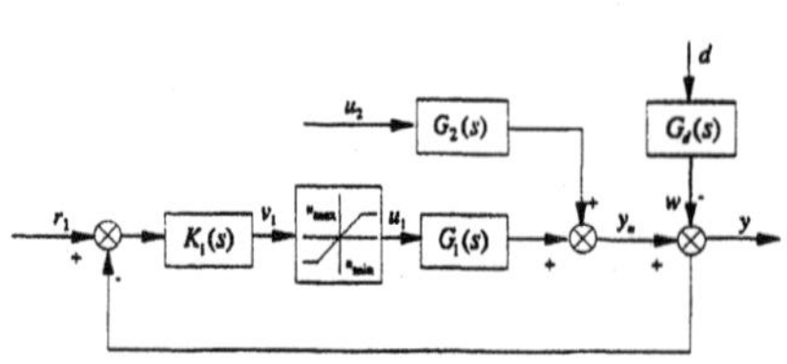

Fig. 1. Control system with two-input and one-output

Combining (1) and (2) yields the reachable range of the output $y(\infty)$ as

$$f_{\min} + G_2(0)u_2 - w(\infty) \leqslant y(\infty) \leqslant f_{\max} + G_2(0)u_2 - w(\infty),$$

where

$$f_{\min} = \min[\text{sgn}(G_1(0))u_{\min},\ \text{sgn}(G_1(0))u_{\max}],$$
$$f_{\max} = \max[\text{sgn}(G_1(0))u_{\min},\ \text{sgn}(G_1(0))u_{\max}],$$

and $\text{sgn}(x)$ is defined as $\text{sgn}(x) = 1$ for $x > 0$, $\text{sgn}(x) = -1$ for $x < 0$.

Since $y(\infty) = r_1(\infty)$ is required, the condition of the output reachability is

$$f_{\min} + G_2(0)u_2 - w(\infty) \leqslant r_1 \leqslant f_{\max} + G_2(0)u_2 - w(\infty). \quad (3)$$

$w(\infty)$ is needed for checking (3), but it is not available since the disturbance $w(t)$ is not measurable. It has to be estimated in some way.

3. DISTURBANCE ESTIMATION

Fig.1 indicates that the collective load disturbance $w(t)$ at the output can be estimated as the difference between the output measurement $y(t)$ and the predicted output $y_u(t)$ without the disturbance:

$$\hat{w}(t) = y_u(t) - y(t), \quad (4)$$

where $y_u(t)$ can be calculated from

$$Y_u(s) = G_1(s)U_1(s) + G_2(s)U_2(s), \quad (5)$$

for the given inputs $u_1(t)$ and $u_2(t)$. The estimation would be accurate if there is no modelling error for G_1 and G_2.

Any type of disturbance and any number of disturbances can always be put equivalently as a collective disturbance $w(t)$, as shown in Fig.1. On the other hand, such an equivalence may lose some information which can otherwise be used for better estimation of the disturbance. Suppose, for example, that a process has a major disturbance, but other disturbances are negligible, and the relation between that disturbance and output is well known. Then, we can model this type of disturbance as

$$W(s) = G_d(s)D(s), \quad (6)$$

with the known $G_d(s)$. One can estimate the true disturbance $d(t)$ by

$$\hat{d}(t) = \int_0^t \bar{g}_d(t-\tau)\hat{w}(\tau)d\tau, \quad (7)$$

where $\bar{g}_d(t)$ is the impulse response of

$$\bar{G}_d(s) = \frac{G_d^{-1}(s)}{(\alpha s + 1)^m},$$

m is chosen such that $\bar{G}_d(s)$ is just proper and thus physically realizable, and α should be much smaller relative to the dominant or equivalent time constant of $G_d(s)$. The advantage of using (7) instead of (4) is that $\hat{d}(t)$ will reach a steady state earlier than $\hat{w}(t)$ and this will enable us to correctly infer whether or not a change in u_2 is needed at an early stage.

To unify two cases in (4) and (7), (7) will be always used for disturbance estimation with $G_d(s)$ defined by

$$G_d(s) = \begin{cases} G_d(s), & \text{if it is available;} \\ 1, & \text{otherwise.} \end{cases}$$

At any particular time $t = T$, knowing no information of future disturbance, a natural estimate for $d(\infty)$ is

$$\hat{d}(\infty) = \hat{d}(T). \quad (8)$$

Then $\hat{w}(\infty)$ is obtained as

$$\hat{w}(\infty) = G_d(0)\hat{d}(\infty). \quad (9)$$

4. ACTIVATING ALGORITHM

In engineering practice, u_2 is normally operated near the design or optimum point. Its change is not welcome for operation optimization and simplicity, unless this change is justified to be necessary. Its frequent and large change should be avoided. Thus, the activating algorithm to be developed in this section has to observe the following basic requirements:

(1) u_2 is activated only when it is necessary, i.e., a true output unreachability may occur;
(2) Times of activation is kept as few as possible;
(3) The change of u_2 is made as small as possible.

It can be noted that these may be in conflict to some extent, and have to be compromised when designing our activating algorithm.

We substitute (9) and (8) into (3) to give the following checkable condition on the output reachability:

$$\frac{1}{G_d(0)}[f_{\min} + G_2(0)u_2 - r_1] \leqslant \hat{d}(T) \leqslant \frac{1}{G_d(0)}[f_{\max} + G_2(0)u_2 - r_1], \quad (10)$$

where $\hat{d}(T)$ is estimated and updated with (4) and (7). Consequently, a violation of (10) forms the condition on output unreachability or activation of u_2. But such an activating criterion may often cause unnecessary or fault activation. This is because the disturbance transient $d(T)$ varies with time T, and can go up and down quickly. It may violate (10) at this moment but meet it shortly afterwards. Obviously, fault activation is undesirable according to requirement (i). In order to reduce it, an observing period T_0 is introduced such that the action will not be taken until (10) is violated for a time period of T_0, that is,

$$\hat{d}(T-\tau) > \frac{1}{G_d(0)}[f_{\max} + G_2(0)u_2 - r_1], \tau \in [0, T_0]; \tag{11a}$$

or

$$\hat{d}(T-\tau) < \frac{1}{G_d(0)}[f_{\min} + G_2(0)u_2 - r_1], \tau \in [0, T_0] \tag{11b}$$

is used as the actual activating criterion. The selection of T_0 is essentially a tradeoff between fast response and few fault actions. A small T_0 will make the algorithm sensitive to $\hat{d}(T)$ and cause frequent activation or large change of u_2, most of which might be unnecessary. A large T_0 requires a long time for the algorithm to verify necessity of activation and may slow down the output response. From extensive simulation, a guideline for selection of T_0 is obtained as

$$T_0 \in [0.2T_p,\ 1.5T_p], \tag{12}$$

where T_p is the time constant of the process. The default value is recommended to be $T_0 = 0.5T_p$.

A natural question which arises after determining activating criterion is how much change in u_2 should be made to restore output reachability. The exact amount of Δu_2 is easily found, by substituting $u_2 = u_2(0) + \Delta u_2$ to (10), as

$$\Delta u_2 = \frac{1}{G_2(0)}[r_1 + G_d(0)\hat{d}(T) - f_{\max}] - u_2(0) \tag{13a}$$

in case of (11a); or

$$\Delta u_2 = \frac{1}{G_2(0)}[r_1 + G_d(0)\hat{d}(T) - f_{\min}] - u_2(0) \tag{13b}$$

in case of (11b). However, this Δu_2 almost always turns out to be too conservative for various reasons. Firstly, since $\hat{d}(T)$ now meets (10) exactly but it keeps changing, a further change in $d(T)$ may cause another activation. Such Δu_2 is simply too small and gives no tolerance for any further change of disturbance in the previous direction, which is likely to occur. Secondly, if Δu_2 merely meets (13), the resultant response for $y(t)$ could be very slow because u_1 will still take its limit value and has no room to regulate y at all. Thirdly, there are always modelling errors in the gains, $G_1(0)$, $G_2(0)$ and $G_d(0)$, due to measurement inaccuracy, material/equipment variations, process nonlinearity and so on. Δu_2 calculated based on these gains could be insufficient. In view of the above observations, Δu_2 is modified with a margin $\beta > 1$ as

$$\Delta u_2 = \frac{1}{G_2(0)}\{\beta[r_1 + G_d(0)\hat{d}(T)] - f_{\max}\} - u_2(0) \tag{14a}$$

for (11a); or

$$\Delta u_2 = \frac{1}{G_2(0)}\{\frac{1}{\beta}[r_1 + G_d(0)\hat{d}(T)] - f_{\min}\} - u_2(0), \tag{14b}$$

for (11b), to reduce times of change in u_2 and also keep u_2 near its optimal point. Through extensive simulations, β is recommended to be $1.05 \sim 1.5$, and the default value is recommended to be $\beta = 1.2$. The above development is summarized as the following algorithm.

Activating algorithm for auxiliary input
Given data: $G_1(s)$, $G_2(s)$, $G_d(s)$, and measurements: $y(t)$, $u_1(t)$, $u_2(t)$.

Step 1: Estimate the disturbance $d(T)$ from (4) and (7) and predict $\hat{w}(\infty)$ from (8) and (9).

Step 2: Check the condition (11). If it holds, calculate Δu_2 from (14) and change u_2 to $u_2 + \Delta u_2$; Otherwise, make no change in u_2;

Step 3: Update $y(t)$, $u_1(t)$ and $u_2(t)$ and go to Step 1.

5. SIMULATION AND REAL-TIME TESTING

Example 1 Consider a system (Zervos *et al.*, 1988) with

$$G_1(s) = \frac{1}{(17s+1)(6s+1)}e^{-6s},$$
$$K_1(s) = 1.21 + \frac{0.06}{s} + 2.86s.$$

For our simulation, G_2 and G_d are taken as

$$G_2(s) = \frac{1}{(s+0.5)(2s+1)}e^{-0.1s},$$
$$G_d(s) = \frac{2}{5s^2+s+1}.$$

The primary manipulated variable is supposed to be constrained to $u_1 \in [1,\ 5]$. The system is initially in its steady state with $u_1(0) = 4$, $u_2(0) = 0$, and $y(0) = 4$. In order to see performance of different activating schemes, two impulse disturbances in w with amplitude of 1.2 and −0.8, and a step disturbance in d with amplitude of 1 are introduced at $t = 10$, $t = 20$ and $t = 30$, respectively. For a conventional trial-and-error activating scheme, the auxiliary manipulated variable u_2 is changed to $u_2 = 0.2$, $u_2 = 0.4$ and $u_2 = 0.6$ when a nonzero steady state error is observed at $t = 100$, $t = 130$ and $t = 150$, respectively. For the proposed scheme, assume first that G_d is unknown. $T_0 = 0.5T_p$ and $\beta = 1.2$ are selected with $T_p \approx 25$. The proposed method without any observing period, i.e., $T_0 = 0$ is also included to show the usefulness of T_0. The output responses are shown in Fig.2. They indicate that the proposed method yields much fast output response with appropriate change in u_2. In the case of no observing period, u_2 is changed too often and too large though the resultant output response is almost the same. By introducing an observing period, the system is more robust to impulse disturbances and unnecessary changes in u_2 are avoided, as expected.

To address the effects of the observing period T_0 and the margin β, the proposed algorithm is simulated with different values of them. The results are omitted here due to space limitation. It is shown that T_0 is reversely proportional to the system speed, but a small T_0 tends to cause fault activations while a large one results in sluggish responses. A small β leads to a small change in u_2 but it may necessitate more times of change in u_2. A large β makes large change in u_2, which may be unnecessary. The effectiveness of those recommended default values in T_0 and β are also observed.

Suppose now that an estimate of $G_d(s)$ is available as

$$\hat{G}_d(s) = \frac{2}{4.8s^2 + s + 1},$$

and it is then used for disturbance estimation $\hat{d}$ in (7) with

$$\bar{G}_d(s) = \frac{G_d^{-1}(s)}{(0.5s+1)^2}.$$

The resultant system performance with the proposed method is shown in Fig.3. Obviously, such a disturbance model helps activate u_2 earlier and speed up the output response.

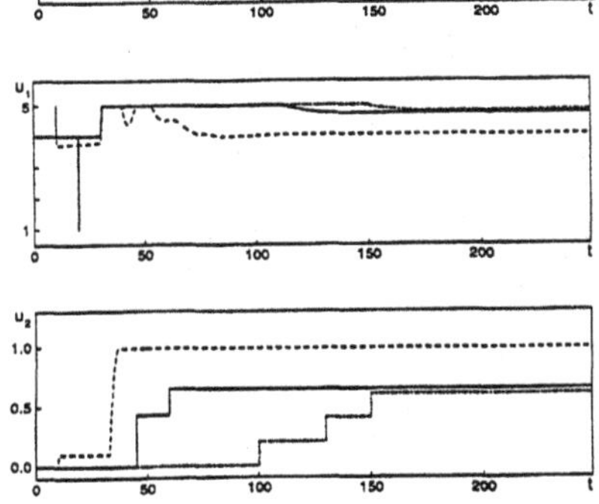

Fig. 2. System performance under different activating schemes (- · - · - : for conventional; - - - : for proposed without observing period; —— : proposed)

To further see the robustness of the proposed algorithm under model mismatch, $G_1(s)$ is supposed to be estimated as

$$\hat{G}_1(s) = \frac{0.95}{(15s+1)(6s+1)}e^{-5.8s},$$

and $\hat{G}_1(s)$ and $G_2(s)$ are then used in (5) for disturbance estimation $\hat{w}$. The resultant system performance is shown in Fig.4 with dash-dot line. Alternatively, if $G_2(s)$ is replaced with its model:

$$\hat{G}_2(s) = \frac{1}{(s+0.6)(2.5s+1)}e^{-0.3s},$$

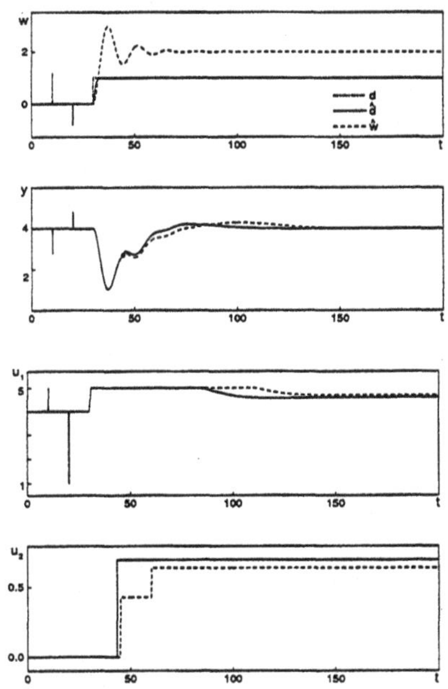

Fig. 3. System performance with disturbance model (- - - : using $\hat{w}$; —— : using $\hat{d}$)

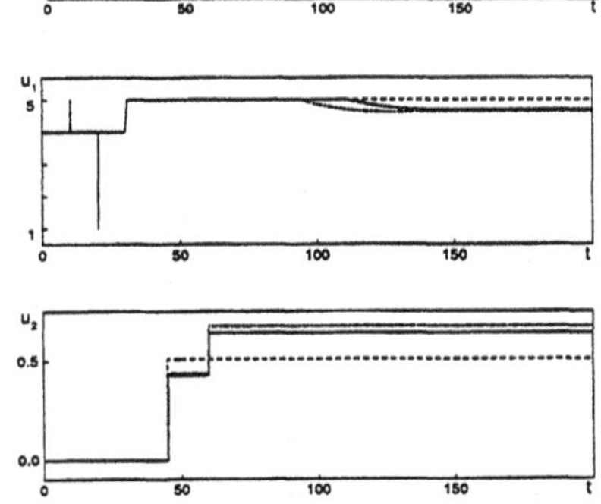

Fig. 4. System performance with models' uncertainty (- · - · - ; for $\hat{G}_1$; —— : without models' uncertainties; - - - : for $\hat{G}_2$)

with $G_1(s)$ in perfect match, then the result is also shown in Fig.4 with dashed line. It is clear that with the proposed algorithm, responses under model mismatches are quite close to that for perfect match (solid line). Introducing T_0 and β thus improves the robustness of the proposed method.

Example 2 The proposed co-operative control strategy was tested on a pilot HVAC system. The system is shown in Fig.5. In this system, supply air to be delivered to the room, is a mixture of return air and outside air. Exhaust air is the air ejected from the room to outside to balance the intake of outside air. A CCU is used for cooling supply air through a heat exchanger with the cooling water. By regulating the opening (u_1) of the room damper, the room is controlled to keep the air temperature (y) at a desired value (r_1). Suppose that there is a big set-point decrease for the room air temperature, but the temperature is still higher than the set-point even when the damper opens to its maximum. In such a case, either off-coil air temperature or fan velocity can be used as an auxiliary manipulated variable. Since regulating the fan velocity is much easier for implementation, it is chosen as the auxiliary manipulated variable u_2. In the experiment, the initial conditions were $y(0) = 19.5^oC$, $u_1(0) \approx 30\%$, and $u_2(0) = 30\%$. The process models and the controller were obtained with the relay auto-tuning (Bi, 1997) as

$$G_1(s) = -\frac{1}{2680.35s + 22.2}e^{-23.158s},$$
$$G_2(s) = \frac{1}{1.37s^2 + 1.47s + 1.23}e^{-1.25s},$$
$$K_1(s) = -(58.6 + \frac{0.48}{s}).$$

The process operation bears constraints to meet indoor air quality and safety requirements. They lead to $u_1 \in [30\%, 70\%]$. In the real-time test, the observing period and the margin were set as $T_0 = 0.5T_p \approx 70s$ and $\beta = 1.1$, respectively. For a new set-point 18.3^oC, the responses of the proposed co-operative control and the conventional activating scheme with 5% increment are shown in Fig.6. The proposed method gave a much faster output response.

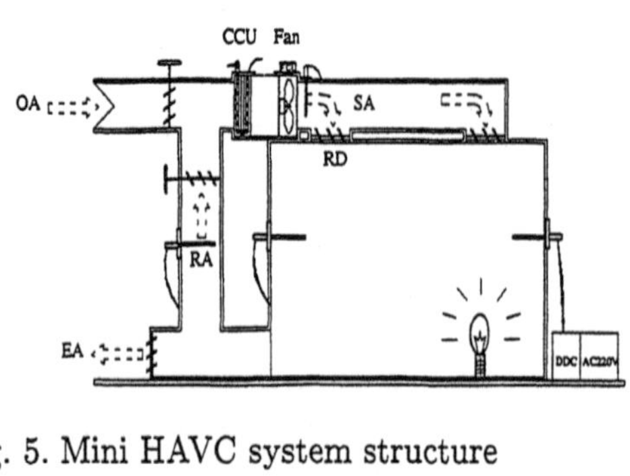

Fig. 5. Mini HAVC system structure

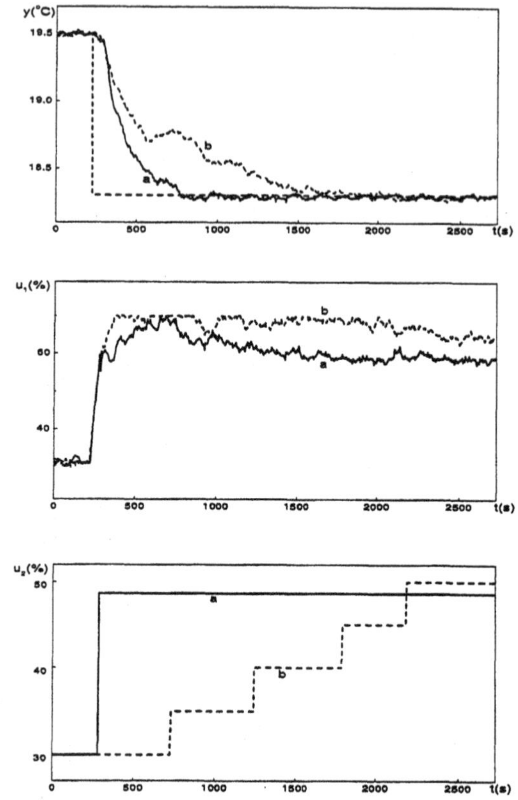

Fig. 6. Experiment results on HVAC system (a: proposed; b: conventional; - - - : setpoint)

6. CONCLUSION

In this paper, a new control algorithm has been proposed for activating auxiliary manipulated variables when necessary. The robustness to unpredictable disturbances and model uncertainties is addressed. The simulation and experiment on a mini-HVAC system show that compared with the conventional strategy, the new method yields much improved control performance with fewer activations.

7. REFERENCES

Bi, Q. (1997). Frequency domain adaptive control with relay feedback. PhD thesis. National University of Singapore. Singapore.

Luyben, W. L. (1990). *Process Modeling, Simulation and Control for Chemical Engineers.* McGraw-Hill International Editions. New York.

Shinskey, F. G. (1981). *Controlling Multivariable Processes.* Instrument Society of America. North Carolina.

Zervos, C., P. R. Belanger and G. A. Dumont (1988). On PID controller tuning using orthonormal series identification. *Automatica* **24**(2), 165–175.

14th Triennial World Congress, Beijing, P.R. China

DESIGN OF 2-DEGREE OF FREEDOM CONTROLLER AND ITS APPLICATION IN INDUCTION MOTORS

Wang Xunxian * **Tan Qing** # **Ding Gang** * **Han Zengjin** *

*Automation Department of Tsinghua University , Beijing 100084, P.R.China
Email: wangmotor@163.net
The First Light Industry Research Institute of Beijing, Beijing 100009, P.R.China

Abstract: Despite 2-degrees of freedom controller can have good servo and disturbance rejection performances in the same time, there is not a standard design method for it until now. In this paper a new design method in which the H_∞ theory has been used is proposed and applied in the speed regulation system of the induction motors. The simulation experiment shows the result is satisfied. *Copyright 1999 IFAC*

Keywords decoupling problems; robust control ; nonlinear control; induction motors; decentralized control.

1. INTRODUCTION

A central concept in automation control is the idea that a physical quantity can be made to behave in a prescribed way by using the difference between the actual value and the desired or reference value to drive the two into correspondence. This idea gives rise to the time-honored unity feedback or error feedback servomechanism depicted in Fig .1a. Since there is only one input to the controller, this classical control structure is also known as one –degree of freedom controller (1-DOFC). However the closed loop system controlled by it can not obtain the good servo and disturbance rejection performance at the same time. Compared to 1-DOFC, two-degrees of freedom controller (2-DOFC) (Fig.1b) has the advantage of higher servo performance and better disturbance rejection performance at the same time. However, there is not exist a standard design method to determine the structure and the parameters of 2-DOFC until now. In this paper H_∞ theory which is developed for improving the robust of the closed loop system is used to design the 2-DOFC .

Induction motors have been widely used for the advantages of lower price, higher reliability and smaller volume compared to DC motors. In order to improve both its speed regulation precision and the power efficiency, some methods such as differential geometry feedback linearization approach and nonlinear adaptive control approach are applied in speed regulation system of induction motors. With the feedback linearization approach, induction motors' mathematical model can be converted into two decoupled subsystems—speed subsystem and rotor flux subsystem in the condition of the rotor flux being observed precisely. However, the coupling effect between the two subsystems will still remain due to the error between the observed and real rotor flux. An effective way to improve the precision of the speed regulation system is to reduce the coupling effect between the two subsystems. In this paper 2-DOFC is used to deal with the problem. The paper is divided into four parts: the design approach of the 2-DOFC is given in the second part, the application of the controller in the induction motors is shown in the third part and the simulation results are displayed in the last part.

2. THE DESIGN METHOD OF 2-DOFC

Fig. 2a is the construction diagram of 2-DOF be used in this paper (Vilanova, R. and I. Serra. ,1997). Where $P(s)$ is the transfer function of the controlled object , $P_0(s)$ is the nominal transfer function of $P(s)$,d is the disturbance, $K_1(s), K_2(s)$ are adjusters. In (Vilanova, R. and I. Serra. ,1997) although all the resolutions of $K_1(s), K_2(s)$ that can stabilize the closed loop system are displayed, but the two adjusters' designing method are not given. Here a new method is proposed to design the adjusters according to the desired performance of the closed system. To do this, Fig. 2a must be translated into Fig.2b and Fig.2c. at first

2.1 The design of servo-adjuster

Suppose $P(s) = P_0(s)$, $d = 0$,then fig 2b can be obtained and $K_1(s)$ is called servo-adjuster for it is used to adjust the servo performance of the system here. There are a lot of methods that can be used to design it and they are not mentioned now. It should be pointed out that controlled object here is the nominal one.

2.2 The design of disturbance rejection adjuster

Set $r = 0$,then fig 2a is converted to fig 2c and alternatively $K_2(s)$ is called disturbance rejection adjuster for it is used to adjust the disturbance reject performance here. In order to conquer the perturbation of the model and the outside disturbance in the same time, the following H_∞ theorem has been used to design it.

Theorem 1 Let x denotes state variable, ξ denotes disturbance, u denotes input variable, z denotes performance index ,A,B,C,D, E, F_a, F_b are the known matrix , $\Delta A, \Delta B$ are the perturbation matrix, Σ is the unknown matrix and $\Sigma \in \Omega$, while $\Sigma^T\Sigma \le I$, I is the identity matrix. and have the quality that

$$[\Delta A, \Delta B] = E\Sigma[F_a, F_b] \tag{1a}$$

The system is given by

$$\begin{cases} \dot{x} = (A + \Delta A)x + B_1\xi + (B_2 + \Delta B)u \\ z = Cx + Du \end{cases} \tag{1b}$$

where $D^T[D \quad C] = [I \quad 0]$ is satisfied. If and only if exist $\lambda > 0$ and positive matrix X which make the RICCATI inequality

$$\begin{aligned} &A^T X + XA + X(B_1B_1^T/\gamma^2 + \lambda^2 EE^T)X \\ &+ C^T C + F_a^T F_a/\lambda^2 - \\ &(XB_2 + F_a^T F_b/\lambda^2)R^{-2}(B_2^T X + F_b^T F_a/\lambda^2) \\ &< 0 \end{aligned} \tag{2}$$

satisfied, where $R^2 = I + \lambda^{-2}F_b^T F_b$ .then there exist the state feedback controller with the form of

$$u = Kx \tag{3}$$

$$(K = -R^{-2}(B_2^T X + \lambda^{-2}F_b^T F_a))$$

which can make the following H^∞ performance satisfied

(a) For $\forall\ \Sigma \in \Omega$, the closed loop system is inner stable.

(b) For $\forall\ \Sigma \in \Omega$,the transfer function $T_{z\xi}(s)$ which is from ξ to z satisfy the following condition

$$\left\| T_{z\xi}(s) \right\| < \gamma \tag{4}$$

Proof (see Shen T. 1996 pp.187-188)

3. THE REAL LINEARIZED MODEL OF IM

3.1 I/O linearisation of MIMO systems

Definition 1: **Lie derivative:** The Lie derivative of a smooth scalar function h, with respect to a vector f ,is denoted as

$$L_f h(X) = \frac{\partial h}{\partial X} f(X) \tag{5a}$$

Higher-order Lie derivatives can be defined as

$$L_f{}^k h(X) = L_f L_f{}^{k-1} h(X) \tag{5b}$$

Definition 2: Relative degree: For m-input ,m-output nonlinear system it is said that the system has the relative degree ($r_1, r_2, \cdots r_m$) at X^0 if

$$\begin{cases} L_{g_i} L_f{}^k h_j(X) = 0 \\ L_{g_i} L_f{}^{r_j - 1} h_j(X^0) \ne 0 \end{cases}$$

$$k < r_j - 1, i = 1,2,\cdots,m \tag{6}$$

satisfied for all X in a neighborhood of X^0.
According to (Isidori A. 1989),the I/O mapping is of the from

$$\begin{bmatrix} y_1^{r_1} \\ y_2^{r_2} \\ \vdots \\ y_m^{r_m} \end{bmatrix} = B(X) + A(X)\begin{bmatrix} u_1 \\ u_2 \\ \vdots \\ u_m \end{bmatrix} \tag{7}$$

The matrix is defined as

$$A(X) = \begin{pmatrix} L_{g_1}L_f^{(r_1-1)}h_1(X) & \dots & L_{g_m}L_f^{(r_1-1)}h_1(X) \\ \dots & \dots & \dots \\ L_{g_1}L_f^{(r_m-1)}h_{mi}(X) & \dots & L_{g_m}L_f^{(r_m-1)}h_m(X) \end{pmatrix} \tag{8a}$$

$$B(X) = \begin{pmatrix} L_f^{(r_1)}h_1(X) \\ \vdots \\ L_f^{(r_m)}h_m(X) \end{pmatrix} \tag{8b}$$

Then if A(X) is nonsingular in all x, by the controller

$$(u_1, u_2, \cdots u_m)^T = A(X)^{-1}(v_1, v_2, \dots v_m)^T - B(X) \tag{9}$$

the linearized model is as following

$$y_i^{r_i} = v_i . \quad i = 1,2,\cdots m \tag{10}$$

3.2 Mathematical model of IM

The mathematical model of current-fed induction motors in a fixed stator reference frame can be described by the Eq. (11) (Marino R., S. Peresada. and P. Valigi.1993):

$$\dot{X} = f(X) + \sum_{i=1}^{2} g_i(X)i_i \tag{11a}$$

$$y_i = h_i(X) \ , \ i = 1,2 \tag{11b}$$

where

$$X = \begin{pmatrix} \omega \\ \psi_{ra} \\ \psi_{rb} \end{pmatrix}, f(X) = \begin{pmatrix} -\frac{T_l}{J} \\ -\alpha\psi_{ra} - n_p\omega\psi_{rb} \\ n_p\omega\psi_{ra} - \alpha\psi_{rb} \end{pmatrix},$$

$$g_1(X) = \begin{pmatrix} -\mu\psi_{rb} \\ \alpha M \\ 0 \end{pmatrix}, g_2(X) = \begin{pmatrix} \mu\psi_{ra} \\ 0 \\ \alpha M \end{pmatrix},$$

$h_1(X) = \omega$, $h_2(X) = \psi_{ra}^2 + \psi_{rb}^2$, ω is the rotor speed , $i_1, i_2, \psi_{ra}, \psi_{rb}$ are the variables of the stator current and rotor flux respect to a,b axis, n_p is the pole pairs, M is the mutual inductance , L_r is the rotor self inductance , R_r is rotor resistance , R_s is stator resistance , L_s is stator self inductance , J is the moment of inertia of the rotor and of any tool attached to it , T_l is load torque , $\mu = (n_pM / JL_r), \alpha = R_r / L_r$, $\sigma = 1 - M^2 / L_sL_r, \beta = M / (\sigma L_sL_r)$ $, \gamma = \alpha M^2 / (\sigma L_sL_r) + R_s / (\sigma L_s)$, y_1, y_2 is the controlled variables of the system .From now on the subscripts s and r will be dropped .

3.3 The real linearised model of induction motors

The relative degree of the system described by the Eq.(11) is (1,1), so that

$$\begin{pmatrix} \dot{y}_1 \\ \dot{y}_2 \end{pmatrix} = \begin{pmatrix} L_fh_1(X) \\ L_fh_2(X) \end{pmatrix} + \begin{pmatrix} L_{g_1}h_1(X) & L_{g_2}h_1(X) \\ L_{g_1}h_2(X) & L_{g_2}h_2(X) \end{pmatrix}\begin{pmatrix} i_1 \\ i_2 \end{pmatrix}$$

$$= \begin{pmatrix} -\frac{T_l}{J} \\ -2\alpha(\psi_a^2 + \psi_b^2) \end{pmatrix} + \begin{pmatrix} -\mu\psi_b & \mu\psi_a \\ 2\alpha M\psi_a & 2\alpha M\psi_b \end{pmatrix}\begin{pmatrix} i_a \\ i_b \end{pmatrix} \tag{12}$$

When ψ_a, ψ_b are not equal to zero at the same time ,

$$\det\begin{pmatrix} -\mu\psi_b & \mu\psi_a \\ 2\alpha M\psi_a & 2\alpha M\psi_b \end{pmatrix} = 2\alpha\mu M(\psi_a^2 + \psi_b^2) \neq 0 \tag{13}$$

define

$$\begin{pmatrix} i_a \\ i_b \end{pmatrix} = \begin{pmatrix} -\mu\psi_b & \mu\psi_a \\ 2\alpha M\psi_a & 2\alpha M\psi_b \end{pmatrix}^{-1} \bullet \left\{ \begin{pmatrix} \phi_1 \\ \phi_2 \end{pmatrix} - \begin{pmatrix} -\frac{T_l}{J} \\ -2\alpha(\psi_a^2 + \psi_b^2) \end{pmatrix} \right\} \tag{14}$$

Substitute Eq. (14) into Eq.(12),equ. (15) cab be obtain

$$\begin{cases} \dot{y}_1 = \phi_1 \\ \dot{y}_2 = \phi_2 \end{cases} \tag{15}$$

It is seen from Eq. (15) that the linear and decoupled model can be obtained if all the value of the parameters in the right side of the Eq.(14) are been known correctly. It should be point out that the condition (13) can be satisfied easily in the real system .

However, some value of the parameters

especially ψ_a,ψ_b and α, T_l can not be measured or observed precisely. Define the nominal value of ψ_a,ψ_b and α T_l as ψ_{a0},ψ_{b0} and α_0 T_{l0}, the Eq. (14) changes to

$$\begin{pmatrix} i_a \\ i_b \end{pmatrix} = \begin{pmatrix} -\mu\psi_{b0} & \mu\psi_{a0} \\ 2\alpha_0 M\psi_{a0} & 2\alpha_0 M\psi_{b0}\end{pmatrix}^{-1} \bullet$$

$$\left\{ \begin{pmatrix}\phi_1\\ \phi_2\end{pmatrix} - \begin{pmatrix} -\dfrac{T_{l0}}{J} \\ -2\alpha_0(\psi_{a0}^2+\psi_{b0}^2)\end{pmatrix}\right\} \qquad (14')$$

Correspondingly, Eq. (15) changes to

$$\begin{cases} \dot{y}_1 = A_1\phi_1 + A_2\phi_2 + A_3 \\ \dot{y}_2 = B_1\phi_1 + B_2\phi_2 + B_3 \end{cases} \qquad (16)$$

where

$$A_1 = \frac{\psi_{a0}\psi_a + \psi_{b0}\psi_b}{\psi_{a0}^2 + \psi_{b0}^2},$$

$$A_2 = \frac{\mu}{2\alpha_0 M}\frac{\psi_a\psi_{b0} - \psi_{a0}\psi_b}{\psi_{a0}^2 + \psi_{b0}^2}$$

$$A_3 = \frac{\psi_{a0}\psi_a + \psi_{b0}\psi_b}{\psi_{a0}^2 + \psi_{b0}^2}\frac{T_{l0}}{J} - \frac{T_l}{J} + \frac{\mu}{M}(\psi_a\psi_{b0} - \psi_{a0}\psi_b)$$

$$B_1 = \frac{2\alpha M}{\mu}\frac{\psi_{a0}\psi_b - \psi_a\psi_{b0}}{\psi_{a0}^2 + \psi_{b0}^2},$$

$$B_2 = \frac{\psi_{a0}\psi_a + \psi_{b0}\psi_b}{\psi_{a0}^2 + \psi_{bo}^2}\frac{\alpha}{\alpha_0}$$

$$B_3 = 2\alpha_0(\psi_{a0}\psi_a + \psi_{b0}\psi_b) - 2\alpha(\psi_a^2 + \psi_b^2) + \frac{2\alpha M}{\mu}\frac{\psi_{a0}\psi_b - \psi_a\psi_{b0}}{\psi_{a0}^2 + \psi_{b0}^2}\frac{T_{l0}}{J}$$

It should be pointed out that the two subsystems still remain decoupled if the observed flux are precise although the parameters are not known correctly.

3.4 The estimate of the ranges of the coefficients in the real linearised model

The parameters of the motor are : $L_s = L_r = 0.0699$, $M = 0.068$, $R_S = 0.18$, $R_r = 0.15$, $R_{r0} = 0.25$, $p = 1$, $J = 0.0586$, $T_l = 27.0$, $T_{l0} = 29.0$ while all the units are in standard form. Suppose that the real and the observed rotor flux are both sine wave with constant speed and constant and equal frequency in the condition that the motor is running in the stable state.

Define

$$\psi_a = r\cos(\omega_0 t + \theta_1)$$
$$\psi_b = r\sin(\omega_0 t + \theta_1)\text{,}$$
$$\psi_{a0} = \hat{r}\cos(\omega_0 t + \theta_2)$$
$$\psi_{b0} = \hat{r}\sin(\omega_0 t + \theta_2)\text{,}$$
$$\frac{|r-\hat{r}|}{r} \le 10\%, |\theta_1 - \theta_2| \le 15^0,$$

Then the ranges of the coefficients in Eq. (16) can be calculated as following

$$A_1 = (0.88,1.11), A_2 = (-16.34,16.34),$$
$$A_3 = (-130.4,134.0), B_1 = (0.52,1.11),$$
$$B_2 = (-0.0084,0.0084), B_3 = (4.08,7.60) \qquad (17)$$

4. THE APPLICATION OF 2-DOFC IN IM

Only the speed subsystem is shown how to design the 2-DOFC .The P controller is used as $K_1(s)$ for $P_0(s) = 1/s$ according to Eq. (15).

To design $K_2(s)$, the first function in Eq. (16) can be written as

$$\dot{y}_1 = A_1 u + \xi_1 \qquad (18)$$

Where $\xi_1 = A_2\phi_2 + A_3$, $u = \phi_1$

In order to get a stable and non-constant error speed regulation system, the closed-loop system shown in fig.3 is constructed to reduce the coupling effect between the subsystems. Here the extended speed subsystem is described as

$$\begin{pmatrix}\dot{y}_1\\ \dot{y}_3\end{pmatrix} = \begin{pmatrix}0 & 0\\ 0 & 1\end{pmatrix}\begin{pmatrix}y_1\\ y_3\end{pmatrix} + \begin{pmatrix}.5\\ 0\end{pmatrix}(2\xi_1) + \left(\begin{pmatrix}1\\ 0\end{pmatrix} + \begin{pmatrix}A_1 - 1\\ 0\end{pmatrix}\right)u \qquad (19a)$$

Define the performance index as

$$z = C(y_1, y_3)' + Du \qquad (19b)$$

Define

$C = \begin{pmatrix} q & 0 & 0\\ 0 & 0 & 0\end{pmatrix}^T$, $D = (0 \quad 0 \quad 1)^T$, Then

$D^T(D \quad C) = (1 \quad 0 \quad 0)$, and

$E = I_2$, $F_a = 0_2$, $F_b = \begin{pmatrix}1\\ 0\end{pmatrix}$, $R^2 = (1+\lambda^{-2})$

By resolving the RICCATI inequality, K can be obtained .

Fig 4 shows the simulation result. The initial reference values of rotor flux and load torque are 1.3WB and 64 NM respectively. The parameters of the controller for the speed subsystem are $k_1 = 500, k_{2p} = 50, k_{2i} = 5$. At the same time the PI controller with $k_p = 30$ and $k_i = 10$ are used in the flux subsystem. Equ. (20) is the rotor flux observer be used in the simulation where T is the sample cycle

$$\begin{cases} \psi_{a0}(k) = (-T\alpha_0 + 1)\psi_{a0}(k-1) \\ \qquad - T\omega\psi_{b0}(k-1) + M\alpha_0 Ti_a(k-1) \\ \psi_{b0}(k) = (-T\alpha_0 + 1)\psi_{b0}(k-1) \\ \qquad + T\omega\psi_{a0}(k-1) + M\alpha_0 Ti_b(k-1) \end{cases} \tag{20}$$

In order to see the advantage of 2-DOF, the simulation result of the closed loop system constitute by 1-DOFC shown in Fig. 1a with $k_p = 120, k_i = 10$ is given in fig 4 also. In instant t_0 the reference values of flux and load torque leap up to 1.6 WB and 84 NM. From Fig 4 the advantage of 2-DOF is easy to be seen.

5. CONCLUSIONS

The advantage of 2-DOFC controlled system proposed in the paper can not only improve the servo performance but also reduce the effect of the disturbance. For used in the induction motors the other advantage of it is that the 2-DOFC controller is so simple to realize that the high performance and expensive microprocessor is not needed to embedded in the system and it can reduce the cost of the whole system

Reference

Hagiwara T. and T. Yamasaki (1996). Two-degree-of-freedom method of LQI servo system disturbance rejection by constant state feedback Int.j.control,vol.63,NO.4,pp.703—719

Vilanova, R. and I. Serra. (1997) Realization of two-degree-of freedom compensators IEE Pro.-Control theory appl. Vol.144,No.6, pp.589-595

Howze, J. W.and S.P. Bhattacharyya.(1997) Robust tracking error feedback, and two-degree-of-freedom controllers. IEEE Trans.A.C ,Vol 42,No7, pp.980—983

Marino R., S. Peresada. and P. Valigi.(1993) Adaptive Input— Output Linearizing Control of Induction Motors IEEE Trans.A.C. ,vol.38, no.2, pp. 208—221

Isidori A. (1989) Nonlinear Control Systems, Springer Verlag,Berlin.

Shen T. (1996) H^{∞} control theory and its application. Press of Tsinghua University (in Chinese)

Han Z.J. (1996) Adaptive control. .Press of Tsinghua University. (in Chinese)

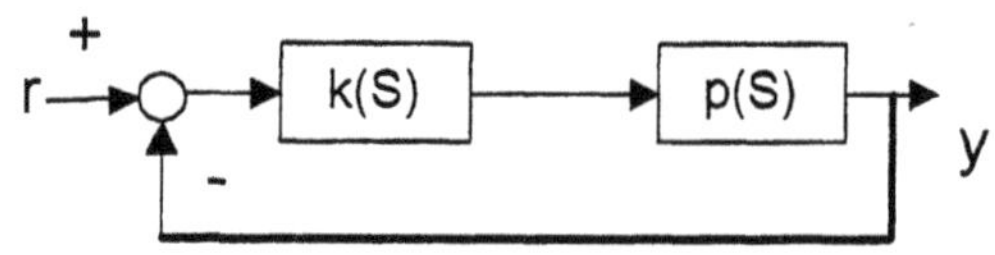

Fig 1a:The closed loop system
controlled by 1-DOFC

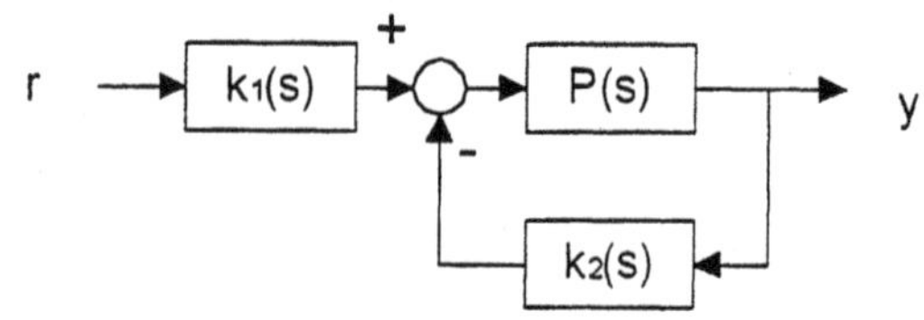

Fig 1b:The closed loop system
controlled by 2-DOFC

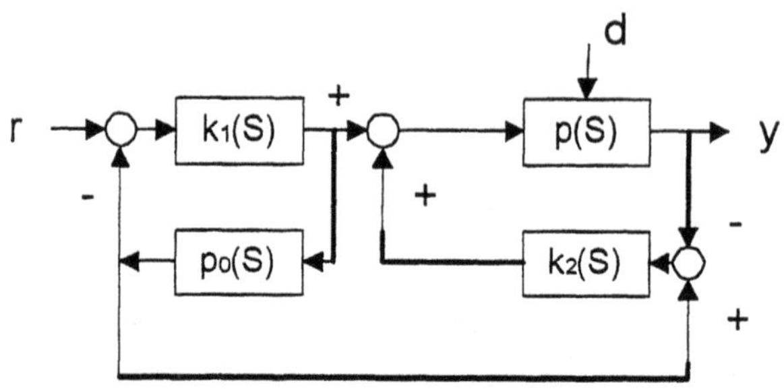

Fig.2a: 2-degrees of freedom controller

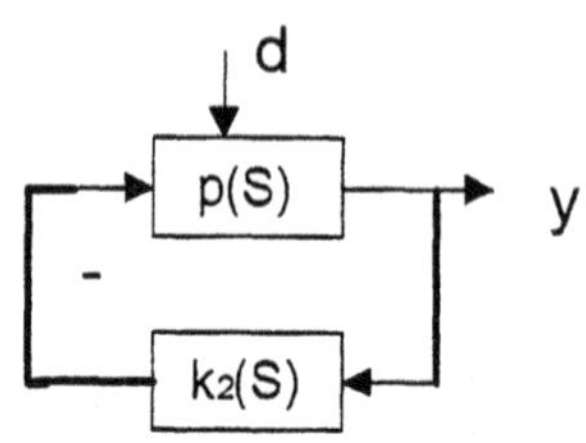

Fig 2c the diagram for designing $K_2(s)$

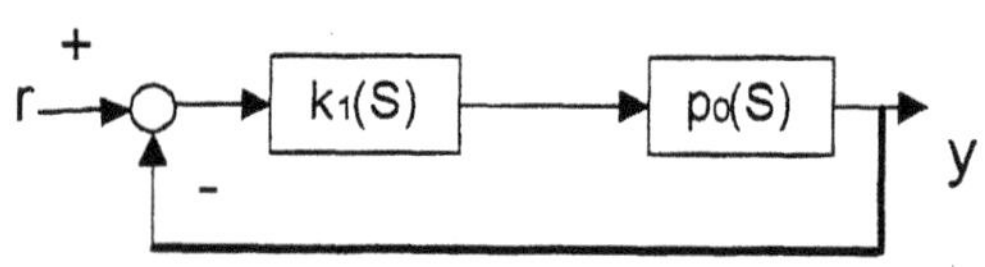

Fig 2b the diagram for designing $K_1(s)$

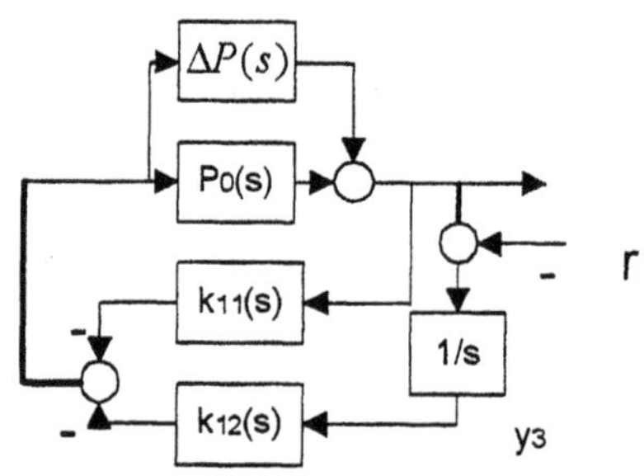

Fig.3:the design diagram of the H^∞ controller

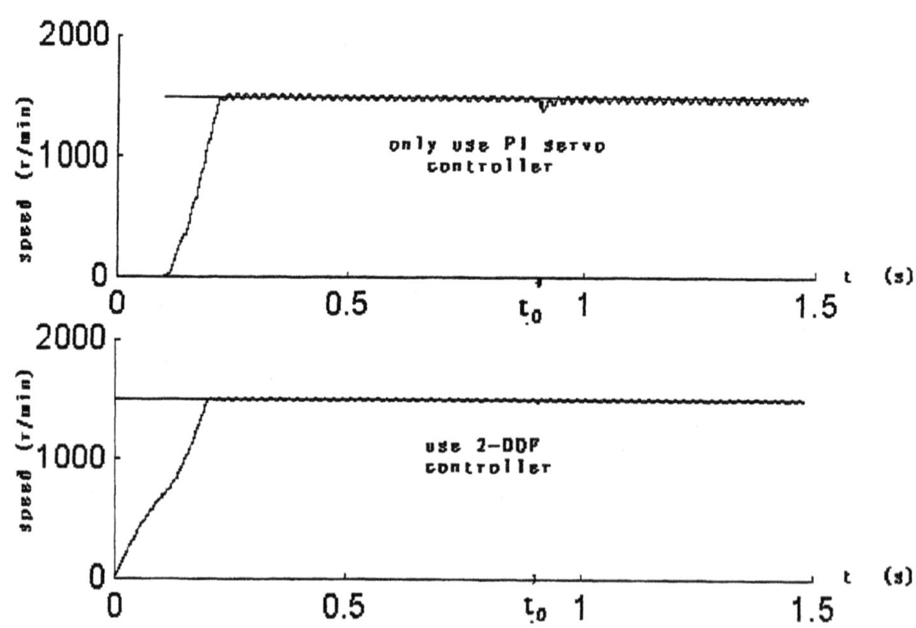

Fig 4: The simulation results of the closed system

Copyright © 1999 IFAC
14th Triennial World Congress, Beijing, P.R. China

SUBOPTIMAL POLE ASSIGNMENT CONTROL FOR A NONMINIMUM PHASE PLANT WITH WHITE NOISE

Wataru Kase and Hideki Okuno

Department of Electrical Engineering, Osaka Institute of Technology.
5-16-1, Omiya, Asahi-ku, Osaka, 535-8585, JAPAN.
Fax: +81-6-6957-2133, E-mail: kase@ee.oit.ac.jp.

Abstract: Recently, a design of model matching and adaptive control systems under the presence of measurement noise is reported by employing a nonminimal order state observer. But this method can be only used for the case where the plant is stably invertible. In this paper, it will be shown that the idea is also applicable to the case of pole placement. The proposed method can be implemented by simple calculation comparing with the design of Kalman filter or H_∞ control. So the proposed method is adequate for an adaptive control in which the controllers must be calculated on line. *Copyright 1999 IFAC*

Key Words: polynomial methods, discrete-time systems, pole assignment, non-minimum phase systems, measurement noise.

1. INTRODUCTION

Recently, a simple design of exact model matching (EMM) (Wolovich, 1974; Ichikawa, 1985) for a plant with measurement noise is shown (Mutoh and Nikiforuk, 1994; Kase and Mutoh, 1997). This method uses the free parameters which arises by employing a non-minimal order observer. For its simplicity, the method is also applicable to model reference adaptive control. In order to extend the result of Mutoh and Nikiforuk (1994) to a nonminimum phase plant, there exist two problems. The most difficult problem is the evaluation of variance. In the case of EMM, the variance of output can be obtained by calculating square sum of controller's parameters. So it is easy to solve the problem if the controller is well parameterized. But in the case of nonminimum phase plant, not only controller parameter but also the parameters of the plant should be considered to minimize the effect of the measurement noise. The other problem is a parameterization of controller. In Mutoh and Nikiforuk (1994), the EMM controller is parameterized by using Markov parameters of the plant. This can be done only for the minimum phase plant. So it must be employed another parameterization.

In this paper, the extended result of Mutoh and Nikiforuk (1994) to the nonminimum phase case is presented, i.e., the suboptimal pole placement design for a nonminimum phase plant with measurement noise will be shown. For such an extension, two major problems which mentioned in the above must be solved. In order to solve these problems, the result of extended division (Kase, 1997) will be used to obtain a new controller parameterization. The controller which minimizes the effect of measurement noise can be determined by using Moore-Penrose pseudoinverse (Albert, 1972) like the previous works. Therefore, the proposed method is simple and is applicable to an adaptive control.

Section 2 provides a statement of problem and derives a norm minimization problem to solve the original problem. Then, a new parameterization of controller, which is the constraint condition of minimization problem in the previous section, is presented in Section 3. The solution of minimization problem is also shown in Section 3. Numerical examples of the pole placement control are then presented in Section 4.

2. PROBLEM STATEMENT

Consider the plant

$$y(t)=\frac{N(z)}{D(z)}u(t) \tag{1}$$

where $u(t)$ is an input, $y(t)$ is an output, z is a shift operator, and t is an integer. In the above plant, it is supposed that the measured output signal $\bar{y}(t)$ is the sum of the real output signal $y(t)$ and the measurement noise $w(t)$, i.e.,

$$\bar{y}(t)=y(t)+w(t). \tag{2}$$

The purpose of the design is to determine the controller which makes the closed-loop system be internally stable and also minimizes the variance of $y(t)$. For this purpose, some assumptions are made as follows:

1. $D(z)$ is monic polynomial of degree n.
2. $N(z)/D(z)$ is strictly proper.
3. A desired closed-loop polynomial $D_*(z)$ and an observer $H(z)$ are stable polynomials of degree n and $d-1\geq n-1$ respectively.
4. $w(t)$ is a white noise and satisfies the following properties:

$$\begin{aligned}&E[w(t)]=0,\\&E[w^2(t)]=\sigma^2,\\&E[w(t)w(t+k)]=0,\quad k\neq 0\ (integer)\end{aligned} \tag{3}$$

where $E[\cdot]$ denotes the expectation operator.

If $w(t)\equiv 0$, it is well known (Wolovich, 1974; Ichikawa, 1985) that the pole placement can be attained by

$$u(t)=\frac{X(z)}{H(z)}u(t)+\frac{Y(z)}{H(z)}y(t)+r(t) \tag{4}$$

where $r(t)$ is a reference input and $X(z)$ and $Y(z)$ are the controller polynomials to be determined so

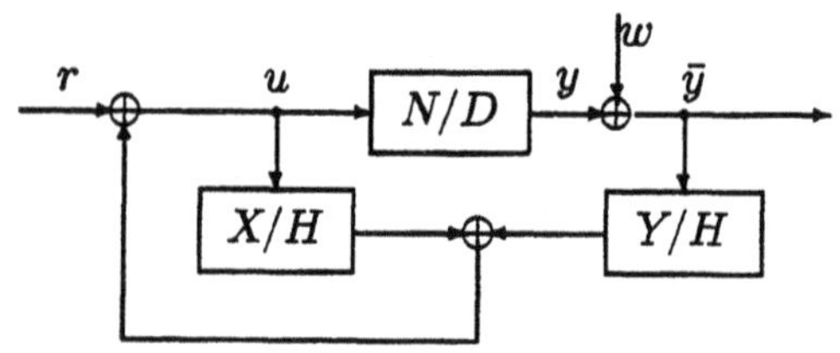

Figure 1: Pole Assignment Control Systems with Measurement Noise.

as to satisfy the following Diophantine equation (Fig.1):

$$X(z)D(z)+Y(z)N(z)=H(z)[D(z)-D_*(z)]. \tag{5}$$

In the case where $w(t)\neq 0$, the control input is given by

$$u(t)=\frac{X(z)}{H(z)}u(t)+\frac{Y(z)}{H(z)}\bar{y}(t)+r(t). \tag{6}$$

From eqs.(5) and (6), the input-output relation from $w(t)$ to $y(t)$ becomes

$$y(t)=\frac{Y(z)N(z)}{H(z)D_*(z)}w(t). \tag{7}$$

Let

$$\begin{aligned}&Y(z)=Y_{n-1}z^{n-1}+\cdots+Y_0=\mathcal{Y}_{d-1}S^{d-1}(z),\\&H(z)=z^{d-1}+H_{d-2}z^{d-2}+\cdots+H_0\\&\quad=\mathcal{H}_{d-1}S^{d-1}(z),\\&\mathcal{Y}_{d-1}=[Y_0\ \cdots\ Y_{n-1}],\\&\mathcal{H}_{d-1}=[H_0\ \cdots\ H_{n-2}\ 1],\\&S^{d-1}(z)=[1\ z\ \cdots\ z^{n-1}]^T,\end{aligned} \tag{8}$$

and

$$\frac{N(z)}{H(z)D_*(z)}=\sum_{i=d}^{\infty}V_{i-d+1}z^{-i}. \tag{9}$$

Then,

$$\frac{Y(z)N(z)}{H(z)D_*(z)}=\mathcal{Y}_{d-1}\mathcal{V}\begin{bmatrix}z^{-1}\\z^{-2}\\\vdots\end{bmatrix},$$
$$\mathcal{V}:=\begin{bmatrix}0&\cdots&0&V_1&\cdots&V_p&\cdots\\0&\cdots&V_1&V_2&\cdots&V_{p+1}&\cdots\\\vdots&&\vdots&\vdots&&\vdots&\\V_1&\cdots&V_{d-1}&V_d&\cdots&V_{p+d-1}&\cdots\end{bmatrix}. \tag{10}$$

Note that the number of columns in $\mathcal{V}$ is infinite. This implies that the variance of $y(t)$ becomes

$$E[y^2(t)]=\sigma^2||\mathcal{Y}_{d-1}\mathcal{V}||^2. \tag{11}$$

Therefore, the purpose of the design is achieved by minimizing $||\mathcal{Y}_{d-1}\mathcal{V}||$. In the next section, it will be derived solved the constraint condition of the above minimization problem and solved the problem.

3. SUBOPTIMAL POLE PLACEMENT FOR A PLANT WITH MEASUREMENT NOISE

In this section, it makes clear the degrees of freedom in $\mathcal{Y}_{d-1}$, and minimize the variance given in eq.(11) using such freedom.

There are many methods to solve eq.(5). In this paper, a method using state space parameters is presented. At first, the following lemma concerning with an extended division holds.

Lemma 1 Assume that $D(z)$ is a monic polynomial of degree n and $N(z)/D(z)$ is strictly proper. Then, for a given polynomial $Y(z)$ of degree $d-1$, there exist polynomials $Q(z)$ and $R(z)$ such that

$$\begin{array}{l} Y(z)N(z) = Q(z)D(z) + R(z), \\ R(z)/D(z) : \text{strictly proper.} \end{array} \tag{12}$$

Let $(A,\ b,\ c)$ be the controllability canonical realization of $N(z)/D(z)$. Then, $Q(z)$ and $R(z)$ are given by the following equations:

$$\begin{array}{l} Q(z) = \mathcal{Y}_{d-1}\mathcal{T}_{d-2}S^{d-2}(z), \\ R(z) = \mathcal{Y}_{d-1}\mathcal{O}_{d-1}S^{n-1}(z), \end{array} \tag{13}$$

where

$$\mathcal{O}_{d-1}(c) = \begin{bmatrix} c \\ cA \\ \vdots \\ cA^{d-1} \end{bmatrix},$$

$$\mathcal{T}_{d-2} = \begin{bmatrix} 0 & 0 & \cdots & 0 \\ cb & 0 & \cdots & 0 \\ cAb & cb & \cdots & 0 \\ \vdots & \vdots & \ddots & \vdots \\ cA^{d-2}b & cA^{d-3}b & \cdots & cb \end{bmatrix} \in \boldsymbol{R}^{d\times(d-1)}. \tag{14}$$

(Proof). Since $(A,\ b,\ c)$ is the realization of $N(z)/D(z)$,

$$\begin{array}{l} zx(t) = Ax(t) + bu(t), \\ y(t) = cx(t) \end{array}$$

holds. From the above equations,

$$S^{d-1}(z)y(t) = \mathcal{O}_{d-1}(c)x(t) + \mathcal{T}_{d-2}S^{d-2}(z)u(t). \tag{15}$$

Let $x_1(t)$ denote the first element of $x(t)$. Then, from the structure of the controllability canonical form, the following relations hold:

$$\begin{array}{l} x(t) = S^{n-1}(z)x_1(t), \\ D(z)x_1(t) = u(t), \\ N(z)x_1(t) = y(t). \end{array}$$

Substituting the above equations into eq.(15),

$$\begin{aligned} S^{d-1}(z)N(z)x_1(t) &= \mathcal{O}_{d-1}(c)S^{d-1}(z)x_1(t) \\ &\quad + \mathcal{T}_{d-2}S^{d-2}(z)D(z)x_1(t). \end{aligned}$$

Premultiplying both side by $\mathcal{Y}_{d-1}$ and noticing that the above equation is valid for an arbitrary $x_1(t)$,

$$\begin{aligned} Y(z)N(z) &= \mathcal{Y}_{d-1}\mathcal{O}_{d-1}(c)S^{d-1}(z) \\ &\quad + \mathcal{Y}_{d-1}\mathcal{T}_{d-2}S^{d-2}(z)D(z). \end{aligned}$$

Since $\mathcal{Y}_{d-1}\mathcal{O}_{d-1}S^{n-1}(z)/D(z)$ is strictly proper from the definition of $S^{n-1}(z)$, eq.(13) holds from the uniqueness of the quotient and remainder (Kailath, 1980).

From eq.(5),

$$\begin{array}{l} X(z) + Y(z)N(z)/D(z) \\ = H(z)[D(z) - D_*(z)]/D(z). \end{array} \tag{16}$$

Let $(A,\ b,\ f)$ denote the controllability canonical realization of $[D(z) - D_*(z)]/D(z)$. Since the remainder of $Y(z)N(z)/D(z)$ coincedes with the one of $H(z)[D(z) - D_*(z)]/D(z)$, we have

$$\mathcal{Y}_{d-1}\mathcal{O}_{d-1}(c) = \mathcal{H}_{d-1}\mathcal{O}_{d-1}(f). \tag{17}$$

Substituting $\mathcal{Y}_{d-1}$ satisfying eq.(17) into eq.(16),

$$X(z) = (\mathcal{H}_{d-1}\mathcal{T}_{d-2} - \mathcal{Y}_{d-1}\mathcal{T}_{d-2})S^{d-2}(z). \tag{18}$$

There exist infinite solution of eq.(17), from the definition of d. Therefore, eqs.(17) and (18) give a parameterization of all pole placement controllers. See Kase (1997) for detail.

From the above discussions, the problem of determining a pole placement system for a plant with measurement noise can be stated as the following simple optimization:

$$\text{minimize } ||\mathcal{Y}_{d-1}\mathcal{V}||^2 \tag{19}$$

$$\text{subject to } \mathcal{Y}_{d-1}\mathcal{O}_{d-1}(c) = \mathcal{H}_{d-1}\mathcal{O}_{d-1}(f). \tag{20}$$

Using a free parameter $\lambda_d \in \boldsymbol{R}^{1\times d}$, all solutions satisfying eq.(20) are given by

$$\begin{aligned} \mathcal{Y}_{d-1} &= \mathcal{H}_{d-1}\mathcal{O}_{d-1}(f)\mathcal{O}_{d-1}^{\dagger}(c) \\ &\quad - \lambda_d(I - \mathcal{O}_{d-1}(c)\mathcal{O}_{d-1}^{\dagger}(c)), \end{aligned} \tag{21}$$

where $\mathcal{O}_{d-1}^{\dagger}(c)$ denotes the Moore-Penrose pseudoinverse (Albert, 1972) and is given by

$$\mathcal{O}_{d-1}^{\dagger}(c) = (\mathcal{O}_{d-1}^{T}(c)\mathcal{O}_{d-1}(c))^{-1}\mathcal{O}_{d-1}^{T}(c) \tag{22}$$

in this case. Substituting eq.(21) into eq.(19),

$$\begin{array}{l} J_d := \min_{\lambda_d} ||\lambda_d\mathcal{F}_d - \mathcal{G}_d||^2, \\ \mathcal{F}_d := (I - \mathcal{O}_{d-1}(c)\mathcal{O}_{d-1}^{\dagger}(c))\mathcal{V}, \\ \mathcal{G}_d := \mathcal{H}_{d-1}\mathcal{T}_{d-2}\mathcal{O}_{d-1}^{\dagger}(c)\mathcal{V}. \end{array} \tag{23}$$

Let $\mathcal{F}_d^{\dagger}$ denote the psuedoinverse of $\mathcal{F}_d$. Then, λ_d^{opt} which minimizes the above norm is given by (Albert, 1972)

$$\lambda_d^{opt} = \mathcal{G}_d\mathcal{F}_d^{\dagger}. \tag{24}$$

Therefore, $\mathcal{Y}_{d-1}$ can be determined by substituting eq.(24) into eq.(21), and the $X(z)$ can be determined by eq.(18).

However, since the number of columns in $\mathcal{V}$ is infinite, the above procedure can not be acheived. In order to calculate an approximate optimal value λ_d^{opt}, define

$$\mathcal{V} := \begin{bmatrix} 0 & \cdots & 0 & V_1 & \cdots & V_p \\ 0 & \cdots & V_1 & V_2 & \cdots & V_{p+1} \\ \vdots & & \vdots & \vdots & & \vdots \\ V_1 & \cdots & V_{d-1} & V_d & \cdots & V_{p+d-1} \end{bmatrix} \quad (25)$$

for some integer $p \geq n$. Then, calculated minimum value $J_{d,p}$ of eq.(23) using $\mathcal{V}_p$ instead of $\mathcal{V}$. Simiraly, calculate $J_{d,p+1}$ using $\mathcal{V}_{p+1}$. If $|J_{d,p+1} - J_{d,p}|$ is small enough, then $J_d = J_{d,p+1}$ and adopt the minimizing value λ_d^{opt} for $J_{d,p+1}$. If not, then p must be increased. Since the optimal value can be attained as $p \to \infty$, the above process will end at finite interval. More explicit discussion will be given in appendix.

In the case where $D_*(z) = z^n$ and $H(z) = z^{d-1}$, the above discussion does not need, since $V_i = 0$ for all $i \geq n$ and the number of columns in $\mathcal{V}$ is finite. Moreover, in that case, eq.(17) simplifies as follow:

$$\mathcal{Y}_{d-1}\mathcal{O}_{d-1}(c) = fA^{d-1}. \quad (26)$$

It should be noted that $E[y^2(t)]$ can be reduced by increasing the value of d and using the corresponding suboptimal pole placement system. Define $\mathcal{Y}_d$ by

$$\mathcal{Y}_d = [0 \ \mathcal{Y}_{d-1}^{opt}], \quad (27)$$

where $\mathcal{Y}_{d-1}^{opt}$ can be obtained from eq.(21) setting $\lambda_d = \lambda_d^{opt}$. The above $\mathcal{Y}_d$ satisfies eq.(20) when d is replaced by $d+1$ since

$$\mathcal{Y}_d\mathcal{O}_d = \mathcal{Y}_{d-1}\mathcal{O}_{d-1}A = fA^d.$$

Using $\mathcal{Y}_d$,

$$J_{d+1} \leq ||\mathcal{Y}_d\mathcal{V}||^2 = ||[0 \ \mathcal{Y}_{d-1}]\mathcal{V}||^2 = J_d. \quad (28)$$

4. EXAMPLE

Consider the following plant:

$$y(t) = \frac{1.5z + 2}{z^3 + 1.5z^2 + 0.5z + 0.04}u(t).$$

In this case, $D_*(z) = z^3$ and $f = [0.04 \ 0.5 \ 1.5]$. Four different controllers are considered in the following. A pulse train function is used as a reference input.

Case 1 (minimal order controller $d = 3$): In this case, $H(z) = z^2$. The controllability canonical realization for this plant is given by

$$A = \begin{bmatrix} 0 & 1 & 0 \\ 0 & 0 & 1 \\ -0.04 & -0.5 & -1.5 \end{bmatrix}, \quad b = \begin{bmatrix} 0 \\ 0 \\ 1 \end{bmatrix},$$
$$c = [2 \ 1.5 \ 0].$$

From the relation of the extended division, the following equation holds:

$$\mathcal{Y}_2\mathcal{O}_2 = \mathcal{Y}_2 \begin{bmatrix} 2 & 1.5 & 0 \\ 0 & 0 & 1 \\ -0.04 & -0.75 & -0.25 \end{bmatrix}$$
$$= [0.07 \ 0.815 \ 1.915] = fA^2.$$

Thus, $\mathcal{Y}_2$ has the unique solution in this case, which is given by

$$\mathcal{Y}_2 = [0.180 \ 2.080 \ 4.818],$$

from which it follows that $J_3 = 15.32$.

On the other hand, $\mathcal{X}_1$ is given by

$$\mathcal{X}_1 = [fAb \ fb] - \mathcal{Y}_2 \begin{bmatrix} 0 & 0 \\ cb & 0 \\ cAb & cb \end{bmatrix}$$
$$= [-8.978 \ 1.500].$$

Fig.2 shows the plant output response when the measurement noise added.

Case 2 (suboptimal controller with $d = 5$): In this case, $H(z) = z^4 I$. The suboptimal controller parameters become

$$\mathcal{Y}_4 = [0.054 \ 0.547 \ 0.559 \ -1.542 \ 1.497],$$
$$\mathcal{X}_3 = [-2.720 \ 4.602 \ -3.996 \ 1.500]$$

from which it follows that $J_5 = 3.961$. Fig.3 shows the output response of the plant with same measurement noise as in case 1.

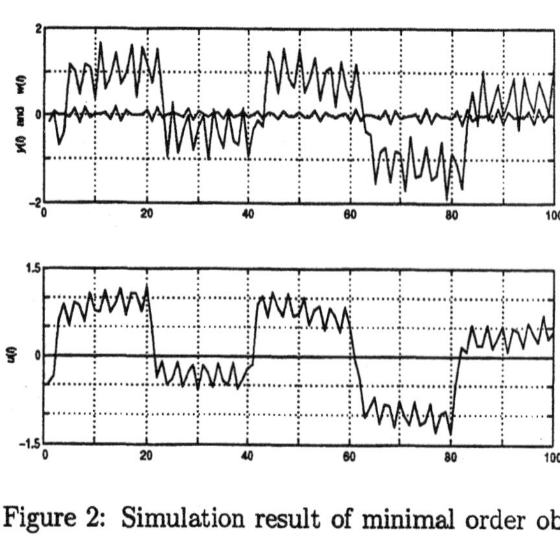

Figure 2: Simulation result of minimal order observer.

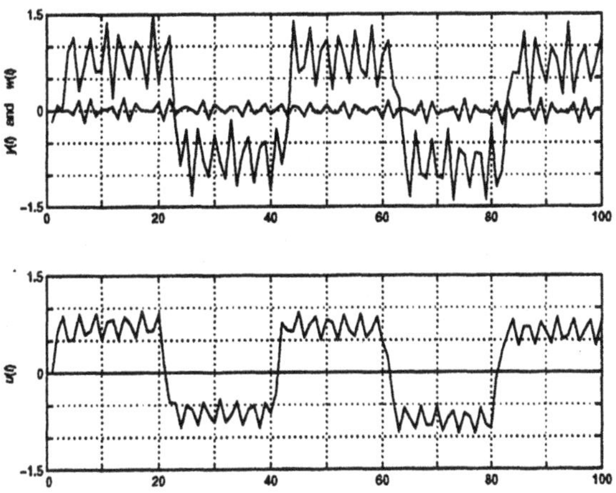

Figure 3: Simulation result of proposed method ($d = 5$).

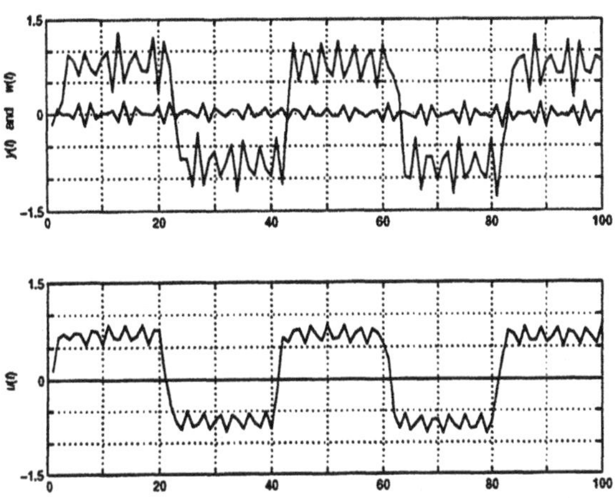

Figure 4: Simulation result of poposed mthod ($d = 7$).

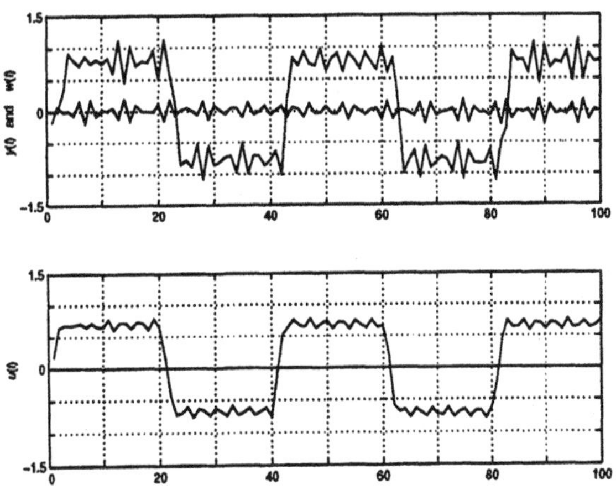

Figure 5: Simulation result of proposed method ($d = 9$).

Case 3 (suboptimal controller with $d = 7$): The suboptimal controller parameters become

$$\mathcal{Y}_6 = [0.0025\ 0.237\ 0.164\ -0.658\ 0.974\ -1.010\ 0.707],$$
$$\mathcal{X}_5 = [-1.232\ 2.607\ -3.507\ 3.607\ -2.810\ 1.500]$$

from which it follows that $J_7 = 2.061$. Fig.4 shows the output response of the plant with same measurement noise as in case 1.

Case 4 (suboptimal controller with $d = 9$): The suboptimal controller parameters become

$$\mathcal{Y}_8 = [0.014\ 0.133\ 0.75\ -0.361\ 0.594\ -0.734\ 0.764\ -0.668\ 0.425],$$
$$\mathcal{X}_7 = [-0.700\ 1.593\ -2.399\ 2.974\ -3.211\ 3.024\ -2.388\ 1.500]$$

from which it follows that $J_9 = 1.405$. Fig.4 shows the output response of the plant with same measurement noise as in case 1.

5. CONCLUSIONS

A design method of pole placement control systems which will reduced the effect of measurement noise on the plant output is presented in this paper. This method uses the freedom of the controller parameters due to increasing the degree of the observer polynomial. For some fixed degree of the observer polynomial, the controller for this pole placement controller is obtained as a solution of a simple optimization problem. It is also shown that the effect of the measurement noise on the plant output becomes smaller as the degree of the observer polynomial is increased. This method will be applicable for a multivariable plant (Kase, 1998) and for a plant with input noise (Kase and Mutoh, 1998). For the choice of stable polynomials $D_*(z)$ and $H(z)$, it maybe useful to consider the LQ optimal control, or the canonical interactor (Kase *et al.*, 1998).

REFERENCES

Albert, A. (1972). *Regression and the Moore-Penrose Pseudoinverse.* Academic Press, NY.

Ichikawa, K. (1985). *Control system design based on exact model matching techniques*, Springer-Verlag, Berlin.

Kailath, T. (1980). *Linear systems*, Prentice-Hall, NJ.

Kase, W. (1997). A solution of polynomial matrix equations using extended division algorithm, *Proc. 2nd Asian Control Conference*, **2**, pp.123-126 (also see its extended version which will appear in *Int. J. Systems Science*).

Kase, W. (1998). Suboptimal dead beat controller for a discrete-time multivariable plant with measurement noise, *18th American Control Conference* (under submission).

Kase, W. and Mutoh Y. (1997), Suboptimal exact model matching for multivariable systems with measurement noise, *IEEE Trans. Automat. Contr.* (under submission).

Kase, W. and Mutoh, Y. (1998). Suboptimal dead beat control for a nonminimum phase plant with input and measurement noise, *5th European Control Conference* (under submission).

Kase, W., Mutoh, Y. and Teranishi, M. (1998) The canonical interactor matrix and its application, *38th IEEE Conference on Disicion and Control* (under submission).

Mutoh, Y. and Nikiforuk, P. N. (1994). Suboptimal perfect model matching for a plant with measurement noise and its application to MRACS, *IEEE Trans. Automat. Contr.*, **39**, pp.422-425.

Wolovich, W. A. (1974). *Linear multivariable systems*, Springer-Verlag, Berlin.

APPENDIX

Let $(A_v,\ b_v,\ c_v)$ denote a minimal realization of $N(z)/H(z)D_*(z)$. Then, V_i are given by

$$V_i = c_v A_v^{i+d-2} b_v. \tag{29}$$

Since A_v is Hurwitz, there exist a positive number a and μ $(0 < |\mu| < 1)$ such that

$$||A_v^k|| \leq a|\mu|^k, \quad \forall k \geq 1. \tag{30}$$

Therefore,

$$\begin{gathered}\left\|\begin{bmatrix} V_{p+1} \\ \vdots \\ V_{p+d} \end{bmatrix}\right\|^2 \leq C_1 |\mu|^{2(p+d-1)}, \\ C_1 := a^2 C_2 ||c_v||^2 ||b_v||^2, \\ C_2 := 1 + |\mu| + \cdots + |\mu|^d - 1 \end{gathered} \tag{31}$$

holds.

Let $\mathcal{Y}_{d-1}^p$ denote the value which minimizes $J_{d,p}$. Then,

$$\begin{aligned} J_{d,p+1} &\leq ||\mathcal{Y}_{d-1}^p \mathcal{V}_{p+1}||^2 \\ &= ||\mathcal{Y}_{d-1}^p \mathcal{V}_p||^2 + ||\mathcal{Y}_{d-1} \begin{bmatrix} V_{p+1} \\ \vdots \\ V_{p+d} \end{bmatrix} ||^2 \\ &\leq J_{p,d} + C_1 ||\mathcal{Y}_{d-1}^p||^2 |\mu|^{p+d-1}. \end{aligned} \tag{32}$$

Using the above relation, it can be obtained

$$J_d \leq J_{d,p} + \frac{C_1 ||\mathcal{Y}_{d-1}^p||^2 |\mu|^{p+d-1}}{1 - |\mu|}. \tag{33}$$

14th Triennial World Congress, Beijing, P.R. China

A DESIGN ALGORITHM OF STATIC OUTPUT FEEDBACK CONTROL FOR EIGENSTRUCTURE ASSIGNMENT

Chia-Chi Tsui

743 Clove Road, NY 10310, USA, 718-727-0670

Abstract: Eigenvectors determine the sensitivity and robustness properties of the corresponding eigenvalues. Hence their assignment is as important as eigenvalue assignment. The algorithm of this paper assigns eigenvectors in a most explicit and therefore the best form -- assigning the linear combination of the basis vectors of each eigenvector. Compare the most recent and similar algorithms which assign the eigenstructure in two consecutive groups, the algorithm of this paper has three more advantages: 1) the explicit and analytical solution of a key matrix equation pair is derived; 2) a very complicated similarity transformation and numerical operation at assigning the second group of eigenvalues/vectors, is eliminated; and 3) the duality of the whole problem, which is fully reflected by the duality of two matrix equation pairs, is fully revealed in the actual design solution and computation. *Copyright © 1999 IFAC*

Keywords: Eigenstructure assignment, Static output feedback

1. INTRODUCTION

Given the state space model ($A \in R^{n \times n}$, $B \in R^{n \times p}$, $C \in R^{q \times n}$) of an irreducible system. The dynamic matrix of the corresponding static output feedback (SOF) system is $A - BK_yC$. Thus SOF control can be considered as a constrained state feedback (SF) control with a constrained SF gain $K = K_yC$, and as a generalized SF control -- the SF is a special case of SOF under the special condition rank(C) = n.

The distinct advantage of SOF over SF is that the loop transfer function of an SOF system is guaranteed to be

$$L(s) = -K(sI - A)^{-1}B \quad \forall s \text{ and } K\ (= K_yC) \tag{1}$$

while (1) cannot be achieved for most (A,B,C)'s if K is designed based on (A,B) only and if the system state is not directly and fully available (Tsui, 1996a). Because loop transfer function determines the critical sensitivity and robustness properties of the corresponding feedback system and (1) is the loop transfer function of the direct SF system, the full realization of the robustness properties of SOF control is guaranteed while that of an SF control is not possible.

The obvious disadvantage of SOF is that it can be much weaker than the SF control. This is because the constraint $K = K_yC$ can be very severe when rank(C) ≪ n. Fortunately, a recently developed dynamic output feedback compensator whose definition uniquely satisfies the necessary and sufficient condition of (1), and which is valid for most (A,B,C), can generate some linear combinations of system state (Tsui, 1996a,b, 1998). These additional signals are linearly independent of the system output signal and can be used as additional system output. Thus this compensator can generate an equivalent SOF control signal whose corresponding rank(C) is *substantially increased.*

Because eigenvectors determine the sensitivity and robustness properties of the corresponding eigenvalues (Wilkinson, 1965), their assignment is *as important as* eigenvalue assignment. Because system poles most directly determine system response/performance, eigenvalue/vector assignment can most effectively improve feedback system performance and robustness (versus the direct design of L(s) of (1)).

This design problem has been studied for over twenty years. The exact pole assignment algorithms have been reported in (Kimura, 1975) and (Wang, 1996), under the general condition p + q > n and the generic condition p x q > n, respectively. A promising approximate pole assignment algorithm is also reported in (Chu, 1993). The references are representative but not exhaustive.

Although the condition $p \times q > n$ is more generous than $p + q > n$, its associated algorithms involve with very complicated numerical search computation and thus unable to consider the critical eigenvector assignment. In addition, the above mentioned substantial increase of rank(C) (or parameter q) has made the generosity of condition $p \times q > n$ less critical. Hence this paper is concerned only with the algorithms under condition $p + q > n$ **and** with actual eigenvector assignment capability. The most recent of such algorithms seems to be (Syrmos et al, 1993).

This algorithm assigns the eigenstructure in two separate groups -- n-q eigenvalues and their corresponding left eigenvectors first and then the q remaining eigenvalues and their corresponding right eigenvectors. An advantage of this two-group eigenvector assignment scheme is that it unifies with the SF design -- if q = n then the algorithm automatically reduces to its second step of assigning all n eigenvalues and n right eigenvectors (identical to the SF design). Another advantage of this design is that it reflects the duality property of the original problem -- the problem can be formulated as to find K_y such that the dual equations

$$T(A - BK_yC) = \Lambda T \qquad (2.a)$$
$$(A - BK_yC)V = V\Lambda \qquad (2.b)$$

are satisfied, where the Jordan form matrix Λ contains all assigned eigenvalues of matrix $A - BK_yC$, and T and V are the corresponding left and right eigenvector matrix, respectively (TV = I).

This paper presents an algorithm which is in the same basic form of (Syrmos et al, 1993). However, a *new solution* of (2) *and* TV = I, which is based on a different solution of (2) in (Tsui, 1993), is used in this paper. This new solution of (2) and TV = I makes the assignment of T and V in the form of assigning the linear combinations of the basis vectors of T and V. Because linear combination of the basis vectors is the most basic and explicit form of the freedom of the corresponding eigenvector, this form of eigenvector assignment should be most effective. In fact, the most effective and general existing robust eigenstructure assignment methods (Kautsky, et al, 1985) are based on this form. In addition, this new solution of (2) and TV = I also makes the above duality property much more relevant in actual design computation, and enables the elimination of a complicated two-step similarity transformation and numerical operation at the assignment of the second group of eigenstructure (Syrmos et al, 1993).

2. THE ALGORITHM AND ITS ANALYSIS

The design algorithm is aimed at partially satisfying (2.a) and then (2.b) (and TV = I). Because (2.a) and (2.b) are redundant, this partial satisfaction of (2.a) and (2.b) also implies the complete satisfaction of (2.a) and (2.b), as will be evident at Step 2 of the algorithm..

Step 0: Divide the n eigenvalues of Λ of (2) into two groups of n-q and q eigenvalues each. The eigenvalues in each group must be either real or complex conjugate. Let Λ_{n-q} and Λ_q be the real Jordan form matrixes which contain the eigenvalues of these two groups, respectively.

Step 1: Compute solution matrix $T_{n-q} \in R^{(n-q)\times n}$ of equation

$$T_{n-q}A - \Lambda_{n-q}T_{n-q} = LC \qquad (3)$$

while matrix $[T_{n-q}':C']'$ has full row rank.

Step 1.a: For each r-dimensional Jordan block J_i of Λ_{n-q}, compute the corresponding matrixes $D_i \in R^{rq \times rn}$ and $E_i \in R^{rq \times rq}$ such that

$$[D_i:E_i][(I_r \otimes A - J_i' \otimes I)':-(I_r \otimes C)']' = 0 \qquad (4.a)$$

where $\otimes$ stands for the Kronecker product (Graham, 1981) and I_r is an r-dimensional identity matrix. If r = 1 and $J_i = \lambda_i$ which is a distinct and real eigenvalue of Λ_{n-q}, then (4.a) becomes

$$[D_i:E_i][(A - \lambda_i I)':-C']' = 0 \qquad (4.b)$$

Because the dimension of the second or the given matrix of (4.a), is $r(n + q) \times rn$, the two solution matrixes with linearly independent rows always exist $\forall i$ (Tsui, 1993).

Step 1.b: Compute $\mathbf{c}_i \in R^{1 \times rq}$ such that matrix $[T_{n-q}':C']'$ has full row rank, where the block of r rows of T_{n-q} corresponding to J_i is

$$T_{n-q,i} = [(\mathbf{c}_iD_{i,1})':(\mathbf{c}_iD_{i,2})':\ldots:(\mathbf{c}_iD_{i,r})']' \quad \forall i \qquad (5.a)$$

and where $D_i = [D_{i,1}:\ldots:D_{i,r}]$ and $E_i = [E_{i,1}:\ldots:E_{i,r}]$ $\forall i$. If the dimension (r) of J_i of (4.a) equals one, then (5.a) becomes

$$T_{n-q,i} = \mathbf{c}_i D_i \qquad (5.b)$$

Comparing (4) and (3), this T_{n-q} together with L where the block of r rows of L corresponding to J_i is $[(\mathbf{c}_i E_{i,1})':(\mathbf{c}_i E_{i,2})':\ldots]'$, satisfy (3). However, only T_{n-q} but not L is needed in the remaining steps of our algorithm.

Because (3) and the full rank of $[T_{n-q}':C']'$ are the necessary and sufficient conditions of the well known state observer, the solution $\mathbf{c}_i$ always exists and is non-unique if $q > 1$. Because (3) and (2.a) show that T_{n-q} would be the left eigenvector matrix of the feedback system dynamic matrix corresponding to Λ_{n-q}, the non-unique selection of $\mathbf{c}_i$ constitutes the assignment of the left eigenvectors of T_{n-q}. The left eigenvector assignment under formulation (5) is most explicit and has the most effective and general decoupling/robust assignment algorithm (Tsui, 1996b, 1993, Kautsky et. al., 1985). Notice that the solution of (3) that is recommended by (Syrmos et al, 1993) is **not** in form (4)-(5).

If $q = n$ (SF case), then Step 1 is unnecessary and $T_{n-q} = 0$.

Step 2: Compute the full column rank solution matrix $V_q \in R^{n \times q}$ of

$$AV_q - V_q\Lambda_q = BK_q \qquad (6.a)$$
$$\text{while} \quad T_{n-q}V_q = 0. \qquad (6.b)$$

This equation pair together with (3) obviously imply that (2) is fully determined and satisfied ($\Lambda = \text{diag}\{\Lambda_{n-q}, \Lambda_q\}$, T_{n-q} will be formed by the first n-q left eigenvectors of T and V_q will be formed by the last q right eigenvectors of V), when K_y of (2) is computed from K_q of (6.a) by an appropriate similarity transformation (as will be done in Step 3 of this algorithm).

It is very interesting to notice that this equation pair to a different system $(A,B,C \triangleq T_{n-q})$, is in perfect dual to the equation pair of (3) and $T_{n-q}B = 0$ to system (A,B,C).

The execution of this step is different for $p + q > n$ and $p + q \not> n$. We will discuss the case $p + q > n$ first.

Step 2.a: For each r-dimensional Jordan block J_i of Λ_q, compute the corresponding matrixes $D_i \in R^{rn \times rp}$ and $E_i \in R^{rp \times rp}$ such that

$$[I_r \otimes A - J_i' \otimes I : -I_r \otimes B][\,D_i' : E_i'\,]' = 0. \qquad (7.a)$$

If $r = 1$ and $J_i = \lambda_i$ which is a distinct and real eigenvalue of Λ_q, then (7.a) becomes

$$[A - \lambda_i I : -B][\,D_i' : E_i'\,]' = 0 \qquad (7.b)$$

Similar to Equation (4) of Step 1.a, because the dimension of the given matrix of (7.a) is $rn \times r(n + p)$, the two solution matrixes with linear independent columns always exist $\forall i$.

Step 2.b: Compute $\mathbf{c}_i \in R^{rp \times 1}$ such that $T_{n-q}V_q = 0$ is satisfied, or

$$[(T_{n-q}D_{i,1})':\ldots:(T_{n-q}D_{i,r})']'\mathbf{c}_i = 0 \quad \forall i \qquad (8.a)$$

or $(T_{n-q}D_i)\mathbf{c}_i = 0$ (if parameter r equals one) (8.b)

where in (7.a) $D_i = [D_{i,1}':\ldots:D_{i,r}']'$ and $E_i = [E_{i,1}':\ldots:E_{i,r}']'$ $\forall i$. This is because the block of r columns of V_q corresponding to J_i of (7.a) is assigned as

$$V_{q,i} = [D_{i,1}\mathbf{c}_i : D_{i,2}\mathbf{c}_i : \ldots : D_{i,r}\mathbf{c}_i] \quad \forall i \qquad (9.a)$$

or $V_{q,i} = D_i\mathbf{c}_i$ (if parameter r equals one).

Similar to Step 1.b, the comparison between (7.a) and (6.a) shows that this V_q together with K_q where the block of r columns of K_q corresponding to J_i of (7.a) is

$$K_{q,i} = [E_{i,1}\mathbf{c}_i : E_{i,2}\mathbf{c}_i : \ldots : E_{i,r}\mathbf{c}_i] \quad \forall i \qquad (9.b)$$

or $K_{q,i} = E_i\mathbf{c}_i$ (if parameter r equals one)

satisfy (6.a). Furthermore, because the dimension of the given matrix of (8) is $r(n - q) \times rp$, the nonzero solution $\mathbf{c}_i$ of (8) always exists $\forall i$ if $(n - q) < p$ or if $p + q > n$.

If $p + q > n + 1$, then the solution $\mathbf{c}_i$ of (8) is not unique. This freedom of $\mathbf{c}_i$, which is still expressed in $\mathbf{c}_i$ itself, can be considered as the freedom of assigning the eigenvectors of V_q of (9.a). We recall that in the well known special SF case where $q = n$, eigenvector assignment of V_q is possible whenever $p > 1$, which is equivalent to this current general condition $p + q > n + 1$.

Because T_{n-q} and V_q will be respectively formed by the first n-q left eigenvectors and the last q right eigenvectors of the feedback system dynamic matrix, and because matrix $[T_{n-q}':C']'$ is full row rank, to make the actual eigenvector matrixes T and V as well conditioned as possible so that the eigenvalues can be as robust as possible (Kautsky, et al, 1985), V_q may be assigned such that matrix CV_q is as well conditioned as possible. The most effective and general existing numerical algorithm for this assignment is based on the same form of (9.a) (Kautsky, et al, 1985).

Finally, if $p + q \not> n$, and if $\mathbf{v}_i \in R^{n\times 1}$ and $\mathbf{k}_i \in R^{p\times 1}$ are the column of V_q and K_q respectively and are corresponding to λ_i of Λ_q, then satisfying (6) implies, that $\forall i$

$$\begin{bmatrix} A - \lambda_i I : & -B \\ T_{n-q} \quad : & 0 \end{bmatrix} \begin{bmatrix} \mathbf{v}_i \\ \mathbf{k}_i \end{bmatrix} = 0 \qquad (10)$$

Equation (10) further implies that λ_i must be a transmission zero of a system with state space model (A, B, T_{n-q}). This understanding is important because of two reasons. First the left eigenvectors in T_{n-q} can be assigned if $q > 1$ (see Step 1.b) and second, the λ_i's do not really need to be given arbitrarily in practice.

Step 3: The comparison between (2.b) and (6.a) shows that

$$K_y = K_q (CV_q)^{-1}. \qquad (11)$$

The inverse of matrix CV_q always exists because $[T_{n-q}':C']'$ is full row rank (see Step 1) and (6.b) holds $(T_{n-q}V_q = 0)$.

3. EXAMPLES

Let $(A, B, C) =$

$$\left(\begin{bmatrix} -4 & 0 & -2 \\ 0 & 0 & 1 \\ 1 & -1 & -2 \end{bmatrix}, \begin{bmatrix} 4 & 2 \\ 0 & -2 \\ 0 & 1 \end{bmatrix}, \begin{bmatrix} 0 & 1 & 0 \\ 0 & 0 & 1 \end{bmatrix} \right).$$

Let the assigned eigenvalues be -1, -2, and -3.

Step 0: We arbitrarily select $\Lambda_{n-q} = -1$ and $\Lambda_q = \text{diag}\{-2, -3\}$.

Step 1.a: Equation (4.b) implies that

$$\underbrace{\begin{bmatrix} 1 & 0 & 3 \\ 0 & 1 & 0 \end{bmatrix}}_{D_1} : \underbrace{\begin{bmatrix} -3 & -5 \\ 1 & 1 \end{bmatrix}}_{E_1} \begin{bmatrix} -3 & 0 & -2 \\ 0 & 1 & 1 \\ 1 & -1 & -1 \\ \hline 0 & -1 & 0 \\ 0 & 0 & -1 \end{bmatrix} = 0$$

Step 1.b: We arbitrarily select $\mathbf{c}_1 = [1 \quad 0]$ so that

$$T_{n-q} = \mathbf{c}_1 D_1 = [1 \quad 0 \quad 3]$$

is linearly independent of the rows of C. This T_{n-q} together with $L = \mathbf{c}_1 E_1 = [-3 \quad -5]$ satisfy equation (3). Selecting $\mathbf{c}_1$ is equivalent of assigning the first left eigenvector T_{n-q}.

Step 2.a: From equation (7.b), to assign $\lambda_2 = -2$ of Λ_q, we have

$$\begin{bmatrix} -2 & 0 & -2 : -4 & -2 \\ 0 & 2 & 1 : 0 & 2 \\ 1 & -1 & 0 : 0 & -1 \end{bmatrix} \begin{bmatrix} -1 & 0 \\ 0 & -1 \\ 2 & 0 \\ \hline 0 & -1/2 \\ -1 & 1 \end{bmatrix} \begin{matrix} \}D_2 \\ \\ \}E_2 \end{matrix} = 0$$

To assign $\lambda_3 = -3$ of Λ_q, we have

$$\begin{bmatrix} -1 & 0 & -2 : -4 & -2 \\ 0 & 3 & 1 : 0 & 2 \\ 1 & -1 & 1 : 0 & -1 \end{bmatrix} \begin{bmatrix} -4 & -3 \\ -1 & 0 \\ 3 & 2 \\ \hline -1/2 & 1/4 \\ 0 & -1 \end{bmatrix} \begin{matrix} \}D_3 \\ \\ \}E_3 \end{matrix} = 0$$

Step 2.b: From equation (8), because

$T_{n-q}D_2 = [5 \quad 0]$ and $T_{n-q}D_3 = [5 \quad 3]$,

we have

$\mathbf{c}_2 = [0 \quad 1]'$ and $\mathbf{c}_3 = [-3 \quad 5]'$

so that $T_{n-q}D_i\mathbf{c}_i = 0$ for $i = 2, 3$.

Because $p + q = 4 \not> n + 1$, the solution $\mathbf{c}_i$ $(i = 2, 3)$ is unique and there is no assignment freedom for the two right eigenvectors of V_q, in this example.

Now from Equations (9.a) and (9.b):

$$V_q = [D_2\mathbf{c}_2 : D_3\mathbf{c}_3] = \begin{bmatrix} 0 & -3 \\ -1 & 3 \\ 0 & 1 \end{bmatrix},$$

and $K_q = [E_2\mathbf{c}_2 : E_3\mathbf{c}_3] = \begin{bmatrix} -1/2 & 11/4 \\ 1 & -5 \end{bmatrix}$.

It can be verified that both (6.a) and (6.b) are satisfied.

Step 3: From Equation (11),

$$K_y = K_q (CV_q)^{-1} = \begin{bmatrix} 1/2 & 5/4 \\ -1 & -2 \end{bmatrix}$$

The eigenvalues of the corresponding matrix $A - BK_yC$ are indeed the desired $\{-1, -2, -3\}$.

Now suppose a different set of eigenvalues $-1 \pm j$ are to be assigned at Step 2. Then

$$\Lambda_q = \begin{bmatrix} -1 & 1 \\ -1 & -1 \end{bmatrix}.$$

Step 2.a: From (7.a) we have

$$\left[\begin{array}{ccc|ccc|cc|cc} -3 & 0 & -2 & 1 & 0 & 0 & -4 & -2 & 0 & 0 \\ 0 & 1 & 1 & 0 & 1 & 0 & 0 & 2 & 0 & 0 \\ 1 & -1 & -1 & 0 & 0 & 1 & 0 & -1 & 0 & 0 \\ \hline -1 & 0 & 0 & -3 & 0 & -2 & 0 & 0 & -4 & -2 \\ 0 & -1 & 0 & 0 & 1 & 1 & 0 & 0 & 0 & 2 \\ 0 & 0 & -1 & 1 & -1 & -1 & 0 & 0 & 0 & -1 \end{array}\right] \left[\begin{array}{cc|cc} -3 & -2 & 1 & 0 \\ 7 & 3 & 7 & 2 \\ -6 & -3 & 0 & 1 \\ \hline 1 & 0 & 3 & 2 \\ 7 & 2 & -7 & -3 \\ 0 & 1 & 6 & 3 \\ \hline 7.5 & 3.5 & 0 & 0 \\ -4 & -1 & 0 & 0 \\ \hline 0 & 0 & -7.5 & -3.5 \\ 0 & 0 & 4 & 1 \end{array}\right] \begin{array}{l} \left.\vphantom{\begin{array}{c}1\\1\\1\end{array}}\right\} D_2 \\ \left.\vphantom{\begin{array}{c}1\\1\\1\end{array}}\right\} D_3 \\ \left.\vphantom{\begin{array}{c}1\\1\end{array}}\right\} E_2 \\ \left.\vphantom{\begin{array}{c}1\\1\end{array}}\right\} E_3 \end{array} = 0$$

Step 2.b: From Equation (8.a), because

$$\begin{bmatrix} T_{n-q}D_2 \\ T_{n-q}D_3 \end{bmatrix} = \begin{bmatrix} -21 & -11 : & 1 & 3 \\ 1 & 3 : & 21 & 11 \end{bmatrix},$$

we have $\mathbf{c}_2 = [1 \quad 0 \quad -4.5 \quad 8.5]'$ and $\mathbf{c}_3 = [0 \quad 1 \quad -2.5 \quad 4.5]'$ so that

$$T_{n-q}D_2\mathbf{c}_i = T_{n-q}D_3\mathbf{c}_i = 0 \text{ for } i = 2,\ 3.$$

Because of the special forms of (8.a) and Λ_q, this solution $\mathbf{c}_i$ (i=2,3) is actually unique (only the sequence i of $\mathbf{c}_i$ is nonunique).

Now from Equations (9.a) and (9.b):

$$V_q = [D_2\mathbf{c}_2 : D_3\mathbf{c}_3] = \begin{bmatrix} -7.5 & 1.5 \\ -7.5 & 6 \\ 2.5 & -0.5 \end{bmatrix},$$

$$\text{and } K_q = [E_2\mathbf{c}_2 : E_3\mathbf{c}_3] = \begin{bmatrix} 7.5 & 3 \\ -4 & -5.5 \end{bmatrix}.$$

Step 3: From Equation (11),

$$K_y = K_q(CV_q)^{-1} = \begin{bmatrix} 1 & 6 \\ -1.4 & -5.8 \end{bmatrix}.$$

The eigenvalues of the corresponding matrix $A - BK_yC$ are indeed the desired $\{-1, -1 \pm j\}$.

4. CONCLUSION

The problem of assigning the eigenvalues **and** eigenvectors to the dynamic matrix of an SOF system is both very important and very challenging. This paper presents an algorithm which assigns the eigenvalues/vectors exactly and in two separate groups in sequence. The algorithm is very simple in light of its tasks, as demonstrated by the very simple, real, and hand computation of the numerical example. The analysis of this algorithm shows that if q + p > n, then arbitrary eigenvalues can be assigned with the only exception of all complex conjugate eigenvalue pairs and an odd q (when Step 0 cannot be implemented). This restriction is similar to that of the existing eigenstructure assignment methods such as (Syrmos et al, 1993). However, as described in Section 1, it is *now* possible to substantially increase q such that q is even and that q + p > n is satisfied *more generally*.

The main advantage of our algorithm is however in the eigenvector assignment aspect, which is as critical as the eigenvalue assignment aspect. In our algorithm, the first n-q left eigenvectors are assignable if q > 1, and the remaining q right eigenvectors are assignable if q + p > n + 1. Eigenvector assignment is not possible at the condition q x p > n only.

In this design algorithm, the dual solutions (5) and (9) of the dual equations (3) and (6.a) respectively, are used. This solution enables the eigenvector assignment to be in the dual forms of (5) and (9.a), where the D_i matrix contains the basis vectors of the corresponding eigenvectors. This form of eigenvector assignment is based on the most explicit form of the freedom of the eigenvectors and enables uniquely the most effective and general existing numerical design algorithms of (Kautsky, et al, 1985). The solution (9) also satisfies equations (6.a) and (6.b) of Step 2 without the very complicated two-step similarity transformation and numerical operation of (Syrmos et al, 1993) on (A,B). To the best of author's knowledge, these three *important features* are *not* shared by the existing algorithms.

REFERENCES

Tsui, C.C., (1996a). "A new design approach of unknown input observers" IEEE Trans. Automatic Control, **AC-41**, 464-468.

Tsui, C.C., (1998) "The first general dynamic output feedback compensator that can implement state feedback control," Int. J. Systems Science, **29**, 1, 49-55.

___, (1996b). Robust Control System Design -- Advanced State Space Techniques. NY: Marcel Dekker.

Wilkinson, J.H. (1965). Algebraic Eigenvalue Problem. London: Oxford University Press.

Kimura, H. (1975). "Pole assignment by gain output feedback," IEEE Trans. Automatic Control, **AC-20**, 509-516.
Wang, X.A. (1996). "Grassmannian, central projection, and output feedback pole assignment of linear systems," IEEE Trans. Automatic Control, **AC-41**, 786-794.
Chu, E.W.E., (1993). "Approximate pole assignment," Int. J. Control, **58**, 471-484.
Syrmos, V.L. and F.L. Lewis, (1993) "Output feedback eigenstructure assignment using two Sylvester equations," IEEE Trans. Automatic Control, **AC-38**, 495-499.
Graham, A. (1981). Kronecker Products and Matrix Calculus with Applications. Chichester, Eng land: Ellis Horwood.
Tsui, C.C. (1993), "On the solution to matrix equation TA - FT = LC and its applications," SIAM J. Matrix Anal. Appl., **14**, 1, 33-44.
Kautsky, J., N.K. Nichols and P. Van Dooren, (1985). "Robust pole assignment in linear state feedback," Int. J. Control, **41**, 1129-1155.

14th Triennial World Congress, Beijing, P.R. China

NON-LINEAR BOUNDS FOR THE GENERALIZED EIGENSYSTEM OF A MATRIX PENCIL WITH DISTINCT EIGENVALUES ⋆

Mihail Mihaylov Konstantinov*
Vera Angelova Angelova Ivan Petkov Popchev****

* *University of Architecture & Civil Engineering*
1 Hr. Smirnenski Blvd., 1421 Sofia, Bulgaria
e-mail: mmk_fte@uacg.acad.bg
** *Institute of Information Technologies, Bulgarian Academy of Sciences, Ak. G. Bonchev Str., Bl. 2, 1113 Sofia, Bulgaria*
e-mail: vangelova@iit.acad.bg, ipopchev@iit.acad.bg

Abstract: In this paper a non-local sensitivity analysis of the generalized eigensystem (eigenvectors and eigenvalues) of an $n \times n$ matrix pencil $A - \lambda B$ with pairwise distinct eigenvalues is made. By rewriting the perturbed problem as an operator equation and using the technique of Lyapunov majorants non-local non-linear bounds are established. An example demonstrates the effectiveness of the estimates proposed.

Keywords: Eigenvalues problems, Generalized eigenvectors, Generalized eigenvalues, Spectral perturbation analysis, Error analysis

1. INTRODUCTION

Many problems from linear system theory come down to the generalized eigensystem of matrix pencils, e.g. computational algorithms for studying the controllability and observability, pole/zero structure, null space structure of linear system (Van Dooren, 1981; Kagström, 1986). The pencil approach is used to study the algebraic Riccati equation arising in optimal control and filtering (Pappas *et al.*, 1980; Lancaster and Rodman, 1995).

Several techniques have been developed for checking the sensitivity of the eigensystem of various classes of matrices, see e.g. (Kahan *et al.*, 1982; Meyer and Stewart, 1988; Stewert and Sun, 1990; Konstantinov *et al.*, 1995). In this work the results from (Konstantinov *et al.*, 1987; Konstantinov *et al.*, 1997) to the generalized eigensystem of a matrix pencil with distinct eigenvalues are extended.

The paper begins with a definition of the generalized eigensystem problem. Next, a non-linear non-local sensitivity analysis of the problem is done. To avoid the solution of an n-th degree algebraic equation when obtaining the perturbation bounds approximations are proposed as well. A detailed consideration of the case $n = 2$ is done. It is shown that for a class of perturbations the proposed bounds are relatively sharp.

The following abbreviations are used: $\mathcal{R}$ and $\mathcal{C}$ – the fields of real and complex numbers, $\mathcal{R}_+ = [0, \infty)$, $\mathcal{C}^{m \times n}$ – the space of $m \times n$ matrices over $\mathcal{C}$, $\mathcal{C}^n = \mathcal{C}^{n \times 1}$, I_n – the unit $n \times n$ matrix, 0 – the zero matrix, $A^T = [a_{ij}] \in \mathcal{C}^{m \times n}$ and $A^H = \bar{A}^T = [\bar{a}_{ij}] \in \mathcal{C}^{m \times n}$ – the transpose and complex conjugate transpose of $A = [a_{ij}] \in \mathcal{C}^{m \times n}$, $\|.\|$ – the Euclidean vector norm or the spectral matrix norm.

⋆ Partially supported by the National Fund on Scientific Research, Bulgaria, Grant I 805/98.

Throughout the whole paper i is a fixed integer between 1 and n, the index k varies from 1 to n and j varies from 1 to n with exception of i. The notation " := " stands for "equal by definition".

2. SCOPE

Let A and B be $n \times n$ matrices. The set of all matrices of the form $A - \lambda B$, for arbitrary $\lambda \in \mathcal{C}$, is referred to as pencil. The generalized eigensystem problem consists in determining those values of λ and vectors $x \neq 0$ which satisfy the equation

$$Ax = \lambda Bx. \tag{1}$$

Such values of λ are called eigenvalues of the pencil $A - \lambda B$ or generalized eigenvalues. The solution x of (1) is an eigenvector of $A - \lambda B$.

Consider the generalized eigensystem for the pencil $A - \lambda B$; $A, B \in \mathcal{C}^{n\times n}$ with pairwise distinct eigenvalues $\lambda_1, \ldots, \lambda_n \in \mathcal{C}$ and corresponding eigenvectors $x_1, \ldots, x_n \in \mathcal{C}^n$.

The round-off errors in the solution and the parameter uncertainties in the data are represented by equivalent perturbations in the data: $A \to A + E$; $B \to B + F$ with $\|E\| = \varepsilon$ and $\|F\| = \varphi = q\varepsilon$, $q \geq 0$. Denote by $\lambda_k + \Delta\lambda_k \in \mathcal{C}$ and $x_k + \Delta x_k \in \mathcal{C}^n$ the eigenvalues and eigenvectors of the perturbed pencil $A + E - \lambda(B + F)$.

The assumption that $A - \lambda B$ has distinct eigenvalues is dictated by the requirement to have full collection of linearly independent eigenvectors, see (Konstantinov *et al.*, 1987).

For ε small enough the eigenvalues of the matrix pencil $A + E - \lambda(B + F)$ are also pairwise distinct. Moreover, the perturbations $\Delta\lambda_k$ and Δx_k are analytic functions of E and F:

$$\Delta\lambda_k = \Delta\lambda_k(E, F),\ \Delta x_k = \Delta x_k(E, F)$$

and $\Delta\lambda_k(0) = 0$, $\Delta x_k(0) = 0$.

Let i $(1 \leq i \leq n)$ be a fixed integer. Our aim is to derive a quantity ε_0 and two analytic not-decreasing functions $f, g : [0, \varepsilon_0) \to \mathcal{R}_+$, satisfying $f(0) = g(0) = 0$ and such that the perturbation bounds

$$|\Delta\lambda_i| \leq g(\varepsilon),\ \varepsilon \in [0, \varepsilon_0)$$

and

$$|\Delta x_i| \leq f(\varepsilon),\ \varepsilon \in [0, \varepsilon_0)$$

hold.

The quantity $\|\Delta x_i\|$ itself may not be a relevant measure for the eigenvector perturbation. On the other hand it would be such a relevant measure if additional requirements such as $\|x_i\| = 1$ or $\|x_i + \Delta x_i\| = 1$ are imposed. In the considerations below it is assumed that $\|x_i\| = 1$, see e.g. (Konstantinov *et al.*, 1997).

3. MAIN RESULTS

Denote by $y_1, \ldots, y_n \in \mathcal{C}^n$ the right eigenvectors of the matrix pencil $A^H - \bar{\lambda}_k B^H$, corresponding to the eigenvalues $\bar{\lambda}_1, \ldots, \bar{\lambda}_n \in \mathcal{C}$ i.e. $A^H y_k = \bar{\lambda}_k B^H y_k$ and suppose that the right and left eigenvectors x_k and y_k are normed as $\|x_k\| = \|y_k\| = 1$. One has $Ax_i = \lambda_i Bx_i$ and $y_j^H A = \lambda_j y_j^H B$. Hence $y_j^H Ax_i = \lambda_i y_j^H Bx_i$, $y_j^H Ax_i = \lambda_j y_j^H Bx_i$ and $y_j^H Bx_i = 0$, $j \neq i$.

Denote $u := \Delta x_i$, $\gamma := \Delta\lambda_i$. The generalized eigenstructure relation

$$(A + E)(x_i + \Delta x_i) = (\lambda_i + \Delta\lambda_i)(B + F)(x_i + \Delta x_i)$$

may be rewritten as

$$\gamma(B + F)(x_i + u) = (A - \lambda_i(B + F))u + E(x_i + u) - \lambda_i F x_i,$$

wich yields

$$\gamma y_i^H (B + F)(x_i + u) = y_i^H (E - \lambda_i F)(x_i + u).$$

and

$$\gamma(y_j^H Bu + y_j^H F(x_i + u)) = y_j^H (A - \lambda_i B)u + y_j^H (E - \lambda_i F)(x_i + u).$$

Since the vectors $x_1, \ldots, x_n$ are linearly independent, the perturbation u may be represented as

$$u = \sum_{k=1}^{n} \alpha_k x_k, \tag{2}$$

where the quantities $\alpha_1, \ldots, \alpha_n \in \mathcal{C}$ shall be determined later on. It follows from (2) that

$$y_j^H Bu = \alpha_j y_j^H Bx_j$$

and

$$Au = \sum_{k=1}^{n} \alpha_k Ax_k = \sum_{k=1}^{n} \alpha_k \lambda_k Bx_k. \tag{3}$$

Now (3) yields

$$y_j^H Au = \alpha_j \lambda_j y_j^H Bx_j.$$

Set $\nu_k = y_k^H Bx_k$ and observe that $\nu_k \neq 0$. Then for a fixed value of i, one has

$$\gamma\nu_i(1 + \alpha_i) + \gamma y_i^H F(x_i + u) = y_i^H (E - \lambda_i F)(x_i + u) \tag{4}$$

and

$$\alpha_j \nu_j (\lambda_i - \lambda_j + \gamma) \tag{5}$$

$$= y_j^H(E - (\lambda_i + \gamma)F)(x_i + u).$$

Since (5) consists of $n-1$ equations, both (4) and (5) in view of (2) constitute a system of n equations for determining the $n+1$ unknowns $\gamma, \alpha_1, \alpha_2, \ldots, \alpha_n$. Thus one additional restriction is imposed $\alpha_i = 0$. Therefore $u = \sum_{j\neq i} \alpha_j x_j$. It must be stressed that this representation of u is always possible if the eigenvector of $A + E - (\lambda_i + \Delta\lambda_i)(B + F)$ does not lie in the subspace

$$\mathrm{span}\{\{x_1, x_2, \ldots, x_n\}\backslash\{x_i\}\}.$$

For ε sufficiently small this is obviously fulfilled.

Thus one has the equations

$$\gamma = \frac{y_i^H(E - \lambda_i F)(x_i + u)}{\nu_i + y_i^H F(x_i + u)} \qquad (6)$$

and

$$\alpha_j = \frac{y_j^H(E - (\lambda_i + \gamma)F)(x_i + u)}{\nu_j(\lambda_i - \lambda_j + \gamma)}, \quad j \neq i. \quad (7)$$

For ε small in a first order approximation one get from (6) and (7)

$$\gamma = \frac{y_i^H(E - \lambda_i F)x_i}{\nu_i} + O(\varepsilon^2), \ \varepsilon \to 0$$

and

$$\alpha_j = \frac{y_j^H(E - \lambda_i F)x_i}{\nu_j(\lambda_i - \lambda_j)} + O(\varepsilon^2), \ \varepsilon \to 0.$$

Since

$$||u|| \leq \sum_{j\neq i} ||\alpha_j x_j|| \leq \sum_{j\neq i} |\alpha_j|$$

then the well known linear asymptotic bounds are obtained

$$|\Delta\lambda_i| \leq \frac{||(E - \lambda_i F)x_i||}{|y_i^H B x_i|} + O(\varepsilon^2), \ \varepsilon \to 0$$

and

$$||\Delta x_i|| \leq ||E - \lambda_i F|| \sum_{j\neq i} \frac{1}{|y_j^H B x_j||\lambda_i - \lambda_j|} + O(\varepsilon^2), \ \varepsilon \to 0.$$

The above linear bounds are valid only asymptotically. This means that the perturbations ε in the data must be small enough to ensure sufficient accuracy of the bounds.

Denote the right-hand side of (6) as $\omega(u, E, F)$. Then

$$u = \Pi(u, E, F) := \qquad (8)$$

$$\sum_{j\neq i} \frac{y_j^H(E - (\lambda_i + \omega(u, E, F))F)(x_i + u)x_j}{\nu_j(\lambda_i - \lambda_j + \omega(u, E, F))},$$

which is an operator equation for u.

Setting

$$c_k := |\nu_k|, \ s_j := |\lambda_i - \lambda_j|, \ \rho := ||u||, \ c := c_i = |\nu_i|$$

one has

$$|\omega(u, E, F)| = \left| \frac{y_i^H(E - \lambda_i F)(x_i + u)}{\nu_i + y_i^H F(x_i + u)} \right| \qquad (9)$$

$$\leq \frac{\varepsilon(1 + |\lambda_i| q)(1 + \rho)}{c - \varepsilon q(1 + \rho)}.$$

Now relations (8) and (9) give

$$||\Pi(u, E, F)|| \leq \qquad (10)$$

$$\sum_{j\neq i} \frac{c\varepsilon(1 + |\lambda_i| q)(1 + \rho)}{c_j(cs_j - \varepsilon(1 + (|\lambda_i| + s_j)q)(1 + \rho))}.$$

In order for (10) to be fulfilled it must be fulfilled

$$\varepsilon < b := \frac{cs}{1 + (|\lambda_i| + s)q},$$

$$s := \min\{s_j : j \neq i\}.$$

Set

$$h(r, \varepsilon, \varphi) := c\varepsilon(1 + |\lambda_i| q)(1 + r) \qquad (11)$$

$$\times \sum_{j\neq i} \frac{1}{c_j(cs_j - \varepsilon(1 + (|\lambda_i| + s_j)q)(1 + r))}$$

and rewrite inequality (10) as

$$||\Pi(u, E, F)|| \leq h(||u||, \varepsilon), \qquad (12)$$

where $\varepsilon \geq 0$ is (small) parameter. Together with (12) consider the corresponding majorant equation

$$r = h(r, \varepsilon). \qquad (13)$$

For $\varepsilon < b$ the function $h(., \varepsilon) : [0, d) \to \mathcal{R}_+$, where

$$d = d(\varepsilon) := \frac{b}{\varepsilon(1 + (|\lambda_i| + s_j)q} - 1$$

is well defined via (11). Moreover, $h(., \varepsilon)$ is real analytic, strictly increasing, convex and

$$\lim_{r\to d-0} h(r, \varepsilon) = +\infty.$$

Having in mind the above properties of the function $h(., \varepsilon)$ and using the method of Lyapunov majorants (Grebenikov and Ryabov, 1979; Konstantinov *et al.*, 1995) it may be shown that there exists a positive quantity $\varepsilon_0 < b$ such that:

- for $\varepsilon < \varepsilon_0$ equation (13) has two positive roots $r_1, r_2 \in [0, d)$,

$$0 \leq r_1 = r_1(\varepsilon) < r_2 = r_2(\varepsilon) < b$$

such that

$$0 < h'_r(r_1, \varepsilon) < 1, \ h'_r(r_2, \varepsilon) > 1;$$

– for $\varepsilon = \varepsilon_0$ equation (13) has a double root

$$r_0 = r_1(\varepsilon_0) = r_2(\varepsilon_0) \in (0, d)$$

i.e., $h'_r(r_0, \varepsilon_0) = 1$;

– for $\varepsilon_0 < \varepsilon < b$ equation (13) has no roots in the interval $[0, d)$.

The pair (r_0, ε_0) is a solution to the system

$$r = h(r, \varepsilon), \quad 1 = h'_r(r, \varepsilon).$$

Set

$$f(\varepsilon) := r_1(\varepsilon), \; \varepsilon \in [0, \varepsilon_0). \tag{14}$$

For each $\varepsilon < \varepsilon_0$ and $u, v \in \mathcal{C}^n$ with $||u||, ||v|| \leq f(\varepsilon)$ one has $||\Pi(u, E, F)|| \leq f(\varepsilon)$ and

$$||\Pi(u, E, F) - \Pi(v, E, F)|| \leq h'_r(f(\varepsilon), \varepsilon) < 1.$$

Hence the operator $\Pi(., E, F)$ maps the set

$$M_\varepsilon := \{u : ||u|| \leq f(\varepsilon)\}$$

into itself and is a contraction. According to the Banach fixed point principle there exists a unique solution

$$u_0 = u_0(E, F) \in M_\varepsilon$$

of the operator equation $u = \Pi(u, E, F)$, i.e. $||u_0(E, F)|| \leq f(\varepsilon)$.

Hence the desired inequalities are

$$||\Delta x_i|| = ||u_0(E, F)|| \leq f(\varepsilon)$$

and

$$|\Delta\lambda_i| = |\omega(u_0(E, F), E, F)| \leq g(\varepsilon), \tag{15}$$

$$g(\varepsilon) := \frac{\varepsilon(1 + |\lambda_i|q)}{c - \varepsilon q(1 + f(\varepsilon))}(1 + f(\varepsilon)).$$

4. APPROXIMATE PERTURBATION BOUNDS

Consider an asymptotic expansion in ε for the solution (14) of equation (13), (11).

The expression $h(r, \varepsilon)$ may be represented in a power series in ε as follows:

$$h(r, \varepsilon) = \sum_{m=1}^{\infty} h_m (1 + r)^m \varepsilon^m, \tag{16}$$

where

$$h_m = \frac{(1 + |\lambda_i|q)}{c^{m-1}} \times \sum_{j \neq i} \frac{(1 + (|\lambda_i| + s_j)q)^{m-1}}{c_j s_j^m}.$$

The solution (14) of (13) when $\varepsilon < \varepsilon_0$ is a real analytic function of ε vanishing for $\varepsilon = 0$. Hence the solution $r_1 = f(\varepsilon)$ satisfying $f(0) = 0$ in the form of power series is chosen

$$f(\varepsilon) = \sum_{m=1}^{\infty} f_m \varepsilon^m. \tag{17}$$

Substituting (17) and (16) in (13) the relations are obtained

$$\begin{aligned}
f_1 &= h_1, \\
f_2 &= h_1^2 + h_2, \\
f_3 &= h_1^3 + 3h_1 h_2 + h_3, \\
f_4 &= h_1^4 + 6h_1^2 h_2 + 4h_1 h_3 + 2h_2^2 + h_4, \\
&\ldots\ldots\ldots
\end{aligned}$$

Thus an asymptotic approximation to the perturbation bound $f(\varepsilon)$ is determined. Having in mind (15) the asymptotic expansion for the eigenvalue perturbation bound is

$$g(\varepsilon) = \sum_{m=1}^{\infty} g_m \varepsilon^m$$

with

$$g_1 = \frac{1 + |\lambda_i|q}{c},$$

$$g_m = \frac{1 + |\lambda_i|q}{c} f_{m-1} + \frac{q}{c} \sum_{k=1}^{m-1} g_{m-k} f_{k-1}, \quad k \geq 2,$$

$$f_0 = 1.$$

Look for an upper bound $f^*(\varepsilon)$ for $f(\varepsilon)$ replacing $h(r, \varepsilon)$ in(13) by some greater quantity $h^*(r, \varepsilon)$.

Denote

$$\theta := c_i \sum_{j \neq i} \frac{1}{c_j(1 + (|\lambda_i| + s_j)q)}.$$

Hence

$$h(r, \varepsilon) \leq h^*(r, \varepsilon) := \frac{\varepsilon\theta(1 + |\lambda_i|q)(1 + r)}{b - \varepsilon(1 + r)}.$$

If

$$\varepsilon < \varepsilon_0^* := \frac{b}{(1 + \sqrt{(1 + |\lambda_i|q)\theta})^2}$$

the equation

$$r = h^*(r, \varepsilon)$$

has two roots r_1^*, r_2^* in the interval $[0, \varepsilon_0^*)$ and

$$f^*(\varepsilon) := r_1^*(\varepsilon),$$

$$r_1^*(\varepsilon) = \frac{2\theta(1 + |\lambda_i|q)\varepsilon}{v(\varepsilon) + \sqrt{v^2(\varepsilon) - 4\theta(1 + |\lambda_i|q)\varepsilon^2}},$$

where

$$v(\varepsilon) := b - (1 + (1 + |\lambda_i|q)\theta)\varepsilon.$$

Thus

$$||\Delta x_i|| \leq f^*(\varepsilon), \; \varepsilon \in [0, \varepsilon_0^*).$$

5. EXAMPLES

For $n = 2$ equation (13) becomes

$$r = \frac{c\zeta\varepsilon(1+r)}{c_j(c|\lambda_1-\lambda_2| - \varepsilon\eta(1+r))} = \frac{c\zeta(1+r)}{c_j\eta(d(\varepsilon)-r)},$$

where

$$\zeta := 1 + |\lambda_i|q,$$
$$\eta := 1 + (|\lambda_i| + |\lambda_1-\lambda_2|)q,$$
$$d(\varepsilon) := \frac{c|\lambda_1-\lambda_2|}{\eta\varepsilon} - 1 > 0.$$

The quantity ε_0 is determined from the condition that equation (13) has a double root, i.e. its discriminant is zero. A quadratic equation for ε is obtained and two subcases are considered.

- If $c_1 = c_2 = c$ then
$$\varepsilon_0 = \frac{c|\lambda_1-\lambda_2|}{(\sqrt{\zeta}+\sqrt{\eta})^2}.$$
- If $c_1 \neq c_2$ then for ε_0 the smaller root corresponding to the sign minus is taken
$$\varepsilon_0 = \frac{c_1c_2|\lambda_1-\lambda_2|}{(\sqrt{c_i\zeta}+\sqrt{c_j\eta})^2}.$$

After some computation one obtains

$$f(\varepsilon) = \frac{2\zeta c_i\varepsilon}{\omega(\varepsilon)+\sqrt{\omega(\varepsilon)^2 - 4\zeta\eta c_1c_2\varepsilon^2}},$$

where

$$\omega(\varepsilon) = \eta c_1c_2|\lambda_1-\lambda_2| - (\zeta c_i + \eta c_j)\varepsilon.$$

Consider now the matrix pencil $A - \lambda B$ with

$$A := \begin{bmatrix}\lambda_1 & 1\\ 0 & \lambda_2\end{bmatrix};\quad B := \begin{bmatrix}1 & \beta\\ 0 & 1\end{bmatrix} \in \mathcal{R}^{2.2}$$

with $s := |\lambda_1-\lambda_2| > 0$.

One has

$$x_1 = \begin{bmatrix}1\\0\end{bmatrix},\quad x_2 = \frac{1}{\sigma}\begin{bmatrix}1\\ \lambda_2-\lambda_1\end{bmatrix},$$
$$y_1 = \frac{1}{\sigma}\begin{bmatrix}\lambda_1-\lambda_2\\ 1\end{bmatrix},\quad y_2 = \begin{bmatrix}0\\1\end{bmatrix}$$

and $c_1 = c_2 = s/\sigma$, where $\sigma = \sqrt{1+s^2}$. Thus

$$\varepsilon_0 = \frac{s^2}{\sigma(\sqrt{\zeta}+\sqrt{\eta})^2}$$

and

$$f(\varepsilon) = (2\zeta\xi)/\left(((\sqrt{\zeta}+\sqrt{\eta})^2\eta - (\zeta+\eta)\xi\right.$$
$$\left. + \sqrt{((\sqrt{\zeta}+\sqrt{\eta})^2\eta - (\zeta+\eta)\xi)^2 - 4\zeta\eta\xi^2)}\right),$$
$$\xi := \frac{\varepsilon}{\varepsilon_0}.$$

The perturbations are taken as

$$E := \begin{bmatrix}0 & 0\\ -\varepsilon & 0\end{bmatrix};\quad F := \begin{bmatrix}0 & 0\\ -q\varepsilon & 0\end{bmatrix},\quad \varepsilon \geq 0$$

i.e. $\|E\| = \varepsilon$ and $\|F\| = q\varepsilon$. Then the perturbed matrix pencil $A + E - \lambda(B+F)$ has eigenvalues

$$\lambda_1^* = \lambda_1 + \Delta\lambda_1,\quad \lambda_2^* = \lambda_2 + \Delta\lambda_2$$

with

$$\Delta\lambda_i = (\lambda_j - \lambda_i)\psi(\varepsilon) + \mu_i(\varepsilon),$$
$$\mu_i(\varepsilon) = \frac{\varepsilon(\beta + q - 2\beta q\lambda_i)}{2(1+\beta q\varepsilon)},$$
$$\psi(\varepsilon) = \frac{1}{2(1+\beta q\varepsilon)}\left(1 - (1 - \frac{4\varepsilon}{s^2}\right.$$
$$+(\varepsilon^2(\beta-q)^2 + 2\varepsilon(\beta+q)(\lambda_1+\lambda_2)$$
$$\left.-4\beta q\lambda_1\lambda_2\varepsilon)/s^2)^{\frac{1}{2}}\right)$$

and corresponding eigenvectors

$$x_1^* = x_1 + \Delta x_1 = \begin{bmatrix}1\\ \dfrac{\Delta\lambda_1}{1-\beta(\lambda_1+\Delta\lambda_1)}\end{bmatrix},$$

$$x_2^* = x_2 + \Delta x_2 = \begin{bmatrix}1\\ \dfrac{\lambda_2-\lambda_1+\Delta\lambda_2}{1-\beta(\lambda_2+\Delta\lambda_2)}\end{bmatrix}.$$

Let

$$\lambda_1 = 0,\ \lambda_2 = 0.1,\ \beta = 1,\ q = 1.$$

Then $s = 0.1$, $\sigma = 1.00499$. Hence

$$|\Delta\lambda_1| = |(\lambda_2-\lambda_1)\psi(\varepsilon) + \mu_1(\varepsilon)|$$
$$= 10\varepsilon + 800\varepsilon^2 + O(\varepsilon^3),$$
$$|\Delta\lambda_2| = |(\lambda_1-\lambda_2)\psi(\varepsilon) + \mu_2(\varepsilon)|$$
$$= 8.1\varepsilon + 801.9\varepsilon^2 + O(\varepsilon^3)$$

while the bounds proposed above give

$$g_1(\varepsilon) = (1 + |\lambda_1|q)\frac{\sigma}{s}\varepsilon$$
$$+(1+|\lambda_1|q)(1+(|\lambda_1|+s)q)\frac{\sigma^2}{s^3}\varepsilon^2 + O(\varepsilon^3)$$
$$= 10.05\varepsilon + 1111\varepsilon^2 + O(\varepsilon^3),$$
$$g_2(\varepsilon) = (1 + |\lambda_2|q)\frac{\sigma}{s}\varepsilon$$
$$+(1+|\lambda_2|q)(1+(|\lambda_2|+s)q)\frac{\sigma^2}{s^3}\varepsilon^2 + O(\varepsilon^3)$$
$$= 11.06\varepsilon + 1333.2\varepsilon^2 + O(\varepsilon^3).$$

The sines between eigenvectors and perturbed eigenvectors are

$$\tau(x_i, x_i^*) = \sqrt{1 - \frac{|x_i^H x_i^*|^2}{\|x_i\|^2\|x_i^*\|^2}}.$$

Hence

$$\tau(x_1, x_1^*) = \frac{|\Delta\lambda_1|}{\sqrt{\Delta\lambda_1^2 + (1-\beta(\lambda_1+\Delta\lambda_1))^2}}$$
$$= 10\varepsilon + 900\varepsilon^2 + O(\varepsilon^3)$$

and

$$\tau(x_2, x_2^*) = \sqrt{1 - \frac{\Gamma^2(\frac{1}{\sigma} + \frac{\Upsilon(\lambda_2-\lambda_1)}{\Gamma\sigma})^2}{\Upsilon^2 + \Gamma^2}}$$

$$= 9.88\varepsilon + 900.35\varepsilon^2 + O(\varepsilon^3),$$

$$\Upsilon = \lambda_2 - \lambda_1 + \Delta\lambda_2, \quad \Gamma = 1 - \beta(\lambda_2 + \Delta\lambda_2).$$

On the other hand

$$f(\varepsilon) = 10.05\varepsilon + 1212\varepsilon^2 + O(\varepsilon^3)$$

for $\tau(x_1, x_1^*)$ and

$$f(\varepsilon) = 11.05\varepsilon + 1455.41\varepsilon^2 + O(\varepsilon^3)$$

for $\tau(x_2, x_2^*)$.

Now let $\lambda_1 = 0$, $\lambda_2 = 0.1$ and $\beta = 0$, $q = 1$ or $\beta = 1$, $q = 0$. Then

$$|\Delta\lambda_1| = |\Delta\lambda_2| = 10\varepsilon + 900\varepsilon^2 + O(\varepsilon^3)$$

while the bound proposed above give

$$g(\varepsilon) = 10.05\varepsilon + 1111\varepsilon^2 + O(\varepsilon^3)$$

when $\beta = 0$, $q = 1$, and

$$g(\varepsilon) = 10.05\varepsilon + 1010\varepsilon^2 + O(\varepsilon^3)$$

when $\beta = 1$, $q = 0$.

The sines between eigenvectors and perturbed eigenvectors are

$$\tau(x_i, x_i^*) = \frac{|\Delta\lambda_i|}{\sqrt{1 + |\Delta\lambda_i|^2}} = 10\varepsilon + 900\varepsilon^2 + O(\varepsilon^3)$$

when $\beta = 0$, $q = 1$ and

$$\tau(x_i, x_i^*) = 10\varepsilon + 1000\varepsilon^2 + O(\varepsilon^3)$$

when $\beta = 1$, $q = 0$.

The bound for the sines between eigenvectors and perturbed eigenvectors gives

$$f(\varepsilon) = 10.05\varepsilon + 1212\varepsilon^2 + O(\varepsilon^3)$$

when $\beta = 0$, $q = 1$ and

$$f(\varepsilon) = 10.05\varepsilon + 1111\varepsilon^2 + O(\varepsilon^3)$$

when $\beta = 1$, $q = 0$.

6. REFERENCES

Grebenikov, E., Yu. Ryabov (1979). *Constructive Methods for the Analysis of Nonlinear Systems.* Nauka, Moscow (in Russian).

Kagström, B. (1986). RGSVD - An algorithm for computing the Kronecker structure and reducing subspaces of singular $A-\lambda B$ pencils. *SIAM J. Sci. Stat. Comput.*, **7**, 185-211.

Kahan, W., B. Parlett and E. Jiang (1982). Residual bounds on approximate eigensystems of non-normal matrices. *SIAM J. Numer. Anal.*, **19**, 470-484.

Konstantinov, M., P. Petkov, N. Christov (1987). Perturbation analysis of linear control problems. *Prepr. 10 IFAC Congress*, Munich, 9, 16-21.

Konstantinov, M., P. Petkov, G. Gu, I. Postlethwaite (1995). *Perturbation Techniques for Linear Control Problems.* LUED Rpt. 95-7, Dept. of Eng., Leicester Univ., UK.

Konstantinov, M., P. Petkov, G. Gu, I. Postlethwaite (1997). Nonlocal sensitivity analysis of the eigensystem of a matrix with distinct eigenvalues. *Numer. Funct. Anal. and Optimiz.*, **18** (3&4), 367-382.

Lancaster, P., L. Rodman (1995). *Algebraic Riccati Equations.* Clarendon press, Oxford.

Meyer, C. and G. Stewart (1988). Derivatives and perturbations of eigenvectors. *SIAM J. Numer. Anal.*, **25**, 679-691.

Pappas, T., A.J. Laub, and N.R. Sandell (1980). On the numerical solution of the discrete time algebraic Riccati equation. *IEEE Trans. Autom. Control*, **AC-25**, 631-641.

Stewert, G. and Ji-Guang Sun (1990). *Matrix Perturbation Theory.* Academic Press, Boston.

Van Dooren, P. (1981). The generalized eigenstructure problem in linear system theory. *IEEE Trans. Autom. Control* **AC-26**, 111-129.

14th Triennial World Congress, Beijing, P.R. China

Robust Output Feedback Stabilization Using Genetic Algorithm

Hei Ka Tam **James Lam**

Department of Mechanical Engineering, The University of Hong Kong
E-mail: phktam@hkusua.hku.hk Fax: (852) 2858 5415

Abstract: The Robust Output Feedback Stabilization (ROFS) problem is tackled using Genetic Algorithm (GA) by finding an output feedback gain matrix such that a fitness function is maximized. The fitness function incorporates the real stability radius of the closed-loop system matrix and performance indices characterizing the assignability of the set of closed-loop poles within certain stable regions in the complex plane. The nature of GA renders it suitable for optimizing nonsmooth objectives such as that involving computation of the real stability radius. Constraints on the elements of the feedback gain matrix can also be catered naturally in the GA parameter encoding procedure. *Copyright © 1999 IFAC*

Key Words: output feedback, robust stabilization, real stability radius, genetic algorithm

1. Introduction

Consider a finite-dimensional linear time-invariant system with q ($q > 1$) inputs and p outputs described by

$$\dot{x} = Ax + Bu \quad (1)$$
$$y = Cx \quad (2)$$

where $x \in \mathbb{R}^{n\times 1}$, $u \in \mathbb{R}^{m\times 1}$, and $y \in \mathbb{R}^{p\times 1}$ represent the state, input, and output vectors respectively. $A \in \mathbb{R}^{n\times n}$, $B \in \mathbb{R}^{n\times m}$, and $C \in \mathbb{R}^{p\times n}$ are respectively the state, input, and output matrices. It is assumed that B and C are of full rank. By applying a static output feedback law

$$u = Ky \quad (3)$$

there results the closed-loop system given by

$$\dot{x} = (A + BKC)x \quad (4)$$

Under the assumption that (A, B) and (C, A) are completely controllable and observable respectively and that $m + p > n$ [8] or $mp > n$ [17] , the spectrum of the closed-loop state matrix

$$A_c = A + BKC$$

can be assigned almostly at any set of self-conjugate complex numbers by proper choice of the feedback gain matrix $K \in \mathbb{R}^{m\times p}$, so as to ensure closed-loop stability, modify dynamic response, and/or improve various robustness properties. Different approaches have been proposed to exploit the design flexibility offered by state feedback pole assignment in general multi-input systems to enhance closed-loop robustness properties [7, 10, 5, 6] . In this paper, a design procedure based on the use of Genetic Algorithm (GA) for robust stabilization via output feedback will be proposed. Specifically, the real stability radius $r_{\mathbb{R}}$ of the closed-loop system matrix will be maximized during the search of feedback gain solutions with the closed-loop spectrum assigned within an open stable region or a trapezoidal region in the s-plane. Elegant procedures for computing $r_{\mathbb{R}}$ have been suggested recently [14, 15, 18] . Still, little work have been done in maximizing this stability robustness index under the output feedback pole assignment framework.

2. Problem Formulation

The problem of *Robust Output Feedback Stabilization (ROFS)* is formulated as an optimization task in which an output feedback matrix K is searched such that the real (Hurwitz) stability radius $r_{\mathbb{R}}$ of A_c defined as

$$r_{\mathbb{R}}(A_c, F, H) = \inf\{\|E\|_2 \mid E \in \mathbb{R}^{p\times \ell} \text{ and } \operatorname{spec}(A_c + FEH^T) \not\subset \mathbb{C}^-\} \quad (5)$$

is maximized with the set of closed-loop poles assigned within an open stable region

$$\mathcal{R}_s(\eta) = \{z \in \mathbb{C} \mid \operatorname{Re}(z) < \eta\} \quad (6)$$

or a trapezoidal region

$$\mathcal{R}_t(\eta, \tau, \theta) = \{z \in \mathbb{C} \mid \tau < \operatorname{Re}(z) < \eta \text{ and } |\operatorname{Im}(z)| < -\tan\theta \cdot \operatorname{Re}(z)\} \quad (7)$$

in the s-plane where $\eta < 0$ and $0 < \theta < \pi/2$ (see Figures 1 and 2). E in (5) is a disturbance matrix with $\|E\|_2 = \sqrt{\lambda_{\max}(E^*E)}$ denoting the spectral norm of E while $F \in \mathbb{R}^{n\times p}$ and $H \in \mathbb{R}^{n\times \ell}$ are constant matrices defining the perturbation structure.

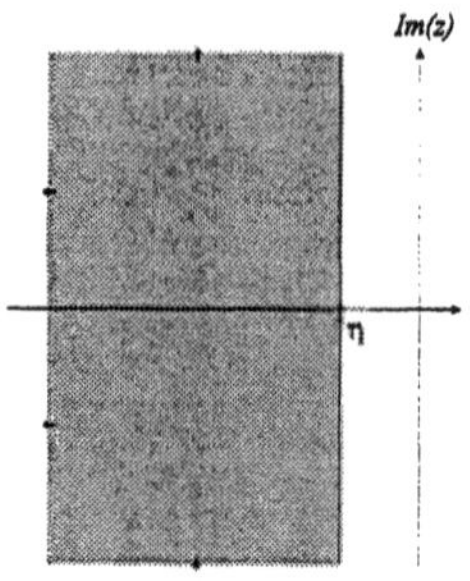

Figure 1: Open Stable Region ($\mathcal{R}_s$)

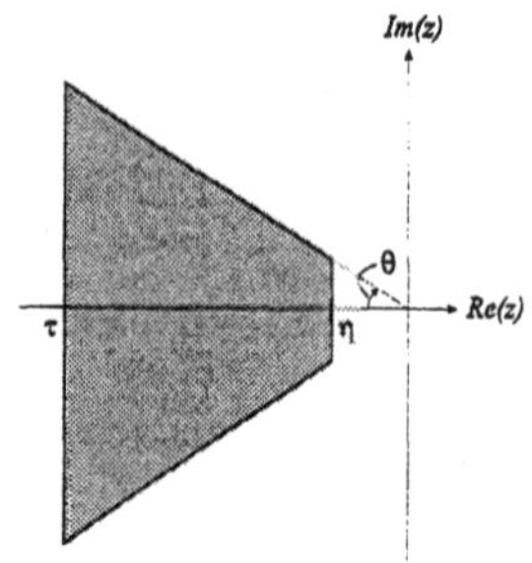

Figure 2: Trapezoidal Region ($\mathcal{R}_t$)

It is apparent that closed-loop system with greater value of $r_{\mathbb{R}}$ exhibits stronger ability to preserve its stability under the particular class of structured perturbations. Pole assignment within the trapezoidal region $\mathcal{R}_t$ characterizes the minimum decay rate η, the minimum damping ratio $\varsigma = \cos\theta$, and the maximum undamped natural frequency $\omega = \tau\tan\theta$ of the closed-loop time-domain response.

The ROFS problem is now tackled via maximization of the fitness function $\mathcal{F}(K)$ with the setting

$$\underset{K}{\text{maximize}}\ \mathcal{F}(K) = r_{\mathbb{R}} - \rho\cdot\mathcal{C} \qquad (8)$$

in which the performance index

$$\begin{aligned}\mathcal{C} \;=\;& \max\left[0, \left(\max_i \mathrm{Re}(\lambda_i(A_c)) - \eta\right)\right] \\ &+ \delta\cdot\left\{\max\left[0, \left(\tau - \min_i \mathrm{Re}(\lambda_i(A_c))\right)\right] + \right.\\ &\left.\max\left[0, \max_i\left(\tan\theta\cdot\mathrm{Re}(\lambda_i(A_c)) + |\,\mathrm{Im}(\lambda_i(A_c))|\right)\right]\right\}\end{aligned}$$

is combined with $r_{\mathbb{R}}$ in $\mathcal{F}(K)$ using a weighting factor $-\rho$ ($\rho > 0$) in order that constraint on pole assignability over $\mathcal{R}_s$ or $\mathcal{R}_t$ (which corresponds to $\mathcal{C}_{\delta=0}$ or $\mathcal{C}_{\delta=1}$ respectively) can be forced to be satisfied during the maximization process. Apparently, $\mathcal{C} = 0$ when the closed-loop spectrum is confined within the desired region. In (8), $r_{\mathbb{R}}$ for a particular K can be evaluated using the computation procedure proposed in Qiu *et al.* [14] as

$$\begin{aligned}& r_{\mathbb{R}}(A+BKC, F, H)^{-1} \\ =\;& \underset{\omega\in[\omega_m,\omega_M]}{\text{maximize}}\ \mu_{\mathbb{R}}[H^T(j\omega I_n - A - BKC)^{-1}F]\end{aligned} \qquad (9)$$

where for any $M \in \mathbb{C}^{m\times p}$

$$\mu_{\mathbb{R}}(M) = \inf_{\gamma\in(0,1]} \sigma_2\left(\begin{bmatrix}\mathrm{Re}(M) & -\gamma\,\mathrm{Im}(M)\\ \gamma^{-1}\,\mathrm{Im}(M) & \mathrm{Re}(M)\end{bmatrix}\right) \qquad (10)$$

The lower and upper bounds of the finite frequency range $[\omega_m, \omega_M]$ in (9) can be computed conveniently in the ways described in Qiu *et al.* [15] .

2.1 GA based ROFS Scheme

Genetic Algorithm (GA) is a stochastic direct search optimization technique based on the mechanics of natural genetics. Several advantages of GA make it superior to other optimization methods. Firstly, the probability for the optimization parameters being trapped into local optimum during the search process is greatly reduced since GA searches from a population of points rather than a single point [3] . Secondly, the intrinsic parallelism of GA makes it readily implementable on parallel computers. Last but not least, constraints on the optimization parameters, such as the elements k_{ij} of K in our ROFS problem, can be catered naturally in the parameter encoding procedure of GA. In fact, GA has been successfully applied to solve many difficult problems in the control community, including model reduction [19, 20] , H_∞ control design [13, 11] , and robust pole assignment [9] . It has been shown recently that the state feedback stabilization by bounded controller is NP-hard [1] , which motivates the use of GA in tackling the much more difficult problem of output feedback stabilization with incorporating feedback gain constraints and the stability robustness property. In the following, a GA based ROFS scheme will be proposed.

Given a controllable pair (A, B) and an observable pair (C, A). Choose an appropriate $\mathcal{R}_s$ or $\mathcal{R}_t$. Supply a ρ to force the set of closed-loop poles satisfying the corresponding pole assignment constraint.

1. **Parameter Encoding:**
   Each feedback gain matrix $K = (k_{ij})_{m\times p}$ (phenotype) in the search space is converted into a row vector $\mathbf{w} \in \mathbb{R}^{1\times mp}$ in a way that

   $$K \mapsto \mathbf{w} = (\ k_{11}\ \ \ldots\ \ k_{1p}\ \ k_{21}\ \ k_{22}\ \ \ldots\ \ k_{mp}\ )$$

   Each element k_{ij} in $\mathbf{w}$ is then coded as a binary string of length $l_{k_{ij}}$. Notice that each coding may have its own string length and range of k_{ij}. Suppose $\underline{k}_{ij}$ and $\overline{k}_{ij}$ is the lower and upper bound of the gain element k_{ij}, then the precision of the linearly-mapped coding $p_{k_{ij}}$ is given by

   $$p_{k_{ij}} = \frac{\overline{k}_{ij} - \underline{k}_{ij}}{2^{l_{k_{ij}}} - 1}$$

   These strings are then concatenated to produce a chromosome (an individual) $\mathbf{x}$ of length $\sum l_{k_{ij}}$ (genotype), that is,

   $$\mathbf{w} \mapsto \overbrace{\underbrace{10110}_{k_{11}}\ \underbrace{010010}_{k_{12}}\ \ldots\ \underbrace{110101}_{k_{22}}\ \ldots\ \underbrace{101}_{k_{mp}}}^{\text{chromosome } \mathbf{x} \text{ of } \sum l_{k_{ij}} \text{ bits}}$$

2. **Population Initialization:**
   Randomly generate an initial population of P chromosomes $\mathbf{x}_r$, for $r = 1, \ldots, P$. Notice that the population size P must be an even number.

3. **Tournament Selection:**

The selection scheme mimics nature's survival-of-the-fittest mechanism. Sets of chromosomes are randomly selected from the current population. The fittest chromosomes within these sets (i.e. with highest $\mathcal{F}$) are then selected and evolved into the new population. The advantage of tournament selection is that the amount of selection pressure can be controlled by the size of tournament s_t. In this way, the convergence speed of maximization can be controlled as well [4] .

4. **Uniform Crossover:**
The crossover operation generates new points in the search space [16] by exchanging genetic materials between "good" chromosomes and hopefully produces offspring that outperform their high fitness parents (i.e. with higher $\mathcal{F}$). The newly selected chromosomes (parents) in the new population are randomly paired together. In each pair of chromosomes, the bits are probabilistically and independently swapped at each bit position with a crossover probability p_c to produce pairs of new chromosomes (offspring).

5. **Bit Mutation:**
The mutation operation simply flips each bit (changing a 1 to a 0 and vice versa) in the population of chromosomes with a small mutation probability p_m. It is essentially a random walk process within the search space which reduces the chance for the optimization parameters to be fooled by local maximum (the chance for premature convergence is reduced).

6. **Elitist Strategy:**
Elitism guarantees the best chromosome in the population always survives and be retained in the next generation. This guarantees asymptotic convergence for the maximization process [2] .

7. **Check for Convergence and Pole Assignment Accuracy:**
Steps 3 to 6 corresponds to a generation. The evolution is repeated for N generations until the search process converges. The best chromosomes (i.e. with highest $\mathcal{F}$) is decoded using Step 1 to produce back the optimal output feedback gain matrix $\tilde{K}$. If the pole assignment criterion is not satisfied, repeat Steps 1 to 6 with a larger value of ρ. On the contrary, the pole assignment constraint can be relaxed by setting a smaller value of ρ in order to further enhance the system robustness property. It is emphasized that the length of encoded binary strings of k_{ij} (i.e. $l_{k_{ij}}$) and the imposed lower and upper bounds on k_{ij} (i.e. $\underline{k}_{ij}$ and $\overline{k}_{ij}$) play an important role on the successful search of a "good" optimal solution.

3. Numerical Example

Consider a 4-state, 2-input, 3-output system [12] given by

$$A = \begin{pmatrix} 1.3800 & -0.2077 & 6.7150 & -5.6760 \\ -0.5814 & -4.2900 & 0 & 0.6750 \\ 1.0670 & 4.2730 & -6.6540 & 5.8930 \\ 0.0480 & 4.2730 & 1.3430 & -2.1040 \end{pmatrix}$$

$$B = \begin{pmatrix} 0 & 0 \\ 5.6790 & 0 \\ 1.1360 & -3.1460 \\ 1.1360 & 0 \end{pmatrix}, C^T = \begin{pmatrix} 1 & 0 & 2 \\ 2 & 1 & 1 \\ 1 & 2 & -1 \\ -1 & 0 & 1 \end{pmatrix}$$

with

$$\mathrm{spec}(A) = \{-8.6659, -5.0566, 0.0635, 1.9910\}$$

To achieve robust stabilization under unstructured perturbations on A_c ($F = H = I_4$), the GA based ROFS scheme is employed with $\mathcal{F}(K)$ as the fitness function to be maximized ($\rho = 1$). The use of two different regions for the assignment of closed-loop poles is now illustrated. In both cases, the GA parameters used are $l_{k_{ij}} = 16$, $P = 100$, $s_t = 8$, $p_c = 0.4$, $p_m = 0.05$, and $N = 300$.

Case 1: Closed-loop spectrum is assigned within $\mathcal{R}_s$ ($\eta = -0.01$). A feedback gain constraint

$$||K||_{\max} < 3$$

is imposed on the GA search space. $||K||_{\max} = \max_{i,j} |k_{ij}|$ denotes the maximum norm of K. The optimal feedback gain matrix is computed as

$$K_s = \begin{pmatrix} 0.7390 & -2.8835 & 0.0152 \\ 0.0770 & -1.4103 & 0.8513 \end{pmatrix}$$

which gives the closed-loop spectrum

$$\mathrm{spec}(A + BK_sC) = \{-2.9392 \pm 3.7433j, -4.4630 \pm 9.9768j\}$$

and the optimal real stability radius

$$r_{\mathbb{R}} = 1.7449$$

where $(\omega, \gamma) = (7.5850, 0.7470)$. In the GA evolution, the maximum value of $\mathcal{F}$ in the population actually increases from 0.2179 to 1.7449. The smallest destabilizing unstructured perturbation on A_c corresponds to K_s is found to be

$$E_s = \begin{pmatrix} 0.9559 & 0.1437 & -0.3483 & -1.1615 \\ -0.2315 & -0.1296 & 0.1047 & 0.1825 \\ -0.6226 & -1.3572 & 0.4973 & -0.5608 \\ -0.2774 & 0.3601 & 0.0151 & 0.7559 \end{pmatrix}$$

which means that closed-loop stability will always be preserved under any unstructured perturbations E on A_c whenever $||E||_2 < ||E_s||_2 = 1.7449$.

Case 2: Closed-loop spectrum is assigned within $\mathcal{R}_t$ ($\eta = -1$, $\tau = -4$, $\theta = \pi/3$). A more stringent feedback gain constraint

$$||K||_{\max} < 1$$

is imposed on the GA search space. The optimal feedback gain matrix is computed as

$$K_t = \begin{pmatrix} 0.8732 & -0.8766 & -0.3605 \\ -0.5072 & 0.7471 & 0.9957 \end{pmatrix}$$

which gives the closed-loop spectrum

$$\begin{aligned} \text{spec}(A + BK_tC) &= \{-2.2765 \pm 3.9394j, \\ &\quad -3.0933 \pm 5.2826j\} \end{aligned}$$

and the optimal real stability radius

$$r_{\mathbb{R}} = 1.4587$$

where $(\omega, \gamma) = (4.5119, 0.5597)$. The maximum value of $\mathcal{F}$ in the population actually increases from -4.0021 to 1.4587 during the GA evolution. The large negative starting value indicates that a subset of the initial closed-loop poles lies far outside $\mathcal{R}_t$. In this case,

$$E_t = \begin{pmatrix} 0.5630 & -0.0343 & 0.0111 & -0.6743 \\ 0.6493 & 0.7398 & -0.3127 & 0.5034 \\ -0.4699 & -0.5801 & 0.2450 & -0.4378 \\ -0.7727 & 0.0133 & -0.0012 & 0.8698 \end{pmatrix}$$

which is the smallest destabilizing unstructured perturbation on A_c corresponding to K_t.

4. Conclusions

In this paper, the problem of ROFS is tackled using GA. The idea is to maximize the fitness function $\mathcal{F}(K)$ defined via the real stability radius $r_{\mathbb{R}}$ and a performance index C characterizing the assignability of the closed-loop spectrum within certain regions in the left-half complex plane in order that the ability for the closed-loop system to preserve its stability under perturbations can be maximized. Apart from the nice nature of GA which renders it capable of optimizing the nonsmooth objective $r_{\mathbb{R}}$, constraint on the feedback gain matrix in the maximum norm sense can also be considered naturally in the GA search process.

References

[1] V. Blondel and J. N. Tsitsiklis. NP-Hardness of some linear control design problems. *SIAM Journal on Control Optim.*, 35(6):2118–2127, 1997.

[2] D. B. Fogel. *Evolutionary Computation: Toward a New Philosophy of Machine Intelligence.* IEEE Press, 1995.

[3] D. E. Goldberg. *Genetic Algorithms in Search, Optimization, and Machine Learning.* Addison-Wesley, 1989.

[4] J. Horn, N. Nafpliotis, and D. E. Goldberg. A niched pareto genetic algorithm for multiobjective optimization. *Proc. 1st IEEE Conference on Evolutionary Computation*, pages 82–87, 1994.

[5] T. Hu and J. Lam. On optimizing performance indices with pole assignment constraints. *J. Systems & Control Engg.*, 1998. to appear.

[6] T. Hu and J. Lam. On parametric stability margin maximization using state feedback pole assignment. *Proc. ACC*, pages 2797–2802, 1998.

[7] J. Kautsky, N. K. Nichols, and P. Van Dooren. Robust pole assignment in linear state feedback. *Int. J. Control*, 41:1129–1155, 1985.

[8] H. Kimura. A further result on the problem of pole assignment by output feedback. *I.E.E.E. Trans. Automatic Control*, 25(3):458–463, 1977.

[9] J. Lam and H. K. Tam. On the use of genetic algorithm for robust approximate pole assignment. *Proc. 2nd Asian Control Conference*, pages 155–158, 1997.

[10] J. Lam and W. Yan. A gradient flow approach to robust pole-placement problem. *Int. J. of Robust and Nonlinear Control*, 5:175–185, 1995.

[11] K. F. Man, K. S. Tang, S. Kwong, and W. A. Halang. *Genetic algorithms for control and signal processing.* Springer, 1997.

[12] M. Oh, D.-W. Gu, and S. K. Spurgeon. Robust pole assignment in a specified region using output feedback. *Optimal Control Applications and Methods*, 14:57–66, 1993.

[13] R. J. Patton and G. P. Liu. Robust control design via eigenstructure assignment, genetic algorithm and gradient-based optimisation. *IEE Proc.-Control Theory Appl.*, 141:202–208, 1994.

[14] L. Qiu, B. Bernhardsson, A. Rantzer, E. J. Davison, P. M. Young, and J. C. Doyle. A formula for computation of the real stability radius. *Automatica*, 31(6):879–890, 1995.

[15] L. Qiu, A. L. Tits, and Y. Yang. On the computation of the real Hurwitz-stability radius. *I.E.E.E. Trans. Automatic Control*, AC-40:1475–1476, 1995.

[16] G. Syswerda. Uniform crossover in genetic algorithms. *Proc. 3rd International Conference on Genetic Algorithms*, pages 2–9, 1989.

[17] X. Wang. Pole placement by static output feedback. *Journal of Math. Systems, Estimation, and Control*, 2(2):205–218, 1992.

[18] W.-Y. Yan and J. Lam. On computation of the stability radius for nonlinearly structured perturbations. *Systems & Control Letters*, 34(5):273–280, 1998.

[19] Z. J. Yang, T. Hachino, and T. Tsuji. Model reduction with time delay combining the least-square method with genetic algorithm. *IEE Proc.-Control Theory Appl.*, 143(3):247–254, 1996.

[20] L. F. Yeung, M. Zhang, J. Lam, and D. Ho. Input/output structure reduction for multivariable control systems by genetic algorithm. *Proc. 4th International Conference on Control, Automation, Robotics and Vision*, pages 2119–2123, 1996.

14th Triennial World Congress, Beijing, P.R. China

CONSTANT GAIN STABILIZATION WITH A SPECIFIED DAMPING RATIO AND DAMPED NATURAL FREQUENCY [1]

Ming-Tzu Ho [*] Aniruddha Datta [**] S.P. Bhattacharyya [**]

[*] *Ritek Corporation, No. 42, Kuangfu N. Road, Hsin Chu Industrial Park, Houko 30316, Taiwan*
[**] *Department of Electrical Engineering, Texas A & M University, College Station, TX 77843-3128, U.S.A*

brutoho0@ms19.hinet.net; datta@eesun1.tamu.edu; bhatt@eesun1.tamu.edu

Abstract: This paper considers the problem of stabilizing a given linear time-invariant plant with a specified damping ratio and damped natural frequency using a constant gain compensator. We show the problem to be solved can be decomposed into two sub-problems. Thereafter, an appropriately generalized version of the Hermite-Biehler Theorem is presented and shown to be useful in providing a solution to this problem. A complete *analytical* characterization of all stabilizing feedback gains is provided as a closed form solution. We also show that the results of (Ho *et al.*, 1997*a*) are a special case of this paper. *Copyright ©1999 IFAC*

Keywords: Damping ratios, Hurwitz polynomial, Natural frequencies, Pole assignment, Stabilization methods

1. INTRODUCTION

A minimum requirement of any control system design is that closed loop stability be guaranteed, i.e., for the continuous-time linear system case, the roots of closed loop characteristic polynomial must lie in the open left half plane. In addition to possessing closed loop stability, a control system must often be designed to meet certain performance requirements. In classical control design, some of these performance requirements are typically stated in terms of time domain specifications such as the maximum overshoot, settling time, rise time, etc. of step response of the closed loop system . These time domain specifications are usually translated into desired closed loop pole locations using certain rules of thumb. By far the most common description of the desired pole locations in the complex plane is given in terms of the *damping ratio* and the *damped natural frequency* of the dominant pair of complex poles. This is because it is well known that these parameters play a predominant role in shaping the time domain behaviour of a system. Thus the problem of designing a constant gain compensator for which the roots of the closed loop characteristic equation satisfy the damping ratio and damped natural frequency requirements arises quite often in classical control design.

In a recent paper (Ho *et al.*, 1997*a*), we presented a generalization of the classical Hermite-Biehler Theorem (Gantmacher, 1959) for real polynomials. Subsequently, we showed that this generalization could be used for analytically characterizing the set of all stabilizing feedback gains for a given plant as well as for providing a computational characterization of all stabilizing proportional-integral (PI) and proportional-integral-derivative

[1] This work was supported by the National Science Foundation under Grants ECS-9210726 and ECS-9417004.

(PID) controllers (Ho *et al.*, 1997*a*; Ho *et al.*, 1997*b*). In this paper, we use a similar methodology to solve constant gain stabilization problem with a specified damping ratio and damped natural frequency. It shows that instead of a generalization of the Hermite-Biehler Theorem for real polynomials, we need a complex version of the generalized Hermite-Biehler Theorem for solving such a constant gain stabilization problem.

The paper is organized as follows. In Section 2, we introduce notations of different stability regions. We also show the stabilization problem to be solved can be decomposed into two sub-problems and the complex version of the generalized Hermite-Biehler Theorem is needed for solving one of such sub-problems. In Section 3, we present an appropriate generalization of the Hermite-Biehler Theorem for complex polynomials. In Section 4, we consider the problem of stabilizing a given plant with respect to a rotated Hurwitz stability region using a constant gain compensator. A complete analytical characterization of all stabilizing gain values is provided. Subsequently, the constant gain stabilization problem with a specified damping ratio and damped natural frequency can be solved immediately. Section 5 contains some concluding remarks.

2. NOTATIONS AND PROBLEM FORMULATION

In this section we first define notations of different Hurwitz stability regions.

Given a complex polynomial $\delta(s)$ of degree n, we say $\delta(s)$ to be Hurwitz stable if $\delta(s)$ has all its roots in the open left half plane. Consider the region S in the left half plane corresponding to a particular damping ratio and damped natural frequency as illustrated in Fig. 1. Given a real polynomial $\delta(s)$ of degree n, we say $\delta(s)$ to be S Hurwitz stable if $\delta(s)$ has all roots in S. Furthermore, the region S can be characterized using three straight lines and their corresponding left half regions:

(I) shifting the imaginary axis parallel to itself by an appropriate number of units α to the left. We define the *shifted Hurwitz* stability region $S_{-\alpha}$ by

$$S_{-\alpha} := \{s : s \in C, Re[s] < -\alpha\}.$$

Now given a real polynomial $\delta(s)$ of degree n, we say $\delta(s)$ to be $-\alpha$ Hurwitz stable if $\delta(s)$ has all roots in $S_{-\alpha}$.

(II) rotating the imaginary axis by a fixed positive angle ϕ about the origin. We define the *rotated Hurwitz* stability region S_ϕ by

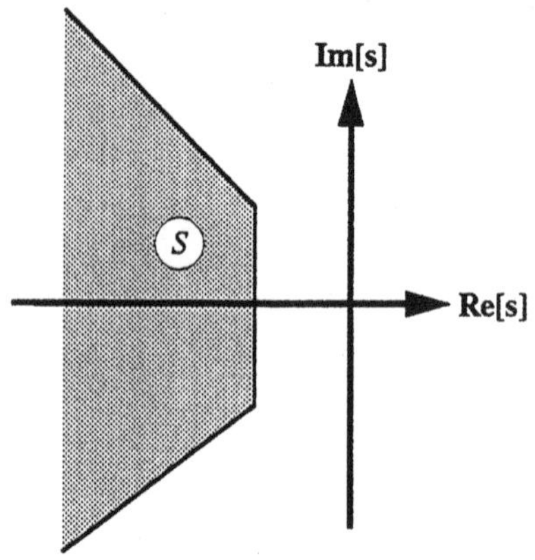

Fig. 1. The region S with a specified damping ratio and damped natural frequency.

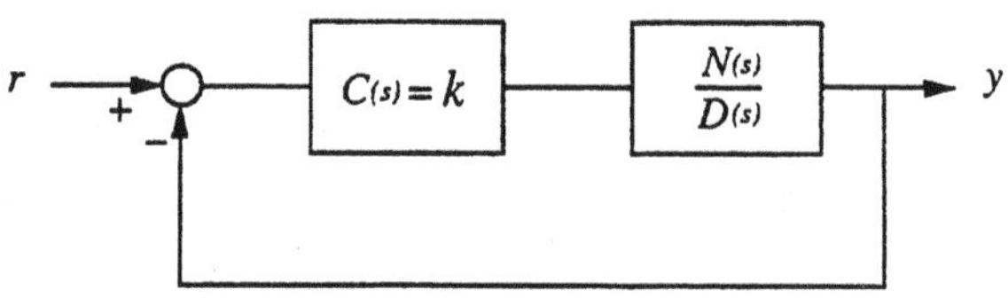

Fig. 2. Feedback control system.

$$S_\phi := \{s : s \in C, Re[se^{-j\phi}] < 0\}.$$

Now given a real polynomial $\delta(s)$ of degree n, we say $\delta(s)$ to be ϕ Hurwitz stable if $\delta(s)$ has all roots in S_ϕ.

(III) rotating the imaginary axis by a fixed negative angle $-\phi$ about the origin. We define the *rotated Hurwitz* stability region $S_{-\phi}$ by

$$S_{-\phi} := \{s : s \in C, Re[se^{j\phi}] < 0\}.$$

Now given a real polynomial $\delta(s)$ of degree n, we say $\delta(s)$ to be $-\phi$ Hurwitz stable if $\delta(s)$ has all roots in $S_{-\phi}$.

It is clear that

$$S = S_{-\alpha} \cap S_\phi \cap S_{-\phi}. \tag{1}$$

Consider the feedback control system shown in Fig. 2.

Here r is the command signal, y is the output, $\frac{N(s)}{D(s)}$ is the plant to be controlled, $N(s)$ and $D(s)$ are coprime polynomials, and $C(s)$ is the controller for stabilizing the closed loop system. Here $C(s)$ is chosen to be a scalar k, i.e. the controller is of zeroth order. Then the closed loop characteristic polynomial $\delta(s, k)$ is given by

$$\delta(s, k) = D(s) + kN(s).$$

Our objective is to determine those values of k, if any, for which the closed loop system is S

Hurwitz stable. The problem to be solved can be decomposed into three parts. First determine the set $K_{-\alpha}$ of those values of k, for which the closed loop system is $-\alpha$ Hurwitz stable. Next determine the sets K_ϕ and $K_{-\phi}$ of those values of k, for which the closed loop system is ϕ and $-\phi$ Hurwitz stable, respectively. From (1), we have the set K of all values of k for which the closed loop characteristic polynomial $\delta(s,k)$ is S Hurwitz stable is given by

$$K = K_{-\alpha} \cap K_\phi \cap K_{-\phi}.$$

Now we set $s = s_1 - \alpha$ and write

$$\begin{aligned}\delta(s,k) &= \delta_{-\alpha}(s_1,k)\\ &= D(s_1-\alpha) + kN(s_1-\alpha).\end{aligned}$$

Then the $-\alpha$ Hurwitz stability of $\delta(s,k)$ is equivalent to the Hurwitz stability of $\delta_{-\alpha}(s_1,k)$. Thus by the generalized Hermite-Biehler Theorem for real polynomials, one can use the results of (Ho *et al.*, 1997*a*) to analytically characterize the set $K_{-\alpha}$ of $-\alpha$ Hurwitz stabilizing gain values.

Next let us consider two rotated Hurwitz cases. For the case of ϕ Hurwitz stabilization problem, we set $s = s_1 e^{j\phi}$ and write

$$\begin{aligned}\delta(s,k) &= \delta_\phi(s_1,k)\\ &= D(s_1 e^{j\phi}) + kN(s_1 e^{j\phi}).\end{aligned}$$

Then the ϕ Hurwitz stability of $\delta(s,k)$ is equivalent to the Hurwitz stability of $\delta_\phi(s_1,k)$. Finally, for the case of $-\phi$ Hurwitz stabilization problem, we substitute $s = s_1 e^{-j\phi}$ into $\delta(s,k)$ and write

$$\begin{aligned}\delta(s,k) &= \delta_{-\phi}(s_1,k)\\ &= D(s_1 e^{-j\phi}) + kN(s_1 e^{-j\phi}).\end{aligned}$$

Then the $-\phi$ Hurwitz stability of $\delta(s,k)$ is equivalent to the Hurwitz stability of $\delta_{-\phi}(s_1,k)$. Now for a given k', we let any s' be a root of $\delta_\phi(s_1,k')$ and denote s'^* to be the conjugation of s'. Since

$$D(s' e^{j\phi}) + k' N(s' e^{j\phi}) = 0$$

and $N(\cdot)$ as well as $D(\cdot)$ are real polynomials, we have

$$D(s'^* e^{-j\phi}) + k' N(s'^* e^{-j\phi}) = 0.$$

It follows that for a given k, $\delta_\phi(s_1,k)$ is Hurwitz if and only if $\delta_{-\phi}(s_1,k)$ is Hurwitz. Hence $K_\phi = K_{-\phi}$. It gives that

$$K = K_{-\alpha} \cap K_\phi.$$

Therefore, the stabilization problem to be solved becomes to determine those values of k, if any, for which the $-\alpha$ and ϕ Hurwitz stabilization problem are simultaneously solvable.

In general, $\delta_\phi(s_1,k)$ and $\delta_{-\phi}(s_1,k)$ are complex polynomials. Hence, the results of (Ho *et al.*, 1997*a*) can not be used since the results of (Ho *et al.*, 1997*a*) are based on a generalized Hermite-Biehler Theorem for real polynomials. They are not applicable to complex cases. This motivate us to derive an appropriate generalization of Hermite-Biehler Theorem for complex polynomials.

3. A GENERALIZATION OF THE HERMITE-BIEHLER THEOREM FOR COMPLEX POLYNOMIALS

In this section, we state a generalization of the Hermite-Biehler Theorem for complex polynomials. To do so, we first establish some notations. The standard signum function $sgn : \mathcal{R} \to \{-1,0,1\}$ is defined by

$$sgn[x] = \begin{cases} -1 & \text{if } x < 0 \\ 0 & \text{if } x = 0 \\ 1 & \text{if } x > 0. \end{cases}$$

Now consider a polynomial $\delta(s)$ of degree n with real leading coefficient:

$$\begin{aligned}\delta(s) &= \delta_0 + \delta_1 s + \delta_2 s^2 + \ldots + \delta_n s^n,\\ &\quad \delta_i \in \mathcal{C},\ i = 0,1,\ldots,n-1,\ \delta_n \in \mathcal{R},\ \delta_n \neq 0.\end{aligned}$$

Define the signature of the polynomial $\delta(s)$ denoted by $\sigma(\delta)$, as

$$\begin{aligned}\sigma(\delta) :=\ &\text{number of open left half plane roots of}\\ &\delta(s) - \text{ number of open right half plane}\\ &\text{roots of } \delta(s).\end{aligned}$$

Now for every frequency $\omega \in \mathcal{R}$, $\delta(j\omega)$ is a point in the complex plane. Let $p(\omega)$ and $q(\omega)$ be two functions defined pointwise by $p(\omega) = Re[\delta(j\omega)]$, $q(\omega) = Im[\delta(j\omega)]$.
With this definition, we have

$$\delta(j\omega) = p(\omega) + jq(\omega)\ \forall \omega \in \mathcal{R}.$$

Furthermore, define

$$\delta_f(j\omega) = p_f(\omega) + jq_f(\omega)$$

where

$$p_f(\omega) = \frac{p(\omega)}{f(\omega)},\ q_f(\omega) = \frac{q(\omega)}{f(\omega)};$$
and $f(\omega) = (1+\omega^2)^{\frac{n}{2}}$.

Now let

$$\omega_1 < \omega_2 < \cdots < \omega_{m-1}$$

be the real, distinct finite zeros of $q_f(\omega)$ with odd multiplicities. Also define $\omega_0 = -\infty$ and $\omega_m = +\infty$.

Using the above notations, we state the following Theorem concerning $\sigma(\delta)$.

Theorem 1. Let $\delta(s)$ be a given complex polynomial of degree n with real leading coefficient. Let $\omega_1 < \omega_2 < \cdots < \omega_{m-1}$ be the real, distinct finite zeros of $q_f(\omega)$ with odd multiplicities. Also define $\omega_0 = -\infty$ and $\omega_m = +\infty$. Then

$$\sigma(\delta) = \begin{cases} \frac{1}{2}\{\text{sgn}[p_f(\omega_0)] \cdot (-1)^{m-1} \\ +2\sum_{i=1}^{m-1} \text{sgn}[p_f(\omega_i)] \cdot (-1)^{m-1-i} \\ -\text{sgn}[p_f(\omega_m)]\} \cdot \text{sgn}[q(\infty)] \\ \qquad \text{if } n \text{ is even} \\ \\ \frac{1}{2}\{2\sum_{i=1}^{m-1} \text{sgn}[p_f(\omega_i)] \cdot (-1)^{m-1-i}\} \\ \cdot\text{sgn}[q(\infty)] \\ \qquad \text{if } n \text{ is odd} \end{cases} \tag{2}$$

4. ROTATED HURWITZ STABILIZATION USING A CONSTANT GAIN

Consider the feedback control system shown in Fig. 2. Assume that $G(s) = \frac{N(s)}{D(s)}$ is the plant to be stabilized. Then the closed loop characteristic polynomial $\delta(s, k)$ is given by

$$\delta(s, k) = D(s) + kN(s).$$

In this section, we consider the constant gain ϕ Hurwitz stabilization problem. First let us set $s = s_1 e^{j\phi}$ and write

$$\begin{aligned}\delta(s, k) &= \delta_\phi(s_1, k) \\ &= D(s_1 e^{j\phi}) + kN(s_1 e^{j\phi}).\end{aligned}$$

Then ϕ Hurwitz stability of $\delta(s, k)$ is equivalent to the Hurwitz stability of $\delta_\phi(s_1, k)$. Our objective now becomes to determine those values of k, if any, for which $\delta_\phi(s_1, k)$ is Hurwitz stable.

To this end, we consider

$$\begin{aligned}N(s_1 e^{j\phi}) &= (a_0 + jb_0) + (a_1 + jb_1)s_1 + \cdots \\ &\quad +(a_m + jb_m)s_1^m,\ a_m + jb_m \neq 0 \\ D(s_1 e^{j\phi}) &= (c_0 + jd_0) + (c_1 + jd_1)s_1 + \cdots \\ &\quad +(c_n + jd_n)s_1^n,\ c_n + jd_n \neq 0\end{aligned}$$

where $n \geq m$. Let

$$\begin{aligned}N'(s_1) &= \frac{1}{c_n + jd_n} N(s_1 e^{j\phi}) \\ &= (a_0' + jb_0') + (a_1' + jb_1')s_1 + \cdots \\ &\quad +(a_m' + jb_m')s_1^m \\ &\quad \text{where } a'_i = \tfrac{a_i c_n + b_i d_n}{c_n^2 + d_n^2} \\ &\quad \text{and } b'_i = \tfrac{-a_i d_n + b_i c_n}{c_n^2 + d_n^2},\ i = 0, 1, \cdots, m\end{aligned}$$

and

$$\begin{aligned}D'(s_1) &= \frac{1}{c_n + jd_n} D(s_1 e^{j\phi}) \\ &= (c_0' + jd_0') + (c_1' + jd_1')s_1 + \cdots + s_1^n \\ &\quad \text{where } c'_i = \tfrac{c_i c_n + d_i d_n}{c_n^2 + d_n^2} \\ &\quad \text{and } d'_i = \tfrac{-c_i d_n + d_i c_n}{c_n^2 + d_n^2}, \\ &\quad i = 0, 1, \cdots, n-1.\end{aligned}$$

Also we denote

$$\begin{aligned}\delta'(s_1, k) &= \frac{1}{(c_n + jd_n)} \delta_\phi(s_1, k) \\ &= D'(s_1) + kN'(s_1).\end{aligned} \tag{3}$$

Dividing $\delta_\phi(s_1, k)$ by $(c_n + jd_n)$ results in $\delta'(s_1, k)$ of a real leading coefficient and also does not alter the roots of of $\delta_\phi(s_1, k)$.

Motivated by the results of (Ho *et al.*, 1997*a*), we first construct a polynomial , said $D^*(s_1)$, such that $\delta'(j\omega, k)D^*(j\omega)$ have that the roots of imaginary part of $\delta'(j\omega, k)D^*(j\omega)$ are independent of k. Then Theorem 1 can be used to determine the values of k, if any, for which $\delta'(s_1, k)$ in (3) is Hurwitz stable. To do so, we now write

$$\begin{aligned}N'(s_1) &= N_R'(s_1) + N_I'(s_1) \\ D'(s_1) &= D_R'(s_1) + D_I'(s_1)\end{aligned}$$

where

$$\begin{aligned}N_R'(s_1) &= a_0' + jb_1' s_1 + a_2' s_1^2 + jb_3' s_1^3 + \cdots \\ N_I'(s_1) &= jb_0' + a_1' s_1 + jb_2' s_1^2 + a_3' s_1^3 + \cdots \\ D_R'(s_1) &= c_0' + jd_1' s_1 + c_2' s_1^2 + jd_3' s_1^3 + \cdots \\ D_I'(s_1) &= jd_0' + c_1' s_1 + jd_2' s_1^2 + c_3' s_1^3 + \cdots.\end{aligned}$$

Define

$$D^*(s_1) = D_R'(s_1) - D_I'(s_1).$$

Now, multiplying $\delta'(s_1, k)$ by $D^*(s_1)$ and examining the resulting polynomial, we obtain

$$\sigma(\delta'(s_1, k)D^*(s_1)) = \sigma(\delta'(s_1, k)) + \sigma(D^*(s_1)).$$

Now, $\delta'(s_1, k)$ of degree n is Hurwitz if and only if $\sigma(\delta'(s_1, k)) = n$. Therefore we have the following.

Lemma 2. $\delta'(s_1, k)$ is Hurwitz if and only if

$$\sigma(\delta'(s_1, k)D^*(s_1)) = n + \sigma(D^*(s_1)). \tag{4}$$

Our task now is to determine those values of k for which (4) holds. Substituting $s_1 = j\omega$, we obtain

$$\begin{aligned}\delta'(j\omega,k) &= [D_R'(j\omega) + D_I'(j\omega)] \\ &\quad + k[N_R'(j\omega) + N_I'(j\omega)] \\ D^*(j\omega) &= D_R'(j\omega) - D_I'(j\omega)\end{aligned}$$

where $D_R'(j\omega)$ and $N_R'(j\omega)$ are purely real, while $D_I'(j\omega)$ and $N_I'(j\omega)$ are purely imaginary. Therefore

$$\delta'(j\omega,k)D^*(j\omega) = p(\omega,k) + jkq(\omega)$$

where

$$\begin{aligned}p(\omega,k) &= p_1(\omega) + kp_2(\omega) \\ p_1(\omega) &= D_R'^{\,2}(j\omega) - D_I'^{\,2}(j\omega) \\ p_2(\omega) &= D_R'(j\omega)N_R'(j\omega) - D_I'(j\omega)N_I'(j\omega) \\ q(\omega) &= \frac{1}{j}[D_R'(j\omega)N_I'(j\omega) - D_I'(j\omega)N_R'(j\omega)].\end{aligned}$$

Since $\delta'(s_1,k)D^*(s_1)$ has degree $2n$ then we define

$$\begin{aligned}p_f(\omega,k) &= \frac{p(\omega,k)}{(1+\omega^2)^n}, \\ q_f(\omega) &= \frac{q(\omega)}{(1+\omega^2)^n}.\end{aligned}$$

We now assume $k \neq 0$ so that the zeros of the imaginary part of $\delta'(j\omega,k)D^*(j\omega)$ are independent of k. For the case of $k = 0$, it is trivial that $\delta'(s_1,k)$ is Hurwitz if and only if $D'(s_1)$ is Hurwitz. We first note that for those values of k for which $\delta'(s_1,k)$ is Hurwitz, the polynomial $\delta'(s_1,k)D^*(s_1)$ satisfies the hypothesis of Theorem 1 so that the latter can be used to determine $\sigma(\delta'(s_1,k)D^*(s_1))$. Since $\sigma(D^*(s_1))$ is known, the stabilizing values of k can be determined from (4).

The formal statement of our main result on constant gain stabilization involves certain *strings* of the real numbers 0, 1 and -1. These strings are used to essentially capture all the different possibilities for the sign of $p_f(\omega,k)$ at the real zeros of $q_f(\omega)$ (with odd multiplicities) and all the possible signs of k. Among these, we are interested in only those possibilities, which when substituted into the first expression of the right hand side of (2) yields a value for the signature of $\delta'(s_1,k)D^*(s_1)$ for which (4) holds.

For clarity of presentation, we first introduce these strings and motivate some definitions before formally stating the result.

Definition 4.1. Let $q_f(\omega)$ be as already defined. Let $\omega_1 < \omega_2 < \cdots < \omega_{m-1}$ be the real, distinct finite zeros of $q_f(\omega)$ with odd multiplicities[2]. Also define $\omega_0 = -\infty$ and $\omega_m = +\infty$. Define

$$A := \{i_0, i_1, \cdots, i_m, i_{m+1}\}$$

where

$$i_t \in \begin{cases}\{-1,1\} & \text{if } j\omega_t \text{ is not a } j\omega\text{-axis root of } D^*(s_1) \\ \{0\} & \text{otherwise}\end{cases}$$
$$\text{for } t = 0,1,\cdots,m$$
$$\textit{and}$$
$$i_{m+1} \in \{-1,1\} \tag{5}$$

In other words, for $t = 0,\cdots,m$, i_t is the set of all possible strings of 1's, 0's and -1's, whose element is equal to -1 or 1, if corresponding $j\omega_t$ is not a $j\omega$-axis root of $D^*(s_1)$. If $j\omega_t$ is a $j\omega$-axis root of $D^*(s_1)$ then the corresponding element is equal to 0.

Next we introduce the set $A(\gamma)$ of strings in A with a prescribed "signature" γ. To do so, we first need to define the "signature" $\gamma(\mathcal{I})$ associated with any element $\mathcal{I} \in A$. This definition is motivated by Theorem 4 to follow.

Definition 4.2. Let $q(\omega)$ and $q_f(\omega)$ be as already defined. Let $\omega_1 < \omega_2 < \cdots < \omega_{m-1}$ be the real, distinct finite zeros of $q_f(\omega)$ with odd multiplicities. Also define $\omega_0 = -\infty$ and $\omega_m = +\infty$. For each string $\mathcal{I} = \{i_0, i_1, \cdots, i_m, i_{m+1}\}$ in A, let $\gamma(\mathcal{I})$ denote the "signature" associated with the string $\mathcal{I}$ defined by

$$\begin{aligned}\gamma(\mathcal{I}) := \frac{1}{2}\{&i_0 \cdot (-1)^{m-1} + 2\sum_{r=1}^{m-1} i_r \cdot (-1)^{m-1-r} \\ &- i_m\} \cdot \mathrm{sgn}[i_{m+1} \cdot q(\infty)].\end{aligned} \tag{6}$$

Remark 3. Note that since $\delta'(s_1,k)D^*(s_1)$ has degree $2n$, if we make the identification $i_t = sgn[p_f(\omega_t,k)]$ for $t = 0,1,\cdots,m$ and $i_{m+1} = sgn[k]$, then the signature of $\delta'(s_1,k)D^*(s_1)$ as determined from the first expression of (2) is the same as the quantity $\gamma(\mathcal{I})$ defined above. Hence, referring to $\gamma(\mathcal{I})$ as the "signature" of $\mathcal{I}$ is appropriate terminology.

Definition 4.3. The set of strings in A with a prescribed signature $\gamma = \psi$ is denoted by $A(\psi)$. We also define the set of feasible strings for the constant gain stabilization problem as

$$\begin{aligned}F^* = \{\mathcal{I} | \mathcal{I} = \{&i_0, i_1, \cdots, i_m, i_{m+1}\} \\ &\in A(n + \sigma(D^*(s_1)))\} \text{ such that}\end{aligned}$$

[2] Note that these zeros are independent of k.

(i) if $p_2(\omega_t) = 0$ for some $t = 0, 1, \cdots, m$, then $i_t = \text{sgn}[p_1(\omega_t)]\}$. (7)

In other words F^* is the subset of strings in $A(n + \sigma(D^*(s_1)))$ which also satisfy (i).

We are now ready to state the main result of this section.

Theorem 4. The constant gain feedback ϕ Hurwitz stabilization problem is solvable for a given plant with transfer function $G(s)$ if and only if the following conditions hold:
(I) $G(s_1 e^{j\phi}) = \frac{N(s_1 e^{j\phi})}{D(s_1 e^{j\phi})}$ is a Hurwitz stable plant where $N(s_1 e^{j\phi})$ and $D(s_1 e^{j\phi})$ are as already defined. In this case, the corresponding stabilizing constant gain K_0 is given by $K_0 = \{0\}$
or
(II)
(1) F^* is not empty where F^* is as already defined, i.e. at least one feasible string exists
and
(2) There exists a string $\mathcal{I} = \{i_0, i_1, \cdots, i_m, i_{m+1}\} \in F^*$ such that
(a) if $i_{m+1} = -1$ then for $t = 1, 2, \cdots, m$

$$\max_{i_t \in \mathcal{I}, i_t \cdot \text{sgn}[p_2(\omega_t)]=1} \left[-\frac{1}{G(j\omega_t e^{j\phi})}\right] < \min\left[0, \min_{i_t \in \mathcal{I}, i_t \cdot \text{sgn}[p_2(\omega_t)]=-1} \left[-\frac{1}{G(j\omega_t e^{j\phi})}\right]\right].$$

In this case, the set of the corresponding stabilizing gains is given by

$$K_- = \left(\max_{i_t \in \mathcal{I}, i_t \cdot \text{sgn}[p_2(\omega_t)]=1} \left[-\frac{1}{G(j\omega_t e^{j\phi})}\right], \min\left[0, \min_{i_t \in \mathcal{I}, i_t \cdot \text{sgn}[p_2(\omega_t)]=-1} \left[-\frac{1}{G(j\omega_t e^{j\phi})}\right]\right]\right)$$

(b) if $i_{m+1} = 1$ then for $t = 1, 2, \cdots, m$

$$\max\left[0, \max_{i_t \in \mathcal{I}, i_t \cdot \text{sgn}[p_2(\omega_t)]=1} \left[-\frac{1}{G(j\omega_t e^{j\phi})}\right]\right] < \min_{i_t \in \mathcal{I}, i_t \cdot \text{sgn}[p_2(\omega_t)]=-1} \left[-\frac{1}{G(j\omega_t e^{j\phi})}\right].$$

In this case, the set of the corresponding stabilizing gains is given by

$$K_+ = \left(\max\left[0, \max_{i_t \in \mathcal{I}, i_t \cdot \text{sgn}[p_2(\omega_t)]=1} \left[-\frac{1}{G(j\omega_t e^{j\phi})}\right]\right], \min_{i_t \in \mathcal{I}, i_t \cdot \text{sgn}[p_2(\omega_t)]=-1} \left[-\frac{1}{G(j\omega_t e^{j\phi})}\right]\right)$$

where $p_2(\omega)$ and $\omega_0, \omega_1, \omega_2, \cdots, \omega_m$ are as already defined. Furthermore, if the above condition (II.2.a) or (II.2.b) is satisfied by the feasible strings $\mathcal{I}_1, \mathcal{I}_2, \cdots, \mathcal{I}_s \in F^*$, then the set of all stabilizing gains is given by $K_\phi = \cup_{r=0}^{s} K_r$ where K_0 is given by condition (I) and K_r is the set of stabilizing gains corresponding to the feasible strings $\mathcal{I}_r$ for $t = 1, 2, \cdots, s$.

Note that as $\phi = 0$, Theorem 4 becomes the analytical solution of the constant gain Hurwitz stabilization problem. The results of (Ho *et al.*, 1997*a*) are, therefore, a special case of Theorem 4.

5. CONCLUDING REMARKS

In this paper, we have provided the analytical solution to the problem of stabilizing a given linear time-invariant plant with a specified damping ratio and damped natural frequency using a constant gain compensator. Although the problem is motivated from our earlier constant stabilization results (Ho *et al.*, 1997*a*), the stabilization problem considered in this paper presents an additional set of unique challenges. The generalized Hermite-Biehler Theorem required for solving such a stabilization problem has to be applicable to complex polynomials and not just real polynomials. The extension of these results to PID controllers is yet to be carried out, and is a topic for further investigation.

6. REFERENCES

Gantmacher, F. R. (1959). *The Theory of Matrices*. Chelsea Publishing Company. New York.

Ho, M.T., A. Datta and S.P. Bhattacharyya (1997*a*). A new approach to feedback design part I: Generalized interlacing and proportional control. *Department of Electrical Engg., Texas A&M Univ., College Station, TX, Tech. Report TAMU-ECE97-001-A.*

Ho, M.T., A. Datta and S.P. Bhattacharyya (1997*b*). A new approach to feedback design part II: PI and PID controllers. *Department of Electrical Engg., Texas A&M Univ., College Station, TX, Tech. Report TAMU-ECE97-001-B.*

14th Triennial World Congress, Beijing, P.R. China

C-2a-10-6

AN ALTERNATIVE SOLUTION TO THE PROBLEM OF POLE ASSIGNMENT BY STATIC OUTPUT FEEDBACK

A. Eisinberg, G. Franzé, P. Muraca, N. Salerno

Università della Calabria
Dip. di Elettronica, Informatica e Sistemistica, (D.E.I.S.)
87030 Rende (Cs) – Italy
tel: 0984 494722 fax: 0984 494713
E-mail: muraca@unical.it

Abstract: This note discusses the problem of pole placement by static output feedback from a geometric point of view. It is shown, without any assumption on the genericity of the system, that the pole placement can be solved by choosing the closed-loop eigenvectors almost freely. We propose a numerical algorithm that allows to fix the desired set of all closed-loop eigenvalues with a small relative error. Moreover, such algorithm seems to be simpler than those known from literature.

Keywords: Linear Multivariable Systems, Output Feedback, Geometric Approach.

1. INTRODUCTION

The eigenvalue assignment problem for linear time-invariant multi-input multi-output systems using static output feedback has been widely investigated in the past years. The first results have been published by Davidson and Wang (1975) and Kimura (1975; 1977). In each of this papers, the authors give a sufficient condition on the number of the closed-loop eigenvalues which can be placed in prescribed points of the complex plane. Kimura uses the properties of the *characteristic subspaces*, while Davidson and Wang propose a solution taking into account the controllability and observability properties of the open-loop system, with the assumption that the closed-loop eigenvalues are different and distinct from any ones of the open loop system. Kwon and Youn (1987) generalize the entire eigenstructure assignment removing the assumption that the eigenvalues of the closed-loop system are distinct and different from any eigenvalues of the open-loop system. Such approach leads to formalize the eigenstructure assignment by output feedback in terms of two coupled Sylvester equations. Alexandridis and Paraskevopoulos (1996) proved that the complexity of assignment problem can be reduced solving it in two steps: the first one assigns the eigenvalues belonging to a set Λ_1 by determining the gain matrix given by the first Sylvester equation, the second step assigns the remaining eigenvalues belonging to Λ_2 by forcing the free parameters in such matrix to satisfy a reduced-order observer equation.

In this paper, using as a basis for the state space the right open-loop eigenvectors, we show that the pole placement problem can be reformulated in terms of a set of n^2 equations that simplify the structure of the eigenvalue equation for the case in which $m + p > n$, where n, m and p are the numbers of state variables, inputs and outputs, respectively.

The result reported in paper, allow to obtain an efficient numerical procedure to syntetize the output feedback gain matrix.

[1] Paper supported by MURST Project *"Identification and control of industrial systems"*

2. PROBLEM STATEMENT AND PRELIMINARY RESULTS

Let us consider the linear time-invariant (LTI) multivariable system

$$\begin{cases} \dot{x} = Ax + Bu, \\ y = Cx, \end{cases} \qquad (1)$$

where $x \in \mathbf{R}^n$, $u \in \mathbf{R}^m$, e $y \in \mathbf{R}^p$. We shall suppose that (A, B, C) is controllable and observable with B and C of full rank and $m + p > n$. For this class of linear systems, has been proved in (Kimura 1975) that it is possible to assign the dynamical behaviour of the closed-loop system under static output feedback. If the control law

$$u = Ky, \qquad (2)$$

is applied to (1), the closed-loop equation is

$$\dot{x} = (A + BKC)\,x. \qquad (3)$$

Problem: *Fixed a symmetric set of n complex numbers* $\Lambda = \{\lambda_1, \ldots, \lambda_n\}$, *find an* $m \times p$ *real matrix* K *such that the eigenvalues of* $A + BKC$ *are equal to* Λ.

In the sequel,

$$\begin{aligned} U &= [u_1, u_2, \ldots, u_n], \\ V &= [v_1^T; v_2^T; \ldots; v_n^T], \\ V &= U^{-1}, \\ \Omega &= [\omega_1, \omega_2, \ldots, \omega_n] \end{aligned}$$

are $n \times n$ matrices, which denote the right and left eigenvectors of A and the right eigenvectors of $A + BKC$, respectively; $\Pi = diag\{\pi_1, \ldots, \pi_n\}$ is the diagonal matrix of the open-loop eigenvalues. The following result gives an alternative formulation of the eigenvalue assignment problem.

Proposition 1. The problem of assigning Λ as the spectrum of the closed-loop system $A + BKC$ is equivalent to finding K and γ_{si}, $i, s = 1, \ldots, n$, such that the system of n^2 nonlinear equations

$$\gamma_{si}\pi_s + \sum_{k=1}^{n} \gamma_{ki} v_s^T BKC u_k = \gamma_{si}\lambda_i \qquad i, s = 1, \ldots, n. \qquad (4)$$

has at least a solution with K real.

Where γ_{si} are the entries of the closed-loop eigenvectors w_i in the basis U:

$$\Omega = U\Gamma$$

$$[\Gamma]_{si} = \gamma_{si}, \quad i, s = 1, \ldots, n.$$

Verifying the Proposition 1 is a straightforward computation.

Equations (4) seem to be a simple way to define the pole placement problem via static output feedback; moreover, they emphasize the relationship between the open-loop and closed-loop eigenvalues and eigenvectors. Before to discuss the main results of this note, the following definitions are introduced:

Definition 1. Given a complex number λ and the triplet (A, B, C), the *characteristic subspaces* $\mathcal{F}(\lambda)$ and $\mathcal{W}(\lambda)$ are defined as

$$\mathcal{F}(\lambda) = \{f \in \mathcal{C}^n : (\lambda I - A)\, f \in \mathrm{Im}(B)\} \qquad (5)$$

$$\mathcal{W}(\lambda) = \{\xi \in \mathcal{C}^{n'} : \xi\,(\lambda I - A) \in \mathrm{Im}(C')\} \qquad (6)$$

Definition 2. Given the triplet (A, B, C) and a complex number λ, representing an eigenvalue of A with algebraic multiplicity $\mu > 1$, there is a corresponding right characteristic subspace chain defined by

$$\begin{aligned} \mathcal{F}_j(\lambda) &:= (\lambda I - A)^{\dagger}\, Im(B) + \\ &\quad (\lambda I - A)^{\dagger}\, v_{j-1} \\ j &= 1, \ldots, \mu. \end{aligned} \qquad (7)$$

where $v_0 = 0$ and $(\lambda I - A)^{\dagger}$ denotes the *Moore-Penrose generalized inverse* of such matrix.

Proposition 2. If the right closed-loop eigenvectors are chosen in a way such that:

$$w_i \in \mathcal{F}_j(\lambda_i)$$

$$i = 1, \ldots, l, \; j = 1, \ldots, \mu_i$$

then the set of equations (4) is a parametric form of the eigenvalue equation with ν free parameters.

$$\nu = mn + \sum_{i=1}^{l} (\mu_i - 1), \qquad (8)$$

where

$$\begin{aligned} &\det(\lambda I - (A + BKC)) = \\ &(\lambda - \lambda_1)^{\mu_1}(\lambda - \lambda_2)^{\mu_2} \cdots (\lambda - \lambda_l)^{\mu_l} \\ &\mu_1 + \mu_2 + \cdots + \mu_l = n. \end{aligned} \qquad (9)$$

The proof is given in the Appendix.

Hence, we shall describe the two main results that enable to solve the posed problem. First, we define the conditions under which there exists a real matrix K such that it is possible to assign exactly p eigenvalues of the closed-loop dynamic matrix. Then, we determine the structure of the subset of $m \times p$ equations for calculating the feedback gain matrix.

Let $\Lambda = \{\lambda_1, \ldots, \lambda_n\}$ be an assignable and symmetric set of complex numbers. In (Champetier and Magni 1991) have shown that, if

$$\omega_k \in \mathcal{F}(\lambda_k)$$
$$k = 1, 2, \ldots, p$$

are independent and

$$M = [C\omega_1 | \ldots | C\omega_p]$$

is nonsingular, then there exists K such that $\lambda_1, \ldots, \lambda_p$ are eigenvalues of $A + BKC$;

However, it does not seem that the conditions under which M is actually nonsingular have been printed out. In order to investigate this issue, assume $p \geq m$, $G(\lambda) = C(\lambda I - A)^{-1} B$ the $p \times m$ transfer function of the system and $g_1(\lambda), \ldots, g_m(\lambda)$ its columns, $g_i(\lambda) \in \mathcal{R}^p$. For the sake simplicity we shall consider only the case of $\lambda_i \notin$ spectrum of A.

Proposition 3. Let $\{\lambda_1, \ldots, \lambda_p\}$ be the closed-loop eigenvalues distict from the eigenvalues of A: if there exist $i_1, \ldots, i_p$ such that

$$\det\left[g_{i_1}(\lambda_1) | g_{i_2}(\lambda_2) | \ldots | g_{i_p}(\lambda_p)\right] \neq 0$$

then for almost all $\omega_1 \in \mathcal{F}(\lambda_1), \ldots, \omega_p \in \mathcal{F}(\lambda_p)$ the matrix $M = [C\omega_1 | \ldots | C\omega_p]$ is nonsingular.

Proof: Under the stated assumptions

$$\mathcal{F}(\lambda_k) = \left\{\omega_k \in \mathcal{R}^n | \omega_k = (\lambda_k I - A)^{-1} B u_k, \; u_k \in \mathcal{R}^m\right\}.$$

Then, M can be written as

$$M = [G(\omega_1) | \ldots | G(\omega_p)] \begin{bmatrix} u_1 & 0 & \ldots & 0 \\ 0 & u_2 & & 0 \\ \vdots & \vdots & \ddots & \vdots \\ 0 & 0 & & u_p \end{bmatrix}$$

It can be shown that

$$\det[M] = \sum d_{i_1 i_2 \ldots i_p} u_{1,i_1} \cdots u_{p,i_p}$$

where

$$d_{i_1 i_2 \ldots i_p} = \det\left[g_{i_1}(\lambda_1) | g_{i_2}(\lambda_2) | \ldots | g_{i_p}(\lambda_p)\right]$$

$$\begin{cases} i_1 \in \{1, \ldots, m\} \\ \vdots \quad \vdots \quad \vdots \\ i_p \in \{1, \ldots, m\} \end{cases}$$

and the summation is over all m^p combinations of $\{i_1, \ldots, i_p\}$.

Now, either $\det[M] = 0$ on a manifold of dimension $mp - 1$, and hence $\det[M] \neq 0$ for almost all choices of $u_1, \ldots, u_p$, or $\det[M] = 0$ for all $u_1, \ldots, u_p$.
But $\det[M] = 0$ for all $u_1, \ldots, u_p$ if and only if all the minors $d_{i_1 i_2 \ldots i_p}$ vanish, which contradicts the hypothesis.

This proves the "genericity" of the condition $\det[M] \neq 0$, under the mild assumption on the transfer function, evaluete at the candidate eigenvalues.
□

Obviously, in the case $m > p$ the Proposition 3 can be written in a dual form by exchanging B, C, m and p by C^T, B^T, p and m.

Proposition 4. Let $\{\lambda_1, \ldots, \lambda_n\}$ be a symmetric set of complex numbers: for at least a subset of indices $\{i_1, \ldots, i_m\}$ and for any subset of indices $\{j_1, \ldots, j_p\}$, the $m \times p$ equations.

$$\begin{cases} v_{i_1}^T BKCw_{j_1} = \gamma_{j_1 i_1}(\lambda_{j_1} - \pi_{i_1}) \\ v_{i_2}^T BKCw_{j_1} = \gamma_{j_2 i_1}(\lambda_{j_1} - \pi_{i_2}) \\ \quad \vdots \quad = \quad \vdots \\ v_{i_m}^T BKCw_{j_1} = \gamma_{j_m i_1}(\lambda_{j_1} - \pi_{i_m}) \\ v_{i_1}^T BKCw_{j_2} = \gamma_{j_1 i_2}(\lambda_{j_2} - \pi_{i_1}) \\ v_{i_2}^T BKCw_{j_2} = \gamma_{j_2 i_2}(\lambda_{j_2} - \pi_{i_2}) \\ \quad \vdots \quad \vdots \quad \vdots \\ v_{i_m}^T BKCw_{j_2} = \gamma_{j_m i_2}(\lambda_{j_2} - \pi_{i_m}) \\ \quad \vdots \quad \vdots \quad \vdots \\ \quad \vdots \quad \vdots \quad \vdots \\ v_{i_1}^T BKCw_{j_p} = \gamma_{j_1 i_p}(\lambda_{j_p} - \pi_{i_1}) \\ v_{i_2}^T BKCw_{j_p} = \gamma_{j_2 i_p}(\lambda_{j_p} - \pi_{i_2}) \\ \quad \vdots \quad \vdots \quad \vdots \\ v_{i_m}^T BKCw_{j_p} = \gamma_{j_m i_p}(\lambda_{j_p} - \pi_{i_m}) \end{cases} \tag{10}$$

exctracted from (4), have a solution for almost all choices of

$$\begin{cases} \omega_{j_1} \in \mathcal{F}(\lambda_{j_1}) \\ \vdots \quad \vdots \quad \vdots \\ \omega_{j_p} \in \mathcal{F}(\lambda_{j_p}) \end{cases}$$

□

Proof: The above system consists of p blocks of m equations:

$$V_m BKC\omega_{j_k} = \begin{bmatrix} (\lambda_{j_k} - \pi_{i_1})\gamma_{j_k i_1} \\ \vdots \\ (\lambda_{j_k} - \pi_{i_1})\gamma_{j_k i_m} \end{bmatrix}, \tag{11}$$
$$k = 1, \ldots, p.$$

where

$$V_m = [v_{i_1}^T; \, v_{i_2}^T; \, \ldots; \, v_{i_m}^T].$$

Since $\text{Rank}(B) = m$ and $B^* = VB$ has also rank m, m rows of V can be choosen such that $V_m B$ is nonsingular; then, each block gives:

$$KC\omega_{j_k} = (V_m^T B)^{-1} z_{j_k}$$

where z_{j_k} is the r.h.s. of (11).

Since

$$\begin{cases} \omega_{j_1} \in \mathcal{F}(\lambda_{j_1}) \\ \vdots \quad \vdots \quad \vdots \\ \omega_{j_p} \in \mathcal{F}(\lambda_{j_p}) \end{cases}$$

can be choosen such that $[C\omega_{j_1}|\ldots|C\omega_{j_p}]$ is nonsingular, K can be computed, in priciple, as

$$K = (V_m^T B)^{-1} [z_{j_1}|\ldots|z_{j_p}] [C\omega_{j_1}|\ldots|C\omega_{j_p}]^{-1}.$$

Note that by changing $\omega_{j_1}, \ldots, \omega_{j_p}$ in the appropiate subspaces, the value of K is changed, but the eigenvalues $\lambda_{j_1}, \ldots, \lambda_{j_p}$ are always assigned. This allows to use a search procedure to assign, or approximate, the remaining $n-p$ eigenvalues.

□

3. ALGORITHM

Proposition 3 and Proposition 4 allow us to write a procedure to compute the gain output feedback matrix K in order to fix p eigenvalues. Moreover, the remaining "free" $n-p$ eigenvalues of the closed-loop system can be placed by a heuristic procedure taking into account that the roots of the characteristic polynomial are continuos functions of its coefficients (Wilkinson 1965). Indeed, we have observed from numerical experiments that only a "little" subset of parameters of (10) affect the $n-p$ eigenvalues. Starting from these considerations, we propose the following algorithm.

Step 1 :
Choose a desired symmetric set of complex numbers Λ^*

Step 2 :
Fix the subsets $\{i_1, \ldots, i_m\}$ and $\{j_1, \ldots, j_p\}$

Step 3 :
Determine the right eigenvectors w_{i_j} of $(A+BKC)$ such that

$$w_{i_j} \in \mathcal{F}_j(\lambda_i), \ \ i = 1, \ldots, l, \ j = 1, \ldots, \mu_i$$

solve the $m \times p$ equations in the unkown $k_{11}, k_{12}, \ldots, k_{mp}$ defined as (10)

Step 4 :

a) Determine numerically the sensitivity of the $n-p$ eigenvalues to the free parameters of (10).
b) Change the basis of Im(B) with respect to such parameters and repeat *Step 3* until

$$\max_{i=1,\ldots,n} |\lambda_i(A+BKC) - \lambda_i^*| \leq \epsilon,$$

where ϵ is a small positive number.

Comment: We have observed from numerical experiments that only a "little" subset of such parameters is effective.

Actually, **Step 4** is more cumbersome, because it implies a trial and error technique. But, it is evident that the equations (10) are multivariable polynomials, this allows to develop a numerical procedure to evaluate the sensitivity of the $n-p$ eigenvalues with respect to the free parameters of the gain matrix.

Moreover, it is evident from (10) that such a step can be solved by a non linear optimization algorithm. This approach is under investigation.

4. NUMERICAL EXAMPLES

In order to put in evidence the performance of the previous algorithm, two numerical examples are reported. The first one point out the capability of the algorithm to place in the same point more than one closed-loop eigenvalues; the second one is taken from (Alexandridis and Paraskevopoulos 1996) in order to compare the performance of the algorithm.

4.1 *Example 1:*

Consider the controllable and observable system given by the following matrices

$$A = \begin{bmatrix} 3 & 1 & 0 & 0 \\ 1 & 0 & 1 & 0 \\ 5 & 0 & 0 & 1 \\ -1 & 0 & 0 & 0 \end{bmatrix}, \ B = \begin{bmatrix} 1 & 0 \\ 0 & 0 \\ 0 & 1 \\ 0 & 0 \end{bmatrix},$$

$$C = \begin{bmatrix} 1 & 0 & 0 & 0 \\ 0 & 1 & 0 & 0 \\ 1 & 7 & 1 & -3 \end{bmatrix}.$$

The eigenvalues of A are :

$eig(A) = \{3.6332, 0.1891, -0.4111 \pm j1.1344\}$

In order to analyze the case where multiple closed-loop eigenvalues are assigned, Λ (the desired closed-loop eigenvalues) is selected as :

$$\Lambda = \{-4, -1, -1, -1\}$$

Divide Λ into the subsets $\Lambda_1 = \{-4, -1, -1\}$ and $\Lambda_2 = \{-1\}$.

After choosing a set of $m \cdot p$ equations from (4) using the structure (10), the output feedback matrix is determined via *Step 4* as

$$K = \begin{bmatrix} -1.00002 & -6.99997 & 0.999995 \\ -34.9999 & 60.9998 & -9.99997 \end{bmatrix},$$

with the following closed-loop eigenvectors

$$\omega_1 = [0.0373831 \ \ 0.0654206 \\ -0.299065 \ \ 0.00934577],$$

$$\omega_{2,1} = \left[-\frac{2}{9} \; -\frac{1}{9} \; \frac{1}{3} \; -\frac{2}{9} \right],$$

$$\omega_{2,2} = \left[-\frac{137}{360} \; \frac{7}{72} \; \frac{17}{60} \; -\frac{137}{360} \right],$$

$$\omega_{2,3} = \left[-\frac{808}{4077} \; -\frac{355}{4077} \; \frac{1457}{4077} \; -\frac{808}{4077} \right],$$

belonging to the subspaces

$$\omega_1 \in F(-4) = Span \left\{ \begin{bmatrix} -\frac{64}{453} & -\frac{4}{453} \\ -\frac{5}{453} & \frac{28}{453} \\ \frac{28}{151} & -\frac{36}{151} \\ -\frac{16}{453} & -\frac{1}{453} \end{bmatrix} \right\},$$

$$\omega_{2,1} \in F_1(-1) = Span \left\{ \begin{bmatrix} -\frac{1}{9} & -\frac{1}{9} \\ -\frac{5}{9} & \frac{4}{9} \\ \frac{2}{3} & -\frac{1}{3} \\ -\frac{1}{9} & -\frac{1}{9} \end{bmatrix} \right\},$$

$$\omega_{2,2} \in F_2(-1) = Span \left\{ \left[\begin{array}{cc|c} -\frac{1}{9} & -\frac{1}{9} & \frac{98}{4077} \\ -\frac{5}{9} & \frac{4}{9} & \frac{220}{4077} \\ \frac{2}{3} & -\frac{1}{3} & -\frac{175}{1359} \\ -\frac{1}{9} & -\frac{1}{9} & \frac{251}{4077} \end{array} \right] \right\},$$

$$\omega_{2,3} \in F_3(-1) = Span \left\{ \left[\begin{array}{cc|c} -\frac{1}{9} & -\frac{1}{9} & \frac{8783}{163080} \\ -\frac{5}{9} & \frac{4}{9} & \frac{29}{8154} \\ \frac{2}{3} & -\frac{1}{3} & -\frac{4493}{27180} \\ -\frac{1}{9} & -\frac{1}{9} & \frac{17711}{163080} \end{array} \right] \right\}.$$

In *table 1* are reported the results. Two eigenvalues are placed exactly, whereas on the remaining two the error is in the order of 10^{-5}.

4.2 *Example 2*

This second example is taken from (Alexandridis and Paraskevopoulos 1996). It represents the linearized model of the CCV-type aircraft which is analytically described in (Syrmos and Lewis 1993). The triplet (A, B, C) is the following

$$A = \begin{bmatrix} -1.3410 & 0.9933 & 0 & -0.1689 & -0.2518 \\ 43.223 & -0.8693 & 0 & -17.251 & -1.5766 \\ 1.3410 & 0.0067 & 0 & 0.1689 & 0.2518 \\ 0 & 0 & 0 & -20 & 0 \\ 0 & 0 & 0 & 0 & -20 \end{bmatrix},$$

$$B = \begin{bmatrix} 0 & 0 \\ 0 & 0 \\ 0 & 0 \\ 20 & 0 \\ 0 & 20 \end{bmatrix},$$

$$C = \begin{bmatrix} 0 & 0 & 1 & 0 & 0 \\ 47.76 & -0.268 & 0 & -4.56 & 4.45 \\ 0 & 0 & 0 & 1 & 0 \\ 0 & 0 & 0 & 0 & 1 \end{bmatrix}$$

Let $\Lambda = \{-20, -19.5, -1, -5.6 + \jmath 4.2, -5.6 - \jmath 4.2\}$ be the desired closed-loop eigenvalues that allow to satisfy the flight specifications.

We split the set Λ into two subset, namely $\Lambda_1 = \{-20, -19.5, -5.6 - \jmath 4.2, -5.6 + \jmath 4.2\}$ and $\Lambda_2 = \{-1\}$. We assign exactly the eigenvalues of Λ_1 solving the equations (10); the remaining eigenvalue is placed using the procedure described in *Step 4*. Consequently, the output feedback gain matrix K is

$$K = \begin{bmatrix} -101.787 & 0.390098 & 2.76459 & -17.2291 \\ -15.6485 & 0.0406955 & 0.248894 & -1.64292 \end{bmatrix}.$$

In *table 2* are reported the results. The absolute errore is on the order of 10^{-2} and only on the fast eigenvalues. In the numerical results, reported in (Alexandridis and Paraskevopoulos 1996), the error on the placement of the closed loop eigenvalues is on the more slow ones.

5. CONCLUSIONS

The pole assignment problem in LTI systems using output feedback has been considered in this paper. The approach here used is based on the geometrical properties of eigenspaces of the closed-loop dynamics.

Using the results of Proposition 1 and Proposition 4, it is possible to reduce the complexity of the procedure to solve the pole placement problem.

An algorithm is proposed to find the output gain matrix K. Such algorithm has been tested on two

Table 1. Results of the example 1

Desired eigenvalues	Computed eigenvalues
-4.0	−4.0
-1.0	−1.0
-1.0	−1.000000010750261
-1.0	−0.999994910893678

Table 2. Reults of the example 2

Desired eigenvalues	Computed eigenvalues
−20.0	−20.032
−19.5	−19.5
−1.0	−1.0
$-5.6 \pm \jmath 4.2$	$-5.6 \pm \jmath 4.2$

numerical examples, the second one is the same proposed in (Alexandridis and Paraskevopoulos 1996).

Moreover, the numerical algorithm seems to be simpler than those known from literature.

APPENDIX

Proof of Result 2:
Equations (4) can be rewritten as

$$f_{si}(K,\gamma_{11},\dots\gamma_{nn}) = 0, \quad i,s = 1,\dots,n.$$

where

$$f_{si}(K,\gamma_{11},\dots\gamma_{nn}) = \sum_{k=1}^{n}\gamma_{ki}v_s^T BKCu_k + \gamma_{si}(\pi_s - \lambda_i), \quad i,s = 1,\dots,n.$$

Let $F(\lambda) = (\lambda I - A)^{-1}B$, then the generalized eigenvectors $\omega_{i,1},\dots,\omega_{i,\mu_i}$ can be written as

$$\omega_{i,1} = \sum_{r=1}^{m}\vartheta(i)_{r1}\operatorname{col}[F(\lambda_i)]_r, \quad i = 1,\dots,l$$

$$\omega_{i,j} = \sum_{r=1}^{m}\vartheta(i)_{rj}\operatorname{col}[F(\lambda_i)]_r + \vartheta_{m+1j}\omega_{i,j-1}, \quad i = 1,\dots,l,\ j = 2,\dots,\mu_i.$$

Then, each generalized right eigenvector chain dependent on ν_i parameters

$$\nu_i = m\mu_i + (\mu_i - 1), \quad i = 1,\dots,l.$$

Then, Ω is a function of $\sum_{i=1}^{l}\nu_i$ parameters, and consequently the matrix Γ too.

This proves the corollary. □

6. REFERENCES

Alexandridis, A.T. and Paraskevopoulos P.N. (1996). A New Approach Assignment by Output Feedback. *IEEE Trans. on Automatic Control* **AC-41**, 1046-1050.

Champetier, C. and Magni J.F. (1991). On Eigenstructure Assignment by Gain Output Feedback. *SIAM J. Control Optim.* **Vol. 29 No. 4**, 848-865.

Davidson, E.J. and S.H. Wang (1975). On Pole Assignement in Linear Multivariable Systems using Output Feedback. *IEEE Trans. on Automatic Control* **AC-20**, 516-518.

Kimura, H. (1975). Pole Assignement by Gain Output Feedback. *IEEE Trans. on Automatic Control* **AC-20**, 509-516.

Kimura, H. (1977). A Further Results on the Problem of Pole Assignment by Output Feedback. *IEEE Trans. on Automatic Control* **AC-22**, 458-463.

Kwon, B.-H. and Wang M.-J. (1987). Eigenvalue-Generalized Eigenvector Assignment by Output Feedback. *IEEE Trans. on Automatic Control* **AC-32**, 417-421.

Syrmos, V.L. and Lewis F.L. (1993). Output Feedback Eigenstructure Assignment using two Sylvester Equations. *IEEE Trans. on Automatic Control* **AC-38**, 495-499.

Wilkinson, J.H. (1965). *The Algebraic Eigenvalue Problem.* Clarendon Press - Oxford.

14th Triennial World Congress, Beijing, P.R. China

C-2a-11-1

ROBUST OPTIMAL CONTROLLERS (MIXED H_2 / H_∞ PROBLEM) VIA GALLERKIN's METHOD

Marcos Azevedo da Silveira **Roberto Ades**

DEE / PUC-Rio, CP 30.630, 22452-900, Rio de Janeiro, RJ., Brazil
E-mail: marcos@ele.puc-rio.br rades@ele.puc-rio.br

Abstract: This paper presents a quadratic optimal controllers design under constraints of quantitative robustness (mixed H_2/H_∞ problem) using Gallerkin's method, where the appropriate space base is chosen on each step by optimizing a quadratic cost-functional related to the problem data. *Copyright © 1999 IFAC*

Keywords: robust control, optimal control, quadratic performance indices, mixed H_2/H_∞ problem, Gallerkin's method.

1. INTRODUCTION

The controllers design of linear systems solving the servomechanism problem under robustness constraints leads naturally to the minimization of quadratic functionals (H_2 problem) with H_∞ constraints. From the parametrization of stabilizing controllers by stable, proper, real-rationals functions, Youla et al. have shown how to solve the linear quadratic optimal control problem in the frequency domain for one and two parameter compensators and for general linear systems (Youla, et al., 1976; Youla, and Bongiorno, 1985; Park, and Bongiorno, 1989). It was called the Wiener-Hopf approach. This method was extended to the servomechanism problem, considering asymptotic tracking, as well as disturbance signal rejection of "known dynamics" (i.e. signals in the linear span of a given finite set of real-rational functions) (Silveira, and Corrêa, 1992) and related to the transient response of the controled system (Corrêa, and Silveira, 1995) by means of a detailed set-up of criteria to be minimized.

The search of robust controllers leads to a H_∞-norm minimization of certain transfer functions associated to the control loop (see (Doyle, et al., 1992) for example).

The conjunction of both problems, in such a way that the quadratic optimization problem is considered under robustness constraints, was used in several papers in the last decade, almost always replacing the original problem by the minimization of an upper norm bound well adapted, as in (Bernstein, and Haddad, 1989; Zhou, et al., 1994).

These modified problems, having the advantage of leading to mathematically tractable formulation, are based upon the intuitively plausible idea that minimizing these upper bound should also reduce the actual objective function. Unfortunately, this may not be the case (conservative solution).

The exact solution of the original problem has been hardly ever treated, in spite of interest in conserve the properties of quadratic criterion (relations with the transient behaviour), as discussed in (Corrêa, and Silveira, 1995).

In (Corrêa, et al., 1997) is presented an approach, without modifications on the criterion, where the H_∞ constraints are approximated by larger sets defined by H_2 functionals. This approach leads to a sequence of approximating problems, whose solutions converges to the original problem solution. Another direct approach is the Gallerkin's method, where the approximations are generated by the projection of solution over the beginning terms of a base, conveniently chosed in the space solution.

A pre-established orthonormal base in H_2, called Laguerre's functions, is used in (Sznaier, et al., 1997). Due to the fact that this basis do not consider nor the cost-functional nor the H_∞ constraint, it can leads to a large number of terms in the base to have a good approximation of the optimal solution.

The work (Scherer, 1993) is more ambicious, proposing the choice of a base by the optimization of a functional related to the Hankel-norm of solution. The new optimization problem is not convex and the proposed algorithm to solve it has only theoretical value.

The present paper propose the choice of each new basis vector by a previous minimization of a functional related to problem data and to the former vectors basis. The obtained optimization problem is not convex, but it shows sufficient properties to be solved by non convex and non differential optimization methods in Hilbert spaces.

In the next section, it will be shown the mixed H_2/H_∞ problem to be treated and its mathematical formulation, being proved the existence and uniqueness of its solution. In the third section, it will be discussed how can be constructed the basis functions from Runge's theorem by an optimization algorithm. In the fourth section, it will be shown an example of this method to achieve the optimal solution.

Notation: $C_+^o(C_-^o)$ denotes the set of complex numers with strictly positive (negative) real part; jR denotes the imaginary axis; L_2 denotes the usual $L_2(jR)$, i.e. the set of square-integrable functions on jR; $H_2^+(H_2^-)$ denotes the subspace of functions in L_2 with analytic continuation in $C_+^o(C_-^o)$; H_∞^+ (H_∞^-) denotes the subspace of functions in L_∞ with a bounded analytic continuation in $C_+^o(C_-^o)$. The inner-product and the norm in L_2 will be denoted by:

$$<F,G>_2 = \int_{-\infty}^{\infty} F^*(s)\, G(s)\Big|_{s=j\omega} d\omega \qquad (1)$$

$$\| F \|_2 = \left[<F,F>_2\right]^{\frac{1}{2}} \qquad (2)$$

where $F^*(s) = F^t(-s)$. For $F \in L_2$, $[F]_+([F]_-)$ denotes the orthogonal projection of F in H_2^+ (H_2^-). The H_∞-norm will be denoted by:

$$\| F \|_\infty = \sup_{\omega \in R} | F(j\omega) | \qquad (3)$$

For any set S of functions defined on C, $\Re S$ denotes the proper real-rational functions in S; for $\Gamma \in \Re S$, $\partial_r \Gamma$ is the relative degree of Γ. Γ is para-hermitian, if $\Gamma = \Gamma^*$. In this case, the Shaked extension for a known Youla's theorem says that $\Gamma = \Phi^*.\Phi$ (Wiener-Hopf factorization), where Φ is a proper real-rational function with poles and zeros in C_-^o or in jR. Obviously, $\partial_r \Gamma = 2\partial_r \Phi$ and Φ will have poles and zeros in jR if and only if Γ has them.

2. THE MIXED H_2/H_∞ PROBLEM

The problem itself is the minimization of a quadratic cost-functional given by:

$$J(K) = \int_{-\infty}^{\infty} [K^*(s)\, \Gamma(s)\, K(s) - 2\, K^*(s)\, \gamma(s)]\Big|_{s=j\omega} d\omega \quad (4)$$

under robustness constraints:

$$\|A(s)\, K(s)\, B(s) + C(s)\|_\infty \le \lambda \qquad (5)$$

where $\Gamma(s)$, $\gamma(s)$, A(s), B(s) and C(s) are propers real-rationals functions related with problem data, $\Gamma(s)$ para-hermitian and λ a real value such that:

$$\lambda \ge \lambda_0 = \inf_{K(s) \in \Re H_\infty^+} \|A(s)\, K(s)\, B(s) + C(s)\|_\infty \qquad (6)$$

and K(s) is real-rational, proper, stable function.

For the problem resolution, it will be necessary to extend the J domain from $\Re H_\infty^+$ to a well defined space.

Lemma 1: If $\Gamma \in \Re L_2$ is para-hermitian, without zeros on jR, $\partial_r \Gamma \ge 2$, then

$$<K,G>_\Gamma = \int_{-\infty}^{\infty} K^*(s)\, \Gamma(s)\, G(s)\Big|_{s=j\omega} d\omega \qquad (7)$$

define an inner-product in $\Re H_\infty^+$, with associated norm $\| K \|_\Gamma = \left[<K,K>_\Gamma\right]^{\frac{1}{2}}$.

The proof of lemma 1 is consequence of the integrability of $\Gamma = \Phi^*.\Phi$ and also by

$$< K, G >_\Gamma = < \Phi K, \Phi G >_2 \qquad (8)$$

The space $H^+_{2\Gamma}$ defined as the completeness of $\Re H^+_\infty$ by the norm $\| \cdot \|_\Gamma$ is a Hilbert space, with the same inner-product mentioned in lemma 1. This construction can be repeated to Γ with zeros in jR, when $H^+_{2\Gamma}$ will contain unstable functions with poles at the same place of Γ zeros.

Lemma 2: If $\gamma \in \Re L_2$, the functional $L(K) = \int_{-\infty}^{\infty} K^*(s)\gamma(s)\big|_{s=j\omega} d\omega$ is continuous in $H^+_{2\Gamma}$ if and only if $\gamma(s)$ doesn't have poles in jR and $\partial_r\gamma \geq \partial_r\Gamma$.

The proof of lemma 2 results from the fact that the linear functional $L(K) = < K, \gamma >_2$ will be continuous in $H^+_{2\Gamma}$ if and only if there is a function $\tilde{\gamma}$ in $H^+_{2\Gamma}$ such that $< K, \gamma >_2 = < K, \tilde{\gamma} >_\Gamma$ (Schwartz, 1970, p.296). However, $< K, \tilde{\gamma} >_\Gamma = < K, \Gamma\tilde{\gamma} >_2$ and then $\gamma = \Gamma\tilde{\gamma}$. So, $\tilde{\gamma} \in \Re L_\infty$ and $\partial_r\tilde{\gamma} \geq 0$, which ensure the conditions of lemma 2 on γ.

The functional J(K) can be rewrited in $H^+_{2\Gamma}$ as follows:

$$J(K) = \| K \|^2_\Gamma - 2\,L(K) \qquad (9)$$

i.e., a functional strictly convex which admits an unique minimum.

Lemma 3: Let the assumptions of lemmas 1 and 2 hold. The unique minimum of J(K) in $H^+_{2\Gamma}$ is given by $\hat{K} = \Phi^{-1}[(\Phi^*)^{-1}\gamma]_+$.

The proof of lemma 3 can be met in (Silveira, and Corrêa, 1992) by completing squares or in (Youla, et al., 1976) by variational arguments. Basically,

$$J(K) = < K, \Gamma K >_2 - 2 < K, \Gamma(\Gamma^{-1}\gamma) >_2 \qquad (10)$$

But $\tilde{\gamma} = \Gamma^{-1}\gamma = \Phi^{-1}\{(\Phi^*)^{-1}\gamma\}$, with $(\Phi^*)^{-1}\gamma \in L_2$ by hypothesis. Moreover,

$$(\Phi^*)^{-1}\gamma = \left[(\Phi^*)^{-1}\gamma\right]_+ + \left[(\Phi^*)^{-1}\gamma\right]_- \qquad (11)$$

Now, note that $< K, \Gamma\Phi^{-1}[(\Phi^*)^{-1}\gamma]_- >_2 = 0$. Rewriting the cost-functional J(K) as:

$$J(K) = < K, \Gamma K >_2 - 2 < K, \Gamma\Phi^{-1}[(\Phi^*)^{-1}\gamma]_+ >_2 \qquad (12)$$

and let $\hat{K} \overset{\Delta}{=} \Phi^{-1}[(\Phi^*)^{-1}\gamma]_+ \in H^+_{2\Gamma}$. It can be proved by completing squares in J(K), that:

$$J(K) = < K, \Gamma K >_2 - 2 < K, \Gamma\hat{K} >_2 + \\ + < \hat{K}, \Gamma\hat{K} >_2 - < \hat{K}, \Gamma\hat{K} >_2$$

$$J(K) = \| K - \hat{K} \|^2_\Gamma - \| \hat{K} \|^2_\Gamma \qquad (13)$$

Thus, J(K) admits unique minimum at $K = \hat{K}$.

Lemma 4: Suppose A(s), B(s), C(s) $\in \Re H^+_\infty$. The set Ω is convex and closed in $H^+_{2\Gamma}$, where $\Omega = \left\{ K \in H^+_{2\Gamma} : \| A(s)K(s)B(s) + C(s) \|_\infty \leq \lambda \right\}$.

To prove this lemma note first that Ω is a level set of convex functions (the composition of H^+_∞ norm with affine functions). It can be assumed that $C = 0$ without loss of generality. Now, let any $G \notin \Omega$. Then, there is an $\varepsilon > 0$ such that $\| AGB \|_\infty > \lambda + \varepsilon$. For any $K \in \Omega$:

$$\| A(G-K)B \|_\infty \geq \| AGB \|_\infty - \| AKB \|_\infty > \varepsilon \qquad (14)$$

Let $\| AB \|_\infty = M$, where M is finite because A, B are bounded on jR. Then, if $I_1 \subset jR$ is a set such that $|A[G-K]B\,(j\omega)| \geq \varepsilon$ for each $\omega \in I_1$ and if $I \subset I_1$ is a subset with measure $m(I) > 0$ and $|\Phi(j\omega)| > \theta > 0$ for each $\omega \in I$ (remember that $\Phi(s)$ has isolated zeros as a meromorphic function):

$$\begin{aligned} \| G-K \|^2_\Gamma &= \int_{-\infty}^{\infty} (G-K)^*\Phi^*\Phi(G-K)\big|_{s=j\omega} d\omega \\ &\geq \int_I (G-K)^*\Phi^*\Phi(G-K)\big|_{s=j\omega} d\omega \\ &> \theta^2 \int_I (G-K)^*(G-K)\big|_{s=j\omega} d\omega \\ &\geq \frac{\theta^2}{M^2} \int_I [A(G-K)B]^*[A(G-K)B]\big|_{s=j\omega} d\omega \\ &\geq \frac{\theta^2}{M^2}\varepsilon^2 m(I) = \delta > 0 \end{aligned} \qquad (15)$$

Therefore $\forall G \notin \Omega$, there exists a $H^+_{2\Gamma}$ neighborhood of G (the $H^+_{2\Gamma}$ ball centered in G with radius δ) without intersection in Ω, which implies the closure of Ω.

Theorem 1: Under the assumptions of lemmas 1, 2 and 4, the minimization problem of J(K) for $K \in \Omega \subset H^+_{2\Gamma}$ has an unique solution.

The proof can be obtained from theorem 1 in (Luenberger, 1969, p.69).

3. $H_{2\Gamma}^{+}$ BASIS AND THE GALLERKIN's METHOD

Riesz has shown that for all positive real numbers "a", the Laguerre's functions:

$$L_n = \sqrt{2a}\,(s-a)^{n-1}/(s+a)^n \ , \ n = 1,2,\ldots \qquad (16)$$

form a orthonormal basis of H_2^{+}. A right away generalization for $H_{2\Gamma}^{+}$, where $\varphi(s)$ denotes the denominator of $\Phi(s)$ (spectral factor of $\Gamma(s)$), is given by the basis:

$$\left\{ L_n^{\Gamma} \ , n = 0,1,\ldots \right\} \qquad (17)$$

where $L_n^{\Gamma} = \varphi^{*}(s)\ (s-a)^{(n-1)v}/(s+a)^{nv}$ for $n \geq 1$, $L_0^{\Gamma} = 1$ and v is the degree of $\varphi(s)$.

This basis is not suitable because its poles are stored at the same position ($s = -a$), resulting in the necessity of a large number of terms to approximate a given function.

Another basis for $H_{2\Gamma}^{+}$ with poles in different positions a_f ($a_f \in R$, $a_f > 0$, $f = 1,2,\ldots$) is given by $\left\{ M_n^{\Gamma} \ , n \geq 0 \right\}$, where:

$$M_0^{\Gamma} = 1, \ M_1^{\Gamma} = \frac{\varphi^{*}(s)}{(s+a_1)^{v}} \ \text{and} \qquad (18)$$

$$M_n^{\Gamma} = \frac{\varphi^{*}(s)}{(s+a_n)^{v}} \prod_{f=1}^{n-1} \frac{(s-a_f)^{v}}{(s+a_f)^{v}} \ \text{ for } n = 2,3,\ldots \qquad (19)$$

Unfortunately, the number of poles increases too fast, generating higher order solutions.

Non orthogonal basis can be obtained from Runge's theorem (Luecking, and Rubel, 1984), if we think that an expansion of an analytical function in C_{+}^{o} (in the topology of uniform convergence in all compact set) will converge in $H_{2\Gamma}^{+}$ if the limit is also in this set.

The Gallerkin's procedure consists in, given a basis for the space of solutions, project the problem in the subspace spanned by the first "n" vectors and solve the finite dimensional problem defined. In the present case, the problem must be adapted such that the approximated problem becomes the minimization of J(K) under the constraints:

$$K \in \Omega \ \text{ and } \ K = \sum_{m=0}^{n} \alpha_m F_m \qquad (20)$$

where the truncated basis is denoted by $\left\{ F_0, F_1, \ldots, F_n \right\}$. The new optimization problem, defined on the parameter set $\left\{ \alpha_0, \alpha_1, \ldots, \alpha_n \right\}$, remain strictly convex with a closed convex constraint. So, there exists a unique solution to the problem, which will be denoted by $\hat{K}_n$.

Lemma 5: If $\left\{ F_m, m = 0,1,\ldots \right\}$ is a basis for $H_{2\Gamma}^{+}$, the $\hat{K}_n$ sequence converges to the unique optimal solution in theorem 1.

A modification of Gallerkin's method is construct a $H_{2\Gamma}^{+}$ basis specially adapted to the problem, searching the next base function in accordance to the following procedure: once the first (n+1) functions $\left\{ 1, \frac{1}{s+a_1}, \frac{1}{s+a_2}, \ldots, \frac{1}{s+a_n} \right\}$ of the basis are known and the optimal solution $\hat{K}_{n+1}$ generated in the span of the above set, find the next basis function minimizing the functional:

$$J(\hat{K}_{n+1} + \alpha_{n+1} \frac{1}{s+a_{n+1}}) \qquad (21)$$

in relation to the couple (α_{n+1}, a_{n+1}). This procedure can be enlarged in order to consider complex conjugate poles.

The minimization problem in (α_{n+1}, a_{n+1}) is non convex, being similar to the optimization problem find in the "output error identification method". It is necessary use an heuristic well adapted, as the ones which will be treated in next section.

4. NUMERICAL EXAMPLE

Consider the block-diagram shown in figure 1.

$$\begin{bmatrix} y \\ z \end{bmatrix} = \begin{bmatrix} P_1 & \overline{P}_1 \\ P_2 & \overline{P}_2 \end{bmatrix} \begin{bmatrix} u \\ d \end{bmatrix} \ \text{ and } \ u = C \begin{bmatrix} r \\ z - v \end{bmatrix} + w \qquad (22)$$

where $C = \left[C_1 \ \vdots \ -C_2 \right]$ denotes a controller; r, u, y, z, d, w, v denote, respectively, the reference signal, plant input, controlled output, measurement signal, a general disturbance, input disturbance and sensor noise. In this case, the plant / sensor model will be given by:

$$P_1(s) = P_2(s) = \frac{s+2}{(s+1)(s-3)} \text{ and } \overline{P}_1 = \overline{P}_2 = 1 \quad (23)$$

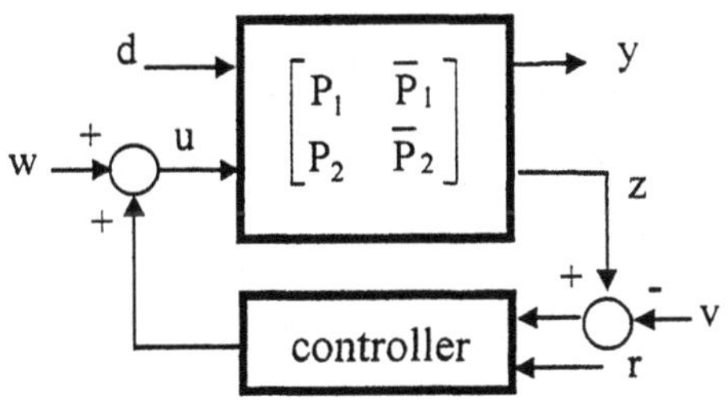

Fig. 1: Closed loop system.

Our aim in this problem is reject the asymptotic effect of disturbance and noise in y and u from d and v. Moreover, it is desired that the optimal stability margin to additive perturbations has a maximum degradation of 10%. With regard to this aim, we define the quadratic cost-functional (performance criterion) as follows:

$$J = \|(\overline{P}_1 - P_1 Q_2 \overline{P}_2)\phi_d\|_2^2 + \|Q_2 \overline{P}_2 \phi_d\|_2^2 + \\ + \|P_1 Q_2 \phi_v\|_2^2 + \|Q_2 \phi_v\|_2^2 \quad (24)$$

where Q_2 is the Youla parameter ($Q_2(K)$ parametrizes all stabilizing controllers for $K \in \Re H_\infty^+$, $C_2 = (I - Q_2 P_2)^{-1} Q_2$) and ϕ_d, ϕ_v can be seen as ponderation filters.

Indeed, $\phi_d = \frac{1}{s+1}$, $\phi_v = \frac{1}{s+2}$ and

$$Q_2 = \frac{16}{5}\frac{(s-3)}{(s+1)^2} - \frac{(s-3)^2}{(s+1)^2} K \quad (25)$$

From (Sales, 1994) the optimal stability margin for this problem is $\|Q_2(K_{H\infty})\|_\infty = 4.8094$. Therefore, relaxing 10% of the optimal margin as discussed above: $\lambda = 4.8094 \times 1.1 \approx 5.2903$. The proposed problem will be:

$$\min_{K \in \Re H_\infty^+} J(K) \quad (26)$$
$$\text{under constraint } \|Q_2(K)\|_\infty \le \lambda = 5.2903$$

The approximate solution for the above problem given in (Corrêa, et al., 1997; Sales, 1994) is a $K_{ml} \in \Re H_\infty^+$ such that:

$$J(K_{ml}) = 16.8480 \text{ and } \|Q_2(K_{ml})\|_\infty = 5.2903 \quad (27)$$

Further, a lower bound for J was supplied as J_{lb} = 16.5697 by the method in (Corrêa, et al, 1997).

Using the technique described in the last section, extended for complex conjugate poles (by adding terms as $\alpha_1 + \frac{\alpha_2 s + \alpha_3}{s^2 + a_1 s + a_2}$), the following results were obtained:

$$K_0 = 0 \quad (28)$$

$$K_1 = -0.74111 + \frac{-1649.46\,s + 174.62}{s^2 + 1235.2\,s + 3341.6} \quad (29)$$

$$K_2 = 0.01289 + \frac{-0.08531\,s - 0.1392}{s^2 + 1.05252\,s + 4.09519} \quad (30)$$

Table 1: Obtained results.

K=	J(K)	$\|Q_2(K)\|_\infty$
K_0	43.96667	9.59999
$\sum_{m=0}^{1} K_m$	16.70000	5.30085
$\sum_{m=0}^{2} K_m$	16.63607	5.29025

Remark 1: Reduction model techniques based on Hankel norm can be applied to reduce the order of the approximate solutions and consequently redefine the truncated basis.

Remark 2: In spite of algorithm obtains a local optimal for the non-convex optimization problem, by the use of lemma 5, it is possible to show that the obtained solutions go to the optimal solution when $n \to \infty$.

Remark 3: To solve the non convex problem, it has been used the Powell non-differential optimization method (Fox, 1971) applied to a new cost-functional, formed adding the constraint as follows:

$$J_\pi = J + \pi . \max(0, \|Q_2\|_\infty - \lambda) \quad (31)$$

where max(. , .) is the maximum value between the arguments, $0 < \pi \in R$.

Remark 4: Basis with multiple poles at the same position, as Laguerre's functions, become numerically ill-conditioned faster than the proposed basis. However, the problem become convex if the poles are pre-established.

5. CONCLUSIONS

In this work, a numerical method to approximately solve the original mixed H_2/H_∞ problem has been derived. The employed method is based on optimization and basis expansion. It has been presented a method to build a optimized basis, adapted to the problem data, such that the solution were formed by a sequence of finite dimensional approximating problems. Moreover, in comparison with other basis as Laguerre's functions (Scherer, 1993; Sznaier, et al., 1997), the convergence was obtained with fewer terms. The disadvantage of this method is the non convexity of the approximating problems. It has been noted that in higher order problems can occur numerical ill-conditioning related to pole-zero cancellation. The next step in our research will be explore new optimization methods more adapted to the studied problem, taking advantage of its special structure.

6. REFERENCES

Bernstein, D. S. and W.N. Haddad (1989). LQG control with an H_∞ performance bound: a Riccati equation approach. *IEEE Transactions on Automatic Control*, **34**, 293-305.

Corrêa, G.O., D.M. Sales and T.M. Soares (1997). Approximate solutions to H_2/H_∞ problems via sequences of H_2-cost/ H_2-constraint optimization problems. *International Journal of Control*, **67**, 333-353.

Corrêa, G.O. and M.A. Silveira (1995). On the design of servomechanisms via H_2-optimization. *International Journal of* Control, **61**, 475-491.

Doyle, J.C., B.A. Francis and A.R. Tannenbaum (1992). *Feedback control theory*. Macmillan Publishing Company.

Fox, R.L. (1971). *Optimization Methods for Engineering Design*, Addison-Wesley, California.

Luecking, D.H. and L.A. Rubel (1984). *Complex analysis - a functional analysis approach.* Springer-Verlag.

Luenberger, D.G. (1969). *Optimization by vector space methods*. John Wiley, NY.

Park, K. and J. Bongiorno (1989). A general theory for the Wiener-Hopf design of multivariable control systems. *IEEE Transactions on Automatic Control,* **34**, 619-626.

Sales, D.M. (1994). *Mixed H_2/H_∞ control - approximate solutions based in sequence of quadratic problems*. M. Sc. Thesis, IME, RJ.

Schwartz, L. (1970). *Analyse*. Herman, Paris.

Scherer, C.W. (1993). Multiobjective H_2/H_∞ control In: *Selected topics in identification, modelling and control*. Vol. 6. Delft University Press.

Silveira, M.A. and G.O. Corrêa (1992). H_2-optimal control of linear systems with tracking / disturbance rejection constraints. *International Journal of Control*, **55**, 1115-1139.

Sznaier, M., H. Rotstein, J. Bu and A. Sideris (1997). An exact solution to continuous-time mixed H_2/H_∞ control problems. *AACC*.

Youla, D., A.H. Jabr and J. Bongiorno (1976). Modern Wiener-Hopf design of optimal controllers. Part II: The multivariable case. *IEEE Transactions on Automatic Control*, **21**, 319-338.

Youla, D. and J. Bongiorno (1985). A feedback theory of two-degree-of-freedom optimal Wiener-Hopf design. *IEEE Transactions on Automatic Control*, **30**, 652-665.

Zhou K., K. Glover, B. Bodenheimer and J.C. Doyle (1994). Mixed H_2 and H_∞ performance objectives I: robust performance analysis. *IEEE Transactions on Automatic Control,* **39**.

14th Triennial World Congress, Beijing, P.R. China

C-2a-11-2

ADDITIVE FEEDFORWARD CONTROL WITH NEURAL NETWORKS

OLE SØRENSEN

Aalborg University
Inst. of Electronic Systems, Dept. of Control Engineering
Fredrik Bajers vej 7C, DK-9220 Aalborg Ø, Denmark
phone +45 96 35 87 48, telefax +45 98 15 17 39
E-mail os@control.auc.dk

Abstract: This paper demonstrates a method to control a nonlinear, multivariable, noisy process using trained neural networks. The basis for the method is a trained neural network controller acting as the inverse process model. A training method for obtaining such an inverse process model is applied. A suitable 'shaped' (low-pass filtered) reference is used to overcome problems with excessive control action when using a controller acting as the inverse process model. The control concept is Additive Feedforward Control, where the trained neural network controller, acting as the inverse process model, is placed in a supplementary pure feedforward path to an existing feedback controller. This concept benefits from the fact, that an existing, traditional designed, feedback controller can be retained without any modifications, and after training the connection of the neural network feedforward controller to the feedback controlled process may happen gradually and controlled. *Copyright © 1999 IFAC*

Keywords: Neural Network, NARMAX, Nonlinear Control, Feedforward Control.

1. INTRODUCTION

When the goal is to control a process, P, it is tempting to try to find a controller, C, which represents the inverse model of the process, $C = P^{-1}$. The serial connection of the controller and the process produces in this way a unity gain transfer function from the input of the controller, which is the reference (the desired output from the process), to the output from the process.

This method implies that three problems arise.

- The first problem concerns stability of the controller itself, which is accomplished, when limiting oneself to consider processes, which are of minimum phase type.
- The second problem concerns causality of the controller, which can be accomplished, when the reference signal is known temporal in advance.
- The third problem is to avoid excessive control action, which is accomplished by a suitable low-pass filtering of the reference signal.

In (Hunt *et al.*, 1992) and (Hunt and Sbarbaro, 1991) different control concepts based on a trained neural network, representing the inverse process model, are mentioned, e.g. Direct Inverse Control and Internal Model Control.

In this paper, however, a trained neural network, representing the inverse process model, is applied in a control structure including an additive feedforward control to the existing feedback control, the principle of which is shown in fig.1.

The starting point of this concept is an *existing*, but not satisfactory functioning, feedback controller. This offers the following advantages

- Collection of data for the training session of the neural network feedforward controller is

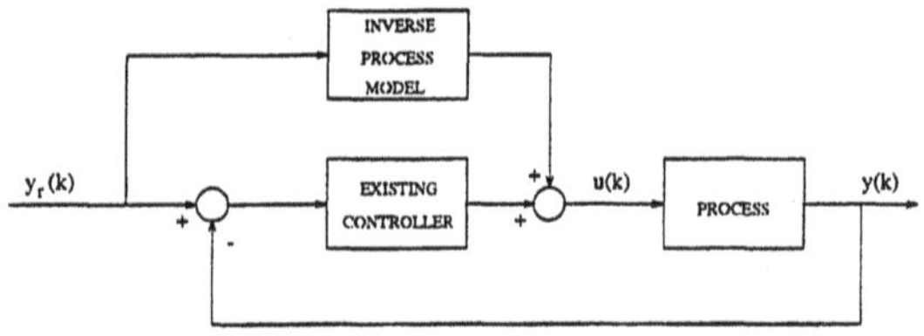

Fig. 1. *The principle of additive feedforward control*

performed from the existing closed loop, thus facilitating the problem of collecting data which are sufficiently representative for the process.

- During training, as well as during control, there is no need for opening the existing control loop. Also, there is no need for retuning the existing feedback controller. Furthermore, all the equipment for supervision and security can be retained.
- After training, the connection of the neural network feedforward controller to the controlled process may happen gradually and controlled.

Section 2 describes, in principle, how to train a neural network controller acting as the inverse process model. Section 3 describes the process, which is a nonlinear, multivariable and noisy thermal mixing process. In section 4 a neural network forward process model is trained and evaluated, and in section 5 a neural network inverse process model for control is found. Finally, in section 6 the trained neural network inverse process model is placed in a supplementary pure feedforward path to the existing traditional controller and a test confirms the improvements of the existing control performance.

2. GENERAL AND SPECIALIZED TRAINING

Fig.2 illustrates two different architectures for training a neural network to represent the inverse of the process. The two methods are named *general* and *specialized* training, respectively. The weights in the neural network are updated during training, based on the error E_u in general training, and on the error E_c in specialized training.

The two methods are described by many researchers applying neural networks for control, e.g. Psaltis and others (Psaltis *et al.*, 1988), K.J.Hunt and D.Scarbano (Hunt and Sbarbaro, 1992) and K.J.Hunt and others (Hunt *et al.*, 1992).

In *general training*, a process input U is selected and applied to the process to obtain a corresponding output Y, and the network is trained, based on the error E_u, to reproduce U at its output from Y. If the training succeeds, the network should then be able to take a desired response Y_r as

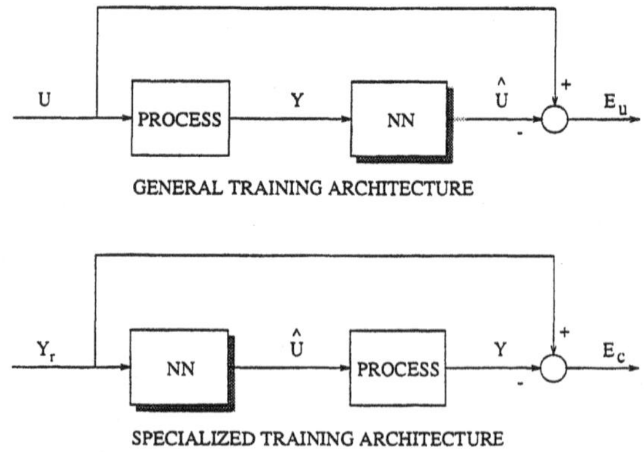

Fig. 2. *General and specialized training architecture*

input and produce the appropriate $\hat{U}$, making the actual process output Y approach Y_r. This method is appealing because the error E_u, used for training, is generated directly on the output of the neural network. Unfortunately the method has two drawbacks. The first drawback is that the method is not 'goal directed', which means that choosing different values of U for training does not ensure that Y is trained in those regions, which are of particular interest for succeeding control application. This is strongly related to the general concept of *persistent excitation*. The second drawback is that an incorrect inverse can be obtained, if the non-linear process mapping is not one-one.

In *specialized training*, the desired output Y_r is used as input to the neural network and the network is trained to find the process input $\hat{U}$, which drives the process output Y to the desired output Y_r. This is accomplished by using the error E_c between the desired and the actual output of the process to update the weights in the network. This method is appealing, because the architecture allows training of the network in those regions for Y, which are of particular interest for succeeding control application, thus ensuring persistent excitation, but also because a particular inverse model will be found, even if the non-linear process mapping is not one-one. The only drawback is that the training becomes more complicated, since the error E_c, measured on the output side of the process, has to be transformed into an equivalent error on the input side of the process, in order to be applied to training the neural network. Transformation of the error E_c, measured on the output side of the process, to an equivalent error on the input side of the process, requires knowledge to the Jacobian of the process. However, the actual value of this Jacobian is easily calculated from another trained, *forward*, neural network model, substituting the process in the specialized training architecture.

Psaltis and others (Psaltis *et al.*, 1988) propose a combination of the two training methods, first performing general training to learn the approximate behavior of the process, followed by specialized training to fine-tune the network in the operating regions of the process.

To avoid excessive control action the desired output Y_r is produced by filtering a reference R with a suitable 'shaping' filter F

$$Y_r(q^{-1}) = F(q^{-1})\ R(q^{-1}) \qquad (1)$$

where q denotes the time shift operator.

3. THE NONLINEAR MULTIVARIABLE PROCESS

The simulated MIMO process is a thermal mixing process, containing several typical practical difficulties. The process is characterized by

- it is dynamic.
- it is multivariable, MIMO.
- it contains measurement noise.
- it is nonlinear.
- it contains a non-measurable state.
- it contains external disturbances.
- the actuators saturate.
- a transducer has a considerable time constant.
- the outputs are mutually coupling.

3.1 *Description of the process*

The process is sketched in fig.3. It is a thermal mixing process. The purpose of the process is to supply water to a user at a specified temperature and water level, regardless of the actual amount of water tapped from the V-shaped tank.

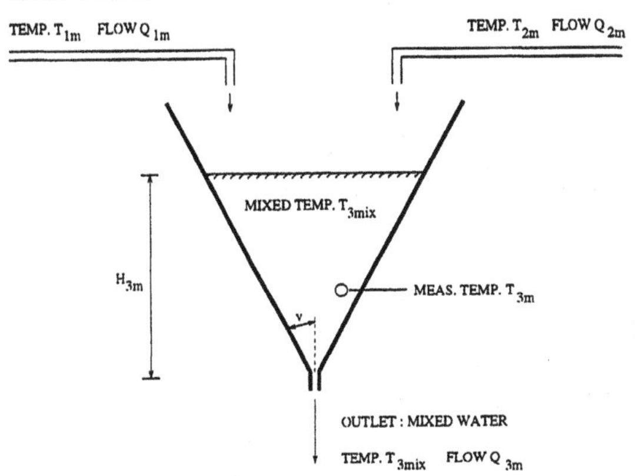

Fig. 3. *A draft of the thermal mixing process*

The inlet flows are the cold water supply Q_{1m} with temperature T_{1m}, and the hot water supply Q_{2m} with temperature T_{2m}. The disturbing outlet flow is Q_{3m} with the mix temperature T_{3mix}. The water level H_{3m} is measured by a fast transducer, not shown in the figure, but unfortunately the mix temperature T_{3mix} is measured by a relatively slow transducer with a time constant τ, where $\tau = 10\ sec$. This measured temperature is denoted T_{3m}. The angle, determining the volume of the tank, is denoted v, where $v = \pi/10\ rad$.

It is assumed that the valves are controlled by fast local flow feedback loops, and further, a pump arrangement is introduced to ensure good mixing in the tank.

Physical constrains imply, that the valves saturate downward at 0 m^3/sec and upward at 0.0005 m^3/sec, and that the water level saturates downward at 0.05 m and upward at 1.00 m.

The sampling rate is chosen to $T_s = 5\ sec$.

The measurement noise is normal distributed, white noise with a standard deviation 0.05 cm for H_{3m} and 0.05 C for T_{3m}.

The vector signals for the process are naturally organized as follows

$$\begin{aligned} &Input \quad U_m(k) &= \{Q_{1m}(k), Q_{2m}(k)\}^T \\ &Disturb.\ D_m(k) &= \{Q_{3m}(k), T_{1m}(k), T_{2m}(k)\}^T \\ &Output \quad Y_m(k) &= \{H_{3m}(k), T_{3m}(k)\}^T \end{aligned}$$

3.2 *How to collect training data*

In traditional linear system identification it is well known, that the *frequency* spectrum of the input signal must be sufficiently wide in order to identify the process. When using a neural network to make a nonlinear system identification, this demand is valid too, but a further demand must now be added, namely that the *amplitude* spectrum of the input signal must be sufficiently wide, i.e. it is necessary, that the network is trained with various amplitudes in order to learn the nonlinearities of the process.

Regarding the thermal mixing process, it is very difficult, in advance, to determine how to manipulate the two inlet valves, in such a way that the water level, as well as the temperature, contain sufficiently varied amplitudes, without frequent saturation of the valves, or without completely filling or emptying the tank. As a consequence, the data forming the training and test sets are collected in a closed loop, controlled by two traditional PI-controllers. After training, the neural network model is tested by placing the network in the closed loop instead of the process.

Fig.4 shows the arrangement for collecting data. It is arbitrarily chosen to let the flow of the cold water control the water level, and to let the flow of the hot water control the temperature, thus neglecting the cross couplings in this multivariable

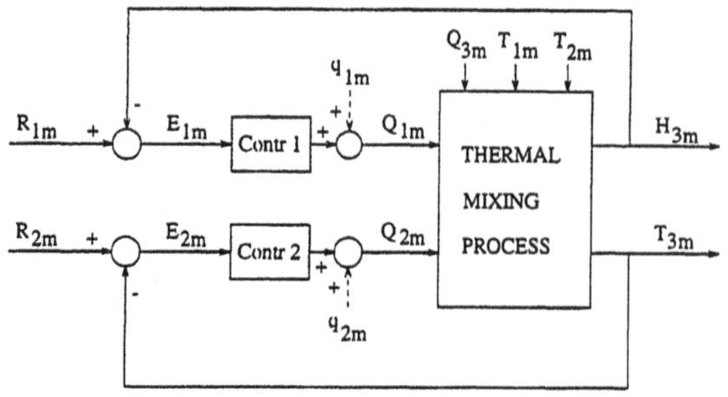

Fig. 4. *The arrangement around the thermal mixing process of two individually tuned control loops to collect data for the training and test set*

process. Using this concept the two discrete time PI-controllers are tuned individually

$$\begin{aligned} Q_{1m}(q) &= 0.002\left(1+0.05\frac{q^{-1}}{1-q^{-1}}\right)E_{1m}(q) \\ Q_{2m}(q) &= 0.00002\left(1+0.05\frac{q^{-1}}{1-q^{-1}}\right)E_{2m}(q) \end{aligned} \qquad (2)$$

For the training set the purpose was to obtain *combinations and variations* in order to produce a sufficiently rich input, concerning frequency and amplitude. Therefore the references were frequently changed. Another reason for frequently changing the references is to prevent, that the input correlates with the output caused by the closed loop. This is always a problem, when identification is performed based on data collected in a closed loop. Here, this problem is avoided simply by adding a small random signal to the input, q_{1m} and q_{2m} in fig.4. Both q_{1m} and q_{2m} are chosen to Pseudo Random Binary Sequences (PRBS), with heights between ± 0.05 l/sec, and with widths between 5 and 10 samples.

The purpose of the data collection arrangement in fig.4 is to obtain *persistent excitation*, where the exciting signals, q_{1m} and q_{2m}, are 'carried' around in the relevant operating area by the controller outputs.

The collected training set is not shown here but the test set is shown in fig.5, and the hidden couplings and nonlinearities of the multivariable process are clearly observed in fig.5.

For a physical process the individual elements of the input and output vectors are measured in physical units, which often are of quite different orders of magnitude. For this reason a scaling is performed. It is here chosen to scale all individual inputs and desired outputs in such a way that, based upon the training set, the mean value becomes zero and the standard deviation becomes one. In the rest of this paper all measured signals are given a subscript 'm', contrary to the scaled signals.

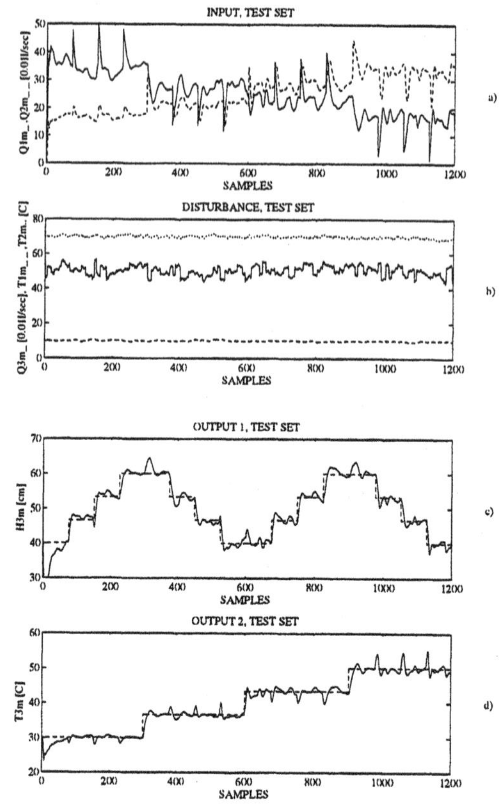

Fig. 5. *Test set collected from the thermal mixing process.*

4. FORWARD NEURAL NETWORK PROCESS MODEL

With U denoting the scaled input vector, Y the scaled output vector and D the scaled disturbance vector, a second order NARMAX (Nonlinear ARMAX) model for the multivariable nonlinear thermal mixing process is introduced

$$\begin{aligned} \hat{Y}(k) &= \mathcal{F}\,(Y(k-1), Y(k-2), U(k-1), U(k-2), \\ &\qquad D(k-1), E(k-1), E(k-2), \theta) \\ Y(k) &= \hat{Y}(k) + E(k) \end{aligned} \qquad (3)$$

where E is the scaled prediction error vector.

In (Ljung, 1987) the ARMAX model is described for linear systems, while (Billings *et al.*, 1990) and (Chen *et al.*, 1990) describe NARX and NARMAX models implemented with neural networks.

The neural network configuration for this model is shown in fig.6.

All the vectors U, Y and E have two elements, while D has three elements, so the total number of inputs to the network is 15, and the number of outputs from the network is 2. The number of hidden neurons is chosen to 10.

After having trained the network with the Back Propagation Error Algorithm (BPEA), with the gradient factor η gradually decreasing from 0.02

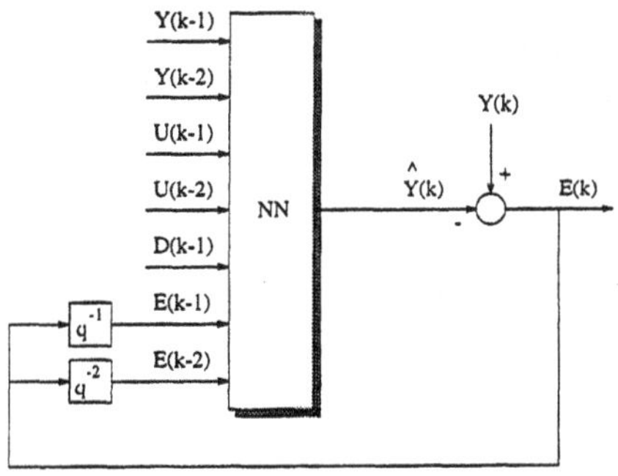

Fig. 6. *Nonlinear ARMAX model (NARMAX) for the thermal mixing process*

to 0.001, examinations of the whiteness of the prediction error plus some relevant correlation analysis showed that the neural network had the potential to form a forward prediction model of the thermal mixing process.

5. INVERSE NEURAL NETWORK PROCESS MODEL

Assume, that the thermal mixing process is described by a forward second order NARX model.

$$\hat{Y}(k+1) = \mathcal{F}(Y(k), Y(k-1), U(k), U(k-1), D(k))$$
$$Y(k+1) = \hat{Y}(k+1) + E(k+1) \tag{4}$$

where $\mathcal{F}$ is a nonlinear vector function, $D(k)$ is the disturbance and $E(k)$ is the output prediction error.

In K.J. Hunt and D. Sbarbaro (Hunt and Sbarbaro, 1991) it is proofed that, if $\mathcal{F}$ is monotonic with respect to $U(k)$, then (4) is invertible, and the desired inverse process model is

$$\hat{U}(k) = \mathcal{G}(Y(k+1), Y(k), Y(k-1), U(k-1), D(k))$$
$$U(k) = \hat{U}(k) + E_u(k) \tag{5}$$

where $\mathcal{G}$ is a nonlinear vector function, $D(k)$ is the disturbance and $E_u(k)$ is the input prediction error.

In fact, this is true only for a process containing no further delay, i.e. $\frac{\partial \hat{Y}^T(k+1)}{\partial U(k)} \neq 0$. In order to handle also processes containing an extra delay, and also in order to obtain a more smoothed control action, it is preferred to assume, that the output is known two samples in advance, giving

$$\hat{U}(k) = \mathcal{G}(Y(k+2), Y(k+1), Y(k), Y(k-1), U(k-1), D(k))$$
$$U(k) = \hat{U}(k) + E_u(k) \tag{6}$$

The neural network configuration for general training is shown in fig.7, and the training is performed by BPEA.

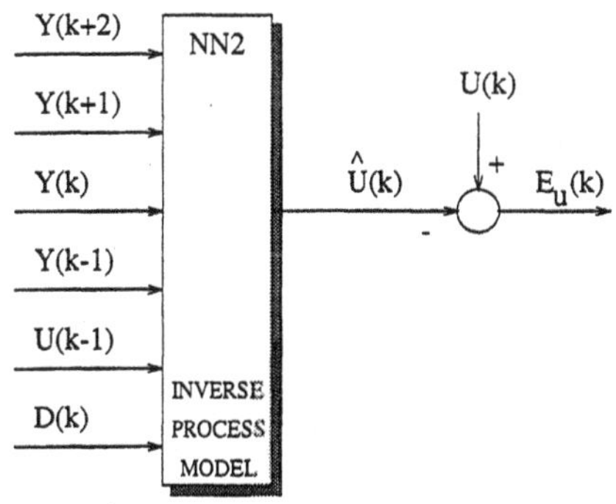

Fig. 7. *General training of the inverse process model NN2*

Based on (6) a neural network can now be trained to act as the inverse process model for controlling the process. The *general* training concept is first applied, using the sequences of the input U, disturbance D and output Y from the *training* set in section 3.2 and afterwards the *specialized* training is continued to fine-tune the controller in all relevant operating areas.

6. ADDITIVE PURE FEEDFORWARD CONTROL

The trained inverse process model (6) is now applied as a 'pure' feed-forward controller

$$U_f(k) = \mathcal{G}(Y_r(k+2), Y_r(k+1), Y_r(k), Y_r(k-1), U_f(k-1), D(k)) \tag{7}$$

If the training succeeded, this a minor approximation, since $Y \sim Y_r$.

The implementation of the neural network based additive pure feed-forward control is shown in fig.8.

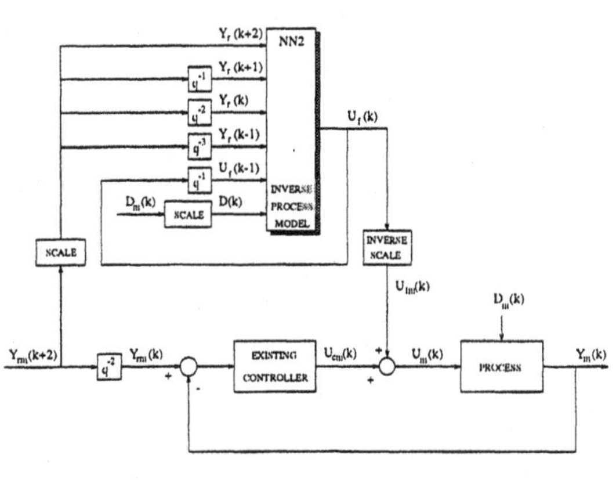

Fig. 8. *Implementation of pure additive feedforward control*

The 'shaping' reference filter has a DC-gain of one and a double-pole in 0.9.

$$\frac{Y_r(q^{-1})}{R(q^{-1})} = \frac{0.01}{1 - 1.80q^{-1} + 0.81q^{-2}} \qquad (8)$$

When applying the reference and disturbance sequences for the *test* set from section 3.2, the result of the control action is shown in fig.9. Sub-figures a and b show the control signals and the disturbances to the process, while sub-figures c and d show the controlled outputs, the water level and the temperature.

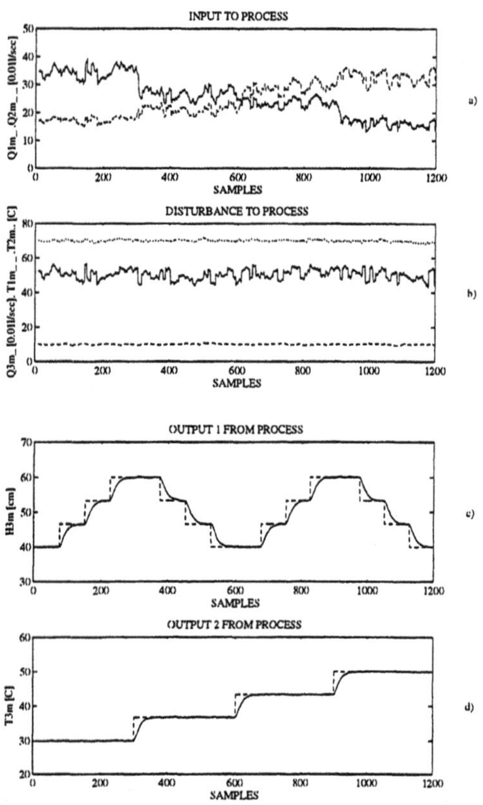

Fig. 9. *Additive pure feed-forward control, inputs to and outputs from the process*

It is obvious, that the training has produced an additive controller, which highly improves the performance of the existing controller, compare the outputs in fig.9 with the outputs in fig.5. Now, the outputs track the 'shaped' references, and the disturbances are well suppressed.

The resulting control signal U_m is composed of two contributions, U_{cm} from the existing feedback controller and U_{fm} from the additive neural network feedforward controller. Depending on the success of the training session, the feedforward controller takes over a major part of the resulting control signal.

This fact is illustrated in fig.10, which shows that the components of the control signal from the additive feedforward controller are totally dominating, while the components of the control signals

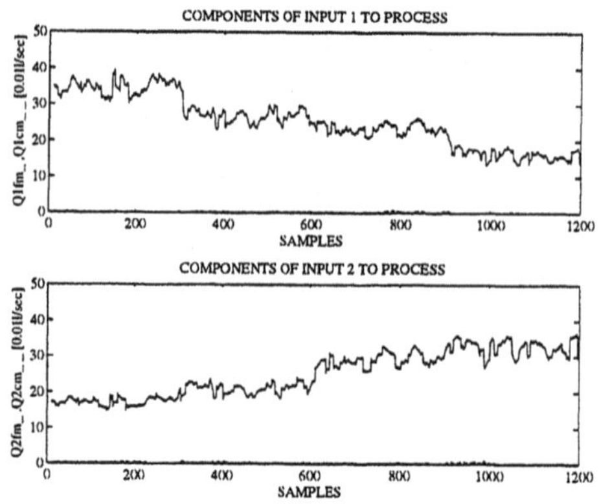

Fig. 10. *Additive pure feedforward control, the components of the control signal*

from the existing controller are very small and hardly visible in the figure, indicating a successful training.

7. CONCLUSION

This paper has demonstrated a method to control a nonlinear, multivariable and noisy process using an additive feedforward control signal from a trained neural network controller, while an existing, traditional designed, feedback controller can be retained without any modifications.

8. REFERENCES

Billings, S. A. et al. (1990). Properties of neural networks with application to modelling non-linear dynamical systems. *Int. J. Control* **1**, 193–224.

Chen, S. et al. (1990). Non-linear system identification using neural networks. *Int. J. Control* **6**, 1191–1214.

Hunt, K. J et al. (1992). Neural networks for control systems - a survey. *Automatica* **6**, 1083–1120.

Hunt, K.J. and D. Sbarbaro (1991). Neural networks for nonlinear internal model control. *IEE Proceedings-D, Control theory and applications* **138**(5), 431–438.

Hunt, K.J. and D. Sbarbaro (1992). Studies in neural network based control. In: *Neural networks for control and systems*. pp. 94–122. Peter Peregrinus Ltd.,London, United Kingdom.

Ljung, Lennart (1987). *System identification, theory for the user*. first ed.. Prentice-Hall.

Psaltis, Demetri et al. (1988). A multilayererd neural network controller. *IEEE Control System Magazine* **8**(3), 17–21.

14th Triennial World Congress, Beijing, P.R. China

C-2a-11-3

ROBUST CONTINUOUS-TIME TRACKING FOR MULTI-RATE DIGITAL CONTROL SYSTEMS [1]

Osvaldo Maria Grasselli* Laura Menini* Paolo Valigi**

* *Dipartimento di Informatica, Sistemi e Produzione*
Università di Roma Tor Vergata
via di Tor Vergata 110, 00133 Roma, Italy
Fax. +39 06 72597430 Tel. +39 06 72597430.
** *Università di Perugia - Istituto di Elettronica*
Via G. Duranti, 93, I-06131 Perugia, Italy
Fax. +39 075 585 2654 Tel. +39 075 585 2684.
e-mail: {grasselli, menini, valigi}@disp.uniroma2.it

Abstract: Rhe robust ripple-free tracking and regulation problem is solved for multirate sampled-data systems, whose state-space description depends on some unknown "physical" parameters. The used control system includes a periodic discrete-time subcompensator and, possibly, a continuous-time subcompensator. The conditions for the existence of a solution are found for two different convergence requirements: a prescribed rate of continuous-time exponential convergence of the error response and the free state response (to be obtained at least in a neighborhood of the nominal parameters), or a continuous-time dead-beat error and free state response at the nominal parameters (together with a continuous-time exponential convergence with any rate of convergence in some neighborhood of the nominal parameters). For both of them, the necessity of a wholly continuous-time internal model of the exogenous signals is proven, and a design procedure is proposed. *Copyright 1999 IFAC*

Keywords: Tracking systems, output regulation, robustness, multirate sampled-data control, multivariable feedback control systems.

1. INTRODUCTION

In view of its great importance, the problem of robust tracking and regulation has been widely investigated both for linear continuous-time systems and for discrete-time ones (see, e.g., (Francis and Wonham, 1975; Davison, 1976; Francis and Wonham, 1976; Francis, 1977; Kucera and Sebek, 1984; Grasselli and Longhi, 1991*c*; Grasselli *et al.*, 1993), and the references therein). If for such a problem a digital control system is used for a continuous-time plant, special care has to be devoted in order to avoid undesirable ripple between sampling instants, which may become unacceptable if the sampling rates are small, so that several contributions appeared about this point, especially for single-rate sampled-data systems (Franklin and Emami-Naeini,1986; Urikura and Nagata, 1987; Yamamoto, 1994; Sanchis and Albertos, 1995; Grasselli *et al.*, 1996, 1997).

For multirate sampled-data systems the tracking problem was studied in (Grasselli *et al.*, 1995), showing that a ripple-free dead-beat convergence of the error response of the closed-loop control system can be achieved if a complete internal model of the class of allowed reference signals is contained in the continuous-time plant. In (Scattolini and Schiavoni, 1993) square plants, subject to parameter perturbations, with multirate output sam-

[1] This work was supported by MURST.

plers and a single-rate input updating mechanism, were considered. The robust tracking and regulation problem for parameter dependent multirate sampled-data systems, was studied in (Grasselli *et al.*, 1994), for (possibly, nonsquare) plants satisfying the Davison condition (Davison, 1976) at the nominal parameters; a continuous-time internal model of the exogenous signals was used as a part of the hybrid control system, as in (Grasselli *et al.*, 1996).

Because of the significance of multirate sampled-data control systems (see e.g., (Dahleh *et al.*, 1992; Grasselli *et al.*, 1995) and the references therein), in this paper the same kind of problem is considered for the case when the Davison condition is not assumed to hold. Namely, a multirate sampled-data hybrid control system is used for obtaining a robust ripple-free asymptotic tracking and disturbance rejection for exponential-sinusoidal reference signals and disturbance functions affecting the (possibly non-square) continuous-time plant to be controlled, which has a state space description whose matrices depend in a known way on some "physical" parameters (as in (Grasselli and Longhi, 1991*c*; Grasselli *et al.*, 1993, 1996)) and, possibly, do not satisfy the Davison condition.

In Section 2, for the multirate control scheme of interest, two different robust tracking and regulation problems are formulated, corresponding to two different convergence requirements (namely that of exponential stability with a prescribed rate of decay and that of dead-beat convergence of the free state response of the closed-loop control system at the nominal values of the parameters) and the necessity of a wholly continuous-time internal model of the exogenous signals is proven. In Section 3 necessary and sufficient conditions for the existence of hybrid control systems that solve the two proposed multirate control problems are given. Such control systems include multirate samplers and holders, a continuous-time subcompensator, whose role is to enlarge the continuous-time internal model of reference signals and disturbance functions possibly contained in the plant, and a discrete-time periodic subcompensator, whose role is to satisfy either of the two convergence requirements. A procedure is proposed for the design of the continuous-time and discrete-time subcompensators.

2. PRELIMINARIES

Consider the linear time-invariant plant P described by

$$\dot{x}(t)=A(\beta)x(t)+B(\beta)u(t)+\sum_{i=1}^{\mu}M_i(\beta)d_i(t), \quad (1a)$$
$$y(t)=C(\beta)x(t)+\sum_{i=1}^{\mu}N_i(\beta)d_i(t), \quad (1b)$$

where $t \in \mathbb{R}$, $x(t) \in \mathbb{R}^n$, $u(t) \in \mathbb{R}^p$ is the control input, $d_i(t) \in \mathbb{R}^{m_i}, i = 1,2,...,\mu$, are the unmeasurable and unknown disturbance inputs, $y(t) \in \mathbb{R}^q$ is the output to be controlled — which is assumed to be measurable — and $A(\beta)$, $B(\beta)$, $C(\beta)$, $M_i(\beta)$, $N_i(\beta)$, $i = 1,2,...\mu$, are matrices with real entries depending on a vector $\beta \in \Omega \subseteq \mathbb{R}^h$ of parameters, which are subject to variations and/or uncertain, and play the role of the "physical" parameters of the plant. The nominal value β_0 of β is assumed to be an interior point of the bounded set Ω. It is assumed that the first $\overline{q}$ components of $y(t)$ must track the corresponding components of the reference vector $r(t) \in \mathbb{R}^{\overline{q}}$, $\overline{q} \leq q$. Then, the error signal $e(t) \in \mathbb{R}^q$ is defined by

$$e(t) := Vr(t) - y(t), \quad V := \begin{bmatrix} I & 0 \end{bmatrix}', \quad (2)$$

where I is the identity matrix. Denote by $\mathbb{Z}^+$ the set of nonnegative integers. The classes $\mathcal{R}$, of reference signals $r(\cdot)$ to be asymptotically tracked, and $\mathcal{D}_i$, of disturbance functions $d_i(\cdot), i = 1,2,...,\mu$, to be asymptotically rejected, are of sinusoidal-exponential type and defined as follows:

$$\mathcal{R} := \mathcal{R}_1 \oplus \mathcal{R}_2 \oplus \cdots \oplus \mathcal{R}_{\overline{\mu}},$$
$$\mathcal{R}_i := \Big\{ r(\cdot) : r(t) = \delta e^{\alpha_i t} + \delta^* e^{\alpha_i^* t}, \quad \forall t \geq 0, \delta \in \mathbb{C}^{\overline{q}} \Big\}, \quad i = 1,2,...,\overline{\mu}, \quad (3)$$
$$\mathcal{D}_i := \Big\{ d_i(\cdot) : d_i(t) = \delta e^{\alpha_i t} + \delta^* e^{\alpha_i^* t}, \quad \forall t \geq 0, \delta \in \mathbb{C}^{m_i} \Big\}, \quad i = 1,2,...,\mu, \quad (4)$$

for some positive integer $\overline{\mu} \leq \mu$ and some known $\alpha_i \in \mathbb{C}, i = 1,2,...,\mu$, where * means complex conjugate. Notice that the pairs $(M_i(\cdot), N_i(\cdot))$ and $(M_j(\cdot), N_j(\cdot))$ may actually coincide for some $i,j,i \neq j$. The α_i's, $i = 1,2,...,\mu$, are assumed to be all distinct, and to have all non-negative real and imaginary parts, while the pair $(M_i(\cdot), N_i(\cdot))$ is assumed to be non-zero for each $i \in \{\overline{\mu}+1, \overline{\mu}+2,...,\mu\}$. For convenience, the above defined exogenous signals $r(\cdot) \in \mathcal{R}$, $d_i(\cdot) \in \mathcal{D}_i$, $i = 1, 2, \ldots, \mu$, can be equivalently considered as output free motions of a properly defined exosystem G, whose state will be denoted by $z(t) \in \mathbb{R}^l$, $z(t) = e^{Ft}z(0)$, according to $r(t) = L_0 z(t)$, $d_i(t) = L_i z(t)$, $i = 1, 2, \ldots, \mu$, so that, viceversa, $z(0) \in \mathbb{R}^l$ yields $r(\cdot) \in \mathcal{R}$, $d_i(\cdot) \in \mathcal{D}_i$, $i = 1, 2, \ldots, \mu$.

When a digital control system is used for system P, it is still needed for the error $e(t)$, at least for all values of β belonging to some neighborhood of β_0, not only the convergence to zero at the sampling times, but the stronger continuous-time convergence to zero, i.e., $\lim_{t\to+\infty} e(t) = 0$.

In the special case of single-rate systems, in order to achieve a ripple-free dead-beat convergence of the error response, for $\beta = \beta_0$, a continuous-time internal model of the exogenous signals $r(\cdot)$ and $d_i(\cdot)$, $i = 1, \ldots, \mu$, is needed, in general (Grasselli *et al.*, 1996). Therefore, it seems reasonable to use a continuous-time subcompensator also here, where a multirate control system is considered, in order to provide a continuous-time internal model of disturbance functions and reference signals, by enlarging

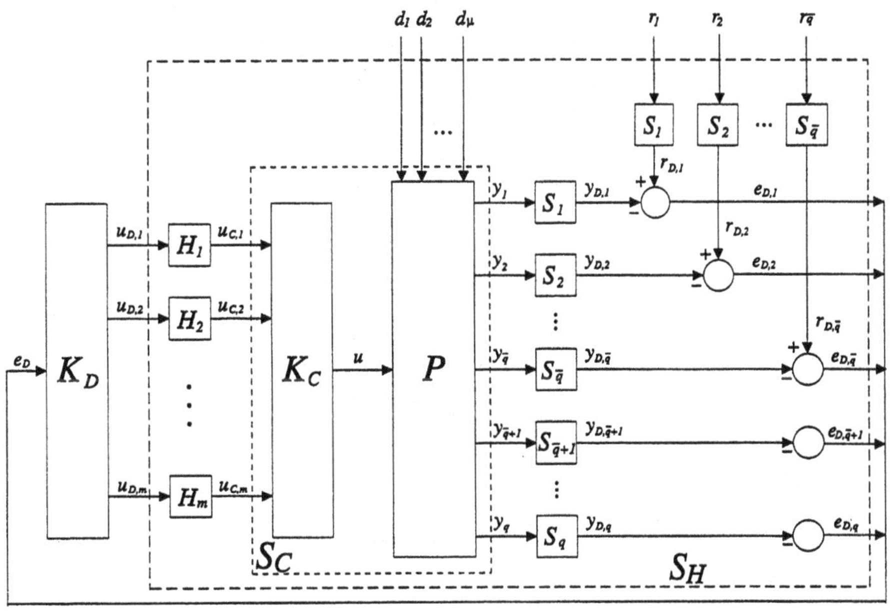

Fig. 1. The hybrid control system Σ.

the internal model that is possibly contained in P, if it is not complete. This motivates the choice of the control scheme depicted in Fig. 1, where K_C is a continuous-time time-invariant subcompensator, whose input $u_C(t) \in \mathbb{R}^m$, and K_D is a discrete-time periodic subcompensator. Each block H_i represents a zero-order holder, having holding period equal to $Q_i\,\delta_T$, $Q_i \in \mathbb{Z}^+$, $Q_i > 0$, $i = 1, 2, \ldots, m$, $\delta_T \in \mathbb{R}$, $\delta_T > 0$; it is described by:

$$u_{C\,i}(t) = u_{D\,i}(k\,Q_i), \quad \forall t \in (k\,Q_i\,\delta_T,\ (k+1)Q_i\,\delta_T],$$
$$u_{C\,i}(t),\ u_{D\,i}(k) \in \mathbb{R},\ t \in \mathbb{R},\ k \in \mathbb{Z}^+.$$

Each block S_i represents a sampler, having sampling interval equal to $Z_i\,\delta_T$, $Z_i \in \mathbb{Z}^+$, $Z_i > 0$, which allows the i-th component $y_{D\,i}(k)$ of $y_D(k)$ (or $r_{D\,i}(k)$ of $r_D(k)$) to be obtained, as sampling of the i-th component $y_i(t)$ of $y(t)$, $i = 1, \ldots, q$ (or $r_i(t)$ of $r(t)$, $i = 1, \ldots, \bar{q}$); the sampler S_i is described by:

$$y_{D\,i}(j\,Z_i) = y_i(j\,Z_i\,\delta_T), \quad j \in \mathbb{Z}^+,$$
$$y_{D\,i}(k) = 0, \quad k \in \mathbb{Z}^+, k \neq j\,Z_i, \forall j \in \mathbb{Z}^+,$$
$$r_{D\,i}(j\,Z_i) = r_i(j\,Z_i\,\delta_T), \quad j \in \mathbb{Z}^+,$$
$$r_{D\,i}(k) = 0, \quad k \in \mathbb{Z}^+, k \neq j\,Z_i, \forall j \in \mathbb{Z}^+.$$

It is assumed that all the holders and samplers are synchronized at time $t = 0$. It is also assumed, without loss of generality, that the positive integers Q_i, $i = 1, 2, \ldots, m$, and Z_i, $i = 1, 2, \ldots, q$, have 1 as their greatest common divisor. Denoting by ω their least common multiple, the period of the subcompensator K_D is chosen to be equal to ω; then, K_D will be said to be ω-periodic.

The following control problems will be studied.

Problem 1. Find, if any, a linear discrete-time ω-periodic subcompensator K_D, and a linear continuous-time time-invariant subcompensator K_C, so that the hybrid control system Σ, depicted in Fig. 1, satisfies the following two requirements for $\beta = \beta_0$:

(a_1) for a prescribed $\alpha \in \mathbb{R}$, $\alpha \geq 0$, for each initial state $x_\Sigma(0)$ of Σ, the free state response of Σ satisfies $\|x_\Sigma(t)\| < c\,e^{-\overline{\alpha}\,t}$ for all $t \geq 0$, for some positive real constants c, $\overline{\alpha}$, $\overline{\alpha} > \alpha$,

(b_1) for each disturbance function $d_i(\cdot) \in \mathcal{D}_i$, $i = 1, 2, \ldots, \mu$, for each reference signal $r(\cdot) \in \mathcal{R}$, and for each initial state $x_\Sigma(0)$ of Σ, $\lim_{t\to+\infty} e(t) = 0$,

and, moreover,

(c_1) there exists a neighborhood $\Psi \subseteq \Omega$ of β_0 such that conditions (a_1) and (b_1) hold for all $\beta \in \Psi$.

Notice that, for $\alpha = 0$, (a_1) reduces to the mere asymptotic stability. In addition, by the periodicity of all the time-varying subsystems appearing in Fig. 1 and by the form of the signals $r(\cdot) \in \mathcal{R}$ and $d_i(\cdot) \in \mathcal{D}_i$, $i = 1, 2, \ldots, \mu$, a solution of Problem 1 guarantees requirements similar to (a_1), (b_1) and (c_1) to be satisfied for any (nonzero) initial time.

Problem 2. Find, if any, a linear discrete-time ω-periodic subcompensator K_D, and a linear continuous-time time-invariant subcompensator K_C, so that the hybrid control system Σ, depicted in Fig. 1, satisfies the following two requirements for $\beta = \beta_0$:

(a_2) for each initial state $x_\Sigma(0)$ of Σ, the free state response of Σ is zero for all $t \geq \hat{t}$, for some $\hat{t} \in \mathbb{R}$, $\hat{t} > 0$,

(b_2) for each disturbance function $d_i(\cdot) \in \mathcal{D}_i$, $i = 1, 2, \ldots, \mu$, for each reference signal $r(\cdot) \in \mathcal{R}$, and for each initial state $x_\Sigma(0)$ of Σ, $e(t) = 0$ for all $t \geq \bar{t}$, for some $\bar{t} \in \mathbb{R}$, $\bar{t} > 0$,

and, moreover,

(c_2) for each $\alpha \in \mathbb{R}$, $\alpha \geq 0$, there exists a neighborhood $\Psi_\alpha \subseteq \Omega$ of β_0 such that conditions (a_1) and (b_1) of Problem 1 hold for all $\beta \in \Psi_\alpha$.

Both Problems 1 and 2 will be studied under two technical assumptions.

Assumption 1. There exists a closed neighborhood $\Psi_a \subseteq \Omega$ of β_0 in which all the entries of $A(\beta)$, $B(\beta)$, $C(\beta)$ are continuous functions of β and such that

$$\text{rank}\begin{bmatrix} A(\beta)-\alpha_i I & B(\beta) \\ C(\beta) & 0 \end{bmatrix} =$$
$$\text{rank}\begin{bmatrix} A(\beta_0)-\alpha_i I & B(\beta_0) \\ C(\beta_0) & 0 \end{bmatrix}, \forall \beta \in \Psi_a, i = 1, ..., \mu.$$

In connection with this, the following notation will be used throughout the paper, for each $i = 1, ..., \mu$:

$$\text{rank}\begin{bmatrix} A(\beta_0)-\alpha_i I & B(\beta_0) \\ C(\beta_0) & 0 \end{bmatrix} - n =: \rho_i. \quad (5)$$

Denote by $\sigma(A(\beta))$ the set of the eigenvalues of $A(\beta)$ and define $\Gamma(\beta) := \sigma(A(\beta)) \cup \{\alpha_1, \alpha_1^*, \alpha_2, \alpha_2^*, ..., \alpha_\mu, \alpha_\mu^*\}$.

Assumption 2. For each element λ of $\Gamma(\beta_0)$, none of the values $\lambda + \dfrac{\jmath 2\pi i}{\omega\,\delta_T}$, $i \neq 0$, $i \in \mathbb{Z}$, is an element of $\Gamma(\beta_0)$, where $\jmath$ is the imaginary unit.

Remark 1. Assumption 2 is standard (see, e.g., (Longhi, 1994)). It implies similar assumptions in (Francis and Georgiou, 1988), i.e. Assumption 2 rewritten with $\omega = 1$, and that none of the values $\dfrac{\jmath 2\pi i}{\delta_T}$, $i \neq 0$, $i \in \mathbb{Z}$, is an element of $\Gamma(\beta_0)$. □

Proposition 1. (Grasselli *et al.*, 1993) If there exists a neighborhood $\overline{\Psi}_a \subseteq \Omega$ of some $\overline{\beta}_0 \in \Omega$ such that all the entries of $A(\beta)$, $B(\beta)$, $C(\beta)$ are continuous functions of β in $\overline{\Psi}_a$, then there exists an interior point β_0 of $\overline{\Psi}_a$ such that Assumption 1 holds.

Notice that a proper choice of δ_T allows Assumption 2 to be satisfied, and, by Proposition 1, also Assumption 1 can be satisfied, possibly by a proper choice of some β_0 near the initial choice $\overline{\beta}_0$ of it, provided that the matrices involved are continuous in some neighborhood $\overline{\Psi}_a$ of $\overline{\beta}_0$.

Now call S_C the series connection of K_C and P (see Fig. 1), and denote its state by $x_C(t)$. Call S_H the hybrid system constituted by the connection of the holders H_i, $i = 1, 2, \ldots, m$, the subcompensator K_C, the plant P, the scalar comparison devices and all the samplers S_i, $i = 1, 2, \ldots, q$ (see Fig. 1). Further, call S_D the discrete-time state-space model of S_H for $r(\cdot) = 0$ and $d_i(\cdot) = 0$, $i = 1, \ldots, \mu$ (Longhi, 1994), i.e., the system described by a pair of equations of the following type:

$$x_D(k+1) = A_D(\beta,\, k)x_D(k) + B_D(\beta,\, k)u_D(k),$$
$$e_D(k) = C_D(\beta,\, k)x_D(k),$$

where $u_D(k) := (u_{D\,1}(k)\ \ldots\ u_{D\,m}(k))'$, $e_D(k) := (e_{D\,1}(k)\ \ldots\ e_{D\,q}(k))'$ (see Fig. 1), and the matrices $A_D(\beta, k)$, $B_D(\beta, k)$, and $C_D(\beta, k)$, are ω-periodic for each $\beta \in \Omega$ (Longhi, 1994).

In the following lemma $\eta_{S_C}(t, \beta, x_C(0), u_C(\cdot), z(0))$ denotes the $y(t)$ response, at time t, of the continuous-time system S_C, from its initial state $x_C(0)$, to the control function $u_C(\cdot)$ and to the initial state $z(0)$ of the exosystem G, for the actual value of vector β, while $\delta_{-1}(t)$ denotes the unit step function.

Lemma 1. If the hybrid control system Σ represented in Fig. 1. satisfies either requirement (c_1), or requirement (c_2), then for each $\beta \in \Psi$, or for each $\beta \in \Psi_\alpha$, and for each initial state $\overline{z} \in \mathbb{R}^l$ of the exosystem G:

(i) there exists an initial state $(\overline{x}'_C\ \overline{w}'_D)'$ of Σ such that, if $\overline{u}_C(t)$ denotes the corresponding response of Σ to $\overline{z}$ in the $u_C(t)$ variable, then the following relation is satisfied:

$$V L_0 e^{Ft}\overline{z} - \eta_{S_C}(t, \beta, \overline{x}_C, \overline{u}_C(0^+)\delta_{-1}(t), \overline{z}) = 0,$$
$$\forall t \geq 0,\, t \in \mathbb{R}; \quad (6)$$

(ii) function $\overline{u}_C(t)$ is periodic of period $\omega\delta_T$; in addition, under Assumption 2, if $\alpha_i \neq 0$ for all $i = 1, 2, \ldots, \mu$, or, more generally, if $\overline{z}$ is such that $z(t) = e^{Ft}\overline{z}$ does not contain the constant mode, then $\overline{u}_C(t)$ is zero for all $t > 0$.

Corollary 1. If $\omega = 1$, then $\overline{u}_C(t)$ is constant.

The proofs of Lemma 1 and Corollary 1 are omitted. See (Grasselli *et al.*, 1996, 1997) for special cases of these results for $\omega = 1$ (*i.e.*, for single-rate sampled-data control systems).

Remark 2. By Lemma 1, if the given plant P does not contain a continuous-time internal model of all the reference signals and disturbance functions for all β in some neighbourhood of β_0, the use of a continuous-time subcompensator K_C (in addition to a discrete-time periodic one K_D) is actually needed (except for the constant mode, if any) in order to satisfy requirement (c_1) or (c_2), *i.e.*, the prescribed robust ripple-free tracking and regulation requirements of Problems 1 and 2. This justifies the use of K_C in the control scheme of Fig. 1. □

3. MAIN RESULTS

The following theorem gives the solvability conditions of Problem 1.

Theorem 1. Under Assumptions 1 and 2, two subcompensators K_D and K_C that constitute a solution to Problem 1 exist if and only if the following conditions are satisfied:

(i) system P is $(-\alpha)$-stabilizable and $(-\alpha)$-detectable (Grasselli and Longhi, 1991*c*) for $\beta = \beta_0$, i.e.

$$\operatorname{rank}\left[\, A(\beta_0) - \lambda I \quad B(\beta_0) \,\right] = n, \quad \forall \lambda \in \mathbb{C},\ \operatorname{re}(\lambda) \geq -\alpha,$$

$$\operatorname{rank}\begin{bmatrix} A(\beta_0) - \lambda I \\ C(\beta_0) \end{bmatrix} = n, \quad \forall \lambda \in \mathbb{C},\ \operatorname{re}(\lambda) \geq -\alpha;$$

(ii) there exists a neighborhood $\Psi_b \subseteq \Omega$ of β_0 such that

$$\operatorname{Im}\begin{bmatrix} M_i(\beta) & 0 \\ N_i(\beta) & -\theta_i V \end{bmatrix} \subseteq \operatorname{Im}\begin{bmatrix} A(\beta) - \alpha_i I & B(\beta) \\ C(\beta) & 0 \end{bmatrix}, \quad \forall \beta \in \Psi_b, i = 1, 2, \ldots, \mu,$$

where $\theta_i := 1, i = 1, \ldots, \overline{\mu}, \theta_i := 0, i = \overline{\mu}+1, \ldots, \mu$.

The proof of Theorem 1 is omitted.

A design procedure for the two subcompensators K_C and K_D, constituting a solution to Problem 1, will now be given, in which the continuous-time subcompensator K_C is obtained as the series connection of μ subcompensators K_μ, $K_{\mu-1}$, ..., K_1 (see Fig. 2). In the following, for each $j = 1, 2, \ldots, \mu$, the series connection of K_j, $K_{j-1}, \ldots$, K_1 and P, having the signal $u_j(t)$ as control input and $y(t)$ as output, will be denoted by P_j, and described by:

$$\dot{x}_j(t) = A_j(\beta)x_j(t) + B_j(\beta)u_j(t) + \textstyle\sum_{i=1}^{\mu} M_{j\,i}(\beta)d_i(t), x_j(t) \in \mathbb{R}^{n_j},\ u_j(t) \in \mathbb{R}^p, \quad (7a)$$

$$y(t) = C_j(\beta)x_j(t) + \textstyle\sum_{i=1}^{\mu} N_{j\,i}(\beta)d_i(t). \quad (7b)$$

Notice that system P_μ coincide with system S_C.

Procedure 1 (which is valid under Assumptions 1 and 2 and conditions (i) and (ii)).

Step 1. Denote system P by P_0 and use the notations in (7) with $j = 0$ for P_0.

Step 2. For each $j = 1, 2, \ldots, \mu$, if $\rho_j = 0$, let K_j be described by $u_{j-1}(t) = u_j(t)$. Else, if $\rho_j > 0$ and $\alpha_j \in \mathbb{R}$, let K_j be described by

$$\dot{w}_j(t) = \alpha_j w_j(t) + [0 \quad I]u_j(t), \quad w_j(t) \in \mathbb{R}^{\rho_j},\ u_j(t) \in \mathbb{R}^p,$$

$$u_{j-1}(t) = E_j w_j(t) + [\tilde{E}_j \quad 0]u_j(t), \quad u_{j-1}(t) \in \mathbb{R}^p,$$

where $\tilde{E}_j \in \mathbb{R}^{p\times(p-\rho_j)}$, E_j and $\tilde{E}_j$ are real matrices such that, using the notations in (7) for the subsystem P_{j-1}:

$$\operatorname{rank}\begin{bmatrix} A_{j-1}(\beta_0) - \alpha_j I & B_{j-1}(\beta_0)E_j \\ C_{j-1}(\beta_0) & 0 \end{bmatrix} = n_{j-1} + \rho_j,$$

$$\text{(i.e.full column rank)} \quad (8a)$$

$$\det[E_j \ \tilde{E}_j] \neq 0. \quad (8b)$$

Otherwise, if $\rho_j > 0$ and $\alpha_j \notin \mathbb{R}$, let K_j be described by

$$\dot{w}_j(t) = W_j w_j(t) + [0 \ \tilde{U}_j]u_j(t), \quad w_j(t) \in \mathbb{R}^{2\rho_j},\ u_j(t) \in \mathbb{R}^p,$$

$$u_{j-1}(t) = \hat{E}_j w_j(t) + [\tilde{E}_j \ 0]u_j(t), \quad u_{j-1}(t) \in \mathbb{R}^p,$$

where $\hat{E}_j$ is given by

$$\hat{E}_j = [e_j^1 \ 0 \ e_j^2 \ 0 \ \ldots \ e_j^{\rho_j} \ 0],$$

in which e_j^i represents the i-th column of E_j, and E_j, $\tilde{E}_j$, are defined through (8) just as in the case of $\alpha_j \in \mathbb{R}$, W_j is defined by

$$W_j = \operatorname{blockdiag}\left\{ \begin{bmatrix} \sigma_j & \omega_j \\ -\omega_j & \sigma_j \end{bmatrix}, \ldots, \begin{bmatrix} \sigma_j & \omega_j \\ -\omega_j & \sigma_j \end{bmatrix} \right\},$$

in which σ_j and ω_j are the real and the imaginary part, respectively, of α_j (with $\omega_j \neq 0$), and $\tilde{U}_j \in \mathbb{R}^{2\rho_j \times \rho_j}$ is defined by

$$\tilde{U}_j = \operatorname{blockdiag}\left\{ \begin{bmatrix} 0 \\ 1 \end{bmatrix}, \ldots, \begin{bmatrix} 0 \\ 1 \end{bmatrix} \right\}.$$

Step 3. Compute the matrices $A_D(\beta_0, k)$, $B_D(\beta_0, k)$, $C_D(\beta_0, k)$, according to (Longhi, 1994).

Step 4. By means of a suitable design technique (e.g., the one reported in (Grasselli and Longhi, 1991*b*, *d*)) find a discrete-time ω-periodic subcompensator K_D such that all the characteristic multipliers of the discrete-time ω-periodic model Σ_D of the closed-loop system Σ are smaller than γ in modulus, for $\beta = \beta_0$, where $\gamma = e^{-\alpha\,\delta_T}$.

Also the proof of the following theorem is omitted.

Theorem 2. Under Assumptions 1 and 2, two subcompensators K_D and K_C that constitute a solution to Problem 2 exist if and only if condition (ii) of Theorem 1 is satisfied, and, in addition:

(i) system P is reachable and observable for $\beta = \beta_0$.

A design procedure for the design of K_C and K_D constituting a solution to Problem 2 can be obtained from Procedure 1 by leaving unchanged Steps 1, 2 and 3, and by modifying Step 4 in order to obtain that, for $\beta = \beta_0$, the free state response of Σ_D is dead-beat convergent to zero (see, e.g., (Grasselli and Lampariello, 1981) or (Grasselli and Longhi, 1988)).

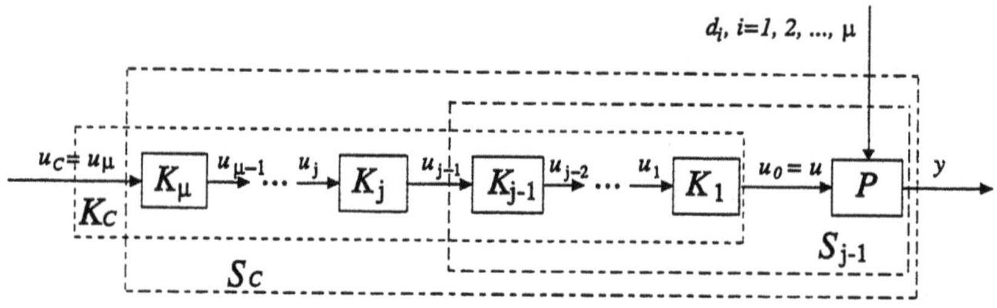

Fig. 2. Structure of subcompensator K_C.

4. REFERENCES

Dahleh, M. A., P. G. Voulgaris and L. S. Valavanis (1992). Optimal and robust controllers for periodic and multirate systems. *IEEE Trans. Automatic Control* **37**, 90–99.

Davison, E. J. (1976). The robust control of a servomechanism problem for linear time-invariant multivariable systems. *IEEE Trans. Aut. Control* **AC-21**, 25–34.

Francis, B. A. (1977). The linear multivariable regulator problem. *SIAM J. Control* **15**, 486–505.

Francis, B. A. and T. T. Georgiou (1988). Stability theory for linear time-invariant plants with periodic digital controllers. *IEEE Trans. Aut. Control* **AC-33**, 820–832.

Francis, B. A. and W. M. Wonham (1975). The internal model principle for linear multivariable regulators. *Appl. Math. Opt.* **2**, 170–194.

Francis, B. A. and W. M. Wonham (1976). The internal model principle of control theory. *Automatica* **12**, 457–465.

Franklin, G. F. and A. Emami-Naeini (1986). Design of ripple-free multivariable robust servomechanism. *IEEE Trans. Automatic Control* **AC-31**(7), 661–664.

Grasselli, O. M. and F. Lampariello (1981). Dead-beat control of linear periodic discrete-time systems. *Int. J. Control* **33**(6), 1091–1106.

Grasselli, O. M. and S. Longhi (1988). Disturbance localization by measurement feedback for linear periodic discrete-time systems. *Automatica* **24**(3), 375–385.

Grasselli, O. M. and S. Longhi (1991*a*). Finite zero structure of linear periodic discrete-time systems. *Int. J. of Systems Science* **22**(10), 1785–1806.

Grasselli, O. M. and S. Longhi (1991*b*). Pole placement for non-reachable periodic discrete-time systems. *Mathematics of Control, Signals and Systems* **4**, 439–455.

Grasselli, O. M. and S. Longhi (1991*c*). Robust output regulation under uncertainties of physical parameters. *Systems and Control Letters* **16**, 33–40.

Grasselli, O. M. and S. Longhi (1991*d*). Robust tracking and regulation of linear periodic discrete-time systems. *Int. J. Control* **54**(3), 613–633.

Grasselli, O. M., L. Jetto and S. Longhi (1995). Ripple-free dead-beat tracking for multirate sampled-data systems. *Int. Journal of Control* **61**(6), 1437–1455.

Grasselli, O. M., L. Menini and P. Valigi (1997). Robust continuous-time asymptotic tracking and regulation for sampled-data systems with structured parametric uncertainties. In: *Preprints of the 2nd IFAC Symp. on Robust Control Design*. Budapest (Hungary). pp. 359–364.

Grasselli, O. M., S. Longhi, A. Tornambè and P. Valigi (1996). Robust ripple-free regulation and tracking for parameter dependent sampled-data systems. *IEEE Trans. Automatic Control* **41**(7), 1031–1037.

Grasselli, O. M., S. Longhi and A. Tornambè (1993). Robust tracking and performance for multivariable systems under physical parameter uncertainties. *Automatica* **29**(1), 169–179.

Grasselli, O. M., S. Longhi and A. Tornambè (1994). Robust continuous-time tracking and regulation for multirate sampled-data systems. *Kybernetika* **30**(6), 725–743.

Kucera, V. and M. Sebek (1984). On dead-beat controllers. *IEEE Trans. Automatic Control* **29**, 719–722.

Longhi, S. (1994). Structural properties of multirate sampled-data systems. *IEEE Trans. Automatic Control* **39**(3), 692–696.

Sanchis, R. and P. Albertos (1995). Design of ripple-free controllers. In: *Proceedings of the 3rd European Control Conference*. Rome, Italy. pp. 3660–3664.

Scattolini, Riccardo and Nicola Schiavoni (1993). On the output control of multirate systems subject to arbitrary exogenous signals. *IEEE Trans. on Automatic Control* **38**(4), 643–646.

Urikura, S. and A. Nagata (1987). Ripple-free deadbeat control for sampled-data systems. *IEEE Trans. Automatic Control* **AC-32**, 474–482.

Yamamoto, Y. (1994). A function space approach to sampled data control systems and tracking problems. *IEEE Trans. Automatic Control* **39**(4), 703–713.

14th Triennial World Congress, Beijing, P.R. China

PRACTICAL TRACKING WITH SETTLING TIME: BOUNDED CONTROL FOR ROBOT MOTION

Annemarie KÖKÖSY

LAMPS - ENI, BP 525, 90016 Belfort, France
Phone: (33)-3-84-58-23-97
Fax: (33)-3-84-58-23-42
e-mail: annemarie.kokosy@enibe.fr

Abstract: This paper presents a new criterion for a practical tracking with a prespecified vector settling time. A new bounded control is synthesised in order to guarantee the requested tracking performance of the robot motion. The model of the robot includes unknown friction forces. They are identified by using a multi-layer neural network with sigmoidal nonlinearities. The control is applied in simulations to a rotational three-degree-of-freedom robot. The simulation results excellently verify the theory. *Copyright © 1999 IFAC*

Keywords: bounded control, nonlinear systems, practical tracking, robot control, settling time

1. INTRODUCTION

Practical tracking concept was introduced by Grujić in (Grujić, 1986). The practical tracking performance of the system motion is guaranteed over a prespecified finite time interval for any initial errors belonging to a prespecified set of initial errors. After the settling time has elapsed the errors rest in a prespecified final error set until the final time. Vector settling time means that every robot segment can be assigned and will achieve an appropriate prespecified finite settling time independently of other segments.

The practical tracking performance is defined over a finite time interval, whereas tracking in the Lyapunov sense refers to an infinite time interval (Grujić, 1993). In the practical tracking concept the set of any permitted initial errors, the set of instantaneous errors over the whole finite time interval and the set of the final errors after the settling time are usually prespecified. Otherwise, they should be exactly determined. In the Lyapunov sense tracking concepts theses sets are not required to be exactly determined, they can be just estimated.

The criteria for the practical tracking concepts were presented in (Grujić, 1988) for a scalar aggregation function. An extension of these results to a vector aggregation function is given in (Lazić, 1995).

In this paper a new criterion will be established for a practical tracking with a prespecified vector settling time. The aggregation function candidate used is more general than in (Lazić, 1995) where a Lurie-type function was utilised. A new bounded control will be synthesised in order to guarantee the requested tracking performance. An illustrative example shows an application of the synthesised control to a real robot (Stoten, 1990). A robot model includes the friction forces. Since the friction nonlinearities are not well known, the controller contains a multi-layer neural network in order to obtain sufficiently accurate information about their responses.

2. NOTATION PRELIMINARIES

$\mathbf{e} = \mathbf{q}_d - \mathbf{q}$ - the q-error vector relative to a desired $\mathbf{q}_d$, $\mathbf{e} = [e_1\ e_2\ \dots\ e_k]^T$, $\mathbf{e}^{(1)} = d\mathbf{e}/dt$,
$\mathbf{e}^1(\cdot)\ :\ \Re^k \times \Re^{2k} \times \Re^d \times \Re^r \times \Re^k \rightarrow \Re^{2k}$ - $\mathbf{e}^1(\cdot) = [\mathbf{e}^T(\cdot)\ \mathbf{e}^{(1)T}(\cdot)]^T$; for short $\mathbf{e}^1(\mathbf{T}) \equiv \mathbf{e}^1$,
$\mathbf{E}_A \subset \Re^k$ - the set of all admitted errors on $\Im^{\tau}$; compact connected neighbourhood of $\mathbf{e} = \mathbf{0}$,
$\mathbf{E}_F \subset \Re^k$ - the set of all admitted final errors on $\Im^{\tau}_{\mathbf{T}_S}$; compact connected neighbourhood of $\mathbf{e} = \mathbf{0}$,
$\mathbf{E}_I \subset \Re^k$ - the set of all admitted initial errors; compact connected neighbourhood of $\mathbf{e} = \mathbf{0}$,
$\Im^{\tau}=[0, \tau[$, $\Im^{\tau}_{\mathbf{T}_S} = [\mathbf{T}_S, \tau[$,
$m \in \Re$ - payload mass,
$\mathbf{q} \in \Re^k$ - general coordinate vector called the robot (internal) dynamics vector,
$\mathbf{q}^{(i)}(\mathbf{T}) \in \Re^k$ - i-th derivative of $\mathbf{q}(\cdot)$ at time $\mathbf{T}$; for short $\mathbf{q}^{(i)}(\mathbf{T}) = \mathbf{q}^{(i)}$, $\mathbf{q}(0) = \mathbf{q}_0$,
$\mathbf{q}_d(\cdot) : \Re^k \rightarrow \Re^k$ - desired robot motion, $\mathbf{q}_d(\mathbf{T}) \in C^{(1)}(\Re^k)$; for short $\mathbf{q}_d(\mathbf{T}) = \mathbf{q}_d$, $\mathbf{q}_d(0) = \mathbf{q}_{d0}$,
$sign(\cdot)\ :\ \Re \rightarrow \Re$ - sign nonlinearity, $sign(e) = e|e|^{-1}$ for $e \neq 0$ and sign(0)=0,
$S(\cdot) : \Re^k \rightarrow \Re^{k\times k}$ - matrix sign function, $S(\mathbf{e}) = diag\{sign(e_1)\ \dots\ sign(e_k)\}$,
$S_{q_d} \subset \Re^k$ - set of all admitted desired motions $\mathbf{q}_d(\cdot)$ on $\Im^{\tau}$, $\mathbf{q}_d(\cdot) \in S_{q_d}$ implies that $\mathbf{q}_d(\mathbf{T}) \in C(\Im^{\tau}, \Re^k)$,
$S_d \subset \Re^d$ - set of all admitted disturbances $\mathbf{d}$ on $\Im^{\tau}$,
$\mathbf{T} \in \Re^k$ - vector time, $\mathbf{T} = [t\ t\ \dots\ t]^T$,
$\mathbf{T}_S \in \Re^k$ - vector settling time,

$$\mathbf{T}_S = [\tau_{S1}\ \ \tau_{S2}\ \dots\ \tau_{Sk}]^T,$$

$T_S \in \Re^{k\times k}$ - $T_S = diag\{\tau_{S1}\ \ \tau_{S2}\ \dots\ \tau_{Sk}\}$,
$\tau \in \Re^k$ - vector final time, $\tau = [\tau_1\ \dots\ \tau_k]^T$,
$\mathbf{v}^{\partial}_{m(\cdot)} = \inf\ [\mathbf{v}(\mathbf{e})\ :\ \ \mathbf{e} \in \partial\mathbf{E}_{(\cdot)}]$, $\mathbf{v}^{\partial}_{m(\cdot)} = [v^{\partial}_{1m(\cdot)}\ \dots\ v^{\partial}_{km(\cdot)}]^T$,
$V^{\partial}_{m(\cdot)} \in \Re^{k\times k}$ - $V^{\partial}_{m(\cdot)} = diag\{v^{\partial}_{1m(\cdot)}\ \dots\ v^{\partial}_{km(\cdot)}\}$,
$\mathbf{v}_{\overline{M}(\cdot)} = \sup\ [\mathbf{v}(\mathbf{e})\ :\ \ \mathbf{e} \in \overline{\mathbf{E}}_{(\cdot)}]$, $\mathbf{v}_{\overline{M}(\cdot)} = [v_{1\overline{M}(\cdot)}\ \dots\ v_{k\overline{M}(\cdot)}]^T$, where $(\cdot)$=A or $(\cdot)$=F or $(\cdot)$=I,
$V_{\overline{M}(\cdot)} \in \Re^{k\times k}$ - $V_{\overline{M}(\cdot)} = diag\{v_{1\overline{M}(\cdot)}\ \dots\ v_{k\overline{M}(\cdot)}\}$.

3. DEFINITION AND CRITERION

Considered robots are described by the nonlinear vector differential equation (1),

$$A(\mathbf{d}, \mathbf{q})\mathbf{q}^{(2)} + C(\mathbf{d}, \mathbf{q}^1)\mathbf{q}^{(1)} + \mathbf{g}(\mathbf{d}, \mathbf{q}) +$$

$$+\mathbf{f}(\mathbf{q}^{(1)}) = B\mathbf{u}, \tag{1}$$

where $A(\cdot)\ :\ \Re^d \times \Re^k \rightarrow R^{k\times k}$ is the robotic system inertia matrix function, $C(\cdot) : \Re^d \times \Re^{2k} \rightarrow \Re^{k\times k}$ is a nonlinear matrix function describing Coriolis and centrifugal forces or their torques, $\mathbf{g}(\cdot) : \Re^d \times \Re^k \rightarrow \Re^k$ is a nonlinear vector function containing gravitation forces, $\mathbf{f}(\mathbf{q}^{(1)})$ is the $k \times 1$ vector containing the friction forces, $\mathbf{d}$ is the $d \times 1$ perturbation vector, $\mathbf{u}$ is the $r \times 1$ control vector, and B is a $k \times r$ elementwise constant matrix describing parametric characteristics of actuators transmitting a control action on the robot mechanism.

Let τ and $\mathbf{T}_S$ be given. Note that their entry values are finite and prespecified.

DEFINITION 3.1. *System (1) exhibits the practical tracking with vector settling time $\mathbf{T}_S$ with respect to $\{\tau, \mathbf{E}_I, \mathbf{E}_A, \mathbf{E}_F, S_d, S_{q_d}\}$ iff*

$$(\mathbf{q}_{d0} - \mathbf{q}_0, \mathbf{d}, \mathbf{q}_d) = (\mathbf{e}_0, \mathbf{d}, \mathbf{q}_d) \in \mathbf{E}_I \times S_d \times S_{q_d},$$

implies that:

i) $\mathbf{q}_d(\mathbf{T}) - \mathbf{q}(\mathbf{T}; \mathbf{q}_0; \mathbf{d}, \mathbf{u}) = \mathbf{e}(\mathbf{T}; \mathbf{e}_0; \mathbf{d}, \mathbf{u}, \mathbf{q}_d) \in \mathbf{E}_A$,
$$\forall \mathbf{T} \in [0, \mathbf{T}_S[,$$

and

ii) $\mathbf{q}_d(\mathbf{T}) - \mathbf{q}(\mathbf{T}; \mathbf{q}_0; \mathbf{d}, \mathbf{u}) = \mathbf{e}(\mathbf{T}; \mathbf{e}_0; \mathbf{d}, \mathbf{u}, \mathbf{q}_d) \in \mathbf{E}_F$,
$$\forall \mathbf{T} \in [\mathbf{T}_S, \tau[.$$

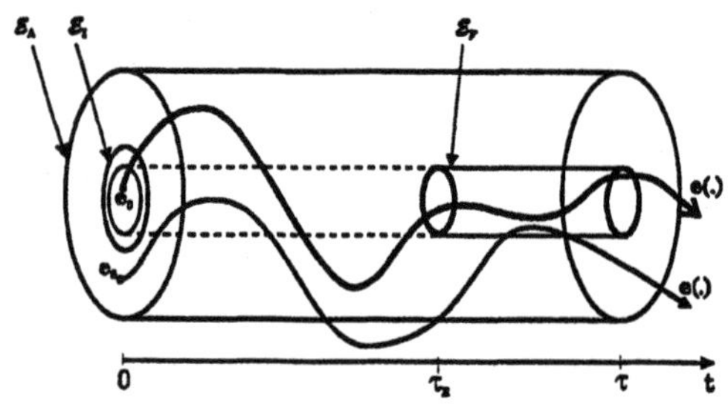

Fig. 1. Practical tracking with settling time

Note. The vector relationships hold elementwise.

Let $\mathbf{v}(\cdot)$ be a vector aggregation function:

$$\mathbf{v}(\cdot) : \mathbf{E}_A \rightarrow \Re^k,\ \ \mathbf{v}(\mathbf{e}) \in C(\mathbf{E}_A),$$

$$\mathbf{v}(\mathbf{e}) = [v_1(e_1)\ \ v_2(e_2)\ \dots\ v_k(e_k)]^T. \tag{2}$$

Let

$$\int_0^{\mathbf{T}} \Psi(v)\mathrm{d}v = \left[\int_0^t \psi_1(v_1)\mathrm{d}v_1 \dots \int_0^t \psi_k(v_k)\mathrm{d}v_k\right]^T,$$
$$\forall t \in \Im^{\tau_i},\ \ \forall i = 1, 2, \dots, k.$$

THEOREM 3.1. In order for system (1) controlled by a control $\mathbf{u}(\cdot)$ to exhibit practical tracking with the vector settling time $\mathbf{T}_S$ with respect to $\{\tau, \mathbf{E}_I, \mathbf{E}_A, \mathbf{E}_F, S_d, S_{q_d}\}$, it is sufficient that there exist a function $\mathbf{v}(\cdot)$ defined by (2), and a vector function $\psi(\cdot)\ :\ \Im^{\tau} \rightarrow \Re^k$, which is integrable over $\Im^{\tau}$, where $\Im^{\tau} = \Im^{\tau_1} \times \Im^{\tau_2} \times \dots \times \Im^{\tau_k}$, so that the following condition hold:

$$D^+\mathbf{v}(\mathbf{e}; \mathbf{d}; \mathbf{u}; \mathbf{q}_d) \leq \psi(\mathbf{T}),$$

$$\forall(\mathbf{T}, \mathbf{e}, \mathbf{d}, \mathbf{q}_d) \in \Im^{\tau} \times \mathbf{E}_A \times S_d \times S_{q_d}, \tag{3}$$

$$\int_0^{\mathbf{T}} \Psi(v)dv \leq \mathbf{v}_{mA}^{\partial} - \mathbf{v}_{\overline{MI}}, \ \forall \mathbf{T} \in \Im^{\tau}, \qquad (4)$$

and

$$\int_0^{\mathbf{T}} \Psi(v)dv \leq \mathbf{v}_{mF}^{\partial} - \mathbf{v}_{\overline{MI}}, \ \forall \mathbf{T} \in \Im^{\mathcal{T}}_{\mathbf{T}_S}. \qquad (5)$$

4. SYNTHESIS OF TRACKING CONTROL

4.1 *Control problem statement*

The control problem is to synthesise a bounded control $\mathbf{u}(\cdot)$ such that the robot system (1) exhibits the practical tracking with the vector settling time $\mathbf{T}_S$ with respect to $\{\tau, \mathbf{E}_I, \mathbf{E}_A, \mathbf{E}_F, \mathcal{S}_d, \mathcal{S}_{q_d}\}$.

4.2 *Control problem solution*

In order to synthesise the control algorithm, the following assumptions are accepted:

Assumption 1
The matrix B has the following property:

$$det(BB^T) \neq 0.$$

Assumption 2
The inertia matrix $A(\mathbf{d}, \mathbf{q})$, the centrifugal and Coriolis forces matrix $C(\mathbf{d}, \mathbf{q}^1)$ and the gravitational forces vector $\mathbf{G}(\mathbf{d}, \mathbf{q})$ are known, and the general coordinnates vector $\mathbf{q}$ of the robot as well as its first and second derivatives are measurable.

The friction forces vector $\mathbf{f}(\mathbf{q}^{(1)})$ is unknown. It will be identified by using a multi-layer neural network (NN) with sigmoidal type nonlinearities. The use of NN for robot dynamics identification is presented in (Kumbla and Jamshidi, 1997). The identified friction forces vector is denoted by $\hat{\mathbf{f}}(\mathbf{q}^{(1)})$.

Assumption 3
The identified friction forces vector is sufficiently close to the real friction forces vector, hence $\hat{\mathbf{f}}(\mathbf{q}^{(1)}) \cong \mathbf{f}(\mathbf{q}^{(1)})$.

This assumption has been well verified by simulating the NN response relative to the real $\mathbf{f}[\mathbf{q}^{(1)}(t)]$ response.

The acceleration is constrained as follows:

$$\mathbf{q}^{(2)} = \begin{cases} -\mathbf{q}_M^{(2)}, & \mathbf{q}^{(2)} \leq -\mathbf{q}_M^{(2)}, \\ \mathbf{q}^{(2)}, & \mathbf{q} \in]-\mathbf{q}_M^{(2)}, \mathbf{q}_M^{(2)}[, \\ \mathbf{q}_M^{(2)}, & \mathbf{q}^{(2)} \geq \mathbf{q}_M^{(2)}. \end{cases} \qquad (6)$$

Let the sets be defined by (7):

$$\mathbf{E}_A = \{\mathbf{e} : |\mathbf{e}| \leq \mathbf{e}_{MA}\}, \ \mathbf{E}_F = \{\mathbf{e} : |\mathbf{e}| \leq \mathbf{e}_{MF}\},$$

$$\mathbf{E}_I = \{\mathbf{e} : |\mathbf{e}| \leq \mathbf{e}_{MI}\}, \qquad (7)$$

with $\mathbf{e}_{MI} < \mathbf{e}_{MA}$, $\mathbf{e}_{MF} < \mathbf{e}_{MA}$. The sets $\mathbf{E}_{(\cdot)}$ are compact and bounded, $\mathbf{E}_{(\cdot)} = \overline{\mathbf{E}}_{(\cdot)}$.

The vector aggregation function which describes these sets is defined by:

$$\mathbf{v}(\mathbf{e}) = |\mathbf{e}|. \qquad (8)$$

The sets described by (7) can be redefined by using (8):

$$\mathbf{E}_A = \{\mathbf{e} : \mathbf{v}(\mathbf{e}) \leq \mathbf{e}_{MA}\},$$
$$\mathbf{E}_F = \{\mathbf{e} : \mathbf{v}(\mathbf{e}) \leq \mathbf{e}_{MF}\},$$
$$\mathbf{E}_I = \{\mathbf{e} : \mathbf{v}(\mathbf{e}) \leq \mathbf{e}_{MI}\},$$

so that $\mathbf{v}_{mA}^{\partial} = \mathbf{e}_{MA} \geq \mathbf{e}_{MI} = \mathbf{v}_{\overline{MI}}$.

The robot geometric and kinematic characteristics determine the set $\mathcal{Q}_c^1$:

$$\mathcal{Q}_c^1 = \left\{\mathbf{q}^1(\cdot) : |\mathbf{q}| \leq \mathbf{q}_M, \ |\mathbf{q}^{(1)}| \leq \mathbf{q}_M^{(1)}\right\},$$

where the values of $\mathbf{q}_M$ and $\mathbf{q}_M^{(1)}$ are well defined and $\mathbf{q}^1(\cdot) = [\mathbf{q}^T(\cdot) \ \mathbf{q}^{(1)T}(\cdot)]^T$.

THEOREM 4.1. In order for system (1) obeying assumptions 1, 2 and 3 to exhibit the practical tracking with the vector settling time $\mathbf{T}_S$ with respect to $\{\tau, \mathbf{E}_I, \mathbf{E}_A, \mathbf{E}_F, \mathcal{S}_d, \mathcal{S}_{q_d}\}$ it is sufficient that the control $\mathbf{u}(\cdot)$ obeys:

$$\mathbf{u} = B^T(BB^T)^{-1}\left\{A(\mathbf{d}, \mathbf{q})\mathbf{w}(\mathbf{q}^{(2)}) + \right.$$
$$+ C(\mathbf{d}, \mathbf{q}^1)\mathbf{q}^{(1)} + \mathbf{g}(\mathbf{d}, \mathbf{q}) + \hat{\mathbf{f}}(\mathbf{q}^{(1)})$$
$$\left. + N(\mathbf{q}^{(2)})\left[\mathbf{e}^{(1)}(\mathbf{T}) + S[\mathbf{e}(\mathbf{T})]\mathbf{r}\right]\right\}, \qquad (9)$$

where

$$n_i(q_i^{(2)}) = \begin{cases} 1, & |q_i^{(2)}| < (q_{Mi}^{(2)} - \epsilon_i) \\ 0, & |q_i^{(2)}| \geq (q_{Mi}^{(2)} - \epsilon_i) \end{cases}, \ i \in \{1, \ldots k\},$$

$$N(\mathbf{q}^{(2)}) = diag\{n_1 \ldots n_k\},$$
$$\mathbf{q}^{(2)} = [q_1^{(2)} \ldots q_k^{(2)}]^T, \ \mathbf{q}_M^{(2)} = [q_{M1}^{(2)} \ldots q_{Mk}^{(2)}]^T,$$
$$\epsilon = [\epsilon_1 \ldots \epsilon_k]^T,$$
$$\mathbf{r}(\mathbf{T}) = \begin{cases} T_S^{-1}\Lambda(\mathbf{v}_{mF}^{\partial} - \mathbf{v}_{\overline{MI}}), & \forall \mathbf{T} \in \Im^{\mathbf{T}_S} \\ \gamma, & \forall \mathbf{T} \in \Im^{\mathcal{T}}_{\mathbf{T}_S} \end{cases},$$
$$\forall [\mathbf{e}_0, \mathbf{q}_d(\cdot), \mathbf{d}] \in \mathbf{E}_I \times \mathcal{S}_{q_d} \times \mathcal{S}_d,$$
$$T_S = diag\{\tau_{S1} \ldots \tau_{Sk}\}, \ \Lambda = diag\{\lambda_1 \ldots \lambda_k\},$$
$$\lambda_i \geq 1, \ \gamma = [\gamma_1 \ldots \gamma_k]^T, \ \gamma_i \geq 0, \ \forall i = 1, \ldots, k. \qquad (10)$$

The function $\mathbf{w}(\cdot)$ is defined by the following relationship:

- *if* $|\mathbf{q}^{(2)}| < (\mathbf{q}_M^{(2)} - \epsilon)$

$$\mathbf{w}(\mathbf{q}^{(2)}) = \mathbf{q}^{(2)}, \qquad (11)$$

- *if* $|\mathbf{q}^{(2)}| \geq (\mathbf{q}_M^{(2)} - \epsilon)$

$$\mathbf{w}(\mathbf{q}^{(2)}) = \begin{cases} \delta(\mathbf{q}_M^{(2)} - \epsilon), & (\mathbf{e} \geq 0, \ \mathbf{q}_d^{(2)} \geq 0), \\ \rho_1 \mathbf{q}_d^{(2)}, & (\mathbf{e} \geq 0, \ \mathbf{q}_d^{(2)} < 0) \\ & \text{or } (\mathbf{e} < 0, \ \mathbf{q}^{(2)} \geq 0), \\ \rho_2 \mathbf{q}_d^{(2)}, & (\mathbf{e} < 0, \ \mathbf{q}_d^{(2)} < 0), \end{cases} \qquad (12)$$

where $0 < \epsilon \lll 1$, $\delta = 0.999\ldots$, $\rho_1 \in]0,1[$, $\rho_2 > 1$ and $-0.8\mathbf{q}_M^{(2)} \le \mathbf{q}_d^{(2)} \le 0.8\mathbf{q}_M^{(2)}$.

The control is bounded and its bound can be calculated from (13):

$$\mathbf{u}_M = \max_{\substack{\mathbf{q}^1 \in \mathcal{Q}_c^1, \\ \mathbf{q}^{(2)} \in \{-\mathbf{q}_M^{(2)}, \mathbf{q}_M^{(2)}\}}} \Big\{ B^T(BB^T)^{-1}\Big[A(\mathbf{d},\mathbf{q})\mathbf{q}^{(2)} + C(\mathbf{d},\mathbf{q}^1)\mathbf{q}^{(1)} + \mathbf{g}(\mathbf{d},\mathbf{q}) + \mathbf{f}(\mathbf{q}^{(1)}) + \mathbf{e}_M^{(1)} + \mathbf{r}_M\Big]\Big\}. \quad (13)$$

Maximum velocity and maximum control values calculation

Substituting (9) in (1) and by using (10), it results:

$$|\mathbf{e}^{(1)}(\mathbf{T})| = -|S[\mathbf{e}(\mathbf{T})]\mathbf{r}(\mathbf{T})| \le |\mathbf{r}(\mathbf{T})| =$$

$$= |T_S^{-1}\Lambda(\mathbf{v}_{mF}^{\partial} - \mathbf{v}_{\overline{M}I})| \le \mathbf{e}_M^{(1)}. \quad (14)$$

From (14) and $\boldsymbol{\lambda} = [\lambda_1 \ \ldots \lambda_k]^T$, $\lambda_i \ge 1$, $\forall i = 1, \ldots k$, it is easy to obtain:

$$\mathbf{1} \le \boldsymbol{\lambda} \le |V_{mF}^{\partial} - V_{\overline{M}I}|^{-1} T_S \mathbf{e}_M^{(1)}. \quad (15)$$

The value of $\mathbf{e}_M^{(1)}$ must be chosen so that the relation (16) holds:

$$\mathbf{e}_M^{(1)} \ge T_S^{-1}|V_{mF}^{\partial} - V_{\overline{M}I}|\mathbf{1}. \quad (16)$$

After $\mathbf{e}_M^{(1)}$ has been adopted, a value of $\boldsymbol{\lambda}$ is chosen so that (15) holds. Therefore, the maximum of $\mathbf{r}(\mathbf{T})$, (10), is

$$\mathbf{r}_M = T_S^{-1}|V_{mF}^{\partial} - V_{\overline{M}I}|\boldsymbol{\lambda} \le \mathbf{e}_M^{(1)}. \quad (17)$$

Knowing $\mathbf{e}_M^{(1)}$, $\mathbf{r}_M$, $\mathbf{q}_M^{(2)}$ and the set $\mathcal{Q}_c^1$ determined by robot geometric and kinematic characteristics, it is possible to determine the maximum control value by using (13).

5. DESIGN EXAMPLE

A rotational three-degree-of-freedom robot is selected for the application (figure 2). Stoten presented the robot model in (Stoten, 1990), (pp. 2.7), which is used in what follows. The payload mass m represents a disturbance variable $\mathbf{d}$ in this example. We assume that the payload mass can vary between 0 and 5 kilogrammes. Hence, $S_d = \{m : 0 \le m \le 5\}$.

The robot inertia matrix is given (Stoten, 1990) by:

$$A(m,\mathbf{q}) = \begin{bmatrix} a_{11} & 0 & 0 \\ 0 & a_{22} & a_{23} \\ 0 & a_{23} & a_{33} \end{bmatrix}, \quad (18)$$

where:

$a_{11} = J_2^* s^2 q_2 + J_{2p} c^2 q_2 + J_{q_3} s^2(q_2+q_3) + J_{3p} c^2(q_2+q_3) + J_3 + 2m_3L_2l_3 sq_2 s(q_2+q_3)$,

$a_{22} = J_2^* + J_{q_3} + 2m_3L_2l_3cq_3$,

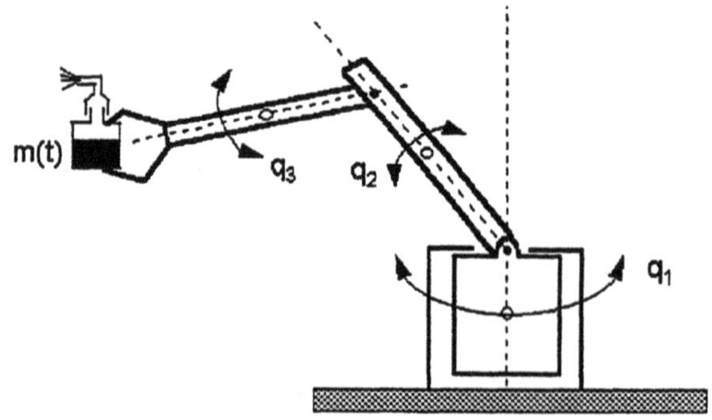

Fig. 2. Functional scheme of the rotational three-degree-of-freedom robot

$a_{23} = J_{q_3} + m_3L_2l_3cq_3$, $a_{33} = J_{q_3}$,
$J_{q_3} = J_3 + m_3l_3^2$, $J_2^* = J_2 + m_2l_2^2 + m_3L_2^2$,

together with $sq_i = sin(q_i)$ and $cq_i = cos(q_i)$, $i = 1,2,3$.

The load mass variations affect the position of the mass centre l_3, the principal moment of inertia of the third link J_3 and a value of the total mass m_3 (mass of the third link m_{3link} and payload m):

$$m_3 = m_{3link} + m, \quad (19)$$

$$l_3 = (m_{3link} \cdot l_{3link} + m \cdot l_{23})/m_3, \quad (20)$$

$$J_3 = J_{3link} + m \cdot l_{23}^2. \quad (21)$$

The centripetal and Coriolis forces matrix is determined from (Stoten, 1990) in the following form:

$$C(m,\mathbf{q}^1) = \begin{bmatrix} c_{11} & c_{12} & c_{13} \\ -c_{12} & c_{22} & c_{23} \\ -c_{13} & c_{32} & 0 \end{bmatrix}, \quad (22)$$

where:

$c_{11} = q_2^{(1)}(J_2^* - J_{2p})sq_2cq_2 + \left[q_2^{(1)} + q_3^{(1)}\right](J_{q_3} - J_{3p})s(q_2+q_3)c(q_2+q_3) + q_2^{(1)}m_3L_2l_3s(2q_2+q_3) + q_3^{(1)}m_3L_2l_3sq_2c(q_2+q_3)$,

$c_{12} = q_1^{(1)}\left[(J_2^* - J_{2p})sq_2cq_2 + m_3L_2l_3s(2q_2+q_3) + (J_{q_3} - J_{3p})s(q_2+q_3)c(q_2+q_3)\right]$,

$c_{13} = q_1^{(1)}\left[(J_{q_3} - J_{3p})s(q_2+q_3)c(q_2+q_3) + m_3L_2l_3sq_2c(q_2+q_3)\right]$,

$c_{22} = -q_3^{(1)}m_3L_2l_3sq_3$,

$c_{23} = -m_3L_2l_3sq_3\left[q_2^{(1)} + q_3^{(1)}\right]$,

$c_{32} = q_2^{(1)}m_3L_2l_3sq_3$.

The gravitational vector is represented (Stoten, 1990) by:

$$\mathbf{g}(m,\mathbf{q}) = -g\begin{bmatrix} 0 \\ (m_2l_2 + m_3L_2)sq_2 + m_3l_3s(q_2+q_3) \\ m_3l_3s(q_2+q_3) \end{bmatrix}. \quad (23)$$

The vector of friction forces considered is modelled as follows:

$$f_i(q_i^{(1)}) = c_i q_i^{(1)} + d_i[q_i^{(1)}]^2 sign(q_i^{(1)}),\ c_i, d_i \in \Re,$$

$$\mathbf{f} = [f_1\ f_2\ f_3]^T.$$

The matrix B is diagonal matrix, $B = diag\{b_1\ b_2\ b_3\}$, where the coefficients $b_i \in \Re^+$, i=1,2,3, are known. The positive diagonality of the matrix B implies its full rank. Hence, the assumption 1 is valid.

The parameter values are the following:

$L_2 = L_3$=1m, $l_2 = l_{3link}$=0.5m, l_{23}=1m,
$m_2 = m_{3link}$=10kg,
$J_2 = J_{3link}$=3kgm^2, $J_{2p} = J_{3p}$=0.01kgm^2,
$b_1 = b_2$=40NmV^{-1}, b_3=20NmV^{-1}, g=9.81ms^{-2},
$c_1 = 1Nms$, $c_2 = 75Nms$, $c_3 = 10Nms$,
$d_1 = 5Ns^2$, $d_2 = 50Ns^2$, $d_3 = 10Ns^2$.

Let:

$$S_{q_d} = \left\{ \mathbf{q}_d(\cdot) : \mathbf{q}_d(\mathrm{T}) = \begin{bmatrix} 30^\circ + 20^\circ sin(t) \\ 10^\circ\,(1 - e^{-t}) \\ 20^\circ + 30^\circ cos(2t) \end{bmatrix} \right\},$$

for all $\mathrm{T} \in \Im^{\mathcal{T}}$,

$$\mathbf{E}_I = \left\{ \mathbf{e} : \begin{bmatrix} -60^\circ \\ -30^\circ \\ -25^\circ \end{bmatrix} \le \mathbf{e} \le \begin{bmatrix} 60^\circ \\ 30^\circ \\ 25^\circ \end{bmatrix} \right\}, \quad (24)$$

$$\mathbf{E}_A = \left\{ \mathbf{e} : \begin{bmatrix} -120^\circ \\ -75^\circ \\ -45^\circ \end{bmatrix} \le \mathbf{e} \le \begin{bmatrix} 120^\circ \\ 75^\circ \\ 45^\circ \end{bmatrix} \right\}, \quad (25)$$

and

$$\mathbf{E}_F = \left\{ \mathbf{e} : \begin{bmatrix} -12^\circ \\ -7.5^\circ \\ -4.5^\circ \end{bmatrix} \le \mathbf{e} \le \begin{bmatrix} 12^\circ \\ 7.5^\circ \\ 4.5^\circ \end{bmatrix} \right\}. \quad (26)$$

By using relations (24), (25), (26) and the definitions of $\mathbf{v}_{\overline{M}I}$, $\mathbf{v}^{\partial}_{mA}$ and $\mathbf{v}^{\partial}_{mF}$, we find:

$$\mathbf{v}_{\overline{M}I} = [60^\circ\ 30^\circ\ 25^\circ]^T,\quad \mathbf{v}^{\partial}_{mA} = [120^\circ\ 75^\circ\ 45^\circ]^T,$$

$$\mathbf{v}^{\partial}_{mF} = [12^\circ\ 7.5^\circ\ 4.5^\circ]^T.$$

Let:

$$\tau = [5\ 5\ 5]^T[s],$$

$$\mathbf{T}_S = [2\ 2.5\ 3.5]^T[s],\quad \mathbf{T}_A = [3.5\ 4\ 4.5]^T[s].$$

The initial angles of the robot are arbitrarily chosen to be:

$$\mathbf{q}_0 = [-20^\circ\ 25^\circ\ 20^\circ]^T.$$

By using this choice, the definition of S_{q_d} and $\mathbf{e} = \mathbf{q}_d - \mathbf{q}$, we find:

$$\mathbf{e}_0 = [50^\circ\ \ -25^\circ\ 30^\circ]^T \in \mathbf{E}_I.$$

The maximal value of the acceleration is chosen as $\mathbf{q}_M^{(2)} = [100\ 75\ 150]^T[deg/s^2]$. Let $\epsilon = [10^{-6}\ 10^{-6}\ 10^{-6}]^T$, $\rho_1 = 0.99$, $\rho_2 = 1.01$ and $\delta = 0.999999$.

The friction forces vector is identified by using a multiple-layer network and nonlinear differentiable transfer function (Kökösy, 1998). Networks with biases, a sigmoid layer and a linear output layer are capable of approximating any function

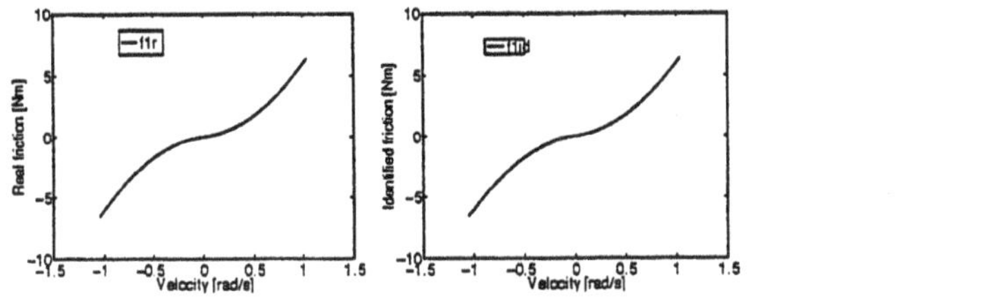

Fig. 3. Real friction forces and identified by NN

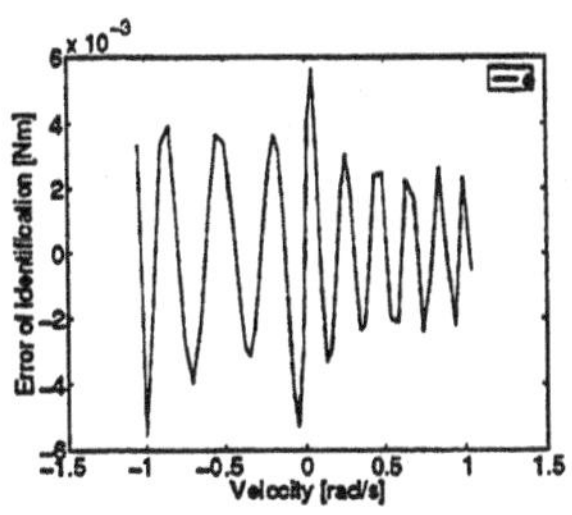

Fig. 4. Error of identification

with finite number of discontinuities. The neural network uses a backpropagation learning rule who was created by generalising the Widrow-Hoff learning rule (Demuth and Beale, 1995).

We propose to use a NN with two layers for each element of friction forces vector identification. The first layer has ten sigmoid neurons and the second one, one linear neuron. The real friction forces and the output of the NN, which represents the identified friction forces, are illustrated in figure 3. The identification error is shown in figure 4. It is in $[-6 \cdot 10^{-3}\ \ 6 \cdot 10^{-3}]$.

The value of $\mathbf{e}_M^{(1)}$ is equal to $\mathbf{e}_M^{(1)} = [25\ \ 10\ \ 7]^T$ $[deg/s]$. Using relation (15), the value of the vector λ is chosen as $\lambda = [1.04\ \ 1.11\ \ 1.19]^T$. The maximal value of $\mathbf{r}(\mathrm{T})$, $\forall \mathrm{T} \in \Im^{\mathcal{T}}$ can be calculated by using relation (17): $\mathbf{r}_M = [1.74\ \ 1.08\ \ 1.56]^T$. Let the set of all admissible angles given by robot geometric characteristics

$$\mathcal{Q}_c = \{\mathbf{q} : |\mathbf{q}| \le [140\ 90\ 60]^T[deg]\}$$

and the set of all admissible velocities given by robot kinematic characteristics

$$\mathcal{Q}_c^{(1)} = \left\{\mathbf{q}^{(1)} : |\mathbf{q}^{(1)}| \le [60\ 60\ 60]^T[deg/s]\right\}.$$

Hence, the set $\mathcal{Q}_c^1$ is now well defined

$$\mathcal{Q}_c^1 = \left\{\mathbf{q}^1 :\ \mathbf{q} \in \mathcal{Q}_c^{(1)},\ \mathbf{q}^{(1)} \in \mathcal{Q}_c^{(1)}\right\}.$$

By knowing the values of $\mathbf{r}_M$, $\mathbf{e}_M^{(1)}$, $\mathbf{q}_M^{(2)}$ and the set $\mathcal{Q}_c^1$, we can calculate the maximal value of the control $\mathbf{u}$, by using relation (13):

$$\mathbf{u}_M = [4.69\ 24.1\ 10.7]^T[V].$$

The simulation results are shown in figures 5, 6, 7 and 8. We can observe in figures 5, 6 and 7 that the initial errors belong to the set $\mathbf{E}_I$ (represented in the figures by the bold vertical segment). The errors belong to the set $\mathbf{E}_A$ all the time, reach the set $\mathcal{E}_F$ after the settling time $\mathbf{T}_S$

has elapsed and belong to it for all $\tau \in [\tau_S, \tau[$ (the bounds of the set $\mathbf{E}_F \times [\tau_S, \tau[$ are shown in the figures by two horizontal dashed lines).

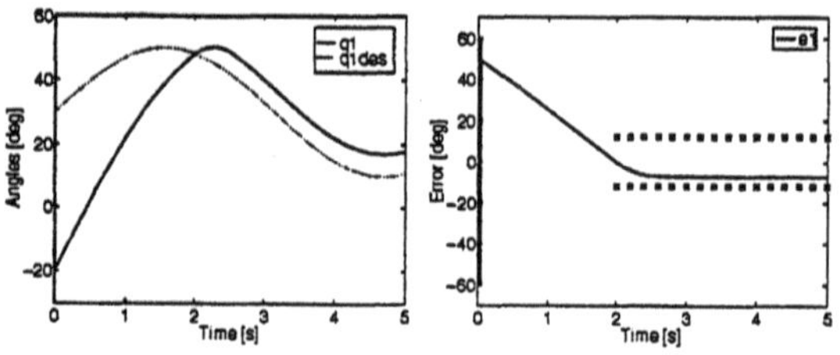

Fig. 5. Real and desired motions and their error of the first link

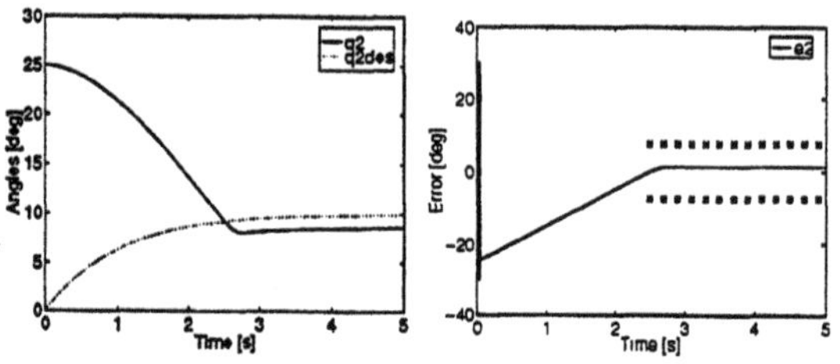

Fig. 6. Real and desired motions and their error of the second link

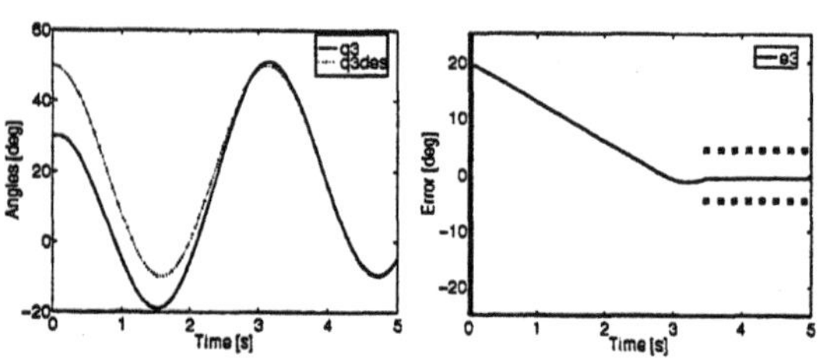

Fig. 7. Real and desired motions and their error of the third link

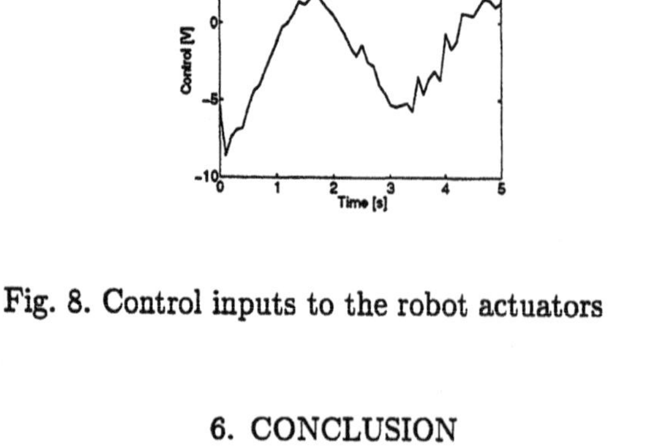

Fig. 8. Control inputs to the robot actuators

6. CONCLUSION

A new criterion for a practical tracking with a prespecified vector settling time is established. A synthesised robot control guarantees the requested tracking performance. A control bound was calculated by using information about the maximal value of the acceleration and about the set $\mathcal{Q}_c^1$, which describes the robot geometric and kinematic characteristics. The results of simulations well verify the theoretical results.

ACKNOWLEDGEMENT

The author is grateful to Professor Lyubomir T. Gruyitch, with the Ecole Nationale d'Ingénieurs de Belfort, France, for his invaluable comments, remarks and suggestions.

7. REFERENCES

Demuth, Howard and Mark Beale (1995). *Neural Network Toolbox.* The Math Works Inc.

Grujić, Lj.T. (1986). On the non-linear tracking systems theory: Ii-phenomena, concepts and problems via state-space. *Automatika* **27**, 9–16.

Grujić, Lj.T. (1993). On the tracking problem for nonlinear systems. In: *Applied Control. Current Trends and Modern Methodologies* (S.G. Tzafestas, Ed.). pp. 325–340. Marcel Dekker Inc.

Grujić, Ljubomir T. (1988). Tracking versus stability: theory. In: *12th IMACS World Congress on Scientific Computation.* Paris, France. pp. 319–327.

Kökösy, A. (1998). Robot control: practical tracking with reachability time. In: *Advances in Systems, Signals, Control and Computers.* Vol. 3. Durban, South Africa. pp. 186–190.

Kumbla, K. K. and M. Jamshidi (1997). Neural network based identification of robot dynamics used for neuro-fuzzy controller. In: *IEEE Conference in Robotics and Automation.* Albuquerque, New-Mexico. pp. 1118–1123.

Lazić, D. (1995). Analysis and synthesis of automatic practical tracking control. PhD thesis. Belgrade, Serbie. In Serbian.

Stoten, D.P. (1990). *Model reference adaptive control of manipulators.* Research Studies Press Ltd.. Somerset, England.

FINITE TIME OUTPUT TRACKING CONTROL OF NONLINEAR DYNAMIC SYSTEMS WITH NON–MINIMUM PHASE

Xinghuo Yu*, Yuqiang Wu, Man Zhihong*****

**Faculty of Informatics and Communication, Central Queensland University, Rockhampton Qld 4702 Australia. E-mail: X.Yu@cqu.edu.au*

***Department of Automation, Qufu Normal University, Qufu, Shandong, People's Republic of China*

****Department of Electrical Engineering and Computer Science, University of Tasmania, Hobart 7001 Australia. E-mail: Zhihong.Man@utas.edu.au*

Abstract. In this paper the terminal sliding mode control technique is used for solving the finite time output tracking problem for non-minimum phase nonlinear systems. A novel idea of stabilizing zero dynamics of the non-minimum phase system is proposed. The terminal sliding mode control is employed to realize the finite time convergence. *Copyright © 1999 IFAC*

Key Words. Sliding mode; variable structure systems; stability; nonlinear systems; non–minimum phase

1. INTRODUCTION

Sliding mode control systems have been studied extensively and used in many applications (Utkin 1992). The sliding mode is attained by designing the control laws which drive the system state to reach and remain on the intersection of a set of prescribed switching surfaces. When in the sliding mode, under the matched condition, the system exhibits invariance properties, such as robustness to certain internal parameter variations and external disturbances. Most commonly used switching surfaces are linear hyperplanes. Such hyperplanes would guarantee the asymptotical stability of the sliding mode. Finite time reachability of a target using sliding modes is desirable in practice. Certain nonlinear switching surfaces may give rise to a finite time sliding mode in which the system state will reach the system equilibrium in finite time. The so called *terminal sliding mode* (TSM) control is such a finite time controller (Man et al 1994; Yu and Man 1996a; Yu and Man 1996b).

In this paper we deal with the finite time output tracking using the TSMs for the SISO nonlinear non-minimum phase system which has not been fully addressed yet. We propose a novel idea for handling the zero dynamics using TSMs. In doing so, the nonlinear non-minimum phase system can be controlled to exhibit finite time convergence and global stability.

2. THE NONLINEAR DYNAMIC SYSTEM

We consider the nonlinear smooth SISO system

$$\begin{aligned} \dot{x} &= f(x) + g(x)u \\ y &= h(x) \end{aligned} \qquad (1)$$

where f and g are the smooth vector fields on $\mathcal{R}^n$, h is the scalar smooth field on $\mathcal{R}^n$, and $u \in \mathcal{R}^1$.

Before we proceed, we introduce several notations. The Lie derivative of h with respect to f is defined as the directional derivative $\mathcal{L}_f h = \nabla h\ f$ where $\nabla h = (\partial h/\partial x)$ representing the gradient of h. Higher order Lie derivatives can be defined recursively as $\mathcal{L}_f^i h = \nabla(\mathcal{L}_f^{i-1} h) f$ for $i = 1, 2, \ldots$ with $\mathcal{L}_f^0 h = h$. The Lie bracket of f and g is defined as $[f, g] = \nabla_g f - \nabla_f g$ which is commonly written as $ad_f g$. Higher order Lie brackets can be defined recursively as $ad_f^i g = [f, ad_f^{i-1} g]$ for $i = 1, 2, \ldots$ with $ad_f^0 g = g$. For details about the Lie derivative and Lie bracket, see (Isidori 1989; Slotine 1991).

Assumption 1 The SISO system (1) has relative degree r in a region $\Omega \in \mathcal{R}^n$, i.e.

$$\mathcal{L}_g \mathcal{L}_f^i h = 0 \quad 0 \le i < r-1 \qquad (2)$$

$$\mathcal{L}_g \mathcal{L}_f^{r-1} h \ne 0 \qquad (3)$$

With Assumption 1, in Ω, the nonlinear system (1) can be transformed to the following normal form (Isidori 1989)

$$\left.\begin{array}{rcl} \dot{z}_i &=& z_{i+1} \quad i=1,\ldots,r-1 \\ \dot{z}_r &=& a(z,\eta)+b(z,\eta)u \\ \dot{\eta} &=& q(z,\eta),\ y=z_1 \end{array}\right\}$$

This normal form (4) will be used in the following discussion.

In this paper, we assume that the nonlinear function $q(z,\eta)$ depends only on the output $y = z_1$, that is

$$\dot{\eta} = q(z_1,\eta) = q(y,\eta), \tag{4}$$

and the zero dynamics

$$\dot{\eta} = q(0,\eta) \tag{5}$$

is unstable, that is, the nonlinear system (1) is not minimum phase. Such a class of systems have been studied extensively (Saberi et al 1990; Sussmann and Kokotovic 1991; Byrnes and Isidori 1991).

Solving globally asymptotical stabilization problems for system (1) usually requires that the zero dynamics (5) is globally asymptotically stable, that is, the nonlinear system (1) is minimum phase. The formation of zero dynamics is due to the "unobservable" part of the nonlinear system. Usually, the "unobservable" part is not controllable directly via the actual control $u(t)$ which can only control the "observable" part. However, if we think that $y(t)$ as a virtual control and design it to stabilize the zero dynamics, we are able to design the control $u(t)$ which enables the ideal virtual control $y(t)$ to be tracked, and hence the entire system is globally stable.

In this paper, we develop a finite time output tracking control to make the output $y(t)$ track a specified signal which stabilize the zero dynamics (5). Our task is to use the TSMs to realize the accurate output tracking in finite time. In order to realizes this control objective, the following assumption is required.

Assumption 2 For the dynamics (4), there exists a virtual feedback control $y(t)$ which is defined by

$$\dot{\xi} = \psi(\xi,\eta),\ y = \phi(\xi,\eta)$$

such that the system

$$\dot{\eta} = q(\phi(\xi,\eta),\eta)$$

is asymptotically stable and as η tends to zero $\xi(t)$ and $\phi(\xi,\eta)$ tend to zero as well.

Now for system (4), if we design the control law $u(t)$ such that the output $y(t)$ accurately tracks $\phi(\xi,\eta)$ in finite time and the state $z(t)$ tends to zero, then the overall system (4) can be guaranteed to be globally asymptotically stable. Hence the main task is to make the output $y(t)$ tracking $\phi(\xi,\eta)$ in finite time. In order to realize this goal, we define the dynamical output tracking error e_1 as

$$e_1 = y - \phi(\xi,\eta) = z_1 - \phi(\xi,\eta). \tag{6}$$

Applying system (4), we obtain

$$\left.\begin{array}{rcl} \dot{e}_1 &=& z_2 - \frac{\partial\phi}{\partial\eta}q(z_1,\eta) \\ \dot{z}_j &=& z_{j+1},\ j=2,\ldots,r-1. \\ \dot{z}_r &=& a(z)+b(z)u \\ \dot{\eta} &=& q(z_1,\eta) \end{array}\right\}.$$

Using the following transformation

$$e_1 = z_1 - \phi(\xi,\eta),\ \ldots, e_r = z_r - \phi^{(r-1)}(\xi,\eta), \tag{7}$$

We obtain

$$\left.\begin{array}{rcl} \dot{e}_1 &=& e_2 \\ \dot{e}_j &=& e_{j+1},\ j=2,\ldots,r-1. \\ \dot{e}_r &=& a(z) - \phi^{(r-1)}(\xi,\eta) + b(z)u \end{array}\right\}.$$

In the following sections, we shall first look at the mechanism of TSMs and relevant results for control design. We then use the TSMs for finite time tracking control.

3. TERMINAL SLIDING MODES

In this section, we first introduce the TSM concept for SISO linear systems (Yu and Man 1996a). This scheme will be used in the next section for the design of finite time output tracking controller for system (1).

Consider the following controllable n-order SISO system in canonical form

$$\left.\begin{array}{rcl} \dot{x}_1 &=& x_2 \\ \dot{x}_2 &=& x_3 \\ \cdots & \cdots & \cdots \\ \dot{x}_{r-1} &=& x_r \\ \dot{x}_r &=& -a_1x_1 - a_2x_2 - \ldots - a_rx_r + u \end{array}\right\} \tag{8}$$

where a_i, $(i = 1,\ldots,r)$ are known. The control target is to achieve finite time convergence, i.e. x_i $(i = 1,\ldots,r)$ reach zero in finite time.

The following recursive TSM structure can be used for

the TSM control of SISO systems (Yu and Man 1996a).

$$\left.\begin{array}{rcl} s_0 & = & x_1 \\ s_1 & = & \dot{s}_0 + \beta_1 s_0^{q_1/p_1} \\ s_2 & = & \dot{s}_1 + \beta_2 s_1^{q_2/p_2} \\ s_3 & = & \dot{s}_2 + \beta_3 s_2^{q_3/p_3} \\ \cdots & \cdots & \cdots \\ s_{n-1} & = & \dot{s}_{r-2} + \beta_{r-1} s_{r-2}^{q_{r-1}/p_{r-1}} \end{array}\right\}, \tag{9}$$

where p_i, q_i $(i = 1, \ldots, r-1)$ are odd positive integers and $q_i < p_i$ $(i = 1, \ldots, r)$, and β_i $(i = 1, \ldots, r-1)$ are certain positive constants. One can easily see that the recursive structure relies on a building block

$$s_{i+1} = \dot{s}_i + \beta_{i+1} s_i^{q_{i+1}/p_{i+1}} \quad i = 1, \cdots, r-1. \tag{10}$$

When $s_{i+1} = 0$ is reached, the dynamics becomes

$$\dot{s}_i + \beta_{i+1} s_i^{q_{i+1}/p_{i+1}} = 0 \tag{11}$$

It can be proved that the solution $s_i(t)$ for (11) will reach zero in finite time because $q_{i+1} < p_{i+1}$ and $\beta_{i+1} > 0$.

If a controller can guarantees that the system state approaches the switching surface $s_{r-1} = 0$ in finite time and confine the state on the surface, $s_{n-2} = 0$ will be reached in finite time according to the definition. Successively s_{n-3} will reach zero in finite time so on so forth, finally $s_1 = 0$ and $s_0 = 0$ (that is $x_1 = 0$) will reach zero in finite time. The following lemmas present some existing results on TSM control.

Lemma 1 (Yu and Man 1996a) For the system (8), if the switching manifolds (9) is chosen, and the control law is taken as

$$u(t) = u_{eq}(t) - K sgn(s_{r-1}), \tag{12}$$

where $K > 0$ and $u_{eq}(t)$ is as

$$u_{eq}(t) = \sum_{i=1}^{r} a_i x_i(t) - \sum_{i=0}^{r-2} \beta_{i+1} \frac{d^{r-i-1}}{dt^{r-i-1}} (s_i^{q_{i+1}/p_{i+1}}), \tag{13}$$

then the system state trajectory starting from any initial state will reach the sliding mode $s_{r-1} = 0$ in finite time.

Lemma 2 (Yu and Man 1996a) If

$$\frac{q_{k+1}}{p_{k+1}} > \frac{r-k-1}{r-k} \tag{14}$$

then when $s_k \to 0$ sequentially from $k = r-2$ to $k = 0$, the control u is bounded.

It is worth noting that in the process of the state trajectory approaching the switching surface $s_{r-1} = 0$, there is a possibility that the trajectory firstly crosses $s_j = 0 (0 \leq j < r-1)$ before reaching $s_{r-1} = 0$ if the initial state is not chosen carefully. This will incur a singularity when the control law in Lemma 1 is used. In order to overcome the singularity problem, an effective means is to prohibit the trajectory from reaching the switching surface $s_i = 0$ $(i = 0, 1, ..., j-1)$ before $s_j = 0$ is reached (for some j). The idea is to find a region in the state space such that any trajectory starting from this region will not incur the singularity problem and the set of switching manifolds $s_{r-1} = 0, ..., s_1 = 0, s_0 = 0$ are reached sequentially. For this purpose, we define a set as

$$\begin{aligned} \Omega &= \{x : s_0 > 0\} \cap \{x : s_1 > 0\} \cap \ldots \\ & \quad \cap \{x : s_{r-1} > 0\}. \end{aligned} \tag{15}$$

Lemma 3 (Wu, Yu and Man 1998) The set Ω is unempty open set, that is $\Omega \neq \phi$. Furthermore, if the initial condition $x(0) \in \Omega$, then the control law will make the trajectory $x(t)$ first reach $s_{r-1} = 0$ and then $s_{r-2} = 0, ..., s_0 = 0$ sequentially. The state $x(t)$ will reach zero in finite time and the control $u(t)$ is bounded.

The above lemmas do not guarantee the boundedness of the control outside Ω. The two phase control was proposed in (Wu, Yu and Man 1998) to remedy this situation.

Define the controllability grammian as

$$G_c(t_0, t_f) \triangleq \int_{t_0}^{t_f} e^{-A\tau} b b^T e^{-A^T \tau} d\tau. \tag{16}$$

The controllability means that the matrix $G_c(t_0, t_f)$ is nonsingular for all $t_f > 0$. Since the solution to the system (8) is given by

$$x(t) = e^{A(t-t_0)} x(t_0) + \int_{t_0}^{t} e^{A(t-\tau)} b u(\tau) d\tau, \tag{17}$$

it follows that the controllability grammian-based control input

$$\begin{aligned} u(t) &= b^T e^{-A^T t} G_c^{-1}(t_0, t_f)[e^{A t_f} x(t_f) \\ & \quad - e^{-A t_0} x(t_0)] \end{aligned} \tag{18}$$

transfers any initial state vector $x(t_0)$ to any preset final state $x(t_f)$ in finite time $t_f > t_0$.

Combining this control with the TSM control we will end up with a nonsingular control strategy stated as follows:

- Select a recursive TSM structure as (9); define

properly the parameters $\beta_i > 0$ and p_i, q_i $(i = 1, ..., r-1)$ such that the conditions in Lemma 2 hold.

- In the open set Ω, choose an arbitrary point x_f.
- For the system (8) and any initial state $x(t_0)$, first choose a proper time t_f such that the state $x(t_f) = x_f$ is reached using the control (18). This step guarantees that in t_f time the trajectory will reach x_f.
- Once x_f is reached, the control is immediately switched to the TSM control law (13). Since $x_f \in \Omega$, lemma 4 and lemma 5 guarantee that the state $x(t)$ will reach zero in finite time and the TSM control law is bounded.

4. FINITE TIME OUTPUT TRACKING CONTROL

Now we look at the finite time control design for system (7) using TSMs. We choose the first switching function s_0 as $s_0 = e_1$ and the hierarchical structure of the switching manifolds is as (9). Because the form of the system (8) is the same as the controllable canonical form in (8) except for the last equation, the TSM control part is easily followed by employing (9). The difficult part is how to design the first phase control (18) as done in Section 3 to bring a state into the region Ω. The approach given in Section 3 is not directly applicable because one can not construct a known controllability grammian matrix $G_c(t_0, t_f)$ for nonlinear dynamics (8). This implies one can not obtain an accurate control signal to move the trajectory from any initial state to a fixed state $x_f \in \Omega$. However, From the discussion in Section 3, as long as the trajectory starts from Ω, the singularity will not take place. This implies that as long as the control can take the trajectory from any initial state into a small neighborhood of a preset $x_f \in \Omega$, the finite time reachability is realized. In fact, since Ω is an open set, a proper small neighborhood can be contained in Ω. Based on this consideration, we firstly choose an arbitrary $x_f \in \Omega$, set $e_f = x_f$, and define

$$\bar{e}(t) = e(t) - \bar{z}(t), \tag{19}$$

where $e(t) = [e_1, ..., e_r(t)]^T$ and $\bar{z}(t)$ is generated by the following dynamics

$$\left.\begin{array}{rcl} \dot{\bar{z}}_1 & = & \bar{z}_2 \\ \dot{\bar{z}}_2 & = & \bar{z}_3 \\ \cdots & \cdots & \cdots \\ \dot{\bar{z}}_{r-1} & = & \bar{z}_r \\ \dot{\bar{z}}_r & = & -a_1\bar{z}_1 - \ldots - a_r\bar{z}_r + v \end{array}\right\}, \tag{20}$$

where a_j $(j = 1, ..., r)$ are known and $v(t)$ is as

$$\begin{aligned} v(t) &= b^T e^{-A^T t} G_c^{-1}(t_0, t_f)[e^{At_f}\bar{z}(t_f) - \\ &\quad e^{-At_0}\bar{z}(t_0)], \end{aligned} \tag{21}$$

which is the same as (18). Such $v(t)$ will make state $\bar{z}(t)$ reach $\bar{z}(t_t) = e_f = x_f$ in time t_f which can be chosen arbitrarily. The dynamics (8) together with (20) yields

$$\left.\begin{array}{rcl} \dot{\bar{e}}_1 & = & \bar{e}_2 \\ \dot{\bar{e}}_2 & = & \bar{e}_3 \\ \cdots & \cdots & \cdots \\ \dot{\bar{e}}_{r-1} & = & \bar{e}_r \\ \dot{\bar{e}}_r & = & \sum_j^r a_j\bar{z}_j + v(t) + a(z) \\ & & -\phi^{(r-1)}(\xi,\eta) + b(z)u \end{array}\right\}. \tag{22}$$

Select a switching manifold for system (22) as

$$s = c_1\bar{e}_1 + \ldots c_{r-1}\bar{e}_{r-1} + \bar{e}_r, \tag{23}$$

where c_j $(j = 1, ..., r)$ are the coefficients of the Hurwitz polynomial $f(\lambda) = \lambda^r + c_{r-1}\lambda^{r-1} + \ldots + c_1$. A sliding mode control to drive the state error $\bar{e}(t)$ to the switching surface $s = 0$ is taken as

$$\begin{aligned} u(t) &= \frac{1}{b(z)}[-\sum_{j=1}^{r} c_j e_{j+1} + \phi^{(r-1)}(\xi,\eta) \\ &- \sum_j^r a_j\bar{z}_j - v(t) - a(z)] - \frac{1}{b(z)} s^{q_0/p_0}. \end{aligned} \tag{24}$$

where $p_0 > q_0 > 0$ and both are odd positive integers, Applying (24) to (23) and (22) yields

$$s\dot{s} = -s^{(q_0+p_0)/p_0}. \tag{25}$$

This implies that the error dynamics $\bar{e}(t)$ will reach the switching manifold $s = 0$ in finite time and will stay there forever. Along $s = 0$, since $f(\lambda)$ is Hurwitz, $\bar{e}(t)$ will tend to zero exponentially. This means that the trajectory $e(t)$ will enter a small neighborhood of the trajectory $\bar{z}(t)$ after a certain period of time T_0 which is independent of $e(t)$, $\bar{z}(t)$ and the choice of $v(t)$.

Choose the time $t_f = T_0$. Then at the time t_f, we get $\bar{z}(t_f) = x(t_f)$. Therefore, $e(t_f)$ is in a small neighborhood of $x(t_f)$. Because $x(t_f) \in \Omega$ and Ω is an open set, and the small neighborhood can be chosen to belong to Ω, we conclude that $e(t_f)$ is also in the open set Ω. This shows that controller (24) will transfer the any initial state $e(0)$ to the $e(t_f) \in \Omega$.

After a state in Ω is reached, the control can switch to the TSM controller and the finite time reachability is realized, as shown in Sections 3. Omitting tedious details of the design procedure, the TSM controller for system (8) is given as

$$u(t) = \frac{1}{b(z)}[-\sum_{i=0}^{r-2}\beta_{i+1}\frac{d^{r-i-1}}{dt^{r-i-1}}(s_i^{q_{i+1}/p_{i+1}})] \\ -a(z) + \phi^{(r-1)}(\xi,\eta) - (s_{r-1})^{q_0/p_0}]. \quad (26)$$

Similar to the analysis shown in Section 3, it follows that $e(t)$ will tend to zero in finite time which implies that the output tracking $e_1(t)$ reaches zero in finite time. Therefore after certain time we have $z_1(t) = \phi(\xi(t), \eta(t))$.

From Assumption 2, we can conclude that even though the zero dynamic is not stable, the state $\eta(t)$ will tend to zero. It can also be obtained that $y(t) = z_1(t)$ tends to zero. From the recursive structure of the switching manifolds $s_0,...,s_{r-1}$ and finite time convergence of $e(t)$, the state $z_2(t),...,z_r(t)$ will also tend to zero. All the analysis above show that the original non-minimum nonlinear phase system (4) is globally asymptotically stable under the two phase control law (24) and (26).

5. CONCLUSION

In this paper, the terminal sliding mode concept has been used in designing the finite time output tracking controller for SISO nonlinear non-minimum phase systems. A novel idea in handling the zero dynamics has been proposed that enables the unstable zero dynamics to be stabilized. The controller is superior to the conventional global stabilizing controllers in that the non-minimum zero dynamics can be indirectly controlled so that the entire system is guaranteed to be asymptotically stable.

6. REFERENCES

B. Fernandez and J. K. Hedrick (1987). Control of multivariable non-linear systems by the sliding mode method, *International Journal of Control*, **46**, 1019–1040.

A. Isidori (1989). *Nonlinear Control Systems*, Springer-Verlag, Berlin, Heidelberg.

Z. Man, A.P. Paplinski and H.R. Wu (1994). A robust MIMO terminal sliding mode control for rigid robotic manipulators, *IEEE Transactions on Automat. Control*, **39**(12), 2464–2468.

H. Sira-Ramirez (1989). Sliding regimes in general nonlinear systems: A relative degree approach, *International Journal of Control*, **50**, 1487–1506.

J.-J. E. Slotine and W. Li (1991). *Applied Nonlinear Control*, Prentice-Hall, Englewood Cliffs, NJ.

V.I. Utkin (1992). *Sliding Modes in Control Optimization*, Springer-Verlag, Berlin, Heidelberg.

X. Yu and Z. Man (1996a). Adaptive model reference following control using terminal sliding modes, *International Journal of Control*, **64**, 1165–1176.

X. Yu and Z. Man (1996b). Finite time output tracking problem using terminal sliding modes, *Proceedings of 1996 IMACS-IEEE MultiConfs: Symposium on Control, Optimization and Supervision*, **1**, 28-31, Lille, France, July 1996b.

Y. Wu, X. Yu and Z. Man (1998). Terminal sliding mode control design for uncertain dynamical systems, in press, *Systems & Control Letters.*

A. Saberi, P.V. Kokotovic, and H.J. Sussmann (1990). Global stabilization of partially linear composite systems, *SIAM J. Control and Optimization*, **28**(6), 1491–1503.

H. J. Sussmann and P. V. Kokotovic (1991). The peaking phenomenon and the global stabilization of nonlinear systems, *IEEE Trans. Automat. Control*, **36**(4), 424–439.

C. I. Byrnes and A. Isidori (1991). Asymptotic stabilization of minimum phase nonlinear systems, *IEEE Trans. Automat. Control*, **36**(10), 1122–1137.

ABSOLUTE OUTPUT NATURAL TRACKING CONTROL: MIMO LURIE SYSTEMS

Lyubomir T. Gruyitch* and William Pratt Mounfield, Jr.**

* *Laboratoire d'Automatique, Mécatronique, Productique et Systémique*
Ecole Nationale d'Ingénieurs, P.O. Box 525, 90016 Belfort Cedex, France.
ljubomir.grujic@enibe.fr
** *M & M Technologies, Inc., P.O. Box 399, Lexington, SC 29071-0399 USA.*
mmtech@netside.com

Abstract: Time-invariant Lurie type nonlinear systems with multiple unknown nonlinearities and unknown internal dynamics are assumed subjected to unknown unpredictable time variable external disturbances. Their outputs are linearly dependent on their states and the disturbances. The goal to synthesize a control ensuring elementwise exponential output absolute tracking under such uncertainties is resolved by developing an algorithm of a natural tracking control. It is based on the introduced concepts of natural absolute trackability, elementwise exponential output absolute tracking and natural tracking control. The necessary and sufficient conditions are established for natural absolute trackability of the Lurie systems and for natural tracking controls to guarantee elementwise exponential output absolute tracking. A simulation example is worked out. *Copyright © 1999 IFAC.*

Keywords: Control theory, nonlinear systems, natural trackability, tracking systems, robust control.

1. INTRODUCTION

The Lurie nonlinear systems representing an important class of nonlinear objects to be controlled have been mainly studied from the stability point of view, (Gantmacher and Yakubovich, 1965) since Lurie and Postnykov (1944). Their output responses were not considered and external disturbances were ignored. Three crucial exact approaches were developed: the algebraic S-procedure (Gantmacher and Yakubovich, 1965), the Popov frequency method (Popov, 1973) and the matrix Lyapunov-like approaches (Grujić, 1981a,b).

Studies of output absolute tracking of the Lurie systems and controller design for them started much later, (Grujić and Porter, 1980, Porter and Grujić, 1980). Such studies were based on the Popov frequency criterion and opened effective investigations on tracking of Lurie nonlinear systems (Porter, 1981). The developed control algorithms were of the classical PID type and were effective to guarantee state-bounded stable output tracking provided the internal dynamics of the linear part of the system was completely known and for a special class of Lurie nonlinearities. Besides, desired outputs and external disturbances were assumed constant. A further development of the absolute tracking of Lurie systems was achieved by using the matrix Lyapunov-like approaches in the framework of time-varying MIMO Lurie systems with multiple unknown nonlinearities, but with known linear parts, when subjected to unknown time-varying inputs (Grujić, 1988). The conditions on the nonlinearities were the same as in Grujić and Porter, (1980), Porter and Grujić, (1980). The structure of the controllers synthesized was predetermined. They were accepted to be Lurie nonlinear PI controllers.

The Lurie nonlinear models have been important for allowing impreciseness in modeling system nonlinearities. It is therefore reasonable to allow impreciseness, hence uncertainties also in modeling their linear parts. Besides, real external disturbances are time varying. Forms of their variations and their real values are usually unknown and unpredictable. In addition, system desired output responses are generally time varying. These are features of the Lurie systems treated in the paper, which impose the problem of designing a controller to guarantee to them global output tracking under the uncertainties. Such a tracking should be with a prespecified quality that is accepted to be an exponential independent convergence of each of the output error responses to the zero error. The convergence rates are prespecified independently of the Lurie system real features (being unknown).

Two new control concepts are introduced in order to solve the posed controller design problem. One is the new qualitative system and control concept called natural absolute trackability. Several natural absolute trackability properties are defined. In general, natural absolute trackability is a system property enabling the existence of a tracking control that can be synthesized without using any information about the system internal dynamics and which can force the real output to follow the desired output in a prespecified way and to reach it in a finite time for any initial output error and under the described uncertainties. Necessary and sufficient conditions are established for the natural absolute trackability.

The natural absolute trackability criteria established in the paper constitute a basis for the controller design. A controller generates a natural absolute tracking control without using information about the system internal dynamics and about the disturbances, and ensures simultaneously the requested tracking quality. The necessary and sufficient conditions are proved for such a controller synthesis.

An illustrative simulation example is worked out, which completely verifies the theory.

2. SYSTEM DESCRIPTION

A description of a MIMO Lurie system with multiple internal nonlinearities to be studied is given by (1),

$$dx/dt = Ax + Ff(x) + Bu + Dd, \qquad (1a)$$

$$y = Cx + Ed, \qquad (1b)$$

where $f(x)=[f_1(x_1)\ f_2(x_2)\ \dots\ f_n(x_n)]^T$, all vectors, matrices and vector functions have appropriate dimensions. Lurie vector nonlinearity f(.) is assumed to belong to a functional family S_{fl} composed of differentiable vector functions obeying the following usual Lurie sector condition and vanishing at the origin (Lurie and Postnykov, 1944; Gantmacher and Yakubovich, 1965):

$$f_i(x_i)/\,x_i \in [0, \infty[\text{ for } x_i \neq 0, \forall i = 1, 2, ..., n,$$
$$f_i(0) = 0, \forall i = 1,2,...,n.$$

External disturbance vector function d(.) and desired output response vector function $y_d(.)$ are from the corresponding families S_{dl} and S_{yl} of bounded differentiable vector functions.

3. NATURAL ABSOLUTE TRACKABILITY

The natural trackability concept was introduced and developed in the framework of time-invariant linear systems in Grujić and Mounfield (1992, 1993), Mounfield and Grujić (1992). It will be broadened to Lurie nonlinear systems as follows:

Definition 1. The Lurie system described by (1) is:

(a) *absolute trackable* over $S_{fl}xS_{yl}$ if and only if for every $(y_0,f,y_d)\in R^m xS_{fl}xS_{yl}$ and every $\sigma\in]0,+\infty[$ there is a control u(.) such that the system real output response $y(t;y_0;A,d,f,u)\equiv y(t)$ equals its desired output response $y_d(t)$ at time σ and on:

$$y(t;y_0;A,d,f,u)=y_d(t)\,,\ \forall(t, y_0)\in[\sigma,+\infty[xR^m. \quad (2)$$

(b) *natural absolute trackable over* $AxS_{dl}xS_{fl}xS_{yl}$ if and only if both it is absolute trackable over $S_{fl}xS_{yl}$ for every $(A,d)\in AxS_{dl}$ and the control u(.) can be generated without using information about $(A,d,f)\in AxS_{dl}xS_{fl}$.

(c) *natural absolute trackable over* $R^{nxn}xS_{dl}x\, S_{fl}xS_{yl}$, for brief *over* $S_{dl}xS_{fl}xS_{yl}$, if and only if both it is absolute trackable over $S_{fl}xS_{yl}$ for every $(A,d)\in R^{nxn}xS_{dl}$ and the control u(.) can be generated without using information about $(A,d,f)\in R^{nxn}xS_{dl}xS_{fl}$.

Absolute trackability generalizes and implies in the framework of Lurie systems the concepts of controllability (Kalman, 1961), output functional controllability (Bertram and Sarachik, 1960; Brockett and Mesarovic, 1965), output reproducibility (Bertram and Sarachik, 1960) and disturbance rejection, but versa does not hold.

Absolute trackability is therefore important as the property of a Lurie type system expressing its potential to permit existence of a control that will force the system real output to follow asymptotically any of its desired outputs from the family S_{yl} for any disturbance vector function in the family S_{dl}, for any nonlinearity f(.) belonging to the family S_{fl} and for arbitrary initial state vector x_0. However, absolute trackability does not show that the object permits the existence of a control that can be synthesized without using information about the linear part of the system.

If the system is to be absolutely trackable and simultaneously should enable the existence of a control that can be synthesized without using information about the linear part of the system then it is also natural absolute trackable. This means that natural absolute trackability of the system ensures existence of a totally robust control relative to the complete uncertainties in the system internal dynamics, external perturbations and arbitrary initial state vector.

Since natural absolute trackability is defined in the output space and the system dynamics is described in the state space, it appears interesting to discover the relevant necessary and sufficient conditions. This is achieved in the next theorem.

Theorem 1. **In order for the Lurie system described by (1) to be natural absolute trackable over $S_{dI}xS_{fI}xS_{yI}$ it is both necessary and sufficient that**

$$\mathbf{Det(G_\infty)\neq 0,\ G_\infty = \lim\{s[G(s)]: s\to\infty\}=CB.} \quad (3)$$

This condition is simple linear systems like algebraic condition. It demands the knowledge of G_∞, or equivalently of the product CB, but not of B or C separated. If the condition (3) is not satisfied then the class of systems should be reduced. Such a reduction concerns mainly a set of allowable linear parts of the systems. Let

$$A_1 = \{(A,B,C) : G_0 = CA^{-1}B \in R^{mxm}\}.$$

Theorem 2. **If det $(G_\infty) = 0$ then in order for the Lurie system described by (1) to be natural absolute trackable over $AxS_{dI}xS_{fI}xS_{yI}$ for $A\subseteq A_1$ it is both necessary and sufficient that**

$$A = \{\mathbf{(A,B,C): det(G_0)\neq 0,\ G_0 = CA^{-1}B}\}. \quad (4)$$

It should be noted that the natural trackability conditions (3) and (4) are necessary and sufficient for the absolute natural trackability of nonlinear Lurie systems described by (1), and that they are expressed in terms of the first two Markov parameters CB and $CA^{-1}B$ of the transfer matrix function $G(s)=[C(sI-A)^{-1}B]$ of the system linear part relative to the control u(.). However, the criteria do not require knowledge of separate matrices A, B, C, but only of terminal matrix values G_∞ and G_0 of G(s).

4. ELEMENTWISE EXPONENTIAL ABSOLUTE TRACKING QUALITY

The widely used tracking property is asymptotic tracking. It means that a system real output y(t) asymptotically approaches its time-varying desired output $y_d(t)$ only as time t tends to infinity. However, systems to be controlled are designed to operate over finite-time intervals rather than over infinite-time intervals. If for theoretical and mathematical reasons it is more convenient to design a control and controller to well control an object (process) over infinite time interval [0, ∞[, or if a real operation time interval can be fairly long, then the controller should ensure sufficiently fast convergence of real object output response to its desired output response over a finite time interval. Such a tracking quality can be requested possibly even independently of object properties. Besides, different nature, meaning and roles of output variables can demand different speeds of convergence of their real values to their desired values. This leads to elementwise independent tracking properties:

Definition 2. The Lurie system described by (1) and controlled by control u(.) *exhibits elementwise exponential absolute tracking over $AxS_{dI}xS_{fI}xS_{yI}$* if and only if there are positive diagonal matrices $L\in\Re_+^{mxm}$, $M\in\Re_+^{mxm}$ and $N\in\Re_+^{mxm}$, L is not less elementwise than the identity matrix I, $L\geq I$, such that also elementwise

$$|y_d(t)-y(t;y_0;A,d,f;u)|\leq L\exp(-Mt)|y_{d0}-y_0|,$$

$$\forall(t,y_0,A,d,f,y_d)\in R_+xR^mxAxS_{dI}xS_{fI}xS_{yI}.$$

What follows is aimed to design a controller generating a tracking control that will ensure (5) in which the output error response is denoted by $\mathbf{e}(t;e_0;A,d,f,u,y_d)\equiv\mathbf{e}(t)$,

$$K_1\mathbf{e}^{(1)}(t;e_0;A,d,f,u,y_d)+K_0\mathbf{e}(t;e_0;A,d,f,u,y_d)=0,$$

$$\forall(t,e_0,A,d,f,y_d)\in R_+xR^mxAxS_{dI}xS_{fI}xS_{yI}, \quad (5)$$

or equivalently:

$$\mathbf{e}(t;e_0;A,d,f,u,y_d)=[\exp(-tK_1^{-1}K_0)]e_0\equiv\mathbf{e}(t;e_0). \quad (6)$$

This shows that such a control will ensure elementwise exponential tracking.

5. NATURAL TRACKING CONTROL

The concept of the natural control is inspired by a control generated by Nature that does not use any information about a mathematical model of the controlled system (Grujić and Mounfield 1992, 1993, 1995; Mounfield and Grujić, 1992, 1994). In addition, the natural control will be requested to guarantee tracking that is the reason to call it natural tracking control.

Definition 3. A control u(.) is a *natural tracking control for the Lurie system described by (1)* if and only if:

a) it is continuous in time $t\in\Re_+$,

b) the controller can generate it without using any information about A, C, D, F, E, d(.), and f(.);

and

c) it forces the system to exhibit tracking.

The time continuity of the natural tracking control permits to consider, as usual in the theory of continuous-

The time continuity of the natural tracking control permits to consider, as usual in the theory of continuous-time control systems without delay, the ideal case that there is no delay of the signal. This means that the signal transmission delay $\varepsilon=0$. Since in reality it is positive but very small then we will write either $\varepsilon=0^+$ or $\varepsilon\to 0^+$. In this sense it will be written t^- instead of $t-\varepsilon|_{\varepsilon=0^+}$ or $t-\varepsilon|_{\varepsilon\to 0^+}$.

The left hand side of (5) will be denoted simply by $T[e(t),e^{(1)}(t)]$:

$$T[e(t),e^{(1)}(t)]=K_1e^{(1)}(t;e_0;A,d,f,u,y_d)+K_0\, e(t;e_0;A,d,f,u,y_d),$$

$$\forall(t,e_0,A,d,f,y_d)\in R_+xR^m xAxS_{d1}xS_{f1}xS_{y1}. \qquad (7)$$

Theorem 3. Let the elementwise exponential absolute tracking algorithm determined by (7), (8),

$$\mathbf{T[e(t),e^{(1)}(t)]=0,\ \forall t\in \mathfrak{R}_+,} \qquad (8)$$

be demanded over $R^{nxn}xS_{d1}xS_{f1}xS_{y1}$. Then, in order for the Lurie system described by (1) and controlled by a natural tracking control u(.) to exhibit the requested tracking it is necessary and sufficient that both

$$\mathbf{det(G_\infty)\neq 0,} \qquad (9)$$

and

$$\mathbf{b[u(t)]=b[u(t^-)]+(G_\infty)^{-1}T[e(t),e^{(1)}(t)],}$$

$$\mathbf{e(t)\equiv e(t;e_0;A,d,f,u,y_d),}$$

$$\forall(t,e_0,A,d,f,y_d)\in R_+xR^m xR^{nxn}xS_{d1}xS_{f1}xS_{y1}. \qquad (10)$$

An inherent characteristics of the natural tracking control and natural tracking controller is the use of the local unit positive feedback in the control u(.) from the controller output to its input. Although such a local unit positive feedback may not be switched on when it is separated from the other controller parts, it becomes extremely effective and useful in combination with the global negative feedback in the system output vector and its derivative from its output to the controller input. The former is to compensate for all uncertainties, and the latter is to ensure the requested tracking quality.

If the condition (9) does not hold then the next result becomes useful.

Theorem 4. Let det $(G_\infty)=0$ and the elementwise exponential absolute tracking algorithm determined by (7), (8) be demanded over $AxS_{d1}xS_{f1}xS_{y1}$. Then, in order for the Lurie system described by (1) and controlled by a natural tracking control u(.) to exhibit the requested tracking it is necessary and sufficient that both

$$A=\{(A,B,C):G_0=CA^{-1}B\in R^{mxm},\ det(G_0)\neq 0\} \qquad (11)$$

and

$$\mathbf{b[u(t)]=b[u(t^-)]+(G_0)^{-1}T[e(t),e^{(1)}(t)],}$$

$$\mathbf{e(t)\equiv e(t;e_0;A,d,f,u,y_d),}$$

$$\forall(t,e_0,A,d,f,y_d)\in R_+xR^m xAxS_{d1}xS_{f1}xS_{y1}. \qquad (12)$$

It is interesting to note that the first two matirx Markov parameters CB and $CA^{-1}B$ are important for the form of the natural tracking control and for the class A of the matrices A, B and C.

6. ROBUST TRACKING CONTROL SYNTHESIS

In order to illustrate the theory the following data were accepted for simulations:

$$A=\begin{bmatrix}0.0 & 1.0 & 0.0\\ -1.0 & -1.0 & 0.0\\ 0.0 & 0.0 & -2.0\end{bmatrix},\ B=\begin{bmatrix}0.1 & 0.2\\ -0.3 & 0.4\\ -0.1 & 0.3\end{bmatrix},$$

$$C=\begin{bmatrix}0.0 & -1.0 & 1.0\\ 3.0 & 0.0 & 1.0\end{bmatrix},\ D=\begin{bmatrix}0.0 & 0.0\\ 1.0 & 0.0\\ 0.0 & 1.0\end{bmatrix},$$

$$E=\begin{bmatrix}0.2 & 0.1\\ 0.3 & -0.1\end{bmatrix},\ F=\begin{bmatrix}1.0 & 1.0\\ 0.0 & 1.0\\ 0.0 & 1.0\end{bmatrix},\ f(y)=\begin{bmatrix}f(y_1)\\ f(y_2)\end{bmatrix},$$

$$x_0=\begin{bmatrix}-0.75\\ 0.26\\ 1.2\end{bmatrix}\Rightarrow y_0=\begin{bmatrix}0.45\\ -1.25\end{bmatrix},$$

where the form of nonlinearities is shown in Fig.1, and those of disturbances in Fig. 2.

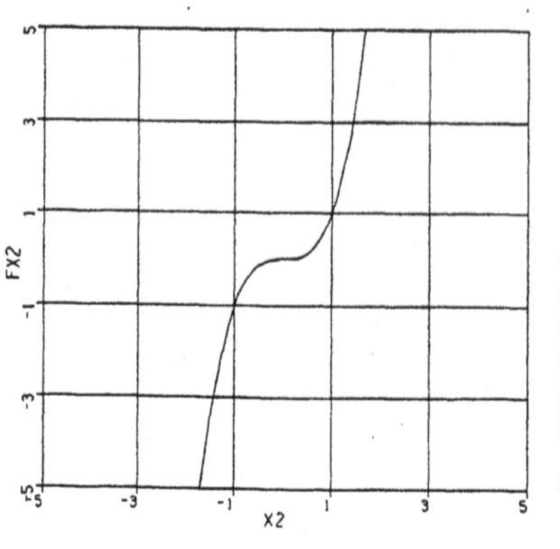

Fig. 1. A representative nonlinearity $f(x_2)$ [FX2].

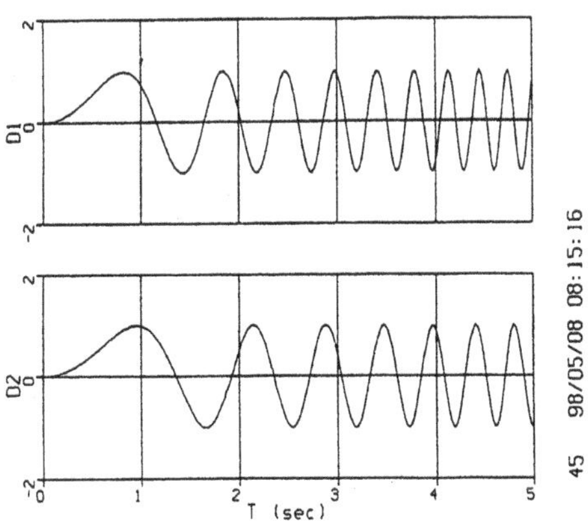

Fig. 2. **Disturbances $d_1(t)$ [D1] and $d_2(t)$ [D2].**

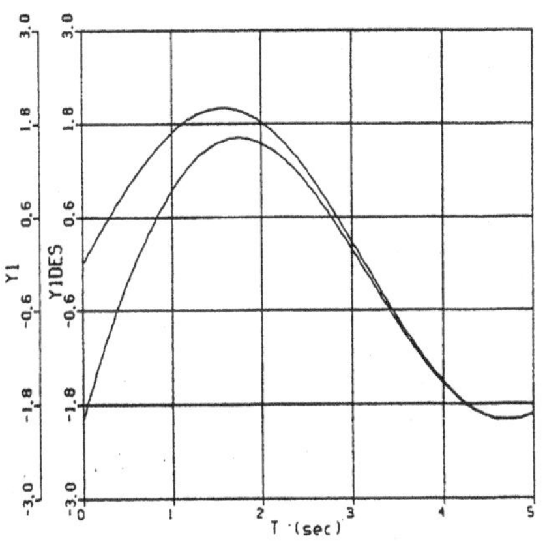

Fig. 3. **Desired output $y_{d1}(t)$ [Y1DES] and actual output $y_1(t)$ [Y1]; elementwise absolute exponential tracking.**

The real versus desired output responses are presented in Fig. 3 and Fig. 4, while their errors are shown in Fig. 5. All they verify the high quality robust elementwise tracking without oscillations in spite of highly oscillatory (unknown to the controller) disturbances, the frequency

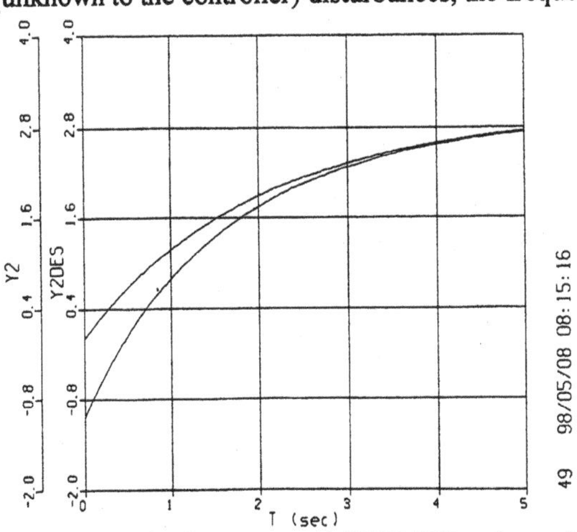

Fig. 4. **Desired output $y_{d2}(t)$ [Y2DES] and actual output $y_2(t)$ [Y2]; elementwise absolute exponential tracking.**

of which is time varying and increasing in time. The control variables are shown in Fig. 6. They are smooth, bounded and their values are technically easily realizable.

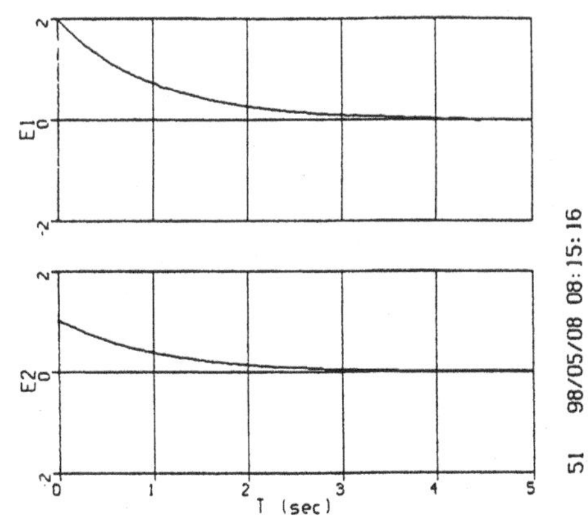

Fig. 5. **Tracking errors $e_1(t)$ [E1] and $e_2(t)$ [E2]; elementwise absolute exponential tracking.**

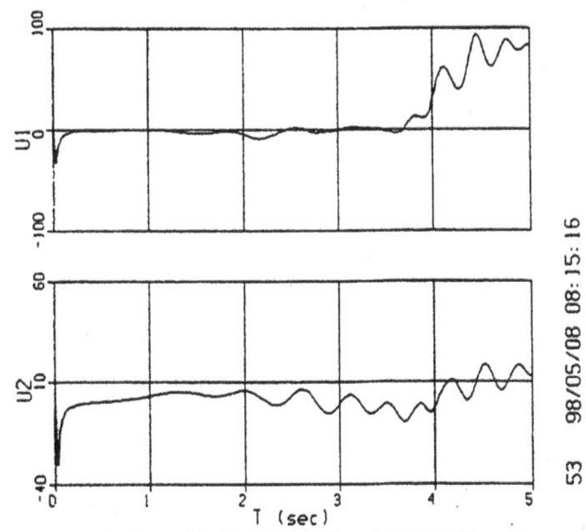

Fig. 6. **Control $u_1(t)$ [U1] and $u_2(t)$ [U2]; elementwise absolute exponential tracking.**

7. CONCLUSION

The concepts of the natural elementwise absolute trackability and of the natural absolute tracking control are introduced in the framework of MIMO Lurie nonlinear systems with unknown multiple internal nonlinearities, unknown linear parts and subjected to unknown external disturbances.

The necessary and sufficient condition for the natural absolute trackability of the Lurie system described by (1) is established in this paper. It is simple, algebraic and can be expressed in terms of the terminal matrix values of the transfer matrix function of the linear part of the system. It is used to design a natural tracking controller that will ensure excellent output error to control channeling, which is important for MIMO systems.

Natural tracking control algorithms guarantee a prespecified elementwise exponential absolute tracking quality which is completely robust relative to the system internal dynamics and external disturbances.

A simulation example verifies the theory applied to a third order MIMO Lurie system with two (unknown to the controller) internal Lurie type nonlinearities. An exponential tracking convergence rate was prespecified independently for every output variable and without using any information about the internal dynamics of the Lurie system.

REFERENCES

Bertram, J.E., and Sarachik, P.E., (1960). On Optimal Computer Control, *Proc. First Int. Congress of IFAC*, Moscow, **1**, pp. 419-422.

Brockett, R.W., and Mesarovic, M.D. (1965). The Reproducibility of Multivariable Systems, *J. of Math. Analysis and Applications*, **1**, pp. 548-563.

Gantmacher, F. R. and V. A. Yakubovich (1965). Absolute stability of non-linear control systems. In: *Analytic Mechanics, Stability of Motion and Space Balistics*, (in Russian), pp. 30 – 63, Nauka, Moscow.

Grujić, Lj. T. (1981a). On absolute stability and the Aizerman conjecture, *Automatica*, **17**, No. 2, pp. 335 - 349.

Grujić, Lj. T. (1981b). Liapunov-like solutions for stability problems of the most general stationary Lur'e-Postnikov systems, *Int. J. Systems Sci.*, **12**, No. 7, pp. 813 - 833.

Grujić, Lj. T. (1988). Algebraic conditions for absolute tracking of Lurie systems, *Int. J. Control*, **12**, No. 2, pp. 729-754.

Grujić, L.T. and Mounfield, W.P., Jr. (1992). Natural Tracking PID Process Control for Exponential Tracking, *AIChE Journal*, **38**, No. 4, pp. 555-562.

Grujić, L.T. and Mounfield, W.P., Jr. (1993). PD-Control for Stablewise Tracking with Finite Reachability Time: Linear Continuous Time MIMO Systems with State-Space Description, *International Journal of Robust and Nonlinear Control*, **3**, No. 4, December, pp. 341-360.

Grujić, L.T. and Mounfield, W.P., Jr. (1995). Natural Tracking Controller, United States Patent, Patent Number 5,379,210, January 3, pp. 1 – 61.

Grujić, Lj. T. and B. Porter (1980). Continuous-time tracking systems incorporating Lur'e plants with single non-linearities, *Int. J. Systems Sci.*, **11**, No. 2, pp. 177-189.

Kalman, R.E. (1961). On the General Theory of Control Systems, *Proc of the First International Congress of the Federation of Automatic Control*, Butterworths, London, pp. 481-492.

Lurie, A. I. and V. N. Postnykov, (1944). On the stability theory of control systems, (in Russian), *Prikl. Mat. i Meh.*, **8**, pp. 246 - 248.

Mounfield, W.P., Jr. and Grujić, L.T. (1992). High-Gain PI Natural Control for Exponential Tracking of Linear Single-Output Systems, *Automatique, Productique Informatique Industrielle (APII)*, vol. 26, No. 2, pp. 125-146.

Mounfield, W.P., Jr. and Grujić, L.T. (1994). High-Gain PI Natural Tracking Control for Exponential Tracking of Linear MIMO Systems with State-Space Description, *I. J. of Systems Science*, Vol. 25, No. 11, November, pp. 1793-1817.

Popov, V. M. (1973). *Hyperstability of Control Systems*, Springer Verlag.

Porter B. (1981). High-gain tracking systems incorporating Lur'e plants with multiple switching non-linearities, *Proc. 8th IFAC World Congress*, Kyoto, Japan, pp. 73 – 78.

Porter B. and Lj. T. Grujić (1980). Continuous-time tracking systems incorporating Lur'e plants with multiple non-linearities, *Int. J. Systems Sci.*, **11**, No. 7, pp. 827-840.

APPLICATION OF OPTIMAL CONTROL THEORY IN REHABILITATION ENGINEERING

K. J. Hunt * M. Munih ** N. de N. Donaldson ***

* *Centre for Systems and Control, Department of Mechanical Engineering, University of Glasgow, Glasgow G12 8QQ, Scotland., UK Email: k.hunt@mech.gla.ac.uk*
** *Faculty of Electrical Engineering, University of Ljubljana, Ljubljana, Slovenia. Email: marko@robo.fer.uni-lj.si*
*** *Department of Medical Physics and Bioengineering, University College London, London, England., UK Email: nickd@medphys.ucl.ac.uk*

Abstract: We are concerned with developing assistive control technologies to aid in the rehabilitation of people with spinal cord injury. Optimal control theory has been applied in feedback systems which allow paraplegic patients to stand without the support of their hands.
This is achieved by combining feedback control with functional electrical stimulation of the paralysed muscles. The approach has been used successfully in experimental trials. Previous work is reviewed in the paper and the current technical and practical problems are discussed. *Copyright ©1999 IFAC*

Keywords: Biomedical control and systems, Control applications, Optimal control, Polynomial methods, Cascade compensation

1. INTRODUCTION

This paper describes work which is concerned with the rehabilitation of people with disabilities, in particular paraplegics. The aim of the paper is to describe the way in which optimal control theory has been applied in feedback systems which allow paraplegic patients to stand without the support of their hands. Of special concern has been the question of providing clinical users of the system, who have little background in control theory, with methods and software which can be straightforwardly applied.

The overall aim of the work is to develop techniques and equipment which will allow paraplegics to stand without support of their hands and arms, utilising the powers of both their paralysed lower-limb muscles, which will be activated by electrical stimulation, and the voluntary actions of their neurologically-intact upper bodies. The equipment is ultimately intended for use in the rehabilitation clinic.

This is achieved by combining two key areas of technology: *feedback control engineering* and *functional electrical stimulation* (FES). Following appropriate training (Kralj and Bajd, 1989), FES of the paralysed muscles can provide useful functional movements; advanced feedback control methods are required in order to provide accurate, stable and safe regulation of forces generated and body posture.

Previous work has considered: methods for empirical modelling of electrically-stimulated muscle (Gollee and Hunt, 1997; Hunt *et al.*, 1998*a*); design of optimal muscle moment controllers (Hunt *et al.*, 1998*b*), and their application in paraplegic subjects (Munih *et al.*, 1996*b*; Munih *et* *al.*, 1996*a*); the development and application of optimal controllers for unsupported standing in paraplegic subjects (Hunt *et al.*, 1997; Munih *et al.*, 1997).

2. PREVIOUS RESULTS

Initial work aimed to provide control of unsupported standing in a restricted situation in which all joints above the ankle were fixed by strapping the patient into a body brace, leaving only the ankles free to move (Hunt *et al.*, 1997; Munih *et al.*, 1997). Thus, the remaining natural motor control abilities of the intact upper body were neglected, and stabilisation was achieved by stimulation of the ankle plantarflexor muscles using surface electrodes.

The overall structure of the feedback control system in these initial experiments is shown in figure 1: two inner loops, of relatively high bandwidth, provide control of the moments generated at each ankle by the electrically-stimulated muscles. An outer loop controls the angle of inclination.

These experiments were carried out by standing the subject in a purpose-built laboratory apparatus know as the 'Wobbler' (Donaldson *et al.*, 1997). This apparatus provided the appropriate measurements required by the artificial controller (see figure 1). It also allowed sinusoidal motion to be applied to the feet, so that the controllers could be tested when the subject was not balancing, thus providing more data.

This initial work has shown that successful periods of unsupported paraplegic standing can be achieved using these techniques.

3. CURRENT WORK

The work described above did not allow interaction from voluntary activity of the upper body. The work concluded that there are major issues to be addressed if the techniques and apparatus are to move towards clinical acceptability. Most important is the fact that paraplegics retain the ability to provide corrective stabilising action by use of the intact upper body (e.g. trunk muscles, arm and head movement) - it is therefore essential to integrate the natural control activity with the artificial control system: what we shall call Integrated Voluntary Control (IVC). We intend to investigate IVC controllers.

These considerations lead to the primary aims of current work in this area:

(1) design, build and test new apparatus for experimental evaluation of unsupported standing, allowing integration of the control actions of the intact upper body. It is essential that the apparatus is easy to use (getting the patient in and out) and has appropriate safety features;
(2) experimentally evaluate the method and apparatus, first using neurologically intact subjects, and finally using paraplegic subjects.

This work implies the need to extend the control techniques for stimulation of the paralysed muscles to take account of upper body movement.

The main research questions which are being investigated are:

(1) how can the voluntary motor control actions of the upper body be integrated with the artificial stimulator-controller system (i.e. development of an appropriate feedback control strategy)?
(2) how should the new apparatus be designed and built such that it fulfils the following requirements:
    - it must allow upper-body movement, and measurement of upper-body motion;
    - the apparatus should be transportable;
    - it should be suitable for use in the rehabilitation clinic, i.e. it should be as easy as possible for patients to use, and satisfy the appropriate safety requirements.
(3) what kinds of functional tasks (e.g. quiet standing, reaching for objects, lifting objects, exercising) are desirable for paraplegics while standing in the apparatus, and how well are these performed in the experimental trials?

Progress in each of these areas will be reported on in the conference presentation.

4. REFERENCES

Donaldson, N., M. Munih, G. F. Phillips and T. A. Perkins (1997). Apparatus and methods for studying artificial feedback-control of the plantarflexors in paraplegics without interference from the brain. *Medical Engineering & Physics* **19**(6), 525–535.

Gollee, H. and K. J. Hunt (1997). Nonlinear modelling and control of electrically stimulated muscle: a local model network approach. *Int J. Control* **68**(6), 1259–1288.

Hunt, K. J., M. Munih and N. Donaldson (1997). Feedback control of unsupported standing in paraplegia. Part I: optimal control approach. *Trans. IEEE on Rehabilitation Engineering* **5**(4), 331–340.

Hunt, K. J., M. Munih, N. Donaldson and F. M. D. Barr (1998*a*). Investigation of the

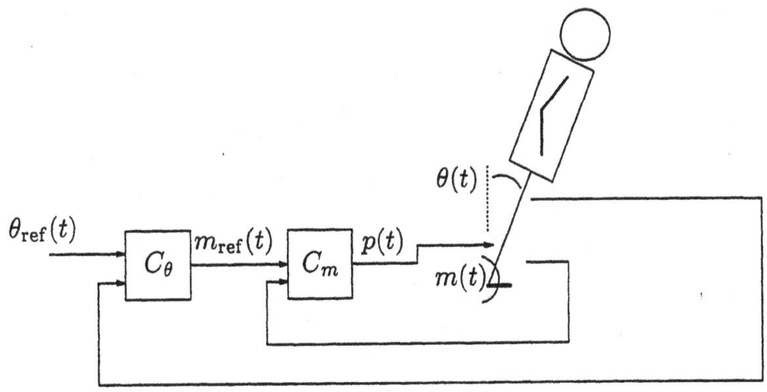

Fig. 1. Feedback control of unsupported standing (body braced above the ankles in pilot study): θ - measured ankle angle; m - measured ankle moment; p - muscle stimulation (constant amplitude and frequency, variable width pulses); θ_{ref} - angle set-point; m_{ref} - moment set-point; C_θ - angle controller; C_m - moment controller

Hammerstein hypothesis in the modelling of electrically stimulated muscle. *Trans. IEEE on Biomedical Engineering* **45**(8), 998–1009.

Hunt, K. J., M. Munih, N. Donaldson and F. M. D. Barr (1998*b*). Optimal control of ankle joint moment: toward unsupported standing in paraplegia. *Trans. IEEE on Automatic Control* **43**(6), 819–832.

Kralj, A. and T. Bajd (1989). *Functional Electrical Stimulation: Walking and Standing After Spinal Cord Injury*. CRC Press. Florida.

Munih, M., K. J. Hunt and N. Donaldson (1996*a*). LQG control for the ankle joint moment. In: *Proceedings ICMMB Conference, Ljubljana, Slovenia.*

Munih, M., K. J. Hunt, N. Donaldson and F. M. D. Barr (1996*b*). LQG moment control in the paraplegic's ankle joint. In: *Proc. 18th Int. Conf. of the IEEE Eng. in Medicine and Biology Society, Amsterdam, the Netherlands.*

Munih, M., N. Donaldson, K. J. Hunt and F. M. D. Barr (1997). Feedback control of unsupported standing in paraplegia. Part II: experimental results. *Trans. IEEE on Rehabilitation Engineering* **5**(4), 341–352.

14th Triennial World Congress, Beijing, P.R. China

HOMOGENEOUS DYNAMIC SYSTEMS THEORY [1]

Clyde Martin* E. P. Loucks Bijoy K. Ghosh*****

* *Department of Mathematics, Texas Tech University*
Lubbock, TX 79409, U.S.A.
Phone: 1-806-7422566, Email: gqcfm@ttacs.ttu.edu

** *Mallinckrodt Inc., Mallinckrodt and Second Streets*
P. O. Box 5439, St. Louis, MO 63147, U.S.A.
Phone: 1-314-5391321, Email: ted.loucks@mkg.com

*** *Department of Systems Science and Mathematics*
Washington University, St. Louis, MO 63130, U.S.A.
Phone: 1-314-9356039, Email: ghosh@zach.wustl.edu

Abstract: In this paper we introduce homogeneous dynamical systems and study many of its important properties. Specifically, we introduce *controllability* problems and show that a homogeneous dynamical system is controllable in between two states, iff the two states are in the same orbit of a *Riccati Flow*. We consider *observability* problems and generalize earlier known 'Popov Belevitch Hautus' rank test for this class of problems. We also obtain observability criterion of a homogeneous dynamical system in the presence of a control input. We introduce *realizability* problems and derive a suitable generalization of the Hankel-Rank type condition, to test for the minimality of a realization. Finally we introduce parameter *identification* problems and explicitly derive identifiability criterion for an important class of homogeneous system. *Copyright © 1999 IFAC*

Keywords: controllability, observability, realizability, Riccati dynamics, perspective, identification

1. INTRODUCTION TO A PERSPECTIVE DYNAMICAL SYSTEM

The class of problem we study in this paper has to do with the problem of controlling and observing the orientation of a state vector in $\mathbb{R}^n$. Of course, the problem of "orientation-control" has been a subject of study in the non-linear control literature for at least the last two decades (Brockett, 1981). A typical example of such a control problem is "satellite orientation control" by active means such as gas jets, magnetic torquing etc. see (Abzug, 1963). In recent years, an important example of the orientation control problem arises in Biology and is known as the 'gaze-control' problem. The 'gaze-control' problem has close connection with the problem of "eye-movement" (see (Robinson, 1968; Robinson, 1964) and (Clark and Stark, 1974)), where the problem is to orient the eye so that a target is in the field of view. An example of the dynamical system we study in this paper is described as follows

[1] This research was supported in part by DOE under grant # $DE-FG02-90ER14140$ Contract No.

$$\begin{pmatrix} \dot{x}_1 \\ \dot{x}_2 \\ \dot{x}_3 \end{pmatrix} = \begin{pmatrix} a_{11} & a_{12} & a_{13} \\ a_{21} & a_{22} & a_{23} \\ a_{31} & a_{32} & a_{33} \end{pmatrix} \begin{pmatrix} x_1 \\ x_2 \\ x_3 \end{pmatrix} + \begin{pmatrix} b_1 \\ b_2 \\ b_3 \end{pmatrix} u \tag{1}$$

$$y = \frac{c_{11}x_1 + c_{12}x_2 + c_{13}x_3}{c_{21}x_1 + c_{22}x_2 + c_{23}x_3} \tag{2}$$

where the scalar output $y(t)$ may be considered to be the slope of the line spanned by the vector $(y_1, y_2)^T$ where

$$\begin{pmatrix} y_1 \\ y_2 \end{pmatrix} = \begin{pmatrix} c_{11} & c_{12} & c_{13} \\ c_{21} & c_{22} & c_{23} \end{pmatrix} \begin{pmatrix} x_1 \\ x_2 \\ x_3 \end{pmatrix} \tag{3}$$

A dynamical system of the form (1), (2) would occur in 'orientation and gaze control problems' where one is interested in controlling only the direction of a vector. To generalize the control problem, we consider a linear dynamical system

$$\begin{aligned} \dot{x} &= Ax + Bu \\ Y &= Cx \end{aligned} \tag{4}$$

in continuous time where we assume that $x \in \mathbb{R}^n$, $n > 1$, $u \in \mathbb{R}^m$ and $Y \in \mathbb{R}^p$, $p > 1$. We do not assume that the vector Y is observed. The observation function $y(t)$ is defined as follows.

$$\begin{aligned} y : \mathbb{R}^p - \{(0,0,\dots,0)\} &\to \mathbb{RP}^{p-1} \\ Y &\longmapsto [Y] \end{aligned} \tag{5}$$

where $[Y]$ is the homogeneous line spanned by the non-zero vector $Y \in \mathbb{R}^p$. Note that $\mathbb{RP}^{p-1}$ denotes the '$p-1$'-dimensional projective space. The pair (4), (5) is a linear dynamical system with a homogeneous observation function and has been introduced in (Ghosh *et al.*, 1994; Ghosh and Loucks, 1995; Ghosh and Loucks, 1996) as an example of a perspective control system. The following two problems would initiate some of the main issues discussed in this paper.

Problem 1: (Gaze Control Problem) Let $[\xi_1]$ and $[\xi_2]$ be two distinct elements of $\mathbb{RP}^{n-1}$, does there exist a $T > 0$ and $u(t)$, $t \in [0, T]$ such that

$$[\xi_2] = [e^{AT}\xi_1^* + \int_0^T e^{A(T-\tau)}Bu(\tau)d\tau] \tag{6}$$

where ξ_1^* is a non-zero vector in $\mathbb{R}^n$ such that $[\xi_1^*] = [\xi_1]$.

Problem 2: (Gaze Observation Problem) Let $[\xi_1]$ and $[\xi_2]$ be two distinct elements of $\mathbb{RP}^{n-1}$, does there exist $T > 0$ and $u(t)$, $t \in [0, T]$ such that

$$\begin{aligned} &[Ce^{At}\xi_1^* + \int_0^t Ce^{A(t-\tau)}Bu(\tau)d\tau] \neq \\ &\quad [Ce^{At}\xi_2^* + \int_0^t Ce^{A(t-\tau)}Bu(\tau)d\tau] \end{aligned} \tag{7}$$

for all $t \in [T_1, T_2]$ where $0 \leq T_1 < T_2 \leq T$, and where ξ_1^* and ξ_2^* are two non-zero vectors in $\mathbb{R}^n$ such that $[\xi_1^*] = [\xi_1]$, $[\xi_2^*] = [\xi_2]$.

The two problems 1 and 2 refer to the extent state vector $x(t)$ can be controlled and observed upto its direction provided that the function $Y(t)$ is observed only upto direction.

Perspective control problems have been originally motivated from problems in motion and shape estimation, well known in computer vision (Kanatani, 1990). In fact, when $u(t) \equiv 0$, observability problems related to the uncontrolled dynamical system (4), (5) has already been studied in (Dayawansa *et al.*, 1995; Ghosh and Rosenthal, 1995) and explicit condition to verify observability has been obtained. In particular, the result obtained generalizes the well known (Hautus, 1969), Popov-Belevitch-Hautus test for ascertaining observability of a linear dynamical system. We have also obtained explicit criterion to identify parameters of a homogeneous dynamical system in (Ghosh *et al.*, n.d.). Our results sharpen many wellknown results in motion and shape estimation.

2. PERSPECTIVE CONTROL PROBLEM

We shall say that the dynamical system (4) is "gaze controllable" if problem 1 has an affirmative answer for almost every pair of points $[\xi_1]$, $[\xi_2] \in \mathbb{RP}^{n-1} \times \mathbb{RP}^{n-1}$ in the usual product topology. Define

$$H_i = \text{column span}\{\xi_i, B, AB, \dots, A^{n-1}B\} \tag{8}$$

for $i = 1, 2$. The following theorem characterizes the main result in gaze control.

Theorem 3: The dynamical system (4) is gaze controllable iff for almost every pair of points $[\xi_1], [\xi_2]$ in $\mathbb{RP}^{n-1}$, the following two conditions are satisfied

(1) The subspaces H_1 and H_2 are such that

$$dim\ H_1 = dim\ H_2 \tag{9}$$

(2) There exists a real number $T > 0$ such that

$$e^{AT}\ H_1 = H_2. \tag{10}$$

Note that theorem 3 may be viewed as a criterion for checking gaze controllability of a homogeneous system (4), in between the two directions $[\xi_1]$ and $[\xi_2]$. The following is an important corollary for gaze control.

Corollary 4: The dynamical system (4) is gaze controllable iff

$$\dim H \geq n - 1. \tag{11}$$

Remark 5: In this remark we point out that the perspective controllability problem has connection with a flow on an associated Grassmannian which is described by a Riccati equation. Let us assume that

$$\dim H_1 = \dim H_2 = n_1 \tag{12}$$

and let us represent by Grass (n_1, n), the space of all n_1 dimensional homogeneous planes in $\mathbb{R}^n$. Clearly the equation

$$\dot{x} = Ax \tag{13}$$

defines a flow in Grass (n_1, n) described by

$$\begin{aligned} \chi : \text{Grass}(n_1, n) \times \mathbb{R} &\to \text{Grass}(n_1, n) \\ (H, T) &\longmapsto e^{AT} H. \end{aligned} \tag{14}$$

The flow (13) can also be described via a Riccati Equation as follows. Let us denote

$$H = \text{span of } \begin{bmatrix} \Theta_0 \\ \Theta_1 \end{bmatrix}. \tag{15}$$

Where Θ_0 is a $n_1 \times n_1$ nonsingular matrix and where Θ_1 is a $(n - n_1) \times n_1$ matrix. Writing

$$X = \begin{pmatrix} X_0 \\ X_1 \end{pmatrix}, \quad A = \begin{pmatrix} A_{11} & A_{12} \\ A_{21} & A_{22} \end{pmatrix} \tag{16}$$

Where X is a $n \times n_1$ matrix, and defining $W = X_1 X_0^{-1}$, we have the equation

$$\dot{W} = A_{21} + A_{22} + W - WA_{11} - WA_{12}W, \tag{17}$$

$$W(0) = \Theta_1 \Theta_0^{-1}.$$

It is trivial to verify that the Riccati Flow (17) is an equivalent representation of the homogeneous flow (14) upto time t when $X_0(t)$ is singular. The differential equation for the homogeneous flow (14) can be written as

$$[\dot{X}] = A[X] \tag{18}$$

where $[X]$ is the column span of the $n \times n_1$ matrix.

Theorem 6: The dynamical system (4) is gaze controllable in between the two directions $[\xi_1]$ and $[\xi_2]$ in $\mathbb{RP}^{n-1}$ iff the subspaces H_1 and H_2 are in the same orbit of the homogeneous flow (14) where n_1 = dimension of H_1 = dimension of H_2.

Remark 7: Theorem 6 is a restatement of theorem 3.

3. PERSPECTIVE OBSERVABILITY PROBLEM

Perspective observation problem or the gaze observation problem has already been introduced in the introduction as Problem 2. The essential question is that if the state $x(t)$ of the dynamical system (4), (5) is not homogeneously observable from the output function $y(t)$ in (5), with $u(t) = 0$, can it be made observable by a proper choice of a control signal $u(t)$. It should be noted that for a linear dynamical system, observability criterion is independent of the control signal. However, (4), (5) is a linear system with homogeneous observation function and we have the following result.

Lemma 8: Consider the dynamical system (4), (5). The pair of points $[\xi_1]$ and $[\xi_2]$ in $\mathbb{RP}^{n-1}$ are perspectively unobservable iff

$$\dim \text{ span } Ce^{A\sigma}(H_1 + H_2) \leq 1 \tag{19}$$

for almost every σ in $\mathbb{R}$.

We are now in a position to state and prove one of the main theorems of this section.

Theorem 9: The dynamical system (4), (5) is gaze unobservable over the base field $\mathbb{C}$ iff there exists a pair of eigenvalues λ_0, λ_1 of A, such that for all pairs of complex numbers μ_0, μ_1 in the set $\{\lambda_0, \lambda_1, \delta_1, \ldots, \delta_s\}$ one has

$$\text{rank} \begin{pmatrix} (A - \mu_0 I)(A - \mu_1 I) \\ C \end{pmatrix} < n \tag{20}$$

where $\delta_1, \ldots, \delta_s$ is the set of eigenvalues of A such that the subspace spanned by the corresponding eigenvectors (generalized eigenvectors) of A is H.

Remark 10: The dynamical system (4), (5) is gaze observable, over the base field $\mathbb{C}$ iff for every pair of eigenvalues λ_0, λ_1 of A, there exist some pair μ_0, μ_1 in the set $\{\lambda_0, \lambda_1, \delta_1, \ldots, \delta_s\}$ such that

$$\text{rank} \begin{pmatrix} (A - \mu_0 I)(A - \mu_1 I) \\ C \end{pmatrix} = n \tag{21}$$

Corollary 11: Assume that $\dim H \geq 2$, the dynamical system (4), (5) is gaze observable over the base field $\mathbb{C}$ iff

$$\text{rank}(CB, CAB, \ldots, CA^{n-1}B) \geq 2. \tag{22}$$

Assume that $\dim H = 1$, the dynamical system (4), (5) is gaze observable over the base field $\mathbb{C}$ iff for every pair of eigenvalues μ_0, μ_1 of A with corresponding eigenspace (generalized eigenspace) containing H, we have (21).

A special case of the gaze observability problem 2 is to consider the case when $u(t) \equiv 0$. This is identical to the case when $B = 0$ and the subspace H is empty. In this case, theorem 9 reduces to the following well-known (Dayawansa *et al.*, 1995) result.

Proposition 12: The dynamical system (4), (5) is gaze un-observable (when $B = 0$) over the base field $\mathbb{C}$ iff there exists a pair of eigenvalues λ_0, λ_1 of A such that

$$\text{rank} \begin{pmatrix} (A - \lambda_0 I)(A - \lambda_1 I) \\ C \end{pmatrix} < n \tag{23}$$

Over the base field $\mathbb{R}$, the above condition (23) is necessary but not sufficient. In fact it is sufficient if all eigenvalues of A are real.

In the general case, when the eigenvalues of A are possibly complex, it is unclear what a suitable necessary and sufficient condition is over the base field $\mathbb{R}$. The following theorem is in this direction. In order to state the theorem we define the set $\mathcal{S}$ of linearly independent pairs of vectors in $\mathbb{R}^n$ as follows

$$\mathcal{S} \triangleq \{(v_1, v_2) \in \mathbb{R}^n \times \mathbb{R}^n : v_1 \text{ and } v_2$$

are linearly independent and we can write

$$v_1 = \sum_{i=1}^{r} \alpha_i u_i, v_2 = \sum_{i=1}^{s} \beta_i u_i, r \leq s$$

where $u_1, \ldots, u_r, \ldots, u_s$ are generalized eigenvectors of A corresponding to eigenvalues

$$\lambda_1, \ldots, \lambda_r, \ldots, \lambda_s,$$

and let p_i be the order of u_i, with the property that the generalized eigenvectors can be arranged so that $\lambda_r + \lambda_s$ is real and if

$$\begin{array}{ll} \lambda_i + \lambda_j = \lambda_r + \lambda_s & 1 \leq i \leq r, 1 \leq j \leq s \\ \text{then } \lambda_i = \lambda_r \text{ and } \lambda_j = \lambda_s. & \end{array} \tag{24}$$

and if $\lambda_i + \lambda_j = \lambda_r + \lambda_s$ $p_i + p_j = p_r + p_s$ then $p_i = p_r$ and $p_j = p_s\}$.

We now have the following theorem:

Theorem 13: The following two conditions are equivalent, over the base field $\mathbb{R}$.

(1) There exists a pair of real or complex conjugate eigenvalues λ_1, λ_2 of A such that

$$\text{rank}\begin{pmatrix} (A - \lambda_1 I)(A - \lambda_2 I) \\ C \end{pmatrix} < n. \tag{25}$$

(2) The dynamical system

$$\begin{aligned} \dot{x} &= Ax \\ y &= [Cx] \end{aligned} \tag{26}$$

does not observe a pair of vectors v_1, v_2 that belong to $\mathcal{S}$.

4. PERSPECTIVE REALIZATION PROBLEM

In this section we propose to answer the question—When is a rational, exponential function minimal? Of course, such a function can be realized in the form of a homogeneous dynamical system

$$\begin{aligned} \dot{x} &= Ax \\ Y_1 &= C_1 X \\ Y_2 &= C_2 X \\ Y &= Y_1/Y_2 \end{aligned} \tag{27}$$

Where $X \in \mathbb{R}^n$ and $Y_1, Y_2 \in \mathbb{R}$. The basic question of importance is, when is n minimal. In general, consider a homogeneous dynamical system

$$\dot{X} = AX, X(0) = X_0 Y = CX \tag{28}$$

where $X \in \mathbb{RP}^{n-1}$, $[X_0] \in \mathbb{RP}^{n-1}$, $Y \in \mathbb{RP}^{p-1}$, $n > 1$, $p > 1$. The following result summarizes an important result of this section.

Theorem 14: A sufficient condition for the homogeneous dynamical system (28) to be minimal is that the vectors

$$\{CX_0, CAX_0, \ldots, CA^{n-1}X_0\} \tag{29}$$

are all linearly independent.

In order to prove theorem 14 we need the following lemma.

Lemma 15: The vectors

$$CX_0, CAX_0, \ldots, CA^{n-1}X_0$$

are linearly independent if and only if there exists T such that the vectors

$$CX_0, Ce^{AT}X_0, \ldots, Ce^{(n-1)AT}X_0$$

are linearly independent.

We remark at this point that the sufficiency condition in theorem 14 is good provided $p \geq n$, i.e. provided the dimension of the output is at least the dimenion of the state. This leaves the following question:

"Under what condition, is a homogeneous system (4), (5) minimal when $p < n$?"

Presently, there does not exist a condition similar to theorem 14 for this case. Fortunately, the condition of minimality can be checked via a "Hankel-type" rank condition which we now describe in the following examples.

Example 16: We consider a problem where $p = 2$ and $n = 3$. Assume that the output $y(t)$ of the homogeneous dynamical system is given by the rational, exponential function

$$y(t) = \frac{2 + 3e^{4t} + 6e^{6t}}{1 + 3e^{4t} + 7e^{6t}} \tag{30}$$

We would like to know if the representation (30) for the function $y(t)$ is minimal. Expanding (30) in power series as

$$y(t) = h_0 + h_1 e^{2t} + h_2 e^{4t} + h_3 e^{6t} + h_4 e^{8t} + \ldots \tag{31}$$

we obtain via cross multiplication

$$\begin{aligned} &h_0 = 2, h_1 = 0, 3h_0 + h_2 = 3, 7h_0 + 3h_1 + h_3 = 6 \\ &\quad 7h_1 + 3h_2 + h_4 = 0 \\ &h_j + 3h_{j-2} + 7h_{j-3} = 0 \quad j = 4, 5, 6, \ldots. \end{aligned} \tag{32}$$

Solving (32) we obtain

$h_0 = 2, h_1 = 0, h_2 = 3 - 6 = -3.$ $h_3 = 6 - 14 = -8$

and

Therefore $y(t) = 2 - 3e^{4t} - 8e^{6t} + h_4 e^{8t} + h_5 e^{10t} + \ldots$.

If the matrix

$$\begin{pmatrix} h_1 & h_2 & h_3 & h_4 \\ h_2 & h_3 & h_4 & h_5 \\ h_3 & h_4 & h_5 & h_6 \\ h_4 & h_5 & h_6 & h_7 \end{pmatrix}$$

has rank 3, and if

$$\begin{pmatrix} h_1 & h_2 & h_3 \\ h_2 & h_3 & h_4 \\ h_3 & h_4 & h_5 \end{pmatrix} \tag{33}$$

also has rank 3, it would follow that the representation (30) is unique. In this example, it is easy to check that the matrix (33) is indeed non-singular. Hence (30) is also minimal.

Example 17: Let us now consider $y(t)$ in the following form

$$y(t) = \frac{2 + 3e^{4t} + 6e^{6t} + e^{\pi t} + e^{3\pi t}}{1 + 3e^{4t} + 7e^{6t} + 2e^{\pi t} + 3e^{3\pi t}} \tag{34}$$

Expanding (34) in power series as

$$z(t) = \sum_{i,j=0}^{\infty} h_{ij} e^{i2t} e^{j\pi t} \tag{35}$$

we have

$$h_{00} = 2, h_{10} = 0, 3h_{00} + h_{20} = 3,$$
$$7h_{00} + 3h_{10} + h_{30} = 6$$
$$h_{j0} + 3h_{j-2,0} + 7h_{j-3,0} = 0 \quad j = 4,5,6,\ldots.$$

Likewise we have

$$h_{00} = 2, h_{01} = 2h_{00} = 1, h_{02} + 2h_{01} = 0,$$
$$h_{03} + 2h_{02} + 3h_{00} = 1$$
$$h_{0j} + 2h_{0,j-1} + 3h_{0,j-3} = 0 \quad j = 4,5,6,\ldots.$$

In general

$$h_{ij} + 3h_{i-2,j} + 7h_{i-3,j} + 2h_{i,j-1} + 3h_{i,j-3} = 0 \tag{36}$$

Writing (36) as follows

$$\begin{pmatrix} h_{i-2,j} & h_{i-3,j} & h_{i,j-1} & h_{i,j-3} \\ h_{i-3,j+1} & h_{i-4,j+1} & h_{i-1,j} & h_{i-1,j-2} \\ h_{i-1,j-1} & h_{i-2,j-1} & h_{i+1,j-2} & h_{i+1,j-4} \\ h_{i,j-2} & h_{i-1,j-2} & h_{i+2,j-3} & h_{i+2,j-5} \\ h_{i-4,j+2} & h_{i-5,j+2} & h_{i-2,j+1} & h_{i-2,j-1} \end{pmatrix} \begin{pmatrix} 3 \\ 7 \\ 2 \\ 3 \end{pmatrix} = \begin{pmatrix} h_{ij} \\ h_{i-1,j+1} \\ h_{i+1,j-1} \\ h_{i+2,j-2} \\ h_{i-2,j+2} \end{pmatrix} \tag{37}$$

If the matrix in (37) is of rank 4, it would follow that the co-efficients 3, 7, 2, 3 are unique. In turn this would imply that the representation (34) is unique and hence minimal. The rank condition on the $\{h_{ij}\}$ matrix, is the "hankel-rank" type condition that would guarantee minimality of a homogeneous system.

Note in particular that in Example 16, we have precisely a Hankel matrix. On the other hand in Example 17 we have a 2 dimenional recursion and the associated matrix is not exactly a Hankel matrix.

5. TWO EXAMPLES

In the last section of this paper, we consider two illustrative examples. Our first example pertains to the notion of observability as discussed in section 3. The second example relates to the circle of ideas in parameter identification.

Example 18: Consider a homogeneous dynamical system

$$\dot{X} = AX, \quad Y = CX \tag{38}$$

where

$$A = \begin{pmatrix} 1 & 1 & 0 & 0 \\ -1 & 1 & 0 & 0 \\ 0 & 0 & 2 & 2 \\ 0 & 0 & -2 & 2 \end{pmatrix}, C = \begin{pmatrix} 1 & -1 & 1 & 0 \\ 1 & 1 & 0 & 1 \end{pmatrix}. \tag{39}$$

The above dynamical system has already been introduced in (Dayawansa *et al.*, 1995) as an example of a homogeneous system which is unobservable over the base field $\mathbb{C}$. In fact it is easy to choose a pair of complex eigenvalues λ_1, λ_2 of the matrix A such that

$$rank \begin{pmatrix} (\lambda_1 I - A)(\lambda_2 I - A) \\ C \end{pmatrix} < 4. \tag{40}$$

It can also be checked that for every pair of complex conjugate eigenvalues λ_1, λ_2 of A, the rank condition (25) (or (40)) is not satisfied. It follows from theorem 13 that every pair of vectors in $\mathcal{S}$ (defined in 24) is observable over the base field $\mathbb{R}$. It is shown in (Dayawansa *et al.*, 1995), using a separate calculation that, not just pairs of vectors in $\mathcal{S}$, but rather every pair of vectors in $\mathbb{R}^4$ is observable for the homogeneous system (38), (39).

Example 19: In this example, we consider a homogeneous dynamical system of the form (38) where

$$A = \begin{pmatrix} -a_{11} & -a_{21} & 0 \\ -a_{12} & -a_{22} & 0 \\ -b_1 & -b_2 & 0 \end{pmatrix} \tag{41}$$

and

$$C = \begin{pmatrix} 0 & -b_1 & a_{12} \\ -b_1 & b_2 & a_{11} - a_{22} \\ b_2 & 0 & -a_{21} \\ 0 & 0 & 1 \end{pmatrix}. \tag{42}$$

Note the existence of 6 unconstrained parameters a_{11}, a_{12}, a_{21}, a_{22}, b_1, b_2 in the pair of matrices A and C. The problem that we would like toconsider in this example is the following: "To what extent is it possible to identify the six parameters from the observation function $Y(t)$?

Let us assume first of all that the dynamical system has been initialized at $X(0) = X_0$. It follows from theorem 14 that the homogeneous system has a minimum state dimension if

$$rank\ (CX_0, CAX_0, CA^2X_0,) = 3, \quad (43)$$

which is satisfied for a generic choice of parameters

$$(a_{11}, a_{12}, a_{21}, a_{22}, b_1, b_2, X_0).$$

For a minimal homogeneous dynamical system, it is known (Ghosh and Loucks, 1995) that the parameters can be identified upto orbit of a perspective group $\mathcal{G}$ on the triplet (C, A, X_0), whose action is described as follows.

$$\begin{aligned}(C, A, X_0), P &\longmapsto (CP, P^{-1}AP, P^{-1}X_0)\\ (C, A, X_0), \lambda &\longmapsto (C, A + \lambda I, X_0)\\ (C, A, X_0), \mu_1, \mu_2 &\longmapsto (C\mu_1, A, X_0\mu_2).\end{aligned} \quad (44)$$

In general, however, the perspective group $\mathcal{G}$ does not preserve the structure of the matrix pair (A, C). It is relatively easy to see that the matrix P which preserves the structure of C is given by

$$P = \begin{pmatrix} p_{11}I & \Theta \\ 0 & p_{33} \end{pmatrix},$$

where $\Theta = (p_{13}, p_{23})^T$. In order that $P^{-1}AP + \lambda I$ has the same structure as that of the matrix A, it is easily verifiable that Θ is either an eigenvector of

$$\begin{pmatrix} a_{11} & a_{21} \\ a_{12} & a_{22} \end{pmatrix} \quad (45)$$

with the eigenvalue given by $\frac{1}{p_{33}}(b_1p_{13} + b_2p_{23})$ or $\Theta = 0$.

6. CONCLUSION

To conclude, we have introduced, in this paper, various problems pertaining to a homogeneous dynamical system. These include obtaining a verifiable condition to test for controllability and observability. The controllability condition has connection with the Riccati Flow on an associated Grassmannian whereas the observability condition generalizes the well-known Popov Belevitch Hautus rank condition on testing the observability of a linear dynamical system. Finally we discuss condition under which the state dimension of a homogeneous dynamical system is minimal. In general, we show, that this problem has connection with a 'Hankel Type Rank Condition'. Overall, we show, that a homogeneous dynamical system has a rich algebraic structure comparable to that of a linear dynamical system.

7. REFERENCES

Abzug, M. J. (1963). On the stability of a class of discontinuous attitude control systems. *AIAA J.* **1(8)**, 1910–1911.

Brockett, R. W. (1981). Nonlinear systems and nonlinear estimation theory. *Stochastic Systems: The Mathematics of Filtering and Identificaiton and Applications, M. Hazewinkel and J. C. Willems (eds.)* pp. 441–477.

Clark, M. R. and L. Stark (1974). Control of human eye movements; i. modelling of extraocular muscles; ii. a model for extraocular plant mechanism; iii. dynamic characteristics of the eye tracking mechanism. *Math Biosc.* **20**, 191–265.

Dayawansa, W. P., B. K. Ghosh, C. Martin and X. Wang (1995). A necessary and sufficient condition for the perspective observability problem. *Systems and Control Letters* **25**, 159–166.

Ghosh, B. K. and E. P. Loucks (1995). A perspective theory for motion and shape estimation in machine vision. *SIAM J. Control and Optim.* **33(5)**, 1530–1559.

Ghosh, B. K. and E. P. Loucks (1996). A realization theory for perspective systems with applications to parameter estimation problems in machine vision. *IEEE Trans. on Automatic Control* **41(12)**, 1–17.

Ghosh, B. K. and J. Rosenthal (1995). A generalized popov-belevitch-hautus test of observability. *IEEE Trans. on Automatic Control* **40(1)**, 176–180.

Ghosh, B. K., H. Inaba and S. Takahashi (n.d.). Identification of Riccati Dynamics under perspective and orthographic Observation. *submitted to IEEE Trans. on Automatic Control.*

Ghosh, B. K., M. Jankovic and Y. T. Wu (1994). Perspective problems in system theory and its application to machine vision. *J. Math. Systems Estim. Control* **4**, 3–38.

Hautus, M. L. J. (1969). Controllability and observability conditions of linear autonomous systems. *Ned. Akad. Wetenschappen, Proc. Ser. A* **72**, 443–448.

Kanatani, K. (1990). *Group-Theoretical Methods in Image Understanding.* Springer Verlag.

Robinson, D. A. (1964). The mechanics of human saccadic eye movement. *J. of Physiol.* **174**, 245–264.

Robinson, D. A. (1968). The oculomotor control system: A review. *Proc. IEEE* **56**, 1032.

14th Triennial World Congress, Beijing, P.R. China

OCULAR DYNAMICS AND CONTROL

Clyde F. Martin *,[1] **Lawrence Schovanec** **,[2]

* *Department of Mathematics and Statistics*
Texas Tech University
Lubbock, TX, USA
** *Department of Mathematics and Statistics*
Texas Tech University
Lubbock, TX, USA
schov@math.ttu.edu

Abstract: Eye movement systems are ideal for studying human control of movement since they are of relatively low dimension and easier to control than other neuromuscular systems. In this paper, a model of the ocular motor plant is developed for the purpose of simulating saccadic eye movement. The system that is presented incorporates muscle mass, general nonlinear musculotendon dynamics and activation dynamics that couple neural controls that are appropriate for saccadic movements to the muscle mechanics. Eye position, velocity and muscle tension are illustrated by numerical simulation of the model. By scrutinizing the trajectories of the eye movements it is possible to infer the effects of motoneuronal activity, deduce the central nervous system's control strategy, and systematically observe the effects of perturbations in the controls. *Copyright © 1999 IFAC*

Keywords: Biocontrol, Biomedical Systems, Nonlinear control systems

1. INTRODUCTION

The eye movement system can be regarded as a nonlinear feedback system that consists of four subsystems concerned with the control of accommodation, vergence, illumination and version. This paper is concerned with a specific movement, horizontal saccades. The purpose of saccadic eye movement is to position the high-resolution fovea, the central part of the retina, on the important features of a scene. Motivation for this study is provided by the anticipation that analysis of saccadic movements will facilitate the study of eye movements and more generally, other human movement systems. Because of the relative simplicity of this horizontal eye system, an analysis of the trajectories, velocities, accelerations, and forces of simulated movements provides information into the role of motoneuronal activity, the central nervous system's control strategy, and effects of pathology and system disturbances on the performance of the system. The insight gained by such analysis should provide insight into more complicated neuromuscular systems.

In human binocular vision the movement of each eye is controlled by a set of six muscles. These muscles work in three agonist-antagonist pairs to exert forces on the eye causing it to rotate. When the eyes are fixed on an object two things occur. First, the eye rotates so that the image of the object formed by the eyes' lens system is projected onto the fovea of the retina. Secondly, the eye lenses adjust to bring the object into focus. The brain and central nervous system process information obtained by the retina, and then transmit signals to the extraocular muscles. This paper is concerned with the first part of this process: how the rotation develops and how it is controlled.

The model proposed herein builds upon earlier models of the oculomotor system(see (Clark and Stark, 1974)

[1] Supported by NSF grants ECS-9720357, DMS-9628558
[2] Supported by NSF grants ECS-9720357, DUE-9813056

and references therein). However, the model presented here incorporates more general formulations of the musculotendon dynamics and allows for the inclusion of muscle mass. There are several reasons that a meaningful model of the horizontal saccadic movement system can be developed. Experimental data has been gathered that provides values for many of the parameters that arise in the description of the model. Most importantly in this regard is information regarding the time course of innervation that is specific to saccadic movement. Experimental data provides fairly conclusive information regarding the nature of the inputs, or controls, that are appropriate to saccades. More specifically, the input into the oculomotor plant is derived from experimental records of raw EMG corresponding to the firing rate of motor units. It is accepted as 'sufficiently realistic as to be useful' (Robinson, 1991) that the rate of discharge corresponding to a saccadic movement can be described by filtered EMG that corresponds to a rectangular pulse followed by a step. Furthermore, because saccades are among the fastest muscle movements that are performed, vision is suppressed during this motion and it is generally accepted that the the oculomotor system operates in an open-loop mode when performing a saccadic movement.

A major objective of this work is the incorporation of physiologically realistic actuation and musculotendon dynamics into the plant model of the eye. In particular the muscle model that is utilized is a two-component version of a Hill type model that consists of a passive and active contractile components. The approach that is presented here allows for the inclusion of very general force-velocity and force-length characteristics in the active component. Furthermore, the model that is developed allows one to distinguish between the elastic and viscous effects in the muscle actuator that are neurologically controlled and those that are due to the passive structure of the actuator.

2. MUSCULOTENDON DYNAMICS

Muscles are the actuators of the neuro-musculo-skeletal control system that produce movement. In the analysis of a control system as complex as that governing movement it is essential to have a clear understanding of the physical nature of the actuators and a tractable mathematical representation of their dynamics. The muscle models utilized in this investigation are referred to as Hill-type models. A phemenological representation of the musculotendon complex as idealized mechanical objects is presented in Figure 1. This model has been shown to incorporate enough complexity while remaining computationally practical.

The muscle of length l_m is in series and off-axis by a pennation angle α with the tendon of length l_t. The total pathlength of the musculotendon complex is denoted by l_{tm}. The muscle is assumed to consist

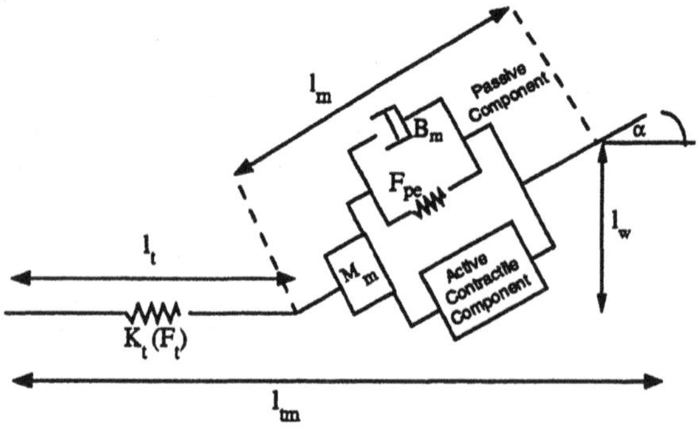

Fig. 1. Model of musculotendon complex

of two components: an active force generator and a parallel passive component. The passive component includes a parallel elastic element (F_{pe}) that describes the passive muscle elasticity and a damping component which corresponds to the passive muscle viscosity (B_m). The model for the active contractile component is based on the generally accepted notion that the active muscle force is the product of three factors: (1) a length-tension relation $f_l(l_m)$, (2) a velocity-tension relation $f_v(\dot{l}_m)$, and (3) the activation level $a(t)$. In this paper, analytical expressions that capture the qualitative properties of the curves will be used. The parameters that appear in these expressions will be determined by imposing smoothness conditions in combination with a fit of experimental data. The parameters needed for each musculotendon group include: (1) maximal isometric active muscle force F_o, (2) optimal muscle length, l_o, (3) pennation angle α_o when $l_m = l_o$, and (4) tendon slack length l_{ts}. In developing nondimensional representations for these curves the approach of (Zajac, 1991) is implemented and all forces and lengths are scaled as $\tilde{F} = F/F_o$ and $\tilde{l}_m = l_m/l_o$. Another quantity used to specify a muscle specific force-velocity relation is the maximum speed of shortening defined as $v_o \equiv l_o/\tau_c$. This quantity scales time and varies for fast and slow muscles.

Muscle force is easily measured at various lengths under isometric conditions to produce force-length relationships. The curve produced when muscle is not stimulated is the passive force-length curve, $F_{pe}(l_m)$. When muscle is activated the curve that results represents both passive and active contributions. The difference in these two curves is the active force-length relation, $f_l(l_m)$. The length at which the maximum active muscle force, F_o, is developed is called the optimal muscle length, l_o. A theoretical explanation for the active f_l relation is based on the microscopic nature of muscle and is explained by the Sliding Filament Theory. This theory offers an explanation for the generally accepted notion that when a muscle is completely tetanized, the active force displays a parabolic dependence on length in a nominal region, $.5 \le \tilde{l}_m \le 1.5$ with a maximum value of F_o when $l_m = l_o$. At less than full activation, the force-length dependence is obtained by scaling the fully activated

f_l curve (Zajac, 1991). In this paper, an analytical model of the force length effect is utilized. Several approaches have been suggested, for example, but a simple normalized form that is utilized here is

$$f_l(\tilde{l_m}) = F_o\left(1 - ((\tilde{l_m} - 1)/0.5)^2\right).$$

The nonlinear passive dependence of muscle force on length is described by the function $F_{pe}(l_m)$. A commonly used form for F_{pe} and that which is employed here is given in (He, *et al.*, 1991) and is expressed as

$$F_{pe}(l_m) = \begin{cases} \left(\dfrac{k_{ml}}{k_{me}}\right)\left[e^{(k_{me}(l_m - l_{ms})} - 1\right], \\ \qquad l_{ms} \le l_m < l_{mc} \\ k_{pm}(l_m - l_{mc}) + F_{mc}, \\ \qquad l_m > l_{mc}. \end{cases} \tag{1}$$

The passive muscle slack length is l_{ms} and corresponds to a length at which no force is generated. The transition length from the linear to nonlinear regime is l_{mc} corresponding to a force of F_{mc}. Theses parameters as well as the stiffness and shape parameters k_{me}, k_{pm}, k_{ml} are determined from experimental data and smoothness conditions imposed on the passive force length curve.

Active muscle force is also dependent on muscle velocity. When a muscle actively shortens, it produces less force than it would under isometric conditions. In contrast to a concentric contraction, when a muscle is actively lengthening it is able to produce forces above the maximal isometric force. An analytical model of the force velocity curve f_v is utilized and due to the way the curve is utilized in the formulation of the dynamics, it is convenient to express this relation in terms of the inverse $f_v^{-1}(F)$. The output of the inverse is normalized with respect to v_o, the maximum speed of shortening. In particular if $\tilde{F}$ denotes a normalized active muscle force,

$$\tilde{v}^m = v^m/v^o = f_v^{-1}(\tilde{F}).$$

There is evidence to suggest that total active force generation is best described by a force-length-velocity relationship which is usually quantified as the product of the force-length and force-velocity curves (Zajac, 1991) where the resulting surface is scaled by muscle activation. Consequently, it is convenient to visualize force generation as a collections of surfaces described in terms of nondimensionalized force velocity and force length curve,

$$F_{act} = a(t)F_0 f_l(\tilde{l_m}) f_v(\tilde{v_m}).$$

Muscle activation, $a(t)$, is related to the neural neural input, $n(t)$, by a process known as contraction dynamics. Both quantities, $n(t)$ and $a(t)$, can be related to experimental data. In particular $n(t)$ is related to rectified EMG while $a(t)$ is related to filtered, rectified EMG. The process through which neural input is transformed into activation is known to be mediated through a calcium diffusion process and is represented by the first order differential equation

$$\frac{da(t)}{dt} + \left[\frac{1}{\tau_{act}}\,(\beta + (1-\beta)u(t))\right] a(t) = \frac{1}{\tau_{act}}\,u(t)$$

where $0 < \beta < 1$ and τ_{act} is an activation time constant that varies with fast and slow muscle.

The series elastic element in Figure 1 corresponds to the muscle tendon. More precisely, the tendon is assumed to behave non-linearly under minimal extension and then to become linear with stiffness constant k_s beyond a given length l_{tc} associated with a particular level of resisting force, F_{tc}. A common approach to tendon dynamics (see, for example (He, *et al.*, 1991)) is to assume a model of the form

$$\dot{F}_t = K_t(F_t)\dot{l}_t \tag{2}$$

where

$$K_t(F_t) = \begin{cases} k_{te}F_t + k_{tl} & 0 \le F_t < F_{tc} \\ k_s & F_t \ge F_{tc} \end{cases}.$$

From Figure 1 it readily follows that the total force of a muscle is the sum of the passive and the active forces,

$$F_m = F_{pe} + F_{act} + B_m\dot{l}_m.$$

Muscle is known to maintain a constant volume and so l_w is constant. With this observation and since

$$l_{mt} = l_t + l_m\cos\alpha,$$

it follows

$$\dot{\alpha} = -\frac{\dot{l}_m}{l_m}\tan\alpha.$$

and

$$\dot{l}_t = \dot{l}_{mt} - \frac{\dot{l}_m}{\cos\alpha}.$$

The tendon dynamics may now be expressed as

$$\dot{F}_t = K_t(F_t)(\dot{l}_{tm} - \dot{l}_m/\cos\alpha). \tag{3}$$

The equation of motion for the muscle mass is

$$M_m\frac{d^2(l_m\cos\alpha)}{dt^2} = F_t - [F_{act} + F_{pe} + B_m\dot{l}_m]\cos\alpha$$

and with some simple manipulations, the muscle dynamics take the form

$$M_m\ddot{l}_m = F_t\cos\alpha - \cos^2\alpha[F_{act} + F_{pe} + B_m\dot{l}_m] + \frac{M_m\dot{l}_m^2\tan^2\alpha}{l_m}. \tag{4}$$

Two state variables are required to describe the contraction dynamics of the musculotendon actuator as given by (3) and (4). For multiple muscle systems such as that needed to describe three dimensional eye movement, it is desirable to reduce the system dimension. This can be achieved by eliminating the muscle mass. In this case, (4) becomes a statement of force balance

$$F_t = F^m \cos\alpha = \Big(F_o a(t) f_l(\tilde{l}_m) f_v(\tilde{v}^m) + F_{pe}(\tilde{l}_m) + B_m \dot{l}_m\Big)\cos\alpha.$$

If it is assumed that the passive muscle viscosity effect is small, then $\dot{l}_m$ can be computed from

$$\dot{l}_m = v^m = v_o f_v^{-1}\left(\frac{(F_t/\cos\alpha) - F_{pe}(\tilde{l}_m)}{F_O a(t) f_l(\tilde{l}_m)}\right).$$

It now follows from (3) and that the differential equation describing the contraction dynamics of the musculotendon is

$$\frac{dF_t}{dt} = \frac{F_0}{l_{ts}} K^t(\tilde{F}_t)\Big[\dot{l}_{mt} - \frac{v_o}{\cos\alpha} f_v^{-1}\left(\frac{(F_t/\cos\alpha) - F_0 f_p(\tilde{l}_m)}{F_O a(t) f_l(\tilde{l}_m)}\right)\Big].$$

3. THE PLANT MODEL OF THE EYE

The three muscle pairs consist of the medial and lateral recti, the superior and inferior recti, and the superior and inferior obliques. The lateral and medial recti produce primarily horizontal rotation. The superior and inferior recti work mainly to control vertical rotation. The superior oblique controls the intorsional rotation (toward the nose) of the eye while the inferior oblique controls mainly extorsional rotation (toward the temple). The eye has three degrees of freedom but experimental evidence shows that for any horizontal rotation (θ) and vertical rotation (ϕ), the amount of torsional rotation (ψ) is determined by a phenomena known as Listing's Law,

$$\psi = \arccos\left(\frac{\sin\theta\sin\phi}{1+\cos\theta\cos\phi}\right).$$

The motion of the eye is a result of moments produced by the six extraocular muscles and a passive moment due to the orbit which restrains the rotation of the globe. If ω_x, ω_y, ω_z denote the components of angular velocity with respect to a fixed inertial reference frame with coordinates x, y, z, then the equation of motion for the globe is of the form

$$\begin{pmatrix}\dot{\omega}_x\\ \dot{\omega}_y\\ \dot{\omega}_z\end{pmatrix} = \frac{1}{J_G}\big[(\boldsymbol{r}_{lr}\times\boldsymbol{F}_{lr}) + (\boldsymbol{r}_{mr}\times\boldsymbol{F}_{mr}) + (\boldsymbol{r}_{sr}\times\boldsymbol{F}_{sr}) + (\boldsymbol{r}_{ir}\times\boldsymbol{F}_{ir}) + (\boldsymbol{r}_{so}\times\boldsymbol{F}_{so}) + (\boldsymbol{r}_{io}\times\boldsymbol{F}_{io}) + (\boldsymbol{r}_p\times\boldsymbol{F}_p)\big].$$

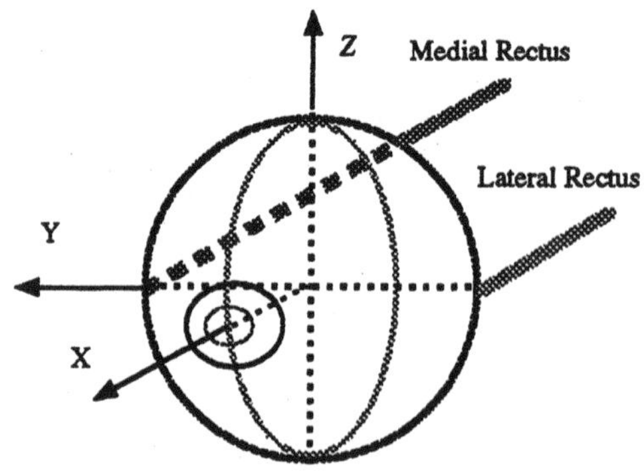

Fig. 2. Left globe with medial and lateral rectus

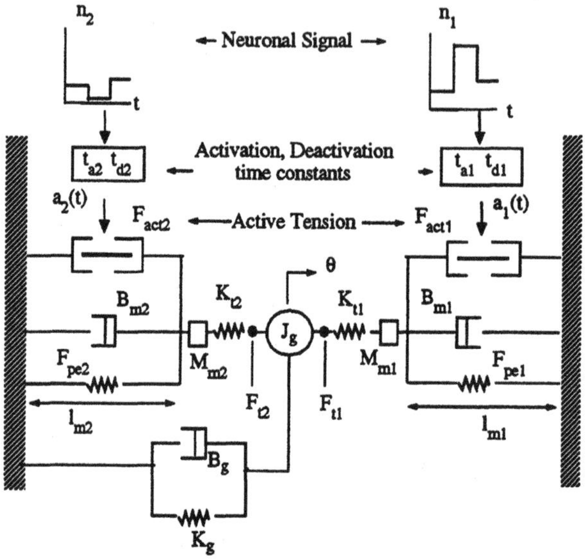

Fig. 3. Model of the eye plant

Here $\boldsymbol{r}_{lr}$ and $\boldsymbol{F}_{lr}$ denote the moment arm and force associated with the lateral rectus, with the obvious interpretation of the other terms. Preliminary three dimensional simulations support the claim that horizontal eye movement may be accurately modeled by including only the medial and lateral rectus muscles. We will assume that the points of attachment are such that the moment these muscles generate is in the direction of the z axis (see Figure 2).

Figure 3 provides a mechanical representation of the eye plant model in which the elements of the plant are displayed as if the muscles and the globe are undergoing linear motion. One must distinguish between the lateral rectus (the agonist) and the medial rectus (the antagonist). All model components that pertain to the agonist will be denoted as $F_{t1}, l_{tm1}, B_{m1}, M_{m1}$, etc, while the corresponding quantities for the agonist are indicated by $F_{t2}, l_{tm2}, B_{m2}, M_{m2}$, etc.

Because of conventions adopted in the ocular literature, it is convenient to report rotations of the eye in degrees (°) and to write the equation of motion for the globe and the musculotendon dynamics in terms of pure tendon force in units of grams tension. To this end let J_G, B_G, and K_G denote the parameters for globe inertia, globe viscosity, and globe elasticity, respectively, in terms of cgs units and radians. Let r

denote the radius of the globe and define

$$J_g = \frac{J_G}{980r(180/\pi)}, \quad \text{in gt} \cdot \text{s}^2/^\circ$$

and similarly for B_g, K_g. If F_{t1} and F_{t2} are reported in gt, then equation of motion for the globe, when expressed in degrees and units of gt, becomes

$$J_g\ddot{\Theta} + B_g\dot{\Theta} + K_g\Theta = F_{t1} - F_{t2}.$$

If pennation effects are ignored, the equation of motion for the muscle mass of the agonist and antagonist, previously expressed in (4), now takes the form for $i = 1, 2$

$$\frac{M}{980}\ddot{l}_{mi} + B_{pm}\left(\frac{180}{\pi r}\right)\dot{l}_{m1} = F_{ti} - F_{acti} - F_{pe1}(l_{mi}).$$

In a similar fashion, the tendon dynamics and the description of the passive muscle elasticity must be amended to account for the change in units. In particular, the modified forms of (1), (2) are

$$F_{pe}(l_m) = \begin{cases} \left(\dfrac{k_{ml}}{k_{me}}\right)\left[e^{(k_{me}(\frac{180}{\pi r})(l_m - l_{ms}))} - 1\right] \\ \qquad l_{ms} \le l_m < l_{mc} \\ \\ k_{pm}\left(\dfrac{180}{\pi r}\right)(l_m - l_{mc}) + F_{mc} \\ \qquad l_m > l_{mc} \end{cases}$$

and

$$\dot{F}_t = K_t(F_t)\left(\frac{180}{\pi r}\right)\dot{l}_t.$$

For the agonist, the tendon dynamics become,

$$\dot{F}_{t1} = K_t(F_{t1})\left[-\dot{\Theta} - \left(\frac{180}{\pi r}\right)\dot{l}_{m1}\right]$$

while for the agonist,

$$\dot{F}_{t2} = K_t(F_{t2})\left[\ \dot{\Theta} - \left(\frac{180}{\pi r}\right)\dot{l}_{m2}\right].$$

If a state vector is selected as

$$\mathbf{x}^\mathsf{T}(t) = [\,\Theta\ \dot{\Theta}\ l_{m1}\ \dot{l}_{m1}\ l_{m2}\ \dot{l}_{m2}\ F_{t1}\ F_{t2}\ a_1\ a_2\,],$$

the state equations for the system are given by

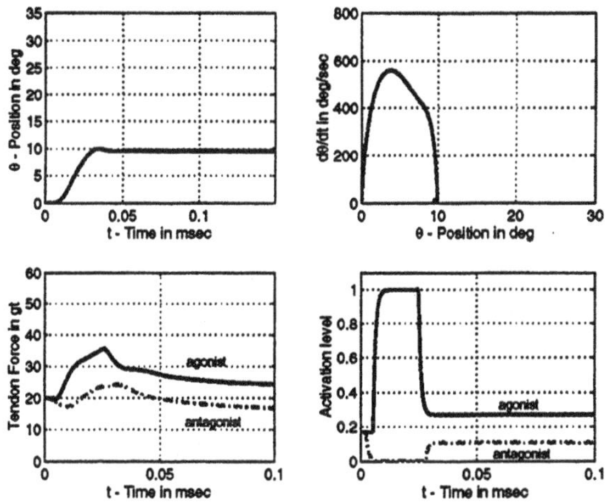

Fig. 4. Model simulation of 10 degree saccade

$$\dot{\mathbf{x}}(t) = \begin{bmatrix} x_2 \\ \frac{1}{J_g}(x_7 - x_8 - B_g x_2 - K_g x_1) \\ x_4 \\ \frac{980}{M}(x_7 - F_{act}(x_3, x_4, x_9) - F_{pe}(x_3) - \\ B_{pm}(\frac{180}{\pi r})x_4) \\ x_6 \\ \frac{980}{M}(x_8 - F_{act}(x_5, x_6, x_{10}) - F_{pe}(x_5) - \\ B_{pm}(\frac{180}{\pi r})x_6) \\ K_t(x_7)\left[-x_2 - \left(\frac{180}{\pi r}\right)x_4\right] \\ K_t(x_8)\left[\ x_2 - \left(\frac{180}{\pi r}\right)x_6\right] \\ \frac{1}{\tau_1(t)}[n_1(t) - x_9] \\ \frac{1}{\tau_2(t)}[n_2(t) - x_{10}] \end{bmatrix}$$

4. SIMULATIONS AND CONCLUSIONS

The simulations illustrated in originate from the primary position and the initial conditions for muscle length, tendon force, and activation are taken from (Robinson, 1991). The vector of initial data is then $\mathbf{x}^\mathsf{T}(0) = [0\ 0\ 4\ 0\ 4\ 0\ 20\ 20\ .17\ .17]$.

The simulated trajectory, velocity, and tendon force are in close agreement with the empiric data of (Robinson, 1991). The results of illustrate the effects of pulse-width and pulse-height mismatch in glissadic overshoot and undershoot. The results show that the pulse-height errors are associated with unusually large peak velocities. In contrast, pulse width mismatches produce saccades with reasonable predictions of velocities. These results are in qualitative agreement

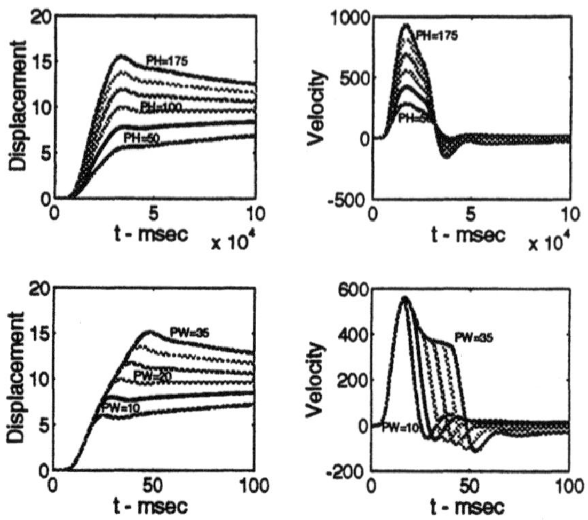

Fig. 5. Glissadic overshoot due to pulse width and height errors

with experimental results of (Baloh, 1974) which suggest that when glissadic overshoot is associated with low peak velocities, the error is caused by the erroneous neural input in computing the pulse width, not the height.

5. REFERENCES

Baloh, R. W. (1974). Internuclear Opthalmoplegia. *Archives of Neurology*, **35**, 484-493.

Clark, M.R. and Stark, L. (1974). Control of human eye movements I. Modeling of extraocular muscles II. A model for the extraocular plant mechanism III. Dynamic characteristics of the eye tracking mechanism. *Mathematical Biosciences*, **20**, 191-265.

He, J.H., Levine, W.S. and Loeb, G.E. (1991). Feedback gains for correcting small perturbations to standing posture. *IEEE Transaction on Automatic Control*, **36**, 322-332.

Robinson, D.A. (1964). The mechanics of human saccadic eye movement. *Journal of Physiology*, **174**, 245-264.

Zajac, F. (1989). Muscle and tendon: Properties, models, scaling and application to biomechanics and motor control. *Critical Reviews in Biomedical Engineering*, **17**, 359-441.

14th Triennial World Congress, Beijing, P.R. China

C-2a-12-4

REFERENCE FRAMES IN LEARNING AND CONTROL

N. Kirupaharan * **W. P. Dayawansa** **,[1]

* *Department of Mathematics and Statistics*
Texas Tech University
Lubbock, Tx 79409, USA
nkirupa@math.ttu.edu
** *Department of Mathematics and Statistics*
Texas Tech University
Lubbock, Tx 79409, USA
daya@math.ttu.edu

Abstract: This article is aimed at introducing the theory of reference frames in mathematical psychology to the control theory community and to point out its relevance in reducing the complexity of control systems design. As an example we discuss the balancing problem for a multi-segment pendulum. *Copyright © 1999 IFAC*

Keywords: Reference Frames, Stabilization

Keywords: reference frames, balancing, human posture control.

1. INTRODUCTION

It is becoming increasingly clear that there are many lessons to be learnt from biological systems, for they are adept at fault tolerance, adaption, learning and extrapolation and many other capabilities which still live in the imagination of designers of engineering systems. For example human brain is capable of breaking down a complex concept, or a task, into several simpler tasks, each of which is familiar to us, thereby making it possible to solve a novel and complex problem by solving several simpler familiar problems. Examples of such occur in reading, facial recognition, learning, sports, and almost everything else we do. On the other hand engineering systems are designed to carry out certain specific tasks in a very orderly manner. Baring a few exceptions they are incapable of learning from their past experiences, and, do not rely upon a hierachical organization as a problem solving methodology. This realization has been the impetus for a renewed and a vigorous call to enumerate a set of fundamental principles which will render biological systems less misterious and even more importantly, will allow human designers to design engineering systems to behave "intelligently".

Our aim here is to point out some relevant concepts from cognitive psychology. They have been interested in developing models of cognition and perception rather than asking how these are carried out at a cellular level. The most fundamental questions from this viewpoint are what do we mean by cognition and perception and how are these mechanisms organized in our psychy. Only a few minutes of thought is required to convince the relevance of the following basic question: How does our brain represent the outside world.

Classically the cognitive system was thought to act as a dynamical system with a very high dimensional state-space representing events in the outside world as points in this state-space. This notion was challenged in 1982 in a series of landmark papers by Michael Leyton (see (?), (?), (?)). He argued that the brain represent each outside object and event as a hierarchy of machines or dynamical systems. To ellaborate, each

[1] Research was funded through NSF Grants ECS 9707927 & 9720357.

event is associated with several simpler events which in turn are associated with even simpler events and so on. For example, each sentence can be broken down into a set of phrases which in turn are broken down into a set of "ideas" etc., and machine descriptions of these arise from rules in linguistic constructions. In this framework cognition of a new and complex object boils down to an organization of a set of links to an existing hierarchy of machines and via them constructing a new machine. Leyton argues forcefully that these structural features are present in every aspect of human cognition, perception, learning and control. Our main point of contention here is that if we are to understand control theory from the stand point of biological systems, then we need to be cognizant of the need to model our systems as a hierarchical structure and device control design tools to design control algorithms which can be built up from the lowest level to the highest level. Unfortunately such methodologies are scarce and the need to develop such has not received due attention. For example we have no tools comparable to wavelet analysis in communication theory and multigrid methods in numerical analysis.

This paper is organized as follows. In section 2 we will briefly describe the theory of reference frames in cognition from the viewpoint of Michael Leyton's work in (?), (?), (?) etc. This will be illustrated via an example dealing with the cognition of geometric figures. In section 3 we will discuss as an example, the problem of balancing a multi-segmented pendulum. The control law derived here is motivated by the reference frame theory. In section 4 we will give some concluding remarks.

2. LEYTON'S VIEW OF THE THEORY OF REFERENCE FRAMES.

In a series of landmark papers (?), (?), (?) Michael Leyton questioned the classical notion of a human *cognitive system* perceived of as a computer delegated with the task of carrying out computations about sorrounding objects. His proposal was to replace this notion by that of *cognitive representation* of the outside world as consisting of dynamical systems. He defined the *cognition of* a stimulus set as the description of the latter as a dynamical system. (In his terminology a dynamical system is the same as a discrete time control system for us. We will continue to use the term dynamical system so a snot to confuse it with some control tasks to be discussed later).

Perhaps the simplest example to illustrate this point is to consider a geometric object, e.g. a parallelogram. Classical viewpoint would say that our cognitive system would form an image of a parallelogram. Let us think a bit harder about how we think of a parallelogram. We are very aware (almost unconciously) that a parallelogram will remain a parallelogram under rotations, translations and certain other deformations. To explain this awareness classical notion would require us to mentally carry out these transformations, i.e. the cognitive system does the computations, and test and see whether the transformed parallelogram matches up with a new geometric figure. In contrast Leyton's theory says our cognitive representation of parallelograms consists of a state-space of parallelograms and transformations which act upon it. This viewpoint obviates the need for a *cognitive system* to carry out computations to match objects. Instead, the natural symmetries of objects under consideration are *stored in* together with the objects in our cognitive representation.

A crucial aspect of the cognitive representation is the notion of reference frames. Eventhough the notion of references has an old history in psychology, Leyton seems to be the first person to mathematically define this notion. Each dynamical system in the outside world is associated with several levels of simpler versions of it. For example, a sentence is associated with a certain set of rules (i.e. dynamics). In linguistics phrases are at a simpler level than sentences, and the rules they have to obey are simpler than those for sentences. Theory of reference frames states that in cognitive representation each dynamical system is associated with a set of simpler dynamical systems, which in turn are associated with even simpler ones and so forth.

To keep things simple in this section let us (as Leyton does) keep our discussion restricted to a very simple class of dynamical systems; those of group actions. Thus we have an object represented as :

a state-space which is a manifold M

an input-space which is a Lie group G

its dynamics :

$$G \times M \rightarrow M,$$

a left action of the group G on M.

Let us represent this object as (G,M).

A reference frame for the object would be another dynamical system (H,N) where N is a submanifold of M and H is a subgroup of G.

To illustrate this let us consider the cognitive representation of parallelogram objects. The description below is a bit more complex than Leyton's description, but is similar in spirit. The state-space in our cognitive representation is an abstract space M. However we can describe the manifold structure of M as follows. Take a parallelogram and identify it as a point in $\Re^2 \times \Re^2 \times \Re^2$ by choosing a vertex and two vectors representing the sides meeting at the vertex. Any nearly point can be identified with a nearly point in $\Re^2 \times \Re^2 \times \Re^2$. This now sets up local coordinates on the manifold M. Input group of the parallelograms representation is $GL(2, \Re) \times \Re^2$ which acts by "translating" a parallelogram and "deforming" it in an obvious way in accordance with its action on $\Re^2$. Thus we have the cognitive representation of parallelograms

$$(GL(2,\Re) \times \Re^2) \times M \to M.$$

To motivate the discussion of reference frames let us ask ourselves the following question. If we were asked to draw a parallelogram which of the parallelograms in figure 1 would be closest to ours?

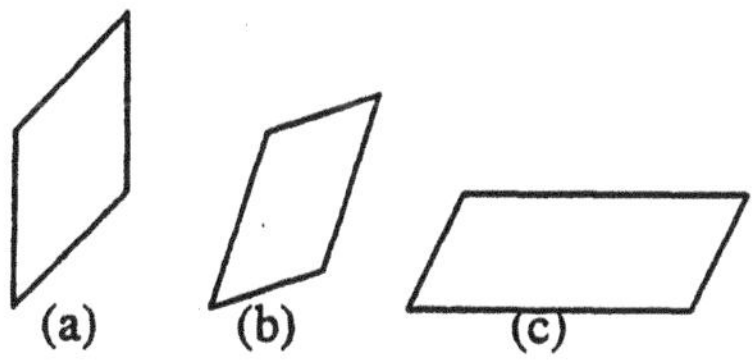

Fig. 1. Parallelogram Objects

It is our contention that chances are very high that it would be (c). Thus, eventhough we know parallelograms in general we have a "preference" to those that "lie flat". Such parallelograms are cognitively represented as a dynamical system,

$$G \times N \to N$$

where N is a submanifold of M, and $G = H \times \Re^2$, where H is the group of "shears" on $\Re^2$. Thus parallelograms that lie "flat" form a reference frame to all parallograms in our cognitive representation. In fact this "referencing" can be carried out further to get rectangle and square objects with their associated dynamics. In general one can think of a cognitive representation of an object as a representation of it as a control system and a reference frame for the cognitive representation as a control system derived from the original one by restricting the input and the state-space appropriately.

3. BALANCING OF A MULI-LINK PENDULUM

Here we ask: What can reference frame theory offer in the form of new insight into a standard control problem.As an example to illustrate a possible use of the idea of reference frames in contol theory let us consider the problem of balancing an inverted multi-segmented pendulum with multiple control inputs. We take the pendulum to consist of n segments and the number of control torques is taken to be n. Here we disregard friction. Let us assume that the system consists of k subsystems and denote by S_i, the ith subsystem. We group control torques acting at joints in S_i, together and denote it by S_i. Subsystems S_{i-1} and S_i are joined at joint P_i, and P_1 is assumed to be fixed. This is schematically represented in Figure 2.

We wish to remark here that the ideas outlined here can be extended to study the human posture control problem. In this case S_1 would consist of the two legs, S_2 would be the lower torso, S_3 would be the upper torso and S_4 would be the head.

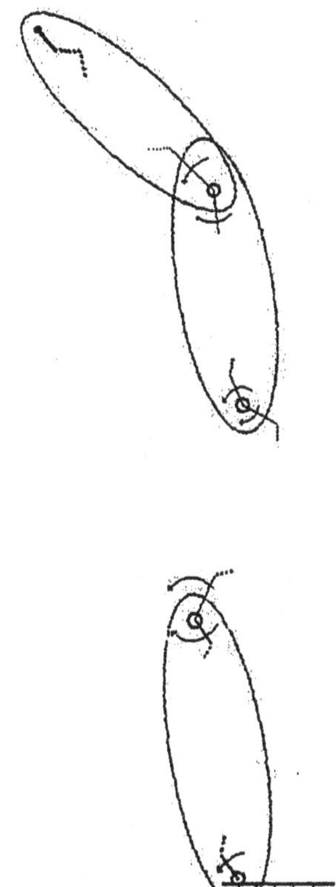

Fig. 2. Multi-link Inverted Pendulum

Our objective here is to study the balancing problem for the multi segment pendulum from the view point of the theory of reference frames. Thus we wish to consider all subsystems below S_i as a reference frame to the collection of subsystems above S_i. More precisely, we aim at the following. Let us first consider the top most subsystem S_k. Suppose we were able to find a control law (for τ_k) that would balance it in the presence of motion of subsystems below it. We wish to employ it to derive a balancing law (for τ_{k-1}) for the subsystem S_{k-1} with minimal modifications of its dynamics. This in turn would be used to derive a balncing law for τ_{k-2} etc, until we end up with balancing control laws τ_1, $\tau_2, \ldots \tau_k$ for the entire system.

The usefulness of a strategy of this type is quite abvious. We stress that at no point in our contol design we will have to deal with the entire systm, thus reducing the complexity quite considerably. In spite of tremendous advances made over the past few years in computer technoloy, still we aren't close to being able to control a biped or a many degrees of freedom robot arm with precision. It is our hope that control design with reference frames in mind will shed some new light into these hard problems.

Getting back to the control problem at hand, let us start from subsystem S_k. Let us represent its state vector by $z_k = \begin{bmatrix} \theta_k \\ \omega_k \end{bmatrix}$ where θ_k are the angles each link make to the vertical direction and ω_k are the angular velocities of the links.

Let us note that the equation of motion for S_k depends on the velocity and acceleration of joint P_k. Thus we have

$$\begin{aligned}\dot{\theta}_k &= \omega_k \\ \dot{\omega}_k &= f_k(z_k, \dot{x}_k, \ddot{x}_k) + G_k(z_k, \dot{x}_k, \ddot{x}_k)\tau_k. \quad (1)\end{aligned}$$

Now, by the assumption that there are as many control torques as links it follows that G_k is invertible, and therefore, we can find a control law, $\tau_k = \alpha_k(z_k, \dot{x}_k, \ddot{x}_k)$, which would stabilize S_k at its vertical equilibrium configuration.

Now let us fix α_k, and consider the stabilization problem for S_k and S_{k-1} together. The crucial point of the whole idea presented here from the viewpoint of control theory is that the nature of τ_k renders the dynamics of $\S_k \cup \S_{k-1}$ block upper triangular. Thus, the stabilization problem reduces to stabilizing $\S_k \cup \S_{K-1}$ while assuming that S_k is constrained to be vertical at all times. Let us represent the state vector of S_{K-1} by $z_{k-1} = \begin{bmatrix} \theta_{k-1} \\ \omega_{k-1} \end{bmatrix}$ where θ_{k-1} and ω_{k-1} and the angular positions and angular velocities of its link. The new contol problem is of dimension $2d_{k-1}$, instead of $2(d_{K-1} + d_k)$ as in the convensional sense. This explains the reduction of compexity offered by this method.

Now, let us suppose that the equation of motion of S_{k-1}, while keeping S_k vertically constrained, is

$$\begin{aligned}\dot{\theta}_{k-1} &= \omega_{k-1} \\ \dot{\omega}_{k-1} &= f_k(z_{k-1}, \dot{x}_{k-1}, \ddot{x}_{k-1}) \\ &\quad + G_k(z_{k-1}, \dot{x}_{k-1}, \ddot{x}_{k-1})\tau_{k-1}.\end{aligned}$$

As for S_k, a control law can be found which will stabilize ((?)) of the form, $\tau k - 1 = \alpha_{k-1}(z_{k-1}, \dot{x}_{k-1}, \ddot{x}_{k-1})$. Continuing itteraively, at the ith step we have a d_i dimensional stabilization problem, instead of a $2(\sigma_{j=i}^{k} d_j)$ problem in the convensional sense.

The control laws, $\alpha_1, \cdots, \alpha_k$ will now provide a set of stabilizing control laws for the system.

3.1 Simulation results

To illustrate this approach we include below some simulation results for a two segment pendulum with controls at both joints. Each link is considered as a seperate subsystem. Simulations were carried out using the Simulink package.

Here, S_1 consists of a link of two meters in length with a mass of one Kg concentrated at its center, and S_2 is a link one meter in length with a cencentrated mass of one Kg at its end. Control laws employed cancel out all $\dot{x}_i$ and $\ddot{x}_i$ terms and make the subsystems feedback linearized. In order to model any computational delay in the feedback loop we take velocity and acceleration measurements of x_i with a 0.2 second transmission delay.

Figure 3 ilustrate simulation results. Note that the control torque for the upper link is much smaller in comparision with the lower one, as it ought to be in any realistic situation. This is in view of the fact that the control law only ask for the uper link to "take care" of its own dynamics.

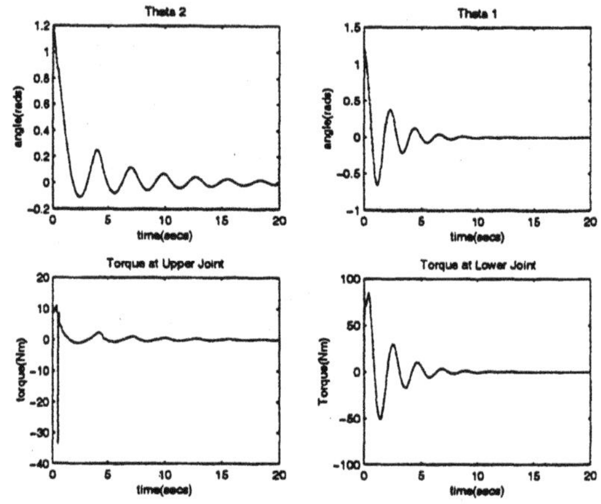

Fig. 3. Balancing a Two-link pendulum

4. CONCLUDING REMARKS

Our aim in the paper was to point out that thinking along the lines of the theory of reference frames bring some new insight into how one may approach the problem of reducing the complexity of controlling multi-link mechanisms. Admittedly, it doesn't produce any new algorithms, but it indicates new guidelines for coming up with such algorithms. It is clear that this line of thinking will only produce a crude solution to a control problem comparable to the unsteady walk of an infant, and undoubtedly much more sophisticated processing takes place after generating an initial solution. Unfortunately, the theory of reference frames doesn't answer the question of how one may go about understanding this post processing process. What natural structures are there to facilitate such processing, i.e. go down the ladder of reference frames rather than going up as in here, is a challenging research problem.

5. REFERENCES

[1] M. Leyton, "Principles of Information Structure Common to Six Levels of the Human Cognitive Systems," *Information Sciences*, **Vol. 38**, 1-120, 1986.

[2] M. Leyton, "A Theory of Information Structures: I. General Principles," *J. Mathematical Psychology*, **Vol. 30**, 103-160, 1986.

[3] M. Leyton, "A Theory of Information Structures: II. Perceptual Organization," *J. Mathematical Psychology*, **Vol. 30**, 257-305, 1986.

[4] T. R. Kane and D. A. Levinson, **Dynamics: Theory and Applications**, *McGraw-Hill*, 1985.

14th Triennial World Congress, Beijing, P.R. China

C-2a-12-5

RECURSIVE ESTIMATION OF MOTION AND SHAPE PARAMETERS UNDER PERSPECTIVE PROJECTION

Hiroyuki Kano * **Hajime Kanai** **

* *Department of Information Sciences, Tokyo Denki University, Hatoyama, Hiki-Gun, Saitama 350-0394, Japan*
** *Nippon Telephone and Telegraph Corp., 2-1-1 Sakuragi-cho, Omiya, Saitama 331-0852, Japan*

Abstract: The problem of estimating motion and shape of a moving rigid object has been considered from a time series of image data obtained via perspective projection. First it is shown that, for a class of motion parameters, the shape parameters are asymptotically decoupled from the optical flow dynamics, hence become unobservable, asymptotically. For other choices of motion parameters, shape parameters remain asymptotically obseravble. Secondly we show that, if the surface is almost planar, then uncertainty in the shape of the surface propagates as an optical flow equation with additive state dependent uncertainty. Finally, we develop a recursive estimation algorithm, and present the numerical simulation results. *Copyright © 1999 IFAC*

Keywords: Image processing, Motion parameters, Asymptotic analysis, Parameter identification, Noisy images, Extended Kalman filters

1. INTRODUCTION

The problem of estimating motion and shape of objects under motion from their corresponding image data observed over time is of paramount importance in various branches of engineering and medical sciences. In general, the solution to the associated estimation problem is ill-posed and is therefore quite susceptible to noise in the observed image data. In order to alleviate the noise problem, various recursive estimation schemes have been proposed, the most important of which is probably Extended Kalman Filtering. Recently a new Implicit Extended Kalman Filtering has also been studied by Soatto (Soatto, 1996). The basic estimation problem is described as follows.

Assume that we have a rigid surface that is undergoing a rigid motion parametrized by a finitely many parameters. Assume that a set of feature points on the surface has been projected perspectively on an image plane and that the image data has been observed over an interval of time. The problem is to simultaneously estimate the shape of the surface and the corresponding motion parameters.

In general, if the feature points are observed exactly, then the problem of computing the motion and shape parameters of a planar surface undergoing a rigid motion has been considered in the literature for over the last 15 years. Many of these techniques are based on solving a set of nonlinear algebraic equations and are therefore nonrecursive (Kanatani, 1990), (Subbarao and Waxman, 1985), (Waxman and Ullman, 1985). These algebraic methods yielded feasible results but were far from providing an acceptable computational algorithm.

In order to obtain algorithms that are robust with respect to noisy image data, recursive algorithms have been introduced and studied. Two questions that are important in this regard are described as follows: *What is a suitable parametric class of motion dynamics, that could yield robust recursive algorithms ?* and *How does noises in the shape propagate under perspective projection ?* The first question is of course connected with the question of parameter identifiability, using recursive algorithms. In fact, if the model class of the motion dynamics is too general, the motion parameters may not be identifiable, using image sequence from a single camera. For the second question, on the other hand, it is shown that the noises in the planar surface give rise to a stochastic optical flow equation.

2. RIGID MOTION DYNAMICS IN $\mathbf{R}^3$

We consider a parametrized class of dynamics that has been studied in great details in the literature (Ghosh and Loucks, 1995), (Kanatani, 1990), pertaining to the problem of estimating the motion parameters. In particular, we discuss the extent to which the parameters can be estimated using asymptotic and recursive algorithms.

Let $\mathcal{X} = (X, Y, Z)^T$ denote a set of co-ordinates in $\mathbf{R}^3$ and consider a motion dynamics as follows:

$$\dot{\mathcal{X}} = \Omega\mathcal{X} + \mathbf{b} \tag{1}$$

where,

$$\Omega = \begin{pmatrix} 0 & \omega_1 & \omega_2 \\ -\omega_1 & 0 & \omega_3 \\ -\omega_2 & -\omega_3 & 0 \end{pmatrix}, \quad \mathbf{b} = \begin{pmatrix} b_1 \\ b_2 \\ b_3 \end{pmatrix}. \tag{2}$$

As one can see, (1) is described by six motion parameters, which are assumed to be time invariant. The dynamics (1) has been studied in the literature as a model for rigid motion and of course it is known classically that any rigid transformation can be represented by (1) if one allows the motion parameters to be time varying.

To get an insight to the geometric structure for the phase portrait of (1), we make the following transformation,

$$\mathcal{X}_1 = \mathcal{X} + \alpha. \tag{3}$$

with $\mathcal{X}_1 = (X_1, Y_1, Z_1)^T$ and $\alpha = (\alpha_1, \alpha_2, \alpha_3)^T$. Note that the skew symmetric matrix Ω in (2) has rank 2 in general. Let us define

$$\mathbf{P} \triangleq \mathrm{Im}\,\Omega \tag{4}$$

where $\mathbf{P}$ is a homogeneous 2-plane in $\mathbf{R}^3$. We now decompose the vector $\mathbf{b}$ as follows: $\mathbf{b} = \mathbf{b}_1 + \mathbf{b}_2$, where $\mathbf{b}_1 \in \mathbf{P}$ and $\mathbf{b}_2 \in \mathbf{P}^\perp$. Note that this decomposition is unique. Let us now choose α in such a way that $\Omega\alpha = \mathbf{b}_1$. Using (3) we get

$$\dot{\mathcal{X}}_1 = \Omega\mathcal{X}_1 + \mathbf{b}_2 \tag{5}$$

where $\mathbf{b}_2$ is an eigenvector corresponding to the eigenvalue 0. It follows that there exists a scalar β such that $\mathbf{b}_2 = \beta\mathbf{v}$ with $\mathbf{v} = (\omega_3, -\omega_2, \omega_1)^T$. We then conclude that there exists an orthogonal matrix R such that if

$$\mathcal{X}_2 = R\mathcal{X}_1 \tag{6}$$

with $\mathcal{X}_2 = (X_2, Y_2, Z_2)^T$, we obtain

$$\dot{\mathcal{X}}_2 = \hat{\Omega}\mathcal{X}_2 + \hat{\mathbf{b}} \tag{7}$$

$$\hat{\Omega} = \begin{pmatrix} 0 & \omega & 0 \\ -\omega & 0 & 0 \\ 0 & 0 & 0 \end{pmatrix}, \quad \hat{\mathbf{b}} = \begin{pmatrix} 0 \\ 0 \\ \gamma \end{pmatrix}. \tag{8}$$

where $\omega = \sqrt{\omega_1^2 + \omega_2^2 + \omega_3^2}$. The equation (7) can be solved easily as follows:

$$\begin{pmatrix} X_2(t) \\ Y_2(t) \end{pmatrix} = \exp\left\{\begin{pmatrix} 0 & \omega \\ -\omega & 0 \end{pmatrix} t\right\} \begin{pmatrix} X_2(0) \\ Y_2(0) \end{pmatrix}$$
$$Z_2(t) = \gamma t + Z_2(0) \tag{9}$$

Thus the magnitude of the vector $\mathcal{X}_2$ is linearly increasing to "infinity" while the direction is asymptotically the direction of the vector $(0\ 0\ 1)^T$ in the $(X_2, Y_2, Z_2)^T$ co-ordinate. It would follow that in co-ordinate $(X_1, Y_1, Z_1)^T$, the asymptotic direction of the vector is along the direction of $(\omega_3\ -\omega_2\ \omega_1)^T$. The conclusion would be same in the co-ordinate system $(X, Y, Z)^T$ as well. Thus we have the following theorem.

Theorem 1. Consider the dynamical system (1) where we assume that $(\omega_1\ \omega_2\ \omega_3) \neq 0$. If $\mathbf{b} \in \mathrm{Im}\,\Omega$, then the phase portrait of (1) is circular and periodic with axis of rotation given by $\beta\mathbf{v}+\alpha$, where α is a vector such that $\Omega\alpha = \mathbf{b}$ and where β is an arbitrary scalar. If on the other hand $\mathbf{b} \notin \mathrm{Im}\,\Omega$, then the magnitude of the vector $\mathcal{X}$ linearly increases to infinity and

$$\lim_{t\to\infty} [\mathcal{X}(t)] = [\mathbf{v}] = [\omega_3, -\omega_2, \omega_1]^T \tag{10}$$

3. SHAPE DYNAMICS

We assume that the points $\mathcal{X} = (X\ Y\ Z)^T$ are actually on a planar surface (see Fig. 1):

$$pX + qY + r = Z \tag{11}$$

In the $(X_2, Y_2, Z_2)^T$ co-ordinate, let

$$p_2X_2 + q_2Y_2 + r_2 = Z_2 \tag{12}$$

be the plane, then we have

$$(p_2\ q_2\ -1\ r_2) \begin{pmatrix} X_2 \\ Y_2 \\ Z_2 \\ 1 \end{pmatrix} = 0. \tag{13}$$

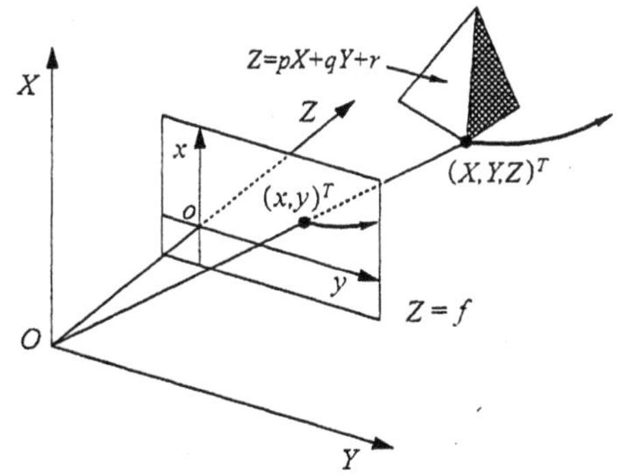

Fig. 1. Planar surface and perspective projection

Differentiating (13) and using (7) and (8) yield

$$\frac{d}{dt}\begin{pmatrix} p_2 \\ q_2 \\ -1 \\ r_2 \end{pmatrix} = \begin{pmatrix} 0 & \omega & 0 & 0 \\ -\omega & 0 & 0 & 0 \\ 0 & 0 & 0 & 0 \\ 0 & 0 & -\gamma & 0 \end{pmatrix} \begin{pmatrix} p_2 \\ q_2 \\ -1 \\ r_2 \end{pmatrix} \quad (14)$$

or, letting $p_2 = \frac{\bar{p}_2}{\bar{s}_2}$, $q_2 = \frac{\bar{q}_2}{\bar{s}_2}$, $r_2 = \frac{\bar{r}_2}{\bar{s}_2}$, we have

$$\frac{d}{dt}\begin{pmatrix} \bar{p}_2 \\ \bar{q}_2 \\ \bar{s}_2 \\ \bar{r}_2 \end{pmatrix} = \begin{pmatrix} 0 & \omega & 0 & 0 \\ -\omega & 0 & 0 & 0 \\ 0 & 0 & 0 & 0 \\ 0 & 0 & \gamma & 0 \end{pmatrix} \begin{pmatrix} \bar{p}_2 \\ \bar{q}_2 \\ \bar{s}_2 \\ \bar{r}_2 \end{pmatrix}.$$

Hence we get $\begin{pmatrix} \dot{\bar{p}}_2 \\ \dot{\bar{q}}_2 \end{pmatrix} = \begin{pmatrix} 0 & \omega \\ -\omega & 0 \end{pmatrix} \begin{pmatrix} \bar{p}_2 \\ \bar{q}_2 \end{pmatrix}$, $\dot{\bar{r}}_2 = \gamma \bar{s}_2$, $\dot{\bar{s}}_2 = 0$, and $\bar{s}_2(t) = \bar{s}_2(0)$, $\bar{r}_2(t) = \gamma \bar{s}_2(0)t + \bar{r}_2(0)$,

$$r_2(t) = \gamma t + r_2(0). \quad (15)$$

Since $\dot{p}_2 = \omega q_2$, $\dot{q}_2 = -\omega p_2$, it follows from (15) that asymptotically the plane (12) is approaching the plane, $Z_2 = \gamma t$. i.e the homogeneous vector, $[p_2\ q_2\ r_2\ 1]$ is approaching $[0\ 0\ \gamma t\ 1]$ in $\mathbf{RP}^3$. It follows that in the original coordinate system, the plane (11) would appoach the plane,

$$\omega_3 X - \omega_2 Y + \gamma t = \omega_1 Z. \quad (16)$$

Thus we have the following theorem.

Theorem 2. If $\mathbf{b} \in \operatorname{Im}\Omega$, then the solution to the shape dynamics is periodic and is given by

$$Z_2 = p_2 X_2 + q_2 Y_2 + r_2(0)$$

where $\dot{p}_2 = \omega q_2$, $\dot{q}_2 = -\omega p_2$. If on the other hand $\mathbf{b} \notin \operatorname{Im}\Omega$, then the solution to the shape dynamics approaches the linearly receding plane (16), receding linearly to infinity.

4. OPTICAL FLOW DYNAMICS

Assume that the point $(X\ Y\ Z)^T$ is observed via perspective projection (Fig. 1 with $f = 1$):

$$x = \frac{X}{Z}, \ y = \frac{Y}{Z}. \quad (17)$$

Differentiating (17) and using (1) we get

$$\dot{x} = (\omega_1 y + \omega_2) + x(\omega_2 x + \omega_3 y) + \frac{b_1}{Z} - x\frac{b_3}{Z}. \quad (18)$$

From (11) we have $\frac{1}{Z} = \frac{1-px-qy}{r}$. Hence

$$\dot{x} = \omega_2 + \omega_1 y + \omega_2 x^2 + \omega_3 xy + (c_1 - xc_3)(1 - px - qy) \quad (19)$$

where $c_i = b_i/r$, $i = 1, 2, 3$. Similarly, we obtain

$$\dot{y} = \omega_3 - \omega_1 x + \omega_2 xy + \omega_3 y^2 + (c_2 - c_3 y)(1 - px - qy). \quad (20)$$

The equations (19) and (20) denote the optical flow equation. In particular when $t \to \infty$

$$p(t) \to \frac{\omega_3}{\omega_1}, \quad q(t) \to -\frac{\omega_2}{\omega_1}, \quad r(t) \to \frac{\gamma t}{\omega_1}$$
$$c_i \to \frac{b_i \omega_1}{\gamma t}, \quad i = 1, 2, 3.$$

Thus we have the following theorem.

Theorem 3. When $\mathbf{b} \notin \operatorname{Im}\Omega$, the coefficients of the optical flow dynamics (19), (20) approaches the coefficients of the following dynamics,

$$\begin{aligned} \dot{x} &= \omega_2 + \omega_1 y + \omega_2 x^2 + \omega_3 xy \\ &\quad + \frac{1}{\gamma t}(b_1 - b_3 x)(\omega_1 - \omega_3 x + \omega_2 y), \\ \dot{y} &= \omega_3 - \omega_1 x + \omega_2 xy + \omega_3 y^2 \\ &\quad + \frac{1}{\gamma t}(b_2 - b_3 y)(\omega_1 - \omega_3 x + \omega_2 y). \end{aligned} \quad (21)$$

If $\mathbf{b} \in \operatorname{Im}\Omega$ on the other hand, the optical flow equation asymptotically is given by

$$\begin{aligned} \dot{x} &= \omega_2 + \omega_1 y + \omega_2 x^2 + \omega_3 xy \\ &\quad + \frac{1}{r(t)}(b_1 - b_3 x)(1 - px - qy), \\ \dot{y} &= \omega_3 - \omega_1 x + \omega_2 xy + \omega_3 y^2 \\ &\quad + \frac{1}{r(t)}(b_2 - b_3 y)(1 - px - qy), \end{aligned} \quad (22)$$

where $p(t)$ and $q(t)$ are periodic functions satisfying the Riccati equation,

$$\begin{aligned} \dot{p} &= \omega_1 q - \omega_2 - \omega_2 p^2 - \omega_3 pq, \\ \dot{q} &= -\omega_1 p - \omega_3 - \omega_2 pq - \omega_3 q^2, \\ \dot{r} &= b_3 - b_1 p - b_2 q - r(\omega_3 q + \omega_2 p). \end{aligned} \quad (23)$$

5. APPROXIMATE OPTICAL FLOW DYNAMICS

Consider the motion dynamics (1) and perspective projection (17). Thus (X, Y, Z) is perspectively projected to (x, y) where we assume that

$$Z^* = pX + qY + r = (px+qy)Z + r, \quad (24)$$

i.e Z^* is not exactly same as Z. Let us define

$$Z - Z^* = \varepsilon \quad (25)$$

which yields

$$Z = \frac{r+\varepsilon}{1-px-qy}. \quad (26)$$

From (17), (1) and (26), we then have

$$\begin{aligned}\dot{x} &= \frac{Z\dot{X} - X\dot{Z}}{Z^2} = \frac{\dot{X}}{Z} - x\frac{\dot{Z}}{Z} \\ &= \left(\omega_1 y + \omega_2 + \frac{b_1}{Z}\right) - x\left(-\omega_2 x - \omega_3 y + \frac{b_3}{Z}\right) \\ &= \frac{1}{r+\varepsilon}\left[\omega_1(r+\varepsilon)y + \omega_2(r+\varepsilon) + \omega_2(r+\varepsilon)x^2\right. \\ &\quad + \omega_3(r+\varepsilon)xy + b_1 - b_3 x - pb_1 x + pb_3 x^2 \\ &\quad \left. - qb_1 y + qb_3 xy\right]. \quad (27)\end{aligned}$$

Hence we get

$$\begin{aligned}(r+\varepsilon)\dot{x} &= (\omega_2(r+\varepsilon) + b_1) - (b_3 + pb_1)x \\ &\quad + (\omega_1(r+\varepsilon) - qb_1)y + (\omega_2(r+\varepsilon) + pb_3)x^2 \\ &\quad + (\omega_3(r+\varepsilon) + qb_3)xy. \quad (28)\end{aligned}$$

Likewise we have

$$\begin{aligned}(r+\varepsilon)\dot{y} &= [\omega_3(r+\varepsilon) + b_2] - [\omega_1(r+\varepsilon) + pb_2]x \\ &\quad - (qb_2 + b_3)y + [\omega_2(r+\varepsilon) + pb_3]xy \\ &\quad + [\omega_3(r+\varepsilon) + qb_3]y^2. \quad (29)\end{aligned}$$

The equations (28) and (29) give a description of the approximate optical flow equation which is of the following form,

$$\begin{pmatrix} d_9 & 0 \\ 0 & d_9 \end{pmatrix}\begin{pmatrix} \dot{x} \\ \dot{y} \end{pmatrix} = \begin{pmatrix} d_1 \\ d_2 \end{pmatrix} + \begin{pmatrix} d_3 & d_4 \\ d_5 & d_6 \end{pmatrix}\begin{pmatrix} x \\ y \end{pmatrix} + \begin{pmatrix} d_7 x^2 + d_8 xy \\ d_8 y^2 + d_7 xy \end{pmatrix}, \quad (30)$$

where

$d_1 = (b_1 + \omega_2 r) + \omega_2\varepsilon$ $\quad d_6 = -(qb_2 + b_3)$
$d_2 = (b_2 + \omega_3 r) + \omega_3\varepsilon$ $\quad d_7 = (pb_3 + \omega_2 r) + \omega_2\varepsilon$
$d_3 = -(b_3 + pb_1)$ $\quad d_8 = (qb_3 + \omega_3 r) + \omega_3\varepsilon$
$d_4 = (\omega_1 r - qb_1) + \omega_1\varepsilon$ $\quad d_9 = r + \varepsilon$
$d_5 = -(pb_2 + \omega_1 r) - \omega_1\varepsilon$

6. ESTIMATION ALGORITHM AND SIMULATION STUDIES

Motion and shape parameters are estimated from noisy image data by employing extended Kalman filter. For convenience of numerical computation, the filter is designed in discrete-time settings.

6.1 *Time-Discretized Motion and Shape Dynamics*

First, the continuous-time motion dynamics (1) is discretized with sampling time T as

$$\mathcal{X}_{k+1} = A\mathcal{X}_k + \mathbf{g} \quad (31)$$

Here $\mathcal{X}_k = \mathcal{X}(kT) = (X_k,\ Y_k,\ Z_k)^T$, and $A = [a_{ij}]$ and $\mathbf{g} = (g_1,\ g_2,\ g_3)^T$ are obtained as

$$A = e^{\Omega T}, \quad \mathbf{g} = \int_0^T e^{\Omega(T-\tau)}\mathbf{b}\,d\tau \quad (32)$$

Motion parameters are now a_{ij} and g_i, $i, j = 1, 2, 3$. On the other hand, the equation of planar surface is given by

$$Z_k = p_k X_k + q_k Y_k + r_k \quad (33)$$

where p_k, q_k and r_k are the shape parameters. The shape dynamics corresponding to (23) is then given (Loucks, 1994) as

$$\begin{aligned} p_{k+1} &= -\frac{\Delta_{11}p_k + \Delta_{12}q_k - \Delta_{13}}{\Delta_{31}p_k + \Delta_{32}q_k - \Delta_{33}}, \\ q_{k+1} &= -\frac{\Delta_{21}p_k + \Delta_{22}q_k - \Delta_{23}}{\Delta_{31}p_k + \Delta_{32}q_k - \Delta_{33}}, \end{aligned} \quad (34)$$

and

$$\begin{aligned} r_{k+1} = g_3 &- \frac{r_k - g_1(\Delta_{11}p_k + \Delta_{12}q_k - \Delta_{13})}{\Delta_{31}p_k + \Delta_{32}q_k - \Delta_{33}} \\ &+ \frac{g_2(\Delta_{21}p_k + \Delta_{22}q_k - \Delta_{23})}{\Delta_{31}p_k + \Delta_{32}q_k - \Delta_{33}}. \end{aligned} \quad (35)$$

Here Δ_{ij} denotes the cofactor of a_{ij} of rotation matrix A.

Under perspective projection, a feature point $\mathcal{X}_k$ is observed as a point (x_k, y_k) on the image plane by $x_k = \frac{X_k}{Z_k}$ and $y_k = \frac{Y_k}{Z_k}$ yielding the following optical flow dynamics.

$$\begin{aligned} x_{k+1} &= \frac{d_{11,k}x_k + d_{12,k}y_k + d_{13,k}}{d_{31,k}x_k + d_{32,k}y_k + d_{33,k}}, \\ y_{k+1} &= \frac{d_{21,k}x_k + d_{22,k}y_k + d_{23,k}}{d_{31,k}x_k + d_{32,k}y_k + d_{33,k}}. \end{aligned} \quad (36)$$

Here $d_{ij,k}, i, j = 1, 2, 3$, called essential parameters, are defined as

$$\begin{aligned} d_{i1,k} &= a_{i1} - c_{i,k}p_k, \quad d_{i2,k} = a_{i2} - c_{i,k}q_k, \\ d_{i3,k} &= a_{i3} + c_{i,k}, \quad i = 1, 2, 3. \end{aligned} \quad (37)$$

and $c_{i,k}, i = 1, 2, 3$, defined by

$$c_{i,k} = \frac{g_i}{r_k} \quad (38)$$

are governed by the following recusive equations.

$$c_{i,k+1} = \frac{(\Delta_{31}p_k + \Delta_{32}q_k - \Delta_{33})c_{i,k}}{\bar{c}_k}, \tag{39}$$

$$\begin{aligned}\bar{c}_k = &(\Delta_{11}p_k + \Delta_{12}q_k - \Delta_{13})c_{1,k} \\ &+(\Delta_{21}p_k + \Delta_{22}q_k - \Delta_{23})c_{2,k} \\ &+(\Delta_{31}p_k + \Delta_{32}q_k - \Delta_{33})c_{3,k} - 1\end{aligned} \tag{40}$$

6.2 *Recursive Estimation Algorithm*

Since the drift vector g and r_k are identifiable only in the form of the ratios, namely as $c_{i,k}$ in (38) (Ghosh and Loucks, 1995), the parameters to be estimated are now $a_{i,j}$, $c_{i,k}$ and p_k, q_k.

Assume that, at each time instant k, a set of n positions (x_{ik}, y_{ik}) corresponding to n feature points $\mathcal{X}_{ik}, i = 1, 2, \cdots, n$ on the surface are measured. Hence as a state vector θ_k we take

$$\begin{aligned}\theta_k = (&a_{11,k}, a_{12,k}, \cdots, a_{33,k}, c_{1,k}, c_{2,k}, c_{3,k}, \\ &p_k, q_k, x_{1k}, y_{1k}, \cdots, x_{nk}, y_{nk})^T\end{aligned} \tag{41}$$

with $\theta_k \in \mathbf{R}^{2n+14}$. The state equation

$$\theta_{k+1} = f(\theta_k) \tag{42}$$

then follows readily from $a_{ij,k+1} = a_{ij,k}$ for $i, j = 1, 2, 3$, (39), (34), and (36).

Now let $(x_{ik}, y_{ik}), i = 1, 2, \cdots, n$ be observed with additive noises, and let $z_k \in \mathbf{R}^{2n}$ be the observation vector. Then

$$z_k = H\theta_k + w_k \tag{43}$$

where $H = [0_{2n,14}\ I_{2n}]$. $\{w_k\}$ is assumed to be a Gaussian white noise with

$$\mathrm{E}\{w_k\} = 0, \quad \mathrm{E}\{w_k w_l^T\} = R\delta_{kl} \tag{44}$$

where δ_{kl} denotes Kronecker delta.

We now employ the extended Kalman filter, as summarized below. Letting $\hat{\theta}_{k|k}$ and $\hat{\theta}_{k|k-1}$ be the estimates of θ_k given the data $\{z_0, z_1, \cdots, z_k\}$ and $\{z_0, z_1, \cdots, z_{k-1}\}$ respectively, the filter equation is given by

$$\begin{aligned}\hat{\theta}_{k|k} &= \hat{\theta}_{k|k-1} + K_k[z_k - H\hat{\theta}_{k|k-1}], \\ \hat{\theta}_{k+1|k} &= f(\hat{\theta}_{k|k}), \quad \hat{\theta}_{0|-1} = \bar{\theta}_0\end{aligned} \tag{45}$$

with the Kalman gain

$$K_k = P_{k|k-1}H^T[HP_{k|k-1}H^T + R]^{-1}. \tag{46}$$

The error covariance matrix $P_{k|k}$ for $\hat{\theta}_{k|k}$, and $P_{k+1|k}$ for $\hat{\theta}_{k+1|k}$, are then given by

$$\begin{aligned}P_{k|k} &= P_{k|k-1} - K_k H P_{k|k-1}, \\ P_{k+1|k} &= F_k P_{k|k} F_k^T, \quad P_{0|-1} = \Sigma_0 \\ F_k &= \left.\frac{\partial f(\theta)}{\partial \theta^T}\right|_{\theta=\hat{\theta}_{k|k}}.\end{aligned} \tag{47}$$

6.3 *Simulation Results*

We first show the motion and shape parameters used in our simulation studies. Ω is set by

$$(\,\omega_1\ \omega_2\ \omega_3\,) = (\,-0.2\ 0.1\ -0.1\,), \tag{48}$$

and we consider the following two cases for $\mathbf{b}$ depending on whether or not $\mathbf{b} \in \mathrm{Im}\Omega$:

$$\text{Case 1}: \ \mathbf{b} = (\,0.1\ 0.1\ -0.1\,)^T \in \mathrm{Im}\Omega \tag{49}$$

$$\text{Case 2}: \ \mathbf{b} = (\,0.1\ 0.1\ 0.1\,)^T \notin \mathrm{Im}\Omega \tag{50}$$

The other parameters are set as follows: $T = 0.1$ in (31); $(p_0, q_0, r_0) = (0.1, 0.1, 2.0)$ in (33); $n = 3$ in (41); $R = \sigma^2 I_6$ with $\sigma = 0.01$ in (44); $\Sigma_0 = 100 I_{20}$ in (47). $\bar{\theta}_0 \in \mathbf{R}^{20}$ is set as the vector with 1's in its 1st, 5th, and 9th positions and 0's elsewhere.

Figure 2 shows a sample noisy trajectory on the image plane used in simulation. Note that in the following Figures 3-6, estimated and true values are plotted in solid and dotted lines, respectively.

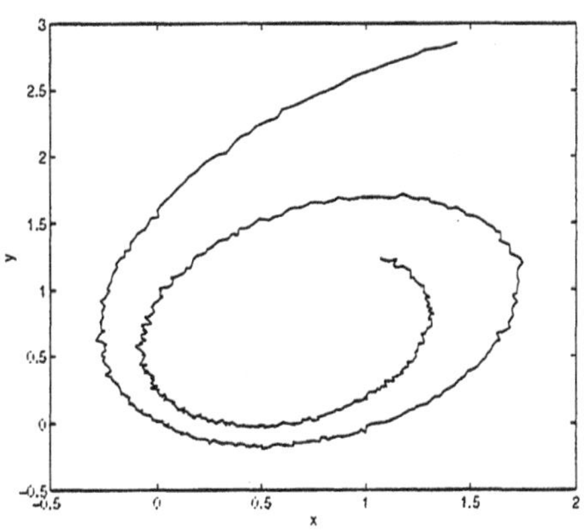

Fig. 2. A noisy image data for the Case 2

First, we present the typical simulation results for the Case 1, i.e. the case b $\in$ ImΩ. As for the motion parameter in Ω, the estimates converged quite fast to the true values (see Fig. 3 for ω_1).

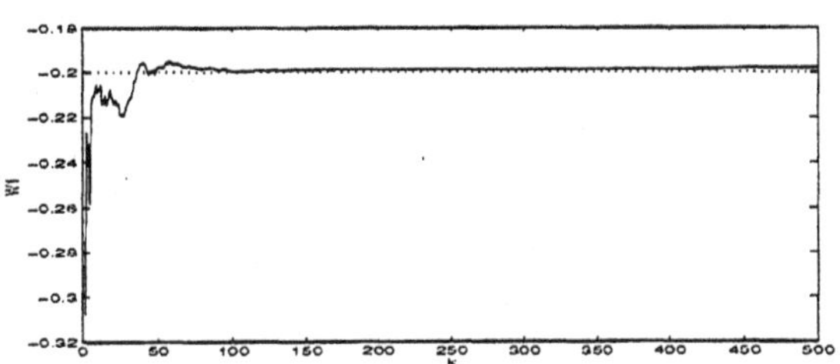

Fig. 3. Estimation of ω_1 for the Case 1

The same is true for the parameters $c_{1,k}$ in (39) as is seen in Fig. 4, in which case the associated error variance decreased to the order of 10^{-7}.

The performances for the shape parameters p_k and q_k are not quite satisfactory (see Fig.5 for p_k).

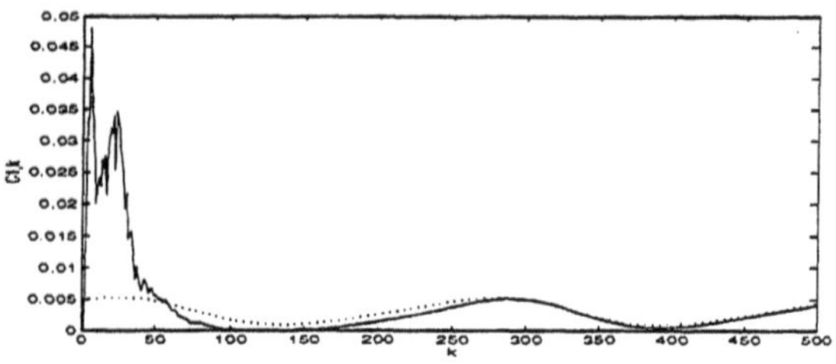

Fig. 4. Estimation of $c_{1,k}$ for the Case 1

In fact the errors grow wherever r_k is large. This phenomenon become more evident in the Case 2.

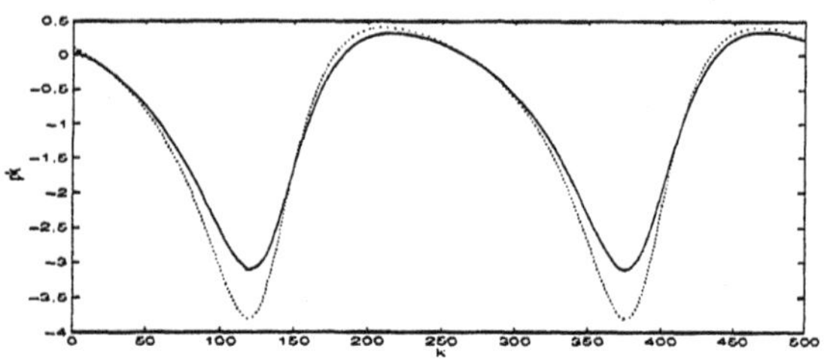

Fig. 5. Estimation of p_k for the Case 1

As for the Case 2 in (50), equally good results as in Case 1 were obtained for the estimation of ω_i and $c_{i,k}, i, j = 1, 2, 3$. The performances for p_k and q_k, however, are much worse than the Case 1 (see Fig.6 for p_k). In order to study further,

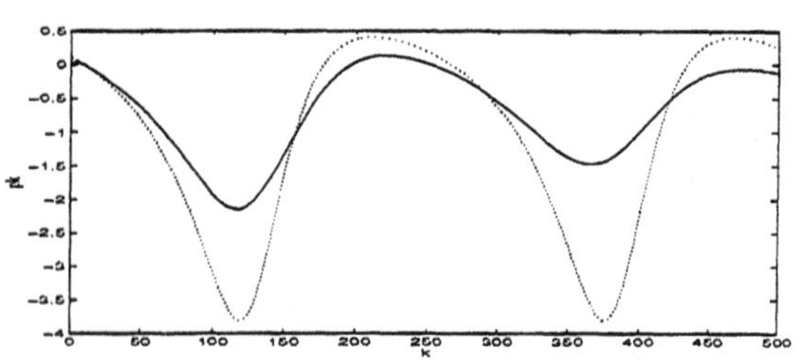

Fig. 6. Estimation of p_k for the Case 2

the following normalized estimation error ϵ_k for p_k and q_k is introduced:

$$\epsilon_k = \frac{\sqrt{(\hat{p}_k - p_k)^2 + (\hat{q}_k - p_k)^2}}{\sqrt{p_k^2 + q_k^2 + 1}} \tag{51}$$

In Fig.7, ϵ_k is plotted in vertical axis against r_k (horizontal axis) with $k = 1, 2, \cdots, 500$ being the parameter. Clearly the larger r_k results in larger

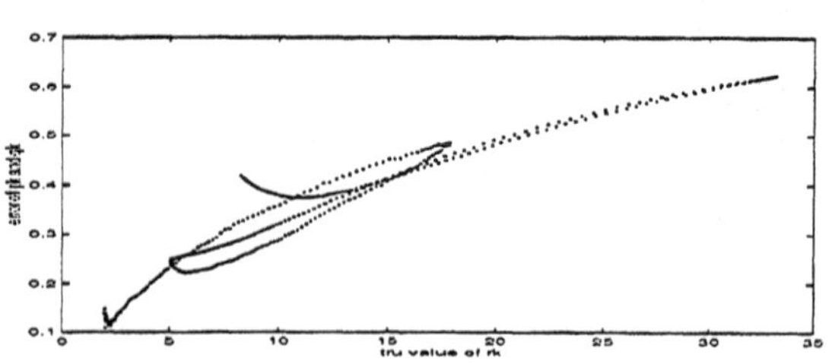

Fig. 7. Normalized error ϵ_k for the Case 2

errors in ϵ_k, and we cannot expect the estimates of p_k and q_k to improve as time proceeds (see Theorem 3).

7. CONCLUDING REMARKS

We considered the problem of estimating motion and shape parameters of moving rigid body from noisy image data under perspective projection.

For a planar surface undergoing a rigid motion, it was shown that, for a generic choice of the motion parameters namely $\mathbf{b} \notin \mathrm{Im}\Omega$, the shape parameters were asymptorically decoupled from the optical flow dynamics. This led to the conclusion that asymptotically the shape parameters were invisible from the image data and hence recursive shape estimation would not be possible under perspective projection for a generic class of motion parameters. By developing a recursive estimation algorithm based on the extended Kalman filter, we confirmed this conclusion via simulations. We also analyzed a suitably rich, non-generic class of motion dynamics, i.e. $\mathbf{b} \in \mathrm{Im}\Omega$, for which "shape-parameters" remain asymptotically observable. Once again, using simulation we showed how the parameters were recoverable.

On the other hand, the motion parameters were estimated accurately in either cases, and the algorithm were shown to be robust to image data with additive observation noises.

Acknowledgement The authors would like to thank Professor Bijoy K. Ghosh, Washington University, St. Louis, for his valuable comments and discussions.

8. REFERENCES

Ghosh, B.K. and E.P. Loucks (1995). A perspective theory for motion and shape estimation in machine vision. *SIAM J. Control and Optimization* **33**, 1530–1559.

Kanatani, K. (1990). *Group-Theoretical Methods in Image Understanding.* Springer Verlag.

Loucks, E.P. (1994). A perspective systems approach to motion and shape estimation in machine vision. *Ph.D. Thesis, Washington University.*

Soatto, S. (1996). A geometric framework for dynamic vision. *Ph.D. Thesis, California Institute of Technology.*

Subbarao, M. and A.M. Waxman (1985). On the uniqueness of image flow solutions for planar surfaces in motion. *Technical Report 114, College Park, University of Maryland, Center for Automation Research.*

Waxman, A.M. and S. Ullman (1985). Surface structure and 3-d motion from image flow : Kinematic analysis. *International J. Robotics Research* **4**, 72–94.

14th Triennial World Congress, Beijing, P.R. China

C-2a-12-6

REAL-TIME PLANNING AND CONTROL FOR ROBOT MANIPULATOR IN UNKNOWN WORKSPACE [1]

Di Xiao[2] Bijoy K. Ghosh[3] Ning Xi[4] Tzyh Jong Tarn[3] Zhenyu Yu[5]

[2] *FANUC Robotics North America, Inc. 3900 W. Hamlin Rd., Rochester Hills, MI 48309 USA*
E-mail: xiaodx@frc.com
[3] *Department of Systems Science and Mathematics, Washington University, St.Louis, MO3130 USA*
[4] *Department of Electrical Engineering, Michigan State University, East Lansing, MI 48824 USA*
[5] *DSP Digital Control Systems Applications, Texas Instrument, Stafford, TX 77477, USA*

Abstract: This paper addresses real-time planning and control for robot manipulator in unknown workspace. The task under our consideration is to control a robot to follow a path on an unknown surface. In order to successfully accomplish the task, a novel strategy for integrating real-time planning and control is proposed to deal with the uncertainties in the environment by means of multisensor fusion. In the proposed scheme, a new planner is designed to complete the path-following task without apriori knowledge of the path, while the hybrid position/force control strategy is utilized to maintain the contact between the robot and the unknown surface. In this paper, to achieve intelligent robot manipulation in an unstructured workspace, complementary sensory data from force-torque and visual sensors is integrated simultaneously during robot constrained motion. *Copyright ©1999 IFAC*

Keywords: Sensor fusion, Robot vision, Force control, Path planning

1. INTRODUCTION

As robotic applications have been extended to flexible manufacturing systems, outspace and undersea, robots are required to work in poorly structured environments. The planning and control issues of robots in the uncertain workspaces become challenging problems. Based on sensory data, this paper proposes a new strategy of integrating planning and control for a typical task with a robot manipulator in an unstructured workspace, in order to increase the level of "intelligence" for a robot in the sense that it can handle uncertainties in the environment without explicit intervention or re-programming. Concretely, the task under consideration is to control a robot such that the tip of a tool grasped by the end-effector of the robot follows a curve on an unknown surface with a camera fixed somewhere above the workcell, as shown in Fig. 1. It is assumed that the robot is equipped with a force-torque sensor and that both the curve and the robot can be observed by the vision system. Also, it is assumed

[1] Partially supported by the Department of Energy under grant number DE-FG02-90ER14140 and by National Science Foundation under grant numbers IRI-9706160 and IRI-9703149, and by Sandia National Laboratories Contract No. AC 3752-C.

that the relationship between the robot and the vision system is *unknown*. Many tasks in manufacturing engineering, such as welding, cutting materials along a curve, scribing parts etc, can be boiled down to this category. To fulfil a path-following task, it is natural to localize the precise position of the path. However, it is impossible to describe exactly the curve in the case where the contact surface is unknown. On the other hand, exact knowledge about the path is not enough to guarantee the completion of this kind of tasks because minute position control error would result in either a large reaction force or no contact between the robot and the surface. To compensate for the uncertainties, a sensor-based integration of real-time planning and control is designed.

As is well-known, a force-torque sensor and a vision system are disparate sensors. Nelson and Khosla (1996) proposed resolvability concept for assimilating those two sensors and utilized the sensors in different stages *separately* during contact transitions based on their resolvability. This paper integrates information from the force-torque sensor and the vision system *simultaneously* to ensure the completion of the task. Real-time planning is developed by projecting the error in the image plane onto the tangent plane, while a hybrid controller is used to execute the task. In this way, real-time planning and control is integrated. The proposed scheme has a hierarchical structure. The lower level is a hybrid position/force control with nonlinear feedback. The upper level is a planner which generates a desired motion based on the information from multiple sensors. The relation between the constrained motion of the robot on the tangent plane and its projection can be continuously updated in time with the measurements from the vision system and the force-torque sensor.

Robot control with visual information in feedback loop has been studied in visual servoing field. The approaches used so far on visual servoing can mainly be classified into two categories: position-based and image-based methods, see (Hutchinson, *et al.*, 1996). To complete the path-following task, the contact force and the motion constrained on the unknown surface should be controlled simultaneously with no conflict. In this case, neither position-based nor image-based approach is suitable, if not impossible. Motion servoing on the tangent plane of the contact surface turns out a better way to design control. In order to utilize visual information, most of previous work on visual servoing needed precise calibration of the camera with respect to the robot. Calibration is a time-consuming procedure and makes the implementation expensive or even impossible for many cases. This paper introduces a new sensor-based motion planning approach for the constrained motion of the robot, which is robust against the uncertainties in the pose of the camera(s).

When the robot contacts with the environment, force control is needed to maintain a proper contact, see (Hogan, 1982; Yoshikawa, 1986). This paper adopts the hybrid position/force control approach, though our proposed scheme differs from previous work in the force control community. In our case, the contact surface is assumed to be unknown. Therefore, the desired trajectory for constrained motion in the hybrid position/force control scheme is unknown a priori and has to be planned in real-time.

2. HYBRID CONTROL

For a robot manipulator contacting with its environment, the dynamics can be written as

$$D(\mathbf{q})\ddot{\mathbf{q}} + c(\mathbf{q}, \dot{\mathbf{q}}) + p(\mathbf{q}) = \tau + \mathbf{F}, \qquad (1)$$

where $\mathbf{q}$, $\dot{\mathbf{q}}$ and $\ddot{\mathbf{q}}$ are the joint angle vector, joint velocity vector and joint acceleration vector, respectively. $D(\mathbf{q})$ is the inertia matrix of the robot while $c(\mathbf{q}, \dot{\mathbf{q}})$ stands for Coriolis and Centrifugal terms. $p(\mathbf{q})$ is the term caused by gravity. τ represents the joint torque vector. $\mathbf{F}$ is the constraint force/torque vector in the joint space.

For our task, the trajectory of the robot should be on a path to be followed. This leads to a tracking nature of the task. Since the robot would touch an unknown surface, force control is needed also. Let us utilize a hybrid position/force control scheme with nonlinear feedback control law (Xiao, *et al.*, 1998), namely,

$$\begin{aligned} \tau &= c(\mathbf{q}, \dot{\mathbf{q}}) + p(\mathbf{q}) + DJ^{-1}\mathbf{u} - DJ^{-1}\dot{J}\dot{\mathbf{q}}, \quad (2)\\ \mathbf{u} &= T[\ddot{\mathbf{x}}_c^d + K_v\dot{\mathbf{e}}_c + K_p\mathbf{e}_c] + \dot{T}\dot{\mathbf{x}}_c \\ &\quad - M^T(\lambda^d + K_f\int_{t_0}^{t} e_f dt), \end{aligned}$$

where J is the Jacobian of the robot. $e_f = \lambda^d - \lambda$ and λ denotes the value of the reaction force. e_c $(=x_c - x_c^d)$ is the error vector of constrained motion with x_c^d being desired constrained motion. The matrix T only depends upon the orientation of the tangent plane of the contact surface at the contact point, which can be measured via the force/torque sensor. $\dot{T}$ reflects the change of the orientation. The matrix M can also be known from the measurements of the force/torque sensor and the encoders of the robot. The proposed control scheme is simple and consists of three parts: one is for constrained motion, one is used to compensate the change of the orientation and the other is for force regulation.

To complete the path-following task, it is necessary for the robot to contact the surface. Hence, maintenance of contact force is very important during the execution of the task. On the other hand, since the contact surface is unknown, the path to be followed is unknown either. Actually, it is very difficult to estimate the path, especially when only one uncalibrated camera is employed. Therefore, the hybrid position/force control scheme cannot directly be applied to the case where the desired motion is unknown. The good news is that it is not necessary to estimate the path at all. On the image plane of an uncalibrated camera, if the error between the tip of the robot and a point p on the path is zero, then in 3-D space the tip of the robot must be on the line of sight starting from the optical center of the camera to the point p. Additionally, if at the same time the tip of the robot is on the unknown surface, one can conclude that the tip of the robot must exactly be on the path to be followed since the point p is indeed the intersection between the line of sight and the surface. More importantly, the above invariant property has nothing to do with the poses of the surface and the camera. Thus, the path-following task can be successfully accomplished if and only if (1) the robot contacts the surface and (2) there is no deviation of the robot from the path in the image plane of an uncalibrated camera. Now the problem becomes how to control the robot from an initial position to achieve the goals (1) and (2). To maintain the contact, force control is used to keep a constant reaction force. However, to use the proposed hybrid position/force scheme, it is needed to develop a planner which can generate the desired motion in real-time. In this paper, real-time planning and control is successfully integrated, as shown in Fig. 2. The sensory data is fed back to a planner which can generate a desired trajectory in real-time for the robot, instead of directly being fed back to the controller like most previous work in robotics.

3. REAL-TIME PLANNING

It can be shown that if the robot is given the desired trajectory the hybrid control can be used to track the trajectory. Since the constrained surface is unknown, neither is the to-be-followed path on the surface. This section shall propose a planning scheme for the constrained motion of the robot based on fusion of the visual data from an uncalibrated camera and additional information from encoders and force/torque sensor.

3.1 *Projection of constrained motion*

Suppose that $O_cX_cY_cZ_c$ is the camera frame with the origin being at the optical center of the camera and Z-axis perpendicular to the image plane. It is true that

$$\begin{pmatrix} x^c \\ y^c \\ z^c \end{pmatrix} = R \begin{pmatrix} x \\ y \\ z \end{pmatrix} + T, \tag{3}$$

where (x^c, y^c, z^c) and (x, y, z) are the coordinates of a point p in the camera frame and the base frame of the robot, respectively. R is the rotation matrix and $T \in \mathbb{R}^3$ is a translation vector.

From a pin-hole model of the camera with focal length f, the coordinates of the image of the point p are given by

$$X = f\frac{x^c}{z^c}, \quad Y = f\frac{y^c}{z^c}. \tag{4}$$

Differentiating both sides of (4) with respect to time t and using (3) gives rise to the following

$$\begin{pmatrix} \dot{X} \\ \dot{Y} \end{pmatrix} = \frac{1}{z^c} \begin{bmatrix} f & 0 & -X \\ 0 & f & -Y \end{bmatrix} R \begin{pmatrix} \dot{x} \\ \dot{y} \\ \dot{z} \end{pmatrix} \tag{5}$$

When the motion of the end-effector of the robot is constrained on a surface $z = z(x, y)$ (note that $z(x, y)$ is unknown), from the force/torque sensor mounted on the end-effector, the robot can locally estimate the surface. More precisely, the tangent plane to the unknown surface at the contact point is computed by using the force/torque sensor. It is easily seen that $\mathbf{f}^T\dot{\mathbf{x}} = 0$, and if $f_z \neq 0$,

$$\dot{z} = -\frac{f_x}{f_z}\dot{x} - \frac{f_y}{f_z}\dot{y} = \alpha\dot{x} + \beta\dot{y},$$

where $\mathbf{f} = (f_x, f_y, f_z, 0, 0, 0)$. Therefore, the motion of the end-effector on the constrained surface and its perspective projection on the image plane of camera are related by

$$\begin{pmatrix} \dot{X} \\ \dot{Y} \end{pmatrix} = \frac{1}{z^c} \underbrace{\begin{bmatrix} f & 0 & -X \\ 0 & f & -Y \end{bmatrix} R \begin{bmatrix} 1 & 0 \\ 0 & 1 \\ \alpha & \beta \end{bmatrix}}_{Q} \begin{pmatrix} \dot{x} \\ \dot{y} \end{pmatrix}$$
$$= \frac{1}{z^c} Q \begin{pmatrix} \dot{x} \\ \dot{y} \end{pmatrix} \tag{6}$$

In (6), z^c and R are unknown, while α and β can be measured by the force/torque sensor. Since both the camera and the base of the robot are fixed, hence R is constant. Once R is recovered, Q can be known via sensory information from the vision system and the force/torque sensor. It is easy to develop a recursive least square algorithm based on Eq. (6) to recover R up to a scalar multiple. This procedure provides an adaptation scheme, wherein it is not needed to determine the rotation matrix R exactly.

3.2 *Motion planning*

With an uncalibrated camera, it is hard to recover exactly the path on the unknown surface. It is assumed that the path and the end-effector of the robot are both visible. Hence it is easy to know the difference between the end-effector and the trajectory on the image plane. Denote the difference on the image plane by $\mathbf{e}$. The desired motion of the robot can be written as

$$\begin{pmatrix} \dot{x}^d \\ \dot{y}^d \end{pmatrix} = -kQ^{-1}\mathbf{e} \qquad (7)$$

where

$$k = \begin{cases} \dfrac{V_{max}}{\|Q^{-1}\mathbf{e}\|_2}, & \text{if } \|Q^{-1}\mathbf{e}\|_2 \geq V_{max} \\ 1, & \text{if } V_{min} \leq \|Q^{-1}\mathbf{e}\|_2 < V_{max} \\ \dfrac{V_{max}}{V_{min}}, & \text{if } \|Q^{-1}\mathbf{e}\|_2 < V_{min}. \end{cases}$$

where V_{max} denotes the upper bound of the speed which the robot can generate within the limited control torque and V_{min} the lower bound to be determined by the task.

It should be pointed out that the control is robust against perturbations in Q. Suppose that $\overline{Q}$ is an estimate of Q. Let $\overline{Q}^{-1} = Q^{-1} + \Delta$. As long as the system $\dot{\mathbf{e}} = -(I + Q\Delta)\mathbf{e}$ is asymptotically stable, the proposed planning and control scheme guarantees that $\mathbf{e} \to 0$ as $t \to \infty$.

3.3 *Multi-camera system*

Suppose that there are m cameras fixed above the workcell. One can expect to benefit from this multi-camera system since any one camera can be easily obstructed. Again, it is assumed that the cameras are not calibrated with respect to the workcell. This section develops an algorithm to apply the proposed control scheme by fusion information from the multiple cameras.

Let (X_i, Y_i) be the image coordinates of the tip of tool grasped by the robot on the image plane of the i-th camera, $(i = 1, \cdots, m)$. Follow the same vein in the previous section, one can obtain

$$\begin{pmatrix} \dot{X}_i \\ \dot{Y}_i \end{pmatrix} = \frac{1}{z_{ci}} Q_i \begin{pmatrix} \dot{x} \\ \dot{y} \end{pmatrix}, \qquad (8)$$

where z_{ci} is the z-coordinate of the tip in the i-th camera frame and

$$Q_i = \begin{bmatrix} f_i & 0 & -X_i \\ 0 & f_i & -Y_i \end{bmatrix} R_i \begin{bmatrix} 1 & 0 \\ 0 & 1 \\ \alpha & \beta \end{bmatrix},$$

where R_i is the rotation matrix of the i-th camera with respect to the base frame of the robot and f_i denotes the focal length of the i-th camera. Using the same algorithm as proposed one above, R_i can be recovered. Eq. (8) can be rewritten as

$$\begin{pmatrix} \dot{X}_1 \\ \dot{Y}_1 \\ \vdots \\ \dot{X}_m \\ \dot{Y}_m \end{pmatrix} = \underbrace{\begin{bmatrix} \frac{1}{z_{c1}} & 0 & 0 \\ 0 & \ddots & 0 \\ 0 & 0 & \frac{1}{z_{cm}} \end{bmatrix}}_{Z} \underbrace{\begin{bmatrix} Q_1 \\ \vdots \\ Q_m \end{bmatrix}}_{W} \begin{pmatrix} \dot{x} \\ \dot{y} \end{pmatrix}.$$

Let

$$\begin{pmatrix} \dot{x}^d \\ \dot{y}^d \end{pmatrix} = -(W^TW)^{-1}W^T \begin{pmatrix} k_1\mathbf{e}_1 \\ k_2\mathbf{e}_2 \\ \vdots \\ k_m\mathbf{e}_m \end{pmatrix}, \qquad (9)$$

where $\mathbf{e}_i \in \mathbf{R}^2$ $(i = 1, \cdots, m)$ are the projections of the difference between the observed point on the tool grasped by the end-effector and the trajectory on the i-th image plane. Since W is a matrix of full column rank, hence W^TW is invertible. k_i's are positive scalars to be determined by users. Similar to the case when only one camera is used, one can simply choose $k_1 = k_2 = \cdots = k_m = k$ and k is given by

$$k = \begin{cases} \dfrac{V_{\max}}{E}, & \text{if } E \geq V_{\max} \\ 1, & \text{if } V_{\min} \leq E < V_{\max} \\ \dfrac{V_{\max}}{V_{\min}}, & \text{if } E < V_{\min} \end{cases}$$

$\mathbf{e} = (\mathbf{e}_1^T, \cdots, \mathbf{e}_m^T)^T$ and $E = \|(W^TW)^{-1}W^T\mathbf{e}\|_2$.

Follow the same vein as for the single camera case, one can prove that under our proposed control law, the errors on each image plane will vanish as time increases. Therefore, the trajectory-following task can be fulfilled. If some cameras are obstructed for some reason, one can simply take the errors on the image planes of the cameras as zero. The desired velocity for the robot still can be obtained based on other cameras. In this sense, the multi-camera system possesses more robustness than a single camera. Also, as shown above, our method can handle redundancy of the measurement very easily, which is indeed an approach of integrating redundant sensory information. Unlike previous work in visual servoing area, using our proposed approach for robot control with a multi-camera system no problems such as observability and controllability need to be addressed.

4. EXPERIMENTS

The experimental system consists of one PUMA 560 manipulator, a vision system and a table on which there is an unknown surface. The computer vision system consists of a CCD camera (fixed above the workspace) with digital image resolution of 256 × 256 and the Intelledex Vision processor based on a 16 MHz Intel 80386 CPU. The focal length of the camera is 0.0125 m. Visual measurements are sent to an SGI 4D/340 VGX via a parallel interface. The robot is controlled by a UMC controller which also interfaces with SGI through memory mapping.

In the experiments, the relative poses between the camera, table and the robot are assumed to be unknown and an unknown surface is randomly placed on the robot workspace. The path to be followed by the end-effector of the robot is characterized by five markers. The proposed control scheme controls the robot successfully such that the trajectory-following task is completed in a robust manner. In the experiments, it is assumed that the end-effector initially contacts with the unknown curved surface (at the beginning of execution, manually move the end-effector of the robot to touch the surface). Force control is used to maintain the contact. A recursive least-square algorithm is employed to recover the motion relation between the tangent plane and the image plane. As soon as the real-time planner generates desired motion with the proposed real-time planning scheme, the robot is then automatically controlled to follow the curve on the unknown curved surface without knowing the path in the base frame of the robot. Planning and control are integrated as a whole procedure without human intervention during the execution. A number of experiments, without changing any parameters in the controller and planner, have been done for different surfaces (locations, shapes and materials are different) and different camera orientations. The path-following tasks have been successfully accomplished.

Some experimental results are, for example, illustrated in Figs.3 and 4. In Fig. 3 the solid line denotes the trajectory of the end-effector of the robot on the image plane. The actual trajectory of the end-effector of the robot in the task space is depicted in Fig. 4. In Figs. 3 and 4, the points labeled by upper letters and lower letters are the markers on the curved surface and their corresponding images on the image plane, respectively. Although the curved surface is deformable and the accuracy of the force sensor is low (see Fig. 4), the task is still accomplished well.

5. CONCLUSIONS

This paper has proposed a sensor integration scheme for real-time planning and control problem of constrained motions of a robot manipulator in unknown workspace. At the contact point, the wrist force/torque sensor of the robot provides local information about an unknown surface. The vision system with camera(s) is used to get global information about the deviation of the end-effector of the robot from the path to be followed. The measurements of the sensors have been fed back to a planner which generates a desired motion on the tangent plane to the unknown surface at the contact point for the robot with a force control loop, using a hierarchical structure. The force control assures the contact between the tool grasped by the end-effector and the unknown surface, while the visual servoing guarantees to vanish the difference between the images of the end-effector and the curve. With the propsed planning and control strategy, the robot is actually controlled in a closed-loop at task-level and the performance is more robust than those in a closed-loop only at joint-level. Therefore, the trajectory-following task is guaranteed to be accomplished.

The experiments with primitive equipments have demonstrated the feasibility of the method. The proposed method is simple to implement and robust against uncertainty in the configuration among camera(s), the robot and the surface with an adaptive scheme. Because the proposed image-based motion planning scheme is implemented on the tangent plane of the contact surface, one can easily develop a fusion algorithm to utilize the scheme for the case where multiple uncalibrated cameras are available.

6. REFERENCES

N. Hogan (1982) Programmable impedance industrial manipulators, Proc. Conf. CAD/CAM Tech. In Mech. Eng., Cambridge, MA, M.I.T.

S. Hutchinson, G.D. Hager and P.I. Corke (1996). A tutorial on visual servo control, IEEE Trans. on Robotics and Automation, Vol. 12, No. 5, pp.651-670.

Bradlely J. Nelson and Pradeep K. Khosla (1996). Force and vision resolvability for assimilating disparate sensory feedback, IEEE Trans. on Robotics and Automation, Vol. 12, No. 5, pp. 714-731.

Di Xiao, B.K. Ghosh, N. Xi and T.J. Tarn (1998). Intelligent robotic manipulation with hybrid position/force control in an uncalibrated workspace, Proc. of 1998 IEEE Int. Conf. on Robotics and Automation, Belgium, May 16-20.

T. Yoshikawa (1986). Dynamic hybrid position/force control of robot manipulators–Description on hand constraint and calculation of joint driving force, Proc. of 1986 IEEE Int. Conf. on Robotics and Automation, pp.1393–1398.

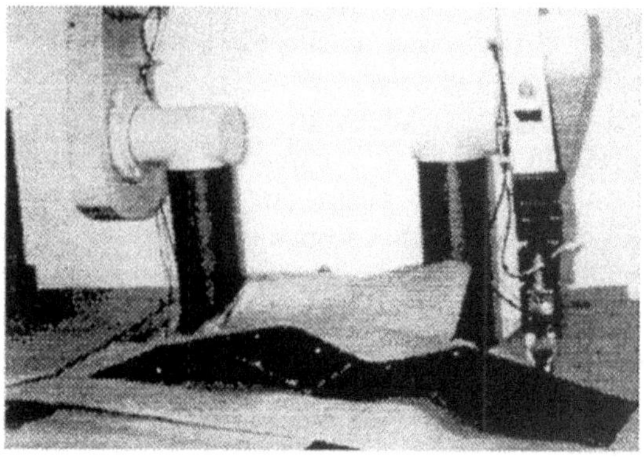

Figure 1: A typical path-following task.

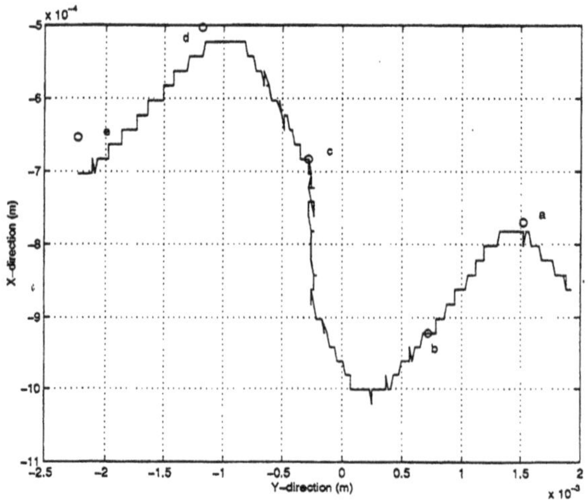

Figure 3: Projections of the desired and actual trajectories of the end-effector on the image plane.

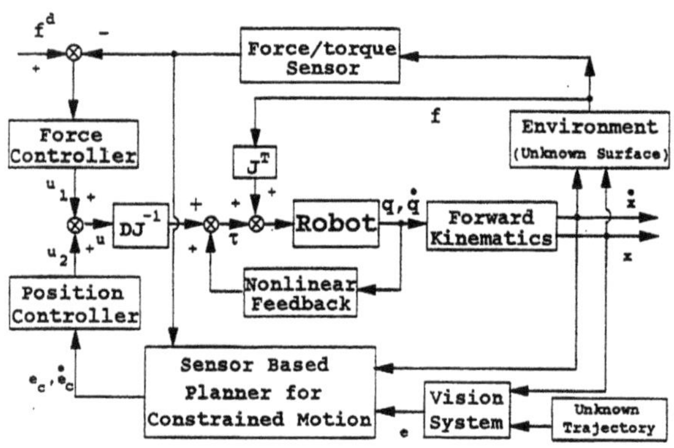

Figure 2: A novel planning and control structure

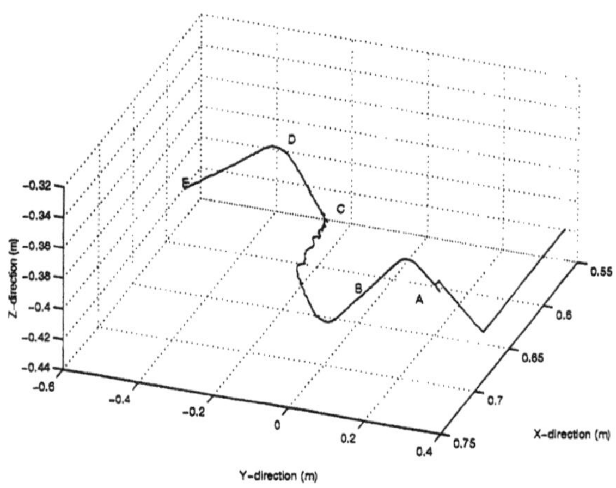

Figure 4: Actual trajectory of the end-effector of the robot in the task space.

14th Triennial World Congress, Beijing, P.R. China

ILC FOR CONSTRAINED NONLINEAR SYSTEMS WITH UNCERTAINTIES AND DISTURBANCES

Wen Chen * **Yeng-Chai Soh** * **Chao-Wan Yin** **

* *School of EEE, Nanyang Technological University Nanyang Avenue, Singapore, 639798.*
** *Robotics Laboratory, Shenyang Automation Institute, Shenyang, Liaoning, 110015, P.R.China*

Abstract: In this paper, we present an ILC method based on the analysis of the constraint, uncertainties and disturbances to achieve precise tracking control of a class of constrained nonlinear systems with uncertainties and disturbances over a finite time interval, using two sorts of matching conditions and relative degrees. The learning is done in a feedback configuration formed by the constraint, and the learning law updates the reference input from the plant input of the previous trail. A sufficient condition which guarantees the convergence of the learning is given. *Copyright © 1999 IFAC*

Keywords: Learning Control, Nonlinear Systems

1. INTRODUCTION

Since the introduction of the concept of iterative learning control by Arimoto et al. (1984), many researchers [Ahn (1993), Arimoto (1994), Wang et al(1995)]. Ahn (1993) et al. proposed an iterative learning control method using the relative degree of the system, and it can guarantee the convergence for nonlinear systems with relative degrees more than one. A sufficient condition under which a uniform convergence of the output to the desired output is also derived. Arimoto et al. (1994) proposed a recursive algorithm of P-type learning control with a forgetting factor for improving robot tasks with geometric end-point constraint. Further, a new concept of exponential passivity with a quadratic margin was introduced for a class of residual error dynamics of robots with or without geometric-point constraint. A rigorous proof of the uniform boundedness of motion trajectories and a rough sketch of the convergence to a neighborhood of the desired motion were presented on the basis of various passivity properties concerning residual error dynamics and difference dynamics of robots. Wang et al. (1995) developed a learning control algorithm for constrained motion control and force control of constrained robot systems. The robustness of the algorithm has been analyzed in the presence of state disturbances, output measurement noise and errors in the initial conditions. It has been shown that the motion and force trajectories converge to neighborhoods of the desired trajectories and these neighborhood tend to zero as the state disturbances, output measurement noise and errors in the initial conditions tend to zero.

For nonlinear systems, Jang et al. (1995) have recently proposed an iterative learning control method for a class of nonlinear systems without uncertainties and disturbances to achieve precise tracking control. The learning is done in a feedback configuration and the learning law updates the feedforward input from the plant input of the previous trial. However, in practice, all constrained nonlinear systems have uncertainties and disturbances. So, it is necessary to discuss the case of constrained nonlinear systems with uncertainties and disturbances in which the iterative

learning law is applied.
In this paper, we propose a simple learning control algorithm, which is different from the common method adopted by many researchers, for the precise tracking control of a class of constrained nonlinear systems with uncertainties and disturbances over a finite time interval. The rest of this paper is organised as follows. In section 2, based on the model of a class of constrained nonlinear systems with uncertainties and disturbances , we discuss the constraint relative degree and output relative degrees, two sorts of matching conditions and give the feedback configuration relative to th Lie derivative of constraint and their pseudo-inverse. In section 3, we propose an iterative learning control method for a class of constrained nonlinear system and establish the convergence to a neighborhood of the desired input. It is shown that the feedback configuration and uncertainties have effects on the convergence condition. An example is presented in section 4 to verify the effectiveness of this learning control method on constrained nonlinear systems modeled in this paper. The concluding remarks are given in section 5.

2. PRELIMINARIES

Consider a class of constrained nonlinear dynamic systems with uncertainties and disturbances defined by the following differential equations:

$$\dot{x} = f(x) + \Delta f(x) + \sum_{j=1}^{m}(\Delta g_j(x) + g_j(x))u_j(t); \quad x \in U \subset R^n,$$
$$z_i = k_i(x) = 0; \qquad i = 1, \ldots, l,$$
$$y_i = h_i(x); \qquad i = 1, \ldots, p, \tag{1}$$

where $f(x), g_j(x), k_i(x)$ and $h_i(x)$ are analytic functions. For convenience we denote that $G(x) = [g_1(x), \ldots, g_m(x)]', \Delta G = [\Delta g_1(x), \ldots, \Delta g_m(x)]'$, $k(x) = [k_1(x), \ldots, k_l(x)]', h(x) = [h_1(x), \ldots, h_p(x)]'$, and $u = [u_1, \ldots, u_m]'$. Suppose that for all $x \in U, g_1(x), \ldots, g_m(x)$ are linearly independent vector fields, $[dk_1(x), \ldots, dk_l(x)]$ and $[dh_1(x), \ldots, dh_p(x)]$ are each linearly independent sets of covector fields, and $m = l + p$. Further, $\Delta f(x)$ and $\Delta g_j(x); j = 1, \ldots, m$, are smooth vector fields which represent the disturbances and modeling uncertainties. The reason why we adopt this affine nonlinear system model is that there are many practical systems which can be modeled as affine nonlinear systems, such as robot manipulators.
Now we state some notations which will be used later.

The Lie derivative of a scalar function $\phi(x)$ along a vector $f(x) = [f_1(x), \ldots, f_n(x)]'$ is denoted as

$$L_f\phi(x) = \frac{\partial \phi}{\partial x} f(x) = \sum_{i=1}^{n} \frac{\partial \phi}{\partial x_i} f_i(x) \tag{2}$$

where $x = [x_1, \ldots, x_n]'$, and the Lie derivative of $\phi(x)$ taken first along $f(x)$ and then along a vector $g(x)$ is

$$L_g L_f \phi(x) = \frac{\partial(L_f\phi)}{\partial x} g(x) \tag{3}$$

If $\phi(x)$ is differentiated j times along $f(x)$, the notation $L_f^j\phi(x)$ is used with $L_f^0\phi(x) = \phi(x)$.
For the constrained nonlinear system with uncertainties and disturbances defined above, we define the relative degree of output y_i with respect to input u, r_i, as the smallest integer for which there exists a $j \in \{1, 2, \ldots, m\}$ such that

$$L_{g_j} L_f^{r_i - 1} h_i(x) \neq 0, \qquad \text{for all} \quad x \in U$$

It is straightforward to show that, if the above system has a (vector) output relative degree $(r_1, \ldots, r_p)$, then the relative degree is unchanged by the addition of perturbations if

$$\Delta f(x) \quad \text{and} \quad \Delta g_j(x) \in ker[dh_i, dL_f^2 h_i, \ldots, dL_f^{r_i-2} h_i]$$
$$\text{for} \quad i = 1, \ldots, p, \quad j = 1, \ldots, m$$

This is called the "matching condition". It assures that the $\Delta f(x)$ and ΔG terms do not appear in derivatives of y_i of order less than r_i. We have the expression of r_i order derivative of y_i according to the matching condition

$$y_i^{r_i} = L_f^{r_i} h_i(x) + L_{\Delta f} L_f^{r_i-1} h_i(x) + \sum_{j=1}^{m} [L_{g_j} L_f^{r_i-1} h_i(x) + L_{\Delta g_j} L_f^{r_i-1} h_i(x)] u \tag{4}$$

Differentiating the constraint relations and suppose that input variables explicitly appear after r_i^c times differentiation of the ith constraint relation, $1 \leq i \leq l$

$$\frac{dz_i}{dt} = L_f k_i(x) = 0$$
$$\vdots$$
$$\frac{d^{r_i^c-1} z_i}{dt^{r_i^c-1}} = L_f^{r_i^c-1} k_i(x) = 0$$
$$\frac{d^{r_i^c} z_i}{dt^{r_i^c}} = L_f^{r_i^c} k_i(x) + \sum_{j=1}^{m} L_{g_j} L_f^{r_i^c-1} k_i(x) u_j = 0$$
$$d(x) + C(x)u = 0 \tag{5}$$

where $C(x) = [c_{ij}(x)]_{l \times m} = [L_{g_j} L_f^{r_i^{c-1}} k_i(x)]_{l \times m}$ and $d(x) = [L_f^{r_i^c} k_1(x), \ldots, L_f^{r_l^c-1} k_l(x)]'$. We say Equation (1) has the constrained relative degree vector $[r_1^c, \ldots, r_l^c]$ on U if the functions

$L_{g_j}L_f^{\rho}k_i(x) = 0, 1 \leq j \leq m, 0 \leq \rho \leq r_i^c - 2$, for all $x \in U$, and the matrix $C(x) = [c_{ij}(x)]_{l\times m}$ has full row rank for $x \in U$.

Now, we extend the concept of matching condition to the constraint terms, i.e. the constraint relative degree is unchanged by adding perturbations if

$$\Delta f(x) \text{ and } \Delta g_j(x) \in ker[dk_i, dL_f^2 k_i, \ldots, dL_f^{r_i-2}k_i]$$
$$\text{for} \quad i, = 1, \ldots, l \quad j = 1, \ldots, m$$

We call this " constraint matching condition". It assures that the $\Delta f(x)$ and ΔG terms do not appear in derivatives of k_i of order less than r_i^c. We have the expressions of r_i^c order derivative of k_i according to the constraint matching condition:

$$L_f^{r_i^c} k_i(x) + L_{\Delta f} L_f^{r_i^c - 1} k_i(x)$$
$$+\sum_{j=1}^{m}(L_{\Delta g_j} L_f^{r_i^c-1} k_i(x) + L_{g_j} L_f^{r_i^c-1} k_i(x))u_j = 0$$

i.e.

$$d(x) + \Delta d(x) + (C(x) + B(x))u = 0 \quad (6)$$

where
$d(x) = [L_f^{r_1^c} k_1(x), \ldots, L_f^{r_l^c} k_l(x)]'$,
$\Delta d(x) = [L_{\Delta f} L_f^{r_1^c} k_1(x), \ldots, L_{\Delta f} L_f^{r_l^c} k_l(x)]$,
$C(x) = [C_{ij}(x)]_{l\times m} = [L_{g_j} L_f^{r_i^c-1} k_i(x)]_{l\times m}$,
$B(x) = [B_{ij}(x)]_{l\times m} = [L_{\Delta g_j} L_f^{r_i^c-1} k_i(x)]_{l\times m}$,
$C(x)$ and $B(x)$ have full row rank for all $x \in U$. Letting $A(x) = C(x) + B(x)$, and $b(x) = d(x) + \Delta d(x)$, we have the solution of (6) written as a feedback law

$$u = -A^+(x)b(x) + (I - A^+(x)A(x))\hat{u}$$
$$\hat{u} \in R^m \quad (7)$$

where $A^+(x) = A'(x)(A(x)A'(x))^{-1}$ is the pseudo-inverse of $A(x)$, and $\hat{u}$ is a reference input, which will be updated by previous input in the learning law.

Remark 1:
Since $A(x)$ has full row rank, we can rearrange input $u' = [w', v'], w \in R^l, v \in R^{m-l}$ such that $A(x) = [A_1(x), A_2(x)]$, where $A_1(x)$ has linearly independent column vectors. In order to impose the constraint, the inputs can be chosen as $w = -A_1(x)^{-1}(b(x) + A_2(x)v)$. But the lineary independent column vectors in $A(x)$ might change as x varies in U. Using the terminology of (McClamroch, 1990), the partition into control inputs and constraint inputs can change as x varies. By avoiding such a partition, the feedback law (7) imposes the constraints without being sensitive to such changes. Therefore the feedback law (7) can impose the constraint in a large region.

Throughout this paper, we assume that systems (1)-(7) satisfy:

(A1) Both the constraint relative degree vector $[r_1^c, \ldots, r_l^c]$ with respect to (f, G, k) and the output relative degree vector $[r_1, \ldots, r_p]$ with respect to (f, G, k) exit.

(A2) For system (1), let R denotes a mapping from $(x(0), u(t), t \in [0,T])$ to $(x(t), t \in [0,T])$ and S be a mapping from $(x(0), u(t), t \in [0,T])$ to $(y(t), t \in [0,T])$. Then, for each fixed $x(0)$, the state map R and the output map S are one-to-one.

(A3) The functions $f(\cdot), g(\cdot)$ and $h(\cdot)$ are Lipschitz in $x(t)$ for $t \in [0,T]$, i.e. there exist constants f_0, g_0, h_0 such that

$$\begin{aligned} \|f(x_1) - f(x_2)\| &\leq f_0\|x_1 - x_2\| \\ \|G(x_1) - G(x_2)\| &\leq g_0\|x_1 - x_2\| \\ \|h(x_1) - h(x_2)\| &\leq h_0\|x_1 - x_2\| \end{aligned} \quad (8)$$

(A4) The desired output $y_d(t)$ is r times continuously differentiable.

(A5) $L_g L_f^{r-1}h(x), A^+(x), b(x), \Delta g_j$ and $\Delta f(x)$ are bounded on R^n.

From (A2), it is guaranteed that given a realizable output and an appropriate initial condition, there exist unique control input and state that yield the output. For practical systems, such as robot manipulators, their models would satisfy the above assumptions.

Our iterative control problem can be stated as follows:

Given a desired output $y_d(x), t \in [0,T]$ and a tolerance error bound ϵ for a class of constrained nonlinear systems (1), the iterative learning control is to find a control input $\hat{u}, \quad t \in [0,T]$ such that $\|y_d(t) - y(t)\| \leq \epsilon$ for $t \in [0,T]$, where $\|\cdot\|$ denotes a vector norm defined as $\|f\| = \max_{1\leq i\leq n} \mid f_i \mid$ for a vector $f = (f_1, f_2, \ldots, f_n)'$ and the matrix norm, induced by the vector norm, as $\|w\| = \max_{1\leq i\leq n}(\sum_{j=1}^m \mid w_{ij} \mid)$ for an $n \times m$ matrix $w = \{w_{ij}\}$. We also define the $\lambda-$ norm for a function $q : [0,T] \to R^m$ as

$$\|q(t)\|_\lambda = \sup_{t\in[0,T]} e^{-\lambda t}\|q(t)\| \quad (9)$$

3. LEARNING CONTROL LAW AND ITS CONVERGENCE

From the constrained nonlinear systems (1)-(7), we propose an input update law for the iterative learning control as the following

$$\hat{u}_{k+1}(t) = u_k(t) + L(t)[y_d^r - y_k^r] \quad (10)$$

where $L(t) : [0,T] \to R^{m\times p}$ is a bounded function, $r = [r_1, \ldots, r_p]$ is the output relative degree, $y_d^r = [y_{d_1}^{r_1}, \cdots, y_{d_p}^{r_p}]'$ is the r order derivative of desired functions, $y_k^r = [y_{k_1}^{r_1}, \cdots, y_{k_p}^{r_p}]'$ is the r order derivative of the output of the kth iteration,

$\hat{u}$ is the reference input from equation (7).
Note that the operator $L(t)$ is an $m \times p$ matrix rather than $m \times m$ one due to the existence of constraint $k_i(x)$ making the dimension of output y less than that of input u.
The iterative learning law adopted by many researchers updates the plant input from the previous trial of plant input itself, which is different from our learning control law. From equation (10), we know that our learning law updates the reference input from the plant input of previous trail. The convergence condition of the proposed learning law is given by the following theorem.
For clarity of the remaining discussion, function parameters will be shown in subscript notation. In particular $\hat{u} \triangleq \hat{u}(t)$, $u_d \triangleq u_d(t)$, $h_k \triangleq h(x_k)$, $A_k^+ \triangleq A^+(x_k)$, $A_k \triangleq A(x_k)$, $b_k \triangleq b(x_k)$.

Theorem 1:
Consider the constrained nonlinear system (1)-(7)with the control law (10) applied iteratively under the assumptions (A1)-(A5). Given a desired output trajectory $y_d(t)$, $t \in [0,T]$, and an initial state $x_d(0)$, which are achievable, and the initial state error is zero for all k. Then with a bounded matrix $L(t) \in R^{m\times p}$ which satisfies the inequality

$$\|(I - L(t)(L_{\Delta g}L_f^{r-1}h_k + L_gL_f^{r-1}h_k))(I - A_k^+A_k)\| \le \rho < 1, \forall t \in [0,T] \quad (11)$$

where $A(x) = C(x) + B(x)$, the error between $\hat{u}$ and u_d is bounded as $k \to \infty$. In addition, the state and output asymptotic error are bounded.

Proof
The update law error for the $k+1$ iteration can be written as

$$u_d - \hat{u}_{k+1} = u_d - u_k - L(t)[L_f^r h_d + L_gL_f^{r-1}h_d u_d - L_f^r h_k - L_{\Delta f}L_f^{r-1}h_k - (L_gL_f^{r-1}h_k + L_{\Delta g}L_f^{r-1}h_k)u_k] \quad (12)$$

where
$L(t) \in R^{m\times p}$, $L_f h_d$ denotes $[L_f^{r_1}h_{d_1}, \ldots, L_f^{r_p}h_{d_p}]'$

$$L_gL_f^{r-1}h_d = \begin{bmatrix} L_{g_1}L_f^{r_1-1}h_{d_1}, & \ldots & , L_{g_m}L_f^{r_1-1}h_{d_1} \\ \ldots & \ldots & \ldots \\ L_{g_1}L_f^{r_p-1}h_{d_p}, & \ldots & , L_{g_m}L_f^{r_p-1}h_{d_p} \end{bmatrix}$$

and $L_f^r h_k$, $L_gL_f^{r-1}h_k$, $L_{\Delta g}L_f^{r-1}h_k(x)$, $L_{\Delta f}L_f^{r-1}h_k(x)$ have similar expressions.
Equation (12) can be further expanded as

$$\begin{aligned} u_d &- \hat{u}_{k+1} \\ &= u_d - u_k - L(t)[L_f^r h_d - L_f^r h_k] \\ &\quad + L(t)[L_{\Delta f}L_f^{r-1}h_k] - L(t)[L_gL_f^{r-1}h_d u_d - \\ &\qquad (L_gL_f^{r-1}h_k + L_{\Delta g}L_f^{r-1}h_k)u_k] \\ &= (u_d - u_k) - L(t)[L_f^r h_d - L_f^r h_k] \\ &\quad - L(t)[L_gL_f^{r-1}h_d u_d - (L_gL_f^{r-1}h_k + \\ &\quad L_{\Delta g}L_f^{r-1}h_k)u_k + L_{\Delta f}L_f^{r-1}h_k \\ &\quad - (L_gL_f^{r-1}h_k + L_{\Delta g}L_f^{r-1}h_k)u_d + \\ &\qquad (L_gL_f^{r-1}h_k + L_{\Delta g}L_f^{r-1}h_k)u_d] \\ &= (u_d - u_k) - L(t)[L_f^r h_d - L_f^r h_k] \\ &\quad + L(t)[L_{\Delta f}L_f^{r-1}h_k] - L(t)[L_{\Delta g}L_f^{r-1}h_k \\ &\quad + L_gL_f^{r-1}h_k](u_d - u_k) - L(t)[L_gL_f^{r-1}h_d \\ &\quad - (L_{\Delta g}L_f^{r-1}h_k + L_gL_f^{r-1}h_k)]u_d \\ &= (u_d - u_k)[I - L(t)(L_{\Delta g}L_f^{r-1}h_k + L_gL_f^{r-1}h_k)] \\ &\qquad - L(t)[L_f^r h_d - L_f^r h_k] \\ &\quad + L(t)[L_{\Delta f}L_f^{r-1}h_k] - L(t)[L_gL_f^{r-1}h_d \\ &\quad - L_gL_f^{r-1}h_k]u_d + L(t)[L_{\Delta g}L_f^{r-1}h_k]u_d \quad (13) \end{aligned}$$

Substituting equation (7) into (13), we have

$$\begin{aligned} u_d &- \hat{u}_{k+1} \\ &= [u_d + A_k^+(x)b_k(x) - (I - A_k^+(x)A_k(x))\hat{u}_k] \cdot \\ &\quad [I - L(t)(L_{\Delta g}L_f^{r-1}h_k + L_gL_f^{r-1}h_k)] \\ &\quad - L(t)[L_f^r h_d - L_f^r h_k] + L(t)[L_{\Delta f}L_f^{r-1}h_k] \\ &\quad - L(t)[L_gL_f^{r-1}h_d - L_gL_f^{r-1}h_k]u_d \\ &\quad + L(t)[L_{\Delta g}L_f^{r-1}h_k]u_d \\ &= [(I - A_k^+A_k)(u_d - \hat{u}_k) + A_k^+b_k + A_k^+A_k u_d] \cdot \\ &\quad [I - L(t)(L_{\Delta g}L_f^{r-1}h_k + L_gL_f^{r-1}h_k)] \\ &\quad - L(t)[L_f^r h_d - L_f^r h_k] + L(t)[L_{\Delta f}L_f^{r-1}h_k] \\ &\quad - L(t)[L_gL_f^{r-1}h_d - L_gL_f^{r-1}h_k]u_d \\ &\quad + L(t)[L_{\Delta g}L_f^{r-1}h_k]u_d \\ &\quad = [(I - A_k^+A_k)(I - L(t)(L_{\Delta g}L_f^{r-1}h_k \\ &\quad + L_gL_f^{r-1}h_k)](u_d - \hat{u}_k) \\ &\quad + [I - L(t)(L_{\Delta g}L_f^{r-1}h_k \\ &\quad + L_gL_f^{r-1}h_k)](A_k^+b_k + A_k^+A_k u_d) \\ &\quad - L(t)[L_f^r h_d - L_f^r h_k] + L(t)[L_{\Delta f}L_f^{r-1}h_k] \\ &\quad - L(t)[L_gL_f^{r-1}h_d - L_gL_f^{r-1}h_k]u_d \\ &\quad + L(t)[L_{\Delta g}L_f^{r-1}h_k]u_d \quad (14) \end{aligned}$$

Since $L_f^r h(x)$ and $L_gL_f^{r-1}h(x)$ are continuous in x, there exist constants m_1 and m_2 such that

$$\|L_f^r h(x_d) - L_f^r h(x_k)\| \le m_1\|x_d - x_k\|$$

$$\|L_gL_f^{r-1}h(x_d) - L_gL_f^{r-1}h(x_k)\| \le m_2\|x_d - x_k\| \quad (15)$$

Taking norms on both sides of (14) and defining $\delta\hat{u}_{k+1} = u_d - \hat{u}_{k+1}$, $\delta\hat{u}_k = u_d - \hat{u}_k$, we get

$$\begin{aligned}
&\| \delta\hat{u}_{k+1}\| \\
&\quad \le \|(I - L(t)(L_{\Delta g}L_f^{r-1}h_k + L_gL_f^{r-1}h_k)) \\
&\quad (I - A_k^+A_k)\|\|\delta\hat{u}_k\| \\
&\quad + \|I - L(t)(L_{\Delta g}L_f^{r-1}h_k + L_gL_f^{r-1}h_k)\| \\
&\quad \|A_k^+b_k + A_k^+A_ku_d\| \\
&\quad + \|L(t)\|\|u_d\|m_2\|\delta x_k\| + \|L(t)\|m_1\|\delta x_k \\
&\quad + \|L(t)\|\|L_{\Delta f}L_f^{r-1}h_k\| \\
&\quad + \|L(t)\|\|u_d\|\|L_{\Delta g}L_f^{r-1}h_k\| \\
&\le \rho\|\delta\hat{u}_k\| + (M_LM_dm_2 + M_Lm_1)\|\delta x_k\| \\
&\quad + M_{gf}M_k + M_L \cdot M_{\Delta f} + M_L \cdot M_d \cdot M_{\Delta g} \quad (16)
\end{aligned}$$

where $M_L, M_d, M_{gf}, M_k, M_{\Delta f}, M_{\Delta g}$ are the norm bounds for $L(t), u_d, (I-(L_{\Delta g}L_f^{r-1}h_k+L_gL_f^{r-1}h_k))$, $(A_k^+b_k + A_k^+A_ku_d), L_{\Delta f}L_f^{r-1}h_k, L_{\Delta g}L_g^{r-1}h_k$, respectively.

Defining $k_1 \triangleq M_LM_dm_2 + M_Lm_1, k_3 \triangleq M_{gf}M_k + M_LM_{\Delta f} + M_LM_dM_{\Delta g}$. we have

$$\|\delta\hat{u}_{k+1}\| \le \rho\|\delta\hat{u}_k\| + k_1\|\delta x_k\| + k_3 \quad (17)$$

With the assumption that $x_k(0) = x_d(0)$ for all $k \ge 0$, the states of the system (1)-(7) satisfy

$$\begin{aligned}
&x_d - x_k \\
&= \int_0^t [(f_d - f_k) + (G_du_d - (G_k + \Delta G_k)u_k - \Delta f_k)]d\tau \\
&= \int_0^t [(f_d - f_k) + (G_du_d - (G_k + \Delta G_k)[-A_k^+b_k + (I - A_k^+A_k)\hat{u}_k] - \Delta f_k)]d\tau \\
&= \int_0^t [(f_d - f_k) + (G_du_d + A_k^+b_k(G_k + \Delta G_k) \\
&\quad + (u_d - \hat{u}_k)(G_k + \Delta G_k)(I - A_k^+A_k) \\
&\quad - (G_k + \Delta G_k)(I - A_k^+A_k)u_d - \Delta f_k)]d\tau \\
&= \int_0^t [(f_d - f_k) + (G_d - G_k)u_d \\
&\quad - \Delta G_ku_d + (G_k + \Delta G_k)(A_k^+b_k + A_k^+A_ku_d) \\
&\quad + (u_d - \hat{u})(G_k + \Delta G_k)(I - A_k^+A_k) \\
&\quad - \Delta f_k]d\tau \quad (18)
\end{aligned}$$

Taking norms of (18), we obtain

$$\begin{aligned}
\|\delta x_k\| \le \int_0^t &[(f_0 + M_dg_0)\|\delta x_k\| + M_g\gamma\|\delta\hat{u}_k\| \\
&+ M_{kd} + M_gM_k + M_{\Delta f_k}]d\tau \quad (19)
\end{aligned}$$

where M_{kd}, M_g and $M_{\Delta f_k}$ are the norm bounds for $\Delta G_ku_d, G_k + \Delta G_k$ and Δf, respectively.

Defining $k_2 \triangleq f_0 + M_dg_0, d \triangleq M_{kd} + M_gM_k + M_{\Delta f_k}, \rho_1 \triangleq M_g\gamma$ and using a basic integral inequality, we have

$$\|\delta x_k\| \le \int_0^t e^{k_2(t-\tau)}(\rho_1\|\delta\hat{u}_k\| + d)d\tau \quad (20)$$

Combining (17) and (20) yields

$$\begin{aligned}
\|\delta\hat{u}_{k+1}\| &\le \rho\|\delta\hat{u}_k\| + k_1(\int_0^t e^{k_2(t-\tau)}(\rho_1\|\delta\hat{u}_k\| \\
&\quad + d)d\tau) + k_3 \\
&\le \rho\|\delta\hat{u}_k\| + k_1\rho_1\int_0^t e^{k_2(t-\tau)}\|\delta\hat{u}_k\|d\tau \quad (21) \\
&\quad + k_1d\int_0^t e^{k_2(t-\tau)}d\tau + k_3
\end{aligned}$$

Multiplying (21) by $e^{-\lambda t}$, defining $k_m = \max(k_1\rho_1, k_2)$, and assuming $\lambda > k_m$, we have that

$$\begin{aligned}
e^{-\lambda t}\|\delta\hat{u}_{k+1}\| &\le \rho e^{-\lambda t}\|\delta\hat{u}_k\| + \int_0^t e^{-\lambda t}\|\delta\hat{u}_k\|e^{(k_m-\lambda)(t-\tau)}d\tau \\
&\quad + k_1d\int_0^t e^{-\lambda t}e^{(k_2-\lambda)(t-\tau)}d\tau + k_3e^{-\lambda t} \quad (22)
\end{aligned}$$

Noticing that the integrals are strictly increasing and that for a constant $\|k_m\|_\lambda = k_m$, we obtain

$$\begin{aligned}
\|\delta\hat{u}_{k+1}\|_\lambda &\le [\rho + \frac{k_m}{\lambda - k_m}(1 - e^{(k_m-\lambda)T})]\|\delta\hat{u}_k\|_\lambda \\
&\quad + \frac{k_1d}{\lambda - k_2}(1 - e^{(k_2-\lambda)T}) + k_3 \quad (23)
\end{aligned}$$

Defining

$$\bar{\rho} = \rho + \frac{k_m}{\lambda - k_m}(1 - e^{(k_m-\lambda)T})$$

and

$$k_4 = k_3 + \frac{k_1d}{\lambda - k_2}(1 - e^{(k_2-\lambda)T}) \quad (24)$$

we have

$$\|\delta\hat{u}_{k+1}\|_\lambda \le \bar{\rho}\|\delta\hat{u}_k\|_\lambda + k_4 \quad (25)$$

Since $\rho < 1$, we can find a $\lambda > k_m$ which makes $\bar{\rho} < 1$. Thus $\hat{u}_k$ converges to the neighborhood of u_d of radius $(\frac{k_4}{1-\bar{\rho}})$ with respect to the λ-norm. This implies that

$$\lim\sup_{k\to\infty} \|\delta\hat{u}_k\|_\lambda \le (\frac{k_4}{1 - \bar{\rho}})$$

Using (20) and similar manipulations, we also have

$$\|\delta x_k\|_\lambda \le \int_0^t e^{(k_2-\lambda)(t-\tau)}\|\delta \hat{u}_k\|_\lambda d\tau + \frac{d}{\lambda - k_2}(1 - e^{(k_2-\lambda)T})$$
$$\le \frac{1}{\lambda - k_2}(1 - e^{(k_2-\lambda)T})\|\delta \hat{u}_k\|_\lambda + \frac{d}{\lambda - k_2}(1 - e^{(k_2-\lambda)T})$$

So

$$lim \sup_{k\to\infty} \|\delta x_k\|_\lambda \le \frac{1}{\lambda - k_2}(1 - e^{(k_2-\lambda)T})(\frac{k_4}{1-\bar{\rho}}) + \frac{d}{\lambda - k_2}(1 - e^{(k_2-\lambda)T})$$

To obtain the result for y_i, we use the fact that $h(x)$ is Lipschitz in x.

Remark 2: From the derivations we can observe that the bounds of the asymptotic errors are dependent on the constraint condition $k_i(x)$, uncertainties, disturbances and their Lie Derivatives. These asymptotic errors can't be eliminated due to the existence of the constraint, which is different from other nonlinear systems such as those discussed by Jang et al.(1995) and Ahn et al. (1993).

Remark 3:

From the convergence condition, we know that although there are uncertainties $\Delta g(x)$ and disturbances $\Delta f(x)$ in the constraint nonlinear systems, only the uncertainties have an influence on the convergence condition.

4. AN EXAMPLE

To show the effectiveness of the proposed learning control law for constrained nonlinear systems with uncertainties and disturbances, a simulation is conducted for a constrained nonlinear system model.

Consider system (1) with

$$f(x) = \begin{bmatrix} -2x_1 + x_1x_2 \\ 0 \end{bmatrix} \quad g(x) = \begin{bmatrix} 0 & 1 \\ 1 & 0 \end{bmatrix} \quad (26)$$

$$k(x) = x_1 - c = 0 \qquad y = h(x) = x_2 \quad (27)$$

where c is a constant, and let the desired trajectory $t \in [0, 10]$ be given

$$y_d(t) = sin(3t) + cos(t)$$

Figure (1) illustrates the outputs of the iterative learning control strategy. Clearly, the output converges quickly to the desired output. The proposed iterative learning law can indeed work very well for this kind ofonstrained nonlinear systems with uncertainties and disturbances.

5. CONCLUSION

In this paper, we have proposed an iterative learning control law, based on the analysis of the constraint terms and two sorts of matching conditions, for a class of constrained nonlinear systems with uncertainties and disturbances over a finite time interval. From the derivation of the convergence, we can identify that the bounds of the asymptotic errors are influended by the existence of the constraint term and uncertainties. The simulation example verifies that the proposed iterative learning control law is effective for the class of constrained nonlinear systems discussed in this paper.

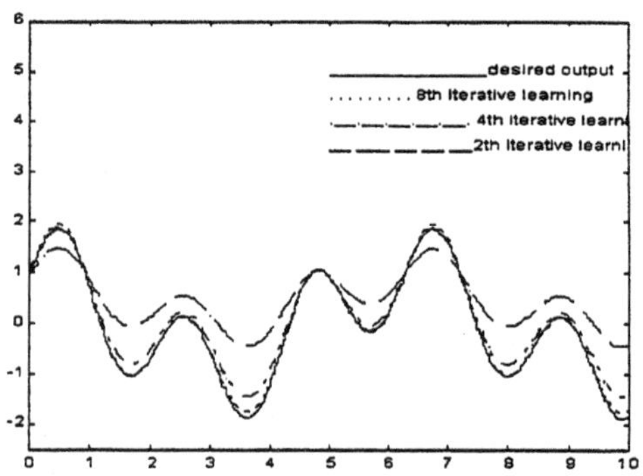

Fig. 1 Simulation results of ILC for constrained nonlinear systems with uncertainties and disturbances

6. REFERENCES

Ahn,H-S., C-H.Choi and K.B.Kim (1993). Iterative learning control for a class of nonlinear systems. *Automatica*, **19**, 1575-1578.

Danwei Wang, Y. C.Soh and C.C.Cheah (1995). Robust motion and force control of constrained manipulation by learning. *Automatica*, **31**, 257-262.

S. Arimoto and Tomohide Naniwa (1994). Learning control for robot motion under geometric end-point constraint. *Robotica*, **12**, 101-108.

Jang, T.-J., et al. (1995). Iterative learning control in feedback systems. *Automatica*, **31**, 243-248.

CMAC NEURAL NETWORKS-BASED FEEDBACK LINEARIZATION FOR NONLINEAR SYSTEMS

You-An Zhang, Ping-Yuan Cui, Xu Yang, and Di Yang

Harbin Institute of Technology, P.O.Box.137, Harbin,150001,P.R.China
E-mail: lian@mail.hr.hl.cn

Abstract: Based on the nominal model of the plant, a CMAC neural network-based controller that feedback-linearizes a class of SISO/MIMO continuous-time nonlinear systems is presented. A stability proof is given strictly in the sense of Lyapunov. It is shown that all the signals in the closed-loop system are uniformly ultimately bounded. Simulation example is also given. *Copyright © 1999 IFAC*

Keywords: CMAC, neural networks, feedback linearization, nonlinear systems, stability analysis

1. INTRODUCTION

In recent years, neural networks (NNs) have been used for feedback linearization of nonlinear systems. The multilayer NN has been used for feedback linearization of a class of SISO continuous-time nonlinear systems (Yesildirek and Lewis, 1995). The multilayer NN is a kind of fully connected NNs, since all the weights are updated during each learning cycle, the learning is essentially global in nature. This global nature of the weight updating does not utilize the information on local NN structure and thus, slows down the speed of learning. The fully connected NNs suffer from an additional drawback in the sense that the function approximation is sensitive to the learning data.

The cerebella model articulation controller (CMAC) NN has advantages over fully connected NNs due to its increased structure, which results in better computational speed and faster learning. The CMAC NN has been proposed for control of complex dynamical systems in (Lane, *et al.*, 1992; Jagannathan, *et al.*, 1998; Commuri and Lewis, 1997). The CMAC is a non-fully connected perceptron-like associative memory network that computes a nonlinear function over a domain of interest. The CMAC takes advantage of the input-redundance by using distributed storage. The contents of these memory locations are referred to as the weights and the output of the CMAC NN is a linear combination of these weights. Feedback linearization of a class of continuous-time nonlinear systems given in affine form using CMAC NNs has been presented by Jagannathan, *et al.* (1998).

In many cases, the nominal model of a plant is known. If the nominal model of a plant can be incorporated into the control law of the adaptive feedback linearization, the burden for the learning of the NNs will be alleviated, and the learning speed of the NNs will be increased. In this paper, based on the nominal model of the plant, the CMAC NN is used to feedback-linearize a class of SISO/MIMO continuous-time nonlinear systems given in state-space affine form.

2. PROBLEM STATEMENT

First, consider a class of SISO state-feedback linearizable system having a state-space affine form

$$\dot{x}_1 = x_2, \dot{x}_2 = x_3 \ , \cdots,$$

$$\dot{x}_n = f(x) + g(x)u + d, y = x_1 \quad (1)$$

where $x = [x_1\ x_2\ \cdots\ x_n\ ]^T$ is the state vector. It is assumed that $d(t)$ is an unknown disturbance with known upper bound b_d so that $|d(t)| \le b_d$, and $f, g: R^n \to R$ are two unknown smooth functions, where $|g(x)| \ge \underline{g} > 0$, $\forall x$, and $\underline{g}$ is a known upper bound. Suppose the nominal models of f and g are $\bar{f}$ and $\bar{g}$ respectively, and define their differences as $\tilde{f} = f - \bar{f}$ and $\tilde{g} = g - \bar{g}$.

Feedback linearization will be used to perform output tracking: Given the desired output y_d, find a control action u, such that the plant follows the desired trajectory with an acceptable accuracy (i.e., bounded-error tracking), while all the states and control remain bounded.

Define the desired states vector as $x_d(t) = [y_d\ \dot{y}_d \cdots y_d^{(n-1)}]^T$, the state error vector as $e = x - x_d$, and a filtered error as $r = [\Lambda^T 1]e$, where $\Lambda = [\lambda_1\ \lambda_2\ \cdots\ \lambda_{n-1}]^T$ is an appropriately chosen coefficient vector such that

$$s^{n-1} + \lambda_{n-1}s^{n-2} + \cdots + \lambda_1$$

is Hurwitz, then, the derivative of r can be written as

$$\dot{r} = (\bar{f} + \tilde{f}) + (\bar{g} + \tilde{g})u + d + y_d \quad (2)$$

where $y_d \equiv -y_d^{(n)} + [0\ \Lambda^T]e$.

Assumption 1: The desired trajectory vector $x_d(t)$ is continuous and available for measurement. $\|x_d(t)\| \le q$ with q a known bound, where $\|\cdot\|$ denotes a norm, taken as Frobenius norm for convenience in the following discussion.

3. THE CMAC NEURAL NETWORK

The CMAC neural networks for control have been discussed in detail by Commuri and Lewis (1997). Here we only give a brief outline of the networks. When a CMAC neural network is used to manufacture a continuous function vector

$$f(x) = [f_1(x) f_2(x) \cdots f_m(x)\ ]^T$$

where $x \in R^n$, and f: $R^n \to R^m$, it is seen that the nonlinear function $f(x)$ produced by the CMAC NN is composed of two primary functions

$$R: X \Rightarrow A,\ P: A \Rightarrow Y \quad (3)$$

where X is the continuous n-dimensional input space, A is the N_A -dimensional association space, and Y is the m-dimensional output space. The function R is fixed and maps each input space onto an association vector $\alpha = R(x)$ in the association space A. The function P computes an output $f \in Y$ by projecting the association vector determined by $R(x)$ onto a vector of adjustable weights W such that $f = P(\alpha) = W^T\alpha$. $R(x)$ in (3) is the multidimensional receptive field function, which assigns an activation value to each point in the input space X.

Definition 1: Let $\mu_{k,j_l}(x_k)$ be the activation value of the j_lth receptive field function along the input dimension k, then the multidimensional receptive field function can be defined as

$$R_{j_1,j_2,\cdots,j_n}(x) = \frac{\mu_{1,j_1}(x_1)\mu_{2,j_2}(x_2)\cdots\mu_{n,j_n}(x_n)}{\sum_{j_n=1}^{N_n}\cdots\sum_{j_2=1}^{N_2}\sum_{j_1=1}^{N_1}\prod_{i=1}^{n}\mu_{i,j_i}(x_i)} \quad (4)$$

The receptive field functions so defined are said to have the normalization property if

$$\sum_{j_n=1}^{N_n}\cdots\sum_{j_2=1}^{N_2}\sum_{j_1=1}^{N_1} R_{j_1,j_2,\cdots,j_n}(x) = \sum_{j_n=1}^{N_n}\cdots\sum_{j_2=1}^{N_2}\sum_{j_1=1}^{N_1}\prod_{i=1}^{n}\mu_{i,j_i}(x_i) = 1 \quad (5)$$

Suppose $x_i \in [x_{i_{min}}, x_{i_{max}}]$, $i = 1,2,\cdots,n$, is divided into $N_i - 1$ equal parts, the partition points satisfy the following relationship

$$x_{i_{min}} = x_{i,1} < x_{i,2} < \cdots < x_{i,N_i} = x_{i_{max}}$$

The triangle receptive field functions are defined as

$$\mu_{i,1}(x_i) = \lambda(-\infty, x_{i,1}, x_{i,2})(x_i),$$
$$\mu_{i,j}(x_i) = \lambda(x_{i,j-1}, x_{i,j}, x_{i,j+1})(x_i), 1 < j \le N_i,$$
$$\mu_{i,N_i}(x_i) = \lambda(x_{i,N_i-1}, x_{i,N_i}, \infty)(x_i),$$
$$\lambda(a,b,c)(y) = \begin{cases} (y-a)/(b-a), a \le y \le b (= 1, \text{if } a = -\infty) \\ (c-y)/(c-b), b \le y \le c (= 1, \text{if } c = \infty) \\ 0, \quad \text{otherwise} \end{cases}$$

By choosing triangle functions with overlap factor of 2 as the one-dimensional receptive field functions, and using (4) as the multi-dimensional receptive field functions, it is pointed out that the multi-dimensional receptive field functions such defined have the normalization property, and take the simplified form

$$R_{j_1,j_2,\cdots,j_n}(x)=\mu_{1,j_1}(x_1)\mu_{2,j_2}(x_2)\cdots\mu_{n,j_n}(x_n)\quad(6)$$

Lemma 1: Any C^1 function $h(x)$ with $h\le L, L>0$ can be expressed as

$$h(x)=w_h^{\mathrm{T}}\Gamma_h(x)+\varepsilon_h \quad (7)$$

where ε_h is the function estimation error and $w_h=\left[w_{h_1}\ w_{h_2}\ \cdots\ w_{h_{N_A}}\right]^{\mathrm{T}}$ is a vector containing the set of weights, and $\Gamma_h(x)$ is a vector of the receptive field activation values.

Assumption 2: $\varepsilon_h \le \varepsilon_{hN}$, with ε_{hN} a known bound.

Assumption 3: The ideal CMAC neural network weight vector w_h is bounded, i.e., $w_h \le W_{h\max}$, with $W_{h\max}$ a known bound.

4. THE CMAC NN CONTROLLER

The control law is chosen as (Yesildirek and Lewis, 1995)

$$u=u_{\mathrm{c}}+u_{\mathrm{d}}$$

$$=u_{\mathrm{c}}+\begin{cases}0.5(u_{\mathrm{r}}-u_{\mathrm{c}})e^{\gamma(u_{\mathrm{c}}-s)} & ,\text{if } u_{\mathrm{c}}\le s\\ (u_{\mathrm{r}}-u_{\mathrm{c}})\left(1-0.5e^{-\gamma(u_{\mathrm{c}}-s)}\right) & ,\text{if } u_{\mathrm{c}}>s\end{cases} \quad (8)$$

$$u_{\mathrm{c}}=\left[-\left(f+\hat{\tilde{f}}\right)+v\right]/\left(g+\hat{\tilde{g}}\right) \quad (9)$$

$$v=-k_{\mathrm{v}}\,r-y_{\mathrm{d}} \quad (10)$$

$$u_{\mathrm{r}}=-\mu\,u_{\mathrm{c}}\ \mathrm{sgn}(r) \quad (11)$$

where k_{v} a positive constant. s>0, $\gamma<(\ln 2)\ s$ and μ are design parameters. Two CMAC NNs are used to estimate $\tilde{f}$ and $\tilde{g}$, i.e.,

$$\hat{\tilde{f}}=\hat{w}_f^{\mathrm{T}}\,\Gamma_f(x),\ \ \hat{\tilde{g}}=\hat{w}_g^{\mathrm{T}}\,\Gamma_g(x), \quad (12)$$

Note that the subscribes f, g in (12) should be $\tilde{f}, \tilde{g}$, but for concise the superscript ~ are omitted. The weight update rules are provided by

$$\dot{\hat{w}}_f=M_f\,\Gamma_f(x)r-kM_f\,r\,\hat{w}_f \quad (13a)$$

$$\dot{\hat{w}}_g=M_g\,\Gamma_g(x)ru_c-kM_g\,r\,u_c\,\hat{w}_g \quad (13b)$$

where M_f, M_g, k are positive constants. The weight estimation errors are defined as $\tilde{w}_f=w_f-\hat{w}_f$ and $\tilde{w}_g=w_g-\hat{w}_g$.

5. STABILITY ANALYSIS

Theorem 1: For system (1), assume the control law is given by (8)-(12) with the weight update rules provided by (13), and assumptions 1-3, lemma 1 hold, when μ satisfies (24), then the filtered tracking error $r(t)$, the CMAC NN weight estimations errors $\tilde{w}_f, \tilde{w}_g$, and the control input u are UUB (Yesildirek and Lewis, 1995). with specific bounds given by (26), and by increasing k_{v}, the filtered tracking error $r(t)$ can be made arbitrarily small.

Proof: Let the Lyapunov function candidate be

$$V=0.5r^2+0.5\left(\tilde{w}_f^{\mathrm{T}}M_f^{-1}\tilde{w}_f+\tilde{w}_g^{\mathrm{T}}M_g^{-1}\tilde{w}_g\right) \quad (14)$$

Consider the derivative of (14) in two mutually exclusive and exhaustive regions.

Region 1: $u_{\mathrm{c}}\le s$. The filtered tracking error system (2) can be rewritten by adding and subtracting v, and using (9) and (10) for v as

$$\dot{r}=-k_{\mathrm{v}}\,r+\left(\tilde{f}-\hat{\tilde{f}}\right)$$
$$+(\tilde{g}-\hat{\tilde{g}})u_{\mathrm{c}}+d+(g+\tilde{g})u_{\mathrm{d}} \quad (15)$$

By using lemma 1, one can obtain

$$\tilde{f}(x)=w_f^{\mathrm{T}}\Gamma_f(x)+\varepsilon_f,\ \tilde{g}(x)=w_g^{\mathrm{T}}\Gamma_g(x)+\varepsilon_g \quad (16)$$

Substituting (12) and (16) into (15), yields

$$\dot{r}=-k_{\mathrm{v}}\,r+\tilde{w}_f^{\mathrm{T}}\,\Gamma_f(x)$$
$$+\tilde{w}_g^{\mathrm{T}}\Gamma_g(x)u_{\mathrm{c}}+(\varepsilon_f+\varepsilon_g u_{\mathrm{c}}+d)+gu_{\mathrm{d}} \quad (17)$$

The time derivative of (14) is

$$\dot{V}=r\dot{r}+\tilde{w}_f^{\mathrm{T}}M_f^{-1}\dot{\tilde{w}}_f+\tilde{w}_g^{\mathrm{T}}M_g^{-1}\dot{\tilde{w}}_g \quad (18)$$

Using (13), (17) and the weight estimation error definitions, (18) can be rewritten as

$$\dot{V}=-k_{\mathrm{v}}\,r^2+r(\varepsilon_f+\varepsilon_g u_{\mathrm{c}}+d)+rgu_{\mathrm{d}}$$
$$+k\,r\,\tilde{w}_f^{\mathrm{T}}\hat{w}_f+k\,r\,u_{\mathrm{c}}\,\tilde{w}_g^{\mathrm{T}}\hat{w}_g \quad (19)$$

Using $\tilde{w}^{\mathrm{T}}\hat{w}\le\tilde{w}\,(w-\tilde{w})$, assumption 1-3, and (8), and with simple manipulation, one can write the final form by completing the squares

$$\dot{V}\le-r\ [k_{\mathrm{v}}\,r-D_1+k(\tilde{w}_f-0.5W_{f\max})^2$$

$$+ks(\|\tilde{w}_g\| - 0.5W_{g\max})^2]$$
$$-0.5|r|g[\mu|u_c| + \mathrm{sgn}(r)u_c]e^{\gamma(|u_c - s|)} \quad (20)$$

where

$D_1 = \varepsilon_{fN} + \varepsilon_{gN}s + b_d + 0.25k(W_{f\max}^2 + sW_{g\max}^2)$

a positive constant. Note that the last term in the right of (20) is always negative when $\mu \geq 1$, and one can therefore deduce from (20) that if $|r| > \delta_{r_1}$, or $\|\tilde{w}_f\| > \delta_{f_1}$, then $\dot{V} < 0$. Where

$$\delta_{r_1} = D_1 / k_v, \ \delta_{f_1} = 0.25W_{f\max} + \sqrt{D_1 / k} \quad (21)$$

By solving (13b), and using the boundedness of $|r|$, $\|\Gamma_g(x)\| \leq 1$, one can obtain $\|\tilde{w}_g\| \leq \delta_{g_1}$, where

$$\delta_{g_1} = \|\tilde{w}_g(0)\| + W_{g\max} + 1/k \quad (22)$$

Since $|u_c| \leq s$, $|u_r|$ is bounded, and $|u|$ is bounded.

Region 2: $|u_c| > s$. Similarly, one can obtain

$$\dot{V} \leq -|r|[k_v|r| - D_2 + k(\|\tilde{w}_f\| - 0.5W_{f\max})^2$$
$$+k|u_c|(\|\tilde{w}_g\| - 0.5W_{g\max})^2] + L_T \quad (23)$$

where $D_2 = \varepsilon_{fN} + b_d + 0.25kW_{f\max}^2$,

$L_T = |r||u_c|\varepsilon_{gN} + 0.25k|r||u_c|W_{g\max}^2 + rgu_d$.

From $|u_c| > s$, $\mu \geq 1$, $g(x) \geq \underline{g}$ and (8), one can obtain

$$rgu_d \leq -\underline{g}|r||u_c|(\mu - 1)(1 - 0.5e^{\gamma s})$$

$$L_T \leq -\underline{g}|r||u_c|[(\mu - 1)(1 - 0.5e^{\gamma s})$$
$$-(\varepsilon_{gN} + 0.25kW_{g\max}^2)/\underline{g}].$$

Since $\gamma < (\ln 2)/s$, when μ satisfies (24), one can obtain $L_T \leq 0$. Where

$$\mu \geq 1 + (\varepsilon_{gN} + 0.25kW_{g\max}^2)/[\underline{g}(1 - 0.5e^{\gamma s})] \quad (24)$$

Thus, from (23), the bounds δ_{r_2} of $|r|$, the bounds δ_{f_2} of $\|\tilde{w}_f\|$ and the bounds δ_{g_2} of $\|\tilde{w}_g\|$ are, respectively

$$\delta_{r_2} = D_2/k_v, \ \delta_{f_2} = 0.5W_{f\max} + \sqrt{D_2/k},$$
$$\delta_{g_2} = 0.5W_{g\max} + \sqrt{D_2/(ks)} \quad (25)$$

Further, by adjusting the accuracy of $\bar{g}$ and $\hat{g}$, and the adaptation gains, the boundedness of $|u_c|$ is ensured, and hence $|u_r|$ and $|u|$ are bounded.

Reprise: Combining the results from region 1 and region 2, when μ satisfies (24), $|r|$、$\|\tilde{w}_f\|$ and $\|\tilde{w}_g\|$ are all bounded, with the bounds given by (26), and $|u|$ is bounded.

$$\delta_r = \max\{\delta_{r_1}, \delta_{r_2}\}, \delta_f = \max\{\delta_{f_1}, \delta_{f_2}\}$$
$$\delta_g = \max\{\delta_{g_1}, \delta_{g_2}\} \quad (26)$$

6. EXTENSION TO THE MIMO CASE

Consider the following MIMO state-feedback linearizable system having a state-space affine form

$$\begin{aligned}
&\dot{x}_1 = x_2, \quad \dot{x}_2 = x_3, \ \cdots, \\
&\dot{x}_{n_1} = f_1(x) + g_1(x)u + d_1, \\
&\dot{x}_{n_1+1} = x_{n_1+2}, \quad \dot{x}_{n_1+2} = x_{n_1+3}, \ \cdots, \\
&\dot{x}_{n_1+n_2} = f_2(x) + g_2(x)u + d_2, \cdots, \\
&\dot{x}_{n_1+n_2+\cdots+n_{m-1}+1} = x_{n_1+n_2+\cdots+n_{m-1}+2}, \\
&\dot{x}_{n_1+n_2+\cdots+n_{m-1}+2} = x_{n_1+n_2+\cdots+n_{m-1}+3}, \cdots, \\
&\dot{x}_n = \dot{x}_{n_1+n_2+\cdots+n_m} = f_m(x) + g_m(x)u + d_m
\end{aligned} \quad (27)$$

with the output equation given as

$$y = \left[x_1 \ x_{n_1+1} \ \cdots \ x_{n_1+n_2+\cdots+n_{m-1}+1}\right]^T \quad (28)$$

It is assumed that $d(t) = [d_1(t) \ d_2(t) \ \cdots \ d_m(t)]^T$ is an unknown disturbance vector with known upper bound b_d, so that $\|d(t)\| \leq b_d$, and $f, g: R^n \to R^m$ are two unknown smooth function vectors

$$f^T(x) = [f_1(x) f_2(x) \cdots f_m(x)],$$
$$g^T(x) = [g_1(x) g_2(x) \cdots g_m(x)].$$

where $g_i(x) \geq \underline{g}$, $i = 1,2,\cdots,m$. and $\underline{g}$ is a known upper bound.

Suppose the nominal models of f_i and g_i are $\bar{f}_i$ and $\bar{g}_i$ respectively, and define their differences as

$$\tilde{f}_i = f_i - \bar{f}_i, \ \tilde{g}_i = g_i - \bar{g}_i,$$

and hence

$$\tilde{f} = f - \bar{f}, \ \tilde{g} = g - \bar{g}.$$

Define the desired output vector as

$$y_d(t) = [y_{d_1}(t)\, y_{d_2}(t) \cdots y_{d_m}(t)]^T,$$

the control action vector as $u = [u_1\ u_2\ \cdots\ u_m]^T$, and the desired states vector as

$$x_d = [x_{d_1}^T\ x_{d_2}^T \cdots x_{d_m}^T]^T,$$

$$x_{d_i} = [y_{d_i}\ \dot{y}_{d_i} \cdots y_{d_i}^{(n_i-1)}]^T,\ i = 1,2,\cdots,m.$$

The states vector $x(t)$ can be further expressed as

$$x = \left[x_1^T\ x_2^T \cdots x_m^T\right]^T,$$

$$x_i = \left[x_{Z_i+1}\ x_{Z_i+2} \cdots x_{Z_i+n_i}\right]^T,$$

$$Z_i = \sum_{j=0}^{i-1} n_j,\ i = 1,2,\cdots,m,\ n_0 = 0, Z_1 = 0 \quad (29)$$

Define the state error vector as

$$e = x - x_d,\ e = \left[e_1^T\ e_2^T \cdots e_m^T\right]^T,$$

$$e_i = [e_{i,1}\ e_{i,2} \cdots e_{i,n_i}]^T, e_{i,k} = x_{Z_i+k} - y_{d_i}^{(k-1)},$$
$$i = 1,2,\cdots,m,\ k = 1,2,\cdots,n_i,$$

where Z_i are the same as in (29), and the filtered error vector

$$r = \text{diag}\{[\Lambda_i^T 1]\}e,\ r = [r_1\ r_2 \cdots r_m]^T,$$

$$\Lambda_i = [\lambda_{i,1}\ \lambda_{i,2} \cdots \lambda_{i,n_i-1}]^T, i = 1,2,\cdots,m,$$

is an appropriately chosen coefficient vector such that

$$s^{n_i-1} + \lambda_{i,n_i-1}s^{n_i-2} + \cdots + \lambda_{i,1}$$

is Hurwitz, then, the derivative of r can be written as

$$\dot{r} = f(x) + Gu + d + Y_d \quad (30)$$

where

$$G = \text{diag}\{g_i\} \in R^{m\times m}$$
$$Y_d \equiv -Y_{dd} + \text{diag}\{[0\ \Lambda_i^T]\}e,$$
$$Y_d = [Y_{d_1}\ Y_{d_2} \cdots Y_{d_m}]^T,$$
$$Y_{dd} = [y_{d_1}^{(n_1)}\ y_{d_2}^{(n_2)} \cdots y_{d_m}^{(n_m)}]^T,$$
$$Y_{d_i} \equiv -y_{d_i}^{(n_i)} + [0\ \Lambda_i^T]e_i, i = 1,2,\cdots,m.$$

The control law is chosen as

$$u = u_c + u_d$$
$$= u_c + \begin{cases} 0.5(u_r - u_c)e^{\gamma(\|u_c\|-s)} & ,\text{if } \|u_c\| \le s \\ (u_r - u_c)(1 - 0.5e^{-\gamma(\|u_c\|-s)}), & \text{if } \|u_c\| > s \end{cases} \quad (31)$$

$$u_c = \left(\bar{G} + \hat{\tilde{G}}\right)^{-1}\left[-\left(\bar{f} + \hat{\tilde{f}}\right) + \nu\right] \quad (32)$$

$$\nu = -K_v\ r - Y_d,\ \nu = [\nu_1\ \nu_2 \cdots \nu_m]^T \quad (33)$$

$$u_r = -\mu\|u_c\|\text{sgn}(r) \quad (34)$$

where $K_v = \text{diag}\{k_{v_i}\} \in R^{m\times m}$, a positive definite matrix. $\text{sgn}(r) \equiv [\text{sgn}(r_1)\,\text{sgn}(r_2) \cdots \text{sgn}(r_m)]^T$. μ satisfies (42) as defined later.

The estimates $\hat{\tilde{f}}$ and $\hat{\tilde{G}}$ will be constructed by the CMAC NNs. The design parameters are: s>0, $\gamma < (\ln 2)/s$, and μ.

From lemma 1(vector form), one can obtain

$$\tilde{f}(x) = W_f^{\ T}\Gamma_f(x) + \varepsilon_f, \|W_f\| \le W_{f\max} \quad (35)$$

$$\tilde{g}_i(x) = w_{g_i}^{\ T}\Gamma_{g_i}(x) + \varepsilon_{g_i}, \quad (36)$$
$$\|w_{g_i}\| \le W_{g_i\max},\ i = 1,2,\cdots,m,$$

Two CMAC NNs are used to estimate $\tilde{f}$ and $\tilde{G}$

$$\hat{\tilde{f}}(x) = \hat{W}_f^T\ \Gamma_f(x) \quad (37)$$

$$\hat{\tilde{g}}_i(x) = \hat{w}_{g_i}^T\ \Gamma_g(x),\ i = 1,2,\cdots,m \quad (38)$$

with the weight update rules provided by

$$\dot{\hat{W}}_f = \Gamma_f(x)r^T M_f - k\|r\|\hat{W}_f M_f \quad (39)$$

$$\dot{\hat{w}}_{g_i} = \Gamma_g(x)r_i u_{c_i} M_{g_i} - k\|r\|\|u_c\|M_{g_i}\hat{w}_{g_i} \quad (40)$$
$$i = 1,2,\cdots,m$$

where, M_f is a symmetric definite matrix, M_{g_i}, k are positive constants.

The weight estimation errors are defined as $\tilde{W}_f = W_f - \hat{W}_f$, $\tilde{w}_{g_i} = w_{g_i} - \hat{w}_{g_i}, i = 1,2,\cdots,m$.

Theorem 2: For system (27), (28), assume the control law is given by (31)-(34) with the weight update rules provided by (39), (40), and assumption 1-3, lemma 1 hold (in vector or matrix form), when μ satisfies (42), then the filtered tracking error vector $r(t)$, the CMAC NNs weight estimation errors $\tilde{W}_f$, $\tilde{w}_{g_i}$, $i = 1,2,\cdots,m$, and the control input u are UUB with specific bounds given by (41a,b,c), and by increasing $k_{v\min}$, the filtered tracking error vector $r(t)$ can be made arbitrarily small. Where

$$k_{v\min} = \min\{k_{v_1}, k_{v_2}, \cdots, k_{v_m}\}$$

$$\delta_r = \max\{\delta_{r_1}, \delta_{r_2}\} \quad (41a)$$

$$\delta_f = \max\{\delta_{f_1}, \delta_{f_2}\} \quad (41b)$$

$$\delta_{g_i} = \max\{\delta_{g_i,1}, \delta_{g_i,2}\},\ i = 1,2,\cdots,m \quad (41c)$$

where

$$\delta_{r_1} = D_1 / k_{\mathrm{vmin}},$$

$$\delta_{f_1} = 0.5W_{f\max} + \sqrt{D_1 / k},$$

$$\delta_{g_i,1} = \|\tilde{w}_{g_i}(0)\| + W_{g_i\max} + 1/k, i = 1,2,\cdots,m$$

$$D_1 = \varepsilon_{fN} + s\sum_{i=1}^{m}\varepsilon_{g_iN} + b_{\mathrm{d}} + 0.25k(W_{f\max}^2 + s\sum_{i=1}^{m}W_{g_i\max}^2)$$

$$\delta_{r_2} = D_2 / k_{\mathrm{vmin}},$$

$$\delta_{f_2} = 0.5W_{f\max} + \sqrt{D_2 / k},$$

$$\delta_{g_i,2} = 0.5W_{g_i\max} + \sqrt{D_2/(ks)}, i = 1,2,\cdots,m$$

$$D_2 = \varepsilon_{fN} + b_{\mathrm{d}} + 0.25kW_{f\max}^2$$

$$\mu \geq 1 + \max\{D_{3,1}, D_{3,2}, \cdots, D_{3,m}\} / \left[g(1 - 0.5e^{\gamma s})\right] \quad (42)$$

$$D_{3,i} = \varepsilon_{g_iN} + 0.25kW_{g_i\max}^2$$

Proof: it is similar to that of theorem 1, and hence it is omitted here for concise.

7. SIMULATION EXAMPLE

Consider the following SISO nonlinear system for simplicity

$$\dot{x}_1 = x_2$$

$$\dot{x}_2 = n_f\left[-x_1 + 2x_2\exp(-x_1^2 - x_2^2) - 0.1x_2\right] + n_g u$$

$$y_1 = x_1.$$

with $n_f = 10$, $n_g = 10$ as the true values, their nominal values are supposed to be $\hat{n}_f = 15$, $\hat{n}_g = 5$. $x_1 \in [-2,2], x_2 \in [-2,2]$. The desired output is $y_{d_1} = 0.5\cos(\pi t/2)$. Take $N_1 = N_2 = 17$, $k_v = 5$, $\mu = 1$, $\gamma = 0.1$, s=5, $\lambda_1 = 5$, $M_f = 100$, $M_g = 100$, $\hat{w}_f(0) = 0$, $\hat{w}_g(0) = 0$. Both the simulations, i.e., with and without neural network adaptation, is conducted. In the simulation with adaptation, the tracking errors decay gradually, but in the simulation without adaptation, the tracking errors are relatively large and don't decay gradually. For the sake of saving space, only the simulation results with adaptation are given, as shown in Fig.1. It is seen that the actual outputs can track the desired outputs very quickly and very accurately.

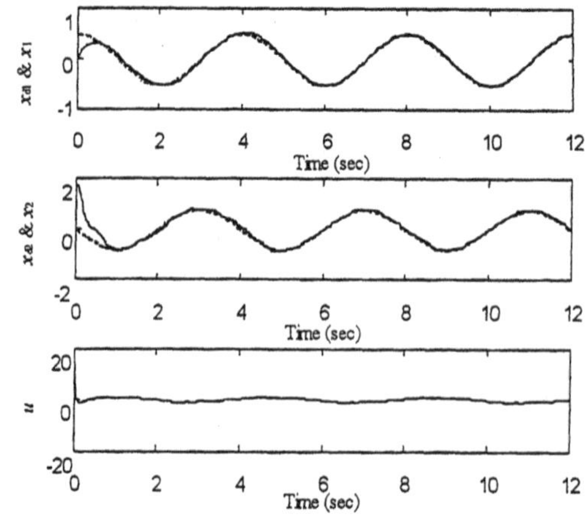

Fig.1. Simulation results (--desired, —actual)

8. CONCLUSIONS

Based on the nominal model of the plant, a CMAC NNs controller for feedback linearization of a class of SISO/MIMO nonlinear systems in state-space affine form has been developed. The burden for the learning of the NNs is thus alleviated, and the learning speed of the NNs is increased. By studying the derivative of Lyapunov function in the mutually exclusive and exhaustive regions, the stability analysis is given strictly and clearly under mild assumptions. Simulation results have shown the rightness and effectiveness of the proposed scheme.

ACKNOWLEDGMENT

This work was supported by the National Natural Science Foundation of China, No.19572114, and the Natural Science Foundation of Heilongjiang Province for Outstanding Youth.

REFERENCES

Commuri, S. and F.L. Lewis (1997). CMAC neural networks for control of nonlinear dynamical systems: structure, stability and passivity. *Automatica*, **33**, 635-641.

Jagannathan, S., S. Commuri and F.L. Lewis (1998). Feedback linearization using CMAC neural networks. *Automatica*, **34**, 547-557.

Lane, S.H., D.A. Handelman and J.J. Gelfand (1992). Theory and development of higher-order CMAC neural networks. *IEEE Control Systems Magazine*, 23-30.

Yesildirek A. and F.L. Lewis (1995). Feedback linearization using neural networks. *Automatica*, **31**, 1659-1664.

14th Triennial World Congress, Beijing, P.R. China

NEW METHOD OF DYNAMIC SYSTEM IDENTIFICATION AND NEURAL NETWORKS LEARNING

Ping-Yuan Cui*, You-An Zhang*, and Masayuki Suzuki**

**Harbin Institute of Technology, P.O.Box.137, Harbin,150001,P.R.China*
***Department of Aerospace Engineering, Nagoya University, Nagoya 464-01, Japan*
E-mail: cuipy@mail.hr.hl.cn

Abstract: A new approximating sensitivity algorithm in prediction error methods is derived for a class of continuous nonlinear dynamic system identification including training multi-layer neural networks. The algorithm can be used to approximate the gradient of output with respect to unknown parameter in a wide class of continuous-discrete nonlinear systems. The comparison between new and conventional algorithm, and simulated example are included to demonstrate the effectiveness of the new algorithm. *Copyright © 1999 IFAC*

Keywords: neural networks, nonlinear systems, prediction error methods, identification, sensitivity

1. INTRODUCTION

System identification and parameter estimation techniques are often used to determine the parameters for the continuous-discrete system. The prediction error (PE) method is a commonly used and effective one. In this method, the loss function is optimized, and the parameters which minimize the loss function will be the estimated results. There are two schemes for the optimization of loss function, i.e., direct search scheme and gradient scheme. The search scheme is attractive since it avoids the sensitivity calculation which causes most part of the computational burden in the gradient scheme. Avoiding the sensitivity calculation also provides a lot of versatility, and it allows the technique to be applied to any model structure without reformulating the algorithm. But the calculation process is rather slow. In the gradient scheme, the convergence speed of calculation is quicker than that of direct search scheme due to the analytical property of the cost function. More attention is paid to getting the derivative of the cost function. The Modified Newton-Raphson algorithm (MNR) is often used in the gradient scheme. But in this method it is necessary to solve sensitivity equation. This accounts for most of the computational efforts since a n-state system with p unknown parameters requires np sensitivity equations plus n-state equations to be integrated at each iteration. For this reason, an optimization scheme referred to as Modified Newton-Raphson with Estimated Sensitivity (MNRES) method is given, which removes the sensitivity equation integrations as in the MNR algorithm (Murphy, 1984).

The MNRES method determines sensitivities without the analytical form of sensitivity equations and can thus do it more effectively than the finite-difference method (Murphy, 1984). Therefore, it can be easily applied to any model structure, and the computational amount of MNRES is much less than that of MNR. This method has the advantages of both the search method and the gradient method. However, because the MNRES algorithm requires the inversion of $p \times p$ matrix, as p gets larger, the $p \times p$ matrix becomes more difficult to invert and, unfortunately, the $p \times p$ matrix is made up of very small numbers as the optimization process converges. Therefore a new algorithm is proposed by using an orthogonal table to calculate the inverse matrix, which overcomes the difficulty of calculating the inverse of initial matrix in the MNRES method, and keeps all the advantages of the MNRES method (Cui, 1995;

Cui, *et al.*, 1997).

Since neural networks have good approximation capabilities and inherent adaptive features, they have been widely used in identification and control of nonlinear system. Chen *et al.* gave a prediction error method in training neural network for discrete-time nonlinear system, and derived batch and recursive algorithm (Chen, *et al.*, 1990; Chen and Billings, 1992). When the prediction error method is used to train neural networks for continuous-time nonlinear dynamic system, however, the problem of the sensitivity calculation which is the same as the conventional parameter estimation is arisen, and the method of sensitivity calculation needs to be used.

The aim of the present paper therefore is to derive an approximating sensitivity algorithm for training neural networks in continuous-time nonlinear dynamic system based on the MNRES method and orthogonal table, which can be used to aerodynamic coefficients estimation using neural networks and keeps all the advantages of the method given by Cui, *et al.* (1997).

2. PREDICTION ERROR METHOD

Consider the following nonlinear continuous-discrete system

$$\begin{aligned} &\dot{x} = f(x,u,\theta),\ x(0) = x_0, \\ &y = g(x,u,\theta) \\ &z_i = y_i + v_i,\ i = 1,2,\cdots,N \end{aligned} \tag{1}$$

where $E(v_i) = 0$, $E(v_i v_j^T) = R\delta_{ij}$, $x \in R^n$, $u \in R^r$ and $y \in R^m$ are the state, input and output vectors, respectively. $\theta \in R^p$ is the unknown parameter vector. z_i, y_i and v_i are the measurement, output and measurement noise vector at $t = t_i$, respectively. R is a diagonal measurement noise covariance matrix. N is the number of data points. δ_{ij} is a Kronecker delta. Then the prediction error algorithm can be given below (Chen, *et al.*, 1990).

The predictor and the prediction error are given by (2), (3). The loss function is given by (4).

$$y_i = g_i(x,u,\theta) \tag{2}$$

$$\varepsilon_i(\theta) = z_i - y_i(\theta) \tag{3}$$

$$J(\theta) = \frac{1}{2N}\sum_{i=1}^{N}\varepsilon_i^T(\theta)\Lambda^{-1}\varepsilon_i(\theta) \tag{4}$$

The minimization of Eq. (4) can be performed efficiently using the following Gauss-Newton algorithm

$$\theta(k+1) = \theta(k) + \eta(k)d(\theta,\delta) \tag{5}$$

where $\eta(k)$ is a scalar $(0 < \eta < 1)$ and

$$d(\theta,\delta) = -[H(\theta,\delta)]^{-1}\Delta J(\theta) \tag{6}$$

is the optimizing direction vector, and

$$\begin{aligned} \Delta J(\theta) &= -\frac{1}{N}\sum_{i=1}^{N}G_i^T(\theta)\Lambda^{-1}\varepsilon_i(\theta) \\ H(\theta,\delta) &= \frac{1}{N}\sum_{i=1}^{N}G_i^T(\theta)\Lambda^{-1}G_i(\theta) + \delta I \end{aligned} \tag{7}$$

are the gradient and the approximate hessian of $J(\theta)$, respectively. $G(\theta)$ is the $m \times p$ sensitivity matrix of $y(\theta)$ with respect to θ

$$G(\theta) = \frac{dy(\theta)}{d\theta} \tag{8}$$

δ is a nonnegative small scalar and I is the identity matrix with appropriate dimension.

3. RECURSIVE SENSITIVITY CALCULATION

In the prediction error method, the sensitivity matrix, $G(\theta)$, plays an important role in determining θ, and calculation of sensitivity is time-consuming in continuous nonlinear dynamic system. There are three methods for the sensitivity computation, i.e., integrating the sensitivity equations (analytical), finite-difference (numerical) and recursive computation (approximate), and integrating the sensitivity equations and finite-difference methods need about the same computational burden (Cui, *et al.*, 1997).

3.1 Fundamental Algorithm (Murphy, 1984)

A Taylor series expansion of y with respect to θ_0 truncated to first order is

$$y(\theta) = y(\theta_0) + \frac{\partial y}{\partial \theta}\bigg|_{\theta_0}\Delta\theta \tag{9}$$

where $\Delta\theta = \theta - \theta_0$. The sensitivities for recursive computation are determined by assuming that they are approximately equal to the slopes of a surface which has been fit to $y(\theta)$. The surface approximates to $y(\theta)$ near the series expansion point of Eq.(9). Considering the simplest form by using a linear surface approximation, $y(\theta)$ is expanded as a first degree polynomial in θ for each point in time and at $p+1$ different points in the p-dimensional parameter space

$$y_{ki}^j(\theta^j) = s_{k0} + s_{k1}\theta_1^j + \cdots + s_{kp}\theta_p^j \quad (10)$$

where i indicates the i th point in time; k indicates the k th element of the output vector, $y(\theta)$ and j indicates one of the $p+1$ sample points used to fit Eq.(10) to $y(\theta)$. Note that $y_k^j(\theta^j) = y_k(\theta)$ at each of the p+1 points. The sample points are chosen by allowing a small perturbation of each parameter around the point where the sensitivities are desired. s_{k1} to s_{kp} are the desired slopes or sensitivities, $(\partial y_k / \partial \theta_j)_i$. s_{k0} is the value of $y_k(\theta)$ evaluated at the series expansion point of Eq.(9). Consider the matrix representation of Eq.(10) for the first element of Y and for the $p+1$ sample points at time "i"

$$Y_{1i} = XS_{1i}^0 \quad (11)$$

Note that y_k is the k th element of the output vector Y, and y_k^j is the j th element of the surface fitting vector Y_k. Since s_{10} is a known point, Eq.(11) can be simplified. The first line in Eq.(11) can be eliminated by subtracting it from the other p equations. Thus,

$$\Delta Y_{1i} = \Delta X S_{1i} \quad (12)$$

where $\Delta Y_{1i} = (y_{1i}^1 - y_{1i}^0, y_{1i}^2 - y_{1i}^0, \cdots, y_{1i}^p - y_{1i}^0)^T$,

$\Delta X = \{\Delta\theta_k^{\dot{\alpha}}\}_{p\times p}$, $\Delta\theta_k^j = \theta_k^j - \theta_k^0$,

$j,k = 1,2,\cdots,p$

Then, at time "i", the sensitivities for the first element in Y are given by

$$S_{1i} = [\Delta X]^{-1} \Delta Y_{1i} \quad (13)$$

In similar manner, the initial values of sensitivity are obtained. When using the iterative least squares approach, the recursive computing algorithm of sensitivity is given by

$$P_0 = [X^T X]^{-1} \quad (14)$$

$$P_{r+1} = [P_r^{-1} + \varphi_r^T \varphi_r - (\varphi_r^0)^T \varphi_r^0]^{-1} \quad (15)$$

where the row vector φ_r corresponding to a new set of θ which produces the smallest value of $J(\theta)$ (i.e., $\hat{\theta}_{k+1}$) is included in X_r, φ_r^0 corresponding to the old set of θ which produces the highest value of $J(\theta)$ is eliminated from X_r. Dropping the k subscript, the recursive equation for S is

$$S_{r+1} = S_r - P_{r+1}\{(\varphi_r^0)^T y_r^0 - \varphi_r^T y_r + [\varphi_r^T \varphi_r - (\varphi_r^0)^T \varphi_r^0] S_r\} \quad (16)$$

where y_r^0 and y_r are the outgoing and incoming elements of Y_k respectively at the r th iteration. With the new sensitivities determined, the algorithm proceeds as before. The recursive relations are derived by Murphy (1984).

3.2 Properties of the Fundamental Algorithm

The first property is that it reduces the computational burden in comparison with finite-difference or integrating sensitivity equations. During the initialization of the algorithm, the recursive algorithm must integrate $n+np$ differential equations too. After the start-up, however, it integrates only n state equations at each iteration, and integrating sensitivity equations or finite-difference algorithm continues to integrate n state plus np sensitivity equations or $n(p+1)$ state equations. It appears that as p gets larger the advantages of recursive algorithm become more obvious. The second property is that it determines sensitivities without the analytical form of the sensitivity equations and removes the need to formulate the sensitivity equations as well as finite-difference.

However, a limited factor in fundamental algorithm is that ΔX in Eq.(11) must be inverted. This $p\times p$ matrix becomes more difficult to invert as p increase and, unfortunately, ΔX is composed of very small numbers as the optimization process converges (Murphy, 1984). In addition, in Eq.(14), P_0 must also be inverted, and its numerical difficulty is also great. Therefore, a modified algorithm was given by Cui (1995).

3.3 Modified Algorithm

The modified algorithm is presented by the orthogonal table with binary and bipolar values to avoid the calculation of the inverse matrix in the fundamental algorithm.

The Properties of the Orthogonal Table with Binary and Bipolar Values. Assume that an orthogonal table with binary and bipolar values $L_{2^u}(2^{q-1})$ consists of elements a_{ml} which are either positive one or negative one, where $m = 1, 2, \cdots, q$; $l = 1, 2, \cdots, q-1$; $q = 2^u$, u is any positive integer. such as $L_{2^2}(2^3)$ below (Cui, *et al.*, 1997).

Table 1. Orthogonal table with binary and bipolar values (u=2)

	1	2	3
1	1	1	1
2	1	-1	-1
3	-1	1	-1
4	-1	-1	1

Let A indicate the matrix $\{a_{ml}\}$. Then the orthogonal table with binary and bipolar values has the following properties (Cui, *et al.*, 1997).

$$\sum_{\alpha} a_{\alpha m} a_{\alpha l} = 0 \quad (m \neq l) \tag{17}$$

$$\sum_{\alpha} a_{\alpha m} a_{\alpha l} = q \quad (m = l) \tag{18}$$

Calculation of Inverse Matrix by the Orthogonal Table. At the first iteration, we must obtain y_{ki}^{j} $p+1$ times near the initial value θ_0. Selecting perturbations of the unknown parameter according to an orthogonal table, the parameters θ_i^j of the matrix X of Eq.(11) could be got through the following formulas.

$$\theta_i^j = \theta_i^0 + a_{ji}\Delta\theta_i \tag{19}$$

where, $i = 1,2,\cdots,p$, $j = 0,1,\cdots,q$, $p \leq q \leq 2p-1$. $a_{ji} = 0$ when $j = 0$, $\Delta\theta_i$ is the perturbation of θ_i relative to θ_i^0. Here will be $q+1$ times of calculation to get y_{ki}^j according to Eq.(19). The elements $\Delta\theta_i^j$ of the matrix ΔX of Eq.(12) are

$$\Delta\theta_i^j = a_{ji}\Delta\theta_i \tag{20}$$

Substituting Eq.(20) in Eq.(12), a new equation can be got as follows

$$\Delta Y_{1i}^{*} = \Delta X^{*} S_{1i}^{*} \tag{21}$$

where $\Delta Y_{1i}^{*} = (y_{1i}^1 - y_{1i}^0,\ y_{1i}^2 - y_{1i}^0,\ \cdots,\ y_{1i}^q - y_{1i}^0)^T$,

$$\Delta X^{*} = \{a_{ml}\}_{q\times p},$$

$$S_{1i}^{*} = \left[\Delta\theta_1 s_{11}, \Delta\theta_2 s_{12}, \cdots, \Delta\theta_p s_{1p}\right]_i^T$$

and the matrix ΔX^{*} is correspondence with matrix A in Eq.(18). Therefore one can obtain

$$S_{1i}^{*} = \left[(\Delta X^{*})^T \Delta X^{*}\right]^{-1} (\Delta X^{*})^T \Delta Y_{1i}^{*} \tag{22}$$

$$s_{1k}^{*} = \Delta\theta_k s_{1k} = \frac{1}{q}\sum_{j=1}^{q} a_{jk}(y_{1i}^j - y_{1i}^0),\quad k = 1,2,\cdots,p \tag{23}$$

From Eq.(23), the initial values of sensitivities are got as follows.

$$S_{1i} = \left[s_{11}, s_{12}, \cdots, s_{1p}\right]_i^T \tag{24}$$

where $s_{1k} = s_{1k}^{*} / \Delta\theta_k$, $k = 1,2,\cdots,p$.

For the same reason, in Eq.(14), the inversion of matrix P_0 is required in recursive calculation. The inverse operation of P_0 is also built according to Eqs.(12) and (20) as follows. According to Eq.(12), for the kth element of the output vectors, dropping the i subscript and writing the matrix form gives

$$\Delta Y_k = \Delta X S_k \tag{25}$$

Thus, the Least-square solution to the sensitivity vector for Eq.(25) is

$$S_k = \left[(\Delta X)^T (\Delta X)\right]^{-1} (\Delta X)^T (\Delta Y_k) \tag{26}$$

Now define a recursive relation for the r th iteration

$$P_r = \left[(\Delta X_r)^T (\Delta X_r)\right]^{-1} \tag{27}$$

and the updating equation is

$$P_{r+1} = \left[P_r^{-1} + (\Delta\varphi_r)^T(\Delta\varphi_r) - (\Delta\varphi_r^0)^T(\Delta\varphi_r^0)\right]^{-1} \tag{28}$$

where $\Delta\varphi_r$ corresponding to a new set of $\Delta\theta$ is the row vector included in ΔX_{r+1}; $\Delta\varphi_r^0$ corresponding to the set of $\Delta\theta$, which corresponds to θ of producing the highest value of loss function, is the row vector eliminated from ΔX_{r+1}. Dropping the k subscript, the modified recursive equation for sensitivities can be written as

$$\begin{aligned} S_{r+1} = S_r - P_{r+1}\{&(\Delta\varphi_r^0)^T(\Delta y_r^0) - (\Delta\varphi_r)^T(\Delta y_r) \\ &+ \left[(\Delta\varphi_r)^T(\Delta\varphi_r) - (\Delta\varphi_r^0)^T(\Delta\varphi_r^0)\right]S_r\} \end{aligned} \tag{29}$$

where Δy_r (corresponding to $\Delta\varphi_r$) and Δy_r^0 (corresponding to $\Delta\varphi_r^0$) are the incoming and outgoing elements of ΔY_k at the r th iteration.

In the recursive formulas composed of Eqs.(28) and (29) (the demonstration process is similar to that of Eqs.(15) and (16)), the initial value S_0 is given by Eqs.(22), (23), and the initial value P_0 is given as follows.

$$P_0 = \left[(\Delta X_0)^T(\Delta X_0)\right]^{-1} \tag{30}$$

In terms of the method in which $\Delta\theta_i^j$ is selected in Eq. (30), ΔX_0 is given as follows

$$\Delta X_0 = \{a_{ml}\Delta\theta_l\}_{q\times p} \tag{31}$$

then

$$P_0 = \left[(\Delta X_0)^T(\Delta X_0)\right]^{-1} = \frac{1}{q}\operatorname{diag}\left\{\frac{1}{\Delta\theta_1^2} \quad \cdots \quad \frac{1}{\Delta\theta_p^2}\right\} \tag{32}$$

is a diagonal matrix. Now define the matrix Q_{r+1} by

$$Q_{r+1} = \left[P_r^{-1} + (\Delta\varphi_r)^T(\Delta\varphi_r)\right]^{-1} \tag{33}$$

then

$$P_{r+1} = \left[Q_{r+1}^{-1} - (\Delta\varphi_r^0)^T(\Delta\varphi_r^0)\right]^{-1} \tag{34}$$

In terms of the matrix inversion lemma, one can obtain the following results

$$Q_{r+1} = P_r - P_r(\Delta\varphi_r)^T[1+(\Delta\varphi_r)P_r(\Delta\varphi_r)^T]^{-1}(\Delta\varphi_r)P_r \quad (35)$$

$$P_{r+1} = Q_{r+1} + Q_{r+1}(\Delta\varphi_r^0)^T[1-(\Delta\varphi_r^0)Q_{r+1}(\Delta\varphi_r^0)^T]^{-1}(\Delta\varphi_r^0)Q_{r+1} \quad (36)$$

Thus, the modified algorithm is composed of Eqs.(22), (29), (32), (35) and (36).

3.4 Properties of the Modified Algorithm

The modified algorithm is an improved technique for the fundamental algorithm presented by Murphy (1984), and all the advantages of fundamental algorithm are retained. The advantages of the modified algorithm over fundamental algorithm are that the former one eliminates the inverse of $p \times p$ matrices ΔX and P_0, and overcomes the difficulty of obtaining the inverse of the matrices. At the same time, it realizes the optimal choice of the perturbation of the initial parameters, and possesses the numerical robustness characteristics. The disadvantage is that the computational burden may be increased from $p+1$ to $q+1$ ($p \le q \le 2p-1$) at the initialization. In the best case when $q=p$, the computational load is not increased at all, and in the worst case when $q=2p-1$, the computational load is doubly increased. However, this case occurs if and only if it is at the initialization.

4. LEARNING IN NEURAL NETWORKS

4.1 The Multi-Layer Neural Networks

The approximating sensitivity algorithm can also be used to train multi-layer (ML) neural networks, which be used to replace the nonlinear functions f and g in the nonlinear continuous system (1). The prediction error algorithm were introduced into the ML neural network learning in nonlinear discrete system by Chen and Billihngs (Chen *et al.*, 1990, Chen and Billings, 1992). The nonlinear functions f and g can be approximated by the ML networks with one hidden layer, which have the following form (Chen, *et al.*, 1990)

$$\hat{f}(x,u,\theta_f) = NN_f = [\hat{f}_1,\cdots,\hat{f}_n]^T \quad (37)$$

where

$$\hat{f}_k = \sum_{i=1}^{n_{hf}} w_{ki}^{(of)} o_{hfi}(t)$$
$$= \sum_{i=1}^{n_{hf}} w_{ki}^{(of)} a\left(\sum_{j=1}^{n_f} w_{ij}^{(hf)} x_{fj}(t) + \mu_i^{(hf)}\right), 1 \le k \le n \quad (38)$$

and

$$\hat{g}(x,u,\theta_g) = NN_g = [\hat{g}_1,\cdots,\hat{g}_m]^T \quad (39)$$

where

$$\hat{g}_k = \sum_{i=1}^{n_{hg}} w_{ki}^{(og)} o_{hgi}(t)$$
$$= \sum_{i=1}^{n_{hg}} w_{ki}^{(og)} a\left(\sum_{j=1}^{n_g} w_{ij}^{(hg)} x_{gj}(t) + \mu_i^{(hg)}\right), 1 \le k \le m \quad (40)$$

where θ_f, θ_g weights and thresholds of the networks NN_f, NN_g

n_f, n_g number of network inputs

n_{hf}, n_{hg} number of hidden nodes

$x_{fj}(t)$, $x_{gj}(t)$ the j th input of the networks

$\mu_i^{(hf)}$, $\mu_i^{(hg)}$ threshold of i th hidden node

$w_{ij}^{(hf)}$, w_{ij}^{hg} connection weight from j th input to i th hidden node

$o_{hfi}(t)$, $o_{hgi}(t)$ output of i th hidden node

$w_{ki}^{(of)}$, $w_{ki}^{(og)}$ connection weight from i th hidden node to k th output node

Without the loss of generality, the activation function $a(\cdot)$ will be chosen as

$$a(z) = 1/[1+\exp(-z)] \quad (41)$$

4.2 Learning Algorithm

Consider using ML networks to approximate nonlinear functions f and g in Eq.(1) to yield

$$\dot{x} = \hat{f}(x,u,\theta_f) = NN_f$$
$$y = \hat{g}(x,u,\theta_g) = NN_g \quad (42)$$
$$z_i = y_i + v_i, i = 1,2,\cdots,N$$

The predictor and prediction error is given as usual by

$$y = \hat{g}(x,u,\theta_g) \quad (43)$$

$$\varepsilon = y - \hat{g}(x,u,\theta_g) \quad (44)$$

The network can be trained using above PE algorithm. The first is to select values for n_f, n_g, n_{hf}, n_{hg}, and the next is to determine values of all weights and thresholds or to estimate θ_f, θ_g by PE algorithm. The gradients, $\frac{d\hat{g}}{d\theta_f}$, $\frac{d\hat{g}}{d\theta_g}$ of the output with respect to θ_f, θ_g is derived as follows:

$$\left[\frac{d\hat{g}(x,u,\theta_g)}{d\theta_g}\right]_{ik} = \frac{d\hat{g}_k}{d\theta_i}$$

$$
= \begin{cases} o_{hgj}(t), & if\ \theta_i = w_{kj}^{(og)}, \\ & 1 \le j \le n_{hg} \\ w_{kj}^{(og)} o_{hgj}(t)(1-o_{hgj}(t)), & if\ \theta_i = \mu_j^{(hg)}, \\ & 1 \le j \le n_{hg} \\ w_{kj}^{(og)} o_{hgj}(t)(1-o_{hgj}(t)) x_l(t), & if\ \theta_i = w_{jl}^{(hg)}, \\ & 1 \le j \le n_{hg}, 1 \le l \le n_g \\ 0 & otherwise \end{cases} \quad (45)
$$

$$
\frac{d\dot{x}}{d\theta_f} = \frac{\partial \hat{f}}{\partial x}\frac{dx}{d\theta_f} + \frac{\partial \hat{f}}{\partial \theta_f}, \quad \left.\frac{dx}{d\theta_f}\right|_{t=0} = 0 \quad (46)
$$

$$
\frac{d\hat{g}}{d\theta_f} = \frac{\partial \hat{g}}{\partial x}\frac{dx}{d\theta_f} \quad (47)
$$

The calculation of the gradient, $\frac{d\hat{g}}{d\theta_f}$ is very time consuming by the integrating sensitivity equations or the finite-difference, and modified approximating sensitivity algorithm can be used to calculate $\frac{d\hat{g}}{d\theta_f}$.

5. SIMULATION EXAMPLE

The state equation and observation equation of the simplified longitudinal models of an aircraft are given as follows:

$$
\begin{aligned} \dot{\alpha} &= -2\alpha + \omega_z \\ \dot{\omega}_z &= M_\alpha \alpha + M_{\omega_z}\omega_z + M_{\delta_z}\delta_z \\ \omega_{zm} &= \omega_z + v \end{aligned} \quad (48)
$$

Here $\alpha, \omega_z, \delta_z$ are angle of attack, pitch rate and elevator deflection, respectively; $M_\alpha, M_{\omega_z}, M_{\delta_z}$ are aerodynamic moment coefficients; ω_{zm} is the measurement value of ω_z; v is a Gaussian noise with zero mean and variance 0.01. Table 2 gives the estimated results by the new approximating sensitivity algorithm (in this example, 5 sec of data (500 data points) and Table 1 ($L_4(2^3)$) are used, and matrices ΔX_0 was constituted according to Eq.(23)).

Table 2. The estimation results of moment coefficients

moment coefficient	true value	initial value
M_α	5	4.5
M_{ω_z}	2	1.5
M_{δ_z}	5	4.5

(continued)

initial perturbation value	estimated value	relative error(%)
0.35	4.963	0.74
0.05	1.997	0.15
0.10	5.020	0.40

6. CONCLUSION

A new algorithm of approximating sensitivity has been introduced for the prediction error method in nonlinear continuous-discrete system. The new algorithms provide a computationally efficient implementation for a class of parameter estimation of nonlinear continuous-discrete systems including training multi-layer neural networks. The algorithms avoid the inverse calculation of the initial matrices ΔX and P in the MNRES algorithm by increasing appropriate sample points (the $p+1$ sample points change to $q+1$ sample points, where $p \le q \le 2p-1$, q is the number of row in the orthogonal table with binary and bipolar values). At the same time, the optimal choice of perturbation of initial parameters and numerical robustness is obtained.

ACKNOWLEDGMENT

This work was supported by the National Natural Science Foundation of China, No.19572114, and the Natural Science Foundation of Heilongjiang Province for Outstanding Youth.

REFERENCES

Chen, S., S.A. Billings and P.M. Grant (1990). Non-linear system identification using neural networks. *Int. J. Control*, **51**, 1191-1214.

Chen, S. and S.A. Billings (1992). Neural networks for nonlinear dynamic system modelling and identification. *Int. J. Control*, **56**, 319-346.

Cui, P.Y. (1995). The implementation for maximum likelihood estimation algorithm of recursive computing sensitivities. *Acta Automatica Sinica*, **21**, 82-86(in Chinese).

Cui, P.Y., N.G. Cui and M. Suzuki (1997). A modified MNRES method of approximating sensitivities for maximum likelihood estimation. In: *The Second Chinese World Congress on Intelligent Control and Intelligent Automation*, Xi'an, P.R.China, June 23-27.

Murphy, P.C. (1984). An algorithm for maximum likelihood estimation using an efficient method for approximation sensitivities. *NASA* TP2311, June.

CONTROL OF NONLINEAR SYSTEMS BASED ON A PROBABILISTIC LEARNING NETWORK

Jinglu Hu * **Kotaro Hirasawa** * **Chunzhi Jin** *
Takuya Matsuoka *

* *Department of Electrical and Electronic Systems Engineering, Graduate School of Information Science & Electrical Engineering, Kyushu University, Higashiku, Fukuoka 812-8581 Japan*
TEL: (+81)92-642-3959, FAX: (+81)92-642-3962
E-mail: jinglu@ees.kyushu-u.ac.jp

Abstract: This paper presents a control design scheme for nonlinear systems based on a probability learning network (ProNet) that has a capability dealing with stochastic signals. Plant and its controllers are described by using a set of related equations and form a unified learning network– ProNet where disturbances are considered as its external inputs. Then controller design is transfered to ProNet learning. By including an effort to reduce variance of ProNet output in training, the trained ProNet has different sensitivity to signals of different frequencies. By taking this advantage, a ProNet control system is designed to increase its robustness against disturbance.

Keywords: Nonlinear system, nonlinear control, probability learning network, robustness

1. INTRODUCTION

Neural networks have attracted much interest in control community because of their ability to learn any nonlinear mapping, see e.g., (MillerIII *et al.*, 1990; Narendra and Parthasathy, 1990). A commonly-trained neural network usually has an identical sensitivity to signals of all frequencies. When such neural networks are used as controller, the control system may not have good robustness against disturbances. To increase robustness of a network-based control system, it is highly motivated to construct a learning network which can be trained to have different sensitivities to signals of different frequencies. A probability learning network (ProNet) proposed by Hirasawa *et al.* (1998*b*) may be considered for this purpose. ProNet is an extension of common learning networks to ones equipped with a capability to deal with stochastic signals. In a ProNet, it is possible to control the variances as well as the means of ProNet output. In this paper, we apply the ProNet to nonlinear control system design.

Any dynamic system can be expressed by a set of equations derived from physical law governing system (Isermann *et al.*, 1997). Consider the case where these equations are fully known; if there are unknown elements in the equations, system identification must be carried out, in which the unknown equations are re-parameterized and the unknown parameters are estimated (Hu *et al.*, 1998*b*). Controllers for the system are also expressed by a set of equations. These equations are usually unknown, hence each unknown equation is re-parameterized using a neural network or a neurofuzzy network. On the other hand, ProNet consists of two kinds of element– nodes and branches. By corresponding nodes to the equations and branches to their relations, the equations described system and its controller form a learning network– ProNet. In the ProNet, system disturbances are considered to be external inputs.

Then controller design in a ProNet-based control system is transfered into ProNet learning. A learning algorithm containing higher order derivatives is introduced for ProNet learning, which is realized by minimizing a criterion function including a minimum squared error control cost and an effort to reduce variance of the controlled system output. The effort leads to a controller with lower sensitivity to disturbances than to signals, so that the obtained control system is expected to have good robustness against disturbances. Computer simulations on a nonlinear crane system control are carried out to demonstrate the effectiveness of the proposed ProNet control scheme.

The paper is organized as follows: Section 2 briefly introduces the ProNet. Section 3 discusses control system design based on ProNet in detail. Section 4 conducts a computer simulation demonstrating the effectiveness of ProNet control. Section 5 presents a conclusion.

2. PROBABILISTIC LEARNING NETWORK

A probabilistic learning network (ProNet) is a learning network equipped with a capability of dealing with stochastic signals. It has a similar structure to deterministic one, but it propagates stochastic signals and controls both means and covariances of propagating signal.

2.1 *ProNet Structure*

ProNet consists of two kinds of element- nodes and branches. Nodes are assigned appropriate node functions including a function realized by a subnetwork, e.g., a sigmoid neural network or a neurofuzzy network, and are interconnected with multi-branches that have arbitrary- positive, zero, or even negative- time delays. A ProNet has learning ability, provided with some adjustable parameters in the nodes or on the branches. For a certain given representation ability, the more sophisticated the node functions are, the less complexity the network will be. One can take this advantage to incorporate prior knowledge about systems as much as possible in order to obtain a simple learning network. For example, any dynamic system can be described by using a set of related equations derived from from physical law governing system (Isermann *et al.*, 1997). Depending on prior knowledge available about system, equations may be fully known, partly known, or totally unknown. Such dynamic system can then be modeled using a ProNet by corresponding nodes to the equations and branches to their relations.

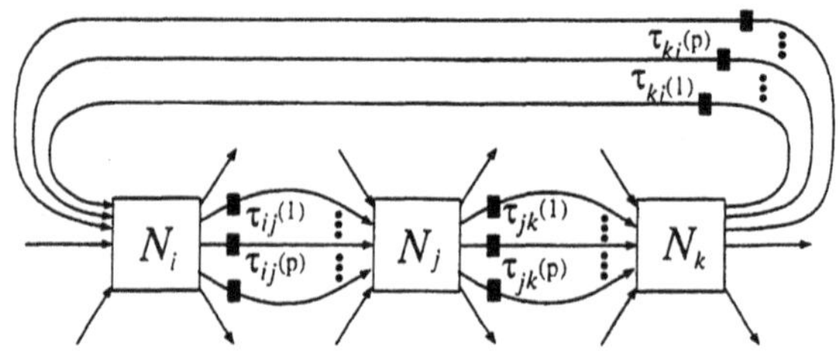

Fig. 1. ProNet Architecture

Figure 1 shows ProNet architecture. It is clear that when sigmoid node functions are used in all nodes, the ProNet reduces to one with similar structure to a common sigmoid neural network.

2.2 *Dynamics Description*

Let N_j $(j \in J)$ be the current node, N_i $(i \in J_F(j))$ and N_k $(k \in J_B(j))$ denote input nodes and output nodes of the current node, where J is the set of total nodes, $J_F(j)$ and $J_B(j)$ of input nodes and output nodes of node j. The ProNet dynamics is described locally as follows

$$h_j(t) = f_j\Big[\{h_i(t-\tau_{ij}(p)|i \in J_F(j), p \in B(i,j)\}$$
$$\{u_n(t)|n \in N(j)\}, \{\beta_q|q \in M(j)\}\Big] \quad (1)$$
$$j \in J \quad t \in T$$

where f_j is the transfer function of node j, $h_j(t)$ the output value of node j at time t, $u_n(t)$ the external input of node j, β_q the adjustable parameter, T the set of sampling instants, $\tau_{ij}(p)$ the time delay of pth branch from node i to node j, $B(i,j)$ the set of branches from node i to node j, and $N(j)$, $M(j)$ the set of external inputs and adjustable parameters related to the node j, respectively. The local description Eq.(1) serves as the node j zoom lens. Moving this zoom lens around the network for all nodes yields a global ProNet dynamics.

However, ProNet is distinctive to common learning networks in that its propagating signal is not deterministic but stochastic one. ProNet dynamics must finally be described by both means and variances of propagating signal. For simplicity, we calculate the means and the variance of propagating signal only for ProNet output nodes. Let N_o $(o = 1, 2, ..)$ be the oth output node. Then its output at time t $(t = 1, 2, ...)$, $h_o(t)$, is described by

$$h_o(t) = \text{ProNet}\,(x_1, x_2, ... x_\ell) \quad (2)$$

where x_i $(i = 1, ..., \ell)$ is the pseudo-external input, introduced to deal with a dynamic network similarly to a static network. For a recurrent (dynamic) ProNet, x_i for instance denotes the jth external input of ProNet at sampling time t_k ($t_k < t$), while for a layered (static) ProNet it denotes a real external input. The mean and the variance of $h_o(t)$ are calculated as follows:

(1) **Means**. The mean of $h_o(t)$ is defined as its ensemble average $E[h_o(t)]$. Applying Taylor series expansion to *Eq.*(2) around the point of $(E[x_1], \cdots, E[x_\ell])$ and recording the results up to second-order term (Hahn and Shapiro, 1967; Ranaweera *et al.*, 1996), we have

$$\begin{aligned} h_o(t) &\simeq \text{ProNet}(E[x_1)], E[x_2)], \cdots, E[x_\ell]) \\ &+ \sum_i \frac{\partial^\dagger h_o(t)}{\partial x_i}(x_i - E[x_i]) \\ &+ \frac{1}{2}\sum_i \sum_j \frac{\partial^{\dagger 2} h_o(t)}{\partial x_i \partial x_j}(x_i - E[x_i])(x_j - E[x_j]) \end{aligned} \quad (3)$$

$$\begin{aligned} E[h_o(t)] &\simeq \text{ProNet}(E[x_1], E[x_2], ..., E[x_\ell]) \\ &+ \frac{1}{2}\sum_i \sum_j \frac{\partial^{\dagger 2} h_o(t)}{\partial x_i \partial x_j}\text{Cov}[x_i, x_j] \end{aligned} \quad (4)$$

where $\frac{\partial^\dagger h_o(t)}{\partial x_i}$ and $\frac{\partial^{\dagger 2} h_o(t)}{\partial x_i \partial x_j}$ are ordered derivatives evaluated at $(E[x_1], \cdots, E[x_\ell])$ (Werbos, 1974).

(2) **Variances**. Using Eq.(3) and Eq.(4), the variance of $h_o(t)$, $\text{Var}[h_o(t)]$, is derived as follows

$$\begin{aligned} \text{Var}[h_o(t)] &= E[(h_o(t) - E[h_o(t)])^2] \\ &\simeq \sum_i \sum_j \frac{\partial^\dagger h_o(t)}{\partial x_i}\frac{\partial^\dagger h_o(t)}{\partial x_j}\text{Cov}[x_i, x_j] \\ &+ \sum_i \sum_j \sum_k \frac{\partial^{\dagger 2} h_o(t)}{\partial x_i \partial x_j}\frac{\partial^\dagger h_o(t)}{\partial x_k}\mu_3[x_i, x_j, x_k] \end{aligned} \quad (5)$$

where $\mu_3[\cdot,\cdot,\cdot]$ is the third-order central moment. In Eq.(4) and Eq.(5), $E[x_i]$, $\text{Cov}[x_i, x_j]$ and $\mu_3[x_i, x_j, x_k]$ are assumed to be known. However, calculations of first-order and second-order derivatives, $\frac{\partial^\dagger h_o(t)}{\partial x_i}$, $\frac{\partial^{\dagger 2} h_o(t)}{\partial x_i \partial x_j}$, are not trivial, referred to (Hirasawa *et al.*, 1998*a*) for their calculations.

3. CONTROL SYSTEM DESIGN

3.1 *Control System Structure*

The systems to be controlled are assumed to be described by a set of related equations derived from physical laws governing system

$$\begin{aligned} h_i(t) &= f_i(S, h_j(t-1); j = 1, 2, ..., n_s + n_c) \quad (6) \\ i &= 1, 2, ..., n_s \end{aligned}$$

where S is system parameter vector, h_{n_s+k} ($k = 1, ..., n_c$) are control inputs discussed later. The equations $f_i(\,)$ ($i = 1, ..., n_s$) and parameter vector S are assumed to be known. If some equations or parameters are unknown, system identification must be carried out, in which the unknown equations are re-parameterized and the unknown parameters are estimated (Hu *et al.*, 1998*b*).

The system (6) is controlled by one or more controllers that are described by the following unknown equations

$$\begin{aligned} &h_{n_s+k}(t) = \\ &f_{n_s+k}(\Theta, h_j(t-1); j = 1, 2, ..., n_s + n_c) \\ &k = 1, 2, ..., n_c \end{aligned} \quad (7)$$

where unknown equations $f_{n_s+k}(\,)$ are supposed to be parameterized by using subnetworks such as neural networks or neurofuzzy networks. $\Theta = [\beta_1, \beta_2, ...]^T$ is the unknown parameter vector of controllers.

In this way, control system including plant and its controllers is described by $n_s + n_c$ related equations. Introducing nodes for equations and branches for their relations, we have a ProNet composed of $n_s + n_c$ nodes, in which control reference and sensor noise are considered as external inputs. It is clear that the controller design becomes a ProNet learning problem.

3.2 *Learning Algorithm*

Consider a criterion function consisting of two parts

$$L_{MV} = L_M + L_V \quad (8)$$

Part L_M is a usual term for realizing a minimum squared error control.

$$L_M = \frac{1}{2}\sum_o \sum_t Q_{no}(h_o(t) - h_o^*(t))^2 \quad (9)$$

where $h_o^*(t)$ is the control reference and Q_{no} the weight coefficients. And the second part L_V is an extra term that is an effort to make the variances of system output desirable

$$\begin{aligned} &L_V = \\ &\frac{1}{2}\sum_o \sum_t Q_{\mu o} \cdot (\text{Var}[h_o(t)] - Vh_o^*(t))^2 \end{aligned} \quad (10)$$

where $Vh_o^*(t)$ is the desired variances, and $Q_{\mu o}$ the weight coefficients. It is clear that, when the desired variances are set to be 0, the second term is an effort to reduce the variances of ProNet output. This effort makes lower the sensitivity of control system to noise which increases the robustness of control system. Then ProNet learning, that is, controller design is performed by minimizing the criterion function described by (8)

$$\Theta = \min_{\Theta} \arg\{L_{MV}\} \tag{11}$$

which can be realized by using a random search method, e.g., RasID (Hu *et al.*, 1998*a*) or a gradient method

$$\beta_m \leftarrow \beta_m - \eta \frac{\partial^{\dagger} L_{MV}}{\partial \beta_m} \tag{12}$$

where β_m is the adjustable parameter and η the learning coefficient assigned an appropriate positive value. Ordered derivative $\frac{\partial^{\dagger} L_{MV}}{\partial \beta_m}$ is obtained as follows,

$$\frac{\partial^{\dagger} L_{MV}}{\partial \beta_m} = \sum_{o}\sum_{t} Q_{no}(h_o(t) - h_o^*(t)) \frac{\partial^{\dagger} h_o(t)}{\partial \beta_m} + \sum_{o}\sum_{t} Q_{\mu o}(\mathrm{Var}[h_o(t)] - Vh_o^*(t)) \frac{\partial^{\dagger} \mathrm{Var}[h_o(t)]}{\partial \beta_m} \tag{13}$$

where $\frac{\partial^{\dagger} \mathrm{Var}[h_o(t)]}{\partial \beta_m}$ is obtained by differentiating (5)

$$\frac{\partial^{\dagger} \mathrm{Var}[h_o(t)]}{\partial \beta_m} = \sum_{i}\sum_{j} \left[\frac{\partial^{\dagger 2} h_o(t)}{\partial x_i \partial \beta_m} \frac{\partial^{\dagger} h_o(t)}{\partial x_j} + \frac{\partial^{\dagger} h_o(t)}{\partial x_i} \frac{\partial^{\dagger 2} h_o(t)}{\partial x_j \partial \beta_m} \right] \cdot \mathrm{Cov}[x_i, x_j] + \sum_{i}\sum_{j}\sum_{k} \left[\frac{\partial^{\dagger 3} h_o(t)}{\partial x_i \partial x_j \partial \beta_m} \frac{\partial^{\dagger} h_o(t)}{\partial x_k} + \frac{\partial^{\dagger 2} h_o(t)}{\partial x_i \partial x_j} \frac{\partial^{\dagger 2} h_o(t)}{\partial x_k \partial \beta_m} \right] \cdot \mu_3[x_i, x_j, x_k] \tag{14}$$

where $\mathrm{Cov}[x_i, x_j]$ and $\mu_3[x_i, x_j, x_k]$ are assumed to be known. The first-order derivative $\frac{\partial^{\dagger} h_o(t)}{\partial x_j}$, the second-order derivative $\frac{\partial^{\dagger 2} h_o(t)}{\partial x_j \partial \beta_m}$, and the third-order derivative $\frac{\partial^{\dagger 3} h_o(t)}{\partial x_i \partial x_j \partial \beta_m}$ are calculated by using an iterative algorithm (Hirasawa *et al.*, 1998*a*).

On the other hand, since it is so time-consuming to calculate the third-order derivatives, it is recommended to use a random search method, e.g. RasID (Hu *et al.*, 1998*a*) for ProNet learning, in which only the derivatives up to second-order must be used.

3.3 *Computing Higher Order Derivatives*

Higher order derivatives play an important role in the proposed robust control scheme because computing the means and the variance of ProNet output involves the second-order derivative computations, see (4) and (5), while computing the gradient of criterion function (8) involves the third-order derivative computations. Computing higher order derivatives in a learning network is rather complicated. We therefore omit the computing algorithm here. The readers are invited to see (Hirasawa *et al.*, 1998*a*) for details.

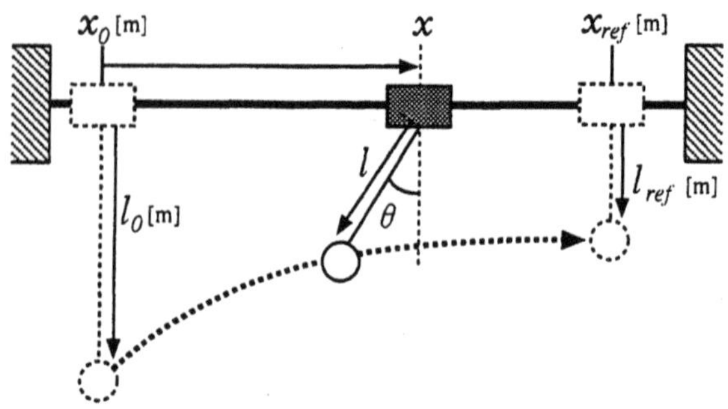

Fig. 2. Nonlinear crane system

4. COMPUTER SIMULATIONS

This section discusses computer simulations where the control scheme presented previously is applied to a nonlinear crane control system.

4.1 *Crane Control System*

Figure 2 shows a crane system. The position of crane stand, angle between rope and vertical line and the position of load and their first and second order derivatives w.r.t t are denoted by $x(t)$, $\theta(t)$, $\ell(t)$, $\dot{x}(t)$, $\dot{\theta}(t)$, $\dot{\ell}(t)$ and $\ddot{x}(t)$, $\ddot{\theta}(t)$, $\ddot{\ell}(t)$. Based on physical law governing the crane system, crane dynamics is described by

$$\ddot{x}(t) = -\frac{mg}{M}\theta(t) - \frac{D+G}{M}\dot{x}(t) + \frac{G}{M}u_1(t) \tag{15}$$

$$\ddot{\theta}(t) = -\frac{M+m}{\ell M}g\theta(t) - \frac{D+G}{\ell M}\dot{x}(t) + \frac{G}{\ell M}u_1(t) \tag{16}$$

$$\ddot{\ell}(t) = -\frac{C+G_m}{m}\dot{\ell}(t) + \frac{G_m}{m}u_2(t) \tag{17}$$

where $u_1(t)$ is the input for moving the crane stand and $u_2(t)$ the input for rolling up the load, $m = 2.0[kg]$ the load mass, $g = 9.8[m/sec^2]$ the gravity coefficient, and $C = 0.42[kg/sec]$, $G = 700[N/V]$, $G_m = 0.98[N/V]$, $D = 300[kg/sec]$, $M = 40[kg]$ the system parameters.

Introducing $h_1(t) = x(t)$, $h_2(t) = \dot{x}(t)$, $h_3(t) = \theta(t)$, $h_4(t) = \dot{\theta}(t)$, $h_5(t) = \ell(t)$ and $h_6(t) = \dot{\ell}(t)$, (15)-(17) can be transformed into a discrete form described by

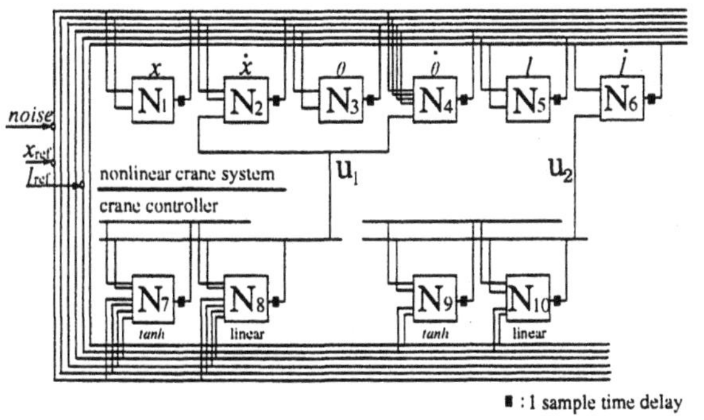

Fig. 3. Treating crane control system as a ProNet

$$h_1(t) = a_{11}h_1(t-1) + a_{21}h_2(t-1) \tag{18}$$

$$h_2(t) = a_{22}h_2(t-1) + a_{32}h_3(t-1) + b_1u_1(t-1) \tag{19}$$

$$h_3(t) = a_{33}h_3(t-1) + a_{43}h_4(t-1) \tag{20}$$

$$h_4(t) = a_{24}\frac{h_2(t-1)}{h_5(t-1)} + a_{34}\frac{h_3(t-1)}{h_5(t-1)} + a_{44}h_4(t-1) + \frac{b_1}{h_5(t-1)}u_1(t-1) \tag{21}$$

$$h_5(t) = a_{55}h_5(t-1) + a_{65}h_6(t-1) \tag{22}$$

$$h_6(t) = a_{66}h_6(t-1) + b_2u_2(t-1) \tag{23}$$

where $t = 1, 2, ...$ and $a_{11} = 1$, $a_{21} = \Delta t$, $a_{22} = 1 - \Delta t(D+G)/M$, $a_{32} = -mg\Delta t/M$, $a_{33} = 1$, $a_{43} = \Delta t$, $a_{24} = -(\Delta t(D+G)/M$, $a_{34} = -\Delta t(M+m)g/M$, $a_{44} = 1$, $a_{55} = 1$, $a_{65} = \Delta t$, $a_{66} = 1 - \Delta t(C+G_m)/m$, $b_1 = \Delta tG/M$, $b_2 = \Delta tG_m/m$, and $\Delta t = 0.002[sec]$ is the sampling time.

Introducing one node for each equation, a network with six nodes (N_1 to N_6) is used to represent the crane system described by (18)-(23). Two controllers, each of which is a recurrent network with two nodes (N_7, N_8 or N_9, N_{10}), are used to control the crane system, in which nodes N_7 and N_9 have sigmoid node functions, while N_8 and N_{10} have linear node functions. The networks for crane system, with known parameters, and for controllers, with unknown parameters, form a unified network, treated as a ProNet (Fig.3), in which white Gaussian disturbance ($e(t) \in N(0,1)$) is supposed to be added to $x(t)$. Sampling time for disturbance is chosen to be $0.33[sec]$ as to simplify computation. The total control time is $10[sec]$, which gives 30 pseudo-external inputs $x_1 \sim x_{30}$ for the ProNet.

The objective of control is to move the load from an initial position ($x(0) = 0[m]$, $\ell(0) = 1[m]$) to a target position ($x_{ref} = 1[m]$, $\ell_{ref} = 0.5[m]$) with $\dot{x}(t_f)$, $\dot{\ell}(t_f)$, $\theta(t)$ and $\dot{\theta}(t)$ as small as possible. In the ProNet crane control system, controller design becomes ProNet training, conducted by minimizing the criterion function L_{MV} (8) where L_M and L_V are defined by

$$L_M = \frac{1}{2}\sum_{s\in T}\{Q_{n1}(x_{ref} - x(s))^2 + Q_{n3}\theta(s)^2 + \theta_{n4}\dot{\theta}(s)^2 + Q_{n5}(\ell_{ref} - \ell(s))^2 + Q_{n7}u_1(s)^2 + Q_{n8}u_2(s)^2\} + \frac{1}{2}\left\{Q_{n2}\dot{x}(t_f)^2 + Q_{n6}\dot{\ell}(t_f)^2\right\} \tag{24}$$

$$L_V = \frac{1}{2}\sum_{s\in T}\Big\{Q_{\mu1}\mathrm{Var}[x(s)] + Q_{\mu2}\mathrm{Var}[\dot{x}(s)] + Q_{\mu3}\mathrm{Var}[\theta(s)] + Q_{\mu4}\mathrm{Var}[\dot{\theta}(s)] + Q_{\mu5}\mathrm{Var}[\ell(s)] + Q_{\mu6}\mathrm{Var}[\dot{\ell}(s)]\Big\} \tag{25}$$

where t_f is final control time and the weighting coefficients Q_{ni} and $Q_{\mu i}$ are determined experientially as follows: $Q_{ni} = 1.0$ ($i = 1, 2, ..., 6$), $Q_{n7} = Q_{n8} = 0.001$, $Q_{\mu1} = Q_{\mu3} = 20.0$, $Q_{\mu2} = Q_{\mu4} = 4.0$ and $Q_{\mu5} = Q_{\mu6} = 0$ are chosen because the disturbance added in $x(t)$ does not affect $\ell(t)$, see Fig.3. When the gradient method is used for this minimization, third-order derivatives must be used because computing L_V along needs derivatives up to the second order, see (5). To simplify calculation, we used random search method RasID (Hu *et al.*, 1998*a*; Hu *et al.*, 1998*c*) in our simulations. The ProNet was trained for 100,000 steps. In the training stage, $x(t)$ was supposed to be corrupted by a white disturbance sequence $e(t) \in N(0,1)$. Namely, $E[e(t)] = 0$ and $\mathrm{Var}[e(t)] = 1$ were used in the training stage. After training, control is conducted by actually adding $e(t) \in N(0,1)$ to variable $x(t)$. Figure 4 shows the control results of $\ell(t)$, $x(t)$ and $\theta(t)$ (solid line). For comparison, dashed lines show the results when only a usual criterion (L_M) was used in the training. Simulation results show that a large disturbance may make the performance of a usual controller poor (dashed line), but this problem is solved by using ProNet with including an effort to reduce the variance of ProNet output. Next, the controls are conducted for the cases where values of disturbance are smaller than those considered in the training stage. Figure 5 shows that the control results of disturbance free case although $e(t) \in N(0,1)$ was considered in the training. We can see that both a usual control and ProNet control have good performance in such cases. Comparing the results in Fig.4 and Fig.5, confirms that poor performance of usual control is caused by the disturbances and ProNet control is robust to the disturbances. However, simulation results also indicate that the ProNet control takes longer time to reach their targets, which is a cost for robustness.

5. CONCLUSIONS

A trained neural network usually has an identical sensitivity to signals of all frequencies. For a

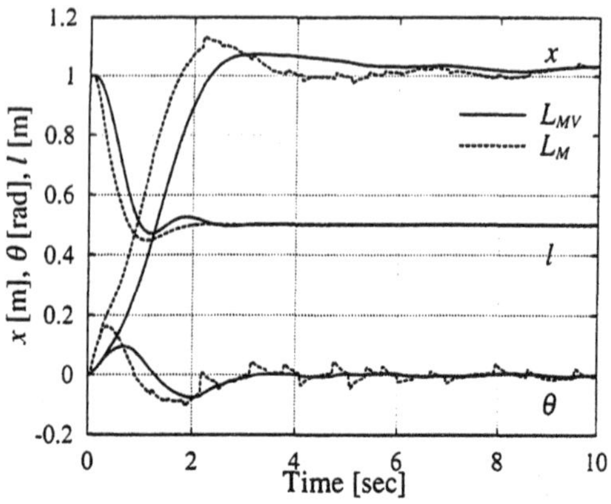

Fig. 4. Control results conducted for the case $x(t)$ is corrupted by $e(t) \in N(0,1)$, solid line denotes ProNet control (L_{MV}); dashed line the usual control (L_M)

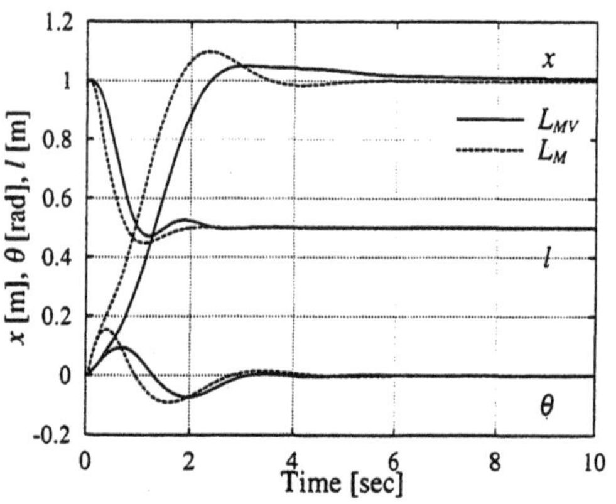

Fig. 5. Control results conducted for disturbance free case although disturbance $e(t) \in N(0,1)$ was considered in the training, solid line denotes ProNet control (L_{MV}); dashed line the usual control (L_M)

neural-network-based control system, it is highly motivated to train neural network such that the system has different sensitivities to signals and disturbances so as to increase its robustness against disturbances.

A ProNet is introduced, which has a capability to propagate stochastic signals. And a ProNet control scheme is proposed, in which the plant and its controllers form a unified network and is treated as a ProNet. Controller design becomes ProNet learning. By including an effort to reduce the variances of system output in the criterion function of ProNet learning, it is possible to construct a ProNet control system that has a lower sensitivity to disturbances than to signals. Computer simulations show that ProNet control has better robustness to disturbances than a usual control.

6. REFERENCES

Hahn, G.J. and S.S. Shapiro (1967). *Stability Models in Engineering*. Wiley, New York.

Hirasawa, K., J. Hu, M. Ohbayashi and J. Murata (1998*a*). Computing higher order derivatives in universal learning networks. *Journal of Advanced Computational Intellignce* **2**(2), 47–53.

Hirasawa, K., M. Ohbayashi, J. Murata and J. Hu (1998*b*). Propabilistic universal learning network theory (in Japanese). *Trans. IEE of Japan* **118-C**(2), 224–230.

Hu, J., K. Hirasawa and J. Murata (1998*a*). RasID– Random search for neural networks training. *Journal of Advanced Computational Intelligence* **2**(4), 134–141.

Hu, J., K. Hirasawa, J. Murata, M. Ohbayashi and K. Kumamaru (1998*b*). Identification of nonlinear black-box systems based on universal learning networks. In: *Proc. of IEEE International Joint Conference on Neural Networks (Alaska)*. pp. 2465–2470.

Hu, J., K. Hirasawa, J. Murata, M. Ohbayashi and Y. Eki (1998*c*). A new random search method for neural network learning -RasID-. In: *Proc. of IEEE International Joint Conference on Neural Networks (Alaska)*. pp. 2346–2351.

Isermann, R., S. Ernst and O. Nelles (1997). Identification with dynamcic neural networks - Architectures, Comparisons, Applications - . In: *Proc. of the 11th IFAC Symposium on System Identification*. Vol. 3. pp. 997–1022.

MillerIII, W.T., Sutton, R.S. and Werbos, P.J., Eds.) (1990). *Neural Networks for Control*. The MIT Press, Massachussetts.

Narendra, K.S. and K. Parthasathy (1990). Identification and control of dynamical systems using neural networks. *IEEE Trans. on Neural Networks* **1**(1), 4–27.

Ranaweera, D.K., George.G. Karady and R.G. Farmer (1996). Effect of probabilistic inputs on neural network-based electric load forecasting. *IEEE Trans. on Neural Networks* **7**(6), 1528–1532.

Werbos, P. (1974). Beyond Regresssion: New Tools for Prediction and Analysis in the Behavior Science. PhD thesis. Harvard University.

COMPARATIVE STUDIES ON NEURO-ASSISTED ITERATIVE LEARNING CONTROL SCHEMES

Xu Jian-Xin[1] and Wong Wai Yen

National University of Singapore, Department of Electrical Engineering
10 Kent Ridge Crescent, Singapore 119260.
[1] *E-mail. Elexujx@nus.edu.sg*

Abstract : In this paper, the comparative studies on the performances of BSNN, RBF and MLP neural networks, as well as the two types of neuro-assisted iterative learning control schemes are presented. A new neural-based iterative learning scheme is developed using BSNN as the feedforward neural network. It is shown from the simulations that the generalization ability of the BSNN is enhanced with capability of achieving small initial tracking errors. *Copyright © 1999 IFAC*

Keywords: BSNN, BFN, MLP neural network, neuro-assisted iterative learning control scheme

1. INTRODUCTION

This paper discusses two different types of learning control schemes. They are namely the repetitive form of learning and the pattern form of learning. A classical example of the former is the iterative learning control scheme which is capable of achieving precise tracking performance through repetitive trials. The latter is through the use of neural network which serves to generalize to the other new unlearned patterns after it has been trained and learned the dynamics of the system or plant based on the given set of training data.

Pure neuro-based form of learning control is useful in learning for new patterns as the trained neural network is able to generalize to the other unlearned patterns. However, this form of pattern based learning is less accurate, as the generalized patterns are only approximate to the ideal ones. The accuracy of the generalized patterns is very much dependent upon the generalization ability of the neural network. Hence, in pure neurocontrol, it may be expected that the errors between the estimated and desired control signals for these unlearned trajectories may be large.

On the other hand, iterative learning control is capable of achieving an exact tracking performance in the control system which is operating repetitively over a fixed time interval (D.H Hwan, *et al.*, 1991; Jim Young Choi and Hyun Joo Park, 1997). A very strong key advantage in using iterative learning is that the output trajectory can be made to converge precisely to the desired trajectory. However, the problem of pure iterative learning is that the learned information will be of no use if the desired trajectory changes. Hence, control inputs for the other trajectories have to be relearned again.

In view of this, an endeavour was made in this project to incorporate neural network with iterative learning, exploiting the desired features of the neural network and iterative learning. Firstly, a comparative study was made among the three types of neural networks, namely the Multilayer Perceptron (MLP), Radial Basis Function (RBF) and B-spline Neural Network (BSNN). The performances of these three types of neural networks are compared in terms of their training and interpolation results.

Secondly, a comparative study was made to analyze the two different forms of neuro-assisted iterative learning. The first type of neuro-assisted iterative learning is a simple scheme which can compensate for the initial trial using the initial control information provided by the trained neural network. In this case, the trained BSNN is only being used for assistance in iterative learning at the initial trial. On the other hand, the other type of neuro-assisted iterative learning scheme is slightly more complicated. In this proposed scheme, the iterative learned control information is transferred to the BSNN by training when a new pattern arises. This scheme ensures precise tracking performance to be achieved in iterative learning as well as the control information obtained by the iterative learning stage can be transferred to the BSNN for accumulation in the neural memory, which is useful for future generalization to the other unlearned trajectories.

In this paper, the simulation results of the above studies made are presented in the subsequent sections.

Problem Formulation and Preliminaries:
A discrete time system dynamics of the form given by equation (1) will be used in the simulation for learning the inverse dynamics of the system as well as in the neural-assisted iterative learning. The next state of the system, $y(t+1)$, is related to the current state, $y(t)$, and current control input, $u(t)$, by the following expression which is given as :

$$y(t+1) = y(t) \{ 1 + 0.01 \sin[y(t)] \} + 0.01\, u(t) \qquad (1)$$

For simulation purpose, the desired system output, $y_d(t)$ is given by $y_d(t) = A\sin(Awt)$, where $1 \leq A \leq 3$, $\pi \leq w \leq 2\pi$, and t is the time index, $t \in [0,200]$. The sampling interval is set to be 0.01 seconds.

2. COMPARATIVE STUDY OF BSNN WITH THE MLP AND RBF NEURAL NETWORKS

First, the second order B-spline basis function of the BSNN was implemented and the activation function of each neuron in the BSNN will be of the second order type of basis function. The BSNN was then presented with a set of training patterns when learning for the inverse dynamics of the system function. Upon training, interpolation analysis was performed on the BSNN using a set of interpolation patterns. The training and interpolation results are then compared with the ones obtained earlier, where the similar system function as given by equation (1) was studied using the Multilayer Perceptron and Radial Basis Functions neural networks.

Based on Table 1, it is observed that the average training errors of the MLP neural network are the lowest among the three networks. The maximum error observed is only about 0.022 for training pattern [T1]. However, the maximum absolute error observed for the same pattern is slightly greater, about 0.072, was obtained by the B-spline neural network.

In general, the overall training results of the MLP neural network is better than the B-spline neural network. As observed from Table 1, the maximum absolute training errors of the MLP neural network are lower than that of the B-spline neural network for this system function. On the other hand, the interpolation ability of the B-spline neural network is better than the MLP network, with lower interpolation errors as observed from Table 2.

Table 1 Maximum Absolute Training Errors observed

Training pattern	MLP	RBF	BSNN
T1	0.022	0.450	0.072
T2	0.025	0.250	0.058
T3	0.026	0.230	0.055
T4	0.025	0.230	0.085
T5	0.030	0.065	0.080
T6	0.045	0.250	0.091
T7	0.050	0.250	0.120
T8	0.080	0.110	0.170
T9	0.050	0.120	0.190

Table 2 Maximum Absolute Interpolation Errors observed

Interpolating Pattern	MLP	RBF	BSNN
S1	0.280	8.0	0.120
S2	0.250	7.5	0.130
S3	0.300	15.0	0.140
S4	0.300	32.0	0.141

Based on the results obtained on the RBF, it can be easily seen that the RBF neural network exhibits quite poor training and interpolation performances. The RBF neural network is too local and it is important to have more centers around the clusters of the training data to ensure good training performance. Due to poor generalization ability, the RBF neural network is not suitable for application in neural based iterative learning control problems.

The results have shown that the performance of the B-spline neural network is relatively good for this simple system. With shorter training time required to achieve a reasonable level of performance, this makes the B-spline neural network a better choice than the MLP for this simple system as the training time of LP is long and the weights of the MLP network can be trapped into the local minima during the training process. On the other hand, BSNN which is being trained using the Least Mean Square Weights Adaptation Algorithm, does not have the local minima problem. Nevertheless, the maximum absolute errors are estimated by observing the various plots given and the simulations in this project. This comparison will serve to present an approximate idea on the performance of the two types of neural network on learning the inverse dynamics of the system. Having discussed the training and interpolated results which were obtained by using the

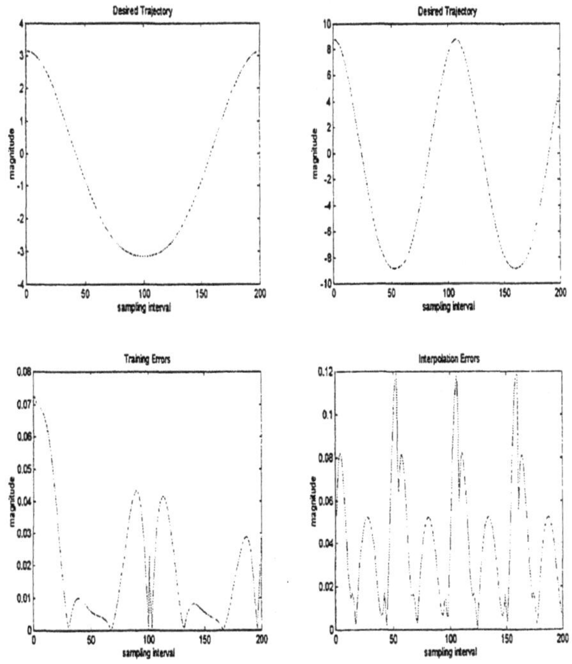

[T1] $A = 1$, $w = \pi$ [S1] $A = 1.5$, $w = 1.25\pi$

Fig.1 Trained and interpolated trajectories using BSNN to learn for the system inverse

B-spline neural network for learning of the system inverse, the next section focuses on the application of the BSNN in iterative learning control.

3. COMPARATIVE STUDY OF NEURO-ASSISTED ITERATIVE LEARNING CONTROL

This section presents a study of the two different forms of neuro-assisted iterative learning control schemes. A simple way of incorporating the trained neural network with iterative learning at the initial stage of learning is examined in section 3.1, and a proposed neuro-assisted iterative learning scheme which combines both iterative learning with neural training at every stage of learning is presented in section 3.2.

3.1 Simulation Results Using Initial Neuro-Assisted ILC

First of all, a BSNN was used to model the inverse dynamics of the given plant and the system function was denoted by equation (1). It was shown from the simulation results that the inverse plant dynamics was modelled well by the BSNN with low training and interpolation errors. This can be clearly illustrated by the trajectory plots of the trained and interpolated patterns in Fig. 1.

Having learned the system inverse, the BSNN was then applied to the iterative learning controller, for providing the initial control information at the first stage of iterative learning. Subsequently, the BSNN was disconnected and the iterative learning controller took over the learning process. A general iterative learning model whereby the BSNN serves to provide for initial neuro-assisted iterative learning is shown in Fig. 2.

With the aid of the trained BSNN, the initial tracking errors of the interpolated patterns are low and it takes not more than 5 iterations for the tracking errors to converge to a value below 0.03. At the fifth iteration, the final tracking errors of the interpolated patterns were again noted. The results of the iterative learning with initial control information provided by the BSNN are summarized in Table 3.

Table 3 Summary of results of iterative learning with BSNN attached during the initial stage of learning

Interpolated pattern	Initial tracking error	Tracking error at the 5th trial
S1	0.0687	0.0150
S2	0.0536	0.0151
S3	0.1008	0.0260
S4	0.0934	0.0253

This is a simple way of neuro-assisted iterative learning control scheme which compensates the initial trial in terms of the interpolation property of the BSNN. It can reduce the initial tracking error and hence fasten the subsequent iterative learning convergence. However, the neural network is not updated with the iterative learned control information for these newly learned patterns. Hence, if a similar pattern arises the next time, the process of iterative learning is repeated again.

3.2 Simulation Results Using Proposed Neuro-Assisted ILC

A simulation of the proposed neuro-assisted iterative learning scheme was performed using the neuro-assisted iterative learning algorithm. This is a more general form of neuro-assisted iterative learning control scheme where the BSNN is used as the feedforward neurocontroller to accumulate the iteratively learned control inputs for the various reference trajectories. Precise tracking performance is ensured by the iterative learning controller. After each iterative learning, the learned control information is transferred to the feedforward neural network by training the neural network for long term memory storage and generalization to any trajectory tracking. In order to prevent any possibility of divergence of the trajectory error in the initial stage of learning, a feedback controller is incorporated into the learning system as shown in Fig. 3. to stabilize

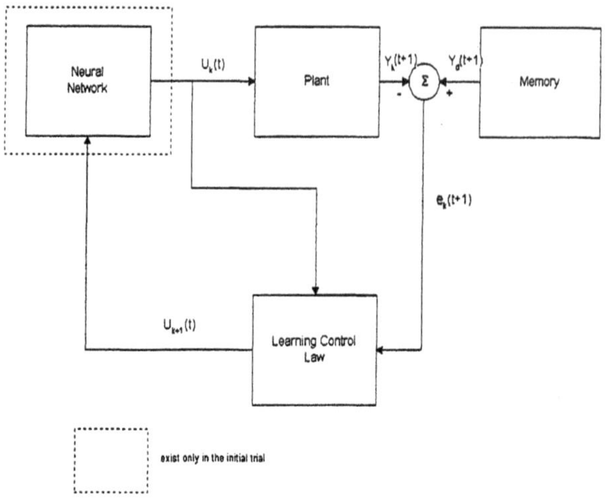

Fig.2 A General Initial Neuro-Assist Iterative Learning Model

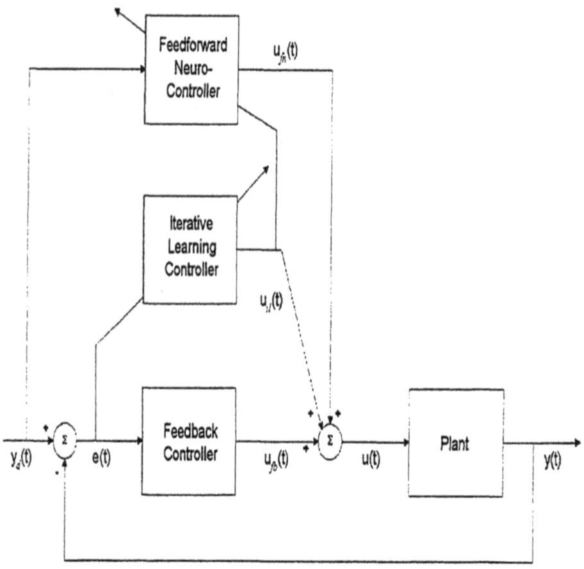

Fig.3 Overall structure of the proposed neural-assisted iterative learning setup

the system at the initial stages of learning.

The setup in Fig. 3 was presented with the set of nine training patterns ([T1], [T2], ... , [T9]) with the corresponding values of A and w are shown in Table 4. Initially, when the setup was presented with the first pattern, that is pattern [T1], 35 iterations were needed for the tracking error to converge to the value smaller than the predefined error threshold of 12 x 10^{-4}. However, after the feedforward neural network had been trained with this pattern, the number of iterations needed for the neural network to be trained with the next neighbouring trajectories, which are patterns [T2] and [T3], were very much lesser than the first pattern [T1]. In this case , about 2 and 14 iterations were required for patterns [T2] and [T3] respectively.

In addition, the maximum absolute tracking errors for the neighbouring patterns at the initial iteration were also lower than that of the previous learned trajectory. This can be seen from Table 4 that the initial tracking errors for patterns [T2] and [T3] are lower than the very first pattern [T1], learned by the feedforward neural network.

This is because the weight vectors, which are associated with these new trajectories, have already been modified during the previous training cycle when the first pattern [T1] was being learnt. As the neighbouring trajectories are very similar to the previous trajectory which the neural network has been trained with, lesser number of times of adjustment for the similar sets of weights, which are adjusted during the previous learning, will be required to achieve the predefined tracking error threshold at this time. Hence, this also means that the number of iterations for the trajectory tracking of these new trajectories will be lesser than the previously learned one. This can be easily seen from the Table 4, that the number of iterations required for the patterns [T2] and [T3], are significantly lesser than the previously learned pattern [T1].

However, if the new trajectory, which is very different from those trajectories learned previously, is presented, the maximum absolute tracking error at the initial trial will be large. In addition, the number of iterations requires for the tracking error to converge has also increased. As from Table 4, the maximum absolute tracking errors for patterns [T4] and [T7], are very much larger than the previous learned patterns [T1], [T2], and [T3]. After the two patterns, [T4] and [T7], have been learned by the learning system, the initial tracking errors of the other remaining patterns such as patterns [T5], [T6], [T8] and [T9], are very much smaller than the number of iterations required by [T4] and [T7].

Table 4 Summary of training results

Training pattern	Iteration	Max absolute tracking error at the initial iteration	Max absolute tracking error at the final iteration
T1	35	0.3311	0.0010
T2	2	0.0013	0.0012
T3	14	0.0302	0.0008
T4	57	0.6868	0.0010
T5	34	0.0175	0.0010
T6	31	0.0095	0.0010
T7	70	0.8070	0.0010
T8	53	0.0670	0.0010
T9	40	0.0317	0.0010

The simulation plots for some of the training patterns are shown in Fig. 4 where the desired plant output and the tracking error, which is measured by the absolute difference between the desired and actual plant outputs for the training pattern [T2] and [T4] are plotted. From the plots in Fig. 4, it was observed that the tracking errors are small and they are in the order of 10^{-3}. Thus, by incorporating the iterative learning controller with the feedforward neurocontroller, the plant output trajectory can track the desired reference trajectory more precisely. In addition, it was noted that as more training patterns were presented to this learning system, or in other words, the feedforward neurocontroller were trained with more training patterns, the control performance became better.

Having trained the feedforward neural network, the entire learning system, depicted in Fig.3, was then presented with the 4 interpolating patterns ([S1], [S2], [S3] and [S4]). As an illustration, the initial tracking errors of the plant output for the interpolated patterns [S1] and [S4] are about 0.0036 and 0.0064 respectively. At the fifth iteration, the final tracking errors of the interpolated patterns were noted. The results are tabulated in Table 5.

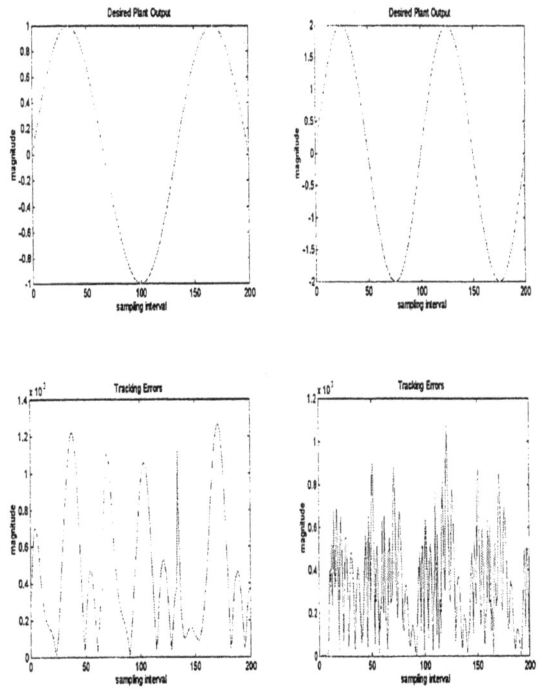

[T2] : $A = 1$, $w = 1.5\pi$ [T4] : $A = 2$, $w = \pi$

Fig.4 Desired and training error trajectories for training patterns [T2] and [T4]

Table 5 Summary of iterative learned results using the Proposed Neuro-Assisted ILC scheme

Interpolated pattern	Initial tracking error	Tracking error at the 5th trial
S1	0.0036	0.0009
S2	0.0031	0.0009
S3	0.0064	0.0018
S4	0.0064	0.0020

As the initial tracking error is low, it takes only a few iterations for the tracking error to converge to a value lower than the predefined error threshold if an iterative learning controller is attached to the learning system. For example, it takes about 5 iterations for the tracking error of the pattern [S1] to converge to a value of 0.0009 from its initial tracking error of 0.0036. The graph of the maximum absolute tracking error vs the number of trials for each of the interpolated patterns using the proposed neural-based iterative learning strategy is shown in Fig. 5.

Fast tracking can also be achieved by the proposed method of neural-based iterative learning even if the BSNN has not been trained at all initially. As observed from the plots in Fig. 6, some form of learning has taken place even at the end of the first trial. It shows that the plant output is able to track the reference trajectory reasonably well even though the initial tracking error is large when the untrained feedforward neurocontroller is used. By the second trial, the tracking performance has improved from the initial ones. Tracking errors have been very much reduced than the previous trial as the learned trajectories follow the desired reference trajectories more precisely than before. However, when a trained feedforward neurocontroller is used, the plant output trajectory is able to follow the desired reference trajectory very precisely even at the initial trial.

Based on the results obtained, small tracking errors and fast tracking performance was achieved using the proposed neural-based iterative learning method. From Table 5, it can be clearly seen that about 5 trials are needed to track any of the four unlearned plant output trajectories within the tracking error threshold of 2×10^{-3}. The overall improved performance using this method will be readily seen if some comparisons are made with the results obtained by the conventional method of neuro-assisted iterative learning at the initial trial only. Upon comparing the simulation results from Table 3 and Table 5, the initial tracking errors obtained by the conventional method of initial neuro-assisted iterative learning are larger than the corresponding ones obtained by the proposed neural-based iterative learning. In addition, it was also noted that the initial tracking errors of the interpolated trajectories were already very much lower than the predefined error threshold of 0.03 for the proposed neural-based iterative learning scheme. Conclusively, the two sets of simulations have very distinctly shown that the proposed neural-based iterative learning is able to achieve fast convergence.

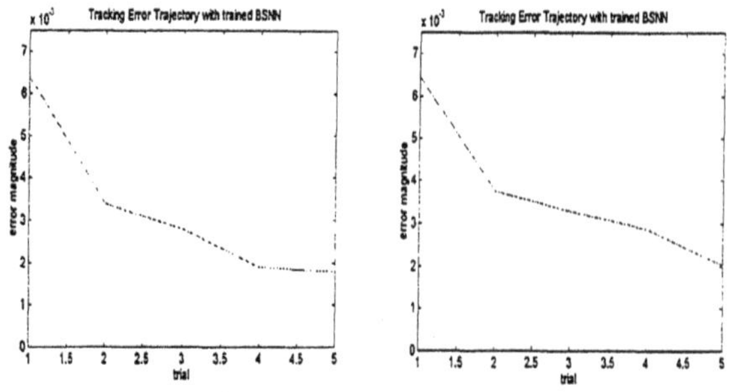

[S3] $A = 2.5$, $w = 1.25\pi$ [S4] $A = 2.5$, $w = 1.75\pi$

Fig.5 Maximum absolute tracking errors vs the number of trials with the trained BSNN and iterative learning controller

However, this scheme is only feasible if sufficient neural training time is available between two consecutive trials, or some parallel computing sources are available. Another problem of this form of learning scheme is that high number of switching operations may have incurred in the system which switches between the consecutive trials of iterative learning and neural training if the parallel computing source is unavailable. Nevertheless, the merit of applying this form of neuro-assisted iterative learning scheme allows the feedforward neurocontroller to be accumulated with the newly learned patterns. It was also found that the generalization ability of the system was enhanced if more training patterns were presented in the learning process.

4. CONCLUSION

In this paper, comparative studies were made on the performances of the 3 different types of neural networks, namely the MLP, RBF and BSNN as well as the two different forms of neuro-assisted iterative learning control schemes. In the first case, it was found from the simulations that BSNN best meets the requirement of good training and interpolation results. It is more flexible to use BSNN as it has good interpolation property which is suitable for the particular problem studied. Two forms of neuro-assisted iterative learning schemes were also examined and the results of the simulations were plotted. It was shown from the simulations that the iterative learning time was much reduced by incorporating the BSNN with iterative learning control.

5. REFERENCES

D.-H. Hwang, Z. Bien, S.-R. Oh (1991). Iterative learning control method for discrete-time dynamic systems, *IEE Proceedings-D*, **1.138**(2), pp.139-144.

Jim Young Choi and Hyun Joo Park (1997). Neural-Based Iterative Learning Control for Unknown Systems, *Proceedings of the 2nd Asian Control Conference*, **2**, pp. 243-246.

Kelvin L. Moore (1993). Iterative Learning Control for Deterministic Systems, Springer-Verlag, Germany.

M. Brown and C.J. Harris (1994). Neurofuzzy Adaptive Modelling and Control, chapter 8, pp. 296-337, Prentice Hall, New York.

Simon Haykin (1994). Neural Networks : A Comprehensive Foundation, chapter 5, pp. 121-132, Prentice Hall, New York.

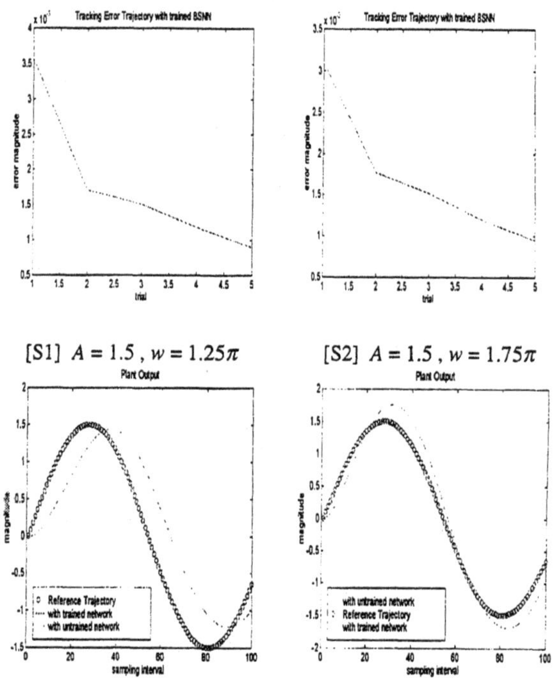

at the end of the first trial at the end of the second trial

Fig. 6. The learned and reference plant output trajectories of interpolated pattern [S1] at the end of the first and second trial.

14th Triennial World Congress, Beijing, P.R. China

MODEL BASED ITERATIVE LEARNING CONTROL (MILC) FOR UNCERTAIN DYNAMIC NON-LINEAR SYSTEMS

Muhammad Arif, Tadashi Ishihara and Hikaru Inooka

Graduate School of Information Sciences, Tohoku University, Japan
arif,ishihara,inooka@control.is.tohoku.ac.jp

Abstract: Model based Iterative Learning Control (MILC) is proposed for a class of time varying nonlinear uncertain systems. Convergence of MILC is analyzed and the uniform boundedness of tracking error is obtained in the presence of uncertainty and disturbances. It is shown that the learning algorithm not only guarantees the robustness, but also improves the learning rate despite of the presence of disturbances and slowly varying desired trajectories in succeeding iterations. The effectiveness of the proposed MILC is presented by simulations. *Copyright © 1999 IFAC*

Keywords: Iterative learning control, Nonlinear control, Uncertain systems, Model based control

1. INTRODUCTION

Various type of learning control methods have been proposed to deal with the uncertainties of the dynamic systems in which the past experience history is utilized to improve the control quality. However if the control task is of repetitive nature as in many robotics applications in the industry, then iterative learning control is found to be effective. Such kind of learning control method is proposed by Uchiyama[1] and is elaborated as a formal theory by Arimoto[2] in 1984. Since then, there has been a great deal of efforts by the researchers to synthesize a better iterative learning control scheme[3,4]. Although sufficient conditions are derived for the convergence of the learning process, the convergence rate is often slow in most of the proposed control laws.

Chen et. al. [5] has described a current iteration tracking error assisted iterative learning control method. They have used the static extrapolation method to predict the current error. However, the effect of the error in the prediction of the current iteration error data has not been incorporated in the convergence analysis of the control algorithm.

In this paper, we present a new model based iterative learning control (MILC) scheme for uncertain time varying nonlinear systems. In our view, it is better to use a nonlinear model of the system for the prediction of error. Future error can be predicted by using the model of the system. This model of the system can be obtained by using the nonlinear system identification technique. However, if the nonlinear system can be linearized around some operating point, we can identify the linearized system by the identification technique for linear systems. It gives us the advantage of using more simpler and easily implementable identification technique. In this way, we can restore the simplicity of the learning algorithm by using an approximate linear model of the system to predict the future error data. In the simulations section, it is shown that the linearized model can be used to predict the error and we can still get a better convergence. Necessary and sufficient conditions are stated for the convergence of the proposed method in the presence of disturbances. It is also shown that the convergence rate of the learning process can be increased using MILC technique. To illustrate the applicability of MILC, we have demonstrated an example of nonlinear system.

2. PROBLEM FORMULATION

Consider a nonlinear time varying dynamic system subjected to the disturbances described as,

$$\dot{x}_i(t) = f(x_i(t),t) + B(t)u_i(t) + d_i(t)$$
$$y_i(t) = C(t)x_i(t) + v_i(t) \tag{1}$$

where i corresponds to the ith iteration of the system; $x_i(t) \in R^{n\times 1}, u_i(t) \in R^{m\times 1}$ and $y_i(t) \in R^{r\times 1}$ are the states, control input and output of the system respectively for $t \in [0,T]$; $d_i(t)$ and $v_i(t)$ are the unknown but bounded disturbances to the system and to the output respectively; $f(x_i(t),t) : R^n \times [0,T] \mapsto R^n$ is a nonlinear, piecewise continuous function of $x_i(t)$ and t.

$f(.)$ satisfies the Lipschitz condition as follows

$$\|f(x_2(t),t) - f(x_1(t),t)\| \le L_f \|x_2(t) - x_1(t)\| \tag{2}$$

where $L_f > 0$ is the Lipschitz constant. Let $y_d(t)$ be the desired trajectory, which is achievable, and continuously differentiable on the finite time interval $t \in [0,T]$, such that for bounded desired trajectory $y_d(t)$, there exists a unique bounded input $u_d(t)$ for $t \in [0,T]$, for which system has unique bounded states $x_d(t)$ and $y_d(t) = C(t)x_d(t)$ at $t \in [0,T]$.

The problem can be stated as follows: For a desired trajectory $y_d(t)$ starting with some initial control input $u_0(t)$, modify the control input $u_i(t)$ sequentially for i^{th} iteration such that for $i \to \infty$, $y_i(t) \to y_d(t)$ $\forall t \in [0,T]$, in the presence of bounded disturbances.

Definition 1. Norms of function $g(t)$, introduced in this paper, are as follows.

$$\|g(t)\|_\lambda = \sup_{t\in[0,T]} e^{-\lambda t}\|g(t)\| \tag{3}$$
$$\|g\| = \max_{1\le i\le n} |g_i| \tag{4}$$
$$\|g(t)\|_\infty = \sup_{t\in[0,T]} \|g(t)\| \tag{5}$$

3. MODEL BASED ITERATIVE LEARNING CONTROL (MILC)

For a nonlinear time varying system with the disturbances $d_i(t)$ and $v_i(t)$ as given in (1), a new model based iterative learning control law MILC is presented in this paper. MILC is defined as

$$u_i(t) = u_{i-1}(t) + \Gamma_1\, \dot{e}_{i-1}(t) + \Gamma_2\, \dot{e}_i^*(t) \tag{6}$$

where

$$\dot{e}_{i-1}(t) = \dot{y}_d(t) - \dot{y}_{i-1}(t),$$
$$\dot{e}_i^*(t) = \dot{y}_d(t) - \dot{y}_i^*(t) \tag{7}$$

The matrices $\Gamma_k \in R^{m\times r}$ $(k = 1,2)$ are the learning matrices. In the above equation, $y_i^*(t)$ is defined as the output of the nonlinear system's model which is identified by using the input-output data of $i-1^{th}$ iteration. Identification of the system can be done on-line in each iteration to ensure the tracking of the parameter drift of the system during the operation.

3.1 *Non Linear System Identification*

The main issue in the system identification is to search for a suitable model structure, within which a good model can be found. Parameter estimation, once the structure is known is a lesser problem. To estimate the model, experimental data $z(t) \in R^{1\times(p+m)}$ can be arranged as,

$$z(t) = [y_1(t)\ldots\ldots\; y_p(t)\; u_1(t)\ldots\ldots\ldots\; u_m(t)] \tag{8}$$

It is very important that the data should reflect all important features of the underlying system. The excitation signal $u(t)$ should carefully be chosen, so that the system dynamics show up in the output signal $y(t)$.

Selection of a model structure M is also an important aspect of system identification. The selection comprises of type and size of model. Once the size and type is defined, one has to search for the model in the model set M^*. The next step is to parameterize the model set. Let the members of M^* is parameterized by a finite dimensional vector $\theta \in D_M \subset R^{d\times 1}$. The model structure is defined by the mapping

$$M : D_M \ni \theta \to M(\theta) \in M^* \tag{9}$$

where $M(\theta)$ is a particular model corresponding to θ.

We denote a family of functions parameterized by θ as follows,

$$y'(t,\theta) = h(\varphi(t),\theta) \tag{10}$$

where

$$\varphi(t) = \varphi(y^{t-1}, u^{t-1}) \tag{11}$$

We call $\varphi(t)$ as regression vector. Regression vector can itself be parameterized:

$$\varphi(t,\eta) = \varphi(y^{t-1}, u^{t-1}, \eta) \tag{12}$$

In practice, measured and model output differs and do not match exactly.

$$\varepsilon(t,\theta) = y(t) - y'(t,\theta) \tag{13}$$

where $\varepsilon(t,\theta)$ is an error due to the unmodelled dynamics and noise. An error criterion can be defined as,

$$V_\varepsilon(\theta, Z_N) = \frac{1}{N}\sum_{t=1}^{N}\varepsilon^2(t,\theta) \tag{14}$$

The problem of identifying a nonlinear dynamic system is a vast field of research and is not in the scope of this paper. As long as the error criterion is minimum, any technique can be used for the identification of the nonlinear system. Instead of using identification technique for nonlinear systems, the system identification techniques for linear systems can also be used for the prediction provided that the prediction error is bounded. It is because of the fact that predicted error at i^{th} iteration will be corrected in the $i+1^{th}$ iteration of the operation. A nonlinear system can be written as,

$$\frac{d}{dt}\underline{x}(t) = \underline{f}(\underline{x}, \underline{u}, t) + B\underline{u}(t) \tag{15}$$

It can be linearized around some operating states. For the nonlinear system described in the above equation, expressing the dependent state variables and independent forcing functions as deviations from some operating state, $\underline{x}_0$ and $\underline{u}_0$, gives

$$\underline{x}(t) = \underline{x}_0(t) + \delta\underline{x}(t) \tag{16}$$
$$\underline{u}(t) = \underline{u}_0(t) + \delta\underline{u}(t) \tag{17}$$

Substituting these expression in equation (15) and writing the nonlinear terms in a first order taylor series expansion about the operating point, we get the expression that represents a linearization of the original nonlinear system.

$$\frac{d}{dt}\underline{\delta x}(t) = J_x(\underline{x}_0, \underline{u}_0)\underline{\delta x}(t) + B\underline{\delta u}(t) \tag{18}$$

where

$$J_x = \left.\begin{bmatrix} \frac{\partial f_1}{\partial x_1} & \frac{\partial f_1}{\partial x_2} & \cdot\cdot & \frac{\partial f_1}{\partial x_n} \\ \frac{\partial f_2}{\partial x_1} & \frac{\partial f_2}{\partial x_2} & \cdot\cdot & \frac{\partial f_2}{\partial x_n} \\ & \cdot & & \\ & \cdot & & \\ \frac{\partial f_n}{\partial x_1} & \frac{\partial f_n}{\partial x_2} & \cdot\cdot & \frac{\partial f_n}{\partial x_n} \end{bmatrix}\right|_{\underline{x}_0, \underline{u}_0} \tag{19}$$

Once the nonlinear system is linearized, we can apply some simple system identification technique for linear system to predict the error.

3.2 *Analysis of Convergence*

MILC is a modified version of standard D type ILC technique to increase the convergence rate. In this technique, we have used predicted error to speed up the convergence. Predicted error is not supposed to be accurate but bounded. The effect of the uncertainty in the prediction will be compensated as the actual error will be available in the next iteration.

Assumptions:

A1: System (1) is stable for $\forall t \in [0,T]$ and $\forall i$.

A2: The disturbances $d(t)$ and $v(t)$ are bounded such that for $\forall t \in [0,T]$ and $\forall i$

$$\begin{aligned} \|d_i(t) - d_{i-1}(t)\|_\lambda &\le b_d \\ \|v_i(t) - v_{i-1}(t)\|_\lambda &\le b_v \end{aligned} \tag{20}$$

A3: Prediction error is also bounded to a certain constant b_m such that,

$$\|e_i^* - e_i\|_\lambda \le b_m \tag{21}$$

where

e_i^* = Predicted error.

e_i = Actual error at i^{th} iteration.

$b_m \equiv$ Modelling error bound.

Remark 1. Prediction error bound b_m is a measure to represent the deviation of e_i^* from e_i, which means that the higher the b_m value, the poorer will be the identification of the system, causing the wrong estimation of the error.

A4: $[\|I_r + CB\Gamma_2\|_\infty]^{-1}$ exists for $\forall t \in [0,T]$ and $\forall i$.

A5: $y_d(t)$ is the desired trajectory which is reachable such that for bounded $y_d(t)$, there exists a bounded $u_d(t)$, for which system has unique bounded states $x_d(t)$ and $y_d(t) = Cx_d(t)$ at $t \in [0,T]$.

A6: $f(x_i(t),t) : R^n \times [0,T] \mapsto R^n$ is a piecewise continuous function and satisfies the Lipschitz condition. Furthermore $C(t)$ and $B(t)$ are uniformly bounded for all $t \in [0,T]$ and for all iterations $i \in [0,\infty]$, such that

$$\begin{aligned} b_c &= \sup_{i\in[0,\infty]}\sup_{t\in[0,T]}\|C(t)\| \\ b_b &= \sup_{i\in[0,\infty]}\sup_{t\in[0,T]}\|B(t)\| \end{aligned} \tag{22}$$

Theorem 1. For the nonlinear time varying uncertain system given in (1), with the assumptions A1 to A6 hold, by using the model based iterative learning control law mentioned in equation (6) incorporating the condition,

$$\gamma\rho < 1 \; for \; \forall t \in [0,T] \tag{23}$$

where

$$\gamma = [\|I_r + CB\Gamma_2\|_\infty]^{-1} \tag{24}$$

$$\rho = \|I_r - CB\Gamma_1\|_\infty \tag{25}$$

provided $[\|I_r + CB\Gamma_2\|_\infty]^{-1}$ exists for $\forall t \in [0,T]$ and $\forall i$, the final tracking error will be bounded for sufficiently large λ, such that

$$\|e_i\|_\lambda \le \gamma\rho \|e_{i-1}\|_\lambda + \gamma(\|CB\Gamma_2\|_\infty b_m + b_v) \tag{26}$$

PROOF.

Let

$$\begin{aligned} e_i(t) &= y_d(t) - y_i(t) \\ &= y_d(t) - y_{i-1}(t) - (y_i(t) - y_{i-1}(t)) \\ &= e_{i-1}(t) - C(t)[x_i(t) - x_{i-1}(t)] \\ &\quad -[v_i(t) - v_{i-1}(t)] \end{aligned} \tag{27}$$

Now integrating equation (1) and combining with equation (6), we get

$$\begin{aligned} x_i(t) - x_{i-1}(t) &= \int_0^t [f(x_i(\tau),\tau) - f(x_{i-1}(\tau),\tau)]d\tau \\ &+ \int_0^t [d_i(\tau) - d_{i-1}(\tau)]d\tau + \int_0^t B(\tau)\Gamma_2 \, \dot{e}_i^* \, (\tau)d\tau \\ &+ \int_0^t B(\tau)\Gamma_1 \, \dot{e}_{i-1} \, (\tau)d\tau \end{aligned} \tag{28}$$

Integrating by parts the last two terms containing $\dot{e}\,(\tau)$, multiplying both side with $C(t)$, putting in equation (27) and rearranging we get,

$$\begin{aligned} [I_r + CB\Gamma_2]\, e_i(t) &= [I_r - CB\Gamma_1]\, e_{i-1}(t) \\ -CB\Gamma_2[e_i^*(t) - e_i(t)] &- C\int_0^t [f_i(\tau) - f_{i-1}(\tau)]d\tau \\ -C\int_0^t [d_i(\tau) - d_{i-1}(\tau)]d\tau &+ C\int_0^t h_1(\tau)e_{i-1}(\tau)d\tau \\ +C\int_0^t h_2(\tau)e_i^*(\tau)d\tau &- [v_i(t) - v_{i-1}(t)] \end{aligned} \tag{29}$$

where $h_{1,2}(\tau) = \frac{\partial}{\partial\tau}(B(\tau)\Gamma_{1,2})$.

Taking the $\lambda - norm$ of the above equation and assuming that A2 and A3 holds, we get

$$\begin{aligned} \|I_r + CB\Gamma_2\|_\infty \|e_i\|_\lambda \le [\|I_r - CB\Gamma_1\|_\infty]\, \|e_{i-1}\|_\lambda \\ +b_c a_3(b_d + L_f(\|x_i - x_{i-1}\|_\lambda + b_{h1}\|e_{i-1}\|_\lambda \\ +b_{h2}\|e_i^*\|_\lambda) + \|CB\Gamma_2\|_\infty b_m + b_v \end{aligned} \tag{30}$$

where

$$a_3 = \max_{0\le t\le T} \frac{(1 - e^{-\lambda T})}{\lambda}$$

$$b_{h1,2} = \left\| \frac{\partial}{\partial\tau}(B(\tau)\Gamma_{1,2}) \right\|_\infty$$

Taking the $\lambda - norm$ of equation (28), we get

$$\begin{aligned} \|x_i(t) - x_{i-1}(t)\|_\lambda \le \frac{1}{[I_n - a_3 L_f]}[a_3 b_d + (a_3 b_{h1} \\ + \|B\Gamma_1\|_\infty)\|e_{i-1}\|_\lambda + (a_3 b_{h2} + \|B\Gamma_2\|_\infty)\|e_i^*\|_\lambda] \end{aligned} \tag{31}$$

Putting the above equation in equation (30),

$$\begin{aligned} \|e_i\|_\lambda \le \gamma\rho\|e_{i-1}\|_\lambda + \gamma[\|CB\Gamma_2\|_\infty b_m + b_v] \\ +\gamma a_3 b_c[b_d + b_{h1}\|e_{i-1}\|_\lambda + b_{h2}\|e_i^*\|_\lambda \\ +\frac{L_f}{[I_n - a_3 L_f]}(a_3 b_d + (a_3 b_{h1} + \|B\Gamma_1\|_\infty)\|e_{i-1}\|_\lambda \\ +(a_3 b_{h2} + \|B\Gamma_2\|_\infty)\|e_i^*\|_\lambda)] \end{aligned} \tag{32}$$

Assuming that λ is sufficiently large, we can neglect the terms with a_3 in equation (32),

$$\|e_i\|_\lambda \le \gamma\rho\|e_{i-1}\|_\lambda + \gamma(\|CB\Gamma_2\|_\infty b_m + b_v) \tag{33}$$

□

Remark 2. The convergence condition for standard D type iterative learning control is elaborated in [2] as

$$\rho < 1 \; for \; \forall t \in [0,T]$$

where ρ is given in equation (25). According to the above convergence condition, we can not select the gain matrix Γ_1 beyond certain upper limit. By incorporating the modified condition mentioned in equation (23) and including an additional parameter γ which is the defined in equation (24), it is possible to increase the gain matrix Γ_1 further to get faster convergence.

If the system (1) is not subjected to the disturbances, that is b_d and b_v are zero, then the equation (33) reduces to the following expression.

$$\|e_i\|_\lambda \le \gamma\rho\|e_{i-1}\|_\lambda + \gamma\|CB\Gamma_2\|_\infty b_m \tag{34}$$

According to the above expression, absolute convergence of the error is not guaranteed due to the presence of b_m, which is the bound of the prediction error. Although the error will converge uniformly to the neighborhood of the desired trajectory in the absence of disturbances, the absolute convergence of the error can not be guaranteed. If the prediction of the error is very accurate, which means that b_m is negligible, it is possible to neglect the last term of equation (34),

$$||e_i||_\lambda \leq \gamma\rho\, ||e_{i-1}||_\lambda \qquad (35)$$

The above expression guarantees the absolute convergence of error to zero as $i \to \infty$. In the case, when the prediction error bound b_m is not negligible, MILC can be applied only in the initial iterations to obtain improved convergence rate. For the rest of iterations, Γ_2 can be set to zero, eliminating the prediction error bound b_m from equation (34) and reducing the model base iterative learning control to the standard D type iterative learning control. In this way, the absolute convergence of the error can be obtained.

4. APPLICABILITY OF MILC FOR SLOWLY VARYING DESIRED TRAJECTORY

In the previous section, desired trajectory $y_d(t)$ is fixed for each iteration and does not change in succeeding iterations. In this section, we are going to address the case when the desired trajectory $y_d(t)$ is allowed to vary in succeeding iterations. Saab et al. [6] has addressed this problem for D type, PD type and PID type of iterative learning control algorithms. Theorem 1 of Saab's paper can be extended to our proposed algorithm.

Following assumptions are made for $y_d(t,i)$.

D1: The desired trajectory $y_d(t,i)$ is differentiable over $t \in [0,T]$ and $\forall i$.

D2: The desired trajectory $y_d(t,i)$ is allowed to vary for each succeeding iteration by a small deviation such that

$$y_d(t,i) - y_d(t,i-1) = \delta_i(t) \qquad \forall i \qquad (36)$$

where

$$||\delta_i(t)||_\lambda \leq b_\delta \qquad t \in [0,T] \qquad (37)$$

for small $b_\delta > 0$.

Theorem 2. For the nonlinear time varying uncertain system given (1), with assumptions A1-A6 and D1-D2 hold, model based iterative learning control law given in (6) will generate the input $u_i(t)$ for $t \in [0,T]$ such that the tracking error will be bounded for sufficiently large λ, provided that the condition

$$\gamma\rho < 1 \; for \; \forall t \in [0,T] \qquad (38)$$

is fulfilled.

5. NUMERICAL ILLUSTRATIONS

In this section, the effectiveness of the proposed MILC is demonstrated. For this purpose, simulation of a nonlinear model of single link direct joint driven manipulator subjected to disturbance has been done.

The dynamic equation of the system is

$$I\,\ddot{\theta} + (\frac{1}{2}m + M)gl\sin\theta = \tau + \tau_d \qquad (39)$$

where $\theta(t)$ is the angular position of the manipulator, $\tau(t)$ is the driving torque and $\tau_d(t)$ is the disturbance torque. m, M and l are the mass, tip load and length of the manipulator respectively. I is the moment of inertia with respect to the joint and defined as $I = Ml^2 + \frac{1}{3}ml^2$. Let $m = 1.0\ Kg$, $M = 2.0\ Kg$, $l = 0.5\ m$ and $g = 9.8\ m/s^2$ be chosen as system parameters and sampling period is 0.01 seconds.

As discussed in the section 3.1, we will linearize the system around the stable point $\theta = 0$ and $\dot{\theta} = 0$. For the identification of the linearized system, Augmented UD Identification (AUDI) technique [7] is used. This technique gives quite accurate results for the identification of the linear systems.

$\theta_d(t)$ is the desired trajectory specified as,

$$\theta_d(t) = \theta_b + (\theta_f - \theta_b)(15t^4 - 6t^5 - 10t^3) \qquad t \in [0,1] \quad (40)$$

where $\theta_b = 0^0$ and $\theta_f = \frac{\pi}{8}$.

To evaluate the performance of MILC, performance index J is defined as follows,

$$J = \max_{0<t\leq T} |e(t)| = \max_{0<j\leq N} |e_j| \qquad (41)$$

Three cases have been considered for the simulation.

Case 1: $\tau_d = 0$

Figure 1 shows the results of simulation using MILC and the standard D type ILC. It can be observed that the convergence is very fast as compared to the standard D type ILC. For comparison, it can be found out that at 10th iteration, the value of J is 0.599 for MILC as compared to 3.930 for D-type ILC which shows the improvement of about 6 times.

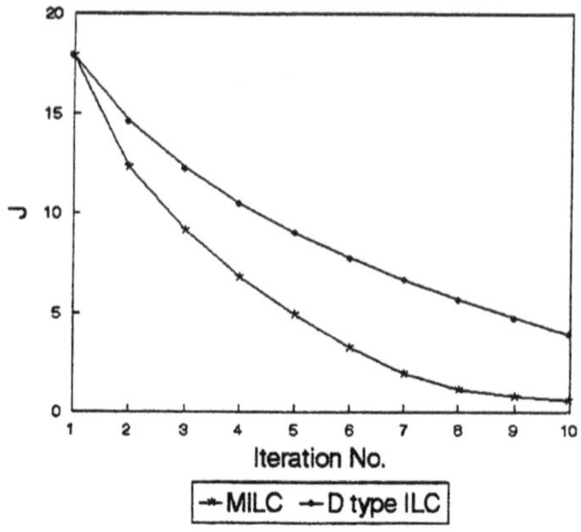

Fig. 1. Performance index J for case 1

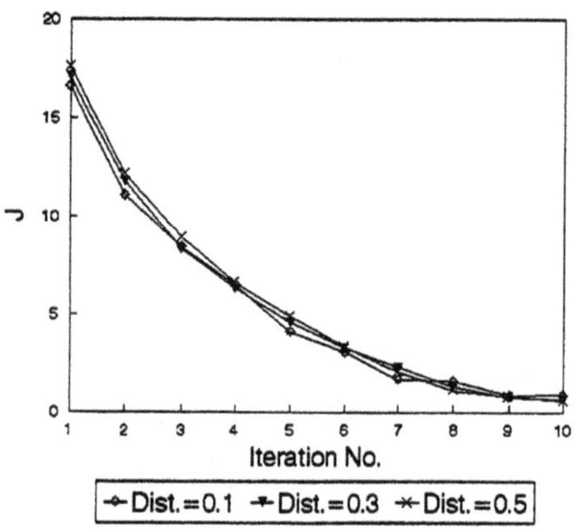

Fig. 2. Performance index J for case 2

Case 2: $\tau_d \neq 0$

τ_d is assumed to be a white noise with zero mean and variance b_e. Simulations are done for three level of variances i.e. $b_e = 0.1, 0.3$ and 0.5 . Figure 2 shows that effect of the level of disturbances are not prominent and the disturbance does not effect the convergence rate considerably.

Case 3: Adaptability

The mass of tip load M is changed after 7th iteration and its effect has been demonstrated. Figure 3 shows the results for 50%, 100%, and 150% increase in M after 7th iteration. It can be observed that parameter change effects the system tracking performance but MILC restores the system to the normal operation after about 5 further iterations depending upon the change in the parameter of the system, ensuring the adaptability of the system.

6. CONCLUSIONS

In this paper, a new iterative learning control based on prediction of future data has been proposed for a class of uncertain time varying nonlinear dynamic systems. Prediction is done by using the system identification technique with the assumption that the prediction error is bounded.

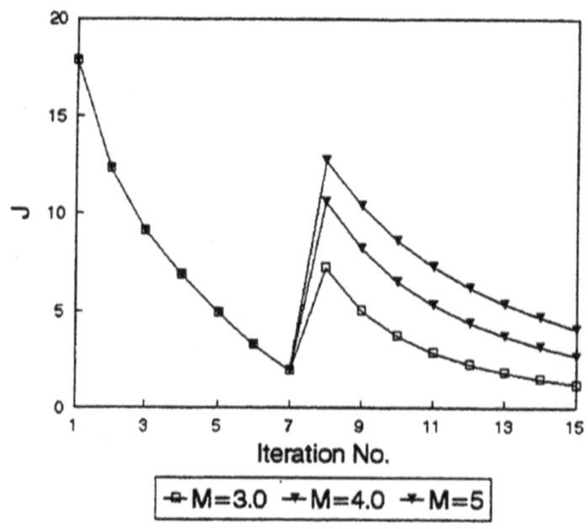

Fig. 3. Performance index J for case 3

It is shown that under certain specified assumptions and conditions on the learning rule, the convergence of the system output to some desired achievable trajectory can be achieved and the final tracking error will be bounded for the uncertain system subjected to bounded disturbances. It is apparent from the simulation results that model based iterative learning control not only offers the improved learning rate, but also provides a better adaptability of the system to its parameter drifts during the operation.

References

[1] M. Uchiyama, Formulation of high speed motion pattern of a mechanical arm by trial, *Trans. SICE (Soc. Instrum. Contr. Eng.)* **14(6)** (1978) 706–712.

[2] S. Arimoto, S. Kawamura, and F. Miyazaki, Bettering operation of robots by learning, *J. Robotic Systems* **1(2)** (1984) 123–140.

[3] P. Bondi, G. Casalina, and L. Gambardella, On the iterative learning control theory for robotic manipulators, *IEEE J. of Robotics and Automation* **4** (1988) 14–21.

[4] S. S. Saab, On the P type learning control, *IEEE Trans. of Automatic Control* **39(11)** (1994) 2298–2302.

[5] C. Chien, Current Iteration Tracking Error Assisted Iterative Learning Control of Uncertain Nonlinear Discrete-time Systems, *Proc. of 35th conference on Decision and Control* (Japan, 1996) 3038-3043.

[6] S. Saab, W. Vogt, and M. Mickle, Learning Control Algorithms for Tracking "Slowly" Varying Trajectories, *IEEE Trans. on SMC* **27(4)** (1997) 657–669.

[7] S. Niu, D. G. Fisher, and D. Xiao, An augmented identification algorithm, *International Journal of Control* **56** (1992) 193–211.

A PENDULATION CONTROL SYSTEM FOR OFFSHORE LIFTING OPERATIONS

D. Fragopoulos◊, M. P. Spathopoulos* and Y. Zheng

System Dynamics and Control, Department of Mechanical Engineering
University of Strathclyde, 75 Montrose str., Glasgow G1 1XJ, Scotland, UK
**tel.: 0044-141-5482326,fax: 0044=141-5525105,e-mail: mps@mecheng.strath.ac.uk.*
◊dimos@spark.net.gr

Abstract: This paper presents a control system that reduces pendulation of suspended loads in offshore lifting operations. The kinematic model of the ship and crane together with the designed anti-pendulation arm are presented. Linear control design techniques for the linearised model are proposed aiming at controlling pendulation in the presence of sea disturbances. These techniques involve LGQ and Generalised predictive control. The use of a dynamic model for the vessel dynamics and the sea wave disturbances in the control design improve the behaviour of the controlled system. Finally simulation results and comparisons are given. *Copyright © 1999 IFAC*

Key words: Active pendulation control system, non-linear system, LQG and Generalised Predictive control, control design

1. INTRODUCTION

The vast majority of crane systems used offshore have no means to compensate for the motions induced on the suspended load caused by the sea, wind and other external disturbances. These induced motions take the form of heave (vertical motion) and pendulation of the load. Previous efforts on the compensation for disturbances induced on the suspended load have resulted in passive systems incorporating damping elements tuned to compensate for specific conditions of load mass and sea state. These however have a limited success on the crane.

The proposed active control system aims at maintaining the position of the suspended load within a sphere of one meter radius relative to an inertial frame of measurement. The control system uses two different actuators to achieve this aim. In the vertical plane defined by the crane boom, actuation is provided by a swinging arm (*anti-pendulation arm*) pivoted at the top of the boom. A hydraulic piston is used to force this arm to move. The cable holding the load is hanging from the lower end of this arm. Thus the boom is not used (in *luffing*) for pendulation control as normally the boom is a slow moving device whose effectiveness is further limited at high boom inclinations. In the lateral direction the whole crane is rotated around a vertical axis (*slewing*). The second actuation corresponding to the crane slewing motion, rotates the base of the crane about a vertical axis.

In terms of measurements the instantaneous position of the load has to be calculated from the measurements of ship motions, suspension line inclination and boom angle, and from the geometry of the crane structure as well as the length of the line. In terms of the ship motions heave , roll and pitch are assumed measurable for this work.

Consider the 3D model of a pendulum consisting of a point mass attached to a light inelastic cable with accelerating cable-top, moving in space. When this model is simplified by the assumption of small angle oscillations, the dynamics become decoupled so that the system may be described by the dynamics of two mutually perpendicular vertical planes. The 3D motion of the pendulum may then be described as the Cartesian combination of the motion in these two planes. Now since our objective is to contain pendulation, it is indeed desirable that the cable angle stays close to vertical, within a few degrees. This observation justifies considering the pendulation problem in one plane at a time.

In this paper linear control design methods, namely LQG and GPC, are considered for the linearised system. For a non-linear analysis including Liapunov and feedback linearization techniques see Fragopoulos and Zheng (1998). Simulation results involving these non-linear techniques and comparisons with the linear ones will appear elsewhere.

In terms of practical considerations, an important aspect is the saturation of the actuator both in angle and angular velocity. Another consideration is that the system needs to operate with different cable lengths. A scheduling scheme to cover various cable lengths has been suchesfully used in Fragopoulos and Zheng (1998). Finally for control purposes on the actuation side, the controller is require to provide angular velocity signals for the slewing and for the motion of the anti-pendulation arm.

2. MODELLING OF THE CRANE MOTION IN THE VERTICAL LUFFING PLANE

The crane consists of a control arm attached with one end on the boom-tip, with the other end attached on a cable. At the lower end of the cable hangs a point mass (load). The cable is modelled like a light string. The whole system is effectively a double pendulum. Moreover the attachment of the double pendulum (boom-tip) is accelerating in the vertical plane. Thus our objective is to use the actuation on the control arm in order to damp the pendulations in the load.

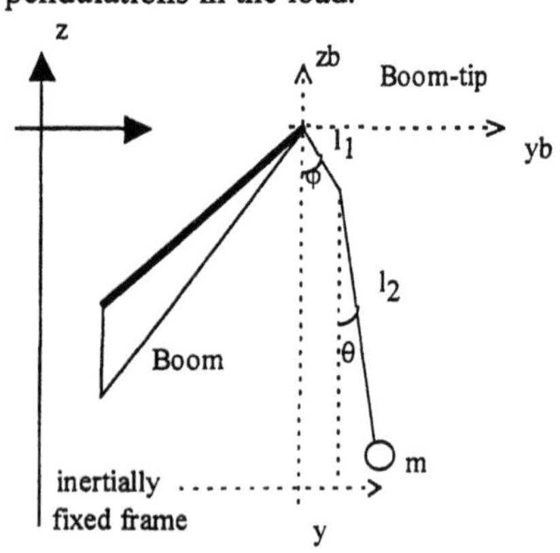

Fig.1. Pendulum in luffing plane

Newton's laws of motion for a point mass, applied to the mass of the pendulum, yield

$$\begin{aligned} m\ddot{y} &= -F_R \sin\theta \\ m\ddot{z} &= -mg + F_R \cos\theta \end{aligned} \tag{1}$$

where F_R is the tension of the cable. Also

$$\begin{aligned} y &= l_1 \sin\phi + l_2 \sin\theta + y_b \\ z &= -l_1 \cos\phi - l_2 \cos\theta + z_b \end{aligned} \tag{2}$$

where y_b, z_b are the co-ordinates of the boom-tip. Taking the first and second order time derivatives of the above equations and substituting in (1) we obtain

$$\begin{aligned} & l_2\ddot{\theta} + l_1 \cos(\phi - \theta)\ddot{\phi} - l_1 \sin(\phi - \theta)\dot{\phi}^2 + \\ & \cos\theta \ddot{y}_b + \sin\theta \ddot{z}_b + \sin\theta g + 2\theta \dot{l}_2 = 0 \end{aligned} \tag{3}$$

The above relation is a kinematic one and is independent of the mass, *m*. Introduce the control as

$$\dot{\phi} = u \tag{4}$$

Equations (3) and (4) are the governing equations of motion for the pendulum. The above non-linear equation is linearised around $\phi = 0, \theta = 0, \dot{y} = 0, \dot{z} = 0$ and $\ddot{y}_b = \ddot{z}_b = u = 0$ to obtain

$$\dot{x} = Ax + B_1\dot{y}_b + B_2 u \tag{5}$$

with $x = [\phi \quad \theta \quad \dot{y}]^T$ and

$$[A \mid B_1 \mid B_2] = \left[\begin{array}{ccc|c|c} 0 & 0 & 0 & 0 & 1 \\ 0 & 0 & l_2^{-1} & -l_1 l_2^{-1} & -l_2^{-1} \\ 0 & -g & 0 & 0 & 0 \end{array}\right] \tag{6}$$

The eigenvalues of this model are at $s = \{0, \pm j\omega_n\}$ where $\omega_n = \sqrt{g/l_2}$. The above state-space system may be written in transfer function form

$$\begin{bmatrix} \phi \\ \theta \\ \dot{y} \end{bmatrix}(s) = \begin{bmatrix} s^2 + g/l_2 \\ -l_1 l_2^{-1} s^2 \\ g l_2^{-1} l_1 s \end{bmatrix} \frac{u(s)}{s(s^2 + g/l_2)} + \begin{bmatrix} 0 \\ -l_2^{-1} s \\ g l_2^{-1} \end{bmatrix} \frac{\dot{y}_b(s)}{(s^2 + g/l_2)} \tag{7}$$

In the actual crane there are restrictions in ϕ and $\dot{\phi}$.

3. MODELLING OF THE CRANE IN LATERAL MOTION

It may be shown that the dynamics of the lateral motion when linearised around an equilibrium point are identical to those of the luffing plane (Fragopoulos and Zheng, 1998). Thus they will not be repeated here. The linearised model holds for small angles of slewing and cable inclinations.

4. STATE-SPACE LQG DESIGN

In this section an LQG (Linear Quadratic Gaussian) optimal control is designed. Two cases are considered. As a first attempt uncoloured (white) noise sources are used to model disturbances. Subsequently a dynamic disturbance model is used for modelling the effect of the waves.

4.1 LQG design with white disturbance model

As a result of the *separation principle* the LQG design is naturally decomposed into LQ optimal state-feedback control and Kalman filtering. We proceed next by designing the LQ control part.

LQ state-feedback design

Consider the state-space model (5). Assume the disturbance, $\dot{y}_b$, to be a white, zero mean, unit variance noise. Construct the error output vector, z,

$z := [l_1\phi \quad l_2\theta \quad 1/\omega_n \dot{y} \quad l_1/\omega_n \dot{\phi} \quad l_2/\omega_n \dot{\theta}]^T$.

Select the weighting matrix, *W*, and then minimise the least square error criterion

$$\min_{\bar{u}} \lim_{T\to\infty} E \int_0^T z(t)^T W^T W z(t) dt \tag{8}$$

From Bode plots it appears that control effort is used in the frequency range it is needed most:

around ω_n. Trying to be economical with control action is an important consideration given the stringent saturation limits on the actuator. This however may result in poor performance in terms of the amplitude of y (disturbance rejection). To improve the disturbance rejection of the system, one may decrease or remove the $\dot{\theta}$ weighting in W. (The $\dot{\theta}$ weighting has the effect of reducing the gains and effectively de-tuning the controller). Alternatively one may decrease the control weighting in W, both for ϕ and $\dot{\phi}$, and/or increase the $\dot{y}$ weighting, bearing in mind that weighting a derivative (e.g. $\dot{\phi}$) takes an effect at higher frequencies than simply weighing ϕ.
Take

$$W = \mathrm{diag}\{1.2 \quad 1 \quad 2.88 \quad 0.74 \quad 0\}.$$

The problem yields a solution

$$\overline{K} = [-1.621 \quad 1.820 \quad -3.696]$$

Reverting back to the original plant parameters by using:

$$K = \omega_n / l_1 \overline{K} \mathrm{diag}\{l_1 \quad l_2 \quad 1/\omega_n\} \tag{9}$$

which for $l_1 = 3.25$, $l_2 = 14$, $g = 9.81$ with $\omega_n = 0.837$ yields closed loop poles -0. 5046, -1.4687 ± 1.0279 i.

Kalman filter

Consider only white noise inputs and white disturbances. An advantage of this design is that it is amenable to normalisation since white noise preserves its statistical properties with a scaling in time.

Consider the state-space system

$$\dot{x} = Ax + B_1 w + B_2 u \tag{10}$$

with zero mean, white noise disturbance $w = [\dot{y}_b \quad u_{err}]$ of covariance matrix, Q, and B_1. Redefined so that the control signal u now has disturbance u_{err} add on to it, resulting in the system description below

$$[A \mid B_1 \mid B_2] = \left[\begin{array}{ccc|cc|c} 0 & 0 & 0 & 0 & 1 & 1 \\ 0 & 0 & l_2^{-1} & -l_1 l_2^{-1} & -l_2^{-1} & -l_2^{-1} \\ 0 & -g & 0 & 0 & 0 & 0 \end{array}\right]$$

The introduction of u_{err} is to excite the integrator associated with the ϕ state, and thus to render (A, B_1) stabilisable. The observations, y, are defined as

$$y = Cx + v \tag{11}$$

with $C = \begin{bmatrix} 1 & 0 & 0 \\ 0 & 1 & 0 \end{bmatrix}$ and v being the measurement noise: a zero mean, white noise vector of covariance R. Note that (A, C) is detectable. The measurement noise intensity is related to inertial measurement error and other angle measurements error. Thus

$$R = 10^{-4} \begin{bmatrix} 1 & 0 \\ 0 & 1 \end{bmatrix}$$

corresponds to standard deviation of 0.6 degrees. The covariance of the process disturbance, Q, is taken

$$Q = \begin{bmatrix} 2 & 0 \\ 0 & 0.01 \end{bmatrix}$$

with the standard deviation of $\dot{y}_b$ being 1.41 m/s, a realistic value and u_{err} having a standard deviation of 6 degrees. This last term may be thought to include a departure from the perfect actuator dynamics. Its presence is also necessary for the well posedness of the problem. The objective of the Kalman filter is

$$\min_{\hat{x}} E(x - \hat{x})^T (x - \hat{x})$$

where $\hat{x}$ is the state estimate.
The filter which is in *state-observer* form, is

$$\dot{\hat{x}} = (A - HC)\hat{x} + B_2 u_s + Hy \tag{12}$$

with H being the observer gain and u_s being the saturated control signal. Thus in our case $u_s = \dot{\phi}$, the actual angular velocity of the actuator as measured directly or inferred from ϕ. This implementation structure is in effect taking care of *integrator windup*. The Kalman filter is calculated using the Control Toolbox of Matlab (routine lqe.m) for $l_2 = 14, l_1 = 3.25, g = 9.81$ yielding an observer gain

$$H = \begin{bmatrix} 9.934 & -1.143 \\ -1.143 & 10.30 \\ 1.143 & -0.0656 \end{bmatrix}$$

and the eigenvalues of $(A - HC)$ are -0.072, -8.921, -11.24 which reveal one dominant pole of time constant of 14 s. This gives the time taken before the filter estimates 'track' the actual state of the system. In particular, this is relevant for $\dot{y}$, since ϕ and θ are measured with little error. Thus once the filter is started with no prior knowledge of the initial value of $\dot{y}$, it will take at least 14s before a useful estimate may be obtained. After this tuning-in period the error is reduced to a ripple.

4.2 *LQG with dynamic disturbance model*

A dynamic model W_d, with state-space description:

$$\begin{aligned} \dot{x}_d &= A_d x_d + B_d \xi \\ \dot{y}_b &= C_b x_d \end{aligned} \tag{13}$$

is used to generate the disturbance, $\dot{y}_b$, from a white noise process, ξ. This model the disturbance process as a frequency spectrum approximation resulting from the ship motion when excited by waves. The above disturbance model may then be combined with the crane model (10) to obtain an augmented model of the form:

$$\dot{x} = Ax + B_{1_1}C_d x_d + B_{1_2}u_{err} + B_2 u$$
$$\dot{x}_d = A_d x_d + B_d \xi \qquad (14)$$
$$y = Cx + v$$

or

$$\dot{x}_a = A_a x_a + B_{1a} w_a + B_{2a} u$$
$$y = C_a x_a + v$$

where

$$[A_a \mid B_{1a} \vdots B_{2a}] = \begin{bmatrix} A & B_{1_1}C_d & 0 & B_{1_2} & B_2 \\ 0 & A_d & B_d & 0 & 0 \end{bmatrix},$$

$C_a = [C \quad 0]$ and $x_a^T = [x^T \quad x_d^T]$, $w_a^T = [\xi \quad u_{err}]$, $B_1 = [B_{1_1} \quad B_{1_2}]$, $v = [v_\phi \quad v_\theta]^T$.

Now and due to the disturbance model involved the equations above cannot be normalised so the controller is parametrised by the cable length l_2.

The model of the disturbance process that affects the top of the boom of the crane is given by the model of the vessel under consideration in connection with the wave model. The wave model is parametrised by the wave height and the frequency of the wave peak energy (Fragopoulos and Zheng, 1998).

For example, a second order approximation of the power spectrum of the vessel named Belos plus the wave model for wave height Hs=2.4 m is given by:

$$W_d = \frac{0.17}{s^2 + 0.198s + 0.9801} \qquad (15)$$

The frequency response of the approximated wave/ship spectrum and the actual disturbance model are shown in the Figure 2.

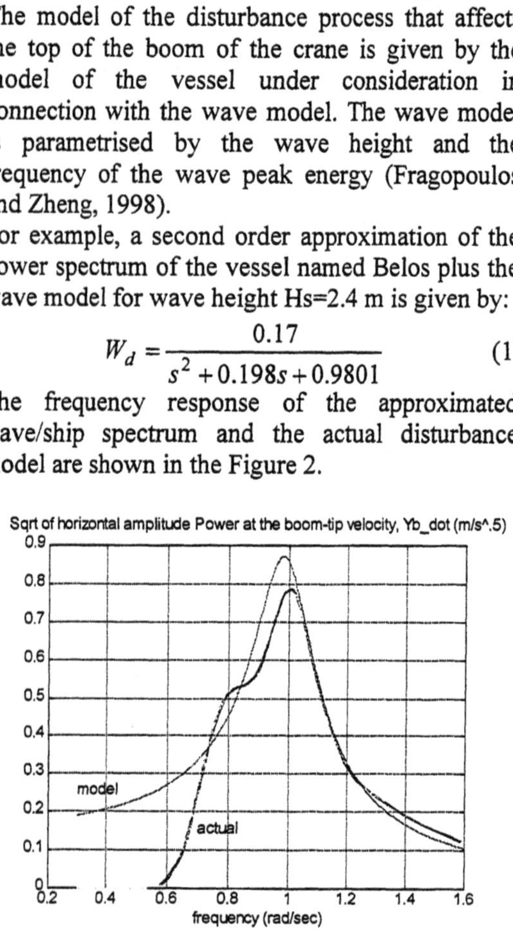

Fig. 2. Frequency response of disturbance: actual and W_d.

For the LQ (linear quadratic) control problem the same weights are used as before. For the Kalman filter the covariance matrices used are

$$Q = \begin{bmatrix} 100 & 0 \\ 0 & 10^{-2} \end{bmatrix}, \quad R = 10^{-4}\begin{bmatrix} 1 & 0 \\ 0 & 1 \end{bmatrix}.$$

The filter modes are: -10.5060, -1.1618 ± 2.4017i, -2.2581, -0.0580.

4.3 Higher-order approximation of the Disturbance model

Since the LQG control with the inclusion of disturbance (wave) model was found to be the more successful we are going to develop it further so that it may be used for the entire operating region. The operating region is defined in terms of cable length and of wave peak frequency which vary in the region of 1-30 m and 0.52-1.16 rad/s (periods of 12.1-5.4s) respectively. The way we decide to cover this operating range is by scheduling in terms of cable length and by designing a disturbance model which is in effect a band pass filter covering the whole of the operating frequency region. This kind of single band-pass disturbance model may result is some loss of performance compared to a disturbance model designed for the appropriate case. It has the advantage however that there is no need to continuously monitor the wave pattern and then schedule the controller accordingly as a function of the wave frequency. Thus we can save one scheduling parameter. The approximated disturbance model used is the following:

$$W_d = \frac{(0.9s+1)}{7.5} \frac{\prod_{i=1}^{3} s^2 + 2\zeta_i \omega_i s + \omega_i^2}{\prod_{j=1}^{4} s^2 + 2\zeta_j \omega_j s + \omega_j^2} \qquad (16)$$

where

Table 1

Poles		Zeros	
ω_n (rad/s)	ζ	ω_n (rad/s)	ζ
.52	.125	.59	.25
.72	.15	.88	.30
.94	.15	1.05	.30
1.16	.11		

The frequency response of the disturbance model is shown in Figure 3.

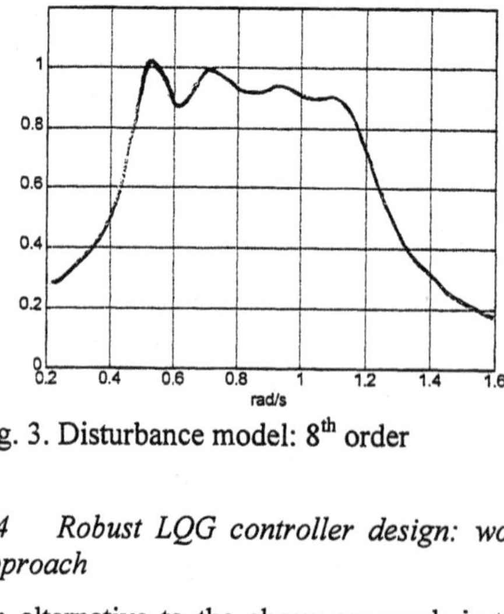

Fig. 3. Disturbance model: 8th order

4.4 Robust LQG controller design: worst case approach

An alternative to the above approach is to design the controller under the worst case disturbance model. The sea wave model (disturbance model) involves parameters, such as the sea wave height H_s and the frequency of maximum of sea wave w_p, that are unknown but bounded. Therefore we can use game theory to obtain the optimal controller under the worst case. The main feature of this robust controller design is that the order of disturbance model is reduced.

Consider again the state-space model (10) and assume that the disturbance $\dot{y}_b$ is white, zero mean, unit variance noise. Construct the error output vector z. The performance criterion is given by (8). Selecting the appropriate weighting coefficient results in deriving the optimal control $u^* = K^* X$ with K^* a constant row vector. Let S_w denote the second order approximated disturbance model (A_d, B_d, C_d) in (13) (similar to (15) for Hs=2.4). The parameters of the model are bounded by $(0.5m \le H_s \le 3.5m,\ 0.5r/s \le w_p \le 1.14r/s)$.

Let U be a control, S_w^* the worst sea wave model and u^* the optimal control. Take

$$S_w^*, u^* : \max_{S_w} \min_{U} J(S_w, U) \tag{17}$$

S_w^* is in a sense the worst case sea wave disturbance model. This model is in effect a band pass filter that is 'optimal' for the bounded operating frequency region. This disturbance model results in a non-optimal performance when compared to the disturbance model designed for the appropriate wave frequency. However, it has the advantage that there is no need to continuously monitor the sea wave pattern and to schedule the controller accordingly as a function of the sea wave frequency. The disturbance model order is reduced from an 8th order approximation derived in the previous section to an 2nd order. Thus the design of the controller is simplified.

After performing optimization (17) we derived the worst sea wave to be: ($H_s = 3m$ and $w_p = .85$ rad/s). A second order approximation of the power spectrum of the ship (Belos) plus worst sea wave model is

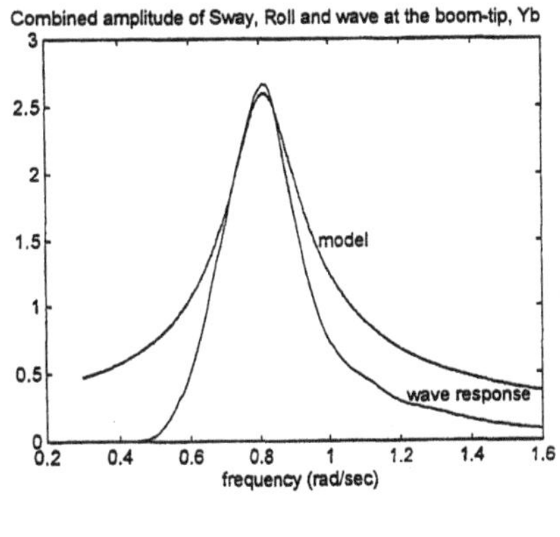

Fig. 4 Worst sea wave model

$$S_w^* = \frac{0.41(s+0.6)}{s^2 + 0.196s + 0.67} \tag{18}$$

The frequency response of the worst wave model is shown in Figure 4

5 GENERALISED PREDICTIVE CONTROL (CGPC) DESIGN

5.1 CGPC design with white disturbance model

Similarly to the LQG design, Continuous-time GPC design (Clark et al., 1987; Gawthrop et al., 1998) is decomposed into CGPC optimal state-feedback control and Kalman filter. The linearised crane pendulation model (5) is a non-square system with more outputs (three) than inputs (one). Since it is only possible to satisfy set-points on one output, the one input three outputs (SI3O) system is transformed into a single input single output (SISO) system by selecting

$$y = k_\phi x_1 + k_\theta x_2 + k_{\dot{y}} x_3 \tag{19}$$

where $[k_\phi\, k_\theta\, k_{\dot{y}}] = [l_1\, l_2\, \omega_n]$. The predictive controller is given by (Fragopoulos and Zheng, 1998; Clark et al., 1987; Gawthrop et al., 1998)

$$u = k_\omega \omega - k_x \hat{x} - k_y y \tag{20}$$

The CGPC gain of (5) is given by

$k_x = [-0.2806\ \ 12.9395\ \ -0.5092]$ and $k_y = 0$

The pendulation study of the crane is a regulation problem instead of a tracking problem, so $\omega = 0$ and k_ω is not to be used.

5.2 CGPC design with dynamic disturbance model

Working as with the LQG, combine the crane linearised model (5) with the dynamic disturbance model to obtain the augmented model similar to (14). The CGPC gains of the system (5) with the approximated second order disturbance model, eighth order model and worst case model are

$K_x = [-0.0262\ \ 73.1305\ \ -0.9511\ \ -0.0190\ \ 0.0684]$

$K_x = [-.0356\ \ 81.2\ \ -1.045\ \ -.063\ \ -.0806$
$-.0624\ \ .0441\ \ -.0395\ \ -.01\ \ -.0062\ \ -.00182]$

and

$K_x = [-0.0603\ \ 91.1383\ \ -1.2076\ \ 0.0901\ \ 0.0642]$

respectively, and $k_y = 0$.

Tuning:

The following parameters have been used in the CPGPC control design, (Fragopoulos and Zheng, 1998).

Prediction range: $N_\phi = 10$. The control results are almost unchanged when $N_\phi = 10$.

Control horizon: The control horizon was set to $N_u = 0$ which means that the role of control horizon is very small.

Time horizon: $t_1 = 0$, $t_2 = 0.8$ with white disturbance and $t_1 = [000\ ]^T$, $t_2 = [.8 .8 .8\ ]^T$ with disturbance models. The crane pendulum has a fast response. The time horizon can not be selected too big.

The parameter matrix P is selected as: $p = [7\quad 1]$ with white disturbance and

$$p = \begin{bmatrix} 7 & 0 & 0 & 1 & 0 & 0 \\ 0 & 7 & 0 & 0 & 1 & 0 \\ 0 & 0 & 7 & 0 & 0 & 1 \end{bmatrix}$$

with dynamic disturbance models.

6. SIMULATION RESULTS: COMPARISONS

Having obtained several controllers from LQG and GPC, the non-linear closed loop system is simulated using realistic ship and wave models. The Belos ship data are used from the Marintek report together with the JONSWAP wave model (Atkins R&D, 1977) for significant wave height Hs=2.4m. The response of the system is simulated for motion in the vertical plane. Moreover the boom tip is assumed to be moving horizontally. The following tables provide the standard deviation (sdv) and maximum values of the various system variables. Note that the angle ϕ is limited ±0.44 rad while $\dot{\phi}$ is limited to ±0.35 rad/s. The first 30s of data are discarded. The simulation results for the LQG and the GPC designs are given in Table 2 where +wh, +2m, +8m or +wo correspond to designs with white noise disturbance, second order disturbance model, eighth order disturbance model and worst case disturbance model, respectively.

The objective of the control design is to keep the pendulation keep the sdv and maximum (max) of y as small as possible. From Table 2, we can see:

1) The control results of LQG and CGPC with disturbance models are better than the ones with white noise disturbance.
2) The control results of CGPC are overall slightly better than that of LQG. We arrive in this conclusion when, in particular, the max value of y and the values of the angle θ are taking into consideration.
3) The control results of LQG and CGPC +8M are better than that of LQG and CGPC +2M and +wo. In practice, the use of the worst case disturbance model is preferable since it leads to simpler controllers.

7. CONCLUSIONS

The problem of pendulation control for a crane for off shores operations has been studied using linear techniques such as LQG and CGPC. The use of approximated dynamic models for the wave and ship disturbances improved considerably the results. A simple controller design for both approaches is proposed using for the worst wave consideration.

REFERENCES

Fragopoulos, D. and Y. Zheng (1998) 'Pendulation control of an Offshore crane' Technical report, Dept. of Mechanical Eng., University of Stathclyde.

Fliess M., J. Levine, P. Rouchon (1991) 'A Simplified approach of crane control via a generalized state-space model' Proc. Conf. Decision Control, Brighton.

Atkins Research & Development (1977). *Dynamics of Marine Structures*, CIRIA Underwater Engineering Group, Report UR8.

Clark, D.W., Mohtadi, C., and P.S. Tuffs (1987) 'Generalised predictive control Part I:the basic algorithm and Part II; extensions and Interpretations, *Automatica* 23(2), pp. 137-160.

Gawthrop, P.J., Demircioglou, H. and I.I. Siller-Alcala, (1998) 'Multivariable Continuous-time Generalised Predictive Control: A state space approach to Linear and Non-linear systems CSC-98001 report.

Table 2: Simulation results with l1=3.25m, l2=14m, Run time: 1800s, Hs=2.4m, ω_p=1.02rad/s.

Design	y_b (m)		Y(m)		$\dot{\phi}$(rad/s)		T (kNm)		P (kW)		ϕ (rad)		θ(rad)	
	Sdv	max	sdv	max	sdv	max	Sdv	max	sdv	max	sdv	max	sdv	max
LQG+wh	.42	1.55	.120	.580	.120	.40	19.7	67.24	2.56	21.86	.124	.440	.011	.040
LQG+2M	.42	1.55	.116	.480	.125	.40	19.8	69.89	2.67	23.25	.125	.440	.010	.038
LQG+8M	.42	1.55	.063	.577	.137	.40	22.1	72.70	3.08	25.89	.138	.442	.008	.047
LQG+wo	.42	1.55	.062	.591	.138	.40	22.4	72.25	3.32	26.41	.139	.440	.006	.044
CGPC+wh	.42	1.55	.136	.557	.126	.40	21.36	69.31	2.71	19.83	.13	.43	.012	.051
CGPC+2M	.42	1.55	.079	.351	.132	.40	21.71	68.98	3.15	25.05	.14	.43	.004	.025
CGPC+8M	.42	1.55	.049	.317	.134	.40	22.51	67.18	3.12	21.11	.14	.43	.002	.009
CGPC+wo	.42	1.55	.080	.384	.142	.40	23.69	68.80	3.47	26.12	.15	.43	.002	.012

14th Triennial World Congress, Beijing, P.R. China

C-2a-14-2

CONTROL OF MAGNETIC SUSPENSION SYSTEMS BY MAINTAINING CONTINUOUS EQUILIBRIUM POINTS

Authors: **B.V. Jayawant** (Corresponding Author)*, **T.C. Yang, S. Töllner**

** School of Engineering, University of Sussex*
Falmer, Brighton BN1 9QT, England
Telephone: +44-1273-606755
Fax: +44-1273-678452
email: b.v.jayawant@sussex.ac.uk

Abstract: The paper describes a technique for enhancing the performance of nonlinear control systems in general but magnetic suspension systems in particular. The main consideration in determining controller structure is based on a heuristic idea of "maintaining continuous equilibrium points". For the case investigated here in addition to the PID Control an adaptive reference current is shown to improve markedly response to external force disturbances. The controller designed is simple, robust and tunable.

Keywords: Magnetic suspension, controllers, electromagnetic devices.

1. INTRODUCTION

Industrial applications of magnetic suspension systems with controlled dc electromagnets are now wide ranging in the field of magnetic bearings, however, they are often in extremely demanding and at times in hazardous environments. Reliability is, thus, at a premium. Wide-spread interest now exists in exploring modern control techniques such as variable structure control (Kanemitsu, et al, 1994) or sliding mode control (Schweitzer, et al, 1994) wherein implementing digital controllers may enhance the reliability and operating range.

The theory of another technique "Maintaining Continuous Equilibrium Points" (MCEP), a control technique which shows promise in nonlinear control systems, has been given by one of the present authors (Yang, 1996). It is shown in this paper that the theory is applicable to the design of controllers of magnetic suspension systems using controlled dc electromagnets. The objectives of this paper are to give a brief review of the theory of MCEP and demonstrate the applicability of the theory by simulation and experiments on a purpose built rig which can be subjected to force disturbances over a range of frequencies.

2. CONTROLLED DC ELECTROMAGNET SYSTEMS AND DESCRIPTION OF TEST RIG

Ferromagnetic objects in the vicinity of permanent magnets or electromagnets with constant current display force-distance characteristics as shown in **Fig. 1(a)** which obviously demonstrate the inherent instability. The requisite force-distance characteristics for converting the configuration into a stable one are also shown in **Fig. 1(a)**. This is achieved by a feedback of position or gap between the object and the magnet, as shown in block schematic **Fig. 1(b)**. The purpose of this paper is to show the advantages of an additional feedback of current as well and set up a framework for the design of more versatile and improved digital controllers (Jayawant, et al, 1996).

Suspension systems for magnetic bearings must exert control over five degrees of freedom. However, the authors have evolved an effective way of developing controllers by conducting tests on a rig shown in **Fig. 2** which restricts the movement of the suspended object along one (vertical) direction only.

The rig consists of a rectangular steel beam hinged at one end and an electromagnet at the other which exerts electromagnetic forces on the top face of the beam. Disturbances forces can be imparted to the free end of the "cantilever" via a spring attached to a motor driven eccentric.

3. MCEP AND ELECTROMAGNETIC SUSPENSION

Block schematic of **Fig. 1(b)** implies that the gap between the magnet and the object to be suspended is controlled by controlling the voltage applied to the coil of the electromagnet i.e. to maintain $x = x^*$ where x^* is the demanded position of the object. If a random force causes it to move away from its demanded position such that the dynamics of the system is then significantly different, a design based on linearisation only is no longer valid. However, if the electromagnetic force $F_m = F_d + mg$, where F_d is the disturbance force, through some effective control for the whole range of x, the control would be much more robust.

Now the electromagnetic force F_m is a function of both the coil current and x i.e.

$$F_m = f(i, x) \qquad (1)$$

It is assumed in the first instance that the only load on the cantilever is that due to its mass i.e. $F_m = mg$. This is later modified to take into account the external load disturbance F_d. Solving eqn. (1) with this assumption gives

$$i = f_{ix}(x) \qquad (2)$$

Eqn. (2) can be interpreted as saying that, for each $x \in [x_{min}\ x_{max}]$, there is a value of $i = f_{ix}(x)$ and if the current can be adjusted to that value, the upward force F_m will be equal to the downward force mg i.e. each x becomes an (unstable) equilibrium function and if in the process of moving from one position of balance $\hat{x}$ to x^* the current is made to follow the relationship of eqn. (2), the whole process can be considered to consist of a series of small changes around equilibrium points. It would then be possible to design an additional simple controller to regulate x for the whole process. This modifies the controller structure to that of **Fig. 3**.

In this figure, i_{ref} is not a constant. It can be set as $i_{ref}(t) = f_{ix}(t)$ and design a current controller to force i to follow the equilibrium function. Setting i_{ref} in this manner, however, regulates i based only on x at current time. We have assumed that control is effected by controlling voltage which leads the current. It is, therefore, essential to incorporate a predictive element to maintain continuous equilibrium points. Hence

$$i_{ref} = f_{ix}(x, t, \hat{k}\dot{x}) \qquad (3)$$

where $\hat{k}$ is a predictive factor and this signal may be termed an adaptive reference input.

4. MATHEMATICAL MODEL AND CONTROLLER DESIGN

4.1 *Modelling*

The mathematical model is based on the geometry shown in **Fig. 4** and coil voltage treated as the independent input to the plant.

The electromagnetic equations are

$$\ddot{x} = \frac{mg - F_m}{m} \qquad (4.a)$$

$$\psi = n\phi = nBA = \frac{n^2 \mu_o A i}{2x + \ell_{Fe}/\mu_r}$$

$$= \frac{k_1 i}{2x + k_2} \qquad (4.b)$$

$$F_m - \frac{\psi^2}{k_1} \qquad (4.c.)$$

and $$V = iR + L\frac{di}{dt} + i\frac{dL}{dt} = iR + \frac{d\psi}{dt} \qquad (4.d)$$

where $k_1 = n^2 \mu_o A$ and $k_2 = l_{Fe}/\mu_r$

Through a process of differentiation and appropriate substitutions one obtains

$$\ddot{x} = g - \frac{k_1}{m(2x + k_2)^2} i^2 \qquad (5.a)$$

$$\dot{i} = \frac{2x + k_2}{k_1}(V - Ri) + \frac{2}{2x + k_2}\dot{x} \qquad (5.b)$$

and $$\dot{x} = \dot{x} \qquad (5.c)$$

Expressed as a function of state variable and control variable

$$\dot{X} = f(X, U) \qquad (6)$$

with the output

$$y = x_1 = x \qquad (7)$$

The system may be linearised around a quiescent point $X = Q$ implying that at that point the system is considered to be in equilibrium and hence

$$\dot{X}_Q = 0 \qquad \dot{I}_Q = 0$$

and $$\Delta\dot{X} = \left.\frac{\partial A}{\partial X}\right|_Q \Delta X + \left.\frac{\partial B}{\partial U}\right|_Q \Delta U$$

Eqns. (5) are a state space representation of the system which could also be written down in matrix notation

$$\left.\begin{aligned} \dot{X} &= AX + BU \\ X &= [x\dot{x}I]^T \\ U &= [V] \end{aligned}\right\} \quad (8)$$

For any equilibrium point $x \in [x_{min}\ x_{max}]$ with $\dot{X} = 0$, the equilibrium function is

$$\left.\begin{aligned} i_e &= f_{ix}(x) = \frac{\left(2x + \frac{\ell_{Fe}}{\mu_r}\right)}{n}\sqrt{\frac{mg}{\mu_o A}} \\ \text{and} \quad V_e &= i_e R \end{aligned}\right\} \quad (9)$$

Eqns. (9) can be derived directly from eqns. (5) with $\dot{\psi} = 0$ and $\ddot{x} = 0$.

Eqns. (9) conform to the notion that the current I_e required to maintain the equilibrium point is a linear function of the gap x. Currents for three gaps, x_{max}, maximum when the cantilever is against the bottom stop, midpoint or when the gap is $x/2$ and when $x = 0$ are those of interest and can be determined experimentally.

4.2 *Controller Design*

The controller configuration incorporating the MCEP control concept is shown in **Fig. 5**. In this configuration a PID controller is chosen for the position control part.

With the assumption that the MCEP controller can maintain the equilibrium at (unstable) points for the whole range of values of x, a linearised small signal plant model around $x = x_{mid}$ is taken as the starting point for the controller design. Applying this to eqns. (5), (6) and (7)

$$\Delta\dot{X} = A\Delta X + B\Delta U \quad (10)$$

and the open loop system becomes

$$\begin{bmatrix} \Delta\dot{x} \\ \Delta\ddot{x} \\ \Delta\dot{I} \end{bmatrix} = \begin{bmatrix} 0 & 0 & 0 \\ \frac{4k_1 I^2}{mk_3^3} & 0 & -\frac{2k_1}{mk_3^2 I} \\ 0 & \frac{2}{k_3} & -R\frac{k_3}{k_1} \end{bmatrix} \begin{bmatrix} \Delta x \\ \Delta\dot{x} \\ \Delta I \end{bmatrix} + \begin{bmatrix} 0 \\ 0 \\ \frac{k_3}{k_1} \end{bmatrix} \Delta V \quad (11)$$

where $k_3 = (2x_Q + k_2)$.

The idea of state-variable feedback is to employ a feedback signal in the form of a constant vector F multiplied by the state vector. Inserting the state feedback i.e. making the input vector $U = V + FX$, the closed loop system equation becomes

$$\dot{X} = [A_o + BF]X + BV \quad (12)$$

The signal to the plant or the suspension system consists of three PID signals V_p, V_d, V_i plus the MCEP signal V_c. In the closed loop the resulting input is

$$\begin{aligned} V = &\left(V_x(x) - V_{ref}\right)k_p + \dot{V}_x(x)k_d\dot{x} - \\ &\left(i - i_{ref}\right)k_c + k_i\int\left(V_x(x) - V_{ref}\right)dt \end{aligned} \quad (13)$$

$V_x(x)$ is the sensor signal voltage at position x and V_{ref} denotes the constant reference voltage for the demanded position. The last term in the eqn. (13) represents the integrator part of the system and is not associated with the state feedback. In the linearised equation V_{ref} does not appear as it is constant, whereas i_{ref} changes adaptively with position and is, therefore, considered separately as Δi_{ref} in the linearisation.

Then

$$\begin{aligned} \Delta V = &k_p \frac{\partial V_x}{\partial_x}\bigg|_Q \Delta x + k_d \frac{\partial V_x}{\partial x}\bigg|_Q \Delta\dot{x} \\ &- k_c\left(\Delta I - \Delta i_{ref}\right) + \Delta V_i \end{aligned} \quad (14)$$

The slope of the sensor voltage at the quiescent point is assumed to be k_m. Thus we get

$$\Delta V = k_p k_m \Delta x + k_d k_m \Delta\dot{x} - k_c\left(\Delta I - \Delta i_{ref}\right) + V_i \quad (15)$$

Δi_{ref} is neither a state variable nor an external input and so it can be expressed in terms of the existing set of state variables.

i_{ref} can be derived as

$$i_{ref}(x) = \left\{2\left(x + k_f\dot{x}\right) + k_2\right\}\sqrt{\frac{mg}{k_1}} \quad (16)$$

Where k_f is a predictive factor.

Linearising

$$\begin{aligned} \Delta i_{ref}(\Delta x, \Delta\dot{x}) &= 2\sqrt{\frac{mg}{k_1}}\Delta x + 2k_f\sqrt{\frac{mg}{k_1}}\Delta\dot{x} \\ &= v_p \Delta x + v_d \Delta\dot{x} \end{aligned} \quad (17)$$

Now that Δi_{ref} has been expressed in terms of state feedback variables, the whole feedback voltage eqn. (15) can be rewritten

$$\begin{aligned} \Delta V(\Delta x, \Delta\dot{x}, \Delta i) = &\left(k_m k_p + k_c V_p\right)\Delta x \\ &+ \left(k_m k_d + k_c V_d\right)\Delta\dot{x} - k_c \Delta i \end{aligned} \quad (18)$$

This supplies the state feedback and the state-feedback vector is

$$F = \left[k_p k_m + k_c V_p + k_d k_m + k_d V_d - k_c\right] \quad (19)$$

As per the requirement of eqn. (12), feedback is now included into the system matrix to yield a description of the closed loop system. Hence

$$\begin{bmatrix} \Delta\dot{x} \\ \Delta\ddot{x} \\ \Delta\ddot{I} \end{bmatrix} = \begin{bmatrix} 0 & 1 & 0 \\ \frac{4k_1 I^2}{mk_3^3} I^2 & 0 & -\frac{2k_1}{mk_3^2 I} \\ \left(k_p k_m + k_c v_p\right)\frac{k_3}{k_1} & \left(k_d k_m + k_c v_d\right)\frac{k_3}{k_1} + \frac{2}{k_3} & -\left(k_c + R\right)\frac{k_3}{k_1} \end{bmatrix} \times \begin{bmatrix} \Delta x \\ \Delta\dot{x} \\ \Delta I \end{bmatrix} + \begin{bmatrix} 0 \\ 0 \\ \frac{k_3}{k_1} \end{bmatrix} \Delta V \quad (20)$$

v_p, v_d and v_c have now been included into the system matrix. Hence V_i is the only component of V that remains as an input to the system. The block diagram takes the form of **Fig. 5**.

Integrator: The block schematic of **Fig. 5** does not have an integrator. Incorporation of an integrator which forms an outer loop is shown in **Fig. 6**.

A suitable gain k_i for the integrator is estimated using the root locus method, the starting points being the poles of the system matrix in eqn. (20). As the state-variable feedback allows the poles to be set arbitrarily, for each set of closed loop system poles there is a set of control parameters k_p, k_d and k_c. Comparing the coefficients in the characteristic polynomial and the system matrix and noting that only the elements of the last row of the A matrix depend on control parameters gives the following result.

$$\left.\begin{aligned} a_{31} &= \left(\underline{k_p} k_m + \underline{k_c} v_p\right)\frac{k_3}{k_1} = k_a \frac{k_3}{k_1} \\ a_{32} &= \left(\underline{k_d} k_m + \underline{k_c} v_d\right)\frac{k_3}{k_1} + \frac{2}{k_3} = k_\beta \frac{k_3}{k_1} + \frac{2}{k_3} \\ a_{33} &= -\left(\underline{k_c} + R\right)\frac{k_3}{k_1} = k_\gamma \frac{k_3}{k_1} \end{aligned}\right\} \quad (21)$$

The parameters which are adjusted are underlined in eqn. (21). It is to be noted that a_{33} depends on the k_c only. Once this has been fixed only k_p is left to be adjusted in a_{31} and only k_d in a_{32}. This means that only one set of control parameters exists corresponding to each set of poles. The relevant equations can and have been compiled into a computer programme which calculates k_p, k_d and k_i for a given set of poles.

The state variables of the system so far cannot express V_i in the system of **Fig. 6**. An additional state variable V_i is now introduced. Block diagram in this figure implies $V_i = k_i \int e\,dt$. Hence

$$\dot{V}_i = k_i e = k_i \left(k_m x - V_{ref}\right) \quad (22.a)$$

Linearising for constant V_{ref} gives

$$\Delta\dot{V}_i = k_i k_m \Delta x \quad (22.b)$$

As the derivative of the new state variable depends solely on Δx it adds only one element to the new 4×4 system matrix.

$$\begin{matrix} \Delta\dot{x} \\ \Delta\ddot{x} \\ \Delta\dot{I} \\ \Delta\dot{V} \end{matrix} = \begin{bmatrix} 0 & 1 & 0 & 0 \\ \frac{4K_1 i^2}{MK_3^3} & 0 & -\frac{2K_1}{MK_3^2 i} & 0 \\ k_a \frac{k_3}{k_1} & k_\beta \frac{k_3}{k_1} + \frac{2}{k_3} & -k_\gamma \frac{k_3}{k_1} & \frac{k}{k_1} \\ k_i k_m & 0 & 0 & 0 \end{bmatrix} \begin{bmatrix} \Delta x \\ \Delta\dot{x} \\ \Delta I \\ \Delta V_i \end{bmatrix} \quad (23)$$

A disturbance force F_d can now be incorporated into the model as an external input. The force means an additional acceleration F_d/m to the suspended mass. This can be added in eqn. ... which affects the second row of the matrix yielding acceleration. Thus

$$\dot{X} = AX + \begin{bmatrix} 0 \\ F_d/m \\ 0 \\ 0 \end{bmatrix} X \quad (24)$$

5. SIMULATION AND EXPERIMENTAL VERIFICATION

5.1 *Simulation Details*

The block schematic of the MATLAB/SIMULINK simulation is shown in **Fig. 7(a)**. This represents the nonlinear model of the plant with provision for external force disturbance which can be applied to the cantilever as shown in **Fig. 2**.

The two blocks, one for calculating ψ^2 and the other for computing the current i from the reluctances, are the nonlinear elements in this block diagram. The other two elements which need explanation are the time delay and representation of the actuator.

The time delay t_d is, on the one hand, to compensate for delays in the op-amps, multipliers and sensors paths. The other reason is the behaviour of the simulation tool itself which facilitates a numerical

The actuator is included in the form of a virtual resistance R_V and the sum of other resistances R_o including that for the power transistors.

The block diagram of **Fig. 7(a)** is condensed as "suspension plant" and inserted into the full control circuit **Fig. 7(b)**. The MCEP controller contains the block "reference generator" that computes the adaptive reference from x and $\dot{x}$. The relationship for the reference current is implemented directly in the form of an equation in this block.

The characteristic line of the sensor output is modelled as a look-up table that contains measured values. Positions outside the permissible range are precluded by the addition of extra elements which "generate" forces that counter transgression beyond ± x_{max}. The elements constitute, so to speak, end stops.

5.2 *Experimental verification and comparisons with PID controller*

The tests carried out on the cantilever rig were for the purpose of comparing the performance of the MCEP plus PID controller with that of the PID control of the electromagnet on the one hand and with the simulation of Sec. 5.1 on the other.

Although various tests were performed results of only two are reported here. The second one, in particular in which the cantilever is subjected to force disturbance via a spring and eccentric arrangement (**Fig. 2**) at three different frequencies, is considered the most onerous and, therefore, significant.

The first test is to determine the speed of response of the cantilever to a demand from the unenergised state i.e. pick-up response. **Fig. 8(a)** shows the correlation for this between experiment and simulation with MCEP. **Fig. 8(b)** shows the correlation between experiment and simulation for PID. These two sets of figures show the considerable improvement in the response achieved by MCEP. The relevant pole placements and various parameter settings are also given on the figures.

The second set of results, as mentioned, is the frequency response to load disturbance applied through the spring-eccentric drive at 5 Hz, 10 Hz and 15 Hz for PID and PID + MCEP. Again it can be clearly seen that (**Fig. 9**) while the PID controller on its own is satisfactory, addition of MCEP leads to a significant improvement, in terms of maximum departure from demanded gap at all frequencies up to 15 Hz. The PID controller, in fact exhibits a resonance at 10 Hz. The maximum variation in position for the 5 Hz disturbance for MCEP is 40 μm peak to peak with the imposed load of 7.92 N. For a small suspension, where the operating flux density in the gap and the generated forces are limited, this is considered highly satisfactory.

6. CONCLUSIONS

It has been demonstrated that adding the approach of maintaining continuous equilibrium points improves the performance of a highly nonlinear system such as magnetic suspension using controlled dc electromagnets. The adaption of digital controllers for straightforward PID systems has already been described elsewhere (Jayawant, et al (1996). Inclusion of MCEP, which is essentially a digital system, makes it now possible to meet the criteria of reliability and flexibility to load disturbances set out in the introduction to this paper. It is also claimed that the MCEP has other applications besides the very practical one described here.

7. REFERENCES

Jayawant, B.V., Dawson, B.E., Whorlow, R.J. and Peyton Jones, J.C. (1996), Digitally Controlled Transducerless Magnetic Suspension System, *Proceedings of the IEE SMT, **143**,* 1, pp. 47-51.

Kanemitsu, Y., Ohsawa, M. and Marui, E. (Aug. 1994), Comparison of Control Laws for Magnetic Levitation, *Proceedings of Fourth International Symposium on Magnetic Bearings,* Zurich, pp. 13-18.

Schweitzer, G., Bleuler, H. and Traxler, A. (1994), Active Magnetic Bearings, *Hochschulverlag AG an der ETH,* Zurich. ISBN 3 7281 21320.

Yang, T.C. (1996), A Case Study of Magnetic Suspension Control, *Proceedings of Int. Conf. on Control,* Exeter UK, pp. 603-607.

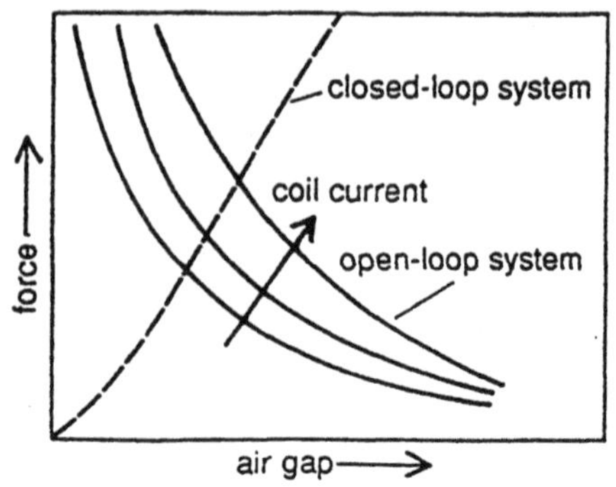

Fig 1(a). Force-distance characteristics for open loop and closed loop.

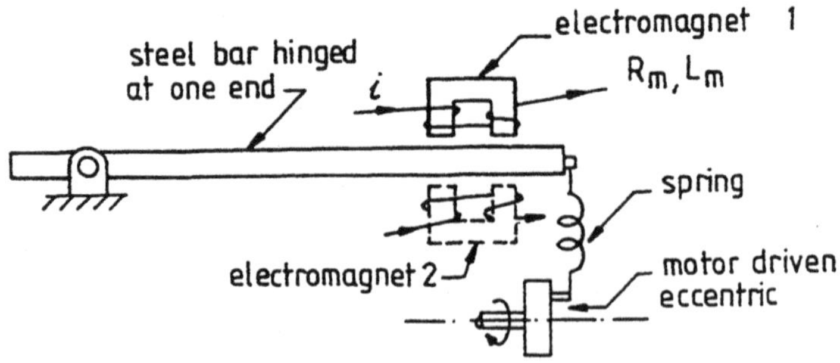

Fig. 2. Cantilever test rig with capability of applying force disturbances.

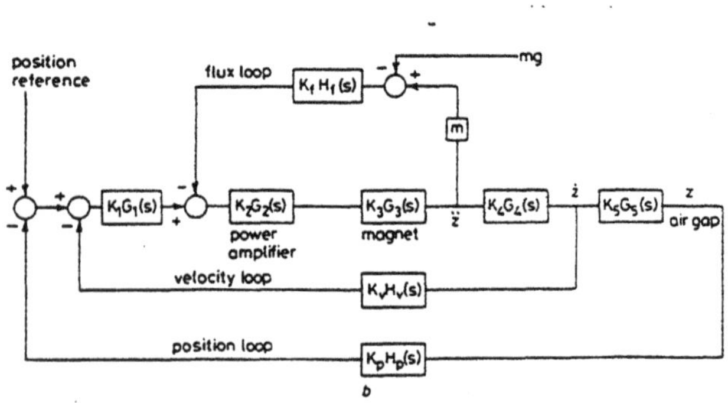

Fig. 1(b). Closed loop control block schematic for suspension.

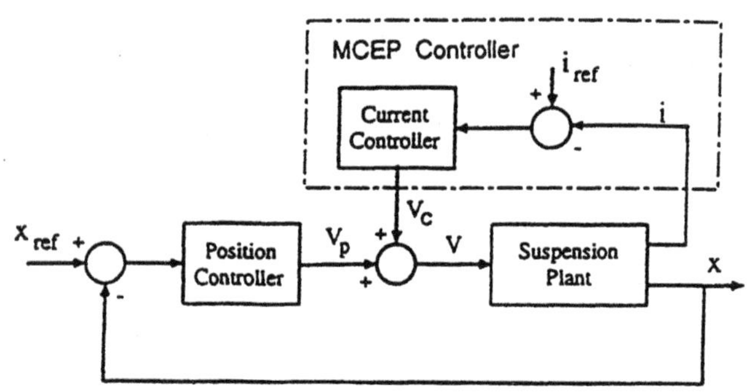

Fig. 3. Modified MCEP controller.

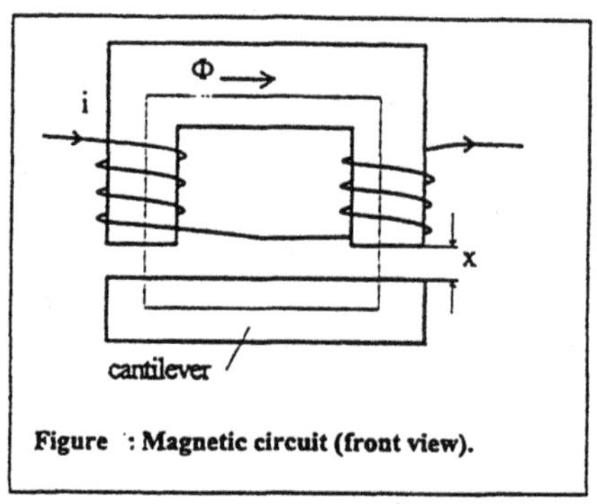

Fig. 4. Magnetic circuit configuration.

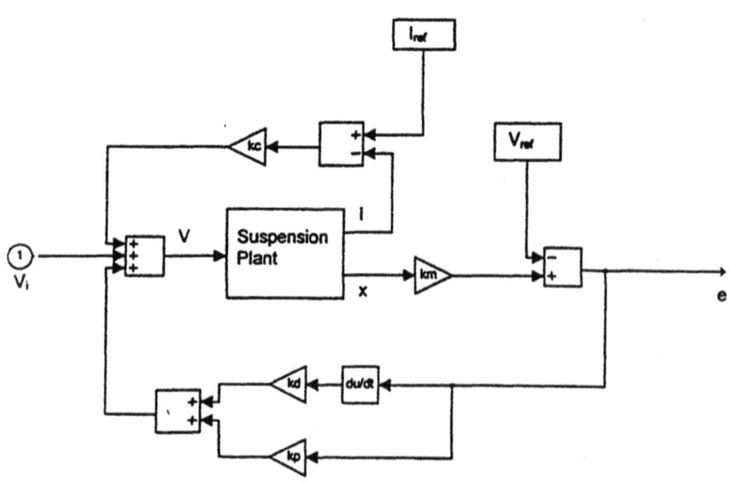

Fig. 5. Closed loop controller without integrator.

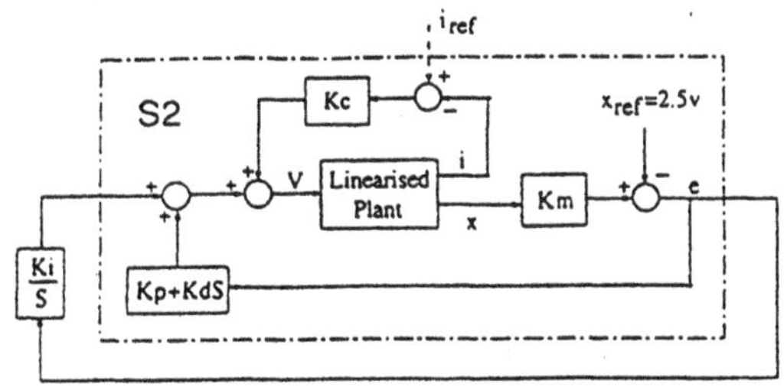

Fig. 6. Closed loop controller with outer loop integrator.

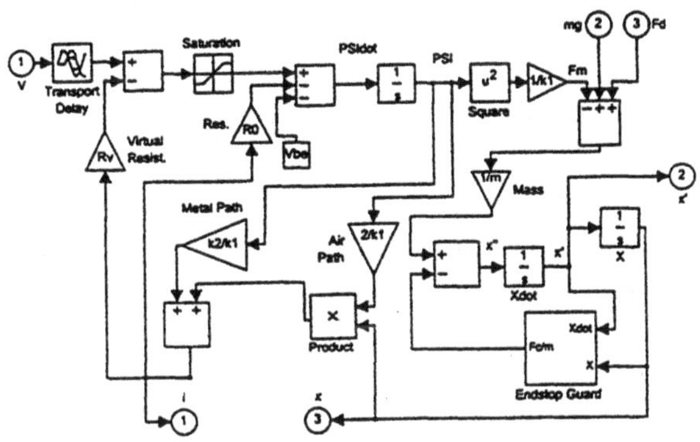

Fig. 7(a). Suspension plant block diagram for simulation.

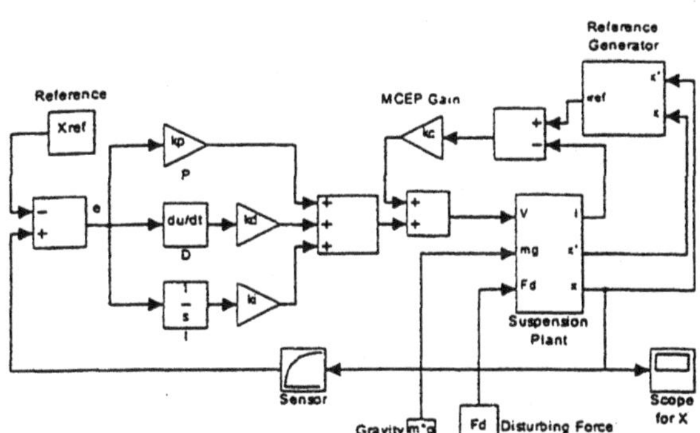

Fig. 7(b). Full control system block diagram (simulation).

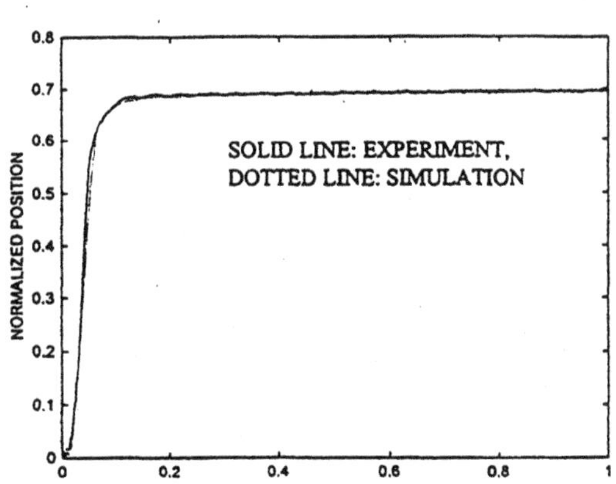

Fig. 8(a). Simulation and experimental pick-up responses for MCEP.

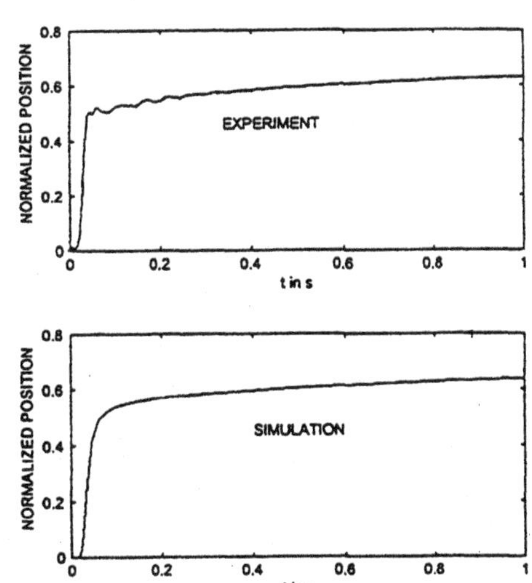

Fig. 8(b). Simulation and experimental responses for PID control.

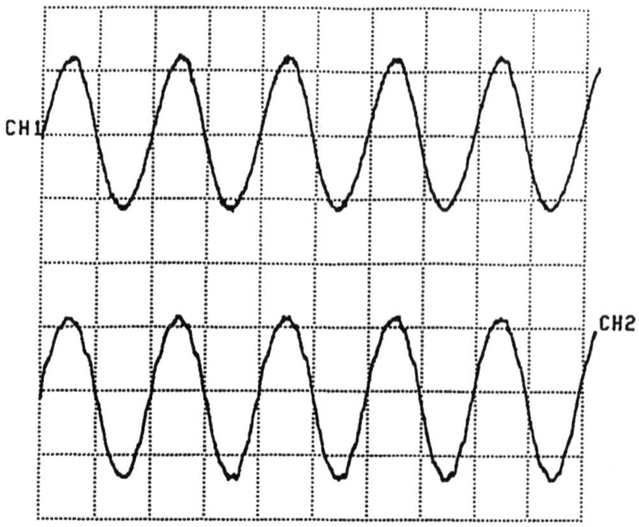

i) 5 Hz; 0.2 V/div; Ch 1 PID, Ch 2 MCEP

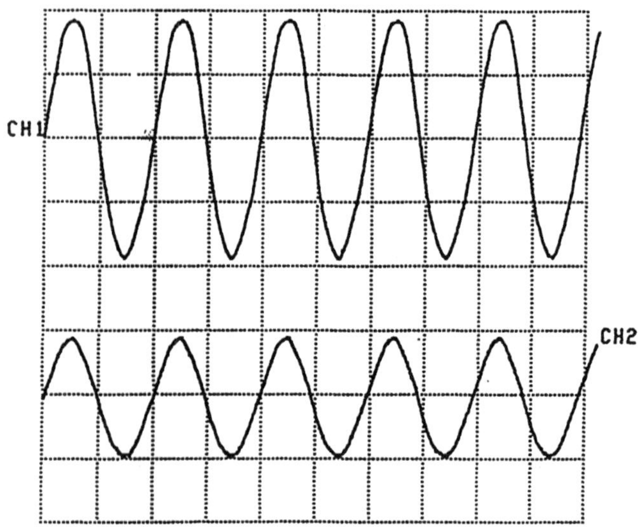

ii) 10 Hz; 0.5 V/div; Ch 1 PID, Ch 2 MCEP

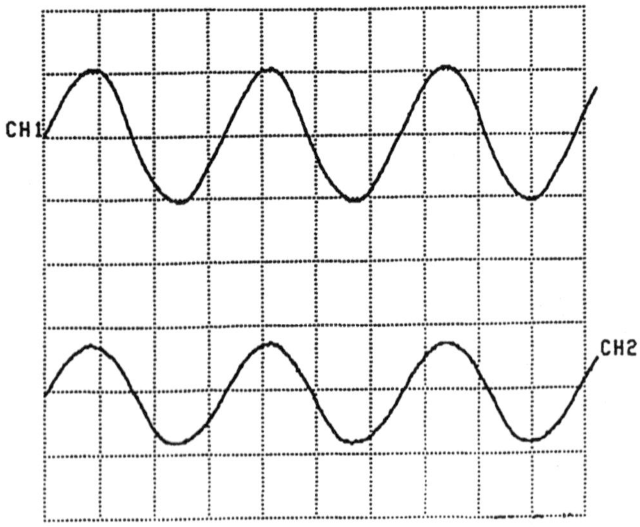

iii) 15 Hz; 0.5 V/div; Ch 1 PID, Ch 2 MCEP

Fig. 9. Comparison of PID and MCEP controllers subjected to force disturbance on cantilever (one dimension).
Vertical scale of displacement
1V = 0.1 mm
suspension magnets pole face area
= 50 mm × 8 mm per pole

operating gap	= 0.6 mm
no. of turns on coil	= 120
coil resistance	= 0.95 Ω
coil inductance	= 0.56 mH
gap flux density	= 0.32 T

DESIGN OF A WIDE ENVELOPE CONTROLLER FOR A STOVL GAS TURBINE ENGINE

A. J. Chipperfield[†], C. M. Fonseca[‡], H. C. Betteridge[*] and P. J. Fleming[†]

[†]*Department of Automatic Control and Systems Engineering*
University of Sheffield, Mappin Street, Sheffield S1 3JD England, UK
[‡]*Unidade de Cieneias Exactos e Humanas, Universidade do Algarve*
Campus de Gambelas, 8000 Faro, Portugal
[*]*Advanced Controls, Rolls-Royce plc*
PO Box3, Filton, Bristol, BS12 7QE England, UK

Abstract: This paper describes a novel approach to the design of a control system for a short take-off, vertical landing (STOVL) aircraft gas turbine engine. A multi-level multiobjective genetic algorithm (MOGA) is employed to design controllers at both individual operating points, using system linearisations and small signal response characteristics, and over the full-flight envelope, using a nonlinear model and large signalresponses. The proposed approach should allow the selection of smoother controller parameters over the flight envelope and ensure that more consistent control demands are made at off-design operating points. Copyright © 1999 IFAC

Keywords: CACSD, gain-scheduling, evolutionary algorithms, aerospace, optimization.

1. INTRODUCTION

Gas turbine engine control is a demanding task that requires the satisfaction of many, often competing, performance measures for stability, sensitivity, safety, etc. While the design of a control system for a conventional propulsion systems poses few hard problems for the control engineer, there may be many candidate solutions available and the choice of the 'correct' controller is paramount. Furthermore, new concepts in aircraft engines, such as the variable cycle engine, are likely to include more controllable elements and will almost certainly require the application of advanced control techniques if they are to realise their true potential.

In this paper, after an introduction to gas turbine engine control and MOGAs, an example is presented demonstrating how a multi-level MOGA may be employed to design a single controller for a wide range of operating points. At a low-level, a set of independent MOGAs are employed to design control schemes for a number of operating points chosen to cover a wide range of the flight envelope. At the high-level, a further MOGA is utilised to design the full controller by selecting sets of the low-level controllers and scheduling them over the operating range. This approach is analogous to the collaborative optimization methods adopted in the multidisciplinary optimization of complex systems.

2. GAS TURBINE ENGINE CONTROL

The mechanical layout of the Rolls-Royce Pegasus short take off, vertical landing (STOVL) gas turbine engine is shown in Fig. 1. The engine contains two spools, low pressure (LP) and high pressure (HP), each comprised of a number of compressor and turbine stages aero-thermodynamically coupled to the other. Air is drawn into the fan (or LP compressor) and compressed. The air is then further compressed by the HP compressor before being mixed with fuel (F), combusted and expelled through the HP and LP turbines. A portion of the air from the fan exit may bypass the HP compressor and turbines and be mixed with the combusted air/fuel mixture before being ejected through the jet pipe and the rear nozzle to produce thrust. Bypass airflow may also be directed through the front nozzle to produce a differential thrust.

The characteristics of operation of such a fixed cycle gas turbine engine, such as specific thrust and specific fuel consumption, are fundamental to the engine design. Dry-engine (without reheat) control of a conventional engine is normally based on a single closed-loop control of fuel flow for thrust rating, engine idle and maximum limitary, and acceleration control. The closed-loop concept provides accuracy and repeatability of control of defined engine parameters under all operating conditions, and compensates automatically for the effects of engine and fuel system ageing.

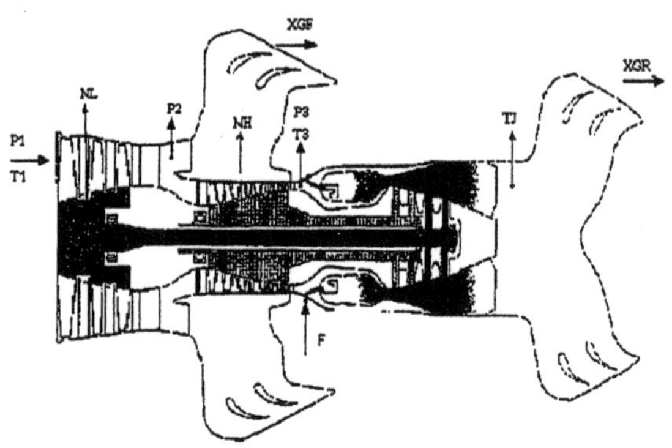

F	fuel flow
NH	high-pressure spool speed
NL	low-pressure spool speed
P3	high-pressure delivery pressure
P2	low-pressure delivery pressure
T3	high-pressure delivery temperature
TJ	jet pipe temperature
XGF	front nozzle thrust
XGR	rear nozzle thrust

Figure 1: The Pegasus turbofan engine

Aircraft gas turbine engines are highly nonlinear over their full operating envelope, exhibiting different characteristics in terms of operational parameters such as aero-thermodynamic stability, sensitivity and robustness, at individual operating points and are also subject to increasingly strict performance requirements. It is usual when designing a controller for such systems to select a number of representative operating points (idle, cruise, loiter, full-power, etc.) and to design controllers that meet the design specifications at each of these points separately. A controller for the full-flight envelope is then achieved by scheduling the controller parameters through these operating points to achieve satisfactory operation.

At each design point, the controller is required to meet a set of performance objectives. These include, but are not limited to, steady-state accuracy, transient accuracy, disturbance rejection, stability, stall margin, structural integrity protection and engine degradation. The control of the engine over its entire operating range must then ensure that their is a smooth transition between these operating points such that actuator demands are consistent and that large-scale transient accuracy is achieved. Additionally, the overall control strategy will usually be chosen to provide some compromise between optimizing the thrust specific fuel consumption, speed of response and minimizing the rate of thermal changes (to reduce thermal fatigue).

Future concepts in aero-engine design, such as the variable cycle engine (VCE), will be expected to operate with greater fuel economy, over a wider flight envelope and have extended mission capabilities. This type of engine will have more active internal variable geometry devices and sensors to meet the perceived operational requirements. These include, but are not limited to, variable and/or multiple bypass ducts, compressor guide vanes, staged combustion and variable nozzle area(s). Efficient operation will thus only be achieved through the accurate control of the variable geometry components to ensure the compressors and turbines run at the design conditions for airflow and pressure ratios. As the design conditions vary with operating point, e.g. cruise, combat, etc., it appears that the approaches adopted for conventional engine control selection and design will be unsuited to the VCE, Garwood and Baldwin (1992).

The application of advanced control techniques, such as multivariable control, to VCEs is likely to provide the extra degrees of freedom necessary to enable cost-effective variation of the engine cycle according to the mission requirements, see, for example, Shuttler and Betteridge (1994). Thus, a wider range of mission capability should be achievable with increased aircraft agility. However, the projected type of VCE design for both conventional and STOVL aircraft applications, and for compound helicopter power plants, shows a significant increase in the number, and type, of variable geometry devices. With this in mind, Chipperfield *et al.* (1997) have demonstrated a MOGA-based approach to the control configuration design (feedback and controller structure) and selection of suitable controller parameters in previous work for a single operating point. In the following sections we describe how we are extending this work to cover a wider range of the full flight envelope.

3. MULTI OBJECTIVE GENETIC ALGORITHMS

The use of multiobjective optimization (MO) recognises that most practical problems require a number of design criteria to be satisfied simultaneously, viz:

$$\min_{\boldsymbol{x} \in \Omega} \quad \boldsymbol{F}(\boldsymbol{x})$$

where $\boldsymbol{x} = [x_1, x_2, \ldots, x_n]$ and Ω define the set of free variables, $\boldsymbol{x}$, subject to any constraints and $\boldsymbol{F}(\boldsymbol{x}) = [f_1(\boldsymbol{x}), f_2(\boldsymbol{x}), \ldots, f_n(\boldsymbol{x})]$ are the design objectives to be minimised.

Clearly, for this set of functions, $\boldsymbol{F}(\boldsymbol{x})$, it can be seen that there is no one ideal 'optimal' solution, rather a set of Pareto-optimal solutions for which an improvement in one of the design objectives will lead to a degradation in one or more of the remaining objectives. Such solutions are also known as non-inferior or non-dominated

If goal and/or priority information is available for the design objectives then it may be possible to differentiate between some non-dominated solutions. For example, if degradations in individual objectives still allow those goals to be satisfied whilst also allowing improvements in other objectives that do not already satisfy their design goals, then these degradations should be accepted. In cases where different levels of priority may be assigned to the objectives then, in general, it is only important to improve the high priority objectives, such as hard constraints, until the corresponding design goals are met, after which improvements may be sought in the lower priority objectives.

These considerations have been formalised by Fonseca (1995) in terms of a transitive relational operator, *preferability*, based on Pareto-dominance, which selectively excludes objectives according to their priority and whether or not the corresponding goals are met. For simplicity only one level of priority is considered here. Consider two objective vectors **u** and **v** and the corresponding set of design goals, **g**. Let the smile $\underset{\smile}{\mathbf{u}}$ denote the components of **u** that meet their goals and the frown $\underset{\frown}{\mathbf{u}}$ those that do not. Assuming minimization, one may then write

$$\mathbf{u}^{\underset{\smile}{\mathbf{u}}} \leq \mathbf{g}^{\underset{\smile}{\mathbf{u}}} \quad \wedge \quad \mathbf{u}^{\underset{\frown}{\mathbf{u}}} > \mathbf{g}^{\underset{\frown}{\mathbf{u}}}$$

where the inequalities apply component wise. This is equivalent to

$$\forall i \in \underset{\smile}{\mathbf{u}}\; u_i \leq g_i \quad \wedge \quad \forall i \in \underset{\frown}{\mathbf{u}}\; u_i > g_i$$

where u_i and g_i represent the components of **u** and **g** respectively. Then, **u** is said to be preferable to **v** given **g** if and only if

$$(\mathbf{u}^{\underset{\frown}{\mathbf{u}}} \;{}_{p}\!\!< \mathbf{v}^{\underset{\frown}{\mathbf{u}}}) \vee \{ (\mathbf{u}^{\underset{\frown}{\mathbf{u}}} = \mathbf{v}^{\underset{\frown}{\mathbf{u}}}) \wedge [(\mathbf{v}^{\underset{\smile}{\mathbf{u}}} \not\leq \mathbf{g}^{\underset{\smile}{\mathbf{u}}}) \vee (\mathbf{u}^{\underset{\smile}{\mathbf{u}}} \;{}_{p}\!\!< \mathbf{v}^{\underset{\smile}{\mathbf{u}}})] \}$$

where $\mathbf{a} \;{}_{p}\!\!< \mathbf{b}$ is used to denote that **a** dominates **b**. Hence **u** will be preferable to **v** if and only if one of the following is true:

1. The violating components of **u** dominate the corresponding components of **v**.
2. The violating components of **u** are the same as the corresponding components in **v**, but **v** violates at least one other goal.
3. The violating components of **u** are equal to the corresponding components of **v**, but **u** dominates **v** as a whole.

Multiobjective ranking is based upon the dominance of an individual, i.e. how many individuals outperform it in objective space, Fig. 2. From Fig. 2, it may also be seen that multiobjective ranking is non-unique, a number of individuals are ranked 0, i.e. non-dominated. Ranking may also be combined with goal and/or priority information to discriminate between non-dominated solutions. For example, a solution in which all the goals are satisfied can be considered superior, or *preferable*, to a non-dominated one in which some components go beyond the goal boundaries. Similarly, an improvement in an objective component such that the objective value lies inside the goal boundary should be accepted even when this causes a degradation in another objective component provided that the corresponding goal is satisfied. In cases where objectives are assigned different priorities, higher priority objectives are optimized in a Pareto fashion until their goals are met at which point the remaining objectives are optimized (see Fonseca and Fleming (1995)). In the example presented here, the objectives are all assigned the same priority

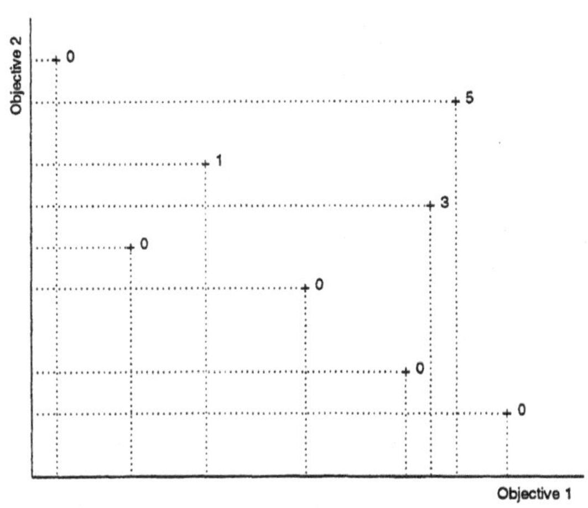

Figure 2: Pareto ranking

All preferred individuals thus achieve the same fitness, however the number of actual offspring may differ due to the stochastic nature of the selection mechanism. Thus, an accumulation of the imbalances in reproduction can lead the search into an arbitrary area of the trade-off surface. This phenomenon described by Goldberg and Sergest (1987) is known as genetic drift and can drastically reduce the quality and efficiency of search. Goldberg and Richardson (1987) proposed a solution to the problem genetic drift, fitness sharing. This penalizes the fitness of individuals in popular neighbourhoods in favour of more remote individuals of similar fitness. A *niche count* is used to derive the density of individuals in genotypic, phenotypic or objective space. This value is derived from a *sharing function* that determines the contribution made by individuals to niches. Raw fitness values may then be adjusted in response to niche counts and the total fitness in the population redistributed over the individuals. However, the use of fitness sharing has been restricted by difficulties in determining appropriate niche sizes.

The approach adopted in the example presented later overcomes this problem by using techniques developed from kernel density estimation methods in statistical analysis, in particular the Epanechnikov kernel [8]. A constant factor is used in computing the sharing function and has been shown by Silverman (1986) to be appropriate for performing sharing in Euclidean decision variable space.

Recombining arbitrary pairs of non-dominated individuals from the trade-off surface can result in the production of an unacceptably large number of unfit offspring, or lethals. A further refinement to the MOGA is therefore to bias the manner in which individuals are paired for recombination, often termed mating restriction (see Deb and Goldberg (1989). This is generally achieved by restricting reproduction to pairs of individuals that are within some given distance of each other. A common practice is to use the niche radius from the sharing function and the approach here employed the smoothing parameter from the kernel density function as the mating restriction limit. The interested reader is referred to Fonseca and Fleming (1995) where a full account of the development of MOGAs may be found.

4. DESIGN PROBLEM

The study presented here is based on models of the Rolls-Royce Pegasus gas turbine engine. The layout of the engine and the physical meaning of the associated variables have already been described in Fig. 1. The nonlinear engine model reproduces dynamic characteristic variations for the fuel thrust range over the complete flight envelope while linearisations may be used to approximate these characteristics around particular operating points.

The full model of the system described by Hancock (1992), shown in Fig. 3, includes the fuel system, stepper motor input and a digital controller for the fan speed. The low pressure spool-speed demand (NLDEM) is the only control input to the system, while the others inputs allowing external perturbations to be applied to the system are not used in this example. The actual speed of the low-pressure spool is compared with NLDEM and used to produce a demand signal for the high-pressure spool speed (NHDEM) which in turn is used to compute the desired position of the stepper motor (THSTD) by comparing NHDEM with the actual high-pressure spool speed (NH).

The two blocks of the controller, NLLOOP and NHLOOP, are each characterised by four parameters - a real-valued gain and three time constants. The structure of the controller had previously been determined by Rolls-Royce based on integration requirements of the engine with other systems on the aircraft (see Shuttler (1991)). These controller parameters may be scheduled against NL and NH respectively to map the controller to the operating conditions.

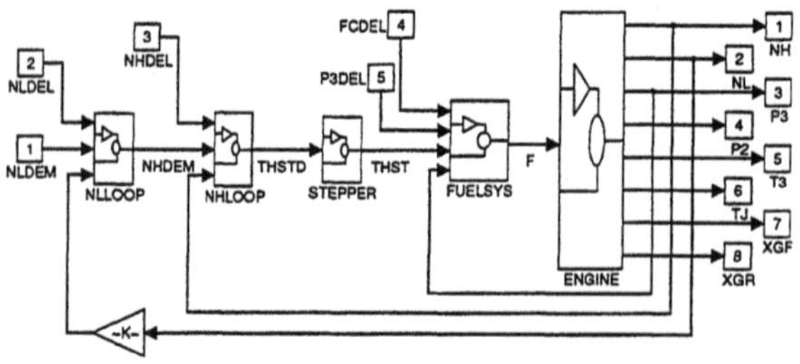

Figure 3: The SIMULINK model of the Pegasus engine and control system

In this example, a sea-level static controller is to be designed to operate from 60% NL to 100% NL. The controller design is based on five operating points, 60%, 70%, 80%, 90% and 100% NL. At each operating point the system is required to meet the following performance objectives:

(1) Maximum pole magnitude ≤ 1
(2) Gain margin ≥ 3.53 dB
(3) Phase margin ≥ 50°
(4) 70% rise-time ≤ 0.59s
(5) 10% settling time ≤ 1.08s
(6) Overshoot ≤ 10%
(7) Error ≤ 0.1%

Objective (1) is a measure of the closed-loop stability of the controller at each operating point and is calculated from the maximum pole value of the discrete-time closed-loop system. Objectives (2) and (3) are measures of robustness calculated by opening the loop after the controller input. Both measures are maximized by minimizing their complement. Objectives (4), (5) and (6) are the dynamic response of the system to a 3% change in NL demand at each operating point. From steady state stability requirements that NL should not fluctuate by more than 0.1% NL, objective (7) calculates the output error over the simulation.

The wide envelope controller (WEC), covering the NL range from 60% to 100%, is designed by scheduling the parameters for NLLOOP and NHLOOP against NL and NH respectively. The WEC should meet the following performance objectives

(i) 70% rise-time ≤ 1.8s
(ii) 10% settling time ≤ 3.6s
(iii) Overshoot ≤ 10%
(iv) Error ≤ 0.15%
(v) TJ ≤ 1030°K

Where objectives (i), (ii) and (iii) are for a step change in NLDEM fro 70% to 95%. Objective (iv) is the output error which is relaxed compared to the small-signal models to allow a shorter simulation time. To ensure that the engine integrity is maintained, the maximum temperature of the jet pipe is not allowed to exceed 1030°K during the manoeuvre, objective (v).

5. IMPLEMENTATION

The structure of the MOGA is shown in Fig. 4. At each design point a separate MOGA, with a population of 50 binary coded individuals, attempts to optimize the controller parameters to find individuals that satisfy design objectives (1, ..., 7). New individuals are generated using the reduded surrogate shuffle crossover of Caruana *et al.* (1989) applied with probability 0.7 and independant bit mutation applied with a probability set according to Harvey (1992) to the selective pressure and chromosome length. When satisfactory controllers are found for each operating point, i.e. ones that satisfy all of the design goals, the real-valued parameters are made available to the higher-level MOGA through lists of acceptable parameters. Therefore, before a new set of parameters can be added to this list by any one of the low-level MOGAs, the corresponding set of objective values must be examined to ensure that only preffered solutions are available to the WEC MOGA.

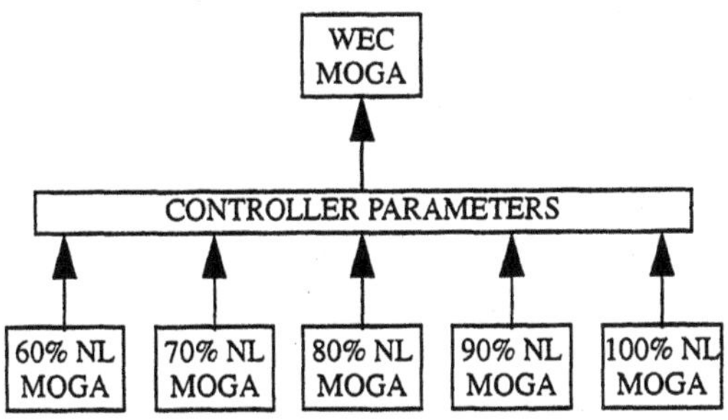

Figure 4: MOGA structure

In the WEC MOGA, controllers are generated by sampling the parameters from the design point controllers using discrete recombination described by Mühlenbein and Schlierkamp-Voosen (1993) and building a look-up table for these values in the controller blocks of the model. The corresponding controllers are then tested for the prescribed manoeuvre. The population size for the WEC MOGA thus varies with the populations of the low-level MOGAs. However, by maintaining families of Pareto-optimal solutions from the low-level MOGAs, it is possible to allow the WEC MOGA to explore a potentially wider range of controllers than those currently available at each generation

The MOGAs and a user interface where implemented using the MATLAB CACSD package. The engine and controller where modelled in SIMULINK with system linearisations used to compute objectives (1), (2) and (3). The remaining objectives where calculated using nonlinear SIMULINK models of the system.

6. RESULTS

To illustrate the approach, consider the trade-off graph for the 70% NL controller design shown in Fig. 5. Each line shows a preferred solution found by the MOGA. That is, solutions that are both Pareto-optimal and satisfy all of the design objectives. The *x*-axis shows the design objectives as described previously and the *y*-axis shows the performance of controllers in each objective domain. Cross-marks represent the design goals and all the lines are therefore below these marks.

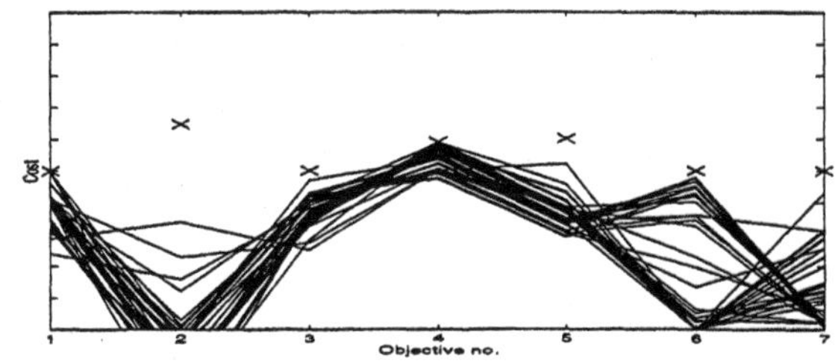

Figure 5: Trade-off graph for 70% NL design point

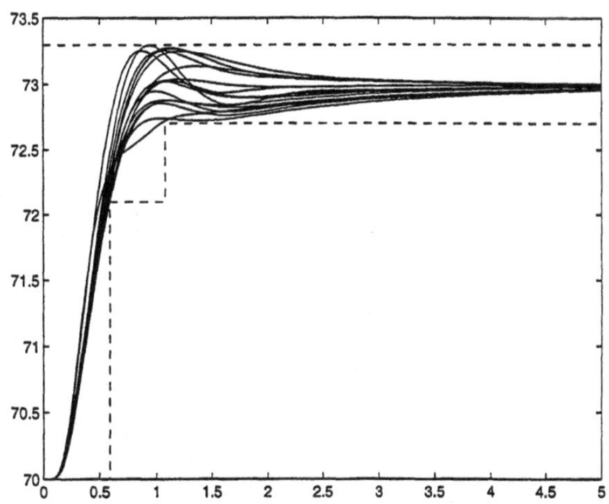

Figure 6: Small-signal responses about 70% NL

In Fig.5, trade-offs between adjacent design objectives result in the crossing of lines between them whereas non-crossing lines indicate that objectives do not compete with one another. For example, improving the phase margin (objective (3)) usually results in a slower rise time (objective (4)) and vice-versa. Similarly, the settling time and overshoot (objectives (5) and (6)) clearly compete with one another and generally it is observed that a faster settling time is achieved at the expense of a greater overshoot. This is confirmed in Fig. 6 where the step responses for a subset of the solutions of Fig. 5 are shown.

The trade-off graph for the WEC is shown in Fig. 7. Note that in this case the position of the design objectives along the *x*-axis have been re-ordered to allow different trade-offs to be examined. The new order of the objectives is now i, v, ii, iv and iii. In this way it can be clearly seen that controllers that offer the fastest rise-time generally have a higher maximum jet pipe temperature but settle to their final value more quickly. However, a good controller offering fast rise and settling times with a low maximum value of TJ has been found but at the expense of output error. The large-signal step responses for these controllers are shown in Fig. 8.

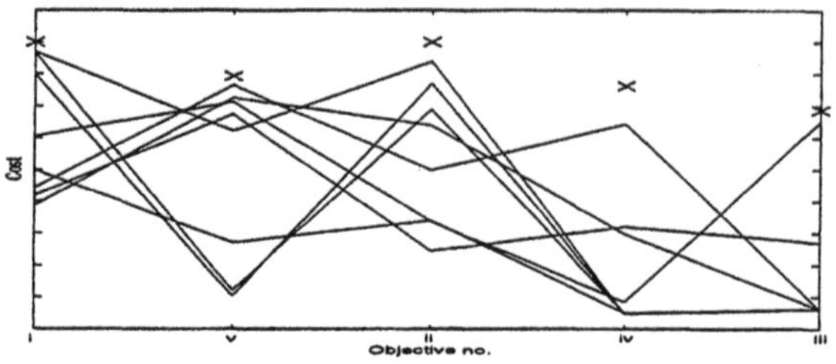

Figure 7: Trade-off graph for wide envelope controller

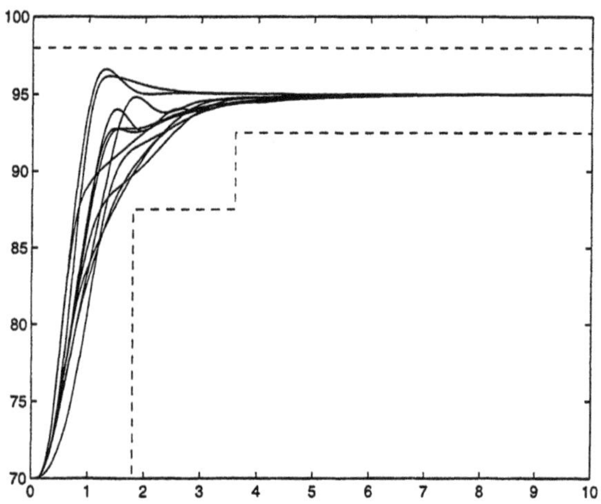

Figure 8: Large signal responses

7. CONCLUDING REMARKS

This paper has demonstrated an approach to the design of a wide envelope controller for a gas turbine engine based on the multiobjective optimization of controller parameters for individual operating points and the selection of suitable subsets of these to implement the final controller. While the design objectives for the individual operating points and their corresponding level of priority have all been the same in the example employed here, it is expected that it will be feasible to allow the objective to be set to reflect the particular concerns at the operating points. For example, at high NL speeds, the limiting of internal temperatures becomes increasingly important. Similarly, the example considered only sea-level static conditions and work is underway to extend this to the full flight envelope.

REFERENCES

Caruana, R.A., L.J. Eshelman and J.D. Schaffer (1989). Representation and hidden bias II: Eleminating defining length bias in genetic search via shuffle crossover, In *Proc. 11th Int. Joint Conf. on Artificial Intelligence*, N. S. Sridharan (Ed.), Morgan Kaufmann, 750-755.

Chipperfield, A.J., H.C. Betteridge and P.J. Fleming (1997). Control mode analysis for advanced concept aero-engines, In *Proc. AIAA/ASME/SAE/ASEE Joint Propulsion Conf.*, Seattle, WA, AIAA 97-3190.

Deb, K. and D.E. Goldberg (1989). An investigation of niche and species formation in genetic function optimization, In *Proc 3rd Int. Conf. on Genetic Algorithms*, J. D. Schaffer (Ed.), Morgan Kaufmann, 42-50.

Fonseca, C.M. (1995). *Multiobjective Genetic Algorithms with Application to Control Engineering Problems*, PhD Thesis, University of Sheffield, UK.

Fonseca, C.M. and P.J. Fleming (1995). An overview of evolutionary algorithms in multiobjective optimization, In *Evolutionary Computing*, **3**(1), 1-16, 1995.

Garwood, K.R. and D.R. Baldwin (1992). The emerging requirements for dual and variable cycle engines, In *10th Int. Symp. on Air Breathing Engines*, RAE-TMP-1220.

Goldberg, D.E. and J. Richardson (1987). Genetic algorithms with sharing for multimodal function optimization, In *Proc. 2nd Int. Conf. on Genetic Algorithms*, J. J. Grefenstette (Ed.), Lawrence Erlbaum, 41-49.

Goldberg, D.E. and P. Segrest (1987). Finite Markov chain analysis of genetic algorithms, In *Proc. 2nd Int. Conf. on Genetic Algorithms*, J. J. Grefenstette (Ed.), Lawrence Erlbaum, 1-8.

Hancock, S.D. (1992). *Gas Turbine Engine Controller Design Using Multi-Objective Optimization Techniques*, PhD Thesis, University of Wales, Bangor, UK.

Harvey, I. (1992). Evolutionary robotics and SAGA: The case for hill crawling and tournament selection, *Cognitive Science Research Paper 222*, University of Sussex, UK.

Mühlenbein, H. and D. Schlierkamp-Voosen (1993). Predictive models for the breeder genetic algorithm, In *Evolutionary Computation*, **1**(1), 25-49

Shuttler, A.G. (1991). A case study problem for the advanced controls technology club, *Internal Report CSAN 2790*, Rolls-Royce plc, Bristol, UK.

Shuttler, A.G. and H.C. Betteridge (1994). Definition, design and implementation of control laws for variable cycle gas turbine aircraft engines, In *The Design & Control of the Next Generation of Civil & Military Engines: Do Variable Cycle Engines Have a Role?*, RAe.

Silverman, B.W. (1986). *Density Estimation for Statistics and Data Analysis*, Chapman and Hall.

A NOVEL QUALITATIVE CONTROL METHOD TO INVERTED PENDULUM SYSTEMS[1]

Deyi Li Hui Chen* Jianhua Fan* and Chengzhi Shen#

The Institute of Electronic System Engineering
No. 6 Wan Shou Road, Beijing, China, 100036
ziqin@public2.bta.net.cn
**The Nanjing Communications Engineering Institute, China*
#Beijing University of Aeronautics and Astronautics, China

Abstract: A new mathematical representation of qualitative concepts is presented by a cloud model in this paper. With the new model, a novel imitating-human control mechanism for inverted pendulum systems driven by single motor is proposed not only serving as foundations of qualitative control engine, but also integrating fuzziness and randomness in uncertainty reasoning. A case study is given to explore a good comprehensibility of human-intelligence control methods . The architecture and object-oriented programming of such a control engine to inverted pendulum systems based on cloud generators and qualitative rule constructors show the advantages in implementations. The physical experimental results with robustness are given and evaluated. *Copyright © 1999 IFAC*

Keywords: Modeling , Qualitative Control , Uncertainty , Robustness.

1. INTRODUCTION

It is worthwhile to model human knowledge and emulate human behavior in real-time control systems (H.meier, *et al.*,1990;K.Furuta, *et al.*,1992; J.Eker, *et al.*,1996). Many researchers who are interested in fuzzy logic, intelligence automation, neural networks and soft computing focus their attention on fuzziness and randomness to attack uncertainty in human thinking and human control (C.Anderson,1989; H.Tai and Shenoi,1994; D.Saez and Cipriano, 1997; L.K.Wong, *et al.*,1997). Human being can well control many unstable objects by simply using their approximate, qualitative knowledge and experiences, such as riding a bicycle, making a long flexible pole upright on his forehead. Therefore, to stabilize an inverted pendulum system has become a very challenging and hot subject in imitating such a human- intelligence and automation for about thirty years. The kernel of inverted pendulum control systems is very much the same as robot control, space aircraft control, and many other unstable systems. An inverted pendulum is also very often used as an experimental instrument on which many control approaches can be examined to explore new control theories. As stated elsewhere this experiment is 'the jewel in the crown of every control department'. Many papers on stabilizing single, double and triple link inverted pendulum systems rely only on computer simulations to validate their approaches. As discussed later in this paper, physical uncertainties not included in the simulation models could cause substantial difference between experimental and simulation results. It is well known (H.meier,*et al.*,1990;T.Yamakawa,1989; L.K.Wong, *et al.*,1997) that the successful control of a triple inverted pendulum using a single motor has not been well solved yet. The powerful combination of advanced computer technologies with state-of-the-art experimental hardware would make it possible if only modern control theory can take into account the uncertainty in physical systems. This paper deals with

This research is supported by the Natural Science Foundation of China under contract 69775016.

a novel qualitative model accomplished in this area, and intends to model and test the actual uncertainties in physical inverted pendulum systems.

2. The QUALITATIVE CONTROL STRATEGY OF INVERTED PENDULUM SYSTEMS

2.1 Problems with Conventional Mathematical and Fuzzy Control Methods

Inverted pendulum control system has a very simple structure. A pole is hinged to a motor-driven cart that moves on rail tracks to its right and left. The pole has only one degree freedom (rotation about the hinge point). The primary control tasks are to keep the multi-link inverted pendulum vertically balanced and keep the cart within the rail track boundaries. The experimental block diagram is shown in Fig.1, and the dynamics of the inverted pendulum system are characterized by eight state variables:

x	position of the cart on the rail;
$\dot{x}$	velocity of the cart;
θ_1, θ_2, θ_3	angle with the lower, middle and upper pole to the vertical axis respectively;
$\dot{\theta}_1$, $\dot{\theta}_2$, $\dot{\theta}_3$	angle velocity with the lower, middle and upper pole to the vertical axis respectively.

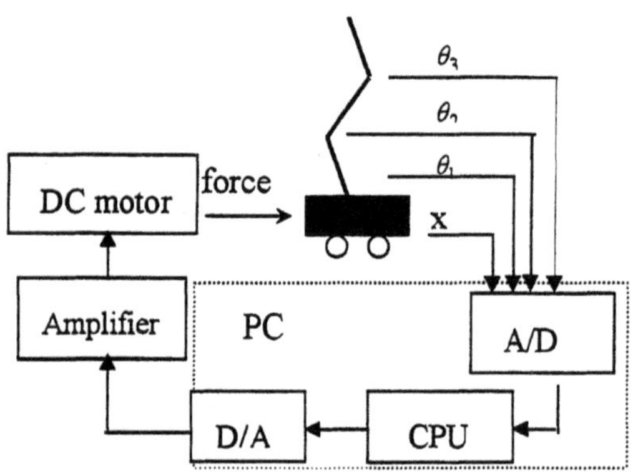

Fig.1 Experimental block diagram

Such an inverted pendulum is nonlinear, and all physical systems are nonlinear to a certain extent. Conventional nonlinear control schemes often require a prior knowledge on mathematical structure and accurate parameters, which are very often not accessible. For example, according to control theory, the mathematical model for the pendulum can be constructed using Lagarange method under some assumptions(D.Saez and A.Cipriano,1997)The nonlinear system equation resulting from the Lagarange method can then be written in the following form:

$$\frac{d}{dt}\left(\frac{\partial T}{\partial \dot{q}_j}\right) - \frac{\partial T}{\partial q_j} + \frac{\partial V}{\partial q_j} + \frac{\partial D}{\partial \dot{q}_j} = Uj + Fj \qquad (j=1..4) \quad (1)$$

$$\mathbf{q} = [\mathbf{x} \quad \theta_1 \quad \theta_2 \quad \theta_3]^T$$

$$F_j = -sign(q_j)u_j N_j \qquad \mathbf{U} = [\mathbf{Ks\ U\ 0\ 0\ 0}]^T$$

where u (in V) is the control input from the computer to the analogue amplifier, Ks (in N/V) is the overall electronic systems input conversion gain and Fj is the coulomb friction term in the jth coordinate direction. The nonlinear system equation resulting from the Lagrange method can then be written in the following form. To find a stabilizing controller for the pendulum, the equation of motion has to be linearized about the vertical position. The linearized model, which then becomes the basic foundation for all stabilization methodologies, is represented in state space form as follows.

$$\mathbf{F(q)}\,\ddot{\mathbf{q}} + \mathbf{G(q,\dot{q})\dot{q}} + \mathbf{H(q)} = \mathbf{L(q,u)} \quad (2)$$

As a matter of fact, a physical multiple-link inverted pendulum is a multivariate, nonlinear, fast reaction and unstable system with a lot of uncertainties under an environment. And there is no accurate mathematical model to fully describe it. However, human experts in theses cases may well achieve the control by control rules which are squeezed out from their long experience and represented by intuitive natural language. The natural languages play a very important role in representing the uncertainty in human control and reasoning.

As an alternative for conventional control theory and methods, fuzzy set theory and fuzzy control techniques have been very flourishing for many years. However, as a foundation of fuzzy set theory, the concept of "membership function" proposed by L. A. Zadeh in 1965 faces some challenges and criticisms. Does there really exist a unique value to which a particular element satisfies the property that characterizes the fuzzy set? How do we set up, measure, and judge the shape of a membership function? It is also argued(Deyi Li, *et al.*,1997) that the membership function frequently takes on fuzzy values itself. This type of fuzzy set is called ultra fuzzy set, level 2 fuzzy set, or interval-valued fuzzy set. Many applications of fuzzy set theory use triangular functions for simplicity, but other types of membership functions are also used in practice, such as sine function, trapezoid, and bell function. Therefore, fuzzy logic theory has been challenged starting from the concept of membership functions. Almost all results in fuzzy control systems are claimed to be dependent on the original membership functions chosen. Besides, the amount of precise computation involved in constructing membership functions can be formidable, especially when they are nonlinear. The product-summation composition and center-of-gravity calculation are most popular ways of defuzzification in the design of fuzzy logic controllers, Unfortunately there is no a clear theoretical interpretation of these operations. Furthermore, the fuzzy inference mechanism has no room to allow inheritance of fuzziness(Deyi Li,1998). After the mapping associated with a membership function, the uncertainty of a fuzzy concept becomes very certain, fixed to that degree for ever. The fuzzy characteristics of the original linguistic term are not passed on the next step of inference at all. That does not keep with the fuzzy thinking of humans.

2.2 Qualitative Representation

So, a very fundamental question arises: "How mathematically natural are qualitative models to represent the uncertainty in human thinking?'' The powerful mathematical tools to represent uncertainty in human thinking and human control are fuzzy set theory and probability theory. The both should enrich each other to model uncertainty. That is our qualitative control strategy. Our point of departure in the qualitative control strategy is to represent qualitative knowledge in linguistic terms.

3. MAPPING BETWEEN QUALITATIVE AND QUANTITATIVE KNOWLEDGE

3.1 Cloud Models

Following the important characteristics of linguistic variables and terms, we define a new concept of cloud models to represent linguistic terms. Let U be the set, U = {u}, as the universe of discourse, and T, a linguistic term associated with U. The membership degree of u in U to the linguistic term T, C_T (u), is a random variable with a probability distribution. C_T(u) takes values in [0,1]. A membership cloud is a mapping from the universe of discourse U to the unit interval [0,1].

The concept of membership clouds is often pictured as two-dimensional graphs. The geometry of membership clouds is a great aid in understanding the uncertainty. It is important to see the properties of the clouds. First of all, the mapping from all u in U to the interval [0,1], is an one-point to multi-point transition, producing a membership cloud, rather than a membership curve. Secondly, any particular drop of the cloud may be paid little attention to, however, the total shape of the cloud, which is visible, elastic, boundless and movable, is most important. That is why we use the terminology ``cloud" to name it. Thirdly, the mathematics expected curve (MEC) of a membership cloud may be considered as its membership function from the fuzzy set theory point of view. Finally, the definition has effectively integrated the fuzziness and randomness of a linguistic term in a unified way. In the cloud, fuzziness lies at the center, and there may be nothing to do with probability, but there is a probability adhered on the fuzziness from the statistical point of view. We can see the integrated uncertainty of fuzziness and randomness and the convergent properties of the cloud model. The cloud concept provides a means of both qualitative and quantitative characterization of linguistic terms.

The bell-shaped clouds, called normal clouds are most fundamental and useful in representing linguistic terms. Since some people may get used to the concept of conventional membership functions, we could also use the normal membership function to represent the mathematical expected curve (MEC) of the cloud model. The digital parameters of a normal cloud characterizes the quantitative meaning of a linguistic atom. The Gaussian distribution transformation is used in a very effective way in characterizing normal clouds. A normal cloud shown in Fig.2 is described with only three digital characteristics, expected value (Ex), entropy (En) and hyper entropy (He).The description of three digital characteristics can be seen in (Deyi Li ,1997).

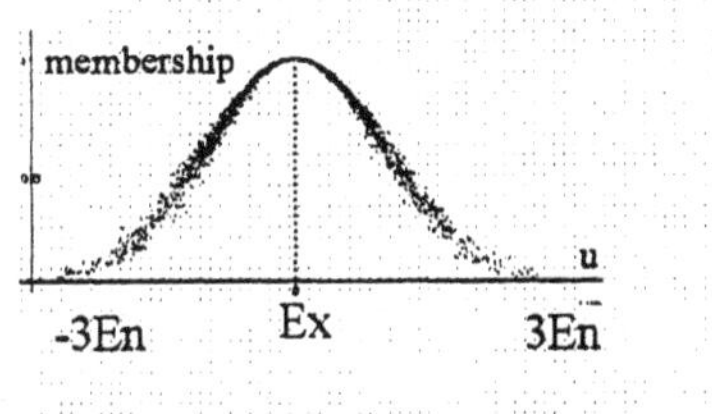

Fig.2 A normal cloud with Ex,En,He.

3.2 Cloud Generators

Given three digital characteristics Ex, En, and He, to represent a linguistic term, a set of cloud drops may be generated by the following algorithm:

1) Produce a random value x which satisfies with the normal distribution probability of mean = Ex, and standard error = En;
2) Produce a random value En' which satisfies with the normal distribution probability of mean = En, and standard error = He;
3) calculate $y = e^{\frac{-(x-E_X)^2}{2(E_N')^2}}$;
4) let (x, y) be a cloud drop in the universe of discourse;
5) repeat 1-4 until the number of drops required all generated.

The idea of using only three digital characteristics to generate a cloud is creative. A series of linguistic term generators have been implemented both in hardware and software and are a patented invention in China. The generator could produce as many drops of the cloud as you like. This kind of generators is called a forward cloud generator. Cloud-drops may also be generated upon conditions. Figure 3 shows the generator producing drops under a given numerical value u in the universe of discourse U. It is easy to set up a half-up or half-down normal cloud generator with the similar strategy, if there is a need to represent such a linguistic term.

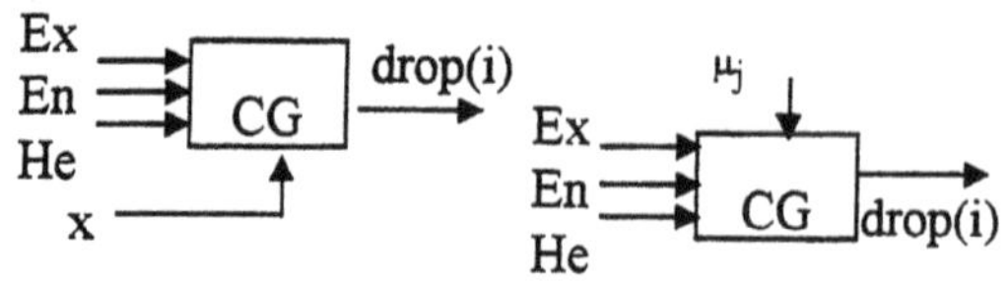

Fig. 3 Generators on condition

It is natural to think about the generator mechanism in an inverse way. Given a number of drops, as samples of a normal cloud, the three digital characteristics Ex, En, and He could be obtained to represent the corresponding linguistic term. This kind of cloud generators may be called backward cloud generators. Since linguistic terms are represented by the cloud model, the forward and backward cloud generators can be served interchangeably to bridge the gap between quantitative and qualitative knowledge.

4 CONSTRUCTION OF QUALITATIVE RULES BY USING ClOUD MODELS

4.1 Qualitative Rule Constructor Implemented in Hardware

We may immediately use two forward cloud generators to construct a qualitative rule, ``If A then B," if the digital characteristics of the linguistic terms A and B in that rule are given. See Figure 4, in which, the membership degree, mu, produced by an input x to the generator CG_A represents the activated strength of the rule which goes to control the generator CG_B to produce a set of drops y quantitatively. The output uncertainty in responding to a same input value is a feature which actually happens with a human controller.

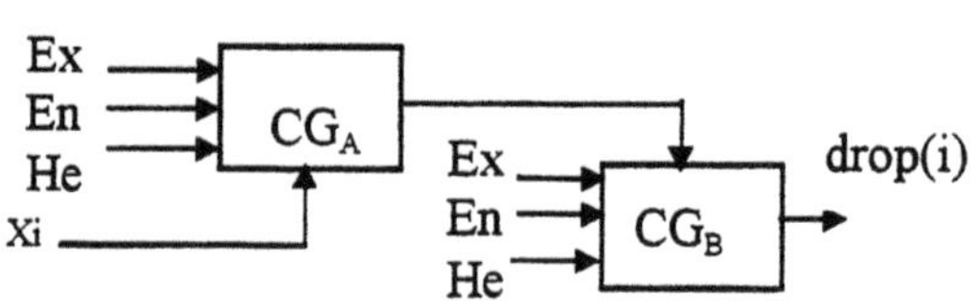

Fig. 4 A qualitative rule implemented by cloud generators

4.2 The Inference Engine with Rule Constructors

Generally speaking, there are a set of rules in a qualitative controller. The rule set is summed up by human experts from their experiences. These specified rules very often put the focus on the nonlinear characteristics, where necessary, in a complex control system. Therefore, all the activated rules will make contributions to a particular input, if more than one rule is fired by an input value in the controller.

5. QUALITATIVE CONTROL MECHANISM IN INVERTED PENDULUM SYSTEMS

5.1 The Imitating-human Control to an One-link Inverted Pendulum System

As to the control of a one-link inverted pendulum, the human qualitative knowledge could be represented in a 5-rule set as following:

1. IF (θ is Negative-Large) THEN (*f* is Negative-Large)
2. IF (θ is Negative-Small) THEN (*f* is Negative-Small)
3. IF (θ is Zero) THEN (*f* is Zero)
4. IF (θ is Positive-Small) THEN (*f* is Positive-Smallθ)
5. IF (θ is Positive-Large) THEN (*f* is Positive-Large)

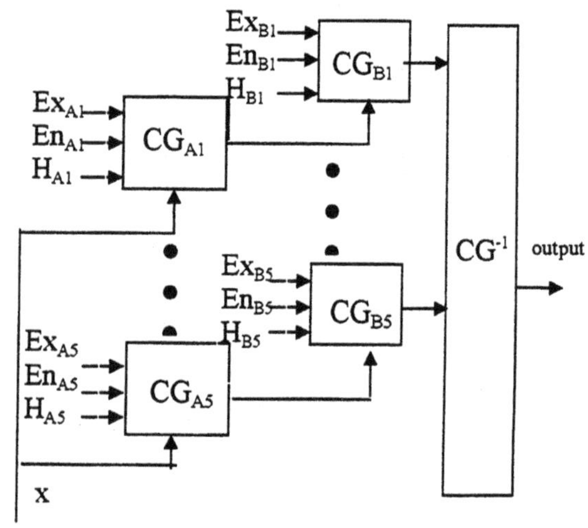

Fig. 5 The construction of a rule set

in which the underlined terms are all linguistic terms implemented in cloud generators. The 5-rule set can be easily implemented in hardware by using the rule construction technology described above. This kind of 5-rule set, named RS(θ), may be treated as a basic objet in programming,. However, if there is a requirement that the cart position must be kept in more or less the center of the rail while balancing the pendulum, another 5-rule set, RS(x), may be added. And at this time, the human control knowledge says that if the pendulum is basically upright, please pay more attention to the position of the cart being at the center of the rail. As a matter of fact, the qualitative knowledge "pay more attention" shows the control priority of θ is higher than x with uncertainty. The control strategy to take account of this knowledge is realized by using linguistic terms, soft-zeroes ($SZ_x, SZ_\theta, SZ_{\dot\theta}, SZ_{\dot x}$). The linguistic term, soft-zero, is very much interesting. The entropy of soft-zero, SZ_θ, can effectively control the probability of activating RS(θ) in control cycles. Once again, soft-zero is also implemented in the cloud model. If the human-intelligence control takes the dθ and dx into account as well, and the priority of activating RS(θ), RS(x), RS($\dot{\theta}$), and RS($\dot{x}$) is given in the same sequence, then the control strategy is also determinated by the digital characteristics of the linguistic terms, soft-zeros which make the mapping between the qualitative and quantitative values. The control block diagram is given in Fig.6.

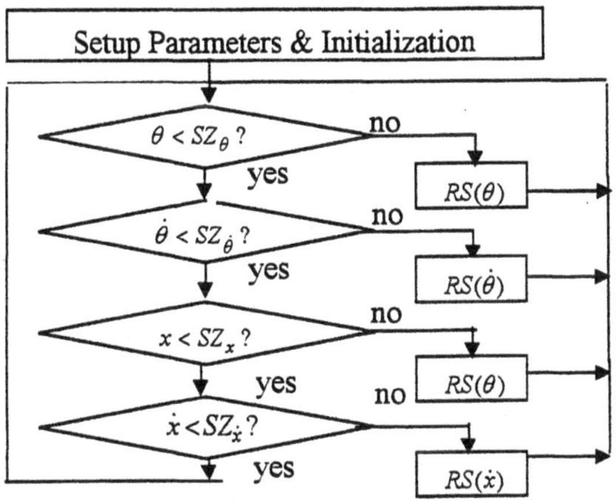

Fig.6 Flow chart of the control mechanism of an one-link inverted pendulum

5.2 The Concept of an Equivalent Cart to Multi-link Inverted Pendulum Systems

In order to be able to make a good understanding and representation of human intelligence on a multi-link inverted pendulum control mechanism, we may set up a new concept, called "an equivalent cart". That is to say that an n-link inverted pendulum is considered as an one-link inverted pendulum supported by an equivalent cart, which is an n-1 link inverted pendulum itself. Again the n-1 link inverted pendulum is another one link inverted pendulum supported by an equivalent cart which is an n-2 link inverted pendulum itself, and so on. Therefore, the qualitative control strategy becomes to pay more attention to keep the one link inverted pendulum first while balancing the equivalent cart.

5.3 The Imitating-human Control to a Double or Triple-inverted Pendulum System

Based on the control strategy of one-link inverted pendulum and the concept of an equivalent cart, the qualitative control process for a double and triple inverted pendulum system becomes much simple. The flow chart of a triple inverted pendulum is given in Fig.7.

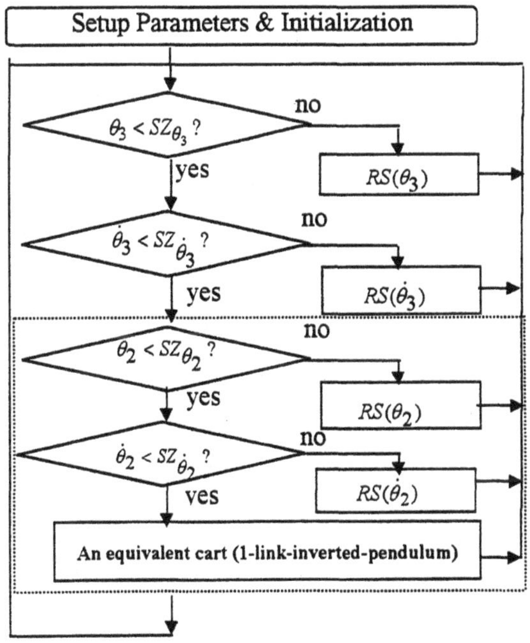

Fig.7 Flow chart of operations of a triple inverted pendulum control system.

6. EXPERIMENTS AND EVALUATIONS

6.1 The Physical Experiments and Testing Environments

For evaluating the qualitative control method, a physical experimental inverted pendulum system driven by single motor is shown in Fig.8.The stabilization of a double inverted pendulum has been succeeded by using the cloud model.

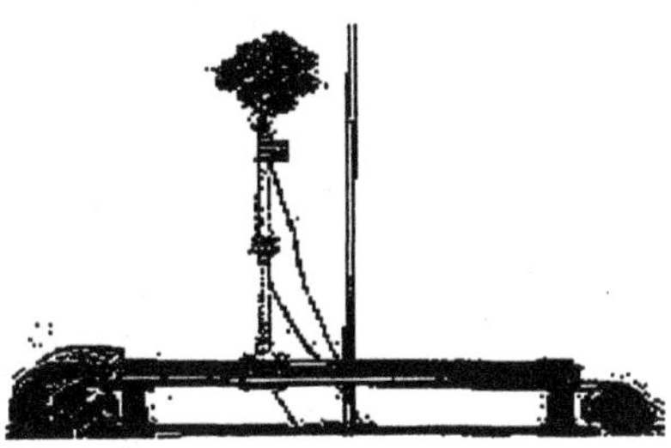

Fig.8 A real testing bed showing the robustness of the system

The sensor for detecting x is a rotating resistance meter WXD7 with resistance of 10kΩ, the sensors for detecting θ1, θ2, and θ3 are the same bearing-rotated resistance meters WDD35D1 with resistance of 2kΩ and rotating range of 360 degrees. The rail effective length is 80cm, and the DC motor is driven between -5V and+5V. An A/D D/A board is used to connect the pendulum instrument to a personal computer with 166MHz under Windows95.

6.2 Test Results and Evaluations

In this section we present the results of the experimental performance study on a double inverted pendulum control system. To make the study more meaningful, all the pre-definitions related to the digital characteristics of linguistic terms used in rule sets RS(x), RS(θ_1), RS($\dot{x}$) and RS($\dot{\theta}_1$) are made as similar as possible. A unified rule set is given , in which the digital characteristics of clouds for input and output are both in the range of [-100,100]. The Ex, En, and He of clouds are represented in the form (Ex, En, He) , as shown in table 1. Then, it is mapped to particular physical scales of RS(x), RS(θ_1), RS($\dot{x}$) and RS($\dot{\theta}_1$) respectively. By using the cloud model, the double inverted pendulum is successfully controlled, which can be stabilized over a long period of time (several hours!).The variation of θ_1 and θ_2 is shown in Fig.10(a). The cart of the double inverted pendulum can even move forward or backward as designed just like a walking robot as shown in Fig.10(b). Experiments also show the strong robustness that the control method is not sensitive to the change of pole length, cart mass and other physical factors. We can even make a bundle of flowers stand upward with the pole without changing any parameters at all. The

Table 1 Digital Characteristics of Clouds in the Unified Rule Set

Linguistic Terms	Unified Values
Negative-Large	(-100,20.7,0.023)
Negative-Small	(-38,12.6,0.015)
Zero	(0,7.6,0.01)
Positive-Small	(38,12.6,0.015)
Positive-Large	(100,20.7,0.023)

successful results encourage us to use these novel methods to balance triple-inverted pendulum, which are undergoing at the time of writing. Multi-dimensional clouds generators are being developed.

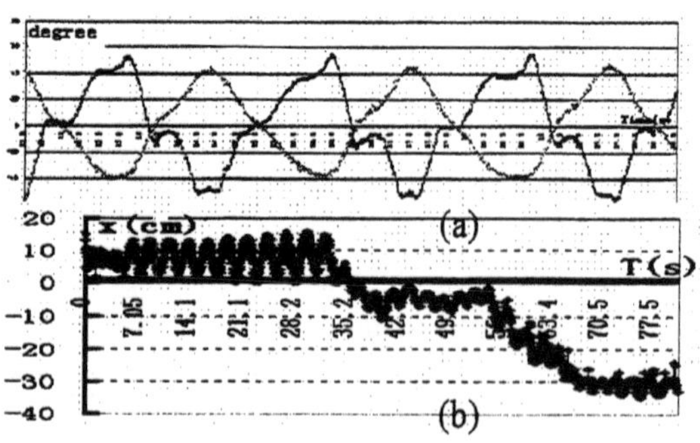

Fig.10 the position of cart ,$\theta_1$and θ_2 of double inverted pendulum when it is stabilized

7. CONCLUSION

A new uncertainty representation, cloud model, is presented integrating randomness and fuzziness with three digital characteristics Ex, En, and He. A soft inference mechanism has been developed to capture the qualitative knowledge contained in the human intelligence. The fuzziness and randomness are not only complementary, but also inseparable. And the experiments to imitating human control mechanism in inverted pendulum control systems are made showing the advantages in large scale qualitative control implementation.

However, to develop sound foundations of knowledge representation and human reasoning in knowledge base systems will require a lot of work. This paper will lead to comparative studies, something that has been lacking in previous reports. We have passed the stage from mathematical computing to information processing by using computers. And we are stepping on the stage from information processing to logic reasoning. By means of soft computation, computing with words, we may jump up from hard reasoning, as the conventional logic does, to a new stage, human thinking.

REFERENCES

C.Anderson (1989), Learning to Control an Inverted Pendulum Using Neural Network, *IEEE Control Systems Magazine*, 31-36.

Deyi Li (1997). On Representing Uncertainty in Commonsense Knowledge, *PACES/SPICIS'97 Conference Proceedings 291-298*, Singapore, 24-27 .

Deyi Li(1998). Uncertainty Reasoning Based on Cloud Models in Controllers, *Computers and Mathematics with Applications*, *Elsevier Science*, **No.3**, **Vol.35**, 99-123,.

D.Saez and A.Cipriano(1997), Design of Fuzzy Model based Predictive Controllers and its Application to an Inverted Pendulum, *Fuzzy-IEEE'97*, 915-919.

H.meier, Z.Farwig and H. Unbehauen(1990). Discrete Computer Control of a Triple-inverted Pendulum, *Optimal Control Applications and Methods*, **Vol.11** 157-172.

H.Tai and S.Shenoi(1994). *Robust Fuzzy Controllers*, 85-90.

J.Eker and J.Astron(1996). A Nonlinear Observer for the Inverted Pendulum, *Proceedings of the 1996 IEEE Int. Conference on Control Applications*, 332-337.

K.Furuta and M.Yamakita(1992). Swing-up Control of Inverted Pendulum Using Pseudo-state Feedback, *J. of Systems and Control Eng*. 263-269.

L.K.Wong, F.Leung & P.Tam (1997), The Design of Stable Fuzzy Controllers with Combination of Converntional Controllers, *Proceedings of ISIE'97*, Portugal, 993-997.

T.Yamakawa (1989), Stabilization of Inverted Pendulum by a high-speed Fuzzy Logic Controller Hardware System, *Fuzzy Sets and Systems*, **Vol.32,** 161-180.

AN ITERATIVE METHOD FOR TUNING DECENTRALIZED PID CONTROLLERS

F. Vázquez, F. Morilla, S. Dormido

fmorilla@dia.uned.es. Dpto de Informática y Automática.Facultad de Ciencias.UNED.
C/ Senda de Rey s/n 28040 Madrid (SPAIN)

Abstract: In this paper a method of tuning decentralized PID controllers for multivariable systems is presented. It is based on an iterative numeric algorithm, and it uses information coming from the frequency response of the full matrix of the system transfer functions. Then, the effects of the interaction are included in the design because the off-diagonal elements in this matrix are taken into account. The aim is to meet the design specifications that were gain margins, phase margins or a combination of both in the different experiments carried out. Three examples are shown: two obtained from the literature and one from a real plant. The results are compared with another type of MIMO tuning.

Keywords: PID controllers, multivariable control systems, gain and phase margins.

1. INTRODUCTION

Although numerous techniques of advanced control exist (robust, predictive, adaptative control, etc) the decentralized control is still frequently used in industry as a strategy of multivariable control, as a mere control technique or used in the most internal control loop and with some of the previous ones in a superior level. The sacrifice that supposes the invariable deterioration of the benefits of a decentralized control structure when is compared with a full multivariable control strategy, is compensated with certain advantages like design and hardware simplicity or easiness of use.

However, the use of this type of strategies is not always possible due to the interaction effects. A severe interaction will produce a loss on closed-loop performance and stability. In these cases it will be necessary to give up the use of these techniques and of think in another type of strategies commented previously. In order to study these interaction effects numerous methods exist, one of most often used in industrial applications being the relative gain matrix (RGA) (Bristol,1966) with their numerous variants, due to their calculation simplicity. The use of this matrix is not only limited to the steady state analysis to solve the pairing problem among variables. Furthermore, it is supplemented with other calculations to provide indexes of stability, or interaction dynamic analysis. In this line, methods like the direct and inverse Nyquist's arrays (DNA and INA), (Rosenbrock,1979; Maciejowsky,1989), supplemented with Gershgorin's bands, and other calculations like the condition numbers, Niederlinsky's index, dynamic RGA, etc, are frequently used.

Once this type of decentralized techniques has been selected for use, either because the interaction is not severe or because only some of the system outputs are needed, and once the pairing problem has been solved, the following step will be choosing the controller and its tuning. Most of the papers published in this field use PID controllers (with their different possibilities) for a multitude of well-known reasons: industrial implantation, robustness, employment easiness... Some methods found in the literature could be classified under the following headings that show the interest that the topic has raised in the last years:

Tuning methods based on heuristic formulas: These formulas indicate the direction in which the parameters have to be detuned to compensate the interaction effects when all the loops are closed. The initial parameters have been previously calculated using some SISO technique, that is, making a design with only the elements of the main diagonal (the g_{ii}) of the system transfer function matrix. (Shinskey,1995; McAvoy, 1983) works are in this line.

Designs based on the relay method: These techniques are able to obtain the different parameters of the controllers from the frequency and amplitude of a maintained oscillation, achieved with the relay when closing the loops progressively. Numerous iteration algorithms exist, like those of (Wang et al, 1996 and 1997; Menani and Koivo, 1996; Halevy et al, 1996; Shiu and Hwang, 1998) to mention some.

True multivariable tuning methods: In this case, in order to calculate the controller parameters, these methods do not use only a part of the information of the transfer function matrix but rather in the design they already take into account the interaction of the off-diagonal elements. One of the most outstanding is the work of Ho et al (1996), that provides on-line tuning formulas to obtain the PID controllers for two by two systems. This is a method that uses Gershgorin's band to provide a correction to the specifications of phase and gain margins and then they enter in a SISO formulation.

The present work could be framed within this last group. In it, an iteration algorithm is proposed to achieve the PID parameters for any multivariable system, paying special attention to those of two inputs and two outputs. First, the tuning method is exposed presenting the notation and the used algorithm. Later the method is applied to three examples: two taken from literature and one developed by the authors.

2. ITERATIVE METHOD FOR TUNING DECENTRALIZED CONTROLLERS

Let suppose a MIMO system with two inputs and two outputs. The use of a decentralized control strategy has been decided and the pairing problem of variables has been solved. It is aimed to control this system by means of two controllers k_1 and k_2. The controller k_i closes a loop between the controlled variable y_i and the manipulated variable u_i. If the controller k_2 has already been tuned by some method and it is sought to make the same thing with the k_1, as can be seen in Figure 1, between the controlled variable y_1 and the manipulated variable u_1 there is something else than the transfer function g_{11}, because an additive action appears. This is due to the existence of one of the hidden loops described in (Shinskey,1995). The effect of this additive action could be represented as the block a_1 shown in Figure 2. In the same figure, the combination of g_{11} and a_1 is denominated $\tilde{g}_{11}$, following the notation of Zhu and Chiu (1998) (in this paper a study of the interaction and stability characterization of this type of systems can be found).

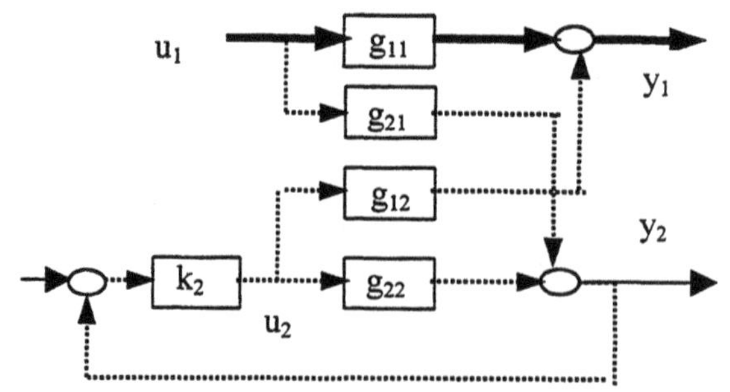

Fig. 1: Interaction of the hidden loop on the loop 1.

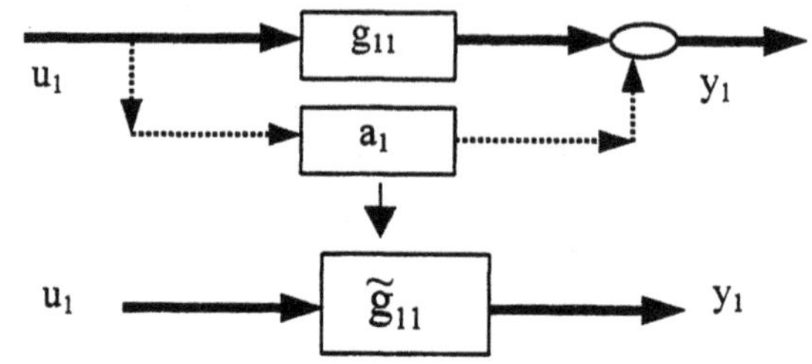

Fig. 2: Effect of the interaction and the final result

With this scheme in mind, the proposed tuning approach instead of calculating a controller for g_{11} (that would be when there is not interaction) will proceed to tune k_1 for $\tilde{g}_{11}$ which already incorporates the effect of this interaction. Nevertheless, the modification of k_1 will have its effect on the other loop, a new tuning of k_2 becoming necessary. This operation could be made successively until the producing changes in k_1 and k_2 are small and each loop meets the design specifications. The calculation could be carried out using some SISO tuning method adapted, as it will be seen later on.

3. TUNING APPROACH FOR PID CONTROLLERS

Let suppose that k_1 and k_2 are two PID controllers described by

$$k_1^i = Kp_1^i\left(1+\frac{1}{Ti_1^i s}+Td_1^i s\right)$$

$$k_2^i = Kp_2^i\left(1+\frac{1}{Ti_2^i s}+Td_2^i s\right) \qquad (1)$$

where the superscript indicates the number of the iteration, and the subscript the loop to which the calculation refers. The algorithm starts with k_1^o and k_2^o obtained exclusively for g_{11} and g_{22} by means of a SISO method or an arbitrary controller that could be Kp=1, Ti=9999, Td=0. The effect of this election has little repercussion in the number of iterations and in the convergence. The following iterative algorithm is proposed: from k_1^o, $\tilde{g}_{11}^o$ is obtained, and from k_2^o, $\tilde{g}_{22}^o$ is obtained, following the expressions:

$$\tilde{g}_{11}^i = g_{11} + a_1^i \qquad \tilde{g}_{22}^i = g_{22} + a_2^i \qquad (2)$$

$$a_1^i = -\frac{k_2^i g_{12} g_{21}}{1 + k_2^i g_{22}} \qquad a_2^i = -\frac{k_1^i g_{12} g_{21}}{1 + k_1^i g_{11}} \qquad (3)$$

If the couples k_1^i $\tilde{g}_{11}^i$ and k_2^i $\tilde{g}_{22}^i$ meet the design specifications the iteration concludes. Otherwise the iteration number increases.

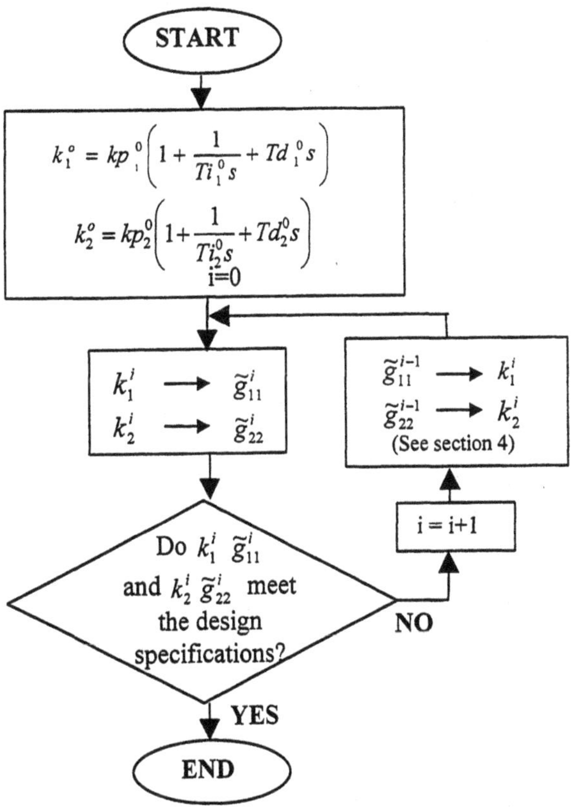

Fig. 3: Flow diagram of the algorithm

The controller k_1^i is then calculated for the model $\tilde{g}_{11}^{i-1}$ and the controller k_2^i for the model $\tilde{g}_{22}^{i-1}$. With the new controllers k_1^i and k_2^i, $\tilde{g}_{11}^i$ and $\tilde{g}_{22}^i$ are recalculated. With these data a new iteration begins. This process is shown in Figure 3.

4. OBTAINING THE k_j^i FROM THE $\tilde{g}_{jj}^i$

In order to obtain the k_j^i from the $\tilde{g}_{jj}^i$ a lot of SISO methods can be employed. However the calculation of the $\tilde{g}_{jj}^i$ from a_j^i is not immediate, due to certain problems: the presence of time delays in the functions g_{ij} can make the analytic calculation complicated. Not less important, the variety of types of models that can take place when making the quotient given in the expression of a_j^i makes it impossible to use a particular formula among the existing ones for models of first order plus delay, two time constant plus delay,..., (Ho et al, 1996). To solve these problems, the proposed method uses an adaptation of a generic tuning method allowing a numeric calculation instead of an analytic one, that is to say, not expressing the g_{ij} as a transfer function but as the array of frequency response of the transfer function. This operation mode is much easier and quicker and the $\tilde{g}_{jj}^i$ will be other response arrays, without an analytic expression in the form of a concrete transfer function.

An adaptation of the analytic method of Morilla and Dormido (1998) has been chosen. This is an extension of the Aström and Hägglund (1984) method consisting in a generalization of the Ziegler-Nichols formulas to move a point of Nyquist's diagram of the open loop transfer function from a position A (in controller's absence) until another position B (in controller's presence), as Figure 4 shows. If the controller is an ideal PID controller given by the following expression

$$k(s) = \mathrm{Kp}\left(1 + \frac{1}{\mathrm{Ti}\ s} + \mathrm{Td}\ s\right) \qquad (4)$$

the determination of its parameters can be summarized in the equations

$$\mathrm{Kp} = \frac{r_b \cos(\phi_b - \phi_a)}{r_a} \qquad (5)$$

$$\mathrm{Ti} = \frac{1}{2\alpha\omega_c}\left(\mathrm{tg}(\phi_b - \phi_a) + \sqrt{4\alpha + \mathrm{tg}^2(\phi_b - \phi_a)}\right) \qquad (6)$$

$$\mathrm{Td} = \alpha\ \mathrm{Ti} \qquad (7)$$

where ω_c is the frequency corresponding to the point A elected in the Nyquist's diagram, r_a and ϕ_a are the gain and the angle of this same point, r_b and ϕ_b the destination point B gain and angle, and α the ratio between the derivative and integral time constants, that should be specified.

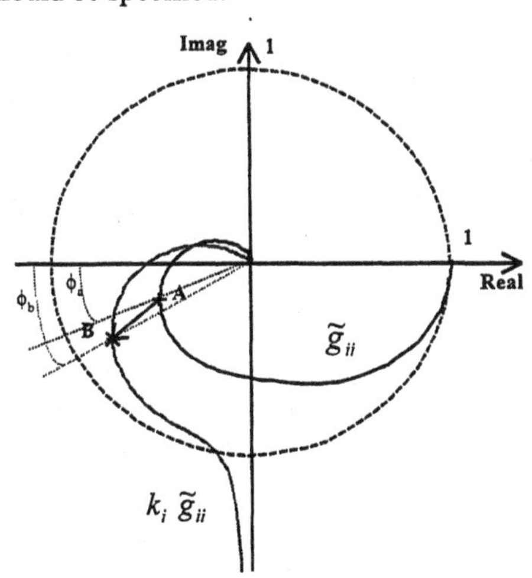

Fig. 4: Process and process plus controller Nyquist's diagram and circle unit. The controller is PID and it has been tuned to move the point A to the point B.

As Morilla and Dormido (1998) show, its design admit three design possibilities: phase margin (PM), gain margin (GM) and a combination of both. Due to space limitations, only the design for phase margin will be described. First of all, the frequency range ω_c that admits solution for a certain controller is calculated, that is to say, the set of points A, keeping in mind that the process included in the algorithm is a frequency response array, $\tilde{g}_{ii}$. From this array, the point A gain and angle, r_a and ϕ_a, are obtained. From the PM specifications, ϕ_b is calculated. Because point B is located on the unit circle, then $r_b = 1$. With these data, Kp and Ti are calculated from expressions (5) to (6) for each point A with possible solution. In order to choose one among all of the solutions, a selection approach is needed. For example, if it is sought to minimize the integral of the absolute value of the error (IAE), the maximum ratio Kp/Ti will be looked for. Parameter α can also be chosen to minimize some other optimization approach.

It is also necessary to notice that it is possible to use any other SISO tuning method whenever it is adapted to solve the problems commented before.

5. EXAMPLES

Example 1: As a first example, a water-methanol distillation column proposed by Ho et al (1996) is analyzed. The system is described by the following transfer function matrix

$$\begin{pmatrix} \dfrac{12.8}{16.7s+1}e^{-s} & \dfrac{-18.9}{21s+1}e^{-3s} \\ \dfrac{6.6}{10.9s+1}e^{-7s} & \dfrac{-19.4}{14.4s+1}e^{-3s} \end{pmatrix} \quad (8)$$

If a PM=20° and a GM=2 for both loops on Ho's tuning formulas are specified, the following parameters are obtained: Kp_1=0.57, Ti_1=20.7, Td_1=0, Kp_2=-0.11, Ti_2=12.88 Td_2=0; while with specifications of PM=30° and GM=3, for both loops, they are Kp_1=0.38, Ti_1=21.64, Td_1=0, Kp_2=-0.07, Ti_2=14.8 Td_2=0. However, if an analysis of these tunings is made the obtained results show that the values are very far from the initial specifications:

Tuning	Loop 1	Loop 2
1th	PM=51.6° GM=3	PM=94° GM=2.2
2th	PM=65° GM=5.3	PM=103° GM=3.8

This is because any design that uses Gershgorin's bands will provide very conservative tunings when the interaction is appreciable. The main reason is that in most of the cases, these bands are excessively wide and they produce a correction in the initial phase and gain margins. Gershgorin's bands for this system with the first tuning can be seen in the graph of Figure 5, superimposed with direct Nyquist array (DNA). In this graph $k_1\tilde{g}_{11}$ and $k_2\tilde{g}_{22}$ (on continuous line) are also superimposed, both being quite far from the band borders, which shows that any design that uses these bands in its calculation process will lead inevitably to extra-conservative tunings. A more exhaustive description of Gershgorin's bands and direct Nyquist's arrays can be found in Rosenbrock (1979) or Maciejowsky (1989).

Using the algorithm proposed in this work with specifications of PM_1=45 and PM_2=45 and beginning with Kp_1=1, Ti_1=9999, Td_1=0 y Kp_2=-1, Ti_2=9999, Td_2=0 the results of Table 1 are obtained for each iteration. It is observed that five iterations have been enough to meet the specifications in the two control loops.

Table1: Results obtained in each iteration for example 1 with specifications of PM_1=PM_2=45°

It.	PM_1	PM_2	GM_1	GM_2	Kp_1	Ti_1	Kp_2	Ti_2
1	0.08	0.09	7.55	1	0.22	2.57	-0.11	4.14
2	37	48	2.39	2.16	0.75	3.33	-0.04	1.67
3	42	40	2.63	1.41	0.70	4.25	-0.09	3.17
4	44	48	2.47	1.51	0.73	3.51	-0.09	3.10
5	45	45	2.48	1.46	0.73	3.56	-0.09	3.11

In Table 2 other tunings are showed, carried out for the PI controllers in example 1 with different sets of specifications: only PM, only GM or combinations of both.

Table 2: Tuning with different specifications for example

Specifications				Performance				Control parameters	
PM_1	*PM_2*	*GM_1*	*GM_2*	PM_1	PM_2	GM_1	GM_2	Kp_1, Ti_1, Td_1	Kp_2, Ti_2, Td_2
40	40	-	-	40	40	0.12	1.36	0.69, 2.88, 0	-0.08, 2.78, 0
40	60	-	-	40	60	2.43	1.43	0.73, 2.88, 0	-0.10, 4.41, 0
60	90	-	-	60	82	2.54	3.32	0.78, 15, 0	-0.06, 7.37, 0
-	-	4	4	20	55	4	4	0.32, 1.64, 0	-0.006, 0.6, 0
-	-	2	5	16	56	2	5	0.66, 1.66, 0	-0.015, 6.15, 0
45	80	4	3	46	80	4	3	0.47, 6.58, 0	-0.06, 6.15, 0
30	65	3	4	30	65	3	4	0.56, 2.49, 0	-0.029, 3.04, 0

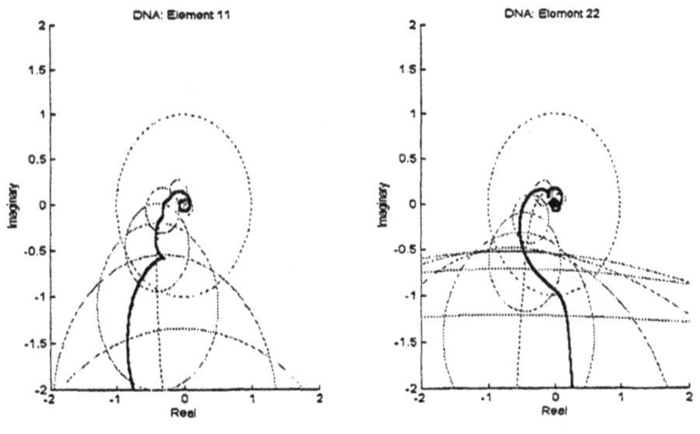

Fig.5: DNA with Gershgorin's bands superimposed for example 1 with the first tuning of Ho's work. $k_1\tilde{g}_{11}$ and $k_2\tilde{g}_{22}$ on continuous line and $k_1 g_{11}$ and $k_2 g_{22}$ in discontinuous

Example 2: It is a four-coupled tanks system, also proposed by Ho et al (1996) whose transfer functions matrix is given by

$$\begin{pmatrix} \frac{4.3}{383s+1}e^{-40s} & \frac{1.8}{383s+1}e^{-140s} \\ \frac{1.2}{281s+1}e^{-80s} & \frac{2.5}{281s+1}e^{-40s} \end{pmatrix} \quad (9)$$

By means of Ho's tuning formulas, if a PM_1=40° and a GM_1=4 for the first loop and a PM_2=30° and a GM_2=3 for the second are specified, the controllers have the following parameters: Kp_1=0.62, Ti_1=217, Td_1=0, Kp_2=0.94, Ti_2=193 Td_2=0. If these tunings are analyzed, the obtained results (PM_1=67°, GM_1=5.35 for the first loop and PM_2=61.5°, GM_2=4.5 for the second) are both far from the specifications, although not as much as in the previous example because Gershgorin's bands are not as wide as in the previous example.

Using the iterative algorithm, designs shown in Table 3 can be obtained, where a great proximity between the specifications and the results can be observed, either in tunings for PM, by GM or combined of PM and GM, and using PI or PID controllers.

Example 3: Lastly, in Table 4 some tunings for the control of a heat exchanger (Morilla and Vázquez, 1997) are shown. The process transfer function matrix is given by the following expression, and its scheme is shown in Figure 6.

$$\begin{pmatrix} T_1 \\ T_2 \end{pmatrix} = \begin{pmatrix} \frac{0.28}{638s+1}e^{-200s} & 0.021\cdot\frac{6000s+1}{1400s+1} \\ \frac{0.49}{824s+1}e^{-200s} & \frac{-0.19}{250s+1}e^{-80s} \end{pmatrix} \begin{pmatrix} Q \\ N_2 \end{pmatrix} \quad (10)$$

Due to the width of Gershgorin's bands (as it can be seen in Figure 7), it was impossible to tune the controllers using Ho's tuning formulas. However, Table 4 shows that the iterative algorithm allows multiple possibilities, most of them tested later in real experiences. Here the difference between the different tuning could be seen.

Table 3: Tuning for different specifications of example 2.

Specifications				Performance				Control parameters	
PM₁	*PM₂*	*GM₁*	*GM₂*	PM_1	PM_2	GM_1	GM_2	Kp_1, Ti_1, Td_1	Kp_2, Ti_2, Td_2
40	30	-	-	40	30	2	1.28	0.88, 101, 0	3.25, 240, 0
67	61	-	-	67	61	5.6	4.9	0.59, 209, 0	0.85, 173, 0
45	45	3	3	45	45	3	3	1.07, 208, 0	1.39, 192, 0
40	30	4	3	40	30	3.8	4.2	0.19, 48, 0	1.24, 91, 0
-	-	6	6	10	10	6	6	0.56, 55, 9.44	0.71, 54, 9.14

Table 4: Tunings for different specifications of example 3.

Specifications				Performance				Control parameters	
PM₁	*PM₂*	*GM₁*	*GM₂*	PM_1	PM_2	GM_1	GM_2	Kp_1, Ti_1, Td_1	Kp_2, Ti_2, Td_2
45	45	-	-	45	45	1.97	4.64	4.8, 436, 0	-6.13, 426, 0
30	70	-	-	30	70	1.85	9.64	6.38, 426, 0	-2.19, 280, 0
70	30	-	-	70	30	2.02	2.55	3.89, 627, 0	-10.1, 243, 0
-	-	3	3	35	27	3	3	0.76, 91, 0	-5.85, 97, 0
-	-	2	5	32.6	27.3	2	5	1.78, 175, 0	-3.71, 111, 0
45	45	-	-	43	47	2.15	3.45	7.1, 554, 77	-2.3, 132, 75
45	45	3	3	45	46	2.89	2.99	2.63, 263, 0	-9.91, 321, 0
45	45	3	3	45	43	3.12	3	2.73, 230, 12	-13.1, 231, 24
60	35	-	-	60	36	1.75	2.36	6.31, 528, 53	-9.47, 271, 27
75	60	5	6	74.7	61	4.8	6.06	2.1, 585, 0	-4.5, 250, 0

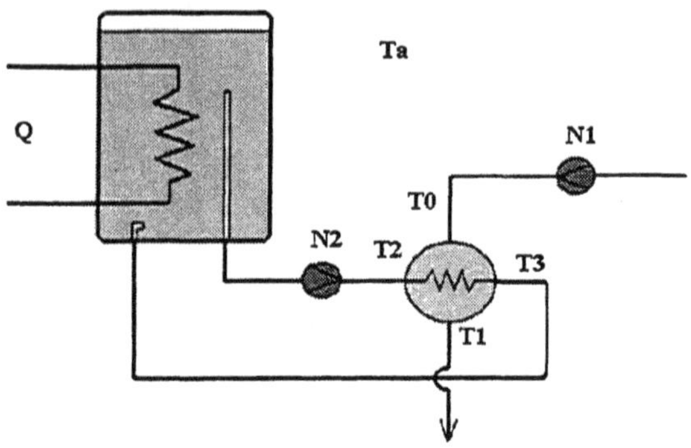

Fig. 6: Process scheme of example 3.

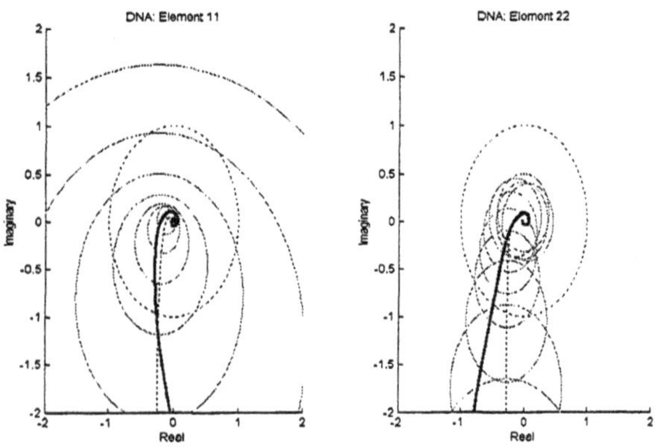

Fig. 7: DNA with Gershgorin's bands superimposed for example 3

In Figure 8 a graph of the real temporal response of the two temperatures is shown, where the set points of T_1 (output process temperature) were changed. The control parameters correspond to a conservative design, with the characteristics of the last row of Table 4.

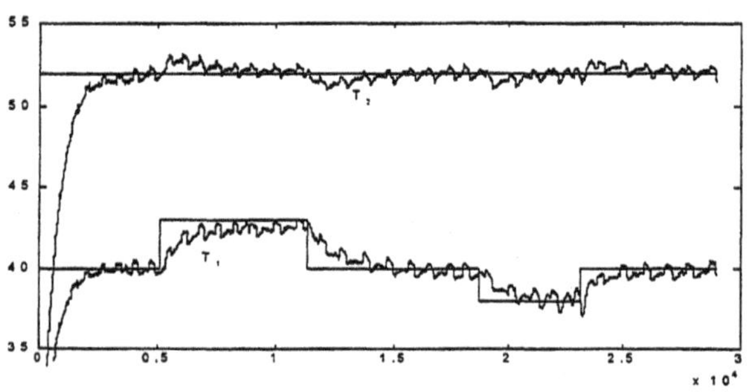

Fig. 8: Time response of the system of example 3 under control

CONCLUSIONS

In this paper, a new multiloop controller tuning method has been presented. The method, based on an iteration algorithm, has been compared with some of the examples used in the literature of MIMO tuning, achieving designs nearer to the specifications. Inside the algorithm any generic method of SISO controllers tuning could be used but with some variation. The difficulty that present some design methods that use Gershgorin's bands has also been shown, because with these, very conservative designs appear in the best of the cases, since in other they don't obtain a solution. In contrast, the method has been successfully applied to tuning the PID controllers of several real systems, like a heat exchanger.

ACKNOWLEDGEMENTS

This work was supported by the CICYT (Comisión Interministerial de Ciencia y Tecnología) under Grant TAP 96-0404.

REFERENCES

Aström, K. J., T. Hägglund (1984). Automatic tuning of simple regulators with specification on phase and amplitude margins. *Automatica*, **20**, 645-651.

Halevy, Y, Palmor, Z., Efrati,T. (1997): Automatic tuning of decentralized PID controllers for MIMO processes, *J. Proc. Cont.* **Vol. 7 N° 2,** pp 119-128 Elsevier Science, Ltd.

Ho, W.K., T.H. Lee, O.P. Gan (1996): Tuning of multiloop PID controllers based on gain and phase margins specifications. *13th IFAC World Congress,* pp 211-216.

Maciejowski, J.M. (1989). In: *Multivariable feedback design.* Addison-Wesley

McAvoy,T. (1983). In: *Interaction analysis: principles and applications.* Instrument Society of America.

Menani, S., Koivo, H. (1996): Relay tuning of multivariable PI controllers, *13th IFAC World Congress,* pp 139-144

Morilla, F., Dormido, S. (1998): Tuning PID controllers using specifications of frequency response. *Internal Report.* Dpto. de Informática y Automática.UNED

Morilla, F., Vázquez, F. (1997): The pasteurizer plant PCT-23. *Internal Report.* Dpto. de Informática y Automática.UNED

Shinskey F. G. (1995). In: *Process Control Systems.* N.Y.: McGraw-Hill.

Shiu, S.J., Hwang,S. (1998): Sequential design method for multivariable decoupling and multiloop PID controllers, *Ind. Eng. Chem. Res.,* **Vol. 37, N° 1,** pp 107-119

Rosenbrock, H.H. (1979). In: *Computer-aided control system design.* N.Y. Academic Press

Wang, Q., Hang, C. and Zou,B. (1996): A frequency response approach to auto-tuning of multivariable PID controllers, *13th IFAC World Congress,* pp 295-300

Wang, Q., Zou, B., Lee, T., and Bi, Q. (1997): Auto-tuning of multivariable PID controllers from decentralized Relay Feedback, *Automatica,* **Vol. 33 N° 3** pp 319-330

Zhu,Z.X., Chiu, M. (1998): Dynamic analysis of decentralized 2x2 control systems in relation to loop interation and local stability. *Ind. Eng. Chem. Res.,* **Vol. 37, N° 2,** pp 464-473

14th Triennial World Congress, Beijing, P.R. China

EVOLUTIONARY COMPUTATION IN DESIGNING A PID CONTROL SYSTEM

Yasue Mitsukura*, Toru Yamamoto*, Masahiro Kaneda* and Kenzo Fujii**

* *Department of Communication Engineering, Okayama Prefectural University 111 Kuboki, Soja, Okayama 719-1197, Japan*

** *Tokuyama Plant, Idemitsu Kosan Co. Ltd., 1-1 Miyamae-Cho, Tokuyama, Yamaguchi 745-8613, Japan*
yama@c.oka-pu.ac.jp

Abstract: PID control schemes have been widely used in most of process control systems represented by chemical processes for a long time. However, it is still a very important problem how to determine or tune the PID parameters, because these parameters have a great influence on the stability and the performance of the control system. On the other hand, in the last twenty years, a genetic algorithm(GA) is attracted as one method which gives us suitable answers for search, optimization and machine learning problems. In this paper, a new genetic tuning algorithm of PID parameters is proposed, in which the search area of PID parameters is reduced sharply by considering an effective parameters' area from the viewpoint of the control engineering. Finally, the proposed scheme is numerically evaluated, and discuss the statistical properties. *Copyright © 1999 IFAC*

Keywords: PID control, Genetic algorithms, process control, Minimum variance control, Nonlinear systems

1. INTRODUCTION

In the field of the control theory, due to the marvelous development of the computer technology, many studies of advanced complex control schemes have been carried out vigorously for a long time, and are continued even now. However, in the field of the process control, for example, various chemical processes, the conventional PID control scheme is used mainly even nowadays. Now, since it plays a great role in determining the behavior of the control system how to tune the PID parameters, a lot of tuning methods have been studied up to now(Zieglar and Nichols, 1942; Chien, *et al.*, 1952; Åström and Hägglund, 1988; Åström, *et al.*, 1992). The authors also have proposed a tuning method(Yamamoto and Kaneda, 1998) of PID parameters from the viewpoint of the self-tuning control scheme based on the relationship between the PID control law and the generalized minimum variance control(GMVC) law(Clarke and Gawthrop,1979), and have illustrated the effectiveness of the proposed method by applying it to a chemical process(Fujii, *et al.*, 1997). However, there were some user-specified parameters in it, and it was left how to design those parameters.

On the other hand, the genetic algorithm(GA), which simulated the mechanism of heredity or evolution in the living things, has attracted our attentions as a method which gives us fairly quickly suitable or suboptimal answers for search, optimization or machine learning problems(Goldberg, 1992). Moreover, some searching methods of PID parameters using GA have also been reported(Porter and Jones, 1992; Jones, *et al.*, 1996; Omatsu and Deris, 1996). However, since three PID parameters are directly searched in those methods, it costs fairly long time to get a suitable set of values due to the wide search area.

In this paper, we propose a new tuning method of PID parameters, in which the user-specified parameters included in the tuning method of PID parameters proposed previously are sought using GA, and are transformed into the corresponding PID parameters. That is, it means that a suitable set of the user-specified parameters in the performance criterion of the GMVC law is sought first using GA, and then a set of PID parameters is tuned using the above set as a mediator. Here, since the user-specified parameters are restricted due to the control engineer sense, that is, taking the control performance of the closed-loop system into consideration, its search area can be reduced sharply. Therefore, the time to get a suitable set of PID parameters is shorten drastically. This paper is organized as follows. First, we discuss the describing model of the controlled object and explain briefly the tuning method of PID parameters proposed previously. Next, we propose a new tuning method of PID parameters mentioned above. Finally, we illustrate the effectiveness of the proposed method with a numerical simulation example, and discuss the statistical properties of the newly proposed scheme.

2. PROBLEM FORMULATION

2.1 System description

Let z^{-1} be the backward shift operator, and the following discrete-time model describes the process:

$$\alpha(z^{-1})y(t) = z^{-k}\beta(z^{-1})u(t-1) + f(u,y) \quad (1)$$

where

$$\left.\begin{aligned} \alpha(z^{-1}) &= 1 + \alpha_1 z^{-1} + ... + \alpha_{n_1} z^{-n_1} \\ \beta(z^{-1}) &= \beta_0 + \beta_1 z^{-1} + ... + \beta_{n_2} z^{-n_2}. \end{aligned}\right\} \quad (2)$$

Moreover, $u(t)$ and $y(t)$ are the input and output, respectively. k is the dead time and $f(u, y)$ implies the nonlinear components. In this paper, we call the system model (1) the 'full model', because it explains the behavior of the system as in detail as possible. Process control systems, in general, are described as the models like (1) which are high order and have some nonlinear components. In this paper, though we assume that the full model (1) is given, if it is not so, the full model may be replaced by an appropriate model, for example, a neural network which simulates approximately the behavior of the controlled object. It is used only for calculating the fitness function in GA as shown in the section 4.4.

On the other hand, consider the second-order linear model which approximates the full model (1) by the following equation:

$$A(z^{-1})y(t) = z^{-k}B(z^{-1})u(t-1) \quad (3)$$

where

$$\left.\begin{aligned} A(z^{-1}) &= 1 + a_1 z^{-1} + a_2 z^{-2} \\ B(z^{-1}) &= b_0 + b_1 z^{-1} + ... + b_m z^{-m}. \end{aligned}\right\} \quad (4)$$

In general, since the control law based on the full model (1) seems to be complex and very difficult in implementation, the second-order linear model (3) is often used in designing the process control system. Therefore, we use the model (3) as a 'design-oriented model'. Note that a_2 is set to 0 in (4) if the controlled object is approximately expressed as a first-order system. Moreover, the degree of $B(z^{-1})$, m, should be set large enough in order to cope with the higher components which can not be approximated as the second-order linear system.

2.2 Brief review of PID control law

Consider the velocity-type PID control law(Cameron and Seborg, 1983) described as

$$\Delta u(t) = k_c\{\Delta + \frac{T_s}{T_I} + \frac{T_D}{T_s}\Delta^2\}e(t) \quad (5)$$

where

$$e(t) := w(t) - y(t). \quad (6)$$

Δ denotes the differential operator defined as $\Delta := 1 - z^{-1}$. $w(t)$ is the reference signal and is given by piecewise constant components. k_c , T_I and T_D are the proportional gain, the reset time and the derivative time, respectively. Moreover, T_s denotes the sampling time.

For convenience, let $C(z^{-1})$ be

$$C(z^{-1}) := k_c\{\Delta + \frac{T_s}{T_I} + \frac{T_D}{T_s}\Delta^2\} \quad (7)$$

then, (5) can be rewritten by

$$C(z^{-1})y(t) + \Delta u(t) - C(z^{-1})w(t) = 0. \quad (8)$$

Since it plays a great role in determining the behavior of the closed-loop system how to tune the PID parameters in (5) or (8), a lot of works for PID tuning schemes have been reported up to now. The authors also have proposed a new tuning method of PID parameters based on the relation to the GMVC law. However, some user-specified parameters were involved in the proposed scheme, and it was left how to design them. Then, in this paper, we consider a tuning method of the PID parameters in which a suitable set of the user-specified parameters is sought using GA and then the corresponding PID parameters are calculated using the above user-specified parameters as mediators.

3. PID PARAMETER TUNING

In this section, based on the relationship between the PID control law and the GMVC law, we introduce how to tune PID parameters(Yamamoto and Kaneda, 1998) in brief. Consider the following cost function to derive a GMVC law:

$$J = \{P(z^{-1})y(t+k+1) + \lambda\Delta u(t) - R(z^{-1})w(t)\}^2. \quad (9)$$

λ in (9) is the weighting factor with respect to the control input, and $P(z^{-1})$ is the user-specified design polynomial of the form:

$$P(z^{-1}) = 1 + p_1 z^{-1} + p_2 z^{-2} \quad (10)$$

$$\left.\begin{aligned} p_1 &= -2e^{-\frac{\rho}{2\mu}}\cos\frac{\sqrt{4\mu-1}}{2\mu}\rho \\ p_2 &= e^{-\frac{\rho}{\mu}} \\ \rho &= \frac{T_s}{\sigma} \\ \mu &= 0.25(1-\delta) + 0.51\delta. \end{aligned}\right\} \quad (11)$$

σ and δ in (11) are parameters related to the rise-time and the damping property, respectively(Yamamoto, *et al.*, 1992). $R(z^{-1})$ in (9) is determined based on the relationship between PID control and GMVC laws. The control input minimizing the cost function (9) is given by the following equation(Clarke and Gawthrop, 1979):

$$\begin{aligned} F(z^{-1})y(t) &+ \{E(z^{-1})B(z^{-1}) + \lambda\}\Delta u(t) \\ &- R(z^{-1})w(t) = 0 \end{aligned} \quad (12)$$

where $F(z^{-1})$ and $E(z^{-1})$ are obtained by solving the following Diophantine equation:

$$P(z^{-1}) = \Delta A(z^{-1})E(z^{-1}) + z^{-(k+1)}F(z^{-1}) \quad (13)$$

$$\left.\begin{aligned} E(z^{-1}) &= 1 + e_1 z^{-1} + \cdots + e_k z^{-k} \\ F(z^{-1}) &= f_0 + f_1 z^{-1} + f_2 z^{-2}. \end{aligned}\right\} \quad (14)$$

In order to obtain a control law with a PID structure, consider the following equation which is replaced $E(z^{-1})B(z^{-1})$ by the static gain $E(1)B(1)$:

$$\begin{aligned} F(z^{-1})y(t) &+ \{E(1)B(1) + \lambda\}\Delta u(t) \\ &- R(z^{-1})w(t) = 0. \end{aligned} \quad (15)$$

Here, ν is defined as

$$\nu := B(1)E(1) + \lambda, \quad (16)$$

and then (15) can be rewritten as

$$\frac{F(z^{-1})}{\nu}y(t) + \Delta u(t) - \frac{R(z^{-1})}{\nu}w(t) = 0. \quad (17)$$

Furthermore, if the following relations are satisfied:

$$\left.\begin{aligned} R(z^{-1}) &= F(z^{-1}) \\ C(z^{-1}) &= \frac{F(z^{-1})}{\nu}, \end{aligned}\right\} \quad (18)$$

(17) becomes identical to (8). Therefore, if we choose the parameters, λ, σ and δ, the PID parameters can be calculated based on (8) and (18) as follows:

$$\left.\begin{aligned} k_c &= -\frac{1}{\nu}(f_1 + f_2) \\ T_I &= -\frac{f_1 + 2f_2}{f_0 + f_1 + f_2} T_s \\ T_D &= -\frac{f_2}{f_1 + 2f_2} T_s. \end{aligned}\right\} \quad (19)$$

By using GA, a scheme to search for the suitable values of λ, σ and δ is considered in the following section.

4. EVOLUTIONARY COMPUTATION OF PID PARAMETERS USING GA

4.1 Summary of evolutionary computation of PID parameters

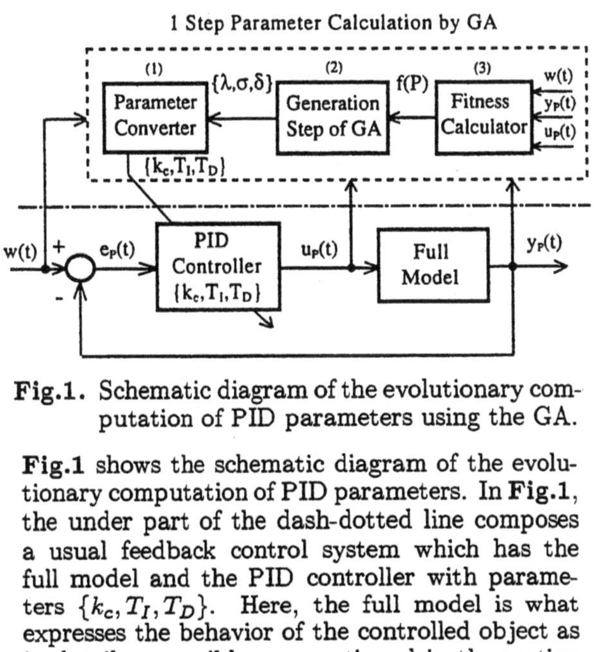

Fig.1. Schematic diagram of the evolutionary computation of PID parameters using the GA.

Fig.1 shows the schematic diagram of the evolutionary computation of PID parameters. In **Fig.1**, the under part of the dash-dotted line composes a usual feedback control system which has the full model and the PID controller with parameters $\{k_c, T_I, T_D\}$. Here, the full model is what expresses the behavior of the controlled object as in detail as possible, as mentioned in the section 2.1.

On the other hand, the upper part of the dash-dotted line is the tuning part of PID parameters during the update of one generation in GA. It is composed of (a) the parameter conversion part in which the user-specified parameters $\{\lambda, \sigma, \delta\}$ sought by GA are converted to the corresponding PID parameters, (b) the updating procedure part of the generation when the fitness function $f(P)$ of each individual is given, and (c) the calculation part of the fitness function of each individual.

The feature of the proposed method can be summarized as follows:

1) As mentioned in the section 3, PID parameters are not directly sought by GA, but they are calculated using the user-specified parameters $\{\lambda, \sigma, \delta\}$ in the performance criterion of the GMVC law, which are sought by GA. Here, since the parameters $\{\lambda, \sigma, \delta\}$ are restricted by taking the control performance of the closed-loop system into consideration, the search area of them is reduced sharply. Therefore, the time to get a suitable set of PID parameters can be shorten drastically.
2) By means of selecting the fitness function appropriately, the response of the closed-loop system can be shaped fairly freely. That is, the rise property, the overshoot, the fluctuation of the control input, and so on can be specified fairly in detail.
3) The full model does not have to be an analytical formulation, but it may be only the model whose input-output property is closely similar to that of the controlled object in a sense. Therefore, as mentioned in the section 2, various neural network models can be used, and it is possible to take any nonlinear property in the model freely.

Now, consider more precisely the tuning part of PID parameters (a), (b) and (c) mentioned above.

4.2 Parameter converter

The calculation flow of the parameter conversion is as follows:

$$\xrightarrow{\text{GA}} \{\lambda, \sigma, \delta\} \xrightarrow{\text{eqn.(11)}} \{\lambda, p_1, p_2\} \longrightarrow$$

$$\xrightarrow{\text{eqns.(13) and (14)}} \{\lambda, e_1, \cdots, e_k, f_0, f_1, f_2\} \longrightarrow$$

$$\xrightarrow{\text{eqn.(16)}} \{\nu, f_0, f_1, f_2\} \xrightarrow{\text{eqn.(19)}} \{k_c, T_I, T_D\}$$

The significant search spans of the parameters λ, σ and δ in the control engineering sense can be considered as follows.

On λ, investigating the stability of the closed-loop system using an appropriate approach, for example, the Bode diagram, the search span of λ can be determined so as to give a suitable gain or phase margin. On σ, since it is said that it is desirable to set it between 1/3 and 1/2 of the system's time constant, it is suggested that the practical search span of σ had better to be set a little larger than the above span. Furthermore, on δ, it is known that, roughly speaking, somehow a desirable response of the closed-loop system can be obtained when δ is set between 0.0 and 2.0(Yamamoto, *et al.*, 1992). Therefore, the practical search span of δ also had better to be a little larger than the above span.

From these control engineering restrictions, the seek span of the corresponding PID parameters can be limited to fairly small area in practice.

4.3 Genetic algorithm

We show the concrete genetic algorithm in order that a suitable set of parameters $\{\lambda, \sigma, \delta\}$ is sought.

Let the geno types of them be G_λ, G_σ, and G_δ, respectively. Each of them is expressed as a n-bits string. Furthermore, gathering them together, the geno type G of each individual in GA is shown as follows:

$$G = [G_\lambda, G_\sigma, G_\delta]. \quad (20)$$

The pheno type P corresponding to G is shown as

follows:

$$P = [P_\lambda, P_\sigma, P_\delta]. \tag{21}$$

Here, the value of each component P_λ, P_σ and P_δ included in P is calculated in the following way: First, regarding each component G_λ, G_σ and G_δ in G as a binary number with n-bits, it is converted into a real number, respectively. Then, each of P_λ, P_σ and P_δ is obtained by scaling the corresponding real number, considering the each seek span of λ, σ and δ mentioned above as the full scale, respectively.

Let the value of the fitness function calculated with the pheno type P be $f(P)$, and then each individual in GA is expressed as follows:

$$I = \{G, P, f(P), p_m\} \tag{22}$$

where p_m denotes the probability of the mutation. By the way, it is necessary to convert the pheno type P into PID parameters $\{k_c, T_I, T_D\}$ through the parameter converter mentioned in the section 4.2, before calculation of the fitness function. Thus, expressing the individual at the i-th generation as I^i, the procedure of GA is shown as follows:

1. Initialization (production of the individual population $\{I^0\}$, setting some parameters in GA and $i = 1$).
2. Calculation of the fitness function $f(P)$.
3. Reproduction of the next generation's population $\{I^i\}$ through selection, crossover and mutation.
4. Update i and return to 2.

4.4 *Fitness function*

As mentioned above, the response of the closed-loop system can be shaped fairly freely by using a suitable fitness function. For example, consider the following as a fitness function:

$$f(P) = \sum_{t=1}^{N} \{y_m(t) - y_P(t)\}^2. \tag{23}$$

$f(P)$ by (23) corresponds to the quadratic performance criterion which is often used in most of the control system design†. $y_m(t)$ denotes the reference model output excited by the reference signal $w(t)$. $y_P(t)$ is the output when the PID controller is employed for the full model, whose PID parameters correspond to the pheno type P of each individual in GA.

Usually, a kind of quadratic performance criteria is mainly used so that it may be convenient to obtain a control law analytically. However, in parameter tuning by GA, it is a feature that the fitness function equivalent to the performance criterion can be freely chosen for the control purpose as mentioned above.

5. SIMULATION EXAMPLE

In this section, the newly proposed scheme is evaluated on a numerical simulation example. Consider the following nonlinear system expressed by a Hammerstein model:

$$y(t) = 0.8y(t-1) - 0.3y(t-2) + 0.4x(t-1) + 0.2x(t-2) \tag{24}$$

where $x(t-1)$ is defined by

$$x(t-1) := u(t-1) + 0.5u^2(t-1). \tag{25}$$

Suppose that the full model is also given by (24) and (25).

Carrying out the parameter estimation by a recursive least squares method using the input-output data of (24) and (25), the design-oriented model was obtained as follows:

$$y(t) = 1.11y(t-1) - 0.47y(t-2) + 1.58u(t-1) + 0.26u(t-2) - 0.47u(t-3). \tag{26}$$

First, for comparison purposes, the control result by Chien, Hrones & Reswick(CHR) method(1952), which is well-known as a PID parameter tuning method, is shown in **Fig.2**. Here, the values of PID parameters were chosen as $k_c = 0.36$, $T_I = 4.07$ and $T_D = 1.25$. **Fig.2** illustrates that there is a some oscillation in the transient state, and the settling time becomes large.

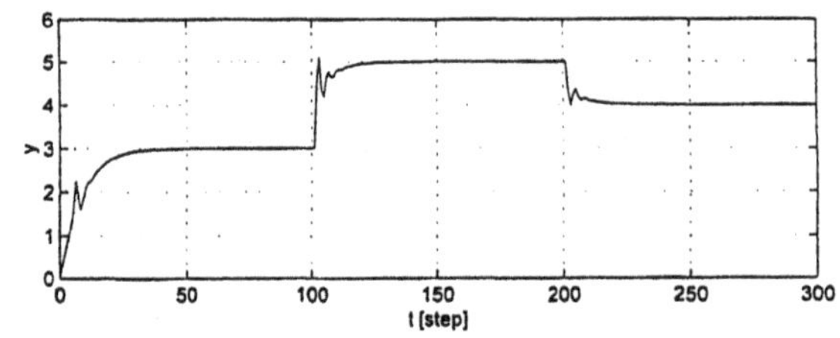

Fig.2. Control result by using CHR type PID control scheme.

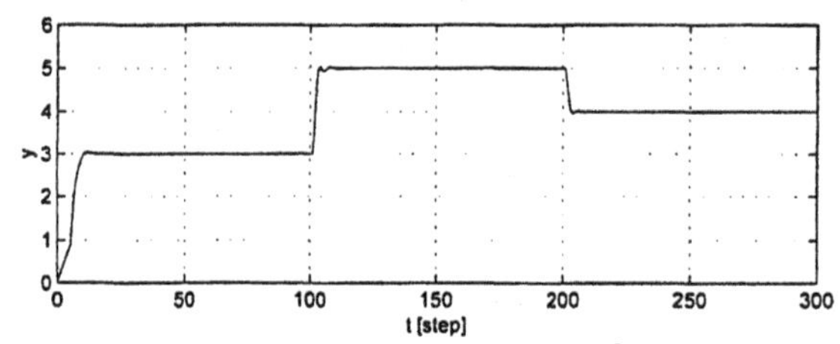

Fig.3. Control result by using the proposed PID control scheme.

Next, the newly proposed scheme was employed, and the control result is shown in **Fig.3**, where the parameters used in GA were set as follows.

- Individual: λ, σ and δ were respectively expressed by 8-bit binary strings, and the geno type of each individual corresponding to (20) was expressed by 24 bits. The number of the population was set to 50.
- Generation: The maximum generation number was set to 20.
- Reproduction & Crossover: The conventional Elite strategy(Goldberg, 1992) was employed, that is, best 10 populations in fitness values were reserved without being changed and 40 populations were reproduced based on the crossover at the random point. The parents in the crossover were also selected at random.
- Mutation: The ratio of the mutation was set to 0.1.
- Seek area: $0.1 \leq \lambda \leq 2.0$, $1.0 \leq \sigma \leq 10.0$, $0.0 \leq \delta \leq 2.0$

† Taking meaning of fitness into consideration, the inverse of the cost function is usually used as a fitness function in GA. However, in order to make the relation to the control theory clear, (23) is defined as a fitness function in this paper.

Moreover, the reference model was designed as:

$$y_m(t) = \frac{z^{-1}0.40}{1 - 0.74z^{-1} + 0.14z^{-2}} w(t). \qquad (27)$$

PID parameters calculated using λ, σ and δ which was sought by GA were the following: $k_c = 0.11$, $T_I = 0.55$ and $T_D = 1.38$, and the sampling interval T_s was set to 1.0[sec].

From **Fig.3**, it is clear that the oscillation is removed, and that the newly proposed scheme works well.

Next, consider the statistical property of the GA's search. Considering the above GA's search as one trial, 100 trials are carried out with changing the initial population. The transitions of the mean value f_a and the best value f_b of the fitness function are shown in **Fig.4**(a) and (b), respectively, which are defined as the following equations:

$$f_a(i) = \frac{1}{100} \sum_{j=1}^{100} \bar{f}(i,j) \qquad (28)$$

$$f_b(i) = \frac{1}{100} \sum_{j=1}^{100} f_{min}(i,j) \qquad (29)$$

where

$$\bar{f}(i,j) = \frac{1}{50} \sum_{k=1}^{50} f(i,j,P_k) \qquad (30)$$

$$f_{min}(i,j) = \min_k f(i,j,P_k), \qquad (31)$$

and $f(i,j,P_k)$ implies the fitness value of the k-th individual of the i-th generation in the j-th trial.

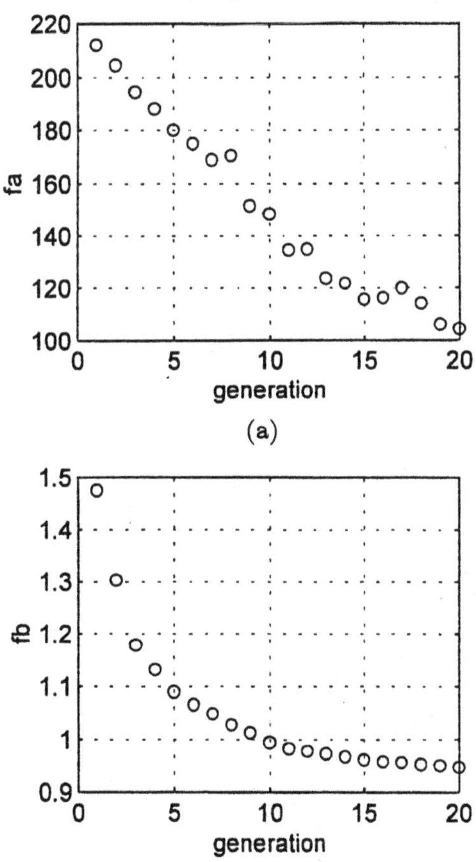

Fig.4. Evaluation of the mean f_a and the best f_b of fitness values over 100 trials.

From **Fig.4**, it is found that both the mean value f_a and the best value f_b are improved desirably without the initial convergence as the generation increases. Especially, it can be thought that the best value f_b has converged in about fifteenth generation.

Furthermore, the distributions of k_c and T_I of PID parameters in the 20th generation, are shown in **Fig.5**, which was obtained in the above trials. Fig.5 indicates that the relationship between k_c and T_I can be regarded to be approximately linear. Furthermore, we obtained the similar results with respect to the relation between k_c and T_D. They are shown in **Fig.6**. Then, it is verified that k_c and T_D are in inverse proportion to each other.

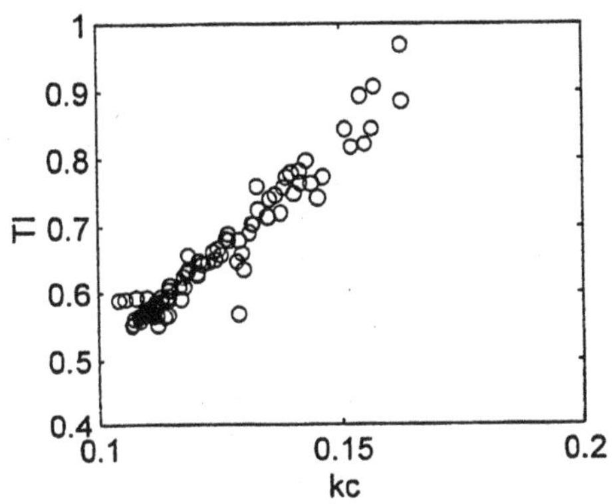

Fig.5 Distributions of k_c and T_I.

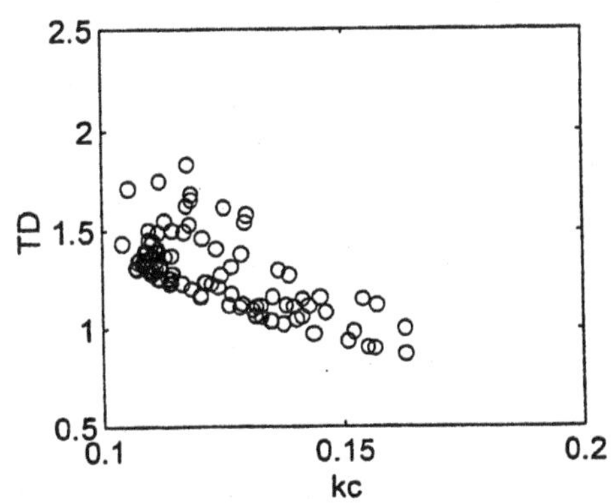

Fig.6 Distributions of k_c and T_D.

By the way, from (5), k_c/T_I and $k_c \cdot T_D$ express the gains of the integral action and the derivative action, that is, the integral gain and the derivative gain, respectively. Though various sets of PID parameters have been obtained in GA's search, it is found that the integral gain and integral gain and the derivative gain converge to certain constants, as the generation goes on, respectively. The proportional gain k_c also converges to a small range enough. Therefore, it can be considered that a suitable set of PID parameters has obtained in a fairly few generations from the aspect of the control action.

As mentioned in the introduction, PID parameters have been searched directly using GA in references(Porter and Jones, 1992; Jones, *et al.*, 1996; Omatsu and Deris, 1996). We will show a comparison between one of those direct GA's search methods and our proposed indirect GA's search method. **Fig.7** shows the convergence behavior of the best fitness value f_b of each method. In this figure, the mark ○ shows f_b of the direct GA's search method, and the mark + shows one of the proposed method. The latter is the same one in **Fig.4**(b). The GA's strategy used in the direct search method is the same as one of the proposed method, except that each individual is expressed as 24 bits binary with 8 bits binary k_c, T_I and T_D. However, we set the search area as $0.12 \leq k_c \leq 1.08$, $1.36 \leq T_I \leq 12.21$ and $0.4 \leq T_D \leq 3.75$, considering the values of the PID parameters by CHR method used in the experiment of **Fig.2**.

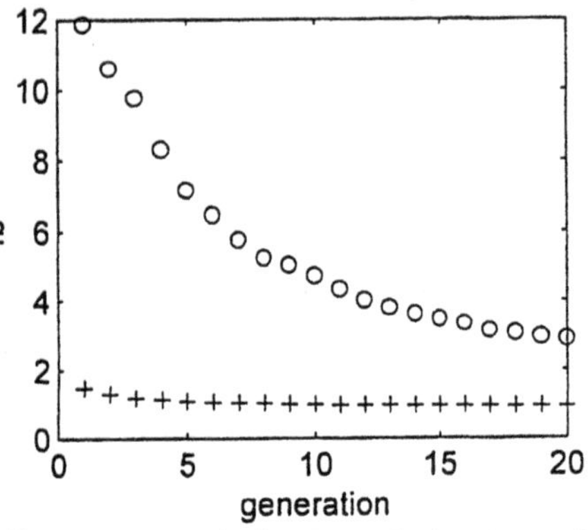

Fig.7 Convergence behavior of the best fitness values f_b.

From **Fig.7**, it is found that the proposed method is superior to the direct GA's search method, because the former has the smaller fitness value than one of the latter in the early generation, and so the required computation time is even less than one of the latter. Furthermore, in the latter, the control system happened to be thrown into the unstable state, and then the calculation of the fitness function happened to be unable when some sets of PID parameters are selected. On the contrary, in the proposed method, such a phenomenon does not happen, because we can set parameter λ so that the control system maintains to be stable. From these reasons, the effectiveness of the proposed method was confirmed.

6. CONCLUSIONS

In this paper, a new genetic tuning method of PID parameters was proposed. Moreover, it was confirmed that the proposed method was useful and effective in a nonlinear simulation example. The newly proposed method is what a suitable set of the user-specified parameters in the performance criterion of the GMVC law is sought by GA, and using the above set as a mediator, PID parameters are calculated in. Therefore, since the parameters which should be sought by GA can be limited by the control engineering restrictions, the seek spans are restricted to a fairly small area, and the time required to seek a suitable set of the parameters can be shortened drastically. And also, the statistical property of the GA's search in the proposed method was investigated. As a result, some valuable and remarkable features were obtained.

Now, in most of design-oriented models, there exist various modeled errors and uncertainties due to the change of the operating conditions. In future, somehow the control scheme taking the uncertainties of the system into consideration will have to be investigated, for example, by fusing the self-tuning control scheme or the robust control scheme with the proposed method. Especially, in the proposed method, since the time to seek parameters is very short, it seems to be possible to tune PID parameters on-line.

REFERENCES

Åström, K.J. and T.Hägglund (1988). *Automatic Tuning of PID Controllers*, Instrument Society of America.

Åström, K.J., C.C.Hang, P.Persson and W.K.Ho (1992). Towards Intelligent PID Control, *Automatica*, **28**, 1-9.

Cameron, F. and D.E.Seborg (1983). A Self-Tuning Controller with a PID Structure, *Int. J. of Control*, **38**, 401-417.

Chien, K.L., J.A.Hrones and J.B.Reswick (1952). On the Automatic Control of Generalized Passive Systems, *Trans. ASME*, **74**, 175-185.

Clarke, D.W. and P.J.Gawthrop (1979). Self-Tuning Control, *Proc. IEE*, **126D**, 633-640.

Fujii, K., T.Yamamoto and M.Kaneda (1997). Self-Tuning Control of a Polybutene Process System, *Proc. of IFAC Symposium on Advanced Control of Chemical Process*, Banff, 107-112.

Goldberg, D.E. (1992). *Genetic Algorithm in Search, Optimization & Machine Learning*, Addison-Wesley.

Jones, A.H., N.Ajlouni and M.Uzam (1996). On-Line Frequency Domain Identification and Genetic Tuning of PID Controllers, *Proc. of IEEE Conference on Emerging Technologies and Factory Automation*, Hawaii, 261-266.

Omatu, S. and S.Deris (1996). Stabilization of Inverted Pendulum by the Genetic Algorithm, *Proc. of IEEE International Conference on Evolutionary Computation*, Nagoya, 700-705.

Porter, B. and A.H.Jones (1992). Genetic Tuning of Digital PID Controllers, *Electronics Letter*, **28**, 843-844.

Yamamoto, T., H.Ishihara, S.Omatu and T.Kitamori (1992). Self-Tuning Controller with Two-Degree-of-Freedom PID Structure and Its Application, *Preprints of IFAC Symposium on Adaptive Systems in Control and Signal Processing*, Grenoble, 527-532.

Yamamoto, T., K.Fujii and M.Kaneda (1998). Design and Implementation of a Self-Tuning PID Controller, *Preprints of IFAC Workshop on Adaptive Control and Signal Processing*, Glasgow, 58-63.

Ziegler, J.G. and N.B.Nichols (1942). Optimum Settings for Automatic Controllers, *Trans. ASME*, **64**, 759-768.

14th Triennial World Congress, Beijing, P.R. China

FORMAL VERIFICATION OF PLC-PROGRAMS

Dr. Th. Filkorn, M. Hölzlein, Dr. P. Warkentin, M. Weiß

Siemens Corporate Technology
81730 Munich, Germany
email: markus.hoelzlein@mchp.siemens.de

Abstract: Siemens Corporate Technology has developed a methodology for formal verification. This methodology allows to prove or falsify the correctness of programs for a PLC (programmable logical controller) *fully automatically* and *exactly* in a strongly mathematical sense. A program is correct, if every possible execution trace of the program fulfils a set of required properties specified by the user. For instance, a property for a traffic light control at some given intersection could be 'traffic lights never all show green at the same time' or 'when a car stops at a red traffic light that traffic light will eventually switch to green'. For this verification method a tool has been developed that checks properties fully automatically and completely. In case a given property is not satisfied by the PLC-program, the verification tool generates an execution trace of the program (a so-called countersequence) that illustrates its erroneous behaviour. In a pilot application, tool machine manufacturers used and evaluated the verification tool with respect to specification, error detection, cost/benefit-estimation and other criteria for verification. The verification tool was received very well. *Copyright © 1999 IFAC*

Keywords: Formal Verification, Control Design, Programming Support, Debugging, Error Detection, Productivity, Quality Control.

1. WHAT IS FORMAL VERIFICATION?

This paper introduces a methodology with the expression 'Formal Verification of PLC-Programs' that allows to give a mathematical proof for the correctness of PLC-programs with regard to a formal specification. Further information about this methodology is given in section 6. Hölzlein *et al* (1998) give a detailed description of the tool. So far formal verification has been applied to instruction list- and HiGraph-programs. HiGraph-programs are graphical state-oriented automation-programs for the PLC SIMATIC S7 by Siemens. Such programs are developed with the CASE-tool HiGraph (Siemens, 1997) by Siemens (the following sections refer to HiGraph mainly). The underlying state diagram methodology is based on results of a German working group consisting of PLC-manufacturers and PLC-applicants. Rath (1997) provides further information on the methodology. Emphasis has to be placed upon the fact that the checking algorithm of the Siemens verification tool works fully automatically which allows an application in the industrial area without problems - not only to HiGraph but also to other automation languages and to SDL for telecommunication software. As a result, the proposed method has the following advantages:

- increase of productivity and cost reduction due to error detection in early development phases by the developer,
- target-oriented debugging of errors by the developer on the basis of fully automatically generated countersequences that illustrate the erroneous behaviour of the program,

Table1 Formal verification versus conventional methods

Method	Advantages	Disadvantages
Formal Verification	+ fully automatic + complete + comfortable + countersequence in case of errors + documentation	- complex algorithmics -new methodology (training required)
Program testing	+ simple + established method	- high costs for intensive testing - incomplete
Simulation	+ illustrative	- high costs for process simulation - long runtime - incomplete
Code-Review/ Inspection	+ documentation	- high costs - incomplete

- safety by complete correctness checking (with regard to specifications),
- regression test after program modifications,
- possible application to certification of automation programs,
- exact specification of properties to be held for the control,
- acceptance of control with program checking by the principal.

Table 1 shows a summary of advantages and disadvantages of various methods for quality management and error detection which are used nowadays. The essential disadvantages of the other methods are the high costs and their incompleteness. Section 2 describes the schematic structure of the verification and introduces a tool machine and its control. In section 3, the user language for properties and assumptions is introduced and examples for the tool machine are given. Section 4 shows how verification is used to detect and debug errors in the control (program) for the tool machine. Section 5 illustrates the support of semantical analysis by the verification tool. In section 6, a methodology for finding a certain class of properties in a systematical manner is introduced. A short description of the technological basis in section 7 and the presentation of our results and experiences in section 8 conclude the paper.

2. ANALYSING PLC-PROGRAMS WITH VERIFICATION

Figure 1 shows the schematic structure of the verification from the user's point of view. The verification tool requires as input the system description and the

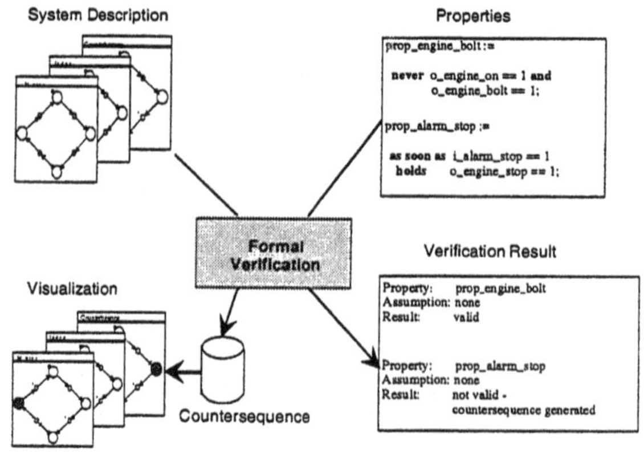

Fig. 1. Verification structure from the user's point of view

properties defined by the user and computes whether the properties are valid or not. In case of invalidity (error case) the tool generates a countersequence that guides the user to the program error. Such a sequence is visualized in the case-tool like HiGraph. Internally, the verification tool works with a finite state machine representation that models the system behaviour. The existing system description given as a HiGraph program is automatically translated into this internal format. Besides HiGraph programs, there are also translations from VHDL (hardware descriptions), SDL (telecommunication protocols) and from instruction list programs (PLC-programs), being used and developed at Siemens, into the internal finite state machine format. In the following the focus is on the verification of HiGraph-programs that are developed with the CASE-tool HiGraph by Siemens. HiGraph generates the machine code MC 7 of the PLC S7 by Siemens.

Based on the example of a tool machine control for a station of a transfer line for motor engine manufacturing the methodology of verification and state diagrams is illustrated. Figure 2 shows the structure of a tool machine unit and its control.

A carriage with two spindles moves back and forth between the rear tool exchange position and the front machining position. In the front position the workpiece, a connecting rod, is machined with the spindles while a cooleant is added. For instance, two holes are drilled into the workpiece. The control is arranged in 3 levels: decision level, coordination level and power level. The levels are realized by various components described as state diagrams (cf. fig. 3). The state diagrams communicate with each other and their environment, i.e. machine process and possibly other control processes, via messages or global variables. For instance, the state diagram „decision" of the decision level releases the carriage to move back by setting the message 're-lease_moving_back'. State diagram 'moving back' of the coordination level reacts to this message by sending the message 'command_spindle_1_off' to the state diagram 'spindle_1' of the power level

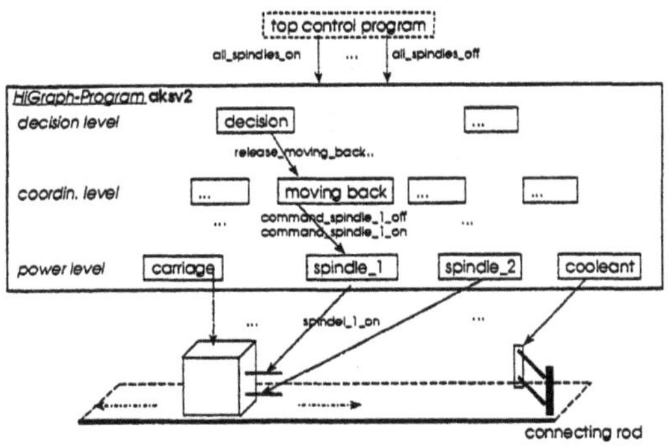

Fig. 2. Program architecture and machine process

which finally turns off spindle_1 by resetting the actuator 'spindle_1_on' to 0. The control program for this station is also influenced by the top control program for the whole transfer line. For example, the top control program sends the signal 'all_spindles_off' to all stations of the transfer line forcing all stations to turn off their spindles.

Figure 3 represents a strongly simplified state diagram 'spindle_1' for the direct control of the spindle_1 on the carriage. It consists of the initial state 0 and the additional states 2, 4 and 6 (bold numbered circles). As can be seen from the diagram, a state transition from state 6 to state 4 occurs if and only if transition condition number 10 holds, i.e. if one of the variables 'command_spindle_1_on' and 'all_spindles_on' has the value 1. In state 4 the action 'N spindle_1_on' is executed as long as the program control remains in state 4. The semantics of this action is such that in state 4 the actuator 'spindle_1_on' is set to 1, but is reset to 0 as soon as state 4 is left! Therefore, in state 6 the actuator has got the value 0 as required. More detailed information about state diagrams and the case-tool HiGraph can be found in the language manual (Siemens, 1998).

3. PROPERTIES AND ASSUMPTIONS

For specifying the desired or required behaviour of control programs, the user formulates properties and assumptions. Properties are required (partial) specifications of the control program that have an exact semantical meaning. Specifications express, for instance, what behaviour is allowed by the control, what is forbidden or what must be guaranteed at any rate. Assumptions, on the other hand, describe specifications about the environment (machine process) the developer has in mind with regard to the interaction with the control program. Both properties and assumptions are formulated using an intuitive user language developed at Siemens that is based on natural language constructs from English or German with an underlying exact semantics. Expressions built with the used program variables are inserted into these patterns. Some constructs of the user

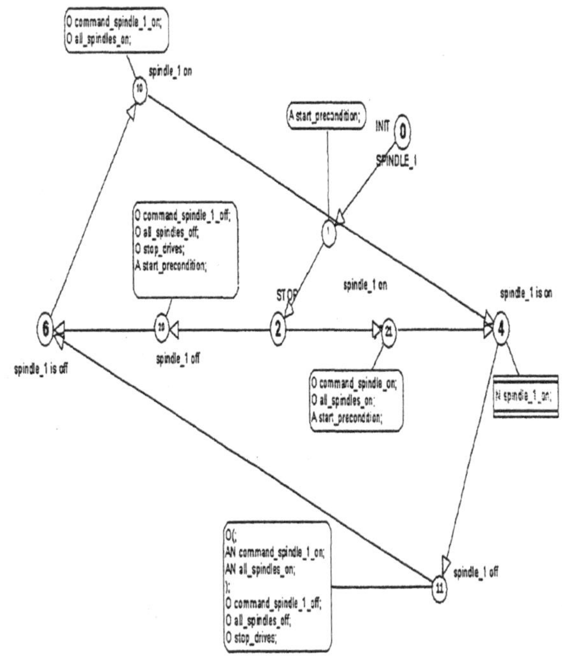

Fig.3. State diagram 'spindle_1'

language that are used frequently include:
ALWAYS HOLDS ...
NEVER HOLDS ...
AS LONG AS ... HOLDS ...
AS SOON AS ... HOLDS ...
AS SOON AS ... THEN AS LONG AS ... UNTIL ...

An example for a property for the tool machine unit described in section 2 is that 'spindle_1_on' receives value 0 within at most 1 PLC cycle after one of the variables 'command_spindle_1_off' or 'all_spindles_off' received the value 1. The formulation of the property in the user language is as follows:

AS LONG AS command_spindle_1_off == 1 or
all_spindles_off == 1
HOLDS AT THE LATEST AFTER 1 CYCLE
(spindle_on == 0) (1)

Given such a property, the verification tool can check whether the property holds for the given program for every possible sequence. In case of an error the tool presents a concrete countersequence starting in an initial state and showing the values of all input, output and program variables for each PLC cycle up to a point in which the computed sequence violates the required property. Such countersequences consist of two parts, the first leading to an error constellation, and the second exhibiting the erroneous behaviour.

It is possible, however, that a generated countersequence does not describe real program errors but represents an unrealistic execution behaviour. In order to understand why this can happen you need to keep in mind that although the verification tool has got the necessary information about the control program itself possibly important information about the environment is lacking. Recall that the environment consists of machine and other control parts that are not part of the control program to be verified, but communicate with the control program (for example via shared memory or via sensors and actuators). Without any additional information about the environment, the verification tool assumes an arbitrary behaviour that might lead to unrealistic executions. With the help of assumptions (interface specifications) the user is in the position to specify environmental behaviour in a more detailed way. Successful verification therefore amounts to showing that the control program fulfils the required properties under the assumptions imposed on the environment.

For this example, one can add the following assumption about the top control program that is not part of the state diagram program under consideration

ALWAYS HOLDS all_spindles_on == 0 or
all_spindles_off == 0 (2)

This assumption expresses that the top control never sends the signals 'all_spindles_on' and 'all_spindles_off' to the stations at the same time. While this assumption seems to be reasonable, it is worth to have a closer look at it. On the one hand this assumption about the behaviour of the top control which is not a verification object in this context reflects a property for the top control which must still be checked. On the other hand the necessity of such an assumption indicates the lack of robustness of the HiGraph-program with regard to "erroneous" behaviour of the top control.

In addition to stating such reasonable requirements on the environment, assumptions are also used to disallow certain behavioural sequences for the verification. For example, if one wants to exclude the manual operation mode of the control from the verification the following assumption about the HiGraph system variable 'om_manual' (operation mode manual) can be added:

ALWAYS HOLDS om_manual == 0 (3)

This effectively focuses the subsequent verification task to those system executions that do not allow manual interaction.

4. APPLICATION OF VERIFICATION

In the following the practised application of the verification method to the example of section 2 is illustrated. The goal is to verify property 1 already introduced in section 3. Without further assumptions

Table 2 Countersequence for property 1

PLC cycle	0 1 2 3 4
[SPINDLE_1,all_spindles_on]	0 0 1 1 0
[SPINDLE_1,all_spindles_off]	0 1 0 0 0
[SPINDLE_1,command_spindle_1_on]	0 0 0 0 0
[SPINDEL_1,command_spindle_1_off]	0 0 0 1 0
[SPINDEL_1,spindle_1_on]	0 0 0 1 1

the verification tool computes a countersequence in which the operation mode is "manual" and in which the top control issues the inconsistent commands 'all_spindles_on==1' and 'all_spindles_off==1' at the same time. Since the property is not required to hold in the operation mode "manual" and since the designer can assume that the top control is implemented correctly - at least with regard to the consistency of the commands 'all_spindles_on' and 'all_spindles_off' sent to the single stations - assumptions 2 and 3 from section 3 are added to the verification task. This approach is typical of verification. Firstly, you try to verify your property without assumptions, i.e. for the most general environment. In case of the output of countersequences assumptions are added successively if necessary. Otherwise the countersequences lead to a program error.

With these assumptions the verification tool determines that the above property is violated, and reports the countersequence as shown in Table 2. In order to interpret such a sequence the program developer concentrates on the end of the sequence where the erroneous behaviour is manifested. The beginning of the sequence just leads into the error constellation. By examining individual columns, the developer gains an overview of the whole constellation. In PLC cycle 4 of the given countersequence 'spindle_1' is turned on although in the preceding PLC cycle a command to turn it off has been issued. A snapshot of the total configuration at PLC cycle 3 is shown in Figure 4. While the carriage is moving back both 'command_spindle_1_off' and 'all_spindles_on' from the top control are set to 1. But the control seems to ignore the command to turn the spindle off. Based on this observation, the next step for the developer consists in putting the behavioural sequence into relationship to the program itself (see Figure 3) in order to detect the program error. The reason for the erroneous behaviour here is that a transition from state 6 to state 4 of state diagram 'spindle_1' is made due to the 'all_spindles_on' command (from top control). During this transition no commands are taken into consideration that could turn off the spindle ('command_spindle_1_off', 'all_spindles_off') although they occur in the countersequence. In order to correct the control program error the developer can now inspect the transition condition of the violating transition from state 6 to state 4 and can

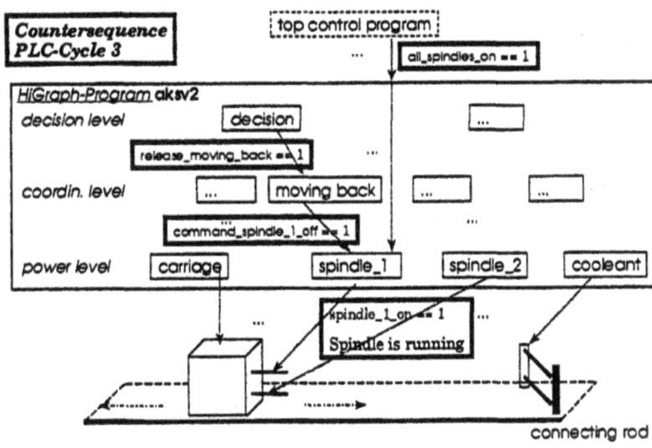

Fig. 4. situation represented in PLC cycle 3 of relating countersequence

modify it in such a way that the transition fires only if there is no command to turn spindle_1 off. In order to achieve this, he simply adds

AN command_spindle_1_off;

AN all_spindles_off;

to the transition 10 ('AN' means 'AND NOT'). The modified program can then be checked by simply running the verification tool again since property and assumptions remain unchanged (only the program itself has been modified). The tool could now report further errors or the validity of the property under these assumptions. In this case verification establishes that the property indeed holds for the modified program under the stated assumptions.

5. SEMANTICAL ANALYSIS

The verification tool does not only check whether properties hold or not, but also produces information based on a semantical analysis of the dynamic behaviour of the control program. Different kinds of information about the program can be extracted that way, among which the following have been found especially useful:

- reachability of the states of the state diagrams,
- indication of transitions that never fire,
- list of all logical input variables, i.e. variables that are only read, but not written to by the program.

The discovery of unreachable states usually hints at a design error. By specifying properties concerning some unreachable state and the transition leading to it the developer can detect errors in his programming logic. A transition that never fires often points to an error in the corresponding transition condition or in the sequence of priorities given to transitions. Finally, the list of all logical input variables can be compared with the actual input variables; differences often point to the mix-up of variables or to faulty mapping of program variables to the address space of the PLC.

6. METHODOLOGY FOR FINDING PROPERTIES

Experience shows that the main job for the user of the verification method is to find suitable properties. In order to support this task, a methodology has been developed that aides the user in achieving this. The basic idea is to determine for each actuator the values of the sensors that requires it to be set or to be reset. More specifically, the proposed method proceeds in the following steps:

1. Build the two lists of all input and output variables
2. For each output variable it is to be determined under which (input-) condition the variable is required to have value 0
3. For each output variable it is to be determined under which (input-) condition the variable is required to have value 1

The formulation of the property in step 2 for the value 0 of the output variable is normally quite easy. This is the case because value 0 normally represents the safe state of an actuator. Therefore, the conditions for making an actuator become "safe" is simple and often very general. Contrary to this, setting an actuator is allowed typically only if neither the reset condition nor some often complex restrictions on the control's environment do hold. Following this methodology the user is guided towards obtaining specifications of the control as complete and exact as possible. The methodology, however, does not help in finding properties about the relationship between different actuators. It is very difficult to support the finding of such properties in a methodological way as these properties typically depend strongly on the control task, at hand. Similarly, properties about the program design such as specifications of the relationship between internal variables (variables that cannot be accessed from outside the program) and input/output variables (interface variables) cannot be derived in a general fashion since such properties typically depend on design decisions made by the programmer. Frequently, a part of these properties can be formulated as invariants. Thus the property

ALWAYS HOLDS

car_park_counter <= car_park_counter_max (4)

specifies an invariant of a control for a multi-storey car park.

7. TECHNOLOGICAL BASIS

Figure 1 not only describes the application of the verification method but also the fundamental system architecture of the verification tool. The system description is translated into a finite state machine representation whose behaviour is represented with the help of Binary Decision Diagrams (Bryant, 1986). The properties are translated into formulae (Emerson, 1990). The actual verification is performed with the techniques for symbolic model checking of temporal logic. Burch *et al* (1994) give a good overview of these techniques. Symbolic model checking allows to verify systems with some hundred (state) variables at enormous amounts of reachable states in the finite state machine.

8. RESULTS

About 80 state diagram programs by tool machine and assembly-line manufacturers have been analysed with the verification tool. Most of them represent programs for a station or top control. The size of them varies between 3 and 24 state diagrams. The average number of states per state diagram is about 7. The average number of variables per program is over 200. The line number of the exclusively textual representation varies from 400 to 3000. In almost all applications errors have been detected with the verification that have led to the modification of the program. Since verification checks every possible run of the control even errors that are rare and very difficult to detect have been found. The cost for formulating properties and the analysis of counter-sequences is rated low in comparison with the effort necessary for error debugging during first system start up or final operation. The time of a verification run per property depends on the program size and the properties and assumptions specified. A typical verification run takes between a few seconds and 120 minutes on a PC with Pentium-processor.

REFERENCES

Bryant, R.E. (1986): Graph-based algorithms for boolean function manipulation.IEEE In : *Transactions on Computers*, **C-35(8)**, 677-691.

Burch, J.R., E.M.Clarke, D.E.Long, K.L.McMillan and D.L.Dill (1994). Symbolic Model Checking for Sequential Circuit Verification. *IEEE Transactions on Computer-Aided Design of Integrated Circuits and Systems*, **Vol. 13, No.4,** 401-424.

Emerson, E.A.(1990). Temporal and Modal Logic In: *Handbook of Theoretical Computer Science*, Chapter 16, pp.996-1072. Elsevier Science Publisher, Oxford.

Hölzlein, M. , Th. Filkorn, P. Warkentin and M. Weiß (1998). Eine Verifikationskomponente für HiGraph In: *GMA-Kongress '98.* VDI-Berichte 1397, pp. 281-290. VDI-Verlag, Düsseldorf.

Rath, G. (1997): Zustandsgraphen - eine neue, aussichtsreiche Programmiermethode. *GMA-Bericht*, **31**.

Siemens (1998) SIMATIC HiGraph for S7-300 and S7-400, Programming State Graphs, Manual, 6ES7-811-3CA02-8BA0, Siemens AG, Nürnberg.

14th Triennial World Congress, Beijing, P.R. China

MODELLING AND ANALYSIS OF CONTINUOUS-DISCRETE SYSTEMS WITH HYBRID PETRI NETS [1]

Mourad Chouikha* Stefan Wegele* Eckehard Schnieder*

* *Institute of Control and Automation Engineering, Technical University of Braunschweig, Langer Kamp 8, D-38106 Braunschweig, Germany, Phone: +49-531-391-7667, Fax: +49-531-391-5197, email:{m.chouikha, s.wegele, e.schnieder}@tu-bs.de*

Abstract: In this paper an overview of existing Petri nets approaches for modelling and analysis of discrete-continuous systems, which are subsequently referred to as hybrid systems, will be given. Focusing on the modelling approach presented in (Chouikha and Schnieder, 1998) an analysis method will be described and applied on a real industrial example of a no-load operation controller. In this context, the analysis deals mainly with the reachability problem. *Copyright © 1999 IFAC*

Keywords: Hybrid Systems, Petri Nets, Differential Equations, Modelling, Analysis

1. MOTIVATION

The subject of hybrid systems has become more and more important in recent years. This trend is due to the increased complexity and volume of modern automation systems in the various technical branches and particularly due to their mixed continuous and discrete characteristics. Hitherto, such systems were developed in modules and following capacious and possibly iterative tests, which despite the high use of capacity did not at all guarantee a faultless functioning in every operational situation. Furthermore, an optimisation of the whole system is still out of question. Therefore the problems of hybrid systems need intensive research works with respect to the integrated design of hybrid system components.

[1] This work is partially sponsered by the german Research Council (Deutsche Forschungsgemeinschaft DFG) within the program KONDISK.

2. OVERVIEW OF PETRI NETS APPROACHES

In this section an overview of the efforts of research teams dealing with the description of hybrid systems is given. The first ideas of hybrid Petri nets were presented by the French researchers David and Alla (David and Alla, 1992). In this first concept the marks were treated as a real quantity by subdividing whole marks in infinitesimally small parts of the marks, the so-called tokens. Thus over the transition there is no more a flow of whole marks but of tokens, hence allowing the markings of the places to take on real quantities. The firing behaviour of the net is determined by the so-called firing speeds of the transitions, which are either constant or depend on the marking of the pre-places of a transition. In this context two models were introduced, namely nets with constant and nets with variable firing speeds. An extension of these nets is made in (Demongodin and Koussoulas, 1998) by defining the differential Petri nets which introduce differential transitions and differential places with negative markings as well as suitable evolution rules.

A similar approach (Hybrid Flow Nets HFN) was presented in (Flaus and Ollagnon, 1997). The HFN were here used for controller design.

An approach combining interpreted Petri nets (IPN) with differential equations was presented in (Nenninger and Krebs, 1998). Here, the discrete dynamic is described by IPN, while continuous parts are modelled by differential equations. The interaction between them is made by C/D and D/C-interfaces. A similar approach was presented in (Chouikha and Schnieder, 1998) describing an integration of Petri nets with differential equations. In this concept differential equations are represented as continuous Petri nets to model continuous systems, so that C/D and D/C-interfaces are no more necessary. Moreover, the description of both discrete and continuous systems by the same description mean has visual as well as formal advantages.

Based on high level Petri nets an approach for modelling hybrid systems was presented in (Decknatel and Schnieder, 1997). Hybrid Petri nets are defined as the limit case of a discrete approximation of the continuous behaviour using timed Coloured Petri nets. In spite of the close relation to discrete nets, there are limitations with regard to the analysis. The classical methods of invariants and occurrence graphs do not agree with the continuous approximation.

3. HYBRID PETRI NETS

In this section the approach made in (Chouikha and Schnieder, 1998) will be briefly described. As already mentioned, the approach expresses the integration of Petri nets and differential equations. Following this approach there is a relation from net structure and behaviour to a system matrix, which represents a differ-ential equation system whose solution is given by a so-called fundamental matrix. In this way it is possible to formulate a basic equation for the description of the continuous net behaviour on the basis of the start marking and the fundamental matrix. The basic equation provides a formal basis of this new description, which is among other things necessary for the formal analysis and verifi-cation of system behaviour. For the formal description of the mixed discrete-continuous behaviour an iterative equation will be given, which describes the hybrid dynamic as an alternation between continuous processes (modes) and discrete events (steps).

Analogous to the papers of (David and Alla, 1992) a representation for hybrid Petri nets can be developed according the occurrence graph of discrete nets. Such a construction is referred to as evolution graph. It will be shown that the evolution graph of a hybrid Petri nets has the same structure as hybrid automata (Alur *et al.*, 1995) known from computer sciences. Representing the net behaviour by means of the evolution graph allows to apply the verification procedures of the hybrid automata. Various tools of analysis can be employed, such as UPPALL (Larsen *et al.*, 1995), KRONOS (Nicollin *et al.*, 1992) or HYTECH (Henzinger *et al.*, 1997). The latter allows parameter and occurrence analysis.

3.1 *Example*

Let a hybrid net be given by the following:

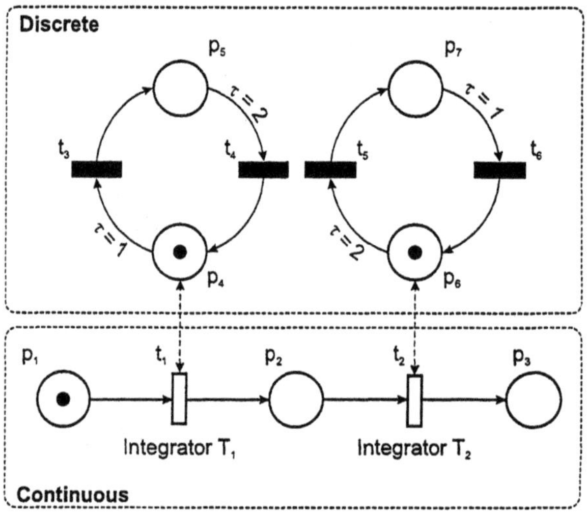

Fig. 1. Example of a hybrid net

After formulating a system of linear differential equations, the obtained solution describing the behaviour of the continuous part can be written as:

$$\begin{pmatrix} m_1 \\ m_2 \\ m_3 \end{pmatrix} = \underbrace{\begin{pmatrix} 1 \\ 0 \\ 0 \end{pmatrix}}_{M_0} + \begin{pmatrix} -1 & 0 \\ 1 & -1 \\ 0 & 1 \end{pmatrix} \cdot \begin{pmatrix} V_1 \\ V_2 \end{pmatrix} \quad (1)$$

where:

$$\begin{aligned} V_1 &= 1 - e^{-t} \\ V_2 &= 1 - (1+t) \cdot e^{-t} \end{aligned} \quad (2)$$

V_1 and V_2 are the firing speeds of t_1 and t_2 respectively. m_1, m_2 and m_3 are the marking of the places p_1, p_2 and p_3 respectively.

the equation 1 assumes that all transitions are fireable at any time. But this is not the case, if e.g. p_4 is no more marked. In this case the transition t_1 cannot fire. The activation of the transition is taken into consideration by multiplying each component of the firing speed vector in equation 1 with a binary function $S(t_i)$ that corresponds to the activity of the affilated transition t_i. The multiplication component by component can be expressed by a diagonal matrix $\mathbf{S}$:

$$\begin{pmatrix} m_1 \\ m_2 \\ m_3 \end{pmatrix} = \begin{pmatrix} 1 \\ 0 \\ 0 \end{pmatrix} + \begin{pmatrix} -1 & 0 \\ 1 & -1 \\ 0 & 1 \end{pmatrix} \cdot \begin{pmatrix} S(t_1) & 0 \\ 0 & S(t_2) \end{pmatrix} \cdot \begin{pmatrix} V_1 \\ V_2 \end{pmatrix}$$

As already mentioned, we can now formulate a recursive equation to describe the entire behaviour. Under the assumption the ith step is continuous and that the end marking of this step is $\mathbf{M}_{i,end}$ the following basic equations are obtained:

$$\begin{aligned} &\text{discrete step:} \quad \mathbf{M}_{i+1} = \mathbf{M}_i + \mathbf{C}\mathbf{V}_d \\ &\text{continuous step:} \quad \mathbf{M}_{i+2}(t) = \mathbf{M}_{i+1} + \mathbf{C}\mathbf{V}_c \end{aligned} \tag{3}$$

Where $\mathbf{C}$ is the incidence matrix of the net. Another form to describe the hybrid behaviour is the evolution graph. Such contruction, like the occurence graph, consists of two elements, i.e. locations and arcs (Fig. 2).

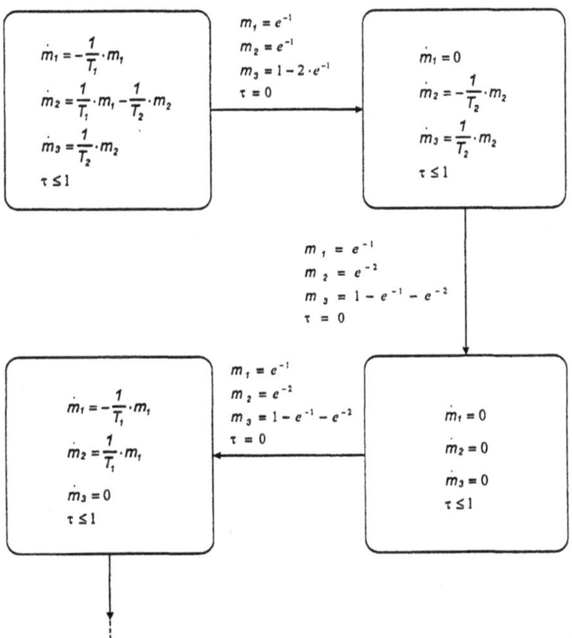

Fig. 2. Evolution graph of the example

The locations represent the modes of the system in which only continuous changes take place. That means, no discrete events occur and hence no discrete transitions can fire. The modes are characterised by the differential equation system that describes the net behaviour. Furthermore, the locations are labeled with the validity conditions of the modes (invariant).

The arcs of the evolution graph are the linking elements between the locations and hence represent the shifts between the modes. The arcs are labelled with the initialised variables (e. g. staring markings). A shift between the modes can, according to the reflections in chapter 3.3, occur when a discrete event occurs, i. e. when a discrete transition fire. On the other hand a change in the modes can happen, when during a modus a marking is occurred which corresponds to a start marking of an already existing location. At this, both locations must have the same dynamic behaviour (differential equations).

4. ANALYSIS

the substantial questions of the analysis of hybrid dynamic systems can be mostly attributed to the problem of the reachability of certain desired or undisired regions in the state space. Same applies to the synthesis of hybrid systems, whose objective is usually to transfer a system from a start mode to a final mode. This means also that the synthesis concerns itself with the question of the reachability.

For the investigation of the reachability in the hybrid state space *search algorithms* have to be defined, which are enable to determine efficiently trajectories in the state space according to the system requirements. Similar algorithms, whose function is also the determination of state trajectories in accordance with an objective function (*functional*), are well-known in mathematics as *optimisation algorithms.*

Due to structural similarity of the reachability and of the optimisation problem, it is obvious to applly the search algorithms from the mathematical optimisation on the reachability problem, i.e. the reachability problem is formulated as optimisation task.

In order to be able to formulate such an optimisation problem, the three constituents of an optimisation task must be defined. These are in detail the objective functional, the system parameters and their validity intervals which can be varied, and the optimisation algorithm as a mathematical instrument.

The parameters which can be varied are usually problem-specific. However some parameters can be given, which must be regarded in most cases. These are the initial conditions, continuous and discrete disturbances, discrete events and often model uncertainties. Additionally validity intervals for these parameters are to be indicated, which are determined by limit values of physical nature, e.g. critical frequencies or - accelerations etc.

The selection of quality functional is a crucial factor for the optimisation task, because the quality of the obtained solutions is evaluated by this *function.* Additionally the time limitation of the optimisation algorithm depends on the quality functional, since it is used as abort criterion for the optimisation procedure. Therefore the quality functional must be a unique criterion for the question of the reachability. The definition of the quality functional for the reachability is graphically illustrated on the basis of Fig. 3. For descriptiveness reasons only two variables of state space x_i and x_j are regarded.

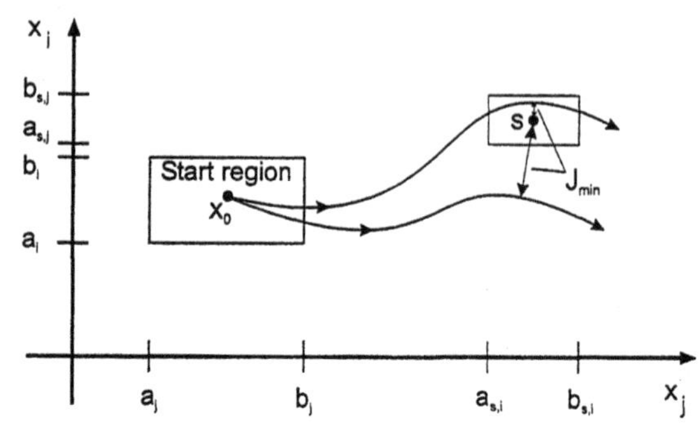

Fig. 3. Definition of a functional

The functional is defined as the distance between the system trajectory in the state space and the centre of gravity of the region that must be reached:

$$
\begin{aligned}
J &= ||\mathbf{X} - \mathbf{S}|| \\
&= \left(x_i - \frac{a_{s,i} + b_{s,i}}{2}\right)^2 + \cdots \\
&+ \left(x_n - \frac{a_{s,n} + b_{s,n}}{2}\right)^2 \rightarrow min \qquad (4)
\end{aligned}
$$

The task of the optimisation algorithm is now to minimise the functional J to obtain the minimal distance to the searched region.

For the solution of discrete-continuous optimisation problems, new procedures e.g. *genetic algorithms* and *Evolution strategies* were developed in more recent time (see (Goldberg, 1989)). These iterative procedures do not make any further requirements to the underlying system and its implementation.

A further procedure, which was 1957 developed by BELLMAN, e.g. (Bellman, 1967), first for the solution of discrete dynamic optimisation problems, so-called *multistage processes*, makes similarly no further requirements against the dynamic system. However the algorithm guarantees a global optimal solution, and it is independent of the selected starting conditions.

This Algorithm of *dynamical optimisation* (or also *dynamical programming*) and the *genetic algorithms* represent thus very general procedures for the solution of optimal optimisation problems, since they do not make requirements to the underlying model and thus for the handling of hybrid systems are best suitable.

5. EXAMPLE

The methodology of the analysis will be now demonstrated by an industrial example (Fig. 4). It is a no-load operation controller. The basic target of the idle regulation, which is active between 700 rpm and 2200 rpm, is it to prevent a going out of the engine under all operating conditions. The two-mass flywheel in the drive train has its natural frequency with approx. 450 rpm, therefore with 500 rpm the quantity supply is interrupted for a certain time, so that the drive train is not damaged. The function of the no-load operation controller is accordingly to that to hold the number of revolutions (no-load speed) over this limit. The parameters of the controller and the palnt depend on different discrete variables, such as gearbox category or transmission mode, and discrete events like air-conditioning on/off or drive clutched/declutcehed. Therefore we deal here with a hybrid system.

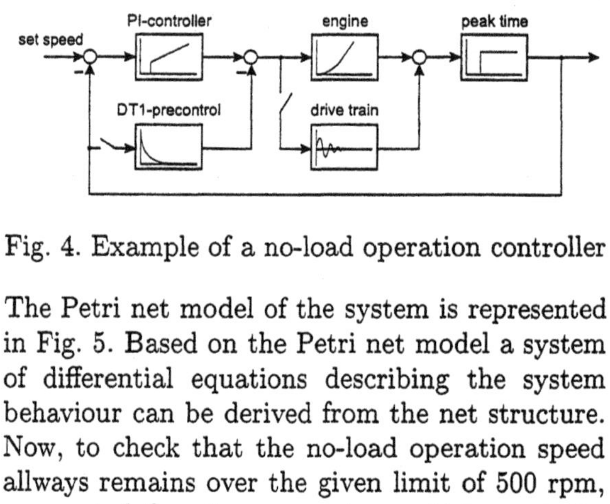

Fig. 4. Example of a no-load operation controller

The Petri net model of the system is represented in Fig. 5. Based on the Petri net model a system of differential equations describing the system behaviour can be derived from the net structure. Now, to check that the no-load operation speed allways remains over the given limit of 500 rpm, we must define the parameters, which are to be varied.

Variation of the disburbances
As model for the disturbance interventions the following function was used:

$$
\begin{aligned}
disturb(t) &= d(t - \tau_1) \cdot A(1 - \exp(10 \cdot (\tau_1 - t))) \\
&\quad + d(t - \tau_2) \cdot 1.5 \sin(\Omega \cdot (t - \tau_2))
\end{aligned}
$$

The disturbance consists of two sections: from a step function and from a sinusoidal oscillation. Since the no load operation speed cannot change precipitously, the step was switched by a first order delay. The function d is a unit step. The parameters of the disturbance have the following meaning:

$\tau_1 \in [0, \text{Simulationtime}]$	step time
$\tau_2 \in [0, \text{Simulationtime}]$	oscillation time
$A \in [-45, 45]$	amplitude of the step
$\Omega \in [0.8, 2] Hz$	oscillation frequency

Variation of initial conditions
The number of revolutions (or speed) and its

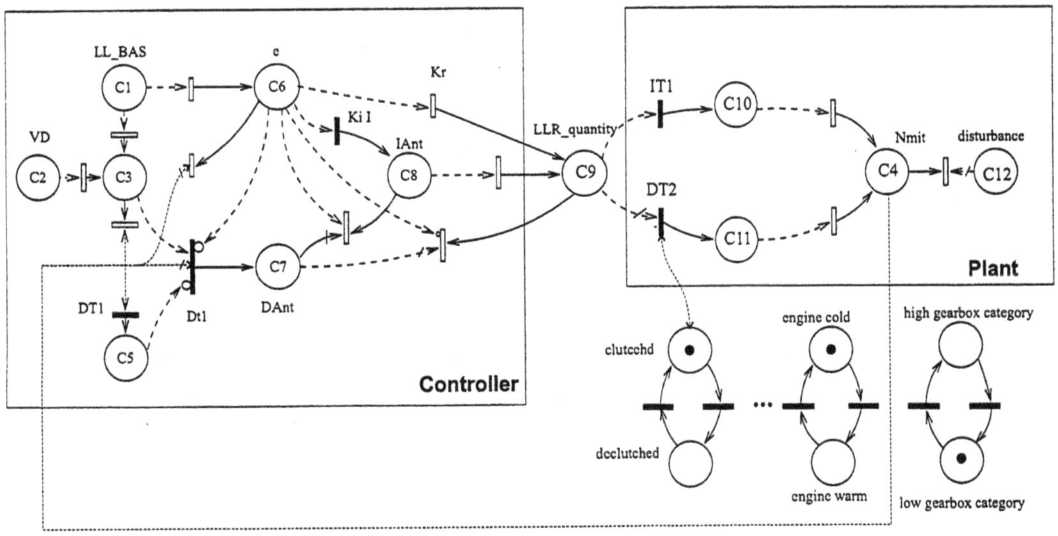

Fig. 5. Petri Net model of the no-load operation controller

temporal derivation can be classified as the most important initial condition. The no-load operation controller is active within the intervall: $700 < N_{mit} < 2200$. From physical considerations, the temporal derivation of the number of revolutions (acceleration) at 700 rpm can be assumed to be between the values of 0 to 300 rpm. Similar for the upper boundary: at 2200 rpm the temporal derivation can take a value from -1000 to 0 (Fig. 6).

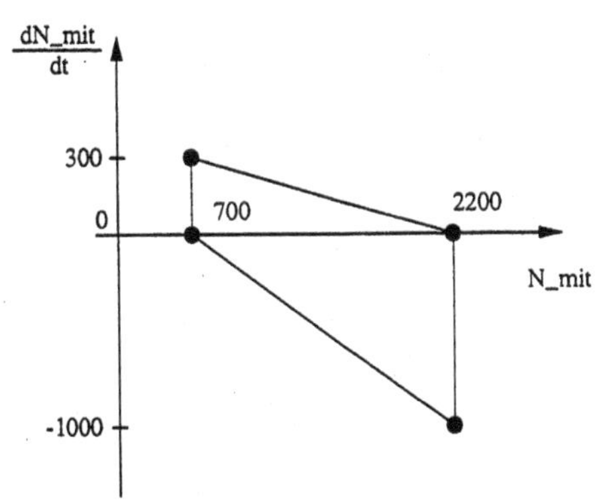

Fig. 6. Initial conditions

Variation of plant parameters

The follwing parameters were determined for example experimentally:

$w \in [9, 18]Hz$ Eigenfrequency of drive train
$K_{IM} = 252 \pm 20\%$ 1/Integration time

Variation of discrete events

The clutch is regarded as discrete event. From physical considerations we assume that the max. frequency, with which the pedal can be operated is smaller than 1 Hz. Therefore this event is approximated by the following function:

$$e = \frac{1}{2}sign(\cos(\omega t + \phi)) + 1$$

if $d = 0$, then the motor is clutched, otherweise declutched. The parameters of the function e are defined in the following intervalls:

$\omega \in [0, 2\pi]s^{-1}$
$\phi \in [0, 2\pi]$

Reachability analysis and discussion

Using genetic algorithms the parameters of the worst case were determined. A *worst-case* simulation is represented in Fig. 7. The dynamical programming gave the same results.

In the first step the homogeneous solution was determined. As result the optimal initial conditions were: $N_{mit}(0) = 2200$ $\frac{d}{dt}N_{mit} = 0$. the functional had the value 476 rpm after this step. In a second step the disturbance function was optimised. As also expected, the largest permitted value for the amplitude of the step function was taken. The functional had the value 463. In the third step the plant parameters were varied. Here the functional decreased relatively strongly, from 463 to 391. That was to be also expected, because the controller was designed based for the plant parameters. After the last step the functional amounted to 387. The transition process with this system runs relatively fast, after 4 seconds it already ran. Therefore the event could occur only once. Thus, after the algorithm was executed, the functional obtained the value 387. A complete search (all variables vary at one time) resulted in the value 376.9. As also expected the algorithm supplied a good approximation for the optimum. The simulation curves for the intermediate results differ little, therefore only the final result is represented (Fig. 7) in this picture, the no-load operation speed, the disturbance and operation of the clutch are represented. All curves were represented full-scale.

The optimisation procedure used the max. amplitude of the disturbance and selected the most favorable point of time to *attack*. The discrete event occurs briefly before the lowest speed was reached, so that the controller was unable to react in time. For this small system we can still imagine plausibly, when and which circumstances can lead to the minimum of the number of revolutions. This is is no more possible for large systems.

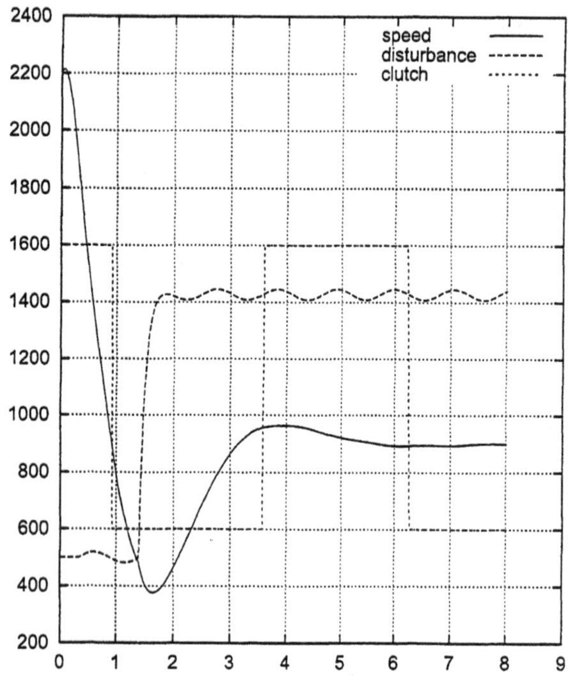

Fig. 7. worst-case simulation

6. CONCLUSIONS

In this paper a new approach for the description and analysis of hybrid systems was developed. The developed description mean is characterised by the fact that it combines discrete petri nets and differential equations as block diagram-similar continuous petri nets. Advantage of this representation is primary the mathematical describability and analysis possibilities.

Moreover a procedure for the analysis of hybrid systems was developed, which is based on formulating the reachability problem as an optimisation task. The procedure can be regarded as an approximation of the mathematically closed direct analysis for linear hybrid automata which can be achieved by model checkers.

The possibilities of the new description mean and the methodology of the analysis approach were demonstrated with the help of a real industrial example. It deals with a no-load operation controller.

7. REFERENCES

Alur, R., C. Coucoubetis, T. A. Henzinger and P. H. Ho (1995). Hybrid automata: An algorithm approach to the specification and verification of hybrid systems. Vol. 736 of *LNCS*. Springer Verlag.

Bellman, R. E. (1967). *Dynamische Programmierung und selbstanpassende Regelprozesse.* R. Oldenbourg Verlag. Munich.

Chouikha, M. and E. Schnieder (1998). Modelling of Continuous-discrete Systems with Hybrid Petri Nets. In: *IEEE: Computational Engineering in Systems Applications* (P. Borne, M. Ksouri and A. El Kamel, Eds.). CESA. IEEE. Hammamet. pp. 606–612.

David, R. and H. Alla (1992). *Petri Nets and Grafcet Tools for Modelling Discrete Event Systems.* Prentice Hall. New York, London, Toronto, Sydney, Tokyo, Singapore.

Decknatel, G. and E. Schnieder (1997). Modelling Railway Systems with Hybrid Petri Nets. In: *Hybrid Dynamical Systems: 3rd Conference on Automation of Mixed Processes* (J. Zaytoon, Ed.). adpm. Reims. pp. 309–315.

Demongodin, I. and N. T. Koussoulas (1998). Differential petri nets: Representing continuous systems in a discrete-event world. *IEEE Transaction on Automatic Control* **43**(4), 573–578.

Flaus, J.-M. and G. Ollagnon (1997). Hybrid Flow Nets for Hybrid Processes Modelling and Control. In: *Hybrid and Real-Time Systems* (O. Maler, Ed.). Vol. 1201 of *LNCS.* Grenoble. pp. 213–227.

Goldberg, D. E. (1989). *Genetic Algorithms in Search, Optimization, and Machine Learning.* Addison-Wesley Publishing Company, Inc.. New York, Tokyo, Singapore, Bonn.

Henzinger, T. A., P. H. Ho and H. Wong-Toi (1997). HyTech: A Model Checker for Hybrid Systems. In: *International Conference on Computer-Aided Verification (CAV 97.* Vol. 1254 of *LNCS.* Springer Verlag. pp. 460–463.

Larsen, K., P. Pettersson and W. Yi (1995). Model-checking for real-time systems. Vol. 965 of *LNCS.* Berlin.

Nenninger, G. and V. Krebs (1998). Analysis of Hybrid Systems using Hybrid Dynamical Models. In: *3rd International Conference on Automation of Mixed Processes.* adpm. Reims. pp. 428–431.

Nicollin, X., J. Sifakis and S. Yovine (1992). Compiling real-time specifications into extended automata. *IEEE: Software Engineering* **18**, 794–804.

14th Triennial World Congress, Beijing, P.R. China

ON CONTROL DESIGN IN SIMULATION OF HUMAN MOTION

Petko Kiriazov, Hyeongseok Ko

Human Animation Center, School of Electrical Engineering, Seoul National University
Seoul 151-742, Korea, email: kiriazov (ko)@graphics.snu.ac.kr

Abstract: Efficient simulation of articulated structures is needed in biomechanical engineering, human animation, virtual reality, and robotics. A conceptual framework for control design in constrained and unconstrained motion tasks is developed. In the first case, decentralized sliding-mode controllers with maximum degree of robustness can be designed. In the other case, simple but consistent with Pontryagin's Maximum principle test functions are employed whose parameters are determined through a fast converging shooting procedure. Besides the analytical guarantees, we verify the efficiency of our approach in dynamic simulation of standing-up motion where both open-loop and closed-loop controllers are implemented.

Keywords: dynamics, open/closed loops, controllability, robustness, optimal learning;

1. INTRODUCTION

Motion simulation of articulated structures such as humans is challenging due to the great number of joints and muscles and the complex multibody and actuator dynamics. A very important question is how that large-scale musculo-skeletal systems are kept controllable in dynamic motion tasks. Numerous models have been developed to give insight to the performance criteria and organization of movement control in humans, (Happee, 1992; Alexander, 1997; Dariush *et al*, 1998).

An optimal control theory is potentially a powerful methodology to evaluate driving forces during movement. It can deliver purely predictive results *independent* of experiments. The predicted time histories of the optimal control forces can be compared with the corresponding measurements and analysed in detail to understand various aspects of multi-joint co-ordination and muscle function. The problem in the aspect of feedforward control is to determine the unique trajectory which yields the best performance based on the selected dynamic model and performance criterion. To solve that problem, there are various control optimization techniques but few of them are appropriate for dealing with the complex human dynamics models.

Classical optimal control theory, based on Pontryagin's Maximum principle (PMP), gives the necessary conditions for optimality and, if dynamic models are complex, the related two-point boundary value-problems (TPBVP) are very difficult or impossible to solve. As the techniques based on this theory involve gradients of functions, the optimal solutions are very sensitive to modelling errors and may not be global ones. Applying PMP to properly simplified models however, can give us useful data regarding the structure and shape of optimal control laws. For global optimization, dynamic programming is sometimes preferred but it is impractical for dynamic systems with more than two degrees of freedom, mainly due to the computation burden.

Control parameter optimization is a unified approach for converting optimal control problems into problems of nonlinear programming where well-developed optimization algorithms exist. It can be applied to inverse dynamics models employing test state trajectories (Dariush *et al*, 1998), or using test control functions applied to direct dynamics models

(Goh and Teo, 1988), (Pandy *et al*, 1992). With inverse dynamics methods, a specific problem is how to define the test state trajectories because the optimal control laws are, in principle, bang-bang (Kirk, 1970). Having experimental data about the kinematics however, the inverse-dynamics method can be efficient, especially, when complex force constraints (e.g. strength, static and dynamic balancing conditions), are to be satisfied (Ko, 1994).

A common problem with most dynamics-based as well as neural nets techniques, is the great number of decision parameters. When applying gradient-based algorithms to large-scale optimization problems, it is very likely to get trapped in local suboptima and we may not be able to find even feasible solutions of the given TPBVP.

The direct dynamics, control synthesis methods have the capability to cope with systems characterised by control as well as dynamics discontinuities. The latter discontinuities are inherent in human locomotion where, in different motion phases, one has to deal with dynamics models having different structures. And the problem here, is how to parameterise the test control functions and what is the minimum number of control parameters. This problem is solved in the cases of a manipulator (Marinov and Kiriazov, 1991), and a five-link biped mechanism (Kiriazov and Schiehlen, 1997).

Feedback controllers are needed for posture or trajectory tracking stabilization. A feedback control design technique based on necessary and sufficient conditions for robust decentralized controllability is developed in (Kiriazov, 1994; Kiriazov *et al*, 1997). For robotic systems satisfying a very weak condition on the transfer matrix, sliding-mode controllers with maximum degree of robustness can be designed.

In simulating, e.g., standing-up, walking, running, constant orientation of the trunk is to be maintained, along with performing optimal point-to-point movements with the legs. It means that, for most locomotion tasks, both open-loop and closed-loop controls have to be applied. Although there are guarantees for the feasibility of our methods for open- and closed-loop control design, it is interesting to see the dynamic performance when both type of controllers are applied at the same time.

The paper is organised as follows. In the next section, we present our concepts in the control design and the necessary mathematical background is briefly described in Section 3. Next, we refer to some physiological results which are very useful for the control structure design and when defining the test control functions. In Section 5, standing-up motion is considered as an illustrative example of combined open-loop and closed-loop control design for a complex dynamic movement. Finally, some concluding remarks are given and directions for ongoing research are outlined.

2. CONCEPTS IN CONTROL DESIGN

A) Simulating human motion means that we have to deal with large-scale, optimal control problems that are difficult or impossible to efficiently solve when full dynamic models are used.

- *Trade-off is always needed between full dynamic modelling, precise parameter identification and optimal system performance*

B) Finding a proper control structure, i.e., input-output pairs, is a problem of principal meaning for a controlled system (Yeung, 1991; Skogestad and Postlethwaithe, 1996). For the controllability, couplings between single input-output pairs have to be reduced as much as possible. Optimal input-output assignments can be done using appropriate dynamic models and design criteria (Kiriazov, 1994; Kiriazov and Schiehlen, 1997).

- *Control structure design with most appropriate input-output pairing;*

C) In voluntary movements, where no space-time or force-time constraints are imposed, it is natural to assume that such movements are controlled in an open-loop manner. Control learning methods, that optimize the performance through repeated forward dynamics simulation and motion evaluation, are to be applied. They should have the following features: *a)* guarantees for existence of feasible solutions; *b)*; motion synthesis done with as small as possible number of decision parameters; *c)* the optimization procedure converges with minimum number of trials.

- *Open-loop control for unconstrained (free) point-to-point movements; control synthesis by learning;*

D) Feedback controllers are to be applied in case of constrained motion, e.g., for posture or trajectory tracking stabilization; They have to be robust and optimal as the feedback controllers of humans are:

- *Robust closed-loop control for movements with space-time or force-time constraints;*

3. MATHEMATICAL BACKGROUND

3.1. Dynamic Modelling

Applying the multi-body system approach (Schiehlen, 1990), and the Lagrange formalism, the dynamic performance of skeletal systems can be, in general, described, by the following system of differential equations

$$\ddot{\mathbf{q}} = \mathbf{M}(\mathbf{q})^{-1}(\mathbf{BF} - \mathbf{C}(\mathbf{q},\dot{\mathbf{q}}) + \mathbf{g}(\mathbf{q})), \qquad (1)$$

where $\mathbf{q}$ is the vector of l generalised co-ordinates (e.g., links' rotation angles or joint angles), $\mathbf{M(q)}$ is the inertia matrix, $\mathbf{C}(\mathbf{q},\dot{\mathbf{q}})$ is the vector of velocity forces, $\mathbf{g(q)}$ stands for friction and gravitation forces, $\mathbf{F}$ is the vector of control forces; matrix $\mathbf{B}$ represents the control force distribution, and $\mathbf{A} = \mathbf{M}(\mathbf{q})^{-1}\mathbf{B}$ is the control transfer matrix (TM);

As the objective in this section is to present the main features of our open- and closed-loop control design methods, the actuator (muscle/tendon) dynamics will not be addressed here. Such dynamic models can be found in (Hill, 1938; Winter, 1990; Dariush *et al*, 1998). That how detailed actuator dynamics models are to be employed should depend on the purpose of human motion simulation: human animation, virtual reality, biomechanical engineering, or robotics. In our numerical considerations, we use the well-known Hill model.

A point-to-point motion task means that the dynamic system (1) has to be transferred from a given initial state $\{\mathbf{q}^0, \mathbf{v}^0\}$ to a required final state $\{\mathbf{q}^f, \mathbf{v}^f\}$. The following TPBVP is to be solved

$$\{\mathbf{q}(t^0) = \mathbf{q}^0,\ \dot{\mathbf{q}}(t^0) = \mathbf{v}^0\} \mapsto \{\mathbf{q}(t^f) = \mathbf{q}^f,\ \dot{\mathbf{q}}(t^f) = \mathbf{v}^f\} \quad (2)$$

such that a performance criterion $J(\mathbf{q},\dot{\mathbf{q}},\mathbf{F})$ (time/energy) is optimised and a set of control and state constraints are satisfied. The control constraints are due to the limited power resources, strength bounds and comfort demands, or due to a task for force interaction with the environment. If we have to pay attention to space or force constraints during motion, we design and apply closed-loop, trajectory tracking controllers. Otherwise, we use open-loop control, and its design is next described.

3.2. Open-loop (feedforward) control design

Unconstrained point-to-point motion tasks are considered as free final-time optimal control problems. At the foundation of our and other studies on human motion control lies the notion that such locomotion patterns are, in general, optimized in time, control effort, or energy expenditure (Happee, 1992; Alexander, 1997). The control functions in simulation of such tasks can be efficiently optimized applying the direct-dynamics control parameter approach (Marinov and Kiriazov, 1991; Kiriazov and Schiehlen, 1997), having the following main steps:

Choose a set of appropriate test control functions: The term "appropriate" concerns the structure, shape, and bounds of the test control functions. Simple linear-spline control functions of "bang-pause-bang" or "bang-slope-bang" types can be used for first-order optimization. Such control laws are consistent with Pontryagin's Maximum principle for linear-in-control dynamic systems and time/energy performance criteria. Parameters p_i, $i = 1,\dots,n$, $n > l$, describing the test control functions, like switch times, slopes, or pause lengths, mostly influence the reached position and the performance criteria.

Define the most relevant input-output pairs: For simplicity, we assume that the end-point velocities v^0, v^f are zero and the generalised co-ordinates q are monotonous functions of time during the trial movements. We define l controlled outputs to be $y_i = y_i(\mathbf{p}) = q_i(t_i^f)$, where $\dot{q}_i(t_i^f) = 0$. Then for each controlled output we assign a control input which mostly influences it. That can be done by a sensitivity analysis using a dynamic model (1) or by engineering intuition. The input-output pairs (p_i, y_i) will be most relevant regarding the controllability if the couplings between these single-input single-output subsystems are minimal.

Solve shooting equations and perform control parameter optimization. With the above input-output pairing, the given TPBVP is transformed into a system of shooting equations $\mathbf{p} \Rightarrow \mathbf{q}^f$ with decision variables p_i, $i = 1,\dots,l$ which is the first level of our control synthesis procedure. At the next level, the other parameters p_i, $i = l+1,\dots,n$ are varied to optimize J. In principle, one should find the minimum time movement first, and then, with given movement execution time, minimise the energy cost.

Following this approach, we have the chance to find a satisfactory suboptimal solution with minimum number of control parameters. Existence of feasible solutions and convergence of bisection algorithms finding them can be guaranteed, provided that a margin of independent parameter controllability (IPC) can be found. The margin (parallelepiped) of IPC is defined by:

Def. 1: $P = \{\mathbf{p}\colon\ p_i \in [p_i^-; p_i^+],\ i = 1,\dots,l\}$ is said to be a margin of IPC iff, for any pair of antipodal points $\underline{p}$ and $\overline{p}$ on the boundary of P, there exists an integer i such that $(y_i(\underline{p}) - q_i^f)\ (y_i(\overline{p}) - q_i^f) < 0$.

The existence of feasible solutions is guaranteed by a Fixed-Point theorem (Kiriazov, 1995), that is an extension of the well-known Borsuk-Ulam theorem for antipodal maps (Todd and Wright, 1980). The dynamic system (1) has an IPC-margin if a set of reasonable control force dominance conditions are satisfied (Kiriazov and Schiehlen, 1997; Marinov and Kiriazov, 1991).

3.3. Closed-loop (feedback) control design

The following reduced model of error dynamics can be used for the purpose of closed-loop control design

$$\ddot{\mathbf{e}} = \mathbf{A}(\mathbf{q})\mathbf{F} + \mathbf{d} \qquad (3)$$

where $\mathbf{e} = \mathbf{q} - \mathbf{q}^{\mathbf{ref}}$, and the vector $\mathbf{d}$ stands here for the model uncertainty and external disturbances. As a measure of tracking precision, we take the absolute value of $\mathbf{s} = \dot{\mathbf{e}} + \lambda\,\mathbf{e}$, $\lambda > 0$.

We consider decentralized controllers which means that each stabilizing control force F_i depends solely on the corresponding controlled output s_i. We use sliding-mode controllers of saturation type (Kiriazov *et al*, 1997), thus avoiding chattering effects.

Def. 2: A decentralized controller is robust against random disturbances $\mathbf{d}$ with known upper bounds $\mathbf{d}^+$ if it gets the local subsystem state $(\dot{\mathbf{q}}, \mathbf{q})$ at each joint to track the desired state $(\mathbf{q}^{\mathbf{ref}}, \dot{\mathbf{q}}^{\mathbf{ref}})$ with maximum allowable absolute values $\mathbf{s}^+$ of errors $\mathbf{s}$.

Theorem (Kiriazov, 1994): Necessary and sufficient condition for a dynamic system (1) to be robustly controlled by a decentralized controller is that matrix A be generalized diagonally dominant (GDD).

When matrix A is GDD, there always exists, (Lunze, 1992), a positive vector $\mathbf{F}^+$ of control function magnitudes solving the following system

$$A_{ii}F_i^+ - \sum_{j \neq i} |A_{ij}| F_j^+ = d_i^+ , \quad i = 1,\dots,n \qquad (4)$$

Eqs. (4) present optimal trade-off relations between the bounds of model uncertainties (and external disturbances) and the control force limits. The robustness of so-designed controllers is proven using Lyapunov theory (Kiriazov, 1994).

4. PHYSIOLOGICAL BACKGROUND

Movement of any segment in the human body requires activation of muscles across several joints, even if the intent is to move a single joint only. This is due to the mechanical interactions among the body segments as well as the muscles spanning more than one joint. We have to take this fact into account whether the movement is voluntary or intended.

4.1. About Control Structure Design in Voluntary, Unconstrained Movements

Bernstein (1967) has suggested that synergies among the muscles are established by the nervous system to reduce the number of degrees of freedom that it must independently control. This fact is in accordance with our understanding of decentralization at the lowest level of the control subsystem.

Further, the patterns of electromyographic activity for a point-to-point movement of a limb are, in general, triphasic (acceleration-pause-deceleration): initial burst of activity in agonist muscles, followed by a pause of agonist activity, and then, activity in antagonist muscles (Gottlieb *et al.*, 1996). The optimal in time/energy, "bang-pause-bang" control laws, mentioned in the previous section, can be considered as a first approximation to such muscle force functions.

Furthermore, there is a linear relationship between the two main joint torques found in arm motion experiments with varied load, direction, speed, angular excursion. Both torques go through extrema and zero crossings almost simultaneously (Gottlieb *et al.*, 1996). Excitation signals are parametrically modified to fit the requirements of different tasks (Bock *et al*, 1996). The control signals for human movements are not calculated anew each time; rather, movements are controlled by stored, pertinent force-time functions, whose parameter values are generated according to the required motion task. The parameters learned by the human are the control magnitudes and switch times (Karniel, and Inbar, 1997), and they are the most important parameters in our open loop control synthesis procedure.

Thus in our direct-search control learning approach, test control functions with shapes/structures similar to that of the human control functions are used.

4.2. Feedback Controllers for Intended (Space or Force Constrained) Movements

The feedback control strategy in this case characterizes the spring-like behaviour of the neuromuscular system in such a way that whenever the limb is disturbed, it will return to its equilibrium posture (Hogan, 1990). The control system stiffness is an important quantity in studies of movement control (Latash, 1993); Dariush *et al*, 1998), because its magnitude determines the resistance to external perturbations.

5. CASE STUDY: CONTROL DESIGN FOR STANDING-UP MOTION

Besides the analytical guarantees given in Section 3 and the successful applications to robotic systems, we now verify the practicability of our approach, considering a specific daily-life motion task of a human: dynamic standing-up motion. In such a task, we have to apply both open-loop and closed-loop controls at the same time. Along with checking the convergence of the open-loop control synthesis algorithm and the robustness of the closed-loop controller, it is interesting to see also if a dynamics-based motion animation looks natural. Such a performance objective is reasonable because the human eye is very sensitive to "errors" and gets disturbed by irregularities in movements that it perceives every day.

In our experiment, human body is modeled as a three-segment (trunk, thigh, and shank), planar linkage, and the ankle is assumed to be hinged to the ground. The dynamics of such an articulated structure is well described in (Winter, 1990). When the human rises from a chair or a static squatting position, the trunk and shank rotations can be considered as space-time constrained motions and, consecutively we have to apply to them closed-loop control. The rotation of the thigh, with the massive trunk hinged to it, is an unconstrained, relatively long-range motion with open-loop control. The structure of the latter control function is triphasic and the control force magnitude in such a voluntary movement is defined according to the Hill model: $F = -a + 1/(b|v|+c)$, where v in our case is taken to be the speed of the thigh rotation (which is proportional to the velocity of shortening in muscle contraction). Varying the control parameters a, b, and c, we can efficiently optimise the movement execution time as well as control energy criteria. In a neighbourhood of the final upward configuration, we apply closed-loop controllers to all the three links, stabilising, in this way, the body at this vertical posture.

The standing-up motion simulation was performed with the numerical data given in (Winter, 1990). With specified values for a, b, and c, the open-loop control synthesis, when using the bisection algorithm, converges within 10 trial movements. For the convergence of this control learning procedure, the robustness of the feedback control of the trunk and shank rotations is of great importance. The time histories of the angular velocities of the three links are depicted in Fig. 1. The stick-diagram of the standing-up motion and the animated figure are shown in Fig. 2. The animation produced from the computed result visualises the standing-up motion in actual speed. The motion picture is available at the web site (ftp://graphics.snu.ac.kr/pub/data/animation/standing-up.mov).

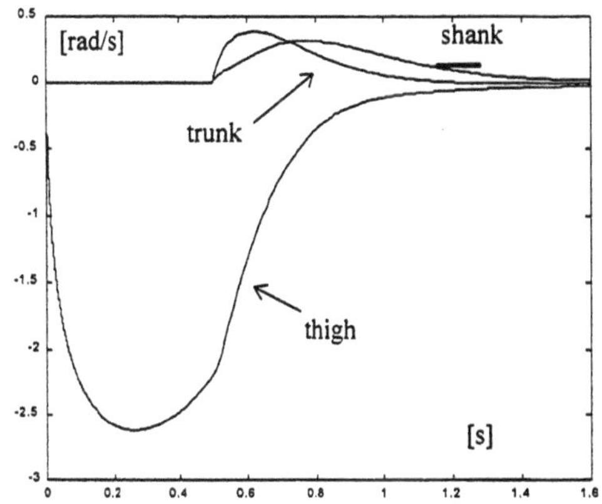

Fig. 1: Velocity time histories

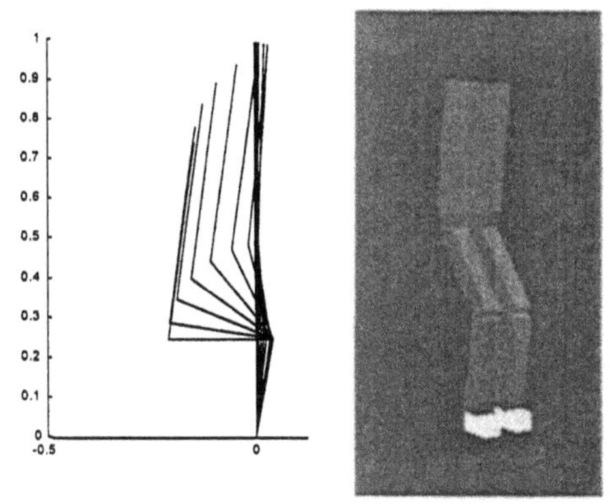

Fig. 2: The stick-diagram and the animated figure

6. CONCLUSION AND ONGOING RESEARCH

A unified approach for dynamics-based simulation of complex articulated structures, like humans or legged robots, has been proposed. Constrained and unconstrained motion tasks have been considered, and accordingly, two types of control strategies are to be applied: closed-loop and open-loop. The first one assigns robust sliding-mode controllers, the other involves simple, but consistent with Pontryagin's Maximum principle, test control functions. Their parameters can be determined through a shooting procedure which converges within minimum number of trial movements.

Despite the analytical guarantees, the feasibility of our approach has been verified on a full dynamic model of the standing-up task in which both, open-loop and closed-loop, controls are applied. The approach can also be efficiently applied in simulating other tasks like walking, running, and grasping which is a subject of ongoing research.

Following the proposed approach, actual problems in biomechanical engineering, human animation, virtual reality, and robotics can be solved. For example, by introducing realistic dynamics models of the muscle activation, new control techniques in functional neuromuscular stimulation of neuroprosthetic systems can be developed.

Knowing how to control certain classes of dynamic motions, we can provide animators/robot operators with a minimum set of movement commands and parameters which completely control the animated figures/robots. The proposed techniques can learn human models or humanoid robots how to locomote automatically, in a way that is inspired by the remarkable ability of real humans to acquire locomotion skills through action and perception.

Acknowledgement: The financial support from the Creative Research Initiatives of the Korean Ministry of Science and Technology is gratefully acknowledged.

REFERENCES

Alexander, R. McN, (1997). A minimum energy cost hypothesis for human arm trajectories. *Biological Cybernetics,* **76**(2), 97-105.

Bernstein, N. (1967). *Coordination and Regulation of Movement.* Oxford: Pergamon Press.

Bock O., D'Eleuterio, G., Lipitkas, J. and Grodski, J. (1996). Parametric motor control: a new approach to the control of point-to-point manipulator movements. *J. Robotic Systems,* **13**(1), 35-40.

Dariush B., Parnianpour, M, and Hemami, H. (1998). Stability and a control strategy of a multilink musculoskeletal model with applications in FES. *IEEE Transactions on Biomedical Engineering,* **Vol. 45**, No.1, 3-14.

Goh, C. and Teo, K (1988) Control parameterization: a unified approach to optimal control problems with general constraints. *Automatica*, **24**: 3-18.

Gottlieb, G., Song, Q., Hong, D., Almeida, G., and Corcos, D. (1996). Coordinating movement at two joints: a principle of linear covariance. *J Neurophysiology*, **75**(5),1760-1764.

Happee, R. (1992). Time-optimality in the control of human movements. *Biol.Cybernetics*, **66**, 357-366.

Hill, A. (1938). The heat of shortening and dynamic constants of muscle. *Proc. Royal Soc. B*, **126**, 136-195.

Karniel, A. and G. Inbar. (1997). A model for learning human reaching movements. *Biological Cybernetics*, **Vol. 77**(3), 173-183.

Hogan, N. (1990). Mechanical impedance of single and multi-articular systems. In *Biomechanics and Movement Organization*, J. Winters and S. Woo Eds., Springer-Verlag, ch. 9, 149-164.

Kiriazov, P. (1995). Controllability of a class of dynamic systems. *ZAMM*, **Vol. 75** SI, 85-86.

Kiriazov, P. (1994). A necessary and sufficient condition for robust decentralized controllability of robot manipulators. In *Proceedings American Control Conference*, Baltimore, MD, 2285-2287.

Kiriazov, P., E. Kreuzer, and F. Pinto (1997). Robust feedback stabilization of underwater robotic vehicles. *Journal of Robotics and Autonomous Systems*. **Vol. 21**, 415-423.

Kiriazov P. and W. Schiehlen. (1997). On direct-search optimization of biped walking. In *CISM Courses,* **Vol. 381**, Eds. A. Morecki *et al.*, Springer-Wien-NewYork, 134-140.

Kirk, D.E. (1970). *Optimal Control Theory: an introduction*, Prentice Hall.

Ko, H. (1994). *Kinematic and Dynamic Techniques for Analyzing, Predicting, and Animating Human Locomotion.* Ph.D. Thesis, University of Pennsylvania, USA.

Latash, M.L. (1993). *Control of Human Movement: Human Kinetics*. Chanpaign, IL.

Lunze, J. (1992). *Feedback Control of Large-Scale Systems*. Prentice Hall, UK.

Marinov P. and Kiriazov, P. (1992). Point-to-point motion of robotic manipulators: dynamics, control synthesis and optimisation. In: *IFAC Symposia Series, Robot Control 1991*, Eds. I.Troch & K.Desoyer, 149-152.

Pandy, M.G., Anderson, F.C., and D.G. Hull. (1992). A parameter optimization approach for the optimal control of large-scale musculoskeletal systems. *Transactions of ASME, Journal of Biomechanical Engineering.*, **114**.

Schiehlen, W. (ed.) (1990) *Multibody Systems Handbook.* Berlin: Springer.

Skogestad, S. and Postlethwaithe, I. (1996). *Multivariable Feedback Control: Analysis and Design*, Wiley.

Todd, M.J. and Wright, A.H. (1980) A variable-dimension symplicial algorithm for fixed-point theorems, *Numerical Functional Analysis and Optimization*, 155-186.

Winter, D. (1990). *Biomechanics and Motor Control of Human Movement*, JohnWiley&Sons, Inc.

Yeung, L.F. (1991). Optimal input-output variable assignments for multivariable systems, *Automatica*, **Vol. 27**, No.4, 733-738.

NONLINEAR DESIGN METHODS FOR CONTROL SYSTEMS

Han Jing-Qing

Institute of Systems Science, Chinese Academy of Sciences, Beijing, 100080, P.R.CHINA
Email:Yao@bamboo.iss.ac.cn
fax: 86-10-62561748

Abstract: Most design approaches are based on the understanding about the models of the plants or are developed by some linear methods. In fact, there are many special characters existing in the nonlinear world. In this paper, some approaches, which take advantage of some nonlinear characters to be independent of the models and the external disturbances, are introduced. *Copyright © 1999 IFAC*

Key Words: Tracking-Differentiator, Extended State Observer, Auto-Disturbances-Rejection Controller, disturbance rejection

1.INTRODUCTION

Making a comprehensive survey of the available design methods and synthesis theories, it can be found that the state space (linear or nonlinear) methods are developed on the basis of the models of the plants. Consequently, strong restrictions associated with these methods make them rather impractical. On the other hand, the classical regulator theory whose kernel is PID and some modern design methods which are developed based on some various recursive approaches (such as the predictive control , the adaptive control) have good advantages in solving practical control problems. But they are mainly linear methods. So efficiencies are limited. In fact, the central point of the control design is to choose a "proper" feedback so as to reject various disturbances, which will obstruct the control objectives. If some feedback law has been designed to stabilize a given plant, the closed-loop system will work normally even if some small disturbances exist. The disturbances that can be rejected by the designed feedback are variances in the system's parameters or the models and the external disturbances. In the view of "rejecting disturbances by the feedback", some properly constructed nonlinear feedback law will be much effective than linear feedback.. It's the reason that some nonlinear characters are used to design control systems, which have strong ability to reject disturbances or even are independent on the model. In Section 2, the ability of nonlinear feedback for disturbance rejection is analyzed. In Section 3, some special nonlinear elements including the nonlinear state observer, the Extended State Observer and the Nonlinear Tracking-Differentiator are introduced. In Section 4, two nonlinear design methods, the Nonlinear PID and the Auto-Disturbances-Rejection Controller are proposed. Some simulation results are shown in Section 5 and some further considerations are concluded in Section 6.

2. THE ABILITY OF NONLINEAR FEEDBACK FOR REJECTING DISTURBANCES

Consider a plant

$$\dot{x} = w(x,t) + u, \quad |w(x,t)| < w_0, \tag{1}$$

where $w(x,t)$ is the total action of the model and the external disturbances. Let's study the ability of the feedback for rejecting the disturbance $w(x,t)$.

Choose a feedback

$$u = -\beta\alpha(x), \tag{2}$$

where $\alpha(x)$ satisfies

$$\alpha(x) = -\alpha(-x), \quad x\alpha(x) > 0, \quad x \neq 0. \tag{3}$$

Assuming that the equation

$$\alpha(x) = \frac{w_0}{\beta} \tag{4}$$

has a root x_1 and $\alpha(x) > \frac{w_0}{\beta}$, when $x > x_1$, there exists

$$x\dot{x} = xw(x,t) - \beta x\alpha(x) < 0, \quad \forall x > x_1.$$

So for some $T > 0$,

$$|x(t)| < x_1, \quad \forall t > T.$$

Hence choosing the feedback (2), which satisfies the condition (3), to make the root of the equation (4) be as small as possible is a way to reduce the system's static error.

If we choose

$$\alpha(x) = |x|^{\alpha}\, sign(x), \tag{5}$$

then the root of the equation (4) is

$$x_1 = |\frac{w_0}{\beta}|^{\frac{1}{\alpha}}.$$

Provided $\beta > w_0$, it's obvious that

$$|\frac{w_0}{\beta}|^{\frac{1}{\alpha_1}} < |\frac{w_0}{\beta}| < |\frac{w_0}{\beta}|^{\frac{1}{\alpha_2}}, \forall 0 \leq \alpha_1 < 1 < \alpha_2. \tag{6}$$

This shows that the smaller α, the stronger the ability of the feedback law determined by (5) for rejecting disturbances. Furthermore, $x_1 = 0$ when $\alpha = 0$, which means the disturbance $w(x,t)$ has been totally suppressed .

This is the basic reason that some nonlinear characters can be used to design a system that has strong ability to reject disturbances and is independent of the models.

3. SOME NONLINEAR ELEMENTS WITH SPECIAL PROPERTIES

3.1 Nonlinear State Observer

First, let's study the linear state observer for the system

$$\begin{cases} \dot{x}_1 = x_2, \\ \dot{x}_2 = f(x_1, x_2, w(t)) + u, \\ y = x_1 \end{cases} \tag{7}$$

where $w(t)$ is external disturbances. If the variation of the function $f(x_1, x_2, w(t))$ is small, the linear observer

$$\begin{cases} \varepsilon = z_1 - y, \\ \dot{z}_1 = z_2 - \beta_{01}\varepsilon, \\ \dot{z}_2 = \quad -\beta_{02}\varepsilon + u \end{cases} \tag{8}$$

can track the state variables of the system (7) quite well. But the tracking result will become remarkably bad when the variation of $f(x_1, x_2, w(t))$ is large.

From the ability of the nonlinear feedback to reject disturbances, which we've analyzed in Section 2, we replace $\beta_{01}\varepsilon$, $\beta_{02}\varepsilon$ in the linear feedback (18) with properly constructed nonlinear functions $\beta_{01}g_1(\varepsilon)$, $\beta_{02}g_2(\varepsilon)$ to get

$$\begin{cases} \varepsilon = z_1 - y, \\ \dot{z}_1 = z_2 - \beta_{01}g_1(\varepsilon), \\ \dot{z}_2 = \quad -\beta_{02}g_2(\varepsilon) + u. \end{cases} \tag{9}$$

Assuming that

$$|f(x_1, x_2, w(t))| < w_0,$$

if we choose

$$g_1(\varepsilon) = \varepsilon, g_2(\varepsilon) = |\varepsilon|^{\alpha}\, sign(\varepsilon),$$

$$w_0 < \beta_{02} < \frac{1}{4}\beta_{01}^2,$$

then the tracking errors $\varepsilon_1(t) = x_1(t) - z_1(t), \varepsilon_2(t) = x_2(t) - z_2(t)$ of the observer (9) can be bounded in

$$|\varepsilon_1(t)| < (\frac{w_0}{\beta_{02}})^{\frac{1}{\alpha}}, |\varepsilon_2(t)| < \beta_{01}(\frac{w_0}{\beta_{02}})^{\frac{1}{\alpha}},$$

though they can't converge to zeros.

This implies that the nonlinear state observer which is independent of $f(x_1, x_2, w(t))$ can be constructed to assure that the tracking errors be small enough when $f(x_1, x_2, w(t))$ is within some bounds. Therefore, the exact information of $f(x_1, x_2, w(t))$ is not necessarily needed here.

3.2 The Extended State Observer (ESO)(Han 1995)

Consider the system (7) again. Define

$$x_3 = f(x_1, x_2, w(t)),$$

i.e. the new state x_3 is being extended for the uncertain value $f(x_1, x_2, w(t))$. Then the system (7) becomes

$$\begin{cases} \dot{x}_1 = x_2, \\ \dot{x}_2 = x_3 + bu, \\ \dot{x}_3 = v(t), \\ y = x_1. \end{cases} \tag{10}$$

where $v(t) = \frac{\partial f}{\partial x_1}\dot{x}_1 + \frac{\partial f}{\partial x_2}\dot{x}_2 + \frac{\partial f}{\partial w}\dot{w}$. Similarly, some properly chosen nonlinear functions $g_1(\varepsilon)$, $g_2(\varepsilon)$, $g_3(\varepsilon)$ and parameters β_{01}, β_{02}, β_{03} can be used to construct a state observer

$$\begin{cases} \varepsilon = z_1 - y, \\ \dot{z}_1 = z_2 - \beta_{01} g_1(\varepsilon), \\ \dot{z}_2 = z_3 - \beta_{02} g_2(\varepsilon) + bu, \\ \dot{z}_3 = \quad - \beta_{03} g_3(\varepsilon), \end{cases} \tag{11}$$

for the system (10). Hence, we can get

$$\begin{cases} z_1 \to x_1(t) = y(t), \\ z_2 \to x_2(t) = \dot{y}(t), \\ z_3 \to x_3(t) = f(x_1(t), x_2(t), w(t)). \end{cases} \tag{12}$$

Here, the third formula $z_3 \to x_3(t) = f(x_1(t), x_2(t), w(t))$ is most important. It shows that $z_3(t)$ can estimate (or track) $f(x_1, x_2, w(t))$ which is the total action of the uncertain models and the external disturbances or the real-time action of the system disturbances.

The $g_i(\varepsilon)$ in the ESO (11) can be chosen as follows

$$g_i(\varepsilon) = \mathrm{fal}(\varepsilon, \alpha_i, \delta) = \begin{cases} |\varepsilon|^{\alpha_i} \operatorname{sign}(\varepsilon), & |\varepsilon| > \delta, \\ \frac{\varepsilon}{\delta^{1-\alpha_i}}, & |\varepsilon| > \delta, \end{cases} \tag{13}$$

where $0 \le \alpha_i \le 1$.

3.3 The Nonlinear Tracking-Differentiator (TD) (Han and Wang, 1994)

It's well known that the classical differentiator is realized by the transfer functions

$$y = \frac{s}{Ts+1}u \quad \text{or}$$

$$y = \frac{1}{T_2 - T_1}\left(\frac{1}{T_1 s + 1} - \frac{1}{T_2 s + 1}\right)u,$$

where the former is sensitive to the noise existing in the input signals while the latter has some filtration on the noise. But the time constants T and $T_1 < T_2$ are all required to be small. For the latter form, its state space realization is

$$\begin{cases} \dot{x}_1 = x_2, \\ \dot{x}_2 = -\frac{1}{T_1 T_2}(x_1 - u) - \frac{T_2 + T_1}{T_1 T_2}x_2, \\ y = x_2. \end{cases}$$

It shows that the smaller $T_1 < T_2$, the quicker $x_1(t)$ tracking $u(t)$. Under these circumstances, is it possible to pick up the differential signal by tracking $u(t)$ quickly enough?

Next we introduce a second order system, which is called the Tracking-Differentiator (Han and Wang, 1994)

$$\begin{cases} \dot{x}_1 = x_2, \\ \dot{x}_2 = -r\,sign(x_1 - u(t) + \frac{x_2|x_2|}{2r}), \end{cases} \tag{14}$$

where, for any input signal $u(t)$, $x_1(t)$ can quickly track it and $x_2 = \dot{x}_1$ can approximate the general differential of $u(t)$ well enough. But the sign function sign(•) induces the chattering when $x_1(t)$ tracks $u(t)$. To avoid this chattering, sign(•) is replaced with a linear saturate function as follows

$$sat(\alpha, \delta) = \begin{cases} sign(\alpha), & |\alpha| > \delta, \\ \frac{\alpha}{\delta}, & |\alpha| \le \delta. \end{cases} \tag{15}$$

Then the following system can be got

$$\begin{cases} \dot{x}_1 = x_2, \\ a = x_1 - u(t) + \frac{x_2|x_2|}{2r} \\ \dot{x}_2 = -r\,sat(a, \delta). \end{cases} \tag{16}$$

By tuning the parameters r and δ, good tracking result and the differential signal can be obtained. Particularly, when *u(t)* is a step function $u_0 = const.$, $x_1(t)$ can quickly track it (the acceleration doesn't exceed r) without any overshoot.

4. TWO NONLINEAR DESIGN APPROACHES

4.1 Nonlinear PID (Han, 1994)

Handling the reference input and the system output by the nonlinear TD respectively, we can obtain a new type of PID---the nonlinear PID (NLPID). Fig.1 shows the structure of the NLPID

The adaptability and the robustness of the NLPID are much better than the classical PID. The contradiction of quick speed and the overshoot no longer exists.

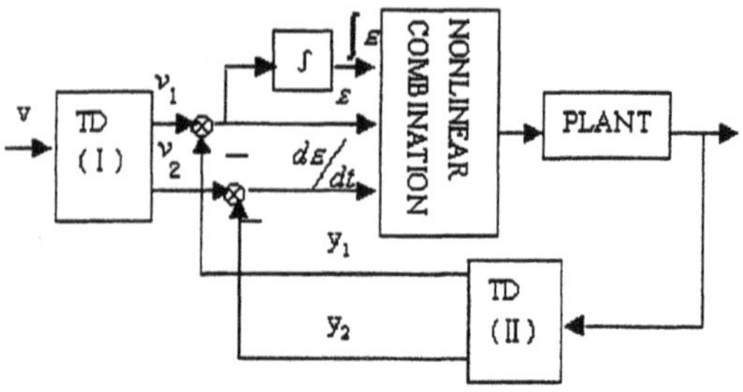

Fig. 1. The structure of NLPID

4.2 *The Auto-Disturbances-Rejection Controller (ADRC)(Han, 1998)*

Combining TD, ESO and nonlinear feedback law, we can get another new type of controller.

The controlled system is

$$\begin{cases} \dot{x}_1 = x_2, \\ \dot{x}_2 = f(x_1, x_2, w(t)) + b(t)u, \\ y = x_1. \end{cases} \qquad (17)$$

where the external disturbance $w(t)$ and $f(x_1, x_2, w(t))$ are unknown, $b(t)$ is uncertain in the neighborhood of b_0=const. Assuming $v(t)$ is the reference input, a new type of controller can be designed according to the following steps:

1. Input the reference signal $v(t)$ to TD to get an arranged transient process v_1 and its differential signal v_2 .
2. Input the system's output $y(t)$ to ESO to get the estimations z_1, z_2 and z_3 for the states x_1, x_2 and $a(t)$--the total action of the system's unknown disturbances.
3. Construct the nonlinear feedback law u_0 by the state errors $\varepsilon_1 = v_1 - z_1$ and $\varepsilon_2 = v_2 - z_2$.
4. Produce the control value

$$u = u_0 - \frac{z_3}{b_0}. \qquad (18)$$

Since $z_3(t)$ is the estimation for the total action of the unknown disturbances, $-\frac{z_3}{b_0}$ in (18) is to compensate the disturbances. So the transfer function from u_0 to y will be in the type of second order cascade integrators. Because this controller can automatically detect the disturbances and compensate it, it is called the Auto-Disturbances-Rejection Controller (ADRC) (Han, 1998). Fig. 2 shows the structure of ADRC.

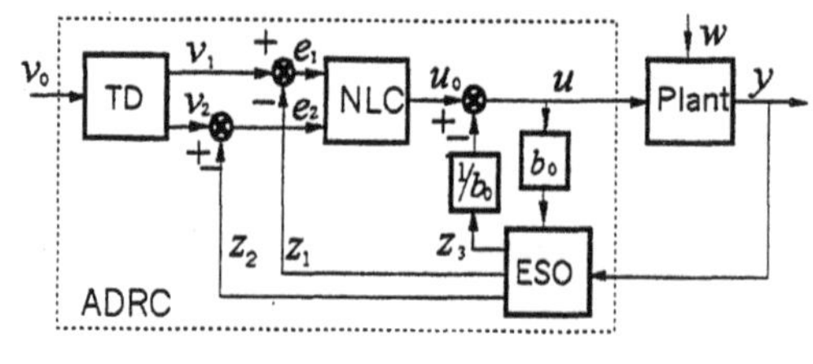

Fig. 2. The structure of ADRC

The algorithm of ADRC is as follows

TD— Arranging the transient process:

$$\begin{cases} \delta_1 = rh, \\ \delta_0 = \delta_0 h, \\ y_1 = v_{1t} - v_t + hv_{2t}, \\ y_2 = v_{2t}, \\ a_0 = \sqrt{\delta_1^2 + 8r \mid y_1 \mid}, \\ a = y_2 + sign(y_1)\dfrac{a_0 - \delta_1}{2}, \\ \text{if } (\mid y_1 \mid < \delta_0); a = y_2 + \dfrac{y_1}{h}; \text{endif} \\ fst2 = -rsat(a, \delta), \\ v_{1t+1} = v_{1t} + hv_{2t}, \\ v_{2t+1} = v_{2t} + hfst2. \end{cases}$$

ESO— The Extended State Observer:

$$\begin{cases} \varepsilon = z_{1t} - y_t, \\ fe_1 = fal(\varepsilon, \dfrac{1}{2}, \delta), \\ fe_2 = fal(\varepsilon, \dfrac{1}{4}, \delta), \\ z_{1t+1} = z_{1t} + h(z_2 - \beta_{01}\varepsilon), \\ z_{2t+1} = z_{2t} + h(z_3 - \beta_{02} fe_1 + b_0 u_t), \\ z_{3t+1} = z_{3t} + h(\quad - \beta_{03} fe_2), \end{cases}$$

NLC— Computing the control value:

$$\begin{cases} \varepsilon_1 = v_{1t} - z_{1t}, \\ \varepsilon_2 = v_{2t} - z_{2t}, \\ fe_1 = fal(\varepsilon_1, \alpha, \delta), \\ fe_2 = fal(\varepsilon_2, \alpha, \delta), \\ uu_{t+1} = \beta_1 fe_1 + \beta_2 fe_2, \\ u_{t+1} = uu_{t+1} - \dfrac{z_{3t+1}}{b_0}. \end{cases}$$

Here the differential equations of TD and ESO are calculated by the Euler recurrence algorithm; the inputs of ADRC are the reference input $v(t) = v_t$ and the system's output $y(t) = y_t$; h is the step length; $r, \beta_{01}, \beta_{02}, \beta_{03}, \delta, \beta_1, \beta_2$ are parameters of ADRC; uu_{t+1} is the control value for the integrator cascaded plant and $\frac{z_{3t+1}}{b_0}$ is the disturbances compensating term.

If the system model can be separated into
$f(x_1, x_2, w(t)) = f_0(x_1, x_2, w_0(t)) + f_1(x_1, x_2, w(t))$

and $f_0(x_1, x_2, w_0(t))$ is known, then $b_0 u$ in ESO can be changed to $b_0 u + f_0(x_1, x_2, w_0(t))$. Hence, $z_3(t)$ will estimate the effect of $f_1(x_1(t), x_2(t), w(t)) + (b(t) - b_0(t))u(t)$. In this case, replacing the control law (18) with

$$u = u_0 - \frac{z_3 + f_0(z_1, z_2, w_0(t))}{b_0},$$

we can get the Model Compensated ADRC. Since the known part of the model has been compensated first, the burdens on ESO can be lightened. Therefore, the estimations can be more exact and the control quality can be improved.

Moreover, according to the information from ESO, the frame of the Model Compensated ADRC can give a self-learning function. First the initial crude estimations $f_0(\bullet)$ and b_0 for $f(\bullet)$ and b are given. So $f(\bullet)$ and b can be expressed as

$$f(\bullet) = f_0(\bullet) + \Delta f, \quad b = b_0 + \Delta b.$$

In this case, the z_3 from ESO will estimate $\Delta f(\bullet) + \Delta bu$. Then modify the original $f_0(\bullet)$ and b_0 by $\Delta f(\bullet)$ and Δb, which are estimated from z_1, z_2, z_3 by some proper identification methods. Repeat this process such that $f_0(\bullet)$ and b_0 approximate to the real values $f(\bullet)$ and b step by step. This process is a self-learning process.

In the ADRC, the model and the external disturbances are all treated as the disturbances. They can be automatically detected and compensated. In the multivariable case, the dynamic coupling effect of the other channels on a certain channel can also be viewed as a kind of external disturbances on this channel. So, when the ADRC is used for the decoupling control, only the static coupling needs to be considered while the dynamic coupling is out of consideration. When the static coupling is in the form of diagonal dominant, it also can be ignored.

Based on the understanding that the time delay system is obtained from some higher order system in which the parts of small time constants have been simplified, the lower order system with time delay can be controlled by a higher order ADRC.

5. SIMULATION EXAMPLE

Given the system

$$\dot{x} = x^m + bu,$$

where b=1 and x^m is unknown.. The control purpose is to track the step signal $r_0(>1)$. Obviously, the open-loop system is unstable. When r_0 and m are large, the system's variables will be sharp and be hard to be controlled.

Next, taking r_0 and m as parameters, the bound of the plant family which can be controlled by a same ADRC is studied.

The algorithm of the ADRC for the first order systems is

Arranging the transient process:

$$\begin{cases} \varepsilon_0 = v_{1t} - r_0, \\ fe_0 = fal(\varepsilon_0, a_0, \delta), \\ v_{1t+1} = v_{1t} - hrfe_0, \end{cases}$$

Estimating states and the action of the disturbances:

$$\begin{cases} \varepsilon_1 = z_{1t} - x_t, \\ fe_1 = fal(\varepsilon_1, a_1, \delta), \\ fe_2 = fal(\varepsilon, \frac{1}{4}, \delta), \\ z_{1t+1} = z_{1t} + h(z_{2t} - \beta_{01}\varepsilon_1 + bu_t), \\ z_{2t+1} = z_{2t} + h(-\beta_{02} fe_1), \end{cases}$$

Computing the control value:

$$\begin{cases} \varepsilon = v_{1t+1} - z_{1t+1}, \\ uu_{t+1} = \beta_1 \varepsilon_1, \\ u_{t+1} = uu_{t+1} - \frac{z_{2t+1}}{b_0}, \end{cases}$$

where the parameters are chosen as:

step length: h=0.001;

the parameters for arranging transient process: $r = 1$, $\alpha = 0.5$ (such that the transient time will be $T_0 = 2\sqrt{r_0}$);

the parameters for the ESO: $a_1 = 0.5$, $\beta_{01} = 600$, $\beta_{02} = 18000$;

the feedback gain: $\beta_1 = 20$;

$\delta = 0.001$.

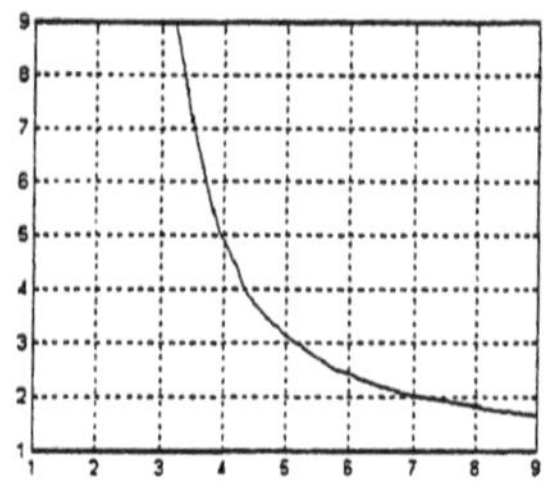

Fig. 3 Bounds of the parameters (m, r_0), m—x-axis, r_0—y-axis

Fig. 3 shows the adaptability and the robustness of the designed ADRC. The systems whose parameters (m, r_0) are the points lying below the curve can be controlled well by the given ADRC.

6. CONCLUSIONS

Summing up what have been studied in this paper, following new knowledge about the problem of the control systems design can be concluded:

1. Some nonlinear characters can be used to construct nonlinear elements with some special properties. So new type of controllers which are independent of the plant can be synthesized.
2. The ADRC is a kind of controllers that is designed based on nonlinear methods. The essential advantage of ADRC is that it can automatically detect the total effect of unknown models and the external disturbances so as to compensate them. This is the sense of "auto-disturbances-rejection".
3. The Model Compensated ADRC can be constructed based on the level of the plant understanding or the requirements of the system's quality.
4. The data obtained from ESO can be used to identify some information about the uncertain plant. Therefore , the self-learning ability can be added to ADRC.
5. The ADRC is suitable for the problems involving

- linear or nonlinear
- time-varying or time-unvarying
- certain or uncertain
- single variable or multivariable
- time-delay or no time-delay

In the nonlinear world, there exists the control mechanism that is independent on the models. So, the concepts of adaptability and the robustness need to be renewed.

Then some problems need further considerations:

- What is the information about the plant that is needed to synthesize new type of controllers by nonlinear methods?
- What is the system's description that is suitable for the new synthesis methods?

REFERENCES

Han J.Q. and W. Wang (1994). Nonlinear Tracking-Differentiator. *Systems Science and Mathematical Sciences*, **14**, 177-183.(in Chinese)

Han J.Q. (1994). Nonlinear PID controller. *ACTA Automatica Sinica*, 487-490.(in Chinese)

Han J.Q. (1995) "Extended State Observer" for a kind of uncertain plants. *Control and Decision*, **10**, 85-88.(in Chinese)

Han J.Q. (1998). The Auto-Disturbances-Rejection Controller and its applications. *Control and Decision*, **14**, 19-23.(in Chinese)

14th Triennial World Congress, Beijing, P.R. China

HIGH-SPEED HIGH-PRECISION TRACKING CONTROL FOR ELECTRONICALLY CONTROLLED WINDING MACHINES

Bagus Mahawan **Zheng-Hua Luo**

Department of Mechanical Engineering, Nagaoka University of Technology, 1603-1, Kamitomioka-machi, Nagaoka, Niigata, 940-2188, JAPAN
Email: mahawan@dragon.nagaokaut.ac.jp, luo@dragon.nagaokaut.ac.jp

Abstract: In this paper, a new type of electronically controlled winding machine is proposed, and a high-speed high-precision tracking control system for the winding machine is designed. To estimate unknown parameters of the plant, a new online identification scheme for general servomechanism is proposed, and the stability of the system is theoretically proved. Furthermore, a disturbance observer is also designed to reject possible disturbances in order to improve the tracking performance of the system. Experimental results show that the identification algorithm guarantees the parameter convergence, and also yields satisfactory transient responses of the identified parameters. It is also observed that the disturbance observer can attenuate the tracking error caused by the disturbances. *Copyright © 1999 IFAC*

Keywords: servomechanism, tracking systems, parameter identification, disturbance rejection, winding machines

1. Introduction

Since electric motors are widely used in industry as well as in daily life, the manufacture and the quality maintenance of motors have become very important. For mass production of high quality electric motors, automatic winding machines have been utilized from an early stage. Most of the existing automatic winding machines, however, are adopting the crank/cam mechanism, and are designed to produce one type of motors. This leads to the following disadvantages:

- The mechanism of the existing winding machines allow only one specific winding trajectory that can be realized,
- The operation speed of the existing winding machines is usually less than 1000 rpm because of the crank/cam mechanism.

Meanwhile, to increase the operation speed, it is highly desireable to develope a new type of electronically controlled winding machines which

- can operate at a speed of 1000-2000 rpm,
- can track any type of specified trajectory with a considerably small tracking error during all the operation time (running time).

It is the purpose of this paper to study the tracking control problem for a prototype of an electronically controlled winding machine that meets the above requirements. Specifically, we propose an online parameter identification scheme for typical servomechanism, which is incorporated into a feedforward controller to realize a high-speed high-precision tracking control system. The asymptotic stability of the closed loop system is proved, as well as the superiority of the convergence rate of the estimated parameters. Experimental results illustrate the applicability of the proposed identification scheme.

Furthermore, we design a disturbance observer to reject the possible existing disturbance and to improve the tracking performance of the system (Onishi, 1993; Iwai *et al.*, 1988). Several experiments are conducted to show that the tracking error arising from the existence of the disturbance and from the parameter error can be attenuated using the disturbance observer.

This paper is organized as follows. In section 2, a

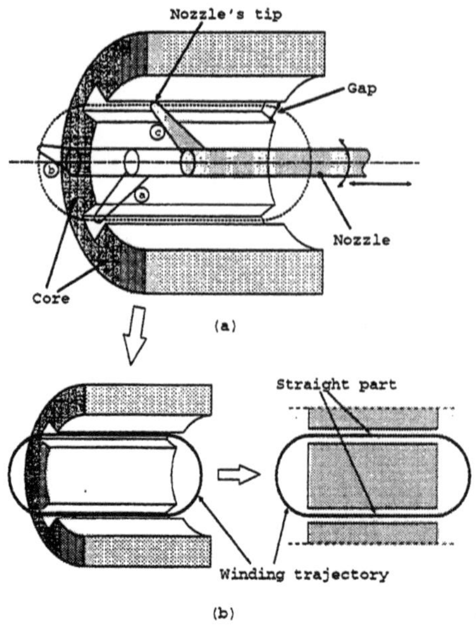

Figure 1: Winding Trajectory

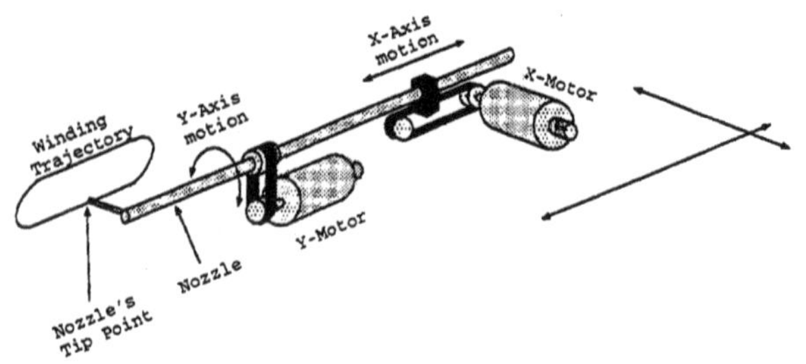

Figure 2: Schematic Diagram of the New Winding Machine

schematic diagram of the new type of winding machine and the configuration of the servomechanism are described. A model of a typical servomechanism is then derived in section 3. In section 4, an on-line parameter identification law is proposed, and the stability of the closed loop system is discussed. In the same section, the experimental results of the identification algorithm are shown. Design of a disturbance observer and the experimental results are discussed in section 5, and finally, conclusions are given in section 6.

2. Problem Statement

Basically, winding machines can be devided into 2 types: those which wind coils into armatures, and those which wind coils into stators. The former are much easier than the latter, so in this paper, we restrict our discussion to the latter, i.e., winding machines for stators.

Direct or indirect methods are currently used to wind coils to stators. In the indirect method, briefly, a coil is first wound into a tool. After it is completely wound, it is then slipped into a stator. In contrast, in the direct method, a coil is wound directly into a stator by passing it through a narrow gap between 2 cores. As depicted in Fig. 1(a), first a nozzle, whose tip leads a wire, moves forward through a gap between the two cores (see ⓐ). After passing the gap, it rotates to the next gap (see ⓑ), and then moves backward passing through the gap (see ⓒ). These ⓐ→ⓑ→ⓒ motions are repeated repetitively until the desired number of winds is reached. Taking notice of the trajectory of the nozzle's tip, we can see the so-called winding trajectory, as shown in Fig. 1(b). The winding trajectory, in general, contains straight parts and curves. In small-sized motors, since the gaps between the two cores have only several milimeters width, high tracking precision of winding trajectory is required, particularly in the straight part of the trajectory.

To realize such a winding trajectory, we design a new type of electronically controlled winding machine, of which schematic diagram is shown in Fig. 2. The new winding machine consists of two servo motors, which are named X-Motor and Y-Motor, respectively. The two motors are placed perpendicularly. A pulley is attached to the free-end of the shaft of each motor. A belt is connected to the pulley of X-Motor with the other part of the belt is mounted to a nozzle in such a way that the rotational motion of X-motor is transmitted to the translational motion of the nozzle. In a similar way, the rotation of Y-Motor is also transmitted to the rotation of the nozzle via belt-pulley transmission. In other words, with X-Motor and Y-Motor, the nozzle has two DOF, i.e., translational and rotational motion. Therefore, by driving both X-Motor and Y-Motor timely, we can realize winding trajectories as shown in Fig. 1(b).

Generally, a desired winding trajectory is specified in advance according to the shape and size of the stators, which is composed from X-Axis trajectory and Y-Axis trajectory, both of which are time functions. Illustrated in Fig. 3(b), for instance, is a desired trajectory for X-Motor and Y-Motor, say $\phi_{dx}(t), \phi_{dy}(t)$ respectively. These two components form a plane desired trajectory shown in Fig. 3(a). An arbitrary desired trajectory can be made by modifying $\phi_{dx}(t)$ and $\phi_{dy}(t)$ separately, which implies that various stators can be wound regardless of the shape of the stators using only one single

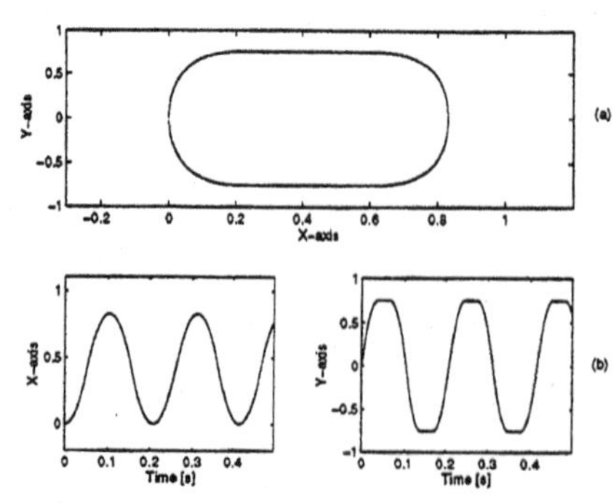

Figure 3: An Example of Desired Trajectory

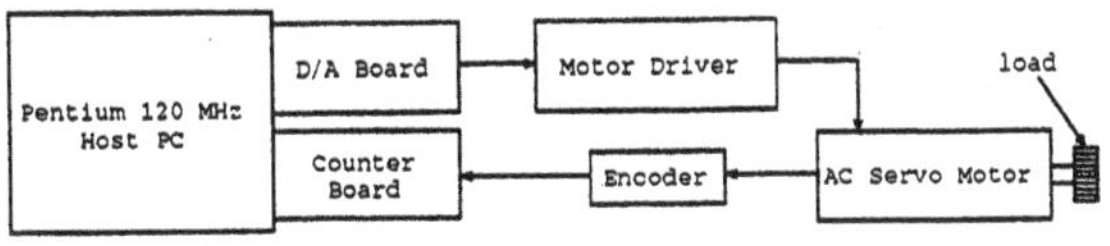

Figure 4: System Configuration of the Servo System

winding machine. This is the most significant advantage of the new type of winding machines, since the conventional winding machines can only generate one kind of winding trajectory.

If the angular position of X-Motor, say $\phi_x(t)$, tracks the given X-axis desired trajectory $\phi_{dx}(t)$, and also if $\phi_y(t)$ tracks $\phi_{dy}(t)$, then it is clear that the nozzle's tip tracks the desired trajectory. Hence, the problem of realizing a winding trajectory shown in Fig. 1(b) can be reduced to a tracking problem of two servomechanisms. Since both X-Motor and Y-Motor are independent, only the tracking problem of X-Motor is considered here, while the tracking problem of Y-Motor can be treated in a similar way.

Fig. 4 shows a system configuration of a servomechanism for the new winding machine. The servomechanism consists of a 3-phase brushless AC servo motor of 7.5KW and a motor driver of velocity reference type. To measure the angular position of the motor shaft, an encoder with 2048 pulse/rev. is attached to the motor shaft. The pulses from the encoder are counted by a counter board and read into a Pentium 120 MHz control computer. According to the control algorithm, which will be mentioned later, the computer calculates the required control voltage which is then sent to the motor driver via a D/A converter.

3. Model of the Servomechanism

Although the structure of AC servo motors is much more complicated than that of DC motors, the commercially available up-to-date drivers for AC motors are well designed so that one can conveniently use them in much the same way as drivers for DC motors. For our winding machine, the motor driver that we chose is of velocity reference type and has a velocity feedback outer loop and a current feedback inner loop. The gain of the current loop is very high and as a result the generating torque τ_m of the motor is proportional to the motor current i. Let ϕ denote the angular position of the motor. Let J and f_d denote, respectively, the inertia of the motor shaft together with the load and the friction

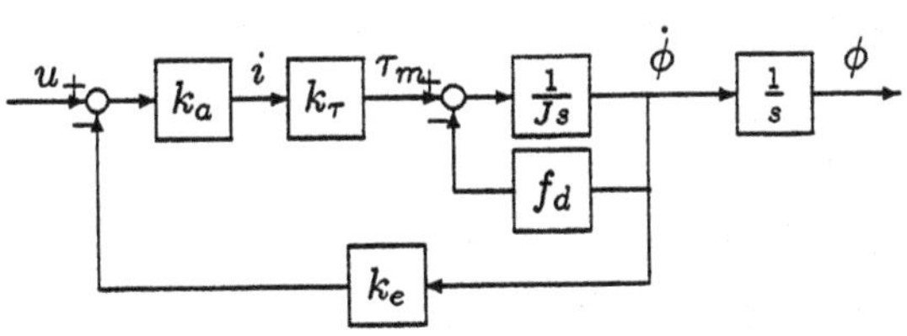

Figure 5: Block Diagram of the Servomechanism

coefficient. Then from the balance of torque, we have

$$\tau_m(t) = J\ddot{\phi}(t) + f_d\dot{\phi}(t).$$

Taking the Laplace transform of this equation, we are able to write the block diagram of the motor and its driver as in Fig. 5 where u denotes the reference input voltage to the motor driver; k_a is the velocity loop feedback gain, and k_τ, k_e are various constants whose meanings are obvious from the block diagram (Yasukawa Denki, 1994).

Now, if we define constants K, T as

$$K := \frac{k_a k_\tau}{f_d + k_a k_e k_\tau} \quad , \quad T := \frac{J}{f_d + k_a k_e k_\tau} \tag{1}$$

then it is seen that the transfer function from $u(s)$ to $\phi(s)$ is given by

$$P(s) = \frac{\phi(s)}{u(s)} = \frac{K}{s(Ts+1)} \tag{2}$$

where K and T are commonly known as gain and time constant respectively. (2) is the transfer function of a typical servomechanism in which all the physical parameters of the motor, motor driver, and load are represented by K and T, which are in general poorly known.

Our goal is to calculate control input $u(t)$ in such a way that for a given achievable desired trajectory $\phi_d(t)$, the angular position of motor $\phi(t)$ converges to the desired trajectory $\phi_d(t)$ with the tracking error $e(t)$ tending to zero as $t \to \infty$. To realize such a high-speed tracking control system, we propose an on-line parameter identification scheme, which will be discussed in the next section.

4. On-line Adaptive Identification

Fig. 6 shows a closed loop system which consists of a predesigned feedback controller $C(s)$ which is to stabilize the plant, a feedforward controller, and the proposed identification law which. The following assumptions are imposed on the class of system shown in Fig. 6.

A1: $\hat{K}^{-1}(t)$ exists

A2: $1, \dot{\phi}_d(t), \ddot{\phi}_d(t)$ are linear independent as functions of t

From Fig. 6, it is clear that if the estimated parameters $\hat{K}(t), \hat{T}(t)$ coincide with the real parameters K, T, then the system output tracks the desired trajectory, i.e., $\phi(t) = \phi_d(t)$, which implies that a perfect tracking performance is achieved. To establish such performance, we propose the following adaptive identification law.

$$\begin{cases} \dot{\hat{T}}(t) = \alpha\left(\dot{\phi}_d(t) - \dot{\phi}(t)\right)\ddot{\phi}_d(t) \\ \dot{\hat{K}}(t) = -\gamma\left(\dot{\phi}_d(t) - \dot{\phi}(t)\right)\hat{K}^{-1}(t) \\ \qquad \times\left(\hat{T}(t)\ddot{\phi}_d(t) + \dot{\phi}_d(t)\right) \end{cases} \tag{3}$$

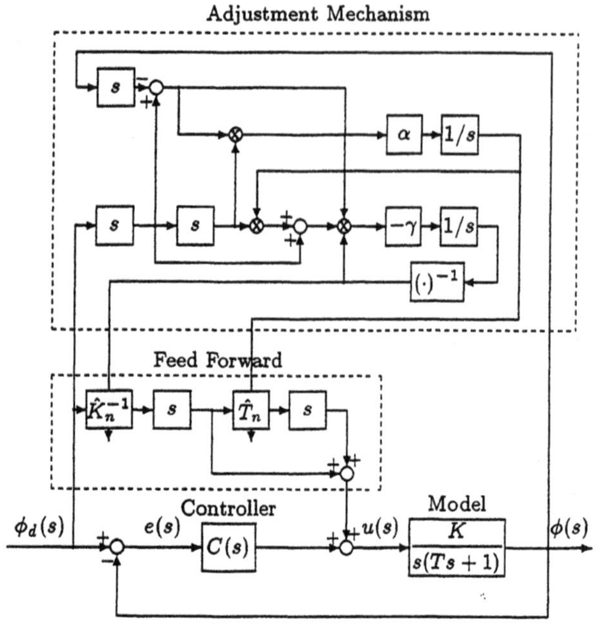

Figure 6: Feedforward Control With Online Adaptive Identification of K and T

where $\hat{T}(0) = T_0, \hat{K}(0) = K_0$ denote initial values of $\hat{T}(t), \hat{K}(t)$. $\alpha > 0, \gamma > 0$ are prespecified positive numbers known as adaptation parameters.

To analyze the stability of our proposed closed loop system, we first rewrite Fig. 6 in the time domain, which yields the following equation

$$\begin{aligned} &T\ddot{\phi}(t) + \dot{\phi}(t) = Ku(t) \\ &u(t) = k_p\left(\phi_d(t) - \phi(t)\right) + \hat{K}^{-1}(t)\left(\hat{T}(t)\ddot{\phi}_d(t) + \dot{\phi}_d(t)\right) \end{aligned} \tag{4}$$

Note that, without loss of generality, we choose the feedback controller $C(s) = k_p$(constant) to simplify our discussion.

We then introduce error variables as follows

$$\begin{cases} e(t) = \phi_d(t) - \phi(t) \\ e_T(t) = \hat{T}(t) - T \\ e_K(t) = \hat{K}(t) - K \end{cases} \tag{5}$$

From (4) and (5), we see that

$$\begin{aligned} T\ddot{e}(t) &= T\ddot{\phi}_d(t) - T\ddot{\phi}(t) \\ &= T\ddot{\phi}_d(t) + \dot{\phi}(t) - Ku(t) \\ &= T\ddot{\phi}_d(t) + \dot{\phi}(t) - K\left[k_p e(t)\right. \\ &\quad \left. + \hat{K}^{-1}(t)\left(\hat{T}(t)\ddot{\phi}_d(t) + \dot{\phi}_d(t)\right)\right] \end{aligned} \tag{6}$$

Thus, some simple calculations lead to the following equation

$$\begin{aligned} &T\ddot{e}(t) + \dot{e}(t) + k_pKe(t) \\ &= T\ddot{\phi}_d(t) + \dot{\phi}_d(t) - K\hat{K}^{-1}(t)\left(\hat{T}(t)\ddot{\phi}_d(t) + \dot{\phi}_d(t)\right) \\ &= e_K(t)\hat{K}^{-1}(t)\left(\hat{T}(t)\ddot{\phi}_d(t) + \dot{\phi}_d(t)\right) - e_T(t)\ddot{\phi}_d(t) \end{aligned} \tag{7}$$

Finally, from (3),(5), and(7), we have the following closed loop equations

$$\begin{cases} T\ddot{e}(t) + \dot{e}(t) + k_pKe(t) \\ \qquad = e_K(t)\hat{K}^{-1}(t)\left(\hat{T}(t)\ddot{\phi}_d(t) + \dot{\phi}_d(t)\right) \\ \qquad - e_T(t)\ddot{\phi}_d(t) \\ \dot{e}_T(t) = \alpha\dot{e}(t)\ddot{\phi}_d(t) \\ \dot{e}_K(t) = -\gamma\dot{e}(t)\hat{K}^{-1}\left(\hat{T}(t)\ddot{\phi}_d(t) + \dot{\phi}_d(t)\right) \end{cases} \tag{8}$$

The closed loop system in (8) is stable if there exists a Lyapunov function, whose time derivative along with the solution of (8) is negative (Khalil, 1996). To prove the stability of the closed loop system, we introduce the following function

$$V(t) = \frac{1}{2}T\dot{e}^2(t) + \frac{1}{2}k_pKe^2(t) + \frac{1}{2\gamma}e_K^2(t) + \frac{1}{2\alpha}e_T^2(t) \tag{9}$$

as the Lyapunov candidate. The time derivative along with the solution of (8) is then

$$\begin{aligned} \dot{V}(t) &= T\dot{e}(t)\ddot{e}(t) + k_pKe(t)\dot{e}(t)\frac{1}{\gamma}e_K(t)\dot{e}_K(t) \\ &\quad + \frac{1}{\alpha}e_T(t)\dot{e}_T(t) \\ &= \dot{e}(t)\left[-\dot{e}(t) - k_pKe(t) - e_T(t)\ddot{\phi}_d(t) + e_K(t)\right. \\ &\quad \left.\times\hat{K}^{-1}(t)\left(\hat{T}(t)\ddot{\phi}_d(t) + \dot{\phi}_d(t)\right)\right] + k_pKe(t)\dot{e}(t) \\ &\quad + \frac{1}{\gamma}e_K(t)\left[-\gamma\dot{e}(t)\hat{K}^{-1}(t)\left(\hat{T}(t)\ddot{\phi}_d(t)\right.\right. \\ &\quad \left.\left. + \dot{\phi}_d(t)\right)\right] + \frac{1}{\alpha}e_T(t)\left[\alpha\dot{e}(t)\ddot{\phi}_d(t)\right] \\ &= -\dot{e}^2(t) \leq 0. \end{aligned} \tag{10}$$

From (10), we see that all error variables $e(t), \dot{e}(t), e_K(t), e_T(t)$ are stable. Furthermore, if $\dot{e}(t) = 0$ then $\ddot{e}(t) = 0, e(t) = C_e, e_T(t) = C_T, e_K(t) = C_K$, where C_e, C_K, C_T are constant variables. Hence, the closed loop system (8) becomes,

$$k_pKC_e = C_1\ddot{\phi}_d(t) + C_2\dot{\phi}_d(t)\ , \ ^{\forall}t > 0 \tag{11}$$

where C_1, C_2 are constants. From assumption A2, (11) holds only if $C_1 = C_2 = C_e = 0$. This implies that

$$\begin{cases} e_K(t) \to 0 & \Longrightarrow \quad \hat{K}(t) \to K \\ e_T(t) \to 0 & \Longrightarrow \quad \hat{T}(t) \to T \\ e(t) \to 0 & \Longrightarrow \quad \phi(t) \to \phi_d(t) \end{cases}$$

when $t \to \infty$. Thus, it is clear that all the error variables are asymptotically stable.

4.1 Experimental Results

The closed loop system shown in Fig. 6 is implemented into the servomechanism described in the previous section, with a sampling frequency of 5[KHz].

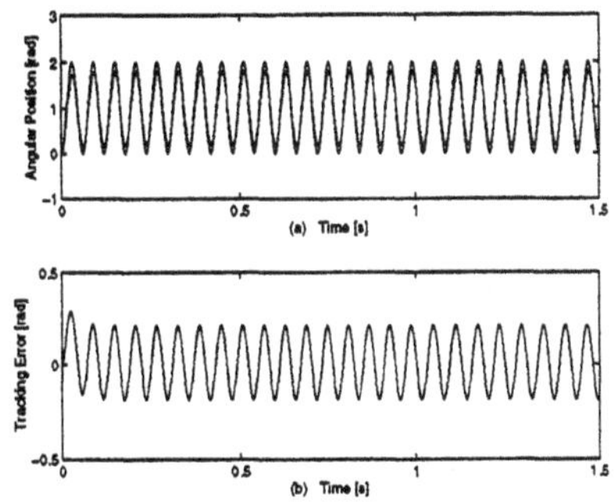

Figure 7: System Output and Tracking Error Without the Proposed Identification Scheme

All control programs are coded with the programming language 'C'. The objective of the experiment is to estimate the plant's parameters K, T, while regulating the control input u such that the angular position of the motor tracks a desired trajectory given by

$$\phi_d(t) = 1 - \cos(\omega_d t). \tag{12}$$

where $\omega_d = 2\pi 1000/60$[rad/sec]. For the closed loop controller $C(s)$, the state-of-the-art PI controller is chosen, with the control gains $k_p = 15.0$, $k_i = 1.0 \times 10^{-3}$, are tuned to achieve the best performance.

In this paper, we show 3 kinds of experimental results. First, an experiment is conducted without using our identification scheme, with parameters in feedforward controller fixed to: $\hat{K}(t) = 40, \hat{T}(t) = 1 \times 10^{-3}$, where the real values $K = 32.1, T = 0.46 \times 10^{-3}$ are obtained from off-line identification. We see from the experimental result that, as shown in Fig. 7, the tracking error remains large for all $t > 0$. This is obvious because of the difference between the estimated parameters and their real values.

We then implement our identification scheme which is incorporated to the feedforward controller, with adaptation parameters and initial values are set as follows

$$\begin{array}{lcl} \gamma = 10 & , & K_0 = 20 \\ \alpha = 2 \times 10^{-4} & , & T_0 = 3 \times 10^{-3} \end{array}$$

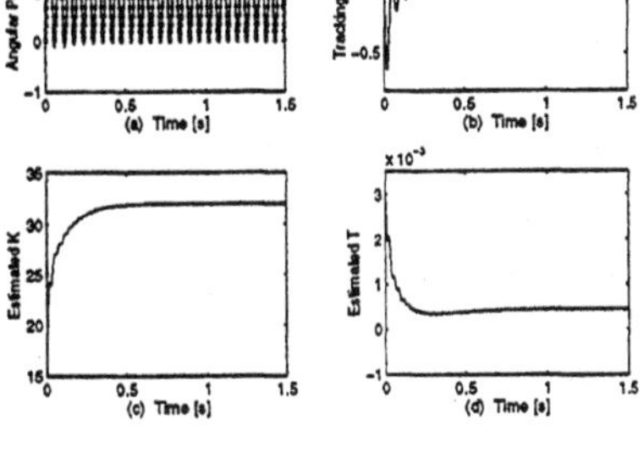

Figure 8: Experimental Result: Tracking Performance With the Proposed Identification Scheme

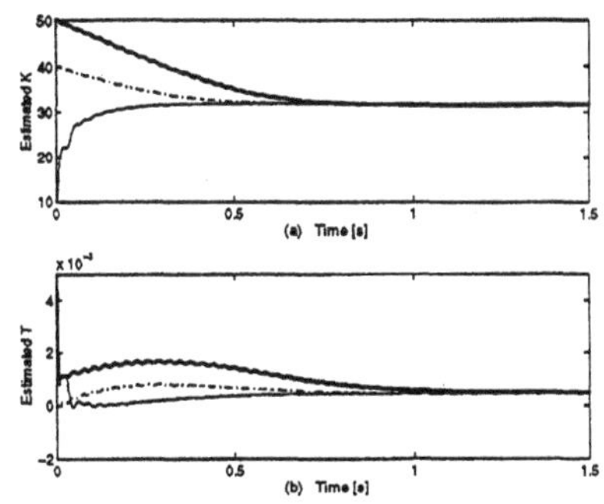

Figure 9: Experimental Results: Parameter Convergence under Various Initial Conditions

The result is shown in Fig. 8. We see that the tracking error $e(t)$ converges to 0 as soon as K, T converge to their real values, in less than 1 second, and the perfect tracking performance is achieved.

The last experiment is to investigate the convergence performance of the estimated parameters when the identification starts with different initial values, which are shown below

bold line :	$K_0 = 50$,	$T_0 = 1 \times 10^{-3}$
dashed line :	$K_0 = 40$,	$T_0 = 0.0$
solid line :	$K_0 = 10$,	$T_0 = 5 \times 10^{-3}$

As shown in Fig. 9, the parameters converge smoothly in all cases. The closer the initial values of the estimated parameters to their real values, the better convergence performance is achieved. Therefore, if there is prior information about the parameters being estimated, then our proposed identification method yields a high-precision tracking control from the beginning of identification, which is not always guaranteed by other identification schemes.

5. Design of Disturbance Observer

In the previous section, it was assumed that any disturbances applying to the system are negligible. However, a disturbance should not be neglected since it usually hinders tracking performance of the system. Thus, in order to get a better tracking performance, it is required to design a disturbance observer which can compensate disturbances.

Fig. 10 shows the system in which disturbance observer is added and the output signal of the disturbance observer $\hat{D}(s)$ is fedback to the control input $u(s)$. From Fig. 10 we are able to write $\hat{D}(s)$ and $\tilde{u}(s)$ as

$$\begin{aligned} \hat{D}(s) &= \frac{w_c}{K_n} \cdot \frac{w_c s}{s + w_c} \cdot \frac{T_n s + 1}{s + w_c} \{\phi(s) \\ &\quad - \frac{K_n}{s(T_n s + 1)} \tilde{u}(s)\} \\ \tilde{u}(s) &= u(s) - \hat{D}(s) \end{aligned} \tag{13}$$

where ω_c denotes a prespecified (large) positive number. K_n, T_n denote the estimated values of the

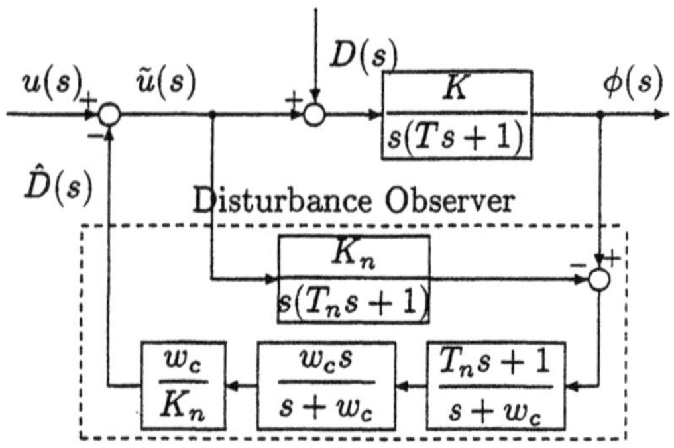

Figure 10: Disturbance Observer

plant's parameters K, T, that, for instance, can be obtained using the proposed identification scheme in the previous section. Assuming that $K_n = K, T_n = T$, and substituting (2) into (13), we get

$$\hat{D}(s) = F(s)^2 D(s), \qquad F(s) = \frac{\omega_c}{s+\omega_c} \tag{14}$$

Hence, the transfer function from $u(s), D(s)$ to $\phi(s)$ becomes

$$\begin{aligned}\phi(s) &= P(s)\left\{u(s) - F(s)^2 D(s) + D(s)\right\} \\ &= P(s)u(s) + P(s)\tilde{F}(s)D(s) \end{aligned} \tag{15}$$

where $\tilde{F}(s) = 1 - F(s)^2 = \frac{\omega_c^2}{s^2 + 2\omega_c s + \omega_c^2}$.

(15) implies that $\tilde{F}(s)$ serves as a high pass filter for $D(s)$. Therefore,If the bandwidth of $D(s)$ is less that ω_c, then it is clear that the gain of $D(s)$ is suppressed, so that $D(s)$ will not affect the system output.

5.1 Experimental Results

The objective of the experiment is to regulate the angular position of the motor shaft to a given desired trajectory, in the presence of disturbance. For the experiment, the disturbance is intentionally given, and is set to $D = 2$ (constant).

For comparison, first, an experiment is conducted without disturbance compensation. Next, the disturbance observer is implemented with $\omega_c = 500.0$ [rad/sec]. Shown in Fig. 11(a) are the tracking error of the system without disturbance compensation (solid line), and with disturbance compensation (bold dotted line). Comparing the results of both experiments, we observe that the tracking error $e(t)$ of the system with a disturbance observer is smaller, which indicates that the effect of the disturbance to the system output can be suppressed by using disturbance observer. Similar experiments with $D = 4, D = 6$ are also conducted. The results are shown in Fig. 11(b),(c), respectively, which are similar to the previous results.

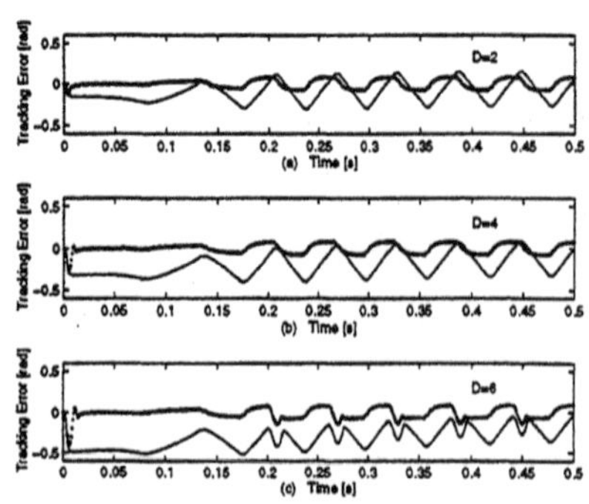

Figure 11: Experimental Results: The Tracking Error

6. Conclusions

In this paper, a new type of electronically controlled winding machine is presented. The new winding machine has the advantage of being able to wind various kinds of stators, since it can generate arbitrary winding trajectories. To realize various winding trajectories, a high-speed high-precision tracking control of a servomechanism is considered.

Since the parameters of the servomechanism are usually unknown, an on-line adaptive parameter identification scheme has been proposed, and the asymptotic stability of the closed loop system using adaptive identification law is theoretically proved. Experimental studies showed that the estimated parameter not only converge to their real values, but also give a satisfactory transient performance. By utilizing these estimated values even to a simple controller such as feedforward controller, the high-speed high-precision tracking control of mechatronic servosystems can be realized.

In addition, to improve the tracking performance of the servomechanism, a disturbance rejection observer is designed, and is implemented on the system. Experimental results showed that the effect of the disturbance is attenuated by the disturbance observer, and that the tracking performance of the system is improved.

REFERENCES

Iwai, Z., A. Inoue and S. Kawaji (1988). *Observers*, Corona-sha, (In Japanese).

Khalil, H.K. (1996) *Nonlinear Systems*, Prentice Hall.

Onishi K. (1993). Robust Motion Control using Disturbance Observer, *Journal of the Robotics Society of Japan*, **11**, **4**, 486-493, (In Japanese).

Yasukawa Denki (1994). *An Introduction to Servo Technology for Mechatronisc*, Nikkan Kogyo Shimbunsha (In Japanese).

THE CONTROL OF FLEXIBLE STRUCTURE SYSTEMS BY DIVISION MOTION METHOD

Victor A. Utkin

Institute of Control Sciences, Profsoyuznaya 65, 117806, Moscow, Russia, E-mail: vicutkin@ipu.rssi.ru

Abstract: The paper presents a new approach to the flexible structure systems synthesis on the base of using a method to dynamic compensation and a division of motion on rate within the framework of systems with large coefficients and discontinuous control, allowing to decompose a problem of large dimensionality to the independently solved a problems of smaller dimensionality (Drakunov at el.. 1990).

Key words: flexible structure, decomposition method, sliding mode, asymptotic observer.

1. INTRODUCTION

The paper presents a new approach to the design of rigid-flexible structure systems, based on a sliding mode technique (Utkin V.I., 1992) and on the method of dynamic compensation (V.A. Utkin and V.I. Utkin, 1983). It is known, that robustness and order reduction are the most important virtue of variable structure control (VSC) with sliding mode. The structure of rigid-flexible system can be presented as two mutually connected subsystems: the first subsystem describes the behavior of rigid coordinates, the second subsystem describes the vibration of flexure coordinates (usually it is expected that flexible variable are described end amount modes) (Balas M.J., 1978). The common approach to the control problem of flexible system, both for a continuous control algorithms (Dong Li, 1994) and for VSC' algorithms (Young K. and Ozguner U., 1990) consists of solution of two problems: the problem of stabilization of rigid subsystem and the restriction of flexible variables excitation. In this paper we suggest the control strategy, which gives a possibility to solve the stabilization problem of flexure subsystem directly. Such possibility appears to account of introduction to sidebar of feedback dynamic element (dynamic compensator or observer), reflect structure to vibrations of flexible arms of robot-manipulator. Greatly, that proposed approach allows realize decompomposition of syntheses problem of large dimensionality on subproblem of smaller dimensianality.

The organization of this paper is as follows. Section 2 provides the and setup problem and background for our work. It introduces the linear model of flexible structure systems and the dynamic compensation method in the invariance system design. Section 3 gives the synthesis procedure for a linear flexible structure systems based on the motion separation method. The application to a nonlinear case is discussed in Section 4.

2. PROBLEM SETUP

The motion for a flexible structure in linear setup can be written as follows:

$$\dot{\theta} = A\theta + Bu, \dot{q} = Wq + Gu, y = D\theta + Pq, \qquad (1)$$

where $\theta \in R^n$ is a rigid-body structure state vector, $q \in R^k$ is a flexible structure state vector, $y \in R^l$ is an output vector and $u \in R^m$ is an input vector. It is supposed that the linear system (1) is controllable and observable.

Known that in the suggestion on controllability and observability system (1) control problem may be solved in the manner of the feedback on the whole state vector , while problem of observing can be a speech with using a theory of asymptotic observer. Difficulty in deciding both problems is their high dimensionality.

Usually in the preceding literature on problems of control flexible systems a subsystem flexible variable is consider as external or singular perturbation to the rigid-body subsystem. Decomposition of problem consists in the independent syntheses of problem of rigid-body subsystem control and following subsystem syntheses flexible variable. Last problem dares by means of introductions to the feedforward of preshaping filter (Khorrami F. at el., 1995), limiting excitement flexible variable or direct control on flexible variable (Aoustin Y. and Chevallereau C., 1993).

In the work is offered decompose an approach to deciding a problem of flexible system control on the system method base with high gain and discontinuous control on sliding modes.

In the proposed approach is also use system division on two subsystems, but instead of preshaping filter is offered to use a flexible subsystem model, but singularity or quick convergence last is ensured to the account of using the system methods with high gain and discontinuous control.

The idea is based on the method of dynamic compensations method, which essence consists in following.

Consider a linear dynamic system, described by equations

$$\dot{\theta} = A\theta + B(u + \Lambda q), \dot{q} = Wq + B_1 u, \quad (2)$$

where $\theta \in R^n, u \in R^m, q \in R^k$-are vectors of the states, controls and disturbances; A, B, B_1, W, Λ - a matrixes of consistent dimensions. To stay the problem to ensure the stabilization to zero the state vector with invariance to external disturbances.

Let us chose the controls as follows:

$$u = -\Lambda z + v, \dot{z} = Wz + B_1 u + Lv, \quad (3)$$

where $z \in R^k$ is the state vector of compensator and $v \in R^m$ is the vector of new control. The motions of closed system (2) and (3) are described by equations

$$\dot{\theta} = A\theta + B\Lambda\varepsilon + Bv, \dot{\varepsilon} = W\varepsilon + Lv, \quad (4)$$

where $\varepsilon = q - z$. The new control is chosen in the class of systems with discontinuous control or high gains

$$v = -M\,\mathrm{sign}(s) \text{ or } v = -ks, \; s = C\theta, \quad (5)$$

where $s \in R^m$ is sliding surface, C is $(m \times n)$ constant matrix, M and k are positive constants. Under sufficiently large M and k, the overall motion of system (4) and (5) is divided into two phases: at the first stage the state vector reaches the sliding surface (or fast motion), whereas at the second stage the sliding (or slow) mode occurs along this surface. According to the equivalent control method the equation of sliding modes (or slow motion) will be described as follows:

$$\begin{aligned} &\dot{\theta} = [A - B(CB)^{-1}CA]\theta, s = C\theta = 0, \\ &\dot{\varepsilon} = -L(CB)^{-1}CA\theta + (W - L\Lambda)\varepsilon. \end{aligned} \quad (6)$$

The characteristic polynomial of the diagonal cell matrix of the system (6) coincides with the product of the characteristic polynomials of diagonal elements. Under assume that the pair (A,B) is controllable and (without loss of generality) the pair (Λ, W) is observable by appropriate choice of matrices C and L it is possible to arbitrarily independently assign (n-m) nonzero eigenvalues of matrix $A - B(CB)^{-1}CA$ and k eigenvalues of matrix $W - L\Lambda$. Note that the system (6) is homogeneous and, hence, the its stability can be assured with the aid of linear control with finite gains.

Also, note that the amplitude of discontinuous control (M) that ensure the fulfillment of conditions of sliding mode occurrence will be nonzero. After finish of transient process the undesirable self-oscillations (named as the chattering phenomena) is occurred. On the other hand, the equivalent control is tended to zero with state variables is tended to zero. Hence, the amplitude of discontinuous control can be diminished to zero, also. Thus, it is possible to eliminate the chattering phenomena.

3. CONTROL SYNTHESIZE OF FLEXIBLE STRUCTURE.

3.1. Control.

Consider a problem of system stabilization (1) in the suggestion, that state vector available for the measurement. The following incremental procedure decides set the problem.

On the first step is interred nonsingular change of flexible variable in the form $\bar{q} = q + M\theta$, where $M = -GB^+$ and B^+ is psevdo-inverse matrix.

Then the second subsystem (1) is written as follows

$$\dot{\bar{q}} = W\bar{q} + \bar{M}\theta, \quad (7)$$

where $\bar{M} = MA - WM$.

Since the system (7) is controllable about the vector θ, regarded as fictitious control there exists a feedback matrix F such that the matrix $W + \bar{M}F$ has any desired eigenvalue distribution. Thereby, the problem to stabilization of system (7) can be solved by means of choice of matrix F in the correlation

$$\bar{\theta}_1 = \theta_1 - F\bar{q} = 0 \quad (8)$$

For ensuring an equality (8) on the second step of the procedure it is necessary to solve the stabilization problem of variable, described on the strength of (7) and (8) by equations

$$\dot{\bar{\theta}} = \bar{A}\bar{\theta} + \bar{W}\bar{q} + Bu, \quad (9)$$

where $\bar{A} = A + F\bar{M}, \bar{W} = FW - A - F\bar{M}$.

The problem to stabilization the system (9) can be solved as problem of ensuring invariance to variable $\bar{q}$ in suggestion on accessories of image of matrixes $Im\,\bar{W} \subset Im\,B$ and organizations of sliding motion on surface s = 0, similarly (5).

Otherwise, the first subsystem (1) nonsingular change variable is bring about block-canonical type

$$\begin{aligned}\dot{\theta}_1 &= A_{11}\theta_1 + A_{12}\theta_2 + W_1\bar{q},\\ \dot{\theta}_2 &= A_{21}\theta_1 + A_{22}\theta_2 + W_2\bar{q} + Bu,\end{aligned} \quad (10)$$

where $\theta_1 \in R^{n-m}, \theta_2 \in R^m, \mathrm{rank}B = m$ and performing a condition $Im\,W_1 \subset Im\,A_{12}$ ensure invariance of subvector θ_1 (V.A.Utkin and V.I. Utkin, 1983). Notice as a rule that invariance condition is executed for mechanical systems when writing models in canonical type, the first equation (10) has of the form of $\dot{\theta}_1 = \theta_2$.

In the case on the first step is interred nonsingular change of flexible variable in the form

$$\bar{q} = q + M_1\theta_1 + M_2\theta_2,$$

where $M_2 = -GB_2^{-1}$, $M_1 = WM_2 - M_2A_{22}$. Then the second subsystem (1) is written as follows

$$\dot{\bar{q}} = W\bar{q} + M\theta_1, \quad (11)$$

where $M = M_2A_{21} - WM_1$. Since the system (11) is controllable about the vector θ_1, regarded as fictitious control there exists a feedback matrix F such that the matrix W+MF has any desired eigenvalue distribution. Thereby, the problem to stabilization of system (11) can be solved by means of choice of matrix F in the correlation

$$\theta_1 = F\bar{q} \quad (12)$$

For ensuring an equality (12) on the second step of the procedure it is necessary to solve the stabilization problem of variable $\bar{\theta}_1 = \theta_1 - F\bar{q}$, described on the strength of (10) and (11) by equations

$$\dot{\bar{\theta}}_1 = (FW + MF)\bar{q} + M\bar{\theta}_1 + \theta_2 \quad (13)$$

Will be considered variables θ_2 in equation (13) as fictitious control and place its equal

$$\theta_2 = -(MF + FM)\bar{q}) - M\bar{\theta}_2 + K\bar{\theta}_1. \quad (14)$$

Then, after substitution of this expression in (13) its stabilization is ensured by suitable choice matrix K:

$$\dot{\bar{\theta}}_1 = K\bar{\theta}_1.$$

Finally, on the third step of procedure necessary to ensure an equality (14) or convergence to zero the variables

$$\bar{\theta}_2 = \theta_2 + (MF + FM)\bar{q} + (M - K)\bar{\theta}_1,$$

which comply with equation of type

$$\dot{\bar{\theta}}_2 = \bar{A}_{12}\bar{\theta}_1 + \bar{A}_{22}\bar{\theta}_2 + \bar{W}\bar{q} + B_2u. \quad (15)$$

The problem to stabilization the system (15) can be solved by the choice of control u in the class of continuous feedback (similarly to second step) or in the class of discontinuous functions by means of introduction of sliding mode on the surface

$$s = \bar{\theta}_2 = 0.$$

Known that in discontinuous systems is ensured invariance to parameters of control object on the strength of that, that conditions of existence of sliding motion be the form of inequalities and, consequently, are to be provided under the known range of changing these parameters. In particular, this fact allows synthesize a stabilization problem for nonlinear system (15) in the suggestion, that are known bounds of nonlinear members, that inherent robot models.

Note that in practice realize moments (u) created by executive devices in class of discontinuous functions is not presented possible. In the case a model of control object must be complement by equations own dynamics of executive devices and intimate above hierarchical procedure must be continue up to real discontinuous controls (Utkin V.A., 1994.) Greatly note that specified decomposition is possible within the framework of discontinuous systems and large coefficients and, besides, is invariant to changing the parameters of control object and external disturbances.

3.2. Observation problem.

As known, if observer is constructed on the basis of equation (1) only, the occurrence of elastic modes leads to observation and control spillover phenomena, which can result in the instability of close-loop system. We below consider the nonspillover effect procedure of observer synthesis, using a full object model and sliding modes control strategy. Let us constructed the observer as follows:

$$\begin{aligned}\dot{\hat{\theta}} &= A\hat{\theta} + Bu + L_1(y - D\hat{\theta} - P\hat{q}),\\ \dot{\hat{q}} &= W\bar{q} + Gu + L_2(y - D\hat{\theta} - P\hat{q}).\end{aligned} \quad (16)$$

Assuming that the sliding mode occurs along the switching surface $s = C\hat{\theta}$ we solve the control problem under assumption $\hat{\theta} = \theta$. Let us write the closed system of equations for variable $\theta, \Delta\theta = \theta - \hat{\theta}$ and $\Delta q = q - \hat{q}$:

$$\begin{vmatrix}\dot{\theta}\\ \Delta\dot{\theta}\\ \Delta\dot{q}\end{vmatrix} = J\begin{vmatrix}\theta\\ \Delta\theta\\ \Delta q\end{vmatrix},$$

$$\text{where } J = \begin{bmatrix}(I - C^*)A & C^*(L_1D - A) & -C^*L_1G\\ 0 & A - L_1D & -L_1G\\ 0 & -L_2D & W - L_2G\end{bmatrix},$$

$$C^* = B(CB)^{-1}C.$$

Note that the equivalent control $u_{eq} = -(CB)^{-1}(A\theta + L_1D\Delta\theta + L_2G\Delta q)$ derived from the equation $\dot{s} = 0$ (Utkin V.I., 1992) is substituted to the equation (16). By virtue of suitable choices of the matrix C and pair (L_1, L_2) independently dimension of the observer (16) causes to a difficulty of the design procedure. To ovecome the obstacle it may be suggested to use the block control principle in the observation problem (Utkin V.A., 1990). The principle bases on a canonical block form representation of observable systems and a hierarchical approach to control; thus the observation problem can be solved by splitting it into independent subproblems of smaller dimension.

Consider the observation problem of flexure coordinates $q \in R^n$ for the linear system

$$\dot{\theta} = A\theta + B(u + \Lambda q), \dot{q} = Wq + B_1 u, \qquad (17)$$

where vector $\theta \in R^n$ is available for measurements and pair $(W, B\Lambda)$ is observable.

Standard procedure of dynamic compensator (observer) syntheses consists in following. Nonsingular change variable is interred $\bar{q} = q + L\theta$ and the second subsystem (17) is wrote about new coordinate:

$$\dot{\bar{q}} = (W + LB\Lambda)\bar{q} + (LA - WL - LB\Lambda L)\theta$$
$$+ (LB + B_1)u.$$

Since pair $(W, B\Lambda)$ is observable, then matrix L exists so, that matrix $(W + LB\Lambda)$ has given distribution own number. Then dynamic device of similar structure

$$\dot{\bar{z}} = (W + LB\Lambda)\bar{z} + (LA - WL - LB\Lambda L)\theta$$
$$+ (LB + B_1)u.$$

allows to get evaluation variable $\bar{q}$ on the strength of system stability, writing comparatively mismatches $\varepsilon = \bar{q} - z$: $\dot{\varepsilon} = (W + LB\Lambda)\varepsilon$.

4. NONLINEAR CASE

In common view a model of multi-link manipulator with flexible arms may be described by the equations

$$\begin{aligned} \dot{\theta}_1 &= \theta_2, \\ \dot{\theta}_2 &= F(\theta_1, \theta_2, q) + B(\theta_1, q)\tau, \\ \dot{q} &= W(\theta_1, \theta_2, q) + B_0(\theta_1, q)\tau, \\ y &= D(\theta_1, q), \end{aligned} \qquad (18)$$

where $\theta_1, \theta_2 \in R^n$ are rigid-body coordinates, the vector of flexure coordinates $q \in R^k$ is modeled by finite number modes, F, W are vector-functions, $B, B_0, (\mathrm{rank} B = n, \mathrm{rank} B_0 \le n)$ are functional matrixes of control distribution, $y \in R^1$ is tip position.

The follows control three steps procedure is suggested for the solution of the stabilization task of the system (18).

1 step.

Let us introduce a nonsingular transformation of coordinates as follows:

$$\bar{\theta}_2 = \theta_2 + K_1\theta_1,$$
$$\bar{q} = q + M\theta_2, M = -B_0 B^{-1}.$$

Thus, the equations about of new flexure coordinates isn't consisted the control variables and stability of own dynamic of variables θ_1 may be supply by choice of matrix K_1:

$$\begin{aligned} \dot{\theta}_1 &= -K_1\theta_1 + \bar{\theta}_2, \\ \dot{\bar{\theta}}_2 &= \bar{F}(\theta_1, \bar{\theta}_2, \bar{q}) + \bar{B}(\theta_1, \bar{\theta}_2, \bar{q})\tau, \\ \dot{\bar{q}} &= \overline{W}(\theta_1, \bar{\theta}_2, \bar{q}) + \bar{B}_0(\theta_1, \bar{\theta}_2, \bar{q}), \\ y &= \bar{D}(\theta_1, \bar{\theta}_2, \bar{q}), \end{aligned} \qquad (19)$$

where $\bar{B}_0 = \overline{M}(\theta_1, \theta_2, \bar{q})\bar{F}(\theta_1, \theta_2, \bar{q})$ and $\bar{F}, \bar{B}, \bar{B}_0, \overline{W}, \bar{D}$ has been get after substitution $q = \bar{q} - M\theta_2$ into F, B, B_0, W, D.

2 step.

In suggestion that system (18) is controllable (or stabilization) for stabilization of flexure variables, described by third equation (18)) the variables $\bar{\theta}_2$ are chosen as the function of variables $\bar{q}$:

$$\bar{\theta}_2 = L(\bar{q}) \text{ and } L(0) = 0.$$

3 step.

Based on the theory of discontinuous control systems, let us introduce the sliding mode along the surface $s = \bar{\theta}_2 - L(\bar{q}) = 0$ by choose control as follows:

$$\tau_i = \begin{cases} M_i^+(\theta_1, \bar{\theta}_2, \bar{q}), s_i > 0, \\ M_i^-(\theta_1, \bar{\theta}_2, \bar{q}), s_i < 0, \end{cases}$$

where functions M_i^+, M_i^- are continuous, $s = (s_1, s_2, \ldots, s_n) \in R^n$. Methods of synthesis of sliding mode in robotics control problems are considered in (Utkin and Izosimov, 1990).

Thus, the motion of system (18) in sliding regime (s=0) will be describe by equations:

$$\begin{aligned} \dot{\theta}_1 &= -K_1\theta_1 + L(\bar{q}), \\ \dot{\bar{q}} &= \overline{W}(\theta_1, L(\bar{q}), \bar{q}) + \bar{B}_0(\theta_1, L(\bar{q}), \bar{q}), \\ \dot{\theta}_2 &= L(\bar{q}), \\ y &= \bar{D}(\theta_1, L(\bar{q}), \bar{q}) \end{aligned} \qquad (20)$$

So far as by choose of matrix K_1 and vector-function $L(\bar{q})$ it is supply the stability of system (9), the problem of tracking tip position may be solved by determination of demanded position of rigid-body generalized coordinates.

As rule, it is need to measure the generalized coordinates in absolute coordinate system. In the case, then it is possible to measure the tip position, it is need to change first equation (20) on the equation, writing about variables y. Then, if a manipulator structure surplus, follows a part of generalized coordinates to refer to second system (20) with according follows procedure.

Let $\text{rank}\{\frac{\partial D(\theta_1, q)}{\partial \theta_1}\} = 1 < n$. Then, according with the implicit function theorem, the vector θ_1 may be divided on the two subvectors $\theta_1^t = [(\theta_1^{'})^t, (\theta_1^{''})^t], \theta_1^{'} \in R^1, \theta_1^{''} \in R^{n-1}$ thus, that $\text{rank}\{\frac{\partial D(\theta_1^{'}, \theta_2^{''}, q)}{\partial \theta_1^{'}}\} = 1$ and, consequently, the vector $\theta_1^{'}$ may be written as a certain function from y: $\theta_1^{'} = D^{+}(y, \theta_1^{''}, q)$. Write the system (18) about variables $y, \theta_1^{''}$:

$$\dot{y} = \{\frac{\partial D}{\partial \theta_1^{'}}\}\theta_2^{'} + \{\frac{\partial D}{\partial \theta_1^{''}}\}\theta_2^{''} + \{\frac{\partial D}{\partial q}\}\dot{q},$$
$$\dot{\theta}_2^{'} = F_1(\theta_1^{'}, \theta_1^{''}, \theta_2^{'}, \theta_2^{''}, q) + B_1(\theta_1^{'}, \theta_1^{''}, q)\tau,$$
$$\dot{\theta}_1^{''} = \theta_2^{''}, \quad (21)$$
$$\dot{\theta}_2^{''} = F_2(\theta_1^{'}, \theta_1^{''}, \theta_2^{'}, \theta_2^{''}, q) + B_2(\theta_1^{'}\theta_1^{''}, q)\tau,$$
$$\dot{q} = W(\theta_1^{'}, \theta_1^{''}, \theta_2^{'}, \theta_2^{''}, q) + B_0(\theta_1^{'}, \theta_1^{''}, q)\tau,$$

The decomposition procedure of stabilization problem of system (21) can look as follows. Supposing the variables $\theta_2^{'}$ in first subsystem as fictitious control, will place its equal

$$s_y = \theta_2^{'} + \{\partial D/\partial \theta_1^{'}\}^{-1}[\{\partial D/\partial \theta_1^{''}\}\theta_2^{''} + \quad (22)$$
$$+ \{\partial D/\partial q\}\dot{q} + K_y y] = 0.$$

Then, by suitable choosing of matrix K_y the first subsystem of (21) will be stability: $\dot{y} = -K_y y$.

On the second step of procedure for performing the correlation (22) the sliding mode is synthesize along the surface $s_y = 0$.

Notice that for making the sliding motion is use by 1 component of control vector. Staying controls are synthesized on the scheme, stated above for system (18). Besides, if components of vector $\theta_2^{''}$ insufficiently for satisfactory stabilization latest subsystem (21), then in surface (22) follows add certain function of flexible variable (q).

5. CONCLUSION

In the paper is explored control problem of robot-manipulator with flexible arms. On the base of method to dynamic compensation is offered new approach to deciding a put problem, allowing realize a stabilization of robot model subsystem, described behavior of flexible variable. Present results on the syntheses both a linear flexible structure models and nonlinear models, inherent robot models with flexible arms.

Offered control algorithms allow realize a decomposition of source syntheses control problem of large dimansionality independently decided subproblems of smaller dimensionality. As follows, the syntheses problems divide into three subproblems - the problem of stabilization of model of flexible variable and two-level control in the space rigid-body coordinates.

REFERENCES

Utkin V.I. (1992). *Sliding modes in control and optimization*. N Y Sprigner Verlag.

Utkin V.A. and. Utkin V.I. (1983). Design of invariant systems by the method of separation of motion. *Automation and remote Control*, **vol. 44, no. 12**, Part 1, pp. 1559-1566.

Bales M.J. (1978). Feedback Control of Flexible systems. *IEEE Transaction on Automatic Control*, **vol. AC - 23, no. 4**, pp. 673 - 679.

Dong Li (1994). Nonlinear control design for tip position tracking of a Flexible manipulator arm. *Int. J. Control*, **vol. 60, no.6**, pp. 1065 - 1082.

Young K. and Ozguner U. (1990). Frequency shaped variable structure control. *Proc. VSS'90*, pp. 181-185, Sarajevo, Yugoslavia,

Aoustin Y. and Chevallereau C. (1993). Some Experemental Results about Singular Perturbation Control of a two Flexible Link. *Proceedings of ECC'93 held in Groningen, June 28-July 1*, **vol. 3,** pp. 904-909, Holland.

Khorrami F., Sandeep J. and Anthony Tzes. (1995). Experimental results on adaptive nonlinear control and input preshaping for multi - link flexible manipulators. *Automatica*, **vol. 31, no. 1**, pp. 83-97.

Drakunov S. V., Izosimov D. B., Lukyanov A. G., Utkin V. A. and Utkin V. I. (1990). The blockcontrol principle. *I. Automatio n and Remote Control*, **vol. 51, no. 5**, Part 1, pp. 601 - 609.

Utkin V. A. (1990). Method of separation of motions in observation problems. *Automation and Remote Control*, **vol. 44, no. 12**, Part 1, pp. 300 - 308.

Utkin V. A. and Izosimov D. B. (1990). Robot - manipulators control based on the method of movements separation. *Proc. VSS' 90, Sarajevo*, March, pp. 86 - 97, Yugoslavia.

SOME ASPECTS ON THE PROPORTIONAL-INTEGRAL OBSERVER IN THE FIELD OF SYSTEM SUPERVISION

Dirk van Schrick and Cumhur Başpınar

Group of Safety Control Engineering
Department of Safety Engineering
University of Wuppertal
D-42097 Wuppertal, Germany
Fon: ++49 (0)202 439 2019
Fax: ++49 (0)202 439 2901
E-Mail: schrick, cumhur@uni-wuppertal.de

Abstract: Mainly in the context of system supervision, the proportional-integral observer (PIO) is discussed in this contribution with respect to different aspects, such as existence conditions, design methods, convergence behaviour, diagnostic purposes or applications. Additionally, this observer type is compared to the unknown-input observer (UIO). Both observer types have different control characteristics and regarding supervision the PIO features fault identification while the UIO features fault isolation. Fulfilling certain requirements, both types of linear observers can be designed as robust variable estimators for control and as robust residual generators for fault detection, isolation and identification (FDII) providing quasi-asymptotic (PIO) or structural (UIO) robustness with respect to unknown inputs for both design cases.

Keywords: System Supervision, Control, PI-Observer, Fault Detection, Isolation, Identification.

1. INTRODUCTION

In the interesting and challenging field *system supervision* an enormous progress in both theory and application can be stated. Nowadays, the entity *diagnosis system (DS)* can be implemented in sensors and actuators, it is no longer just a mechanics-based item for emergency stop, it is a highly integrated intelligent system consisting of hardware and software. Ideally, a DS supports the technical part of *fault management (FM)*, essentially the diagnosis-related functions *monitoring, diagnosis* and *prognosis* that are based on means for *description, design, implementation* and *operation* to fulfil the main diagnosis-oriented purposes *fault detection, fault isolation, fault identication (FDII)* where the purposes *fault isolation, fault identification (FII)* are subsumed by the purpose *fault diagnosis (FD)* (van Schrick, 1997).

With regard to modern quantitative model-based diagnosis procedures, the estimation of both system parameters and system variables is of interest. Control theory provides suitable methods for the design of parameter estimiators and variable estimators. These methods have been adopted and modified for diagnosis purposes such that the resulting estimators represent realizations of an item termed *residual generator (RG)* possessing the RG-property that, in a technical context, the generator's output called *residual* is zero in the fault-free case and reflects the effect of a fault by deviating from zero. With respect to undesired effects existent additionally, e. g. those from allowed parameter variations, this property is termed *robust RG-property (RRGP)*, cf. for details Patton and Chen (1997), for instance.

In this contribution the *variable estimator approach (VEA)* in shape of observer-based FDII residual generators is of interest. From the set of observer types, the *proportional-integral observer (PIO)* (Beale and Shafai, 1989) is considered generally. In the light of system control and supervision the PIO is compared additionally to the *unknown-input observer (UIO)* (Wünnenberg, 1990). One result will is that both the PIO and the UIO can be designed as FDI residual generators where the letter 'I' of the acronym *FDI* symbolizes *identification* with respect to the PIO and *isolation* with respect to the UIO. The representation of a class of engineering systems considered and an outline of both observer types including a sketch of the convergence proof for some PIO-types is followed by a summarizing comparison of both observer types with respect to FDII. Relevant means for comparison are the aspects *residual generator, existence conditions, design methods, purpose, minimal number of measurements, robustness* and *applications.*

2. SYSTEM REPRESENTATION

The class of engineering systems considered is modeled as a linear, time-invariant, deterministic, uncertain, disturbed and faulty system in the state space where the latter three attributes refer to operational effects considered as *uncertainties* caused in the nonoperational phases modeling, design, implementation, and to effects occuring during the operational phase caused by *external disturbances* such as electro-magnetic radiation, *internal disturbances* such as nonlinearities or couplings between subsystems and *faults* to be detected, isolated and identified. More generally, the total index-set of effects, including their derivatives or sums, is divided into the index-sets $\mathcal{Z}$ referring to *effects of interest (EI)* and $\overline{\mathcal{Z}}$ referring to *effects of no interest (EN)* where $\mathcal{Z} \cap \overline{\mathcal{Z}} = \mathcal{O}$ and $\mathcal{Z} \cup \overline{\mathcal{Z}} = \{1, 2, ..., p\}$. This classification is reflected by

$$\dot{\mathbf{x}}(t) = \mathbf{A}\mathbf{x}(t) + \mathbf{B}\mathbf{u}(t) + \mathbf{E}_D\mathbf{s}_D(t) + \overline{\mathbf{E}}_D\overline{\mathbf{s}}_D(t), \quad (1)$$
$$\mathbf{y}(t) = \mathbf{C}\mathbf{x}(t) + \mathbf{D}\mathbf{u}(t) + \mathbf{E}_M\mathbf{s}_M(t) + \overline{\mathbf{E}}_M\overline{\mathbf{s}}_M(t) \quad (2)$$

where the underlying mathematical system representation is based on the assumption that all effects can be represented in a structured fashion, i. e. with known matrices and unknown vectors. The totality of such effects, i. e. uncertainties, disturbances and faults, is termed *unknown inputs.* The n-dimensional vector $\mathbf{x}(t)$ symbolizes the state, the r-dimensional known vector $\mathbf{u}(t)$ the input and the m-dimensional vector $\mathbf{y}(t)$ the output in the system description. The EI are symbolized by the q-dimensional unknown vectors $\mathbf{s}_D(t)$, $\mathbf{s}_M(t)$ and the $\overline{q}$-dimensional unknown vectors $\overline{\mathbf{s}}_D(t)$, $\overline{\mathbf{s}}_M(t)$ describe the EN. The known matrices with appropriate dimensions are represented by $\mathbf{A}, \mathbf{B}, \mathbf{C}, \mathbf{D}, \mathbf{E}_D, \overline{\mathbf{E}}_D$ and $\mathbf{E}_M, \overline{\mathbf{E}}_M$.

With respect to observer design, different *system properties*, cf. Hou (1995) for instance, can be checked by inspecting (1,2):

Observability with EI-Diagnosability

$$\text{rank}\begin{bmatrix} \lambda\mathbf{I}_n - \mathbf{A} & \mathbf{E}_D \\ \mathbf{C} & \mathbf{E}_M \end{bmatrix} = n + \text{rank}\begin{bmatrix} \mathbf{E}_D \\ \mathbf{E}_M \end{bmatrix}, \ \forall\, \lambda \in \mathcal{C}, \quad (3)$$

Observability with EN-Decoupability

$$\text{rank}\begin{bmatrix} \lambda\mathbf{I}_n - \mathbf{A} & \overline{\mathbf{E}}_D \\ \mathbf{C} & \overline{\mathbf{E}}_M \end{bmatrix} = n + \text{rank}\begin{bmatrix} \overline{\mathbf{E}}_D \\ \overline{\mathbf{E}}_M \end{bmatrix}, \ \forall\, \lambda \in \mathcal{C}, \quad (4)$$

EI/EN-Decoupability (Decoupability of EI from EN)

$$\text{rank}\begin{bmatrix} \mathbf{E}_{D_i} & \overline{\mathbf{E}}_D \\ \mathbf{E}_{M_i} & \overline{\mathbf{E}}_M \end{bmatrix} > \text{rank}\begin{bmatrix} \overline{\mathbf{E}}_D \\ \overline{\mathbf{E}}_M \end{bmatrix}, \ \forall\, i \in \mathcal{Z}. \quad (5)$$

Whether or not these conditions have to be met actually depends on the type of observer selected for robust FDII. This is to be seen in the following two sections assuming the system output not to be influenced by effects of both the known inputs and the unknown inputs (EI, EN) resulting in $\mathbf{D} = \mathbf{0}$, $\mathbf{E}_M = \mathbf{0}$ and $\overline{\mathbf{E}}_M = \mathbf{0}$.

3. PROPORTIONAL-INTEGRAL OBSERVER

Consider the linear time-invariant system

$$\dot{\hat{\mathbf{x}}}(t) = (\mathbf{A} - \mathbf{L}_x\mathbf{C})\hat{\mathbf{x}}(t) + \mathbf{B}\mathbf{u}(t) + \mathbf{L}_x\mathbf{y}(t) + \mathbf{L}_E\mathbf{L}_C\hat{\mathbf{v}}(t), \quad (6)$$
$$\dot{\hat{\mathbf{v}}}(t) = \mathbf{V}\hat{\mathbf{v}}(t) + \mathbf{L}_v\mathbf{e}_y(t), \quad (7)$$
$$\mathbf{w}(t) = \mathbf{L}_C\hat{\mathbf{v}}(t) \quad (8)$$

where $\mathbf{e}_y(t) = \hat{\mathbf{y}}(t) - \mathbf{y}(t)$. The symbol $\hat{\mathbf{v}}(t)$ represents the n_v-dimensional vector of estimated fictitious variables to be interpreted purpose-dependently, $\mathbf{w}(t)$ represents the w-dimensional output vector that comprizes approximated EI, for example, and the symbol $\mathbf{V}$ represents the system matrix of the extending system (7). The symbols $\mathbf{V}$ and $\mathbf{L}_x, \mathbf{L}_v, \mathbf{L}_C, \mathbf{L}_E$ represent matrices to be determined.

The idea behind the extended observer (6,7,8) providing $\mathbf{w}(t)$ consists in the approximation of the EI's time behaviour by a fictitious linear dynamic system

$$\dot{\mathbf{v}}(t) = \mathbf{V}\mathbf{v}(t), \quad (9)$$
$$\mathbf{s}_D(t) \approx \mathbf{L}_C\mathbf{v}(t) \quad (10)$$

with $\mathbf{v}(t)$ representing fictitious variables that can, via fixing $\mathbf{L}_C$, be used for both observer-based closed-loop control (CLC) in shape of disturbance rejection control (DRC) (Müller and Lückel, 1977) or state feedback control (SFC) (Beale and Shafai, 1989) and open-loop supervision (OLS) (Söffker, 1996).

System (6,7) is a full-order observer based on an extended system description of order $l = n + n_v$. If the extended system is completely observable, i. e.

$$\text{rank}\begin{bmatrix} \lambda \mathbf{I}_n - \mathbf{A} & -\mathbf{L}_E\mathbf{L}_C \\ \mathbf{0} & \lambda \mathbf{I}_{n_v} - \mathbf{V} \\ \mathbf{C} & \mathbf{0} \end{bmatrix} = n + n_v,\ \forall\, \lambda \in \mathcal{C}, \tag{11}$$

the gains $\mathbf{L}_x, \mathbf{L}_v$ can be fixed to stabilize observer (6,7) quasi-asymptotically with arbitrary eigenvalues in the left half-plane (Müller, 1995). The resulting output error dynamics is

$$\dot{\mathbf{e}}_y(t) = \mathbf{C}(\mathbf{A} - \mathbf{L}_x\mathbf{C})\mathbf{e}_x(t) + \mathbf{C}\mathbf{L}_E\mathbf{L}_C\hat{\mathbf{v}}(t) - -\mathbf{C}\mathbf{E}_D\mathbf{s}_D(t) - \mathbf{C}\overline{\mathbf{E}}_D\overline{\mathbf{s}}_D(t) \tag{12}$$

where $\mathbf{e}_x(t) = \hat{\mathbf{x}}(t) - \mathbf{x}(t)$. Regarding EI-estimation, condition (11) has to be met only. It is important to note, equations (7,8) describe the integral path for $\mathbf{e}_y$-feedback in addition to the proportional one described by $\mathbf{L}_x$. This observer type is therefore termed *proportional-integral observer (PIO)*. The PIO-structure is given in Fig. 1.

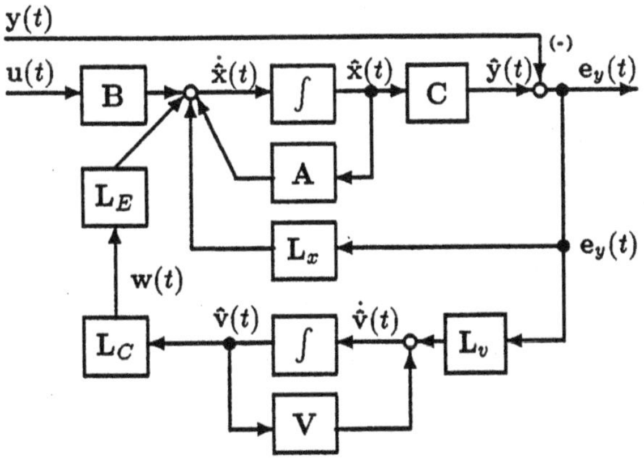

Fig. 1. Proportional-Integral Observer

Note: In opposition to condition (3) implying $\mathbf{x}$-observability, EI-detectability and $q \leq m$, condition (11) implies $\mathbf{x}$-observability, $\mathbf{v}$-detectability, $n_v \leq m$ and, moreover, $\mathbf{v}$-observability with respect to the fictitious output $\mathbf{w}(t)$. With respect to system (1,2), this is due to the extended system underlying the PIO (6,7).

3.1 PIO for Control and Supervision

Regarding the system state and the uncertainties, disturbances and faults, three cases, all aiming at robustness against unknown inputs, can be distinguished:

a) Regarding *parameter uncertainties* $\overline{\mathbf{E}}_D\overline{\mathbf{s}}_D(t) = \Delta\mathbf{A}(t)\mathbf{x}(t) = \mathbf{B}\mathbf{D}_A(t)$ where $\mathbf{D}_A(t)$ is an approximative signal influencing the system via matrix $\mathbf{B}$, the classical PIO-purpose (Kaczorek, 1979) is robust state-estimation for feedback (Beale and Shafai, 1989) in sense of

$$\lim_{t\to\infty}\begin{bmatrix} \hat{\mathbf{x}}(t) - \mathbf{x}(t) \\ \hat{\mathbf{v}}(t) \end{bmatrix} = \mathbf{0} \tag{13}$$

where observer design, i. e. fixing $\mathbf{L}_x$ only, is based on pole-placement (PP), linear quadratic regulation (LQR) etc. with $n_v = m, r = m, \mathbf{V} = \mathbf{0}, \mathbf{L}_v = \mathbf{L}_C = \mathbf{I}_m, \mathbf{L}_E = \mathbf{B}$ yealding $\mathbf{w}(t) \to \mathbf{0}$, cf. e. g. Beale and Shafai (1989) or Shafai et al. (1995) for $\mathbf{V} = \mathbf{I}_m$ and fixing $\mathbf{L}_x, \mathbf{L}_v$ by linear quadratic gaussian (LQG) yealding $\mathbf{w}(t) \to \mathbf{0}$. In both cases, system (6,7) is used to gain robust estimation of the state $\mathbf{x}(t)$.

b) With respect to a class of *external disturbances* $\mathbf{E}_D\mathbf{s}_D(t) = \mathbf{D}_D\mathbf{d}(t)$, disturbance observer (DO) design, i. e. fixing $\mathbf{L}_x, \mathbf{L}_v$, is based on PP, LQR etc. where $n_v = q, \mathbf{L}_E = \mathbf{D}_D$ for a suitable $\mathbf{L}_C$-fixing yealding $\mathbf{w}(t) = \hat{\mathbf{d}}(t)$, e. g. for DRC (Müller and Lückel, 1977) where $\mathbf{L}_C$ depends on the estimated quantities selected for disturbance compensation. With respect to a class of *internal disturbances* $\mathbf{E}_D\mathbf{s}_D(t) = \mathbf{N}_D\mathbf{n}(\mathbf{x}(t), t)$, the nonlinearity observer (NO) design is appropriate to the DO-design with the exception of $\mathbf{L}_E = \mathbf{N}_D$ and $\mathbf{w}(t) = \hat{\mathbf{n}}(\mathbf{x}(t), t)$ (Müller, 1995). For both DO and NO, the purpose is robust $\mathbf{v}$-estimation in sense of

$$\lim_{t\to\infty}\mathbf{w}(t) = \begin{cases} \mathbf{d}(t), & \text{for DO} \\ \mathbf{n}(\mathbf{x}(t), t), & \text{for NO} \end{cases}. \tag{14}$$

c) With respect to *faults* $\mathbf{E}_D\mathbf{s}_D(t) = \mathbf{F}_D\mathbf{f}(t)$, fault observer (FO) design, here residual generator (RG) design, i. e. fixing $\mathbf{L}_x, \mathbf{L}_v$, is based on PP, LQR etc. where $n_v = q, \mathbf{L}_C = \mathbf{I}_q, \mathbf{L}_E = \mathbf{F}_D$ and $\mathbf{w}(t) = \hat{\mathbf{f}}(t)$. The purpose is robust estimation of the extended state in sense of

$$\lim_{t\to\infty}\mathbf{e}_y(t) = \mathbf{0}, \quad \lim_{t\to\infty}\mathbf{w}(t) = \mathbf{f}(t) \tag{15}$$

where $\mathbf{e}_y(t)$ and $\mathbf{w}(t) = \hat{\mathbf{f}}(t)$ can be used as residuals for FDII.

3.2 Remarks on PIO Design

Here, two cases are important:

a) With respect to the design of DO, NO and FO, in general, a suitable choice of the matrices $\mathbf{V}, \mathbf{L}_C$ requires good a priori knowledge of the system behaviour. Approximating the time-depending behaviour by step functions yealds $\mathbf{V} = \mathbf{0}$ (Müller, 1995). According to (8), $\mathbf{L}_C$ is used for selecting suitable variables $\hat{\mathbf{v}}_i(t), i \in \mathcal{Z}$. Considering the output-error dynamics (12) in case of $\mathbf{L}_C = \mathbf{I}_{n_v}$, setting $\mathbf{L}_E = \mathbf{E}_D$ according to the estimation purpose guaranties steady-state $\mathbf{e}_y(t)$-robustness against $\mathbf{C}\mathbf{E}_D(\mathbf{w}(t) - \mathbf{s}_D(t))$. With regard to fixing $\mathbf{L}_x, \mathbf{L}_v$ and robustness to $\overline{\mathbf{s}}_D(t)$, an LQG-approach is given in Shafai et al. (1995). In general, a full-order quasi-asymptotically robust PIO for CLC and OLS purposes exists iff condition (11) holds.

b) With respect to the convergence of the PIO, a proof in sense of tuning $\mathbf{L}_{x_o}, \mathbf{L}_{v_o}$ can be achieved by the substitution

$$\mathbf{L}_x = \rho_x\mathbf{L}_{x_o};\ \ \mathbf{L}_v = \rho_v\mathbf{L}_{v_o} \tag{16}$$

with scalars $\rho_x \to \infty, \rho_v \to \infty$. The ρ-splitting proof is given in Söffker (1996). In this high-gain approach, the values of the scalar quantities depend on both the maximal absolute values of $\dot{\mathbf{s}}_D(t), \bar{\mathbf{s}}_D(t)$ and, referring to the error dynamics (12), the nulling of $\mathbf{s}_D(t), \bar{\mathbf{s}}_D(t)$.

In general, it is difficult to prove the PIO-convergence, i. e. the approximation property $\mathbf{w}(t) \rightarrow \mathbf{s}_D(t)$. In the following, a sketch of a new convergence proof is given. Therein, condition $\mathbf{CE}_D = \mathbf{C\bar{E}}_D = \mathbf{0}$ (Söffker, 1996) is no longer required. The proof relating to the convergence of the PIO-types DO, NO and FO is stated in the frequency domain. With $\mathbf{E} = [\mathbf{E}_D \quad \bar{\mathbf{E}}_D]$, $\mathbf{s}(t) = [\mathbf{s}_D(t) \quad \bar{\mathbf{s}}_D(t)]^T$ and $\mathbf{L}_C = \mathbf{I}_{q+\bar{q}} = \mathbf{I}_p$ the frequency domain equations for the observer errors $\mathbf{e}_x(t) = \hat{\mathbf{x}}(t) - \mathbf{x}(t)$, $\mathbf{e}_s(t) = \hat{\mathbf{s}}(t) - \mathbf{s}(t)$ are

$$\mathcal{E}_x(jw) = (jw\mathbf{I} - \mathbf{A} + \mathbf{L}_x\mathbf{C})^{-1}\mathbf{E}\,\mathcal{E}_s(jw), \quad (17)$$

$$\mathcal{E}_s(jw) = \Big(\Big[jw\mathbf{I} - \mathbf{V} + \mathbf{L}_s\mathbf{C}(jw\mathbf{I} - \mathbf{A} + \mathbf{L}_x\mathbf{C})^{-1}\mathbf{E} \Big]^{-1} [\mathbf{V} - jw\mathbf{I}] \Big)^* \mathcal{S}(jw) \quad (18)$$

where $\mathbf{T}(jw) = (\cdot)^*$ of eqn. (18). Inspecting the equations (17,18) reveals the following:

1. The estimation error $\mathcal{E}_s(jw)$ influences the estimation error $\mathcal{E}_x(jw)$. Consequently, a reduction of $||\mathcal{E}_s(jw)||$ reduces $||\mathcal{E}_x(jw)||$. The reduction of $||\mathcal{E}_s(jw)||$ requires a reduction of $||\mathbf{T}(jw)||$.
2. The parameters $\mathbf{V}$, $\mathbf{L}_x$ and $\mathbf{L}_s$ have to determined by the designer to reduce $||\mathbf{T}(jw)||$.

- Fixing $\mathbf{V}$ does not influence the estimation errors significantly, i. e. this does not reduce $||\mathbf{T}(jw)||$ significantly, because $\mathbf{V}$ appears in the nominator and denominator of the transfer function $\mathbf{T}(jw)$, cf. eqn. (18). Moreover, $\mathbf{T}(jw)$ simplifies to $\mathbf{T}(\infty) = -\mathbf{I}$ for $w \rightarrow \infty$: The observer reacts on fast effects $\mathbf{s}(t)$ with proportional behaviour! Consequently, setting $\mathbf{V} = \mathbf{0}$ is the simplest solution of the design problem, in general with highly satisfying results (Müller, 1995).
- Fixing $\mathbf{L}_x$ arbitrarily for $||\mathbf{T}(jw)||$-reduction requires $\text{rank}\,[\mathbf{C}] = n$, i. e. the state of system (1) must be measurable. Although the expression $(\cdot)^{-1}$ in eqn. (17) is decreased in general, $\text{rank}\,[\mathbf{C}] = n$ does not guarantee a $||\mathcal{E}_x(jw)||$-reduction because increasing $\mathbf{L}_x$ may cause an increasing of $||\mathcal{E}_s(jw)||$.
- Fixing $\mathbf{L}_s$ arbitrarily for $||\mathbf{T}(jw)||$-reduction requires

  $$\text{rank}\,[\mathbf{L}_s\mathbf{C}(jw\mathbf{I} - \mathbf{A} + \mathbf{L}_x\mathbf{C})^{-1}\mathbf{E}] = q + \bar{q} \quad (19)$$

  for an arbitrary reduction of $||\mathcal{E}_s(jw)||$. Therefor,

  $$\text{rank}\,[\mathbf{L}_s] = \text{rank}\left[\mathbf{C}(jw\mathbf{I} - \mathbf{A} + \mathbf{L}_x\mathbf{C})^{-1}\mathbf{E}\right] = q + \bar{q} \quad (20)$$

  has to be fulfilled. Regarding minimal observer errors, the necessary conditions

  $$\text{rank}\,[\mathbf{E}] = q + \bar{q}, \;\; \text{rank}\,[\mathbf{C}] = m \geq q + \bar{q}, \;\; n \geq q + \bar{q} = p \quad (21)$$

  result from condition (19). Under the assumption that w and $||\mathcal{S}(jw)||$ are bounded, the reduction of $||\mathbf{T}(jw)||$ reduces also $||\mathcal{E}_s(jw)||$. Again a tuning proof can be given with

  $$\mathbf{L}_s = \rho_s \mathbf{L}_{s_o} \quad (22)$$

  resulting in

  $$||\mathbf{T}(jw)|| = \frac{1}{\rho_s}||\mathbf{T}_o(jw)|| \quad (23)$$

  where $\mathbf{L}_{s_o}$ symbolizes a matrix fulfilling

  $$\text{rank}\left[\mathbf{L}_{s_o}\mathbf{C}(jw\mathbf{I} - \mathbf{A} + \mathbf{L}_x\mathbf{C})^{-1}\mathbf{E}\right] = q + \bar{q}. \quad (24)$$

  For $\rho_s \rightarrow \infty$, $||\mathbf{T}(jw)||$ can be reduced arbitrarily.
  Consequently, the fullfilment of both the boundedness assumption and the conditions (19,21) make sure the reduction of $||\mathcal{E}_x(jw)||$ to arbitrary small values, but not to zero.

3. The observer reaction on fast effects $\mathbf{s}(t)$ can be improved by increasing the gains $\mathbf{L}_x$ and $\mathbf{L}_s$, depending on $\dot{\mathbf{s}}(t)$. This may cause some implementation problems due to too large gain value. Moreover, the observer stability could not be guaranteed when fixing $\mathbf{L}_s$ with too large values without considering the values of $\mathbf{L}_x$ and vice versa. To avoid this stability problem, the LQ-design is applicable: The Riccati formalism guarantees a stable design of system (6,7) (Müller and Lückel, 1977).

4. UNKNOWN INPUT OBSERVER

Consider the linear time-invariant system

$$\dot{\mathbf{z}}(t) = \mathbf{F}_z\mathbf{z}(t) + \mathbf{J}\mathbf{u}(t) + \mathbf{G}\mathbf{y}(t), \quad (25)$$

$$\mathbf{w}(t) = \mathbf{L}_1\mathbf{z}(t) + \mathbf{L}_2\mathbf{y}(t) \quad (26)$$

where $\mathbf{z}(t)$ symbolizes the t-dimensional UIO state vector and $\mathbf{w}(t)$ the w-dimensional output vector (Wünnenberg, 1990).

The idea behind observer (25,26) providing $\mathbf{w}(t)$ consists in a transformation-based decoupling of the state $\mathbf{z}(t)$ from the EN treated as *unknown inputs*. The output $\mathbf{w}(t)$ can be used for observer-based closed-loop control (CLC) or open-loop supervision (OLS).

In general, system (25,26) is a reduced-order observer with order $l < n$. If system (1,2) provides EN-measurability and EN-decoupability, i. e. conditions

$$\text{rank}\left[\mathbf{C\bar{E}}_D\right] = \text{rank}\left[\bar{\mathbf{E}}_D\right] \quad (27)$$

and (4) hold for CLC, or it provides EI-diagnosability and EI/EN-decoupability, i. e. conditions (3,5) hold for OLS, the observer matrices

$\mathbf{F}_z, \mathbf{J}, \mathbf{G}, \mathbf{L}_1, \mathbf{L}_2$ can be fixed by solving the equations

$$\mathbf{F}_z\mathbf{T} - \mathbf{T}\mathbf{A} + \mathbf{G}\mathbf{C} = \mathbf{0}, \quad \mathbf{F}_z\text{: stable,} \quad (28)$$
$$\mathbf{J} = \mathbf{T}\mathbf{B}, \quad (29)$$
$$\mathbf{T}\overline{\mathbf{E}}_D = \mathbf{0}, \quad (30)$$
$$\mathbf{L}_1\mathbf{T} + \mathbf{L}_2\mathbf{C} = \mathbf{X}, \quad (31)$$

cf. Wünnenberg (1990), Hou (1995) or Chen, Patton and Zhang (1995), where $\mathbf{T}$ symbolizes a transformation matrix fulfilling (30) and, with respect to (31), $\mathbf{X}=\mathbf{I}$ for CLC and $\mathbf{X}=\mathbf{0}$ for OLS. For $\mathbf{e}_z(t) = \mathbf{z}(t) - \mathbf{T}\mathbf{x}(t)$, this results in

$$\dot{\mathbf{e}}_z(t) = \mathbf{F}_z\mathbf{e}_z(t) - \mathbf{T}\mathbf{Y}\mathbf{s}_D, \quad (32)$$

where $\mathbf{Y} = \mathbf{0}$ for CLC and $\mathbf{Y} = \mathbf{E}_D$ for OLS reflecting faults in the dynamics. Regarding UI-decoupling, conditions (4,27) imply observer-state decoupling from EN $\overline{\mathbf{s}}_D(t)$, condition (30) refers to system-state decoupling from EN $\overline{\mathbf{s}}_D(t)$ and condition (31) results in OLS-output decoupling from the system state $\mathbf{x}(t)$. Therefore, this type of observer is termed *unknown input observer (UIO)*. The UIO-structure is given in Fig. 2.

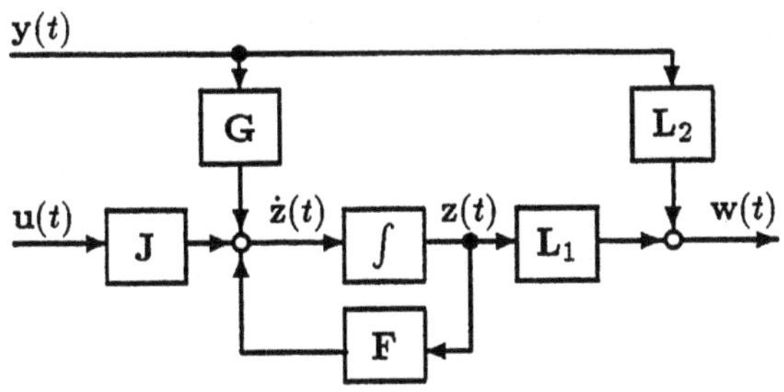

Fig. 2. Unknown Input Observer

In addition to (27), a full-order UIO with order $l=n$ exists iff

$$\text{rank}\begin{bmatrix} \lambda\mathbf{I} - \mathbf{A}_1 \\ \mathbf{C} \end{bmatrix} = n, \ \forall\, \lambda \in \mathcal{C} \quad (33)$$

where $\mathbf{A}_1 = \mathbf{A} - \overline{\mathbf{E}}_D[(\mathbf{C}\overline{\mathbf{E}}_D)^T\mathbf{C}\overline{\mathbf{E}}_D]^{-1}(\mathbf{C}\overline{\mathbf{E}}_D)^T\mathbf{C}\mathbf{A}$ (Chen, Patton and Zhang, 1995), (Hou, 1995).
Note: Condition (27) must hold for EN-measurability, additionally (3) for both EI-isolability and EI-identifiability and (4) for EN-decoupability, and condition (5) allows testing the EI/EN-decoupability necessary for complete EI-isolability and EI-identifiability (Hou, 1995).

4.1 UIO for Control and Supervision

Regarding CLC, the classical UIO-purpose is robust state-estimation for feedback in sense of

$$\lim_{t\to\infty} \mathbf{w}(t) = \mathbf{x}(t) \quad (34)$$

where the observer design is based on a number of design methods for robust UIO, e. g. Hou (1995). With respect to OLS, the UIO-purpose is robust residual generation in sense of

$$\lim_{t\to\infty} \mathbf{w}(t) = \lim_{t\to\infty} \mathbf{r}(t) \quad (35)$$

where $\mathbf{r}(t)$ symbolizes the residual. FDII-observer design, here RG-design, is based on a number of design methods for robust UIO-based RG, cf. Patton and Chen (1997), Chen, Patton and Zhang (1995), Chang and Hsu (1993), Ding (1992).

5. SUMMERIZING COMPARISON FOR FDII

In the field *system supervision*, the set $\mathcal{Z}$ refers to effects of faults to be detected, isolated and identified. All other effects are of no interest and associated to $\overline{\mathcal{Z}}$. For robust FDII, a residual generator has to provide RRGP, i. e. a residual sensitive to EI and robust to EN.

With respect to the RRGP, both system (6,7,8) and system (25,26) represent a RG each: the PIO with residual $\mathbf{r}(t) = \mathbf{w}(t) = \hat{\mathbf{f}}(t)$ for FII or with residual $\mathbf{r}(t) = \mathbf{e}_y(t)$ for early FD; according to (26), the UIO with residual $\mathbf{r}(t) = \mathbf{w}(t) = \mathbf{L}_1\mathbf{e}_z(t)$ for FDI(I). However, the PIO-based RG provides quasi-asymptotic robustness by *high-gain tuning*, in opposition to the UIO-based RG with structural robustness by *T-transformation*.

The PIO-existence condition is weak, only condition (11) must hold for FDII whereas the much harder conditions (3,5) and additionally those, for instance, given in Ding (1992) for complete fault identification, are required for the UIO-existence. If, moreover, complete or perfect decoupling according to (4,27) cannot be achieved optimal decoupling can be the solution to obtain robustness (Patton and Chen, 1997), (Wünnenberg, 1990). No structural decoupling methods are available for robust PIO.

According to condition (11), the PIO allows $q \leq m$ EI to be identified while being robust to $\overline{q} > n - m$ EN, theoretically. Practically, with respect to acceptable values of the gains $\mathbf{L}_x$, $\mathbf{L}_v$, $\overline{q}$ could be restricted to $\overline{q} \leq n - m$ EN, e. g. by suitable gain values depending on the system considered. Practical experiences reveal lower gains for CLC (Müller, 1995) in opposition to OLS. According to condition (3), the UIO allows $q \leq m$ EI to be diagnosed while $\overline{q} \leq n - m$ EN can be decoupled. For identification, additional requirements have to be met, cf. Ding (1992).

In both approaches, PIO and UIO, condition (5) can be used to test the EI/EN-decoupability where, depending on the test result, a bank of observers, such as the generalized or dedicated observer scheme GOS or DOS (Wünnenberg, 1990) providing structured or directional residuals (Chen, Patton and Zhang, 1995) has to be designed, if the number of measurements is too low to cover the EI-totality, cf. van Schrick (1994) for a scheme comparison.

Besides classical control applications, e. g. Beale and Shafai (1989), Shafai et al. (1995), the PIO has been applied to estimation and compensation of nonlinearities in robot position control (Müller, 1995) and to estimation of crack forces in rotors for diagnosis purposes (Söffker, 1996). The UIO has been applied to FDII problems, cf. e. g. Patton and Chen (1997), Isermann and Ballé (1996). Moreover, internal disturbances in form of friction can also be detected, e. g. for monitoring in robot position control (Wünnenberg, 1990).

6. CONCLUSIONS

Mainly with respect to closed-loop supervision, in this paper different aspects of the proportional-integral observer have been shown, explained and put into relation to the unknown input observer. A new aspect of interest is the convergence of the proportional-integral observer and its proof. A first approach to it has been sketched and future work is directed to the complete convergence proof for this type of observer.

In conclusion, the *PIO*-based RG provides "high-gain" ***steady-state state-estimation, quasi-asymptotic robustness*** and features the diagnosis purpose ***fault identification***. The *UIO*-based RG provides ***asymptotic output-estimation, structural robustness (decoupling)*** and features the diagnosis purpose ***fault isolation***.

REFERENCES

Beale, S. and B. Shafai (1989). Robust Control System Design with a Proportional Integral Observer, *Int. J. Contr.*, Vol. 50, pp. 97-111.

Chang, S.-K. and P.-L. Hsu (1993). Fault Detection Observer Design for Linear Systems with Unknown Inputs. In: *Proc. ECC'93*, pp. 1975-1980.

Chen, J.; Patton, R.J. and H.Y. Zhang (1995). Design of Robust Structured and Directional Residuals for Fault Isolation via Unknown Input Observers. In: *Proc. ECC'95*, pp. 348-353. State Observers. In: *Proc. UITAM Symp. Nonlin. Dynamics in Engineer. Systems*, Springer Berlin, pp. 205-215, 1990.

Ding, X. (1992). Frequenzbereichsverfahren zur beobachtergestützten Fehlerentdeckung, *VDI-Fortschritt-Berichte*, Vol. 8, No. 295, VDI-Verlag (in German).

Hou, M. (1995). Descriptor Systems: Observers and Fault Detection, *VDI-Fortschritt-Berichte*, Vol. 8, No. 482, VDI-Verlag.

Isermann, R. and P. Ballé (1996). Trends in the Application of Model Based Fault Detection and Diagnosis of Techical Processes. In: *Proc. IFAC World Congress '96*, Vol. N, pp. 1-12.

Kaczorek, T. (1979). Proportional-Integral Observers for Linear Multivariable Time-Varying Systems, *Regelungstechnik*, Vol. 27, pp. 359-363.

Müller, P.C. and J. Lückel (1977). Zur Theorie der Störgrößenaufschaltung in linearen Mehrgrößenregelsystemen, *Regelungstechnik*, Vol. 25, pp. 54-59 (in German).

Müller, P.C. (1995). Non-Linear Robot Control: Method of Exact Linearization & Decoupling by State Feedback & Alternative Control Design Methods, *Appl. Math. and Comp. Sci.*, Vol. 5, pp. 359-371.

Patton, T. and J. Chen (eds) (1997). *Prepr. SAFEPROCESS'97.*

Shafai, B. et al. (1995). Proportional-Integral Observers for Discrete-Time Systems. In: *Proc. ECC'95*, pp. 520-525.

Söffker, D. (1996). Fault Detection using Proportional-Integral Observer for Application to Elastic Mechanical System. In: *Proc. CESA'96*, pp. 522-527.

van Schrick, D. (1994). A Comparison of IFD Schemes. In: *Proc. CCA'94*, pp. 889-894.

van Schrick, D. (1997). The Diagnostic System - A Terminological Approach. In: *Prepr. SYSID'97*, pp. 637-642.

Wünnenberg, J. (1990). Observer-Based Fault Detection in Dynamic Systems, *VDI-Fortschritt-Berichte*, Vol. 8, No. 222, VDI-Verlag, Düsseldorf.

14th Triennial World Congress, Beijing, P.R. China

C-2a-16-2

CONTROL OF EVOLUTIONARY PROCESSES AND DEFORMATION THEORY

A.G. Aleksandrov

Institute of Control Sciences, Russian Academy of Sciences, Profsojuznaja str. 65, GSP-7, Moscow, 117806 Russian Federation
phone: 7-095-334-8980; fax: 7-095-420-2016
e-mail: alexandr@ipu.rssi.ru

Abstract: A general approach to the study of control of general evolutionary processes is discussed. Such a process is described by a vector field with polynomial, analytic or smooth coefficients in phase space. The corresponding multidimensional systems are investigated by means of perturbation analysis of the associated bifurcation sets and the discriminants. The approach is based essentially on the modern deformation theory, the theory of discriminants and on the theory of integrable connexions associated to deformations of multidimensional systems with singularities. Such a connexion is presented as a Fuchsian system whose coefficients have logarithmic poles along the discriminant of a deformation.

Key words: processes, phase stability, equilibrium, multidimensional systems, perturbation analysis theory, parameters.

1. EVOLUTIONARY PROCESSES

An evolutionary process is described by a vector field in phase space. A point of phase space defines the state of the system. The vector at this point indicates the velocity of change of the state. The points where the vector field vanishes are called equilibrium points or equilibrium positions.

It was shown by H.Poincaré (1879) that the typical phase portraits in the neighborhood of an equilibrium point of generic systems can be classified so that the corresponding list consists of the five simple types: two stable (focus, node) and three unstable (saddle, focus, node).

Real evolutionary processes are usually described by means of generic systems or, in other words, of systems which are in general position. Such systems always depend on parameters that are never known exactly. A small generic change of parameters transforms a non-generic system into a generic one.

Thus, at first sight, more complicated cases might not to be considered since they turn into combinations of the above types after a small generic perturbation of the system.

However, if one is interested not in an individual system but in systems depending on parameters the situation is quite different and more complex. Thus, let us consider the space of all systems divided into domains of generic systems. The dividing sets (hypersurfaces) correspond to degenerate systems. Under a small change of the parameters a degenerate system becomes non-degenerate. A one-parameter family of systems is presented by a curve which can intersect transversely the boundary separating different domains of nondegenerate systems.

Hence, although for each fixed value of the parameter the system can be always transformed by a small perturbation into a nondegenerate one, it is impossible to do this simultaneously for *all* values of the parameter. In fact, every curve closed to the one considered intersects the boundary of the separate hypersurface at a close enough value of the parameter.

Thus, if one studies not an individual system only but the whole family, the degenerate cases are not *removable*. If the family depends on a one parameter than the simplest degeneracies are unremovable one, those represented by boundaries of codimension one (that is, boundaries given by one equation) in the space of all systems. The more complicated degenerate systems, forming a set of codimension two in the space of all systems, may be gotten rid of by a small perturbation of the one-parameter family.

If one analyzes two-parameter families then one needs not consider degenerate systems forming a set of codimension three and so on. Therefore at first it ought to analyze all generic systems, then degeneracies of codimension one, then – two and so on, see (Arnol'd, 1981). Herewith one must not restrict the study of degenerate systems to the picture at the moment of degeneracy, but must also include a description of the reorganizations that take place when the parameter passes through the degenerate value.

If a stable equilibrium state describes the established conditions in some real system then when it merges with an unstable equilibrium state, the system must *jump* to a completely different state: under changing of parameters the equilibrium state in the neighborhood considered suddenly disappears.

2. CONTROL SPACE AND PARAMETERS

Let us consider a family of *smooth* functions

$$f : \mathbf{R}^n \times \mathbf{R}^r \to \mathbf{R},$$

describing a certain process happening in various copies of $\mathbf{R}^n$ governed by the function f and affected by the point in $\mathbf{R}^r$. The coordinate space $\mathbf{R}^n$ is usually called the space of *internal* variables while $\mathbf{R}^r$ the space of *external* variables over which each copy sits. Such terminology is suitable when the variables in $\mathbf{R}^r$ label in physical space as in optics, biology, and so on.

For systems which one alters something and then to observe what happens the variables in $\mathbf{R}^r$ are called the *control* parameters while the variables in $\mathbf{R}^n$ are named the *behavioral* parameters. Accordingly the space $\mathbf{R}^r$ is referred to as the *control* space while $\mathbf{R}^n$ as the *behavior* space. In the strictly mathematical context it natural to call the space $\mathbf{R}^r$ the *deformation space* while its points (or their coordinates) the *parameters of a deformation.* The number r is correspondingly the *external* or *control* dimension, or the *dimension* of deformation.

Suppose that a submanifold $M \subset \mathbf{R}^n \times \mathbf{R}^r$ is given by the equation

$$Df_u(x) = 0,$$

where $f_u(x) = f(x,u)$, $(x,u) \in \mathbf{R}^n \times \mathbf{R}^r$, and D is the usual differential of the image

$$f_u : \mathbf{R}^n \longrightarrow \mathbf{R}.$$

In other words, the manifold M is the set of all *critical points* of all the potentials f_u in the family f. Denote by ξ the restriction to M of the natural projection

$$\pi : \mathbf{R}^n \times \mathbf{R}^r \longrightarrow \mathbf{R}^r, \qquad \pi(x,u) = u.$$

The *critical set* is identified with the subset $C \subset M$ consisting of *singular* points of the image ξ. In other words, C consists of points in which the map ξ is *singular*, that is, the rank of the derivative $D\xi$ is less than r. The image of the critical set $\xi(C) \subset \mathbf{R}^r$ is called the *bifurcation set* $\mathcal{B}$.

It is not difficult to see, by computing $D\xi$, that C is the set of points $(x,u) \in M$, at which $f_u(x)$ has a *degenerate critical* point. It follows that $\mathcal{B}$ is the locus where the number and nature of critical points *change* (that is, it happens jump changes in the state of a control system); for by structural stability of Morse functions such changes can only occur by *passing* through a degenerate critical point. In most applications (for instance, in problems of stability, optimization, in studying caustics, wave fronts and so on) it is the *bifurcation* set that is the most important, for it lies in the control space, hence is "observable", and all delay convention jumps occur in it.

Investigations show that a bifurcation set as variety possesses highly complicated topological, analytic, algebraic and geometric *structures*. Herewith it appears that characteristics of such a variety depend mainly on the structure of its subvariety of *singularities*, which, in turn, also can possess singularities and so on. This observation directly leads to the notion of a *stratification* variety, but in the most general context the study of bifurcation sets is reduced to the study of *stratified* varieties, see (Whitney, 1955).

Remark that in virtue of the known splitting lemma a *smooth* function f can be presented around a points, where it has corank m, in the form:

$$\tilde{f}(x_1,\ldots,x_m) \pm x_{m+1}^2 \pm \ldots \pm x_n^2$$

(perhaps with parameters in $\mathbf{R}^r$ for $\tilde{f}$). Herewith the variables $x_1,\ldots,x_m$ are called *essential* while $x_{m+1},\ldots,x_n$ - *unessential*. Certainly, such presentation is very far from unique. It should be also noted that most singularities met by an r-dimensional family will even when not regular or Morse, have codimension less than r. However, it is possible to write an r-parameter family f, around a point, where it meets transversely a singularity of codimension ν, in a way in which only ν control parameters appear. When one has done so, one may call the coordinates on $\mathbf{R}^r$ that no longer appear, *disconnected* or *dummy* control parameters.

3. DEFORMATIONS

In this work using the modern deformation theory the control of general evolutionary processes is studied. In fact, most such processes can be described with the help of polynomial and analytical functions or by systems of differential equations. In particular, using properties of associated bifurcation sets, or more generally, the discriminants basic properties of the corresponding systems are investigated. The main tool of the investigation is the general notion of the integrable connection associated naturally with any deformation of a system with singularities. Let us shortly discuss basic ideas of the theory. Consider the system of polynomial or analytic equations

$$\begin{cases} f_1(z_1,z_2,\ldots,z_m) = 0 \\ \vdots \\ f_k(z_1,z_2,\ldots,z_m) = 0 \end{cases} \tag{1}$$

given in a neighborhood U of the origin in $\mathbb{C}^m$. For simplicity we shall assume that $k = m-1$ and the set X_0 of the solutions of our system in the neighborhood U is one-dimensional. We shall say that a point, laying on the curve X_0, is *nonsingular* if the differential form $df_1 \wedge \ldots \wedge df_{m-1}$ does not vanish on it. Otherwise, this point (and the curve) refers to as *singular*, or shortly a *singularity*. Without loss of generality one can suppose that X_0 has the only singularity at the origin $\{0\} \in X_0 \subset \mathbb{C}^m$, that is, X_0 is the germ of a reduced *curve*.

We shall assume now that the equations of the system (1) can be *perturbed:*

$$\begin{cases} f_1(z_1,z_2,\ldots,z_m) = t_1 \\ \vdots \\ f_{m-1}(z_1,z_2,\ldots,z_m) = t_{m-1} \end{cases} \tag{2}$$

in such a manner that at each small enough value of parameters $t = (t_1, \ldots, t_{m-1}) \in \mathbb{C}^{m-1}$ in the chosen neighborhood U the set X_t of the solutions of the system (2) is also *one-dimensional.* In other words, we shall consider the *principal (flat) deformation* of the curve singularity X_0 given by the holomorphic map:

$$f: \mathcal{X} \longrightarrow \mathbb{C}^{m-1}. \qquad (3)$$

Let X be the intersection of a ball of a small radius $\varepsilon > 0$ centered at the distinguished point $\{0\} \in X_0$ with $f^{-1}(T)$, where $T \subset \mathbb{C}^{m-1}$ is a punctured ball of a radius $0 < \delta \ll \varepsilon$ centered at the origin $0 = f(0)$. Consider the natural restriction of the mapping (3):

$$f: X \longrightarrow T.$$

For some values of parameters $t \in T$ the fibres X_t are non-singular curve germs, for other ones the fibres may have singular points which is called the *critical points* of mapping f.

4. PERIOD INTEGRALS

Denote by $C \subset X$ the set of critical points of f and by D its image $f(C) \subset T$. Thus, parameters corresponding to the fibres with singularities form the set D which refers to as the discriminant set, or the *discriminant* of the principal deformation X_0. In many important cases the discriminant is the zero-set of the only equation $h(t_1, \ldots, t_k) = 0$, i.e. D is a *hypersurface.* Set

$$T' = T - D, \quad X' = X - C.$$

The restriction $f: X' \rightarrow T'$ is a local trivial differentiable fibre bundle called the *Milnor fibration* of f, that is, the fibres $X_t = f^{-1}(t) \cap X$ (of *real* dimension two) form the smooth fibre bundle over T'. Fix a point $t_0 \in T'$. Then for each *smooth closed path* $\gamma_0 \subset X_{t_0}$, corresponding to the 1-cycle in $H^1(X_{t_0}, \mathbb{C})$, it is possible to construct a family of 1-cycles $\gamma(t) \subset X_t$, $t \in T'$, such that $\gamma(t_0) = \gamma_0$.

If one takes a holomorphic differential form $\omega = g(z)dz_1 \wedge \ldots \wedge dz_m$ of the maximal degree in a neighborhood of the origin in $\mathbb{C}^m$, then using the identity $df_1 \wedge \ldots \wedge df_{m-1} \wedge \psi = \omega$ one can find a differential form ψ, which is the result of the division of ω by $df_1 \wedge \ldots \wedge df_{m-1}$. The form ψ is not determined uniquely, but up to the summands containing differentials of the functions $f_1, \ldots, f_{m-1}$. It is easy to prove, that for all parameters t, rather close to zero, the integral

$$I(t) = \int_{\gamma(t)} \psi = \int_{\gamma(t)} \frac{\omega}{df_1 \wedge \ldots \wedge df_{m-1}} \qquad (4)$$

is determined correctly. Moreover, the integral $I(t)$ is an *analytic* function in t. Integrals of such type are called the *period integrals.*

Replacing the differential form ω with any other one, the integral (4), generally speaking, will also change. However, it is possible to prove that the set of all such integrals contains a finite number of the elements $I_1(t), \ldots, I_\mu(t)$ so that any integral of the type (4) may be expressed by means of these generators as a linear combination with *holomorphic* coefficients. In the present context μ is the *Milnor number* which is a topological invariant of the singularity X_0.

The same observation holds, if one fixes the form ω and takes various families $\gamma(t)$. For definiteness, we shall fix a family of vanishing cycles $\gamma(t)$ and consider μ independent *period integrals* of the following type

$$I_j(t) = \int_{\gamma(t)} \frac{\omega_j}{df_1 \wedge \ldots \wedge df_{m-1}}, \qquad (5)$$

where $1 \leq j \leq \mu$. The period integrals $I_j(t)$ can be differentiated along the parameter t. Between integrals and their derivatives there arose linear relations *(syzygies)* with polynomial coefficients in t. These relations generate a system of differential equations for the integrals $I_j(t)$ that is expressed by a finite number of independent ones.

5. CONNEXION

In such a way a system of differential equations in t may be associated to the singularity X_0 which is determined correctly outside of the discriminant. This system refers to as *Gauss-Manin connexion*, or *Gauss-Manin system*, associated with the principal deformation of X_0. The main problem is to describe a system of differential equations defined on the *whole* space of parameters, which is equivalent to the initial one outside of the discriminant (or, in other words, to *prolong* the initial system to the discriminant set). It is possible to show that the solution of this problem depends mainly on properties of the discriminants as well as on that of the singular fibres.

It turns out that the connexion in question can be represented in a quite elegant form. In order to explain this idea we need the following notion. A meromorphic form ω on S is called the *logarithmic* (along a divisor $D \subset S$) differential form if and only if ω and its differential $d\omega$ have poles along D at worst of the first order. That is, $h\omega$ as well as $hd\omega$ are *holomorphic* differential forms on S where h is a local equation of the hypersurface $D \subset S$.

The $\mathcal{O}_S$-module of logarithmic differential q-forms is usually denoted by $\Omega^q_S(\log D)$. The logarithmic differential forms have many remarkable analytic and algebraic properties.

Following Saito (1977) denote by $\mathrm{Der}_S(\log D)$ the $\mathcal{O}_S$-module of *logarithmic vector fields* along D on S. This module consists of germs of holomorphic vector fields η on S for which $\eta(h)$ belongs to the principal ideal $(h)\cdot\mathcal{O}_S$ (in particular, the vector field η is tangent to D at its smooth points). The inner multiplication of vector fields and differential forms induces a natural pairing of $\mathcal{O}_S$-modules

$$\mathrm{Der}_S(\log D) \times \Omega^q_S(\log D) \longrightarrow \Omega^{q-1}_S(\log D).$$

For $q = 1$ this $\mathcal{O}_S$-bilinear mapping is a *non-degenerate* pairing so that $\mathrm{Der}_S(\log D)$ and $\Omega^1_S(\log D)$ are $\mathcal{O}_S$-dual.

Let $\mathcal{H}$ be a free $\mathcal{O}_S$-module. Then a connexion ∇ on $\mathcal{H}$ with *logarithmic* poles along $D \subset S$ is a morphism satisfying the following conditions:

1) $\nabla(\omega + \omega') = \nabla(\omega) + \nabla(\omega')$,

2) $\nabla(f\omega) = \omega \otimes df + f\nabla(\omega), \quad f \in \mathcal{O}_S.$

Consider the case where $\Omega^1_S(\log D)$ is a *free* $\mathcal{O}_S$-module of rank m. Obviously, in such a case $\Omega^p_S(\log D) = \bigwedge^p \Omega^1_S(\log D)$, $p \geq 1$. It is often said that the divisor D is *free* or, equivalently, D is a *Saito* divisor. The following characteristic property of such divisors was discovered by Saito (1977).

Suppose that there exist m vector fields

$$v_1, \ldots, v_m \in \mathrm{Der}_S(\log D)$$

such that for the $m \times m$-matrix $\mathcal{V}$ whose entries are the coefficients of v_i, $i = 1, \ldots, m$, one has $\det(\mathcal{V}) = ch$, where c is a unit. Then $v_1, \ldots, v_m$ is a *basis* for the free $\mathcal{O}_S$-module $\mathrm{Der}_S(\log D)$. In particular, $\Omega^1_S(\log D)$ is a *free* $\mathcal{O}_S$-module.

For example, $\Omega^1_S(\log D)$ is free when D is the discriminant of the *versal* deformation of the system defined by a function with isolated singularity.

Let us describe the *logarithmic connexion*

$$\nabla_{X/S}\colon \mathcal{H} \longrightarrow \mathcal{H} \otimes_{\mathcal{O}_S} \Omega^1_S(\log D) \qquad (6)$$

on $\Omega^1_S(\log D)$ itself, that is, consider the case where $\mathcal{H} = \Omega^1_S(\log D)$:

$$\nabla\colon \Omega^1_S(\log D) \to \Omega^1_S(\log D) \otimes_{\mathcal{O}_S} \Omega^1_S(\log D).$$

Let $\omega_1, \ldots, \omega_m$ be free generators of the module $\Omega^1_S(\log D)$. Then the connexion ∇ can be expressed in terms of of Christoffel symbols in the following way:

$$\nabla\omega_i = \sum_{j=1}^m \omega_j \otimes \omega_i^j, \quad \omega_i^j = \sum_{k=1}^m \Gamma_i^{jk}\omega_k.$$

The connexion ∇ is called *torsion free* if

$$d\,\omega_i = \sum_{j=1}^m \omega_i^j \wedge \omega_j = \sum_{k,j=1}^m \Gamma_i^{jk}\omega_k \wedge \omega_j,$$

and ∇ is called *integrable* if

$$d\omega_i^j = \sum_{k=1}^m \omega_i^k \wedge \omega_k^j,$$

that is,

$$d\,\nabla = \nabla \wedge \nabla,$$

where $\nabla = \| \omega_i^j \|$ is the *coefficient matrix* of the connexion ∇.

In particular, it means that the composition

$$\mathcal{H} \xrightarrow{\nabla} \mathcal{H} \otimes \Omega^1_S(\log D) \xrightarrow{\nabla} \mathcal{H} \otimes \wedge^2\Omega^1_S(\log D)$$

is zero.

6. FUCHSIAN SYSTEMS

In fact, one can associate to any integrable and torsion free connexion ∇ on the module $\Omega^1_S(\log D)$ the Fuchsian system in the following way.

It is known (Aleksandrov, 1990) that the multiplication on h induces the surjection

$$\Omega^1_S(\log D) \xrightarrow{\cdot h} \operatorname{Tors}\Omega^1_D \longrightarrow 0, \qquad (7)$$

whose kernel is coincide with $\mathcal{O}_S$-module

$$\frac{dh}{h}\mathcal{O}_S + \Omega^1_S.$$

Here Ω^1_S is the module of holomorphic differential 1-forms on S generated by the differentials $dz_1, \ldots, dz_m$ over $\mathcal{O}_S$,

$$\Omega^1_D = \Omega^1_S/(h\,\Omega^1_S + \mathcal{O}_S dh)$$

is the module of regular differential 1-forms on the divisor D, and $\operatorname{Tors}\Omega^1_D$ is the torsion submodule of Ω^1_D. The support of $\operatorname{Tors}\Omega^1_D$ is contained in the singular locus $\operatorname{Sing} D$ of the hypersurface D. The torsion $\mathcal{O}_D$-module $\operatorname{Tors}\Omega^1_D$ has a system of generators containing at least $m-1$ elements.

By definition, the generalized Fuchsian system is a system of linear differential equations on S with *meromorphic* coefficients containing in $\Omega^1_S(\log D)$:

$$d\mathbf{I} = \left(A_0\frac{dh}{h} + \sum_{i=1}^{\ell} A_i\frac{\vartheta_i}{h}\right)\mathbf{I}, \qquad (8)$$

where $\mathbf{I} = {}^{\mathrm{tr}}(I_1, \ldots, I_k)$ is a vector-column of unknown functions, the differential 1-forms $\vartheta_i \in \Omega^1_S$, $i = 1, \ldots, \ell$, correspond via (7) to the non-zero elements of the torsion module $\operatorname{Tors}\Omega^1_D$, and $A_i \in \operatorname{End}(\mathbf{C}^k) \otimes \mathcal{O}_S$, $i = 0, 1, \ldots, \ell$, are coefficient matrices with holomorphic entries satisfying the integrability condition:

$$d\Omega = \Omega \wedge \Omega,$$

where

$$\Omega = A_0\frac{dh}{h} + \sum_{i=1}^{\ell} A_i\frac{\vartheta_i}{h}.$$

It is not difficult to show that one can associate to any integrable and torsion free connexion ∇ on the module $\Omega^1_S(\log D)$ the generalized Fuchsian system (8), see (Aleksandrov, 1998). Moreover, using the Christoffel symbols of such connexion, it is possible to express the integrability condition in terms of commuting relations of the coefficient matrices A_i, $i = 1, \ldots, \ell$.

Under some additional assumption on entries of the coefficient matrices A_i it is possible to investigate the system (8), to describe its explicit solutions and to show that there exist ones which do not have an integral representation of Euler type. Hence they give us examples of functions of unknown type that satisfy systems of differential equations rather close to the hypergeometric type. In fact, such functions are quite useful in describing the control of evolutionary processes and perturbations of multidimensional systems with singularities.

REFERENCES

Aleksandrov, A.G.(1990). Nonisolated hypersurface singularities. *Advances in Soviet Mathematics,* **1**, 211-246.

Aleksandrov, A.G.(1998).Generalized Picard-Fuchs systems in several complex variables. *Journal of Shanghai Jiaotong University,* **E-3**, Sup.(I), 1-9.

Arnol'd, V.I. (1981). *Catastrophe theory.* Znanie, Moscow.

Poincaré, H. (1879). *Sur les propriétés des fonctions définies par les équations aux différences partielles.* Thèse inaugurale, Paris.

Saito, K. (1977). On the uniformization of complements of discriminant loci. *Hyperfunctions and Linear partial differential equations, RIMS Kōkyūroku,* **287**, 117-137.

Whitney, H. (1955). On singularities of mappings of euclidean spaces I. Mappings of the Plane into the Plane. *Annals of Mathematics,* **62**, 374-410.

14th Triennial World Congress, Beijing, P.R. China

C-2a-16-3

CONTROL OF LINEAR MOTOR-BASED ARM MOTION GENERATOR USING ACTIVE IMPEDANCE IMPLEMENTATION

Se-Han Lee and Jae-Bok Song

Dept. of Mechanical Engineering, Korea University, Anam-Dong, Sungbuk-Ku, Seoul 136-701, South Korea, E-mail: jbsong@kuccnx.korea.ac.kr, FAX: +82-2-928-9769

Abstract : In this research a 2-dimensional arm motion generator composed of two linear motors was developed. The inertia, damping and/or stiffness characteristics of the motion generator can be changed on the real-time basis by properly regulating the force generated by the linear motors. That is, active impedance is implemented without actual change in the physical structure of the motion generator. Control of the motor force is carried out by regulating the input currents supplied to the linear motors by using a common voltage-driven driver. In the control system the time delay due to A/D conversion of the current output has an adverse effect on the stability of the system. Furthermore, disturbances caused by characteristics of the motion generator exist. To cope with these difficulties, 2-DOF controller combined with the LQ-servo and H-infinity controller was used. The gains of the controller are selected so that disturbance rejection, stability guarantee and tracking performance may be achieved. This motion generator can be used to measure kinesthetic sense associated with the human arm and thus leads to developing the products for which the kinesthetic sense is taken into account. *Copyright © 1999 IFAC*

Keywords: Arm Motion Generator, Active Impedance, 2-DOF Controller, H-infinity controller, Linear Motor

1. INTRODUCTION

As lifestyles modernize, products with not only function in its mind but products with kinesthetic sense along with its function draw attention to customers. That is, people consider the sensibility of a product as well as its performance. In recent years, there are lots of efforts and considerations to apply this even from the designing and early stages of development.

The object of this research is to develop an arm motion generator necessary for quantification and implementation of kinesthetic sense, which mainly relates to arm movement. In most mechanical systems, motion has close relations with mechanical impedance, which consists of inertia, stiffness, and damping. The arm motion generator developed in this research must measure the kinesthetic sense difference experienced by a person by alternating the impedance or must let the person experience the desired 'sense'. But varying the impedance physically is very tedious, costly, and time demanding. The arm motion generator controls the force outputs of the linear motors as to have the same effect as to physically changing the impedance. And more to that, use of two linear motors makes two-dimensional movement possible.

These kinds of motion generators are used in various flight and vehicle simulators and vigorous research of this continues today. However, these motion generators mainly use pneumatic systems to generate an entire body movement, and thus systems get complicated and manufacturing gets costly. But since the motion generator developed in this research only deals with arm movement, it makes the use of electrical motor possible, which brings system composing and controlling simple.

And the similar concept of active impedance used in this research has been widely studied in robot engineering. In robotics, when a manipulator is in contact with an object, the relation between the force exerted on the object by the manipulator and the motion at the contact point is modeled with impedance such as inertia, damping and stiffness and then the appropriate position/force control is carried out.(Spong, *et al.*, 1989) The concept of active impedance in this research is basically the same with the impedance in robotics, but has many differences in application purpose and method.

Chapter 2 of this paper takes a close look at the concept of active impedance. Chapter 3 deals with hardware structure of the motion generator system and 2-DOF control system including H-infinity controllere for controlling current supply to the motor. In Chapter 4, example of active impedance implementation using the arm motion generator developed in this research is given. Finally, in Chapter 5, conclusion is mentioned.

2. IMPLEMENTATION OF ACTIVE IMPEDANCE

A motion generator capable of implementing two dimensional plane motion is made by combining two linear motors perpendicularly as shown in Fig. 1 As an operator moves the lever with his hand to generate some movement pattern, displacement, velocity, and acceleration involved in this pattern are detected. Based on this information, motor driving current is regulated to get desired active impedance, that is active inertia, active damping, and/or active stiffness. By implementing this kind of active impedance, by varying inertia, stiffness, and/or damping, one can measure the kinesthetic sense most suitable for the operator. These data can be used to develop product with better kinesthetic sense for the user. This chapter will explain concretely about implementation of active impedance using the motion generator mentioned above.

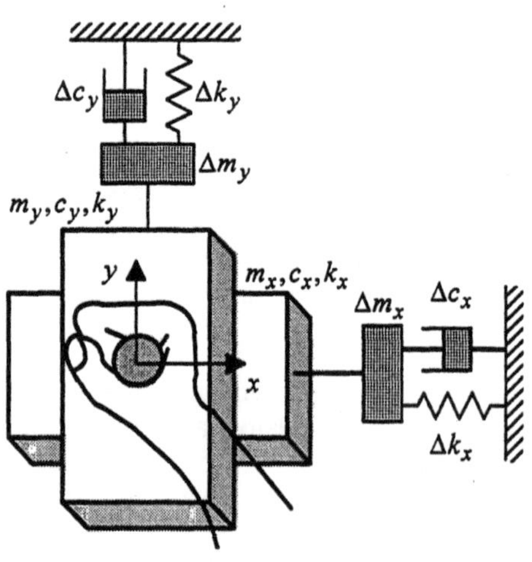

Fig.1. Configuration of motion generator and concept of active impedance

2.1 Active Impedance

Impedance in a mechanical system is defined as a measure of motion response to force applied to the system, and consists of inertia, damping, and stiffness. Active impedance, mentioned in this research, means getting the same effect of physically changing the impedance of the moving part of the linear motor with only generating proper force from the motor.

Although the motion generator is capable of two-dimensional movement in x-axis and y-axis, only one-dimensional movement is explained here because exactly the same formula is used for both linear motors. Let us first consider active inertia. When an operator applies force F to the lever of the linear motor in which moving part has mass of m and no energy is supplied, acceleration of $a_o = F/m$ is experienced. Assuming that we add active inertia of Δm to mass m of actual moving part,

$$M = m + \Delta m \qquad (1)$$

the operator must experience acceleration of $a = F/(m+\Delta m)$ $(< a_o)$ for the same input of F. If appropriate current is supplied to the motor to generate force of $F_1 = \Delta m \cdot a$ to the opposite direction of the force given by the operator, the net force input on the moving part becomes

$$F - F_1 = (m + \Delta m)\,a - \Delta m \cdot a = m \cdot a \qquad (2)$$

and so the acceleration felt by the operator is not a_o but a, which is the acceleration earned from force F applied on active inertia M, even when the operator puts force F on mass m. In this way, active inertia larger than actual inertia can be implemented.

While, given torque constant (force constant in case of linear motor) of the motor K_T and current supply i, the force generated by the linear motor F_m becomes $F_m = K_T\, i$. And supplying the current of

$$i_m = -\Delta m \cdot a / K_T \qquad (3)$$

to the motor could generate the effect of active inertia M. Here, the letter m at the current stands for the current relating to mass. Also, the negative sign means that the force generated by the motor should be in the opposite direction to the force applied by the operator. A plus sign (+) should be used as to get active inertia less than the actual inertia, because the force from the motor should be in the same direction with the force applied by the operator.

Active damping C and active stiffness K can be presented in the same manner,

$$C = c + \Delta c\,, \quad K = k + \Delta k \qquad (4)$$

where c and k are the original damping and stiffness of the moving part of the linear motor, $\Delta\, c$ and $\Delta\, k$ the added damping and stiffness, respectively. Suitable current value for the motor to obtain these active damping and stiffness are,

$$i_c = -\Delta c \cdot v / K_T\,, \quad i_k = -\Delta k \cdot x / K_T \qquad (5)$$

where v and x is the velocity and the position of the moving part of the linear motor.

If active inertia, damping, and stiffness wanted in the Eqn (3) and (4) are set, the values for the added inertia, damping, and stiffness can be calculated easily. But since the actual force input by the operator on the lever is not known, the values for acceleration a, velocity v, and displacement x must be obtained by actual measurements rather than by calculations. The acceleration and velocity values are obtained by numerically differentiates the displacement of the moving part measured by an encoder, which will be explained later.

2.2 Implementation of Complex Active Impedance

Inertia, damping, and stiffness do not exist in simple forms but rather in complex structures in actual mechanical systems. That is why implementing complex active impedance is required to let us reach closer to actual systems. When an operator applies force to the arm motion generator (to the lever connected to the linear motor), the motion generator also exerts the same amount of force to the opposite direction, as noted in the Newton's law of action-reaction. This force is called reaction force. Thus, when the operator generates motion by exerting force on the lever connected to the motor, the operator will experience reaction to inertia, damping, and stiffness all in complex. Expressing this in an equation gives,

$$F_{\text{total}} = F_{\text{iner}} + F_{\text{damp}} + F_{\text{stiff}} = Ma + Cv + Kx \quad (6)$$

When considering a reaction of an instant, the complex form of the inertia, damping, and stiffness would not be known. But by considering that the reaction due to inertia be proportional to acceleration, damping to velocity, and stiffness to position, one can measure the form of time-varying reaction which leads to classification of the each component of the total reaction.

The velocity and acceleration in the figure can be obtained by using the basics of the measurement of the position of the linear motor moving part using the encoder. First, when the values of active impedance M, C, K which are to be implemented are set, additional impedance Δm, Δc, Δk can be calculated by using Eqn (1) and (4). By multiplying these values of additional mass, damping, and stiffness by the acceleration, velocity and displacement obtained above and then adding this value would get the force which the linear motor should generate. At this time, the total reaction felt by the operator is expressed in Eqn (6), but for this to happen, the motor has to generate the following force.

$$F_{\text{motor}} = \Delta m \cdot a + \Delta c \cdot v + \Delta k \cdot x \quad (7)$$

To generate this force, the current that must be supplied to the linear motor is

$$i_{\text{total}} = i_m + i_c + i_k = \frac{-(\Delta m \cdot a + \Delta c \cdot v + \Delta k \cdot x)}{K_T} \quad (8)$$

3. MOTION GENERATOR SYSTEM

3.1 Structure of Motion Generator System

Figure 2 shows an arm motion generator composed of the two linear motors, drive circuit to run linear motors, TMS320 DSP board used as a controller, and parallel port connected PC used to collect and analyze experimental data. Linear motor used here has 3-phase, 4-pole motor part which consists of permanent magnet and drive coil, and an encoder used to track position, combined as one structure. Voltage driven BLDC (Brushless DC) drive, exclusively for motors, is used to drive the linear motor. That is, it detects position of the linear motor with Hall sensor signals and applies voltage to corresponding driving coil in PWM (Pulse Width Modulation) mode.

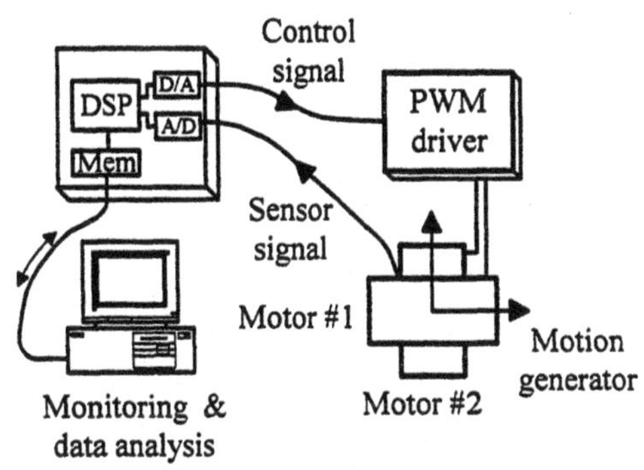

Fig. 2. Schematic of experimental setup

The main controller TMS320 DSP board can be operated at 40MHz speed and it carries out algorithms necessary to control the motor. This board has ±10V range, 12-bit resolution A/D and D/A converters. The A/D converter measures analog current supplied to the motor and digitizes it, then sends it to the DSP, while the D/A converter transforms the output digital signal from the controller to analog voltage to send it to the motor drive. The PC communicates with the controller and takes charge of collecting and analyzing of the experimental data.

3.2 Motion Generator Control System

As mentioned before, in order to implement active impedance, control of the force generated by the linear motor is needed. This force is proportional to the current supplied to the motor. But since most motor drivers are usually voltage-driven, the current is not directly controlled, but it is controlled indirectly by controlling the voltage. That is, to supply necessary current, the voltage of the linear motor must be controlled by the controller on the basis of the difference between the targeted value and the actual value of the current. The relation between the input voltage u and the current i flowing through the motor is commonly known as

$$u - K_E v = L\frac{di}{dt} + Ri \quad (9)$$

where L and R denote the inductance and resistance of the linear motor winding, respectively, and K_E the back emf constant. The back emf $K_E v$ is generated which is proportional to the velocity of the BLDC motor, and is applied in the opposite direction to the supplied voltage to the motor. Since the motion is produced by combination of the force applied by the operator and the force generated by the motor, the displacement, velocity and acceleration of the motion generator depend mostly upon the force arbitrarily applied by the operator. Therefore, the term $K_E v$ on the left-hand side of Eqn (9) varies arbitrarily, and this is treated as disturbance in the control system. In this case the plant can be described by

$$G_P(s) = 1/(Ls + R) \tag{10}$$

The linear motor drive in Fig. 2 provides the current required for force control based on the current command. Since the current output is measured by the A/D converter and fed back to the control system, the time delay of at least the sampling period T_s due to A/D conversion occurs. As for the linear motor, the electric time constant τ_e is about 1,250μs, while the sampling period of the controller is T_s = 400μs. It is generally preferable that the sampling period be less than 1/10 of the plant time constant. Because the time constant of the linear motor cannot be changed without altering the structure, the sampling period of the controller must be decreased to improve stability of the control system. However, it is also difficult because lowering the sampling period leads to deteriorating the resolution of the encoder used for the displacement measurement. Since use of the encoder with higher resolution is costly, robust control scheme to cope with this time delay problem is adopted in this research.

It is interesting to investigate the effect of this time delay on the control system. The z-transform of Eqn (10) with the sampling period of T_s = 400μs yields

$$G_P(z) = \frac{0.10784z^{-1}}{1 - 0.72615z^{-1}} \tag{11}$$

In Fig. 3 showing the root locus plot for Eqn (11), the proportional gain is limited to 16 for the closed-loop system to be stable. Implementation of the PID controller with the P gain greater than the above critical value showed unstable behavior. With the low P gain, however, fast tracking capability could not be achieved. It is noted that infinite gain is allowed in the 1st-order analog plant described by Eqn (10). Lowering the sampling period can increase this limit, but it is quite difficult for the reasons explained above.

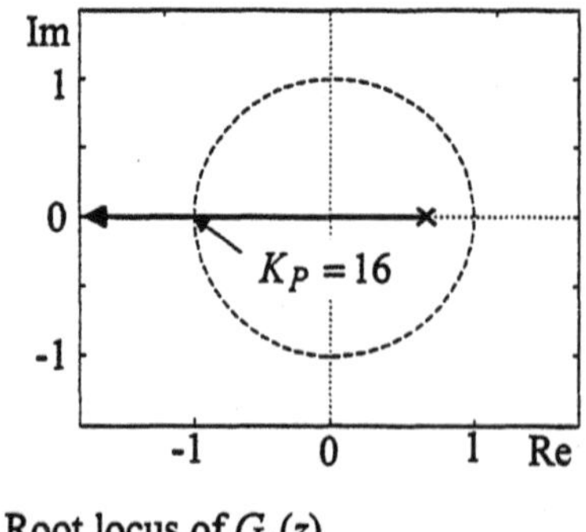

Fig. 3. Root locus of $G_p(z)$

To cope with such a problem, a 2-DOF control system is employed, which includes an internal plant and an H-infinity controller as shown in Fig. 4. The time delay due to sampling is absorbed in the multiplicative modeling error Δ_m.

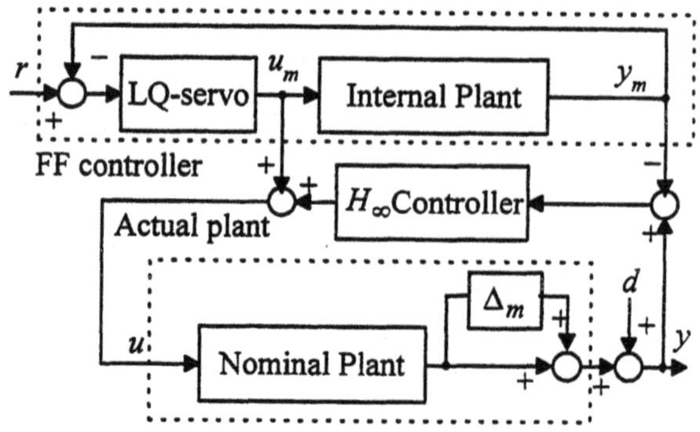

Fig. 4. Block diagram of 2-DOF current controller including internal plant and H-infinity controller.

For the internal plant, an LQ-servo controller was employed as a feedforward controller, and the difference between the internal plant output and the actual plant output was regulated by H-infinity controller which treats the modeling error. The total control input u is considered as the sum of the two outputs of the LQ-servo controller and the H-infinity controller.

3.3 H-infinity control formulation

As explained before, the time delay related to the sampling period T_s was treated as part of multiplicative modeling error Δ_m in Fig. 4. The multiplicative modeling error is generally used to describe sensor-related uncertainty. (Bahram, *et al.*, 1993) It is well known that robust stability is guaranteed provided that the multiplicative modeling error satisfies the following condition:

$$|\Delta_m(j\omega)| = |e^{-j\omega T_s} - 1| < |T(j\omega)|^{-1} \text{ for all } \omega \tag{12}$$

where $T(s)$ is the nominal co-sensitivity function (i.e., closed-loop transfer function). Figure 5 shows the Bode plot of this multiplicative modeling error. Based on this condition, the co-sensitivity weighting function $W_T(s)$ can be selected as an envelope for the multiplicative modeling error (John, *et al.*, 1992)

$$W_T(s) = \frac{2.2 \times T_s \times s}{T_s \times s + 1} \tag{13}$$

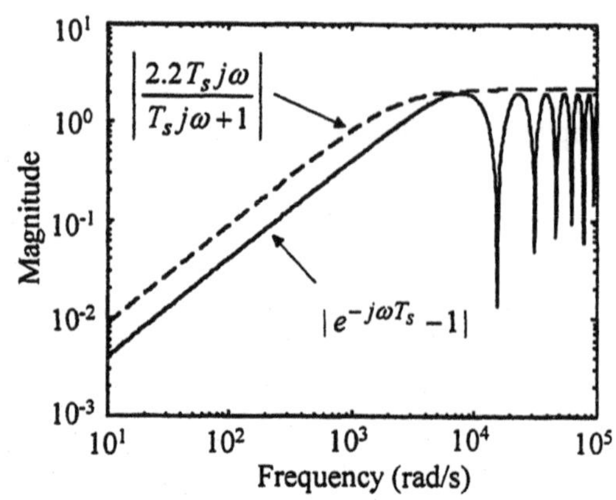

Fig. 5. Bode plots of $\Delta_m(s)$ and $W_T(s)$

Another weighting function to be considered is a sensitivity weighting function as a performance factor. The operator of a motion generator is a human, and the activating frequency may be within 10 to 20Hz. In this research, the activating frequency region is limited to 50Hz, which is large enough to cover the motion. The sensitivity weighting function has cutoff frequency of 50Hz and includes pure integral part to eliminate a steady-state error caused by constant disturbance.

$$W_S(s) = \frac{50 \times 2\pi}{s(s + 50 \times 2\pi)} \tag{14}$$

The weighting functions $W_T(s)$ and $W_S(s)$ are used for a mixed sensitivity H-infinity control problem shown in Fig. 6 as follows:

$$\left\| \begin{matrix} W_S(s)S(s) \\ W_T(s)T(s) \end{matrix} \right\|_\infty < \gamma \tag{15}$$

where $S(s)$ is the sensitivity function and γ is a positive real number.

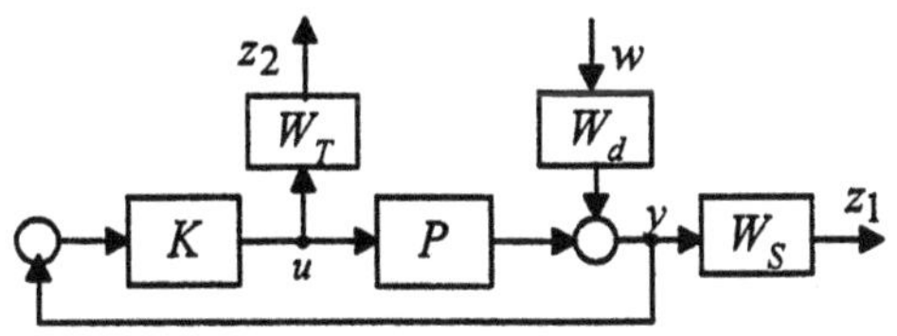

Fig. 6. Mixed sensitivity H-infinity problem with weighting functions

The generalized plant corresponding to Eqn (15) can be represented by

$$G(s) = \left[\begin{array}{c|c} W_S(s)W_d(s) & W_S(s)P(s) \\ 0 & W_T(s) \\ \hline W_d(s) & P(s) \end{array} \right] \tag{16}$$

At $\gamma \approx 0.14$, the controller $K(s)$ corresponding to Eqn (16) can be obtained by (Kemin, *et al.*, 1996)

$$K(s) = \frac{41.5(s + 2.5 \times 10^5)(s + 800)(s + 335.6)}{s(s + 35659)(s + 1156)(s + 314.16)} \tag{17}$$

The sensitivity function and its inverse weighting function with $K(s)$ are shown in Fig. 7. With this plot, it is assured that the condition (15) is satisfied.

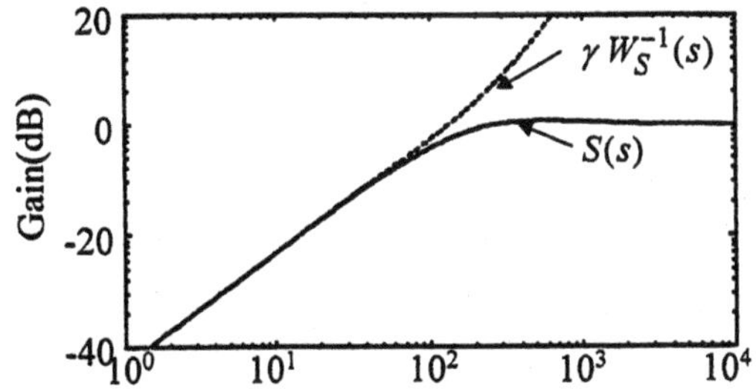

Fig. 7. Sensitivity function and its inverse weighting function

3.4 LQ-servo formulation

An LQ-servo is used to control the internal plant. Though the LQ-servo problem is a kind of trial and error, the LQ-servo has a merit to obtain control gain considering control input constraint. (Ken, *et al.*, 1997) In this research, control input weighting matrix R was chosen not to exceed maximum motor drive output. Overall block diagram for the LQ-servo system is shown in Fig. 8. Figure 9 shows the responses to step change in the target current based on the LQ-servo controller. Even though the input signal to the motor driver has a small overshoot, the consequent current signal to the motor shows smooth and accurate behavior.

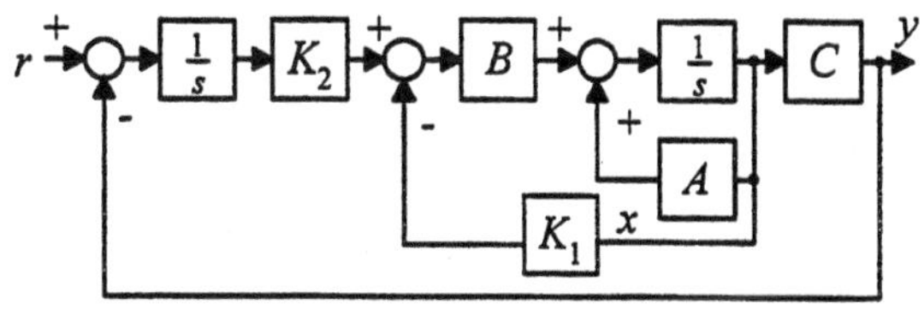

Fig. 8. Block diagram of the LQ-servo controller

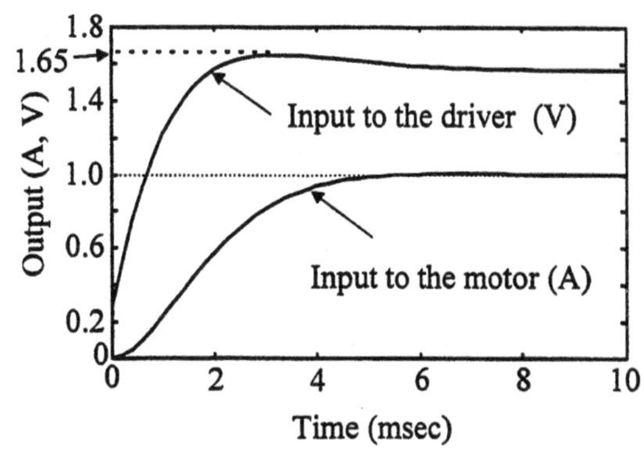

Fig. 9. Step responses with LQ-servo controller

4. PERFORMANCE OF MOTION GENERATOR

Figure 10 shows the experimental results when the operator activates the motion generator with arbitrary reciprocal motion with active stiffness of 500 N/m. The dotted and solid lines represent the reference and actual currents, respectively. When the conventional PI controller is used as shown in (b), the P gain is limited to some low value for stability and thus the tracking performance is also limited. If the higher gain is used, unstable behavior may occur, thus leading to high frequency audible sound. However, combination of the LQ-servo controller and the H-infinity controller provide much better performance as shown in (a). That is, the actual current can track the oscillatory reference value both accurately and quickly. In addition, no stability problem may occur. The similar results were obtained for the cases of active damping and inertia, though omitted here.

Here are some of the test results of the two-dimensional motion generator developed based on the active impedance implementation. Fig. 11 shows the linear motor having moving part mass of m = 6kg, damping constant of $c \approx 20\text{N} \cdot \text{s/m}$, and the stiffness of $k \approx 0$. The active stiffness was set to $K = 300\text{N/m}$ by letting additional stiffness of $\Delta k = 300\text{N/m}$. Then the displacement of the moving part of the motor was measured at each time as the lever was released at the initial displacement of about 8cm. As shown in the Fig. 11, the amplitude decreases gradually because the linear motor originally has damping of its own. All that is to say, implementation of active impedance on the system enabled the motion generator to vibrate as if it were a spring mechanism.

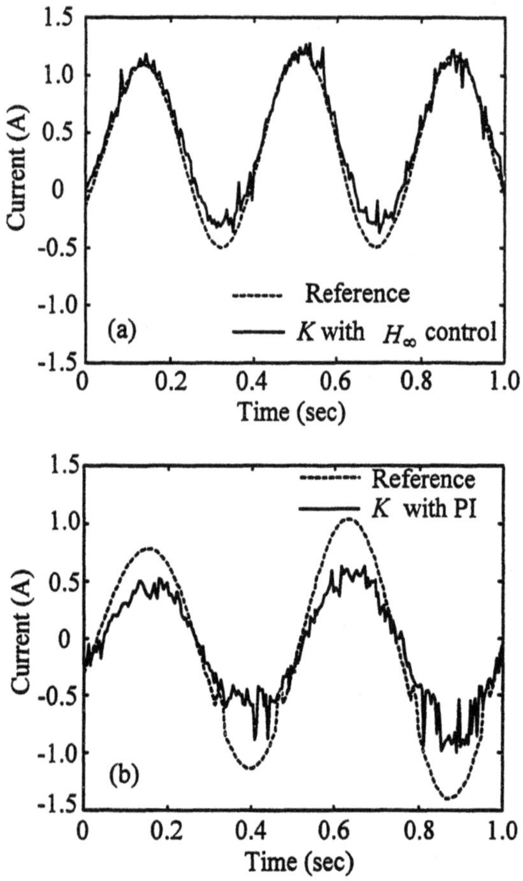

Fig. 10. Current responses of the motion generator with active stiffness implemented, (a) 2-DOF controller with H-infinity controller, and (b) PI controller.

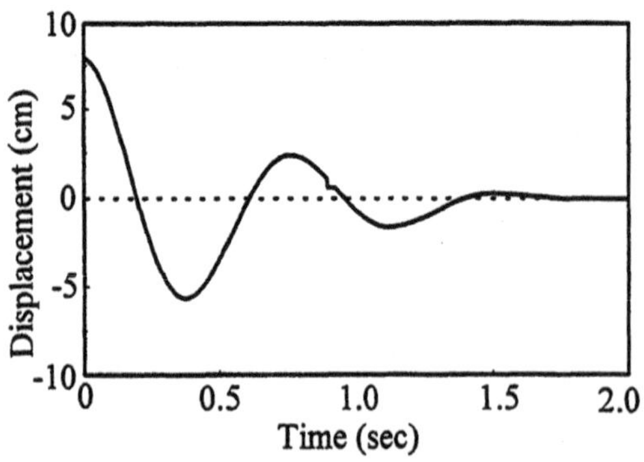

Fig. 11. Displacement response of the motion generator subject to initial displacement with active stiffness implemented

Fig. 12 shows the linear motor having same specifications as the above one. But this time, not only the active stiffness of 300N/m are added as in Fig. 11, but the additional damping of $\Delta c = 60\text{N·s/m}$ are also added. It shows the time-variant displacement of the motor with the same initial conditions. As expected, the vibration fades away more rapidly than before on reasons to the additional damping.

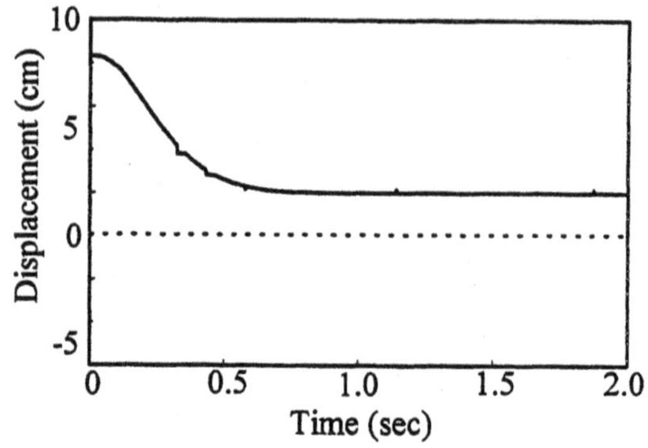

Fig. 12. Displacement response of the motion generator subject to initial displacement with active stiffness and damping implemented

5. CONCLUSIONS

In this research, an arm motion generator capable of two-dimensional movement is made by perpendicularly connecting two linear motors. By controlling the force generated by these motors on the motion generator, active impedance is implemented, which has the same effect as physically changing the inertia, damping, and stiffness of the motion generator. Numerous kinesthetic senses relating to arm movement can be measured due to the ability to change the impedance, such as inertia, damping, and stiffness, on real-time basis.

In this research, a 2-DOF controller was employed to cope with some inherent problems with current control of the linear motor. The control system shows reasonable performance on both disturbance rejection and tracking.

REFERENCES

Bahram, S. and M. Hassel, (1993). *Control System Design Using MATLAB*, pp.404, Prentice-Hall, New Jersey.

John, C. D., B.A. Francis, and A.R. Tannenbaum,. (1992). *Feedback Control Theory*, pp.48–49, Macmillan Publishing Company, New York.

Kemin, Z.H., J.C. Doyle, and K. Glover, (1996). *Robust and Optimal Control*, pp.459-462, Prentice-Hall, New Jersey.

Ken, D., S. Thompson and B. Barraclough, (1997). *The Art of Control Engineering*, pp.603–604, Addison-Wesley, England.

Lewis, F.L., C.T. Abdallah, and D.M. Dawson, (1993). *Control of Robot Manipulators*, Macmillan Publishing Company, New York.

Spong, M.W. and M. Vidyasagar, (1989). *Robot Dynamics and Control*, pp.252-254, Wiley.

14th Triennial World Congress, Beijing, P.R. China

NONLINEAR CONTROL DESIGN OF DISCRETE-TIME STRICT-FEEDBACK SYSTEMS [1]

Jung-Shan Lin

Department of Electrical Engineering
National Chi Nan University
Puli, Nantou, Taiwan 545
e-mail: jslin@ncnu.edu.tw

Abstract: Following the development of backstepping design methodology for a class of continuous-time nonlinear systems, the main attention is shifted to the backstepping control design of *discrete-time* strict-feedback nonlinear systems in this paper. A step-by-step systematic design procedure, which is analogous to the backstepping scheme in continuous-time systems, is developed to guarantee the stability of the resulting closed-loop system in discrete-time cases. The results show that the discrete-time backstepping control design is not only to globally stabilize the plant, but also to drive the system output to zero asymptotically. *Copyright © 1999 IFAC*

Keywords: Discrete-time systems, Nonlinear control, Lyapunov stability

1. INTRODUCTION

In recent years, a systematic design scheme, which is called *backstepping* (Kanellakopoulos *et al.*, 1991; Kanellakopoulos *et al.*, 1992; Khalil, 1996; Krstić *et al.*, 1995), has been developed for the control design of nonlinear feedback systems, either strict-feedback or output-feedback. In nonlinear systems, existing design methods can guarantee global stability only under the condition that all the state variables entering the nonlinearities must be measured (Khalil, 1996; Krstić *et al.*, 1995; Marino and Tomei, 1995). As a result, in many nonlinear control applications, full-state measurement is viewed as necessary to meet the stability and performance requirements. The backstepping design methodology can guarantee global stability, tracking, and transient performance for the broad class of strict-feedback systems. However, these results were all discussed thus far in continuous-time systems, so one may want to see if the concept of backstepping design strategy can be extended to the control design of discrete-time systems.

Since backstepping approach is a Lyapunov-based design scheme, the choice of the *control Lyapunov function* (Krstić *et al.*, 1995) plays an important role for the design of controllers. In Lyapunov stability analysis of continuous-time systems, one must check whether the derivative of the chosen Lyapunov function is negative definite or not, so the normal choice of the control Lyapunov function is *quadratic* terms in continuous-time backstepping design. However, in discrete-time systems one needs to investigate the *forward difference* (Kalman and Bertram, 1960; Kuo, 1992; Ogata, 1987) of the chosen Lyapunov function instead of the derivative. If quadratic terms are still employed as the candidates of control Lyapunov functions in discrete-time cases, it may make the analysis of control design procedure more complicated since the forward difference of quadratic terms must be computed. Hence, one should seek an appropriate control Lyapunov function for the discrete-time backstepping design. In the paper,

[1] This work was supported in part by National Science Council, Taiwan, R.O.C. under the grant NSC 88-2212-E-260-001.

we show that the terms of *absolute value* are suitable to be chosen as a candidate of this control Lyapunov function in discrete-time cases.

In comparison to nonlinear strict-feedback systems (Krstić *et al.*, 1995) in continuous-time cases, let us consider the following discrete-time nonlinear system in the standard *strict-feedback form*:

$$\begin{aligned} x_1(k+1) &= x_2(k) + \varphi_1(x_1(k)) \\ x_2(k+1) &= x_3(k) + \varphi_2(x_1(k), x_2(k)) \\ &\vdots \\ x_{n-1}(k+1) &= x_n(k) + \varphi_{n-1}(x_1(k), \ldots, x_{n-1}(k)) \\ x_n(k+1) &= \beta_0(x(k))u(k) + \varphi_n(x(k)) \\ y(k) &= x_1(k), \end{aligned} \tag{1}$$

where $x(k) \in \mathbb{R}^n$ is the state, $u(k) \in \mathbb{R}$ is the input, $y(k) \in \mathbb{R}$ is the output, and $\beta_0(\cdot)$ and $\varphi_i(\cdot)$, $1 \leq i \leq n$, are known nonlinear functions. In the regulation problem, the control objective is not only to globally stabilize the plant (1), but to also drive the output $y(k)$ to zero under the following assumption:

- $\beta_0(x(k)) \neq 0$, $\forall x(k) \in \mathbb{R}^n$.

Therefore, in this paper a discrete-time backstepping design procedure is developed to deal with the regulation problem of the nonlinear strict-feedback system (1).

The remainder of the paper is organized as follows: In Section 2 the general discrete-time nonlinear backstepping design scheme is developed for the given nth-order strict-feedback system step by step. Then, in Section 3 we investigate the choice of the design constants, which are employed in our discrete-time backstepping design procedure. Section 4 shows an example to illustrate the design procedure and to verify our conclusions through simulation. Finally, some concluding remarks are given in Section 5.

2. NONLINEAR BACKSTEPPING DESIGN

In order to mimic the concepts of continuous-time backstepping design procedure as much as possible, our discrete-time backstepping design procedure for the system (1) is carried out in n steps. At the ith step, an ith-order subsystem is stabilized with respect to a Lyapunov function $V_i(k)$ by the design of a *stabilizing function* $\alpha_i(k)$. The feedback control $u(k)$ is designed at the final step. Hence, the discrete-time backstepping scheme is derived and outlined step by step as follows:

Step 1: Starting with the regulated output $y(k)$ and defining a new variable $z_1(k) = y(k)$ for the consistence of notations, from (1) we have

$$\begin{aligned} z_1(k+1) &= y(k+1) \\ &= x_1(k+1) \\ &= x_2(k) + \varphi_1^{(0)}(k), \end{aligned} \tag{2}$$

where

$$\varphi_1^{(0)}(k) = \varphi_1(x_1(k)). \tag{3}$$

Then, we use the state $x_2(k)$ as the virtual control in (2) and introduce the error variable $z_2(k) = x_2(k) - \alpha_1(k)$, where $\alpha_1(k)$ is the first *stabilizing function* to be chosen later. In general, quadratic terms are the most common choice for the candidates of Lyapunov functions in the stability analysis of continuous-time systems since the derivative of the chosen Lyapunov function must be investigated. However, in Lyapunov stability analysis of discrete-time systems one has to check the *forward difference* of the selected Lyapunov function instead of its derivative, so the other choice of Lyapunov functions may be considered. Let us now consider the following partial Lyapunov function

$$V_1(k) = |z_1(k)|, \tag{4}$$

so the difference between $V_1(k+1)$ and $V_1(k)$ is computed as

$$\begin{aligned} \Delta V_1(k) &= V_1(k+1) - V_1(k) \\ &= |z_1(k+1)| - |z_1(k)| \\ &= \left|z_2(k) + \alpha_1(k) + \varphi_1^{(0)}(k)\right| - |z_1(k)|. \end{aligned} \tag{5}$$

With the choice of the stabilizing function

$$\alpha_1(k) = c_1 z_1(k) - \varphi_1^{(0)}(k), \tag{6}$$

where c_1 is a design constant to be chosen later, we obtain

$$\begin{aligned} \Delta V_1(k) &= |c_1 z_1(k) + z_2(k)| - |z_1(k)| \\ &\leq (|c_1| - 1)|z_1(k)| + |z_2(k)|, \end{aligned} \tag{7}$$

and (2) becomes

$$z_1(k+1) = c_1 z_1(k) + z_2(k). \tag{8}$$

Note that the term $(|c_1| - 1)|z_1(k)|$ in the right hand side of (7) is negative definite if the design parameter c_1 is chosen such that $|c_1| < 1$. If $x_2(k)$ is the actual control, then one can use $x_2(k) = \alpha_1(k)$ as the control law, which yields $z_2(k) = 0$. Hence, $\Delta V_1(k)$ is negative definite, which implies that the asymptotic stability of $z_1(k)$ is guaranteed. However, $x_2(k)$ is not the actual control signal in the plant (1), so the backstepping design procedure must be continued.

Step i ($2 \leq i \leq n-1$): The forward signal of the error variable $z_i(k)$ is

$$
\begin{aligned}
&z_i(k+1) \\
&= x_i(k+1) - \alpha_{i-1}(k+1) \\
&= x_{i+1}(k) + \varphi_i(x_1(k), \ldots, x_i(k)) - c_{i-1}z_{i-1}(k+1) \\
&\quad + \sum_{j=1}^{i-1} \varphi_j^{(i-j-1)}(k+1) - \sum_{j=1}^{i-2} c_j z_j^{(i-j-1)}(k+1) \\
&= x_{i+1}(k) + \sum_{j=1}^{i} \varphi_j^{(i-j)}(k) - \sum_{j=1}^{i-1} c_j z_j^{(i-j)}(k), \quad (9)
\end{aligned}
$$

where

$$
\begin{aligned}
\varphi_i^{(0)}(k) &= \varphi_i(x_1(k), \ldots, x_i(k)) \\
\varphi_j^{(i-j)}(k) &= \varphi_j^{(i-j-1)}(k+1) \quad (\text{for } 1 \leq j \leq i-1) \\
&= \varphi_j^{(i-j-2)}(k+2) \\
&\ \vdots \\
&= \varphi_j^{(0)}(k+i-j) \\
&= \varphi_j(x_1(k+i-j), \ldots, x_j(k+i-j)) \\
&= \varphi_j\Bigg(x_{i-j+1}(k) + \sum_{l=1}^{i-j} \varphi_l^{(i-j-l)}(k), \ldots, \\
&\qquad x_i(k) + \sum_{l=j}^{i-1} \varphi_l^{(i-1-l)}(k)\Bigg) \quad (10) \\
z_{i-1}^{(1)}(k) &= c_{i-1}z_{i-1}(k) + z_i(k) \\
z_j^{(i-j)}(k) &= z_j^{(i-j-1)}(k+1) \quad (\text{for } 1 \leq j \leq i-2) \\
&= z_j^{(i-j-2)}(k+2) \\
&\ \vdots \\
&= z_j^{(1)}(k+i-j-1) \\
&= z_j(k+i-j) \\
&= c_j z_j(k+i-j-1) + z_{j+1}(k+i-j-1) \\
&= c_j z_j^{(i-j-1)}(k) + z_{j+1}^{(i-j-1)}(k).
\end{aligned}
$$

We employ the state $x_{i+1}(k)$ as the virtual control in the $z_i(k+1)$-equation (9), define $z_{i+1}(k) = x_{i+1}(k) - \alpha_i(k)$, and choose the corresponding stabilizing function $\alpha_i(k)$. With the choice of

$$
\alpha_i(k) = c_i z_i(k) - \sum_{j=1}^{i} \varphi_j^{(i-j)}(k) + \sum_{j=1}^{i-1} c_j z_j^{(i-j)}(k), \quad (11)
$$

where c_i is a design constant to be chosen such that $|c_i| < \frac{1}{i}$, the equation (9) becomes

$$
z_i(k+1) = c_i z_i(k) + z_{i+1}(k). \quad (12)
$$

Let us consider the following partial Lyapunov function

$$
V_i(k) = V_{i-1}(k) + i|z_i(k)|, \quad (13)
$$

so the difference between $V_i(k+1)$ and $V_i(k)$ is calculated as

$$
\begin{aligned}
\triangle V_i(k) &= V_i(k+1) - V_i(k) \\
&= \triangle V_{i-1}(k) + i|z_i(k+1)| - i|z_i(k)| \\
&\leq \sum_{j=1}^{i-1} (j|c_j| - 1)|z_j(k)| + (i-1)|z_i(k)| \\
&\quad + i|c_i z_i(k) + z_{i+1}(k)| - i|z_i(k)| \\
&\leq \sum_{j=1}^{i} (j|c_j| - 1)|z_j(k)| + i|z_{i+1}(k)|. \quad (14)
\end{aligned}
$$

Step n: From the definition of the error variable $z_n(k) = x_n(k) - \alpha_{n-1}(k)$, we have

$$
\begin{aligned}
&z_n(k+1) \\
&= x_n(k+1) - \alpha_{n-1}(k+1) \\
&= \beta_0(x(k))u(k) + \varphi_n(x_1(k), \ldots, x_n(k)) \\
&\quad - c_{n-1}z_{n-1}(k+1) + \sum_{j=1}^{n-1} \varphi_j^{(n-j-1)}(k+1) \\
&\quad - \sum_{j=1}^{n-2} c_j z_j^{(n-j-1)}(k+1) \\
&= \beta_0(x(k))u(k) + \sum_{j=1}^{n} \varphi_j^{(n-j)}(k) - \sum_{j=1}^{n-1} c_j z_j^{(n-j)}(k), \quad (15)
\end{aligned}
$$

where

$$
\begin{aligned}
\varphi_n^{(0)}(k) &= \varphi_n(x_1(k), \ldots, x_n(k)) \\
\varphi_j^{(n-j)}(k) &= \varphi_j\Bigg(x_{n-j+1}(k) + \sum_{l=1}^{n-j} \varphi_l^{(n-j-l)}(k), \ldots, \\
&\qquad x_n(k) + \sum_{l=j}^{n-1} \varphi_l^{(n-1-l)}(k)\Bigg) \quad (16) \\
&\qquad (\text{for } 1 \leq j \leq n-1) \\
z_{n-1}^{(1)}(k) &= c_{n-1}z_{n-1}(k) + z_n(k) \\
z_j^{(n-j)}(k) &= c_j z_j^{(n-j-1)}(k) + z_{j+1}^{(n-j-1)}(k) \\
&\qquad (\text{for } 1 \leq j \leq n-2).
\end{aligned}
$$

Since the actual control $u(k)$ appears in (15), we can determine the control law at this final step. With the choice of the control signal

$$
u(k) = \frac{1}{\beta_0(x(k))}\Bigg(c_n z_n(k) - \sum_{j=1}^{n} \varphi_j^{(n-j)}(k) + \sum_{j=1}^{n-1} c_j z_j^{(n-j)}(k)\Bigg), \quad (17)
$$

where c_n is a design constant to be chosen such that $|c_n| < \frac{1}{n}$, the equation (15) becomes

$$
z_n(k+1) = c_n z_n(k). \quad (18)
$$

Hence, the forward difference of the Lyapunov function

$$V_n(k) = V_{n-1}(k) + n|z_n(k)| \qquad (19)$$

can be computed as

$$\begin{aligned}\Delta V_n(k) &= V_n(k+1) - V_n(k)\\ &= \Delta V_{n-1}(k) + n|z_n(k+1)| - n|z_n(k)|\\ &\le \sum_{j=1}^{n-1}(j|c_j|-1)|z_j(k)| + (n-1)|z_n(k)|\\ &\quad + n|c_n z_n(k)| - n|z_n(k)|\\ &\le \sum_{j=1}^{n}(j|c_j|-1)|z_j(k)|\,. \end{aligned} \qquad (20)$$

Clearly, the right hand side of (20) is negative definite with appropriate choice of the design constants $c_1, c_2, \ldots, c_n$ such that $|c_j| < \frac{1}{j}$, $1 \le j \le n$. Hence, from (8), (12) and (18) the resulting closed-loop system becomes

$$z(k+1) = \begin{bmatrix} c_1 & 1 & 0 & \cdots & 0\\ 0 & c_2 & 1 & \ddots & \vdots\\ 0 & 0 & \ddots & \ddots & 0\\ \vdots & \ddots & \ddots & c_{n-1} & 1\\ 0 & \cdots & 0 & 0 & c_n \end{bmatrix} z(k) \triangleq Az(k)\,, \qquad (21)$$

where $z(k) = \begin{bmatrix} z_1(k) & z_2(k) & \cdots & z_n(k)\end{bmatrix}^{\mathrm{T}}$, and it is globally asymptotically stable. Since $z_1(k) = y(k) \to 0$ as $k \to \infty$, the regulation goal is also achieved. Note that the original *nonlinear* system (1) with the control law (17) results in the *linear* closed-loop system (21), where A is a constant matrix with eigenvalues located at c_j, $1 \le j \le n$. Hence, from the Lyapunov stability analysis of linear time-invariant discrete-time systems, there exists another choice of Lyapunov function

$$V_z(k) = z^{\mathrm{T}}(k)Pz(k)\,, \qquad (22)$$

where the matrix $P = P^{\mathrm{T}}$ is positive definite, if and only if the error system (21) is asymptotically stable. In (22), the matrix P must satisfy the Lyapunov equation

$$A^{\mathrm{T}}PA - P = -Q\,, \qquad (23)$$

where $Q = Q^{\mathrm{T}}$ is also positive definite.

3. CHOICE OF DESIGN CONSTANTS c_i's

If we replace the Lyapunov function (19) with the following one:

$$V_d(k) = \sum_{i=1}^{n} d_i|z_i(k)|\,, \qquad (24)$$

where positive constants d_i's are chosen to satisfy the condition $1 = d_1 < d_2 < d_3 < \cdots < d_{n-1} < d_n$, then we can compute the difference between $V_d(k+1)$ and $V_d(k)$ as

$$\begin{aligned}\Delta V_d(k) &= V_d(k+1) - V_d(k)\\ &\le (|c_1|-1)|z_1(k)|\\ &\quad + \sum_{i=2}^{n}[d_i|c_i| - (d_i - d_{i-1})]|z_i(k)|\,. \end{aligned} \qquad (25)$$

In order to make the right hand side of (25) negative definite, we have to choose c_i's such that

$$|c_1| < 1 \text{ and } |c_i| < \frac{d_i - d_{i-1}}{d_i}\,, \quad 2 \le i \le n\,. \qquad (26)$$

For example, (19) is identical to (24) with the parameters $d_i = i$, $1 \le i \le n$, so the design constants c_i, $1 \le i \le n$, must be chosen to satisfy

$$|c_i| < \frac{1}{i}\,, \qquad 1 \le i \le n\,. \qquad (27)$$

Since the choice of the parameters d_i's is just for the stability analysis (not for the design of control law), one can choose them to be large enough to satisfy $d_i >> d_{i-1}$, $2 \le i \le n$, and then (26) becomes

$$|c_i| < 1\,, \qquad 1 \le i \le n\,. \qquad (28)$$

This is to say, all the design constants c_i's in the control law (17) can be chosen arbitrarily with their magnitudes being less than 1. This result agrees with the fact that the resulting closed-loop system (21) whose eigenvalues of matrix A are c_i's, which can be chosen arbitrarily inside the interval $(-1, 1)$ to make the system asymptotically stable.

In similarity to pole placement technique, we can place the poles of the linear error system (21) inside the interval $(-1, 1)$ by choosing appropriate design constants c_i's. If all the design constants c_i's are chosen to be zero, that is, all the poles of the system are placed at zero, then the matrix A defined in (21) becomes

$$A = \begin{bmatrix} 0 & 1 & 0 & \cdots & 0\\ 0 & 0 & 1 & \ddots & \vdots\\ 0 & 0 & \ddots & \ddots & 0\\ \vdots & \ddots & \ddots & 0 & 1\\ 0 & \cdots & 0 & 0 & 0 \end{bmatrix} \triangleq A_0\,. \qquad (29)$$

As one can see, the matrix A_0 is *nilpotent* (Chen, 1984; Kailath, 1980; Ogata, 1987; Rugh, 1996) since $A_0^n = 0$. The nilpotent matrix A_0 plays an important role in discussing *deadbeat response* (Chen, 1984; Kuo, 1992; Ogata, 1987). This implies that the error state $z(k)$ will be brought to zero from any initial state $z(0)$ in "*at most*" n sampling periods, that is,

$$z(k) = 0\,, \quad \forall k \ge q \quad (\text{where } q \le n)\,. \qquad (30)$$

In brief, the response to any initial state $z(0)$ is deadbeat. The concept of deadbeat control is unique to discrete-time systems, but there is no such thing as deadbeat response in continuous-time systems. However, it is not physically possible to have an extremely large control signal without bound to achieve deadbeat control, so we still need to choose the design constants c_i's appropriately in our discrete-time backstepping design procedure.

4. AN ILLUSTRATIVE EXAMPLE

Consider the following third-order discrete-time nonlinear system in strict-feedback form

$$\begin{aligned} x_1(k+1) &= x_2(k) + 2\cos x_1^2(k) \\ x_2(k+1) &= x_3(k) + x_1(k)\sin x_2(k) \\ x_3(k+1) &= u(k) + x_1(k)x_2(k)x_3(k) \\ y(k) &= x_1(k), \end{aligned} \tag{31}$$

which is equivalent to the system (1) with $n = 3$ and the functions $\beta_0(x(k)) = 1$, $\varphi_1(x_1(k)) = 2\cos x_1^2(k)$, $\varphi_2(x_1(k), x_2(k)) = x_1(k)\sin x_2(k)$ and $\varphi_3(x(k)) = x_1(k)x_2(k)x_3(k)$. According to the discrete-time backstepping design procedure derived in the previous sections, the control law $u(k)$ can be obtained as follows:

$$\begin{aligned} u(k) = {}& c_3 z_3(k) - x_1(k)x_2(k)x_3(k) \\ & - \left(x_2(k) + 2\cos x_1^2(k)\right) \times \\ & \quad \sin\left(x_3(k) + x_1(k)\sin x_2(k)\right) \\ & - 2\cos\Big[x_3(k) + x_1(k)\sin x_2(k) \\ & \qquad + 2\cos\left(x_2(k) + 2\cos x_1^2(k)\right)^2\Big]^2 \\ & + c_2\left(c_2 z_2(k) + z_3(k)\right) \\ & + c_1\Big[c_1\left(c_1 z_1(k) + z_2(k)\right) \\ & \qquad + c_2 z_2(k) + z_3(k)\Big], \end{aligned} \tag{32}$$

where

$$\begin{aligned} z_1(k) &= x_1(k) \\ z_2(k) &= x_2(k) - \alpha_1(k) \\ z_3(k) &= x_3(k) - \alpha_2(k) \\ \alpha_1(k) &= c_1 z_1(k) - 2\cos x_1^2(k) \\ \alpha_2(k) &= c_2 z_2(k) - x_1(k)\sin x_2(k) \\ & \quad - 2\cos\left(x_2(k) + 2\cos x_1^2(k)\right)^2 \\ & \quad + c_1\left(c_1 z_1(k) + z_2(k)\right). \end{aligned} \tag{33}$$

In (32) and (33), c_1, c_2 and c_3 are the backstepping design constants. With the choice of the original state initial conditions: $x_1(0) = x_2(0) = x_3(0) = 2$, the simulation results with the different choices of these design constants in the following three cases are shown in Fig. 1–Fig. 3.

- Case 1: $c_1 = 0.7$, $c_2 = -0.4$ and $c_3 = -0.1$. This choice of the design constants is to satisfy those conditions: $|c_1| < 1$, $|c_2| < \frac{1}{2}$ and $|c_3| < \frac{1}{3}$, which have been mentioned in Section 2. From Fig. 1, we can see that all the error states (z-coordinates) converge to zero asymptotically, and the output regulation is achieved (i.e., $y(k) = x_1(k) = z_1(k) \to 0$ asymptotically).
- Case 2: $c_1 = 0.7$, $c_2 = -0.7$ and $c_3 = 0.7$. They are chosen such that $|c_i| < 1$, $1 \le i \le 3$, which have been stated in Section 3. Fig. 2 shows that all the control objectives still can be attained with much more flexible design constant selections.
- Case 3: $c_1 = 0$, $c_2 = 0$ and $c_3 = 0$. This case is to verify the deatbeat response mentioned in Section 3. In Fig. 3, one can see that all the error states ($z_1(k)$,$z_2(k)$ and $z_3(k)$) are driven to zero with only three time units. However, the control effort compared to the previous two cases is increased.

Since the equilibrium point $z(k) = [0, 0, 0]^{\mathrm{T}}$ of the resulting closed-loop system is asymptotically stable, the equilibrium state in the original coordinates (x-coordinates) can be obtained by substituting $z(k) = \mathbf{0}$ in

$$\begin{aligned} x_1(k) &= z_1(k) \\ x_2(k) &= z_2(k) + \alpha_1(k) \\ &= z_2(k) + c_1 z_1(k) - 2\cos x_1^2(k) \\ x_3(k) &= z_3(k) + \alpha_2(k) \\ &= z_3(k) + c_2 z_2(k) - x_1(k)\sin x_2(k) \\ & \quad - 2\cos\left(x_2(k) + 2\cos x_1^2(k)\right)^2 \\ & \quad + c_1\left(c_1 z_1(k) + z_2(k)\right). \end{aligned} \tag{34}$$

Therefore, $x(k) = [0, -2, -2]^{\mathrm{T}}$ is the equilibrium point in the original coordinates, which is verified in Fig. 1–Fig. 3.

5. CONCLUDING REMARKS

We have developed a discrete-time backstepping control design to deal with the regulation problem for nonlinear discrete-time strict-feedback systems. The results show that our discrete-time backstepping design scheme can guarantee the globally asymptotic stability of the plant (1) as well as regulate the output signal asymptotically.

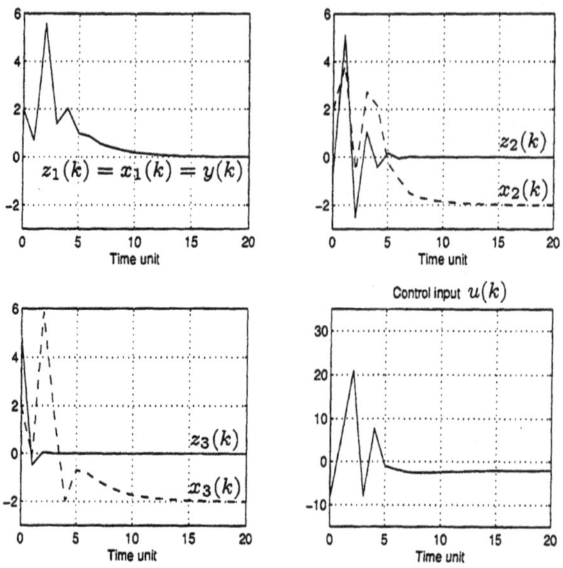

Fig. 1. The design constants are $c_1 = 0.7$, $c_2 = -0.4$ and $c_3 = -0.1$. All the error states converge to zero asymptotically and the output regulation is achieved.

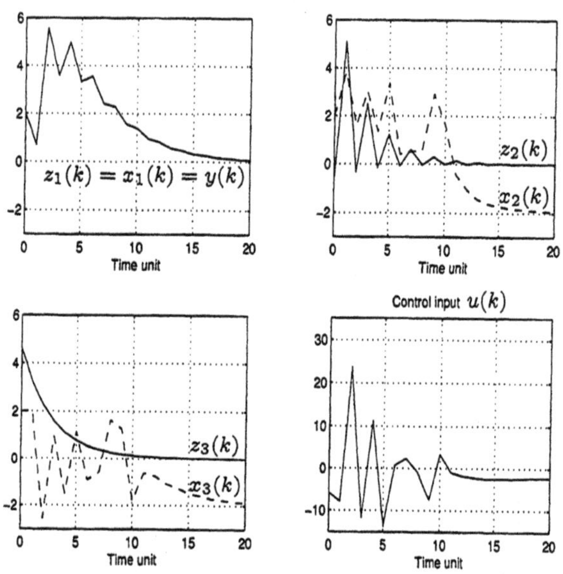

Fig. 2. The design constants are $c_1 = 0.7$, $c_2 = -0.7$ and $c_3 = 0.7$. All the error states converge to zero asymptotically, but the rate of convergence is slower than that in Fig. 1.

However, the class of systems (1), which was considered here, is not the largest class of systems to which our discrete-time backstepping design procedure can be applied. Extensions such as systems with zero dynamics and even pure-feedback systems (Krstić *et al.*, 1995), can be dealt with via slight modifications of this procedure. Similarly, the tracking problem in discrete-time strict-feedback systems also can be done with our discrete-time backstepping design procedure if the reference signal for tracking is well-known.

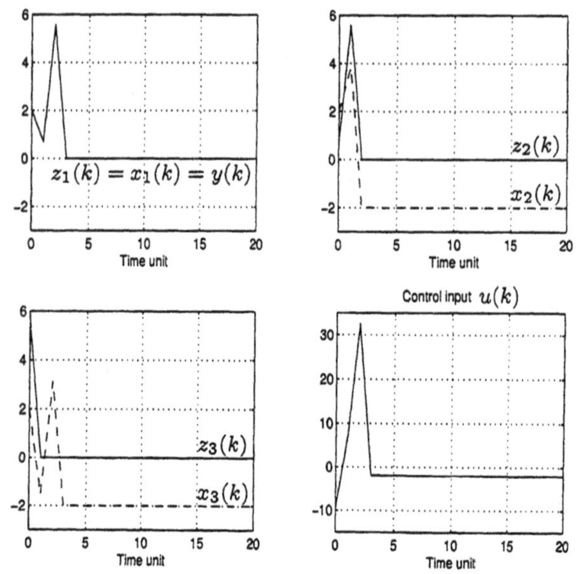

Fig. 3. The design constants are $c_1 = 0$, $c_2 = 0$ and $c_3 = 0$. All the error states converge to zero asymptotically with three time units only (deadbeat control).

REFERENCES

Chen, C.-T. (1984). *Linear System Theory and Design.* Holt, Rinehart and Winston. New York, NY.

Kailath, T. (1980). *Linear Systems.* Prentice-Hall. Englewood Cliffs, NJ.

Kalman, R. E. and J. E. Bertram (1960). Control system analysis and design via the second method of Lyapunov: I. continuous-time systems; II. discrete-time systems. *ASME Journal of Basic Engineering* **82**, 371–400.

Kanellakopoulos, I., P. V. Kokotović and A. S. Morse (1991). Systematic design of adaptive controllers for feedback linearizable systems. *IEEE Transactions on Automatic Control* **36**, 1241–1253.

Kanellakopoulos, I., P. V. Kokotović and A. S. Morse (1992). A toolkit for nonlinear feedback design. *Systems & Control Letters* **18**, 83–92.

Khalil, H. K. (1996). *Nonlinear Systems.* 2nd ed.. Prentice-Hall. Upper Saddle River, NJ.

Krstić, M., I. Kanellakopoulos and P. V. Kokotović (1995). *Nonlinear and Adaptive Control Design.* John Wiley & Sons. New York, NY.

Kuo, B. C. (1992). *Digital Control Systems.* 2nd ed.. Saunders College Publishing. Orlando, FL.

Marino, R. and P. Tomei (1995). *Nonlinear Control Design: Geometric, Adaptive and Robust.* Prentice-Hall International. London, UK.

Ogata, K. (1987). *Discrete-Time Control Systems.* Prentice-Hall. Englewood Cliffs, NJ.

Rugh, W. J. (1996). *Linear System Theory.* 2nd ed.. Prentice-Hall. Upper Saddle River, NJ.

RIESZ BASIS GENERATION OF DISCRETE OPERATORS AND APPLICATION TO A LINEAR THERMOELASTIC SYSTEM

Baozhu GUO *

* *Department of Applied Mathematics, Beijing Institute of Technology, Beijing 100081, China*

Abstract: In this paper, we give a new result of Riesz basis generation property for discrete operators in Hilbert spaces, from which, we show that the generalized eigenfunctions of a one-dimensional linear thermoelastic system with Dirichlet-Dirichlet boundary condition form a Riesz basis in the state Hilbert space. This develops a parallel result which is already known for the same systems with Dirichlet-Neumann or Neumann-Dirichlet boundary conditions [S.W.Hansen, J.Math.Anal.Appl., 167(1992), 429-442]. *Copyright © 1999 IFAC*

Keywords: Riesz basis, spectrum-determined growth condition, eigenvalues.

1. INTRODUCTION

For a linear vibration system, the property that the (generalized) eigenvectors of the system form a Riesz basis in the state Hilbert space is one of the most important features from both theoretical and application point of view. The valid of this property will result in naturally the most concerned problems such as spectrum-determined growth condition and exponential stability of the system. Recall an earlier result (see e.g. Gohberg and Krein, 1969) due to Bari on the Riesz basis perturbation that if $\{\phi_n\}_1^\infty$ is a Riesz basis in a Hilbert space $\mathbf{H}$ and another an ω-linearly independent sequence $\{\psi_n\}_1^\infty$ in $\mathbf{H}$ is quadratically close to $\{\phi_n\}_1^\infty$ in the sense that

$$\sum_{n=1}^{\infty} \|\phi_n - \psi_n\|^2 < \infty$$

then $\{\psi_n\}_1^\infty$ is a Riesz basis itself in $\mathbf{H}$. But unfortunately, it is not easy to completely understand the behavior of all eigenfunctions even for most extensively studied linear systems. The difficulty is that we have no useful (or simple) way to know the number of (generalized) eigenfunctions corresponding to those eigenvalues that are located on bounded region even if we have found a suitable reference Riesz basis $\{\phi_n\}_1^\infty$ as in Bari's theorem and got a clear picture on high eigenmodes by asymptotic technic. This limits greatly the application of Bari's theorem.

The purpose of this paper is to overcome this difficulty. In the next section, an abstract result on Riesz basis generation for discrete operators in Hilbert spaces is developed, which avoids the efforts for the understanding of the low eigenmodes as it was done in many literatures (Conrad and Morgul, 1998; Rao, 1997). To demonstrate its applications, the result is easily used, in section 3, to show the Riesz basis generation for the following one-dimensional linear thermoelastic system with Dirichlet-Dirichlet boundary condition (Liu and Sheng, 1993):

$$\begin{cases} u_{tt}(x,t) - u_{xx}(x,t) + \gamma\theta_x(x,t) = 0, \\ \qquad 0 < x < 1, t > 0, \\ \theta_t(x,t) + \gamma u_{xt}(x,t) - k\theta_{xx}(x,t) = 0, \\ \qquad 0 < x < 1, t > 0, \\ u(i,t) = \theta(i,t) = 0, i = 0,1, t \geq 0 \end{cases} \tag{1}$$

where $\gamma > 0, k > 0$ are constants. This is a parallel result for thermoelastic system with Dirichlet-

Neumann or Neumann-Dirichlet boundary conditions developed in (Hansen, 1992).

2. RIESZ BASIS PROPERTY OF DISCRETE OPERATORS

Lemma 1. Let $\{\phi_n\}_1^\infty$ be a Riesz basis in a Hilbert space **H**. Let $\{\psi_n\}_{N+1}^\infty$ be another sequence in **H**. If there is an $N \geq 0$ such that

$$\sum_{n=N+1}^{\infty} \| \phi_n - \psi_n \|^2 < \infty$$

then there exists a $M \geq N$ such that $\{\phi_n\}_1^M \cup \{\psi_n\}_{M+1}^\infty$ is a Riesz basis in **H**.

Proof. Since $\{\phi_n\}_1^\infty$ is a Riesz basis, there is an invertible operator $T \in \mathcal{L}(\mathbf{H})$, the set of all linear bounded operators on **H**, such that

$$T\phi_n = e_n$$

where $\{e_n\}_1^\infty$ is an orthonormal basis in **H**. By hypothesis, there is a $M \geq N$ such that

$$\| T \|^2 \sum_{n=M+1}^{\infty} \| \phi_n - \psi_n \|^2 < 1.$$

Define operator Q as follows:

$$Qx = \sum_{n=1}^{M} a_n\phi_n + \sum_{n=M+1}^{\infty} a_n\psi_n,$$

$$\text{for } x = \sum_{n=1}^{\infty} a_n\phi_n \in H.$$

It is apparent that Q is well-defined under assumption. Moreover

$$\begin{aligned} \| (I - Q)x \|^2 &= \| \sum_{n=M+1}^{\infty} a_n(\phi_n - \psi_n) \|^2 \\ &\leq \sum_{n=M+1}^{\infty} | a_n |^2 \sum_{n=M+1}^{\infty} \| \phi_n - \psi_n \|^2 \\ &\leq \| T \|^2 \sum_{n=M+1}^{\infty} \| \phi_n - \psi_n \|^2 \| x \|^2 \end{aligned}$$

for all $x \in \mathbf{H}$. This shows that $I - Q \in \mathcal{L}(\mathbf{H})$, $\| I - Q \| < 1$. Therefore, both Q and Q^{-1} belong to $\mathcal{L}(\mathbf{H})$. Consequently, $\{\phi_n\}_1^M \cup \{\psi_n\}_{M+1}^\infty$ is a Riesz basis in **H** because $Q\phi_n = \phi_n$ for $1 \leq n \leq M$ and $Q\phi_n = \psi_n$ as $n > M$.

Theorem 1. Let A be a densely defined discrete operator (that is, there is a $\lambda \in \rho(A)$ such that $R(\lambda, A) = (\lambda - A)^{-1}$ is compact) in a Hilbert space **H**. Let $\{\phi_n\}_1^\infty$ be a Riesz basis in **H**. If there are $N \geq 0$ and an ω-linearly independent subset of generalized eigenvectors $\{\psi_n\}_{N+1}^\infty$ of A corresponding to $\{\sigma_n\}_{N+1}^\infty$ (σ_n may be the same for different n) such that

$$\sum_{N+1}^{\infty} \|\phi_n - \psi_n\|^2 < \infty$$

then there are eigenvectors $\{\psi_n\}_1^N$ of A corresponding to $\{\sigma_n\}_1^N$ such that $\{\psi_n\}_1^\infty$ forms a Riesz basis in **H**. Hence $\sigma(A) = \{\sigma_n\}_1^\infty$ counting by algebraic multiplicity. And σ_n are algebraically simple for sufficiently large n while they are distinct for sufficiently large n.

Proof. First it follows from Lemma 1 that there exists a $M > N$ such that $\{\phi_n\}_1^M \cup \{\psi_n\}_{M+1}^\infty$ is a Riesz basis in **H**.

Let $\{\psi_n\}_{N+1}^\infty \cup \{\psi_\alpha\}$ be the maximal set of ω-linearly independent generalized eigenvectors of A. Then by Bari's theorem, the number of $\{\psi_\alpha\}$ can not exceed N. Let $\mathbf{H}_2$ be the closed subspace spanned by $\{\psi_n\}_{N+1}^\infty \cup \{\psi_\alpha\}$. The proof is complete if we can show $\mathbf{H}_2 = \mathbf{H}$.

Suppose without loss of generality that $0 \in \rho(A)$. Let $B = A^{-1}$. It is well-known that $\lambda \in \rho(A)$ if and only if $\lambda^{-1} \in \sigma(B)$ and the root subspace of A corresponding to λ is the same as the root subspace of B corresponding to λ^{-1} (see Dunford and Schwartz, 1971, pp.2293). Hence $\{\psi_n\}_{N+1}^\infty \cup \{\psi_\alpha\}$ also is the maximal set of ω-linearly independent generalized eigenvectors of B. However, our operator B now, as assumption, is compact on **H**.

Suppose on the contrary: $\mathbf{H}_2 \neq \mathbf{H}$. Then by orthogonal decomposition theorem, there is a subspace $\mathbf{H}_1$, the orthogonal complement of $\mathbf{H}_2$ in **H**, such that

$$\mathbf{H} = \mathbf{H}_1 \oplus \mathbf{H}_2, \mathbf{H}_1 = \mathbf{H}_2^\perp, \mathbf{H}_1 \neq \{0\}.$$

Since $\{\phi_n\}_1^M \cup \{\psi_{M+1}^\infty\}$ is a Riesz basis in **H**, it follows from Bari's theorem that $\mathbf{H}_1$ must be finite dimensional.

Let $\mathbb{P}$ be the orthogonal projector from **H** to $\mathbf{H}_1$. It follows from Lemma 4.2 in (Gohberg and Krein, 1969, pp.17) that $\mathbb{P}B\mathbb{P}$ is a Volterra operator and so is its adjoint $(\mathbb{P}B\mathbb{P})^* = \mathbb{P}B^*\mathbb{P}$, that is, $\mathbb{P}B^*\mathbb{P}$ is compact and has no nonzero eigenvalues.

On the other hand, since $\mathbf{H}_2$ is an invariant subspace of B and so $\mathbf{H}_1$ is an invariant subspace of its adjoint B^*. Note that $\mathbb{P}B^*\mathbb{P}\mathbf{H}_1 = B^*\mathbf{H}_1$, therefore, $B^* |_{\mathbf{H}_1}$, the operator B^* confined to $\mathbf{H}_1$, has no nonzero eigenvalue. Since $\mathbf{H}_1$ is finite dimensional, it must have $0 \in \sigma(B^*|_{\mathbf{H}_1})$, that is, there is a $x_0 \in \mathbf{H}_1, x_0 \neq 0$ such that $B^*x_0 = 0$. This contradicts a well-known fact that $\mathbf{H} = \overline{\mathcal{R}(B)} \oplus \mathcal{N}(B^*)$ and the assumption $\overline{\mathcal{R}(B)} = \overline{D(A)} = \mathbf{H}$, where $\mathcal{R}(\bullet)$ and $\mathcal{N}(\bullet)$ denote respectively, the range and null subspace of operator $\bullet$. Therefore, $\mathbf{H}_2 = \mathbf{H}$. The proof is complete.

3. APPLICATION TO THERMOELASTIC SYSTEM

Now, we return to system (1). The following results were collected from (Liu and Sheng, 1993; Guo and Yung, 1997).

Theorem 2. (i). The system (1) associates with a solution of C_0-semigroup of contractions $T(t) = e^{At}$ on the state Hilbert space $\mathcal{H} = H_0^1(0,1) \times (L^2(0,1))^2$, where $A : D(A) \to \mathcal{H}$ is given by

$$A = \begin{pmatrix} 0 & 1 & 0 \\ D^2 & 0 & -\gamma D \\ 0 & -\gamma D & kD^2 \end{pmatrix}, \text{ with}$$
$$D = \partial/\partial x,$$
$$D(A) = (H^2(0,1) \times H_0^1(0,1) \times H_0^1(0,1)) \cap \mathcal{H}.$$

(ii). A^{-1} exists, and the eigenvalues of A consist of a real sequence $\{\sigma_n\}$ and a sequence of conjugate pairs $\{\lambda_n, \overline{\lambda}_n\}$ with the following asymptotic properties:

$$\begin{cases} \sigma_n = -k(n\pi)^2 + \gamma^2/k + \mathcal{O}(n^{-2}), \\ \lambda_n = -\dfrac{\gamma^2}{2k} + in\pi + \mathcal{O}(n^{-1}) \end{cases} \tag{2}$$

where n is a large positive integer.

The exponential stability of system (1) was first proved in (Liu and Sheng, 1993). For thermoelastic system with Dirichlet-Neumann or Neumann-Dirichlet boundary conditions, it was proved in (Hansen, 1992) that the generalized eigenfunctions of the system form a Riesz basis in the state Hilbert space. Unfortunately, the method used in (Hansen, 1992) can not be applied to treat system (1).

Let A be defined as in Theorem 2. It is known from (Liu and Shneg, 1993; Guo and Yung, 1997) $\lambda \in \sigma(A)$ (it must be $Re\lambda < 0$) if and only if there exists a $(\phi, \psi) \neq 0$ such that

$$\begin{cases} \phi^{(4)}(x) - \lambda(k\lambda + \gamma^2 + 1)\phi''(x) + \lambda^3\phi(x) = 0, \\ \phi(i) = \lambda(k\lambda + \gamma^2)\phi'(i) - k\phi'''(i) = 0, i = 0, 1, \end{cases} \tag{3}$$

with

$$\gamma\lambda\psi(x) = k\phi'''(x) - \lambda(k\lambda + \gamma^2)\phi'(x). \tag{4}$$

Moreover, λ is geometrically simple and the associated eigenfunction is $(\phi, \lambda\phi, \psi)$.

The characteristic equation of (3)

$$ka^4 - \lambda(k\lambda + \gamma^2 + 1)a^2 + \lambda^3 = 0 \tag{5}$$

has four different roots $a_1, a_2, -a_1, -a_2$, where

$$a_1 = \sqrt{\frac{\lambda}{2k}[k\lambda + \gamma^2 + 1 + \sqrt{(k\lambda + \gamma^2 + 1)^2 - 4k\lambda}]},$$
$$a_2 = \frac{\lambda\sqrt{\lambda/k}}{a_1}. \tag{6}$$

For any λ,

$$\begin{aligned} \phi(x) &= (g_2 sha_1 - g_1 sha_2)(cha_1 x - cha_2 x) \\ &\quad -(cha_1 - cha_2)(g_2 sha_1 x - g_1 sha_2 x) \end{aligned} \tag{7}$$

satisfies

$$\begin{cases} k\phi^{(4)}(x) - \lambda(k\lambda + \gamma^2 + 1)\phi''(x) + \lambda^3\phi(x) = 0, \\ \phi(i) = \lambda(k\lambda + \gamma^2)\phi'(0) - k\phi'''(0) = 0, i = 0, 1, \end{cases} \tag{8}$$

where

$$g_i = a_i(k\lambda^2 + \gamma^2\lambda - ka_i^2) \neq 0, i = 1, 2. \tag{9}$$

Moreover, to make $\phi(x)$ defined by (7) to be a solution of (3), the necessary and sufficient condition is ϕ satisfying the last boundary condition $\lambda(k\lambda + \gamma^2)\phi'(1) - k\phi'''(1) = 0$ which deduces that

$$\begin{aligned} & -2[\lambda(k\lambda + \gamma^2)\phi'(1) - k\phi'''(1)] \\ &= 4g_1g_2 - 4g_1g_2 cha_1 cha_2 + 2[g_1^2 + g_2^2] sha_1 sha_2 \\ &= 4g_1g_2 + (g_1 - g_2)^2 ch(a_1 + a_2) \\ &\quad -(g_1 + g_2)^2 ch(a_1 - a_2). \end{aligned} \tag{10}$$

We obtain once again the characteristic equation appeared in (Guo and Yung, 1997).

To simplify the presentations, we use λ instead of $k\lambda$ throughout the analysis, but return to $k\lambda$ when we draw our final conclusions in Theorem 3. The following correspondences are kept temporarily:

$$\lambda \to k\lambda, g_i \to \beta_i, a_i \to \alpha_i, i = 1, 2. \tag{11}$$

From (Guo and Yung, 1997)

$$\begin{cases} \alpha_1 = \dfrac{\lambda}{k} + \dfrac{\gamma^2}{2k} + \dfrac{4\gamma^2 - \gamma^4}{8k}\dfrac{1}{\lambda} + \mathcal{O}(|\lambda|^{-2}), \\ \alpha_2 = \dfrac{\sqrt{\lambda}}{k} - \dfrac{\gamma^2}{2k}\dfrac{1}{\sqrt{\lambda}} + \mathcal{O}(|\lambda|^{-3/2}), \end{cases} \tag{12}$$

and hence

$$\begin{cases} \beta_1 = \alpha_1(\dfrac{\lambda^2}{k} + \dfrac{\gamma^2}{k}\lambda - k\alpha_1^2) \\ \quad = -\dfrac{\gamma^2}{k^2}\lambda[1 + \mathcal{O}(|\lambda|^{-1})], \\ \beta_2 = \alpha_2(\dfrac{\lambda^2}{k} + \dfrac{\gamma^2}{k}\lambda - k\alpha_2^2) \\ \quad = \dfrac{1}{k^2}\lambda^2\sqrt{\lambda}[1 + \mathcal{O}(|\lambda|^{-1})]. \end{cases} \tag{13}$$

After a simple calculation, we write $\phi(x), \psi(x)$ defined by (7) and (4), respectively, to be

$$\begin{aligned} \phi(x) &= (\beta_2 sh\alpha_1 - \beta_1 sh\alpha_2)(ch\alpha_1 x - ch\alpha_2 x) \\ &\quad -(ch\alpha_1 - ch\alpha_2)(\beta_2 sh\alpha_1 x - \beta_1 sh\alpha_2 x) \\ &= \beta_2 sh\alpha_1(1 - x) + \beta_1 sh\alpha_2(1 - x) \\ &\quad +\beta_2 ch\alpha_2 sh\alpha_1 x + \beta_1 ch\alpha_1 sh\alpha_2 x \\ &\quad -\beta_2 sh\alpha_1 ch\alpha_2 x - \beta_1 sh\alpha_2 ch\alpha_1 x \end{aligned} \tag{14}$$

$$\begin{aligned} \psi(x) &= \frac{k^2}{\gamma\lambda}\phi'''(x) - \frac{1}{\gamma}(\lambda + \gamma^2)\phi'(x) \\ &= -\frac{k}{\gamma\lambda}\Big[- \beta_1\beta_2 ch\alpha_1(1 - x) - \beta_1\beta_2 ch\alpha_2(1 - x) \\ &\quad -\beta_1^2 sh\alpha_2 sh\alpha_1 x - \beta_2^2 sh\alpha_1 sh\alpha_2 x \\ &\quad +\beta_1\beta_2 ch\alpha_1 ch\alpha_2 x + \beta_1\beta_2 ch\alpha_2 ch\alpha_1 x\Big]. \end{aligned} \tag{15}$$

It follows from (12) that

$$\begin{aligned} e^{\alpha_1 x} &= e^{(\frac{\lambda}{k}+\frac{\gamma^2}{2k})x}[1+\mathcal{O}(|\lambda|^{-1})], \\ e^{\alpha_2 x} &= e^{\frac{\sqrt{\lambda}}{k}x}[1-\frac{\gamma^2}{2k}\frac{x}{\sqrt{\lambda}}+\mathcal{O}(|\lambda|^{-1})] \end{aligned} \tag{16}$$

$$\begin{aligned} e^{-\alpha_1 x} &= e^{-(\frac{\lambda}{k}++\frac{\gamma^2}{2k})x}[1+\mathcal{O}(|\lambda|^{-1})], \\ e^{-\alpha_2 x} &= e^{-\frac{\sqrt{\lambda}}{k}x}[1+\frac{\gamma^2}{2k}\frac{x}{\sqrt{\lambda}}+\mathcal{O}(|\lambda|^{-1})] \end{aligned} \tag{17}$$

Lemma 2. The eigenfunctions $\{(\phi_n, \sigma_n\phi_n, \psi_n)\}$ associated with real eigenvalues $\{\sigma_n\}$ of A have the following asymptotic expression:

$$-\frac{2\gamma\sigma_n}{k}\frac{e^{\alpha_1}}{\beta_2^2}\begin{pmatrix}\phi_n(x)\\ \sigma_n/k\phi_n(x)\\ \psi_n(x)\end{pmatrix} = \begin{pmatrix}0\\0\\ i\sin n\pi x\end{pmatrix}$$
$$+F_n(x) \quad \text{with } \| F_n \|_{\mathcal{H}} = \mathcal{O}(n^{-1}).$$

Proof. It follows from (2), the real eigenvalue λ satisfies

$$\lambda = -(kn\pi)^2 + \gamma^2 + \mathcal{O}(n^{-2})$$

for large positive integer n, and so

$$\frac{\sqrt{\lambda}}{k} = in\pi + \mathcal{O}(n^{-1}).$$

By (16) and (17),

$$\begin{aligned} sh\alpha_2 x &= i\sin n\pi x + \mathcal{O}(n^{-1}), \\ ch\alpha_2 x &= \cos \pi x + \mathcal{O}(n^{-1}). \end{aligned} \tag{18}$$

Note that $\beta_1/\beta_2 = -\frac{\gamma^2}{\lambda\sqrt{\lambda}} + \mathcal{O}(|\lambda|^{-1})$, we have, by refereeing (15) and (18), that

$$\begin{aligned} &-\frac{2\gamma\lambda}{k}\frac{e^{\alpha_1}}{\beta_1\beta_2}\frac{\beta_1}{\beta_2}\psi(x) \\ &= -\beta_1/\beta_2 e^{\alpha_1 x} + (\beta_1/\beta_2)^2 sh\alpha_2 e^{\alpha_1(1-x)} + sh\alpha_2 x \\ &+\beta_1/\beta_2 ch\alpha_2 x + \beta_1/\beta_2 ch\alpha_2 e^{\alpha_1(1-x)} + \mathcal{O}(n^{-1}) \\ &= sh\alpha_2 x + \mathcal{O}(n^{-1}) = i\sin n\pi x + \mathcal{O}(n^{-1}). \end{aligned}$$

Moreover, since

$$\begin{aligned} \phi'(x) &= -\alpha_1\beta_2 ch\alpha_1(1-x) - \alpha_2\beta_1 ch\alpha_2(1-x) \\ &+\alpha_1\beta_2 ch\alpha_2 ch\alpha_1 x + \alpha_2\beta_1 ch\alpha_1 ch\alpha_2 x \\ &-\alpha_2\beta_2 sh\alpha_1 sh\alpha_2 x - \alpha_1\beta_1 sh\alpha_2 sh\alpha_1 x \end{aligned}$$

we have

$$\begin{aligned} &2\frac{e^{\alpha_1}}{\beta_1\beta_2}\frac{\beta_1}{\beta_2}\phi'(x) = -\alpha_1/\beta_2 e^{\alpha_1 x} \\ &+\alpha_1/\beta_2 ch\alpha_2 e^{\alpha_1(1-x)} + \alpha_2\beta_1/\beta_2^2 ch\alpha_2 x \\ &+\alpha_2/\beta_2 sh\alpha_2 x + \alpha_1\beta_1/\beta_2^2 sh\alpha_2 e^{\alpha_1(1-x)} + \mathcal{O}(e^{-c|\lambda|}) \end{aligned}$$

where $c > 0$ is a constant independent of λ. Hence

$$2\frac{e^{\alpha_1}}{\beta_1\beta_2}\frac{\beta_1}{\beta_2}\phi'(x) = \mathcal{O}(|\lambda|^{-3/2})$$

or

$$-\frac{2\gamma\lambda}{k}\frac{e^{\alpha_1}}{\beta_1\beta_2}\frac{\beta_1}{\beta_2}\phi'(x) = \mathcal{O}(|\lambda|^{-1/2}) = \mathcal{O}(n^{-1}).$$

Similarly,

$$-\frac{2\gamma\lambda}{k}\frac{e^{\alpha_1}}{\beta_1\beta_2}\frac{\beta_1}{\beta_2}\frac{\lambda}{k}\phi(x) = \mathcal{O}(n^{-1}).$$

Therefore,

$$-\frac{2\gamma\lambda}{k}\frac{e^{\alpha_1}}{\beta_2^2}\begin{pmatrix}\phi(x)\\ \lambda/k\phi(x)\\ \psi(x)\end{pmatrix} = \begin{pmatrix}0\\0\\ i\sin n\pi x\end{pmatrix} + F_n(x)$$
$$\text{with } \| F_n \|_{\mathcal{H}} = \mathcal{O}(n^{-1}).$$

This is the result required.

Lemma 3. The eigenfunctions

$$\{(\phi_n, \lambda_n\phi_n, \psi_n), (\overline{\phi_n}, \overline{\lambda_n}\overline{\phi_n}, \overline{\psi_n})\}$$

associated with complex conjugate eigenvalues $\{\lambda_n, \overline{\lambda_n}\}$ of A have the following asymptotic expression:

$$\frac{2e^{-\alpha_2}}{\alpha_1\beta^2}\begin{pmatrix}\phi_n(x)\\ \lambda_n/k\phi_n(x)\\ \psi_n(x)\end{pmatrix} = \begin{pmatrix}f_n(x)\\ i\sin n\pi x\\ 0\end{pmatrix} + F_n(x)$$
$$\text{with } \| F_n \|_{\mathcal{H}} = \mathcal{O}(n^{-1}).$$

where $f_n'(x) = \cos n\pi x + \mathcal{O}(n^{-1})$.

Proof. For the complex conjugate eigenvalues $\{\lambda, \overline{\lambda}\}$ of A, it hold

$$\lambda = -\frac{\gamma^2}{2} + ikn\pi + \mathcal{O}(n^{-1})$$

for large positive integer n and so $\frac{\lambda}{k} + \frac{\gamma^2}{2k} = in\pi + \mathcal{O}(n^{-1})$, which yields

$$\begin{aligned} ch\alpha_1 x &= \cos n\pi x + \mathcal{O}(n^{-1}), \\ sh\alpha_1 x &= i\sin n\pi x + \mathcal{O}(n^{-1}). \end{aligned} \tag{19}$$

Moreover, $\frac{\sqrt{\lambda}}{k} = \frac{1+i}{\sqrt{2}}\sqrt{n\pi/k}[1+\mathcal{O}(n^{-1})]$. From (14)

$$\begin{aligned} \phi'(x) &= -\alpha_1\beta_2 ch\alpha_1(1-x) - \alpha_2\beta_1 ch\alpha_2(1-x) \\ &+\alpha_1\beta_2 ch\alpha_2 ch\alpha_1 x + \alpha_2\beta_1 ch\alpha_1 ch\alpha_2 x \\ &-\alpha_2\beta_2 sh\alpha_1 sh\alpha_2 x - \alpha_1\beta_1 sh\alpha_2 sh\alpha_1 x \end{aligned}$$

and so by (19)

$$\begin{aligned} \frac{2e^{-\alpha_2}}{\alpha_1\beta_2}\phi'(x) &= -\alpha_2\beta_1/(\alpha_1\beta_2)e^{-\alpha_2 x} + ch\alpha_1 x + \\ &\alpha_2\beta_1/(\alpha_1\beta_2)ch\alpha_1 e^{-\alpha_2(1-x)} \\ &-\alpha_2/\alpha_1 sh\alpha_1 e^{-\alpha_2(1-x)} \\ &-\beta_1/\beta_2 sh\alpha_1 x + \mathcal{O}(n^{-1}). \end{aligned}$$

Note that $\frac{\beta_1}{\beta_2} = -\frac{\gamma^2}{\lambda\sqrt{\lambda}}[1+\mathcal{O}(|\lambda|^{-1})]$, $\frac{\alpha_2}{\alpha_1} = \frac{1}{\sqrt{\lambda}}[1+\mathcal{O}(|\lambda|^{-1})]$, we have $\alpha_2\beta_1/(\alpha_1\beta_2) = -\frac{\gamma^2}{\lambda^2}[1+\mathcal{O}(|\lambda|^{-1})] = \mathcal{O}(n^{-2})$, therefore,

$$\frac{2e^{-\alpha_2}}{\alpha_1\beta_2}\phi'(x) = ch\alpha_1 x + \mathcal{O}(n^{-1}) = \cos n\pi x + \mathcal{O}(n^{-1}). \tag{20}$$

Similarly,

$$\begin{aligned} \frac{2e^{-\alpha_2}}{\alpha_1\beta_2}\frac{\lambda}{k}\phi(x) &= \frac{\lambda}{k}\beta_1/(\alpha_1\beta_2)e^{-\alpha_2 x} + \frac{\lambda}{k}/\alpha_1 sh\alpha_1 x \\ &+\frac{\lambda}{k}\beta_1/(\alpha_1\beta_2)ch\alpha_1 e^{-\alpha_2(1-x)} \\ &-\frac{\lambda}{k}/\alpha_1 sh\alpha_1 e^{-\alpha_2(1-x)} \\ &-\beta_1/(\alpha_1\beta_2)ch\alpha_1 x + \mathcal{O}(n^{-1}) \\ &= sh\alpha_1 x + \mathcal{O}(n^{-1}) \\ &= i\sin n\pi x + \mathcal{O}(n^{-1}). \end{aligned} \tag{21}$$

$$\frac{2e^{-\alpha_2}}{\alpha_1\beta_2}\psi(x) = -\frac{k}{\gamma\lambda}\Big[-\beta_1/\alpha_1 e^{-\alpha_2 x} -\beta_1^2/(\alpha_1\beta_2)sh\alpha_1 x - \beta_2/\alpha_1 sh\alpha_1 e^{-\alpha_2(1-x)} +\beta_1/\alpha_1 ch\alpha_1 e^{-\alpha_2(1-x)} + \beta_1/\alpha_1 ch\alpha_1 x + \mathcal{O}(e^{-c\sqrt{n}})\Big] = \mathcal{O}(n^{-1}) \quad (22)$$

where $c > 0$ is a constant independent of λ. Combining (20)-(22), we obtain

$$\frac{2e^{-\alpha_2}}{\alpha_1\beta^2}\begin{pmatrix}\phi(x)\\ \lambda/k\phi(x)\\ \psi(x)\end{pmatrix} = \begin{pmatrix} f(x)\\ i\sin n\pi x\\ 0\end{pmatrix} + F_n(x)$$
$$\text{with } \| F_n \|_{\mathcal{H}} = \mathcal{O}(n^{-1}).$$

where $f'(x) = \cos n\pi x + \mathcal{O}(n^{-1})$. Lemma 3 follows.

Summarizing, we get the estimates for the approximate normalized eigenfunctions which are stated in the following theorem.

Theorem 3. There are two families of approximate normalized eigenfunctions of operator A: one family $\{\Phi_n\}, \Phi_n = (\phi_n, \lambda_n\phi_n, \psi_n)$ associates with real eigenvalues σ_n with the following asymptotic expression:

$$\Phi_n = \begin{pmatrix}\phi_n(x)\\ \sigma_n\phi_n(x)\\ \psi_n(x)\end{pmatrix} = \begin{pmatrix}0\\ 0\\ i\sin n\pi x\end{pmatrix} + F_{1n}(x)$$
$$\text{with } \| F_{1n} \|_{\mathcal{H}} = \mathcal{O}(n^{-1});$$

where $f_n'(x) = \cos n\pi x + \mathcal{O}(n^{-1})$, and another family $\{\Psi_n, \overline{\Psi}_n\}, \Psi_n = (\phi_n, \lambda_n\phi_n, \psi_n), \overline{\Psi}_n = (\overline{\phi}_n, \overline{\lambda}_n\overline{\phi}_n, \overline{\psi}_n)$ correspond to complex conjugate eigenvalues $\{\lambda_n, \overline{\lambda}_n\}$ with the following asymptotic expression:

$$\Psi_n = \begin{pmatrix} f_n(x)\\ i\sin n\pi x\\ 0\end{pmatrix} + F_{2n}(x)$$
$$\text{with } \|F_{2n}\| = \mathcal{O}(n^{-1})$$

where $f_n'(x) = \cos n\pi x + \mathcal{O}(n^{-1})$.

Let A_0 denote the operator A with $\gamma = 0$, that is, there is no coupling between wave and heat equations in (1). It is well-known that A_0 is a self-adjoint operator in $\mathcal{H}$ with compact resolvent. There are two family of eigenvalues of A_0: the real eigenvalues $\lambda_{n0} = -k(n\pi)^2$ with positive integer n, associate with normalized eigenfunctions $(0, 0, i\sin n\pi x)$, and the complex conjugate pairs $\{\lambda_{n0}, \overline{\lambda}_{n0}\}, \lambda_{n0} = in\pi$ with positive integer n, correspond to the normalized eigenfunctions $(\sin n\pi x/(n\pi), i\sin n\pi x, 0)$. Denote

$$\Phi_{n0} = \begin{pmatrix}0\\ 0\\ i\sin n\pi x\end{pmatrix}, \Psi_{n0} = \begin{pmatrix}\sin n\pi x/(n\pi)\\ i\sin n\pi x\\ 0\end{pmatrix}. \quad (23)$$

Then $\{\Phi_{n0}, \psi_{n0}, \overline{\Psi}_{n0}\}_1^\infty$ is the system of maximal ω-linearly independent eigenfunctions of A_0, which forms an orthogonal basis in $\mathcal{H}$. By Theorem 3, there is an $N > 0$ such that

$$\sum_{n>N}^{\infty} [\| \Phi_n - \Phi_{n0} \|_{\mathcal{H}}^2 + \| \Psi_n - \Psi_{n0} \|_{\mathcal{H}}^2 + \| \overline{\Psi}_n - \overline{\Psi}_{n0} \|_{\mathcal{H}}^2] < \infty. \quad (24)$$

By Theorem 1 and (24), we have

Theorem 4. Let A be the operator associated with thermoelastic system (1) by Theorem 1. Then there is a system of generalized eigenfunctions of A forms a Riesz basis in the state space $\mathcal{H}$. Moreover, all eigenvalues with sufficiently large module are algebraically simple. Therefore, the spectrum-determined growth condition holds for system (1).

Acknowledgements: The author would like to thank Dr.K.S.Liu for his contribution to the new proof of Theorem 1 attached here. The support of National Key Project of China is gratefully acknowledged.

REFERENCE

B.Z. Guo and S.P. Yung(1997). Asmptotic behavior of the eigenfrequency of a one dimensional linear thermoelastic system, *J. Math. Anal. Appl.*, **213**, pp.406-421.

B.P.Rao(1997). Optimal energy decay rate in a damped Rayleigh beam, in *"Contemporary Mathematics"* (S.Cox and I.Lasiecka eds.), **209**, pp.211-229.

F.Conrad and O.Morgul(1998). On the stabilization of a flexible beam with a tip mass, *SIAM J.Control & Optim.*, **36**, pp.1962-1986.

I.C.Gohberg and M.G.Krein(1969). Introduction to the Theory of Linear Nonselfadjoint Operators, *Trans.of Math.Monographs* **18**, *AMS Providence.*

N.Dunford and J.T.Schwartz(1971). Linear Operators, Part III, *Wiley-Interscience.*

S.W.Hansen(1992). Exponential energy decay in a linear thermoelastic rod, *J.Math. Anal.Appl.*, **167**, pp.429-442.

Z.Y.Liu and S.M.Zheng (1993). Exponential stability of semigroup associated with thermoelastic system, *Quart.Appl.Math.*, **51**, pp.535-545.

AUTHOR INDEX

KEYWORD INDEX

www.ingramcontent.com/pod-product-compliance
Ingram Content Group UK Ltd.
Pitfield, Milton Keynes, MK11 3LW, UK
UKHW052301280926
15471UKWH00006B/1040

9 780080 432144